GAMMA 2001

The Gamma 2001 Symposium was originally planned and scheduled as the sixth in a series of "Compton" Symposia. The series was initiated in 1992 to highlight scientific results from the highly successful Compton Gamma-Ray Observatory. Since the Compton GRO mission was terminated in June of 2000, the theme was modified to highlight forthcoming gamma-ray astrophysics missions, but the Symposium series was continued. In addition to the Compton Symposia, a separate series of five Gamma-Ray Burst Workshops, which focused on the Compton GRO studies of the enigmatic cosmic gamma-ray bursts.

Most of the five Compton Symposia proceedings, as well as all five Gamma-Ray Burst Workshop proceedings volumes were published by the American Institute of Physics.

Compton Symposia

	Year	Held in	Publisher	ISBN
5th	1999	Portsmouth, New Hampshire	AIP Conf. Proceedings vol. 510	1-56396-932-7
4th	1997	Williamsburg, Virginia	AIP Conf. Proceedings vol. 410	1-56396-659-X
3rd	1995	Munich, Germany	Astronomy & Astrophys. Suppl.	
2nd	1993	College Park, Maryland	AIP Conf. Proceedings vol. 304	1-56396-261-6
1st	1991	St. Louis, Missouri	AIP Conf. Proceedings vol. 280	1-56396-104-0

Gamma-Ray Burst Workshops

	Year	Held in	Publisher	ISBN
5th	1999	Huntsville, Alabama	AIP Conf. Proceedings vol. 526	1-56396-947-5
4th	1997	Huntsville, Alabama	AIP Conf. Proceedings vol. 428	1-56396-766-9
3rd	1995	Huntsville, Alabama	AIP Conf. Proceedings vol. 384	1-56396-685-9
2nd	1993	Huntsville, Alabama	AIP Conf. Proceedings vol. 307	1-56396-336-1
1st	1991	Huntsville, Alabama	AIP Conf. Proceedings vol. 265	1-56396-018-4

Other Related Titles from AIP Conference Proceedings

586 Relativistic Astrophysics: 20th Texas Symposium
Edited by J. Craig Wheeler and Hugo Martel, October 2001, 0-7354-0026-1

579 Radio Detection of High Energy Particles: First International Workshop; RADHEP 2000
Edited by David Saltzberg and Peter Gorham, July 2001, 0-7354-0018-0

566 Observing Ultrahigh Energy Cosmic Rays from Space and Earth: International Wkshp.
Edited by Humberto Salazar, Luis Villaseñor, and Arnulfo Zepeda, May 2001, 0-7354-0002-4

558 High Energy Gamma-Ray Astronomy: International Symposium
Edited by Felix A. Aharonian and Heinz J. Völk, April 2001, 1-56396-990-4

556 Explosive Phenomena in Astrophysical Compact Objects: First KIAS Astrophysics Workshop
Edited by H.-Y. Chang, C.-H. Lee, M. Rho, and I. Yi, March 2001, 1-56396-987-4

To learn more about these titles, or the AIP Conference Proceedings Series, please visit the
webpage http://www.aip.org/catalog/aboutconf.html

GAMMA 2001

Gamma-Ray Astrophysics 2001

Baltimore, Maryland   4–6 April 2001

EDITORS
Steven Ritz
Neil Gehrels
Chris R. Shrader
NASA/Goddard Space Flight Center
Greenbelt, Maryland

Melville, New York, 2001
AIP CONFERENCE PROCEEDINGS ■ VOLUME 587

Editors:

Steven Ritz, Neil Gehrels, and Chris R. Shrader
NASA/Goddard Space Flight Center
Code 661
Greenbelt, MD 20771
USA

E-mail: ritz@milkyway.gsfc.nasa.gov
gehrels@lheapop.gsfc.nasa.gov
shrader@grossc.gsfc.nasa.gov

The articles on pp. 213–217, 281–285, 432–441, 545–554, 570–574, 603–612, 668–672, 706–710, 722–728, 781–790, 848–852, 867–876, and 942–951 were authored by U. S. Government employees and are not covered by the below mentioned copyright.

Authorization to photocopy items for internal or personal use, beyond the free copying permitted under the 1978 U.S. Copyright Law (see statement below), is granted by the American Institute of Physics for users registered with the Copyright Clearance Center (CCC) Transactional Reporting Service, provided that the base fee of $18.00 per copy is paid directly to CCC, 222 Rosewood Drive, Danvers, MA 01923. For those organizations that have been granted a photocopy license by CCC, a separate system of payment has been arranged. The fee code for users of the Transactional Reporting Service is: 0-7354-0027-X/01/$18.00.

© 2001 American Institute of Physics

Individual readers of this volume and nonprofit libraries, acting for them, are permitted to make fair use of the material in it, such as copying an article for use in teaching or research. Permission is granted to quote from this volume in scientific work with the customary acknowledgment of the source. To reprint a figure, table, or other excerpt requires the consent of one of the original authors and notification to AIP. Republication or systematic or multiple reproduction of any material in this volume is permitted only under license from AIP. Address inquiries to Office of Rights and Permissions, Suite 1NO1, 2 Huntington Quadrangle, Melville, N.Y. 11747-4502; phone: 516-576-2268; fax: 516-576-2450; e-mail: rights@aip.org.

L.C. Catalog Card No. 2001094992
ISBN 0-7354-0027-X
ISSN 0094-243X

CD-ROM available separately: ISBN 0-7354-0030-X

Printed in the United States of America

Contents

Preface xvii
Sponsors xix

INTRODUCTION

The Compton Gamma Ray Observatory in Review 3
    N. Gehrels and C. R. Shrader

GALACTIC CENTER

Comparative Studies of Line and Continuum Positron Annihilation
Radiation 11
    P. A. Milne, J. D. Kurfess, R. L. Kinzer, and M. D. Leising
Can INTEGRAL Detect 0.511 MeV Radiation from Giant
Molecular Clouds? 16
    N. Guessoum, P. von Ballmoos, J. Knodleseder, and G. Vedrenne
COMPTEL Observations of a Source in the Direction of the
Galactic Centre 21
    A. W. Strong, W. Collmar, K. Bennett, H. Bloemen, R. Diehl, W. Hermsen,
    A. Iyudin, H. Mayer-Hasselwander, J. Ryan, and V. Schönfelder

NEUTRON STAR X-RAY BINARIES

Propagation of Gamma-Rays in Cen X-3 29
    W. Bednarek
Discovery of Weak EXO 2030+375 Outbursts with BATSE 34
    C. A. Wilson, M. H. Finger, M. J. Coe, and S. Laycock
Monte-Carlo/Fokker-Planck Simulations of Accretion onto
Magnetized Neutron Stars 39
    M. Böttcher and E. P. Liang
Studies of Hard X-ray Tails in Z Sources with HEXTE/RXTE 44
    F. D'Amico, W. A. Heindl, R. E. Rothschild, and D. E. Gruber
Discovery of Two Transient X-Ray Pulsars in the Small
Magellanic Cloud 49
    S. Laycock, R. H. D. Corbet, M. J. Coe, D. Perrodin, F. Marshall,
    and C. Markwardt
Self-Similar Hot Accretion Flow onto a Rotating Neutron Star:
Structure and Stability 54
    M. V. Medvedev and R. Narayan

BLACK HOLE X-RAY BINARIES

XMM-Newton Observation of the Black Hole Microquasar GRS 1758- 258 61
    A. Goldwurm, D. Israël, P. Goldoni, P. Ferrando, A. Decourchelle, F. I. Mirabel, and R. S. Warwick

RXTE Observations of GRS 1915+105. 66
    J. Rodriguez, P. Durouchoux, and M. Tagger

Models of Phase Lags in the Rapid Aperiodic Variability of X-ray Binaries 71
    M. Böttcher

The Temporal and Spectral Properties of Cyg X-1 during a Large X-ray Flare 76
    Y. X. Feng, W. Cui, and S. N. Zhang

X-ray Dips and Orbital Modulation in Cyg X-1 81
    Y. X. Feng and W. Cui

Can INTEGRAL Detect 2.223 MeV Radiation from X-Ray Binary Sources? 86
    N. Guessoum and P. Jean

Wide-band X-ray Variability of GRS 1915+105 Observed with BeppoSAX 91
    P. Casella, M. Feroci, E. Massaro, E. Costa, M. Litterio, G. Matt, T. Belloni, M. Tavani, F. I. Mirabel, A. J. Castro-Tirado, A. Harmon, and G. Pooley

Gamma-Ray Spectral Variability of Cygnus X-1 96
    M. L. McConnell, K. Bennett, H. Bloemen, W. Collmar, W. Hermsen, L. Kuiper, W. Paciesas, B. Phlips, J. Poutanen, J. M. Ryan, V. Schönfelder, H. Steinle, A. W. Strong, and A. A. Zdziarski

Relativistic Effects on X-ray Emissions from Accretion Disks around Black Holes 101
    X. Zhang, S. N. Zhang, and Y. Yao

Studying the Accretion Disks in Black Hole X-ray Binaries with Monte-Carlo Simulations 106
    Y. Yao, S. N. Zhang, and X. Zhang

X-ray Spectra of Accretion Discs with Dynamic Coronae 111
    J. Malzac, A. M. Beloborodov, and J. Poutanen

Detection of the γ-Ray Emission from the X-Ray Nova GRO J0422+32 116
    A. F. Iyudin and F. Haberl

The Effects of a Comptonizing Corona on the Appearance of the Reflection Components in Accreting Black Hole Spectra 121
    P. O. Petrucci, A. Merloni, A. Fabian, F. Haardt, E. Gallo, and J. Malzac

The Compact Jet of the Black Hole Candidate XTE J1550-564 during the 2000 X-ray Outburst 126
    S. Corbel, P. Kaaret, R. K. Jain, C. D. Bailyn, R. P. Fender, J. A. Tomsick, E. Kalemci, V. McIntyre, D. Campbell-Wilson, J. M. Miller, and M. L. McCollough

Accretion-Ejection Instabilities in Black-Hole Binaries. 131
    P. Varnière, S. Caunt, and M. Tagger
A Review of Cygnus X-1 Soft γ-ray Observations. 135
    J. C. Ling

GAMMA RAY BURSTS

Recent Advances in Our Understanding of GRB. 143
    P. Mészáros
Gamma-Ray Burst Spectral Diagnostics in the GLAST Era 153
    M. G. Baring
Energy Outflows in γ-ray Bursts: Discontinuous versus Continuous? 158
    E. Ramirez-Ruiz and A. Merloni
Gamma Ray Bursts as Probes of the First Stars 163
    J. E. Rhoads
GeV and X-ray Inverse Compton and Proton Synchrotron Signatures
in Gamma-ray Burst Afterglows 168
    B. Zhang and P. Mészáros
The Supernova-Collapsar Scenario for the Gamma-Ray Bursts 173
    P. J. T. Leonard
Pulses, Spectral Lags, Durations, and Hardness Ratios in Long GRBs 176
    J. P. Norris, J. D. Scargle, and J. T. Bonnell
Super-LOTIS/LOTIS/LITE: Prompt GRB Followup Experiments. 181
    H. S. Park, E. Ables, S. Barthelmy, M. Bradshaw, T. Cline, N. Gehrels,
    D. Hartmann, K. Hurley, R. Nemiroff, W. Pereira, D. Perez-Ramirez,
    G. G. Williams, and K. Ziock
Number Counts of GRBs in a Vacuum Dominated Universe. 185
    K. Strohmaier and T. Wickramasinghe
Modeling and Implications of the Transient Absorption Feature in
GRB 990705. 190
    M. Böttcher, C. D. Dermer, L. Amati, and F. Frontera
Continuously-fed Fireballs and Signatures in Gamma-ray Burst
Afterglows 195
    B. Zhang and P. Mészáros
Trans-Relativistic Supenovae, Circumstellar Gamma-Ray Bursts,
and Supernova 1998bw 200
    C. D. Matzner, J. C. Tan, and C. F. McKee
Physics of Collisionless GRB Shocks and Their Radiation Properties 205
    M. V. Medvedev
White Hole Model of Gamma Ray Bursts and Recent Observations 210
    S. Ramadurai
GRB Coordinates Network (GCN): A Status Report 213
    S. D. Barthelmy, T. L. Cline, and P. Butterworth
Milagro's Sensitivity to GRB's 218
    E. Hays and D. Noyes for the Milagro Collaboration
Intensity Distributions of Gamma-Ray Bursts 223
    D. L. Band

The Blandford-Znajek Mechanism with Vacuum Breakdown as a
Gamma-ray Burst Progenitor 228
    G. Barbiellini and F. Longo

BLAZARS ACTIVE GALACTIC NUCLEI

Very High Energy Gamma Rays from Blazars 235
    J. H. Buckley

Very High Energy Flaring Activity of Markarian 421 during
2000-2001 as Seen by CAT and CELESTE Telescopes 246
    A. Djannati-Ataï for the CAT and CELESTE Collaborations

The Highest Energy Emission Detected by EGRET from Blazars 251
    B. L. Dingus and D. L. Bertsch

An X-ray Survey of Extragalactic Radio Jets with Chandra 256
    R. M. Sambruna, L. Maraschi, F. Tavecchio, C. C. Cheung,
    C. M. Urry, G. Chartas, R. Scarpa, and J. E. Pesce

Extended X-ray Emission from the Radio Galaxy 3C 390.3 261
    H. Krawczynski and S. Wagner

The Extraordinary Campaigns on MKN 421 of 2000 (SAX) and
2001 (RXTE) 266
    G. Fossati, M. Jordan, and J. Buckley

COMPTEL Observations of the Virgo Blazars 3C 273 and 3C 279 271
    W. Collmar, V. Schönfelder, S. Zhang, H. Bloemen, W. Hermsen,
    M. McConnell, K. Bennett, and O. R. Williams

Shock Structures in Relativistic Jets 276
    P. E. Hardee

Recent Observations of 1ES2344+514 Using the Whipple
Gamma-Ray Telescope 281
    H. M. Badran

Model for the X-ray Emission in the Jets and Hot Spots of Radio
Galaxies 286
    C. D. Dermer and H. Li

The Central Black Hole Masses of Gamma-ray Loud Blazars 291
    J. H. Fan and K. S. Cheng

VHE Observations of Unidentified EGRET Sources 296
    S. J. Fegan for the VERITAS Collaboration

The Effect of the SED Shape on the Gamma-ray vs. Radio Emission
Dependence in AGNs 301
    A. Lähteenmäki and E. Valtaoja

The Nature of the EGRET Source 3EG J1621+8203 304
    R. Mukherjee, J. Halpern, N. Mirabal, D. Stern, and E. V. Gotthelf

High Energy Observations of Blazars—Archival Analysis 309
    G. Nandikotkur, K. M. Jahoda, J. H. Swank, P. Sreekumar,
    and R. M. Sambruna

γ-ray Spectral Changes during Blazar Outbursts 314
    P. Sreekumar, R. C. Hartman, R. Mukherjee, and M. Pohl

Multiwavelength Observations of 3EG J2006-2321 and
3EG J0433+2908 319
    P. M. Wallace, M. Eracleous, J. V. Foreman, J. P. Halpern,
    O. Reimer, and D. J. Thompson

Observations of the BL Lac Object, 1H1426+428 at TeV
Gamma-ray Energies 324
    D. Horan for the VERITAS Collaboration

Optical Identification and Monitoring of High Energy
Gamma-Ray Sources 329
    S. D. Bloom and D. A. Dale

CCD Photometry of Blazars at Abastumani: Progress Report 333
    O. M. Kurtanidze and M. G. Nikolashvili

Colour Variation in BL Lacertae 338
    O. M. Kurtanidze and M. G. Nikolashvili

COMPTEL Observations of the Blazars 3C 454.3 and CTA 102 343
    S. Zhang, W. Collmar, V. Schönfelder, H. Bloemen, W. Hermsen,
    M. McConnell, K. Bennett, and O. R. Williams

Optical Photometric Observations of γ-ray Loud Blazars in the
October of 2000 348
    J. H. Fan, O. M. Kurtanidze, and M. G. Nikolashvili

The Spectral Energy Distribution of Centaurus A (NGC 5128)—
A Summary of All Observations Including All CGRO Results 353
    H. Steinle

The Gamma-Ray Horizon 358
    T. M. Kneiske, K. Mannheim, and D. H. Hartmann

Neutrino-Emission from Active Galactic Nuclei as a Diagnostic Tool 363
    C. Schuster, M. Pohl, and R. Schlickeiser

Space VLBI: Past, Present and Future, and Connections to
Gamma-Ray Astronomy 368
    P. G. Edwards and H. Hirabayashi

SEYFERT GALAXIES

Modeling the R-Γ Correlation in Compact Sources 375
    J. Malzac

BeppoSAX Observations of the Radio-Quiet QSO MR 2251-178 380
    A. Orr, P. Barr, M. Guainazzi, A. Parmar, and A. Young

X-ray Spectra Emitted by a Hot Plasma Containing Cold Clouds 385
    J. Malzac

Thermal Comptonization and Disk Reprocessing in Type 1
Seyfert Galaxies 390
    J. Chiang and O. Blaes

Constraints on the High Energy Source of Seyfert 1 from BeppoSAX
Observations 395
    P. O. Petrucci, L. Maraschi, F. Haardt, P. Grandi, J. Malzac, G. Matt,
    F. Nicastro, L. Piro, G. C. Perola, and A. De Rosa

Mrk 335: A NLSy1 with an Unusual Warm Absorber? 400
    A. Orr

A Quasi-Spherical Inner Accretion Flow in Seyfert Galaxies 404
    J. Malzac

GALAXY CLUSTERS AND DIFFUSE GAMMA-RAY BACKGROUND

Accelerated Particles from Shocks Formed in Merging Clusters of
Galaxies 411
    R. C. Berrington, C. D. Dermer, and S. J. Sturner

The Soft γ-ray Spectrum of Galaxy Clusters 416
    M. Henriksen and J. Chiang

Clusters of Galaxies—The EGRET Observations between
1991 and 2000 422
    O. Reimer and P. Sreekumar

Gamma-Rays from Galaxy Clusters: Preliminary Evidences and
Future Expectations 427
    S. Colafrancesco

The Extragalactic Gamma-Ray Background 432
    F. W. Stecker and M. H. Salamon

Contributions of GRBs and Cen A-like Radio Galaxies to the Cosmic
Gamma-ray Background 442
    K. Watanabe and D. H. Hartmann

PARTICLES AND COSMIC RAYS

The Origin of Cosmic Rays and the Diffuse Galactic Gamma-Ray
Emission 449
    S. W. Digel, S. D. Hunter, I. V. Moskalenko, J. F. Ormes, and M. Pohl

Gould's Belt and the Local Cosmic Ray Electron Spectrum 459
    M. Pohl, C. Perrot, and I. Grenier

A Study of Very High Energy Gamma-Rays and Electrons with
GLAST 464
    R. Terrier, A. Djannati-Ataï, A. Chehktman, J. E. Grove, and
    W. N. Johnson

Excess GeV Radiation and Cosmic Ray Origin 469
    I. Büsching, M. Pohl, and R. Schlickeiser

Channeled Blast Wave Behaviour Based on Longitudinal,
Electrostatic Instabilities 474
    M. Pohl, I. Lerche, and R. Schlickeiser

Gamma-ray Detection of Particle Dark Matter 479
    L. Bergström

NUCLEOSYNTHESIS AND GALACTIC DIFFUSE EMISSION

Gamma-Ray Signatures of Supernovae and Hypernovae 487
    K. Nomoto, K. Maeda, Y. Mochizuki, S. Kumagai, H. Umeda,
    T. Nakamura, and I. Tanihata

Gamma-ray Signatures of Classical Novae 498
    M. Hernanz, J. Gómez-Gomar, and J. José

Study of the Galactic Distribution of Nova-Produced ^{22}Na with COMPTEL 508
    A. F. Iyudin, V. Schönfelder, A. W. Strong, K. Bennett, R. Diehl,
    W. Hermsen, G. G. Lichti, and J. Ryan

TGRS and the 478 keV Line from ^{7}Be in Novae 513
    M. J. Harris, D. M. Palmer, G. Weidenspointner, H. Seifert,
    B. J. Teegarden, T. L. Cline, N. Gehrels, and R. Ramaty

Gamma-Ray Line Emission from Superbubbles in the Interstellar Medium: The Cygnus Region 518
    S. Plüschke, R. Diehl, K. Kretschmer, D. H. Hartmann, and U. Oberlack

Unveiling the True Age of the "Vela Junior" Supernova Remnant 523
    S. Mereghetti and A. Pellizzoni

On the Nature of the Nonthermal Emission from the Supernova Remnant IC 443 528
    S. J. Sturner, O. Reimer, J. W. Keohane, C. M. Olbert, R. Petre,
    and C. D. Dermer

SNR and Fluctuations in the Diffuse Galactic γ-ray Continuum 533
    A. W. Strong and I. V. Moskalenko

EGRET Observations of Diffuse Gamma-Ray Emission in Taurus and Perseus 538
    S. W. Digel and I. A. Grenier

PULSARS

Gamma-Ray Pulsars: At the Tip of the Iceberg 545
    A. K. Harding

The Parkes Multibeam Pulsar Survey and the Discovery of New Energetic Radio Pulsars 555
    N. D'Amico, V. M. Kaspi, R. N. Manchester, F. Camilo, A. G. Lyne,
    A. Possenti, I. H. Stairs, M. Kramer, G. Hobbs, and J. F. Bell

Rotationally-induced Asymmetry in the Double-Peak Lightcurves of the Bright EGRET Pulsars? 560
    J. Dyks and B. Rudak

Resolving the Crab σ - Problem 565
    D. Kazanas and J. Contopoulos

RXTE Observations of the Vela Pulsar: The Pulsar Rosetta Stone 570
    M. S. Strickman, A. K. Harding, C. Gwinn, P. McCulloch, and D. Moffett

BeppoSAX Observations of the γ-ray Pulsars PSR B0656+14 and PSR B1706-44 575
    E. Massaro, T. Mineo, G. Cusumano, B. Sacco, and W. Becker

Galactic Population of Radio and Gamma-Ray Pulsars 580
    P. L. Gonthier, M. S. Ouellette, S. O'Brien, J. Berrier,
    and A. K. Harding

Gamma Ray Emission in Relativistic Pulsar Magnetospheric Plasmas 585
    Q. Luo, D. B. Melrose, and G. Z. Machabeli

Observations of the Crab Nebula with CAT and CELESTE 590
    C. Masterson for the CAT and CELESTE Collaborations

Variable High-Energy γ-Ray Emission from Pulsar Wind Nebulae 595
    M. S. E. Roberts, B. M. Gaensler, and R. W. Romani

SOLAR AND STELLAR FLARES

Solar Gamma-Ray Physics Comes of Age 603
    G. H. Share and R. J. Murphy

**COMPTEL Gamma-Ray Observations of the C4 Solar Flare on
20 January 2000** 613
    C. A. Young, M. B. Arndt, K. Bennett, A. Connors, H. Debrunner,
    R. Diehl, M. McConnell, R. S. Miller, G. Rank, J. M. Ryan,
    V. Schönfelder, and C. Winkler

**X- and Gamma-Ray Observations of the 15 November 1991
Solar Flare** 618
    M. B. Arndt, A. Connors, J. Lockwood, M. McConnell, R. Suleiman,
    J. Ryan, C. A. Young, G. Rank, V. Schönfelder, H. Debrunner, K. Bennett,
    O. Williams, and C. Winkler

Energetic Proton Spectra in the 11 June 1991 Solar Flare 623
    C. A. Young, K. Bennett, A. Connors, R. Diehl, M. McConnell, G. Rank,
    J. M. Ryan, R. Suleiman, V. Schönfelder, and C. Winkler

**Expected Gamma-Ray Fluxes from Interactions of Flare Energetic
Particles with Solar Wind Matter** 628
    L. I. Dorman

**Interactions of Flare Energetic Particles with Stellar Wind Matter:
Expected Gamma Ray Fluxes from Local Stars** 633
    L. I. Dorman

SURVEYS AND POPULATION STUDIES

A Multiwavelength Strategy for Identifying Celestial γ-ray Sources 641
    P. A. Caraveo

**Neutron Star Contribution to the Galactic Unidentified
EGRET Sources** 649
    I. A. Grenier and C. A. Perrot

Population Studies of the Gamma-ray Sources 663
    A. W. Chen, S. Mereghetti, A. Pellizzoni, M. Tavani, and S. Vercellone

Artifact Sources Near Bright EGRET Pulsars 668
    D. J. Thompson, D. L. Bertsch, and R. C. Hartman

Possible New Identifications for Southern EGRET Sources.................. 673
    M. Tornikoski, A. Lähteenmäki, M. Lainela, and E. Valtaoja
A Search for Supernova-Remnant Masers toward Unidentified
EGRET Sources 678
    Z. Arzoumanian, F. Yusef-Zadeh, and T. J. W. Lazio
Galactic Plane EGRET Unidentified Source Distribution.................... 683
    D. Bhattacharya, A. Akyüz, T. Miyagi, J. Samimi, and A. Zych

ANALYSIS TECHNIQUES

Effects of Background Counts in RMS Normalization 691
    W. T. Bridgman
An Observability Study for the Tentatively Identified 3EG Sources
Likely to be Detected by the Next-Generation Cherenkov Telescopes 696
    D. Petry and O. Reimer
Bayesian Multiscale Deconvolution Applied to Gamma-ray
Spectroscopy 701
    C. A. Young, A. Connors, E. Kolaczyk, M. McConnell, G. Rank,
    J. M. Ryan, and V. Schönfelder
EGRET's Detection Efficiency in the Later Phases of the Mission............ 706
    D. L. Bertsch, R. C. Hartman, S. D. Hunter, D. J. Thompson,
    and P. Sreekumar

FUTURE MISSIONS (GLAST/AGILE)

The Gamma-ray Large Area Space Telescope Mission:
Science Opportunities... 713
    P. F. Michelson
Gamma-Ray Large Area Space Telescope (GLAST) Project 722
    S. Lambros
Science with AGILE 729
    M. Tavani, G. Barbiellini, A. Argan, N. Auricchio, P. Caraveo, A. Chen,
    V. Cocco, E. Costa, G. Di Cocco, G. Fedel, M. Feroci, M. Fiorini,
    T. Froysland, M. Galli, F. Gianotti, A. Giuliani, C. Labanti, I. Lapshov,
    P. Lipari, F. Longo, E. Massaro, S. Mereghetti, E. Morelli, A. Morselli,
    A. Pellizzoni, F. Perotti, P. Picozza, C. Pittori, C. Pontoni, M. Prest,
    M. Rapisarda, E. Rossi, A. Rubini, P. Soffitta, M. Trifoglio, E. Vallazza,
    S. Vercellone, and D. Zanello
Gamma-Ray Imaging by Silicon Detectors in Space: Presentation of
the AGILE Reconstruction Method and Kalman Filter Algorithms........... 739
    C. Pittori, A. Giuliani, S. Mereghetti, and M. Tavani
GEANT Simulation of the AGILE Gamma-Ray Imaging Detector 744
    V. Cocco, F. Longo, and M. Tavani
Test Campaign of the Mini-Calorimeter for the AGILE Satellite 749
    N. Auricchio, E. Celesti, G. Di Cocco, M. Galli, F. Gianotti, C. Labanti,
    A. Mauri, M. Malaspina, E. Rossi, J. B. Stephen, A. Traci,
    and M. Trifoglio

The Next Generation of High-Energy Gamma-ray Detectors for
Satellites: The AGILE Silicon Tracker 754
    G. Barbiellini, G. Bordignon, G. Fedel, F. Liello, F. Longo, C. Pontoni,
    M. Prest, and E. Vallazza
AGILE Sky Exposure and Sensitivity Maps.............................. 759
    A. Pellizzoni, A. Chen, A. Giuliani, S. Mereghetti, M. Tavani,
    and S. Vercellone
Imaging of High Energy Sources with AGILE............................ 764
    S. Vercellone, A. W. Chen, V. Cocco, M. Feroci, M. Galli, A. Giuliani,
    I. Lapshov, P. Lipari, F. Longo, S. Mereghetti, A. Pellizzoni, C. Pittori,
    P. Soffitta, and D. Zanello
Super AGILE: The X-ray Monitor on-board of AGILE..................... 769
    I. Lapshov, L. Barbanera, E. Costa, E. Del Monte, M. Feroci,
    G. Porrovecchio, M. Mastropietro, L. Pacciani, A. Rubini, P. Soffitta,
    E. Morelli, M. Rapisarda, G. Barbiellini, F. Longo, M. Prest, E. Vallazza,
    A. Argan, S. Mereghetti, M. Tavani, S. Vercellone, and A. Morselli
The AGILE Scientific Instrument 774
    G. Barbiellini, M. Tavani, A. Argan, N. Auricchio, P. Caraveo, A. Chen,
    V. Cocco, E. Costa, G. Di Cocco, G. Fedel, M. Feroci, M. Fiorini,
    T. Froysland, M. Galli, F. Gianotti, A. Giuliani, C. Labanti, I. Lapshov,
    P. Lipari, F. Longo, E. Massaro, S. Mereghetti, E. Morelli, A. Morselli,
    A. Pellizzoni, F. Perotti, P. Picozza, C. Pittori, C. Pontoni, M. Prest,
    M. Rapisarda, E. Rossi, A. Rubini, P. Soffitta, M. Trifoglio, E. Vallazza,
    S. Vercellone, and D. Zanello

FUTURE MISSIONS (GAMMA RAY BURSTS)

Swift: A Gamma Ray Burst MIDEX 781
    S. D. Barthelmy on behalf of the Swift Team
The Swift Ultra-Violet/Optical Telescope................................ 791
    P. W. A. Roming, S. D. Hunsberger, J. A. Nousek, and K. Mason
Swift Burst Alert Telescope Hard X-Ray Monitor and Survey 796
    H. A. Krimm, L. M. Barbier, S. D. Barthelmy, A. J. Dean,
    A. Eftekharzadeh, E. E. Fenimore, N. Gehrels, D. D. Hullinger, H. Ozawa,
    D. M. Palmer, A. M. Parsons, T. Takahasi, M. Tashiro, J. Tueller,
    and G. Weidenspointner
The GLAST Burst Monitor (GBM)....................................... 801
    R. M. Kippen, M. S. Briggs, R. Diehl, G. J. Fishman, R. H. Georgii,
    C. Kouveliotou, G. G. Lichti, C. A. Meegan, W. S. Paciesas, R. D. Preece,
    V. Schönfelder, and A. von Kienlin

FUTURE MISSIONS (INTEGRAL)

Science with INTEGRAL in Perspective.................................. 809
    V. Schönfelder
Gamma-Ray Polarization Measurements with INTEGRAL/IBIS 816
    J. B. Stephen, E. Caroli, R. C. da Silva, and L. Foschini

Evaluation of the INTEGRAL/IBIS Photons Detectors Efficiencies by
Monte Carlo Simulation 821
    G. De Cesare, C. Ciocca, M. Del Santo, G. Di Cocco, P. Laurent,
    F. Lebrun, G. Malaguti, L. Natalucci, V. Reglero, and P. Ubertini

In-flight Calibration Sources Simulations for the IBIS Telescope 826
    M. Del Santo, A. Bazzano, A. J. Bird, G. De Cesare, P. Laurent,
    G. Malaguti, and L. Natalucci

Results from the SPI Imaging Test Setup 831
    C. B. Wunderer, R. Diehl, R. Georgii, A. von Kienlin, G. G. Lichti,
    V. Schönfelder, A. Strong, P. Connell, F. Sanchez, and G. Vedrenne

First Results on SPI/INTEGRAL Flight-Model Gamma-Camera
Calibration... 836
    P. Paul, L. Bouchet, and J. P. Roques

MISSION CONCEPTS: HIGH-ENERGY GAMMA RAYS

Feasibility Studies of Coded Masks for High-Energy Gamma-ray
Telescopes in Space... 843
    Y. C. Lin, P. F. Michelson, P. L. Nolan, and D. J. Thompson

Design of a Next-Generation High-Energy Gamma-Ray Telescope............ 848
    S. D. Hunter, D. L. Bertsch, and P. Deines-Jones

MISSION CONCEPTS: GAMMA-RAY LINE SPECTROSCOPY

Diffractive/Refractive Lenses—A Revolution in Gamma-Ray
Astronomy? 855
    G. K. Skinner

B-MINE—The Balloon-Borne Microcalorimeter
Nuclear Line Explorer 860
    E. Silver, H. Schnopper, C. Jones, W. Forman, S. Bandler, S. Murray,
    S. Romaine, P. Slane, J. Grindlay, N. Madden, J. Beeman, E. E. Haller,
    D. Smith, M. Barbera, A. Collura, F. Christensen, B. Ramsey, S. Woosley,
    R. Diehl, G. Tucker, J. Fabregat, V. Reglero, and A. Gimenez

MISSION CONCEPTS: ADVANCED COMPTON TELESCOPES

Progress towards an Advanced Compton Telescope 867
    J. D. Kurfess and R. A. Kroeger

The Nuclear Compton Telescope: A Balloon-borne Soft γ-ray
Spectrometer, Polarimeter, and Imager.................................... 877
    S. E. Boggs, P. Jean, R. P. Lin, D. M. Smith, P. von Ballmoos,
    N. W. Madden, P. N. Luke, M. Amman, M. T. Burks, E. L. Hull, W. Craig,
    and K. Ziock

The TIGRE Gamma-Ray Telescope.. 882
    T. J. O'Neill, A. Akyüz, D. Bhattacharya, M. Polsen, J. Samimi,
    and A. Zych

MEGA — A Next Generation Mission in Medium Energy
Gamma-Ray Astronomy...887
    G. Kanbach for the MEGA Collaboration
Development of CdTe Imaging Detectors for a Compton Telescope...........892
    K.-L. Giboni, E. Aprile, and U. Oberlack

MISSION CONCEPTS: HARD X-RAY TELESCOPES

EXIST: The Ultimate Spatial/Temporal Hard X-ray Survey.................899
    J. Grindlay, L. Bildsten, R. Blandford, D. Chakrabarty, M. Elvis, A. Fabian,
    F. Fiore, G. Fishman, N. Gehrels, C. Hailey, F. Harrison, D. Hartmann,
    C. Kouveliotou, T. Prince, B. Ramsey, R. Rothschild, G. Skinner, and
    S. Woosley
The Development of Coplanar CZT Strip Detectors for
Gamma-Ray Astronomy...909
    M. L. McConnell, L.-A. Hamel, J. R. Macri, M. McClish, and J. M. Ryan

GROUND-BASED HIGH-ENERGY ASTRONOMY

Gamma Ray Astronomy with Air Shower Arrays..........................917
    A. I. Mincer
CANGAROO-II and CANGAROO-III.....................................927
    M. Mori for the CANGAROO Collaboration
The Current Status and Future Plans of the STACEE Observatory..........932
    R. Mukherjee, L. M. Boone, D. Bramel, E. Chae, C. E. Covault, P. Fortin,
    D. M. Gingrich, J. A. Hinton, D. S. Hanna, C. Mueller, R. A. Ong,
    K. Ragan, R. A. Scalzo, D. R. Schuette, C. G. Theoret, and D. A. Williams
The Keck Solar Two Gamma-Ray Observatory..........................937
    M. Tripathi, D. Bhattacharya, J. Lizarazo, G. Mohanty, U. Mohideen,
    P. Murray, H. Tom, T. Tümer, G. Xing, and J. Zweerink
The Next Generation of Ground-based Gamma-ray Telescopes...............942
    T. C. Weekes

List of Participants...953
Author Index..963

PREFACE

The *Gamma-Ray Astrophysics 2001 Symposium* was held April 4-6, 2001 in Baltimore Maryland. More than 280 participants attended the meeting hosted by NASA's Goddard Space Flight Center.

The *Gamma-Ray Astrophysics 2001 Symposium* offered the opportunity for participants to discuss important results from the nine years of operations of the Compton Gamma Ray Observatory (CGRO), which ceased operations on June 4, 2000. Discussions also included new results from current missions such as HETE-II, Chandra, and XMM-Newton. In addition, results from ground-based VHE gamma-ray and radio observatories, and other ground-based and space missions related to high energy astrophysical sources were discussed. On the horizon, new gamma-ray space telescopes like GLAST, Swift, INTEGRAL and AGILE will be launched in the next few years. Discussions included mission capabilities and anticipated science from these upcoming missions. With such enormous leaps in capabilities on all fronts, there is every reason to expect this to be a decade of discovery with continued great advances in gamma-ray astrophysics.

A special session entitled "Advanced Compton Telescope (ACT) for Gamma Ray Astrophysics-II" was held the day prior to the start of the symposium. The workshop covered the topics of instrument concepts for an advanced Compton telescope, progress in related detector technologies, as well as simulation activities to support instrumental design, background estimation and reduction, and predict instrumental performance.

Additionally, an education workshop for local middle- and high-school educators was held on the day prior to the opening of the technical sessions. This workshop involved sessions on both hands-on astronomy activities as well as short lectures with emphasis on gamma-ray astrophysics.

The members of the *Gamma-Ray Astrophysics 2001 Symposium*'s Scientific Organizing Committee included N. Gehrels (NASA/GSFC, Co-Chair), S. Ritz (NASA/GSFC, Co-Chair), D. Bertsch (NASA/GSFC), R. Blandford (Caltech), E. Bloom (SLAC), P. Caraveo (IFC/CNR), G. Fishman (NASA/MSFC), P. Fleury (École Polytechnique), I. Grenier (Université Paris VII), K. Hurley (UCB), R. Johnson (UCSC), T. Kamae (SLAC/Hiroshima University), D. Kniffen (NASA Headquarters), J. Kurfess (NRL), R. Lin (UCB), C. Meegan (NASA/MSFC), P. Michelson (Stanford), J. Ormes (NASA/GSFC), L. Piro (Istituto Astrofisica Spaziale/CNR), G. Ricker (MIT), J. Ryan (UNH), V. Schoenfelder (MPE, Garching), C. Shrader (USRA/NASA/GSFC), P. Ubertini (IAS/CNR), G. Vedrenne

(C.E.S.R.), T. Weekes (Harvard-Smithsonian, CfA), C. Winkler (ESA/ESTEC), and S. Woosley (UCSC). In addition, we acknowledge the individuals on the local organizing committee for their efforts: C. Shrader (chair), N. Gehrels, R. Hartman, J. Norris, S. Ritz, S. Barnes, L. Londot, and E. Pentecost.

We especially thank Sandy Barnes from USRA for leading the logistical organization of the Symposium. We also thank Liz Pentacost, Lorna Londot, Cathy Dicks, Lee Mewshaw, Ginny Peles and the staff at USRA for their excellent support of the meeting. Finally, we would like to acknowledge Jerry Bonnell, Seth Digel, Alice Harding, Bob Hartman, Demos Kazanas, Barbara Mattson, Floyd Stecker and Dave Thompson for their assistance in reviewing the papers in this volume.

The Symposium would not have been possible without the greatly appreciated support from the following organizations: NASA, DOE, Italian Space Agency (ASI), TRW, Spectrum Astro, Lockheed Martin, Southwest Research, Orbital Sciences, USRA, and the following missions: GLAST, CGRO, Swift and INTEGRAL.

Steve Ritz
Neil Gehrels
Chris Shrader
July, 2001

We would like to thank all of our generous
sponsors for their contributions.

 Compton Gamma Ray
Observatory

Introduction

The Compton Gamma Ray Observatory in Review

N. Gehrels & C.R. Shrader

Laboratory for High Energy Astrophysics,
NASA Goddard Space Flight Center

Abstract. The Compton Gamma Ray Observatory was de-orbited on 4 June 2000 after 9 highly successful years in orbit. Major discoveries were made every year with *Compton*. We present a retrospective overview of the mission from launch to deorbit,, highlighting some seminal scientific findings.

THE COMPTON GRO MISSION

After nearly a decade in orbit, and numerous scientific discoveries, the Compton Gamma Ray Observatory (*Compton*) mission came to an end in the early morning hours of June 4, 2000. Engineers at the Goddard Space Flight Center in Greenbelt began planning for the Observatory's reentry in April 1999 when one of its gyroscopes, which provide attitude reference position and rate information, first began showing signs of problems. The gyroscope in question failed in December of 1999. Goddard engineers inferred from the telemetry data that its ball bearings had seized causing its spin-motor to stall. Although out of 45 similar gyroscopes with hundred of years of flight time only two other failures have occurred, the loss of an additional gyro would have lead to a more complex reentry plan incurring a greater risk of failure. A number of deorbit scenarios were conceived, but none were simple or fool proof, and given the mass of the observatory and the density of a number of its constituent components safety was the key consideration. Thus, NASA decided to perform a controlled reentry maneuver at the earliest plausible time.

Compton was the second of NASA's Great Observatories. At 15,900 kg, it was the most massive scientific payload ever flown at the time of its launch by the space shuttle Atlantis on April 5, 1991. Compton had four instruments that covered an unprecedented six decades of the electromagnetic spectrum, from 30 keV to 30 GeV [1,2,3]. In order of increasing spectral energy coverage, these instruments were the Burst and Transient Source Experiment (BATSE), the Oriented Scintillation Spectrometer Experiment (OSSE), the Imaging Compton Telescope (COMPTEL), and the Energetic Gamma Ray Experiment Telescope (EGRET). For each of the instruments, an improvement in sensitivity of more than a factor of ten was realized over previous missions. The Observatory was named in honor of Dr. Arthur Holly Compton in 1991.

SCIENCE HIGHLIGHTS

Thus, we will not attempt to present a comprehensive list of the scientific highlights of the mission – those highlights comprise a major portion of these proceedings. Instead we will summarize a few seminal findings.

One of the great achievements of *Compton* was the compilation of the first all-sky map in high-energy gamma rays. Data obtained with EGRET during the first 4 years of the *Compton* mission were recently carefully reanalyzed in a uniform manner using the most up to date instrumental calibration information to compile a comprehensive source catalog. The resulting (3EG) catalog consists of 271 sources (Figure 1): 5 pulsars, 1 solar flare, 66 high confidence blazar identifications, 27 possible blazar identifications, 1 likely radio galaxy (Cen A), 1 normal galaxy (LMC), and 170 unidentified sources (a sixth EGRET pulsar is also detected, but is seen only in pulsed data, and so is not included in the catalog) [4]. Progress is already being made on studying the unidentified source component. For example, archival researchers have recently identified a probable Geminga like "radio quiet" pulsar associated with 3EG J1835+5918 [5]. Other efforts, including the statistical association of a subset of 3EG sources with the Gould's belt, a possible association of high-latitude sources with Galaxy clusters, and on prospects for the future are reviewed in this volume [6,7,8].

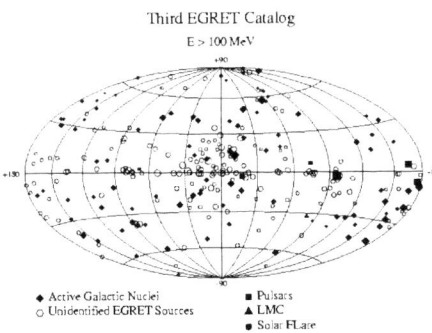

Figure 1. A graphical representation of the source catalog of obtained with the EGRET experiment. Different symbols represent different objet classes as denoted. A large fraction (about 60%) of the sources are unidentified. Work on identifying these sources, drawing heavily upon multi-wavelength study remains an active are of research.

Perhaps the greatest legacy of the COMPTEL experiment is an l-sky map, which illustrates the power of imaging in a narrow band of the radioactive ^{26}Al (Figure 2).

Figure 2. COMPTEL all-sky map of radioactive emission of from 26-aluminum. The inset shows a corresponding gamma-ray spectrum, illustrating the strength of the spectral line. The level of spatial detail, and the association of intensity enhancements with specific regions make this compilation one of *Compton's* great legacies.

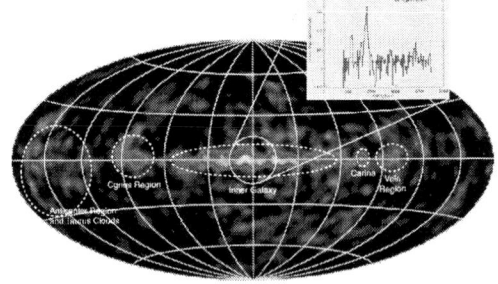

This map reveals unexpectedly high concentrations of this particular isotope in small regions. Supernova or massive, evolved stars are the likely sources of the 26-Al. The spatial and intensity distribution of this radioactive emission has major implications for the chemical evolution of our Galaxy, as the 26-Al isotope has a decay-half lifetime of about 1 million years. The wide longitudinal extent and clumpiness of the emission is seen much more clearly in maps utilizing the complete dataset allowing one to discriminate between different source distribution [9]. It was found that models of the galactic molecular gas distribution and sharply peaked nova distributions provide a relatively poor match to the data (although note reference [10]). Spiral structure in the 26-Al source distribution is also revealed.

OSSE has produced revised maps of 511 keV positron annihilation line and positronium continuum radiation from the Galactic center region. The positrons can result from the decay of radioactive elements produced in explosive objects (supernovae and novae), in energetic processes associated neutron stars and black holes, and from interactions of cosmic rays with the interstellar medium. Differential observations between "source" and "background" measurements using each of the

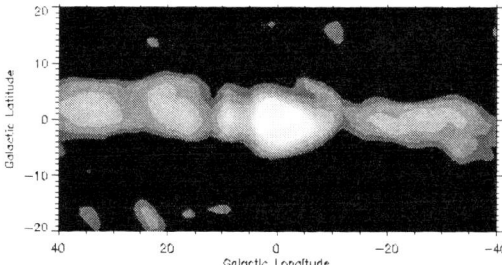

Figure 3. OSSE Ps continuum map of the Galactic center region. The emission is clearly extended, however the evidence for a north-south latitude asymmetry is now considered marginal. Work towards improving this map is ongoing with emphasis on using BATSE to better characterize possible point-

four OSSE detectors are used to map the region. Maps generated using several distinct mapping algorithms represent the most comprehensive gamma ray picture of the Galactic Center region to date [11]. The maps utilize over 20 million seconds of OSSE observing time and draw on previous experiments for comparison. The emission is dominated by an intense, diffuse bulge component, emission along the Galactic plane, and with marginal evidence for excess emission at positive Galactic latitudes. No variability of the 511 keV emission has been observed by OSSE.

BATSE's primary objective was the study of the mysterious phenomenon of gamma-ray bursts BATSE's all sky map of burst positions showed shortly after launch that, unlike Galactic objects which cluster near the plane or center of the Galaxy the bursts are distributed isotropically on the sky (Figure 4). It is now known that the gamma-ray bursts originate from cosmological distances, apparently in star-forming regions of young Galaxies. Work supporting this scenario was presented in this conference [12,13]. Additional work presented here suggests that a distinct subclass of "dim" GRBs can be identified, which can lead to meaningful subsampling of the data and thus to refinement of statistical studies[14]. Aside from its primary goal of studying GRBs, BATSE has contributed significantly to the study of Galactic X-ray binaries, notably the long-baseline characterization of accretion-driven pulsar spin histories and their interpretation [15].

GUEST INVESTIGATOR PROGRAM

In all areas of *Compton* research described here, a rigorous Guest Investigator Program contributed significantly in the observation planning, data analysis, correlative observations, and theoretical interpretations. A total of over 750 scientists from 128 institutions in 23 countries were involved since launch. In considering these numbers, one should bear in mind that prior to *Compton*, the international gamma-ray astrophysics community consisted of perhaps 100-200 indivduals. Clearly, one of *Comtons's* greatest legacies was the infusion of gamma-ray astrophysics into the astronomical mainstream. All observing time on *Compton*, during its final 6 years of operation, was awarded competitively through scientific peer review, during successive annual cycles. Cycle 9 was about one third complete at the time *Compton* was deorbited.

PUBLIC DATA ARCHIVE

All data from the *Compton* mission are available through a public archive managed by the Goddard Space Flight Center. Data were placed in the archive following delivery by the instrument teams and the expiration of their proprietary period (all proprietary data rights have now expired). The data are easily accessible via computer networks using the web-based HEASARC W3Browse search and retrieval facility. Magnetic tape distribution is also available upon request in the case of extremely large data sets and a limited number of BATSE and EGRET CD ROMs are available.

Compton's RETURN TO EARTH

Once the aforementioned spacecraft problems were assessed and a decision was made to terminate the mission, a strategy of lowering *Compton*'s altitude using four deorbit burns was devised. Windows of opportunity, driven by a combination of the required vector orientation at apogee and optimal illumination of the solar arrays, were calculated to occur about ever 45 days. The first plausible window, taking in to account the necessary time for preparation and staff training, was near the end of

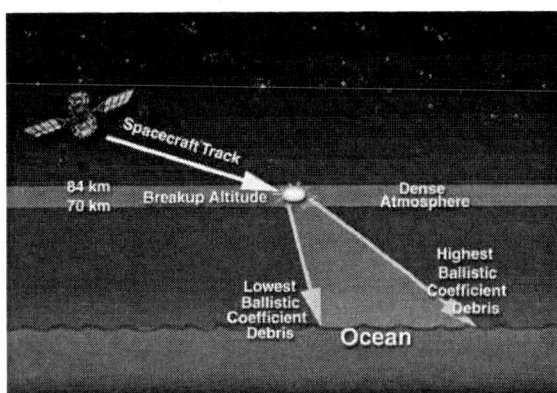

Figure 4. Schematic illustration of *Compton's* reentry path, and the "breakup" altitude at which the telemetry was lost and the spacecraft began to disintegrate. This led to an impact "footprint" of about 700 km centered about 3500-km southeast of Hawaii. The operation was carried out flawlessly, with the spacecraft and flight-operations crew performing commendably until the end.

May, 2000. On May 28, a test burn was performed, and the four science instruments were powered down. Following this, a series of 4 burns of about 30-minutes duration were initialized via a sequence of commands sent to the spacecraft by controllers on the ground. The final two descent burns occurred on June 4, 2000. As the spacecraft reentered the Earth's atmosphere, it began to tumble and heat up as about 60-70% of the mass disintegrated (Figure 5). More than six tons of metal debris, consisting of fragments ranging from a few grams to a few 10's of kilograms, survived the reentry process and reached the surface of the Earth. This debris was spread over a target area defined by a long and narrow corridor in the pacific ocean, southeast of Hawaii.

Thus ended one of NASA's most successful scientific missions. The *Compton* spacecraft and its flight control team performed admirably right up to the end. Nonetheless, it was a bittersweet moment for the many scientists who participated in the mission, and had hoped for further discoveries. As "Murphy's Laws" would dictate, approximately one week subsequent to reentry, a major solar flare occurred.

Looking over the entire mission, it is clear that *Compton* moved the field of gamma-ray astrophysics, which had consisted of studies of a few enigmatic sources, forward into to a new era highlighted by discoveries of source classes with implications for all of astronomy.

REFERENCES

1. Gehrels, N. & Shrader, C., 2000, "The Fifth Compton Symposium.", ed. M. McConnell & J.Ryan, AIP CP-510.
2. Kurfess, J, et al., 1997, "The Fourth Compton Symposium.", ed. C. Dermer, M. Strickman & J. Kurfess, AIP CP-410.
3. Shrader, C. & Gehrels, N. 1995, PASP, **107**, 606
4. Hartman, R., et al., 1999, ApJS, **123**, 79
5. Mirabal, N. & Halpern, J, 2001, ApJ, **547**, L137
6. Genier, I, 2001, (these proceedings)
7. Caraveo, P., 2001, (these proceedings)
8. ColaFrancesco, S., 2001, (these proceedings)
9. Diehl, R., et al., 2001 (these proceedings)
10. Iyudin, A., et al, 2001 (these proceedings)
11. Milne, P., et al., 2001 {these proceedings)
12. Fruchter, A., 2001, (these proceedings)
13. Rhoads, J,m 2001, (these proceedings)
14. Norris, J., et al., 2001, {these proceedings)
15. Wilson-Hodge, C., et al., 2001, (these proceedings)

Galactic Center

Comparative Studies of Line and Continuum Positron Annihilation Radiation

P.A. Milne [1], J.D. Kurfess [2], R.L. Kinzer [2] and M.D. Leising [3]

[1] *NRC/NRL Resident Research Associate, Naval Research Lab,
Code 7650, Washington DC 20375*
[2] *Naval Research Lab, Code 7650, Washington DC 20375*
[3] *Clemson University, Clemson, SC 29631*

Abstract. Positron annihilation radiation from the Galaxy has been observed by the OSSE, SMM and TGRS instruments. Improved spectral modeling of OSSE observations has allowed studies of the distribution of both positron annihilation radiation components, the narrow line emission at 511 keV and the positronium continuum emission. The results derived for each individual annihilation component are then compared with each other. These comparisons reveal approximate agreement between the distribution of these two emissions. In certain regions of the sky (notably in the vicinity of the previously reported positive latitude enhancement), the distribution of the emissions differ. We discuss these differences and the methods currently being employed to understand whether the differences are physical or a systematic error in the present analysis.

I INTRODUCTION

Nine years of observations made with the Oriented Scintillation Spectrometer Experiment (OSSE) on-board NASA's COMPTON observatory (1991-2000) [1], eight years of observations made with the Gamma-Ray Spectrometer on-board the Solar Maximum Mission (SMM) (1980-1989) [2], and two years of observations made with the Transient Gamma-Ray Spectrometer (TGRS) on-board the WIND mission (1995-1997) [3] have been utilized to study the galactic distribution of positron annihilation radiation. The OSSE instrument featured a 3.8° x 11.4° FWHM FoV, a $\sim$3 x 10^{-5} photons cm^{-2} s^{-1} line sensitivity (10^6 s on-source time), and a 45 keV energy resolution at 511 keV. These detector attributes have permitted the first detailed studies of the distribution of annihilation radiation in the inner radian of the Galaxy. The annihilation of positrons with electrons gives rise to two spectral features, a line emission at 511 keV and a positronium continuum emission (which increases in intensity with energy roughly as a power law up to 511 keV and falls abruptly to zero above 511 keV) [4]. The TGRS instrument, which featured a germanium detector with excellent energy resolution, has demonstrated that the integrated flux from the inner radian is best described as a narrow 511 keV line (FWHM ≤ 1.8 keV) and a positronium continuum to 511 keV line ratio of ∼ 3.6 (which corresponds to a positronium fraction of f_{Ps}=0.94) [3].

Purcell et al. (1997) (hereafter PURC97) [6] reported results from OSSE/SMM/TGRS studies of the 511 keV line component of annihilation radiation. They found the 511 keV emission to be comprised of three components; 1) an intense bulge emission, 2) a fainter disk emission, and 3) an enhancement of emission at positive latitudes (hereafter called a PLE). The PLE was also reported by Cheng et al. (1997) [7], and has been interpreted to be an "annihilation fountain" by Dermer & Skibo [8]. PURC97 characterized the emission via mapping, employing the SVD matrix inversion algorithm, and via model fitting, testing the combination of a spheroidal Gaussian bulge, a disk that is flat in longitude to ±40° and Gaussian in latitude (FWHM = 9°), and a spheroidal PLE. The two characterizations differ in the thickness of the Gaussian disk (SVD being narrower) and the extension of the PLE. The enhancement of the PLE differed between the two characterizations, varying from 1.5 x 10^{-4} photons cm^{-2} s^{-1} for the SVD map to 9 x 10^{-4} photons cm^{-2} s^{-1} for the broad 2D Gaussian PLE (FWHM = 16.4°). A parallel study of both line and continuum annihilation radiation along the galactic plane by Kinzer et al. (1996,2001) [9,10] reported that positronium continuum emission is similarly distributed as 511 keV line emission. The Kinzer studies did not investigate the PLE.

We report here updates from our continuing analysis which extends the study of PURC97 (see also Milne et al. (1998,1999) [11,12]). The primary differences between current studies and PURC97 are; 1) the inclusion of more observations, both archival and data collected after PURC97, and 2) reporting maps of the positronium continuum emission in addition to the 511 keV line. To extract the positronium continuum component from the underlying galactic continuum emission, we widened the spectral modeling to include thermal bremsstrahlung and exponentially-truncated power-law models. We also removed high-energy diffuse continuum emission following a prescription from Kinzer et al. (1999), distributing the emission spectrally according to an α = -1.65 power-law and spatially according to a 90° x 5° 2D Gaussian [13]. Two maps of the 511 keV line emission and two maps of the positronium continuum emission are shown in Figure 1.

II DISCUSSION

Although not identical, the two 511 keV maps share certain fundamental features. Both exhibit an intense bulge emission and a fainter planar emission. The regions that appear anomalous in the RL map (relative to symmetrical bulge and disk emissions) are also anomalous in the SVD map. The positronium continuum maps are also dominated by an intense bulge and a fainter disk component. Pairings of bulge and disk components suggest the same families of solutions. Both suggest

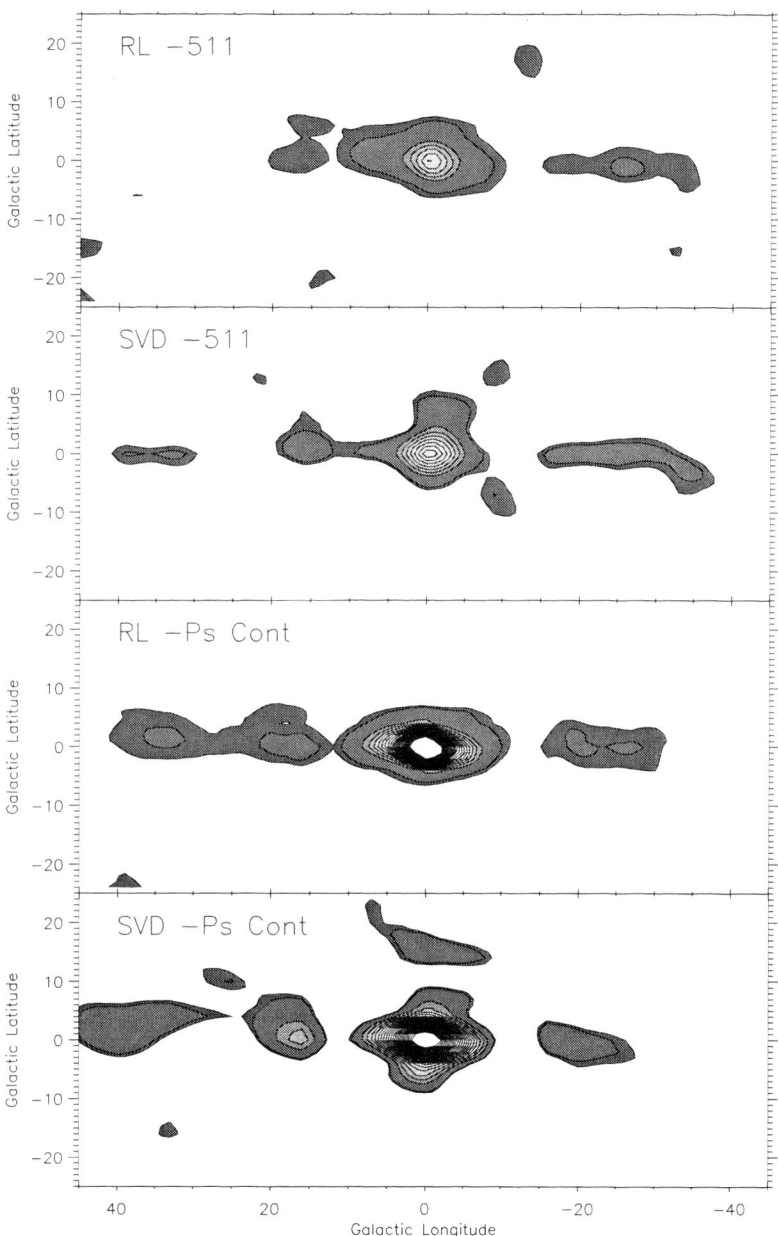

FIGURE 1. Four characterizations of positron annihilation radiation. The upper two panels are Richardson-Lucy and SVD maps of 511 keV line emission. The lower two panels are Richardson-Lucy and SVD maps of positronium continuum emission.

that the bulge-to-disk ratio can vary from 0.2 -3 depending upon whether the bulge component features a halo (which leads to a large B/D). The most noticeable difference between the four maps is how the PLE is characterized.

Both mapping algorithms suggest enhancement of 511 keV line emission from the region reported in PURC97, although at lower flux levels ($\leq 1.0 \times 10^{-4}$ photons cm^{-2} s^{-1}). The broad 2D Gaussian PLE described in PURC97 is *not* an acceptable solution; the quality of fit is much worse than solutions without a PLE. By contrast, there is no suggestion of a PLE at the location reported in PURC97 (l, b = -2°, 7°) in either of the positronium continuum maps. Shown in Figure 2 are 1° wide cuts cuts through the four maps, taken at 70° relative to the negative-longitude galactic plane. It is apparent that the four curves are consistent with the same distribution towards negative latitudes, but differ at positive latitudes. The lower panel of Figure 2 shows the positive latitude cut after subtraction of the mirror negative latitude cut. The enhancement at 511 keV has a corresponding *deficit* for positronium continuum for both mapping algorithms.

An enhancement of 511 keV emission that is due to an additional source of positrons (relative to the Galaxy-wide bulge and disk components) would be expected to feature an enhancement of both annihilation components. For the lowest f_{Ps} value of zero, we would expect the positronium continuum emission to be symmetric and the 511 keV line emission to be enhanced. A positronium continuum *deficit* at positive latitudes would not be expected. If, alternatively, the PLE is a region with no excess of positrons but is where the local f_{Ps} varies from the integrated f_{Ps}=0.94 (suggested both by OSSE analysis [10], and by wide FoV germanium detectors [3]), then the 511 keV line emission could be enhanced and the positronium continuum emission could be deficient. However, variations of the f_{Ps} it do not conserve photon flux. The 1:1 enhancement-to-deficit ratio suggested in Figure 2b would not be expected. A third possibility is that the enhancement is instead due to the influence of gamma-ray sources that corrupt the spectral fitting. Two possible mechanisms for this biasing could be from observations made while a source is exhibiting a hard X-ray flare, and/or if the source exhibits a previously undetected hard tail. The fact that both the line and positronium continuum components peak at 511 keV combined with the 45 keV FWHM energy resolution of the OSSE instrument mean that as many as 30% of the counts at that energy are due to positronium continuum photons. It is unclear whether flaring gamma-ray sources can bias the the spectral fitting to a large enough extent to entirely account for the apparent asymmetry in the annihilation radiation.

A few compact sources which produce this type of biasing have been identified. At the present time, it has not been established which of the three explanations is correct. The CGRO/BATSE instrument made observations of these sources which were nearly simultaneous with the OSSE observations [14]. It is an objective of the current analysis effort to determine whether joint analysis of OSSE & BATSE observations of these sources will permit the unambiguous extraction of annihilation radiation from this complex environment. Fortunately, although other regions of the inner galactic radian may be similarly biased, the majority of the region is not

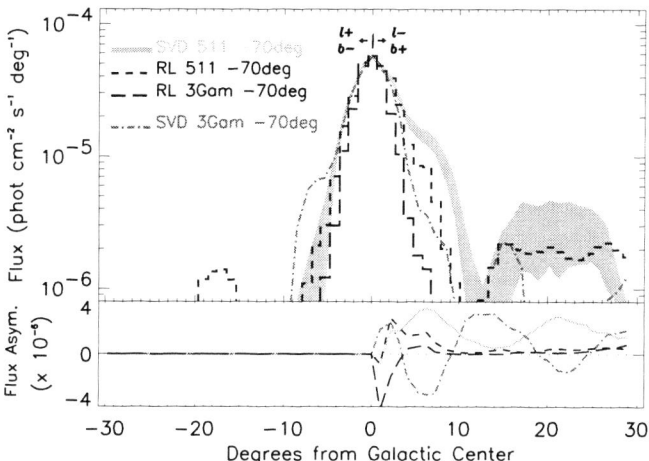

FIGURE 2. Cuts through the positive latitude enhancement taken at 70° relative to the negative-longitude galactic plane displayed on a logarithmic scale. Shown in the upper panel is a 1° wide cut through the RL and SVD maps of line and continuum annihilation radiation. The SVD -511 maps shows 1σ error bars. The lower panel shows the positive latitude portion of the cuts with the mirror negative latitude portion subtracted off.

expected to have been affected. The wealth of information available both from the complete OSSE data-set as well as from the expanding data-set of monitoring of compact sources with the BATSE instrument may permit definitive statements as to the galactic distribution of positron annihilation radiation, particularly of the existence of a PLE.

REFERENCES

1. Johnson, W.N., et al. *ApJS*, **86**, 693 (1993).
2. Share, G.H., et al., *ApJ* **326**, 717 (1988).
3. Harris, M.J., et al., *ApJ* **501**, L55 (1998).
4. Ore, A. & Powell, J.L., *Phys Rev*, **75**, 11 (1949).
5. Gehrels, N., et al. *ApJ*, **375**, L13 (1991).
6. Purcell, W.R., et al., *ApJ* **491**, 725 (1997).
7. Cheng, L.-X., et al., *ApJ* **481**, L43 (1997).
8. Dermer, C.D. & Skibo, J.G., *ApJ* **487**, L57 (1997).
9. Kinzer, R.L., et al., *A &AS* **120**, 317 (1996).
10. Kinzer, R.L., et al., *ApJ -submitted*, (2001).
11. Milne, P.A. et al., *Astro. Lett. & Comm.* **38**, 441 (1998).
12. Milne, P.A. et al., *in Proceedings of the 5th Compton Symposium*, 21 (1999).
13. Kinzer, R.L., Purcell W.R., & Kurfess, J.D., *ApJ* **515**, 215 (1999).
14. Harmon, B.A., et al. *in Proceedings of the 1st Compton Symposium*, 69 (1992).

Can INTEGRAL Detect 0.511 MeV Radiation from Giant Molecular Clouds?

N. Guessoum[1], P. von Ballmoos[2], J. Knodleseder[2], G. Vedrenne[2]

[1] American University of Sharjah, College of Arts &Sciences, Sharjah, UAE
[2] Centre d'Etude Spatiale des Rayonnements, Toulouse, France

Abstract. One of the interesting prospects raised by CGRO but never properly investigated due to the lack of instrumental sensitivity and resolution is the gamma-ray line emission from giant molecular clouds. In this paper, we study the 0.511 MeV annihilation line that is emitted when cosmic rays (CRs) bombarding a cloud produce radioisotopes that decay and produce positrons, which subsequently annihilate either directly or after the formation of Positronium. Several interesting factors come into play to decide whether the flux of this emission is strong enough to be detected by INTEGRAL, among them the intensity of the cosmic rays and their nuclear composition, as well as the mass, distance, and other characteristics of the molecular cloud. We have computed both the flux and the line width of the radiation for four nearby clouds. It is found that the line width can be very small (0.1 keV or less) in the case of a partially ionized cold core. This preliminary result, however, does not take into account Doppler shifts due to various motions of the cloud, as well as field-view effects, which may lead to a modification of the overall measured spectrum. We compare our preliminary results to INTEGRAL's capabilities and conclude with the prospects for a successful detection of such a radiation.

1. INTRODUCTION

Interest in the annihilation of positrons and electrons in giant molecular clouds arose from the (retracted) detection of nuclear gamma-ray lines from the Orion cloud by CGRO's Comptel instrument. The cosmic rays that would have excited the CNO nuclei and led to the production of nuclear rays would also have produced positrons by radioactive nuclei; the positrons would then annihilate with electrons in the cloud, on timescales that depend almost exclusively on the cloud's density: $\tau \approx 10^5$ yrs / n (cm^3).

Ramaty *et al.* (1995), when modeling the presumed nuclear emission from Orion, considered the annihilation possibility very briefly and estimated the flux from the cloud at $\approx 2 \times 10^{-5}$ γ/cm^2/s. Realizing that the special physical conditions of the cloud (very low temperatures in particular, along with the possibility of magnetic shielding of the core), Guessoum *et al.* (1997a) computed the expected spectra of the line and showed that it could very well be an extremely narrow line (≈ 0.1 keV or less).

Since there are other, bigger and closer, clouds, it was realized that the same consideration could be applied to them, first because they would result in higher fluxes and secondly because INTEGRAL's superior sensitivity and resolution give better prospects for detection and spectroscopic analysis.

2. CHARACTERISTICS OF THE GIANT MOLECULAR CLOUDS OF INTEREST

Giant molecular clouds (GMCs) are believed to be embedded in lower density H I gas of total mass $M \approx 10^4 - 10^5 \, M_\odot$. They consist of cold cores of density between 10^5 and 10^6 cm^{-3} and a temperature between 70 and 150 K, and are surrounded by H II regions, which physical parameters have been measured using optical, infrared, and radio spectroscopy; the information obtained regarding the ionized gas has converged toward the following standard values: an average electron density $n_e \approx 5000$ cm^{-3}, and an electron temperature $T_e \approx 8500$ K. Magnetic fields are also known to thread the clouds; for instance, 21-cm measurements of Orion have yielded estimates of 50 to 125 µG in the line of sight, the actual magnetic field strength being perhaps twice as high as the line-of-sight value.

The interface region between the molecular cloud and H II regions is thought to consist mainly of Hydrogen at a temperature in the range of 100-500 K, with a density of at least 2×10^5 cm^{-3} (see Genzel & Stutzki 1989 for references on the Orion cloud complex, which serves as a prototype for GMCs as we consider them here).

Finally, except for the H II region, which is obviously partially ionized – though with an uncertain fraction, the ionization state of the other two regions is unknown, as the cosmic rays (CRs) may or may not produce substantial ionization in the cold cloud cores. We will assume, for definiteness and for the sake of the analysis, that the ionization fraction in the H II region is 50%, while in the other two regions it can range between 0 and say 25%; the non-zero values then imply CR penetration and ionization of the cold regions, the zero value implies magnetic shielding of the core from the cosmic rays. This latter factor will prove important with regard to the profile of the annihilation line, because the core being extremely cold and dense the annihilation therein produces substantial, extremely narrow radiation (see below the details of the annihilation processes and their conditions).

The best survey of GMCs in the Galaxy was conducted by Dame *et al.* (1987) using CO molecular emission. A detailed position map was drawn from the results obtained, as well as a radial velocity map, and a table of interesting clouds was given with parameters. We have selected four GMCs, owing to their total mass and distance to Earth, as presented in the following table:

TABLE 1. GMCs of Interest and Parameters.

GMC	Total Cloud Mass	Distance from Earth
Taurus	$3.0 \times 10^4 \, M_\odot$	140 pc
Aquila Rift	$1.5 \times 10^5 \, M_\odot$	200 pc
Perseus OB2	$1.3 \times 10^5 \, M_\odot$	350 pc
Rho Ophiuchi	$3.0 \times 10^4 \, M_\odot$	165 pc

3. ANNIHILATION PROCESSES, FLUXES AND SPECTRA

The physical processes that a positron can undergo when it enters a cloud are known to a great extent. First the positron, which is normally emitted by a radioactive nucleus with energy of about 1 MeV, loses energy by collisions with neutral atoms (discrete excitations and ionizations) and free electrons (continuous Coulomb losses). During this process and before it reaches an energy of about 100 eV, the positron can form a Positronium atom "in flight", which then decays either into two photons of 0.511 MeV each, if it is in a singlet "para" state, or into three photons of various energies between 0 and 511 keV, if the Positronium is in a triplet "ortho" state. The singlet and triplet states have 25 and 75 % probabilities of formation, respectively. The timescale for such losses has been computed by Bussard *et al.* (1979), who find it to be $\approx 3.5 \times 10^{12} / n \, (cm^{-3})$ s.

Alternatively, the positrons can slowly thermalize with the ambient medium; in this case they can undergo a variety of processes: charge exchange with neutral Hydrogen and Helium (and other elements if they are significantly abundant), which leads to the formation of Positronium, direct annihilation with a bound electron, direct annihilation with a free electron, radiative combination (i.e. capture of a electron), leading to Positronium formation, or capture on dust grains, with subsequent direct annihilation, formation of Positronium, or escape.

The annihilation timescale can then be written as follows (in terms of the rates of annihilation for each process):

$$t_{ann} = \left(R_{ce} n_H + R_{rc} n_e + R_{daf} n_e + R_{dab} n_H + R_{gr} n_{gr} \right)^{-1} \quad (1)$$

where R_{ce}, R_{rc}, R_{daf}, R_{dab}, and $R_{gr}n_{gr}$ are the rates (σv) for annihilation by charge exchange with atomic Hydrogen, radiative combination with free electrons, direct annihilation with free electrons, direct annihilation with bound electrons, and annihilation on grains; n_H and n_e are the number densities of Hydrogen atoms and free electrons, respectively.

Each of the annihilation processes has been studied in detail, although in many cases, experimental measurements of the cross sections is impossible in practice, the process having such a low probability, and one has to rely on theoretical calculations or estimates. In some cases, reevaluations, whether experimental or theoretical, produce results significantly different from the adopted ones that a new analysis of the astrophysical problem is warranted (Guessoum *et al.* 1997b). For all practical purposes, we can state that the following theoretical treatments contain most of the information one needs to carry out any investigation of e^+e^- annihilation in an astrophysical situation: Bussard *et al.* (1979); Zurek (1985); Guessoum *et al.* (1991).

As briefly mentioned above, the line spectra resulting from the annihilation of positrons in the cold core regions depend crucially on the ionization fraction assumed. Let us explain this.

Charge exchange between positrons and Hydrogen atoms has by far the highest reaction rate of all the processes. However, this process has an energy threshold of 6.8 eV, so at low temperatures (less than about 5000 K) it becomes totally negligible. Therefore, at low temperatures (of the order of 100 K), such as in the cold cores of the giant molecular clouds, radiative combination dominates by more than an order of magnitude (see Figure 3 in Bussard *et al.* 1979), that is, if there are any free electrons, i.e. if the ionization fraction is not zero. If the medium is totally neutral, the dominant process is the direct annihilation with bound electrons. So the ionization fraction is obviously crucial, as even 1% of ionized Hydrogen can make a drastic difference.

Another effect of the ionization state of the medium is the fraction of positrons that forms Positronium in flight. Laboratory measurements (Brown, Leventhal, and Mills 1986) have shown that this fraction is about 0.9 for a totally neutral medium, but this fraction drops very rapidly with the ionization fraction: about 0.3 at 10%, 0.045 at 50%, etc. And since charge exchange in flight produces a relatively wide line (FWHM = 6.4 keV), the spectra are significantly affected by the ionization state of the medium.

Moreover, the radiative combination process leads to the formation of Positronium, which gives both a continuum spectrum below 511 keV and a line of width $\Gamma_{rc} = 1.1 \times 10^{-2} \, T^{1/2}$ keV, whereas the direct annihilation with bound electrons gives only a line of width $\Gamma_{daH} = 1.56$ keV (Brown *et al.* 1984; Brown and Leventhal 1986). Therefore, at low temperatures (≈ 100 K) the spectrum will consist of a very narrow line (≈ 0.1 keV) with a low-energy continuum if the medium is partially ionized, while it will consist of just a line – no continuum – of width 1.56 keV if the medium is totally neutral.

The dust grain processes are reasonably well understood, but their relative significance is highly uncertain, as this depends strongly on the physical conditions as well as on the grains themselves (size, electric charge, density, metallicity, etc.). For simplicity, we will take the grains to be neutral, to have normal galactic metallicity, a density of about 1 g/cm^3, and a standard size of 0.1 μm. These parameters correspond to a value of $x_{gr} = 1$ in Guessoum *et al.* (1991), where it was shown that grains make essentially no difference at all in the cold medium, whereas in the partially ionized warm medium their effect is moderate. We include them here for completeness.

Based on these parameters and the physics of the annihilation outlined above, we have determined the fluxes and obtained the spectra of the line from each cloud, based on standard low-energy CR fluxes and 3 different CR compositions ("normal", i.e. solar; "WC Wolf-Rayet", i.e. ≈ 28% H, 24% He, 20% C, 28% O; "Metal-Rich", i.e. 0% H, 25% of He, C, N, and O, each); the fluxes are presented in Table 2:

TABLE 2. Fluxes of Annihilation Radiation from GMCs (in γ/cm^2/sec).

GMC	Normal CR Composition	Wolf-Rayet CR Composition	Metal-Rich CR Composition
Taurus	2.4 x 10^{-6}	3.1 x 10^{-5}	3.5 x 10^{-5}
Aquila Rift	5.8 x 10^{-6}	7.6 x 10^{-5}	8.5 x 10^{-5}
Perseus OB2	1.6 x 10^{-6}	2.1 x 10^{-5}	2.4 x 10^{-5}
Rho Ophiuchi	1.7 x 10^{-6}	2.2 x 10^{-5}	2.5 x 10^{-5}

The results for the spectra are interesting. If the cloud core is totally neutral, the annihilation spectrum is essentially a composite of two Gaussian lines: a very wide (6.4 keV) line due to the annihilation in flight, surmounted by a narrower ($\approx$ 2 keV) one, which itself is the product of the thermal annihilation processes. If the cloud core is partially ionized, the spectrum consists of a very narrow line ($\approx$ 0.1 keV) due to the dominant direct annihilation with free electrons; for the partially ionized warm H II region, the spectrum is a medium-wide ($\approx$ 1.8 keV) line. The overall spectrum that combines the various phases of the cloud, assuming a neutral core or a partially ionized core, can be fitted by a Gaussian of widths 1.6 and 1.4 keV, respectively.

4. CONCLUSION

Can INTEGRAL's Ge-based detectors distinguish between these spectra? There is no doubt that these detectors can distinguish between two lines of width 6.4 and 1.56 keV. With high enough statistics they might even distinguish between a totally broad line and one with a broad base and a narrow tip. This means that these instruments should be able to tell us whether or not the cores of these giant clouds are ionized, that is whether the cosmic rays are penetrating them.

But can such instruments distinguish between a line of width 0.1 keV and a line of width 1.8 keV, and so can the data tell us whether the positrons are annihilating primarily inside the cores or in the warm H II regions? If the instrumental resolution of INTEGRAL's SPI instrument really performs at 1 keV (at 0.511 MeV), then the two lines of width 0.1 keV and 1.8 keV can be distinguished at almost a 2 σ statistical level. If for any reason this instrumental resolution is degraded, say to about 2 keV at 0.511 MeV, then the difference is only about 1 σ.

On the basis of these results (fluxes and spectra) we believe that observational and theoretical investigations of e^+e^- annihilation in giant clouds could yield interesting insights into these environments and open new windows on the Galaxy.

REFERENCES

1. Brown, B. L., et al., *Phys. Rev. Lett.* **53**, 2347-2350 (1984).
2. Brown, B. L., and Leventhal, M., *Phys. Rev. Lett.* **57**, 1651-1654 (1986).
3. Brown, B. L., Leventhal, M., and Mills, A. P., Jr., *Phys. Rev.* **A33**, 2281 (1986).
4. Bussard, R. W., Ramaty, R., and Drachman, R. J., *ApJ* **228**, 928-934 (1979).
5. Dame, T. M. et al., *ApJ* **322**, .706 -720 (1987)
6. Genzel, R., and Stutzki, J., *Ann. Rev. Astron. Astrophys.* **27**, 41-85 (1989).
7. Guessoum, N., Ramaty, R., and Lingenfelter, R. E., *ApJ* **378**, 170-180 (1991).
8. Guessoum, N., Skibo, J. G., and Ramaty, R., "Positron Annihilation in the Orion Cloud" edited by M. S. Potgieter et al., Proc. XXVth ICRC, Durban (South Africa), vol.3, 1997, pp.149—152.
9. Guessoum, N. et al., "Positron Annihilation Processes Update", in *The Transparent Universe*, edited by C. Winkler et al., Proc. 2nd INTEGRAL Workshop, ESA Publ. SP 382, 1997, pp. 113-118.
10. Ramaty, R., Kozlovsky, B., and Lingenfelter, R. E., *ApJ Letters* **438**, 21-24 (1995).
11. Zurek, W. H., *ApJ* **289**, 603-608 (1985)

COMPTEL Observations of a source in the direction of the Galactic centre

A.W. Strong*, W. Collmar*, K. Bennett[†], H. Bloemen**, R. Diehl*,
W. Hermsen**, A. Iyudin*, H. Mayer-Hasselwander*, J. Ryan[‡] and
V. Schönfelder*

*Max-Planck-Institut für extraterrestrische Physik, Postfach 1312, 85741 Garching, Germany
[†]Astrophysics Division, ESTEC, Noordwijk, Netherlands
**Space Research Organization of the Netherlands (SRON), Utrecht, Netherlands
[‡]Space Science Center, University of New Hampshire, Durham, NH 03824

Abstract. During the CGRO mission, the Galactic centre region was well exposed by COMPTEL. There is evidence for a distinct excess near the Galactic centre direction in the COMPTEL energy range 1 – 30 MeV. The analysis for point sources is however complicated by the intense emission from the Galactic ridge. We use data from the full mission to investigate this source, present a multiwavelength spectrum and discuss various possible counterparts, including the microquasar source 1E1740.7 - 2942.

INTRODUCTION

The Galactic centre (GC) region was well observed by COMPTEL in the 1-30 MeV range over the full mission lifetime. Maps of the continuum emission for the entire sky and the Galactic plane have been presented previously [1, 2]; these show intense diffuse emission from the Galactic ridge as well as the principal sources detected by COMPTEL such as Crab, Vela, Cyg X-1, and also apparent source-like excesses at $l = 18^o$ and the Galactic Centre direction. Because of the difficulties associated with the ridge emission, the latter sources have not been the subject of a great deal of attention up to now. All-sky source maps have been presented in [3].

ANALYSIS

The standard COMPTEL source likelihood-analysis tool SRCFIX was used [4], including estimates of the diffuse Galactic emission based on HI, CO maps and inverse-Compton models. However in view of the uncertainty in such models (much of the emission may be unresolved point-sources, see [5, 6]), the analysis is not unique. Especially the GC is problematic because of the large CO peak there which is believed to reflect abnormal conditions in the ISM rather than a large molecular hydrogen mass; hence the CO maps are usually interpolated in some way to remove this peak for use in gamma-ray analysis, and this is also done here, but this procedure is clearly somewhat

TABLE 1. Summary of fitting parameters for source near GC

Energy MeV	l,b of max	$-2ln(L/L_{max})$*	prob (3 df)	σ	Flux 10^{-5} cm^{-2} s^{-1}†
1-3	358.5, -1.5	26.8	6.5 10^{-6}	4.5	5.12 ± 1.0
3-10	359.5, -0.5	3.0	0.39	0.85	0.77 ± 0.49
10-30	359.5, -0.5	12.2	6.7 10^{-3}	2.7	0.60 ± 0.19
1-30	358.5, -0.5	37	4.6 10^{-8}	5.4	6.5 ± 1.1

* at maximum
† at $l=358.5^o$, $b=-0.5^o$

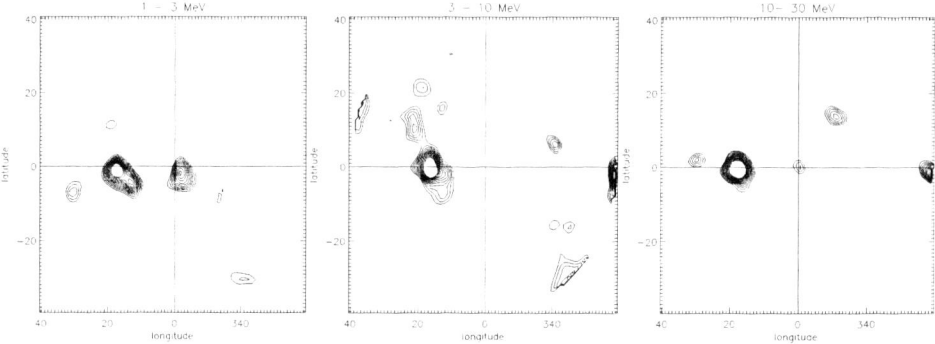

FIGURE 1. Likelihood ratio map of inner Galaxy for 1 – 3, 3 – 10 and 10 – 30 MeV. An estimate of the intense Galactic diffuse emission has been subtracted. The quantity $-2ln L/L_{max}$ is contoured at levels 8, 10, 12 ...

ad-hoc [1].

The analysis was performed for the energy ranges 1-3, 3-10 and 10-30 MeV on data from Cycles 1-6 of the mission. Table 1 summarizes the significances and fluxes. The method uses the likelihood-ratio statistic $-2ln(L/L_{max})$; since we have no *a priori* position for the source we have three degrees of freedom (flux, l, b) so that $-2ln(L/L_{max})$ is distributed as χ_3^2 and this is here used to derive probabilities/significances [2].

Fig 1 shows contours of the likelihood-ratio statistic $-2ln(L/L_{max})$ for the energy ranges 1-3, 3-10 and 10-30 MeV, and in Fig 2 summed for 1 – 30 MeV. While the source at $l = 18^o$ is the most prominent feature on these maps, a source is detected near the GC at 4.5σ for 1-3 MeV and 2.7σ for 10-30 MeV with consistent positions; for 3-10 MeV there is no significant detection at this position, although the upper limit is not incompatible with the other ranges.

---

[1] the CO peak was not removed from the model in [3] which may explain the lack of a significant residual γ-ray GC excess in those maps
[2] strictly we should allow for the trials involved in a full-sky search, but we consider the GC region as of sufficient *a priori* interest for independent reasons.

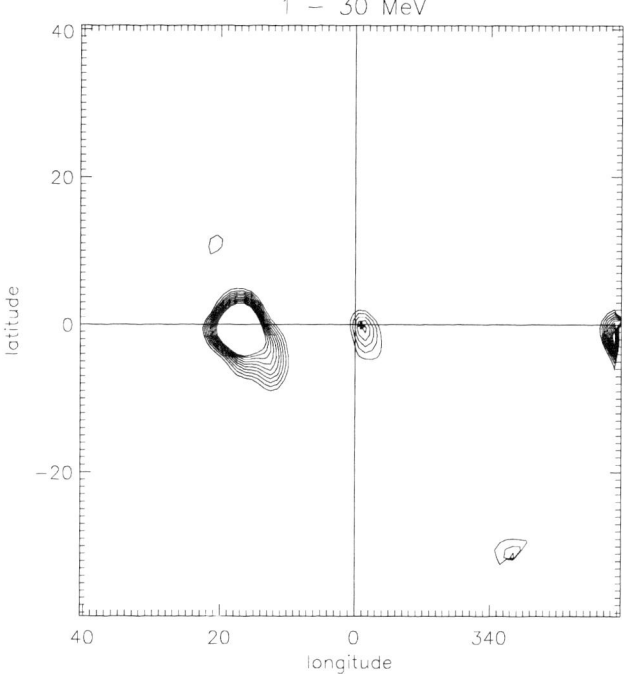

FIGURE 2. Likelihood ratio map of inner Galaxy for 1 – 30 MeV. An estimate of the intense Galactic diffuse emission has been subtracted. The quantity $-2\ln L/L_{max}$ is contoured at levels 20,24,28,32 corresponding to χ_3^2 probabilities $1.7\ 10^{-4}$, $2.5\ 10^{-5}$, $3.6\ 10^{-6}$, $5.2\ 10^{-7}$ The position of 1E1740.7-2942 is marked by '+'.

The maximum value of the total 1–30 MeV lnL-statistic near the GC is 37 corresponding to 5.4σ. While the significances in the individual energy ranges are not very high, the combined ranges add up to a significant detection in 1–30 MeV. The position of the 1–30 MeV maximum is $l = 358.5^o$, $b = -0.5^o$ with an error circle radius about 2° (change in lnL-statistic of 6 corresponding to 95% probability). For consistency the spectrum has been derived for this position.

Fig 3 shows the present COMPTEL estimates of the spectrum together with one from EGRET [7] and various hard X-ray spectra as discussed below.

DISCUSSION

The nature of the COMPTEL source in the direction of the GC is not known, but the spectrum is not inconsistent with a steep continuation of the ASCA/SIGMA/SAX spectra of the BH/microquasar source 1E1740.7-2942 [8, 9, 10, 11, 12]. This is a persistent hard X-ray source very near to the GC direction ($l = 359.15$, $b = -0.12$) and

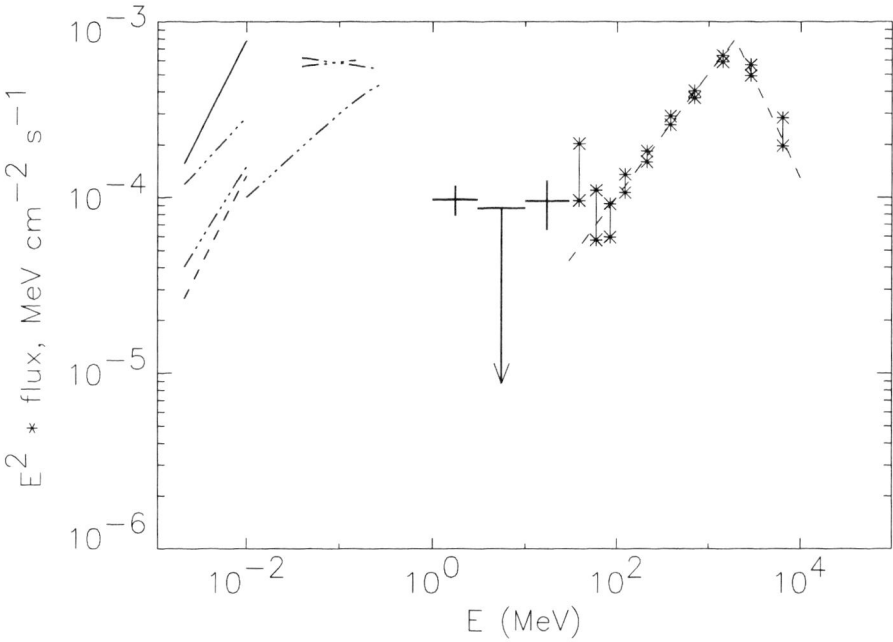

FIGURE 3. Spectrum of sources near to GC. COMPTEL points from Table 1, (3-10 MeV shown as 2σ upper limit). EGRET: [7] and Mayer-Hasselwander, private communication (asterisks). 1E1740.7-2942 (in various states): dash-dot lines, [8, 13] (SIGMA), [11] (ASCA); [10] (ASCA, SIGMA), [12](SAX). X-ray spectra of extended emission in the inner 1° are shown as full lines [15, 16] (ASCA), and dashed lines [17] (SAX).

the position is consistent with the COMPTEL excess, although an identification certainly cannot be claimed. It is the hardest X-ray source in the GC region with a spectrum similar to Cyg X-1 [13]; the latter is known to have a spectrum extending up to a few MeV [14]. The EGRET source position above 300 MeV is not compatible with 1E1740.7-2942 [7], but at lower energies it is consistent; hence there would have to be 2 EGRET sources if one were 1E1740.7-2942 : this is indeed plausible. The COMPTEL and EGRET fluxes are compatible with a continuous spectrum from 1 to 100 MeV. The non-detection in the 3 – 10 MeV band is consistent with the apparent overall minimum in the luminosity spectrum in this range.

Another possible origin of the COMPTEL source is inverse-Compton emission from the electrons in the GC radio arc, as discussed for the EGRET source in [7]. Extended X-ray emission with a hard spectrum has been reported based on ASCA data [15, 16], see also SAX observations [17], which could also be associated with the gamma-ray source.

The possible association with the 1E microquasar is of particular interest because of the proposed identification of the EGRET source at $l=18^o$ with the microquasar

LS5039 [18] and in view of the COMPTEL source at this position [19] which would imply the extension of the emission from such objects into the MeV-GeV range.

ACKNOWLEDGMENTS

The COMPTEL project is supported by the German Ministerium für Bildung und Forschung through DLR grant 50 QV 9096 8.

REFERENCES

1. Strong, A.W. et al. *Astrophys. Lett Comm.* 39, 209 (1999)
2. Bloemen, H. et al. *AIP Conf. Proc.* 510, 586 (2000)
3. Collmar, W. et al. *AIP Conf. Proc.* 510, 591 (2000)
4. Bloemen, H. et al. *ApJS* 92, 419 (1994)
5. Strong, A.W., Moskalenko, I.V., Reimer, O. *ApJ* 537, 763 (2000)
6. Strong, A.W., Moskalenko, I.V., *AIP Conf. Proc.* 510, 291 (2000)
7. Mayer-Hasselwander, H.A. et al. *A&A* 335, 161 (1998)
8. Cordier, B. et al. *A&A* 272, 277 (1993)
9. Laurent P., Paul J. *ApJS* 92, 375 (1994)
10. Vilhu, O. et al. *Proc 4th Compton Symposium, AIP* 410, 887 (1997)
11. Sakano M. et al. *ApJ* 520, 316 (1999)
12. Sidoli, L et al. *ApJ* 525, 215 (1999)
13. Kuznetsov S. et al. *MNRAS* 292, 651 (1997)
14. McConnell, M.L. et al. *ApJ* 543, 928 (2000)
15. Koyama et al. *PASJ* 48, 249 (1996)
16. Tanaka et al. *PASJ* 52, L25 (2000)
17. Sidoli, L, Mereghetti S. *A&A* 349, L49 (1999)
18. Parades J.M. et al. *Science* 288, 2340 (2000)
19. Schönfelder, V. et al. *A&AS* 143, 145 (2000)

Neutron Star X-Ray Binaries

Propagation of gamma-rays in Cen X-3

W. Bednarek

Department of Experimental Physics, University of Łódź, 90-236 Łódź, ul. Pomorska 149/153, Poland

Abstract. We consider the cascade initiated by high energy electrons or γ-rays in the radiation field of the massive star in Cen X-3. The spectra of cascade γ-rays, which escape from the system and collide with a surface of the massive star, are computed for different viewing angles. They have different shapes and intensities. We predict the γ-ray light curves at energies above 100 MeV and 300 GeV from the Cen X-3 system. Photons falling onto the massive star should excite γ-ray lines. However we show that the phase distribution of the γ-ray line and the continuum γ-ray emission is different. This feature can be used as a good test of our model by future γ-ray observations in the MeV, GeV and TeV energy ranges.

INTRODUCTION

The massive X-ray binary system Cen X-3, containing a neutron star with a 4.8 s period in a 2.09 day orbit around an O-type supergiant, has been detected above 100 MeV γ-rays by the EGRET detector on the Compton Observatory [1]. The γ-ray emission at these energies has a form of outbursts with evidences of modulation with a 4.8 s pulsar period. TeV emission from this source has been also reported. It has been localized to a relatively small region between the pulsar orbit and the surface of a massive companion which can be identified with the accretion wake or the limb of the massive star [2]. More recently the Durham group has detected a persistent flux of γ-rays above 400 GeV on a lower level than previous reports [3, 4, 5]. No evidence of correlation with the pulsar or orbital periods has been found and no evidence of correlation with the X-ray flux has been detected so far [4, 5].

It has been already shown that TeV photons, if injected close to the surface of the massive star, can meet problems with escape from the soft radiation field of the massive star in the binary system Cen X-3. The computations of the optical depth for TeV photons in the radiation fields of such type massive stars show that the absorption of TeV photons should be significant [6, 7, 8, 9]. We have performed Monte Carlo simulations of cascades initiated by relativistic electrons via Compton scattering of thermal photons from the massive companion in Cen X-3 and $e^{\pm}$ pair production in photon-photon collisions [9, 10]. The electrons have different energy and angular distributions. Schematic picture of such cascade is shown in Fig. 1. The spectra of escaping γ-ray photons and their light curves and the spectra of cascade photons which fall onto the surface of the massive star have been obtained and discussed in the context of recent GeV and TeV observations. In this paper we discuss further consequences of propagation of high energy γ-rays inside the binary system Cen X-3.

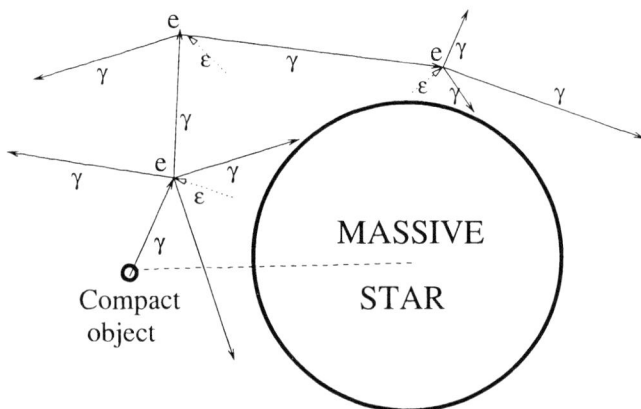

FIGURE 1. Schematic picture of the cascade initiated by primary gamma-ray photon in the radiation field of a massive star in Cen X-3 system. Gamma-ray, injected by the compact object, creates an $e^{\pm}$ pair in the interaction with a soft star photon ε. The secondary $e^{\pm}$ pairs create secondary gamma-rays by Compton scattering soft star photons. These next generation of gamma-rays can: create next $e^{\pm}$ pairs; escape from the system or collide with the stellar surface.

OPTICAL DEPTH FOR GAMMA-RAYS

Let us assume that photon with energy E_γ is injected at the distance r_γ from the center of the massive star at an angle α in respect to the direction determined by the place of the photon injection (e.g. the compact object) and the center of the massive star. We compute the optical depth for different injection parameters of the photon (E_γ, r_γ and α), applying parameters of the massive star in Cen X-3 system: the radius of the O star $r_s = 8.6 \times 10^{11}$ cm, its surface temperature $T_s = 3 \times 10^4$ K, the binary star separation $a = 1.2 \times 10^{12}$ cm [11]. Results of computations for photons with different energies as a function of the angle α are shown in Fig. 2. It is clear that photons with energies in the TeV range, if injected at the distance within a few radii from the massive star, interact with the soft stellar radiation for wide range of injection angles. Only photons injected within a few tens of degrees around the direction defined by the centers of the stars can escape without interaction. For large angles α the optical depth drops since it is calculated up to the moment of photon collision with the surface of the massive star.

GAMMA-RAY SPECTRA FROM ICS CASCADE

We assume that relativistic electrons, with the power law spectrum extending up to 10^7 MeV and spectral index equal to 2, are injected isotropicly by the compact object inside the binary system Cen X-3. Let us consider the cascade initiated in the anisotropic thermal radiation of the massive star which develops through comptonization of soft photons and absorption of secondary γ-rays (the inverse Compton $e^{\pm}$ pair cascade). If the secondary photons reach low enough energies they do not participate further in the cascade process. These photons escape from the binary system or fall onto the surface of the massive star. In Fig. 3 we show the spectra of such secondary cascade photons which escape at different range of angles α. Our simulations show that the highest fluxes of escaping γ-ray photons with different energies should be observed at different observation angles. In Fig. 4 the spectra of secondary γ-rays which fall onto the surface of the massive star are shown for these same range of angles α. The highest fluxes of secondary photons fall onto the surface of the massive star just below the position of the compact object. However significant amount of photons can also fall onto the part of the star which is unvisible from the location of the compact object. Note also that the angular distributions of photons which escape from the system and fall onto the surface of the star are different.

GAMMA-RAY LIGHT CURVES FROM CEN X-3

Since the orbit of the compact object around the massive star in Cen X-3 is almost circular and its plane lays close to direction towards the Earth, the expected γ-ray light curve should be symmetric with evidences of the eclipse. Based on our simulations we obtain the γ-ray light curves at photon energies above 100 MeV and 300 GeV for the case of isotropic injection of electrons with the power law spectrum. The results are shown in Fig. 5. It is interesting that the γ-ray light curves above 100 MeV and 300 GeV show clear anticorrelation. Even during the eclipse of the compact object by the massive star we predict some fluxes of secondary cascade γ-ray photons in the GeV energy range. This is the consequence of the type of cascade considered in this paper. However in the TeV energy range the photon flux drops to zero in the moment of the total eclipse of the compact object by the massive star.

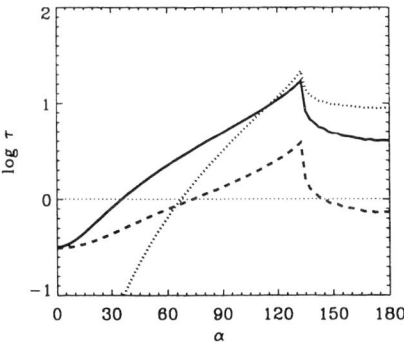

FIGURE 2. The optical depth for γ-ray photons in the soft radiation field of the massive star in Cen X-3 system is shown as a function of the angle of injection of photons α, for different photon energies $E_\gamma = 10^5$ MeV (dotted curve), 10^6 MeV (full), and 10^7 MeV (dashed). Photons are injected at the distance of 1.4 radii from the massive star with the parameters of Cen X-3.

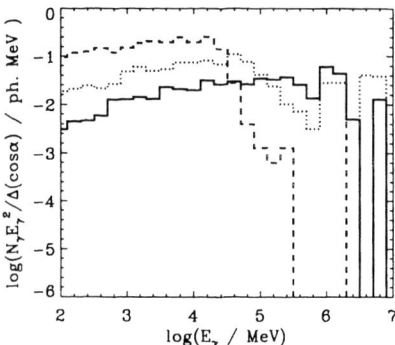

FIGURE 3. The secondary γ-ray spectra which escape from the binary system Cen X-3 inside the range of cosine angles α, measured from the direction defined by the stars, equal to $\cos\alpha = 0.9 \leftrightarrow 1$. (full histogram), $0.5 \leftrightarrow 0.6$ (dashed), and $-0.5 \leftrightarrow -0.4$ (dotted). The primary electrons with a power law spectrum and spectral index -2. are injected by the compact object.

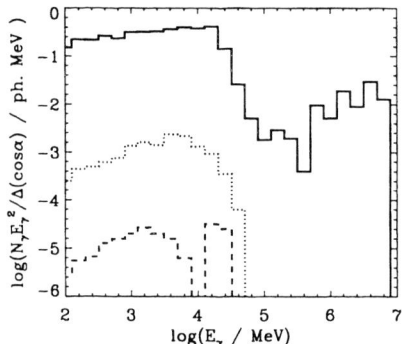

FIGURE 4. The secondary γ-ray spectra which fall onto the surface of the massive star in the binary system Cen X-3 inside the cosine angles α equal to $\cos\alpha = 0.9 \leftrightarrow 1$. (full histogram), $0.5 \leftrightarrow 0.6$ (dashed), and $-0.5 \leftrightarrow -0.4$ (dotted). The primary electrons with a power law spectrum and spectral index -2. are injected by the compact object.

CONCLUSION

We compute the γ-ray light curves and γ-ray spectra at different phases of Cen X-3 binary system assuming that electrons or photons are injected by the compact object into the radiation field of the massive star in Cen X-3. These primary particles interact with the soft radiation of the massive companion initiating inverse Compton $e^{\pm}$ pair cascade. It is found that the γ-ray light curves above 100 MeV and 300 GeV show clear anticorrelation in the case of injection of primary electrons by the compact object in Cen X-3 binary system. The γ-ray spectra depend strongly on the orbital phase of the system. Since

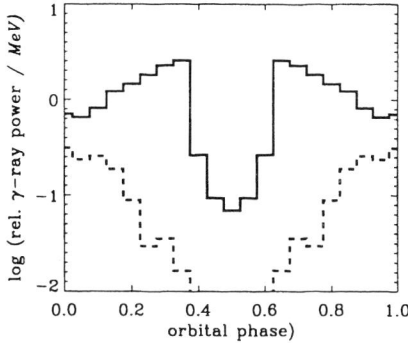

FIGURE 5. The γ-ray light curves observed at the inclination angle of the orbital plane in Cen X-3 (70°) at energies above 100 MeV (full histogram) and 300 GeV (dashed) in the case of isotropic injection of electrons with the power law spectrum and spectral index -2.

escaping photons are produced in the cascade process, we predict that the modulation of γ-ray emission with the pulsar period should not be observed. Significant amount of secondary cascade γ-ray photons fall onto the surface of the massive star. These photons induce the $e^{\pm}$ annihilation line and γ-ray nuclear lines which can be detected by the INTEGRAL Observatory [10].

ACKNOWLEDGMENTS

This work is supported by the Polish KBN grant 2P03C 006 18

REFERENCES

1. Vestrand, W.T., Sreekumar, P., Mori, M. *ApJ* **483**, L49-52 (1997).
2. Raubenheimer, B.C., Smit, H.J. *Astropart.Phys.* **7**, 63-71 (1997).
3. Chadwick, P.M. et al. *ApJ* **503**, 391-395 (1998).
4. Chadwick, P.M. et al., "TeV gamma ray emission from Cen X-3" in *Proc. 26th ICRC*, eds. B.L. Dingus, D.B. Kieda, M.H. Salamon, Salt Lake City, 1999, v. 4 pp. 72-75.
5. Chadwick, P.M. et al. *A&A* **364**, 165-169 (2000).
6. Protheroe, R.J., Stanev, T. *ApJ* **322**, 838-841 (1987).
7. Moskalenko, I.V., Karakuła, S., Tkaczyk, W. *MNRAS* **260**, 681-685 (1993).
8. Bednarek, W. *A&A* **322**, 523-532 (1997).
9. Bednarek, W. *A&A* **362**, 646-654 (2000).
10. Bednarek, W., "Propagation of gamma-rays in massive binary Cen X-3: Interaction of cascade gamma-rays with the massive star" in *Proc. INTEGRAL Workshop*, ed. V. Reglero, Alicante, 2001, in press
11. Krzemiński, W. *ApJ* **192**, L135-138 (1974).

Discovery of Weak EXO 2030+375 Outbursts with BATSE

Colleen A. Wilson*, Mark H. Finger*, M.J. Coe[†] and Silas Laycock[†]

*Space Science Research Center, SD 50, NSSTC, 320 Sparkman Dr., Huntsville, AL 35805
[†]Dept. of Physics and Astronomy, The University, Southampton, SO17 1BJ, England

Abstract. EXO 2030+375 is a 42-second X-ray pulsar orbiting a Be star every 46 days. Previous work using epoch-folded frequency searches over 1-day intervals of BATSE data [1] indicated that EXO 2030+375 underwent a 2.5 year period of quiescence from 1993 September-1996 March. Improvements in the search method that reduced systematic errors due to (1) aperiodic noise from the nearby black hole candidate Cyg X-1 and (2) sources undergoing Earth occultation, have allowed longer time intervals to be searched, hence increasing BATSE's sensitivity. Using the improved method with 4-day intervals, we detect EXO 2030+375 near most of its periastron passages during 1993 September-1996 March and for most periastron passages during BATSE's 9 years in orbit. Earth occultation measurements in the 20-100 keV band, selected when Cyg X-1 was below the Earth's horizon and epoch-folded at EXO 2030+375's orbital period of 46 days, also indicate that EXO 2030+375 was active for most of the mission. We will present histories of EXO 2030+375's pulse frequency and pulsed flux. In addition, we will present results of a pulse timing analysis and evidence that EXO 2030+375's outbursts shift in orbital phase.

INTRODUCTION

EXO 2030+375 was discovered in 1985 with EXOSAT during a bright 80 day outburst [9]. The companion was later identified as a B0 Ve star [4]. The companion star is a Be star, a main sequence star with strong Balmer emission lines. These lines indicate a circumstellar disk of material around the Be star. X-ray outbursts occur when the neutron star interacts with this disk. Be/X-ray binaries typically show two types of outbursts: (i) giant outbursts, characterized by high luminosities and high spin-up rates (e.g., the discovery outburst of EXO 2030+375), and (ii) normal outbursts, characterized by lower luminosities, low spin-up rates (if any), and modulation at the orbital period [12, 1].

From 1991-2000, EXO 2030+375 was monitored with the Burst and Transient Source Experiment (BATSE) on the Compton Gamma Ray observatory (CGRO) [6], using epoch folded pulse frequency searches over 1-day intervals of 20-50 keV BATSE data and using the Earth occultation method. This monitoring revealed a series of 13 consecutive outbursts spaced at approximately 46 day intervals. A few detections of marginal significance preceeded and followed this series. This series of outbursts was used to determine the orbital parameters for EXO 2030+375 [13]. Following these outbursts, the source appeared to be quiescent for 2.5 years before reappearing in BATSE and in the Rossi X-ray Timing Explorer (RXTE) in April 1996 [1, 11].

FREQUENCY SEARCH AND ORBITAL DETERMINATION

In previous studies with BATSE, histories of pulse frequency and pulsed flux for known pulsars were generated using grid searches over a range of candidate frequencies. The best fit frequency was determined using the Z_n^2 statistic [2]. These studies were often limited to 1-day intervals by systematic effects. We have developed an "advanced" pulsar monitor that reduces 3 systematic effects: (1) aperiodic noise from sources in the BATSE field of view, (2) Earth occultations of bright sources during the folding interval, and (3) bright pulses from other pulsars in the BATSE field of view. Aperiodic noise from other sources, primarily Cyg X-1, raises the average power of the data and causes the Z_n^2 statistic and BATSE's sensitivity to be dependent on the noise level if the Poisson level is assumed. Earth occultation steps from bright sources can produce spurious signals at high harmonics of the spacecraft orbital period which can overwhelm the real signal. This effect was greatly reduced by maintaining a database that included sources active on a given day and what level they were active. Depending upon the pulse period of the source, i.e., which spacecraft harmonic it was closest to, we determined which occultation steps should be fitted with the background model. If a harmonic of the pulse frequency of the source of interest happens to be close in frequency to a harmonic of a bright pulsar, e.g. Vela X-1, the bright pulsar can produce spurious signals. This effect can be easily addressed by removing the contaminated harmonic from the test statistic. Reducing these effects has allowed us to combine much longer intervals of data, e.g. 4 days, with near statistical errors.

Using our advanced pulsar monitor, new EXO 2030+375 pulse frequencies and pulsed fluxes were determined from a grid search in pulse frequency using 4-day intervals of BATSE DISCLA 20-50 keV data. Each 4-day interval was split into 300 second segments. The data in each segment were fitted with an empirical background model, which included fits to Earth occultation steps from sources indicated in our database as bright, and a 3 harmonic Fourier expansion in the pulse phase model. Arrival times were corrected to the solar system barycenter and using the orbital parameters from [13]. After this preliminary fit, the harmonic amplitudes are re-fitted for a grid of frequency points. Aperiodic noise from Cyg X-1, caused the variances on the Fourier coefficients to be larger than for Poisson statistics. Corrections to the Poisson variances were estimated by multiplying the variances on the mean Fourier coefficients for each 4-day interval by the reduced χ^2 of a fit to the mean coefficients for each harmonic to those from the 300-s segments within each 4-day interval. The best fit frequency was determined using a modified Z_n^2 statistic which incorporated the corrected variances [5].

Our initial search revealed many more outbursts than had been previously detected, including several during the 2.5 year "quiescent" interval 1993 August - 1996 April (MJD 9200-50175) [1]. However, there was considerable unexpected scatter in the pulse frequency measurements, starting during the so-called "quiescent" interval. These outbursts were occurring at an earlier orbital phase than those used by [13] to determine the orbit. One outburst in 1996 July (MJD 50265-50275) was also observed with the RXTE PCA [10]. We generated pulse phase measurements for each RXTE PCA observation by epoch-folding barycentered Standard 1 data using a pulse phase model consisting of a constant pulse frequency estimated from the BATSE measurements and the orbital parameters of [13]. Phase offsets to this model were generated using a template

TABLE 1. Best Fit Orbital Parameters for EXO 2030+375

Parameter	Value
Orbital Period P_{orb}	46.026 ± 0.003 days
Periastron Epoch τ_p	JD2450133.54 $\pm$ 0.02
Semi-Major Axis $a_x \sin i$	242 ± 4 light-seconds
Eccentricity e	0.414 ± 0.003
Periapse Angle ω	$212.0° \pm 0.6°$
Mass Function $f(M)$	7.0 ± 0.3 M$_\odot$
χ^2/dof	174.9/147

pulse profile selected from the brightest PCA observation. Pulse phase measurements at 1-day intervals were also generated using BATSE data for the 13 outbursts used by [13]. Data from each outburst were epoch-folded using an empirical background model (described earlier) and a 3 harmonic Fourier expansion using a pulse phase model consisting of a constant pulse frequency for each outburst. Phase offsets to this model were generated by cross-correlating individual pulse profiles with a template profile selected from the brightest 4-day interval. Pulse phase measurements from BATSE and RXTE were fitted with a global orbit plus a different quadratic for each outburst using the Levenberg-Marquardt method for χ^2 minimization. Although this model does not fully describe the intrinsic pulse frequency variations, the different orbital phase coverage of the RXTE PCA observations allow us to better decouple the orbital effects from the intrinsic torque effects. Our best fit orbit is given in Table 1. Using this new orbit, we re-did the frequency search and found that the new orbit removed the unexpected scatter in the pulse frequencies. The resulting pulse frequency history from BATSE data is shown in Figure 1c. The 4-day average 20-50 keV rms pulse fluxes generated assuming an exponential energy spectrum with $kT = 15$ keV are shown in Figure 1d for intervals when EXO 2030+375 is detected. Using the new monitor, we now detect 52 outbursts of EXO 2030+375 out of a total of 72 periastron passages during the BATSE mission.

RESULTS AND CONCLUSIONS

Figure 1a shows the K-band magnitudes measured for EXO 2030+375's companion. The diamonds indicate values from Table 1 in [11] and the squares indicate values from the Southampton-Valencia database [3]. The K-band magnitude is a good indicator of the size of the circumstellar disk around the Be star. Prior to MJD 48500 the K-band magnitude remained stable at a value of about 9.8 [11]. From MJD 48700 to 50300, the K-band magnitude increased, indicating that the disk was shrinking. After MJD 50300, the disk appeared to begin to grow again, but it still remains smaller than it was prior to MJD 48700. Figure 1b shows the 20-50 keV pulsed flux measured with BATSE at 4-day intervals. No upper limits are shown. The pulsed flux declined sharply after MJD 9200, suggesting that the reservoir of material (i.e. the Be disk) had declined to the point where the mass accretion rate onto the pulsar dropped significantly. Figure 1c shows the barycentered spin-frequency history of EXO 2030+375. Early in the

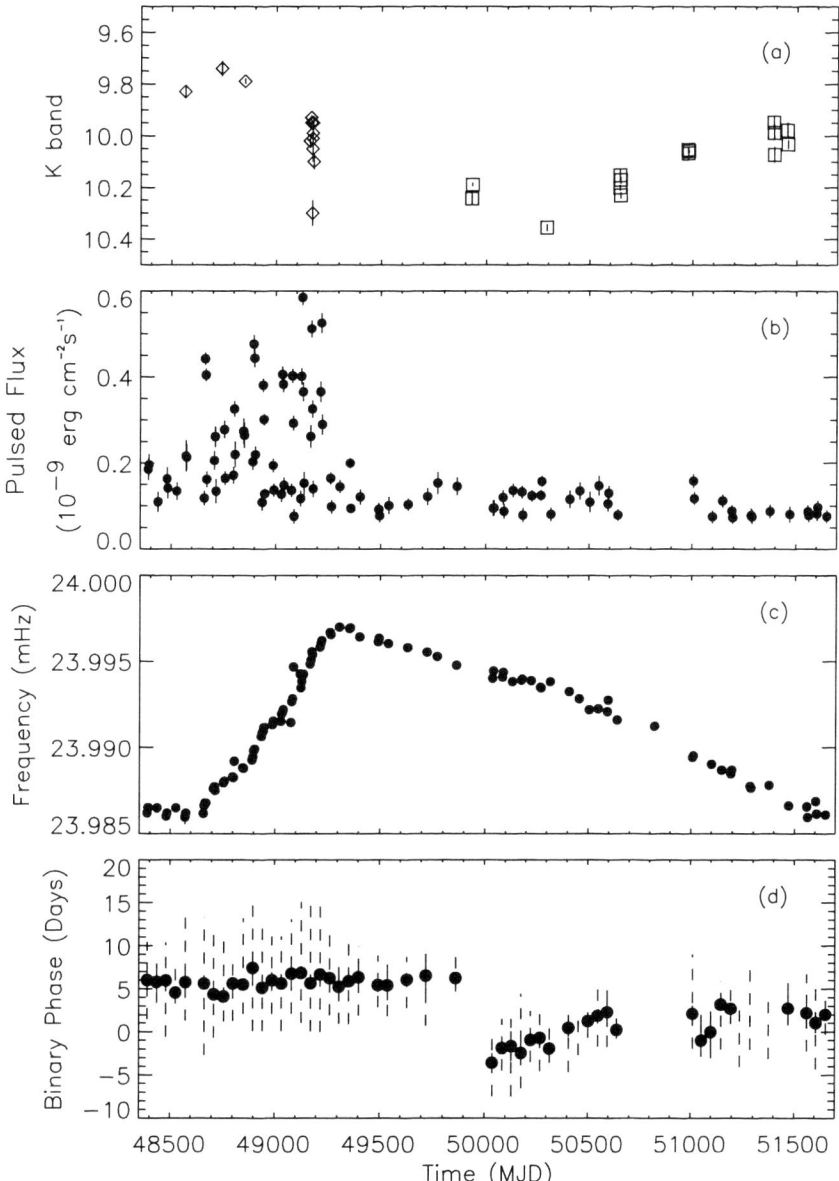

FIGURE 1. (a) K-band magnitude; (b) 4-day average 20-50 keV pulsed flux within intervals of detection; (c) 4-day average barycentered spin frequency; (d) Orbital phase of outburst peaks in days past periastron.

BATSE mission, MJD 48360-48600, the spin frequency measurements are consistent with a constant global trend. From MJD 48600-49300, when the outbursts were the brightest, the average spin-up rate is 1.9×10^{-13} Hz s^{-1}. Shortly after the pulsed flux drops dramatically, the global trend in the pulse frequencies changes to spin-down, at an average rate of -5.3×10^{-14} Hz s^{-1}. This spin-down is consistent with that expected if the pulsar enters centrifugal inhibition of accretion between outbursts, i.e. the propeller effect. This is also consistent with the drop in mass accretion rate indicated by the pulsed fluxes (Figure 1b).

Figure 1d shows the orbital phase of the outbursts versus time. Dashed lines indicate the intervals of orbital phase where the outbursts were detected with BATSE. Filled circles indicate the time of outburst peaks determined by fitting a Gaussian to the 1-day pulsed flux measurements. No filled circle is shown where the error on the Gaussian centroid was larger than half the outburst duration. From MJD 48360-49900, the outbursts occurred at a stable orbital phase, peaking about 6 days after periastron passage. A few outbursts were not detected following this interval, but when the outbursts were again detected after MJD 50000, they peaked about 4 days before periastron passage. The orbital phase of these outbursts slowly recovered to a different stable orbital phase of about 2.5 days after periastron.

A possible explanation of the shift in orbital phase of the outbursts is a density perturbation (global one armed oscillation) in the Be disk. Evidence for these density perturbations is seen in the Hα line profiles for several Be/X-ray binaries (e.g. [7, 8]). When a density perturbation is present, the Hα line is double peaked. The relative sizes of the two peaks changes with a cycle of several years. The density perturbation would produce a non-axially symmetric Be disk. If the pulsar interacts with a region of the disk affected by the perturbation, more material would be available for accretion, possibly causing a shift in outburst phase. From the trend in the orbital phases of the outburst peaks in Figure 1d, we estimate a period of about 12 years for the density pertubation to propogate around the Be disk. No significant change in X-ray intensity was seen when the outbursts shifted in orbital phase.

REFERENCES

1. Bildsten, L. et al. 1997, ApJS, 113, 367
2. Buccheri, R. et al. 1983, A&A, 128, 245
3. Coe, M.J. 2001, private communication
4. Coe, M.J. et al. 1988, MNRAS, 232, 865
5. Finger, M.H. et al. 1999, ApJ, 517, 449
6. Fishman, G.J. et al. 1989, in Proc. of the GRO Science Workshop, ed. W.N. Johnson (Greenbelt:NASA/GSFC), 2
7. Negueruela, I. et al. 2001, A&A, 369, 108
8. Negueruela, I. & Okazaki, A.T. 2001, A&A, 369, 117
9. Parmar, A.N. et al. 1989, ApJ, 338, 359
10. Reig, P. & Coe, M.J. 1998, MNRAS, 294,118
11. Reig, P. et al. 1998, MNRAS, 301, 42
12. Stella, L., White, N.E., & Rosner, R. 1986, ApJ, 308, 669
13. Stollberg, M. et al. 1999, ApJ, 512, 313

Monte-Carlo / Fokker-Planck Simulations of Accretion onto Magnetized Neutron Stars

Markus Böttcher[*,1] and Edison P. Liang[*]

[*]*Physics and Astronomy Department, Rice University, Houston, TX*

Abstract. We discuss the results of coupled Monte-Carlo/Fokker-Planck simulations of the thermal/nonthermal radiation and electron acceleration and cooling in the case of accretion onto a magnetized neutron star. Most of the energy release into thermal/nonthermal electrons is assumed to happen in a thin shell near the Alfvén radius, where the radiation from the neutron star surface is reprocessed into thermal/nonthermal high-energy (hard X-ray — γ-ray) emission. We explore the parameter space defined by the accretion rate, stellar surface field and the level of wave turbulence in the shell. Our results are relevant to the emission from atoll sources, transient X-ray binaries containing weakly magnetized neutron stars, and to recently suggested models of accretion-powered emission from anomalous X-ray pulsars.

INTRODUCTION

Recent observations of weak-field neutron star binaries, such as low-luminosity, X-ray bursters (e.g., Zhang et al. 1996, Barret et al. 1996), bursting soft X-ray transients (e.g., Harmon et al. 1996), or pulsar binary systems (e.g., Tavani et al. 1996) indicate that many of them exhibit soft (photon index $\gtrsim 2$) power law tails extending beyond ~ 100 keV, at least episodically, in addition to the thermal component at temperatures of $\sim$ a few keV, which presumably originates from the stellar surface. The luminosity of this high-energy tail appears to be anti-correlated with the soft X-ray luminosity (Barret & Vedrenne 1994, Tavani & Liang 1996).

The origin of this high energy tail is unexplained at present. It could be due to thermal Comptonization by a hot coronal plasma, or it could be due to nonthermal emission. Tavani & Liang (1996) examined systematically the possible sites of nonthermal emissions and concluded that the Alfvén surface is the most likely candidate since the dissipation of the rotation energy of the disk is strongest there, due to magnetic reconnection and wave turbulence generation. Here we focus on particle acceleration by wave turbulence. We assume that the leptons are energized by Coulomb

---

[1]) Chandra Fellow

collisions with virial ions and accelerated nonthermally by Alfvén and whistler wave turbulence, and cooled by cyclotron/synchrotron, bremsstrahlung, and inverse Comptonization of both internal soft photons and blackbody photons from the stellar surface. We use our coupled Monte-Carlo/Fokker-Planck code (Böttcher & Liang 2001) to solve self-consistently for the resulting thermal/nonthermal equilibrium electron distribution and radiation transport.

The primary focus of the parameter study presented in this paper is the application to weakly magnetized neutron stars with surface magnetic fields of $B_{\rm surf} \lesssim 10^{11}$ G. In this context, however, it is interesting to note that Chatterjee, Hernquist, & Narayan (2000; see also Mereghetti & Stella 1995, Wang 1997) have recently proposed a similar type of accretion-powered emission for anomalous X-ray pulsars (AXPs), as an alternative to models based on magnetic-field decay (Thompson & Duncan 1996) or residual thermal energy (Heyl & Hernquist 1997). According to Chatterjee et al. (2000) the X-ray emission from AXPs (which generally consists of a soft, thermal component with $kT \sim 0.3 - 0.4$ keV plus a hard X-ray tail with photon index $\Gamma \sim 3 - 4$) is powered by accretion of material from the debris of the supernova which had formed the neutron star, onto the surface of the neutron star, which possesses a typical pulsar magnetic field of $B_{\rm surf} \sim 10^{12}$ G. Therefore, we extend our parameter study to parameter values relevant to accreting pulsars. However, we point out that in the case of a magnetic field as high as $B_{\rm surf} \gtrsim 10^{12}$ G, the assumed shell geometry and the quasi-isotropy of the emission from the neutron star surface may be a gross over-simplification. However, although consequently the precise parameter values used in this region of the parameter space should not be taken at face value, our parameter study might still provide interesting insight into the dependence of the equilibrium electron and photon spectra on the various input parameters in the high-magnetic-field case.

MODEL SETUP

In both weakly magnetized neutron stars (atoll sources and soft X-ray transient neutron star binary systems) with $B_{\rm surf} \lesssim 10^{11}$ G and X-ray pulsars (including, possibly, anomalous X-ray pulsars) with $B_{\rm surf} \sim 10^{12}$ G, energy dissipation will be most efficient at the Alfvén radius, where the optically thick, geometrically thin outer accretion disk is disrupted and the dynamics of the accretion flow becomes dominated by the magnetic field. We idealize this region of efficient energy dissipation at the Alfvén radius of disk accretion onto a magnetized neutron star as part of a spherical shell whose distance r_0, magnetic field, and column thickness are fixed by the accretion rate (Ghosh & Lamb 1979a,b). The distance r_0 is given by

$$r_0 = 2 \times 10^8 \, f \, \mu_{30}^{4/7} \, l_*^{-2/7} \, M_*^{-1/7} \, R_6^{-2/7} \; {\rm cm} \qquad (1)$$

where $\mu_{30} = $ (neutron star magnetic moment)$/(10^{30}\,{\rm G\,cm}^3)$, $l_* = L/L_{\rm Edd}$, $M_* = M_{\rm NS}/M_\odot$, and $R_6 = R_{\rm NS}/(10^6\,{\rm cm})$. For the current simulations, for definiteness,

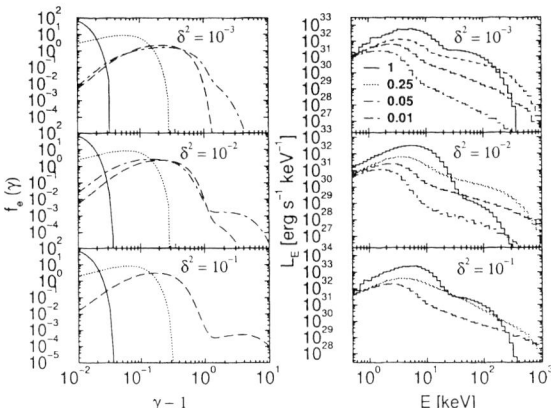

FIGURE 1. Equilibrium electron spectra (left) and photon spectra (right) for fixed pulsar dipole moment $\mu_{30} = 10^{-3}$. The legend in the upper left-hand panel refers to the various values of l_*.

we fix the parameter $f = 0.3$, and set $M_* = R_6 = 1$. Hence, the dipole magnetic field at the Alfvén radius is

$$B_0 = 4.2 \times 10^6 \, l_*^{6/7} \, \mu_{30}^{-5/7} \, M_*^{3/7} \, R_6^{6/7} \, f_{0.3}^{-3} \text{ G}, \quad (2)$$

where $f_{0.3} = f/0.3$. The virial ion temperature at r_0 is

$$kT_i = \frac{2}{3} \frac{G M m_H}{r_0} \approx 9.3 \left(\frac{r_0}{10^7 \text{ cm}}\right)^{-1} M_* \text{ MeV}. \quad (3)$$

The column density of the shell can be estimated using the poloidal accretion rate $\dot{M} \sim 4\pi r_0 \Delta r_0 n_i v_p m_H$, where we assume that the poloidal velocity $v_p \sim v_{\text{ff}}/2$ with v_{ff} being the free-fall velocity. Hence, the radial Thomson depth of the shell is approximately:

$$\tau_T = \Delta r_0 \, n_i \, \sigma_T \sim \frac{\dot{M} \, \sigma_T}{2\pi r_0 \, v_{\text{ff}} \, m_H} \approx 0.97 \, l_*^{8/7} \, \mu_{30}^{-2/7} \, M_*^{4/7} \, R_6^{1/7} \, f_{0.3}^{-1/2}. \quad (4)$$

The neutron star (taken to be a 10 km spherical surface) is assumed to emit a blackbody luminosity at the temperature $kT_{\text{BB}} = 1.78 \, l_*^{1/4}$ keV. In addition the level of wave turbulence is specified by the dimensionless amplitude $\delta^2 = (\Delta B/B)^2$ and the spectral index q. The minimum wavevector k_{min} is set to $2\pi/(\Delta r_0)$, where $\Delta r_0 \sim 0.1 r_0$ is the shell thickness (Ghosh & Lamb 1979a,b). Due to the spherical symmetry of our simulations the magntic field is assumed to be nondirectional in the shell and the synchrotron emissivities and absorption coefficients are angle-averaged.

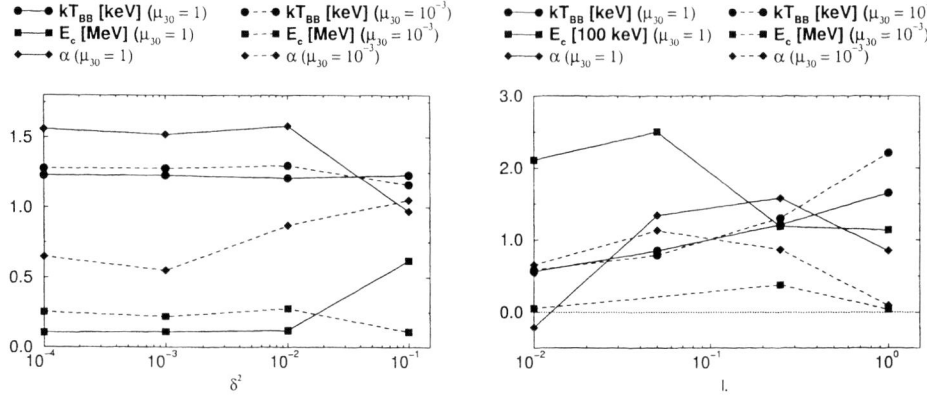

FIGURE 2. Left: Dependence of the spectral fit parameters α, E_c, and kT_{BB} on the input parameter δ^2 for fixed $l_* = 0.25$ and two different values of μ_{30}. Right: Dependence of the fit parameters on l_* for fixed $\delta^2 = 10^{-2}$ and two different vlaues of μ_{30}.

NUMERICAL RESULTS

We have explored the parameter space spanned by different values of the accretion rate — corresponding to a variation of l_* —, the neutron star surface magnetic field — corresponding to a variation of μ_{30} —, and the turbulence level δ^2. Fig. 1 shows some sample results for a surface magnetic field of $B_{\text{surf}} = 10^9$ G and different values of the accretion rate and the turbulence level.

In order to quantify the dependence of the resulting photon spectra on the input parameters, we have fitted the simulated spectra with a phenomenological spectral function consisting of a thermal blackbody + an exponentially cut-off power-law:

$$F(E) = A\,E^3 \left(e^{E/kT_{\text{BB}}} - 1\right)^{-1} + B\,e^{-(3\,kT_{\text{BB}}/E)^2}\,E^{-\alpha}\,e^{-E/E_c}, \tag{5}$$

and investigate the dependence of the spectral parameters α (energy index), E_c, and kT_{BB} as a function of the input parameters.

Fig. 2 illustrates the dependence of the spectral fit parameters on the turbulence level and on the accretion rate, respectively, for both the low and high magnetic field cases. The main results of our parameter study are:

1. As the accretion rate is decreasing, the lepton thermal temperature increases, and the hard X-ray photon spectrum becomes harder, at least for moderate accretion rates ($l_* \lesssim$ a few %). At the same time, the normalization of the hard X-ray power-law, relative to the thermal blackbody from the neutron star surface, becomes smaller.

2. As the turbulence level is increasing, the nonthermal tails in the electron spectra become more dominant and harder. At the same time, in the low-magnetic-

field case the quasi-thermal electron temperature decreases, leading to a softer hard-X-ray spectrum. In contrast, in the high-magnetic field case, an increasing turbulence level leads to a slight hardening of the hard X-ray / γ-ray tails in the photon spectra.

3. For low accretion rates ($l_* \lesssim$ a few %), the photon spectra are only very weakly dependent on the magnetic moment of the neutron star. For higher accretion rates and moderate turbulence levels, an increasing neutron star magnetic moment leads to a softening of the hard X-ray spectrum.

DISCUSSION AND MODEL PREDICTIONS

We have shown that the anti-correlation of the hardness and luminosity of the hard X-ray emission with the soft X-ray luminosity is a natural consequence of the energetics of particle acceleration and cooling near the Alfvén radius. We predict that the nonthermal tails in the hard X-ray spectra of accreting, weakly magnetized neutron stars in the low/hard state may extend up to several hundred keV. We also predict that similar hard X-ray tails should exist in the high/soft state as well, and extend to $\sim$ 100 – 200 keV. A solid detection and the measurement of the cutoff energy of these high-energy tails by the INTEGRAL satellite will provide important constraints on accretion-based models for the hard X-ray emission from accreting neutron stars.

Our model calculations indicate that the hard X-ray spectral slopes of $\alpha \sim$ 2 – 3 observed in some AXPs requires surface magnetic fields of $B_{\mathrm{surf}} \gtrsim 10^{12}$ G and rather low turbulence levels, $\delta^2 \lesssim 10^{-3}$. In that case, we predict a cut-off energy of $E_c \sim$ 100 – 500 keV, which may also be testable with the upcoming INTEGRAL mission.

REFERENCES

1. Barret, D., et al., *A&AS*, **120**, 121 (1996).
2. Barret, D., & Vedrenne, G., *ApJS*, **92**, 505 (1994).
3. Böttcher, M., & Liang, E. P., *ApJ*, **552**, 248 (2001).
4. Chatterjee, P., et al., *ApJ*, **534**, 373 (2000).
5. Ghosh, P., & Lamb, F., *ApJ*, **232**, 259 (1979a).
6. Ghosh, P., & Lamb, F., *ApJ*, **234**, 296 (1979b).
7. Harmon, B. A., et al., *A&AS*, **120**, 197 (1996).
8. Heyl, J. S., & Hernquist, L., *ApJ*, **489**, L67 (1997).
9. Mereghetti, S., & Stella, L., 1995, *ApJ*, **442**, L17 (1995).
10. Tavani, M., et al., *A&AS*, **120**, 221 (1996).
11. Tavani, M., & Liang, E. P., *A&AS*, **120**, 133 (1996).
12. Thompson, C., & Duncan, R. C., *ApJ*, **473**, 322 (1996).
13. Wang, J. C. L., *ApJ*, **486**, L119 (1997).
14. Zhang, N. S., et al., *A&AS*, **120**, 279 (1996).

Studies of Hard X–ray Tails in Z Sources with HEXTE/RXTE

Flavio D'Amico*, William A. Heindl[†], Richard E. Rothschild[†] and Duane E. Gruber[†]

*Center for Astrophysics and Space Sciences, University of California, San Diego, and Instituto Nacional de Pesquisas Espaciais - INPE
Av. dos Astronautas 1758, 12227-010 S. J. dos Campos, Brazil
[†]Center for Astrophysics and Space Sciences, University of California, San Diego
9500 Gilman Dr., La Jolla, CA 92093-0424

Abstract. We report *RXTE* results of spectral analyses of three (Sco X-1, GX 349+2, and Cyg X-2) out of the 6 known Z sources. No hard X–ray tails were found for Cyg X-2 ($< 8.4 \times 10^{-5}$ photons cm^{-2} s^{-1}, 50–100 keV, 3σ) and for GX 349+2 ($< 7.9 \times 10^{-5}$ photons cm^{-2} s^{-1}, 50–100 keV, 3σ). For Sco X-1 a variable hard X–ray tail (with an average flux of 2.0×10^{-3} photons cm^{-2} s^{-1}, 50–100 keV) has already been reported. We compare our results to reported detections of a hard component in the spectrum of Cyg X-2 and GX 349+2. We argue that, taking into account all the results on detections of hard X–ray tails in Sco X-1 and GX 349+2, the appearance of such a component is correlated with the brightness of the thermal component.

INTRODUCTION

The class of Z sources comprises 6 LMXBs (Sco X-1, GX 349+2, GX 340+0, Cyg X-2, GX 5−1, and GX 17+2) in which the primary is a neutron star with a low magnetic field ($\sim 10^{10}$ G) accreting at or near the Eddington limit [1]. They share similar timing properties and are among the most luminous known LMXBs. The designation Z source results from the shape described in a x-ray color × color diagram (CD), with the movement along the Z interpreted in terms of changes in the mass accretion rate ($\dot{M}$, see, e.g., [1]). Apart from Sco X-1 and Cyg X-2, the Z sources are all found near the Galactic mid-plane (i.e., $b = 0°$).

Hard X–ray spectra from both Z and low luminosity atoll sources have already been reported in the literature [2-8]. The production of the hard X–ray tails in atoll sources has been presented in the context of various thermal emission models [9] from which the accretion geometry can be inferred. The situation is less clear for the Z sources, where non–thermal mechanisms are invoked to explain the production of such a component, and little, or nothing, is known about the details of the accretion geometry.

We are currently analyzing all of the Z source observations in the public *RXTE* database which contain long pointings. The aim is to create an uniform database that will allow us to make direct hard X–ray spectra comparisons. From this we expect to better understand the behavior of any non–thermal emission in these sources. We report here the preliminary results of this work, with data from 3 (Sco X-1, GX 349+2, and

Cyg X-2) out of the 6 known Z sources.

DATA SELECTION AND ANALYSIS

We used data from HEXTE [10] to search for hard X–ray tails in the spectrum of Sco X-1, GX 349+2, and Cyg X-2 in the ~ 20–220 keV interval and data from PCA [11] to determine the position of the source in the CD and to study the 2–20 keV spectrum. We selected, from the public *RXTE* database, those subsets of data in which $\gtrsim 5000$ s of HEXTE total on–source time was available, in order to achieve good sensitivity at high energies. Table 1 shows the selected subsets for GX 349+2 and Cyg X-2. The list of selected observations of Sco X-1 is given in [7]. We used XSPEC to analyze the PCA source spectra, using published models for GX 349+2 (a blackbody plus a disk-blackbody and an iron line, see [12]) and Cyg X-2 (an absorbed cutoff power-law plus an iron line, see [13]). A complex multicomponent model (an absorbed blackbody plus a power-law, a Comptonization spectrum, and a Gaussian line) was used to heuristically fit the PCA Sco X-1 spectra. Low enewrgy (20–50 keV) HEXTE spectra were fitted by a simple thermal bremsstrahlung. The hard X-ray component (i.e. $E > 50$ keV), found only in Sco X-1, was modeled as a simple power-law (see [6] for a more detailed description of the instrument and procedure used for data analysis). We carefully verified our background subtraction procedures, specially for GX 349+2, which is located near the Galactic mid-plane, where the diffuse Galactic Plane background up to $E \sim 800$ keV [14] is known to vary in latitude [15]. We took advantage of HEXTE aperture modulation to remove this contribution to the background since HEXTE Cluster A measured the background at the same latitude as the source. Source confusion is also a concern for GX 349+2 due to the presence of 4U 1700−37 (see, e.g., [16]) inside the field of view of one of the regions used by HEXTE Cluster B to measure background (the B$^-$ region). This is easily solved using only the B$^+$ region to measure the background for HEXTE Cluster B. We found no evidence of source confusion/contamination for Cyg X-2 and Sco X-1.

RESULTS

Cygnus X-2 and GX 349+2 were easily detected by HEXTE up to 50 keV. Nevertheless, the detection level was *always* below 3σ in the 50–75 keV band. We show in Fig. 1 a typical spectrum for Cyg X-2 and GX 349+2 together with a detection and a non-detection of a hard X–ray tail in Sco X-1.

All sources show some degree of variability in the 20-50 keV range. From the results in [7], for Sco X-1, a factor of 2 was detected, while it was a factor of 5 for Cyg X-2 and 2 for GX 349+2. Among the three, Cyg X-2 is the least luminous in the 2–20 keV energy range, with an average luminosity of 0.4 L_{Edd} (using d and M_{ns} measurements in [17]), while Sco X-1 and GX 349+2 emit at or above Eddington levels, for $M_{\text{ns}} = 1.4 M_\odot$ (see [18] and [19] for measured distances to Sco X-1 and GX 349+2, respectively). We found no evidence of the presence of a hard X–ray tail in our database for

TABLE 1. Selected *RXTE* observations of GX 349+2 and Cyg X-2

GX 349+2							
OBSID	**MJD**	T_{obs}*	T_{HEX}†	$F_{(2-20)}$**	$F_{(20-50)}$‡	$F_{(50-100)}$§	Z¶
20054-05-01-00	50570	10032	5902	$1.75^{+0.09}_{-0.09}$	$2.03^{+0.43}_{-0.41}$	< 4.64	SA
30042-02-01-01	50822	8688	5492	$2.42^{+0.22}_{-0.17}$	$3.94^{+0.59}_{-0.52}$	< 5.52	FB
30042-02-01-02	50823	10336	6527	$1.98^{+0.06}_{-0.06}$	$3.46^{+0.48}_{-0.48}$	< 6.28	(lower) NB
30042-02-01-07	50823	14160	8850	$1.95^{+0.02}_{-0.02}$	$2.87^{+0.34}_{-0.34}$	< 1.37	SA
30042-02-01-03	50825	13728	8602	$2.50^{+0.25}_{-0.22}$	$3.94^{+0.08}_{-0.83}$	< 4.84	FB
30042-02-01-04	50826	14304	8865	$2.55^{+0.20}_{-0.18}$	$2.56^{+0.28}_{-0.26}$	< 3.71	FB
30042-02-01-08	50826	10368	6318	$2.08^{+0.25}_{-0.25}$	$4.97^{+0.45}_{-0.45}$	< 3.17	FB
30042-02-02-00	50830	9760	5632	$1.72^{+0.07}_{-0.07}$	$2.46^{+0.44}_{-0.44}$	< 5.32	NB-FB
30042-02-02-08	50838	7704	4689	$1.71^{+0.08}_{-0.08}$	$2.10^{+0.48}_{-0.46}$	< 6.49	NB-FB
30042-02-03-01	50842	9216	5684	$2.71^{+0.49}_{-0.32}$	$4.22^{+0.51}_{-0.46}$	< 2.97	FB
Cyg X-2							
10063-10-01-00	50316	8088	5044	$1.16^{+0.35}_{-0.39}$	$8.37^{+4.77}_{-4.44}$	< 6.92	FB
30418-01-05-00	51000	10760	6349	$1.14^{+0.14}_{-0.18}$	$5.62^{+4.22}_{-3.82}$	< 4.12	FB
30046-01-01-00	51009	13376	8180	$1.88^{+0.13}_{-0.13}$	$21.89^{+3.50}_{-3.28}$	< 4.61	SA
30046-01-02-00	51015	14736	9240	$1.65^{+0.07}_{-0.15}$	$15.26^{+3.66}_{-3.66}$	< 4.21	FB
30046-01-03-00	51022	13728	8881	$1.42^{+0.30}_{-0.28}$	$28.03^{+3.64}_{-3.64}$	< 4.88	FB
30046-01-04-00	51029	13584	8157	$1.23^{+0.09}_{-0.10}$	$9.74^{+3.12}_{-3.02}$	< 1.78	FB
30046-01-06-00	51041	15104	9114	$1.76^{+0.18}_{-0.21}$	$15.56^{+3.11}_{-3.11}$	< 4.86	FB
30046-01-07-00	51048	13888	8231	$1.35^{+0.09}_{-0.11}$	$5.38^{+3.34}_{-3.12}$	< 3.02	FB
30046-01-08-00	51055	13920	8566	$1.20^{+0.06}_{-0.06}$	$25.95^{+3.37}_{-3.37}$	< 4.40	NB
30046-01-09-00	51061	16256	9010	$1.30^{+0.08}_{-0.08}$	$6.54^{+3.07}_{-2.88}$	< 3.62	FB
30046-01-10-00	51068	8600	5419	$1.81^{+0.11}_{-0.13}$	$23.97^{+5.03}_{-5.03}$	< 3.03	SA
30046-01-11-00	51078	12512	8098	$1.48^{+0.10}_{-0.12}$	$9.24^{+3.88}_{-3.60}$	< 3.96	FB
30046-01-12-00	51081	14608	9703	$1.37^{+0.01}_{-0.01}$	$29.16^{+3.50}_{-3.50}$	< 7.01	HB

* total *RXTE* source's exposure time, in s
† corrected HEXTE exposure time, in s
** Flux, in 2-20 keV range, in units of 10^{-8} ergs cm^{-2} s^{-1}; uncertainties are given at 90% confidence level
‡ Flux, in 20-50 keV range, in units of 10^{-10} ergs cm^{-2} s^{-1}, for GX 349+2, and 10^{-11} ergs cm^{-2} s^{-1}, for Cyg X-2; uncertainties are given at 90% confidence level
§ 3σ upper limit on power-law Flux, in units of 10^{-11} ergs cm^{-2} s^{-1}, in the 50-100 keV range; power-law index frozen at a value of 2
¶ HB=horizontal branch; NB=normal branch; FB=flaring branch; SA=soft apex

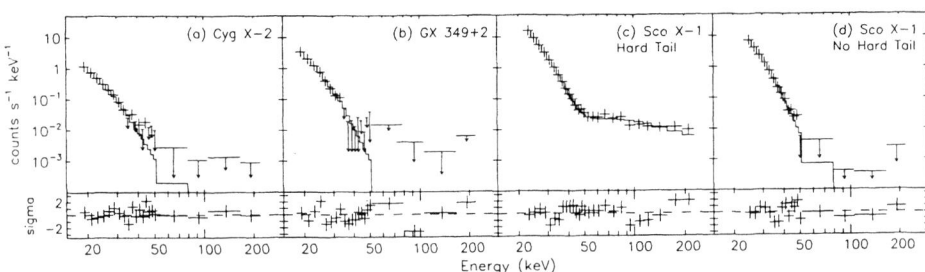

FIGURE 1. Typical HEXTE spectra (upper panels) for (a) Cyg X-2, (b) GX 349+2, (c) a hard X-ray tail detection in Sco X-1, and (d) a non-detection in Sco X-1 (for comparison). Residuals are given in units of standard deviations (lower panels). Upper limits are 2σ.

GX 349+2 or Cyg X-2. The HEXTE 3σ upper limit to 50–100 keV flux from GX 349+2 is 7.9×10^{-5} photons cm^{-2} s^{-1} and for Cyg X-2 is 8.4×10^{-5} photons cm^{-2} s^{-1}. For these two sources, a hard X–ray tail was, however, reported by *BeppoSAX* ([8] and [3], respectively), at a level of 4.6×10^{-4} photons cm^{-2} s^{-1} for GX 349+2 (using the fit parameters given in [8]; for Cyg X-2 it is not possible to estimate the flux from [3]). Our results, thus, can be interpreted in terms of variability in the appearance of this component, as was observed in Sco X-1 [7] on a 4 hour time-scale.

DISCUSSION

Scorpius X-1 remains as a special case among the Z sources. It is the only one in which a hard X–ray tail has been observed more than once, and by two different instruments ([5] and [7]). For Cyg X-2, GX 17+2 and GX 349+2 hard X–ray tails were reported by *BeppoSAX* ([3], [4], and [8], respectively) on one occasion. From our combined HEXTE database, we found the presence of a hard X–ray tail in 8 out of 28 occasions for Sco X-1, and zero out of 10 and 13 observations of GX 349+2 and Cyg X-2, respectively. Fitting our HEXTE data for GX 349+2 and Cyg X-2 with a power-law with indices frozen in the range 1–2 (within the values found for those three sources: see [3], [7-8]), we found a 3σ upper limit on the luminosity of the power-law component, $L^{PL}_{20\text{-}80\,keV} = 6.8 \times 10^{35}$ ergs s^{-1} and $L^{PL}_{20\text{-}80\,keV} = 5.0 \times 10^{35}$ ergs s^{-1} for GX 349+2 and Cyg X-2, respectively. Our HEXTE result (for $\Gamma = 1-2$) for hard X–ray tail detections in Sco X-1 is $L^{PL}_{20\text{-}80\,keV} = 6.7 \times 10^{35}$ ergs s^{-1}. It thus appears that our observations were sensitive enough to detect hard X–ray tails in Cyg X-2 and GX 349+2. As we pointed out in [7] the chance of observing a hard X–ray tail (in Sco X-1) is higher when the thermal component of the spectrum is brighter. From our results here (see Table 1), we have, for GX 349+2 $L^{Thermal}_{20\text{-}50\,keV} = 1.2\text{-}3.1 \times 10^{36}$ ergs s^{-1}, while for Cyg X-2 the results are $L^{Thermal}_{20\text{-}50\,keV} = 0.4\text{-}2.1 \times 10^{36}$ ergs s^{-1}. The same component in Sco X-1, when a hard tail is detected [7], is in the range $L^{Thermal}_{20\text{-}50\,keV} = 4.5\text{-}9.0 \times 10^{36}$ ergs s^{-1}. While comparable values were not given by the *BeppoSAX* results in [3], [4], and [8] (nor by the OSSE/*CGRO* results in [5]), it is possible to extrapolate the results presented in [8] in order to find an estimate of the luminosity of the thermal component. We estimate that the 20–50 keV GX 349+2 luminosity measured by *BeppoSAX* was greater than 5×10^{36} ergs s^{-1}. Thus, one can speculate that the production of a hard X–ray tail in a Z source is a process triggered when the thermal component is brighter than a level of $\sim 4 \times 10^{36}$ ergs s^{-1}.

CONCLUSIONS

We have shown *RXTE* results of broad-band spectral analyses of three Z sources, with emphasis on the hard X–ray spectrum. We found no evidence for a detection of a hard X–ray tail in the spectra of GX 349+2 and Cyg X-2, although one detection of such a component has been reported for each of these sources. We interpret this in terms of variability, which was shown to be as fast as 4 hours in Sco X-1. We found an indication that the production of hard X–ray tails in Z sources is a process triggered when the thermal component brightness is above a value of $\sim 4 \times 10^{36}$ ergs s^{-1}. We are currently creating a uniform HEXTE database including the other three Z sources (GX 17+2, GX 340+0, and GX 5−1), from which we hope to be able to better understand the production of hard X–ray in Z sources.

ACKNOWLEDGMENTS

This research has made use of data obtained through the HEASARC, provided by NASA/GSFC. F.D. gratefully acknowledges FAPESP/Brazil for financial support under grant 99/02352-2. This research was supported by NASA contract NAS5-30720.

REFERENCES

1. van der Klis, M., *X-Ray Binaries*, edited by W. H. G. Lewin, J. van Paradjis, and E. P. J. van den Heuvel, Cambridge University Press, Cambridge, 1995, pp.252–307.
2. Barret, D. et al., *ApJ* **533**, 329-351 (2000).
3. Frontera, F. et al., *Nucl. Phys. B* **69**, 286-293 (1998).
4. Di Salvo, T. et al., *ApJ* **544**, L119-L122 (2000).
5. Strickman, M., and Barret, D. 2000, "Detections of Multiple Hard X-ray Flares from Sco X-1 with OSSE", in *Proceedings of the Fifth Compton Symposium*, edited by M. L. McConnel and J. M. Ryan, AIP Conference Proceedings 510, New York, 2000, pp. 222-226.
6. D'Amico, F. et al., *ApJ* **547**, L147-L150 (2001).
7. D'Amico, F. et al., *Adv. Spa. Res.*, in press (2001) (astro-ph/0101396).
8. Di Salvo, T. et al., *ApJ*, in press (2001) (astro-ph/0102299).
9. Barret, D., *Adv. Spa. Res.*, in press (2001) (astro-ph/0101295).
10. Rothschild, R. E. et al., *ApJ* **496**, 538-549 (1998).
11. Jahoda, K. et al., *Proc. SPIE* **2808**, 59-70 (1996).
12. Christian, D. J., and Swank, J. H., *ApJS* **109**, 177-224 (1997).
13. Kuulkers, E. et al., *A&A* **323**, L29-L32 (1997).
14. Boggs, S. E. et al., *ApJ* **544**, 320-329 (2000).
15. Valinia, A., and Marshall, F. M., *ApJ* **505**, 134-147 (1998).
16. Reynolds, A. P. et al., *A&A* **349**, 873-876 (1999).
17. Orosz, J., and Kuulkers, E., *MNRAS* **305**, 132-142 (1999).
18. Bradshaw, C. F., Fomalont, E. B., and Geldzahler, B. J., *ApJ* **512**, L11-L14 (1999).
19. McNamara, D. H. et al., *Pub. Astr. Soc. Pac.* **112**, 202-216 (2000).

Discovery of Two Transient X-Ray Pulsars In The Small Magellanic Cloud

S. Laycock[*], R. H. D. Corbet[†‡], M. J. Coe[*], D. Perrodin[†‡¶], F. Marshall[†], C. Markwardt[†],

[*]University of Southampton, Dept. of Physics & Astronomy, SO17 3AA, UK
[†]Goddard Space Flight Center, Greenbelt, MD 20771, USA
[‡]Universities Space Research Association
[¶]Case Western Reserve University

Abstract. Recent *RXTE* observations of the SMC have revealed two previously unknown transient X-ray pulsars with pulse periods of 95s and 4.78s. The sources are proposed as Be/neutron star systems on the basis of their pulsations, transient nature and characteristics hard spectra. Optical observations indicate an Hα emission-line star as a candidate optical counterpart. These results add to the emerging picture of the SMC as containing an extremely dense population of transient HMXBs, only a fraction of which are active at any one time.

INTRODUCTION

Observations of the SMC by the current generation of X-ray satellites have detected an increasing number of new transient X-ray pulsars over the last few years. As well as new discoveries, there have been many identifications as pulsars of X-ray sources seen by *Einstein* and *ROSAT*. Compared to all predictions based on its mass and composition, the SMC contains a suprisingly large population of High Mass X-ray binary (HMXB) systems (Coe 2000). This population provides the closest available approximation to a luminosity-limited sample of X-ray pulsars due to the low extinction and small size of the SMC. The depth of the SMC is small relative to its distance from us, placing all of the pulsars at effectively the same distance. Rossi X-Ray Timing Explorer's (*RXTE*) proportional counter array (PCA) is well suited to studying this population since its field of view, sensitivity and timing resolution allow simultaneous monitoring of all sources in a significant fraction of the SMC.

XTE J0052-727

The 4.78s pulsar XTE J0052-727 was first noticed in an *RXTE* Proportional Counter Array (PCA) observation performed on Jan 5th 2001. The field was centered on the co-ordinates α: $0^h53^m53^s$, δ: -72°26'42", in the western wing of the SMC. The source was sufficiently bright for its position to be localized to within 1arcmin by slewing the PCA across the field. Slews were performed in RA and Dec on two separate occasions, combining the results from these measurements resulted in a best-fit position at 99% confidence of α: $0^h52^m19^s \pm 24^s$, δ: -72°19'48" ± 50".

The *RXTE* position enabled an optical observation to be made using the South African Astronomical Observatory (SAAO) 1-meter telescope, while the source was still in X-ray outburst.

X-ray Observations

The PCA (Jahoda et al 1996) Good Xenon data was used to create lightcurves with 100ms timing resolution. From these, the source was determined to have been active between Dec 27th 2000 and Jan 24th 2001. The power spectrum from the discovery observation of XTE J0052-727 is shown in Figure 1, clearly showing the fundamental period and first harmonic. Pulse profiles were then created for the available observations, the first of these is also shown in Figure 1. The pulse profile is double-peaked and exhibits evolution on timescales of a few days, the peak X-ray flux was about 8 mCrab on Jan 9th.

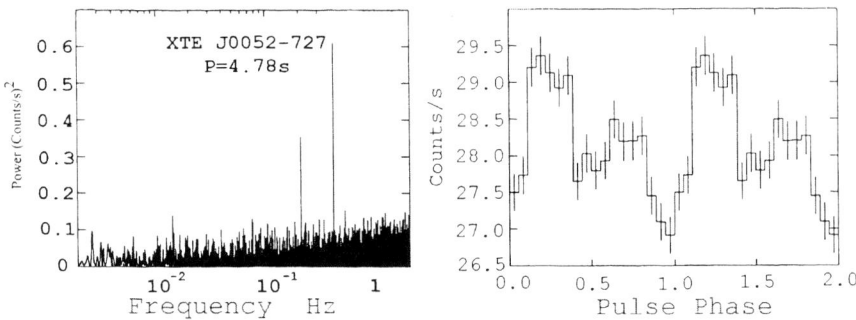

FIGURE 1. The discovery of XTE J0052-727 in an RXTE PCA observation. The power spectrum (left) shows the fundamental and first harmonic.

Optical Photometry

Following the *RXTE* slew measurements, a series of photometric observations were made on Jan 16th 2001, covering the 99% error box. CCD images were taken in Johnson U, B, V, R and Hα. This field is shown in Figure 2 with the *RXTE* error box indicated. The CCD images were analyzed using an elliptical-Gaussian PSF fitting routine and calibrated by aperture photometry. A two-color diagram was then constructed from the reduced photometric data in order to identify any Be stars in the field, this diagram is shown in Figure 3. The B-V color separates stars by their effective temperature, which is directly related to their spectral type. The R-Hα color then identifies those stars that have a significant Hα excess in their red-end continuum. This procedure yielded seven Be stars in the 5 arcmin field, two of which lie in the *RXTE* error box, these stars are labeled in Figures 2 & 3. Star 13 is also known as the emission-line star [MA93]537.

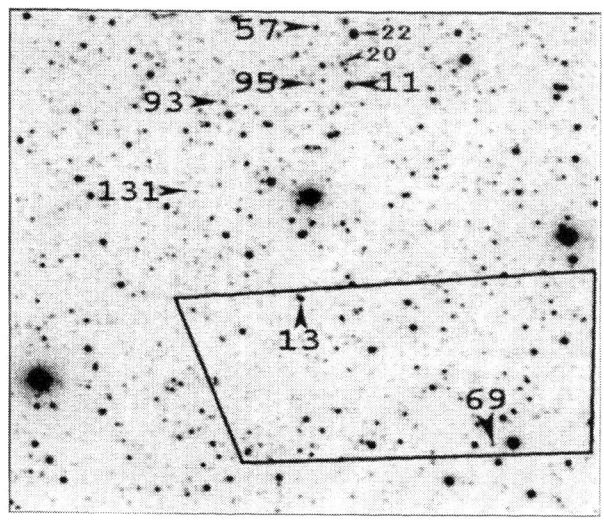

FIGURE 2. CCD image covering the field of XTE J0052-727. Show are the *RXTE* 99% error box and Be stars identified by multicolor photometry (see Figure 3).

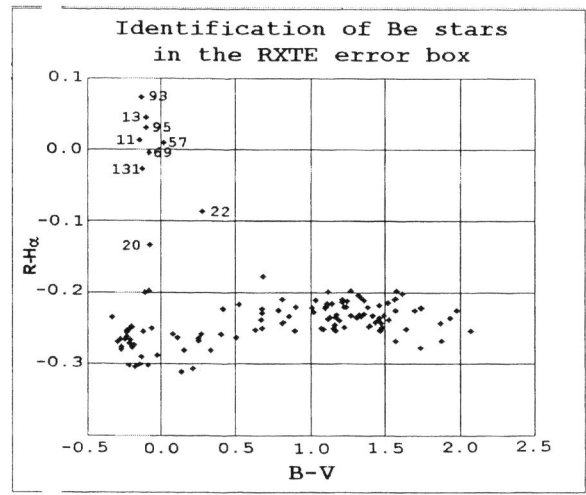

FIGURE 3. Two-color diagram for stars in the field of XTE J0052-727. Be stars are indicated by numbers, which refer to Figure 2.

XTE SMC95

A 95s pulsar was first detected in an *RXTE* PCA observation on 11th Mar 1999. The field was centered on the co-ordinates α: $0^h53^m53^s$, δ: -72°26'42", in the western wing of the SMC. The source was seen again one week later at a similar brightness, and in a third observation two weeks later on Apr 1st, by which time it had faded considerably. No source location was possible in this case, and we have provisionally designated it XTE SMC95. Power spectra and pulse-profiles for discovery observation are shown here in Figure 4. The pulse profile was double-peaked in the first observation and changed significantly after seven days. X-ray spectra were also obtained for XTE SMC95, they are well described by an absorbed power-law model and show no significant variation between the two observations (No results could be derived from the third observation due to an interfering source). The fitted spectral model implies a 2-10 keV luminosity of $\geq 2\times10^{37}$ ergs s^{-1} which is within the range of luminosities seen during normal outbursts in galactic HMXBs (Negueruela 1998).

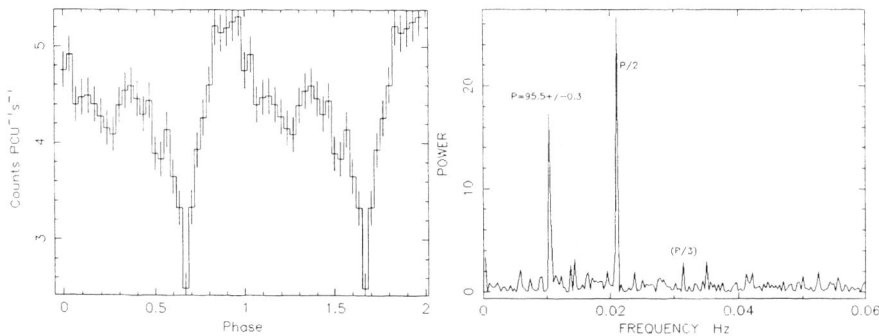

FIGURE 4. 3-25 keV background corrected pulse profile and power spectrum for the first detection of XTE SMC95 on Mar 11th 1999.

CONCLUSIONS

With the addition of these two recent discoveries, there are now 25 binary X-ray pulsars known in the SMC. Of these X-ray sources, the majority have been identified as belonging to Be systems rather than supergiant systems (e.g. Haberl & Saski 2000). Comparing these figures to the population in the Galaxy in which about 70 HMXBs are known, 31 identified with Be optical companions, the SMC emerges as containing a particularly dense population of Be X-ray binaries. If the comparison is scaled by the relative masses of the SMC and Galaxy it is clear that the population density of HMXBs in the SMC is higher even than the Galaxy's most populous regions. This abundance of HMXB's represents a young population of systems, and coupled with other evidence (e.g. Maeder et al 1999, Yokogawa et al 2000), suggests that a major epoch of star formation occurred ~5My ago. The likely instigator of this event was the enormous tidal force acting on the SMC as it neared its closest approach to the LMC (see Gardiner & Noguchi).

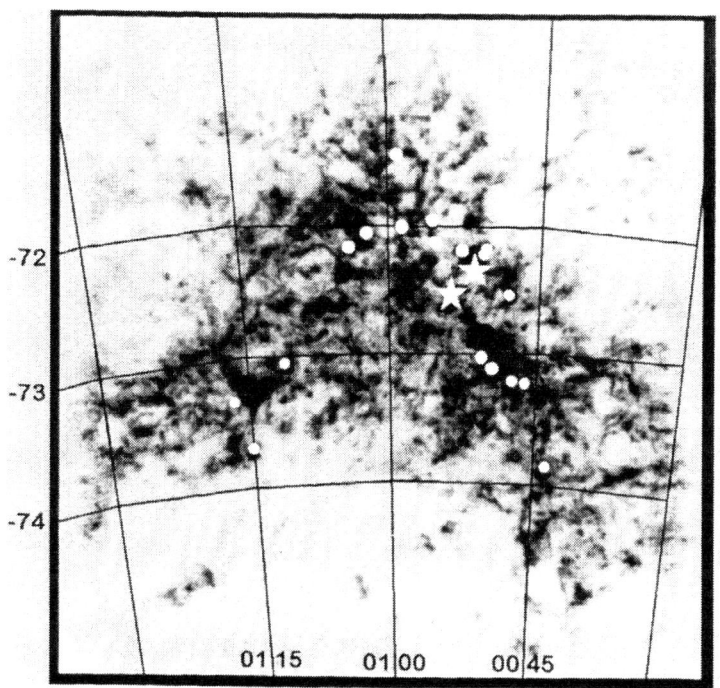

FIGURE 5. Positions of X-ray pulsars in the SMC, stars mark the two new sources. (Superimposed on an Hl radio map of Stavely-Smith et al 1997).

REFERENCES

1. Haberl., F., Saski M., *Astron. Astrophys.* 359, 573-585 (2000)
2. Coe M. J., *ASP Conference Series*, Vol. 214, 656 (2000)
3. Jahoda K., Swank J. H., Stark M. J., Strohmeyer T., Zhang W., Morgan E. H., *Proc.SPIE*, 2808, 59 (1996)
4. Maeder A., Grebel E. K., & Mermilliod J. C., *A & A*, 346, 456 (1999)
5. Negueruela I., A\&A 338, 505 (1998)
6. Stavely-Smith, L., Sault R. J., Hatzidimitriou D., Kesteven M. J., McConnell D., *MNRAS*, 289, 225 (1997)
7. Yokogawa J., Imanishi K., Tsujimoto M., Nishiuchi M., Koyama K., Nagase F., Corbet R. H. D., *ApJS* 128, 491-509 (2000)
8. Gardiner L. T., Noguchi M., *MNRAS*, 278, 191 (1996)

Self-Similar Hot Accretion Flow onto a Rotating Neutron Star: Structure and Stability

Mikhail V. Medvedev[1] and Ramesh Narayan[2]

[1] *CITA, University of Toronto, Toronto, Ontario, M5S 3H8, Canada*
[2] *Harvard-Smithsonian Center for Astrophysics, 60 Garden Street, Cambridge, MA 02138*

Abstract. We present analytical and numerical solutions which describe a hot, viscous, two-temperature accretion flow onto a rotating neutron star or any other rotating compact star with a surface. We assume Coulomb coupling between the protons and electrons, and free-free cooling from the electrons. Outside a thin boundary layer, where the accretion flow meets the star, we show that there is an extended settling region which is well-described by two self-similar solutions: (i) a two-temperature solution which is valid in an inner zone $r \lesssim 10^{2.5}$ (r is in Schwarzchild units), and (ii) a one-temperature solution at larger radii. In both zones, $\rho \propto r^{-2}$, $\Omega \propto r^{-3/2}$, $v \propto r^0$, $T_p \propto r^{-1}$; in the two-temperature zone, $T_e \propto r^{-1/2}$. The luminosity of the settling zone arises from the rotational energy of the star as the star is braked by viscosity. Hence the luminosity and the flow parameters (density, temperature, angular velocity) are independent of $\dot M$. The settling solution described here is not advection-dominated, and is thus different from the self-similar ADAF found around black holes. When the spin of the star is small enough, however, the present solution transforms smoothly to a (settling) ADAF.

We carried out a stability analysis of the settling flow. The flow is convectively and viscously stable and is unlikely to have strong winds or outflows. Unlike another cooling-dominated system — the SLE disk, — the settling flow is thermally stable provided that thermal conduction is taken into account. This strong saturated-like thermoconduction does not change the structure of the flow.

THE SETTLING FLOW

At small mass accretion rates, $\lesssim 10^{-2}$ of the Eddington rate, black holes (BHs) accrete via an ADAF — a hot, two-temperature, radiatively inefficient, geometrically thick, advection-dominated accretion flow [1,2]. In contrast, accretion onto compact stars, e.g., a neutron star (NS) may occur via either an ADAF of a settling solution [3]. The latter corresponds to strongly rotating stars only. In the settling flow the rotational energy is extracted from the star via viscous torques in the boundary layer where the flow meets the star surface. The extracted energy heats the flow and ultimately escapes from the flow as free-free radiation. In addition, viscosity extracts angular momentum from the star as well. In the stationary flow, this angular momentum must be transported though the flow to the outermost radii, where it goes into the ambient medium. It is this huge angular momentum flux $\dot J$ which modifies the entire structure of the accretion flow and makes it drastically different from an ADAF. Note that no angular momentum may be extracted from a BH horizon by viscosity. Thus a viscous settling flow may exist in NS systems and not in BH systems.

The structure of the steady, rotating, axisymmetric, quasi-spherical, two-temperature settling flow has been found analytically and confirmed numerically [3]. We use the height-integrated form of the viscous hydrodynamic equations with the Shakura-Sunyaev-type viscosity parametrized by dimensionless α. We assume viscous heating of protons, Bremsstrahlung cooling of electrons and Coulomb energy transfer from the protons to the electrons; we neglect Comptonization but include thermal conduction in the form discussed in the next section. In the inner zone

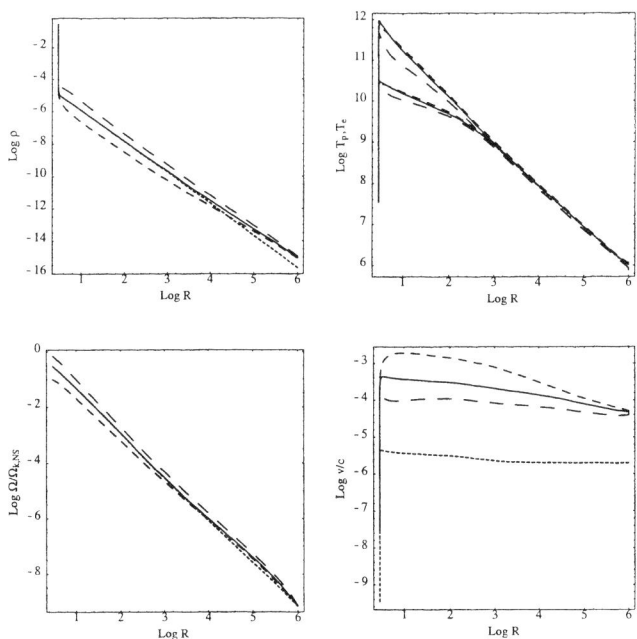

FIGURE 1. Profiles of density ρ (g cm^{-3}), proton temperature T_p (°K), electron temperature T_e (°K), angular velocity Ω (in units of the Keplerian angular velocity at the NS radius R_{NS}), and radial velocity v (in units of c) for accretion flows with $\alpha = 0.1$ and $(\dot m, s) = (0.01, 0.3)$ – solid line, (0.0001, 0.3) – short-dashed line, (0.01, 0.1) – medium-dashed line, (0.01, 0.7) – long-dashed line.

$r < 10^{2.5}$ (r is in Schwarzchild units, $R_S = 2GM/c^2$), the flow is two-temperature with the density, proton and electron temperatures, angular and radial velocities scalings as

$$\rho = \rho_0\, r^{-2}, \quad T_p = T_{p0}\, r^{-1}, \quad T_e = T_{e0}\, r^{-1/2}, \quad \Omega = \Omega_0\, r^{-3/2}, \quad v = v_0\, r^0, \qquad (1)$$

where $\rho_0, T_{p0}, T_{e0}, \Omega_0, v_0$ are functions of M, α and the star spin $s = \Omega_*/\Omega_K(R_*)$, and $\Omega_K(R) = (GM/R^3)^{1/2}$ is the Keplerian angular velocity. In the outer zone $r > 10^{2.5}$, we have $T_e = T_p \propto r^{-1}$ and the same other scalings. This self-similar solution is valid for the part of the flow below the radius r_s related to the mass accretion rate $\dot{m}$ (in Eddington units, $\dot{M}_{\rm Edd} = 1.4 \times 10^{18} m$ g/s, and here $m = M/M_\odot$):

$$\dot{m} < 2.2 \times 10^{-3} \alpha_{0.1}^2 s_{0.3}^2 r_{s,3}^{-1/2}, \qquad (2)$$

where $r_{s,3} = r_s/10^3$, $\alpha_{0.1} = \alpha/0.1$, etc.. The numerical solution of the hydrodynamic equations with appropriate inner and outer boundary conditions is represented in Figure 1. It is in excellent agreement with the self-similar solition (1).

There is a remarkable property of the settling flow: all quantities, e.g., ρ, T_p, etc., except v, are *independent* of the mass accretion rate $\dot{M}$. This happens because the gravitational energy of the accreting gas is much smaller than the energy extracted from the rotating star. It is this energy which dominates the flow luminosity. We should also remark that the settling flow is more similar to a steady, radiative cooling-dominated "atmosphere" rather than to a rapidly infalling flow: the radial infall velocity is constant and is much less then the free-fall velocity $v/v_{\rm ff} \propto r^{1/2} \to 0$ as $r \to 0$. The structure of the settling is very sensitive to the rotation rate of the central star. As the angular velocity decreases below few percents of the Keplerian value, the settling flow smoothly transforms into a conventional ADAF solution, as represented in Figure 2.

It was shown that the settling flow is (i) convectively stable and (ii) may not have strong winds and outflows (the Bernoulli number is negative) if the adiabatic index satisfies

$$\gamma > \frac{3(1 - s^2/2)}{(2 - s^2/2)} \sim 1.5. \qquad (3)$$

Other properties of the settling flow are discussed elsewhere [3,4].

THERMAL STABILITY OF THE SETTLING FLOW

The settling flow is cooling-dominated. Thus it is similar to the Shapiro-Lightman-Eardley solution [5] which is known to be thermally unstable [6]. Hence, our settling solution may be unstable as well. To study the thermal instability in sheared circular accretion flows we use the shearing sheet approximation with the velocity given by

$$\mathbf{V}_0(x) = 2A\,x\,\hat{y}, \qquad (4)$$

where $2A = d\mathbf{V}_0/dx$ is the shear frequency, x, y, z are the radial, azimuthal, and vertical coordinates, and "hat" denotes a unit vector. We assume that there is a Coriolis acceleration due to $\mathbf{\Omega} = \Omega\,\hat{z}$. The vorticity and epicyclic frequency are then $2B = 2A + 2\Omega$, $\kappa_{\text{epi}}^2 = 4\Omega B$. For a Keplerian-type flow, $\Omega \propto R^{-3/2}$, which is the case for both the settling and SLE solutions considered below, one has $2A = -(3/2)\Omega$ and $2B = \Omega/2$. We assume that perturbations have only s-component, which corresponds to axisymmetric perturbations. We ignore motions in z direction. We use hydrodynamic equations with thermoconductive flux.

The settling flow is hot (sub-virial), so that the mean-free-path (of both electrons and protons) is larger than the size of the system. Hence the conventional Spitzer theory fails. Without collisions but in the presence of magnetic fields electrons stream freely along the field lines, therefore the parallel heat flux remains large. In contrast, transverse heat flux is greatly reduced in a magnetic field because electrons are tied to the field lines on the scale of the Larmor orbit and cannot move across the field lines too far. In a tangled field, however, electrons can jump

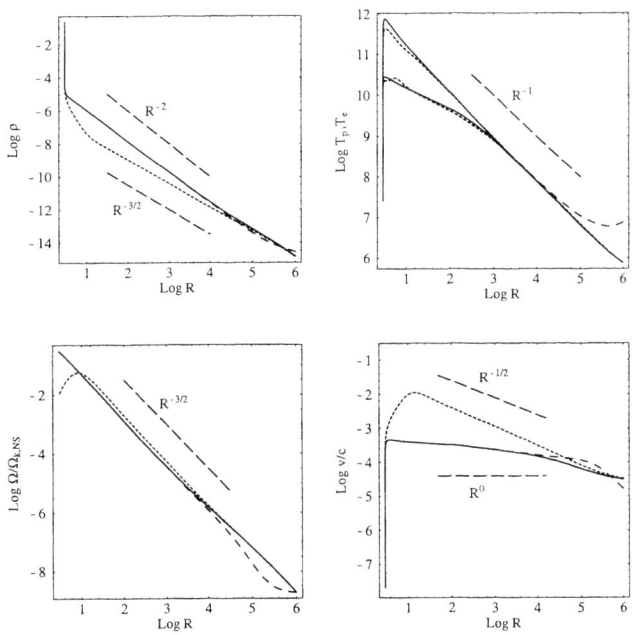

FIGURE 2. Same as in Fig. 1 for $\gamma = 4/3$ and $s = 0.3$ (solid curve) and $s = 0.01$ (dotted curve). The self-similar slopes for an ADAF flow and a settling flow are shown for comparison. The long-dashed curves represent the same solution as the solid curve, but with ten times higher temperature at R_{out}.

from one field line to another and thus transfer heat across the field [7]. The average thermal conductivity in tangled fields is

$$\kappa_B \simeq n k_B v_T l_B \vartheta \simeq 10^{-2} n k_B v_T R \xi_{-1} \vartheta_{-1}, \quad (5)$$

where $l_B = \xi R$ is the correlation scale of the magnetic fields set by flow turbulence and $\vartheta < 1$ takes in to account that only a fraction of all particles may pass though the magnetic barriers. The rest of them will remain trapped in magnetic wells and, hence, will not transport energy to large distances $\gg l_B$. The thermoconductive flux is

$$q_{\text{cond}} = -\alpha_c \frac{\rho c_s^2}{\Omega_K} \frac{dc_s^2}{dx}, \quad \text{where} \quad \alpha_c \simeq \frac{R}{H} \xi \vartheta \, F(e,p) \simeq 10^{-2} \xi_{-1} \vartheta_{-1} \, F(e,p) \quad (6)$$

is the "alpha-prescription" thermo-conductivity coefficient and $F(e,p)$ [$1 \leq F(e,p) \lesssim 15$] takes into account that conduction may be dominated by the protons or the electrons, depending on the flow conditions. Here we used that $v_T \simeq c_{se}$ and $H/R \sim c_s/v_{\text{ff}} \sim c_s/\Omega_K R$, where $H \sim R$ is the accretion disk scale height.

Now, from hydrodynamic equations for the perturbations, we obtain the dispersion relation

$$\omega \left[\frac{\omega}{\gamma - 1} + \frac{i(n-1)}{\tau_{\text{cool}}} + \frac{ik^2 R^2}{\tau_{\text{cond}}} \right] \left(\omega^2 - \kappa_{\text{epi}}^2 - k^2 c_s^2 \right) - \omega \left[\omega + \frac{i(2B/A - 1)}{\tau_{\text{cool}}} \right] k^2 c_s^2 = 0,$$

where $\quad \tau_{\text{cool}} \doteq \dfrac{4}{(9\alpha s^2 + 2\alpha_c)} \Omega_K^{-1} \simeq \dfrac{2}{\alpha_c} \Omega_K^{-1}, \quad \tau_{\text{cond}} = \dfrac{3}{\alpha_c} \Omega_K^{-1}. \quad (7)$

In the large-k limit, (7) yields the growth rate of the thermal mode. The stability criterion $\text{Im}\,\omega < 0$ is (for $\alpha s^2 \lesssim \alpha_c$)

$$kR > \left[\frac{\tau_{\text{cond}}}{\tau_{\text{cool}}} \left(2 - n - 2\frac{B}{A} \right) \right]^{1/2} = \left[\frac{26\alpha_c}{9(9\alpha s^2 + 2\alpha_c)} \right]^{1/2} \simeq \sqrt{\frac{13}{9}} \simeq 1.2, \quad (8)$$

that is, thermal modes with $kR \geq 2$ are stable. Whether the mode $kR = 1$ is stable or not cannot be reliably determined from our local approach. A global stability analysis is necessary to properly account for the effects of geometry and curvature on the eigenmode structure. Finally, we conclude that the settling flow is very likely stable.

REFERENCES

1. Narayan, R., and Yi, I., *Astrophys. J.* **428**, L13 (1994)
2. Narayan, R., and Yi, I., *Astrophys. J.* **452**, 710 (1995)
3. Medvedev, M. V., and Narayan, R., *Astrophys. J.* **554**, 756 (2001)
4. Medvedev, M. V., in Proceedings of the 20th Texas Symposium, eds. J. C. Wheeler and H. Martel (2001)
5. Shapiro, S. L., Lightman, A. P., & Eardley D. M., *Astrophys. J.* , **204**, 187 (1976)
6. Piran, T., *Astrophys. J.* , **221**, 652 (1978)
7. Rechester, A. B., & Rosenbluth, M. N., Phys. Rev. Lett., **40**, 38 (1978)

Black Hole X-Ray Binaries

XMM-Newton Observation of the Black Hole Microquasar GRS 1758-258

A. Goldwurm*, D. Israël*, P. Goldoni*, P. Ferrando*, A. Decourchelle*, F. I. Mirabel* and R. S. Warwick[†]

*Service d'Astrophysique/DAPNIA, CEA-Saclay, F-91191 Gif-Sur-Yvette, France
[†]Dept. of Physics & Astronomy, Leicester University, Leicester, LE1 7RH, U.K.

Abstract. The XMM-Newton X-ray observatory pointed the galactic black hole candidate and microquasar GRS 1758-258 in September 2000 for about 10 ks during a program devoted to the scan of the Galactic Center regions. Preliminary results from EPIC MOS camera data are presented here. The data indicate that the source underwent a state transition from its standard low-hard state to an intermediate state. For the first time in this source the ultra-soft component of the accretion disk, which black hole binaries display in intermediate or high-soft states, was clearly detected and measured thanks to the high spectral capabilities of XMM-Newton.

INTRODUCTION

The source GRS 1758-258 was discovered in 1990 with the SIGMA soft γ-ray telescope at about 5° from the Galactic Center [1]. The hard spectrum extending up to 200-300 keV [2,3], very similar to the Cyg X-1 spectrum, strongly suggests that this source is an accreting black hole in a galactic binary system with a low mass companion star. The source was then observed in radio and two symmetrical radio lobes were detected at 6 cm with the VLA [4] on either sides of a point-like radio source close to the X-ray source. The radio point source position was compatible with both the SIGMA error circle and the much smaller Rosat error circle (10" radius) of GRS 1758-258. In spite of a large drop in hard X-ray flux detected with SIGMA in 1991-1992 and some claims of sporadic appearance of a soft component [5], no spectral transitions have ever been clearly observed from this source. We present here the first convincing detection of an ultra-soft disk emission and of a spectral transition in GRS 1758-258 during a XMM-Newton observation.

XMM-NEWTON OBSERVATIONS AND RESULTS

The source was observed on 19th September 2000 with XMM-Newton for about 10 ks, with EPIC cameras EMOS 1 in timing mode, EMOS 2 in imaging refresh frame store (RFS) mode and PN in small window mode [6]. These modes were selected to avoid as much as possible pile-up effect, expected for such bright source. The medium filter was selected to reduce potential optical loading on the CCDs. We report here preliminary results from the MOS camera data. Data reduction was performed using the XMM SAS (Science Analysis Software), the standard XSPEC, and the XRONOS packages. The observation was contaminated by a large flux of "soft protons" background events; however, thanks to the strength of the source, the signal to noise ratio remained very high.

We have used the MOS 2 data to build an image and a spectrum of the source. Fig. 1 (left) shows the 0.2-10 keV image of the source obtained using the central MOS 2 CCD, which was employed in RFS mode for a effective integration time of 1325 s. The count distribution is compatible with a point-like source positioned at (2000 equinox) $R.A. = 18^h\ 01^m\ 12.5^s\quad Dec. = -25°\ 44'\ 40''$ with an error radius of 5''.

To source spectrum was derived by applying standard cuts to events collected within 30'' from the source center. Background was estimated using offset regions in the same central CCD. The source rate of 11.9 events/frame induced a non negligible pile up [7]. While no attempt has been made at this stage for correcting for it, its effect on the determination of the spectral shape was found to be within the statistical error bars of the derived model parameters. Its influence on the absolute flux is however much more important, and we roughly estimate that it induces a flux loss by a factor of ~ 1.5.

Data were rebinned to reach 20 counts per bin and the derived source count spectrum in the range 0.2-10 keV was compared to several models. As demonstrated in Fig. 2 (left), a simple power law, with a reduced chi-square > 3, does not fit the data, and the residuals indicate the need to include a soft component. The chi-square reaches acceptable values when a soft black body component is included. In Table 1 we report the best fit parameters ($\chi^2_\nu = 1.026$) for a model of a power-law plus a black body and the unfolded data are compared to the model in Fig. 2 (right). The soft component with a temperature of ≈ 0.3 keV is clearly detected, in addition to a power-law with photon index of ≈ 2.0, and reaches a fraction of 15 % of the total absorbed 0.2-10 keV flux, flux which amounts to $3.7\ 10^{-10}$ ph cm^{-2} s^{-1}. The column density is $1.7\ 10^{22}$ cm^{-2} and the derived source luminosity in the 0.2-10 keV band at 8 kpc is $8.7\ 10^{36}$ erg s^{-1}, out of which ≈ 30 % is due to the black-body component. No iron lines or other relevant features were detected.

Fig. 3 (left) reports the source light curve with time bins of 1.75 s obtained using the EMOS 1 data, collected in timing mode for a total exposure of 9865 s. In timing mode the EPIC instruments record the position on one axis only and the arrival time with a 1.75 ms resolution. The power density spectrum (PDS) was built using events in the central part of the CCD, and grouping the light curve in 21 ms bins. The spectrum regrouped in 14 channels after subtraction of the statistical noise is displayed in units of rms^2 Hz^{-1} in Fig. 3 (right). It can be modeled by a broken power-law with flat slope below a break frequency ν_B of 1.48 Hz, and with slope -1.32 above ν_B and normalization of $7.53\ 10^{-3}$ rms^2 Hz^{-1} at ν_B.

TABLE 1. Best-Fit Parameters for a Power Law plus Black Body Model

Parameter	Value (errors at 90% c.l.)
N_H (10^{22} cm^{-2})	1.74 ± 0.07
α_{pl}	1.99 ± 0.09
N_{pl} (ph $keV^{-1}cm^{-2}s^{-1}$)	0.13 ± 0.02
kT_{bb} (keV)	0.32 ± 0.02
N_{bb} (L_{36}/D_{10}^2)	3.95 ± 0.55

COMPARISON WITH PREVIOUS RESULTS

In Fig. 1 (right) the XMM-Newton error circle of GRS 1758-258 is reported on the optical image obtained by Marti et al. [8] and compared to the Rosat error circle of 10″ radius and to the VLA position of the point radio source (named VLA C) found at the origin of the extended radio lobes. The XMM error circle intersects the Rosat circle and includes the VLA C position confirming the identification of GRS 1758-258 with the radio source at the origin of the relativistic jets.

The spectrum obtained with XMM-Newton can be compared to previous X-ray spectra obtained with ASCA [9] or RXTE [10]. In all cases it is clear that a spectral evolution took place, since during the XMM observation the source clearly displayed a significant soft component accounting for more than 15 % of the measured X-ray flux, component which was never clearly detected before. Compared to the ASCA measurement the 1-10 keV flux also increased by a factor > 1.5, and the power-law steepened from 1.7 to 2.0. Variability characteristics of the source also changed. In Fig. 3 (right) we report, on the XMM measured PDS, the model of PDS derived with RXTE data [11]. This plot shows the Belloni-Hasinger effect typical of black hole systems in low-hard state [12]. However the low level of the flat slope and the fact that ν_B reaches values > 1 Hz show that the source state during this observation was strikingly close to the intermediate states of Cyg X-1 [12].

We conclude that the source was probably in an intermediate state considerably different from the standard low-hard state which the source displayed since its discovery. The presence of a soft excess was claimed in the past but not fully demonstrated and the disk parameters could not be derived. A more dramatic state transition was very recently observed with RXTE [13,14] when the source entered a very soft state in March 2001 and in that occasion the black body component was detected with a temperature compatible with the values reported here.

CONCLUSIONS

We have presented preliminary results from a XMM-Newton observation of the microquasar GRS 1758-258 in September 2000. The main findings are the following.
1) The source position is determined with 5″ error radius and is compatible with the position of the radio point source at the center of the 2 radio jets.

 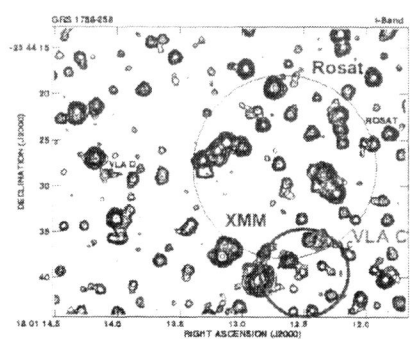

FIGURE 1. XMM Newton 0.2-10 keV image of GRS 1758-258 from the central CCD (11 arcmin size) of EMOS 2 (left). XMM-Newton error circle of GRS 1758-258 (5 " radius) reported on the optical image of the field [8] and compared to the Rosat error circle of the X-ray source and to the VLA error circle of the radio point source (named VLA C) at the origin of the jets (right).

2) The spectrum of GRS 1758-258 cannot be fit by a simple power-law and a soft component (accounting for 15 % of the measured flux) must be included.
3) The soft component can be described by a black body of kT = 0.3 keV while the power-law requires a photon spectral index of 2.0. The total luminosity in 0.2-10 keV band increased by factor > 1.5 with respect to the 1995 ASCA observation.
4) The power density spectrum can be described with the standard flat plus power-law function seen in black hole binaries in low-hard state, but the rms has decreased and break frequency has increased significantly with respect to the values found previously for this source with RXTE.

This observation clearly reveals for the first time in this source the ultra-soft spectral signature of an accretion disk, as seen in many confirmed black hole binary systems. Moreover the strength of the disk emission, the steeper power-law and the different variability characteristics indicate that the source underwent a state transition leaving its standard low-hard state to enter a typical intermediate state, as observed in other black hole binaries like Cyg X-1.

These preliminary results, though need to be confirmed by deeper and more complete analysis, further support the black hole nature of this source and strengthen the link between relativistic jets and black holes.

REFERENCES

1. Mandrou P., *IAUC 5032*, (1990).
2. Gilfanov M., et al., *ApJ* **418**, 844, (1993).
3. Kuznetsov S. I., et al., *Ast.L.*, **25(6)**, 351, (1999).
4. Rodriguez L. F., Mirabel I. F. & Martí J., *ApJ*, **401**, L15 (1992).
5. Mereghetti S., Belloni T., Goldwurm A., *ApJ*, **433**, L21 (1994).
6. Turner M.J.L., et al., *A&A* **365**, L27 (2001).

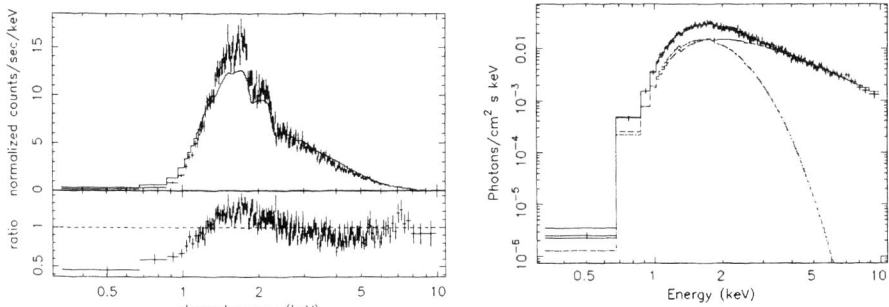

FIGURE 2. GRS 1758-258 folded spectrum obtained from EMOS 2 data compared to the best fit power-law model. Residuals indicate the need to include a soft component (left). The unfolded spectrum compared to the best fit model of a power-law and a black body (right).

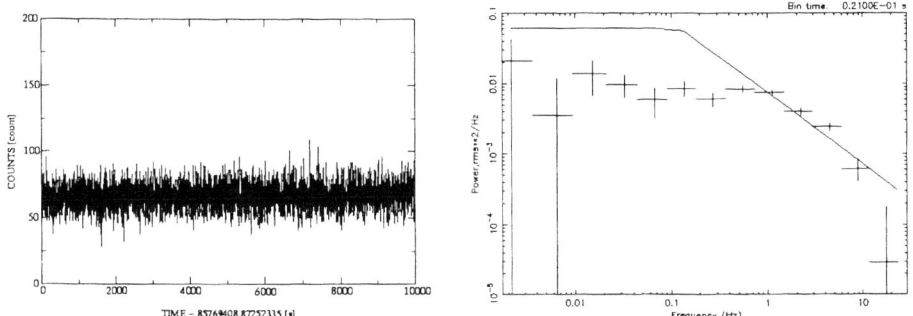

FIGURE 3. GRS 1758-258 light curve from the EMOS 1 data in timing mode with bins of 21 ms (left). Power Density Spectrum (PDS) in units of rms^2 Hz^{-1} derived from the same data and compared to the best fit model of a broken power law found in previous observations of the source (left).

7. Ballet J., *A&A Suppl. Ser.* **135**, 371 (1999).
8. Martí J., et al., *A&A* **338**, L95 (1998).
9. Mereghetti S., et al., *ApJ*, **476**, 829, (1997).
10. Lin D., et al., *ApJ*, **532**, 548, (2000).
11. Smith D.M., et al., *ApJ*, **489**, L51, (1997).
12. Belloni T., et. al., *ApJ*, **472**, L107, (1996).
13. Smith D.M., et al., *IAUC 7595*, (2001).
14. Smith D.M., et al., *ApJL*, submit. (astro-ph/0103381) (2001).

RXTE Observations of GRS 1915+105

J. Rodriguez*, Ph. Durouchoux* and M. Tagger*

*CEA/DSM/DAPNIA/Service d'Astrophysique (CNRS URA 2052), 91191 Gif Sur Yvette Cedex, France

Abstract.
We analyse a set of three RXTE Target of Opportunity observations of the Galactic microquasar GRS 1915+105, observed in April 2000. We concentrate on the timing properties of the source, and examine the properties of a low frequency QPO, with its harmonic, in several energy ranges. The source was found in two different states of the spectro/temporal classification of Belloni *et al.* (2000), and exhibited in the three observations a strong, low frequency QPO together with a strong harmonic. We discuss the properties of the QPO, of its harmonic and of their spectral behaviour in the framework of the Accretion Ejection Instability (AEI) (Tagger & Pellat, 1999; Varnière, Rodriguez & Tagger, 2001; Rodriguez *et al.*, 2001).

INTRODUCTION : QPO'S IN BLACK HOLE BINARIES

X Ray binaries exhibit quasi-periodic behaviors on time scales ranging from millisecond to days or more. Systematic monitoring of these sources, and the use of instruments with high timing resolution such as the Rossi X ray Timing Explorer (RXTE), it has become possible to distinguish several type of QPO's, based on their frequency, and to correlate their properties with the spectral state of the source.
In the case of GRS 1915+105 at least three types of QPO's have so far been detected : a 67Hz one during soft high states (Morgan, Remillard & Greiner, 1997), a 67mHz one, also present during soft high states (Morgan, Remillard & Greiner, 1997), and a 1 − 10Hz variable QPO which appears to be ubiquitous during the low hard state. We focus here only on this ubiquitous QPO, for which several studies have pointed out correlations between the spectral and the temporal parameters, such as the source flux and QPO frequency (Markwardt *et al.*, 1999), or the inner disk temperature and QPO frequency (Muno *et al.*, 1999).
An intriguing result was recently reported by Sobczak *et al.* (1999), who found that in GRO J1655-40 the correlation between the QPO frequency and the disk inner radius was the opposite of that found in XTE J1550-564. In previous work we have confirmed this reversed correlation and compared GRO J1655-40 with GRS 1915+105 (Varnière, Rodriguez & Tagger, 2001; Rodriguez *et al.*, 2001). We showed that it could be explained (see fig 1) if the QPO is identified with the Accretion-Ejection Instability (AEI) of Tagger & Pellat (1999).

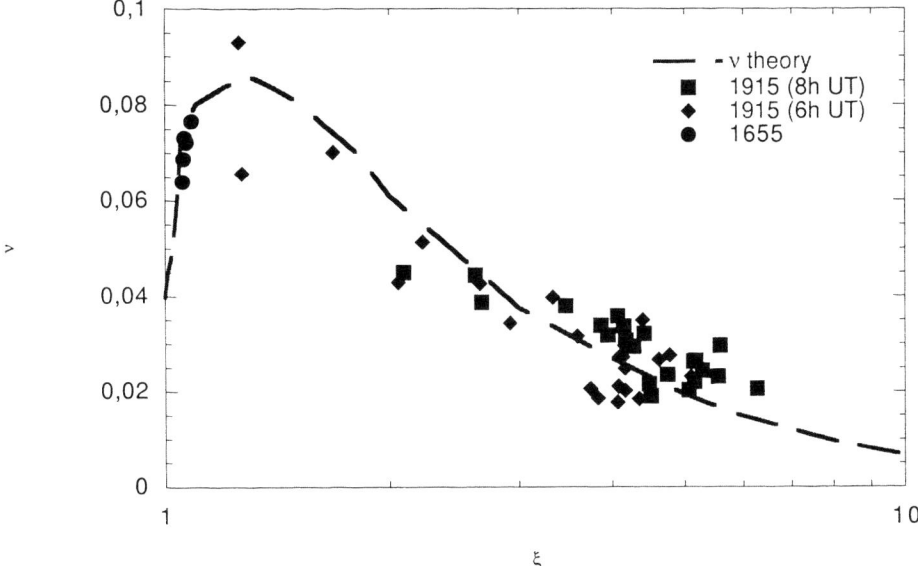

FIGURE 1. Plot of the QPO frequency ν_{QPO} vs. the inner disk radius, r_{int}, for GRS 1915+105 and GRO J1655-40, together with the theorical curve; the X axis is in units of $\xi = r_{int}/r_{Last\ Stable\ Orbit}$, and the Y axis is in units of the keplerian frequency at the last stable orbit. The fit is obtained with the estimated mass of GRO J1655-40, while for GRS 1915+105 it gives a mass of $\sim 20 M_\odot$, compatible with independent studies.

THE SOURCE ON APRIL 17TH

We show on figure 2 dynamical power spectra of the source in five PCA energy bands. The source is in the ν state of Belloni *et al.* (1999), characterized by a ~ 30 mn cycle between a high/soft and a low/hard state. Each interval starts with the source in the high/soft state, showing episodically the QPO. After the "dip" (at $t \simeq 600s.$ and $6000s.$) marking the transition to the low/hard state, the QPO is stronger and shows a prominent harmonic, appearing as a second dark lane in the lower panels of figure 2. A strong X-ray spike, at times $t \simeq 2000s.$ and $7700s.$, marks a return to the high state, with again episodic occurence of the QPO; incidentally the fundamental frequency of the QPO is close to that of the harmonic just before the spike.

Comparing the lightcurves to the variations of the QPO frequency, we find that although the QPO power is stronger in the 5-13 keV band, its frequency variation is better correlated with the softer flux; no QPO is observed above 40 keV; we note that the large dips are smoothed while rising in energy, so that the soft spike corresponds to a sudden decrease of the hard emission; this can be interpreted as the disappearance of a part of the corona, blown away by a sudden ejection coincident with the spike, as already seen during multiwavelength observations of similar states (Eikenberry *et al.*, 1998; Mirabel *et al.*, 1998).

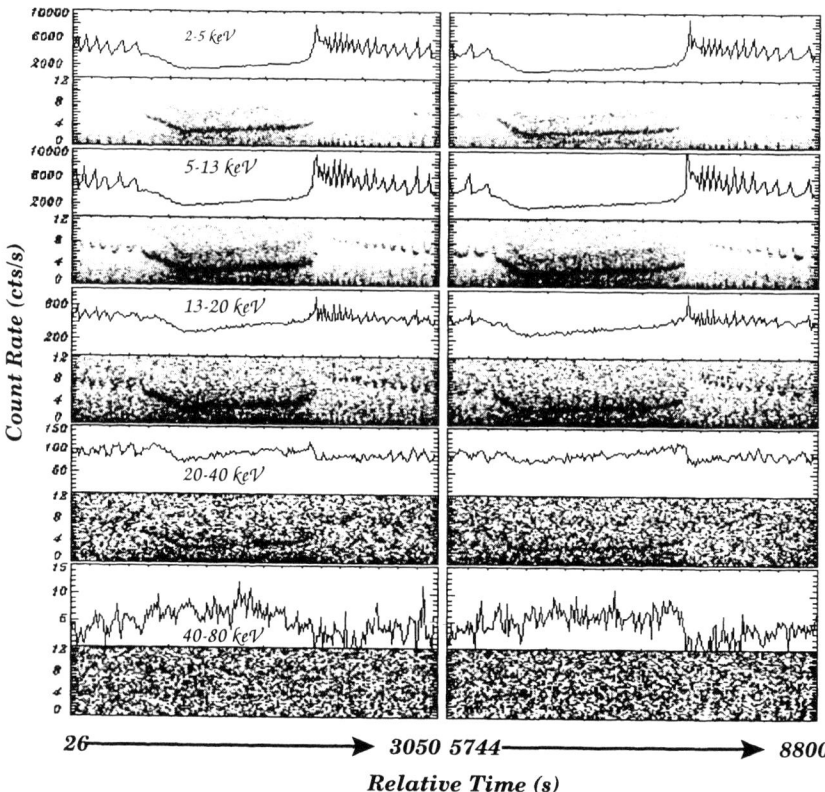

FIGURE 2. Dynamical power spectra of the source on April 17th, in five PCA energy ranges. The X axis is in seconds since the beginning of the time intervals studied. The Y axis is in count/sec. in the upper panels, and in Hz in the lower ones. The gap in the data corresponds to occultations during the orbit

APRIL 22ND AND 23RD

The source is in a low and steady state at these dates, and as we could expect from previous studies (Markwardt et al. , 1999; Muno et al. , 1999; Rodriguez et al. , 2001) the QPO frequency is fairly constant (figure 3), over the whole observations, with a value of ~ 2.15Hz on the first two intervals of April 22, ~ 2.37Hz during the third one, and ~ 2.9Hz on April 23.

We again extracted lightcurves and produced power spectra in 5 PCA energy channels ($\sim 2 - 5$keV, $\sim 5 - 13$keV, $\sim 13 - 20$keV, $\sim 20 - 40$keV, and above 40 keV). power vs. energy, for the QPO and its harmonic, are plotted in figure 4. Only four ranges are plotted since the last one suffers from the lack of flux, which did not allow us to extract QPO parameters. Nevertheless, we find that the QPO power increases with the energy, up to about 30 keV, whereas the harmonic decreases above 10 keV. These distinct spectral behavior thus represent a new challenge for models of the QPO.

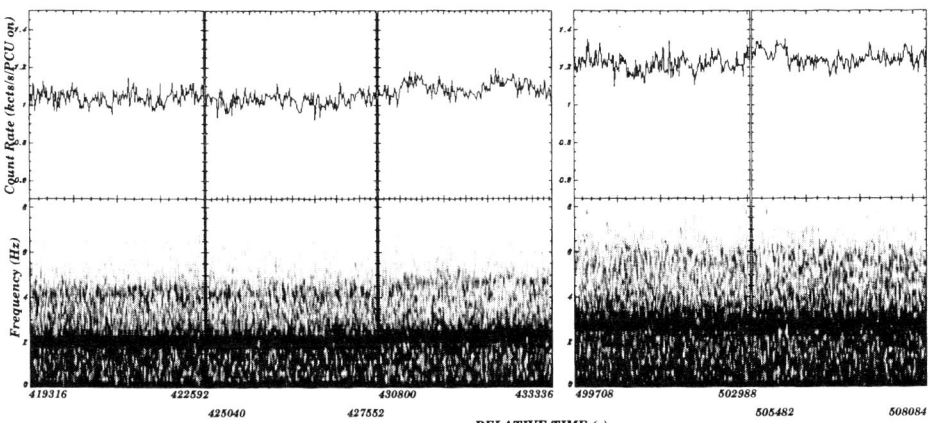

FIGURE 3. Standard lightcurves (upper pannels), and dynamical power spectra of the source (lower pannels) on April 22nd and 23rd. X axis is in second, time 0 is the same as in figure 2.

CONCLUSION

Our study confirms and expands the conclusion (Markwardt *et al.*, 1999; Muno *et al.*, 1999) that the QPO frequency is better correlated with the softer flux, which tends to show that it has its origin in the disk, but that it affects more strongly the flux at higher energies usually considered to be emitted by the corona. In addition figure 2 shows that the presence of the harmonic is correlated to a strong coronal emission; whenever the latter disapear, only the fundamental remains.

Tagger & Pellat (1999) have shown that the inner region of the disk could exhibit what they called an Accretion-Ejection instability, extracting energy and angular momentum from the disk and transferring them to Alfven waves emitted toward the corona. The instability forms a steady spiral pattern, rotating at a frequency compatible with that of the "ubiquitous" QPO (see Varnière and Tagger, these proceedings). In this context, the harmonic can be seen as a diagnostic of the non-linear development of the instability, *e.g.* the formation of a hot point or a spiral shock in the disk (Rodriguez *et al.*, 2001). A useful analogy can be made with the gaseous shock marking the spiral arms of galaxies. In this context the diappearance of the harmonic during the high state could correspond to a weaker instability leading to less sharp non-linear features; the decrease in the power of the harmonic at high energies, contrasting with the behavior of the fundamental, would indicate that the harmonic does not propagate to the corona.

REFERENCES

Belloni T., *et al.*, *Atronomy & Astrophysics*, **355**, 271-290 (2000)
Eikenberry S. S., *et al.*, *The Astrophysical Journal Letters*, **494**, L61 (1998)
Markwardt C. B., Swank, J. H., Taam, R. E., *The Astrophysical Journal*, **513**, L37-L40 (1999)
Mirabel L. F., *et al.*, *Astronomy & Astrophysics*, **330**, L9-L12 (1998)

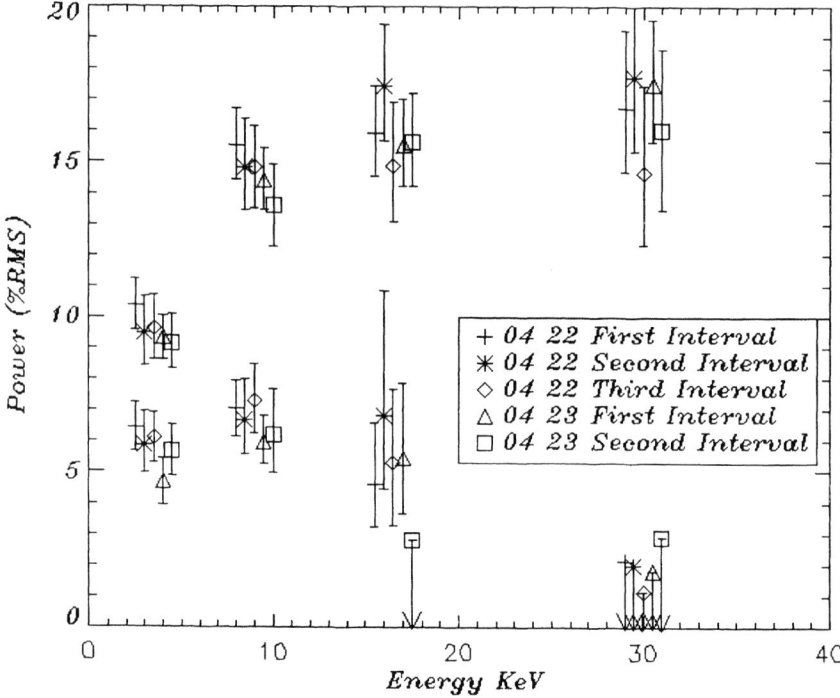

FIGURE 4. Plot of the power(in unit of % RMS) vs. Energy range (keV), for the QPO and its harmonic, in four bands defined in the text; the higher range is not represented since the lack of flux above 40 keV did not permit a detection of a QPO. The lower points correspond to the harmonic, the upper ones to the fundamental.

Morgan E. H., Remillard, R. A., Greiner, J., *The Astrophysical Journal*, **482**, 993-1010 (1997)
Muno M. P., Morgan E. H., Remillard, R. A., *The Astrophysical Journal*, **527**, 321-340 (1999)
Remillard R. A., *et al.* , *The Astrophysical Journal*, **517**, L127-L130 (1999)
Remillard R. A., *et al.* , *The Astrophysical Journal*, **522**, 397-412 (1999)
Rodriguez J., Varnière P., Tagger M., Durouchoux P., accepted for publication in *Astronomy & Astrophysics*
Sobczak G. J. *et al.* ,*The Astrophysical Journal*, **531**, 537-545 (2000)
Tagger M., Pellat R., *Astronomy & Astrophysics*, **349**, 1003-1016 (1999)
Varnière P., Rodriguez J., Tagger M., submitted to *Astronomy & Astrophysics*

Models of phase lags in the rapid aperiodic variability of X-ray binaries

Markus Böttcher[*][1]

Physics and Astronomy Department, Rice University, Houston, TX

Abstract. The most popular models for the complex phase and time lags in the rapid aperiodic variability of Galactic X-ray binaries are based Comptonization of soft seed photons in a hot corona, where small-scale flares are induced by flares of the soft seed photon input (presumably from a cold accretion disc). However, in their original version, these models have neglected the additional cooling of the coronal plasma due to the increased soft seed photon input, and assumed a static coronal temperature structure. In this paper, our Monte-Carlo/Fokker-Planck code for time-dependent radiation transfer and electron energetics is used to simulate the self-consistent coronal response to the various flaring scenarios that have been suggested to explain phase and time lags observed in some Galactic X-ray binaries. It is found that the predictions of models involving slab-coronal geometries are drastically different from those deduced under the assumption of a static corona. However, with the inclusion of coronal cooling they may even be more successful than in their original version in explaining some of the observed phase and time lag features.

INTRODUCTION

The X-ray emission from Galactic X-ray binaries is known to exhibit aperiodic variability on a variety of time scales. Since it is generally believed that the X-ray emission above $\sim 2 - 10$ keV is due to Comptonization of soft photons in a hot, tenuous coronal gas, it is natural to assume that the rapid aperiodic variability of this emission also reveals information about the size scales, dominant physical mechanisms, and geometry of the Comptonizing region.

Theoretical calculations of the expected time-dependent signatures of Comptonization (e.g., Kazanas, Hua & Titarchuk 1997, Hua, Kazanas & Titarchuk 1997, Böttcher & Liang 1998) have been done for several different source geometries under the assumption of a static corona with a constant temperature. While such models, in particular in the case of central injection, representative of a slab-coronal geometry, could successfully reproduce the dominant spectral and timing features of some

---

[1] Chandra Fellow

objects, such as Cyg X-1, they generally required that the corona should extend out to radii of $R \gtrsim 10^4 R_s$, where R_s is the Schwarzschild radius of the accreting compact object, and that out to those radii, the rate of energy dissipation per unit volume would have to decrease only $\propto r^{-1}$, implying that most of the energy is dissipated at large distance from the central object. The required large coronal size scales were implied by the fact that the maximum time lag achievable between hard and soft X-rays corresponds to the difference in photon diffusion time in the process of Compton upscattering and is thus of the order of the light travel time through the corona.

The Comptonization-based models mentioned above had so far been calculated under the assumption of a static corona with a fixed electron temperature, which does not change in response to the increased soft photon input during flares. Since it is generally believed that the coronal temperature — at least in spectral states other than the very-low or off state of transient sources — is determined by the energy balance between various heating mechanisms and cooling dominated by Compton cooling, this may be an unrealistic assumption. The coronal response to accretion-disc flares in the case of a slab-coronal geometry has recently been investigated for two special test cases by Malzac & Jourdain (2000) and in a more systematic parameter study by Böttcher (2001). They found that coronal cooling due to the increased soft photon input leads to rapid cooling of the coronal electron population and a pivoting of the X-ray spectrum around $\sim 10 - 50$ keV for typical parameters. This implies that generally the variability amplitude should decrease with increasing photon energy in the *RXTE* PCA energy range and that flares in the rapid aperiodic variability should be accompanied by spectral softening.

MODEL SETUP

In order to investigate the coronal response to an accretion-disc flare in a slab-coronal geometry (e.g., Liang & Price 1977, Bisnovatyi-Kogan & Blinnikov 1977), we approximate the soft photon input from the disc as a thermal blackbody spectrum of $kT_e = 0.2$ keV, typical of the soft excess in the X-ray spectra of black-hole X-ray binaries in the low/hard state. An accretion disc flare is simulated by increasing the blackbody temperature to 0.5 keV over a limited time interval $\Delta t_{\rm flare}$.

The disc is sandwiched by a tenuous corona of scale height $h = 10^8$ cm and vertical Thomson depth $\tau_{\rm T} = 1$. The distribution function of electrons (i.e. their temperature if the electrons are primarily thermal) is determined by solving the time-dependent Fokker-Planck equation (Böttcher & Liang 2001), including heating through Coulomb interactions with a coronal proton plasma at a fixed temperature of $kT_p = 100$ MeV. Electron cooling through Compton scattering, bremsstrahlung, thermal cyclotron, and nonthermal synchrotron emission is taken into account.

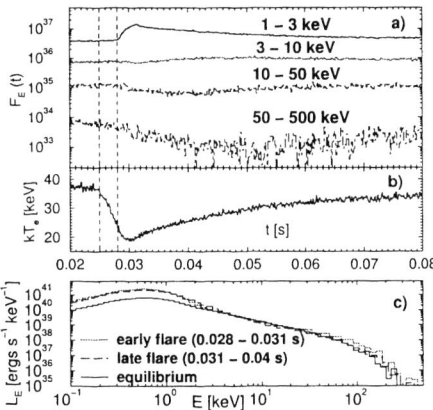

FIGURE 1. Energy-dependent light curves resulting from the coronal response to an accretion disc flare of duration $\Delta t = 3 \times 10^{-3}$ s, within which the disc temperature is increased from 0.2 to 0.5 keV. The vertical dashed lines in panels a) and b) indicate the duration of the flare. The corona (slab geometry) has a Thomson depth $\tau_T = 1$, height $h = 10^8$ cm, and proton temperature $kT_p = 100$ MeV. Panel b) shows the average coronal temperature as a function of time. Panel c) shows snapshot spectra in the early and late flare phases and in equilibrium.

NUMERICAL RESULTS

Fig. 1 shows the energy-dependent light curves, the evolution of the average coronal temperature, and some snapshot energy spectra of the observable X-ray emission for a typical simulation, where $\Delta t_{\text{flare}} = 3 \times 10^{-3}$ s $\approx h/c$. The accretion-disc flare leads to strong cooling of the coronal electrons. Consequently, at hard X-ray energies ($E \gtrsim 3$ keV) a broad dip rather than a flare results.

The simulated hard X-ray light curves can be reasonably well fitted with a function $f(t) = \min\{F_0, F_1(t)\}$, where

$$F_1(t) = F_1^0 \left(e^{-\frac{t-t_0}{\tau_d}} \Theta[t_0 - t] + e^{\frac{t-t_0}{\tau_r}} \Theta[t - t_0] \right). \quad (1)$$

Fig. 2 shows the phase and time lags between the 3 – 10 keV and the 10 – 50 keV bands resulting from three representative simulations with $\Delta t_{\text{flare}} = 10^{-2}$ s, 3×10^{-3} s, and 10^{-3} s, respectively. The figure illustrates that a variety of phase and time lag phenomena can result from this scenario. It is consistent with a Fourier-frequency independent time lag at low frequencies, breaking into a power-law with $\Delta t \propto f^{-\alpha}$, where generally $0.5 \lesssim \alpha \lesssim 1$ (e.g., Cui et al. 1997, Crary et al. 1998).

In the limit $f \ll \tau_r^{-1}, \tau_d^{-1}$, the resulting time lags are dominated by the difference in the fit parameter t_0, which exhibits a photon energy dependence (see Fig. 3) consistent with the logarithmic time lag dependence measured, e.g., for Cyg X-1, GRS 1915+105 (e.g., Cui 1999), and XTE J1550-564 (Wijnands et al. 1999).

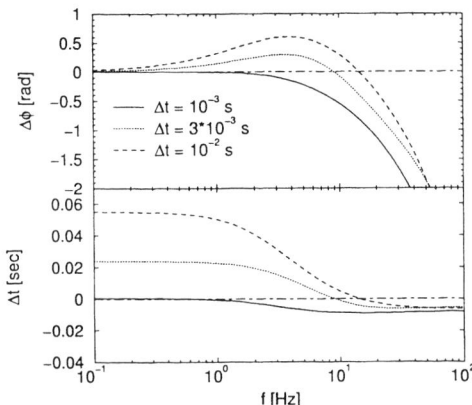

FIGURE 2. Phase lags (upper panel) and time lags (lower panel) between the 3 – 10 keV and the 10 – 50 keV energy bands resulting from the coronal response to accretion disc flares of different durations in the case of slab-coronal geometry. The corona has a Thomson depth $\tau_T = 1$, height $h = 10^8$ cm, and proton temperature $kT_p = 100$ MeV.

Note that this logarithmic energy dependence of the maximum time lag is related to the time scales for Compton cooling and relaxation back to thermal equilibrium due to Coulomb heating rather than due to the difference in photon diffusion time between hard and soft X-rays as in the case of a static corona. This naturally avoids the problem of the size-scale constraint for static-corona models and, instead, yields constraints on the proton temperature and density within the corona, which determine the Coulomb heating time scale. Assuming that both the electron and proton temperatures, $\Theta_{e,p} = kT_{e,p}/(m_{e,p}c^2)$, are non-relativistic, the Coulomb heating time scale (which is expected to be comparable to the maximum time lag between soft and hard X-rays) can be estimated as

$$\tau_{\text{Coulomb}} \sim 3 \times 10^{-3} \, n_{15}^{-1} \frac{\Theta_e}{\Theta_p} (\Theta_e + \Theta_p)^{3/2} \text{ s}. \qquad (2)$$

(Dermer & Liang 1998). Here, n_{15} is the coronal proton density in units of 10^{15} cm^{-3}. Eq. (2) indicates that maximum time lags of order $\sim 10^{-2} - 10^{-1}$ s can naturally occur in such a scenario. Interestingly, Fig. 2 also indicates that over a limited frequency range also negative phase lags (i.e. soft lags) may result.

In this scenario, the onset of an accretion-disc flare marks the onset of the episode of enhanced coronal cooling and thus the onset of the decay of the high-energy light curves. Consequently, considering that in a realistic scenario there will be a rapid succession of such flares occurring throughout the disc, the hard X-ray light curves will exhibit maxima around the onsets of the accretion disc flares, with no appreciable offset between the maxima at different photon energies. This is consistent with the recent result of Maccarone et al. (2000) that the peaks of the

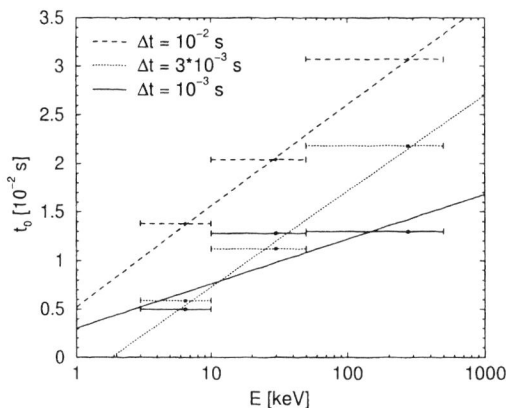

FIGURE 3. Energy dependence of the light-curve fitting parameter t_0, marking the time of the light curve dip, resulting from the coronal response to accretion disc flares of different durations in the case of slab-coronal geometry. The corona has a Thomson depth $\tau_T = 1$, height $h = 10^8$ cm, and proton temperature $kT_p = 100$ MeV. The curves are best fits of a logarithmic dependence, $t_0 = A + B\ln(E)$.

cross-correlation function between light curves at different energy bands observed in Cyg X-1 are consistent with 0 time lag. This would be inconsistent with accretion-disc flaring scenarios in slab-coronal geometry if the observed phase lags were due to the time-dependent Comptonization response in a static corona.

REFERENCES

1. Bisnovatyi-Kogan, G. S., & Blinnikov, S. I., *A&A*, **59**, 111 (1977).
2. Böttcher, M., & Liang, E. P., *ApJ*, **506**, 281 (1998).
3. Böttcher, M., & Liang, E. P., *ApJ*, **511**, L37 (1999).
4. Böttcher, M., & Liang, E. P., *ApJ*, **552**, 248 (2001).
5. Crary, D. J., et al., *ApJ*, **493**, L71 (1998).
6. Cui, W., et al., *ApJ*, **474**, L57 (1997).
7. Cui, W., *ASP Conf. Ser.*, **161**, 97 (1999).
8. Dermer, C. D., & Liang, E. P., *ApJ*, **339**, 512 (1989).
9. Hua, X.-M., Kazanas, D., & Titarchuk, L., *ApJ*, **428**, L57 (1997).
10. Kazanas, D., Hua, X.-M., & Titarchuk, L., *ApJ*, **480**, 735 (1997).
11. Liang, E. P., & Price, R. H., *ApJ*, **218**, 427 (1977).
12. Maccarone, T. J., et al., *ApJ*, **537**, L107 (2000).
13. Malzac, J., & Jourdain, E., *A&A*, **359**, 348 (2000).
14. Wijnands, R., et al., *ApJ*, **526**, L33 (1999).

The Temporal and Spectral Properties of Cyg X-1 during a Large X-ray Flare

Y.X. Feng [*], Wei Cui[*] and S.N. Zhang[†]

[*]Department of Physics, Purdue University, West Lafayette, IN 47907
[†]Physics Department, University of Alabama in Huntsville

Abstract. We present results from a monitoring campaign on Cyg X-1 during a large X-ray flare. During this period, the spectrum of the source evolved significantly: the higher the soft flux the steeper the spectrum. The evolution follows a simple pattern of pivoting around 8-10 keV, which is similar to that observed in the 1996 state transition. The power density spectrum during the flare is typical of the source during a spectral state transition, so do the correlations among temporal properties and the spectral properties. Therefore, an X-ray flare and a spectral state transition appear to be similar phenomenon and the difference seems to be only quantitative.

INTRODUCTION

Cyg X-1 has two distinct spectral states: the low (hard) and the high (soft) states (Oda 1977; Liang & Nolan 1984; Tanaka & Lewin 1995). Usually, Cyg X-1 stays in the low state, but occasionally it undergoes a transition to the high state. In the low state the soft X-ray flux is lower and X-ray spectrum is flatter than those in the high state; the spectrum evolves continuously during a state transition. The continuum component in power density spectrum (PDS) in the low state is composed of a white-noise(flat) component that extends to roughly 0.4-1 Hz then it is cut off with a power law at higher frequencies, while in the high state it is represented by a break power low. During the state transition, its PDS is composed of a low-frequency red-noise (power-law) component, followed by a white-noise (flat) component that extends to roughly 0.4-4 Hz then it is cut off with a power law at higher frequencies (Cui et al. 1997). Recently, it was found that the spectral hardness and soft X-ray flux of the source are positively correlated in the high state and the correlation gradually evolves to a negative one as the source returns to the low state, and vice versa (Li et al. 1999; Wen et al. 2000).

The long-term monitoring of Cyg X-1 shows that it not only undergoes rare spectral state transitions, it also experiences X-ray flares. The duration of the flares varies from less than one day to weeks. Cyg X-1 became very active in 1999. In this paper, we report analysis results of Cyg X-1 monitored with RXTE during a large flare which lasted for about two weeks(Sept.27 - Oct. 11). Fig.1 shows the ASM light curve of Cyg X-1. In the flaring state, there were many short flares with duration less than one day. For comparison between the flaring state and the 1996 state transition, eight observations during the 1996 state transition were analyzed.

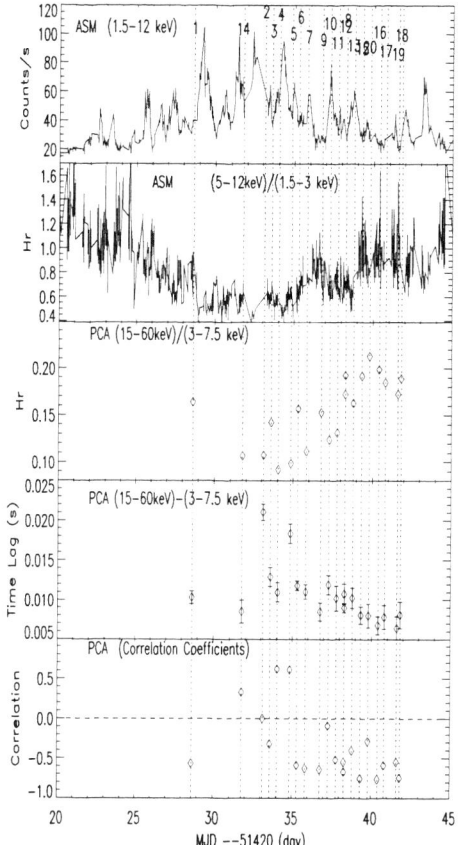
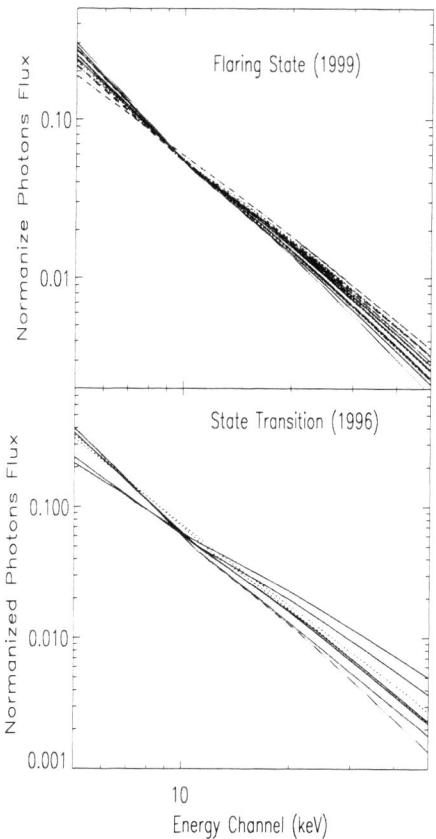

FIGURE 1. Monitoring Cyg X-1 observed with ASM/RXTE in 1999 and the evolution of its parameters. The vertical lines indicated the times of pointing observations of RXTE.

FIGURE 2. The unfolded energy spectra of Cyg X-1. Each of the spectra is normalized to its photons flux in energy range of 3-200 keV.

ANALYSIS AND RESULTS

Energy spectrum in 2-30 keV band was extracted for the first xenon layer of each PCU and the spectrum in 20-200 keV band was extracted from the HEXTE. All the spectra can be represented by the model composed of a multi-color disk, a broken power-law with a high energy cut off, and a Gaussian function. Unfolded spectrum from each observation was normalized to its photon flux (3-200 keV) and plotted in Fig. 2. It shows a pivoting effect around 8-10 keV in the flaring state and around 10 keV in the 1996 state transition.

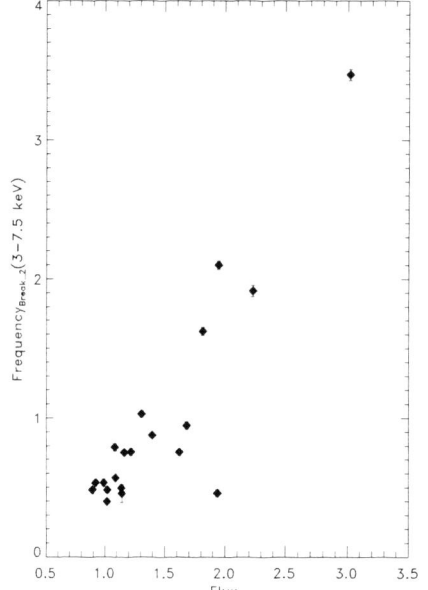
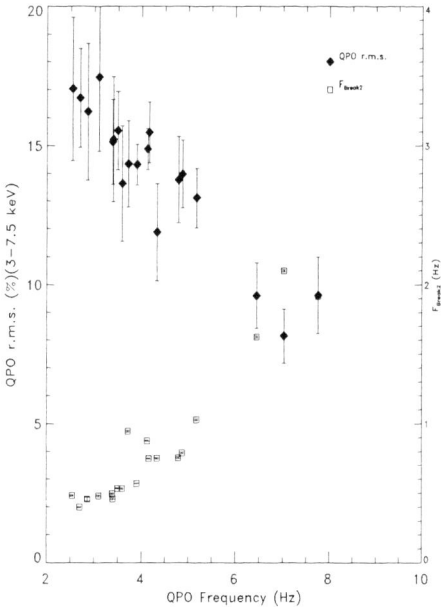

FIGURE 3. The Break$_2$ Frequency and Flux correlation during the flaring state in 1999.

FIGURE 4. The correlations among the break frequency, QPOs frequency, and r.m.s. of the QPOs in 1999.

For each observation, a FFT was performed to obtain a Leahy–normalized power density spectra (PDS) in 3-7.5 keV, 7.5-15 keV and 15-60 keV bands. From the average PDS we subtracted the Poisson noise power corrected for the instrument dead–time effects. Three typical styles of PDS are shown in Fig. 5. To parameterize the power density spectra (PDS), we model the continuum with the following function:

$$PDS(f) = \begin{cases} C(rms_c)^2(\frac{f}{f_{b1}})^{-\alpha_1} & f < f_{b1} \\ C(rms_c)^2 & f_{b1} \leq f < f_{b2} \\ C(rms_c)^2(\frac{f}{f_{b2}})^{-\alpha_2} & f \geq f_{b2}, \end{cases} \quad (1)$$

where C is a constant chosen so that rms_c is the integrated fraction r.m.s. amplitude of the continuum in the frequency range of 0.02–32 Hz. The QPO is modeled by a Lorentzian function. Based upon the FFT, time lags between 3-7.5 keV and 15-60 keV bands were also calculated. The correlation coefficient between hardness ratio (15-60 keV)/(3-7.5 keV) and count rate was calculated in each observation and plotted in Fig. 1.

The low-frequency red-noise component in the PDS disappeared when the spectrum became harder, as shown in Table 1 which indicated also the energy dependent in the continuum of PDS. From Fig.3, it shows that the break frequencies were shifted to higher frequency when X-ray flux increased. From Fig. 4, it could be found that the QPOs

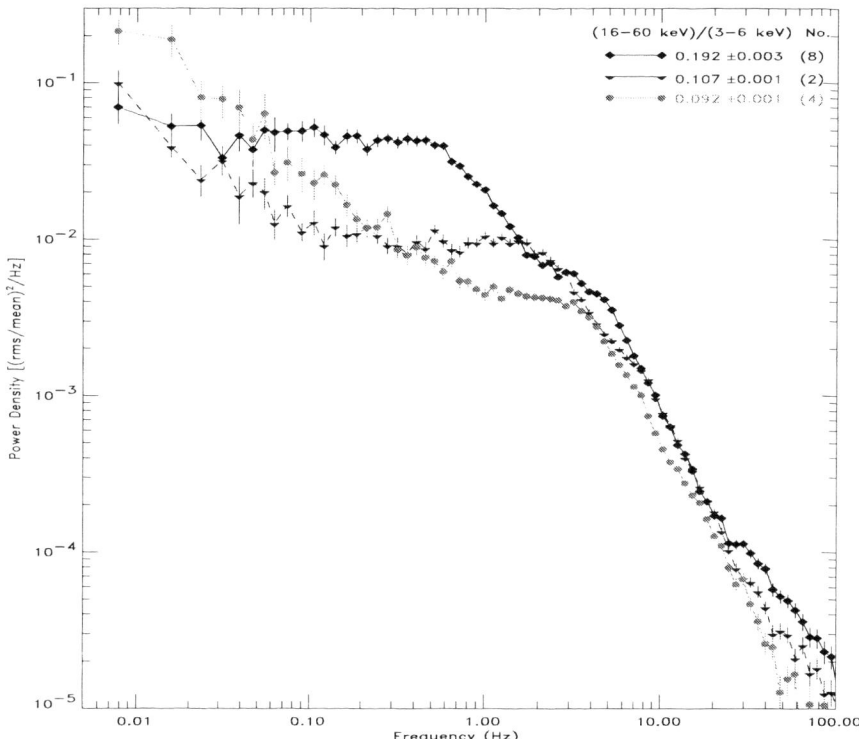

FIGURE 5. The typical PDSs in the flaring state of Cyg X-1 in 1999. The PDSs were obtained from the light curves in the energy band of 3-7.5 keV.

fractional r.m.s. amplitude is anti-correlation with the QPOs frequency. The frequency is correlated with the break frequency. The correlation between the hardness and count rate was positive when the spectrum was soft and became negative when the spectrum was hard, as shown in Fig. 1.

DISCUSSION

The timing and spectral properties and their correlations observed in the flaring state are very similar to those observed in the 1996 state transition except for that Cyg X-1 did not achieve the high state during the flaring state. During observation 4 the flux (3-200 keV) of Cyg X-1 is 5.6×10^{-8} erg cm^{-2}s^{-1} which is higher than that observed in HS during 1996 state transition, but the PDS and time lags showed that Cyg X-1 was in the transition state during the observation 4. It suggested that the flux might be not the only factor to indicate the states of Cyg X-1.

The pivoting energy in the flaring state was about 10 keV which was almost the same as that observed in the 1996 state transition. To qualitatively understand the pivoting effect, compTT model in XSPEC was used to simulate the X-ray emission spectrum from a spherical corona with seed photons in the center of the corona. In this case, output spectrum shows that pivoting effect can be caused by changing the optical depth of the corona alone and higher pivoting energy can be achieved by raising electron temperature in the corona. Therefore, we speculate that the averaged energy of the electrons in the corona during the 1996 state transition was comparable to that during the flaring state in 1999. In phenomemon, the flaring state is a failed state transition. Why Cyg X-1 failed to achieve the high state during the flaring state is still unknown? Other parameters besides the accretion rate may play important roles to control the state transition.

The break frequency increased with X-ray flux has been attributed to decrease in the size of corona which services as a low pass filter for variability in the seed photons (Cui et al. 1997). In this scenario, for a given energy of seed photons, a given optical depth, and a given electron temperature, the higher observed energy band is, the steeper of high frequency cut off in the PDS is. Since the observed cut off was flatter in high energy band than that in low energy band, an extra component with high frequency besides a comptonization component is required in high energy band to explain the flatter cut off.

Acknowledgments: We gratefully acknowledge support from NASA through an ADP grant NAG5-9098 and an LTSA grant NAG5-9998.

REFERENCES

1. Cui, W., Focke, W., & Swank, J. 1996, IAUCIRC 6439
2. Cui, W., Zhang, S., N., Focke, W., & Swank, J. 1997, ApJ 484, 383
3. Li, T. P., Feng, Y. X., & Chen, L. 1999, ApJ, 521, 789
4. Liang, E. P. 1998, Physics Reports, 302, 67
5. McConnell, M. L., Bennett, K., Bloemen, H., Collmar, W., Hermsen, W., Kuiper, L., Phlips, B., Ryan, J. M., Schoenfelder, V., Steinle, H., Strong, A.W. 2001 AIP Conf. Proc., "Gamma Ray 2001"
6. Narayan, R., & Yi, I. 1994, ApJl, 428, L13
7. Ode, M.1977, Space Sci. Rev., 20, 757
8. Pottschmidt, K., Wilms, J., Nowak, M. A., Heindl, W. A., Smith, D. M, et al. 2000, A&A, 357, L17
9. Tanaka, Y., & Lewin, W. H. G. 1995, in "X–ray Binaries", eds. W. H. G. Lewin, J. van Paradijs, & E. P. J. van den Heuvel (Cambridge U. Press, Cambridge) p. 126
10. Wen, L., Cui, W., & Bradt, H. V. 2001, ApJl, 546, L105
11. Zhang, S. N., Cui, W., Harmon, B. A., Paciesas, W. S., Remillard, R. E., & van Paradijs, J. 1997, ApJl, in press (astro-ph/9701027)

X-ray Dips and Orbital Modulation in Cyg X-1

Y.X. Feng* and Wei Cui*

Department of Physics, Purdue University, West Lafayette, IN 47907

Abstract. We observed Cyg X-1 contiguously with RXTE over one 5.6-day binary orbit. Many X-ray dips were detected in the X-ray light curves, which lie mostly between orbital phases 0.8 and 1.2 (with phase 0.0 or 1.0 defined as the times of superior conjunction of the black hole), but dips were also seen at other orbital phases. We discovered that the dips fall into two distinct categories, based on their spectral properties. One (common) type exhibits additional energy-dependent attenuation of X-ray emission at the lowest energies during a dip, which is characteristic of photoelectric absorption, but the other type shows nearly energy-independent attenuation up to at least 20 keV. Moreover, the former seems to occur around superior conjunction but the latter almost at the opposite side of the binary orbit (around phase 0.6), based on limited statistics. Therefore, the first type of dips are likely caused by density enhancement in an inhomogeneous wind of the companion star, while the second type might be due to partial obstruction of an extended X-ray emitting region by an optically thick trailing tidal stream. Such a tidal stream has been shown to exist in hydrodynamic simulations of wind accretion in high-mass X-ray binaries. We also made an attempt to quantifying the varying amount of absorbing material along the line of sight over the orbit. The column density does seem to be higher, on average, around superior conjunction, but large uncertainties in the measurements make it difficult to draw any definitive conclusions.

INTRODUCTION

Cyg X-1 is the first astronomical system that shows strong evidence to contain a stellar-mass black hole (Bolton 1972; Webster & Murdin 1972). It is a binary system with an orbital period of 5.6 days. The most recent radial velocity measurements show that the black hole has a mass about 10 $M_\odot$ (Herrero et al. 1995). The companion star is identified as an O9.7 Iab supergiant (Walborn 1973; Gies & Bolton 1986), with a mass of about 20 $M_\odot$ (Herrero et al. 1995). The X-ray emission from Cyg X-1 is likely powered by accretion of material from the companion star by the black hole. In this case, the accretion flows are thought to follow a pattern that is intermediate between that of Roche-lobe overflow, as in low-mass systems, and that of wind accretion, which is common for high-mass systems (Gies & Bolton 1986). Such an accretion process is sometimes referred to as "focused wind accretion", which can occur when the companion star is very close to filling its Roche lobe (Friend & Castor 1982).

In the hard state, the X-ray intensity of Cyg X-1 is strongly modulated by the orbital motion (Wen et al. 1999; Brocksopp et al. 1999; Priedhorsky, Brandt, & Lund 1995; Holt et al. 1979). Besides the observed global modulation of X-ray emission with orbital motion, X-ray intensity dips are often seen in the light curves of Cyg X-1 (e.g., Pravdo et al. 1980; Kitamoto et al. 1984; Balucinska-Church et al. 1991; Ebisawa et al. 1996). The dips vary in duration from minutes to hours. In this paper, we present results from a long observations of Cyg X-1 with the *Rossi X-Ray Timing Explorer* (RXTE), over one

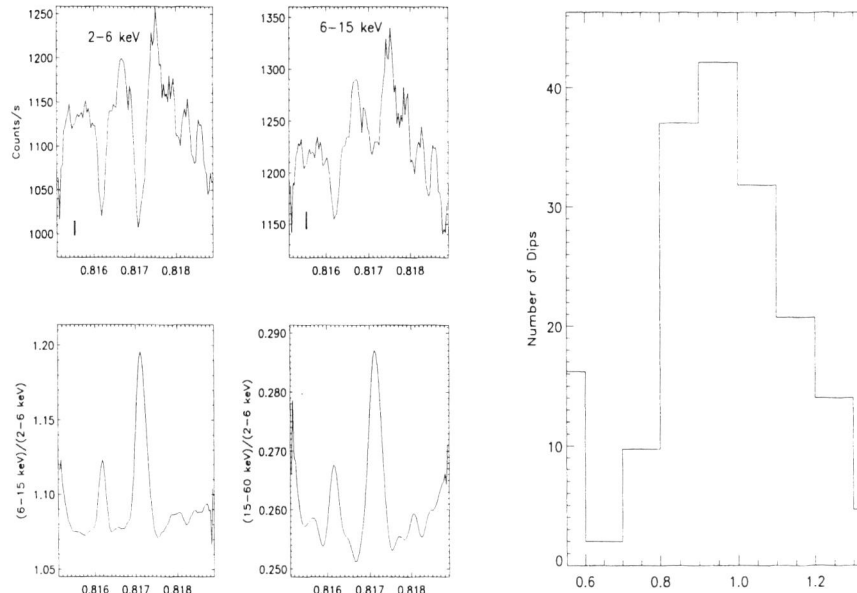

FIGURE 1. X-ray intensity dips of Cyg X-1. The time bin of each data point is 16 s. The light curveshave been smoothed by a running boxcar of width 4 bins. Typical size of error bars is shown for each light curve.

FIGURE 2. Distribution of dips over orbital phases. For phase bins that are not fully covered (due to gaps in the observation), the number of dips has been corrected for coverage fraction (i.e., it represents the number of dips that would be detected if a phase bin is fully covered).

complete 5.6-day orbital cycle.

OBSERVATION

For six consecutive days (Jan 5–11, 2000) Cyg X-1 was observed with the large-area detectors on RXTE, with usual interruptions due to earth occultations and passages of the satellite through the South Atlantic Anomaly. The data was collected for an effective exposure time of about 240 ks. Judging from the long-term ASM/RXTE light curve (flux and hardness ratios) of the source, Cyg X-1 was in its usual hard state when our pointed RXTE observations was made.

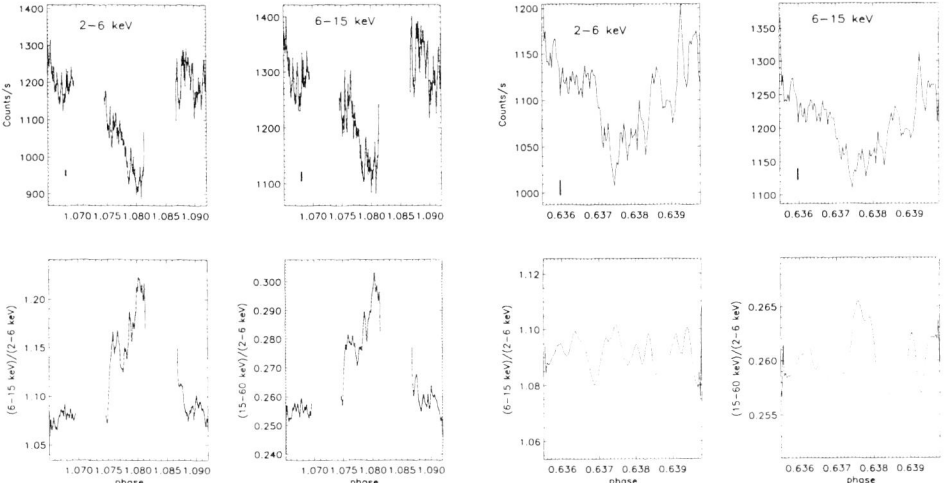

FIGURE 3. Same as Fig. 1, but for a dip of the longest duration. It occurred almost exactly at superior conjunction of the black hole, and lasted for at least 1 hour. Note that there are observation gaps on both sides of the dip.

FIGURE 4. Same as Fig. 1, but for the type B dip observed. The dip occurred near phase 0.6, where a previous study (Blucinska-Church et al. 2000) indicates a secondary peak in the distribution of dips in Cyg X-1.

ANALYSES AND RESULTS

Both spectral and timing analyses were carried out using the *Standard 2* data with time bin size of 16 s. We used *ftools* version 5.0 and made a light curve for each energy bands of 3–6 keV, 6–15 keV, and 15–60 keV, using data from all PCUs (and all xenon layers) that were turned on for a segment of the observation. A corresponding background light curve was then constructed from background models and was subtracted from the overall light curve. Note that we have plotted orbital phases instead of times to emphasize the distribution of dips over the orbital cycle. For that purpose, we have used the most updated ephemeris of Cyg X-1 (Brocksopp et al. 1999).

We identified all the peaks in the time series of hardness ratios. We then verified that each of the peaks actually corresponds to a local minimum in the light curve. As an example, Fig. 1 shows a short segment of the light curve where two dips are clearly visible. A total of 86 dips were selected. Fig. 2 shows their distribution (normalized by exposure time) over the orbit. it shows a peak around phase 0.95 and looks asymmetric about phase 0. It is worth noting that the dip of the longest duration was seen almost exactly at phase 0, as shown in Fig. 3.

Two quite different spectral properties of dips were found. The dips with the different properties were refereed to as type A and type B. Fig. 4 shows an example of a type

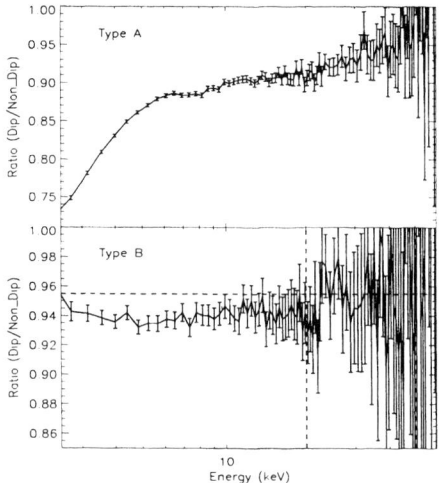
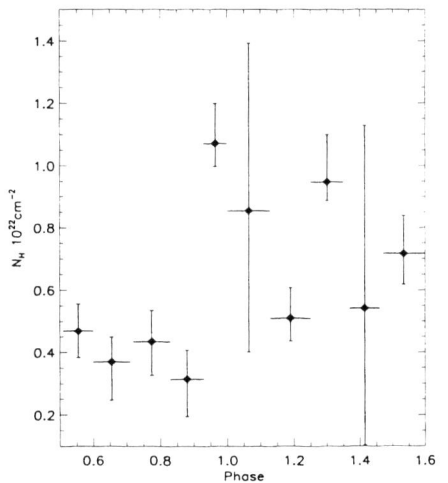

FIGURE 5. Ratio of dip spectrum to non-dip spectrum for the type A dip that is shown in Fig. 3 (upper panel) and the type B dip (lower panel), respectively. Note that the horizontal dashed line in the lower panel represents the averaged value above 20 keV.

FIGURE 6. Observed variation in column density over one orbital cycle.

B dip. The difference between the two kinds of dips can be clearly seen in the ratios between the dip spectrum and non-dip spectrum, as shown in Fig. 5. To study the spectra of the longest dip and non-dip, we fit the non-dip spectrum with a broken power law plus a Gaussian function. The inferred column density is $0.8 \pm 0.4 \times 10^{22}\ cm^{-2}$. We then applied the best-fit model for the non-dip spectrum to the dip spectrum by fixing all model parameters except for the column density. The inferred column density for the dip is $1.7 \pm 0.1 \times 10^{22}\ cm^{-2}$. To investigate the global orbital modulation of X-ray emission previously seen (e.g., Wen et al. 1999), we binned the observation into 10 orbital phases. For each orbital phase, we fit the spectrum with the same model (broken power law plus a Gaussian function). Fig. 6 shows the inferred column density at different orbital phases.

DISCUSSION

The main result of this investigation is the recognition of the existence of two types of dips. While we confirmed (with less statistics) the results of previous studies (Balucinska-Church et al. 2001), including an apparent asymmetry in the orbital distribution of dips, we realized that one dip around phase 0.6 (type B) has quite different spectral properties from those around superior conjunction (type A). The spectrum of type A dips shows energy-dependent reduction at low energies that is characteristic of photoelectric absorption. Detailed spectral modeling shows that the only difference between the spectrum of type A dips and average non-dip spectrum seems to be the presence of additional column density during a dip. We stress, however, that our data is not of sufficient quality to warrant more complicated modeling, as has been done (e.g., Ebisawa et al. 1996). The spectrum of type B dips, on the other hand, shows almost energy-independent reduction . We speculate that this reduction may be caused by partial covering of an extended X-ray emission region by an opaque "screen". In the context of Comptonization models (e.g., review by Liang 1998 and references therein), such an extended emitting region may not be unreasonable for Cyg X-1. The question is what can physically serve as a "screen".

The type A dips are the most common type. They are probably produced by density enhancement in an inhomogeneous wind from the companion star, since the column density changes only moderately during such a dip. They occur more frequently around superior conjunction probably because the line-of-sight follows a longer path through the wind. The inhomogeneities in the wind can be caused by a variety of physical processes (see Balucinska-Church et al. 2001 for a few examples). It remains to be seen whether the observed global orbital modulation of X-ray emission is mostly due to the presence of such dips.

We made an attempt to quantifying how the column density varies with orbital motion by carrying out orbital-phase-resolved spectroscopy. The results seem to indicate that the column density is higher at superior conjunction, although error bars are too large for us to be definitive about it. The issue can be resolved by a similar observation with detectors of much improved low-energy response and spectral resolution, such as those aboard *Chandra* and *XMM*. Much improved spectral resolution of these detectors will also allow the detection of absorption edges, which provides means of determining column density in a manner that is independent of continuum modeling (Schulz et al. 2001).

Acknowledgments: We gratefully acknowledge support from NASA through an ADP grant NAG5-9098 and an LTSA grant NAG5-9998.

REFERENCES

1. Bałucińska-Church, M., Church, M. J., Charles, P. A., Nagase, F., LaSala, J., & Barnard, R. 2000, MNRAS, 311, 861
2. Bałucińska-Church, M., Takahashi, T., Ueda, Y., Church, M. J., Dotani, T., Mitsuda, K., & Inoue, H. 1997, ApJL, 480, L115
3. Blondin, J. M., Stevens, I. R., & Kallman, T. R. 1991, ApJ, 371, 684
4. Bolton, C. T. 1972, Nature, 235, 271
5. Brocksopp, C., Tarasov, A. E., Lyuty, V. M., & Roche, P. 1999, AAP, 343, 861
6. Ebisawa, K., Ueda, Y., Inoue, H., Tanaka, Y., & White, N. E. 1996, ApJ, 467, 419
7. Gies, D. R. & Bolton, C. D. 1986, ApJ, 304, 371
8. Holt, S. S., Kaluzienski, L. J., Boldt, E. A., & Serlemitsos, P. J. 1979, ApJ, 233, 344
9. Kitamoto, S., Miyamoto, S., Tanaka, Y., Ohashi, T., Kondo, Y., Tawara, Y., & Nakagawa, M. 1984, PASJ, 36, 731
10. Pravdo, S. H., White, N. E., Becker, R. H., Kondo, Y., Boldt, E. A., Holt, S. S., Serlemitsos, P. J., & McCluskey, G. E. 1980, ApJL, 237, L71
11. Priedhorsky, W. C., Brandt, S., Lund, N. 1995, AAP, 300, 415
12. Schulz, N. S., Cui, W., Canizares, C. R., Marshall, H. L., Lee, J. C., Miller, J. M., & Lewin, W. H. G. 2001, ApJ, submitted
13. Walborn, N.R., 1973, AJ, 78, 10, 1067
. Wen, L., Cui, W., Levine, A. M., & Bradt, H. V. 1999, ApJ, 525, 968

Can INTEGRAL Detect 2.223 MeV Radiation from X-Ray Binary Sources?

Nidhal Guessoum[1], Pierre Jean[2]

[1]*American University of Sharjah, College of Arts &Sciences, Sharjah, UAE*
[2]*Centre d'Etude Spatiale des Rayonnements, Toulouse, France*

Abstract. We consider the production of 2.223 MeV radiation resulting from the capture of neutrons in the atmosphere of the secondary in an X-ray binary system, the neutrons being produced in the accretion disk around the compact primary star and radiated in all directions. We have considered several accretion disk models (ADAF, ADIOS, SLE, Uniform-Temperature) and a variety of parameters (accretion rate, mass of the compact object, mass, temperature, and composition of the secondary star, distance between the two objects, etc.). The neutron production rates are calculated by a network of nuclear reactions in the accretion disk, and this is handled by a reaction-rate formulation taking into account the structure equations given by each accretion model. The processes undergone by the neutrons in the atmosphere of the companion star are studied in detail, including thermalization, elastic and inelastic scatterings, absorption, escape from the surface, decay, and capture by protons. The radiative transfer of the 2.22 MeV photons is treated separately, taking into consideration the composition and density of the star's atmosphere. We compute neutron production rates in each model and set of parameters. The final flux of the 2.22 MeV radiation that can be detected from earth is calculated taking into account the distance to the source, the direction of observation with respect to the binary system frame, and the rotation of the source, as this can lead to an observable periodicity in the flux. We have produced spectra of the line, where rotational Doppler shift effects can lead to changes in the spectra that are measurable by INTEGRAL's spectrometer (SPI).

1. INTRODUCTION

The capture of a neutron by a proton and the subsequent emission of a photon at 2.223 MeV has been the object of much interest since the dawn of gamma-ray astrophysics because it represents a potentially important spectroscopic tool in the investigation of high-energy environments such as solar flares, neutron star surfaces or accretion disks around compact objects. Several studies have been conducted both from theory and observation, realizing that this line is the most promising candidate of all nuclear gamma-ray processes. With the obvious exception of solar flares no detection has been confirmed and established from any galactic sources to date, despite several previous announcements (Jacobson 1982; McConnell *et al.* 1997). The latter showed an excess emission at (l,b) = 300°,-30° at about 3.7 σ. No obvious counterparts (X-ray binaries or the like) could be linked to that general direction, although a magnetic DA white dwarf has been suggested as a likely candidate for that emission. Van Dijk (1996) has compiled a list of upper-limit 2.22 MeV binary-source radiation fluxes based on COMPTEL observations of 27 black-hole candidates. These upper limits range from 1 to 5×10^{-5} γ/cm^2/s.

On the theoretical front, Guessoum & Dermer (1988) conducted a detailed investigation of compact binary sources, taking Cygnus X-1 as a case-study and pointing out the possibility of emission of a very narrow 2.223 MeV line from the atmosphere of the companion, while Aharonian & Sunyaev (1984) had considered a two-temperature accretion disk with the possibility of emission from the disk itself, where the line would be very broad. Bildsten and co-workers considered the emission at the surface of a neutron star, as a result of bombardment by accreted protons (Bildsten 1991; Bildsten, Salpeter, and Wasserman 1993). Finally, Guessoum & Kazanas (1999), interested in Lithium production by neutron bombardment of X-ray binary companions' atmospheres and using the ADAF model, estimated the flux of 2.22 MeV line emission from a source at 1 kpc and found it very low ($\sim 10^{-7}$ $\gamma/cm^2/s$).

2. NEUTRON PRODUCTION IN THE ACCRETION DISK

Neutrons are produced by breakup of Helium nuclei. Once produced, the neutrons can escape from the disk, provided their kinetic energies allow them to overcome the gravitational binding of the compact object, as they are not affected by the presence of any magnetic fields. So the calculation of the flux of neutrons irradiating the surface of the companion star can be performed, after one takes into account the model-dependent temperature and density distributions of matter in the accretion disk, the nuclear composition of the plasma, and other factors. For general purposes, we assume a mass of 1 $M_\odot$ for the compact object, unless a specific case is considered.

In this work we consider several accretion disk models:
- The Advection-Dominated Accretion Flow (ADAF) model proposed by Narayan and co-workers (Narayan & Yi 1994, and subsequent works);
- The Advection-Dominated Input-Output Solution (ADIOS) proposed by Blandford and Begelman (1999), which consists of a "generalized" ADAF model where only a small fraction of the gas initially supplied ends up falling onto the compact object and where the energy released is transported radially outward and effectively becomes a wind, driving away the rest of the accreted material;
- The two-temperature Shapiro-Lightman-Eardley (1976) (SLE) model;
- The Uniform-Temperature model, where T_i is set by a viscosity condition but is taken to be uniform in the disk, and T_e is chosen arbitrarily (0.1 to 0.5 MeV).

The disk structure equations are given in Jean & Guessoum (2001).

Assuming an initial composition for the accreting plasma (90% H, 10% He in the "normal" case, and 10% H, 90% He in the "He-rich case", the latter corresponding to a secondary star of Wolf-Rayet WN type), we numerically compute the rates of the main nuclear reactions that the various ions can undergo and the values of the abundances of the various nuclei by following the plasma over the accretion timescale. The calculation is performed for various values of the model parameters: $\dot{M} = 10^{-10}$, 10^{-9}, and 10^{-8} $M_\odot/yr$; $\alpha = 0.3$ and 0.1; etc. The escape of the neutrons is accounted for by computing the fraction of neutrons that have velocities sufficient to overcome the gravitational binding to the compact object (see Guessoum & Kazanas 1990).

3. PROPAGATION AND CAPTURE OF NEUTRONS IN THE ATMOSPHERE OF THE SECONDARY

A fraction of the neutron flux emitted by the accretion disk irradiates the atmosphere of the secondary star. Most of these neutrons thermalize; some are captured by a nucleus (H or ^{3}He), the others decay or escape the secondary if after several scatterings their kinetic energy becomes larger than the gravitational potential energy binding them to the star. The 2.22 MeV photons resulting from the capture of neutrons by protons escape the secondary atmosphere if they are not absorbed or Compton scattered in the surrounding gas, so the probability of escape of the photons depends on the depth of their creation site (in the star's atmosphere), on their direction of emission with respect to the surface, and on the composition of the atmosphere.

These processes have all been modeled: the transport of neutrons in the secondary was simulated with the code GEANT/GCALOR, which takes into account elastic and inelastic interactions as well as the neutrons' decay and capture by nuclei; the radiative transfer of 2.22 MeV photons in the atmosphere and the fraction of neutrons falling back onto the secondary were modeled separately, using the results of the previous simulations; the total emissivity at 2.22 MeV is estimated by integrating over the whole secondary surface irradiated by neutrons. The intensity of the radiation that one expects to measure depends on the direction of observation with respect to the binary system frame. The observable photons come only from the area of the secondary that is irradiated by neutrons and is visible to the observer. Therefore we can expect to observe a periodic 2.22 MeV line flux due to the rotation of the binary system. Finally, the rotation of the binary system also leads to a shift in the centroid of the line; this Doppler shift varies with the phase and depends on the observer's direction. Details of the method and effects are presented in Jean & Guessoum (2001).

4. RESULTS

We have considered 2 secondary star prototypes: a normal-composition (90% H) star of 1 solar mass and 1 solar radius, atmospheric density 10^{-4} g/cm^3 and surface temperature 5000 K; a Helium-rich (90% He) star of mass 10 solar masses and 3.6 solar radii, atmospheric density 10^{-5} g/cm^3, and surface temperature 10000 K. These values have been chosen so as to compare neutron capture efficiency in He-enriched and "normal" atmospheres.

Combining the two parts of the work (neutron production in the accretion disk and neutron capture in the secondary's atmosphere), we obtain the final fluxes $F_{2.22}$. In Table 1 we present the "standard" results: for a secondary star of the first kind, an accretion rate $\dot{M} = 10^{-8}$ $M_\odot$/yr, a 2 $R_\odot$ separation between the primary and the secondary, and a distance of 1 kpc for the binary system from Earth.

We have also obtained the 2.22 MeV line profiles for various cases. In Figure 1 we show the spectra for a 3 solar-mass secondary and three values of the star separation.

TABLE 1. Mean 2.22 MeV Flux for Each Accretion Disk Model.

Accretion Disk Model	$\alpha = 0.3$	$\alpha = 0.1$
ADAF	1.1×10^{-5}	2.3×10^{-5}
ADIOS	1.3×10^{-5}	3.4×10^{-5}
SLE	4.8×10^{-6}	1.8×10^{-5}
Accretion Disk Model	$kT_e = 0.5$ MeV	$kT_e = 0.5$ MeV
Uniform-T_i (30 MeV)	1.6×10^{-5}	4.9×10^{-6}
Uniform-T_i (10 MeV)	1.1×10^{-5}	1.7×10^{-6}

The line is broadened and double-peaked due to the rotation of the secondary star.

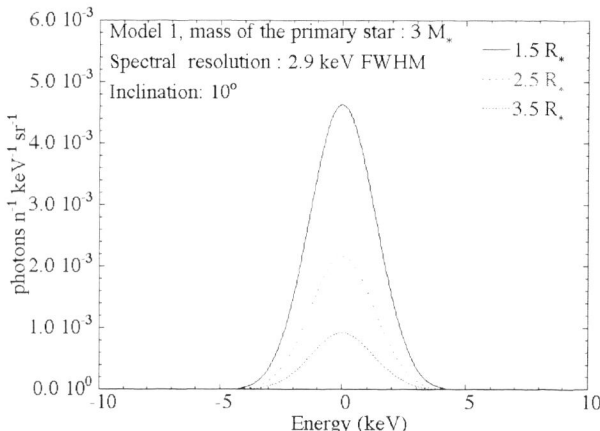

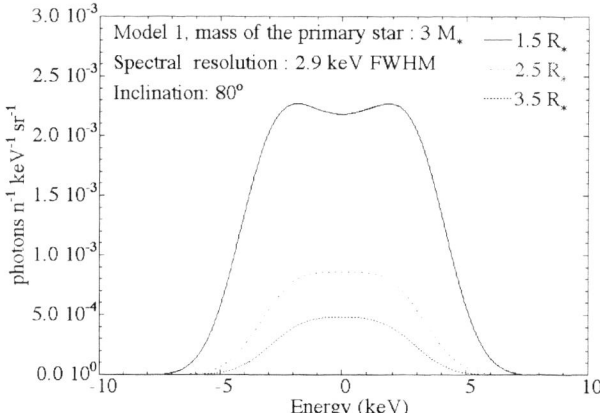

FIGURE 1. Profile of the 2.22 MeV line for a normal-composition secondary star of mass $M = 3\ M_\odot$. The inclination of the binary system with respect to the observer is 10° (above) and 80° (below). Several values of the binary separation are shown.

5. CONCLUSION

We have performed a theoretical investigation of the 2.22 MeV line emission from X-ray binary where neutrons are produced in the accretion disk by nuclear spallation of Helium nuclei. The line intensity was calculated for several accretion disk models, two simple secondary star models, and various X-ray binary geometries.

Once interesting effect we have shown is that, due to the rotation of the secondary, the intensity of the 2.22 MeV radiation varies periodically, and the line centroid is Doppler shifted. Consequently, the spectral and temporal analyses of the 2.22 MeV line flux would provide valuable insights on the characteristics of the binary system (separation, inclination with respect to the observer, masses, accretion rate and disk structure, etc.). If one knows the distance and the geometry of the X-ray binary system, then a measure of a 2.22 MeV line flux from it can set constraints on the neutron production rate and consequently on the accretion disk models (ADAF, ADIOS, SLE, etc.) and on the accretion rate, as the flux depends strongly on $\dot{M}$.

At 2.22 MeV the narrow γ-ray line sensitivity of INTEGRAL's spectrometer SPI is expected to be 7-10 x 10^{-6} γ/cm^2/s for an observation time of 10^6 seconds (Jean et al., 2000). Therefore the 2.22 MeV fluxes estimated in our simple models for various disk models can be measured by SPI (see Table 1) at least in some cases, that is if the accretion rate is large enough ($\dot{M} > 10^{-8}$ M$_\odot$/yr). In case of detection, SPI's spectral resolution (≈ 3 keV at 2.22 MeV) would allow the measurement of the broadening of the line, if the separation and the inclination of the X-ray binary are favorable.

REFERENCES

1. Aharonian F. A. and Sunyaev, R. A., *MNRAS* **210**, 257-277 (1984).
2. Bildsten, L., "Gamma-ray lines from accreting neutron stars" in *Gamma-Ray Line Astrophysics*, edited by P. Durouchoux and N. Prantzos, New York: AIP, 1991, pp. 401-406.
3. Bildsten, L., Salpeter, E. E., and Wasserman, I., *ApJ* **408**, - 3 , (1993).
4. Blandford R. D. and Begelman, M. C., *MNRAS* **303**, 1-5 (1999).
5. Guessoum, N. & Dermer, C.D., "Properties of Hydrogen/Helium Accretion Plasmas" in *Nuclear Spectroscopy of Astrophysical Sources*, edited by N. Gehrels and G. H. Share, New York: AIP, 1988, pp. 332-337.
6. Guessoum, N. and Kazanas, D., *ApJ* **358**, 5 5-53 (1990).
7. Guessoum, N. and Kazanas, D., *ApJ* **512**, 33 -339 (1999).
8. Jacobson, A. S. et al., Gamma-Ray Spectroscopy in Astrophysics (NASA T.M. 79619), 1978, p.228.
9. Jean, P. et al., "The Spectrometer SPI of the INTEGRAL Mission" in *The Fifth Compton Symposium*, edited by M. L. McConnell and J. M. Ryan, New York: AIP, 2000, pp. 708-711.
10. Jean, P. and Guessoum, N., submitted to *A & A*, (2001).
11. McConnell, M. et al., "COMPTEL All-Sky Imaging at 2.2 MeV" in *Proceedings of the Fourth Compton Symposium*, edited by C. D. Dermer et al., New York: AIP, 1997, pp. 1099-1103.
12. Narayan, R. and Yi, I., *ApJ Letters* **428**, L13-16, (1994).
13. Shapiro, S.L., Lightman, A.P., and Eardley, D.M., *ApJ* **204**, 187-199, (1976).
14. Van Dijk, R., PhD Thesis, p. 14, (1996).

Wide-band X-ray variability of GRS 1915+105 observed with BeppoSAX

P. Casella, M. Feroci, E. Massaro, E. Costa, M. Litterio*, G.Matt[†], T. Belloni**, M. Tavani[‡], F.I. Mirabel[§], A.J. Castro-Tirado[¶], A. Harmon[∥] and G. Pooley[††]

Istituto di Astrofisica Spaziale, CNR, Roma, Italy
[†]*Physics Department, Universitá Roma TRE, Roma, Italy*
***Osservatorio Astronomico di Brera, Merate, Italy*
[‡]*Istituto di Fisica Cosmica, CNR, Milano, Italy*
[§]*Service d' Astrophysique, CEA, Saclay, France*
[¶]*Istituto de Astrofisica de Andalusia, Granada, Spain*
[∥]*NASA Marshall Space Flight Center, Hunstville, Alabama, USA*
[††]*Mullard Radio Astronomy Observatory, Cambridge, UK*

Abstract.
The Galactic Microquasar GRS 1915+105 was observed by the Narrow Field Instruments onboard BeppoSAX in two pointings during year 2000. We present the preliminary results of wide-band study of the observed variability, carried out also with the wavelet analysis.

INTRODUCTION

Since its discovery in 1992 ([4]), GRS 1915+105, the prototype of galactic microquasars ([7]), is one of the most observed sources, particularly in the X-ray range. Its light curve shows a high variable behavior, characterized by strong and structured outbursts interrupted by quiescent phases. No evidence of periodic variations, related to orbital motion in a binary system, has been firmly established. It is general opinion, by the analogy with similar sources, that the X-ray emission is originated in an accretion disk rotating around a few stellar mass black hole. The X-ray spectrum is generally fitted by at least two components: a multitemperature disk blackbody and a power law extending to several hundreds keV. During the flares the main spectral parameters of the thermal component show significant variations, which have been interpreted by the emptying and refilling of the inner portion of the accretion disk ([1]).

In a recent paper, Belloni et al. ([2]), on the basis of a large set of RXTE observations, defined 12 different variability modes of the X-ray emission, each of them characterized by a time profile and spectral variability as apparent from the dynamical hardness ratio plots. This classification is potentially useful for the understanding of the physical processes occurring in this exceptional source and will be used in our analysis.

In this contribution we present some preliminary results of a wide band X-ray observations of GRS 1915+105 recently performed with the Italian-Dutch satellite BeppoSAX. In particular, we compare the light curves in the energy bands of the MECS (2-10 keV)

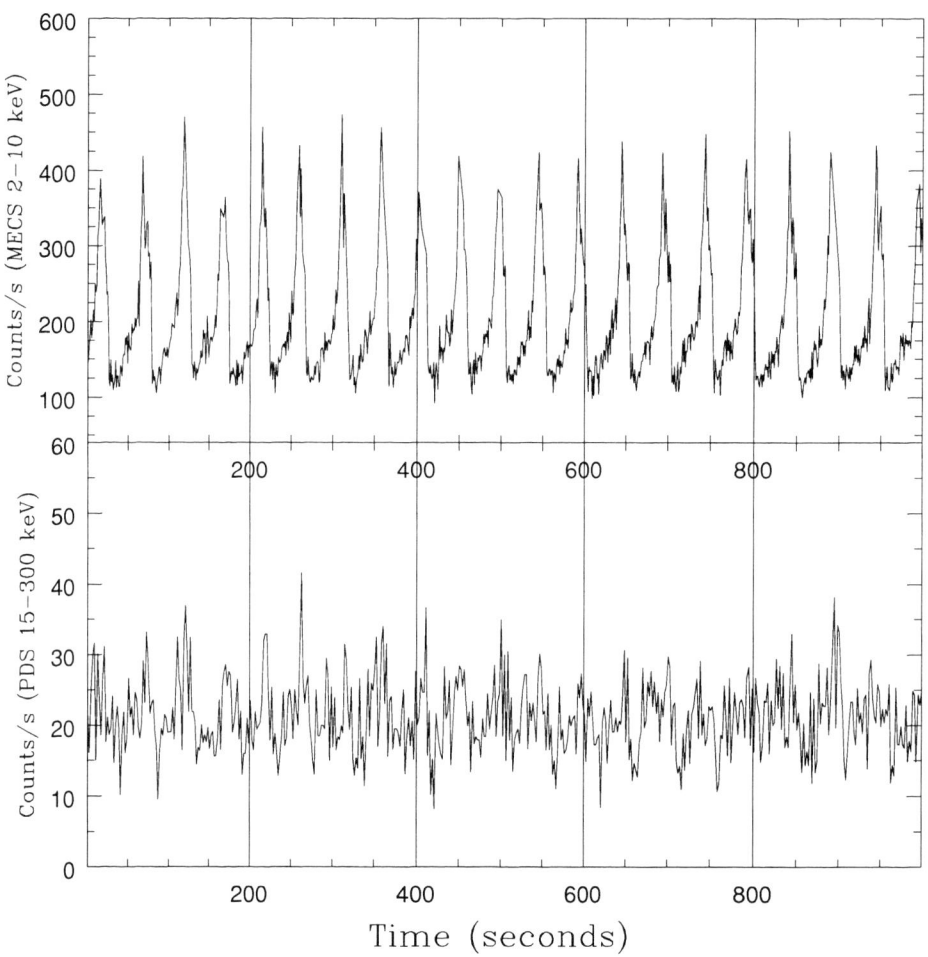

FIGURE 1. Two simultaneous portions of the light curve in the MECS and PDS during a regular ρ mode.

and PDS (13-300 keV) in regular and irregular variabiltiy modes and describe the evolution of short time scale by means of wavelet power spectra.

OBSERVATIONS AND DATA REDUCTION

Recent BeppoSAX observational campaigns on GRS 1915+105 were performed in the period April 1999, April and October 2000. The main parameters of the last three

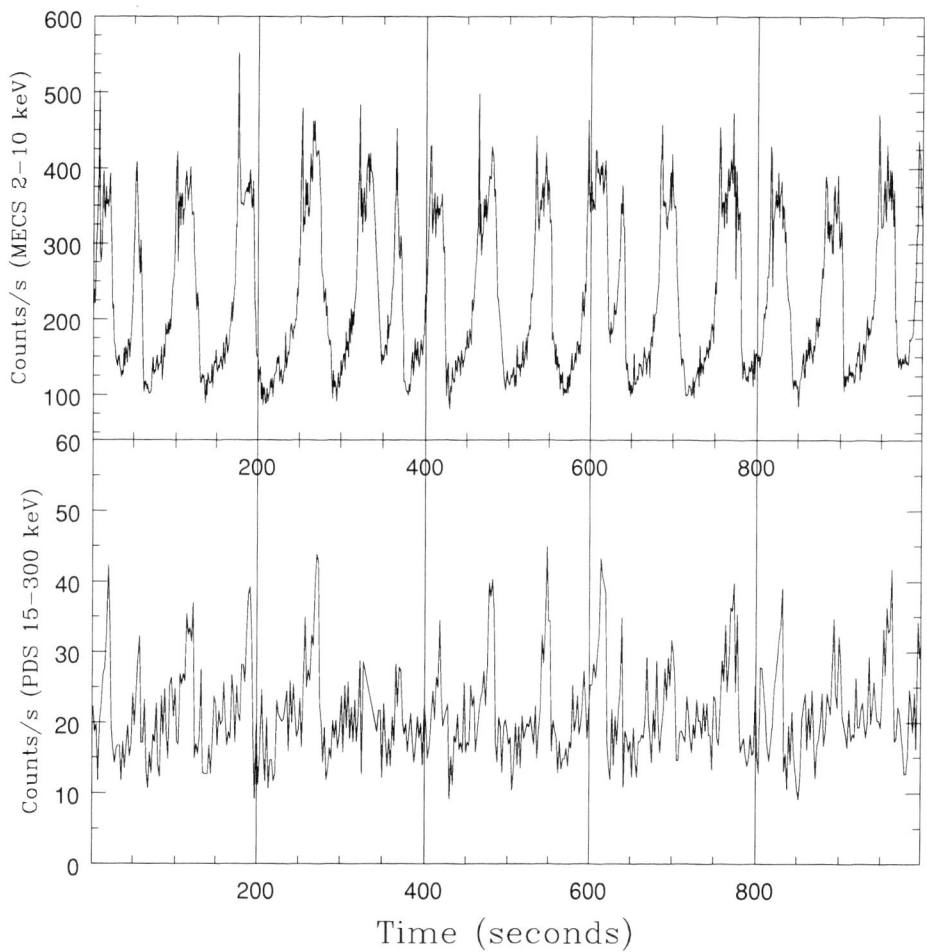

FIGURE 2. Two simultaneous portions of the light curve in the MECS and PDS during a non-regular ρ mode.

observations are reported in Table 1. Standard procedures and selection criteria were applied to the BeppoSAX data to avoid the South Atlantic Anomaly, solar, bright Earth and particle contamination using the SAXDAS v. 2.0.0 package. Source counts for each energy channel were extracted from the LECS and MECS images using standard algorithms and parameters. PDS data were taken in the standard direct mode, with a collimator rocking law with 96 s dwell time in order to observe simultaneously source and background.

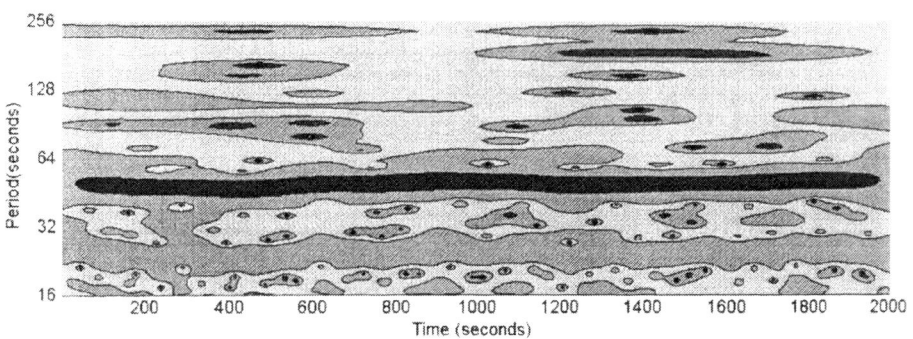

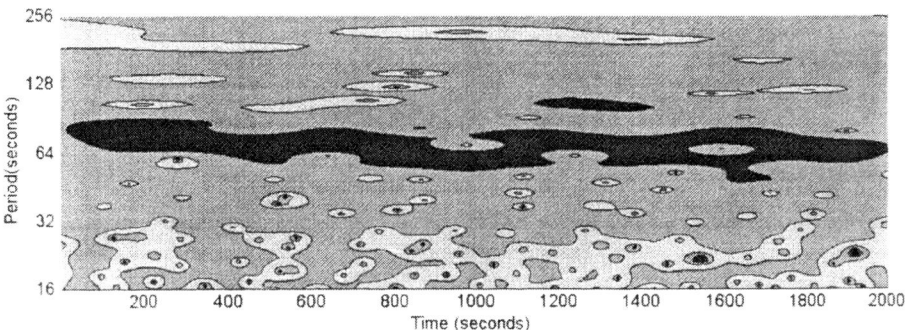

FIGURE 3. Wavelet power spectra of two MECS light curve portions (each 2,000 s long) during regular (Fig. 1 time series, upper panel) and non-regular (Fig. 2 time series, lower panel) ρ mode. The dark bands at 50 s and 70 s, the most significant structures in the two plots, show the evolution of the recurrent time scales in the two data sets.

THE STRUCTURE OF THE LIGHT CURVES

For the entire duration of the April 1999 observation the 2-10 keV X-ray emission of GRS 1915+105 showed the characteristic pulsed modulation of the ρ mode ([2]). Pulses were not well evident in the PDS (13-300 keV) and LECS (0.1-2 keV) bands because of the lower count rate due to the strong interstellar absorption below a few keV and the dominance of the the power law component above 15 keV (for a detailed analysis see e.g. [3], [6]).

The April 2000 pointing covered only about 3 days in the course of a longer (10 days) multifrequency campaign ([10]). In the MECS and PDS time series the X-ray emission of GRS 1915+105 was characterized by a slow variability superimposed to a random noise, whose amplitude was compatible with the Poisson statistical fluctuations. A large flare was detected in the MECS, but not in the PDS. Unfortunately, it was just at the end of the visibility window during the BeppoSAX orbit and therefore its time evolution was not entirely followed.

The observation of October 2000 was particularly important because of its long overall duration, about 10 days, which gave us the possibility to detect the transition between different modes. A stable ρ mode was observed during the first 160,000 s followed by an irregular phase about 25,000 s long. Another and longer not stable phase was in the interval from about 275,000 s to 375,000 s and after the source showed again for a long time a stable ρ mode interrupted by a quiescent state at 615,000 s. Finally, in the final part, a transition to the ν mode occurred. Two pairs of short segments, 1,000 s long, of the MECS and PDS light curves are shown in the Fig. 1 and 2, respectively. The former, corresponds to the regular ρ mode, while the latter to the irregular one, which is characterized by broader pulses with a recurrence time longer than 70 s, and by the apperence of narrower pulses between two main ones. The higher energy time series are also quite different: pulses are barely detectable in Fig. 1, while they are well evident in Fig. 2.

The stability of these two types of behavior of the ρ mode was also studied by means of the wavelet analysis, which gives a description of the time evolution of the periodogram and allow to recognize the occurrence and the stability of the various time scales. Wavelet power spectra were computed using the Morlet wavelet of order 12 ([5]). Two spectra for the MECS time series are presented in the panels of Fig. 3. The spectrum of the regular ρ mode is characterized by a rather stable recurrence time of the peaks of about 50 s, as shown by an uninterrupted dark linear strip (corresponding to the highest power), centered at this period value. In the irregular mode this strip shows meanderings and in some cases more dominant frequencies appear at the same time.

An important finding is that the pulses in the high energy curve occurr in the second half of the corresponding low energy peaks. This effect suggest that there is a delay like the emission would be due to different components. Such behavior is reminiscent of that observed in the so called *plateau* intervals, during which a QPO frequency lower than 2 Hz is observed ([9]), and interpreted by Nobili et al. ([8]) in terms of a comptonization model. Another possibility is that the non-regular ρ mode is associated with a greater oscillation amplitude of the innermost region of the accretion disk where higher temperatures can be reached. A more complete time resolved spectral analysis, however, is necessary to derive further information on the physical conditions of the emitting plasma and its instabilities.

REFERENCES

1. Belloni T. et al. 1997, *ApJ 488, L109*
2. Belloni T. et al. 2000, *A&A 355, 271*
3. Casella P. et al. 2001, *Proc. Integral Workshop, Alicante*
4. Castro-Tirado A.J., Brandt S., Lund N. 1992 *IAUC 5590*
5. Farge M. 1992, *Ann. Rev. Fluid. Mech., 24, 395*
6. Feroci M. et al. 2001, *Proc. Microquasars Workshop, Granada*
7. Mirabel I.F., Rodriguez L.F. 1994, *Nature 371, 46*
8. Nobili L. et al. 2000, *ApJ 538, L137*
9. Reig P. et al. 2000, *ApJ 541, 883*
10. Ueda Y. et al. 2001, *Proc. Microquasars Workshop, Granada*

Gamma-Ray Spectral Variability of Cygnus X-1

M. L. McConnell[1], K. Bennett[2], H. Bloemen[3], W. Collmar[4], W. Hermsen[3], L. Kuiper[3], W. Paciesas[5], B. Phlips[6], J. Poutanen[7], J. M. Ryan[1], V. Schönfelder[4], H. Steinle[4], A. W. Strong[4], and A. A. Zdziarski[8]

[1] Space Science Center, University of New Hampshire, Durham, NH 03824
[2] Space Science Department, ESTEC, Noordwijk, The Netherlands
[3] SRON – Utrecht, Utrecht, The Netherlands
[4] Max Planck Institute for Extraterrestrial Physics, Garching, Germany
[5] University of Alabama, Huntsville, AL
[6] George Mason University, Fairfax, VA
[7] Stockholm Observatory, SE-106 91 Stockholm, Sweden
[8] N. Copernicus Astronomical Center, Warsaw, Poland

Abstract. We have used observations from CGRO to study the variation in the MeV emission of Cygnus X-1 between its low and high X-ray states. These data provide a measurement of the spectral variability above 1 MeV. The high state MeV spectrum is found to be much harder than that of the low state MeV spectrum. In particular, the power-law emission seen at hard X-ray energies in the high state spectrum (with a photon spectral index of 2.6) is found to extend out to at least 5 MeV, with no evidence for any cutoff. Here we present the data and describe our efforts to model both the low state and high state spectra using a hybrid thermal/nonthermal model in which the emission results from the Comptonization of an electron population that consists of both a thermal and nonthermal component.

INTRODUCTION

It has long been recognized that the soft X-ray emission of Cygnus X-1 (~10 keV) generally varies between two discrete levels [1,2]. The 20–100 keV time history of Cygnus X-1, as derived from BATSE occultation data, is shown in the center panel of Figure 1. The top panel of Figure 1 shows the 20–100 keV power-law spectral index, as derived from the BATSE occultation data. These data cover most of the CGRO mission, from the launch in April of 1991 until the end of 1999. During the first few months of the CGRO mission (up until October of 1991), all-sky monitoring data from Ginga (1–20 keV) was available, confirming that the source was in its low X-ray state during this period [3]. From October of 1991 until December of 1995, there were only sporadic pointed X-ray observations of the soft X-ray flux from Cygnus X-1. It was not until the launch of RXTE, in December of 1995, that continuous data on the soft X-ray flux once again became available. The data from the RXTE All-Sky Monitor (ASM) are shown in the lower panel of Figure 1, in the form of the 2–10 keV count rate.

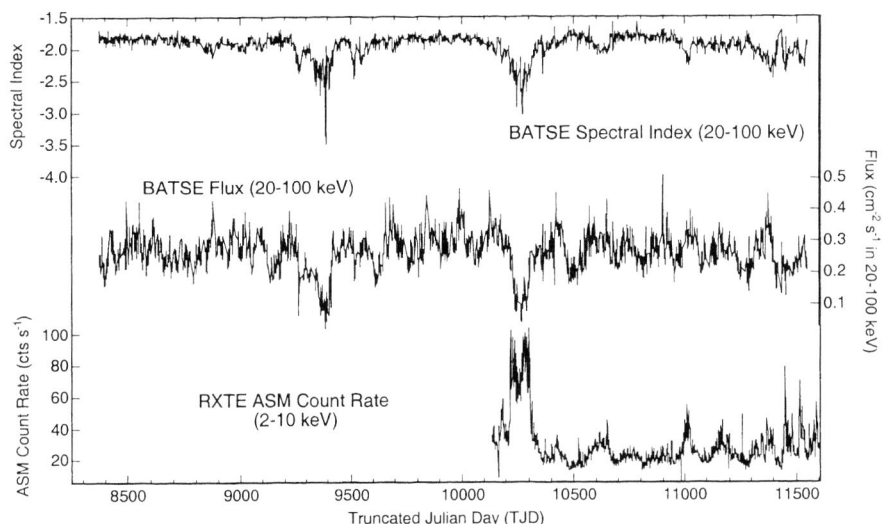

FIGURE 1. Time histories of the X-ray behavior of Cygnus X-1 for a time period encompassing most of the CGRO mission (1991–1999). The top two panels represent data from the BATSE experiment, showing the 20–100 keV flux (middle) and the power-law index for an assumed power-law spectrum in the 20–100 keV energy band (top). The lower panel shows the 2–10 keV count rate from the All-Sky Monitor (ASM) on RXTE.

The data shown in Figure 1 dramatically demonstrate the general X-ray behavior of Cygnus X-1. About 90% of the time, Cygnus X-1 is in its so-called "low" X-ray state (as defined by the soft X-ray emission). In this state, the soft X-ray flux (2–10 keV) is relatively low, while the hard X-ray flux (20–100 keV) is relatively high. The spectral shape in the 20–100 keV energy band is a relatively hard power-law spectrum with a photon spectral index near 1.8. During the so-called "high" X-ray state, the soft X-ray flux (2–10 keV) increases by about a factor of 3–4, while the hard X-ray flux (20–100 keV) decreases by about a factor of 3. This results in a much softer power-law spectrum, with a spectral index between 2.5 and 3.0. Data provided by all four instruments on CGRO provide an unprecedented opportunity to study the variability of the γ-ray emission between these two states.

THE LOW STATE SPECTRUM

In earlier work [4], we assembled a broadband spectrum of the γ-ray emission from Cygnus X-1 by combining COMPTEL data with contemporaneous data from the BATSE [5,6], OSSE [7], and EGRET [8] experiments on CGRO. Due to the small FoV of OSSE (and the near all-sky monitoring capability of BATSE), the selection of data in this case was driven by the availability of contemporaneous OSSE data. The selection of data was also confined to the first three cycles of CGRO observations (1991–1994), based on the quality of the COMPTEL data that was available at the time the study was undertaken. A final selection on the data was imposed based on the level of hard X-ray flux and the spectral index observed by BATSE (Figure 1). This was done to ensure that there were no spectral state changes during the selected observations.

The general form of the spectrum (Figure 2) is that of the "breaking γ-ray state" that is generally associated with the low X-ray state of galactic black hole binaries [10]. The low-state spectrum provides evidence of significant emission out to at least 2 MeV. There is additional evidence for emission bewteen 2–5 MeV, with no evidence for emission above 5 MeV. The data do not provide any evidence for a high-energy cutoff to the spectrum. Also shown in Figure 2 is an estimated upper limit from EGRET data collected during Cycles 1–4 of the CGRO mission, an exposure that is similar to that for the other data included in this analysis [8]. Unfortunately, the EGRET upper limit (based on an assumed E^{-3} source spectrum) does not provide any further constraints on the extrapolated spectrum. Attempts were made to fit the broadband spectrum with both one- and two-component Compton models (based on [9]), and with an exponentiated power-law model. None of these models provided a statistically acceptable fit to the data. They all underestimated the measured flux levels at the highest energies near 1 MeV and above.

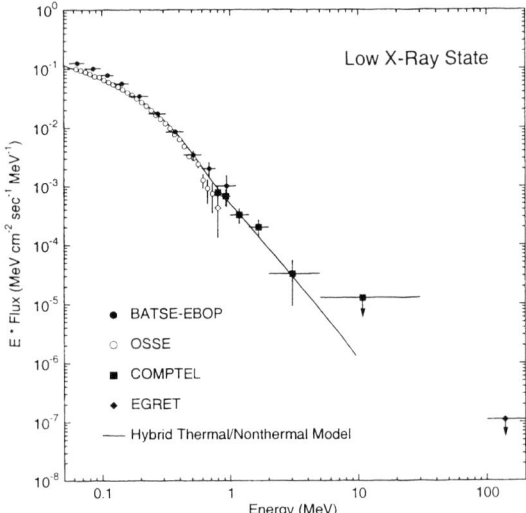

FIGURE 2 - Contemporaneous broadband CGRO spectrum of the low X-ray state of Cygnus X-1. For the sake of clarity, upper limits from OSSE and BATSE are not shown, but these are consistent with the total dataset.

THE HIGH STATE SPECTRUM

During the CGRO mission, Cygnus X-1 spent only about 10% of its time in the high X-ray state. The high X-ray state was clearly observed on only two occasions. In each case, the high state period lasted about 5 months. The high X-ray state was first observed by CGRO in January of 1994, at a time (prior to the launch of RXTE) when there was no soft X-ray monitoring data available. (This transition is clearly seen in Figure 1 near TJD 9400.) A CGRO target-of-opportunity was declared (CGRO viewing period 318.1) so that all four CGRO instruments (not just BATSE) could collect data. Observations by COMPTEL showed no detectable level of emission. This null result, however, was consistent with an extrapolation of the $E^{-2.7}$ power-law spectrum measured at hard X-ray energies by both BATSE [11] and OSSE [7].

The second observation of a high X-ray state took place in May of 1996. The transition was first observed by RXTE, beginning on May 10 [12]. The 2-12 keV flux reached a level of 2 Crab on May 19, four times higher than its normal value. Meanwhile, at hard X-ray energies (20-200 keV), BATSE measured a significant *decrease* in flux [13]. Motivated by these dramatic changes, a ToO for CGRO was declared and observations by OSSE and COMPTEL began on June 14 (CGRO viewing period 522.5). (Unfortunately, the EGRET experiment was turned off during this viewing period, as

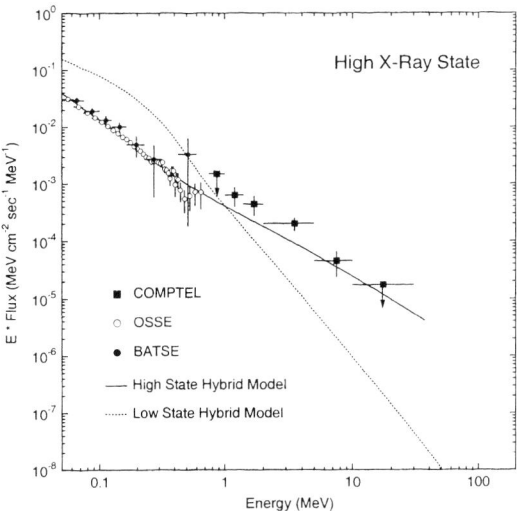

FIGURE 3 - Broadband CGRO spectrum of the high X-ray state of Cygnus X-1 (VP 522.5). For the sake of clarity, upper limits from OSSE and BATSE are not shown, but these are consistent with the total dataset.

part of an effort to conserve its supply of spark chamber gas.) During the ToO, COMPTEL collected 11 days of data (from June 14 to June 25) at a favorable aspect angle of 5.3°.

Whereas the low-state CGRO spectrum (Figure 2) shows the breaking type spectrum that is typical of most high-energy observations of Cyg X-1, the high-state CGRO spectrum (Figure 3) shows the power-law type spectrum that is characteristic of black hole binaries in their high X-ray state [10]. The high state power-law spectral behavior had already been reported for the 1996 transition based on observations with both BATSE [11, 13] and OSSE [7, 14]. An earlier detailed study of the broadband high state spectrum was based on data from ASCA, RXTE and CGRO/OSSE, but did not include the higher energy COMPTEL data [14]. The inclusion of the COMPTEL data in the high state spectrum provides evidence of a continuous power-law (with a photon spectral index of 2.6) extending beyond 1 MeV, up to ~10 MeV. No clear evidence for a cutoff in the power-law spectrum can be discerned from these data.

SPECTRAL MODELING

More detailed physical modeling of the spectrum has been attemped using a hybrid thermal/non-thermal Comptonization model, which describes the photon spectrum resulting from the Compton scattering of low energy photons off a hybrid electron distribution [15]. The total electron distribution includes both a thermal (Maxwellian) distribution plus a non-thermal (power-law) distribution at higher energies. The thermal electron component is defined by a characteristic temperature (kT_e). The non-thermal electron component is characterized as a power law with spectral slope p_e

extending from an electron Lorentz factor γ_{min} (where the Maxwellian transforms to the power-law tail) up to a Lorentz factor of γ_{max}. The electron population is assumed to reside in an accretion disk corona with an optical depth of τ. This particular model is useful in that it permits a quantitative description of the underlying electron distribution based on the observed photon spectrum. We have found that this model can provide very good fits to both the low- and high-state spectra, as can be seen in Figures 2 and 3. However, with the CGRO data alone, we find it difficult to constrain many of the model parameters. In our earlier analysis of the low-state spectrum, we used an energy threshold of 200 keV [4]. Unfortunately, such a high threshold leaves the fitting process very insensitive to some of the most important model parameters, such as kT_e and γ_{min}.

We are continuing the analysis of these data in an effort to provide tighter constraints on the model parameters, especially those (such as p_e and γ_{max}) that are most influenced by the high-energy data. The latest analysis uses a lower threshold energy of 50 keV and (unlike those fits shown in Figures 2 and 3) also allows for independent detector normalizations. In addition, we may also incorporate data from lower energy experiments (such as RXTE and SAX). The results of this on-going analysis will be reported elsewhere [16].

ACKNOWLEDGMENTS

The COMPTEL project is supported by NASA under contract NAS5-26645, by the Deutsche Agentur für Raumfahrtgelenheiten (DARA) under grant 50 QV90968 and by the Netherlands Organization for Scientific Research NWO. This work was also supported by NASA grant NAG5-7745.

REFERENCES

1. Priedhorsky, W.C., Terrell, J., and Holt, S.S., *ApJ* **270**, 233-238 (1983).
2. Ling, J.C., et al., *ApJ*, **275** 307-315 (1983).
3. Kitamoto, S., et al., *ApJ* **531**, 546-552 (2000).
4. McConnell, M.L. et al., *ApJ* **543**, 928-937 (2000).
5. Ling, J. C., et al., *A&AS* **120**, C677-C679 (1996).
6. Ling, J.C., et al., *ApJ Supp.* **127**, 79-124 (2000).
7. Phlips, B., et al., *ApJ* **465**, 907-914 (1996).
8. Hartman, R.C., et al., *ApJ Supp.* **123**, 79-202 (1999).
9. Titarchuk, L., *ApJ* **434**, 570-586 (1994).
10. Grove, J.E., et al., *ApJ* **500**, 899-908 (1998).
11. Ling, J.C., et al., *ApJ* **484**, 375-382 (1997).
12. Cui, W., et al., *ApJ* **484**, 383-393 (1997).
13. Zhang, S.N., et al, *ApJ* **477**, L95-L98 (1997).
14. Gierlinski, M.,et al., *MNRAS* **309**, 496-512 (1999).
15. Poutanen, J., and Svensson, R., *ApJ* **470**, 249-268 (1996).
16. McConnell, M.L. et al., *ApJ*, in preparation (2001).

Relativistic Effects on X-ray Emissions from Accretion Disks around Black Holes

Xiaoling Zhang*, Shuang Nan Zhang* and Yangsen Yao*

University of Alabama in Huntsville and National Space Science and Technology Center, Physics Department, Huntsville, AL 35899, USA

Abstract. Special and general relativistic effects on the X-ray emissions, especially the continuum spectra, from the accretion disks around black holes are investigated using the ray-tracing method. Because both the special and general relativistic effects are more important at distances closer to the black hole, the relativistic modifications to the emitted X-ray spectra are more significant at higher energies. In a simple accretion disk precession model, the relativistic effects can account for the energy dependence of the QPO amplitude and phase-lags observed in several black hole binaries. The dramatic QPO phase-lag transitions only show up when the central black hole is spinning rapidly. The narrow distribution of the observed system inclination angles of black hole binaries may be due to the selection effect caused by the relativistic effects around black holes.

INTRODUCTION

Many black hole binaries have been known and studied in recent years. Because of the strong gravitational field near the black hole, the radiation from the inner portion of the accretion disk experiences strong relativistic modifications before escaping from the system. In order to infer the physical parameters of the accretion disk and the black hole from the observed X-ray radiation, it is necessary to consider the relativistic effects (e.g., Zhang, Cui & Chen 1997; Cui, Zhang & Chen 1998). The exact modifications depend strongly on the angular momentum of the black hole and the inclination angle of the disk, as well as the black hole mass and the accretion rate.

Analytical calculations of the relativistic modifications can only be done in a few very simple cases. For most applications, numerical methods must be adopted. Here we use the ray-tracing method by Fanton *et al.* (1997), assuming that the disk is a thin Keplerian disk. The ray-tracing method takes all special and general relativistic effects into account.

In this paper we calculate the relativistically modified X-ray spectra from black hole X-ray binaries with different angular momenta and disk inclination angles. In particular we investigate the X-ray light curve modulations if the accretion disk is assumed to precess. Our calculations reproduce the observed light curve modulation *rms* (root-mean-squares) and the peculiar phase-lags around QPO peaks from the microquasar GRS1915+105, if the black hole in GRS1915+105 is spinning rapidly. Our results thus suggest that the observed light curve modulations and phase-lags are manifestation of the strong relativistic effects around a rapidly spinning black hole. We also propose that the observed system inclination angles around 60 to 70 degrees for all black hole binaries may be due to the selection effect caused by the relativistic effects around black holes.

MODELS FOR X-RAY SPECTRA AND FLUX MODULATIONS

We follow the widely established model for the X-ray spectra from black hole X-ray binaries, i.e., the two component model consisting of a soft, blackbody-like component and a hard, power-law like component. The soft component is usually approximated by the *diskbb* model in the *XSPEC* package. Here we use the radial temperature profile derived in the Kerr metric for the soft component.

For the hard component, currently there is no widely accepted physical model yet. For simplicity, we assume that the hard component is also produced from very close to the accretion disk. We further assume that the local hard component is of a power-law shape with a low energy cutoff. The local power-law photon index is assumed to be $\alpha = 2.2 + (r/r_g)/40$ in order to mimic the observed power-law shape. The low energy turn-over is the same as that in the Comptonization model (Sunyaev and Titarchuk 1980) i.e., implying that the power-law is produced via (thermal or non-thermal) Comptonization and the seed photons are the local blackbody emission from the disk. The emissivity of the hard and soft component is assumed to be the same, i.e., the gravitational energy release to X-ray radiation is shared equally in the soft and hard component.

We consider all relativistic effects when calculating the X-ray spectra at infinity. First, the rapid Keplerian motion of the disk causes both the Doppler frequency shift and Doppler boosting. Secondly, the strong gravitational field introduces the gravitational redshift and focusing. These effects can be expressed with the transfer function (Cunningham, 1975), calculated here with the ray-tracing method of Fanton *et al.* (1997).

RESULTS OF RAY-TRACING CALCULATIONS

In Fig. 1, we show the calculated X-ray spectra at infinity, for different black hole angular momenta and disk inclination angles. We fixed the black hole mass and mass accretion rate, such that the peak temperature for the extremal Kerr black hole is 2.1 keV; for all other black hole angular momenta, the corresponding peak temperatures are lower accordingly. These spectra may be fitted using the XSPEC package with the RXTE PCA response and *diskbb+powerlaw* model, with reasonable parameters and residuals, demonstrating that our spectral model can indeed reproduce the observed X-ray spectra from black hole X-ray binaries.

In Fig. 2, we show the observed flux in different energy bands as a function of the disk inclination angle. It is clear that significant deviations from the cosine law exist for a rapidly spinning black hole, especially at higher energies. This demonstrates clearly the necessity of taking into account relativistic effects if the black hole is spinning.

In a simple disk precession model, we may assume that the whole inner disk region precesses at a certain period, which is determined by the exact physical mechanism and mode of the disk precession. The observed X-ray flux will then display periodic oscillations. As an example, we assume that the disk between the last stable orbit and 200 r_g undergoes stable precession with inclination angles between 60 and 80 degrees. No other source of flux modulation is assumed to exist. In Fig. 3, the observed light curves in different energy bands are plotted for different black hole angular momenta. It is striking that the low energy and high energy light curves are opposite in phase for a rapidly spinning black hole.

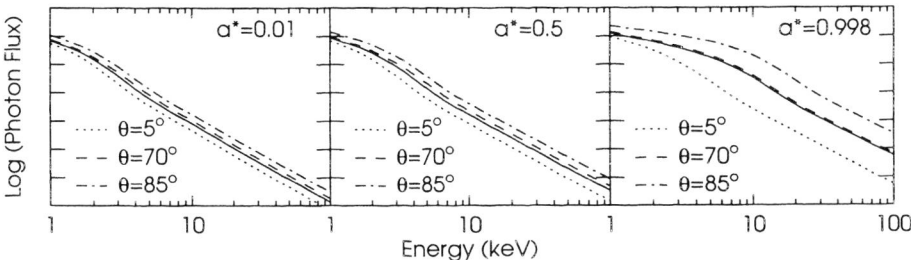

FIGURE 1. Accretion disk spectra around black holes with different angular momenta and viewed with different inclination angles from infinity. The solid lines are the local spectra and the dotted lines are the spectra observed at infinity, after modifications by special and general relativistic effects. For a rapidly spinning black hole system, the observed spectrum at infinity above 10 keV may deviate from the local spectrum significantly.

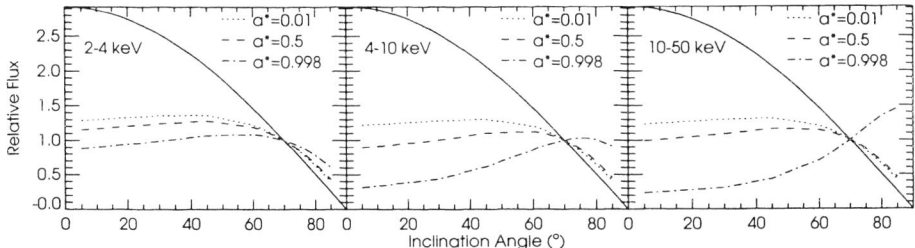

FIGURE 2. Flux at infinity in different energy bands as a function of the inclination angle after the relativistic modifications, as shown by the dotted lines; the solid lines are for the cosine law, appropriate when relativistic effects are negligible. The deviations from the cosine law are very significant for a rapidly spinning black hole, especially in high energy bands. This indicates that the observed light curve modulations in different energy band will not be in phase when the disk is forced to precess.

In Fig. 4, we show two sample power density spectra and the phase-lag spectra between them for different black hole angular momenta, with or without any white noise. Relatively strong harmonics of the precession frequency are observed. When $a^* = 0.5$ no phase-lag is present, because the low and high energy light curves are in phase. However, when $a^* = 0.998$, the low and high energy light curves are always opposite in phase. However, in the existence of a weak white noise, only sharp phase-lag transitions are observed around the peaks of the power density spectra.

Finally in Fig. 5, we show the *rms* of the light curve modulations as a function of energy for different black hole angular momenta. For non- and slowly spinning black holes, the light curve modulation *rms* does not change significantly as a function of energy. However, for the case of an extremal Kerr black hole, the light curve modulation depends strongly on the photon energy.

In summary, when the black hole is spinning rapidly, sharp phase-lag transitions and strong energy dependence of the X-ray light curves are expected. Therefore the X-ray light curves may be used for probing the strong relativistic effects near black holes.

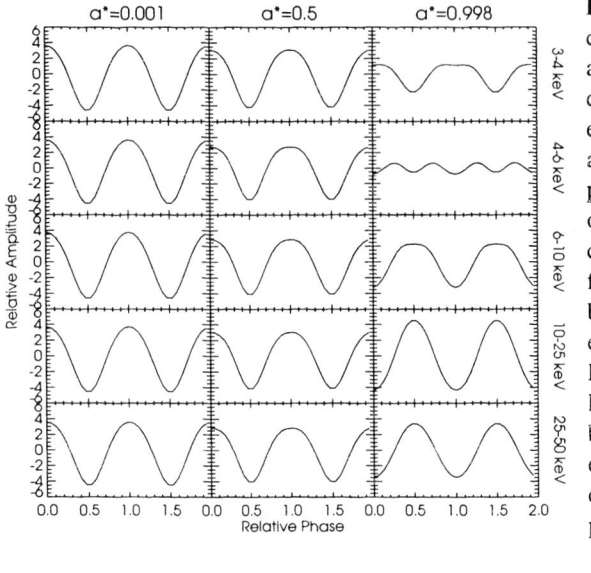

FIGURE 3. Observed light curves in different energy bands at infinity. The whole accretion disk responsible for X-ray emission is assumed to precess at a certain frequency and the precession is assumed to be the only source of the observed light curve modulation. Note that for a non- and slowly spinning black hole, the low and high energy light curves are in phase. However when $a^* = 0.998$, the light curves in different energy bands are very different; low energy and high energy light curves are in fact opposite in phase.

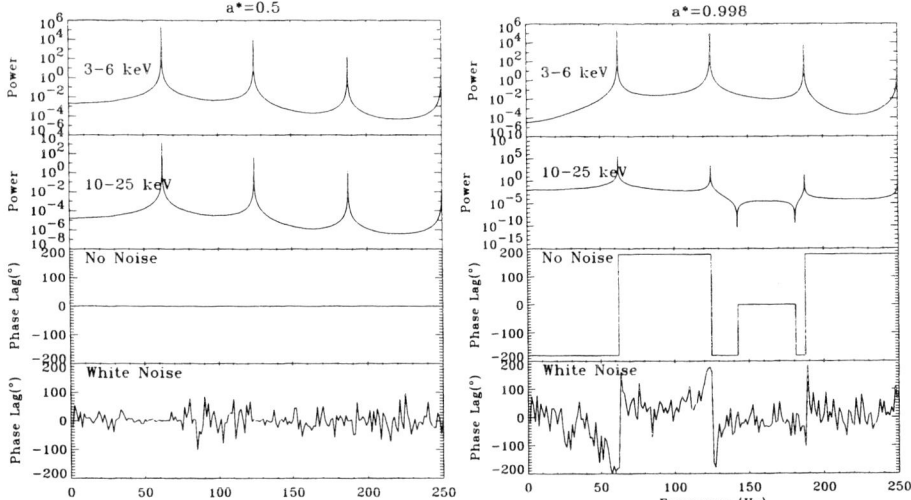

FIGURE 4. Sample power density and phase lag spectra, for two light curves in 3-6 keV and 10-25 keV bands. The white noise is about 0.2% of the signals. For the extremal Kerr black hole ($a^* = 0.998$), sharp phase-lag changes are expected around the peaks of the power density spectrum. The sharp phase-lag changes are purely due to the relativistic effects around a rapidly spinning black hole.

DISCUSSIONS

The most striking result of this investigation is the naturally produced sharp phase-lag transitions around the peaks in the power-density spectra. Regardless the details of

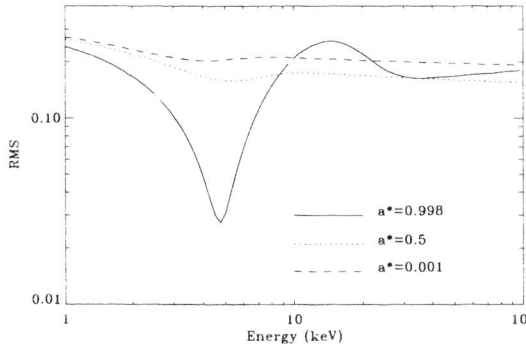

FIGURE 5. The *rms* (root-mean-squares) of the light curve modulations, as a function of energy. For non- and slowly spinning black holes, the light curve modulation *rms* does not change significantly as a function of energy. However, for the case of an extremal Kerr black hole, the light curve modulation depends strongly on the photon energy.

our calculations, this feature is robust and relies essentially on only three assumptions: higher energy photons are produced at distances closer to the black hole, the periodic flux modulations are mainly due to the accretion disk precession, and the black hole is spinning rapidly. This phenomenon has been observed in GRS1915+105 (Cui, 1999; Lin et al., 2000). The increasing trend of the light curve *rms* for an extremal Kerr black hole in the middle energy range, as shown in Fig. 5, has also been detected in GRS1915+105 (e.g., Cui, 1999). Therefore our results confirm that the black hole in GRS1915+105 is spinning rapidly (Zhang, Cui & Chen 1997; Cui, Zhang & Chen 1998). From Fig. 5, we predict that the *rms* decreases at energies above about 20-30 keV for extremal Kerr black hole binaries.

Finally our calculations may shed light on the outstanding puzzle that almost all observed X-ray black hole binaries have inclination angles around 60°-70° (Zhang et al. 1997). The combined effects of higher space density for high inclination angle systems and the relative flux of these high inclination angle systems (see Fig. 2) favor detection of binary systems at inclination angles around 60°-70°. Thus it is likely that the inclination angle distribution of known black hole binaries is simply a selection effect favoring the brightest X-ray sources, manifesting the strong relativistic effects around black holes.

Acknowledgments: We thank Dr. Fanton for providing us with the ray-tracing code. We have also benefited from discussions with Drs. Lev Titarchuk, Wei Cui, Yuxin Feng and Dingxiong Wang. This work was supported in part by NASA MSFC under contract NCC8-200 and by NASA LTSA Program under grants NAG5-7927 and NAG5-8523.

REFERENCES

1. Cui, Wei, Zhang, S.N., Chen, Wan, 1998, ApJ, 492, L53
2. Cui, Wei, 1999, ApJ, 524, L59
3. Cunningham, C. T., 1975, ApJ, 202, 788
4. Fanton, C., et al., 1997, PASJ, 49, 159
5. Lin, D., Smith, I.A., Liang, E.P., and Bottcher, M., 2000, ApJ, 543, L141
6. Makishima, K. et al., 1986, ApJ, 308, 635
7. Sunyaev, R. A., & Titarchuk, L. G., 1980, A&A, 86, 121
8. Zhang, S. N., Cui, Wei, and Chen, Wan, 1997, ApJ, L155
9. Zhang, S. N., et al., 1997, Invited Review of the 4th Compton Symposium Proceedings, AIP Conference Series, 410, 141, eds Charles D. Dermer, Mark S. Strickman, James D. Kurfess.

Studying the Accretion Disks in Black Hole X-ray Binaries with Monte-Carlo Simulations

Yangsen Yao*, Shuang Nan Zhang* and Xiaoling Zhang*

*University of Alabama in Huntsville and National Space Science and Technology Center, Physics Department, Huntsville, AL 35899, USA

Abstract. Understanding the properties of the hot corona is very important for studying the accretion disks in black hole X-ray binary systems. Using the Monte-Carlo technique to simulate the inverse Compton scattering process between disk photons and electrons in the hot corona, we have produced two table models in the XSPEC package. Applying the models to the broad-band BeppoSAX observations of the black hole candidate XTE J2012+381, we demonstrate the power of this table model. Our results indicate that the electron distribution in the corona has a powerlaw shape with a spectral index around 4 and the size of the corona is just several tens of gravitational radius and the inclination angle of the disk is around 60 degrees.

INTRODUCTION

The spectrum of a black hole X-ray binary system usually consists of two components (see Zhang, et al. 1997 for a review and references therein): a blackbody-like component and a powerlaw-like component. The blackbody-like component turns off above 20 keV and is believed to be emitted from the accretion disk. The powerlaw-like component can extend up to 200 keV and is believed to be produced in a hot corona around the black hole through the inverse Compton scattering process between the disk photons and electrons in the corona. This process can be described by the Kompaneets's equation, which is a non-linear diffusion equation and very difficult for analytical solutions.

Analytic solutions have been worked out by Sunyaev and Titarchuk (1980) and Titarchuck (1994) based on these assumptions: 1) uniform corona with either spherical or disk geometry; 2) seed photons are emitted in the center of the corona; 3) the seed photon distribution follows the Wien form; 4) the thermal temperature of electrons in the corona is much higher than the Wien temperature of the seed photons. However, some or all of these assumptions are not appropriate for black hole binary systems.

In the X-ray astronomy community, the multi-color blackbody (Mitsuda et al., 1984; Makishima et al., 1986; *diskbb* in *XSPEC*) plus a powerlaw model has been employed traditionally to fit the energy spectra of the black hole X-ray binary systems; this model is non-physical and unreasonable as we pointed out in our previous paper (Zhang et al., 2001).

In this paper, we use the Monte-Carlo technique to simulate the inverse Compton process in the corona; two powerful table models (as the standard model in *XSPEC* package) have been built up based on our simulation results. Using these table models to fit the data on XTE J2012+381 with BeppoSAX, we estimated several physical parameters of the accretion disk and corona.

MONTE-CARLO SIMULATIONS

In our simulations, we assume that the accretion disk is a Keplerian disk, and during the accretion process, the potential energy loss of the accreted material is radiated away in blackbody radiation. Therefore, the temperature distribution along the accretion disk is $T \propto r^{-3/4}$. The corona may take either spherical or disk geometry; the electron density distribution in the corona may be either uniform or non-uniform (take the powerlaw for non-uniform distribution); and the electron energy distribution in the corona may take either thermal form or non-thermal form (powerlaw).

Our simulation results can be summarized as following: 1) for a given electron temperature, the scattered photon spectrum in the high energy band is determined by the optical depth of the corona (FIGURE 1.a); 2) the shape of the scattered photon spectrum in the low energy band is related to the geometrical size of the corona (FIGURE 1.b); 3) for the same properties of disk and corona, the ratio between the flux of soft component and that of the hard component is related to the inclination angle of the accretion disk (FIGURE 1.c); 4) the high energy portions of the spectra are very different for different electron energy distributions in the corona, even though the low energy portions are quite similar (FIGURE 1.d).

Based on our simulation results, we have built up two different table models for thermal and powerlaw electron energy distributions respectively. The table models consist of the following parameters: temperature of the inner boundary of the accretion disk (T_{in}), thermal electron temperature (KT) or the electron powerlaw index (Γ_e), size of the corona (*Size*), scattering optical depth of the corona (τ), inclination angle of the accretion disk, and normalization parameter (K_{BB}) which is added by the *XSPEC* automatically.

APPLICATION TO XTE J2012+381

The two models are applied to model the BeppoSAX broad-band data on the black hole candidate XTE J2012+381. FIGURE 2 shows the modeling results with *diskbb + powerlaw* model in *XSPEC* and the results with our table models. During the rising phase of the outburst, the powerlaw component was very strong (Sun et al., 2001)(FIGURE 2(a)). For the BeppoSAX observation during the rising phase we have analyzed, the observed spectrum can be fitted reasonably well using *diskbb + powerlaw* model with photon index around 2.4 (FIGURE 2(b)). When fitting with our table model for the thermal electron case, the fit to the high energy data above 10 keV is not acceptable (FIGURE 2(c)) and thus the thermal model is rejected. However, when we fitting with our table model for the powerlaw electron distribution, the result is comparable to that with the *diskbb + powerlaw* model. The powerlaw index for electron energy distribution is around 4, the size of corona is around several tens of gravitational radius and the inclination angle of the disk is around 60 degrees (FIGURE 2(d)).

FIGURE 3 shows the results of inner disk radius inferred with different models. When modeling with *diskbb + powerlaw* model and *diskbb + compTT* model, the inner disk radius inferred directly from the normalization of *diskbb* model are very small compared to the results obtained with our table model. The results after the radiative transfer correction (see Zhang et al. 2001 for detail) are consistent with the results of our table model reasonably well.

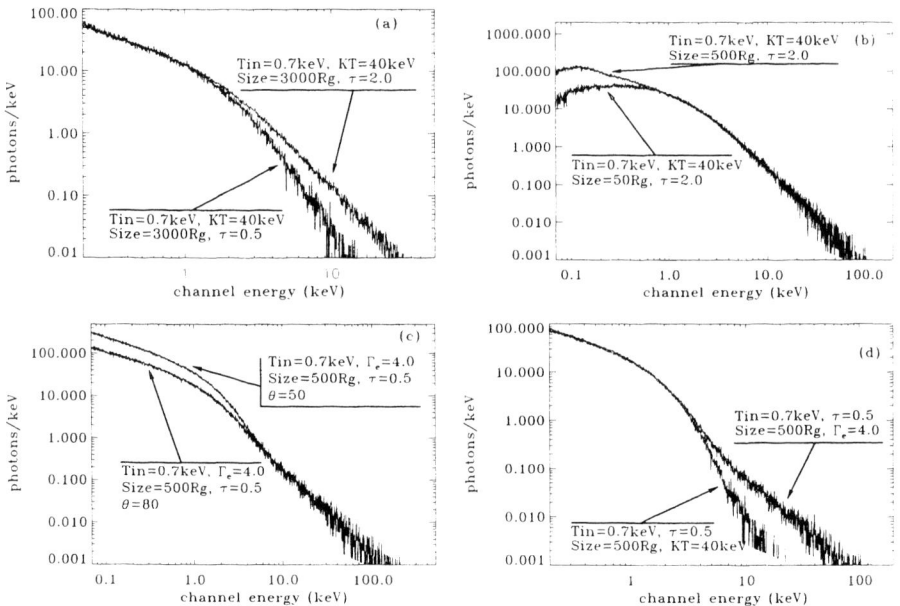

FIGURE 1. X-ray photon spectra for the spherical corona with different physical properties of the accretion disk and the corona. Panel (a) and panel (b) show the photon spectra for the thermal electron energy distribution with a temperature of 40 keV. Panel (a) is for a coronae with different optical depths but with the same size. Panel (b) is for a coronae with different sizes but with the same optical depth. Panel (c) shows different photon spectra with different inclination angles when the electron energy distribution has the powerlaw form. Panel (d) shows the comparison of the photon spectra between thermal and powerlaw electron energy distributions in the corona.

CONCLUSION AND DISCUSSION

According to our simulation results and data analysis, the X-ray spectral shape in the low energy band might be related to the size of the corona and the electron energy distribution may be inferred from the spectral shape in the high energy band. Two powerful table models have been built up based on our simulation results and the physical parameters can be obtained directly when modeling the data with these table models.

According to the fitting results with our table models, the electron energy distribution in the corona of XTE J2012+381 during the rising phase seems to have a powerlaw form rather than a thermal form and the size of the corona is just several tens of gravitational radii and the inclination angle of the disk is around 60 degrees.

The reason that the size of the corona may be determined from the X-ray data is because in our model the seed photons for the inverse Compton scattering come from the disk, on which temperature is a function of the distance from the central black hole. The relatively small size of the corona indicates that most of the hard X-ray photons come from the region very close to the black hole.

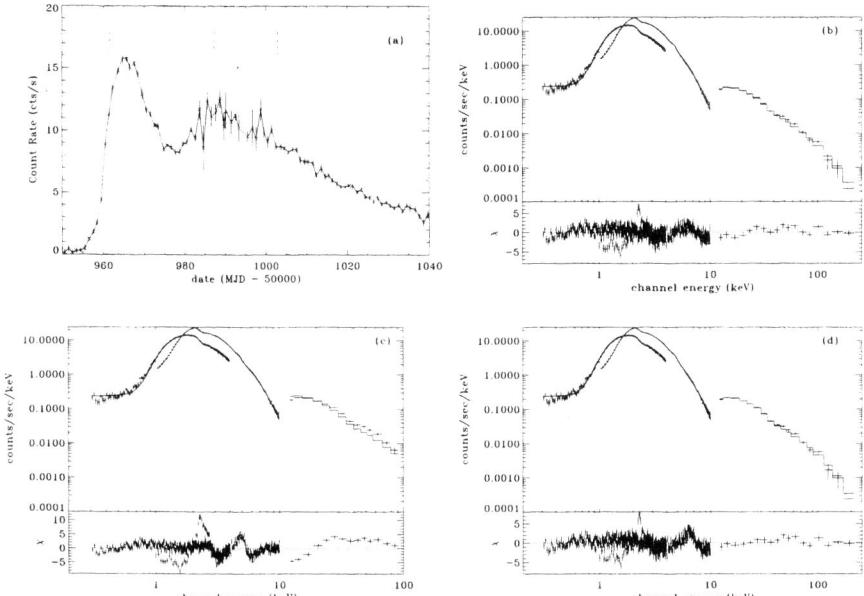

FIGURE 2. Panel (a) shows the ASM light curve of the black hole candidate XTE J2012+381 during its burst in 1998 and the dotted lines in the plot indicate the dates of BeppoSAX observations we have analyzed. The first observation is in the rising phase in which the hard component is relatively strong and the other two observations are in the soft state in which the hard component is very weak. Panel (b) shows the modeling result for the first observation with *diskbb + powerlaw* model. The best fit parameters are: $T_{in} = 0.77 \pm 0.002$ *keV*, $K_{BB} = 887 \pm 10.0$, $\Gamma = 2.4 \pm 0.03$, and the Chi-Squared is 1368 with 553 degrees of freedom. Panel (c) shows the modeling result with our table model for thermal electron distribution in the corona. The best fit parameters are: $T_{in} = 0.6 \pm 0.02$ *keV*, $KT = 52 \pm 6.2$ *keV*, $Size = 19 \pm 1.7 R_g$, $\tau = 0.6 \pm 0.06$, and the Chi-Squared is 3109 with 548 degrees of freedom. Panel (d) shows the modeling result with our table model for powerlaw electron distribution in the corona. The best fit parameters are: $T_{in} = 0.76 \pm 0.002$ *keV*, $\Gamma_e = 4$, $Size = 49 \pm 6.2 R_g$, $\tau = 0.09 \pm 0.003$, $\theta = 59^0 \pm 3.6^0$, and the Chi-Squared is 1564 with 552 degrees of freedom.

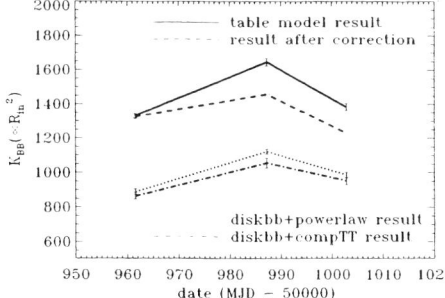

FIGURE 3. Inner disk radius ($R_{in}^2 \propto K_{BB}$) inferred with different models. These three data points correspond to the three observations indicated in FIGURE 2(a). When modeling with *diskbb + powerlaw* model and *diskbb + compTT* model, the results inferred directly from the soft component are quite small compared to the results inferred from our table model and the results after the radiative transfer correction.

Another interesting result is that the spectral fitting is quite sensitive to the disk inclination angle. This is because the observed flux from the disk depends strongly upon the inclination angle, while the hard X-ray flux is almost isotropic, especially for the case of spherical corona. Unfortunately, if the quality of the data is not good enough, the inclination angle and the geometrical factor (opaqueness of the corona) may be coupled together in the spectral fitting. If the inclination angle of the system is determined independently, e.g., with optical photometry observations or radio measurements of jets, the shape of the corona may be constrained reliably with X-ray spectral fitting.

Determining the value of the inner disk radius is very important in understanding the physics of the accretion disk and the black hole angular momentum (Zhang, Cui and Chen 1997). The normalization parameter inferred with our table model, which is proportional to square of the inner disk radius (Makishima et al., 1986), is significantly different from that determined with the simple $diskbb + powerlaw$ or $diskbb + compTT$ model without radiative transfer correction. This is because the corona, though optically thin in most cases, scatters some of the photons emitted from the disk and makes the observed soft component different significantly from the original disk emission.

Therefore modeling broad-band X-ray continuum spectra with physically consistent and accurate models provides a powerful tool in determining the properties of accretion disks and coronae in black hole X-ray binaries. However such studies require high quality and broad-band data, which may be provided by BeppoSAX, Chandra and XMM currently. Future data from Integral, Swift and especially the Constellation X-ray missions will provide significant breakthroughs in this field.

Currently, we have not included non-uniform corona and other geometry in our table models. The general relativistic effects and Doppler effects (Zhang, et al., these proceedings) are not taken into account in our table model either. These will be our future work.

Acknowledgments: Mr. Yongzhong Chen is acknowledged for his initial work on this project during his visit to UAH in 1999-2000. We thank Drs. Lev Titarchuk, Wei Cui and Yuxin Feng for interesting discussions. This work was supported in part by NASA Marshall Space Flight Center under contract NCC8-200 and by NASA Long Term Space Astrophysics Program under grants NAG5-7927 and NAG5-8523.

REFERENCES

1. Makishima et al., ApJ 308, 635, (1986)
2. Mitsuda et al., PASJ, 36, 741, (1984),
3. Sun, X. J, et al., to be submitted to ApJ, 2001
4. Sunyaev, R. A., & Titarchuk, L. G., 1980, A&A, 86, 121
5. Titarchuk, L., 1994, ApJ, 434, 570
6. Zhang, S. N., Cui, Wei, and Chen, Wan, 1997, ApJ, L155
7. Zhang, S. N.,et al. , 1997, Invited Review of the 4th Compton Symposium Proceedings, AIP Conference Series, 410, 141, eds Charles D. Dermer, Mark S. Strickman, James D. Kurfess.
8. Zhang, S. Nan, et al., to be submitted to ApJ, 2001
9. Zhang, Xiaoling, Zhang, S. Nan, & Yao, Yangsen, these proceedings

X-ray spectra of accretion discs with dynamic coronae

Julien Malzac*, Andrei M. Beloborodov[†] and Juri Poutanen[†]

Osservatorio Astronomico di Brera, via Brera, 28, 20121 Milano, Italy
[†]*Stockholm Observatory, SE-133 36 Saltsjöbaden, Sweden*

Abstract. We compute X-ray spectra produced by *non-static* coronae atop accretion discs around black holes. The hot corona is radiatively coupled to the underlying disc (the reflector) and generates an X-ray spectrum which is sensitive to the bulk velocity of the coronal plasma, $\beta = v/c$. We show that an outflowing corona atop a neutral reflector reproduces the hard-state spectrum of Cyg X-1 and similar objects. The dynamic model predicts a correlation between the observed amplitude of reflection R and the X-ray spectrum slope Γ since both strongly depend on β. A similar correlation was observed and its shape is well fitted by the dynamic model.

INTRODUCTION

The hard X-ray spectra of galactic black holes (GBHs) and active galactic nuclei (AGN) indicate the presence of hot plasmas with temperature $kT \sim 100$ keV in the vicinity of accreting black holes (see e.g. Poutanen 1998 for a review). The plasma can be identified with a corona of a black hole accretion disc (e.g. Galeev et al. 1979; see Beloborodov 1999b for a review, hereafter B99b). The corona is likely to form as a result of magnetorotational instabilities in the disc and the buoyancy of the generated magnetic field (Tout & Pringle 1992; Miller & Stone 2000). The coronal plasma is probably heated in flare-like events of magnetic dissipation producing the variable X-ray emission. The observed power-law X-ray spectra are generated by Comptonization process. The X-rays are partly reprocessed into soft radiation by the underlying disc. The soft radiation reenters the source, providing the feedback loop that regulates the temperature of the corona (Haardt & Maraschi 1993). The geometry of the corona can hardly be derived from first principles. It might be a large cloud covering the whole inner region of the disc. It may also be a number of small-scale blobs with short life-times. The resulting X-ray spectrum is however not sensitive to the exact shape of the cloud, its density distribution, and other details. The only important parameter is the effective feedback factor (see e.g. Stern et al. 1995b) that is the fraction of the X-ray luminosity which reenters the source after reprocessing. Previous computations of the disc-corona models all assumed that the corona is static (e.g. Haardt & Maraschi 1993; see also Poutanen 1998). The model was successfully applied to Seyfert 1 AGN, however, it was found to disagree with observations of some black-hole sources in the hard state, for instance, Cyg X-1 (Gierliński et al. 1997). A possible explanation could be that the coronal plasma is moving away from the disc and emits beamed X-rays. Beloborodov 1999 (hereafter B99a) argued that the flaring plasma is likely to acquire a mildly relativistic

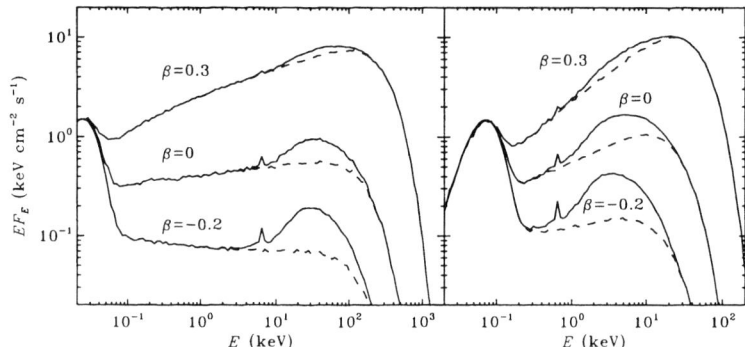

FIGURE 1. Effect of bulk motion on the emitted spectra. Here $h/r = 2$, $\tau_T = 3$, and nearly face-on inclination is assumed, $0.9 < \cos i < 1$. Left panel corresponds to the case of AGN ($kT_{bb} = 5$ eV) and right panel corresponds to GBHs ($kT_{bb} = 150$ eV).

bulk velocity $\beta \lesssim 0.5$ as a result of magnetic dissipation and/or radiative pressure in the corona. The bulk velocity may be directed away or towards the disc (and it may change) with the preferential direction away from the disc. Mildly relativistic bulk motion causes aberration of the X-ray emission and strongly affects both the amplitude of reflection, R, and the spectrum slope, Γ. Here we perform exact computations of the X-ray spectra produced by dynamic coronae. We use a non-linear Monte-Carlo code (Malzac & Jourdain 2000) which is based on the large-particle method (Stern et al. 1995a).

SET UP

For definiteness, consider a cylinder of radius r and height h located atop the accretion disc. The cylinder may be associated with a hot outflow covering the disc or a heated magnetic tube in a compact flare. The plasma in the cylinder is assumed to have a constant density and it is heated homogeneously with a constant rate. The plasma moves through the cylinder with a velocity β directed normally to the disc. The electrons are assumed to have a Maxwellian distribution with a temperature T_e in the plasma rest frame. T_e is calculated from the heating=cooling balance. We assume that reprocessed radiation is the main cooler of the hot coronal plasma and neglect soft radiation generated viscously inside the accretion disc. We assume that the reflecting material of the disc is sufficiently dense so that the ionization parameter $\xi \lesssim 10^3$ and the ionization effects are weak. Then the albedo is small, $a \sim 0.2$, and most of the X-rays impinging the disc are reprocessed. We assume that the reprocessed flux has a quasi-blackbody spectrum (possibly diluted) with a constant temperature T_{bb}. In the simulations, we consider two cases: $kT_{bb} = 150$ eV and $kT_{bb} = 5$ eV, representing the typical temperatures in GBHs and AGN, respectively. We start a simulation from an initial (non-equilibrium) state and follow the evolution of the plasma and radiation in the cylinder until a steady state is achieved. The model has four parameters: (i) Thomson optical depth τ_T (defined along

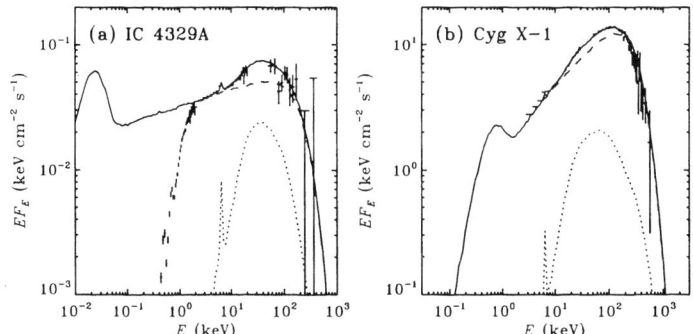

FIGURE 2. Right panel: Spectrum of Cyg X-1 as observed by Ginga and CGRO/OSSE in September 1991 (crosses, set 2 from Gierliński et al. 1997). The solid curve is the model spectrum for $\tau_T = 3$, $h/r = 1.25$, $\beta = 0.3$ at inclination $i = 50°$. Left panel: Spectrum of the Seyfert 1 galaxy IC4329A observed by ROSAT, Ginga, and CGRO/OSSE (crosses, from Madejski et al. 1995). The solid curve is the model spectrum for $\tau_T = 3$, $h/r = 2$, $\beta = 0.1$ at inclination $i = 40°$. In both panels, dotted curves give the reflected components, dashed curves show the intrinsic Comptonized spectra.

the height of the cylinder), (ii) height to radius ratio of the cylinder h/r, (iii) bulk velocity β and (iv) blackbody temperature T_{bb}. Given these parameters the code computes the emitted spectrum as a function of the disc inclination i. Details of the computations are given in Malzac, Beloborodov & Poutanen (2001, hereafter MBP).

THE EFFECTS OF BULK MOTION

Fig. 1 illustrates the effects of bulk motion on the emitted spectra. In the case of $\beta = 0.3$ (plasma moves away from the disc), the observed Comptonized luminosity is enhanced as a result of relativistic aberration. The X-rays are beamed away from the disc, and the reprocessed and reflected luminosities are reduced. The low feedback leads to a hard intrinsic spectrum. In the case of $\beta = -0.2$ (plasma moves towards the disc), the Comptonized luminosity is beamed towards the disc and the reprocessed and reflected components are enhanced. The high feedback leads to a soft intrinsic spectrum. Since τ_T is fixed in Fig. 1, a high (low) feedback leads to low (high) coronal temperature. This causes the shift of the spectral break to lower energies with decreasing β. The dynamic corona model reproduces the broad-band spectra of black hole sources. For illustration, we show in Fig. 2 the model and observed spectra of Cyg X-1 and the bright Seyfert 1 galaxy IC4329A.

REFLECTION AND SPECTRAL INDEX

There are two parameters that are commonly used to quantify the shape of the observed X-ray spectra: Γ, the slope of the primary Comptonisation spectrum in the 2-10 keV range, and R, the amplitude of reflection. The R is defined so that $R = 1$ for an isotropic

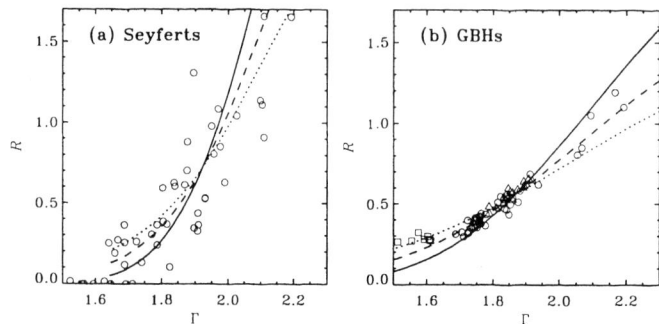

FIGURE 3. Left panel: The observed R–Γ correlation for Seyfert galaxies (data from Zdziarski et al. 1999). The curves show the model with $\mu_s = 0.6$, $\tau_T = 3$, $a = 0.15$ at three different inclinations. Right panel: The $R - \Gamma$ correlation for GBHs (data from Gilfanov et al. 2000). The model curves have $\mu_s = 0.45$, $\tau_T = 3$, $a = 0.15$. In both panels, solid, dashed, dotted curves correspond to $i = 30°, 60°, 70°$.

point source illuminating an infinite slab. Both R and Γ expected from a dynamic corona can be evaluated using a simple analytical model (B99a). We here improve the model by accounting for additional attenuation of reflection by scattering in the corona. The amplitude of reflection is given by

$$R(\mu) = \frac{(1-\beta\mu)^3}{(1+\beta\mu_s)^2} \left\{ \mu_s \left(1 + \frac{\beta\mu_s}{2}\right) + \frac{(1-\mu_s)[1+\beta(1+\mu_s)/2]}{(1+\beta)^2} e^{-\tau_T(1-\mu_s)} \right\}, \qquad (1)$$

where $\mu = \cos i$, i is the inclination angle, and the geometrical parameter $\mu_s \approx (h/2)/\sqrt{r^2 + (h/2)^2}$ describes the geometry of the blob. In equation (1), the reflected luminosity is represented as a sum of two parts: the first one is reflected outside the cylinder base and does not experience any attenuation and the second one is reflected from the base and it is partly attenuated, depending on τ_T. When μ_s approaches unity (or $\tau_T \to 0$), the attenuation is not important and equation (1) becomes equation (3) in B99a. In MBP we show that formula (1) is in good agreement with the results of exact simulations. The spectral index Γ can be evaluated using a simple relation between Γ and the amplification factor A (B99b)

$$\Gamma = C(A-1)^{-\delta}. \qquad (2)$$

Equation (2) is a good approximation for models with kT_e in the range between 50 keV and 300 keV (see Fig. 7 in MBP). From our simulations we find $C = 2.19$, $\delta = 2/15$ for GBHs ($kT_{bb} = 150$ eV) and $C = 2.15$, $\delta = 1/14$ for AGN ($kT_{bb} = 5$ eV). The amplification factor ($A = D^{-1}$ where D is the feedback factor) is determined by the geometrical parameter μ_s, the energy-integrated albedo of the disc, a, and the plasma velocity β (B99a)

$$A = \frac{2}{(1-a)(1-\mu_s)} \frac{\gamma^2(1+\beta)^2(1+\beta\mu_s)^2}{1-\beta^2(1+\mu_s)^2/4}. \qquad (3)$$

Combining equations (2) and (3) we get Γ as a function of the corona parameters.

A correlation between R and Γ was found in individual objects observed at different epochs as well as in a sample of sources (Fig. 3). Equations (1-3) give $R(\beta)$ and $\Gamma(\beta)$ predicted by the dynamic corona model for given μ_s, τ_T, a. Variations in β result in $R - \Gamma$ correlation with a shape similar to the observed correlation (see Fig. 3). Note that the correlations for AGN and GBHs cannot be fitted with the same value of μ_s. It might indicate different geometries of the coronae in GBHs and AGN.

CONCLUSIONS

We performed Monte-Carlo simulations of X-ray production by coronae atop accretion discs. The disc-corona model agrees with the data if the hot plasma moves with a mildly relativistic velocity away from the accretion disc. The spectrum of Cyg X-1 is reproduced by the model with $\beta = 0.3$, confirming the estimate of B99a. The observed $R - \Gamma$ correlation is well explained by varying β. It suggests that β may be the main parameter controlling the X-ray spectrum. The results of the simulations are in good agreement with the analytical description of B99a,b. We improved the analytical model by accounting for the attenuation of the reflection component by the hot plasma atop the disc.

ACKNOWLEDGMENTS

This work was supported by the Italian MURST grant COFIN98-02-15-41 (JM), the Swedish Natural Science Research Council (AMB, JP), the Anna-Greta and Holger Crafoord Fund (JP), and RFBR grant 00-02-16135 (AMB).

REFERENCES

- Beloborodov A. M., 1999a, ApJ, 510, L123 (B99a)
- Beloborodov A. M., 1999b, in High Energy Processes in Accreting Black Holes, ed. J. Poutanen & R. Svensson (ASP Conf. Series), 161, 295 (B99b)
- Galeev A. A., Rosner R., Vaiana G. S., 1979, ApJ, 229, 318
- Gierliński M. et al., 1997, MNRAS, 288, 958
- Gilfanov M., Churazov E., Revnivtsev M., 2000, in Proc. 5th CAS/MPG Workshop on High Energy Astrophysics, in press (astro-ph/0002415)
- Haardt F., Maraschi L., 1993, ApJ, 413, 507
- Madejski G. M. et al., 1995, ApJ, 438, 672
- Malzac J., Jourdain E., 2000, A&A, 359, 843
- Malzac J., Beloborodov A. M., Poutanen J., 2001, MNRAS in press, astro-ph/0102490
- Miller K. A., Stone J. M. 2000, ApJ, 534, 398
- Poutanen J., 1998, in Theory of Black Hole Accretion Disks, ed. M. Abramowicz, G. Björnsson, & J. Pringle (Cambridge University Press), 100
- Stern B. E., Begelman M. C., Sikora M., Svensson R., 1995a, MNRAS, 272, 291
- Stern B. E., Poutanen J., Svensson R., Sikora M., Begelman M. C., 1995b, ApJ, 449, L13
- Tout C. A., Pringle J. E., 1992, MNRAS, 259, 604
- Zdziarski A. A., Lubiński P., Smith D. A., 1999, MNRAS, 303, L11

Detection of the γ-Ray Emission from the X-Ray Nova GRO J0422+32

A.F. Iyudin* and F. Haberl*

Max-Planck-Institut für extraterrestrische Physik, Postfach 1312, 85741 Garching, Germany

Abstract. The evolution of the γ-ray emission from GRO J0422+32 after its main outburst of August 1992 was followed by COMPTEL during 25 observation periods (VPs) between August 1992 and August 1997, where NPer 92 was in the FoV of COMPTEL, and for 3 VPs preceding the nova outburst. It is found that the NPer 92 γ-ray emission is greatly confined to energies between 1.5 and 2.0 MeV. This emission is prominent during observations that correspond to phases between 0.0 and 0.5 of the ~120 days period discovered in the nova optical emission.

This periodical activity of the nova in form of mini-outbursts is supported by the detection of the γ-ray emission from GRO J0422+32 in seven episodes up to ~ 1800 days (August 1997) after the main outburst, and by the All Sky Monitor's (ASM, Rossi XTE) detection of X-ray mini-outbursts (four episodes). In one episode we have succeeded to derive a contemporaneous spectrum based on the measurements by COMPTEL, ASM and BATSE.

The plausible mechanisms of the GRO J0422+32 periodical mini-outbursts and of its characteristic γ-ray emission are discussed. The possibility of the system being a triple is also evaluated.

Introduction

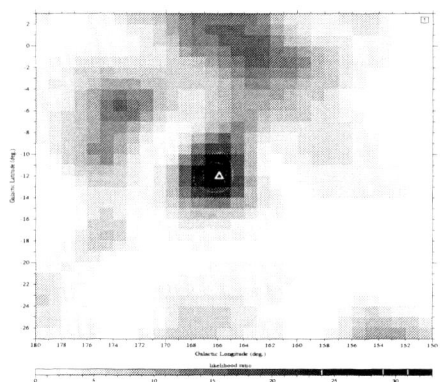

Fig. 1: The GRO J0422+32 image in the 1.6 MeV line emission, for the combined data of 7 viewing periods within phases 0.0-0.5 of the assumed 120,4 day period of the nova recurrent activity.

On August 5, 1992, BATSE detected the X-ray transient GRO J0422+32 [23], also known as Nova Persei 1992 and V518 Per. Following this discovery by BATSE, the X-ray nova GRO J0422+32 (NPer 92) was observed by a number of space and ground observatories during its main and secondary outbursts. The spectral analysis of the bright X-Nova performed for the main outburtst by BATSE [17], OSSE [22], SIGMA [25], MIR-KVANT [29] pointed to the possibility of the NPer 92 being a system containing a Black Hole Candidate (BHC). This conclusion at first glance is supported by the timing analysis of the OSSE data during the main outburst with most of the details provided in paper [13]. Very unusual timing behavior of GRO J0422+32 was observed 42 days after main outburst by ROSAT (HRI) in soft X-rays, namely, a strong flaring on a time scale of ~ 1 sec, and an integrated rms power of ∼ 40% [24], that is higher than for any other X-ray binary [2]. An optical counterpart of NPer 92 was found by Castro-Tirado et al. (1992) inside of an error circle with radius of 0.2° defined on the basis of scanning observations by OSSE combined with the error box of BATSE [16], later confirmed by soft γ-ray observations of SIGMA [12].

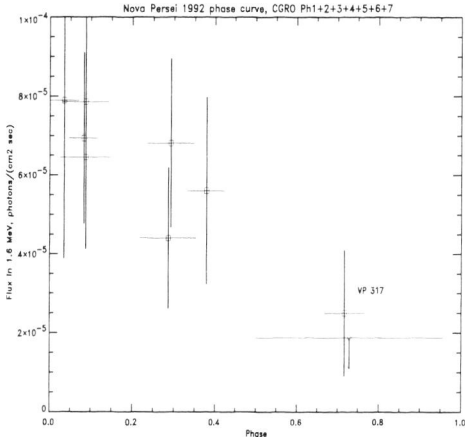

The follow-up optical observations helped to establish that GRO J0422+32 is a low-mass system [7, 8, 26, 30]. A number of optical measurements were performed after the main outburst until the quiescence, that was reached around February 1994. As a result of optical measurements a binary mass-function, orbital period and the secondary spectral type were established. The most reliable optical data have been acquired with the W.M. Keck 10 m telescope by Fillipenko, Matheson and Ho (1995).

Fig. 2: The GRO J0422+32 light curve in the 1.6 MeV line emission, folded with assumed 120,4 days period of the nova recurrent activity. Zero phase corresponds to TJD 8840.500±nP.

The Black Hole Candidate Signature in the γ-Rays

Nova Persei 1992 belongs to the few X-ray novae detected above 100 keV. The follow up measurements of the NPer 92 evolution resulted in the detection of a secondary outburst in optical and X-ray wavelengths and of additional optical mini-outbursts with a recurrence period of about 120 days [6, 8, 9, 14, 18, 19, 20, 27].

COMPTEL detected GRO J0422+32 in the main outburst up to 2 MeV [10]. The higher energy γ-ray emission was not seen neither by COMPTEL nor EGRET. The spectral analysis of the hard X-ray and γ-ray emission from GRO J0422+32 has suggested a possible contribution of a nonthermal mechanism to the nova hard X-ray and gamma-ray emission [10, 25]. By following the evolution of the γ-ray emission of the nova with COMPTEL during 25 observation periods (VPs) between August 1992 and August 1997, where NPer 92 was in the FoV of COMPTEL, and for 3 viewing periods preceding the nova outburst, we have found that the NPer 92 γ-ray emission was confined between 1.5 and 2.0 MeV, and is detectable only during observations that correspond to phases 0.0 - 0.5 of the 120 day period. Finer spectral analysis revealed that the large part of the nova flux in 1.5-2.0 MeV is in fact due to line emission at ∼1.6 MeV (Fig. 1).

It is stunning, that the γ-ray emission is detectable only in viewing periods which correspond to phases between 0.0 and 0.5 of the ∼ 120 day period of the nova mini-outburts detected in visible [9]. The γ-ray emission was detectable by COMPTEL up to ∼ 1800 days (August 1997) after the main outburst. The confirmation of possible periodicity of the nova outbursts has come from the All Sky Monitor (ASM on board of Rossi XTE) detection of the X-ray mini-outburst confined to the time consistent with the nova recurrent activity period of about 120 days [19, 20]. Figure 2, the phase curve of the γ-ray emission convolved with the optical period, clearly demonstrates that only upper limit could be derived at the position of GROJ0422+32 during off-phase VPs, e.g., for observations within phases 0.5-1.0 of the assumed 120.4 day period of the X-nova recurrent activity.

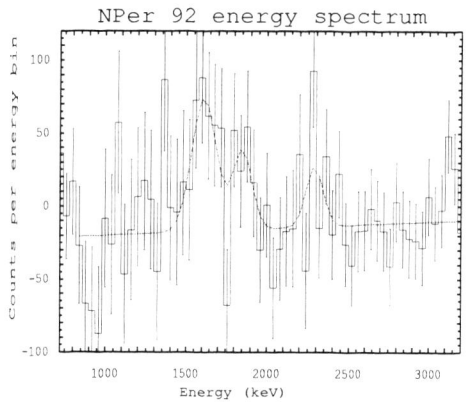

Fig. 3: The GRO J0422+32 on-phase spectrum constructed for the combination of the X-nova recurrent activity periods.

The remarkable recurrent activity of the nova in the 1.6 MeV γ-ray line emission was detected in seven episodes by COMPTEL. In three viewing periods the detection had a significance $\geq 3\sigma$ and in another four it was $\geq 2\sigma$.

The verification of the 1.6 MeV line intensity and significance was made with the use of different background models in imaging and spectral analyses. A spectral analysis of GRO J0422+32 was performed for each of the 28 VPs by selecting events with scatter directions and scatter angles from a circular region of 2° radius centered at the position of source. A spectral analysis was performed with background models, like: - high-latitude observations background spectra; - background generated from the mirror points of the same observation; - and with background spectra generated for the same observation, but from a larger annulus around the source position.

Both imaging and spectral analyses with different background models gave consistent results (difference no more than 25%), in the number of line counts and of flux values.

An example of the COMPTEL background-subtracted spectra from the 2° radius circle centered on the position of GRO J0422+32 is shown in Fig. 3 for a combination of the active periods of GRO J0422+32. Two lines are clearly seen, at ∼1.6 MeV and ∼1.8 MeV, with significances of ∼ 6 σ and of ∼ 4 σ, respectively. There is also a hint for the line at ∼2.3 MeV. For comparison the spectrum (not shown) constructed for the position of GROJ0422+32 with the same background model but for the combination of "off" periods of the X-nova shows no traces of lines. The χ^2-test and F-test performed for two models of the residual spectra of NPer 92 in individual and combined active periods, namely: a) linear background and no lines present; b) linear background plus line at ∼1.6 Mev; strongly support the presence of the line in the spectrum of the nova. Here we will discuss only the strongest line at ∼1.6 MeV of the nova active periods. The implications of other lines detection will be discussed elsewhere.

The detected by COMPTEL γ-ray line emission at 1.6 MeV can be explained as an emission of the hot plasma at the compact object boundary layer or accretion disk (AD), a model that was considered for the X-binaries containing a BHC or neutron star 1, 15]. The temperature limits of the inner AD plasma can be derived from the detection and non-detection of the specific line emission from the source, under reasonable assumption on the relative abundances of the radiating plasma. If the detected 1.6 MeV γ-ray line is produced in the accretion flow in the neutron star atmosphere [3, 4, 5], then it is likely produced as a redshifted 2.2 MeV neutron capture line at the neutron star atmosphere depth, corresponding to the redshift of 0.38, which is a reasonable value for a massive neutron star [5].

For the GRO J0422+32 main outburst later scenario is consistent with the observed total luminosity from the source, evaluated as redshifted neutron capture lines luminosity, assuming a maximum neutron production yield of $\leq 2 \times 10^3$ neutrons/erg, which has to be $\geq 3 \times 10^{37}(D/2.4\ kpc)^2$ erg s^{-1}. Unfortunately, both above mentioned scenarious are not applicable to the mini-outbursts. For both considered processes the ratio of the X-ray luminosity to the γ-ray luminosity are expected to be of the order $L_x/L_\gamma \sim 10^3$-10^4 [1, 5]. But, in the case of GRO J0422+32 mini-outbursts the luminosity ratio is $L_x/L_\gamma \sim 0.1$, if one compares X-ray fluxes: $F^x_{1.3-12 keV} \sim 10^{-12}$ erg cm^{-2}s^{-1} (ASM RXTE) and $F^x_{20-100 keV} \sim 10^{-12}$ erg cm^{-2}s^{-1} (BATSE CGRO), with the γ-ray line luminosity of $F^\gamma_{1.6 MeV} \sim 1.3 \times 10^{-10}$ erg cm^{-2}s^{-1} (COMPTEL) [21].

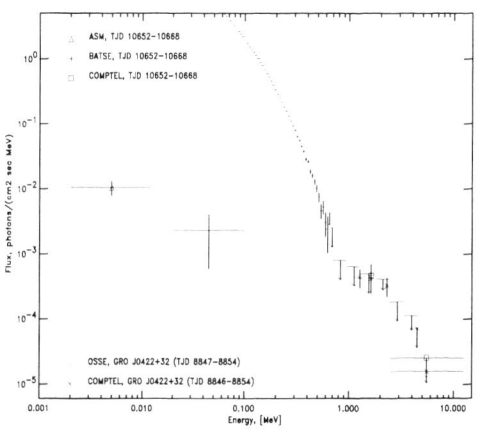

Fig. 4: Spectrum of the GRO J0422+32 mini-outburst detected in X-rays by ASM (RXTE) at 2-10 keV, BATSE at 20-100 keV and by COMPTEL at 1.6 MeV γ-ray line emission, compared to that of the GRO J0422+32 main outburst spectrum as derived by OSSE (Kröger et al. 1996) and by COMPTEL (this work) for the TJD 8847-8854.

A key point to establishing an emission mechanism of the γ-ray line could be a trigger of the periodic mini-outbursts. What kind of mechanism could regularly switch on an accretion (acceleration ?) in the binary system, and after a reasonably well defined time interval to switch it off again? One of the possible hypothesis was put forward in [21] that the NPer 92 is a triple system. Indeed, a well defined periodic activity of the NPer 92 indicates that it could be a triple system composed of a tight binary and of a detached third component, rotating in an eccentric orbit around a common centre of gravity, but does not exclude other possible interpretations.

Alternatively, periodical mini-outbursts could be explained by the accretion disk - neutron star magnetosphere interaction, which could lead to cyclic accretion enhancements [28]. A quasiperiodical solar-like cyclic activity of the secondary can yet be another explanation.

The peculiar poperties of GRO J0422+32 revive the hypothesis that it could contain a neutron star instead of a black hole. What are the reasons for which this idea is still alive?

The detection by COMPTEL of Nova Per'92 in the narrow energy interval, if interpreted as a redshifted and broadened line at 2.2 MeV, during main and multiple (quasi)periodical mini-outbursts, is a strong argument of the compact object being a neutron star. The redshifted 2.2 MeV line emission could be produced as a result of ^{4}He spallation and subsequent neutron capture by protons in the neutron star atmosphere [4]. But where is the associated X-ray emission? And what if there are 2 or more lines?

We leave these and other related questions to be addressed in a forthcoming publication.

REFERENCES

1. Aharonian, F.A. & Syunaev, R.A., MNRAS, 210, 257 (1984).
2. Belloni, T., Hasinger, G., A&A, 227, L33 (1990).
3. Bildsten, L. et al., ApJ, 384, 143 (1992).
4. Bildsten, L. et al., ApJ, 408, 615 (1993).
5. Brecher, K. & Burrows, A., ApJ, 240, 642 (1980).
6. Callanan, P.J. et al., 1995, ApJ, 441, 786 (1995).
7. Castro-Tirado, A.J. et al., IAU Circ. No 5588 (1992).
8. Castro-Tirado, A.J., PhD Thesis, The Amsterdam Univ., (1994).
9. Chevalier, C. & Ilovaisky, S.A., A&A, 297, 103 (1995).
10. van Dijk, R. et al., A&A, 296, L33 (1995).
11. Filippenko, A.V., Matheson, T., & Ho, L.C., ApJ, 455, 614 (1995).
12. Goldwurm, A. et al., IAU Circ. No 5589 (1992).
13. Grove, J.E., et al., in AIP 304 Conf. Proceed., The 2nd COMPTON Symposium, College Park, eds. C.E.Fichtel, N. Gehrels, J. P. Norris, (New-York: AIP), p. 192 (1994).
14. Guarneri, A. et al., IAU Circ. No 6740 (1997).
15. Guessom, N., ApJ, 345, 363 (1989).
16. Harmon, B.A. et al., IAU Circ. No 5584 (1992).
17. Harmon, A.B., et al., in AIP 304 Conf. Proceed., The 2nd COMPTON Symposium, College Park, eds. C.E.Fichtel, N. Gehrels, J. P. Norris, (New-York: AIP), 210 (1994).
18. Harrison, T.E., et al., in AIP 304 Conf. Proceed., The 2nd COMPTON Symposium, College Park, eds. C.E.Fichtel, N. Gehrels, J. P. Norris, (New-York: AIP), p. 319 (1994).
19. Iyudin, A.F. & Haberl, F., IAU Circ. No 6605 (1997a).
20. Iyudin, A.F. & Haberl, F., IAU Circ. No 6738 (1997b).
21. Iyudin, A.F., Haberl, F., and Schönfelder, V., Proc. of the Symposium "*Highlights in X-Ray Astronomy*", eds. B. Aschenbach & M. Freyberg, MPE Report 272, 90 (1999).
22. Kroeger, R.A., et al., A&ASuppl.Ser., 120, 117 (1996).
23. Paciesas, W.C., et al., in AIP 304 Conf. Proceed., The 2nd COMPTON Symposium, College Park, eds. C.E.Fichtel, N. Gehrels, J. P. Norris, (New-York: AIP), 365 (1994).
24. Pietch, W., Haberl, F., Gehrels, N. and Petre, R, A&A, 273, L11 (1993).
25. Roques, J.P., et al., ApJS, 92, 451 (1994).
26. Shrader, C.R., Wagner, R.M., & Starrfield, S.G., IAU Circ. No 5591 (1992).
27. Shrader, C.R. et al., in AIP 304 Conf. Proceed., The 2nd COMPTON Symposium, College Park, eds. C.E.Fichtel, N. Gehrels, J. P. Norris, (New-York: AIP), p. 365 (1994).
28. Spruit, H.C., in "The Lives of the Neutron Stars", eds. Alpar, M.A. et al., (Kluwer Academic Publishers: Dordrecht), 377 (1995).
29. Sunyaev, R.A., et al., A&A, 280, L1 (1993).
30. Wagner, R.M. et al., IAU Circ. No 5589 (1992).

The effects of a comptonizing corona on the appearance of the reflection components in accreting black hole spectra

P.O. Petrucci*, A. Merloni†, A. Fabian†, F. Haardt**, E. Gallo** and J. Malzac*

Osservatorio Astronomico di Brera, Milano, Italy
†*Institute of Astronomy, Cambridge, UK*
**Universitá dell'Insubria, Como, Italy*

Abstract. We discuss the effects of a comptonizing corona on the appearance of the reflection components, and in particular of the reflection hump, in the X-rays spectra of accreting black holes. Indeed, in the framework of a thermal corona model, we expect that part of (or even all, depending on the corona covering factor) the reflection features should cross the hot plasma, and thus suffer Compton scattering, before being observed. We have studied in detail the dependence of these effects on the physical (i.e. the temperature and optical depth) and geometrical (i.e. the inclination angle) parameters of the corona, concentrating on the slab geometry. Due to the smoothing and shifting towards high energies of the comptonized reflection hump, the main effects on the emerging spectra appear mainly above 100 keV. We have also investigated the importance of such effects on the interpretation of the results obtained with the standard fitting procedures. We found that fitting Comptonization models, taking into account comptonized reflection, by the usual cut–off power law + uncomptonized reflection model, may lead to an underestimation of the reflection normalization and an overestimation of the high energy cut–off. We discuss and illustrate the importance of these effects by analyzing recent observational results as those of the galaxy NGC 4258. We also find that the comptonizing corona can produce and/or emphasize correlations between the reflection features characteristics (like the iron line equivalent width or the covering fraction) and the X-ray spectral index similar to those recently reported in the literature. We also underline the importance of these effects when dealing with accurate spectral fitting of the X-ray background.

THE MAIN EFFECTS

In this section we discuss the effect of the Comptonization in the corona on the shape of the reflection component and on the iron line equivalent width varying the optical depth and/or the temperature and/or the inclination of the corona. In the following R_{dir} and R_{comp} correspond to the reflection components computed with or without the comptonization effects respectively. The continuum will be simply called C.

- **Varying τ at fixed coronal temperature:** We will first suppose here that the temperature of the corona is fixed, equal to 50 keV, while τ varies. We have plotted in Fig. 1a the reflection shapes of R_{dir} and R_{comp} (in dashed and solid line respectively) for different values of τ. The increase of the optical depth hardens the X-ray primary spectrum, thus modifying the intrinsic shape of the reflection component. It thus explains the hardening of the uncomptonized reflection R_{dir} between $\tau=0.1$

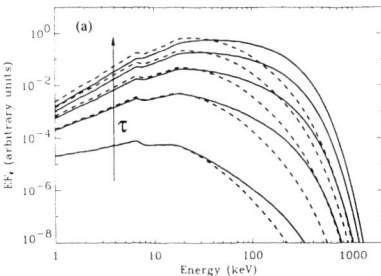
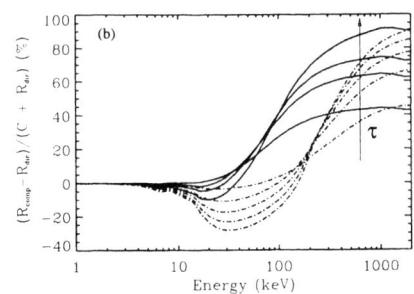

FIGURE 1. (a) The comptonized (solid lines) and uncomptonized (dashed lines) reflection humps, R_{comp} and R_{dir}, for different values of the coronal optical depth, the temperature being fixed to kT_e=50 keV. From bottom to top, τ=0.1, 0.3, 0.5, 0.7 and 0.9. (b) Deviations (in %) between the outgoing spectra. The solid line and dot-dashed lines correspond to kT_e = 50 keV and 350 keV respectively.

and τ=0.9. Increasing τ also magnifies the effect of the Comptonization. The larger τ, the larger the probability of a photon to be comptonized. In this process the reflected photons are shifted towards higher energies and an increasing deviation between $R_{\rm dir}$ and $R_{\rm comp}$ is seen, at energies below and above $\sim kT_e$ =50 keV, for increasing τ.

To estimate quantitatively these effects on the total outgoing spectrum, we have plotted in Fig. 1b (solid lines) the deviations (in %) between the total spectrum $(C+R_{\rm comp})$, expected if all the reflected photons cross the comptonizing corona before being observed, and the total one $(C+R_{\rm dir})$ predicted when the Comptonization of the reflection hump is not taken into account. The most important effects (variations of $>50\%$) occur at very high energies (above 100 keV). They are due to the hardening of $R_{\rm comp}$ for large optical depths. For the larger optical depth case considered ($\tau = 0.9$), a factor of ~ 2 is expected near 1 MeV. At these energies, $R_{\rm comp}$ tends to have the same shape as the primary continuum and the fractional deviations between the two spectra attain a maximum.

The Comptonization produces also smaller ($< 10\%$) differences between $R_{\rm dir}$ and $R_{\rm comp}$ near 10 keV. The larger one is still produced in the high optical depth case, as expected.

- **Varying τ and T_e at fixed Compton parameter:** In the case of a constant Compton parameter, changes in τ will be necessarily accompanied by changes in kT_e, namely a larger optical depth will require a smaller coronal temperature. This produces a hardening of $R_{\rm comp}$ at high energy when τ decreases, oppositely to what we obtained when the temperature is fixed. However the deviations between the outgoing spectra closely resemble those shown in Fig. 1b (cf. Petrucci et al. 2001). It is worth noting that in a corona geometry with a smaller covering factor, only a part of the reflected photons cross the comptonizing plasma before being observed. The effects of Comptonization are thus reduced.

- **Dependence on the inclination angle:** We have plotted in Fig. 2a, the different shapes of $R_{\rm comp}$ and $R_{\rm dir}$ for different values of the viewing angles. Because an

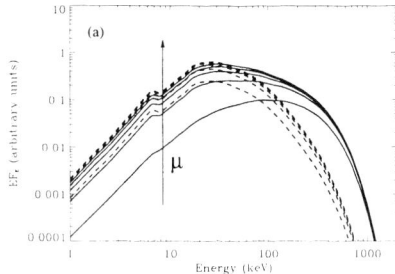 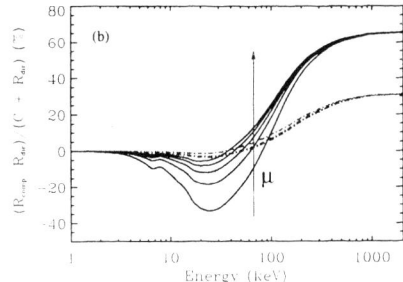

FIGURE 2. (a) The comptonized (solid lines) and uncomptonized (dashed lines) reflection humps, R_{comp} and R_{dir}, for different viewing angle. From bottom to top $\mu=0.1, 0.3, 0.5, 0.7$ and 0.9. The optical depth and temperature of the corona are fixed to 0.35 and 90 keV respectively. (b) Deviations (in %) between the outgoing spectra. In dot-dashes line the case of an hemispherical corona with $\tau=0.35$ and $kT_e=200$ keV.

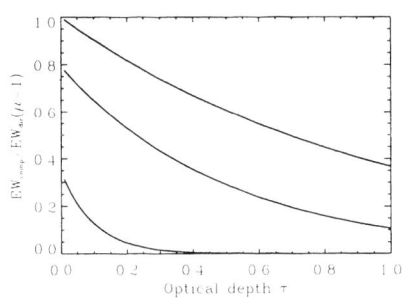

FIGURE 3. Reduced equivalent width $EW_{comp}(\mu)/EW_{dir}(\mu=1)$ of the comptonized iron line in function of the optical depth of the corona for different inclination angles. From top to bottom $\mu=1, 0.5$ and 0.1

increase of i will correspond to an increase of the effective optical depth τ/μ, we observe, for increasing i, the same effects we observed for increasing τ, the main difference is that now the shape of the primary spectrum impinging on the accretion disk is constant.

- **Effects on the iron line:** The Comptonization of the iron line may affect the measurement of its EW as pointed by Haardt et al. (1993) and discuss by Matt et al. (1997). We have plotted in Fig. 3 the reduced equivalent width $EW_{comp}(\mu)/EW_{dir}(\mu=1)$[1] as a function of the optical depth of the corona and for different viewing angles. We see that the effects of Comptonization are very strong, especially at large inclination angles. For $\mu=0.9$ the EW is reduced by a factor ~ 3 when τ varies between 0 and 1 whereas it becomes rapidly negligible for

---

[1] $EW_{dir}(\mu=1)$ corresponds to the face-on EW of the uncomptonized line

$\tau > 0.5$ if $\mu = 0.1$.

Remark: the Comptonization effects on the reflection hump beyond ~ 100 keV may not be easily detected. Firstly because the high energy instruments are generally not very sensitive beyond ~ 100 keV and thus prevent any good constrains on the spectral shape. Secondly, any deviations from the ideal case studied here like, for instance, a stratified temperature corona, may distort the derived reflected shape, especially at high energy, and complicate the spectral analysis. The effects at lower energy (near ~ 20 keV where the instrumental sensitivity is very good) would certainly be more relevant for any estimation of the comptonization effects. We also expect these modifications of the Iron line EW to be a lot more observable than the spectral changes of the reflection hump detailed.

OBSERVATIONAL CONSEQUENCES

- **The case of NGC 4258:** The nearby bright galaxy NGC 4258 possesses a (narrow) iron line near 6.4 keV with an equivalent width of the order of 100 eV (Reynolds et al. 2000, hereafter R00; Fiore et al. 2001, hereafter F01). This is an AGN in a low state with an Eddington accretion rate of 0.0002 (F01). The presence of an ADAF seems however to be ruled out by the recent BeppoSAX observations and the most natural explanation of the X-ray luminosity of this source may be in terms of Comptonization of soft photons in a hot corona (F01). The inclination of the disk is large and very well constrained i=82 ± 1 degrees (Herrnstein et al. 1999). We thus expect the Comptonization effects on the reflection components coming from the disk to be relatively important, especially in the case of a slab corona.

  Since the inclination angle of the disk is large we expect part of the iron line photons to be comptonized in the corona. Thus the EW may be significantly reduced in comparison to the one we expect without a comptonizing corona. Supposing that all iron line photons have to cross the corona before being observed the actual uncomptonized face–on line EW would be well above the average narrow iron line equivalent width observed in Seyfert 1 galaxies. Thus, the narrow iron line must preferentially originate from matter not associated with the accretion disk.

  The Comptonization effects also enable to put less strict constrains than R00 on the EW upper limit of a possible broad line, not seen in the data. Consequently if there were a broad line present at the level seen in most other AGN (~ 100–300 eV, Nandra et al. 1997; Lubinski & Zdziarski 2000 hereafter LZ00), we do not expect to see it. The data thus do not require that the inner part of the accretion disc be truncated as suggested in R00.

- **Correlations between reflection features and X-ray slope:** The Comptonization effects we have discussed so far may influence the apparent correlations observed between the reflection features characteristics (broad Fe line EW, reflection normalization) and the X-ray photon index (LZ00; Zdziarski et al. 1999 hereafter Z99).

  If for example objects with hard spectra (small Γ) have a corona with a large optical depth, correlations between EW–Γ and R–Γ are expected. A possible reason for an anticorrelation between τ and Γ may be the following. Suppose that the corona+disk configuration is the one proposed by Z99 i.e. a central hot plasma + cold disc model, the inner radius of the disk being able to vary. In this interpretation, the increase of the inner radius of the disk, which will produce a decrease of R and Γ, may be due to the evaporation of part of the inner regions of the disk (due to some disk instabilities, Meyer et al. 2000; Menou et al. 2000; Turolla & Dullemond 2000) in the corona, thus resulting in a increase of the corona optical depth. Interestingly, in this case, the comptonization effects would reduce the apparent reflection component and thus could explain the small R values observed at small Γ.

- **Consequences for the X-ray background:** All the spectral models for the XRB assume that the intrinsic AGN spectrum (before any absorption in a putative large scale molecular torus) is the sum of a power-law continuum and an uncomptonized reflection. We expect that the effects described here can be of some relevance for any detailed fitting model for the XRB. Indeed, in all current models (Madau et al. 1994; Wilman & Fabian 1999; Gilli et al. 2001), for an average continuum slope of about 2, the peak of the XRB (in νF_ν) is associated with the peak of the reflection hump. Furthermore, there is evidence that the observed peak of the X-ray background is located at slightly higher energies than those predicted by the standard models that neglect the effect of Comptonization in the corona.

Although many uncertainties come into the exact determination of the possible consequences of such effect on the XRB shape, as the temperature and optical depth distribution in the different sources or the total coronal covering fraction, the work we have presented here should indicate that Comptonization of the reflection component in the corona have to be taken into account when dealing with accurate spectral fitting of the X-ray background.

ACKNOWLEDGMENTS

POP acknowledges a grant of the European Commission (contract number ERBFMRX-CT98-0195, TMR "Accretion onto black holes, compact stars and protostars").

REFERENCES

Fiore F., Pellegrini S., Matt G. et al. 2001, ApJ in press (astro-ph/0102438) (F01)
Gilli, R., Salvati, M., & Hasinger, G. 2001, A&A, 366, 407
Haardt, F., Done, C., Matt, G., & Fabian, A. C. 1993, ApJL, 411, L95
Herrnstein, J. R. et al. 1999, Nature, 400, 539
Lubinski, P. and Zdziarski, A. A. 2000, submitted to MNRAS (astro-ph/0009017) (LZ00)
Madau, P., Ghisellini, G., & Fabian, A. C. 1994, MNRAS, 270, L17
Matt, G., Fabian, A. C., & Reynolds, C. S. 1997, MNRAS, 289, 175
Menou, K., Hameury, J., Lasota, J., & Narayan, R. 2000, MNRAS, 314, 498
Meyer, F., Liu, B. F., & Meyer-Hofmeister, E. 2000, A&A, 361, 175
Nandra, K., George, I. M., Mushotzky, R. F., Turner, T. J., & Yaqoob, T. 1997, ApJ, 477, 602
Petrucci, P. O., Merloni, A., Fabian, A., Haardt, F., & Gallo, E., MNRAS, submitted
Reynolds, C. S., Nowak, M. A., & Maloney, P. R. 2000, ApJ, 540, 143 (R00)
Turolla, R. & Dullemond, C. P. 2000, ApJL, 531, L49
Wilman, R. J. & Fabian, A. C. 1999, MNRAS, 309, 862
Zdziarski, A. A., Lubinski, P. and Smith, D. A. 1999, MNRAS, 303, L11 (Z99)

The compact jet of the black hole candidate XTE J1550−564 during the 2000 X-ray outburst

S. Corbel*, P. Kaaret[†], R.K. Jain**, C.D. Bailyn**, R.P. Fender[‡],
J.A. Tomsick[§], E. Kalemci[§], V. McIntyre[¶], D. Campbell-Wilson[||], J.M. Miller[††] and M.L. McCollough[‡‡]

*Université Paris VII and Service d'Astrophysique, CEA Saclay, F-91191 Gif sur Yvette, France
[†]Harvard–Smithsonian Center for Astrophysics, 60 Garden Street, Cambridge, MA 02138, USA
**Yale University, Department of Astronomy, P.O. Box 208101, New Haven, CT 06520-8101, USA
[‡]Astronomical Institute 'Anton Pannekoek', University of Amsterdam, and Center for High Energy Astrophysics, Kruislaan 403, 1098 SJ Amsterdam, The Netherlands
[§]Center for Astrophysics and Space Sciences, University of California, San Diego, MS 0424, La Jolla, CA 92093, USA
[¶]Australia Telescope National Facility, PO Box 76, Epping NSW 1710, Australia
[||]School of Physics, University of Sydney, NSW 2006, Australia
[††]Center for Space Research, Massachusetts Institute of Technology, 70 Vassar Street, Cambridge, MA 02139, USA
[‡‡]Universities Space Research Association, Hunsville, AL 35812, USA

Abstract.
We report on radio, near-infrared, optical and X-ray observations of the black hole candidate (BHC) XTE J1550−564 performed during its 2000 X-ray outburst. These observations have allowed us to sample the behavior of XTE J1550−564 in the X-ray Low Hard and Intermediate/Very High states. The radio emission in the Low Hard state most likely originates from a compact jet and the synchrotron emission from this jet may extend up to the optical range or beyond, therefore indicating that the total power of the compact jet is a significant fraction of the total luminosity of the system. In the Intermediate/Very High state the radio emission is quenched, implying a suppression of the outflow. We discuss the properties of radio emission in the X-ray states of BHCs.

XTE J1550−564 and its 2000 outburst

The recurrent soft X-ray transient and microquasar XTE J1550−564 was first detected by the All Sky Monitor (ASM) on-board the *Rossi X-ray Timing Explorer (RXTE)* on 1998, September 7. XTE J1550−564 went through all canonical black hole states (Sobczak et al. 2000; Homan et al. 2001) before its return to quiescence in May 1999. Optical and radio counterparts were reported by Orosz, Bailyn and Jain (1998) and Campbell-Wilson et al. (1998).

In April 2000, XTE J1550−564 became active again in soft X-rays and optical. The light curves in soft and hard X-rays have been presented in Figure 1 of Corbel et al. (2001). The 2000 outburst can be understood as follows: an initial Low Hard state followed by a transition to the Intermediate state (or Very High state, see discussion in Homan et al. 2001) on April 26 and then a return to the Low Hard state after May 13. The RXTE/PCA+HEXTE observations performed during this outburst are discussed in

Tomsick, Corbel & Kaaret (2001), Kalemci et al. (2001) and Miller et al. (2001).

ATCA radio continuum observations of XTE J1550−564 were carried out on three different dates in 2000: April 30, May 6 and June 1, they are summarized in Table 1 of Corbel et al. (2001). In this paper, we concentrate on the observation perfomed during the Intermediate/Very High state and during the final Low Hard state. We also present near-infrared and optical YALO observations taken simultaneously with the radio observation on June 1. Before the ATCA observation, the MOST radio telescope performed several observations while XTE J1550−564 was in its initial Low Hard state and detected it at a level of 8–15 mJy at 843 MHz. Full discussion of these observations can be found in Corbel et al. (2001) and Jain et al. (2001).

A powerful compact jet

After the final transition to the Low Hard state, we performed our last ATCA observation. Radio emission is detected at a level of ∼ 1 mJy from 1.4 GHz to 8.6 GHz. The spectrum is slightly inverted with a spectral index of $\alpha = 0.37 \pm 0.10$ for a flux density $S_\nu \propto \nu^\alpha$ (Figure 1). This is reminiscent of the behavior of the few BHCs (transient or persistent) which have been observed at radio frequencies in the Low Hard state (Brocksopp et al. 1999; Corbel et al. 2000; Fender 2001). Such radio properties have been interpreted as arising from a self absorbed compact jet (on milli-arcsec scale), similar to those considered for flat spectrum active galactic nuclei (Blandford & Königl 1979, Hjellming & Johnston 1988). This interpretation has been successfully confirmed with the resolution of a compact jet in Cyg X−1, based on very long baseline interferometric radio observations (Stirling et al. 2001). Nevertheless, compact jet models predict a cut-off to the flat or inverted spectral component at high frequency. Various observations indicated that this might be taking place in the near-infrared/optical range (e.g. Hynes et al. 2000, Fender et al. 2001, Brocksopp et al. 2001). In GX 339−4, a cut-off was clearly observed to fall in the near-infrared range (Corbel & Fender 2001).

In Figure 1, we show the simultaneous radio, near-infrared and optical observations of XTE J1550−564 taken on 2000 June 1. The spectral index in the near-infrared and optical bands is not compatible with the thermal spectrum of an optically thick accretion disk. It is therefore likely that a significant fraction of the optical-near infrared emission in XTE J1550−564 during the Low Hard state is (optically thin) synchrotron emission from the compact jet. The full optical and near-infrared YALO monitoring of the 2000 X-ray outburst of XTE J1550−564 has confirmed such interpretation. Indeed, a secondary maximum was observed when XTE J1550−564 entered the final Low Hard state, it may be linked with non-thermal emission associated with the formation of the compact jet (Jain et al. 2001).

Using the June 1 observations, we find that the minimum total radiative luminosity (for a distance of 2.5 kpc) of the compact jet is of the order of $\sim 2\times 10^{34}\,\mathrm{erg\,s^{-1}}$, several per cent of the 2-20 keV band X-ray luminosity of $5\times 10^{35}\,\mathrm{erg\,s^{-1}}$ measured on the same day (Tomsick, Corbel, & Kaaret 2001). Thus, the compact jet is likely very powerful with a total power that is a significant fraction of the bolometric luminosity of the system (accretion disk and corona).

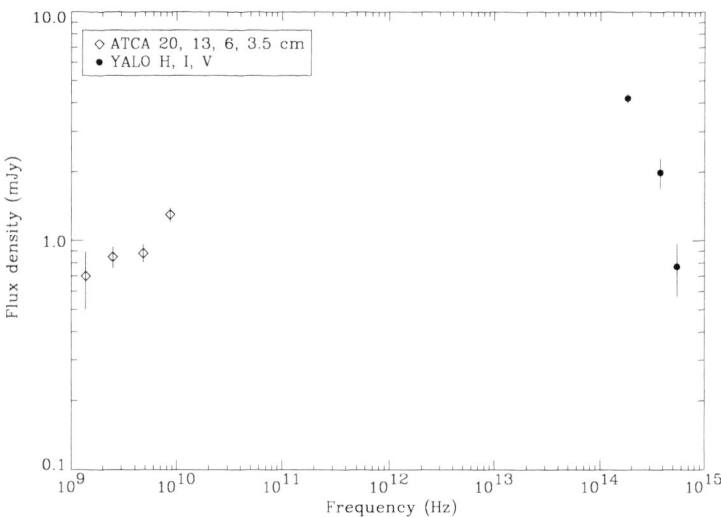

FIGURE 1. Radio, near-infrared (H) and optical (V,I) measurements of XTE J1550−564 on June 1. The optical data originate from the Yale 1-m telescope (dereddened using an optical extinction of 2.2 ± 0.3 mag.) and the radio data from the ATCA radio interferometers. The spectral index in the radio range is +0.37, and near infrared/optical emissions are compatible with the optically thin synchrotron emission contribution from the compact jet. (NB: The H band datapoint has been added in this figure, compared to Figure 2b in Corbel et al. 2001).

X-ray states and radio emission in black hole candidates

Black hole candidates (BHCs) are known to exhibit five distinct X-ray spectral states, distinguished by the presence or absence of a soft blackbody component at $\sim$ 1 keV and the luminosity and spectral slope of emission at harder energies. Systems in the Low Hard state have hard power-law spectra and no (or only weak) evidence of a soft thermal component. On the other hand, High Soft state spectra are dominated by a disk blackbody component (contribution from an optically thick accretion disk) with a characteristic temperature $\sim$ 1 keV. The dominant radiative process in the Low Hard state appears (in our current understanding) to be thermal comptonization of soft accretion disk photons by a hot corona surrounding a black hole. But in some or perhaps all cases, significant hard X-ray synchrotron emission from the compact jet may be observed (e.g. Markoff, Falcke, & Fender 2001). In the Intermediate and Very High states, both emission components (the accretion disk and the corona) are observed. These two states can be viewed as having a similar geometry but with different X-ray luminosity, the Very High state being close to the Eddington luminosity (Homan et al. 2001). In addition, all transient BHCs spend most of their time in a quiescent/off state, characterized by a very low X-ray flux.

The second ATCA observation was performed when XTE J1550−564 was in the Intermediate/Very High state. At that time, XTE J1550−564 was not detected with

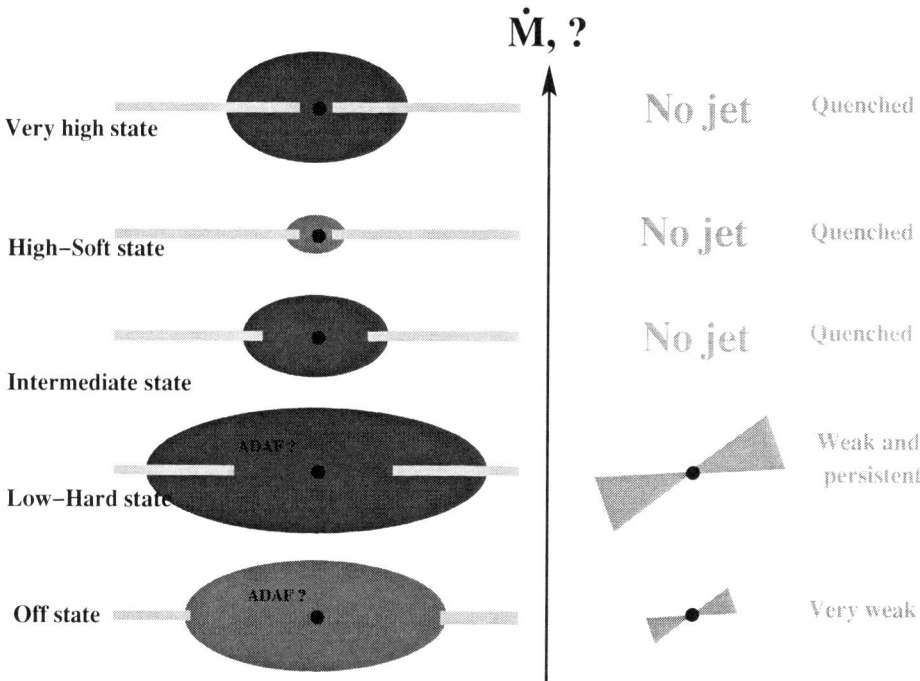

FIGURE 2. A schematic view of black hole states and jet properties (as observed in radio). Variation in the mass accretion rate and/or possibly the size of the corona may lead to state transition, see discussion in Homan et al. (2001).

ATCA at 4800 and 8640 MHz, with a 3 σ upper limit of 0.15 mJy at 8640 MHz, i.e. a reduction of radio emission by a factor greater than 50, if we take into account the MOST detection in the initial Low Hard state. This is the first time a BHC was observed in radio during the Intermediate/Very High state. These observations therefore indicate that the outflow of XTE J1550−564 is suppressed in the Intermediate/Very High state, which is similar to the quenching of the compact jet of GX 339−4 in the High Soft state (Fender et al. 1999).

Based on the data we (and other groups) have been collecting, the scenario represented in Figure 2 seems to emerge regarding the connection between X-ray states of BHCs and radio emission. The Low Hard state (and the Off state to a weak extent, e.g. see Corbel et al. 2000 for GX 339−4) does produce radio emission via a compact jet, whereas the states with a stronger soft component (High Soft and Intermediate/Very High states) lead to the quenching of these compact jets for a reason which is not understood (Corbel et al. 2001).

REFERENCES

1. Blandford, R. D., & Königl, A. 1979, *Astrophys. J.*, **232**, 34
2. Brocksopp C., Fender R. P., Larionov V., Lyuty V. M., Tarasov A. E., Pooley G. G., Paciesas W. S., & Roche P. 1999, *Mon. Not. R. Astron. Soc.*, **309**, 1063
3. Brocksopp C., Jonker, P. G., Fender R. P., Groot, P.J., van der Klis, M., & Tingay, S. J. 2001, *Mon. Not. R. Astron. Soc.*, **323**, 517
4. Campbell-Wilson, D., McIntyre, V., Hunstead, R., & Green, A. 1998, *IAU Circ.*, 7010
5. Corbel, S., Fender, R. P., Tzioumis, A. K., Nowak, M., McIntyre, V., Durouchoux, P., & Sood, R. 2000, *Astron. & Astrophys.*, **359**, 251
6. Corbel, S. et al. 2001, *Astrophys. J.*, accepted, astro-ph/0102114
7. Corbel, S., & Fender, R. P. 2001, *Astrophys. J.*, submitted
8. Fender R. P. et al. 1999, *Astrophys. J.*, **519**, L165
9. Fender R. P., Hjellming, R. M., Tilamus, R. P. J., Pooley, G. G., Deane, J. R., Ogley, R. N., & Spencer, R. E. 2001, *Mon. Not. R. Astron. Soc.*, accepted, astro-ph/0101346
10. Fender R. P. 2001, *Mon. Not. R. Astron. Soc.*, **322**, 31
11. Hjellming R. M., & Johnston K. J. 1988, *Astrophys. J.*, **328**, 600
12. Homan, J., Wijnands, R., van der Klis, M., Belloni, T., van Paradijs, J., Klein-Wolt, M., Fender, R. P., & Méndez, M. 2001, *Astrophys. J. Suppl. Ser.*, **132**, 377
13. Hynes, R. I., Mauche, C. W., Haswell, C. A., Shrader, C. R., Cui, W., & Chaty, S. 2000, *Astrophys. J.*, **539**, L37
14. Jain, R. K., Bailyn, C. D., Orosz, J. A., McClintock, J. E., & Remillard, R. A. 2001, *Astrophys. J.*, accepted, astro-ph/0105115
15. Kalemci, E., Tomsick, J.A., Rothschild, R.E., Pottschmidt, K. & Kaaret, P. 2001, *Astrophys. J.*, submitted, astro-ph/0105395
16. Markoff, S., Falcke, H., & Fender, R.P., 2001, *Astron. & Astrophys.*, accepted, astro-ph/0010560
17. Miller, J.M., Wijmands, R., Homan, J., Belloni, T., Pooley, D., Corbel, S., Kouveliotou, C., van der Klis, M. & Lewin, W.H.G. 2001, *Astrophys. J.*, submitted, astro-ph/0105371
18. Orosz, J. A., Bailyn, C. D., & Jain, R. K. 1998, *IAU Circ.*, 7009
19. Sobczak, G. J., McClintock, J. E., Remillard, R. A., Cui, W., Levine, A. M., Morgan, E. H., Orosz, J. A., & Bailyn, C. D. 2000, *Astrophys. J.*, **544**, 993
20. Stirling, A. M., Spencer, R. E., de la Force, C. J., Garrett, M. A., Fender, R.P., & Ogley, R. N. 2001, *Mon. Not. R. Astron. Soc.*, submitted
21. Tomsick, J.A., Corbel, S. & Kaaret, P. 2001, *Astrophys. J.*, submitted, astro-ph/0105394

Accretion-Ejection Instabilities in Black-Hole Binaries

P. Varnière*, S. Caunt† and M. Tagger*

*Service d'Astrophysique CEA/Saclay France
†Astronomy Division, University of Oulu

Abstract. The Accretion-Ejection Instability has been proposed to explain the "ubiquitous" low frequency QPO of microquasars.

This instability which occurs in the inner region of moderately magnetized disks has the the unique property that it can send toward the corona the energy and angular momemtum from the disk.

I will present recent results on the physics of the instability, and comparison with observations. In particular I will describe how it generates Alfven waves propagating the energy and angular momentum to the corona.

INTRODUCTION

The AEI is a spiral instability, similar to galactic spirals but driven by magnetic stresses rather than self-gravity. It occurs in the inner region of an accretion disk threaded by a vertical magnetic field of the order of equipartition with the gas pressure, $\beta = 8\pi p/B^2 \sim 1$. It forms a quasi-steady pattern rotating in the disk at a frequency of the order of $0.1 - 0.3\ \Omega_{int}$, the rotation frequency at the inner edge of the disk.

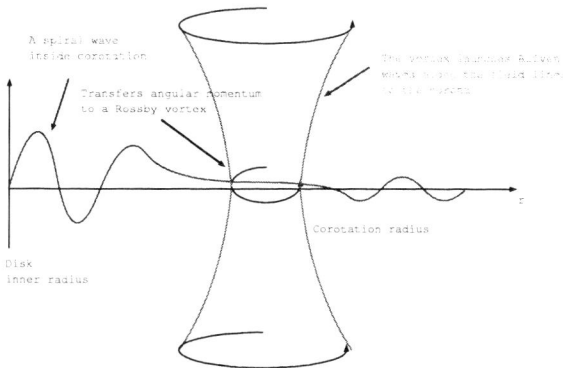

FIGURE 1. The instability is formed of a standing spiral density wave in the inner part of the disk, coupled to a Rossby vortex it excites at its corotation radius. The Rossby vortex in turn generates Alfvén waves propagating toward the corona of the disk.

The spiral grows by extracting energy and angular momentum from the disk, causing accretion, and storing them in a Rossby vortex (analogous to the Great Red Spot of Jupiter) it generates at its corotation radius.

We expect the vortex to "leak" energy and angular momentum as Alfvén waves to the corona, where it might power a wind or a jet. Since the energy and angular momentum are transported by waves, they should not heat the disk as in usual viscous models.

NUMERICAL SIMULATION

We have performed 2D (r,ϕ), non-linear MHD simulations of a thin disk threaded by a vertical magnetic field (Caunt & Tagger 2001). These simulations show growing spiral patterns, in detailed agreement with the linear theory. The initial number of arms of the spiral depends on the initial conditions, but generally, as magnetic flux is advected to the inner region of the disk, it reduces to a 1-armed spiral. Non-linearly, as expected (and as in galaxies), the spiral saturates at a finite amplitude to give a quasi-stationary spiral pattern in the disk, giving a very plausible mechanism for the low frequency ("ubiquitous") Quasi-Periodic Oscillation (QPO) of the X-ray binaries.

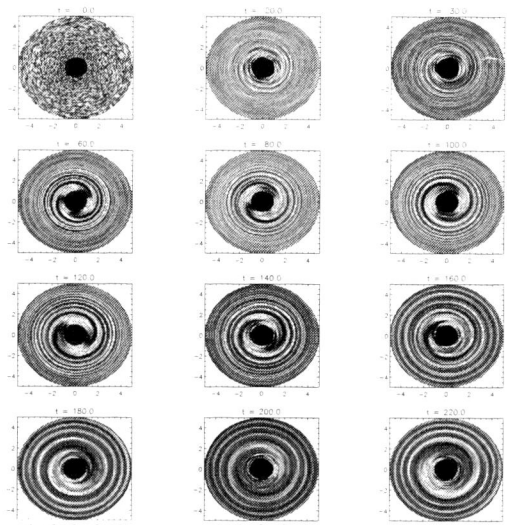

FIGURE 2. MHD simulations show an $m = 1$ spiral forming in the inner region of the accretion disk.

THE ACCRETION-EJECTION INSTABILITY AS A MODEL FOR QPO

The Accretion-Ejection Instability is a good candidate to explain the low-frequency QPO because of several of its characteristics:

- its frequency (0.1-0.3 Ω_{int} where Ω_{int} is the inner radius keplerien frequency) is compatible with the observed ones.
- the quasi-steady spiral pattern (as in galaxies) would be observed as a quasi-periodic feature
- the emission of Alfven waves to the corona would explain the observatin that, although it has its origin in the disk, the QPO strongly affects the corona.

RELATIVISTIC EFFECTS

The propagation of the spiral in the disk gives a special role to the Inner Lindblad Resonance (ILR), defined by the radius where $\omega = m\Omega - \kappa$ (with m the number of arms, Ω the rotation frequency and κ the epicyclic frequency). In a keplerian disk $\Omega = \kappa$ and therefore no ILR is possible for the $m = 1$ mode. Relativistic effects near the last stable orbit decrease κ and thus would allow the existence of an ILR for the $m = 1$ mode. This will then change the properties of the 1-armed spiral (best candidate to explain the QPO), and in particular its frequency.

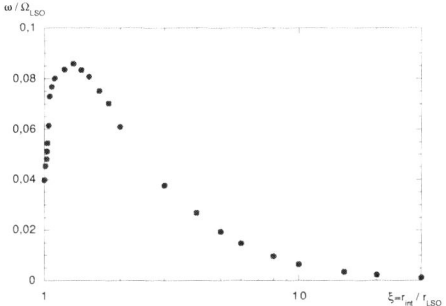

FIGURE 3. Theoretical curve showing the change in correlation between the frequency of the QPO, scaled by the keplerian frequency at the last stable orbit, and the inner radius of the disk, scaled by the radius of the last stable orbit.

We have studied, using a pseudo-newtonian potential which reproduces the relativistic rotation curve, the properties of the instability when the disk inner radius r_{int} approaches the Last Stable Orbit at r_{LSO}.

When r_{int} is large the QPO frequency varies as $\omega \propto r^{-3/2}$; but as r_{int} approaches r_{LSO} the correlation changes and becomes *positive* (the QPO frequency decreasing with a decreasing radius) when $r_{int} < 1.4 r_{LSO}$, *i.e.* when the $m = 1$ mode has an ILR in the disk. We consider this as a possible explanation for the positive correlation found (contrary to other sources) in GRO J1655 by Sobczak *et al.*. See Rodriguez *et al.* for a more detailed comparison.

EMISSION OF ALFVÉN WAVES

The Rossby vortex twists the footpoints of magnetic field lines threading the disk. If the disk has a low density corona this twist will be propagated upward as Alfvén Waves. The energy and angular momentum extracted from the disk and stored in the vortex will thus be transferred to the corona, where they might power a wind or jet. We study this mechanism with a variational form

$$F = \text{[energy of the waves]} \\ + i \text{ (outgoing spiral} + \text{ coupling with the vortex} \\ + k_z \text{ Alfvén Waves)}$$

where imaginary terms correspond to an energy and momentum flux, causing amplification or damping of the instability.

In the variational form the term representing the flux of Alfven waves is singular at the corotation. This results in an efficient transfer of the accretion energy and momentum to the corona. The details of the results show that in practice this fraction will be fixed by non-linear effects.

CONCLUSION

⋆ The AEI provides a possible explanation for the main properties of the QPO: its frequency, the frequency-radius correlation, the amplitude-energy correlation.
⋆ Numerical simulation confirm the expected behavior, namely a spiral wave and a Rossby vortex forming a steady rotating pattern.
⋆ We progress in the detailed theoretical understanding of the AEI and its role in the inner part of the disk.
⋆ Alfven waves can energize the corona efficiently.

REFERENCES

1. Caunt, s. & Tagger, M. Numerical simulations of the Accretion-Ejection Instability in magnetized accretion disks. A&A, **367**, 1095-1111, 2001.
2. J.Rodriguez, P.Varnière & M.Tagger Accretion-Ejection Instability and QPO: II. Observations. to be published by A&A
3. Sobczak, G. J, McClintock, J. E., Remillard, R. A., Cui, W., Levine, A. M., Morgan,E. H., Orosz, J. A., and Bailyn, C. D., 2000, Correlations between Low-Frequency Quasi-periodic Oscillations and Spectral Parameters in XTE J1550-564 and GRO J1655-40 2000, ApJ , **531**, 537.
4. M.Tagger & Pellat Accretion-Ejection Instability in Magnetized Disks. A&A, 2:155–192, 1999.
5. P.Varnière , J.Rodriguez & M.Tagger Accretion-Ejection Instability and QPO: I. Theory. submitted to A&A

A Review of Cygnus X-1 Soft γ-ray Observations

James C. Ling

Jet Propulsion Laboratory
California Institute of Technology

Abstract. Since the first discovery of Cygnus X-1 in the mid 1960's, the source has been the subject of intense soft γ-ray (30 keV – 10 MeV) observations by many balloon and satellite experiments. A large body of spectral and temporal information about the source has been gathered to date. While these results have significantly enhanced our understanding of the Cygnus X-1 system, they have also raised new questions that need to be addressed by future missions. This paper provides a brief summary of some of the important long-term temporal and spectral results obtained over the last thirty years, and discusses the current status and issues that need to be addressed by upcoming missions.

1. INTRODUCTION

Cygnus X-1 is one of the brightest soft γ-ray sources in our Galaxy. For more than thirty years since the source was first discovered in the mid 1960's[1], it has been observed by numerous balloon[2-4] and satellite experiments. The latter include OSO-7[5], OSO-8[6], HEAO-1[7], HEAO-3[8-9], SMM[10], SIGMA[11], OSSE[12], BATSE[13-15], COMPTEL[16-17], ROSSI[18] and BEPPOSAX[19]. This paper gives a brief summary of the long-term spectral and temporal behavior of the source. Specific emphasis is given to the description of three distinct soft γ-ray spectra that were observed when the source was in the standard state (Section 2), high x-ray state (Section 3) and superlow x-ray/high γ-ray state (Section 4), respectively. Important issues that need to be addressed by future missions is also discussed.

2. THE STANDARD STATE

Cygnus X-1 is highly variable with time scale ranging from milliseconds to years. The source stayed most of the time in its standard state, known also as the x-ray "low/hard" state. Under this condition, the soft γ-ray spectrum can be generally described as having two components: a Comptonized[20] shape up to ~300-400 keV with an electron temperature of kT ~30 - 100 keV and optical depth of τ_s ~ 1-3[2,7,8,9,11] and a high energy power-law tail extending from 0.4 to several MeV[12,13,16]. Figure 1 (left panel) shows a composite broadband spectrum measured contemporaneously by COMPTEL, BATSE and OSSE during the first three cycles of the CGRO mission in 1991-1994[10]. The best-fit power-law index of the high energy component (> 0.6

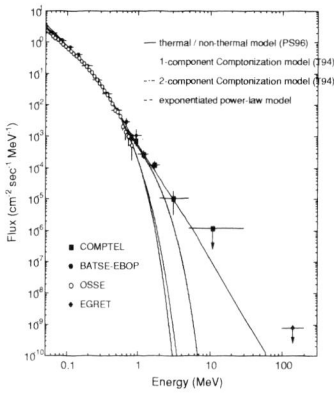
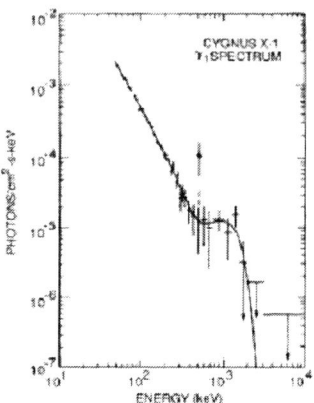

Figure 1. Three very different soft γ-ray spectra associated with the Cygnus X-1 states were observed in the past decades: *Left Panel:* the standard-state spectrum observed by COMPTEL, BATSE and OSSE in 1991-1994[16] consists of two components: a Comptonized shape below 400 keV followed by a power-law tail extended to ~5 MeV with an index of ~3.3. *Middle Panel*- the high X-ray state spectrum observed by BATSE[13] and OSSE[12] in 1994 is power-law like with photon index of ~2.7. *Right Panel::* the superlow X-ray /high γ-ray state spectrum measured by HEAO-3 in 1979 consists of a Comptonized component below 0.4 MeV, a broad γ-ray Gaussian feature (0.4-1.5 MeV) centered at 1 MeV[9] and a weak (~2σ) narrow 511 keV feature[34].

MeV) measured by COMPTEL[10] is ~3.3. The luminosity in the 0.02 - 0.2 MeV, 0.2 - 1 MeV and >1 MeV bands was estimated[21] to be 20.5, 4.8 and 0.6 x 10^{36} erg/s, respectively, with ~80% in the 20 - 200 keV band, assuming a source distance of 2.5 kpc.

The two-component standard spectrum may be interpreted in terms of a thermal model consisting of two interacting regions: a high-temperature core embedded in a lower temperature corona[13,22]. Such a model is qualitatively consistent with the advective accreting model[23], which is quasi-spherical and has a temperature gradient along the radius. The spectrum could also be interpreted in terms of a nonthermal model[24] of free infalling matter onto the BH in the converging flow region. More recently, a hybrid thermal/non-thermal model[25] has been shown to fit both the standard (low-state) spectrum[16] and the high x-ray state spectrum[17] well.

Electron-positron pairs could also exist in such a BH system and produce an annihilation feature[35]. HEAO-1[36] observed a broad feature centered at 500 keV with a flux of 5 x 10^{-3} photons/cm^2-s in 1977-1978 when the source was in the standard state. However, this was not seen in the γ$_2$, or the standard state, spectrum measured by HEAO-3 in 1979-1980, which set a 3σ upper limit of 4 x 10^{-3} photons/cm^2-sec for a broad feature[9] and 1 x 10^{-4} photons/cm^2-sec for a narrow feature[34]. More recently, OSSE[12] also placed an upper limit of 7 x 10^{-5} photons/cm^2-sec for a narrow 511 keV line and 2 x 10^{-4} photons/cm^2-sec for a broad 511 keV line when the source was in its standard state.

3. THE HIGH X-RAY STATE

While Cygnus X-1 stayed most of the time in the standard (low) state, it is also known to make an occasional transition to a high (intensity) x-ray state. Such a transition has been seen at least five times in the last thirty years: in 1970 by Uhuru[26], in 1975 by Ariel 5 All Sky Monitor[27-28] and OSO-8[6], in 1980 by Hakucho[29] and HEAO-3[8-9], in 1994 by BATSE[13], and in 1996 by ROSSI[18], BATSE[15], and COMPTEL/OSSE/BATSE[17]. During a typical "low" to "high" state transition, the 1-10 keV flux increased significantly, while the flux above 10 keV decreased in an anti-correlated fashion, with the spectrum pivoting at around 10 keV[15,30]. The overall soft γ-ray spectrum (>30 keV), however, evolved from the two-component shape (Comptonized + power-law tail) described above to a power-law with index of 2.6-2.7. Such spectral change was seen by both BATSE[13] and OSSE[12] for the 1994 event (see Figure 1, the middle panel), and by COMPTEL/OSSE/BATSE[17] for the 1996 event. It is important to note that in contrast to the softening of the spectrum below 200 keV, the high-state spectrum above 200 keV is actually harder than the standard (low) state spectrum. This is indicated by the power-law spectral index of 2.6-2.7 for the former and 3.3 for the latter. The standard and high x-ray state spectra were shown to intersect at ~1 MeV[17] in addition to that seen at ~10 keV[15].

4. THE SUPERLOW X-RAY/HIGH γ-RAY STATE

HEAO-3 γ-Ray Spectrometer[31] and Ariel 5 All Sky Monitor[27-28] discovered a new low x-ray state in 1979 which was called the "superlow" x-ray state[8-9]. Several unique features associated with this state are:

• Both the soft x-ray (3-6 keV) and hard x-ray (45-140 keV) fluxes were simultaneously low by at least a factor of two compared to those of the standard state, During the recovery phase from the superlow to the standard (low) state, the soft and hard X-ray fluxes tracked each other in a correlated fashion (Figure 2: left half of the figure), in contrast to the anti-correlation seen in a typical standard (low) to high x-ray state transition (e.g. see Figure 2: right half of the figure). The superlow-state spectrum and fluxes in the 50-200 keV range observed by HEAO-3 were independently confirmed by a contemporaneous balloon observation in 1979[3].

• From the standard state to the superlow state, a large fraction of the luminosity below 0.4 MeV was converted to energy above 0.4 MeV, showing a new type of anti-correlation between x rays and γ Rays[9]. The luminosity in the 0.05 – 0.4 MeV and 0.4-1.5 MeV bands are 10.5 and 13.8 x 10^{36} ergs/s, respectively, for the superlow state compared to 17.0 and 0.85 x 10^{36} erg/s for the standard (low) state[9]. Because the γ-ray (0.4 –1.5 MeV) luminosity increased by more than an order of magnitude, the superlow x-ray state could also be considered a high γ-ray state.

• The superlow state spectrum, which was also called the γ_1 spectrum[9] (see Figure 1 right panel), consists of a Comptonized component below 0.4 MeV and a broad

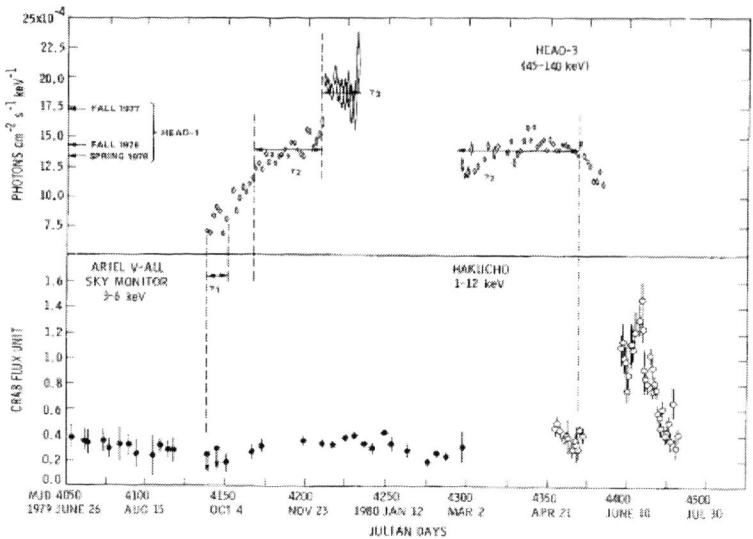

Figure 2. Two contrasting transitions were seen by HEAO-3 (45-140 keV), Ariel V All Sky Monitor (3-6 keV) and Hakucho (1-12 keV) experiments in 1979-1980[8-9]. During the 1979 superlow x-ray state to standard state transition (left half of Fig 2), the soft and hard x-ray fluxes measured by Ariel V and HEAO-3 tracked one another in a correlated fashion. This is in contrast to the standard state to high x-ray state transition seen in 1980 when the soft and hard x-ray fluxes, provided by near contemporaneous data of HAKUCHO and HEAO-3, were anti-correlated (right half of Fig 2).

Gaussian feature centered at 1 MeV with a flux of 1.6×10^{-2} photons/cm^2-sec measured at ~5σ significance. One model suggested[32] that the γ-ray feature was produced in a 4×10^9 K pair-dominated, optically thick cloud near the horizon, while hard x-rays (E < 0.4 MeV) were produced in a cooler, optically thinner, and ion-dominated plasma in the outer disk. Pairs may escape such a system and produced a narrow annihilation feature in the cold surrounding regions[33]. Such a feature was also seen by HEAO-3[34] at the predicted level of 4.4×10^{-4} photons/cm^2-sec, at ~2σ significance.

5. SUMMARY

Cygnus X-1 has displayed two types of transitions over the last 30 years: (1) standard state-to-high x-ray state transition, and (2) standard state-to-superlow x-ray/ high γ-ray state transition. The former has been seen five times, but the latter was seen only once. The spectra for the three states are distinctly different, reflecting complex processes in the system that need to be understood. It is important to note that the shape of the standard state spectrum is harder in regions <200 keV but softer >200 keV, compared to that of the high x-ray state spectrum which is softer <200 keV and harder >200 keV. One of the outstanding unresolved questions since its discovery in 1979 is when will the source make another transition to the superlow x-ray /high γ-ray state? The broad MeV feature and narrow 511 keV feature shown in the superlow state spectrum need to be confirmed. If such a transition could be captured by INTEGRAL during its mission, instruments such as JEX-X, SPI and IBIS onboard the INTEGRAL spacecraft are ideally suited to make sensitive measurements of these features, and to effectively address the triggering mechanism for such an event.

ACKNOWLEDGMENTS

The author wishes to thank M. McConnell and Wm. Wheaton for their comments. This work was carried out by the Jet Propulsion Laboratory, California Institute of Technology, under contract with the National Aeronautics and Space Administration.

REFERENCES

1. Bowyer, S. et al., 1965, Science, 147, 394.
2. Sunyaev, R. A., and Trumper, J., 1979, Nature, 279, 507.
3. Perotti, F., et al., 1986, ApJ., 300, 297.
4. McConnell, M. L., et al., 1989, Ap. J., 343, 317.
5. Baity, W.A., et al., 1973, Nature Phys. Sci. 245, 90.
6. Dolan, J. F., et al., 1979, ApJ., 230, 551.
7. Nolan, P:. L., et al., 1981, Nature, 293, 275.
8. Ling, J. C., et al., 1983, Ap. J., 275, 307.
9. Ling, J. C., et al., 1987, ApJ, 321, L117.
10. Schwartz, R. A., et al., 1991, ApJ 376, 312.
11. Salotti, L., et al., 1992, A&A, 253, 145.
12. Phlips, B. F., et al., 1996, Ap. J. 465, 907.
13. Ling, J. C., et al., 1997, ApJ, 484, 375.
14. Ling, J. C., et al., 2000, ApJS, 127, 79.
15. Zhang, S. N. et al., 1997, ApJ 477, L95.
16. McConnell, M. L., et al., 2000, Ap. J., 543, 928.
17. McConnell, M. L., et al., 2001, These Proceedings.
18. Cui, W., et al., 1997, ApJ 474, L57.
19. Frontera, F., et al., 2001, ApJ, 546, 1027.
20. Sunyaev, R. A., and Titarchuk, L.G., 1980, Astron. Astrophys., 86, 121.
21. Moskalenko, I. V., Collmar, W., and Schonfelder, V., 1998, ApJ, 502, 428.
22. Skibo, J.G. and Dermer, C.D., 1995, Ap. J. 455, L25.
23. Narayan, R., & Yi, I. 1994, ApJ, 428, L13.
24. Chakrabarti S.K. and Titarchuk, L., 1995, Ap.J . 455, 623.
25. Poutanen, J, and Svensson, R., 1996, ApJ., 470, 249.
26. Tananbaum, H., et al., 1972, ApJ, 177, L4.
27. Holt, S. S., et al., 1975, Nature, 256, 108.
28. Holt, S. S., et al., 1979, ApJ, 233, 344.
29. Ogawara, Y., et al., 1982, Nature, 295, 675.
30. Nolan, P. L., 1982, UCSD Ph. D. thesis.
31. Mahoney, W. A., J. C. Ling, A. S. Jacobson, and R. M. Tapphorn, NIM, 178, 363, 1980.
32. Liang, E. P., and Dermer, C. D., 1988, Ap. J., 325, L39.
33. Dermer, C. D., and Liang, E. P., 1988, Nucl. Spec. of Astrophysical Sources Proc., 326.
34. Ling, J. C. and Wm. A. Wheaton, 1989, Ap.J. (Letters), 343, L57.
35. Liang, E. P., 1979, ApJ., 234, 1105.
36. Nolan, P. L., & Matteson, J. L., 1983, ApJ., 265, 389.

Gamma Ray Bursts

Recent Advances in our Understanding of GRB

P. Mészáros[1]

[1] *Pennsylvania State University, 525 Davey, University Park, PA 16802*

Abstract.
We discuss several recent developments in gamma-ray burst and afterglow theory, related in particular to the interaction with the environment and how this bears on the progenitor issue. Pair production induced by the initial gamma-rays in the nearby environment will modify the initial spectrum and the afterglow light curve, and the magnitude of these changes provides a diagnostic for the external density. The illumination of the progenitor remnant and/or the surroundings by the X-ray afterglow continuum can produce substantial Fe K-alpha line and edge emission, with implications for the progenitor model. TeV neutrinos are expected to be produced both by successful (γ-ray detectable) and choked (γ-ray dark) fireball jets as they make their way out of a massive progenitor star, whose fluence is detectable with planned km^3 detectors. Two mechanisms, IC and proton synchrotron, may play a role in afterglow energetics, and would have detectable signatures in the X-ray light curves as well as in the GeV range.

I PAIR PRODUCTION IN GRB ENVIRONMENTS

Gamma-ray burst sources with a high luminosity can produce e$^\pm$ pair cascades in their environment as a result of back-scattering of a seed fraction of their original hard spectrum. New pairs can be made as some of the initial energetic photons are backscattered and interact with other incoming photons. Previous work on this investigated the acceleration of new pairs for a particular fireball model [1,4], the effect of pair formation for a low compactness parameter external shock model of GRB [3], and Compton echoes produced by pairs [2]. Here we discuss a simplified analytical treatment [5] of pair effects from γ-rays arising in internal shocks in a wind; the remaining wind energy drives a blast wave which decelerates as it sweeps up the external medium, and gives rise to the afterglow emission. The γ-rays would propagate ahead of the blast wave, leading to pair production (and an associated deposition of momentum) into the external medium. The pair cascades saturate after the external (pair-enriched) medium reaches a critical bulk Lorentz factor, which is generally below that of the original relativistic wind. For external baryonic densities similar to those in molecular clouds the pairs can achieve scattering optical depths $\tau_{\pm} \lesssim 1$. Even for less extreme external densities the effect of the additional pairs can be substantial, increasing the radiative efficiency of the blast wave and leading to distortions of the original spectrum. This provides a potential tool for diagnosing the compactness parameter of the bursts and thus the radial distance at which shocks can occur. It also provides a tool for diagnosing the baryonic density of the external environment, and testing the association with star-forming regions.

For the maximum Lorentz factor to which an $e^{\pm}$ can be accelerated by scattering, and the maximum Lorentz factor at which back-scattered photons can still make new pairs, one finds two regimes defined by the effective duration of the light pulse seen by the screen of accelerated pairs. At low radii (wind regime) the effective duration is the burst duration t_w; for large radii (impulsive regime), the effective duration is $\Delta t \sim r/c\Gamma_{\pm}^2$. For an incident photon number index $\beta = 2$, in the former $\Gamma_{\pm} \propto r^{-1/3}$ and in the latter $\Gamma_{\pm} \propto r^{-2}$. The critical radius and Lorentz factor for the transition between the wind and the impulsive regimes are [5] $r_c = 5 \times 10^{14} L_{w50}^{2/5} t_{w1}^{3/5}$, $\Gamma_c = 3 \times 10^1 L_{w50}^{1/5} t_{w1}^{-1/5}$. The maximum radius at which pair cascades cut off is $r_\ell \sim (4r_* c t_w/3)^{1/2} \sim 4 \times 10^{15} L_{w50}^{1/2} t_{w1}^{1/2}$ cm.

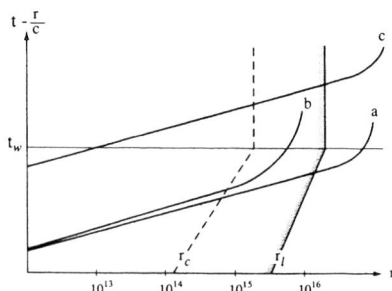

FIGURE 1. Pair enrichment effects on GRB. The axes are logarithmic, t is observer time, being zero when the burst starts. (In this plot, light rays are horizontal lines). a) Low density external medium, a high $\eta = L/\dot{M}c^2$ blast wave; b) Higher external density, deceleration occurs before r_ℓ; c) Lower η case, pair sweeping occurs during afterglow. [5]

Before the pairs start accelerating, assuming they are held back by the environmental protons through magnetic fields, an initial cascade amplification factor $k_p \sim (m_p/m_e)$ is achieved. After the mean mass per scatterer drops to a value comparable to the electron mass, before reaching r_ℓ a further amplification $k_a \sim 2^s \sim 50$ (where $s \sim \log \Gamma_c / \log 2$) is possible, so the total pair amplification factor is [5] $k_c = k_p k_a(r_c) \sim (m_p/m_e)\Gamma_c \sim 5 \times 10^4 L_{w50}^{1/5} t_{w1}^{-1/5}$. The maximum pair optical depth at r_c, which is prevented from exceeding $\tau_{\pm} \sim 1$ by self-shielding, is achieved for external densities $n_p \gtrsim n_{p,c}$, where $n_{p,c} \simeq 10^5 L_{w50}^{-3/5} t_{w1}^{-2/5} \text{cm}^{-3}$.

The external density and the initial Lorentz factor η determine when the outer shock and the reverse shock become important and whether this happens within the radius already polluted with pairs. If $\eta \lesssim r_l/ct_w$ the external shock responsible for the afterglow occurs beyond the region "polluted" by new pairs, and otherwise the afterglow shock may experience, after starting out in the canonical manner, a "resurgence" or second kick as its radiative efficiency is boosted by running into an $e^{\pm}$-enriched gas [5].

Additional effects are expected when $\tau_{\pm} \to 1$, for external baryon density $n_p \gtrsim n_{c,p} \sim 10^5 L_{w50}^{-2/5} t_{w1}^{-3/5} \mathrm{cm}^{-3}$ at radii $r < r_{\ell}$. Such high densities could be expected if the burst is associated with a massive star in which prior mass loss led to a dense circumstellar envelope. The pair optical depth saturates to $\tau_{\pm} \sim 1$ and in addition to an increased efficiency and softer spectrum of the afterglow reverse shock, the original gamma-ray spectrum of the GRB will be modified as well. One of the consequences of such a critical external density leading to $\tau_{\pm} \sim 1$ would be the presence of an X-ray quasi-thermal pulse, whose total energy may be a few percent of the total burst energy [5].

In Figure 1 three schematic cases are illustrated. The criterion for runaway pair production is satisfied within the ambient material out to $r = r_{\ell}$. The associated absorption of momentum would accelerate this pair-enriched material to an r-dependent Lorentz factor. The wind carries more momentum than the gamma rays, and drives a relativistic blast wave that sweeps up the pair-enriched medium. In (a), the external medium has a low density, the blast wave has a high $\eta = L/(\dot{M}c^2)$, sweeping up all the pair-enriched medium before it has decelerated: the effects of the pairs are then observed primarily during the burst itself. In (b), with higher external density, deceleration occurs at radii $< r_{\ell}$, and the blast wave is still moving through pair-enriched material during the afterglow. When the ambient medium is dense, the pairs may provide an optical depth of unity, so the primary burst itself would be reprocessed, and its short time-structure smeared out. Case (c) corresponds to a lower η. The sweeping-up of pairs then occurs during the afterglow (modifying the radiative efficiency of the outer shock) even if the external density is low and there has not (as in case b) already been deceleration.

II FE X-RAY LINES FROM GRB PROGENITORS

Important clues for identifying the nature of the progenitors of the long ($t \gtrsim 2$ s) GRBs may be available from the recent report at a 4.7σ level of X-ray Fe line features in the afterglow after 1.5 days of the gamma-ray burst GRB 991216 [7], as well as similar detections at the 3σ level in 5 other bursts with Beppo-SAX and ASCA. X-ray atomic edges and resonance absorption lines are theoretically expected to be detectable from the gas in the immediate environment of the GRB, and in particular from the remnants of a massive progenitor stellar system [12,11,13].

A straightforward interpretation [7] of the GRB 991216 observation would imply a mass $\gtrsim 0.1-1 M_\odot$ of Fe at a distance of about 1-2 light-days, possibly due to a remnant of an explosive event or supernova which occurred days or weeks prior to the gamma-ray burst itself (a 'supranova', [7,9]). The long time delay is necessary both to get the relatively massive, slow moving ejecta out to few light-day distances (to explain the line appearance at a few days with light travel arguments), and in order to get the initial Ni and Co to decay to Fe ($\sim$ 55 days). This requires a two-step process, in which an initial supernova leads to a temporarily stabilized neutron star remnant, which after weeks collapses to a black hole leading to a canonical burst ([8,9]). It is unclear whether fall-back from the supernova leading to the second collapse to a BH could occur with such a ($\sim$ weeks) long delay (e.g. [15]). Another possibility is that a massive progenitor has previously emitted a copious wind ($\dot{M} \gtrsim 10^{-4} M_\odot/\text{yr}$), which would need to be unusually Fe-rich and highly inhomogeneous ([11]; c.f. [7]).

An alternative, and perhaps less restrictive scenario for such Fe lines [14] involves an extended, possibly magnetically dominated wind from a GRB impacting the expanding envelope of a massive progenitor star. This could be due either to a spinning-down millisecond super-pulsar or to a highly-magnetized torus around a black hole (e.g. [10]), which could produce a luminosity that was still, one day after the original explosion, as high as $L_m \sim 10^{47} t_{day}^{-1.3}$ ergs. An outflow with such a dependence can also be powered by accretion of fall-back material onto a central black hole [15]. This jet luminosity may not dominate the continuum afterglow; but its impact on the outer portions of the expanding stellar envelope at distances $\lesssim 10^{13}$ cm, even with just solar abundances, can be efficiently reprocessed into an Fe line luminosity comparable to the observed value, together with a contribution to the X-ray continuum. Under this interpretation, the dominant continuum flux in the afterglow, even in the X-ray band, is still attributable to a standard decelerating blast wave.

The relativistic magnetized wind from the compact remnant would develop a stand-off shock before encountering the envelope material, and shocked relativistic plasma would be deflected along the funnel walls. Non-thermal electrons will be accelerated behind the standoff shock in the jet material; the transverse magnetic field strength (which decreases as $1/r$ in an outflowing wind) would be of order 10^4 G at 10^{13} cm – strong enough to ensure that the shock-accelerated electrons cool promptly, yielding a power-law continuum extending into the X-ray band. Some of these X-rays would escape along the funnel, but at least half (the exact proportion depending on the geometry and flow pattern) would irradiate the material in the stellar envelope. Pressure balance in the shocked envelope wall implies densities of $n_e = \alpha L_m / 6\pi r^2 ckT \sim 10^{17} \alpha L_{47} r_{13}^{-2} T_8^{-1}$ cm^{-3}, where $\alpha \sim 1$ is a geometric factor, and the recombination time for hydrogenic Fe in the funnel walls photoionized by the non-thermal continuum is $t_{rec} = 6 \times 10^{-6} T_8^{1/2} n_{17}^{-1} \sim 6 \times 10^{-6} \alpha L_{m47}^{-1} r_{13}^2 T_8^{3/2}$ s. Standard cal-

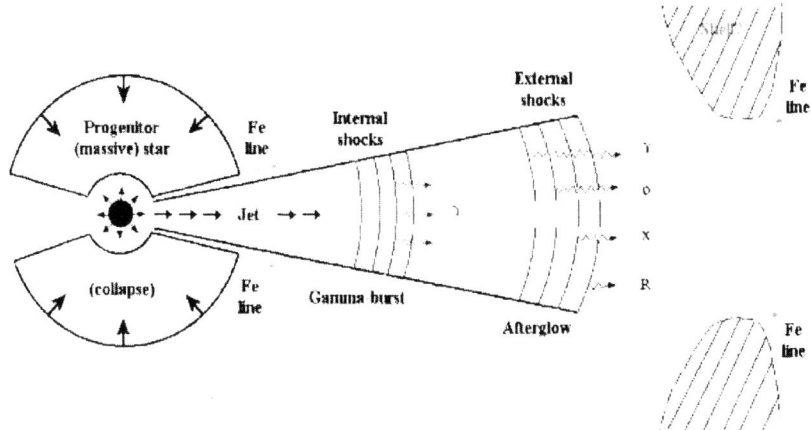

FIGURE 2. Schematic GRB from a massive stellar progenitor, resulting in a relativistic jet which undergoes internal shocks producing a burst of γ-rays and (as it decelerates through interaction with the external medium) an external shock afterglow, leading to γ-rays, X-rays, optical and radio radiation. Iron lines can arise from X-ray illumination of a pre-ejected shell (e.g. supernova remnant or supranova) [7] or from continued X-ray irradiation of the outer stellar envelope [14]

culations of photoionization of optically-thin slabs (e.g. [6]) show that the equivalent width of the Fe K-alpha line, for solar abundances, is about 0.5 keV, or twice as strong if the Fe has ten times solar abundances. These results are applicable provided that the ionizing photons encounter a Fe ion before being scattered by free electrons i.e. provided that $\tau_T = \sigma_T d_i n_e \lesssim 1$. Under these conditions the Fe K-α photon flux is about 0.1 of the X-ray continuum [14], $\dot{N}_{LFe} \sim 10^{54} L_{47} \beta$ ph/s, where $\beta < 1$ is the ratio of ionizing to MHD luminosity. This line luminosity compares well with Fe line luminosity 6×10^{52} ph/s observed $t \sim 1.5$ day after the GRB 991216 burst by [7].

The total amount of Fe needed to explain the observed K-α line flux, arising in a thin layer of the funnel walls of a collapsar model, amounts to a very modest mass of $M_{Fe} \sim 10^{-8} M_\odot$, which could be Fe synthesized in the core. The Fe-enriched core material can easily reach a distance comparable to $r \sim 10^{13}$ cm in 1 day for an expansion velocity below the limit $v \sim 10^9$ cm s^{-1} inferred by [7] from the line widths. Even without this, a solar abundance ($10^{-5} M_\odot$ of Fe) in the envelope is sufficient to explain the observations. The initial, energetic portion of the relativistic jet, with a typical burst duration of 1 − 10 s, will rapidly expand beyond the stellar envelope, leading in the usual way to shocks and a decelerating blast wave. A continually decreasing fraction of energy, such as put out by a decaying magnetar, may continue being emitted for periods of a day or longer, and its reprocessing by the stellar envelope can be responsible for the observed Fe line emission in GRB 991216. Since the energy in this tail can decay faster than t^{-1}, the usual standard shock gamma-ray and afterglow scenario need not be affected, being determined by the first 1-10 s worth of the energy input.

Another way of producing Fe lines in a collapsar model of GRB with a small amount ($\lesssim 10^{-5} M_\odot$) of Fe uses the fact that the cocoon produced by the jet as it burrows its way through the progenitor will lead to a relativistic bubble. This bubble breaks through the stellar envelope several hours to a day after the GRB trigger, leading to conditions suitable for the line production [16].

III TEV NEUTRINOS FROM BURIED JETS IN CHOKED AND SUCCESSFUL GRB

A widely discussed scenario for (long, i.e. $t \gtrsim 2$ s) gamma-ray bursts invokes as its ultimate energy source the core collapse of a massive star, resulting in a relativistic fireball jet which breaks through the stellar envelope. However, for very extended or slowly rotating stars, the fireball may be unable to break through the envelope, resulting in a choked fireball. On the other hand, for "thin skinned" (small enveloped) or fast rotating stars (with a centrifugally much thinned-out spin axis, one expects in this scenario a "successful" GRB, where the jet does break through the envelope, and leads to internal and external shocks in the standard GRB fireball shock scenario. An interesting particle astrophysics aspect of this is that both the penetrating (successful) and choked fireballs will produce, by photo-meson interactions of accelerated protons, a burst of $\gtrsim 5$ TeV neutrinos while propagating in the envelope [17]. The predicted flux, from both penetrating and chocked fireballs, should be easily detectable by planned 1 km^3 neutrino telescopes

The way this occurs is that, while the jet is making its way out through the stellar core and/or the stellar envelope, it advances at a slower speed than the actual bulk Lorentz factor with which the jet is being fed by the central engine (which, in this scenario, would be a black hole resulting from the core collapse, fed by intermittent accretion of a fall-back torus of matter around it, resulting in an ultrarelativistic, super-Eddington jet). The jet thus takes on the order of tens of seconds in drilling through (or getting stuck). There is a termination shock, where the jet is suddenly decelerated from its initial $\Gamma_j \gg 1$ to a much lower value. (This termination shock may be collisional, or radiatively dominated; and the particle and photon densities in it is too high for significant relativistic particle acceleration, which would be stymied by radiative losses). The energy input of the jet heats the interior of the envelope, which produces approximately black-body radiation of temperature $T_x \sim$ keV. These photons cannot penetrate far back into the jet, which has a high Thompson depth in the frame of the advancing jet head, but they create a high photon density layer near the top of the jet, just below the termination shock. Below that, the jet is still highly relativistic and much less dense, and it can undergo internal shocks (which are collisionless, and are permeated by a much lower photon density, so inverse Compton losses are not important in the jet comoving frame). These internal shocks can accelerate protons to very high Lorentz factors. Protons with $\epsilon_p \gtrsim 10^5$ GeV produced in these low density internal shocks, as they proceed upward they meet an increasing density of $T_x \sim$ keV photons, the optical depth against photo-meson interactions becoming large in a skin depth near the jet head. This leads to pions, muons, and muon as well as electron neutrinos.

The energy of these neutrinos is $\epsilon_\nu \gtrsim$ few TeV, and their fluence at Earth, from a typical fireball jet of isotropic equivalent energy $E_{iso} \sim 10^{53} E_{53}$ erg at redshift $z \sim 1$, is $\Phi_\nu \sim 10^{-5} E_{53} D_{28}^{-2}$ erg cm^{-2}, resulting in an upward-directed average number of muons $N_\mu \sim 0.2 E_{53} D_{28}^{-2}$ km^{-2} [17]. For the typical angular resolution ~ 1 degree expected from a km^3 detector such as ICECUBE [18], this is far below the fluence from the neutrino background at these TeV energies, and thus is easily detectable, at least in an aggregate sense, from many bursts. However, the total fireball energy and bulk Lorentz factors will be distributed around the mean values, and fluctuations around the mean in these quantities and in the distance will lead to detections dominated by rare, energetic nearby events, as argued for GRB [19]. Thus, one could expect a significant number of neutrino bursts (few/year) with $N_\mu \gtrsim 10$/km^2 muon events/burst, easily detectable as individual bursts of neutrinos.

For every successful "collapsar" GRB (in which the jet punches through the envelope and produces detectable γ-rays), one may expect at least a comparable number of choked fireball jet events directed at Earth, $\gtrsim 10^3$/year. If jet punch-through requires a fast core rotation rate, e.g. from merging of the massive star with a binary companion in only a fraction of cases, this number could, in fact, be significantly larger. In successful breakthrough collapses, the TeV neutrinos would precede by $t_{cross} \lesssim t_j = 10 - 100$ s the GRB trigger. Their detection would be a test of the collapsar hypothesis for long GRB, and the duration of the neutrino precursor would constrain the dimensions and sound speed of the stellar envelope. In this case the absence (or a much reduced duration) of ν-precursors in short GRB would support the current view that long and short GRB arise from qualitatively different progenitors.

The TeV neutrino signals preceding the γ-rays in GRB, as well as those from choked, γ-ray dark fireballs, would differ significantly from the neutrino signals expected during and after the γ-ray phase of GRB. In the latter, one expects $\gtrsim 100$ TeV neutrinos [20] from internal shock protons interacting with MeV photons in a jet well beyond any stellar envelope, or $\gtrsim 10^{17}$ eV neutrinos [21] from external shock protons interacting with UV photons even further out, and one also expects 2-5 GeV neutrinos [22-24] from inelastic nuclear collisions when neutrons decouple from protons. For the flat $\epsilon_\nu^2 \Phi_\nu$ neutrino spectra considered here, the number of precursor TeV neutrino events per burst is also larger by at least one order of magnitude, relative to those at $\gtrsim 100$ TeV and at GeV expected during and after the γ-bright phase of GRB.

Such GRB precursor or γ-ray dark TeV neutrino burst signals may therefore be the likeliest targets for early detection with planned experiments [18].

IV GEV AND X-RAY SIGNATURES OF IC AND PROTON SYNCHROTRON

Electron inverse Compton emission and proton synchrotron radiation are two high energy radiation mechanisms which may be important for the energy budget, and which may have signatures in the broad-band spectra and in the keV to GeV light curves of gamma-ray burst afterglows. It is advantageous to use a simple analytical approach, allowing also for the effects of photon-photon pair production, in order to explore the conditions under which one or the other of these components dominates [25], with particular attention to potential detection in the X-ray band with *Chandra*, and in the GeV band with *GLAST*.

For the frequency range below the electron's synchrotron cut-off, $\nu < \nu_{u,e}$, there is a competition between the electron synchrotron component on the one hand, and the proton synchrotron component or the electron IC component on the other, which can affect the higher energy bands including X-rays or above. This competition divides the ϵ_e, ϵ_B phase space into three regimes. (Fig.1 of [25]). One interesting consequence is that there is a substantial region (regime III) in which neither IC nor proton synchrotron are important. These bursts, as well as the regime I (proton dominated) bursts may be still detectable by *GLAST* in the prompt afterglow phase due to the electron synchrotron emission if the source is located closer to the earth. Above the electron synchrotron cut-off, the competition is between the electron IC component and the hadron-related photo-meson decay components, which we treated as a reduced extension of the proton synchrotron component. Again, the phase space region where the latter effects are important in the afterglow is small. For the external shock and the afterglow phase the $\gamma - \gamma$ absorption is not important below the TeV range.

The most likely origin for an extended afterglow component at GeV energies is from the electron IC component [25], at least for the values of ϵ_e, ϵ_B most commonly used so far in snapshot spectral analyses. Not only is the phase space region where the IC component dominates much larger than that where the proton component dominates, but also the time scale during which an appreciable flux level is maintained in the GeV band is much longer for the IC component than for the hadronic components. In the parameter regime favorable for the IC emission, this component is observable at and above the X-ray band. In the X-ray band, it will lead to a flattening of the light curve at late times, as long as the medium density is not too low (see also [26]). Above the X-ray band, the time after which IC emission becomes dominant appears earlier, and the IC dominates the GeV-band emission almost from

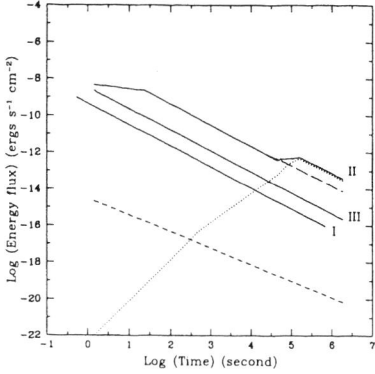

FIGURE 3. X-ray light curves in three regimes of the afterglow dominated by (I) protons synchrotron effects, (II) electron IC effects and (III) electron synchrotron. In particular, the IC hump in curve II may be an example of what is observed in GRB 000926 [25]

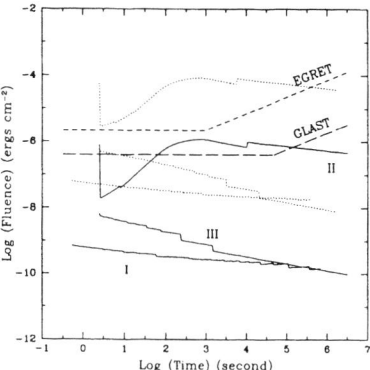

FIGURE 4. GeV light curves in three regimes of the afterglow dominated by (I) protons synchrotron effects, (II) electron IC effects and (III) electron synchrotron [25]. The darker lines are for $z = 1$, while the dotted lines are for $z = 0.1$. Also shown are the EGRET and GLAST sensitivities.

the onset of the afterglow phase. In general, a high external density medium favors the detectability of the IC component. Such an IC component is likely to have been responsible for the GeV photons detected from GRB 940217 with *EGRET*, and similar events should in the future be detectable by *GLAST*.

The proton synchrotron component, as well as the hadron-related photo-meson electromagnetic components in the afterglow radiation are likely to be, in most cases, less important than previous approximate estimates indicated The ϵ_e, ϵ_B phase space region where this component overcomes the electron synchrotron component is very small, unless the medium density is high. Even in the most favorable region of the phase space, the proton component is not expected to show up in the X-ray band, but it may overcome the electron synchrotron component in the GeV band shortly after the onset of the afterglow phase, if the medium density is moderately high. However, the flux level drops with time, and may be only detected by *GLAST* in the very early phase of the afterglow if the source distance is close. Nonetheless, the detection of a proton component is of extremely high interest for the GRB and shock physics. Its detection, if successful, would imply, depending on its strength, fireball and shock parameters which are more extreme than currently commonly assumed. In particular, it would provide a diagnostic for a high ϵ_p, ϵ_B and/or a low ϵ_e.

This research was supported by NASA NAG5-9192, NAG5-9153. I am grateful to M.J. Rees, E. Waxman, B. Zhang and E. Ramirez-Ruiz for valuable discussions and collaborations on these topics.

REFERENCES

1. Madau, P & Thompson, C, 2000 ApJ, 534, 239
2. Madau, P, Blandford, R & Rees, M.J., 2000, ApJ, 541, 712.
3. Dermer, C & Böttcher, M., 2000 ApJ 534, L155
4. Thompson, C & Madau, P, 2000 ApJ, 538, 105
5. Mészáros, P, Ramirez-Ruiz, E & Rees, M, 2001, ApJ in press(astro-ph/0011284)
6. Young, A.J., 1999, Ph.D. thesis, Cambridge University
7. Piro, L, et al., 2000, Science, 290, 955
8. Vietri, M & Stella, L.A., 1998, ApJ 507, L45
9. Vietri, M, Perola, G, Piro, L & Stella, L, 2000, MNRAS 308, L29.
10. Wheeler, J.C, et al., 2000, ApJ 537, 810 (astro-ph/9909293)
11. Weth, C, Mészáros, P, Kallman, T & Rees, M.J, 2000, ApJ 534, 581
12. Mészáros, P & Rees, M.J. 1998, MNRAS, 299, L10
13. Böttcher, M & Fryer, C.L, 2001, ApJ, 547, 338 (astro-ph/0006076)
14. Rees, M.J. & Mészáros, P, 2000, ApJ, 545, L73
15. MacFadyen, A, Woosley, S & Heger, A, 2000, ApJ(astro-ph/9910034)
16. Mészáros, P & Rees, MJ, 2001, ApJL, subm (astro-ph/0104402)
17. Mészáros, P & Waxman, E, 2001, PRL subm (astro-ph/0103275)
18. Halzen, F, 2001, in Intl Symp on High Energy Gamma Ray Astronomy, Heidelberg, June 2000 (astro-ph/0103195)
19. Alvarez-Muniz, J, Halzen, F, Hooper, D, 2000, PRD 62:093015
20. Waxman, E & Bahcall, JN 1997, PRL, 78, 2292
21. Waxman, E & Bahcall, JN 2000, ApJ, 541, 707
22. Bahcall, JN & Mészáros, P, 2000, PRL, 85, 1362
23. Mészáros, P. & Rees, M.J. 2000, ApJ(Lett.), 541, L5
24. Derishev, EV, Kocharovsky, VV & Kocharovsky, Vl.V 1999, ApJ 521, 640
25. Zhang, B & Mészáros, P, 2001, ApJ, 559, in press (astro-ph/0103229)
26. Sari, R., & Esin, A. A. 2001, ApJ, 548, 787

Gamma-Ray Burst Spectral Diagnostics in the GLAST Era

Matthew G. Baring

Department of Physics and Astronomy, MS-108
Rice University,
P. O. Box 1892,
Houston, TX 77251-1892
baring@rice.edu

Abstract. The spectra obtained above 100 MeV by the EGRET experiment aboard the Compton Gamma-Ray Observatory for a handful of gamma-ray bursts has given no indication of any spectral attenuation that might preclude detection of bursts at higher energies. With the discovery of optical afterglows and counterparts to bursts in the last few years, enabling the determination of significant redshifts for these sources, it is anticipated that profound spectral attenuation will arise in the GLAST energy band of 30 MeV–300 GeV for many if not most bursts. An important goal will be to discriminate between such extrinsic absorption, due to the cosmic infra-red background, and that which arises internally in GRBs. This paper explores expectations for the spectral properties in the GLAST band for bursts, in particular how attenuation of photons by pair creation internal to the source modifies the spectrum to produce distinctive signatures. The energy of spectral breaks and the associated spectral indices provide valuable information that can constrain the bulk Lorentz factor of the GRB outflow at a given time. Moreover, distinct temporal behavior is present for internal attenuation, and is easily distinguished from extrinsic absorption. These characteristics define palpable observational goals for both the GLAST mission and ground-based Čerenkov telescopes, and strongly impact the observability of bursts above 1 GeV.

INTRODUCTION

High energy gamma-rays have been observed for six gamma-ray bursts by the EGRET experiment on CGRO. Most conspicuous among these observations is the emission of an 18 GeV photon by the GRB940217 burst [1]. Taking into account the instrumental field of view, these detections indicate that emission in the 1 MeV–10 GeV range is probably common among bursts, if not universal. One implication of GRB observability at energies around or above 1 MeV is that, at these energies, spectral attenuation by two-photon pair production ($\gamma\gamma \to e^+e^-$) is absent in the source. From this fact, Schmidt [2] deduced that a typical burst had to be closer than a few kpc, if it produced quasi-isotropic radiation.

In the aftermath of BATSE's revelation [3] that most if not all bursts are at cosmological distances (where $\tau_{\gamma\gamma} \sim 10^{11}-10^{12}$ for isotropic photons), Fenimore et al. [4] proposed that GRB photon angular distributions are highly beamed and produced by a relativistically moving plasma, a suggestion that has become very popular. This can dramatically reduce $\tau_{\gamma\gamma}$ and blueshift spectral attenuation turnovers above the observed spectral range. Various determinations of the bulk Lorentz factor Γ of the GRB medium have been made in recent years, mostly concentrating [5–7] on cases where the angular extent of the source was of the order of $1/\Gamma$. These calculations generally assume an infinite power-law burst spectrum, and deduce [8] that gamma-ray transparency up to the maximum energy detected by EGRET requires $\Gamma \gtrsim 100-10^3$ for cosmological bursts.

While the power-law source spectrum assumption is expedient, the spectral curvature seen in most GRBs by BATSE [9] is expected to play an important role in reducing the opacity for potential upper GLAST band and TeV emission from these sources (Baring & Harding [10]). Such curvature is patently evident in 200 keV–2 MeV spectra of some EGRET-detected bursts (e.g. [11]), and its prevalence in bursts is indicated by the generally steep EGRET spectra for bursts [1,12,13]. In this paper, the principal properties of pair production opacity that couple to spectral shape in the BATSE/EGRET energy range are exhibited, focusing the work of [10] to identify observational diagnostics for the GLAST mission and future ground-based initiatives such as the VERITAS, MILAGRO, HESS and MAGIC Čerenkov telescopes. These spectral signatures are clearly distinguishable from absorption by background radiation fields, in particular via their time-dependent evolution. Good prospects exist for "mapping" Lorentz factors as functions of time in the brightest bursts, so as to elicit details of the adiabaticity or otherwise of the post-fireball expansion.

γ-γ ATTENUATION AND GLAST SCIENCE

The simplest picture [6,7] of relativistic beaming that is pertinent to opacity calculations has "blobs" of material moving with a bulk Lorentz factor Γ more-or-less toward the observer, and having an angular "extent" $\sim 1/\Gamma$. For an infinite power-law spectrum $n(\varepsilon) = n_\gamma \varepsilon^{-\alpha}$, where ε is the photon energy in units of $m_e c^2$ (a dimensionless convention used throughout), for which the optical depth to pair creation assumes the form $\tau_{\gamma\gamma}(\varepsilon) \propto \varepsilon^{\alpha-1} \Gamma^{-(1+2\alpha+\delta)}$ for $\Gamma \gg 1$. Here $\delta = 0$ for the blob case, however it can assume values like unity, depending on how the size of the emission region couples to the source variability timescale. As noted above, the input source spectrum needs to be modified, to include the effects of a relative depletion of low energy (target) photons in the BATSE range. The simplest approximation to spectral curvature is a power-law broken at a dimensionless energy ε_{B} ($= E_{\text{B}}/0.511\,\text{MeV}$), with spectral indices α_l below and α_h above the break. More gradual spectral curvature can be treated by fitting the GRB continuum with piecewise continuous power-laws, to arbitrary accuracy. The optical depth

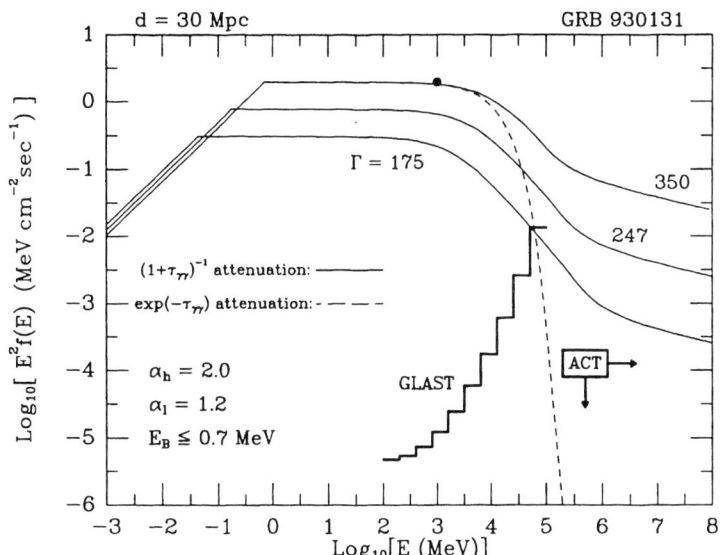

FIGURE 1. An evolutionary sequence for γ-γ attenuation, starting with a GRB930131 data fit ($\Gamma = 350$ case), appropriate for an adiabatically-decelerating blast wave. Notice the separation of the $\sim$ GeV turnover and sub-TeV break energies as Γ declines at later times. The GLAST steady-source differential ($\Delta E/E = 2$ step-function, for a one year all-sky survey) and approximate Whipple integral sensitivities (ACT box; derived from upper limits to later bursts listed) are depicted. A case of exponential attenuation (for the $\Gamma = 350$, dashed line) is illustrated; it would inhibit detections by atmospheric Čerenkov telescopes (ACTs).

determination for such a distribution uses results obtained in [14] for truncated power-laws, and was presented in [10] and [15]:

$$\frac{\tau_{\gamma\gamma}(\varepsilon)}{n_\gamma \sigma_T R} \propto \begin{cases} \dfrac{\varepsilon^{\alpha_h - 1}}{\Gamma^{2\alpha_h}}, & \text{if } \varepsilon \lesssim \Gamma^2/\varepsilon_B \ , \\ \dfrac{\varepsilon^{\alpha_l - 1}}{\Gamma^{2\alpha_l}}, & \text{if } \varepsilon \gtrsim \Gamma^2/\varepsilon_B \ , \end{cases} \qquad (1)$$

This form implies breaks in the absorbed portion of the hard gamma-ray spectrum that "image" the BATSE band break in the seed photons. A variability "size" $R_v = 3 \times 10^7$ cm ($= R/\Gamma$) is chosen here following [8,10], and the observed flux at 1 MeV normalizes the source density coefficient n_γ. The results of the attenuation of the BATSE-type spectra are depicted in [15] and in Fig. 1, indicating marked absorption whose onset energy couples to Γ and the EGRET band spectral index α_h. Above this turnover the immediate spectral index is $1 - 2\alpha_h$, which flattens to $1 - \alpha_h - \alpha_l$ at even higher energies. Precise knowledge of the GRB distance, such as through redshifts of accompanying optical afterglows, would facilitate the determination of tight constraints on Γ.

The potential for observational diagnostics is immediately apparent, given the displayed instrumental sensitivities. First, the extant EGRET data already provides a lower bound to Γ: the dot on Fig. 1 represents the highest energy photon from GRB930131, and clearly suggests that $\Gamma \gtrsim 300$ for $d = 30\,\mathrm{Mpc}$ (or a somewhat higher value of $\Gamma \gtrsim 800$ for a $d = 1\,\mathrm{Gpc}$ case, not depicted). Second, the sensitivity of atmospheric Čerenkov telescopes (ACTs) is easily sufficient to detect relatively nearby bursts even with significant attenuation, so that they could well probe the spectral issues raised here. Yet the Whipple rapid search [16] postdated the EGRET detections, and produced merely upper limits as indicated in Fig. 1, which is not unduly disturbing if one assumes that most bursts are at moderate redshifts. What is more intriguing is the possible detection of the weak BATSE burst GRB 970417 by the MILAGRITO forerunner to MILAGRO [17]. While this is perhaps viewed in a pessimistic light, since it was just one out of more than fifty bursts (spatially and temporally) sampled by MILAGRITO's observing program, it has profound implications if it can be supported by subsequent TeV/sub-TeV detections of nearby bursts by the next generation of ACT experiments. Such consequences include (i) an upper bound for the distance to the nearest burst, which constrains the space density of GRBs and concomitantly their degree of beaming, and (ii) an obvious increase in the minimum Γ from those for EGRET bursts.

Great strides in understanding in the next decade will be precipitated by broadband spectral coverage afforded by simultaneous detection of bursts by GLAST and Čerenkov telescopes. To illustrate some of the diagnostic possibilities, Fig. 1 displays a time-evolutionary sequence of GRB spectra, including the effects of γ-γ attenuation, and compares this with the projected GLAST *steady-source differential* sensitivity, and the potentially-constraining pre-1998 Whipple integral sensitivity threshold (deduced from the results of [16]; this "box" will move somewhat to the lower left for imminent ACT experiments like HESS and VERITAS). The GLAST sensitivity is obtained from simulations (Digel, private communication) of the spectral capability for high latitude, steady sources in a one-year survey, i.e. roughly 8 weeks on source. The real GLAST sensitivity for transient GRBs of duration t_{dur} can be estimated to be roughly $[(8\,\mathrm{weeks})/t_{\mathrm{dur}}]^{1/2}$ times that depicted, i.e about a factor of 30–100 higher. Note that the differential sensitivity is the most appropriate measure for spectral diagnostic capabilities.

Evidently, GLAST and atmospheric Čerenkov telescopes working in concert (and ideally together with Swift) will be able to determine the spectral shape and evolution of bright, flat-spectrum bursts like GRB930131 if the attenuation is no more dramatic than $1/(1+\tau_{\gamma\gamma})$. The particular evolutionary scenario depicted in Fig. 1 is an adiabatic, non-radiative one for blast wave deceleration during the sweep-up phase, where the dependences on time t are $\Gamma \propto t^{-3/8}$, $\varepsilon_{\mathrm{B}} \propto \Gamma^4 \propto t^{-3/2}$, and $\varepsilon_{\mathrm{B}}^2 f(\varepsilon_{\mathrm{B}}) \propto \Gamma^{8/3} \propto t^{-1}$ for the flux at the peak [18]. This specialization is sufficient to illustrate the global properties without loss of generality. The lowering of the break energy in time reflects the fact that the drop in Lorentz factor outweighs the influence of the decline in the density of target photons. Shifts in the turnover energy and sub-TeV break energy, and correlations with BATSE flux and break energy should be discernible in bright sources. Moreover, the actual evolutionary details (i.e. rates at which these breaks evolve) will differ for radiative cooling-dominated GRB expansion scenarios, so that it may prove possible to discriminate such models from their adiabatic competitors with GLAST data.

It must be emphasized that these internal absorption characteristics are easily distinguishable from those of external absorption generated by the cosmological infra-red background along the line of sight [19,20]. Attenuation by such background fields couples to the redshift, not parameters internal to the source nor the shape of the spectrum in the BATSE and EGRET bands. Furthermore, it is always exponential in nature (i.e. of severity equivalent to the dashed curve in Fig. 1) since the emission region is distinct from the location of the soft target photons, and is patently independent of time. The possibility of confusing such with the internal attenuation that forms the focus of this paper seems minimal. Hence, the prospects for powerful spectral diagnostics in bright bursts with the GLAST mission and atmospheric Čerenkov telescopes promise an exciting future for the field of high energy gamma-ray astronomy.

Acknowledgments: I thank Seth Digel for simulating GLAST spectral sensitivities, and Brenda Dingus, Julie McEnery and Alice Harding for many discussions.

REFERENCES

1. Hurley, K. et al. *Nature* **372**, 652 (1994).
2. Schmidt, W. K. H. *Nature* **271**, 525 (1978).
3. Meegan, C., et al. *Ap. J. Supp.* **106**, 65 (1996).
4. Fenimore, E. E., Epstein, R. I. & Ho, C. in *Gamma-Ray Bursts,* eds. Paciesas, W. S. and Fishman, G. J., (AIP, New York) p. 158 (1992).
5. Epstein, R. I. 1985 *Ap. J.* **297**, 555
6. Krolik, J. H. & Pier, E. A. *Ap. J.* **373**, 277 (1991).
7. Baring, M. G. *Ap. J.* **418**, 391 (1993)
8. Baring, M. G. & Harding, A. K. *Ap. J.* **491**, 663 (1997b).
9. Band, D., et al. *Ap. J.* **413**, 281 (1993).
10. Baring, M. G. & Harding, A. K. *Ap. J. Lett.* **481**, L85 (1997a).
11. Schaefer, B. E., et al. 1992 *Ap. J. Lett.* **393**, L51
12. Schneid, E. J., et al. *Astron. Astr. (Lett.)* **255**, L13 (1992).
13. Sommer, M., et al. *Ap. J. Lett.* **422**, L63 (1994).
14. Gould, R. J. & Schreder, G. P. *Phys. Rev.* **155**, 1404 (1967).
15. Baring, M. G. in *GeV-TeV Gamma-Ray Astrophysics Workshop,* ed. B. L. Dingus, M. H. Salamon, & D. B. Kieda (AIP Conf. Proc. 515, New York), p. 238 (2000).
16. Connaughton, V. et al. *Ap. J.* **479**, 859 (1997).
17. Atkins, R., et al. *Ap. J.* **533**, L119 (2000).
18. Dermer, C. D., Chiang, J. & Böttcher, M. *Ap. J.* **513**, 656 (1999).
19. Stecker, F. W. & De Jager, O. C. *Space Sci. Rev.* **75**, 401 (1996).
20. Mannheim, K., Hartmann, D. & Funk, B. *Ap. J.* **467**, 532 (1996).

Energy outflows in γ-ray bursts: discontinuous *versus* continuous?

Enrico Ramirez-Ruiz and Andrea Merloni

Institute of Astronomy, Madingley Road, Cambridge, CB3 0HA, UK

Abstract. The recent realization that energy conversion in γ-ray bursts is most likely via internal shocks rather than via external shocks provides additional information about the inner engine: the relativistic flow must be irregular, it must be variable on a short time scale, it must be able to turn off to a very low level and then turn on again, and it must be active for up to a few hundred seconds and possibly much longer. An acceptable model requires that the central engine evolves into a configuration which is stable enough to survive the violent gravitational instabilities associated with the merging/collapse of compact objects, while still keeping enough binding energy to power the burst and in some cases to be active again after some period of *quiescence*. At present, it is unclear if these separated emission episodes observed in some γ-ray burst light curves are consequences of the same physical process, and if the time separation is due to some intrinsic property of the central source or of its environment. The hypothesis of an intermittent central engine, although intriguing, has to be tested against observations. The feasibility of different models in the production of episodes of quiescence are discussed. Some key theoretical issues are highlighted, along with the types of observations that would determine whether or not the central engine goes dormant for a period of time comparable to the duration of the gaps.

γ-RAY BURSTS IN THE TIME DOMAIN

The study of γ-ray bursts (GRBs) has undergone a revolution since the the first fading sources at X-ray, optical and radio wavelengths were discovered, making them the most powerful photon-emitters known in the Universe (see [4] for a recent review). Although much attention has been devoted to the late afterglow emission since then, the prompt γ-ray emission still has to be understood. The data collected by the *Burst and Transient Source Experiment*, *BATSE*, have provided us with an unprecedented wealth of information. Nonetheless, GRBs are so complicated and diverse in the time domain that, at first sight, their behaviour obeys no simple rule. The durations range from 10^{-2}s to about 10^3s, with a roughly bimodal distribution of short ($< 2s$) and long ($> 2s$) bursts [3], and substructure sometimes down to milliseconds.

Besides this apparent bimodality, γ-ray burst temporal profiles are enormously varied. Many bursts have a highly variable temporal profile with a variability time scale that is significantly shorter than the overall duration, while in a minority of them there is only one peak, with no substructure. Furthermore, some long γ-ray light curves often show multiple episodes of emission, separated by background intervals, or *quiescent times*, of variable durations. In other words, it seems that the emission can turn off to a very low level and then turn on again (Fig. 1).

The work we present here has been done with the twofold aim of understanding the

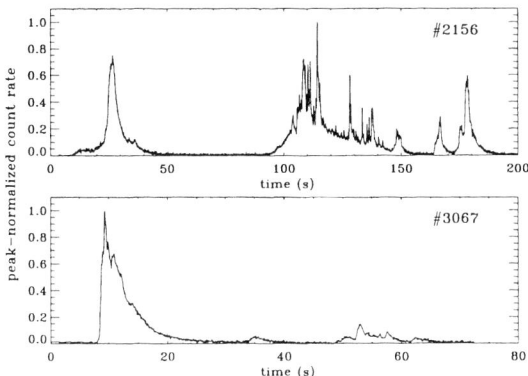

FIGURE 1. Time profiles observed with *BATSE* (at energies > 20keV) that contain periods of quiet emission. Upper panel: *BATSE* burst #2156 have a very strong main burst after a long quiet emission period. Lower panel: *BATSE* burst #3067 have a very strong main burst before the count rate drops to the background level.

nature of such quiescent times and finding possible observational tests that could help us to discriminate between the two main possible explanations for this phenomenon: a turning-off central engine or a continuous relativistic outflow, modulated at the central site or by the interaction with the ambient medium. In the former case, the flow will be discretised in a number of thick shells, each one of size roughly comparable to the duration of the emission episodes observed at the detector. In the latter, instead, the flow can be associated with a single shell whose thickness is related to the total burst duration.

DISCONTINUOUS *VERSUS* CONTINUOUS?

The complicated light-curves can be understood in terms of internal shocks [8] in the ejecta itself, caused by velocity variations in the outflow. In this early phase, the time-scale of the burst and its overall structure follows, to a large extent, the temporal behaviour of the source [2]. In contrast, the subsequent afterglow emerges from the shocked regions of the external medium where the relativistic flow is slowed down; therefore the inner engine cannot be seen directly in the afterglow. A magnetic field configuration capable of powering the bursts is likely to have a large scale structure. Flares and instabilities occurring on the characteristic dynamical time scale would naturally cause intermittency in the overall outflow that would manifest itself in internal shocks [7]. There is thus no problem in principle in accounting for sporadic large-amplitude variability on all time-scales, even in the most long-lived bursts.

The presence of quiescent times could be regarded as an indication of a turning-off of the central site for a period of time. However, if GRBs are produced by internal shocks in relativistic winds, it is possible as an alternative to attribute the quiescent times in the

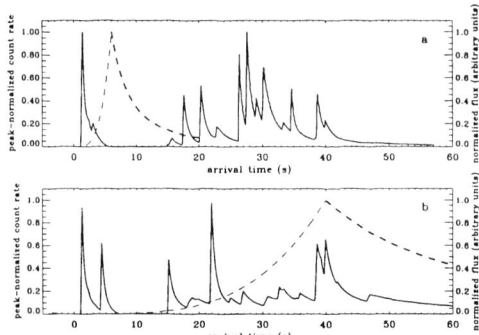

FIGURE 2. Multi-wavelength signature for the two different scenarios described in section 2. The solid curve is the main burst produced by internal shocks. The dashed line is the emission from the reverse shock. This emission terminates once the reverse shock crosses the shell and the cooling frequency drops below the observed frequency. During this period of emission one expects a $\sim t^{-2}$ dependence. For a central engine that turns off (panel a), the emission from the reverse shock peaks much earlier than in the case of an engine in a steady state (panel b).

γ-ray light-curves to a complicated modulation of the ejecta velocities in the relativistic outflow [6]. There are at least two simple mechanisms which might lead to a period of quiescent emission in the observed light curve without postulating any quiet phase in the central engine. The simplest possibility is that the central engine ejects consecutive shells moving with Lorentz factors that are essentially constant over a certain period of time. The second possibility may arise when the Lorentz factors of a series of consecutive shells monotonically decrease during a certain time interval. In both cases, an observed quiescent time interval will be a consequence of the modulation of a continuous wind.

If the central source really turns off for a long period when a quiescent time is observed, the relativistic flow will be discretised in a number of thick shells, each one of size roughly comparable to the duration of the different emission episodes observed at the detector. This will not be the case if such features arise from the modulation of the relativistic ejecta, because all these scenarios envisage the ejection of a continuous outflow over the whole duration of the main event. Clearly, since the width of the shell Δ responsible for sweeping up the external medium varies from one scenario to the other, one expects the reverse shock emission to peak earlier if the central engine turns off since the peak time $\propto \Delta/c$ (Fig. 2). The reverse shock contains, at the time it crosses the shell, and amount of energy comparable to that in the forward one. However, its effective temperature is significantly lower (typically by a factor of Γ).

We can thus conclude that the multi-wavelength signature from the reverse shock emission [6], if measured, would be a clear way of determining whether or not the central engine turns off in order to produce a quiet period in the γ-ray time history.

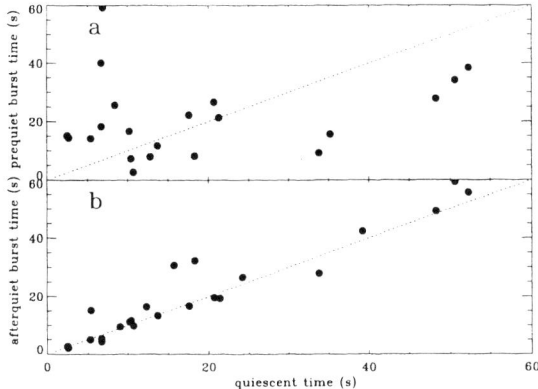

FIGURE 3. Correlations between the temporal properties of the different emission periods. In panel (a) we plot the duration of an emission episode (*prequiet burst time*) against the duration of the following quiescent time. No clear correlation is found in this case. In panel (b) we plot the duration of an emission episode (*afterquiet burst time*) against the duration of the previous quiescent time. There is a clear trend: the longer the quiescent time, the longer the duration of the following emission period.

QUIESCENT TIMES:PERIODS OF FUEL ACCUMULATION?

An observation that attracted much of our attention was the discovery [5] of a strong quantitative proportionality relation between the duration of an emission episode and the quiescent time elapsed since the previous episode; while no clear correlation between the length of an emission episode and that of the following quiescent time was found (Fig. 3). If the duration of an emission episode is proportional to the total energy radiated, the correlation between times we found should simply reflect a correlation between the burst strength and the time elapsed since the previous emission episode. It will then be indicative of an accumulation of fuel.

The tentative correlation found hints at the following general scenario.

The system builds up its energy (via an MHD instability driven dynamo, for example) and reaches a near critical, or *meta-stable* state. Any local instability can, by definition, cause a rapid dissipation of all the stored energy through an avalanche of dissipation events. The system will tend to return to a more stable configuration, characterised by a certain threshold energy E_0, or a sub-critical coronal magnetic field configuration. The source then becomes quiescent. If the lifetime of the accreting torus is long enough and depending on the rate at which the energy is actually extracted from the disc (or from the black hole) and deposited into the external magnetic field, the system can undergo another episode of strong emission. As we can assume E_0 to be fixed by the geometry and by the physical parameters of the black hole–accretion disc system, the longer the quiescent time, the higher will be the stored energy above the threshold available for the next episode. Such a situation will give rise to the observed correlation (Fig. 4).

This is a mechanism different from any relaxation oscillator, in which the threshold energy is an *upper* limit for the system. As soon as the system reaches such a limit it is

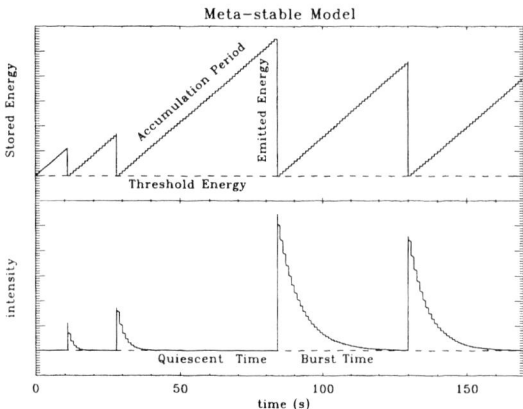

FIGURE 4. Diagram shows how the mechanism responsible for extracting and dissipating the GRB energy has to take place in a meta-stable configuration, such that the longer the accumulation period, the higher is the stored energy available for the next episode.

forced to release energy: in this case, the larger the amount of energy released, the longer will be the time needed to reach the threshold again, and we would obtain a correlation between the burst time and the prequiet time, contrary to what observed.

It is also important to note that the threshold energy E_0 above which the system is in a meta-stable state is not directly related to the intensity of the seed magnetic field inside the neutron torus. It is instead most likely related to the intensity and the topological configuration of the amplified field which emerges from the torus.

Finally, we would like to point out the suggestive analogy with the micro-quasar GRS 1915+105, a galactic black hole candidate which exhibits a strikingly similar (even quantitatively) correlation between the duration of a quiescent time and that of the following burst [1]. This source is believed to accrete at a rate very close to (or maybe higher than) the Eddington limit, which is also the case for the configuration suggested for gamma-ray bursters discussed above.

REFERENCES

1. Belloni T., et al., *ApJ*, **488**, L109-L112 (1997).
2. Kobayashi, S., Piran, T., and Sari, R., *ApJ*, **490**, 92-98 (1997).
3. Kouveliotou, C., et al., *ApJ*, **413**, L101–L104 (1993).
4. Mészáros, P., *Science*, **291**, 79–84 (2001).
5. Ramirez-Ruiz, E. and Merloni, A., *MNRAS*, **320**, L25–L28 (2001).
6. Ramirez-Ruiz, E., Merloni, A. and Rees, M. J., *MNRAS*, in press (2001).
7. Rees, M. J., *A&AS*, **138**, 491–497 (1999)
8. Rees, M. J. and Mészáros, P., *ApJ*, **430**, L93-97 (1994).

Gamma Ray Bursts as Probes of the First Stars

James E. Rhoads

STScI, 3700 San Martin Dr., Baltimore, MD 21210, USA

Abstract. The redshift where the first stars formed is an important and unknown milestone in cosmological structure formation. The evidence linking gamma ray bursts (GRBs) with star formation activity implies that the first GRBs occurred shortly after the first stars formed. Gamma ray bursts and their afterglows may thus offer a unique probe of this epoch, because they are bright from gamma ray to radio wavelengths and should be observable to very high redshift. Indeed, our ongoing near-IR followup programs already have the potential to detect bursts at redshift $z \sim 10$. In these proceedings, we discuss two distinct ways of using GRBs to probe the earliest star formation. First, direct GRB counts may be used as a proxy for star formation rate measurements. Second, high energy cutoffs in the GeV spectra of gamma ray bursts due to pair production with high redshift optical and ultraviolet background photons contain information on early star formation history. The second method is observationally more demanding, but also more rewarding, because each observed pair creation cutoff in a high redshift GRB spectrum will tell us about the integrated star formation history prior to the GRB redshift.

INTRODUCTION

The high redshift frontier of observational cosmology currently stands at redshifts $z \approx 6$. The current redshift record is a quasar at $z = 5.8$, and a few galaxies are known at marginally lower redshift. Beyond $z = 6$, we have yet to identify any individual objects. We do know that hydrogen was predominantly neutral at redshifts $z \gtrsim 30$ based on the observed anisotropy of the cosmic microwave background, which would be smoothed out by Thomson scattering if the free electron density at $z \gtrsim 30$ were too great. The redshift range $6 \lesssim z \lesssim 30$ remains unknown territory. It is a very interesting territory, too, for it should include the formation of the first stars, galaxies, and quasars, and certainly includes the epoch at which hydrogen was reionized.

Searches for starlight (and other rest-frame near ultraviolet tracers) can make incremental progress into the low-redshift end of this period. However, these methods face a practical limit where the Lyman break redshifts out of the optical window to the near-infrared, at $z \approx 7$. At higher redshift, essentially no flux is expected in the optical window (observed wavelengths $0.36\mu m \lesssim \lambda_{obs} \lesssim 1\mu m$). Atmospheric conditions and present detector technologies conspire to make searches at $\lambda_{obs} \gtrsim 1\mu m$ much less efficient. Future instrumentation like the Next Generation Space Telescope (NGST) promise extensions of "conventional" optical methods to the observed near-IR and thus to redshifts $z \gg 6$, but this may be a decade or more away. In the meantime, we expect the upcoming extension of our Large Area Lyman Alpha (LALA) survey (Rhoads et al 2000; Malhotra et al 2001) to $z = 6.6$ to be at or near the practical limit for some years.

We would like to find tracers of $z > 6$ objects that are accessible now. Fortunately, this is possible so long as we are willing to use something besides starlight. In practice, this means higher energy photons (γ and x-rays), since lower energies still face either confusion or sensitivity issues.

Gamma ray bursts (GRBs) are an excellent candidate for detection at high redshift because the bursts and their afterglows are extremely bright at all wavelengths. Two conditions must be met for such a candidate to work well. First, there should be a reasonable expectation that the object exists at high redshift; and second, it should be detectable there.

The best argument that gamma ray bursts should occur at high redshift comes from the growing body of evidence linking GRBs to star formation activity (and hence presumably to the deaths of massive, short-lived stars): GRB host galaxy colors are characteristically blue (Fruchter et al 1999); the spatial distribution of GRBs on their hosts matches expectations for hypernova models (Bloom, Kulkarni, & Djorgovski 2000); and the emission lines of GRB host galaxies are unusually strong (Fruchter et al 2001). Structure formation models yield estimated redshifts $z \sim 15 \pm 5$ for the first stars to form in the universe (cf. Barkana & Loeb 2001). This is supported by studies of heavy element abundances: It has proven extremely difficult to find objects with primordial (i.e., big bang nucleosynthesis) abundances at any redshift currently accessible. The immediate inference is that a substantial generation of stars must have existed at earlier redshifts to produced the ubiquitous metals. The association of GRBs with star formation then implies that the first GRBs also occurred in the redshift range $z \sim 15 \pm 5$.

The detectibility of GRBs at $z \gg 6$ has been considered in detail by Lamb & Reichart (2000), who find that the bright end of the luminosity distribution would be detectable at very high redshifts (though quantitative predictions depend substantially on unknown details of the GRB luminosity function). This applies also to the X-ray and optical afterglows, for which time dilation of the most distant afterglows helps offset the increase in luminosity distance with redshift (Lamb & Reichart 2000; Ciardi & Loeb 2000).

The Lyman break will render afterglows at $z > 7$ invisible to optical detectors, just as it does for galaxies. But the problem here is not so serious. Searches for $z > 7$ galaxies suffer because galaxies at such high redshifts are faint, and the required combination of large solid angles and high sensitivity to find them is not yet practical at near-IR wavelengths. Because GRB afterglows outshine their host galaxies at early times, and because X-ray detectors can determine GRB locations with accuracy comparable to the current near-IR field of a 4m class telescope, an afterglow at this redshift is easier to find than are the galaxies around it. Indeed, published near-infrared afterglow observations (Rhoads & Fruchter 2001) already achieve a sensitivity sufficient to detect afterglows at $z \sim 10$ for several hours following a GRB (cf. figures 2,3 of Lamb & Reichart 2000). The followup program described in Rhoads & Fruchter 2001 is continuing at the NASA Infrared Telescope Facility, and we have a similar program at the National Optical Astronomy Observatory. The observed signature of a $z > 7$ GRB would be a near-infrared afterglow exhibiting a Lyman break at $\lambda_{obs} = 0.1215(1+z)\mu m > 1\mu m$. Such breaks have been used to measure $z = 2.05$ for GRB 000301C (Smette et al 2001) and to estimate $z \approx 5$ for GRB 980329 (Fruchter 1999; see also Reichart et al 1999). Their extension to longer wavelengths is straightforward. Thus, it is reasonable to expect that $z > 7$ GRBs will be detected with current technology.

The prospect of detecting gamma ray bursts at $z > 7$ opens two possible methods of studying star formation activity at these epochs: GRB rate evolution, which should trace star formation activity; and pair production cutoffs in the GeV spectra of bursts, which probe the total optical-ultraviolet background light produced by high redshift stars.

BURST RATE EVOLUTION

The most basic inference from the observed burst rate is that the highest redshift where a burst has been detected $z_{max,grb}$ implies the onset of star formation at some redshift $z_{max,*} > z_{max,grb}$. It is likely that in fact $z_{max,*} \approx z_{max,grb}$: The association of GRBs with star formation tracers requires short progenitor lifetimes ($\ll 10^8$ years), so the redshift difference between the first stars formed and the earliest possible hypernovae is small.

It will be possible to go further by measuring the GRB rate as a function of redshift, $R_{grb}(z)$, and taking it as a surrogate for the star formation rate. Such studies would require a large sample (several tens) of high redshift GRBs, together with an understanding of the selection effects that went into the sample. This method is likely to be limited by at least two systematic factors. First, uncertainties in the GRB luminosity function will introduce uncertain corrections to the inferred total GRB rate and the inferred star formation rate, since the high redshift sample will contain only bright bursts. Second, evolution in the burst progenitor population may influence the burst rate. One plausible example is that the GRB rate could depend on progenitor metallicity, which is likely to be lower in the early universe. Another is that the stellar initial mass function (IMF) may vary, thereby affecting the relation between GRB rate and star formation rate, and perhaps also the shape of the GRB luminosity function. Possible evidence for IMF variations has recently been found in at some high redshift Lyman α emitting galaxies (Malhotra et al 2001).

Overall, these complications suggest that calibration of the GRB rate as an indicator of global star formation might be possible to within a factor of a few. While higher accuracies would be desirable, the present uncertainties with more conventional star formation estimators are not much better. For example, rest ultraviolet continuum measurements are corrected by a factor of ~ 7 for dust absorption, and the uncertainty in this correction could easily be a factor of two given the range of possible dust properties.

PAIR PRODUCTION CUTOFFS IN GRB SPECTRA

The observed spectra of gamma ray bursts sometimes extend to very high photon energies: The EGRET experiment on the Compton Gamma Ray Observatory detected four bursts with unbroken power law tails extending to $E_\gamma > 1 \text{GeV}$, and the Milagrito air shower experiment has tentatively detected one burst at $E_\gamma \gtrsim 1 \text{TeV}$. Photons with such high energies have mean free paths shorter than a Hubble distance due to $\gamma + \gamma \to e^+ + e^-$ interactions with low energy background photons. The threshold for such pair production reactions is $E_\gamma \varepsilon_\gamma > m_e^2 c^4 = (511 \text{keV})^2$, corresponding to the requirement that each photon have the rest mass energy of an electron in their center of momentum frame. (Here E_γ and ε_γ are the two photon energies measured in an arbitrary frame, and $E_\gamma \geq \varepsilon_\gamma$ by convention.) The cross section (for a head-on collision) peaks at $E_\gamma \varepsilon_\gamma = 2 m_e^2 c^4$ and falls asymptotically as $1/(\mathcal{E}_\gamma\varepsilon_\gamma)$ for $\mathcal{E}_\gamma\varepsilon_\gamma \gg 2m_e^2 c^4$.

Pair production cutoffs in the TeV gamma ray spectra of blazars due to interactions with the cosmic infrared background have been predicted (Stecker, De Jager, & Salamon 1992; MacMinn & Primack 1996; Madau & Phinney 1996; Malkan & Stecker 1998) and observed (e.g., De Jager, Stecker, & Salamon 1994; Konopelko et al 1999) for several years now. The extension of the same physics to higher redshifts and lower gamma ray energies has been explored recently by several groups (Salamon & Stecker 1998; Primack et al 2000; Oh 2000).

The observer frame gamma ray energy determines simultaneously the redshift and rest frame energies of the background photons that dominate the pair production optical depth. At low redshifts ($z \ll 1$), the effective absorption coefficient $\alpha(\mathcal{E}_\gamma)$ increases with $\mathcal{E}_\gamma$ and changes relatively little with redshift, so that the relevant physics is simply $\alpha(\mathcal{E}_{cut})d = 1$, with d the distance to the source. However, at $z \gtrsim 1$, redshift effects become important: The threshold energy $\varepsilon_\gamma(z) \propto 1/(1+z)$, and the background radiation field will also evolve with redshift. The optical depth for photons near $\mathcal{E}_{cut}$ is therefore dominated by absorption at high redshift, unless the source redshift is so high as to precede the creation of any substantial optical-IR background. By the time the photon reaches lower redshifts, the threshold for pair creation grows so large that the density of relevant photons is extremely low. Oh (2000) has shown that the highest energy background photons capable of producing optical depth $\tau \approx 1$ over a Hubble distance have energies below the ionization threshold for hydrogen (i.e., $\varepsilon_\gamma < 13.6 \text{eV}$), since hydrogen absorption in stellar atmospheres, galaxies, and the intergalactic medium ensures a strong decrement in background photon number density at 13.6eV.

The most robust observable consequence of the pair creation cutoff is the observer frame gamma ray energy $\mathcal{E}_{cut}(z)$ for which the optical depth $\tau = 1$. Lower pair creation optical depths ($\tau \ll 1$) cannot be measured reliably because of our imperfect knowledge of the intrinsic (i.e., unabsorbed) source spectrum, while at higher optical depths ($\tau \gg 1$) the absorption reduces the flux below detection thresholds of present or near-future instruments. We might measure $\tau(\mathcal{E}_\gamma)$ with reasonable accuracy over the range $1/2 \lesssim \tau \lesssim 2$.

Detailed predictions of $\mathcal{E}_{cut}(z)$ differ from model to model, depending on the theoretical treatment adopted for the earliest star formation (see Primack et al 2000; Oh 2000). For example, the observer frame energy where $\tau = 1$ for redshift $z = 6$ is $4\text{GeV} \lesssim \mathcal{E}_{cut}(6) \lesssim 6\text{GeV}$ for different models in Primack et al (2000), and $10\text{GeV} \lesssim \mathcal{E}_{cut}(6) \lesssim 26\text{GeV}$ for models in Oh (2000). Therein lies the power of this method for learning about the first generations of stars, for these strong differences in predictions allow the models to be distinguished with comparative ease from even a modest data set.

Moreover, if we can observe the GeV cutoffs in spectra of a few GRBs spread over the redshift range $6 \lesssim z < z_{max,*}$, we can infer the evolution of the optical-UV background radiation over the same period with little dependence on models. This follows because the *difference* in pair creation optical depth between two bursts at redshifts z_1 and z_2 ($z_1 < z_2$) is determined only by the background radiation in the range $z_1 < z < z_2$.

DISCUSSION

The two methods of using gamma ray bursts to probe high redshift star formation complement each other in many ways. GRB rate measurements at high redshift are technically easier. They require a GRB monitor plus rapid multiband near-infrared followup. Existing instrumentation and indeed existing observational programs are already adequate for this work. Pair creation cutoffs require one additional observation, namely, a GeV energy spectrum obtained during the GRB. This GeV spectrum will have to come from GLAST or a similar space mission.

The physical assumption behind the GRB rate evolution method is that the bursts are associated with star formation activity. Under this assumption, there will be some systematic uncertainties in converting the GRB rate to the star formation rate (see above). In contrast, the pair creation cutoff method requires only that some high redshift GRBs have GeV spectra that are sufficiently bright and sufficiently smooth for the cutoff to be observed. Beyond this, there is no requirement on the nature of the bursters, which are needed only as beacons to probe the intervening background radiation. The physics of pair creation is then well understood and probes the total background radiation produced by high redshift stars.

Thus, combining the two methods of studying high redshift star formation with GRBs may overcome the physical uncertainties of either method alone. Additional constraints from other techniques using other classes of objects (galaxies observed at infrared wavelengths, or quasars at X-ray wavelengths) will become available over the next few years, and will again have complementary strengths and weaknesses. By adding these to the GRB results, we can reasonably expect to understand star formation at $z \sim 10$ as well as we understand it at $z \sim 3$ today.

REFERENCES

1. Barkana, R., & Loeb, A. 2001, *Physics Reports*, in press
2. Bloom, J. S., Kulkarni, S. R., & Djorgovski, S. G. 2000, submitted to AJ, astro-ph/0010176
3. Ciardi, B., & Loeb, A. 2000, ApJ 540, 687
4. De Jager, O. C., Stecker, F. W., & Salamon, M. H. 1994, *Nature* 369, 294
5. Fruchter, A. S., et al 1999, ApJ 519, L13
6. Fruchter, A. S. 1999, ApJ 512, L1
7. Fruchter, A. S., et al 2001
8. Konopelko, A. K., Kirk, J. G., Stecker, F. W., & Mastichiadis, A. 1999, ApJ 518, L13
9. Lamb, D. Q., & Reichart, D. E. 2000, ApJ 536, 1
10. MacMinn, D., & Primack, J. R. 1996, *Space Science Reviews* 75, 413
11. Madau, P., & Phinney, E. S. 1996, ApJ 456, 124
12. Malhotra, S., et al 2001, in preparation
13. Malkan, M. A., & Stecker, F. W. 1998, ApJ 496, 13
14. Oh, S. P. 2001, to appear in ApJ, astro-ph/0005263
15. Primack, J. R., Somerville, R. S., Bullock, J. S., & Devriendt, J. E. G. 2000, astro-ph/0011475
16. Reichart, D. E., et al 1999, ApJ 517, 692
17. Rhoads, J. E., Malhotra, S., Dey, A., Stern, D., Spinrad, H., & Jannuzi, B. T. 2000, ApJ 545, L85
18. Rhoads, J. E., & Fruchter, A. S. 2001, ApJ 546, 117
19. Salamon, M. H., & Stecker, F. W. 1998, ApJ 493, 547
20. Stecker, F. W., De Jager, O. C., & Salamon, M. H. 1992, ApJ 390, L49

GeV and X-ray Inverse Compton and Proton Synchrotron Signatures in Gamma-ray Burst Afterglows

Bing Zhang* and Peter Mészáros*

Pennsylvania State University, 525 Davey Lab, University Park

Abstract. Two high energy radiation mechanisms, i.e., the electron inverse Compton emission and the proton synchrotron emission, as well as their relative importance with respect to the electron synchrotron emission, are explored in the context of gamma-ray burst afterglow theories, with the focus on the possible signatures of these emission components in the broad-band spectra and in the GeV to keV light curves of gamma-ray burst afterglows. A general conclusion is that the electron inverse Compton emission dominated parameter regime is the most favorable regime for the high energy emission, and the inverse Compton component is detectable by *GLAST* within hours for bursts at typical cosmological distances, and by *Chandra* in days if the ambient density is high.

INTRODUCTION

Gamma-ray burst (GRB) afterglows are believed to be due to the emission of the non-thermal particles from the external shock when a fireball blastwave sweeps up the interstellar medium. The simplest mechanism proven to be quite adequate to interpret the broadband afterglow data from a dozen of GRBs is the synchrotron emission of the relativistic electrons (e.g. [1]), at least in the low-energy regime. In the high energy regime, some other high energy spectral components should exist and may, under certain conditions, become the dominant radiation mechanisms. These include the inverse Compton component of the electrons ([2] [3] [4] [5]) and the proton synchrotron emission ([6] [7] [8]). Previously, these components were usually treated singly, without a clear manifest about their significance with respect to the main electron synchrotron component in various bands or about their relative importance with respect to each other. We have performed a coherent study of the high energy spectral components of GRB afterglows. Here we will outline some of the main findings in this study. A full treatment is presented in [9].

Following notations will be adopted throughout: $\varepsilon_e, \varepsilon_p, \varepsilon_B$ are the equipartition parameters for the shock-accelerated electrons, protons, and the magnetic field, respectively; ζ_e, ζ_p are the injection parameters of the electrons and the protons; $\mathcal{E}_{52}$ is the total energy per solid angle in unit of 10^{52}ergs; n is the interstellar medium density; t_h is the observer's time in unit of hour; z is the redshift of the GRB; $F_\nu(\nu)$ is the spectral flux at the frequency ν; Y_e is the Compton factor; α is the acceleration parameter; and k is the assumed reduction factor of the photo-meson components relative to the proton synchrotron component.

RELATIVE IMPORTANCE OF VARIOUS SPECTRAL COMPONENTS

We consider three spectral components:

Electron Synchrotron Emission: A four-segment broken power law separated by $\nu_{a,e}$ (the self-absorption frequency), $\nu_{m,e}$ (the characteristic frequency of the electrons with the minimum injection energy) and $\nu_{c,e}$ (the cooling frequency), and terminate around $\nu_{u,e}$ (the characteristic frequency of the electrons with the maximum injection energy allowed by shock acceleration). Assuming the injection power index of the electrons to be p, the spectral indices in the four segments are $[2, 1/3, -1/2, -p/2]$ for the fast cooling ($\nu_{a,e} < \nu_{c,e} < \nu_{m,e}$) regime, or $[2, 1/3, -(p-1)/2, -p/2]$ for the slow cooling ($\nu_{a,e} < \nu_{m,e} < \nu_{c,e}$) regime [1].

Proton Synchrotron Emission (and other hadron-related components): The proton synchrotron component has a similar four-segment broken power law shape as the electron synchrotron component, but only in the slow cooling regime. The separation frequencies are $\nu_{a,p}$, $\nu_{m,p}$, $\nu_{c,p}$ and $\nu_{u,p}$. Above $\nu_{u,p}$, there are some other hadron-related components, e.g., synchrotron emission from the positrons produced by π^+ decay and the γ-rays produced directly from π^o decay, with a reduced level [7].

Electron Inverse Compton Emission: A four-segment broken power law (approximately) separated by $\nu_{a,e}^{IC} \sim \gamma_{m,e}^2 \nu_{a,e}$, $\nu_{m,e}^{IC} \sim \gamma_{m,e}^2 \nu_{m,e}$ and $\nu_{c,e}^{IC} \sim \gamma_{c,e}^2 \nu_{c,e}$, and terminate around $\nu_{u,e}^{IC}$ (minimum of $\gamma_{u,e}^2 \nu_{u,e}$ and the Klein-Nishina limit). The spectral indices are $[1, 1/3, -1/2, -p/2]$ for fast cooling ($\nu_{a,e}^{IC} < \nu_{c,e}^{IC} < \nu_{m,e}^{IC}$), and $[1, 1/3, -(p-1)/2, -p/2]$ for slow cooling ($\nu_{a,e}^{IC} < \nu_{m,e}^{IC} < \nu_{c,e}^{IC}$).

The relative importance of the three components could be addressed in two regimes:

1. $\nu < \nu_{u,e}$: competitions are among all the three components. There are three sub-cases:

   Regime I: Proton synchrotron dominated regime; the condition is $F_{\nu,p}(\nu_{u,p}) > F_{\nu,e}(\nu_{u,p})$, which reads (line 1 in Fig.1)

$$(1+Y_e)^{2/3}\varepsilon_B > 594(\varepsilon_e/\varepsilon_p)^{2(p-1)/3}(\zeta_p/\zeta_e)^{2(p-2)/3}\alpha^{2/3} \times E_{52}^{-5/12} n^{-7/12}[t_h/(1+z)]^{-1/12}, \qquad (1)$$

   Regime II: Electron IC dominated regime; the condition is $F_{\nu,e}^{IC}(\nu_{c,e}^{IC}) > F_{\nu,e}(\nu_{c,e}^{IC})$ for the slow-cooling case, or $F_{\nu,e}^{IC}(\nu_{m,e}^{IC}) > F_{\nu,e}(\nu_{m,e}^{IC})$ for the fast-cooling case, both of which can be reduced to (line 2 in Figure 1).

$$(1+Y_e)\varepsilon_B < 3.8(\varepsilon_e/\zeta_e)^{(p-1)}\zeta_e(E_{52}/n)^{(p-2)/8} \times [t_h/(1+z)]^{-3(p-2)/8}. \qquad (2)$$

   Regime III: Electron synchrotron dominated regime; this is the parameter space excluded by eqs.(1) and (2).

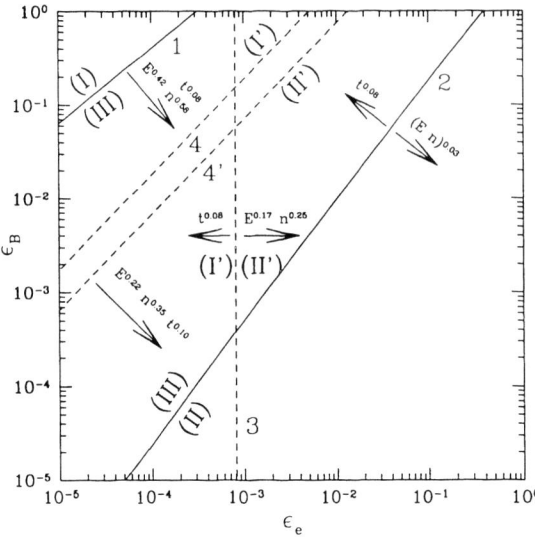

FIGURE 1. Regions in the $\varepsilon_e, \varepsilon_B$ parameter space where the various radiation mechanisms dominate. The parameter dependences of the boundaries are marked by arrows.

2. $\nu > \nu_{u,e}$: competitions are between the hadron-related components and the electron IC component. There are two regimes:
    Regime I': Proton synchrotron and the related photo-meson emission components dominated regime, the condition is $F_{\nu,p}(\nu) > F_{\nu,e}^{IC}(\nu)$.
    For $\nu < \nu_{u,p}$, one has (line 3 in Fig.1).

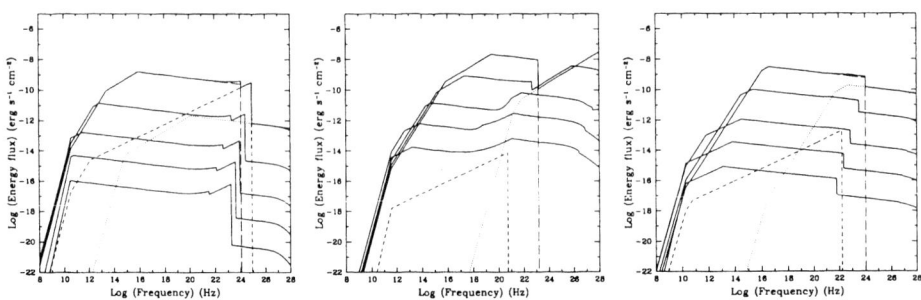

FIGURE 2. Temporal evolution of the broad-band spectra of GRBs for the three parameter regimes (I, II, and III from left to right). Thick solid curves are the final spectra for various observer times, starting from (top) the onset of the afterglow, 1 minute, 1 hour, 1 day to (bottom) 1 month, respectively. For the top curve, contributions from the various radiation components are also plotted. Long dashed are electron synchrotron, short dashed are proton synchrotron, and dotted lines are electron inverse Compton emission.

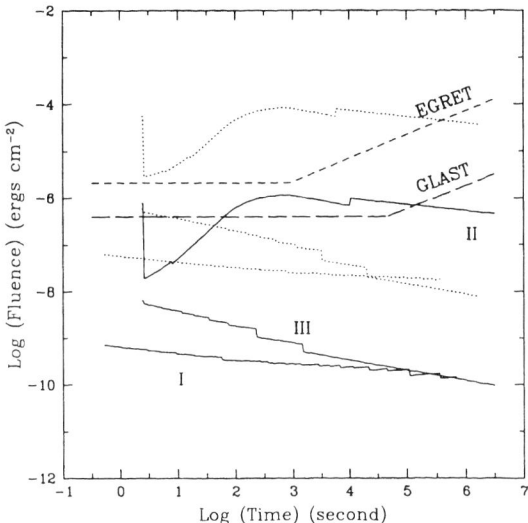

FIGURE 3. GeV $\int_{\nu_1}^{\nu_2} \nu F_\nu d\nu \cdot t$ lightcurves for the three types of bursts, as compared with the *EGRET* and *GLAST* sensitivities. Solid curves are for bursts at $z = 1$ while the dotted curves are for bursts at $z = 0.1$.

$$\varepsilon_e < 8.1 \times 10^{-4} \varepsilon_p^{\frac{1}{2}} \zeta_e^{\frac{p-2}{p-1}} \zeta_p^{-\frac{p-2}{2(p-1)}} \mathcal{E}_{52}^{-\frac{p+1}{16(p-1)}}$$
$$\times n^{-\frac{7-p}{16(p-1)}} [t_h/(1+z)]^{\frac{3p-5}{16(p-1)}}. \quad (3)$$

For $\nu > \nu_{u,p}$, one has ($p = 2.2$ adopted) (line 4 in Fig.1)

$$(1+Y_e)^{0.85} \varepsilon_B > 215 k_{-1}^{-0.43} \varepsilon_e^{1.02} \varepsilon_p^{-0.51} \zeta_e^{-0.17} \zeta_p^{0.09} \alpha^{0.34}$$
$$\times \mathcal{E}_{52}^{-0.22} n^{-0.35} t_h^{-0.10} (1+z)^{0.01} \nu_{26}^{-0.04}. \quad (4)$$

Regime II': Electron IC dominated regime; this is the parameter space excluded by eqs.(3) and (4).

AFTERGLOW SIGNATURES IN THE GEV & X-RAY BAND

The broadband snapshot spectra for different times are presented in Figure 2, for the three different parameter regimes in which the proton synchrotron, electron IC, and electron synchrotron dominate, respectively.

By requiring $F_{\nu,e}^{IC}(\nu) > F_{\nu,e}(\nu)$, the critical time for the electron IC component overtaking the electron synchrotron component at the frequency ν is

$$t_{IC} = 3.4 \text{ days } \varepsilon_{e,.5}^{0.89} \varepsilon_{B,-2}^{0.08} \zeta_e^{1.63} E_{52}^{-0.06} n^{-0.66} (1+z)^{-0.36} \nu_{18}^{-0.68}, \qquad (5)$$

where $p = 2.2$ is adopted. Thus the IC component will show up in the X-ray band within days for a typical regime II burst if the medium density is mildly high (cf. [4][5]). At higher energies, the IC component shows up at an even earlier time. In the GeV band, the IC component will form a bump peaking at hours after the burst trigger, due to the sweeping of the peak energy flux of the IC component. A typical regime II burst is detectable by *GLAST* at a typical cosmological distance, and a close regime II burst should have been detectable by *EGRET* (see Fig.3). This might have been the case of GRB 940217.

By requiring $F_{\nu,p}(\nu) > F_{\nu,e}(\nu)$, the critical time for the proton component overtaking the electron synchrotron component at the frequency ν is

$$t_p = 1.1 \text{ hr } \varepsilon_{B,.5}^{-3} (\varepsilon_{e,-3}/\varepsilon_p)^{4(p-1)} (\zeta_p/\zeta_e)^{4(p-2)} E_{52}^{-1} n_2^{-2} (1+z)^{-3} \nu_{23}^{-2}. \qquad (6)$$

It is impossible for this component to show up in the X-ray band. In the GeV band, for typical regime I parameters, it may overcome the electron component. However, the absolute flux level is low. A typical regime I burst is non-detectable by *GLAST* even at $z = 0.1$, unless the burst has a greater than average energy budget and/or in a very dense environment.

ACKNOWLEDGMENTS

This work is supported by NASA NAG5-9192 and NAG5-9153.

REFERENCES

1. Sari, R., Piran, T., and Narayan, R., *ApJ*, **497**, L17–L20 (1998).
2. Mészáros, P., Rees, M. J., and Papathanassiou, H., *ApJ*, **432**, 181–193 (1994).
3. Dermer, C. D., Chiang, J., and Mitman, K. E., *ApJ*, **537**, 785–795 (2000).
4. Panaitescu, A., and Kumar, P., *ApJ*, **543**, 66–76 (2000).
5. Sari, R., and Esin, A. A., *ApJ*, **548**, 787–799 (2001).
6. Vietri, M., *Phys. Rev. Lett.*, **78**, 4328–4331 (1997).
7. Böttcher, M., and Dermer, C. D., *ApJ*, **499**, L131–L134 (1998).
8. Totani, T., *ApJ*, **502**, L13–L16 (1998).
9. Zhang, B., and Mészáros, P., *ApJ*, **559**, in press (astro–ph/0103229) (2001).

The Supernova-Collapsar Scenario for the Gamma-Ray Bursts

Peter J.T. Leonard

*Science Systems and Applications, Inc.,
10210 Greenbelt Road, Suite 400,
Lanham, MD 20706, USA*

Abstract. We propose that a gamma-ray burst can result from the merger of a neutron star with a massive main-sequence star following the supernova explosion of the primary star of a binary system. The newly-formed neutron star receives a random kick during the explosion, and, in a small fraction of cases, the kick has the appropriate direction and amplitude to remove most of the orbital angular momentum from the post-supernova binary system, which results in an orbit with a pericenter distance smaller than the radius of the secondary. The neutron star rather quickly becomes embedded inside the secondary, and sinks to its center. Here the neutron star experiences runaway accretion due to neutrino losses, and collapses into a black hole, which continues to accrete from the disk that surrounds it, and produces a pair of jets that bore their way out of the merged object. Observers who lie in the direction of either jet will see a gamma-ray burst. This scenario naturally explains the iron lines in gamma-ray burst spectra due to the processed material from the supernova explosion that surrounds the burst site.

INTRODUCTION

The gamma-ray burst problem is one of the most enduring mysteries of modern astrophysics (see [7] for a nice review). It has now been established that these events occur in star forming galaxies at cosmological distances, that each burst releases an amount of energy comparable with the rest mass energy of our sun in a matter of seconds, and that gamma-ray burst lightcurves and spectra are described quite well by the so-called fireball model. There is also growing evidence that at least some gamma-ray bursts appear to be correlated with supernova explosions (e.g., [3]). In addition, there is strong observational evidence for iron lines in gamma-ray burst spectra at X-ray energies (e.g., [8]), which is commonly interpreted as being proof of supernova processed material surrounding the burst sites. All this evidence suggests that gamma-ray bursts likely result from the formation of several-solar-mass black holes when massive stars end their lives. The precise details of how this happens have not yet been established, and here we present one possible scenario.

The Details of the Supernova-Collapsar Scenario

We propose that a gamma-ray burst can result from the merger of a neutron star with a massive main-sequence star following a supernova explosion in a binary system. The sequence of events is based on the scenario outlined by Leonard, Hills and Dewey [4].

The first part of the scenario is fairly standard. Consider a massive close main-sequence binary system and its subsequent evolution. Let us call the component stars the primary and the secondary. The primary is initially more massive than the secondary. The primary evolves more rapidly than the secondary, expands, fills its Roche lobe, and transfers matter to the secondary. The primary is now less massive than the secondary. The core of the primary burns it nuclear fuel until the core can no longer support itself against gravity. The core collapses into a neutron star, and the primary undergoes a supernova explosion. Since the outer layers of the originally more massive star have been transferred to the secondary, then the supernova may be hydrogen deficient.

The next part of the scenario is less standard, but is still rather straightforward. As the primary undergoes the supernova explosion, the newly-formed neutron star receives an asymmetric kick. This velocity kick is an observational fact (e.g., [2]), although the kick mechanism remains poorly understood. Assuming that the velocity kick is randomly oriented and has a Maxwellian distribution of amplitudes, then, in a small fraction of cases (roughly 1%), the kick removes most of the orbital angular momentum from the binary system, and hence the post-explosion orbit can have a periastron distance smaller than the radius of the secondary. If so, then the neutron star rather quickly becomes embedded inside the secondary, and spirals toward the center of the secondary, giving the envelope of the merged object a lot of rotational angular momentum in the process. Leonard, Hills and Dewey [4] estimate the rate of this process in the Galaxy to be roughly 0.06 per square kpc per Myr for secondaries more massive than 15 solar masses. The neutron star also accretes matter from a disk that inevitably forms around it.

Note that the neutron star just misses the secondary on the first pass in a comparable number of cases, but hydrodynamic effects dissipate enough orbital energy to ultimately doom the neutron star to becoming embedded within the companion at a later time. This increases the rate of the process over that estimated by Leonard, Hills and Dewey [4] by roughly a factor of two.

The fate of the merged object has been the source of much speculation, and we shall assume that a collapsar-like scenario results, as described by MacFadyen, Woosley and Heger [5, 6]. That is, the neutron star experiences runaway accretion due to neutrino losses, collapses into a black hole, which continues to accrete, and produces a pair of jets that bore their way out of the merged object. Note that the collapse to a black hole may not happen right away, because roughly one solar mass of material must first be accreted. How long this takes depends upon the accretion rate, which is unknown, and may vary from case to case. Runaway accretion due to neutrino losses can result in fantastic accretion rates − roughly 0.1 to 10 solar masses per year, or higher, are possible. These rates would result in a range of delays between the supernova event and the collapse to a black hole of roughly 10 to 0.1 years, or less, respectively. On the other hand, some researchers (e.g., [1]) argue that the runaway neutrino instability does not occur in this situation, and consequently the resulting accretion rates are too small to allow the neutron star to collapse into a black hole before the neutron star and its disk entirely dissipate the secondary.

Assuming that the collapsar scenario does indeed take place in this situation, then a pair of oppositely directed relativistic particle jets are thrown off from the vicinity of the black hole. These jets bore their way through the envelope of the rapidly rotating merged object along the rotation axis, and then the jets bore through the surrounding supernova remnant into interstellar space. The jets produce gamma rays either by internal shocks, external shocks, pair creation and annihilation, or via some other mechanism. Observers who lie in the direction of either jet will see a gamma-ray burst, in addition to the supernova event which occurs at some earlier time. Roughly 0.01 of supernovae in massive binary systems result in neutron stars that quickly become embedded in the secondaries, and of those which produce black holes, only roughly 0.01 would be observable as gamma-ray bursts, if the jets are beamed into roughly 0.01 of the sky. This scenario naturally explains the iron lines in gamma-ray burst spectra, due to the processed material from the supernova explosion that surrounds the collapsar.

ACKNOWLEDGMENTS

The author is thankful for the continuing support of NASA.

REFERENCES

1. Armitage, P.J., and Livio, M., *The Astrophysical Journal* **532**, 540-547 (2000).
2. Arzoumanian, Z., Chernoff, D.F., and Cordes, J.M., submitted to *The Astrophysical Journal* (2001).
3. Galama, T.J., Vreeswijk, P.M., van Paradijs, J., et al., *Nature* **395**, 670-672 (1998).
4. Leonard, P.J.T., Hills, J.G., and Dewey, R.J., *The Astrophysical Journal* **423**, L19-L22 (1994).
5. MacFadyen, A.I., and Woosley, S.E., *The Astrophysical Journal* **524**, 262-289 (1999).
6. MacFadyen, A.I., Woosley, S.E., and Heger, A., *The Astrophysical Journal* **550**, 410-425 (2001).
7. Meszaros, P., *Science* **291**, 79-84 (2001).
8. Piro, L., Garmire, G., Garcia, M., et al., *Science* **290**, 955-958 (2000).

Pulses, Spectral Lags, Durations, and Hardness Ratios in Long GRBs

Jay P. Norris[*], Jeffrey D. Scargle[†], and Jerry T. Bonnell[*¶]

[*]*Laboratory for High Energy Astrophysics,
NASA/Goddard Space Flight Center, Greenbelt, MD 20771, USA*

[†]*Space Science Division, NASA/Ames Research Center, Moffett Field, CA 94035-1000, USA*

[¶]*Universities Space Research Association, Washington, DC, USA*

Abstract. We analyze BATSE 64-ms data for long gamma-ray bursts ($T_{90} > 2.6$ s), exploring the relationships between spectral lag, duration, and the number of distinct pulses per burst, N_{pulses}. We measure durations using a brightness-independent technique. Within a similarly brightness-independent framework, we use a "Bayesian Block" method to find significant valleys and peaks, and thereby identify distinct pulses in bursts. Our results show that, across large dynamic ranges in peak flux and the N_{pulses} measure, bursts have short lags and narrow pulses, while bursts with long lags tend to have just a few significant, wide pulses. There is a tendency for harder bursts to have few pulses, and these hardest bursts have short lags. Spectral lag and duration appear to be nearly independent – even for those bursts with relatively long lags: wider pulses tend to make up in duration for fewer pulses. Our brightness-independent analysis adds to the nascent picture of an intimate connection between pulse width, spectral lag, and peak luminosity for bursts with known redshifts: We infer that lower-luminosity bursts should have fewer episodes of organized emission – wider pulses with longer spectral lags.

INTRODUCTION

Gamma-ray burst (GRB) time profiles are notoriously heterogeneous – chaotic and unpredictable in appearance – sufficiently so to evade physical modeling attempts. The best demonstration that each GRB time profile is unique and unpredictable comes from burst gravitational lens searches. The results of Marani et al. show that no two BATSE bursts studied have identical temporal and spectral development [1].

One of the first quantitative indications of a global tendency was Nemiroff's "ψ" temporal asymmetry analysis [2], which showed that bursts tend to be asymmetric on all timescales. Even for bursts at one extreme, where spike-like pulses are nearly symmetric at BATSE energies, the burst envelope is often asymmetric. The "Pulse Paradigm" further elucidates burst behavior: Pulses range from narrow and nearly symmetric, to wide and asymmetric, with low energy lagging high energy [3]. Thus, diversity at the level of burst duration masks a determinism at the pulse level: *Individual pulses are organized in time and energy*, and pulses within a given burst tend to exhibit the same degree of asymmetry and spectral lag [3]. In fact, for the few bursts with redshifts there appears to be an important correlation between pulse width, spectral lag, and luminosity [4]. The physical mechanisms giving rise to these correlations are not yet well elucidated. However, recent work by Soderberg and Fenimore shows that a pulse's locus in time in { Intensity, Peak in $\nu \cdot F(\nu)$ } space is inconsistent with pure kinematics of colliding shells, requiring the inclusion of cooling mechanisms to explain pulse profiles [5].

The question we begin to address here is: For a given burst, how is pulse behavior contingent upon global behavior – how are number of pulses, duration, spectral lag, and hardness interrelated?

ANALYSIS

To better inform studies of pulse behavior, we performed the first automated, brightness-independent analysis of the number of pulses in a burst, N_{pulses}, and then examined the relationships of N_{pulses} to hardness ratio, duration, and spectral lag. We used Bayesian Block methods developed by Scargle [6,7] in conjunction with brightness- and noise-equalization methods [8] to estimate N_{pulses} and durations.

The sample comprises 659 long bursts, including all BATSE GRBs with background fits, peak flux (PF) > 1.3 photons cm^{-2} s^{-1}, peak intensity (PI) > 4000 counts s^{-1}, and T_{90} > 2.6 s, where the duration for each burst was measured after equalization of signal to noise (S/N) and PI to average sample threshold levels. Spectral lags between BATSE channels 1 and 3 (100–300 and 25–50 keV) were measured via the cross correlation function (CCF) [4]. Durations were measured as described in reference [8], also with S/N levels equalized to the average threshold level. The burst data files at 64-ms resolution and their associated background fits are available at the COSSC web site: http://cossc.gsfc.nasa.gov/cossc/batse/index.html .

We bootstrapped the data, making 51 time profiles per burst, fitting a cubic to the CCF, and requiring the fit to be concave up for 51 consecutive trials; else, we increased the length of the fitted interval and restarted the set of fits. The CCF peak was taken as the spectral lag measure, using the measurements of individual fits to generate error bars. We applied the Bayesian Block (BB) cell coalescence method [7] to the S/N equalized time profiles, estimating significant peaks and valleys. The BB representation thus yields the number of identifiable pulse peaks per burst (N_{pulses}).

RESULTS

Figure 1 illustrates two scatter plots of peak flux (50–300 keV) versus spectral lag between BATSE energy channels 1 and 3. The obvious general trend, previously reported for a sample with a higher PF threshold [4,9], is that bursts with longer CCF lags tend to occur at lower peak fluxes. Now, extension of the sample down to PF > 1.3 photons cm^{-2} s^{-1} reveals some bursts with even longer lags, up to ~ 4 s. Notice, however, that most lags tend to concentrate near 0 to ~ 100 ms, even at low PF. For the few GRBs with associated redshifts, luminosity and spectral lag appear to be inversely correlated, following an approximate power law with slope ~ -1 [4,9], and therefore most BATSE bursts with PF $\gtrsim$ 1 photons cm^{-2} s^{-1} are probably highly luminous. At lower peak fluxes the situation may be different (see Discussion).

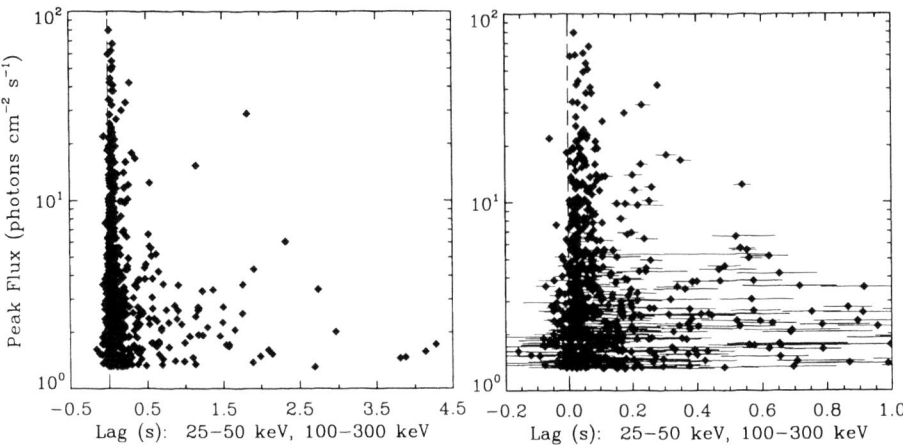

FIGURE 1. Peak Flux versus Spectral Lag. Left panel, showing full spectral lag range, illustrates tendency for lower peak-flux bursts to have longer lags. Right panel: Lag scale is magnified, thereby showing only bursts with lags < 1 s. Lag error bars, larger at lower peak flux, indicate that most long lag determinations are significantly different than the majority of lags, which concentrate $\lesssim$ 100 ms.

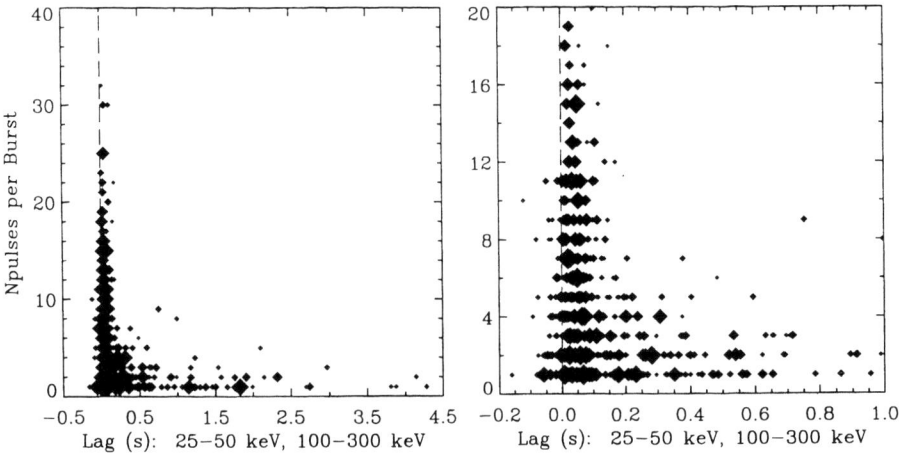

FIGURE 2. Number of Pulses versus Spectral Lag. Symbol size is proportional to ln(PF). Left panel: Full ranges of N_{pulses} and spectral lag are shown. Right panel: Lag and N_{pulses} scales are magnified to show detail. Bursts with long lag and few (wide) pulses tend to have low PF (smaller diamonds).

Scatter plots of N_{pulses} versus lag illustrate (Figure 2) an important extension of the "Pulse Paradigm" [3]. While bursts over a wide range in N_{pulses} have short lags, bursts with long lags tend to have just a few significant, *wide* pulses. The connection between luminosity and lag would therefore imply that these long-lag bursts occupy the low-end tail of the *observed* luminosity distribution; their actual numbers must be far greater than observed, due to brightness selection bias. That the pulses in these bursts are wide, monolithic structures – rather than composites of many narrow pulses – is evidenced by the spectral organization revealed in their long lags. To some degree the appearance of bursts with short lags and few pulses must be due to rendering the S/N to threshold level. However, our analysis of relatively bright bursts performed at original S/N levels yields many short lag bursts with few, narrow pulses.

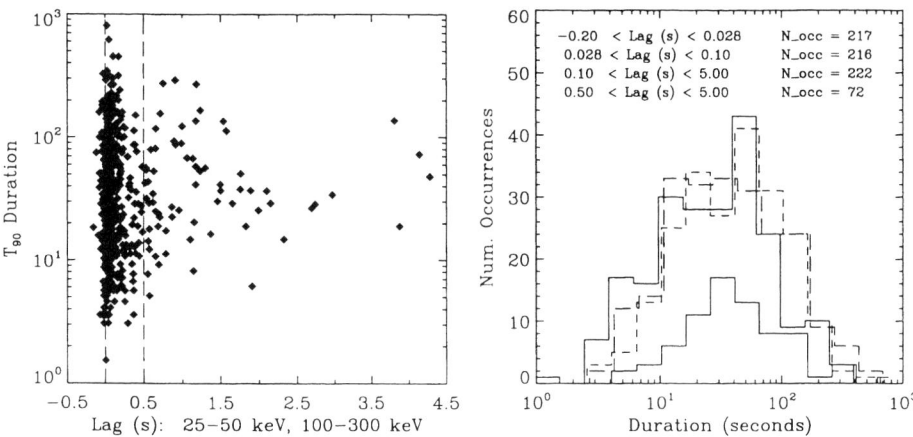

FIGURE 3. Left panel: Brightness-independent Duration versus Spectral Lag. Right panel: Duration histograms for three mutually exclusive ranges of lag, which divide the sample nearly equally, and for lag > 0.5 s (that portion of the sample right of dashed line in left panel).

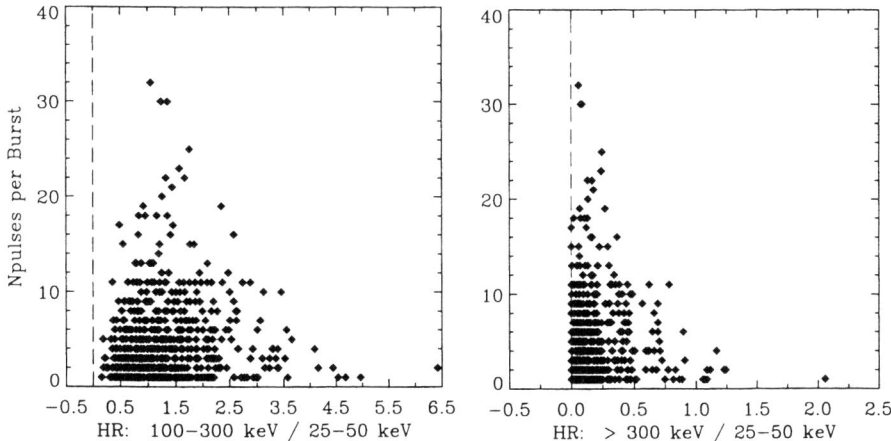

FIGURE 4. N_{pulses} versus Hardness Ratio (HR). Left panel: HR = integral counts above background, (100–300 keV) / (25–50 keV). Right panel: Same for (> 300 keV) / (25–50 keV).

The left panel of Figure 3 illustrates a scatter plot of duration versus spectral lag. The right panel shows durations histogrammed for three lag ranges (solid, dotted, dashed – increasing lag) which span the measured values, and for the lags > 0.5 s (lower solid histogram). No trend is apparent; bursts with few long-lag pulses must tend to lengthen the duration by virtue of their wider pulses, as might be expected. Figure 4 illustrates scatter plots of N_{pulses} versus hardness ratio for channels 3/1 ([100–300 keV] / [25–50 keV]) and channels 4/1 ([> 300 keV] / [25–50 keV]). There is a tendency for the harder bursts to have few pulses. Also, the hardest bursts have shorter lags: For HR 3/1 > 2.5, 36 (10) bursts have lags < (>) 60 ms; for HR 4/1 > 0.5, 31 (7) have lags < (>) 60 ms.

SUMMARY AND DISCUSSION

This work describes the first combined brightness-independent measurements of several parameters for long GRBs ($T_{90} > 2.6$ s): number of pulses per burst, spectral lag, hardness ratio, and duration. Some interesting relationships are quantified, pertaining mostly to GRBs with few pulses: GRBs with long spectral lag tend to have just a few, wide pulses. At the other extreme, short spectral lag, are GRBs also with few pulses, but with the hardest spectra. Spectral lag and duration appear to be nearly independent – even for GRBs with relatively long lags; wider pulses must compensate in duration for few pulses. From the nascent connection [4,9] between pulse width, spectral lag, and peak luminosity for GRBs with known redshifts we infer that *the lower-luminosity GRBs should tend to have fewer episodes of wide-pulsed emission.* Essentially, wide pulses with long spectral lags means lower luminosity. The analysis of average GRB profiles by Stern et al. as a function of PF [10] suggested an admixture of a larger fraction of simple bursts with wide pulses and long lag near the triggered BATSE PF threshold, ~ 0.25 photon cm^{-2} s^{-1}. Since our present sample (PF > 1.3 photon cm^{-2} s^{-1}) evidences the beginning of this trend, we infer that low-luminosity, long-lag, few-pulse bursts may dominate near the BATSE PF threshold.

REFERENCES

1. Marani, G.F., Nemiroff, R.J., Norris, J.P., Kevin, H., and Bonnell, J.T., *ApJ* **512**, L13 (1999).
2. Nemiroff, R.J., et al., *ApJ* **423**, 432 (1994).
3. Norris, J.P., et al., *ApJ* **459**, 393 (1996).
4. Norris, J.P., Marani, G.F., and Bonnell, J.T., *ApJ* **534**, 248 (2000).
5. Soderberg, A.M., and Fenimore, E.E., "The Unique Signature of Shell Curvature in Gamma-Ray Bursts," in *Proc. 2nd Rome Workshop on GRBs in the Afterglow Era* (2001).
6. Scargle, J.D., *ApJ* **504**, 405 (1998).
7. Scargle, J.D., "Bayesian Blocks: Divide and Conquer, MCMC, and Cell Coalescence Approaches," in *MaxEnt99* (2001).
8. Bonnell, J.T., Norris, J.P., Nemiroff, R.J., and Scargle, J.D., *ApJ* **490**, 79 (1997).
9. Norris, J.P., Marani, G.F., and Bonnell, J.T., "Connection between Spectral Lags and Peak Luminosity in GRBs," in *Gamma-Ray Bursts*, edited by R.M. Kippen, R.S. Mallozzi, and G.J. Fishman, AIP Conference Proceedings 526, New York, 2000, p. 87.
10. Stern, B., Poutanen, J., and Svensson, R., *ApJ* **510**, 312 (1999).

Super-LOTIS / LOTIS / LITE: Prompt GRB Followup Experiments

H. S. Park[1], E. Ables[1], S. Barthelmy[2], M. Bradshaw[3], T. Cline[2], N. Gehrels[2], D. Hartmann[4], K. Hurley[5], R. Nemiroff[6], W. Pereira[6], D. Perez-Ramirez[6], G. G. Williams[3], K. Ziock[1]

[1] *Lawrence Livermore National Laboratory, Livermore, CA 94550*
[2] *NASA/Goddard Space Flight Center, Greenbelt, MD 20771*
[3] *Steward Observatory, Tucson, AZ 85721*
[4] *Dept. Of Physics and Astronomy, Clemson University, Clemson, SC 29634*
[5] *Space Science Laboratory, University of California, Berkeley, CA 94720*
[6] *Dept. Of Physics, Michigan Technological University, Houghton, MI 49931*

Abstract. LOTIS (Livermore Optical Transient Imaging System) and Super-LOTIS are automatic telescope systems that measure very prompt optical emission occurring within seconds of the gamma-ray energy release during a Gamma Ray Burst (GRB). Unlike hour-to-days delayed afterglow measurements, very early measurements will contain information about the GRB progenitor. To accomplish this, we developed and have been operating automated telescopes that rapidly image GRB coordinate error boxes in response to triggers distributed by the GRB Coordinate Distribution Network (GCN). LOTIS, located in California, consists of 4 cameras each with a different astronomical filter (B, V, R, open) that can respond to GRB triggers within 5 s. Super-LOTIS can point to any part of the sky within 30 s upon receipt of a GCN trigger and its sensitivity is as deep as V=17~19 depending on the integration times. Since the shutdown of the CGRO, there has been no real-time GRB triggers that enable the LOTIS systems to measure real-time GRB counterpart fluxes as of May 2001. This paper describes performance of these systems. We also present our plan to replace the current optical CCD camera on the Super-LOTIS to a near infrared camera to be able to probe dusty GRB environment.

INTRODUCTION

The dramatic breakthrough in our understanding of GRBs occurred when the high resolution X-ray detector on the Beppo/SAX satellite was able to determine the position of a GRB with sufficient accuracy to enable a large telescope to observe a faint, fading afterglow days later. Optical and radio afterglows now have been observed for many GRBs during the last two years. These long-lasting but faint afterglows have been successfully explained in the "fireball models" as the result of the synchrotron interaction with surrounding material[1].

However, prompt counterparts associated with GRBs are rare. Rapidly slewing and automatic telescopes such as LOTIS[2] and ROTSE[3] have been operating to catch early glimpse of the GRBs detected by the BATSE instrument. After running these telescopes for many years recording real-time images of the GRB error boxes, only one event (GRB990123) has been detected simultaneously with a GRB. But this event was unusual in terms of its total X-ray fluence, peak flux and spectrum[4]. Unlike the observed later-time afterglows, prompt optical measurements (or even stringent constraints on that optical emission) would provide information about the GRB progenitors. In order to detect many more prompt counterparts, we are currently operating two instruments, LOTIS and Super-LOTIS, connected to the Coordinate Distribution by the current and future satellites such as the HETE-2, Rossi-XTE, INTEGRAL, and Swift missions.

SUPER-LOTIS

The Super-LOTIS telescope is a Boller & Chivens 0.6 m f/3.5 reflector telescope. It has superb optical quality and is equipped with computer controllable drives. Its focal array is a Loral 2048 x 2048 pixel 15 x 15 µm/pixel CCD cooled to –30°C with custom-built readout electronics. This focal plane array is placed at the primary focus of the mirror with a coma corrector yielding 0.84 x 0.84° field of view. Super-LOTIS began operation in October 2000 at Kitt Peak National Observatory in Arizona. Figure 1 shows its operation inside a roll-off roof housing.

FIGURE 1. Super-LOTIS in operation. It is housed in a building with a roll-off roof and is fully automated.

Since there were no real-time GRB triggers available until the end of May, 2001, Super-LOTIS has been responding to Beppo/SAX and IPN GRB triggers that usually come many hours after the burst. After this delay, the relativistic shock wave from the GRB has interacted with the interstellar medium producing an afterglow that decays as a power-low. Super-LOTIS imaged one of these afterglows, GRB010222, on Feb. 23, 2001. The image was taken 23.6 hours after the burst and the brightness of the afterglow was R=20.0. The Super-LOTIS image of this afterglow is shown in Figure 2. When it is not imaging the GRB fields, it systematically searches for many other classes of transient objects, such as novae and supernovae. It will also monitor long- and short-period stellar variability.

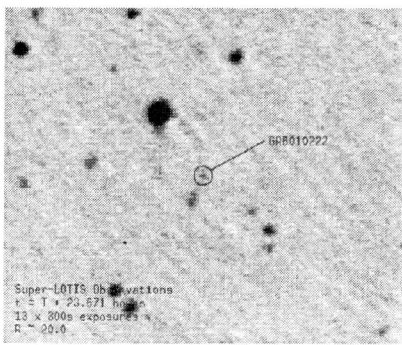

FIGURE 2. Super-LOTIS response to GRB010222.

LOTIS

The LOTIS was originally constructed to respond rapidly to real-time GRB triggers provided by the BATSE which had a 1-σ error of 2~10°. It utilizes commercially available Canon f/1.8 with a 200 mm focal length lenses. With a 2048 x 2048 pixel 15 x 15 µm/pixel CCD array, one camera images 8.8 x 8.8° field-of-view. The mount can point any part of the sky within 5 s after it receives a trigger. We have a 2 x 2 array of such cameras originally configured to have 17.4 x 17.4° field-of-view for BATSE triggers. Results from LOTIS have been published[5]. Recently we collapsed the angular offsets of this system to have all 4 cameras view the same patch of the sky but each with a different astronomical filter, R, V, B or clear filter, for simultaneous color imaging of a GRB counterpart. Its sensitivity is mag~13 to 15 depending on the integration times, weather and filter types. LOTIS is located at a LLNL's test facility, 25 miles east of Livermore, California. The system is fully automated to respond the HETE-2 triggers and is operating every night when the weather permits.

LITE

The LITE (Livermore Infrared Transient Experiment) is our planned upgrade system for the Super-LOTIS. Measured GRB distance scales, recent conjecture of their star-forming environment and theoretical arguments of their high red-shifts[6] suggest that the optical signal may be obscured compared to the infrared signals. In addition, all the future planned GRB missions lack coverage of the infrared signals from a GRB. In order to fill this gap, we plan to upgrade the Super-LOTIS by replacing the current optical CCD camera with an infrared camera. We will install a f/2 secondary mirror in place of the current coma corrector and a HgCdTe 512 x 512 pixel camera for the focal plane array with a filter wheel equipped with J, H and K filters. This system will have 6.7 x 6.7 arcmin field-of-view well-matched to the Swift error box of 4 arcmin. Its expected sensitivity and prompt response capability will be the same as the Super-LOTIS system.

SUMMARY

Table 1 summarizes the parameters for the automatic telescopes discussed in this paper. Prompt optical and infrared counterparts of the GRB will provide important clues to understanding of the GRB progenitors. With our current operating LOTIS and Super-LOTIS and planned LITE systems connected to the real-time GRB triggers from the space-borne GRB detectors, we will measure many early-time optical and infrared fluxes associated with the GRBs.

TABLE 1. LOTIS System Parameters.

Instrument	Super-LOTIS	LOTIS	LITE
Aperture Diameter	0.6 m	0.11 m	0.6 m
Optics Speed	f/3.5	f/1.8	f/8
Imaging Sensor	2048 x 2048 CCD	4 of 2048 x 2048 CCD	512 x 512 HgCdTe
Field-of-View	0.84 x 0.84°	8.8 x 8.8°	6.7 x 6.7 arcmin
Resolution	1.5 arcsec/pixel	15 arcsec/pixel	0.78 arcsec/pixel
Sensitivity	R~17 (5 s integration)	V~14 (10 s integration)	J~17
Filters	B, V, R, clear	B, V, R, clear	J, H, K
Response time	< 30 s	< 5 s	< 30 s

ACKNOWLEDGMENTS

This research is supported under NASA contract numbers S-03975G and S-57797F and under the auspices of the U.S. Department of Energy by University of California Lawrence Livermore National Laboratory under contract No. W-7405-Eng-48.

REFERENCES

1. Wijers, R. A. M J., et al., 1997, MNRAS **288**, L51.
2. Park, H., et al., *Explosive Phenomena in Astrophysical Compact Objects*, edited by H. Chang et al., AIP Conference Proceedings 556, New York: American Institute of Physics, 2000, p. 261.
3. Akerlof, C., et al., ApJ, 2000, **532**, L25.
4. Briggs, M., et al., ApJ, 2000, **524**, 82.
5. Williams, G., et al., *Gamma-Ray Bursts*, edited by M. Kippen et al., AIP Conference Proceedings 526, New York: American Institute of Physics, 2000, p. 250.
6. Lamb, D., et al., ApJ, 2000, **536**, 1.

Number Counts of GRBs in a Vacuum Dominated Universe

K. Strohmaier and T. Wickramasinghe

Department of Physics, The College of New Jersey, Ewing, NJ 08628

Abstract. It is known that the population of GRBs is consistent with a standard cosmological model with $\Omega = 1$. However, recent theoretical and observational studies indicate that the universe could be vacuum dominated. We develop an analytical algorithm, beginning with the luminosity distance, D_L, to analyze bursts in a vacuum dominated universe, $\Omega = \Omega_v + \Omega_m = 1$. Using that result, we then show that the BATSE sample of GRBs is consistent with a universe where $\Omega_v \approx 0.6$.

INTRODUCTION

Our present understanding is that the universe is dominated by vacuum [3]. This in turn means that most of the energy of the universe is in free space. The energy density of this free space known as the vacuum is calculated to be $\rho_{vac} = \lambda c^2/8\pi G$, where λ is the cosmological constant, G the gravitational constant, and c the speed of light [2]. The critical density of the universe is given by $\rho_{cr} = 3H_0^2/8\pi G$ where H_0 is the Hubble constant, the rate of expansion of the universe. Present observations show that $\Omega_v = \rho_{vac}/\rho_{cr} \approx 0.6$; that is, more than 60 percent of mass of the universe is in the form of free space. Already, several projects are under way to measure this energy density using the concept of the luminosity distance [4].

The luminosity distance of an object having a luminosity L is given by $D_L^2 = L/4\pi F_0$ where F_0 is the observed flux of the source [2]. Thus D_L contains information about the nature of the global spacetime of the universe itself. The traditional method to calculate D_L is to implement a series of numerical integrations.

We show that a very efficient analytical treatment can be established to carry out the calculation of D_L. We find that our method is exceedingly efficient and easy to implement and has a negligible error, which even decreases with the distance. In turn, after calculating D_L, we then apply our algorithm to the GRB sample collected by the BATSE satellite. In doing so, we use the previously derived results for D_L and discover that a value of $\Omega_v \approx 0.6$ best fits the observational data.

THEORY

Current observations show that the universe is spatially flat; that is, $\rho_{cr} = \rho_0$ or $\Omega_0 = 1$, where ρ_0 is the present mass density of the universe. Thus, $\rho_0 = \rho_{vac} + \rho_m$, where ρ_m is the matter density. Therefore the observed mass density parameter Ω_0 is given by

$$\Omega_0 = \frac{\rho_0}{\rho_{cr}} = \frac{\rho_{vac}}{\rho_{cr}} + \frac{\rho_m}{\rho_{cr}} = \Omega_v + \Omega_m = 1 \tag{1}$$

At any cosmic epoch t, the distance between any two points in the universe is proportional to the scale factor $S(t)$. In a vacuum-dominated universe, this is given by [2]

$$\left(\frac{dS}{dt}\right)^2 = H_0^2 \Omega_v S^2 + H_0^2 \Omega_m \frac{S_0^3}{S} \tag{2}$$

where $S_0 = S(t_0)$, where the present cosmic epoch $= t_0$. Letting $A = H_0^2 \Omega_v$ and $B = H_0^2 \Omega_m$, we write the foregoing equation as

$$\left(\frac{dS}{dt}\right)^2 = AS^2 + \frac{B}{S} \tag{3}$$

Assuming a big bang model $S \to 0$ and $t \to 0$, the above equation above is readily integrated to

$$t = \int_{S=0}^{S} \frac{dS\sqrt{S}}{\sqrt{AS^3 + B}} \tag{4}$$

Thus,

$$AS^3 = \frac{B}{2}[\cosh(3t\sqrt{A}) - 1] \tag{5}$$

Substituting for A and B, we get

$$\left(\frac{S}{S_0}\right)^3 = \frac{1}{2}\frac{\Omega_m}{\Omega_v}[\cosh(3H_0 t\sqrt{\Omega_v}) - 1] \tag{6}$$

The spacetime of the universe is given by the Robertson-Walker metric [5]

$$ds^2 = c^2 dt^2 - S^2\left[\frac{dr^2}{1 - kr^2} + r^2(d\theta^2 + (\sin\theta d\phi^2)^2)\right] \tag{7}$$

where $k = 0$ represents a flat universe and (r, θ, ϕ) are the comoving coordinates of a point in space. Considering a point at r and taking our vantage to be at $r = 0$, we get from Eq. (4)

$$r = \int_0^r dr = -\int_{t_0}^t \frac{cdt}{S} \tag{8}$$

Eqs. (6) and (8) yield

$$rS_0 = \frac{c}{H_0} \frac{1}{3\Omega_v^{1/6}\Omega_m^{1/3}} \int_x^{x_0} \frac{dx}{[\sinh \frac{x}{2}]^{3/2}} \qquad (9)$$

where $x = x(t) = 3\sqrt{\Omega_v}H_0 t$, $x_0 = x(0)$ and $x_0 > 0$. Letting

$$\Psi(x) = \int_\epsilon^x \frac{dx}{[\sinh \frac{x}{2}]^{3/2}} \qquad (10)$$

where the constant $\epsilon \approx 0$. The luminosity distance $D_L = rS_0(1+z)$ is therefore given by

$$D_L = \frac{c}{H_0} \frac{1+z}{3\Omega_v^{1/6}\Omega_m^{1/3}} [\Psi(x_0) - \Psi(x)] \qquad (11)$$

Eq. (10) can be evaluated analytically [1]. A series expansion yields

$$\Psi(x) = a + (3)2^{2/3}x^{1/3}\left[1 - \frac{x^2}{252} + \frac{x^4}{21060}\right] \qquad (12)$$

where a is a constant. Let us now define a new function T as

$$T(x) = x^{1/3}\left[1 - \frac{x^2}{252} + \frac{x^4}{21060}\right] \qquad (13)$$

Then,

$$\frac{\Psi(x_0) - \Psi(x)}{3 \, 2^{2/3}} = T(x_0) - T(x) \qquad (14)$$

The redshift z is related to S via $1+z = S_0/S$. Now, Eq. (6) with $x = 3H_0 t\sqrt{\Omega_v}$ gives

$$x = \cosh^{-1}\left[2\frac{\Omega_v}{1-\Omega_v}\frac{1}{(1+z)^3} + 1\right] \qquad (15)$$

Combining Eqs. (1), (11), (13), and (15) we obtain the luminosity distance to be

$$D_L = 2^{2/3}\frac{c}{H_0}\frac{1+z}{\Omega_v^{1/6}(1-\Omega_v)^{1/3}}[T(x_0) - T(x)] \qquad (16)$$

where $\Omega_v + \Omega_m = 1$ as previously taken.

Once we have the foregoing formulas to calculate the luminosity distance, as well as the formulas for $T(x)$ and $x(z)$, we are able to apply these formulas to the BATSE sample. Our calculations focused on the $C_{\max}/C_{\min}$ values for each data point. Therefore, they can be understood to be a measure of the relative strength of each burst.

Our first step was plotting the BATSE 3B Catalog samples, as our method involved calculating and plotting $\log(C_{max}/C_{min})$ versus $\log(N/N_\infty)$, where N is the number count of GRBs at the corresponding value of $\log(C_{max}/C_{min})$. However, we first had to remove from the sample any data points that were incomplete, as some had only one of the two values we needed. In addition, we also used points that had a signal to noise ratio about a certain value.

What is known as the brightness statistic distribution, a plot of $\log(N/N_\infty)$ against $\log(C_{max}/C_{min})$, was calculated according the following equations.

$$\frac{N}{N_\infty} = \frac{15}{\sqrt{\Omega_v(1-\Omega_v)}} \int_z^0 [T(x_0) - T(x)]^2 \frac{dT(x)}{dz} \frac{dz}{1+z}$$

$$\frac{C_{max}}{C_{min}} = \Theta \frac{(1+z)^{2-\alpha}}{D_L^2}$$

As can be readily seen, this depends on both the redshift z and the value of Ω_v, as well as the parameters α (the spectral index), N_∞ and the normalization, Θ. These two parameters are constants determining where the right normalization should be obtained.

We were able to fit the theoretical models with the data given from the BATSE sample. Obviously, the best fit comes with $\Omega_v = 0.6$ and $z = 3$, which is shown in Figure (1), for $\alpha = 1$. From the plot of the Einstein De Sitter universe we found that there is agreement at the faint end of the sample but not at the strong end.

Finally, in order to test the accuracy of our model, we first took the best-fit curve (see Fig. 1) and calculated that approximately 84% of the data points agreed with the theory. We also applied the Kolmogrov-Smirnov test to our model. Doing this showed that our model has around a 92 percent confidence level.

RESULTS AND DISCUSSION

Of course, the analytical result in Eq. (16) is an excellent approximation to the more exact numerical result. The percentage error in the analytical result can be expressed as $100(\gamma - 1)$, where

$$\gamma(\Omega_v, z) = \frac{3 \, 2^{\frac{2}{3}} [T(x_0) - T(x)]}{\int_x^{x_0} \frac{dt}{[\sinh \frac{t}{2}]^{2/3}}} \tag{17}$$

The relative error in the analytical expression slightly increases with Ω_v. However, the maximum value for Ω_v is constrained to about 0.6 from other astronomical observations (as well as our own work). Thus, the percentage error in our method is much less than 0.1 percent at any practically significant redshift.

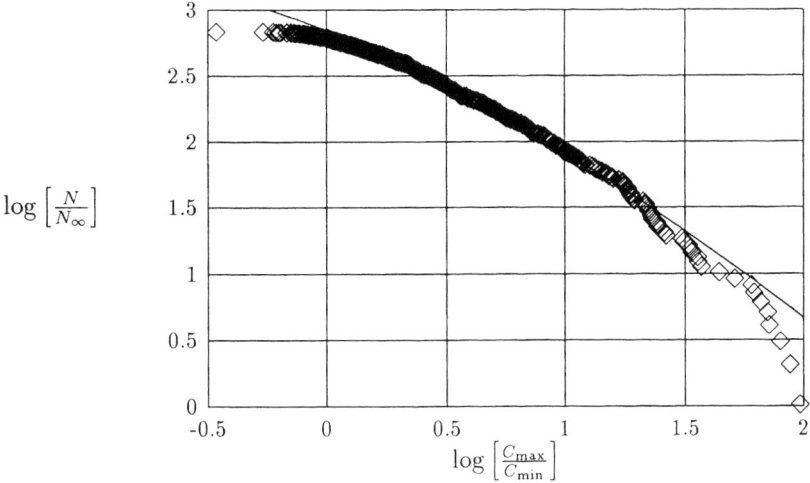

Figure 1: This is the plot of $\log(N/N_\infty)$ vs $\log[C_{max}/C_{min}]$

The analytical result becomes indistinguishable from the numerical result at any redshift larger than unity giving an extremely efficient method to evaluate D_L.

The best fit yields that $\Omega_v \approx 0.6$. This is an excellent agreement with with the observational data available, as well as the results from other investigations. Our model, when challenged with the Kolmogrov-Smirnov test, shows a 92 percent confidence level. Finally, our algorithm has shown that GRBs are not occurring with any substantial frequency beyond $z = 3$. However, the parameter space to which an acceptable fit could be obtained is fairly large. This is why in this preliminary analysis we can only claim that the distribution of GRBs is only consistent with a vacuum dominated universe having $\Omega_v \approx 0.6$.

REFERENCE

1. Arfken G. 1985, Mathematical Methods for Physicists, Academic Press, 748.

2. Narlikar, J.V. 1983, Introduction to Cosmology, Cambridge.

3. Ostriker, J.P., Steinhardt, P.J. 1995, Nature, 377, 600.

4. Perlmutter, S., et al. 1997, ApJ, 483, 565.

5. Peacock, J.A. 1999 Cosmological Physics, Cambridge, 69.

Modeling and Implications of the Transient Absorption Feature in GRB 990705

Markus Böttcher*[1], Charles D. Dermer[†], Lorenzo Amati[‡], and Filippo Frontera[§]

*Physics and Astronomy Department, Rice University, Houston, TX
[†]Naval Research Laboratory, Washington, D.C.
[‡]ITESRE, CNR, Bologna, Italy
[§]Dipartimento di Fisica, Università di Ferrara, Ferrara, Italy

Abstract. A transient Fe absorption edge during the prompt phase of a GRB has recently been observed for the first time. This has been interpreted as evidence for photoionization of the GRB environment by the burst radiation, which allows diagnostics of the density structure and element abundances in the vicinity of the burst. Assuming that the observed absorption feature is caused by photoelectric absorption, we model the time-dependent photoionization and X-ray radiation transport, and deduce constraints on the size and matter distribution of the photoionized region responsible for the absorption edge. If the density of the region is $\lesssim 10^{10}$ cm^{-3}, then we find that the intervening material would have to contain $\sim 44\,\Omega\,M_\odot$ of iron within ~ 1.3 pc of the burst source, assuming the measured best-fit ~ 75-fold overabundance of iron. Alternatively, and more plausibly, the observed absorption feature could be caused by photoelectric absorption in dense clouds of radius $r_c \lesssim 10^{12}$ cm at $\sim 10^{17}$ cm from the burst source, containing $\lesssim 0.7\,\Omega\,M_\odot$ of iron. In such an environment, recombination would compete with photoionization, until the clouds are effectively heated to the Compton equilibrium temperature of the ionizing GRB continuum. We also briefly discuss the recent suggestion that the absorption feature may be due to resonant scattering by Fe XXVI in a highly ionized, clumpy high-velocity outflow.

INTRODUCTION

Recently, Amati et al. (2000) have reported the detection of a transient, redshifted Fe K edge in the prompt X-ray emission from GRB 990705. This is the second time (after GRB 980329; Frontera et al. 2000) that X-ray spectroscopy during the prompt phase of a GRB revealed evidence for excess absorption above the Galactic hydrogen column, and the first time that evidence for a time-dependence

---

[1] Chandra Fellow

of such absorption has been found. The transient nature of the absorption feature has been interpreted as unmistakable evidence for photoionization of the absorber by the prompt burst emission (Amati et al. 2000, Böttcher et al. 2001). The corresponding time-dependent photoionization and radiation transfer problem had been simulated before by Böttcher et al. (1999) who predicted that excess absorption features are either transient on time scales of $\lesssim 1$ minute, or do not change appreciably as the burst evolves, depending primarily on the average distance of the absorbing material from the GRB source.

In this paper, we review the results of detailed modeling of the time-dependent photoionization and radiation transfer problem in quasi-isotropic, intermediate-density media obtained in Böttcher et al. (2001) and in small, dense clumps in which photoionization is balanced by rapid recombination. We also discuss the possibility that the absorption feature is a blue-shifted resonance scattering feature in an inhomogeneous high-velocity outflow, as recently suggested by Lazzati et al. (2001). The implications of these results for GRB progenitor models will be discussed briefly.

PHOTOELECTRIC ABSORPTION IN AN INTERMEDIATE-DENSITY HOMOGENEOUS SHELL

The most straightforward interpretation of the transient absorption feature would be photoelectric absorption in a homogeneous cloud or an isotropic shell of intermediate density, in which recombination is slow compared to photoionization. In this case, we model the absorbing material as a uniform shell with inner radius $r_{\rm in}$, density $n_{\rm sh}$ [cm^{-3}], and radial column density N_H [cm^{-2}], which also determines the thickness of the shell. The element abundances are fixed to the values found through spectral analysis of the BeppoSAX WFC spectrum of GRB 990705 by Amati et al. (2000), i.e. standard solar-system abundances for all elements except iron for which a 75-fold overabundance of was deduced. The material in the shell is assumed to be neutral before the onset of the burst. We are using the code developed in Böttcher et al. (1999) to simulate the time-dependent photoionization and radiation transfer problem in the CBM. The modeling is done by varying the initial column density N_H, the radius $r_{\rm in}$, and the density $n_{\rm sh}$.

The prompt burst emission from GRB 990705 can be reasonably well represented by a FRED-type single-pulse burst. We can thus describe the GRB spectral evolution by the parametrization of Dermer et al. (1999), based on the external shock model for GRBs. GRB 990705 had a γ-ray duration of $t_d \sim 40$ s and a 2 – 700 keV fluence of $(9.3 \pm 0.2) \times 10^{-5}$ erg cm^{-2}. At a redshift of $z = 0.86$, as determined from the transient Fe absorption edge, this corresponds to an isotropic energy output in X- and γ-rays of $E_{X\gamma} \approx 2.2 \times 10^{53}$ ergs. During the early phase of the GRB, the 2 – 700 keV continuum was consistent with a power-law of photon index $\Gamma \approx 1.1$ (i.e. νF_ν spectral index $v = 0.9$). No evidence for a high-energy spectral break or turnover in the early 2 – 700 keV spectrum was found, implying that the νF_ν peak

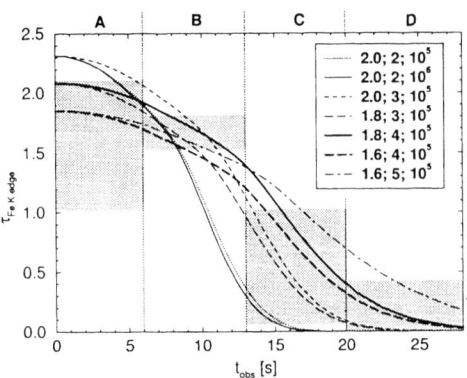

FIGURE 1. Simulated time evolution of the Fe K edge in GRB 990705 for various combinations of column density, N_H, radius, r_{in}, and density, n_{sh}, of the absorbing material. The shaded regions show the values (with 1 σ error) allowed by the measurements of Amati et al. (2000). The curves show the model results, using the parameters quoted in Section 3 and in the legend, where the first number is N_H [10^{22} cm^{-2}], the second number is r_{in} [10^{18} cm], and the third number is n_{sh} [cm^{-3}].

energy is beyond 700 keV. For our modeling, we assume $\epsilon_0 \equiv E_{pk}/(m_e c^2) = 1.5$. All the above observables can be reproduced in the framework of the Dermer et al. (1999) parametrization using $E_0 = 10^{54}$ erg, $\Gamma_0 = 440\, n_0^{-1/8}$, $v = 0.9$, $\delta = 0.5$, $g = 1.8$, $q = 2.5 \cdot 10^{-3}$.

Fig. 1 shows sample results of some of our simulations. We find that the measurements place rather tight constraints on the three fitting values N_H, r_{in}, and n_{sh}. Good agreement is achieved for $N_H = 1.8 \times 10^{22}$ cm^{-2}, $r_{in} = 1.3$ pc, $n_{sh} = 10^5$ cm^{-3}. These values are in agreement with a simple analytical estimate based on the assumption that the decay time scale of the absorption feature, $t_{abs} \sim 10$ s corresponds to the time scale t_{ion} of photoionization of the CBM by the prompt GRB radiation, which yields $t_{ion} = 36/(\Phi_0 E_{thr} \sigma_0)$, where $\Phi_0 \approx (1.2 \times 10^{56}/r_{in}[cm]^2)$ photons cm^{-2} s^{-1} keV^{-1} is the illuminating photon flux at $E = m_e c^2$, $E_{thr} \sim 8$ keV is an average value of the ionization threshold and $\sigma_0 \sim 3.5 \times 10^{-20}$ cm^2 is an average value of the photoelectric absorption cross section at threshold for the various ionization stages of iron. $t_{ion} \sim 10$ s is achieved for $d \sim 1$ pc.

This result implies that the absorber would have to contain a total mass of $M_{CBM} = 365\, \Omega\, M_\odot$, where Ω is the solid angle subtended by the absorber. The amount of iron contained in this cloud would be $M_{Fe} = 44\, \Omega\, M_\odot$. Even if the cloud were strongly anisotropic, $\Omega \lesssim 1$, it appears very unlikely that such an extreme mass and concentration of iron could be produced in the context of any reasonable GRB progenitor model.

PHOTOELECTRIC ABSORPTION IN DENSE CLOUDS

In a medium of moderate density, where recombination is inefficient, each iron atom absorbs ~ 12 ionizing photons (taking into account Auger ionization) before being completely ionized. The amount of iron required to reproduce the observed absorption feature in terms of photoelectric absorption could be reduced if the material is so dense that the recombination time scale is of the order of the ionization time scale, so that each iron atom absorbs more than ~ 12 ionizing photons. To estimate the conditions under which this is the case, we assume that a fraction ζ of the CBM is contained in these clouds, i.e. the total mass in the clouds is $M_c = \zeta M_{\rm ej}$, and we write $M_{\rm ej} = m_{\rm ej} M_\odot$. These clouds are covering a fraction $a_c \sim 1$ of the sky as seen from the burst source. The radius of the clouds is $r_c = 10^{10} r_{10}$ cm, and their distance from the central source is $x = 10^{17} x_{17}$ cm. The cloudy ejecta subtend a solid angle Ω_s around the GRB source. The recombination timescale of fully ionized hydrogen may be estimated as $t_{\rm rec,Fe}^{-1} \sim 2.13 \times 10^3 \, \zeta \, m_{\rm ej}/(T_4^{3/4} x_{17}^2 r_{10} a_c)$ s^{-1}, where T_4 is the electron temperature in units of 10^4 K.

To estimate the photoionization time scale, we assume that during the first ~ 10 s the GRB spectrum is approximately constant in flux and spectral shape. The average spectrum of GRB 990705 during this time interval, at the redshift of the burst, is well characterized by $\Phi(E) = 4 \times 10^{22} x_{17}^{-2} h_{17}^{-2} E_{\rm keV}^{-1.1}$ photons cm^{-2} s^{-1} keV^{-1}, where $E_{\rm keV}$ is the photon energy (in the GRB rest frame) in units of keV. This results in a photoionization time scale for iron of $t_{\rm pi,Fe}^{-1} \sim 97 \, x_{17}^{-2} h_{70}^{-2}$ s^{-1}.

Furthermore, we require that the column density of iron through the cloud layer is $N_{\rm Fe} \sim 2.2 \times 10^{20}$ cm^{-2}, as indicated by the depth of the observed absorption feature. Finally, we assume that the decline of the absorption depth is due to ongoing Compton heating of the electrons in the cloud (note that the recombination rates scales as $t_{\rm rec}^{-1} \propto T_e^{-3/4}$), and estimate the Compton heating time as $t_{\rm heat}^{-1} \sim 0.52 \, x_{17}^{-2} h_{70}^{-2} T_4^{-0.1}$ s^{-1}.

Now, setting $t_{\rm heat} = 10$ s, and $t_{\rm rec,Fe} = t_{\rm ion,Fe}$, we find the following physical conditions for the material in the clouds: $x \sim 2 \times 10^{17} h_{70}^{-1} T_4^{-0.05}$ cm, $r_c \sim 8 \times 10^{11} \zeta/([\zeta_{\rm Fe}/75] T_4^{0.85} a_c)$ cm, $n_c \sim 8 \times 10^{10} T_4^{0.85}$ cm^{-3}, $M_{\rm ej} \sim 3.5 \, \Omega_s h_{70}^{-2} T_4^{-0.1}/(\zeta_{\rm Fe}/75) \, M_\odot$, $M_{\rm Fe} \sim 0.7 \, \Omega_s h_{70}^{-2} T_4^{-0.1} \, M_\odot$. Here, $\zeta_{\rm Fe}$ is the overabundance of iron w.r.t. standard solar-system abundances. As a consistency check, we require that the Thomson depth through ejecta be $\tau_{\rm T} < 1$ in order to avoid a very low transmission efficiency and smearing of the millisecond variability in GRB 990705. We find: $\tau_{\rm T} \sim \sigma_{\rm T} r_c n_c a_c \sim 4.3 \times 10^{-2} \zeta/(\zeta_{\rm Fe}/75)$.

RESONANCE SCATTERING BY FE XXVI

Lazzati et al. (2001) have recently suggested that the transient absorption feature in GRB 990705 could be due to resonant scattering by Fe XXVI in highly ionized clumps in a high-velocity outflow from the burst source. In order to reproduce the observed width of the absorption feature, an outflow velocity of $v_0 \sim 4 \cdot 10^9$ cm s^{-1}, and a velocity dispersion of $\Delta v \sim v_0$ was required in order to produce the observed energy and width of the absorption feature (recall that resonance scattering in a stationary medium would produce a very narrow absorption line).

For this scenario, it is required that the material is highly ionized, but a significant fraction of iron remains in the Fe XXVI state through balance between photoionization and recombination, and that the transient nature of the iron line is due to Compton heating of the plasma. From these requirements, Lazzati et al. (2001) infer the following CBM parameters: $x \sim 2.6 \times 10^{16} L_{51}^{1/2}$ cm, $r_c \sim 2 \times 10^{13} \tau_T^2 (\zeta_{Fe}/10)$ cm, $n_c \sim 8.3 \times 10^{10} T_4^{5/3} \tau_T^{-1} (\zeta_{Fe}/10)^{-1}$ cm^{-3}, $M_{ej} \sim 6.1\, M_\odot$, $M_{Fe} \sim 0.16\, M_\odot$, where L_{51} is the burst luminosity in units of 10^{51} ergs s^{-1} and the Thomson depth through the CBM is restricted to $\tau_T \leq 1$.

SUMMARY AND DISCUSSION

We have investigated different scenarios for the production of the transient absorption feature seen in GRB 990705. We found that an unreasonable amount of iron in the circumburster material (CBM) would be required if the transient nature of the absorption feature was due to photoionization, not effectively balanced by recombination. More plausible CBM parameters are found assuming that the absorption is photoelectric absorption in dense clouds with covering factor ~ 1 around the burst source, although rather extreme clumping ($r_c/x \sim 4 \cdot 10^{-6}$) is required.

As an alternative to photoelectric absorption, Lazzati et al. (2001) have suggested resonant scattering by Fe XXVI in a high-velocity outflow. This model requires less extreme clumping than the photoelectric-absorption scenario only if the Thomson depth through the CBM is assumed to be $\tau_T \sim 1$. Also it appears to require a rather extreme mean outflow velocity of $v \sim 0.13$ c, which also implies a rather huge kinetic energy in the ejecta of $\sim 10^{53} \Omega$ ergs.

Both versions of the clumpy-CBM scenario point towards progenitor models in which the GRB is preceded by a supernova, which ejects several solar masses of highly iron-enriched material into the environment. If photoelectric absorption is the principal absorption mechanism, then the delay between supernova and GRB can be estimated to $\Delta t \sim$ several years, while in the case of resonant scattering by Fe XXVI, $\Delta t \sim$ a few months.

REFERENCES

1. Amati, L., et al., *Science* **290**, 953 (2000).
2. Böttcher, M., et al., *A&A*, **343**, 111 (1999).
3. Böttcher, M., et al., proc. of "Gamma-Ray Bursts in the Afterglow Era II", in press (2001).
4. Dermer, C. D., et al., *ApJ*, **513**, 656 (1999).
5. Frontera, F., et al., *ApJS*, **127**, 59 (2000).
6. Lazzati, D., et al., *ApJ*, in press (2001).

Continuously-fed Fireballs and Signatures in Gamma-ray Burst Afterglows

Bing Zhang* and Peter Mészáros*

Pennsylvania State University, 525 Davey Lab, University Park

Abstract. In some types of the gamma-ray burst central engines, a significant energy input into the fireballs may in principle continue for a longer time scale and pose important impacts on afterglow emissions. If the injection is Poynting-flux-dominated rather than kinetic-energy-dominated, the accumulation of the fireball energy may change the global blastwave dynamics, which brings an achromatic bumping signature in the afterglow lightcurves. We discuss the condition of such a signature. A millisecond magnetar pulsar as the central engine is discussed in detail as an example.

INTRODUCTION

Gamma-ray burst (GRB) afterglows set up when the GRB fireball starts to decelerate due to the interaction with the interstellar medium (ISM). The blastwave deceleration dynamics is well quantified provided the density profile of the ambient medium (a constant medium or a stellar wind) and the beaming geometry. For example, for an isotropic fireball running into a constant density medium, the bulk Lorentz factor decays as $\Gamma \propto R^{-3/2}$, where R is the radius of the fireball in the fixed frame. This simple decay manner, when coupled with shock acceleration and electron radiation physics, generally well delineates the broadband afterglow lightcurves observed in the hitherto known GRBs. However, in some types of the gamma-ray burst central engines, a significant energy input into the fireballs may in principle continue for a longer time scale (e.g. [1] [2] [3]). The injection energy, if in the form of Poynting-flux-dominated flow, may be damped into the fireball and the total energy budget of the fireball will be boosted. Under certain conditions, the global fireball dynamics will be modified due to the injection (see [3] for a detailed treatment). If the injection is rather kinetic-energy-dominated, an additional pair of shocks may, under certain conditions, form at the discontinuity of the leading shell and the trailing shell. The global blastwave dynamics may not be altered, but the shell merging process will bring more complicated signatures on afterglow lightcurves [4]. Here we only focus on the case that the injection is dominated by the Poynting flux. A practical example is that the central engine is a highly-magnetized millisecond pulsar (e.g. [5] [6] [7] [8]). The rotation energy loss of a pulsar can in principle act as a continuous energy injection component. In the magnetar central engine models, some other emission components, such as the dissipation energy of the toroidal magnetic fields (e.g. [7][1]), are invoked to interpret the "impulsive" energy released during the GRB prompt phase (although the rotation energy itself may power the GRB in some models [5]).

CONTINUOUS-INJECTION DYNAMICS

Consider an adiabatic relativistic shell which is collecting materials when expanding into the ISM and which is receiving energy input from the central engine, the differential energy conservation equation in the fixed frame is $d[\Gamma(M_0c^2 + M_{ism}c^2 + U)] = dM_{ism}c^2 + dE_{inj}$, where M_0, and M_{ism} are the masses of the impulsive shell and the swept-up ISM, $U = (\Gamma - 1)M_{ism}c^2$ is the total internal energy of the blastwave, E_{inj} is the received injection energy, and Γ is the bulk Lorentz factor of the blastwave. The integrated energy conservation equation then reads [4]

$$E_{inj} + (\Gamma_0 - \Gamma)M_0c^2 = (\Gamma^2 - 1)M_{ism}c^2. \qquad (1)$$

When E_{inj} is negligible, eq. (1) is the standard equation for an impulsive blastwave evolution [9], which could be reduced to the familiar form of $E_{imp} = (4\pi/3)R^3 nm_pc^2\Gamma^2$ (so that $\Gamma \propto R^{-3/2}$) when $\Gamma_0 \gg \Gamma \gg 1$, where $E_{imp} = \Gamma_0 M_0 c^2$. With the presence of E_{inj}, the blastwave dynamics will change when E_{inj} exceed $(\Gamma_0 - \Gamma)M_0c^2$, or essentially E_{imp}.

Assuming that the intrinsic luminosity of the central engine has a constant temporal index, i.e., $dE_{inj}/dT \propto T^q$ (here T is also the observer's time since the central engine is at rest with respect to the observer), E_{inj} will have a temporal index of $q+1$, so that $q > -1$ is essential for the term E_{inj} to dominate the term E_{imp}. Notice that in terms of fixed frame $t = R/c$, the luminosity law has a different index [3] $q' = q(m+1) + m$, since Γ generally is a function of t, and $dt = 2\Gamma^2 dT$. The equivalent essential condition of changing the dynamics is of the same form, i.e., $q' > -1$. When the blastwave dynamics is dominated by the injection term, the bulk Lorentz factor will then decay as

$$\Gamma \propto R^{-(2-q)/2(2+q)} \propto T^{-(2-q)/8}, \quad R \propto T^{(2+q)/4}. \qquad (2)$$

when $\Gamma_0 \gg \Gamma \gg 1$.

THREE TIME SCALES AND THE INJECTION SIGNATURE

There are three relevant time scales in the continuous-injection problem. These are:

The blastwave deceleration time, T_0: the onset time of the GRB afterglow. For an isotropic fireball running into a constant density ISM, this is

$$T_0 = 2.4 \text{ s } \mathcal{E}_{52}^{1/3} n^{-1/3} (\Gamma_0/300)^{-8/3}. \qquad (3)$$

The critical time, T_c: the time when the continuous-injection energy component dominates the impulsive-injection energy component. For an adiabatic blastwave, assuming a luminosity law of $L(T) = L_0(T/T_0)^q$, this is

$$T_c = \left[(q+1)\left(\frac{E_{imp}}{L_0 T_0}\right)\right]^{1/(q+1)} T_0, \qquad (4)$$

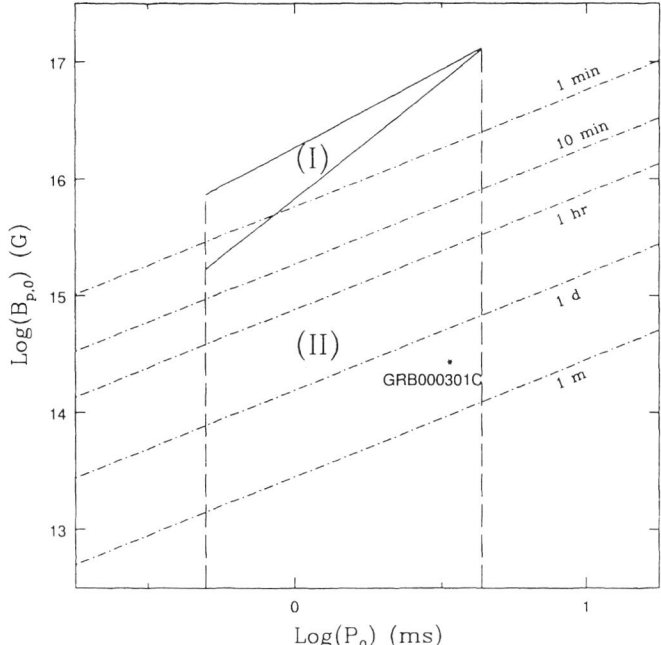

FIGURE 1. A $B_p - P_0$ diagram for the initial parameters of a pulsar born in a GRB. The enclosed areas are the phase spaces where an achromatic pulsar signature is expected in the GRB afterglow lightcurves. Regime I and II are for immediate and delayed injection, respectively. Dotted-dashed lines denote various $\mathcal{T}_{em}$ expected. $E_{imp} = 10^{51} \mathrm{erg\ s^{-1}}$, $\Gamma_0 = 300$, $n = 1 \mathrm{cm}^{-3}$, and $P_0(\min) = 0.5$ ms have been adopted (from [3]).

The characteristic time, $\mathcal{T}$: the time scale when the central engine has an injection index $q > -1$. Both $\mathcal{T}$ and q depend on the properties of the central engine.

The condition for the injection signature to show up in the afterglow lightcurves:

$$\mathcal{T} > \max(T_c, T_0). \tag{5}$$

There are two sub-cases:

1. $T_c < T_0 < \mathcal{T}$, the dynamics is defined by the continuous-injection as soon as the afterglow is set up. The lightcurves will be flat from T_0 to $\mathcal{T}$, and will steepen after $\mathcal{T}$. We call this case as an *immediate injection* (regime I in Fig.1).
2. $T_0 < T_c < \mathcal{T}$, the signature will not show up until T_c. So there are two temporal index changes: the lightcurves will flatten around T_c, and resume the original steepness around $\mathcal{T}$. We call this case as a *delayed injection* (regime II in Fig.1).

The injection-signature is **achromatic** due to the change of the blastwave dynamics. Assuming synchrotron emission from the forward shock, the temporal index at the

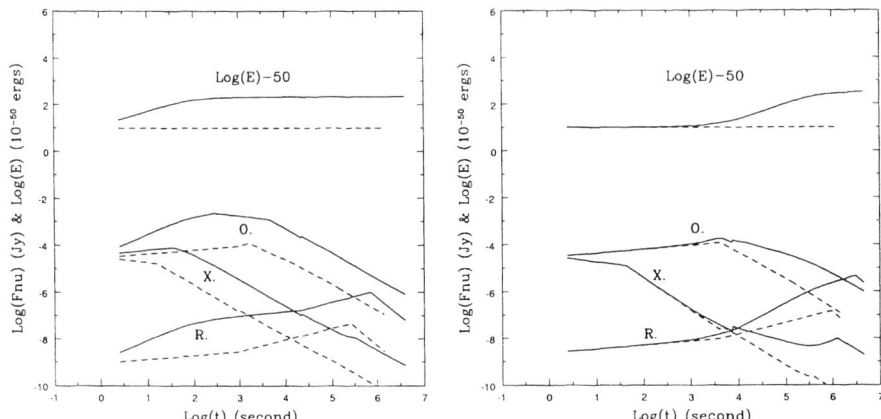

FIGURE 2. Two typical pulsar injection cases. Solid and dashed lines are the cases with or without a continuous energy injection. Top curves indicate the total energy (in unit 10^{50} ergs) contained in the fireball; and the bottom curves are the lightcurves in the X-ray, optical, and radio, respectively. Parameters adopted: $E_{imp,51} = 1$, $n = 1$ cm^{-3}, $\Gamma_0 = 300$, $\varepsilon_e = 0.5$, $\varepsilon_B = 0.01$, and $p = 2.2$. The pulsar parameters are: (1) immediate injection: $B_{p,15} = 10$ and $P_{-3} = 1$; (2) delayed injection: $B_{p,15} = 10$ and $P_{-3} = 0.8$.

injection phase is $\alpha = (1 - q/2)\beta + 1 + q$, as compared with the standard case of $\alpha = 3/2\beta$, where β is the spectral index at a certain band of interest.

HIGHLY MAGNETIZED MILLISECOND PULSAR AS THE CENTRAL ENGINE – AN EXAMPLE

If the central engine is a millisecond magnetar pulsar, after the prompt phase which powers the GRB, a residual dipolar field can lie in a wide range (e.g. [5][7][1]). Depending on the initial pulsar parameters, P_0 (initial period) and B_d (dipole field strength at the surface), the injection may or may not bring signatures in the afterglow lightcurves. When it does, the injection signature could be either "immediate" or "delayed".

During the injection phase, a pulsar engine will have $q = 0$, and the characteristic time scale is

$$T_{em} = \frac{3c^3 I}{B_p^2 R^6 \Omega_0^2} = 2.05 \times 10^3 \text{ s } I_{45} B_{p,15}^{-2} P_{0,-3}^2 R_6^6, \qquad (6)$$

Various conditions for different injection cases can be then worked out using (5) and other conditions. Figure 1 presents the parameter spaces for the injection-dominated regime.

The temporal index at the injection phase is simply $(\beta + 1)$ for the pulsar case. Figure 2 presents the indicative lightcurves in various bands for the immediate and delayed injection, respectively.

ACKNOWLEDGMENTS

This work is supported by NASA NAG5-9192 and NAG5-9153.

REFERENCES

1. Dai, Z. G., and Lu, T., *Phys. Rev. Lett.*, **81**, 4301–4304 (1998).
2. Rees, M. J., and Mészáros, P., *ApJ*, **545**, L73–L76 (2000).
3. Zhang, B., and Mészáros, P., *ApJ*, **552**, L35–L38 (2001).
4. Zhang, B., and Mészáros, P., p. in preparation (2001).
5. Usov, V. V., *Nature*, **357**, 472–474 (1992).
6. Blackman, E. G., and Yi, I., *ApJ*, **498**, L31–L34 (1998).
7. Kluzniak, W., and Ruderman, M., *ApJ*, **505**, L113–L116 (1998).
8. Wheeler, J. C., Yi, I., Höflich, P., and Wang, L., *ApJ*, **537**, 810–823 (2000).
9. Huang, Y. F., Dai, Z. G., and Lu, T., *MNRAS*, **309**, 513–516 (1999).

Trans-Relativistic Supernovae, Circumstellar Gamma-Ray Bursts, and Supernova 1998bw

Christopher D. Matzner*, Jonathan C. Tan† and Christopher F. McKee**

*CITA, 60 St George St., Toronto, ON M5S 3H8, Canada
†Astronomy Department, U.C. Berkeley, Berkeley, CA 94720 USA
**Depts. of Physics and Astronomy, U.C. Berkeley, Berkeley, CA 94720 USA

Abstract. Supernova (SN) 1998bw and gamma-ray burst (GRB) 980425 offer the first direct evidence that supernovae are the progenitors of some GRBs. However, this burst was unusually dim, smooth and soft compared to other bursts with known afterglows. Whether it should be considered a prototype for cosmological GRBs depends largely on whether the supernova explosion and burst were asymmetrical or can be modeled as spherical. We address this question by treating the acceleration of the supernova shock in the outermost layers of the stellar envelope, the transition to relativistic flow, and the subsequent expansion (and further acceleration) of the ejecta into the surrounding medium. We find that GRB 980425 could plausibly have been produced by a collision between the relativistic ejecta from SN 1998bw and the star's pre-supernova wind; the model requires no significant asymmetry. This event therefore belongs to a dim subclass of GRBs and is not a prototype for jet-like cosmological GRBs.

INTRODUCTION

A growing body of indirect evidence links some long duration gamma-ray bursts with recent star formation, and therefore with the core collapse of massive stars (e.g., [1]). The most direct evidence of such a link is provided by the probable association [2, 3] of GRB 980425 with SN 1998bw. However, at the distance of the supernova, this burst was 10^6 times dimmer than the brightest of cosmological bursts (10^{48} ergs [2], not 10^{54}, in γ-ray isotropic equivalent energy). Should this burst be considered the first of a new class of weak, supernova-related GRBs [4], or a cousin of strong cosmological bursts?

Central to this question is the degree of asymmetry needed to understand the supernova and its GRB. Evidence of large-scale asymmetry suggests a jet-like explosion of the core, which is considered a necessary ingredient for cosmological bursts if they involve internal shocks within high Lorentz factor flows [27] from the cores of stars [5]. If instead the event is consistent with spherical symmetry, then it should be inadmissible as evidence of a causal relation between SNe and any model for GRBs requiring a jet.

As a spherical explosion, SN 1998bw possessed about 3×10^{52} erg of kinetic energy, thirty times more than what is typical of supernovae. Höflich, Wheeler & Wang [6] argue that SN 1998bw may have been an asymmetric explosion on the basis that this would allow a lower explosion energy, and Nakamura et al. [7] find evidence for asymmetry of the inner ejecta in the late decay of the supernova light curve. Note, however, that the polarization of light from supernova 1993J suggested asymmetry of its inner ejecta [8], whereas radio emission from its outermost ejecta [9] shows no asymmetry; moreover, SN 1998bw had lower polarization than SN 1993J and most type II supernovae [10].

More compelling would be evidence that the observed GRB originated in a highly asymmetrical event. In this regard, [11], [12], and [13] advocate a scenario in which a beamed, highly relativistic outflow is viewed off-axis to produce GRB 980425.

The competing, more conservative hypothesis construes the burst as the earliest phase of interaction between high-velocity (spherical) stellar ejecta and progenitor star's wind – the same interaction that gave rise to the later radio emission [14]. This possibility is similar to the suggestion of Colgate [15] that GRBs might be due to shock breakout in supernovae. In this model, the energy that emerged as gamma rays was previously locked up in the kinetic energy of expanding ejecta. Even at the distance of 1998bw, a burst of GRB 980425's brightness probably required (mildly) relativistic motion in order to avoid excessive self-opacity. Corroborating evidence comes from the very high mean velocity inferred for the supernova's synchrotron shell ($c/3$ at 12 days; Kulkarni et al. [16]). To assess the viability of a spherical model for GRB 980425, we must:

1. estimate the minimum Lorentz factor needed to produce the GRB;
2. investigate whether a model for the supernova explosion that accounts for the optical emission can simultaneously produce sufficient kinetic energy in material above this Lorentz factor; and
3. determine whether the pre-supernova stellar wind was dense enough to convert the kinetic energy into gamma rays (without absorbing them).

MINIMUM LORENTZ FACTOR OF GRB 980425

Lithwick & Sari [17] considered the minimum possible Lorentz factor for GRB 980425, finding it to be at least 3.8 in order for the burst not be obscured by e^{+-} pairs produced by its radiation field. However, this analysis relied on a power law extrapolation of the observed spectrum to (comoving-frame) energies above $m_e c^2$. [17] considered spectra no steeper than $d\log N_\gamma / d\log e_\gamma = -3$; further, they took $m_e c^2$ as the maximum observed photon energy. In contrast, the BATSE light curve for this burst exhibited a 37-σ detection in the 50-100 keV channel, 20-σ detection in the 100 – 300 keV channel, and no detection (< 1-σ) in the > 300 keV channel – i.e., no evidence for photons with energies above $m_e c^2$. Interpreted as a power law, the higest channels give a slope of -4 or steeper; however, they are more suggestive of a spectral cutoff (likely a dilute Wien spectrum; C. Thompson, private communication, 2001) than a power law.

Another estimate of the minimum Lorentz factor, and one that does not depend on the specifics of the emission mechanism, arises from the requirement that 10^{48} ergs of gamma rays be produced in an interaction between stellar ejecta and the pre-supernova stellar wind. For mean ejecta Lorentz factor $\bar{\Gamma}$, the wind mass must be about $1/\bar{\Gamma}$ of the ejecta mass. This mass of wind must be found in a radius that is roughly $2\bar{\Gamma}^2 c$ times the observed duration of the burst (~ 15 seconds). But, the wind cannot be opaque at this radius. Applied to the parameters of GRB 980425, these considerations (including the difference between the velocity of the ejecta and that of the emitting swept-up shell, and the Klein-Nishina opacity correction) give $\bar{\Gamma} > 1.9$, roughly; see [18] for a more thorough discussion. Both estimates of the minimum Lorentz factor merit further investigation, preferably careful modeling of both the dynamical interaction and the emission mechanism; we shall adopt the latter as the more robust estimate.

RELATIVISTIC EJECTA FROM SN 1998BW

As a supernova explosion engulfs a star's envelope, the velocity of its leading shock front responds to two competing trends: a general deceleration as increasing mass is swept up, and a tendency to accelerate down any sharply declining density gradient (in a manner analogous to the cracking of a whip). Matzner & McKee [19] have shown that these trends can be combined into a single formula that tracks the behavior seen in numerical simulations remarkably well. After the shock emerges from the stellar surface, the shocked material accelerates further as its residual heat is converted into kinetic energy. The highest velocity attained by the ejecta is set by the fact that the shock front spans a finite optical depth; the star must therefore be relatively compact or have an energetic explosion in order to produce any relativistic ejecta. Matzner & McKee determined that a compact Wolf-Rayet star would most likely satisfy this criterion. Although their formulae did not address relativistic motion, they were able to estimate the kinetic energy in relativistic ejecta by evaluating their formulae at a final velocity of c. This estimate illustrated that an explosion like that of SN 1998bw would indeed produce of order 10^{48} erg in relativistic ejecta, roughly enough to power GRB 980425.

Woosley, Eastman & Schmidt [20] considered the production of relativistic ejecta in specific models developed to fit the light curve of SN 1998bw. The most promising of these is the $6.6\,M_\odot$ CO core of a $\sim 25\,M_\odot$ main-sequence star, exploding with 2.8×10^{52} ergs of final kinetic energy. Woosley et al. used the theory of Gnatyk [21] to extrapolate their nonrelativistic simulations into the relativistic regime. They concluded that the supernova could not have powered GRB 980425; however, this conclusion was flawed on several counts. First, Gnatyk's formula (an interpolation between nonrelativistic [22] and relativistic [23] scaling laws) was of unknown validity. Second, and much more importantly, Woosley et al. made an allowance for the postshock acceleration that was valid in the nonrelativistic regime (in which the four-velocity $\Gamma\beta$ of a fluid element increases by a factor 2.5), but did not account for the very different character of this acceleration found by Johnson & McKee [23] for relativistic flow (in which $\log[\Gamma\beta]$ nearly quadruples). This led them to predict a much steeper decline of kinetic energy with increasing Lorentz factor than actually holds. Lastly, Woosley et al. assumed that the minimum Lorentz factor was at least about 5, whereas we have argued above that this value is not supported by observations and the lower value of ~ 1.9 is more appropriate.

To put the theory of this burst on a more solid footing, Tan, Matzner & McKee [18] have considered in detail the evolution of explosions involving a transition from nonrelativistic to relativistic motion. Among the results of this investigation are:
- A relativistic extension of Matzner & McKee's theory for the shock velocity;
- Likewise for the postshock acceleration of fluid elements to their final velocities;
- Formulae for the resulting distribution of kinetic energy among ejecta of different final velocities and Lorentz factors;
- An analysis of what aspects of stellar envelopes enhance the efficiency with which they produce relativistic ejecta;
- Simple formulae for the yield of relativistic ejecta from stars with radiative outer envelopes, in terms of parameters like mass, radius, luminosity and composition;
- Formulae to predict the relativistic ejecta in different directions for numerical simulations of asymmetrical explosions (including ejecta produced by shock acceleration in beamed and jet-like events, which could give rise to GRB precursors [5]);
- Generalization to the collapses of compact objects (e.g., accretion-induced collapse of white dwarfs) in which gravity sets the characteristic ejecta velocities; and

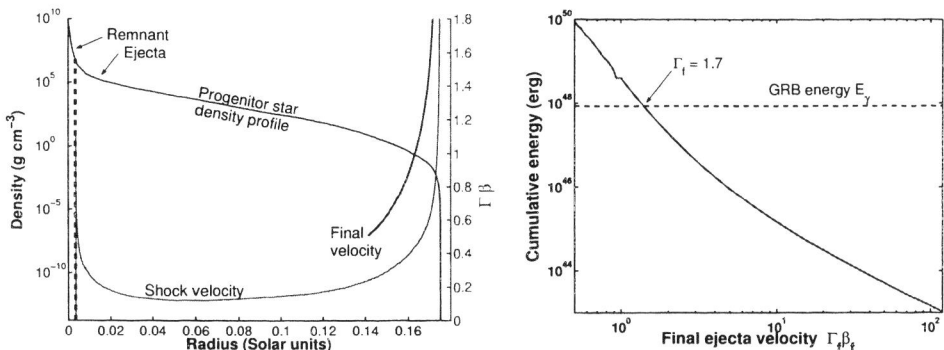

FIGURE 1. 2.8×10^{52} erg explosion of progenitor CO6 of Woosley, Eastman & Schmidt [20] (provided by S. Woosley), a model for SN 1998bw. *Left panel:* Stellar density distribution, and velocities of the shock front and in free expansion, in the theory of Tan et al. [18]. *Right panel:* Kinetic energy contained in ejecta traveling higher than a given final velocity. The observed energy of GRB 980425 is realized in ejecta with final Lorentz factors above 1.7.

- A consideration of "hypernovae" of very high explosion energy.

These analytical results were verified and calibrated by means of well-resolved, relativistic numerical simulations in spherical and planar symmetry.

Tan et al. verified Matzner & McKee's prediction of the kinetic energy in relativistic ejecta, demonstrating that this energy is associated with ejecta moving with $\Gamma_f > 1.41$. Applying their results to Woosley, Eastman & Schmidt's model CO6 (kindly provided by Stan Woosley), Tan et al. find that the energy of GRB 980425 emerged in material whose minimum Lorentz factor was 1.7, for which $\bar{\Gamma} = 2$. Coupled with the minimum Lorentz factor identified above, this confirms Matzner & McKee's prediction that SN 1998bw produced enough energy in relativistic ejecta to have powered GRB 980425. For higher Lorentz factors, Tan et al. predict a decline in kinetic energy roughly as $E_k(>\Gamma_f) \propto 1/\Gamma_f$ because of dramatic postshock acceleration in the relativistic regime. This relatively shallow decline indicates that explosions that can produce any relativistic ejecta also channel significant energy into ultrarelativistic motion.

CIRCUMSTELLAR MATERIAL

As discussed above, the interaction that gives rise to the GRB occurs at a radius that is roughly $2\Gamma^2 c t_{obs}$. Within this radius, a mass $E_\gamma/[c^2 \bar{\Gamma}(\bar{\Gamma}-1)]$ of circumstellar material must be found. The circumstellar material must therefore have a mass per unit radius of $\sim E_\gamma/[2c^3 t_{obs} \bar{\Gamma}^3(\bar{\Gamma}-1)]$. The minimum $\bar{\Gamma}$ thus implies a maximum mass per unit length in the circumstellar material; for a wind, this is the ratio of mass loss rate to wind velocity. Evaluated for GRB980425, the maximum is $(3 \times 10^{-4} M_\odot/\text{yr})/(1000 \text{ km/s})$ – dense, but within the range of Wolf-Rayet (WC subclass [24]) winds.

The radio afterglow from SN 1998bw provides a consistency check on any model in which the GRB arises from an early circumstellar interaction. Applying the theory of Chevalier [25] and Nadyozhin [26] to the collision of the nonrelativistic ejecta with the circumstellar wind, Tan et al. find a mean expansion velocity of $0.35c$ for the first 12 days – in excellent agreement with the value $0.3c$ given by Kulkarni et al. [16].

CONCLUSIONS

We have argued that the GRB and the later radio emission associated with supernova 1998bw can be explained in a spherical model – the same model proposed by Woosley, Eastman, & Schmidt [20] to explain its light curve. The only additional requirement is a relatively dense circumstellar wind, but one within the range observed around Wolf-Rayet stars. The viability of a spherical model casts significant doubt on the hypothesis that GRB 980425 was intimately related to beamed cosmological bursts. Specifically, there is no evidence for a jet of high Lorentz factor material.

Tan et al.'s analysis demonstrates that the fraction of a supernova's energy that winds up in relativistic ejecta is enhanced if the stellar atmosphere is as diffuse as possible compared to its core. Stars whose luminosity is comparable to the Eddington limit are ideal in this regard. A high explosion energy and low envelope mass are even more important, as the energy in relativistic motion scales as $E^{3.6} M_{env}^{-2.6}$. Pre-explosion mass loss therefore enhances the possibility of a GRB both by increasing the amount of energy in relativistic ejecta, and by giving rise to the circumstellar wind necessary for converting this energy into gamma rays in a brief period.

ACKNOWLEDGMENTS

We thank Stan Woosley for kindly providing a model of SN 1998bw's progenitor star. C. D. M. is supported by an NSERC fellowship; the research of J. C. T. and C. F. M. is supported by NSF grants AST- 9530480 and AST-0098365.

REFERENCES

1. Kulkarni, S. R. et al. 2000, astro-ph/0002168
2. Galama, T. J. et al. 1998, Nature, 395, 670
3. Soffitta, P., et al 1998, *IAUC* 6884
4. Bloom, J. S. et al. 1998, ApJ, 506, L105
5. MacFadyen, A., & Woosley, S. E. 1999, ApJ, 524, 262
6. Höflich, P., Wheeler, J. C., & Wang, L. 1999, ApJ, 521, 179
7. Nakamura, T., Mazzali, P. A., Nomoto, K., & Iwamoto, K. 2001, ApJ, 550, 991
8. Höflich, P. 1995, ApJ, 440, 821
9. Bartel, N. et al. 2000, Science, 287, 112
10. Kay, L. E., Halpern, J. P., Leighly, K. M., Heathcote, S., Magalhaes, A. M., & Filippenko, A. V. 1998, IAU Circular, 6969, 1
11. Eichler, D. & Levinson, A. 1999, ApJL, 521, L117
12. Nakamura, T. 1999, ApJL, 522, L101
13. Salmonson, J. D. 2001, ApJL, 546, L29
14. Li, Z. Y., & Chevalier, R. A. 1999, ApJ, 526, 716
15. Colgate, S. A. 1974, ApJ, 187, 333
16. Kulkarni, S. R. et al. 1998, Nature, 395, 663
17. Lithwick, Y. & Sari, R. 2001, ApJ, submitted
18. Tan, J. C., Matzner, C. D., & McKee, C. F. 2001, ApJ, 551, 946
19. Matzner, C. D., & McKee, C. F. 1999, ApJ, 510, 379
20. Woosley, S. E., Eastman, R. G., & Schmidt, B. P. 1999, ApJ, 516, 788
21. Gnatyk, B. I. 1985, Sov. Astron. Lett. 11(5), 331
22. Sakurai, A. 1960, Comm. Pure Appl. Math., 13, 353
23. Johnson, H. M., & McKee, C. F. 1971, PRD, 3, 858
24. Koesterke, L. & Hamann, W. -. 1995, AAP, 299, 503
25. Chevalier, R. A. 1982, ApJ, 258, 790
26. Nadyozhin, D. K. 1985, ApSS112, 225
27. Sari, R., & Piran, T. 1997, MNRAS, 287, 110

Physics of Collisionless GRB Shocks and Their Radiation Properties

Mikhail V. Medvedev

CITA, University of Toronto, Toronto, Ontario, M5S 3H8, Canada
Harvard-Smithsonian Center for Astrophysics, 60 Garden Street, Cambridge, MA 02138

Abstract.
We present a theory of ultrarelativistic collisionless shocks based on the relativistic kinetic two-stream instability. We demonstrate that the shock front is unstable to the generation of small-scale, randomly tangled magnetic fields. These fields are strong enough to scatter the energetic incoming (in the shock frame) protons and electrons over pitch angle and, therefore, to convert their kinetic energy of bulk motion into heat with very high efficiency. This validates the use of MHD approximation and the shock jump conditions in particular. The effective collisions are also necessary for the diffusive Fermi acceleration of electrons to operate and produce an observed power-law. Finally, these strong (sub-equipartition) magnetic fields are also required for the efficient synchrotron-type radiation emission from the shocks.

The predicted magnetic fields have an impact on polarization properties of the observed radiation (e.g., a linear polarization from a jet-like ejecta and polarization scintillations in radio for a spherical one) and on its spectrum. We present an analytical theory of jitter radiation, which is emitted when the magnetic field is correlated on scales smaller then the gyration (Larmor) radius of the accelerated electrons. The spectral power of jitter radiation is described by a sharply broken power-law: $P(\nu) \propto \nu^1$ for $\nu < \nu_j$ and $P(\nu) \propto \nu^{-(p-1)/2}$ for $\nu > \nu_j$, where p is the electron power-law index and ν_j is the jitter break, which is independent of the magnetic field strength but depends on the shock energetics and kinematics. Finally, we present a composite jitter+synchrotron model of GRB γ-ray emission from internal shocks which is capable of resolving many puzzles of GRB spectra, such as the violation of the "line of death", sharp spectral breaks, and multiple spectral components seen in some bursts (good examples are GRB910503, GRB910402, etc.). We stress that simultaneous detection of both spectral components opens a way to a precise diagnostics of the conditions in GRB shocks. We also discuss the relation of our results to other systems, such as internal shocks in blazars, radio lobes, and supernova shocks.

THE STRUCTURE OF COLLISIONLESS SHOCKS

The conventional paradigm of GRBs assumes optically thin synchrotron radiation from ultra-relativistic shocks where the radiation is produced by Fermi-accelerated electrons moving in strong, nearly equipartition magnetic fields. This purely phenomenological model contains several serious assumptions which require justification: (i) standard hydrodynamic shock physics must be valid for these highly collisionless shocks, i.e., one needs effective collisions; (ii) magnetic fields must be generated *in situ* much faster than the dynamical time; (iii) acceleration of electrons requires multiple scatterings (i.e., effective collisions) in the shock.

These problems have successfully been resolved [2,3]. It was shown that magnetic fields are naturally produced via the relativistic two-stream instability operating at the shock front. In essence, this work [2] makes the bridge between the theories non-relativistic collisionless shocks [4] (those observed in the interplanetary space were studied *in situ* in great details by many satellites, and their ultra-relativistic counterparts. Here we briefly describe the main results.

- The two-stream instability operates in both internal and external shocks. The field is produced by both the electrons and protons.

- The generated magnetic field is randomly oriented in space, but always lies in the plane of the shock front.

- The characteristic *e*-folding time in the shock frame for the instability is $\tau \sim \gamma_{\rm sh}^{1/2}/\omega_{\rm p}$ (where $\gamma_{\rm sh}$ is the shock Lorentz factor) which is $\sim 10^{-7}$ s for internal shocks and 10^{-4} s for external shocks. This time is much shorter than the dynamical time of GRB fireballs.

- The characteristic coherence scale of the generated magnetic field is of the order of the relativistic skin depth $\lambda \sim c\bar{\gamma}^{1/2}/\omega_{\rm p}$ (where $\bar{\gamma}$ is the mean thermal Lorentz factor of particles), i.e. $\sim 10^3$ cm for internal shocks and $\sim 10^5$ cm for external shocks. This scale is much smaller than the spatial scale of the source.

- The instability converts a large fraction of the kinetic energy of particles into magnetic energy, hence $[B^2/8\pi]/[mc^2n(\bar{\gamma}-1)] = \eta \sim 10\%$. This agrees well with direct particle simulations.

- Random fields scatter particles over pitch-angle and, thus, provide effective collisions. Therefore MHD approximation works well for the shocks. The magnetic fields communicate the momentum and pressure of the outflowing fireball plasma to the ambient medium and define the shock boundary.

- The instability isotropizes and heats the electrons and protons. Moreover, effective collisions will diffusively further accelerate the electrons to higher energies.

RADIATION FROM SHOCKS

Since the geometry of magnetic fields is not entirely random, it affects the observed properties of radiation. In particular, polarization of radiation will always be radial. Therefore, one expects non-vanishing degree of polarization observed at any wavelength for a non-spherically symmetric explosion, e.g., a jetted geometry. An optical transient of GRB990510 shows linear polarization of the degree $\sim 1-2\%$. For a spherically symmetric case, radio scintillations may reveal the polarization map of the source (afterglow) [5].

The small-scale nature of magnetic fields affects the radiation process as well. In fact, it breaks the conventional paradigm of synchrotron nature of the radiation from shocks. The magnetic field produced in GRB shocks randomly fluctuates on a very small scale of roughly the relativistic skin depth, which is much smaller than the Larmor radius of the ultra-relativistic emitting electron. Therefore, the electron trajectories are not helical, as they would be in a homogeneous field, as in Figure 1. Thus, the theory of synchrotron radiation derived for homogeneous fields is not applicable and the spectrum of the emergent radiation is different. Such a situation has never been considered in the astrophysical literature.

If the magnetic field is randomly tangled and the correlation length is less then a Larmor radius of an emitting electron, then the electron experiences random deflections as it moves through the field. Its trajectory is, in general, stochastic. This is similar to a collisional motion of an electron in a medium. Bremsstrahlung quanta are emitted in every collision. Unlike the bremsstrahlung case, here "collisions" are due to small-scale inhomogeneities of the magnetic field rather than due to electrostatic fields of other charged particles. Since the Lorentz force depends on particle's velocity, the emergent spectrum will be somewhat different from pure bremsstrahlung. There is also an alternative physical interpretation of the process. For an ultrarelativistic electron, the method of virtual quanta applies. In the rest frame of the electron, the magnetic field inhomogeneity with wavenumber $k \sim 1/\lambda$ is transformed into a transverse pulse of electromagnetic radiation with frequency kc. This radiation is then Compton scattered by the electron to produce observed radiation with frequency $\sim \gamma^2 kc$ in the lab frame.

Keeping this general physical picture in mind, we now analyze the problem in more details. Let's consider a nonuniform random magnetic field with a typical

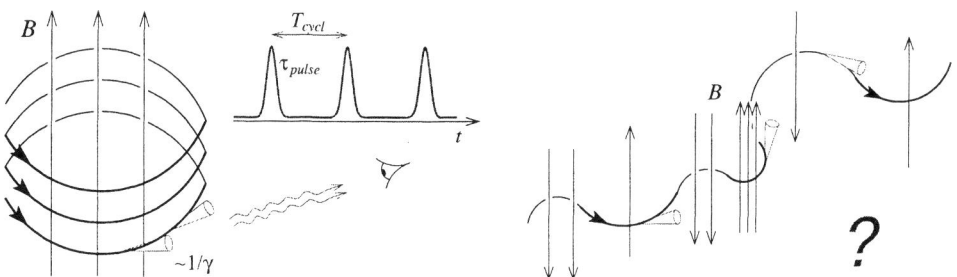

FIGURE 1. Illustration of the process of radiation in homogeneous and inhomogeneous fields.

correlation scale λ_B, the Larmor radius of the electron, $\rho_e = \gamma m_e c^2/eB_\perp$ is less or comparable comparable to λ_B. The emerging spectrum depends on the relation between the particle's deflection angle, α, and the beaming angle, $\Delta\theta$. For ultrarelativistic particles and small deflection angles, the latter is estimated as follows. The particle's momentum is $p \sim \gamma m_e c$. The change in the perpendicular momentum due to the Lorentz force acting on the particle during the transit time $t \sim \lambda_B/c$ is $p_\perp \sim F_L t \sim eB_\perp \lambda_B/c$. The angle α is then $\alpha \sim p_\perp/p \sim eB_\perp \lambda_B / \gamma m_e c^2$. We now define the deflection-to-beaming ratio as follows,

$$\delta \equiv \frac{\gamma}{k_B \rho_e} \sim \gamma \frac{\lambda_B}{\rho_e} \sim \frac{\alpha}{\Delta\theta} \sim \frac{eB_\perp \lambda_B}{m_e c^2}. \qquad (1)$$

It is interesting to note that this ratio is independent of particle's energy (i.e., of γ) and is determined by B and λ_B.

There are two limiting cases, as in Figure 2a,b. First, $\delta \sim \alpha/\Delta\theta \gg 1$; an observer sees radiation coming from short segments ("patches") of the electron's trajectory, the magnetic field is almost uniform but it varies from patch to patch. The radiation is pulsed with a typical duration $\tau_p \sim 1/\omega_c$ as for pure synchrotron. The ensemble-averaged spectrum completely identical to synchrotron radiation from large-scale weakly inhomogeneous magnetic fields. Second, $\delta \sim \alpha/\Delta\theta \ll 1$; the particle moves along the line of sight almost straight and experiences high-frequency jittering in the perpendicular direction due to the random Lorentz force. The emergent spectrum is determined by random accelerations of the particle.

Spectra of jitter radiation were calculated in Ref. [5]. They are well approximated by the sharply broken power-law, as is seen from Figure 2c. The jitter break frequency is

$$\omega_{jm} = 2^{7/4} \gamma_{sh} \gamma_{int} \gamma_{min}^2 \bar{\gamma}_e^{-1/2} \omega_{pe} \propto n_e^{1/2}, \qquad (2)$$

where γ_{sh}, γ_{min}, γ_{int}, $\bar{\gamma}_e$ are the Lorentz factors of the ejecta, internal shock, electron power-law cutoff, and the mean thermal γ-factor of electron in front of the shock.

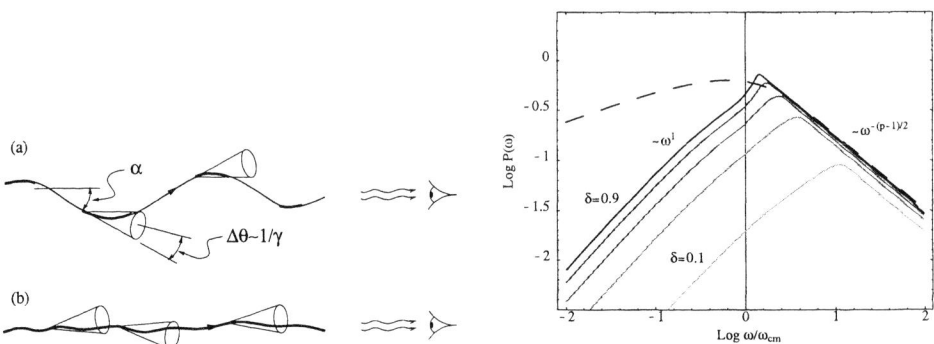

FIGURE 2. (a,b) The emission process in $\delta \sim \alpha/\Delta\theta \gg 1$ and $\delta \ll 1$ regimes. (c) Spectral power of jitter radiation for various δ; synchrotron spectrum is shown by dashed curve.

Note, this break frequency is independent of the magnetic field strength; instead it directly "measures" the density of particles in the shock. Below and above the break, the photon spectra scale as (p is the electron power-law index)

$$F(\omega < \omega_{jm}) \propto \omega^0, \qquad F(\omega > \omega_{jm}) \propto \omega^{-(p+1)/2}, \qquad (3)$$

that is the spectrum is harder than synchrotron ($\propto \omega^{-2/3}$) at low frequencies.

As was mentioned earlier, the magnetic field is produced by both the electrons and the protons. However, only the electron-produced field is small-scale enough to produce jitter radiation, whereas for the proton-produced field $\delta > 1$ and synchrotron spectrum is expected. The ratio of the jitter and synchrotron break frequencies and peak fluxes completely determine both free parameter of the model: the small-to-large scale field strength ratio and the deflection-to-beaming ratio,

$$\frac{\omega_{jm}}{\omega_{sm}} \simeq \frac{2}{3}\frac{\bar{B}_{SS}}{\bar{B}_{LS}}\delta^{-1}, \qquad \frac{F_{J,\max}}{F_{S,\max}} \simeq \delta^2. \qquad (4)$$

This jitter+synchrotron model explains perfectly well the diversity of time-resolved GRB γ-ray spectra and resolves several long-standing puzzles, namely

- the violation of the constraint on the low-energy spectral index called the synchrotron "line of death" in about a third of BATSE and BSAX bursts [6];
- the sharp spectral break at the peak frequency seen in some bursts which is inconsistent with the broad synchrotron bump [7];
- the evidence for two spectral sub-components seen in some GRBs [8];
- possible existence of emission features ("GRB lines") seen in few bursts [9].

All this strongly supports that (i) the proposed jitter radiation mechanism operates in astrophysical objects and (ii) the magnetic field is generated in shocks by the two-stream instability. In general, the detection of both spectral components in GRB spectra would be a powerful and precise tool to investigate the properties of cosmological fireballs.

REFERENCES

1. Piran, T., *Phys. Rep.* **314**, 575 (1999)
2. Medvedev, M. V., and Loeb, A., *Astrophys. J.* **526**, 697 (1999)
3. Medvedev, M. V., in Proceedings of the 20th Texas Symposium, eds. J. C. Wheeler and H. Martel (2001)
4. Moiseev, S. S., and Sagdeev, R. Z., *J. Nucl. Energy C* **5**, 43 (1963)
5. Medvedev, M. V., *Astrophys. J.* **540**, 704 (2000)
6. Preece, R. D., et al., *Astrophys. J. Suppl.* **126**, 19 (2000)
7. Pelaez, F., et al., *Astrophys. J. Suppl.* **92**, 651 (1994)
8. Barat, C., et al., *Astrophys. J.* **538**, 152 (2000)
9. Schaefer, B. E., et al., *Astrophys. J.* **492**, 696 (1998)

White Hole Model of Gamma Ray Bursts and Recent Observations

S.Ramadurai

Astrophysics Group, Tata Institute of Fundamental Research, Homi Bhabha Road, MUMBAI 400 005, INDIA

Abstract. Recent observations have unambiguously established the metagalactic model for the distribution of Gamma Ray Bursts (GRBs). While the interpretation of the afterglow has addressed the question of emission and propagation of radiation, the central engine powering this process is subject to a lot of speculative models. The possibility of a Whitehole as the progenitor object is examined in the context of recent observations of the afterglow. After giving a brief summary of the conclusions from the observations of the GRBs, the early works on the whitehole models are mentioned. The close similarity of the whiteholes to the concept of collapsar or hypernovae models is pointed out. The possibility of establishing the whitehole model as the correct model for at least some of the GRBs using the measurements of the pulse duration at different wavelength bands is indicated. Further work is in progress on the comparison of detailed predictions of the whitehole model with the light curves of the GRBs with multiwavelength observations including the optical bands.

INTRODUCTION

Though the Gamma Ray Bursts were discovered in 1973 from the analysis of the observations made by the surveillance Vela Satellites (see Klebasadel et al.[1]), only after the Compton Gamma Ray Observatory observations, they are established as a new class of extremely interesting sources (see for a summary, Fishman and Meegan [2]). The BATSE observations confirmed this and defined the various important characteristics like 1) isotropy, 2) the spectra are non-thermal 3) the duration ranges from 0.001 to 10000 seconds with a bimodal distribution – long (greater than 2 s) and short (less than 2 s). The gamma ray light-curves have wide variations. The pulse distribution is complex and the time histories at different energies giving indications about the nature of the source regions. The isotropic distribution established unambiguously their metagalactic nature (see Paczynskii[3]).

The first precise prediction of the fireball model of the synchrotron emission from the relativistically expanding blastwave from a compact object was given by Meszaros and Rees[4,5], though the relations between the observable synchrotron spectra and shock parameters have been derived in a number of papers (see for a summary, van Paradjis et al.[6]). With the observations of nearly three dozens of GRB afterglows and with more than 25 objects identified with host galaxies some important characteristics have emerged. These are the most energetic explosions (reaching energies of 10^{52-54} ergs) after the bigbang in the history of the Universe. They result from the ejection of ultrarelativistic matter from a powerful compact energy source and subsequent collision with the environment. While the afterglow is now more or less well-understood, the progenitors are not known well. Suggestions range from a sudden collapse of the cores of massive stars called hypernovae or collapsars through the merger of the neutron star and blackhole binaries, to the formation from a stellar collapse of rapidly rotating neutron star with ultrahigh magnetic field. It is also not clear whether there are more than one class of objects involved in the progenitors to explain the bimodal distribution of long and short GRBs. In this investigation a revival of the model of Whiteholes as the progenitors of GRBs is being attempted (see Narlikar et al.[7] and Narlikar and Appa Rao[8]).

WHITEHOLE MODEL

Whitehole is taken as a spherical object with uniform density and zero pressure in the commoving frame of the outward moving particles. In the simplest version, it is an object exploding from a singularity and obeying Einstein's Equations of gravitation subsequent to the singular event. This may be termed `delayed bang' or `lagging core' in which matter explodes into existence over a limited region of space. Essentially the explosions are due to the bounce of a collapsing massive object. Conventional physics precludes such phenomena. But unusual equations of state (see Novikov and Zeldovich[9]) or the inroduction of negative energy field (see Hoyle and Narlikar[10]) can convert implosions to explosions. In the former case the equation of state must obey the following peculiar properties. The pressure initially increases with increase in the density, it obtains a maximum at certain point and then starts decreasing. This calls for the emergence of new forces at very high densities, causing the energy density to decrease with increasing particle density and the pressure to become negative. Since this violates one of the conditions of the Penrose-Geroch-Hawking theorem, the reversal of the initial contraction to expansion is not ruled out. It is assumed that such a thing is essentially the case in the case of the progenitors of the gamma ray bursts. Essentially this is very similar to the physical situation in the `collapsar' or `hypernova' models (see Paczynskii[11] and Woosley[12]). In these models as a result of either a `failed supernova' or the merger of a pair of neutron stars, a neutron disk or torus spinning around a stellar mass blackhole with a period of about 10^{-3} s, can release about 10^{53-54} ergs. In the case of whiteholes, the energy release is due to the peculiar equation of state at ultrahigh densities of the collapsing cores. So what exactly are the consequences of this peculiar phenomenon of the whiteholes. While the special features in the spectra of high energy radiation from whiteholes have been worked out earlier, in this work it is proposed to concentrate on a single unique prediction about the pulse duration at different wavelength bands so that at least for some of the GRBs the progenitors may be established unambiguously as whiteholes.

The relation between the frequency band of observations and time duration of the pulse has an interesting relation. Suppose the spectrum of radiation inside the whitehole is in the range (f_1, f_2) and the corresponding observed frequency range is (f_a, f_b). Then it can be shown using Einstein's theory (see Narlikar and Appa Rao[8]) that the time corresponding to the beginning of the pulse T_i and the end of the pulse T_f are given by

$$T_i = T_0 (f_1/f_b)^4 \ ; \ T_f = T_0 (f_2/f_a)^4 \qquad (1)$$

If observation is made at another waveband (f_a', f_b') and T_i' and T_f' are the corresponding times to this band, then

$$\frac{T_i - T_f}{T_i' - T_f'} = \frac{(f_1/f_b)^4 - (f_2/f_a)^4}{(f_1/f_b')^4 - (f_2/f_a')^4}$$

If observations are made in three or more bands, then the above relationship can be easily verified. This will be a powerful signature of the whitehole being the progenitor object. With the observations now extending from gamma ray, X-ray, optical and radio wavelength bands for at least a few objects, it is easily possible to verify the above relationship. However, there may be some observational constraints to establish the absolute duration of the pulse due to the presence of a background. This may be especially true with the spectral index of the power spectrum lying between 1 and 2, as observed, will lead to problems at lower frequency bands radiation being stronger and introducing some observational bias. But with the powerful space missions already in the sky like the HEE2 and the planned ones like SWIFT, the prospects for the detection of this nature is quite bright indeed.

WORK IN PROGRESS

At present the details of the spectral signatures of the spectra as well as the pulse structure as well as duration are being investigated for the whitehole model in more complete detail than has been presented here. These will be compared with the available observations to establish the soundness of the whitehole model for at least some of the GRBs.

REFERENCES

1. Klebesadel, R. W., Strong, I. B., and Olson, R.A., Astrophys. J. 182, L 85- 88 (1973).
2. Fishman, C., and Meegan, C., Ann. Rev. Ast. Ap. 33, 415 – 58 (1995).
3. Paczynskii, B., Astrophys. J. 308, L 43- 46 (1986).
4. Meszaros, P., and Rees, M., Astrophys. J. 418, L 59- 62 (1993).
5. Meszaros, P., and Rees, M., Astrophys. J. 476, 232- 37 (1997).
6. van Paradijs Jan, Kouveliotou, C., and Wijers, R.A.M.J., Ann. Rev. Ast. Ap. 38, 379- 425 (2000).
7. Narlikar, J. V., Appa Rao, M.V.K., and Dadich, N., Nature 251, 590-91 (1974).
8. Narlikar, J.V., and Appa Rao, M. V. K., Astrophys. Sp. Sci. 35, 321- 336 (1975).
9. Novikov, I. D., and Zeldovich, Ya. B., Ann. Rev. Ast. Ap. 11, 387- 412 (1973).
10. Hoyle, F., and Narlikar, J.V., Proc. Roy. Soc. A278, 465-80 (1964).
11. Paczynskii, B., Astrophys. J. 494, L 45-48 (1998).
12. Woosley, S.E., Astrophys. J. 405, 273- 77 (1993).

GRB Coordinates Network (GCN): A Status Report

Scott D. Barthelmy, T.L. Cline, P. Butterworth

NASA-GSFC, Greenbelt, MD 20771

Abstract. The GRB Coordinates Network (GCN) continues to deliver locations of GRBs to instruments and observers in real-time (a few seconds) -- while the burst is still bursting -- so that they can make multi-band simultaneous follow-up observations. This was routine during the GRO-BATSE years and has resumed with HETE. This goal was realized with the optical detection of the burst counterpart for GRB990123 by the ROTSE instrument [1]. A brief review of the function and capabilities of the GCN system is given. Complementing the real-time location Notices, the GCN Circulars allow the follow-up observers to share the results of their observations rapidly with the community. A status report on recent improvements to the GCN system and a list of future improvements is given.

INTRODUCTION

The original BATSE Coordinates Distribution Network (BACODINE) started operations in 1993 by distributing GRB locations calculated from the real-time telemetry from the CGRO-BASTE instrument [2]. It has since grown into the GRB Coordinates Network (GCN) by distributing location and light curve information for GRBs detected by all spacecraft capable of detecting GRBs. There are two basic components of the GCN system (the details of GCN are described in Barthelmy et al.[3]). The GCN is a system of computers and programs which: 1) Collect GRB location information from other spacecraft and distribute them to interested parties (the Notices). 2) Collect reports from burst follow-up observers and distribute them to the GRB community (the Circulars).

The GCN system is all encompassing and all automatic. GCN collects all the information on GRB locations from all the various sources into a single point and transmits that information to all the various sites. Each site need only develop and maintain one connection for all their GRB needs. There are no humans involved (within the GCN system proper), so there is minimal delay (1 sec for HETE information and only a 1-60 sec delay after receipt of the information from the other sources (e.g. BeppoSAX, RXTE, ALEXIS, etc.).

The original goal of the GCN (BACODINE) system of observing a GRB in a non-gamma-ray bandpass while the GRB is still busting has been realized. At 9:46:59 UT (T+3sec) on 23 Jan 99, the GCN system detected a GRB in the CGRO-BATSE telemetry stream, calculated a rough location, and distributed that location to the ROTSE instrument located at LANL [1]. The ROTSE instrument is a fully computer-controlled, fast slewing, wide FOV CCD camera instrument connected to the GCN system. ROTSE received the GCN burst location and was on-target and integrating the first of a series of exposures at T+22 sec. During the 90-sec duration of the burst, ROTSE recorded open-filter magnitudes of 11.8, 8.9, and 10.1. The idea of making observations of GRBs in bandpasses other than the gamma-ray while the burst is still bursting has been proven.

NOTICE TYPES and DISTRIBUTION METHODS

Table 1 shows all the Notice types currently available through the GCN system. The time delay after the burst, size of the error box, occurrence rates, and a comment about which types of instruments are suitable to make use of the type is given.

SOURCE	TIME DELAY	ERROR BOX	RATE
HETE	10-30 sec	20-200' dia	30/yr
RXTE-ASM	1-2 hr	4 x 15-150'	8/yr
SAX-WFC	2-3 hr	6-20' dia	8/yr
RXTE-PCA	2-5 hr	6-40'	6/yr
SAX-NFI	12-48 hr	100" dia	4/yr
IPN	0.5-3 day	4 x 4-8 '	2/month
ALEXIS	12 hr	0.6° dia	20/yr

All of the Notice types listed in Table 1 are available via the 5 distribution methods: Internet socket, phone/modem, e-mail, pagers/cellphones, and the GCN web site. The Internet socket connection directly couples the site's instrument control program to the GCN program with a TCP/IP socket connection. It is fast -- 0.1-2 sec. All the software required at the "site"-end of the connection has been developed and is available [4]. The dedicated phone line is slightly faster (0.3 sec) than the socket method, but seldom worth the expense. The e-mail method uses the standard e-mail protocols with the body of the e-mail having a "TOKEN:value" format suitable for human and computer parsing. The delivery times are typically less than 30 sec, which is fine for human destinations. There are several forms of pager and cell-phone methods -- you can get beeped by the Universe. There are normal and terse formats, subject-line-only formats, and decimal degrees vs hh:mm:ss formats. The alpha-numeric pagers are convenient because you can get the RA,Dec,Time,Intensity of the burst no matter your location. They have the same time delays as the e-mail method.

The web site [4] is an archive facility. There are pages within the site that contain all the Notices for each Source instrument (HETE-WXM/-SXC, RXTE-PCA/-ASM, BeppoSAX-WFC/-NFI, IPN, BATSE, COMPTEL, and ALEXIS). There are also pages which contain the lightcurves from the CGRO-BATSE, Wind-KONUS, and NEAR-XGRS instruments in alpha-numeric text, JPEG, GIF, and Postscript formats. All of the various web archive pages are updated within 60 sec of the Notice being distributed to the community. The web site also archives the published GCN Circulars. It also has the detailed technical description of the system.

GCN FILTERING METHODS

There is a wide range of filters available to sites wishing to customize the quantity and quality of the notices they receive. There are 4 main filtering categories and each filter category acts independent of the other 3 categories. 1) Source type -- sites can elect to receive each of the following Source types: HETE, RXTE, BeppoSAX, IPN, ALEXIS. Within each Source type, there are one or more sub-types for the instrument and/or the Alert, Updated, Final, GndAnalysis, Will-observe, Won't-observe, Saw-something, Didn't-see-something, etc sub-types. 2) There is a filter category based on the location of the burst on the sky: ALL, VISIBLE, or NIGHT. The VISIBLE filter requires that the RA,Dec location of the bust have an elevation angle at the site greater than 10degrees. The NIGHT filter further requires nautical twilight at the site. There are also customized versions which can include declination-based, Sun angle, UT window, and proximity to previous GRBs constraints. 3) There is also a filter based on the size of the uncertainty in the locations, and 4) a filter based on the amount of time from burst to Notice availability -- the details for both are discussed in the Improvements section.

SOME GCN STATISTICS

Currently, the GCN system distributes Notices to 204 "sites" involving ~350 researchers. There are 55 locations with 75 instruments (34 optical, 12 radio, 16 gamma-ray, 7 x-ray, 3 gravity wave and 3 neutrino). There are 11 fully automated instruments; the rest are manual. There rest are individuals and teams on telescopes, heads-up, and cross-instrument correlation operations. Approximately 600 burst follow-up observations have been made by the GCN sites on about 450 GRBs. As of 01 Jun 01, 1063 Circulars were distributed to a list of 520 recipients.

IMPROVEMENTS TO THE GCN SYSTEM

The GCN system is an evolving and improving system. It is continuously adding new capabilities and new sources of GRB information to provide a better system to the clients. The GCN system is funded by NASA through the SR&T and HETE Operations programs. It is envisioned that it will have a long and productive future, and therefore can be counted on for the long-range plans of the follow-up community. The following are improvements made within the last 2 years -- the dates in parentheses are when the feature was added to the system.

HETE-2 was launched 09 Oct 2000. The Wide-field X-ray Monitor (WXM) and Soft X-ray Camera (SXC) positions are being distributed via GCN (May 01). There will be about 30/yr and with error box sizes of 0.3-30 arcmin.

In the past the BeppoSAX-WFC/-NFI burst location notices had been distributed semi-automatically, such that the last step in the sequence involved the GCN Operator formulating a message to be imported into the GCN system for distribution. Now this last step has been automated too (Dec 99), resulting in a decrease in the time delay of 0.5 to a few hours.

The GCN system captures the periodic telemetry down-loads from the Wind and NEAR spacecraft and extracts the data from the KONUS (Oct 98) and XGRS (starting Mar 99; ending Feb 01) [5] instruments, respectively. It then scans these data looking for sudden increases in the background counting rates. A 2-minute section of this count-rate lightcurve is extracted and sent to the IPN system operated by Cline [6]. These lightcurves are cross-correlated with the Ulysses burst detections to produce automated IPN annuli and IPN boxes. The IPN solutions are then sent back to the GCN system for distribution to those sites requesting this IPN Notice type. The time delay on these IPN solutions is dependent on the rate of telemetry dumps from the spacecraft; typically 1-25 hours for Wind and 1 hour to 1.1 days for NEAR. There have been 6 IPN solutions using KONUS data and 4 solutions using the XGRS data, including the IPN solution for GRB991208 which resulted in a radio and optical counterpart detections based on the small error-box of the IPN solution [7]. These IPN solutions based on the Wind-KONUS and NEAR-XGRS contributions are significant because they provide the third node in the IPN triangulation technique to yield true error boxes, instead of the annuli segments using only the Ulysses and BATSE data.

The ASM instrument on RXTE scans about 80 percent of the sky every 90 minutes. D. Smith et al. [8] has created a real-time data analysis system to scan the count-rate data from the ASM to look for hard x-ray transients. When a location of detected, the location and error box is sent to the GCN for distribution to those sites requesting this Notice type (Jan 99).

Two new filtering criteria were added (Jan 00) that allow each site to filter based on the time delay between t he GRB and when the Notice is available for distribution and a filter based on the size of the error box (or circle). The time-delay filter is useful for the non-automated sites that have a sensitivity limit such that they would not be able to detect an afterglow N hours after the burst. It is continuously adjustable from 1 sec to 1000 days. The error-box filter is useful for sites that have a small FOV or can not perform large amounts of tiling observations. Since some of the sources produce burst positions with a wide range of error-box sizes, sites can use this filter to get only those positions for which the uncertainty is comparable to their instrument's FOV. The range is adjustable from 1 arcsec to 360 degrees.

When INTEGRAL launches (Apr 2001), the GRB locations imaged by the SPI and ISGRI instruments (20/yr at few arcmin with delays of less than 60 sec) will be distributed by GCN. And when Swift launches in late 2003, the locations from the BAT, XTR, and UVOT instruments will be also distributed (150-300/yr at 4arcmin at T+10sec and 1-2arcsec at T+60sec).

REFERENCES

1) Akerlof, C., et al., Nature, **398**, pp 400, 1999.
2) Barthelmy, S.D., et al.; Huntsville 3rd GRB Workshop, AIP, **384**, pp 580, 1996.
3) Barthelmy, S.D., et al., Huntsville 4th GRB Workshop, AIP, **428**, pp 99, 1998
4) The GCN Web Pages URL: http://gcn.gsfc.nasa.gov/gcn.
5) Barthelmy, S.D., et al., SPIE Conference Proceedings, **3768**, pp 444, 1999
6) Cline, T.L.. et al., Huntsville 5th GRB Workshop, AIP, **526,** pp 726, 1999
7) Hurley, K., et al., GCN Circular 450, 1999.
8) Smith, D.A., et al.; ApJ, **526**, pp 683, 1999.
9) Kippen, R.M., et al.; Huntsville 4th GRB Workshop, AIP, 428, pp 119, 1998

Milagro's Sensitivity to GRBs

Elizabeth Hays and David Noyes for the Milagro Calaboration

University of Maryland (College Park, MD)

Abstract. Milagro is a TeV gamma-ray observatory whose all-sky capability and high duty cycle make it an excellent tool for studying transient phenomena such as GRBs. In this paper we calculate Milagro's sensitivity to GRBs including the possible effects of attenuation due to the scattering of TeV photons by the intergalactic infrared (IR) background. Using a model for the energy dependence of the attenuation, we show that lowering Milagro's energy threshold increases the detector's viewing range allowing the detection of more GRBs.

THE MILAGRO DETECTOR

Located in the Jemez mountains near Los Alamos, New Mexico, Milagro is a large-area water Cherenkov detector sensitive to gamma rays in the energy range from 100 GeV to 10 TeV. The detector lies at an altitude of 2630 m, and consists of a 6 million gallon artificial pond containing 723 photomultiplier tubes (PMTs) [1].

Milagro currently uses a PMT multiplicity trigger. The trigger threshold corresponds to the number of PMTs that register a hit within a 200 ns time window. The current threshold varies between 50 and 70 PMTs and is limited by the maximum data rate of the data acquisition system.

The detector response and air shower development are simulated using GEANT [2] and CORSIKA [3] respectively. The gamma-ray simulations used for this study are thrown on an $E^{-2.4}$ differential spectrum over an area of 10^6 m^2.

MOTIVATING A LOWER TRIGGER THRESHOLD

High energy photons are attenuated by interactions with the intergalactic IR background. Some theories use semi-analytic models to calculate the gamma-ray survival probability for various cosmologies [4]. For our analysis we use predictions for a ΛCDM universe as shown in figure 1. This shows that for a given redshift there is a cutoff energy (from 50 TeV at z=0.02 to 100 GeV at z=1) above which no photons survive. To have a greater chance of seeing GRBs at redshifts less than 1, it is necessary to increase sensitivity below the cutoff.

To see the effect of trigger threshold on Milagro's energy response, we calculate effective areas using Monte Carlo simulations. Milagro's effective area is defined as $A_{core} \times (N_{trig} / N_{total})$, where A_{core} is the area used in Monte Carlo for the distribution of shower cores, N_{trig} is the number of showers above threshold, and N_{total} is the number of showers thrown. Figure 2 shows the effective area for low (30 PMT) and high (70 PMT)

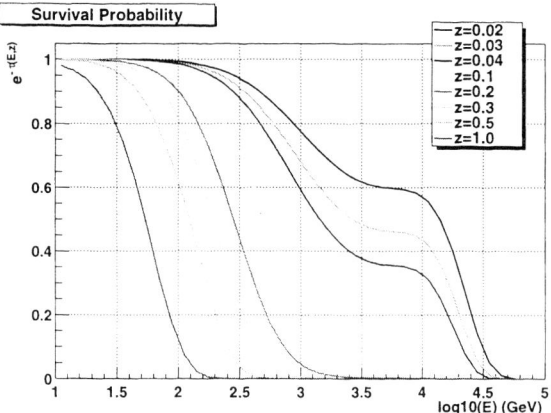

FIGURE 1. Survival probability for photons as a function of energy for various redshifts. The plot is based on semi-analytic models for a ΛCDM cosmology [4]. This indicates that improving sensitivity below 1 TeV will give an increased chance of seeing more GRBs at greater distances.

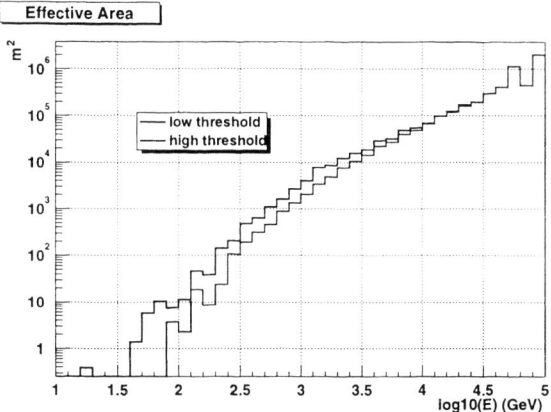

FIGURE 2. Lowering the trigger threshold increases the effective area below 1 TeV. The calculation is based on Monte Carlo simulaitons of gamma-ray showers.

thresholds. The low threshold has a larger effective area in the energy range below the cutoff due to IR attenuation. To see how this effects Milagro's ability to detect GRBs, we must calculate the sensitivity based on some assumptions about the source.

SENSITIVITY TO GRBs

To detect GRBs with Milagro the sky is searched with a square bin of width 3° (optimal bin size from Monte Carlo simulations) on a chosen time scale. This bin contains signal events that are compared to a predicted number of background events to get an event excess. If the Poisson probability that the excess is a statistical fluctuation is less than 5σ, then the GRB can be detected. Because the whole sky is searched on multiple time scales, a trials factor must be included that changes the pre-trials probability requirement to 10^{-20} (9σ).

To calculate a rate from a GRB, we must assume an energy spectrum. Many theories predict TeV emission from GRBs [5], however, the exact form of the spectrum is not known. For our analysis we use a simple power law with an index of -2.4 for the differential energy spectrum and calculate fluences using the spectrum above 1 TeV. Figure 3 shows Milagro's fluence sensitivity at the TeV scale as a function of duration without IR attenuation effects. This is plotted on top of a set of 100 keV scale GRB fluences and durations as measured by BATSE [6].

When we convolve the effective area for some trigger threshold with an $E^{-2.4}$ spectrum modified by the IR attenuation, we get a signal rate dependent on redshift. The background rate is obtained from data taken at the same trigger threshold. By combining these rates with a time scale for GRB emission, we can calculate the minimum isotropic energy as a function of redshift that would need to be released for a 9σ detection probability. Figure 4 shows the fluence threshold as a function of redshift for high and low trigger thresholds.

LOWERING THE THRESHOLD

At low thresholds the data rate of signal events increases but not as rapidly as the rate of background events. Several methods are being developed to reject this background which will reduce the data rate below the limit of the electronics and increase signal sensitivity. Gamma/hadron separation has been used to obtain an improvement in sensitivity of 1.8 [7] at high thresholds and continues to be studied at low thresholds.

The increased background rate at low thresholds is largely due to high-angle muons, or muons with angles greater than 45° from zenith. At a low trigger threshold the light from a muon of this sort can hit enough PMTs as it crosses the pond to trigger the detector. However, the time development of light from a single muon should be slower than that of an air shower. This is because light in the detector travels at speed c/n, where c is the speed of light and n is the index of refraction, while the particles in the air shower travel at speeds near c before interaction. This effect allows rejection of high-angle muons with a rise time cut. This cut requires 10% - 90% of the PMTs hit in an

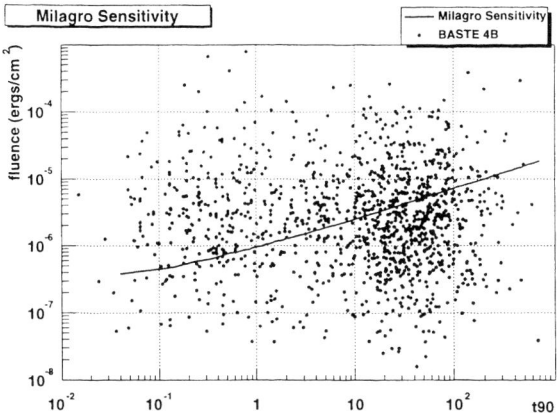

FIGURE 3. Milagro's fluence sensitivity as a funciton of GRB duration is calculated using an $E^{-2.4}$ differential spectrum above 1 TeV. The blue dots show measured BATSE fluences (20-300 keV energy band) and durations [6]. The number of GRBs that Milagro can detect depends on two things: their redshift distribution (they must be close enough for us to see them), and if GRBs emit radiation with comparable fluences in the TeV range as in the keV range.

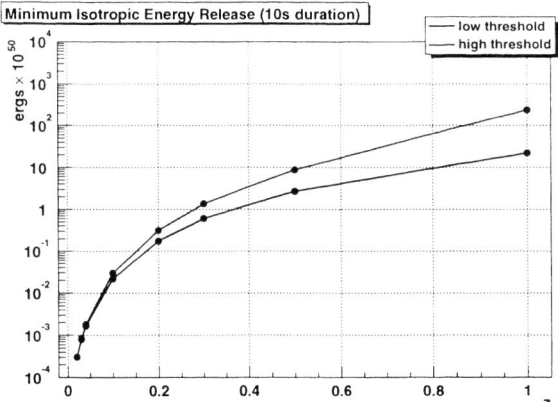

FIGURE 4. If one requires a 9σ probability for a detection, the minimum isotropic energy release can be calculated. Here it is shown for a GRB of 10s duration plotted as a function of redshift. The plot shows that lowering Milagro's trigger threshold moves the redshift horizon to greater distances. For example, a GRB of energy 10^{51} ergs could be detected out to $z = 0.8$ for low threshold but only to $z = 0.5$ for high threshold.

event to occur within a time chosen to keep most of the photon-induced showers while rejecting enough muons to keep the data rate manageable. A new trigger card has been designed to implement this cut in electronics and soon it will allow Milagro to lower its energy threshold.

ACKNOWLEDGMENTS

This work is supported by the Department of Energy Office of High Energy Physics, the National Science Foundation, the LDRD program at Los Alamos National Laboratory, Los Alamos National Laboratory, the University of California, the Institute of Geophysics and Planetary Physics, the Research Corporation, and the California Space Institute. We would also like to recognize the hard work of Scott Delay and Michael Schnieder, without whom these data would not exist.

REFERENCES

1. Sullivan, G., "Status of the Milagro Gamma Ray Observatory", ICRC Conference Proceedings, 2001.
2. CERN Application Software Group, CERN W, 5013, Version 3.21, 1994.
3. Heck, D., et al., CORSIKA: A Monte Carlo Code to Simulate Extensive Air Showers, Forschungszentrum Karlsruhe, Wissenschaftliche Berichte FZKA 6019, 1998.
4. Boone, L.M., Bullock, J.S., Primack, J.R., Williams, D.A., "The Implications of Galaxy Formation Models for the TeV Observations of Current Detectors", in *Snowbird 1999, Toward a Major Atmospheric Cerenkov Detector*, 1999, pp. 100–104.
5. Dermer, C.D., Böttcher, M., Chiang, J., *ApJ*, **537**, 255–260 (1999).
6. Paciesas, W., et al., *ApJS*, **122**, 465 (1999), URL http://gammaray.msfc.nasa.gov/batse/grb/catalog/4b/.
7. Sinnis, C., "Background Rejection in the Milagro Gamma Ray Observatory", ICRC Conference Proceedings, 2001.

Intensity Distributions of Gamma-Ray Bursts

David L. Band

X-2, Los Alamos National Laboratory, Los Alamos, NM 87545 USA

Abstract. Observations of individual bursts chosen by the vagaries of telescope availability demonstrated that bursts are *not* standard candles and that their apparent energy can be as great as 10^{54} erg. However, determining the distribution of their apparent energy (and of other burst properties) requires the statistical analysis of a well-defined burst sample; the sample definition includes the threshold for including a burst in the sample. Thus optical groups need to report the criteria behind the decision to search for a spectroscopic redshift. Currently the burst samples are insufficient to choose between lognormal and power law functional forms of the distribution, and the parameter values for these functional forms differ between burst samples. Similarly, the actual intensity distribution may be broader than observed, with a low energy tail extending below the detection threshold.

INTRODUCTION

Major advances in the study of gamma-ray bursts have resulted both from the construction and analysis of carefully selected burst samples, and from fortuitous discoveries of key properties of individual bursts. The BATSE burst sample, with precisely defined detection thresholds, showed that the peak flux distribution and the isotropic angular distribution were inconsistent with an origin in the Galactic plane[9] and marginally inconsistent with a Galactic halo progenitor population, and thus before 1997 most of the burst community concluded by a process of elimination that bursts occur at cosmological distances. On the other hand, beginning in 1997 the arcsecond localization of a few bursts coincident with faint galaxies became the definitive confirmation that bursts are indeed a cosmological phenomenon; these bursts were chosen largely on the basis of the available telescopes and the willingness of observers with access to these telescopes to devote their precious observing time to following these bursts. (Note that the possibility still remains that some bursts—such as bursts with durations under ~ 1 s— originate from less distant sources.) Thus some of the major advances in the study of bursts have resulted from the statistical analysis of well-constructed burst samples, while others (often the more spectacular breakthroughs) have followed observations of individual bursts selected through the vagaries of the available telescopes. Here I discuss the interplay of statistical analysis and serendipitous discovery in the determination of bursts' intrinsic intensity. I argue that observations of individual bursts established the qualitative characteristics of burst intensities but that determining the quantitative properties requires well-defined criteria for choosing the bursts which should be studied. In particular, a multi-wavelength effort involving different telescopes is required for determining properties such as a burst's intensity and redshift: an X-ray burst detector (e.g., *Beppo-Sax*'s Wide Field Camera) first localizes the burst to ~ 3 arcminute; an X-ray telescope (e.g., *Beppo-Sax*'s Narrow Field Instrument) then localizes the afterglow to about an arcminute; an imaging optical telescope localizes the optical afterglow to less than an arcsecond; and finally, optical spectroscopy determines the burst redshift through the detection of either absorption lines in the afterglow spectrum or emission lines from the host galaxy. The most restrictive intensity threshold for mounting any component of this campaign must be defined and reported to construct a well-defined statistical sample.

CONSTANT CANDLE BURSTS

After the statistical analysis of the BATSE peak flux and spatial distributions favored a cosmological origin, the burst community by and large adopted the "minimal cosmological model"[2] wherein bursts were standard candles and the shape of the intensity distribution (e.g., for the peak photon flux) resulted from the cosmological curvature of space. Although this paradigm was expected to be overly simplistic, it was the basis for Schaefer's conclusion[11] that the predicted host galaxies were missing from the best burst localizations then available. Schaefer's assertion was contested[8], but the statistical analysis of Band and Hartmann[2] verified that the *expected* host galaxies were indeed absent. Band and Hartmann considered the galaxy detection thresholds in each burst error box as well as the presence of the host and unrelated background galaxies. However, testing the cosmological paradigm which motivated Schaefer's search for the host galaxies became moot with the observed coincidence of *faint* galaxies with afterglows localized to an arcsecond. As Schaefer had reported, and Band and Hartmann corroborated, the minimal cosmological model was indeed incorrect, not because bursts are not cosmological, but because the model was too "minimal:" the deviation of the cumulative intensity distribution from the $-3/2$ power law expected for a homogeneous population in three-dimensional Euclidean space does not result solely from the cosmological curvature of space, and bursts are farther than derived by the minimal model. Band, Hartmann and Schaefer[3] used the statistical methodology of Band and Hartmann to find the burst energy for which the host galaxy detections (or upper limits) were consistent with the expected galaxy luminosity function. They found a standard candle total burst energy of $\sim 10^{53}$ erg. The observed redshifts and energy fluences did give some burst energies of this magnitude; however, because a range of burst energies spanning ~ 3 decades was observed, these observations also demonstrated that bursts are *not* standard candles in terms of the apparent total energy.

Thus the statistical analyses of Band and Hartmann[2] and Band, Hartmann and Schaefer[3] were appropriate for answering questions such as whether the minimal cosmological model was valid and what is the standard candle energy. However, these questions are answered more directly by a few observations of randomly selected bursts: only two bursts with significantly different energies demonstrate that bursts are not standard candles; only one bright burst with a faint host galaxy (or with $z > 1$) shows that the intensity distribution cannot be explained solely by cosmological curvature; and a few burst energies give the typical energy scale.

THE INTENSITY DISTRIBUTION

What is the distribution of the total energy emitted by a gamma-ray burst? We only observe the energy emitted in our direction, which can be expressed as the *apparent* total energy, the energy emitted if the burst actually radiated gamma-rays isotropically. The apparent total energy is the actual energy divided by the beaming fraction, the fraction of the sky into which the gamma-rays are actually emitted; both the total energy and the beaming fraction are relevant for understanding burst physics. Indeed, Frail *et al.*[6] calculated the beaming fraction for a burst sample based on afterglow evolution, and found a narrow distribution of the total energy centered around 10^{51} erg.

Different measures of burst intensity have been used, such as the peak energy luminosity, the total gamma-ray energy, the peak photon flux, the total number of photons emitted, or the total afterglow energy. Note that bursts can be standard candles for at most one of these intensity measures. The intensity measure studied is a matter of theoretical prejudice and ease of calculation (the data may not be available to calculate some measures). Many studies have used the peak photon flux because burst detectors, such as BATSE, trigger on the peak count rate, and consequently the detection threshold for the peak flux is fairly sharp. However, I prefer the total energy emitted; the observable is a burst's energy fluence. In the current theoretical scenario (see Piran[10] for a review of current burst theories) the gamma-ray emission results from internal shocks when regions in a relativistic outflow with different Lorentz factors collide. The total gamma-ray energy emitted should be related to the energy of the outflow while the peak luminosity is a consequence of a particular internal shock, which will result from the burst-specific distribution of Lorentz factors within the outflow. Thus I suspect that the emitted total energy is fairly representative of the energy released while the peak luminosity is more contingent on the details of the energy release.

In studying the burst intensity distribution, my collaborators and I have assumed specific functional forms. Thus Jimenez, Band and Piran[7] fit lognormal distributions to the total apparent energy, the peak gamma-ray luminosity, and the total X-ray afterglow energy, while recently I fit both lognormal and simple power law distributions to the total apparent energy.[1] When normalized to unity, these distributions are the probability $p(I)$ that a burst has a given intensity I. However, the probability $p_{\rm obs}(I)$ of *observing* a burst with a given intensity is that part of the intensity distribution above the threshold for including the burst in our sample. For example, a high redshift burst will be detected, and thus included in our sample, only if it was drawn from the high end of the intensity distribution. In both studies likelihood functions were constructed from the probabilities $p_{\rm obs}(I)$ of obtaining each member of the sample. The parameters of the distribution function were determined by maximizing the likelihood, and the parameters' confidence ranges were determined by integrating over the likelihood surface.

These studies considered different burst samples. Both Jimenez *et al.*[7] and my recent study[1] used a sample of 9 bursts with spectroscopic redshifts and BATSE spectra. The energy fluences were calculated by fitting the GRB spectral function[4] to the BATSE spectra, and then integrating the fits over the 20–2000 keV energy band and the burst duration. In my study[1] I also considered the 17 burst sample of Frail *et al.*[6] which adds bursts observed by *Beppo-SAX*, *Ulysses*, KONUS and *NEAR* to the 9 BATSE bursts. For these two samples the detection thresholds are unknown since the criteria for attempting to localize the bursts and determine their redshifts have not been reported. Finally, in my study I used the 220 BATSE bursts with redshifts determined through the conjectured correlation between light curve variability and peak burst luminosity[5]; this sample has a well-defined threshold for including a burst.

Jimenez et al.[7] extended their sample of bursts with spectroscopic redshifts by including bursts with only a host galaxy magnitude. A redshift probability distribution can be derived from a host galaxy magnitude using an empirical galaxy redshift distribution and assuming a model for the rate at which bursts occur in galaxies. Galaxy surveys such as the Hubble Deep Field weight each detected galaxy equally, yet in most burst scenarios bursts occur preferentially in massive or luminous galaxies. Jimenez et al. tested various weighting schemes using a sample of 10 bursts with both host galaxy magnitudes and spectroscopic redshifts. The test consisted of calculating a likelihood using the redshift probability distributions evaluated at the observed spectroscopic redshifts. The redshift probability distribution used in this test should be modified to include only the redshift range over which the observations could have determined the redshift: one of the few spectral lines detectable in the spectra of faint galaxies must have been redshifted into the telescope's bandpass. Once again a detection threshold is required for the statistical analysis of a burst sample. We found that weighting the empirical galaxy redshift distribution (derived from the Hubble Deep Field) by the host galaxy's luminosity at the time of the burst was favored. This result is relevant to progenitor scenarios (e.g., the galaxy luminosity for $z > 1$ may be proportional to the star formation rate, consistent with the progenitors being massive, short-lived stars), although a proper analysis requires sophisticated modeling of a galaxy's luminosity history.

In my study of the burst energy[1] I found that the lognormal and power law distribution functions are both acceptable descriptions of the data because the average of the cumulative probability is consistent with the expected value of 1/2 within the uncertainty resulting from the sample size. The Bayesian odds ratio demonstrated that neither function was favored over the other (odds ratio of ~ 1) for the two small burst samples with spectroscopic redshifts, but the lognormal function is favored (odds ratio of $\sim 10^4$) for the large sample with redshifts derived from the variability-luminosity correlation. The three samples give significantly different best-fit parameter values, which may result from the small sample sizes, the poorly determined detection thresholds for the spectroscopic redshift samples, and the uncertain validity of the variability-luminosity correlation. For example, for the central energy of the lognormal distribution I find: $E_0 = 1.3 \times 10^{53}$ erg (with a 90% confidence range of 0.016–3.2$\times 10^{53}$ erg) for the 9 burst BATSE sample; $E_0 = 5.2 \times 10^{53}$ erg (0.016–1.0$\times 10^{53}$ erg 90% confidence range) for the 17 burst Frail et al.[6] sample; and $E_0 = 0.12 \times 10^{53}$ erg (0.02–0.23$\times 10^{53}$ erg 90% confidence range) for the 220 burst variability-luminosity[5] sample. For all three samples the logarithmic width is nearly the same (the standard deviation of $\ln E$ is ~ 2).

The likelihood contours for the 2 parameters of the lognormal distribution have a ridge of high likelihood running from the peak towards lower central energy E_0 and larger logarithmic width σ. Similarly, the low energy cutoff for a simple power law cannot be determined—we only know that it is below the lowest energy observed. Thus for both distribution functions the likelihood does not rule out the possibility that the actual energy distribution is wider than observed, with the low end unobservable because of the detection threshold. Only a larger burst sample sensitive to fainter bursts will determine the extent of the actual energy distribution function.

Implicit in this statistical methodology are assumptions about the burst sample. I assume that the energy distribution does not evolve with redshift; this can eventually be tested by subdividing a larger sample. Further, I assume that there is no correlation between a burst's energy and the ability to determine its redshift; a burst's intrinsic intensity is assumed to be unrelated to environmental factors which promote or suppress the afterglow necessary to localize the burst.

As mentioned above, an accurate determination of the energy distribution requires the fluence threshold for including a burst in the sample. In general, the statistical determination of a burst property's distribution requires an understanding of the thresholds for all observational steps. In the case of the energy distribution, the most restrictive threshold is the determination of the redshift. A decision was made that: a) the burst was bright enough to attempt an afterglow detection; b) the afterglow was sufficiently well-localized to find a host galaxy; and c) the afterglow or the host galaxy are bright enough for spectroscopic observations. Thus, clear criteria are needed for the systematic ground-based observations which will follow-up the large number of well-localized bursts anticipated from *HETE-II* and *SWIFT*. Only with an understanding of the detection thresholds will burst property distributions be determined quantitatively.

ACKNOWLEDGMENTS

I thank Dieter Hartmann for his comments on this paper. This work was performed under the auspices of the U.S. Department of Energy by the Los Alamos National Laboratory under Contract No. W-7405-Eng-36.

REFERENCES

1. Band, D. L., *Ap. J.*, submitted, (2001).
2. Band, D. L., and Hartmann, D. H., *Ap. J.* **493**, 555–562 (1998).
3. Band, D. L., Hartmann, D. H., and Schaefer, B. E., *Ap. J.* **514**, 862–868 (1998).
4. Band, D., Matteson, J., Schaefer, B., Teegarden, B., Cline, T., Paciesas, W., Pendleton, G., Fishman, G., Meegan, C., Wilson, R., and Lestrade, P., *Ap. J.* **413**, 281–292 (1993).
5. Fenimore, E., and Ramirez-Ruiz, E., *Ap. J.*, in press (2001).
6. Frail, D., *et al.*, *Nature*, submitted (2001).
7. Jimenez, R., Band, D., Piran T., *Ap. J.*, in press (2001).
8. Larson, S. B., McLean, I. S., and Becklin, E. E., *Ap. J. Lett.* **460**, L95–97 (1996).
9. Meegan, C., *et al.*, *Nature* **355**, 143-145 (1992).
10. Piran, T., *Phys. Rep.* **333**, 529–553 (2000).
11. Schaefer, B., in *Gamma-Ray Bursts: Observations, Analyses and Theories*, ed. C. Ho, R. I. Epstein, and E. E. Fenimore (Cambridge: Cambridge University Press), 107-110 (1992).

The Blandford-Znajek mechanism with Vacuum Breakdown as a Gamma-ray Burst Progenitor

Guido Barbiellini* and Francesco Longo[†]

*Department of Physics University of Trieste and INFN, Trieste
[†]Department of Physics University of Ferrara and INFN, Ferrara

Abstract.
The energetics of the long duration GRB phenomenum is compared with the BZ mechanism. A rough estimate of the energy extracted from a rotating Black Hole with the Blandford-Znajek mechanism is evaluated with a very simple assumption: an inelastic collision between the rotating BH and an accreting torus. The GRB energetics requires an high magnetic field that breaks down the vacuum around the BH and gives origin to a $e^{\pm}$ fireball.

PHENOMENOLOGICAL OVERVIEW

Gamma-ray Bursts (GRBs) until a few years ago were largely devoid of any observable counterpart at any other wavelengths. However, a dramatic development in the last several years has been the measurement and localization of fading X-ray signals from some GRBs, lasting typically for days and making possible the optical and radio detection of afterglows, which mark the location of the GRB event. These afterglows in turn enabled the measurement of redshift distances, the identification of host galaxies, and the confirmation that GRB were at cosmological distances (for a recent brief review see [1]). The temporal decay of the emission in different frequencies for several GRBs has been interpreted according to the fireball model and suggested jet beaming with opening angle $\theta \sim 4°$ [2].

Another important discovery made in the last year is the presence of iron lines in the X-ray spectrum of GRBs (for example [3, 4, 5]). This provides a powerful tool to understand the nature and the environment of GRB primary sources [6, 7]. The presence of strong iron lines implies a rich environment located very close to the GRB and it may be an argument in favour of massive-star progenitor models of GRB [8, 9, 10, 11].

The presence of an iron cloud is in favour of the interpretation of GRBs as a second step of the residual of the primary explosion(e.g.[11]). In this interpretation the primary explosion leaves over a compact object that could be a rotating black hole, at the center of the environment consisting of ejecta. In this scenario, it is plausible the hypothesis of energy extraction from a rotating BH (compact object left over from the primary explosion), through the Blandford-Znajek mechanism [12], where the external magnetic field can be supplied by an Fe torus circulating around the BH at a distance R (of the order of R_s).

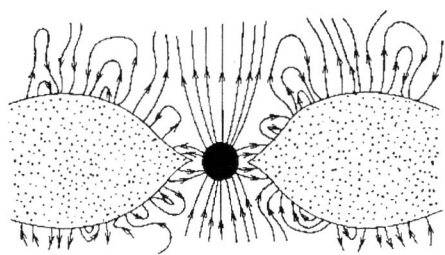

FIGURE 1. Possible configuration of the magnetic field close to the BH surrounded by an accretion disk (reprinted with permission from [13], © 1986 Yale University Press).

GAMMA-RAY BURST PROGENITOR

At cosmological distances the observed GRB fluxes imply energies of order of up to a solar rest-mass ($\sim 10^{54}$ erg), and from causality these must arise in regions whose size if of the order of kilometers in a time scale of the order of seconds. This implies that an $e^{\pm}, \gamma$ fireball must form, which would expand relativistically. The short life time of the GRB and the large distance suggest as the GRB source a compact system undergoing a very energetic transition with high an efficient mechanism for conversion of energy into γ-ray emission. Accreting black holes are so good candidates for being source of GRBs.

Blandford and Znajek have proposed an interaction between a rotating BH and an accretion disk to explain the energetics of Active Galactic Nuclei[12]. The same mechanism could be a good candidate for GRB engines[10, 14]. In the BZ mechanism the magnetic field of the accretion disk acts as a brake on the BH, and the energy output is mainly due to the loss of rotational energy. The rotational energy for a maximally rotating BH, with the rotation parameter $\tilde{a} = \frac{Jc}{M^2 G} = 1$, is $0.29\, Mc^2$. Even with the optimal efficiency the energy available for the BZ mechanism is [14]:

$$E_{BZ} \simeq 0.3 \cdot 10^{54} \left(\frac{M}{M_\odot} \right) \text{erg}$$

In the following considerations it we assume that a dissipative interaction is at work between the BH and the torus surrounding it. The dissipative interaction is due to an internal torque. If the short interaction is treated as an inelastic shock it is possible to apply the angular momentum conservation law (in its classical mechanics approximation).

$$I_{BH}\Omega_{BH} + I_{disk}\Omega_{disk} = I\Omega$$

where I_{BH} and I_{disk} are the momenta of inertia of the BH and the disk, Ω_{BH} and Ω_{disk} are the angular velocities. I and Ω are the same quantities for the resulting BH. In this approximation the loss of kinetic energy is

$$\Delta E = \frac{1}{2} \frac{I_{BH} I_{disk}}{I_{BH} + I_{disk}} (\Omega_{BH} - \Omega_{disk})^2$$

Let us consider as a typical configuration a BH with a mass of $10\,M_\odot$ with a rotating disk of $0.1\,M_\odot$ at the last stable orbit $R_l \sim 3R_s$:

$$I_{disk} = M_{disk} \cdot 9R_s^2 \qquad I_{BH} = \tfrac{2}{5}M_{BH}R_s^2$$
$$I_{disk} < I_{BH} \qquad \Delta E = \tfrac{1}{2}I_{disk}(\Omega_{BH} - \Omega_{disk})^2$$

Assuming that $\Omega_{disk} < \Omega_{BH}$, the energy availaible for the BZ process to power the GRB is:

$$\Delta E = \frac{1}{2}I_{disk}\Omega_{BH}^2 \sim 8 \cdot 10^{53}\left(\frac{M_{disk}}{0.1M_\odot}\right)\,\mathrm{erg}$$

where we have used $I_{disk} = 9M_{disk}R_s^2$ and $\Omega_{BH} \sim \frac{c}{R_s}$.

VACUUM BREAKDOWN

If the environment is sufficiently clean, the GRB fireball could be generated by the vacuum breakdown[15] in the volume close to the polar cap of the BH. The magnetic field required to explain the high luminosity of GRB generates an electric field that could break down the vacuum.

The luminosity of the BZ mechanism is [10]:

$$L_{BZ} \sim 10^{51}\left(\frac{B}{10^{15}\mathrm{G}}\right)^2\left(\frac{M_{BH}}{10M_\odot}\right)^2\,\mathrm{erg\,s^{-1}}$$

The BZ mechanism generates a voltage drop of the order [13]:

$$\Delta V = 10^{23} \cdot \left[\frac{M}{10M_\odot}\right]\left[\frac{B}{10^{15}\mathrm{G}}\right]\,\mathrm{V}$$

Assuming a mass of the BH of the order of $10\,M_\odot$, to this voltage drop corresponds an electric field in the proximity of the BH:

$$E = \frac{\Delta V}{2\pi R} = 10^{23}\frac{\frac{B}{10^{15}\mathrm{G}}}{6\pi \cdot 10^4}\,\frac{\mathrm{V}}{\mathrm{m}}$$

The energy acquired by an electric charge e over the reduced Compton electron wavelenght $\bar{\lambda} = \frac{\hbar}{m_e c}$ under this electric field can exceed $2 \cdot m_e c^2$ if:

$$eE\bar{\lambda} > 1.6 \cdot 10^{-6}\,\mathrm{erg}$$

From this relation it is possible to derive an estimate for the magnetic field:

$$B \sim 5 \cdot 10^{15}\,\mathrm{G}$$

In the proximity of the BH is possible to generate $e^{\pm}$ pairs than could give origin to the GRB fireball, provided a sufficiently clean environment in order to avoid previous electric field discharge.

ENERGETICS, TIME DURATION AND VARIABILITY

In the following we assume as hypothesis that the energy source of the GRB is the gravitational collapse of a torus of 0.1 $M_\odot$ onto a rotating BH of 10 $M_\odot$. The available energy from this inelastic collision derived in a semiclassical approximation is $\Delta E \sim 8 \cdot 10^{53}$ erg. The interaction between the torus and the BH is assumed to be the BZ mechanism with a magnetic field of the order of 10^{15} G.

The accreted material which releases its gravitational energy gives origin to a variable magnetic field that breaks the vacuum in a volume $\sim R_s^3$. The number of $e^\pm$ pairs produced is:

$$N_{e^\pm} = 2\frac{R_s^3}{\lambda^3} = 0.84 \cdot 10^{51}$$

The particle density is $\sim 3 \cdot 10^{31}$ $e^\pm$ cm^{-3}. If we assume that the magnetic energy $\frac{B^2}{2\mu_0} \sim 4 \cdot 10^{28}$ erg cm^{-3} is shared among them, each particle gets an energy $\varepsilon_0 \sim 1.3\eta_{acc} \cdot 10^{-3}$ erg, where η_{acc} is the acceleration efficiency. Each particle has an energy $\gamma_0 m_e c^2$, so it is possible to derive the relativistic Lorentz factor $\gamma_0 = \frac{\varepsilon_0}{m_e c^2} \sim 1.5\eta_{acc} \cdot 10^3$.

After the formation of the plasmoid containing $\sim 10^{51}$ $e^\pm$ pairs in a volume $V \sim R_s^3$, the particles undergo three important processes:

1) Particle acceleration in a time scale $t_{acc} \sim 10^3 \cdot \frac{\lambda}{c}$ to acquire an energy $\sim 10^3 \cdot mc^2$, since each particle acquires on average an energy $\sim mc^2$ every λ.

2) Momenta randomization in a time scale $t_{random} \sim \frac{l}{c}$, where l is the mean free path for $e^\pm$ interaction. $l = \frac{1}{\sigma n} \sim 0.3$ cm, using $\sigma = \frac{87(\text{nb})}{E(\text{GeV})^2}$ and $n = 3.5 \cdot 10^{31}$ cm^{-3}.

3) Single particle collimation by synchrotron radiation. The particles momentum components normal to the magnetic field $p_\perp$ are damped in a time scale $t_{coll} < \frac{\rho}{c}$, where ρ is the curvature radius. $\rho \simeq \frac{0.3 E_{GeV}}{c \cdot B_T}$ m. With the presence of a magnetic field of the order of 10^{15} G, the particles radiate all the energy corresponding to $p_\perp$ on a small fraction of a turn.

The momentum components perpendicular to the residual field line outside the plasmoid for all the particles are damped in a time $\sim \frac{R_s}{c}$ and the plasmoids becomes a stream of particles with velocity parallel to the external field lines with $\gamma \sim \frac{1}{3}\gamma_0$. As a result the plasmoid travels as a parallel stream with bulk Lorentz factor $\Gamma = \gamma \sim 500\eta_{acc}$.

The energy of the particles in the plasmoid before the cooling by synchrotron emission is:

$$E_{particles} = 2\frac{R^3}{\lambda^3} \cdot m_e c^2 \Gamma \sim 3.4\eta_{acc} \cdot 10^{47} \text{erg}$$

The available energy in the overall inelastic collision is $\Delta E \sim 8 \cdot 10^{53}$ erg, so that the emission of plasmoids could happen $N_{plasmoid}$ time where:

$$N_{plasmoid} \sim \eta_B \frac{\Delta E}{E_{plasmoid}} = 2.4\eta_B \cdot 10^6$$

where we have taken into account also the efficiency, η_B, for conversion of mechanical energy into the magnetic field, and $E_{\text{plasmoid}} = E_{\text{particles}}/\eta_{\text{acc}}$ is the total energy available in each plasmoid.

Since the time duration of the engine is smaller or equal to the observation time $t_{\text{obs}} \sim 10^2$s, the average time separation between two consecutive plasmoids is $t_s \sim t_{\text{obs}}/N_{\text{plasmoid}} \sim 4\eta_B^{-1} \cdot 10^{-5}$s. In the fireball model the prompt γ emission of the GRB is due to internal shocks between plasmoids proceeding at different speed.

CONCLUSIONS

This models predicts for long duration GRB a pulsed emission from 10^5 emitted plasmoids (with $\eta_B \sim 0.1$) with an average time separation $\Delta t \sim \frac{t_{\text{obs}}}{N_{\text{plasmoids}}} \sim 10^{-4}$ s, corresponding to a distance in space of $\sim 1 R_s$. The train of "sausage" plasmoids is $\sim 10^{10}$m at the end of engine activity. Its lenght is slightly modified during the internal shock phase so that the overal time duration is dominated by the duration of engine activity, the shortest time variability instead is determined by the plasmoids interactions.

ACKNOWLEDGMENTS

We thank A.Celotti, G.Cocconi and M.Tavani for critical reading and valuable comments and G.Fishman and N.Gehrels for the encouragement.

REFERENCES

1. Mészáros P. (2001), Science, **291**, 79
2. Frail D.A. *et al.* (2001), Nature submitted, `astro-ph/0102282`
3. Amati L. *et al* (2000), Science **290**, 953.
4. Piro L. *et al* (2000), Science **290**, 955.
5. Antonelli L.A. *et al.* (2000), ApJ **545**, L39
6. Rees M.J. & Mészáros P. (2000), ApJ **545**, L73
7. Vietri M. *et al* (2001), ApJ **550**, L43
8. Woosley S.E. (1993), ApJ **405**, 273
9. Paczynski B. (1993), in *Texas/Pascos 92*, Akerlof C.W. and Srednicki M.A. eds., Annals of the New York Academy of Sciences **688**, 321
10. Paczynski B. (1998), ApJ **494**, L95
11. Vietri M. and Stella L. (1998), ApJ **507**, L45
12. Blandford R.D. & Znajek R.L. (1977), MNRAS **179**, 433
13. Thorne K.S., Price R.H. and Macdonald D.A.(1986), *Black holes: the membrane paradigm*, Yale University Press
14. Lee H.K. *et al.* (2000), Physics Reports **325** 83
15. Ruffini R. (1998), `astro-ph/9811232`, in *Black holes and high energy astrophysics*, Sato, H. and Sugiyama, N., eds. Proceedings of the Yamada conference XLIX.

Blazars Active Galactic Nuclei

Very High Energy Gamma Rays from Blazars

James H. Buckley

Washington University, Dept. of Physics, St. Louis, MO, 63130, USA

Abstract.
Very high energy (GeV to TeV) γ-ray observations provide a new spectral window on high energy astrophysical phenomena. VHE gamma-ray observations of the BL Lacs Mrk 421 and Mrk 501 have provided new information about the nature of the relativistic jets in these objects and the origin of the nonthermal emission. These sources also provide probes of the radiation fields in the nuclear region of these active galaxies and in intergalactic space. Observations of sub-hour-scale TeV flares provide constraints on the emission mechanism, and on any energy dependent dispersion in the velocity of light. This paper summarizes the status of VHE γ-ray observations of blazars and highlights some of the most important recent measurements.

INTRODUCTION

The *Energetic Gamma-Ray Experiment Telescope* (EGRET) on board the *Compton Gamma-Ray Observatory* (CGRO) provided evidence for variable γ-ray emission from a number of AGNs. The types of AGN detected at high energies, which include flat spectrum radio quasars (FSRQs) and BL Lacertae (BL Lac) objects, are collectively referred to as *blazars*. Blazar emission is dominated by highly variable, nonthermal continuum emission from an unresolved nucleus. The Whipple Observatory 10m atmospheric Cherenkov telescope provided remarkable evidence for the extension of the emission spectra to TeV energies for some of these objects. The SEDs of these objects have a double-peaked shape (see Fig. 1) with a synchrotron component that peaks in the UV or X-ray band, and a second component typically rising in the X-ray range and peaking at energies between 1 MeV and 1 TeV [1]. The most natural explanation of the second peak is inverse-Compton scattering of ambient or synchrotron photons [2] although other possibilities have not been ruled out [3].

The EGRET catalog includes more than 65 blazars between 30 MeV and 30 GeV[4]. Of the ~50 sources reported by the EGRET team in the second catalog [5, 6] roughly 37 are FSRQs, and 14 are BL Lac objects [6]. The EGRET sources are all radio-loud, flat-spectrum radio sources and are detected over a range of redshifts between 0.03 and 2.28. They are characterized by two component spectra with peak power in the infrared to optical waveband and in the 10 MeV to GeV range. For many of the GeV blazars, the total power output of these sources peaks in the γ-ray waveband.

Whipple observations of the vast majority of EGRET blazars yielded only upper limits [17, 18, 19]; Mrk 421 is the single exception. Subsequent searches for emission from X-ray selected BL Lac objects led to the detection of Mrk 501, and four other as yet unconfirmed sources. The spectral energy distributions (SEDs) observed for these sources show higher energy synchrotron and γ-ray peaks, and comparable power output at the synchrotron and γ-ray peak. These observations are well described by the classification scheme of Padovani and Giommi [7] where the EGRET sources include low energy peaked BL Lacs (LBLs) and the VHE sources belong to the high energy peaked BL Lacs (HBLs).

Detailed studies of two of these sources, Mrk 421 and Mrk 501, have led to significant advances in our understanding of blazars and AGN jets. Because of their time-variable emission and broadband spectra, BL Lacs can best be understood through simultaneous multi-wavelength observations. The sensitivity of VHE telescopes to sub-hour scale variability [8], and the measurement of spectra between 200 GeV and 10 TeV make VHE observations an important new addition to multi-wavelength observations of AGN. Once the source mechanism of γ-ray production is understood, these objects can be used as probes of the low energy radiation fields in the object and filling intergalactic space, and may even provide probes of spacetime itself.

THE VHE BLAZAR CATALOG

Very high energy (VHE) emission extending to TeV energies has been observed at high levels of significance from Mrk 421 (at $z=0.031$) [9] and Mrk 501 (at ($z=0.034$) [10]. Other credible, but as yet unconfirmed detections include that of 1ES 2344+514 ($z=0.044$) [11], 1ES 2155-304 ($z=0.117$) [12], 1ES1959+650 ($z=0.048$) [13] and now a sixth potential VHE source 1H1426+428 ($z=0.13$) announced at this meeting [14]. Table 1 summarizes the properties of these sources.

These objects appear to form a new class, distinct from the EGRET sources. All are high-energy peaked [7] BL Lacs (HBLs) defined as sources with their synchrotron emission peaked in the UV/X-ray band (see, e.g., Fig. 1). The correspondence of the position of the peak of the synchrotron and γ-ray energy is naturally explained in models where the same population of electrons produces both spectral components. Proton induced cascade models [3] might also reproduce the spectra, but require an extra parameter that governs the termination of the proton spectrum.

The sensitivity of EGRET for a one-year exposure is comparable to that of Whipple for a 50 hour exposure for a source with spectral index of 2.2. The failure of ACTs to detect any but the nearest AGNs therefore can not be explained as a lack of sensitivity, but requires a cut-off in the γ-ray spectra of the EGRET sources. This cutoff could be intrinsic to the electron acceleration mechanism, due to absorption off of ambient photons from the accreting nuclear region [20], or caused by absorption by $\gamma-\gamma$ pair production off of the diffuse extragalactic background radiation [21, 22]. This apparent liability can be turned into an asset; high energy BL Lacs can serve as a unique probe of the radiation fields around the nucleus of the active galaxy and filling intergalactic space by virtue of their interaction with low energy (IR-UV) photons to produce electron positron pairs.

While pair absorption off of the diffuse intergalactic IR is the limiting factor that defines the TeV horizon, it does not explain why nearby EGRET sources like BL Lacertae and W Comae have not been detected [15, 23]. Even with a handful of TeV detections and a larger number of upper limits it is possible to put unified models for

TABLE 1. Properties of VHE blazars.

BL Lac Name	Redshift z	EGRET flux[a] ($E >100$ MeV) (10^{-7}cm^{-2}s^{-1})	F_X^b 2 keV (μJy)	Visible Magnitude M_V (mag)	F_R 5 GHz (mJy)
Mrk421	0.031	1.4±0.2	3.917	14.4	722
Mrk501	0.033	3.2±1.3	3.702	14.4	1371
1ES2344+514	0.044	<0.7	1.142	15.5	215
1ES1959+650	0.048		3.645	13.7	252
PKS2155-304	0.116	3.2±0.8	5.746	13.5	310
1H1426+428	0.130		2.678	16.4	38

Source: [b]Radio, optical and X-ray data from Perlman [26], [a][6, 27]

blazars to the test. In the framework of Fossati et al. [24] the low energy peaked EGRET BL Lacs correspond to AGNs with a more luminous nuclear emission component than HBLs. The relatively high ambient photon density in the LBLs is Comptonized to γ-ray energies with an intensity that exceeds the SSC power explaining the dominant emission in the γ-ray band. In these objects inverse-Compton losses dominate over synchrotron losses and therefore limit the maximum electron energy achieved by shock acceleration (this limit occurs at the energy at which the cooling time is equal to the acceleration time [25]). Thus one also obtains a natural explanation for the lower energies of the peak synchrotron and IC power in these objects. In HBLs, the ambient photon fields are presumably weaker and self-Compton emission dominates over Comptonization of external photons (EC). Electrons can reach higher energies by shock acceleration, and the peaks in the SED move to higher energies and have more nearly equal peak power. This model has proved itself to be consistent with the data and has served as a useful paradigm for searching for new VHE sources.

MULTIWAVELENGTH OBSERVATIONS: SPECTRA

The SEDs shown in Fig. 1, and 4 summarize the results of a number of different measurements of the X-ray and VHE spectra of Mrk 421 and Mrk 501, and compare them with simple SSC models [25, 27]. In these Figures we in include the results of the Whipple, CAT [28], and HEGRA [?] experiments. The agreement between the spectral measurements is exceptionally good for Mrk 501. For Mrk 421, the lack of truly simultaneous measurements and rapid variability may account for the larger variation in spectra. The overall trend for both Mrk 421 and Mrk 501 for all experiments can be explained if the VHE spectra get harder when the TeV flux increases, due to a shift in the peak energies of the SEDs.

MULTIWAVELENGTH OBSERVATIONS: VARIABILITY

Mrk 421 Observations, May 1996: Extreme variability on time-scales from minutes to years is the most distinctive feature of the emission from BL Lac objects. Both Mrk 421 and Mrk 501 have been detected in emission states that range over nearly two orders of

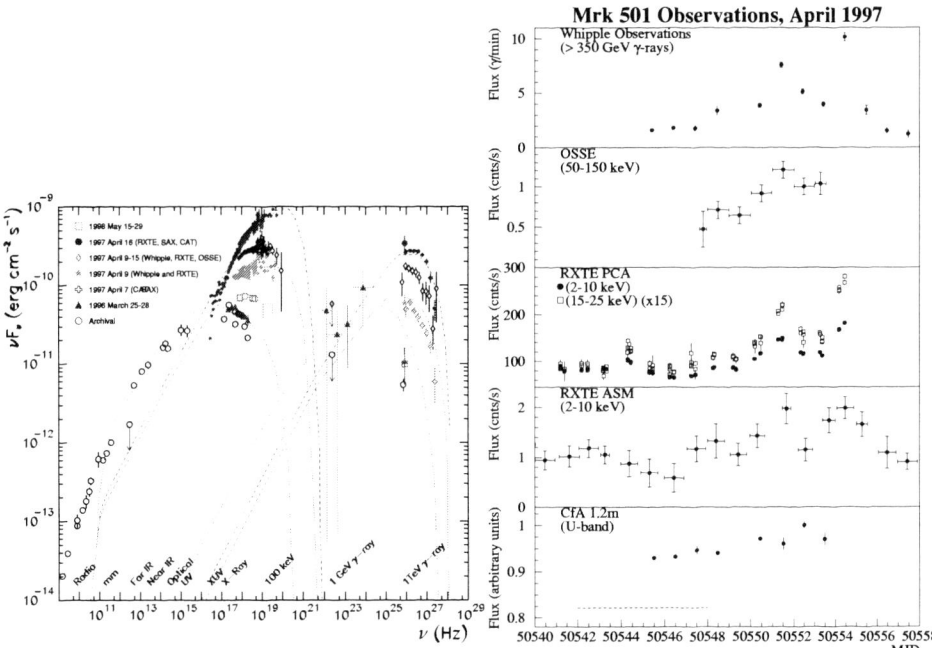

FIGURE 1. *Left:* SED of Mrk 501 from contemporaneous and archival observations. Data points come from Buckley [31] and references therein including the Whipple spectrum, RXTE (PCA and HEXTE) data and OSSE spectrum from Catanese et al. [29]. The CAT spectral points from Tavernet [32] and Djannati-Ataï et al. [28] and simultaneous BeppoSAX data from Pian et al. [28] are also shown. The curves are SSC model fits [27, 25]. *Right:* Multi-wavelength observations of Mrk 501[29]): (a) γ-ray, (b) hard X-ray, (c) soft X-ray, (d) U-band optical light curves during the period 1997 April 2-20 (April 2 corresponds to MJD 50540). The dashed line in (d) indicates the optical flux in 1997 March.

magnitude in flux. Monitoring data taken on Mrk 421 over the years 1995 [33] to 2001 [34] show that the γ-ray emission is characterized by a succession of rapid hour-scale flares with relatively symmetric profiles (e.g., see Figure 2).

Observations of Mrk 421 by the Whipple collaboration in 1996 revealed two dramatic flares. On May 7, a flare with the largest flux observed from any VHE γ-ray source to date was detected [8]. The flux increased over a $\sim$2 hour period, beginning at a rate twice as high as any previously observed flare and reaching a counting rate $\approx$10 times the steady rate observed from the Crab Nebula. The doubling time of this flare was $\sim$1 hour. The second flare, observed eight days later, was remarkable for its very short duration (Fig. 6): the doubling and decay times of the flare were both $\sim$15 minutes, the fastest ever observed in any blazar at any γ-ray energy.

Mrk 501, April 1997: A multi-wavelength campaign conducted on Mrk 501 in 1997 revealed a strong correlation between TeV γ-rays and soft X-rays and with the 50-500 keV band detected by OSSE (Fig. 1). Mrk 421 has never been detected by OSSE, while the Mrk 501 flux was the highest recorded by OSSE from any blazar [35]. Also, unlike

Mrk 421, the amplitude of the TeV γ-ray variations in Mrk 501 are typically larger than the X-rays. A multi-wavelength campaign on Mrk 501 conducted in 1998 during a more quiescent emission state again showed a detectable flux in the OSSE energy range, indicating that even in a relatively low flux state, the hard X-ray spectrum of Mrk 501 is significantly different than Mrk 421.

Mrk 421 March 2001: Multiwavelength observations of Mrk 421 were made during the period March 18, 2001 to April 1, 2001 with the Whipple gamma-ray telescope, the Whipple Observatory's 1.2 m optical telescope, the newly completed 0.5 m Antipodal Transient Observatory (ATO) optical telescope, and with the RXTE PCA detector. To better sample the extremely rapid variability of Mrk421, this campaign included the highest temporal density X-ray observations possible with a nearly continuous >330 ks exposure with RXTE [36]. Numerous ground-based atmospheric Cherenkov and optical observations were scheduled during this period to improve longitudinal/temporal coverage. Frequent correlated hour-scale X-ray and γ-ray flares were observed (see Fig. 2). Intra-day optical variability was also observed, although the details of the temporal correlation is unclear [34].

X-ray data were obtained with the Proportional Counter Array (PCA) aboard the Rossi X-ray Timing Explorer (RXTE). These data will be presented in full in a subsequent paper [36]. Fig. 2 shows a subset of these data showing the close correlation of the well-sampled TeV and X-ray (2-10 keV) lightcurves on March 19, 2001 [34].

The dramatic flaring of Mrk 421 and fortuitous multiwavelength observations provide invaluable new data. While it is too early to draw detailed conclusions, we note that the symmetric flares at keV and TeV energies may indicate comparable acceleration times and synchrotron cooling times as predicted by diffusive shock acceleration models [25]. However, the short optical variability timescales are significantly shorter than the synchrotron cooling time for typical jet parameters (e.g., a 30 to 100 mGauss field, and Doppler factor $\delta \sim 10$).

INTERPRETATION: BLAZAR EMISSION MODELS

The simplest viable model for blazar emission is the one-zone synchrotron self-Compton (SSC) model where energetic electrons in a compact emission region up-scatter their own synchrotron radiation. As shown in Fig. 1, such a model results in very good fits to the Mrk 501 SED. In the SSC model, the intensity of the synchrotron radiation is proportional to the magnetic energy density and the number density of electrons $I_{\text{synch}} \propto n_e$. Since these same electrons up-scatter this radiation, the IC emission scales as $I_{\text{IC}} \propto n_e^2$. Thus we expect $I_{\text{IC}} \propto I_{\text{synch}}^2$.

Krawczynski et al. [37] examined the correlation of TeV γ-ray and X-ray intensity for several strong flares of Mrk 501 in 1997. The results, plotted in Figure 3, show evidence for such a quadratic dependence. (However the possibility of a baseline level of the X-ray emission probably can not be excluded.) These observations provide further support for SSC in Mrk 501.

In the framework of either the EC or SSC models the γ-ray and X-ray data can be used to constrain the Doppler factor δ and magnetic field B in the emission regions of Mrk 421 and Mrk 501. The maximum γ-ray (IC) energy $E_{\text{C,max}}$ provides a lower limit on the

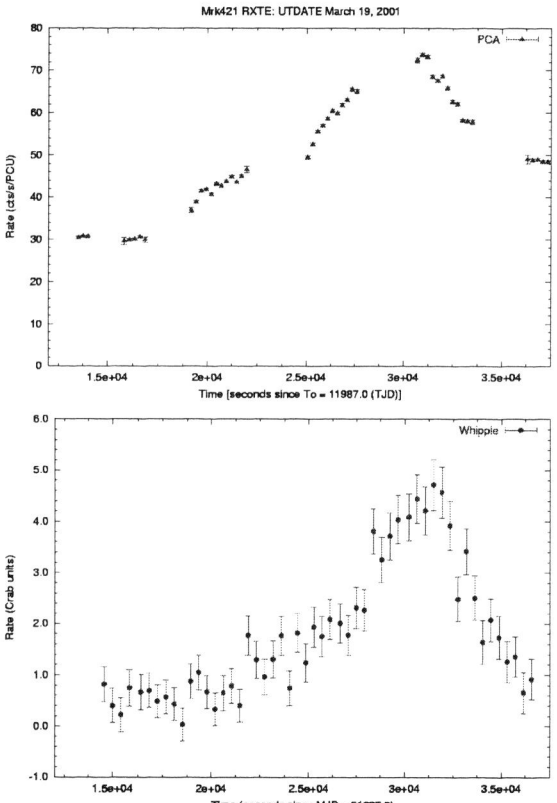

FIGURE 2. Simultaneous X-ray/ γ-ray flare observed on March 19, 2001. The 2-10 keV X-ray light curve was obtained with the PCA detector on RXTE [34, 36]; data points are binned in roughly 4 minute intervals.

maximum electron energy (with Lorentz factor $\gamma_{e,max}$) given by $\delta\gamma_{e,max} > E_{C,max}/m_e c^2$; combining this with the measured cut-off energy of the synchrotron emission $E_{syn,max}$ one obtains an upper limit on the magnetic field $B \lesssim 2 \times 10^{-2} E_{syn,max} \delta E_{C,max}^{-2}$ (where $E_{C,max}$ is in TeV). A lower limit on the magnetic field follows from the requirement that the electron cooling time, $t_{e,cool} \approx 2 \times 10^8 \delta^{-1} \gamma_e^{-1} B^{-2}$s, must be less than the observed flare decay timescale. These limits depend on the Doppler factor of the jet and in some cases cannot be satisfied unless δ is significantly greater than unity [29, 38]. Typically, these arguments lead to predictions of ~100 mGauss fields and Doppler factors $\delta > 10$ to 40 for Mrk 421. Similar values for Mrk 501 but typically with a reduced lower limit on the Doppler factor.

Since Mrk 421 may be a transitional object between the EGRET (LBL) and VHE (HBL) blazars, I will consider an example of a simple model fit to the Mrk 421 SED that includes both SSC and EC components: The model shown in Fig. 4 includes the following features: I assume a self-consistent shock-accelerated electron spectrum with

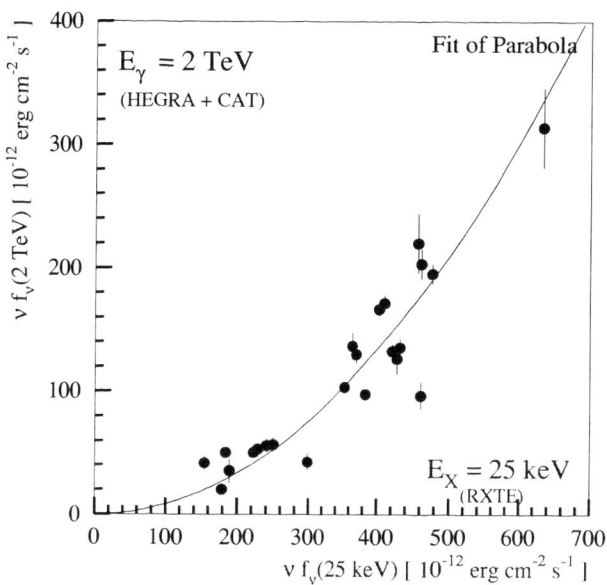

FIGURE 3. Plot of TeV γ-ray flux versus X-ray flux measured with the HEGRA experiment during an intense flare of Mrk 501 (courtesy Henric Krawczynski).

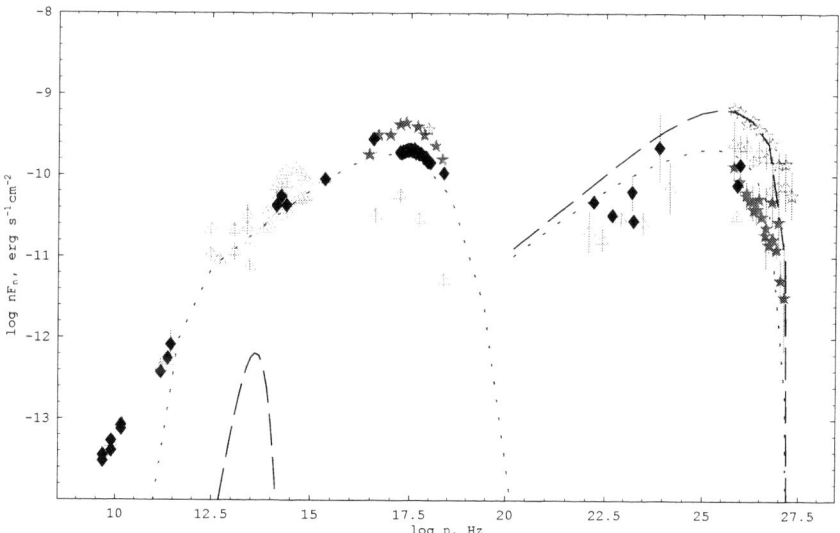

FIGURE 4. Model fit to Mrk 421 SED with both an SSC and external Compton component. The data is taken from Buckley [31] and references therein. Also shown are the Whipple 1996 May 7 and 1996 May 15 spectra from [39] as well as the HEGRA spectrum measured in 1998 April [30]) Also shown are the ASCA spectral points from the high X-ray state measured during 1995 April [40].

its maximum energy limited by the condition that the acceleration time is equal to the shortest cooling timescale (from synchrotron radiation or Comptonization of ambient photons); the full synchrotron kernel is used and self absorption is taken into account; the full Klein-Nishina cross-section is used for the Compton kernel [25]; ambient photons (modeled as an isotropic black-body distribution transformed into the jet frame) are considered as seed photons for inverse-Compton and an absorbing target for $\gamma\gamma$ pair production in the source. I further assume that although the shock is mildly relativistic in the bulk frame of the jet, the diffusive shock acceleration model is still applicable.

The dotted curve shows the synchrotron and self-Compton emission components, and the dashed curve shows the external Compton fit for the γ-ray emission and an adhoc IR black-body distribution of seed photons. (Only a small fraction of photons are Doppler boosted and then Comptonized.) The synchrotron model is fit to the data of [41]. Care is taken to ensure that the self-absorption feature is at sufficiently high energy to avoid overproducing mm-wavelength emission. This condition effectively places an upper limit on the size of the emission region. The external photon flux limits the maximum energy of electron shock acceleration, the intensity of the IC emission, and pair absorption of the γ-ray emission [25]. This model fit results in a magnetic field of $B = 0.08$ Gauss, requires a mean free path for electron scattering a factor of 500 times the Bohm limit for a shock of velocity $u_s = 0.1$ c (in the jet frame), an emission region 30 times the Schwartzschild radius of a 10^8 $M_\odot$ black hole, an electron spectral index of 2.2, and a soft photon field with a peak wavelength of 10 μm in the observer's frame. A Doppler factor of $\delta = 100$ is required to match the observed flux level. This simple model does not attempt to address the important effects of time-variability in the data. Such an extreme value for the Doppler and compact emission region is not *required* by the spectral data. The resulting acceleration and cooling timescales at the maximum electron energy are $\tau_{acc} \approx \tau_{cool} \approx 5$ minutes and the light crossing time is 2.5 minutes. These values are not far from those required to match the shortest observed flares [8] or the rapid optical variability observed in March 2001.

IR BACKGROUND CONSTRAINTS

The diffuse extragalactic background light (EBL) causes distortions in the spectra of distant AGNs due to the pair absorption process $\gamma_{TeV} + \gamma_{IR} \rightarrow e^+e^-$. The optical depth depends on an integral over the EBL spectrum from threshold for pair creation $\varepsilon_{IR} = (2m_ec^2)^2/2E_\gamma$ up to higher energies. Since high energy background photons have a lower number density than low energy photons, low energy γ-rays can travel a larger distance before they interact. This results in a spectral cutoff that moves to lower energies with increasing redshift.

From considering the effect of $\gamma-\gamma$ interactions on the TeV spectra of Mrk 421 and Mrk 501, new robust limits have been set on the density of extragalactic background photons. Biller et al. [42], e.g., derive an upper limit on $\varepsilon^2 n(\varepsilon)$ of 0.0027 eV/cm^3 in the energy interval 0.042 eV $< \varepsilon < 0.167$ eV, where ε is the energy of the IR photon and n is the spectral density of photons. The resulting limits are more than an order of magnitude more restrictive than direct observations in the 0.025-0.3 eV regime [42].

Figure 5 (Adapted from [43] and references therein) shows the status of observations

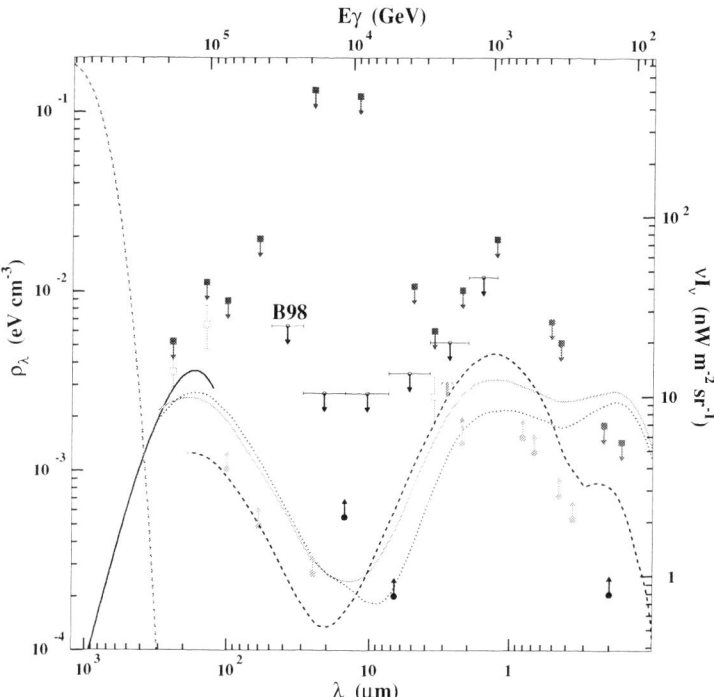

FIGURE 5. Spectrum of the extragalactic background light with initial *measurements* by DIRBE, upper limits by Whipple [42] and lower limits from K-band galaxy counts (adapted from [43]).

including Whipple upper limits [42] (B98), DIRBE measurements (recently retracted), and lower limits come from K-band galaxy counts. Model curves from [44] show sensitivity to the history of star formation (e.g., CHDM vs. SCDM) in the 1-10μm regime and sensitivity to the initial mass function and re-radiation by dust in the interval ($\lambda \lesssim 1\ \mu$m and $\lambda \gtrsim 10\ \mu$m).

LIMITS ON QUANTUM GRAVITY

On the Planck scale, classical gravity must break down and a quantum theory must take it's place. Any theory of quantum gravity is likely to result in a granular or frothy structure of space-time on very small distance scales due to quantum fluctuations. These fluctuations can turn empty space into into a dispersive medium for high energy photons (e.g., [45]). This small dispersion in the velocity of light can best be measured at very high energies for distant sources with short intrinsic variability timescales. In some theoretical frameworks, the relevant energy scale could be significantly less than the Planck mass $m_P \approx 10^{19}$ GeV and the length scale of the quantum fluctuations significantly larger than the Planck length $\approx 10^{-33}$ cm (e.g., $\sim 10^{16}$ GeV; [46]) making it accessible to GeV or TeV measurements of GRB or AGNs.

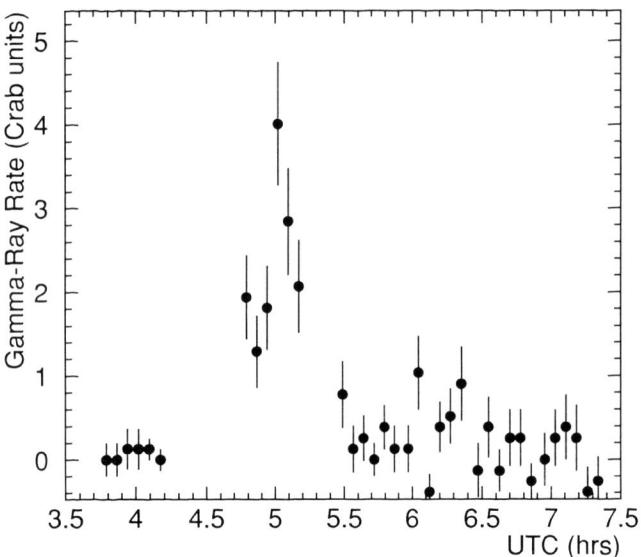

FIGURE 6. Mrk 421 flare observed on May 15, 1996 (reprinted with permission of *Nature* 383, 26 September 1996, pp. 320-321: ©1996 Nature Publishing Group[8].

If the energy dependence contains a first order Lorentz violating term then the propagation delay Δt for photons of energy E traveling a distance L is given by $\Delta t = \xi(E/E_{QG})(L/c)$. The observation of variability in Mrk 421 with $t = 280$ sec in the May 15 flare produces a limit $E_{QG} > 6 \times 10^{16}$ GeV [47]. If this first-order term is ignored, the first Lorentz invariant term $\Delta t \approx (aE/E_{QG})^2 (L/c)$ would require a limit $E_{QG}/a > 6 \times 10^9$ GeV. A VERITAS detection of a 1 minute flare at $E > 5$ TeV from a source at $z = 0.1$ would constrain QG effects within a factor of 5 of the Planck mass.

ACKNOWLEDGMENTS

I would like to thank the VERITAS, CAT and HEGRA collaborations, Giovanni Fossati, Mead Jordan, Arache Djannati-Atai, and Henric Krawczynski This work was supported in part by the U.S. Dept. of Energy.

REFERENCES

1. C. von Montigny, et al., *ApJ* **440**, 525 (1995).
2. M. Sikora, M.C. Begelman, M.J. Rees, *ApJ* **421**, 153 (1994).
3. Mannheim, K. *A&A* **269**, 67 (1993).
4. Hartman, R.C., et al., 1999, *ApJS* **123**, 79 (1999).
5. Thompson, D.J., et al., *ApJS* **101**, 259 (1995).
6. Mukherjee, R., et al., *ApJ* **490**, 116 (1997).
7. Padovani, P. & Giommi, P. *ApJ* **444**, 567 (1995).

8. J.A. Gaidos, et al., *Nature* **383**, 319 (1996).
9. Punch, M., et al., *Nature* **358**, 477 (1992).
10. Quinn, J., et al., *ApJ* **456**, L83 (1996).
11. Catanese, M., *ApJ* **501**, 616 (1998).
12. Chadwick, P.M., et al., *Astropart. Phys* **11**, 145 (1999).
13. Nishiyama, et al., in *26th ICRC (Salt Lake City, Utah)*, unpublished communication (1999).
14. Horan, D., et al., in *Proc. 27th ICRC (Hamburg, Germany)*, in press, (2001).
15. M. Catanese, *ApJ* **480**, 562 (1997).
16. M. Catanese, et al., in *Proc. of the 4th Compton Symp.* C.D. Dermer, M.S. Strickman, J.D. Kurfess, eds., AIP **410**, 1376 (1997).
17. A.D. Kerrick, et al., *ApJ* **452**, 588 (1995b).
18. J. Quinn, et al., in *Proc. of 24th ICRC (Rome)* **2**, 369 (1995).
19. D. Petry, et al., in *Proc. 25th ICRC (Durban)* **3**, 241 (1997).
20. Dermer, C.D., Schlickeiser, R., & Mastichiadis, A. *A&A* **256**, L27 (1992).
21. R.P. Gould, G.P. Schreder, *Phys. Rev.* **155**, 1408 (1967).
22. F.W. Stecker, O.C. De Jager, *ApJ* **415**, L71 (1993).
23. M. Catanese, et al., in *Proc. 25th ICRC* (Durban) **3**, 277 (1997).
24. Fossati, G., Maraschi, L., Celotti, A., Comastri, A., Ghisellini, G., *MNRAS* **299** 433 (1998).
25. Inoue, S., & Takahara, F., *ApJ* **463**, 555 (1996).
26. E.S. Perlman, et al., *ApJS* **104**, 251 (1996).
27. Kataoka, J., et al., *Astropart. Phys.* **11**, 149 (1999).
28. Djannati-Atai, A., et al. *A&A* **35**, 17 (1999).
29. M. Catanese, et al., *ApJ* **487**, L143 (1997a).
30. Aharonian, F., et al., *A&A* **349**, 29 (1999).
31. Buckley, J., *Astroparticle Phys.* **11**, 119 (1999).
32. Tavernet, J.P. et al., in *Proc. 26th ICRC (Salt Lake City, Utah)*, in press (1999).
33. J.H. Buckley, et al., *ApJ* **472**, L9 (1996).
34. M. Jordan, et al., in *Proc. 27th ICRC (Hamburg, Germany)*, in press, (2001).
35. K. McNaron-Brown, et al., *ApJ* **451**, 575 (1995).
36. Fossati, G., et al., in preparation (2001).
37. Krawczynski, H., Coppi, P.S., Maccarone, T., Aharonian, F.A., *A&A* **353**, 97 (2000).
38. J.H. Buckley, et al., in *Proc. of the 4th Compton Symp.* C.D. Dermer, M.S. Strickman, J.D. Kurfess, eds., AIP **410**, 1381 (1997).
39. F. Krennrich, et al., in *Proc. 26th ICRC (Salt Lake City, Utah)*, ed. D.B. Kieda, M.H. Salamon and B.L. Dingus, **3**, 301 (1999).
40. Takahashi, T., Madejski, G., & Kubo, H., *Astropart. Phys.* **11**, 177 (1999).
41. Macomb, D.J., et al., *ApJ* **449**, L99 (1995).
42. Biller, S.D., et al., *Phys. Rev. Lett.* **80**, 2992 (1998).
43. Vassiliev, V.V., *Astroparticle Physics* **12**, 217 (2000).
44. Primack, J.R., Bullock, J.S., Somerville, R.S., MacMinn, D., *Astroparticle Physics* **11**, 93 (1999).
45. Amelino-Camelia et al., *Nature* **393**, 319 (1998).
46. Witten, E., *Nucl. Phys. B* **471**, 135 (1996).
47. Biller, S.D., et al., *Phys. Rev. Letters* **83**, 2108 (1999).

Very High Energy Flaring Activity of Markarian 421 during 2000-2001 as seen by CAT and CELESTE Telescopes

Arache Djannati-Ataï for the CAT and CELESTE collaborations

PCC - Collège de France, 11 pl. Marcelin Berthelot, Paris 75231, France

Abstract.
During the observation seasons 2000-2001, Mkn 421 showed a very strong flaring activity, with variability time-scales sometimes as short as one hour. We report here on the observations made with the CAT and CELESTE Čerenkov telescopes. VHE spectral properties of Mkn 421, as measured by CAT above 250 GeV, are compared and discussed for both seasons.

INTRODUCTION

Since its discovery as a VHE source [10], Markarian 421 has been one of the most studied γ-ray-emitting blazars; it has been monitored by the CAT telescope, at energies above 250 GeV, since the commencement of observations in december 1996 [9]. Figure 1 shows the light curve for the the years 2000-2001: during the 2001 observation season (from December 2000 to April 2001) the mean flux ($\Phi_{>250\,\text{GeV}}=20.4\pm0.6\times 10^{-11}\,\text{cm}^{-2}\,\text{s}^{-1}$) was twice as high as that for the preceding season ($\Phi_{>250\,\text{GeV}}=9.0\pm 0.5\times 10^{-11}\,\text{cm}^{-2}\,\text{s}^{-1}$).

The CAT imaging telescope, operating in the French Pyrénées, has now a young sister, the CELESTE experiment. CELESTE currently consists of 40 heliostats and a system of secondary optics mounted on the 100 m tower of the former solar plant, which now operates as a low energy γ-ray telescope achieving a threshold of 60 GeV [11].

Simultaneous observations by CAT and CELESTE during 2000 period are shown in figure 2. The remarkable correlation between the two sets of measurements definitely validates the experimental method and analysis procedure of CELESTE .

CAT sampling of Markarian 421 in year 2001 was marked by two episodes of strong activity during the months of February and March 2001, with a spectacular variability during the night of March 23/24 where the source flux changed from a level of two to more than six times that of the Crabe nebula. Figure 4 shows the intra-night variability of Markarian 421 for 15 min timing samples during that night.

SPECTRUM MEASUREMENTS

The CAT spetcrum analysis procedure is presented in detail in [9]. It is based on a forward-folding method in which the detector effective area and energy resolution (de-

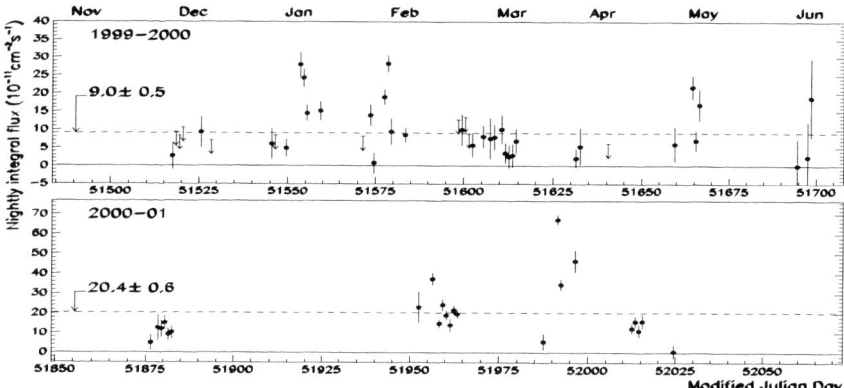

FIGURE 1. Markarian 421 nightly-averaged integral flux above 250 GeV between December, 1999 and May, 2001. The γ-ray effective area has been weighted using a differential index of −2.9, in order to estimate the integral flux for observations far from the Zenith (see [9]). Arrows stand for 2σ upper-limits when no signal was recorded, and dashed line shows the mean flux for each observation year.

termined by detailed simulations) are included. A maximum likelihood method is used to directly determine the relevant spectral parameters from the distributions of estimated energies of ON and OFF-source γ-ray candidates for different bins in zenith angle. The likelihood-function expression relies on the Poissonian distributions of ON and OFF

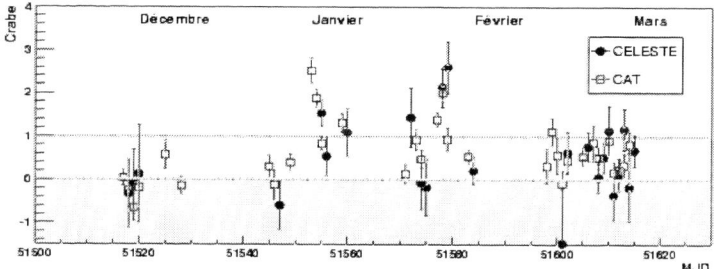

FIGURE 2. Simultaneous observations by CAT and CELESTE during 2000 in Crab units. The remarkable correlation between the two sets of measurements validates the experimental method and analysis procedure of CELESTE. The negative values correspond simply to non detections.

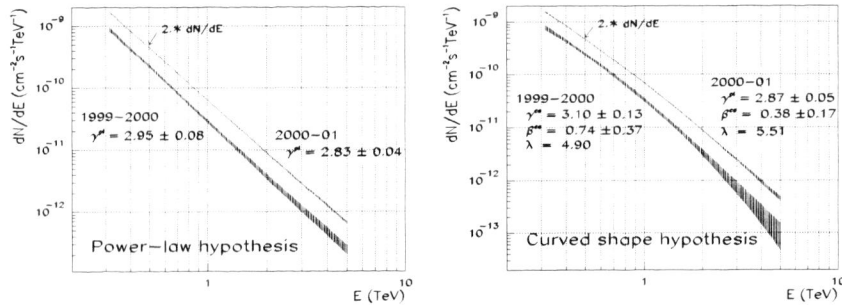

FIGURE 3. Mkn 421 time-averaged spectra between 0.3 and 5.0 TeV in 1999–2000 and 2000–01. The areas show the 68% confidence level contour given by the likelihood method. Left and right hand plots correspond to the power-law hypothesis and the curved shape hypothesis, repectively. The differential flux of 2000–01 is multiplied by a factor of 2 for the clarity of the figures.

events, thus allowing possible low statistics to be treated in a rigorous manner. Two hypotheses are successively considered for the differential γ-ray spectrum $\frac{d\phi}{dE}$:

i) a simple power law, $\phi_0^{pl} E_{TeV}^{-\gamma^{pl}}$ (hyp. H^{pl}), and

ii) a curved shape, $\phi_0^{cs} E_{TeV}^{-(\gamma^{cs}+\beta^{cs}\log_{10}E_{TeV})}$ (hyp. H^{cs}).

The relevance of a given hypothesis with respect to the other one is then estimated from their likelihood ratio (λ).

Fig. 3 shows the spectrum for the 1999–2000 and 2000–01 periods for both hypotheses. The fitted parameters are summarized in Table 1. These spectral shapes are consistent in spectral slope to less than 2σ on statistical errors and are compatible with a power-law, although there is some evidence for curvature: the likelihood ratio values are 4.90 (corresponding to a chance probability of 0.027) and 5.51 (a chance probability of 0.019) for 2000 and 2001 seasons, respectively. Thus, for 2001, the following characterisation can be retained:

$$\frac{d\phi}{dE} = (12.44 \pm 0.81^{stat} \pm 2.49^{syst})10^{-11}\,\text{cm}^{-2}\,\text{s}^{-1}\,\text{TeV}^{-1}$$

$$\times E_{TeV}^{-2.80\pm0.09^{stat}\pm0.06^{syst}-(0.81\pm0.27^{stat}\pm0.03^{syst})\log_{10}E_{TeV}}.$$

The spectrum during the flare of March 23/24, 2001, is given in figure 4. It shows significant curvature with a likelihood ratio of 11.3 (chance probability of 7.7×10^{-4}):

$$\frac{d\phi}{dE} = (12.44 \pm 0.81^{stat} \pm 2.49^{syst})10^{-11}\,\text{cm}^{-2}\,\text{s}^{-1}\,\text{TeV}^{-1}$$

$$\times E_{TeV}^{-2.80\pm0.09^{stat}\pm0.06^{syst}-(0.81\pm0.27^{stat}\pm0.03^{syst})\log_{10}E_{TeV}}.$$

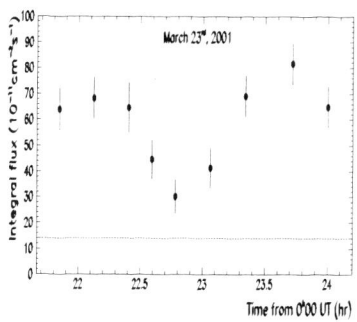 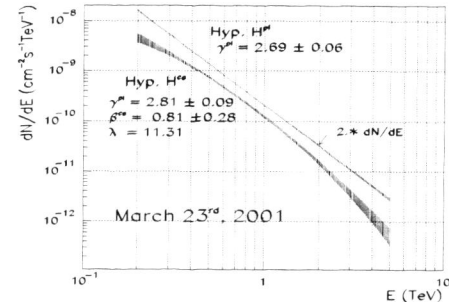

FIGURE 4. On the left: Integral flux above 250 GeV during the night of March 23 rd, 2001 (MJD 51991/92). Each point indicates for a ~15 min observation, the dashed line shows the Crab nebula level emission ($\Phi_{>250\,\mathrm{GeV}} = 14.10 \pm 0.35 \times 10^{-11}\,\mathrm{cm}^{-2}\,\mathrm{s}^{-1}$). The χ^2 per d.o.f corresponding to a constant flux is 5.64. On the right: Mkn 421 time-averaged spectrum between 0.3 and 5.0 TeV in March 23 rd, 2001. The curve labelled *Hyp. H^{pl}* uses a power-law shape, and the one referenced by *Hyp. H^{cs}* uses a curved shape. The differential flux of the power-law hypothesis is multiplied by a factor of two for the clarity of the figure.

TABLE 1. Parametrisation of Mkn 421 spectra as measured by CAT. Are indicated the observation period, the energy range used in the likelihood method, $\Delta \widetilde{E}_\gamma$, the total observed number of γ-ray events, S_γ, the spectral parameters obtained in the H^{pl} and H^{cs} hypotheses, and the likelihood ratio λ. We also quote the decorrelation energy, E_d, in the H^{pl} hypothesis. The energies are given in TeV, and the flux constants in units of $10^{-11}\,\mathrm{cm}^{-2}\,\mathrm{s}^{-1}\,\mathrm{TeV}^{-1}$.

Period	$\Delta \widetilde{E}_\gamma$ (TeV)	S_γ	ϕ_0^{pl}	γ^{pl}	E_d	ϕ_0^{cs}	γ^{cs}	β^{cs}	λ
1999–2000	0.3–5.0	1424±71	2.90±0.18	2.95±0.08	0.63	3.34±0.28	3.10±0.13	0.74±0.37	4.90
2000–01	0.3–5.0	5612±157	3.18±0.10	2.83±0.04	0.68	3.47±0.16	2.87±0.05	0.38±0.17	5.51
23/03/2001	0.3–5.0	1320±50	10.50±0.49	2.68±0.06	0.64	12.44±0.81	2.80±0.09	0.81±0.27	11.3

DISCUSSION

Measurement of spectral properties of blazars for different activity states constitutes one of the best means for the understanding of the underlying physical processes. Curved shape and spectral variability - in terms of hardness-intensity correlations - have already been observed for the other very well studied blazar, Markarian 501 , [4] [3] during its tremendous activity in 1997. Focusing on the recent data taken during March 2001 on Markarian 421 , a clear curvature is seen by CAT (likelihood ratio of 11.3 corresponding to a chance probability of 7.7×10^{-4}) while the hardening of the spectrum correponds to a 2.7 sigma effect when compared to the time averaged 2000 spectrum. One should mention here that during March 2001, the simultaneous X-ray measurements, which were carried out by the Rossi X-Ray Timing Explorer satellite, have also shown exceptional acitivity from Markarian 421 , with a very hard spectrum peaking at several keV [5] as compared to the previous years' X-ray campaign results.

The situation regarding hardness-intensity correlations remains somewhat unclear

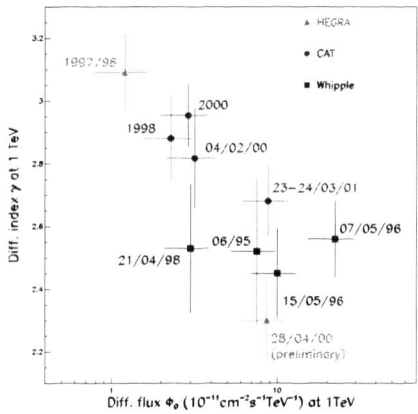

FIGURE 5. Compilation of spectral measurements on Markarian 421, obtained by the HEGRA ([1, 2]; [6]), CAT ([9] and this work) and Whipple ([12]; [7, 8]) experiments between 1995 and 2000. The results are given in the $\{\phi_0^{pl},\gamma^{pl}\}$ plane of spectral parameters in the power-law hypothesis. The error bars take account, for *all* measurements, of both statistical and systematic errors, the latter being conservative if unknown: $(\Delta\phi_0^{pl}/\phi_0^{pl})^{syst}=\pm 30\%$ and $(\Delta\gamma^{pl})^{syst}=\pm 0.10$ were used where these values are not quoted directly.

when comparing the available results from different groups. This is summarized in 5: Whipple data does not indicate any spectral variability for a wide range of fluxes, while HEGRA and CAT measurements (including the March 23/24, 2001 flare), seem to show some. A word of caution is necessary here: combining different experiments' measurements is meaningless if the systematic errors between them are not understood. So more work — and simultaneous data — is needed here to be able to draw any definite conclusion. However, one should also recall that in case of Markarian 501 spectrum measurements were in quite good agreement between the three telescopes.

REFERENCES

1. Aharonian, F.A., et al. 1999a, A&A 350, 757
2. Aharonian, F.A., et al. 1999b, 26th ICRC (Salt-Lake City) 3, 350
3. Aharonian, F.A., et al. 2000, ApJ 539, 317
4. Djannati-Ataï, A., et al. 1999, A&A 350, 17
5. Fossati, G., et al., these proceedings
6. Horns, D. 2001, 36th Rencontres de Moriond on "very high-energy phenomena in the universe" (Les Arcs, France), astro-ph/0103514
7. Krennrich, F., et al. 1999a, ApJ 511, 149
8. Krennrich, F., et al. 1999b, 26th ICRC (Salt-Lake City) 3, 305
9. Piron, F., et al. 2001, accepted for publication in A&A
10. Punch, M., et al. 1992, Nature 358, 477
11. Smith, D.A., et al. 2000, Nucl. Phys. 80B, 163
12. Zweerink, J.A., et al. 1997, ApJ 490, L141

The Highest Energy Emission Detected by EGRET from Blazars

Brenda L. Dingus[1] & David L. Bertsch[2]

(1) Physics Department, University of Wisconsin, Madison, WI 53711
dingus@physics.wisc.edu
(2) NASA Goddard Space Flight Center, Greeenbelt, MD 20771

Abstract. Published EGRET spectra from blazars extend only to 10 GeV, yet EGRET has detected approximately 2000 γ-rays above 10 GeV of which about half are at high Galactic latitude. We report a search of these high-energy γ-rays for associations with the EGRET and TeV detected blazars. Because the point spread function of EGRET improves with energy, only ~2 γ-rays are expected to be positionally coincident with the 80 blazars searched, yet 23 γ-rays were observed. This collection of > 10 GeV sources should be of particular interest due to the improved sensitivity and lower energy thresholds of ground-based TeV observers. One of the blazars, RGB0509+056, has the highest energy γ-rays detected by EGRET from any blazar with $2 > 40$ GeV, and is a BL Lac type blazar with unknown redshift.

INTRODUCTION

EGRET catalogs have been produced in different energy ranges, > 100 MeV [1] and > 1 GeV [2]. However, EGRET also detected approximately 2000 γ-rays above 10 GeV. At these high energies the number of γ-rays detected is small due to both the source flux and EGRET's effective area decreasing with energy. The loss of detection efficiency is caused by higher energy γ-rays producing showers in the calorimeter with particles that propagate backwards and hit the anticoincidence dome, and thus the trigger is self-vetoed. The energy resolution is ~20-50% at these energies because of shower leakage out of the calorimeter. However, the point spread function (psf) dramatically improves with energy going from ~6° at 100 MeV (68% containment) to 0.3° at 10 GeV. The preflight calibration of the psf is shown in Figure 1 and the locations of the 35 >10 GeV γ-rays near the EGRET-detected pulsars are plotted in Figure 2. More information about the Galactic sources is given in an accompanying paper [3]. The subject of this paper is a seach for positional coincidences between the ~1000 γ-rays more than 10 degrees from the Galactic plane and the sample of 77 active galactic nuclei (AGN) detected by EGRET and 3 additional AGN (1ES2344+514, 1ES1959+650, 1ES1426+428) detected only at TeV energies.

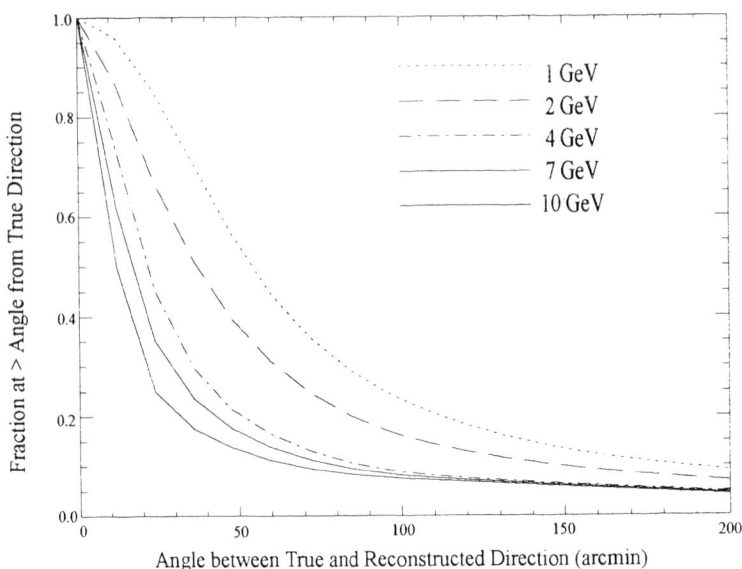

FIGURE 1. The EGRET point spread function for γ-rays of different energies as determined by the pre-flight calibration.

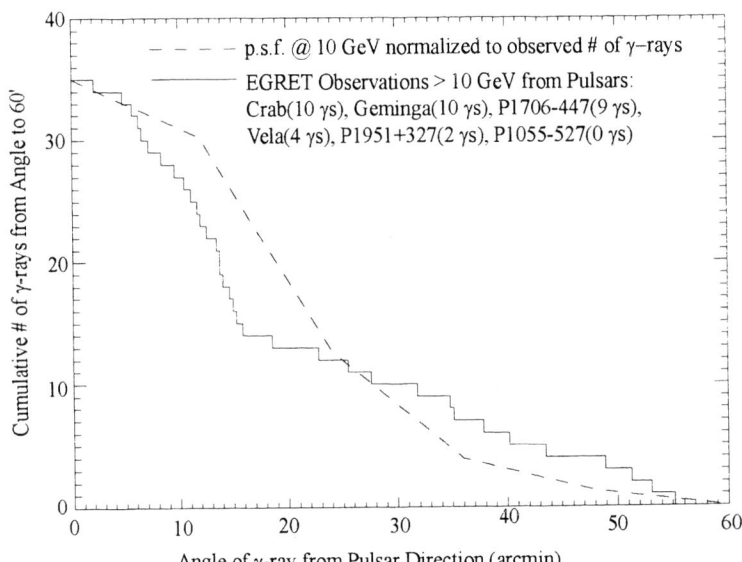

FIGURE 2. The direction of the 35 γ-rays > 10 GeV within 1° of the EGRET-detected pulsars. The dashed line corresponds to the 10 GeV point spread function of Figure 1, but normalized to the number of γ-rays detected.

SEARCH TECHNIQUE & RESULTS

The improved angular resolution and the small number of γ-rays and sources allow a simple search technique. All γ-rays within 30 arcmin of any of the 80 active galactic nuclei known to emit high-energy γ-rays are selected. The search region of 30 arcmin radius is chosen to include a large fraction of the point spread function as shown in Figures 1 and 2. The number of positional coincidences occurring by chance can be estimated by multiplying the solid angle of the search region, 80 x π (30 arcmin^2), by the average density of γ-rays more than 10 degrees from the Galactic plane, ~ 1000 γ / 4π(1-sin10°) sr, which gives ~2 γ-rays expected due to random coincidences. A more detailed calculation using the EGRET exposure as a function of Galactic latitude and longitude plus a model of the Galactic diffuse γ-rays [5] and the isotropic diffuse γ-ray flux [6] normalized to the observed number of γ-rays, yielded a similar number of 2.2 γ-rays expected by chance in the 80 regions surrounding these AGN. Since 23 γ-rays were found to be positionally coincident with the 80 AGN, the probability is <10^{-16} that these γ-rays are chance coincidences.

The 23 selected γ-rays are listed in Table 1. The characteristics of the sources and the γ-rays are listed as well as 3 probabilities. The probability #1 is the chance that a single γ-ray > 10 GeV will be within 30 arcmin of that AGN by random chance. This calculated number depends on the source location and is determined from the EGRET exposure and diffuse models mentioned in the previous paragraph. The probability #2 is not a true a priori chance probability, but is an attempt to identify which sources are most likely to emit higher energy γ-rays. This number is calculated by multiplying the probability #1 by the number of γ-rays with the energy of γ-ray detected or higher divided by the number of γ-rays > 10 GeV, as well as multiplying by the square of the ratio of the distance between the γ-ray detected and 30 arcmin. Thus the higher the energy and the closer the γ-ray is to the source location, the lower this second probability. The probability #3 is given only when more than one γ-ray is associated with a single source. This probability #3 is the chance of getting the combination of the #2 probabilities, α_i for i=1 to N, the number of γ-rays detected from that source, and is determined assuming $-2 \ln \Pi \alpha_i$ is distributed as a $\chi^2(2N)$. The list is sorted such that the 4 sources with more than one γ-ray are listed first, and the remaining sources are ordered by increasing probability #2.

Several of these sources are not extensively studied at lower energies, and some do not have known redshifts. For example, the first source in the table RGB0506+059 [7] has 2 γ-rays above 40 GeV, but is not detected in the 3rd EGRET catalog [1], because it does not emit a significant flux in the energy range above 100 MeV. It is a 4.5 σ detection in the >1GeV catalog [2] with only 23±7 γ-rays detected from the source, and a flux >1 GeV of (1.4 ± 0.4) x 10^{-8} γ/cm^2 sec. The flux above 10 GeV flux is very uncertain due to the statistical error with only 2 γ-rays detected, and due to the systematic errors in calculating the self-veto by the anticoincidence shield.

TABLE 1. All γ-rays >10 GeV within 0.5 degrees of EGRET or TeV detected blazars.

Name	AGN Properties			Gamma-Ray Properties			Chance Probabilities		
	Gal. Long.	Gal. Lat.	z	Date	Energy (GeV)	Arcmin From AGN	#1	#2 See Notes	#3
0506+059	195.40	-19.64	?	Feb-94	45.1	3.0	6.9E-02	4.1E-05	5.0E-08
				Aug-94	44.9	3.6	6.9E-02	5.9E-05	
2155-304	17.73	-52.25	0.116	Jan-98	11.0	7.8	6.2E-03	3.5E-04	2.8E-08
				Feb-93	11.1	12.6	6.2E-03	9.2E-04	
				May-96	11.2	7.2	6.2E-03	3.0E-04	
0430+285	170.52	-12.60	?	Aug-95	15.7	6.0	1.7E-01	3.1E-03	4.5E-05
				Aug-96	14.1	6.6	1.7E-01	4.6E-03	
				Feb-97	29.2	30.0	1.7E-01	2.5E-02	
1406-076	333.88	50.28	1.494	Jan-93	11.2	22.8	1.1E-02	2.7E-03	1.7E-04
				Dec-94	10.1	15.0	1.1E-02	5.2E-03	
1622-253	352.14	16.32	0.786	Dec-91	14.2	14.4	1.6E-01	2.0E-02	2.7E-03
				Sep-93	19.4	16.2	1.6E-01	1.5E-02	
1219+285	201.74	83.29	0.102	Apr-93	27.3	7.2	7.6E-03	7.5E-05	
0208-512	276.10	-61.78	1.003	Mar-96	10.7	11.4	9.7E-03	1.2E-03	
1101+384	179.83	65.03	0.031	Sep-92	14.2	19.8	6.7E-03	1.6E-03	
1730-130	12.03	10.81	0.902	Apr-94	23.7	7.2	1.6E-01	2.0E-03	
1759-396	352.45	-8.43	?	Jun-94	29.6	10.8	2.0E-01	2.0E-03	
1611+343	55.15	46.38	1.401	Nov-92	10.9	19.8	6.2E-03	2.3E-03	
1622-297	348.82	13.32	0.815	Apr-94	13.2	6.0	1.1E-01	2.6E-03	
1633+382	61.09	42.34	1.814	Sep-91	11.0	20.4	7.1E-03	2.8E-03	
0440-003	197.20	-28.46	0.844	Aug-94	15.4	24.0	1.9E-02	5.6E-03	
0219+428	140.14	-16.77	0.444	Oct-98	12.2	26.4	2.1E-02	1.1E-02	
1322-428	309.52	19.42	0.0018	Apr-93	10.0	26.4	3.7E-02	2.9E-02	

Notes:
#1 is chance probability that a diffuse γ-ray > 10 GeV is within 0.5 degrees from that AGN.
#2 is chance probability that a diffuse γ-ray > energy detected is within the distance detected from that AGN.
#3 is combined probability of multiple γ-rays with chance probability #2. (See text for the formula.)

However, a comparison with the Crab observation implies the flux is of order one-tenth the Crab flux at these energies. This blazar is observed optically with a magnitude of 15.3. The radio flux is variable with the flux density as high as 1 Jy at 4.85 GHz and the radio spectrum is flat. ROSAT detected the source with a flux of $(1.74 \pm 0.47) \times 10^{-15}$ W/m^2 in the all sky survey [8].

DISCUSSION

The AGN listed in Table 1 are a better indicator of the highest energy γ-ray emitters than the extrapolations of the spectra from the 3rd EGRET catalog [1]. These spectral fits have large uncertainties and there may be spectral changes before these energies. TeV sources have proven to be weak in the EGRET energy range implying a steepening of the spectral slope at energies above EGRET's detections. In fact, many bright >100 MeV EGRET sources, such as 3C279 or PKS0528+134, are not in this table. Some of the sources in the table have been reported as TeV sources, such as 1101+384 (Mrk 421), PKS2155-304, and 0219+428 (3C66A) [4]. This list points to several AGN that should be detectable by new ground-based observatories which have increased sensitivity and lower energy thresholds of ~100 GeV. GLAST will also have a much larger area and field of view, and will be able to better measure the flux and spectral features such as spectral breaks due to the source or due to pair production on the extragalactic background light.

ACKNOWLEDGMENTS

The author acknowledges support from NASA, NSF and Research Corporation. This research has made use of the NASA / IPAC Extragalactic Database (NED) which is operated by the Jet Propulsion Laboratory, California Institute of Technology, under contract with NASA.

REFERENCES

1. Hartman, R.C. et al. *ApJ Supp* **123** 79 (1999).
2. Lamb, R. C. & Macomb, D. L. *Ap. J.* **488**, 872 (1997).
3. Bertsch, D.L. et al, this proceedings.
4. Weekes, T. C. GeV – TeV Gamma-Ray Astrophysics Workshop , ed. Dingus, B.L., Salamon, M.H., Kieda, D.B. (AIP Conf. Proc. 515: New York) 3 (2000)
5. Bertsch, D.L., et al, *ApJ* **416** 587 (1993)
6. Sreekumar, P. et al, *ApJ* **494** 593 (1998)
7. Punsly, T. *Ap. J.* **516**,141, (1999).
8. Laurent-Muehleisen, S.A. et al, *A&A Supp*, **122** 235 (1997).

An X-ray survey of extragalactic radio jets with Chandra

Rita M. Sambruna[1], L. Maraschi[2], F. Tavecchio[2], C. C. Cheung[3],

C. Megan Urry[4], G. Chartas[5], R. Scarpa[6], and J. E. Pesce[1]

[1] *George Mason University, Dept. of Physics & Astronomy and School of Computational Sciences, MS 3F3, 4400 University Dr., Fairfax, VA, 22030-4444 (rms@physics.gmu.edu);*
[2] *Osservatorio Astronomico di Brera, via Brera 28, 20121 Milano, Italy;* [3] *Physics Department, MS 057, Brandeis University, Waltham, MA 02454;* [4] *STScI, 3700 San Martin Dr., Baltimore 21218, MD;* [5] *Pennsylvania State University, 525 Davey Lab, State College, PA 16802;* [6] *European Southern Observatory, 3107 Slonso de Cordova, Santiago, Chile*

Abstract.
    We are performing a survey of a sample of 17 radio jets with *Chandra* and *HST*, with the aim of finding their X-ray and optical counterparts. In this paper, we present the preliminary *Chandra* results for the first six sources observed and analyzed so far. In four of them an X-ray jet is clearly detected. We compare the X-ray morphologies to the radio ones, using archival radio data and discuss similarities and differences with respect to previously studied cases. Our results show that X-ray jets are no longer rare and exotic phenomena, but qualify as a rather common occurrences of the extragalactic sky.

INTRODUCTION

One of the first surprises from *Chandra* was the detection of a bright X-ray jet in the distant quasar PKS 0637–752 during early calibration phase [3]. The detection of the radio jet at X-rays was unexpected, as no optical counterpart to the radio jet was known. Indeed, the *Chandra* discovery prompted a re-analysis of the archival *HST* image of the quasar, which resulted in the detection of three faint optical knots coincident with the radio and X-ray knots [12]. Canonical synchrotron or synchrotron self-Compton models fail to reproduce adequately the radio-to-X-ray spectral distribution of the jet. The latter is instead well explained by a model invoking inverse Compton scattering of the microwave background photons [13]. This model also succesfully accounts for the shape of the X-ray spectrum, and yields minimal estimates for the jet total power.

The detection of the PKS 0637–752 jet and its unusual properties raise several questions. How common is X-ray and optical emission from extragalactic radio jets? Is PKS 0637−752 a "rare butterfly" or just the tip of the iceberg?

To address these and other questions, we proposed a survey at X-ray and optical wavelengths of a sample of radio jets using *Chandra* and *HST*, aimed at detecting new jets at shorter wavelengths for subsequent deeper follow-up studies. Here we present the *Chandra* results for the first six sources observed as part of the Guest Observer cycle 2 program. Throughout this work we adopt $H_0 = 75$ km s^{-1} Mpc^{-1} and $q_0 = 0.5$. More details will be given in [11].

SAMPLE SELECTION AND OBSERVATIONS

The survey sample was selected from published list of known radio jets [1,4]. Basically, the jets were selected on the basis of their length and surface brightness, to ensure a *Chandra* detection in a modest exposure of 10 ks for plausible values of the radio-to-X-ray spectral index. These criteria resulted in a sample of 17 radio jets, spanning a range of redshift, core and extended radio power, and radio galaxy classification. Here we report the results for the first six sources observed with *Chandra*. For four of them, 0723+679, 1136−135, 1150+497, and 1354+195, an X-ray counterpart of the radio jet was detected for the first time with *Chandra*. Three of them (1136−135, 1150+497, and 1354+195) also have an optical detection with *HST*. The two for which no jet was detected are 1055+018 and 2251+134.

Chandra observed the sources with ACIS-S in the focal plane, with exposures of 10 ks per target. Archival VLA radio images at 5 GHz were also analyzed. The images of 0723+679 and 1150+497 were presented originally by [7] and have been reimaged by us using calibrated (u,v) data kindly provided by F. Owen. Each map was restored with a circular beam with FWHM=0.5″, except for 1354+195 which was restored with a 1″ beam.

I RESULTS

Figure 1 shows the *Chandra* ACIS-S images of the four newly detected jets in the 0.4–8 keV energy band. Overlayed on the X-ray images are the radio contours from the archival VLA data. In all four cases, an X-ray jet is detected. The projected length of the jets varies from 5″ (20 kpc) in 1150+497 to 25″ (134 kpc) in 1354+195. A summary of the X-ray and radio properties is given below for each target.

0723+679 (z=0.846): Two X-ray knots are detected at $\sim$ 10 and 15″ from the nucleus (Fig. 1), coincident with the radio knots in the VLA radio map [7]. The jet appears to bend after the first knot. Faint X-ray emission is also visible from the hot spot of the extended radio counter-lobe. This source also exhibits an interesting X-ray nuclear spectrum, with a possible redshifted Fe Kα emission line and will be the subject of a future separate publication. On the whole the X-ray and radio brightnesses track each other very well.

1136−135 (z=0.554): An X-ray knot is detected at 7.5″ from the nucleus, with very weak radio counterpart [9]. The X-ray and radio brightnesses along the jet behave

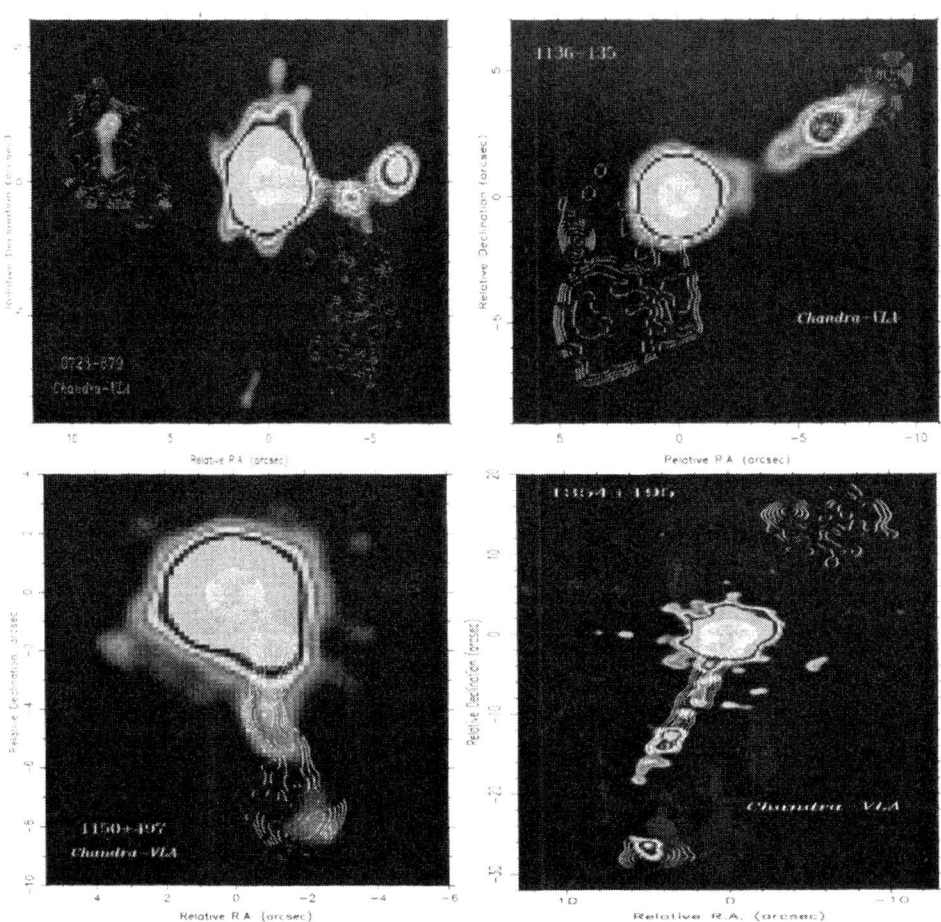

FIGURE 1. *Chandra* ACIS-S images in 0.4–8 keV of the four newly discovered jets from our GO2 survey (color image). Overlayed are the radio contours from archival VLA data. Both the colors and the contours are plotted logarithmically, in steps of factor 2. The *Chandra* images were smoothed with a Gaussian of width $\sigma=0.45''$, yielding a resolution of $1.17''$ FWHM. Each radio map was restored with a circular beam with FWHM=$0.5''$, except for 1354+195 which was restored with a $1''$ beam.

oppositely, with the X-rays peaking at $\sim 6''$ from the core, in correspondence to a radio relative minimum. After this knot, the X-rays fade substantially, while the radio brightness increases.

1150+497 (z=0.334): X-ray emission at $8''$ from the nucleus is visible on the ACIS image (Fig. 1). There are hints for a curved structure of the X-ray jet, tracking rather well the helix-like shape seen in the radio [7]. Also in this case the X-ray and radio morphologies trace each other quite well.

1354+195 (z=0.720): The X-ray jet is long ($25''$) and has a knotty structure, again similar to the radio [6]. Note however that in the second half of the jet a conspicuous X-ray knot is present which is faint or absent in the radio. Weak X-ray emission from the southern radio hot spot is also present. Both the radio and X-ray jets terminate in a bright hot spot, which appears misaligned with respect to the jet axis.

II DISCUSSION AND CONCLUSIONS

We presented the first X-ray results from our ongoing survey of extragalactic radio jets with *Chandra* and *HST*. The X-ray counterpart of the radio jet was detected in four out of the six sources observed and analyzed so far, on physical scales ranging from 20 kpc (1150+497) to 134 kpc (1354+195). We compared the *Chandra* images to archival radio images of the jets.

Based on the comparison between the X-ray and radio morphologies, we suggest that the jets could be classified in two main classes. In the first one (class 1) the X-ray brightness distribution closely resembles the radio one, as in 0723+679, 1150+497, and 1354+195. This is similar to the case of PKS 0637−752. Class 2 contains jets like 1136−135, where the X-rays peak close to the nucleus while the radio peaks at the end of the jet, in a region of weak X-ray emission (Figure 1). The jets of 3C 371 and 3C 273 would also fall into the latter category [8,5,10].

A number of processes can contribute to the production of X-rays from extended jets in AGN. The non-thermal mechanisms include synchrotron radiation from a population of relativistic leptons, or inverse Compton scattering processes, with the seed photons provided by the synchrotron photons themselves (SSC) or by external radiation fields. Recently, [13] and [2] have shown that, if the plasma is still relativistic on large (10 kpc or more) scales, the dominant source of seed photons is the cosmic microwave background (CMB). In fact, the CMB radiation density is amplified in the comoving frame of the emitting plasma by a factor Γ^2, where Γ is the bulk Lorentz factor of the plasma. Inverse Compton scattering on the CMB photons (IC/CMB) explains well the broad-band spectrum of PKS 0637−751, and yields a reasonable estimate of the jet kinetic power. In this framework, the similarity of the X-ray and radio morphologies can be understood, as the same electron population is responsible for the emission at both wavelengths.

By analogy with PKS 0637−751, we can reasonably argue that the most likely candidate mechanism for the production of X-rays in the jets of class 1 (0723+679,

1150+497, and 1354+195) is IC/CMB. Using a similar analysis to PKS 0637–751, for 0723+679 we directly verified that the IC/CMB scenario is consistent with equipartition between magnetic field and emitting particles for a moderately large value of the beaming factor ($\delta \sim 8-10$). SSC requires in general conditions very far from equipartition and larger intrinsic power in the jet. Independent evidence for large bulk velocities on kpc scales in these jets is provided by the radio, which gives large (> 55) ratios of the jet to counterjet flux.

For class 2 we favor a synchrotron origin for the X-rays, requiring electrons of much higher energy than those producing the radio emission. The different X-ray morphology could be related, in this picture, to the short cooling time of X-ray emitting electrons. Electrons producing X-rays cool rapidly after the peak knot, while radio electrons can survive until they reach regions in the jet which are further away. This scenario could hold for 1136–135. Indeed, a preliminary analysis for this source indicates that, assuming equipartition, the cooling time of X-ray electrons in the knot is comparable with the crossing time of the knot, consistent with the morphology. In this case a moderate beaming is also required. In this picture, we would also expect the optical morphology of the jet to be similar to the radio. This will be tested with our *HST* data, whose analysis is still in progress.

In conclusion, the first results from our ongoing *Chandra* GO2 survey of extragalactic radio jets show that X-ray emission from these structures is common. This implies the presence of relativistic plasma on very large scales, at tens to hundreds of kpc from the central black hole. It is safe to say that X-ray jets are no longer "rare butterflies"; they are starting to qualify as a rather common phenomenon of the extragalactic sky.

This project is funded by NASA grants GO1-2110A and HST-GO-08881.01-A.

REFERENCES

1. Bridle, A. H. & Perley, R. A., *ARA&A*, **22**, 319 (1984).
2. Celotti, A., Ghisellini, G., & Chiaberge, M., *MNRAS*, **321**, L1 (2001).
3. Chartas, G. et al., *ApJ*, **542**, 655 (2000).
4. Liu, F. K. & Xie, G. Z., *A&AS*, **95**, 249 (1992).
5. Marshall, H.L. et al., *ApJ*, **549**, L167 (2001).
6. Murphy, D. W., Browne, I. W. A., & Perley, R. A., *MNRAS*, **264**, 298 (1993).
7. Owen, F.L. & Puschell, J. J., *AJ*, **89**, 932 (1984).
8. Pesce, J.E. et al., *ApJ*, in press (astro-ph/0106426)
9. Saikia, D. J., Junor, W., Cornwell, T. J., Muxlow, T. W. B., & Shastri, P., *MNRAS*, **245**, 408 (1990).
10. Sambruna, R. M. et al., *ApJ*, **549**, L161 (2001).
11. Sambruna, R. M. et al., in prep. (2001).
12. Schwartz, D.A. et al., *ApJ*, **540**, L69 (2000).
13. Tavecchio, F., Maraschi, L., Sambruna, R.M., and Urry, C.M., *ApJ*, **544**, L23 (2000).

Extended X-ray Emission From The Radio Galaxy 3C 390.3

H. Krawczynski*, S. Wagner[†]

* *Yale University, P.O. Box 208101, New Haven, CT 06520-8101, USA*
[†] *Landessternwarte, Königstuhl 10, 69117 Heidelberg, Germany*

Abstract. We report on Chandra ACIS observations of the radio galaxy 3C 390.3 which reveals spatially extended X-ray emission from the northern and the southern hotspot regions. We compare the complex X-ray morphology of the hotspot regions with their radio morphology. We argue that Synchrotron emission from electrons accelerated by the highly non-stationary and non-uniform plasma naturally accounts for the observed X-ray emission and radio to X-ray energy spectra. We discuss the implications for the dynamics of the jet fluid and particle acceleration processes.

I INTRODUCTION

Recent Chandra observations have revealed extended X-ray emission from a large fraction of observed radio galaxies (e.g. [1–4]). Although the responsible emission mechanism is still a matter of debate it is obvious that X-ray emission from the radio structure is a powerful diagnostic tool for studying the acceleration of particles by the jet plasma and the dynamics of the jet, which complements the information from radio observations. In this paper we present Chandra X-ray observations of the broad line region radio galaxy 3C 390.3. It is a very nearby ($z=0.0561$) classical double radio-source with (given its nearness) a rather high radio power of $\log_{10} P_{178\,\mathrm{MHz}} = 25.41$ W Hz^{-1} sr^{-1}. At 1.6 GHz its angular diameter of 230" corresponding to a projected extension of 265 kpc (we use $H_0 = 65$ km s^{-1} Mpc^{-1} and $q_0 = 0.5$, one arcsec corresponds to 1.16 kpc). At milli-arcsecond scale superluminal motion of the northern jet component with an apparent pattern speed of $3.5c$ was found [5], indicating a bulk relativistic motion towards the observer with a line of sight angle smaller than 32°. Prieto & Kotilainen [6] reported the detection of X-ray emission from the northern hotspot region based on a 11.5 ksec ROSAT PSPC observation. Using the higher angular resolution of the ROSAT HRI detector Harris, Leighly and Leahy [7] identified the X-ray source (still unresolved) with the compact and radio bright hotspot B (here and in the following we use the radio nomenclature introduced in [8]).

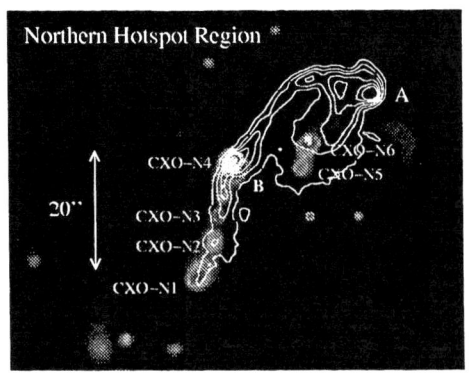 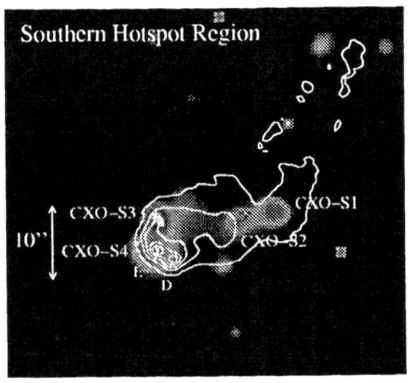

FIGURE 1. X-ray (greyscale) and Radio (contours) images of Northern (left panel) and Southern (right panel) hotspot regions of 3C 390.3. The X-ray image has been smoothed adaptively with the algorithm csmooth [9]. Note that the X-ray features CXO-N2, CXO-N3, and CXO-N5 have individual statistical significances of only $\simeq 2\sigma$, all other X-ray features have significances exceeding 3σ. Radio contours are linearly spaced with a maximum brightness of 0.026 Jy/1".1×1".00 beam (left panel) and 0.1016 Jy/1".1×1".00 beam (right panel). Labels beginning with "CXO" designate X-ray features; labels "A, B, D, E" designate radio features and follow the nomenclature of Leahy & Perley (1995).

II OBSERVATIONS AND RESULTS

We observed 3C 390.3 for 34 ksec on April, 2000, using the ACIS instrument on board the Chandra X-ray Observatory. Motivated by the earlier ROSAT results, we placed the aim-point of the S3 chip near the northern hotspot B. We reduced the data with the CIAO 2.0 analysis software. Rejecting periods of high background we extract a usable exposure time of 29.13 ksec. The image analysis uses data in the energy range from 0.2 keV–8.0 keV. Results of spectral fits will be quoted with 90% confidence interval errors. We correlate the X-ray morphology with the radio morphology using the 5 GHz VLA radio map of 3C 390.3 of [5]. We matched the absolute astrometry of the radio and X-ray image by matching the center of the X-ray emission and the radio emission of the core of 3C 390.3. Since the X-ray image of the core is heavily affected by pile up and pixel saturation the accuracy of our matching procedure is only about 1".

We detect extended X-ray emission with complex morphology from the northern and southern hotspot regions as shown in Fig. 1 together with the 5 GHz radio intensity contours. Features in the northern and southern hotspot region have been labeled with CXO-N1 to CXO-N6 and CXO-S1 and CXO-S4, respectively, to facilitate discussion in the following. Chandra count rates and flux estimates of the X-ray features together with radio and optical flux estimates are given in Table 1.

We focus first on the discussion of the emission from the northern hotspot region. The brightest X-ray feature of the northern hotspot region (CXO-N4) is

TABLE 1. Radio and X-Ray Fluxes of Jet/Hotspot Features

Feature[a]	S(5 GHz)[b]	X-ray Counts[c]	X-ray Flux[d]
CXO-N1	0.0002	20.7±5.8 (3.4)	2.4±0.7
CXO-N2 (J3)	0.0011	10.5±4.4 (2.9)	1.2±0.5
CXO-N3 (J4)	0.0012	8.1±3.5 (1.9)	0.9±0.4
CXO-N4 (Hotspot B)	0.0240	197.3±14.5 (3.5)	23.0±1.7
CXO-N5	0.0014	8.1±3.5 (1.9)	0.9±0.4
CXO-N6	0.0034	74.8±8.6 (2.8)	8.7±1.0
CXO-S1	0.0007	15.9±4.7 (2.5)	2.4±0.7
CXO-S2	0.0029	15.2±4.6 (2.4)	2.3±0.7
CXO-S3	0.0268	23.5±5.3 (2.3)	3.5±0.8
CXO-S4 (Hotspot E)	0.0900	17.6±4.9 (2.5)	2.6±0.7

[a] Brackets give names of associated radio features (names according to Leahy & Perley 1995)
[b] Radio surface brightness (5 GHz, in Jy / 1".1×1".00 beam) at approximate location of X-ray feature
[c] ACIS-S source counts (0.2-8 keV) after an exposure time of 29.13 ksec; value in brackets gives radius of circle used for event-extraction in arcsec
[d] X-ray flux (0.5-8 keV) in (10^{-15} ergs cm^{-2} s^{-1}) computed under the assumption of a powerlaw spectrum with photon index $\Gamma = -2$ and a neutral hydrogen column density $N_H = 4.28 \cdot 10^{20}$ cm^{-2}

clearly associated with radio hotspot B. The X-ray picture shows a structure which is extended perpendicular to the jet-flow. The data suggest two bright maxima separated by 1.5" which could resemble the X-ray emission of a torus seen edge-on. Although the extension of the X-ray feature is highly statistically significant, the dip in the center between the two emission maxima is only marginally significant. The maximum of the radio-emission is offset by about 1" to the north which is still within the systematic uncertainties of the astrometry. Note that the X-ray flux of hotspot B given in Table 1 is consistent with the flux estimate of Harris et al. (1998). The results of fits of power law and thermal bremsstrahlungs models to the X-ray data of hotspot B (including both humps) are given in Table 2. We fitted the data twice, once using all events with energies between 0.3 keV and 3 keV allowing the neutral hydrogen column density N_H to vary, and once with the events from 0.6 keV to 3 keV fixing N_H at its galactic value of $4.28 \cdot 10^{20}$ cm^{-2} (Dickey & Lockman 1990) which excludes the low energy data where the detector response matrix is still affected by substantial systematic uncertainties. Both fits give identical results within statistical errors. The data can be fitted satisfactorily with a power law model and with a thermal bremsstrahlungs model. However, a thermal model predicts a Faraday Rotation measure of more than 1000 rad m^{-2} which exceeds the observed ones by more than one order of magnitude. A thermal origin of the radiation is therefore strongly disfavored.

To the north of hotspot B (downstream) there is little evidence for X-ray emission from the other radio features (N1-N4, F, A of Leahy & Perley 1995) or the diffuse radio emission, except for two features (Fig. 1, CXO-N5 and CXO-N6) which are inconspicuous in the radio image. The features lie on the straight line of the jet

TABLE 2. Hotspot B – Parameters of X-ray Spectral Fits

Range of Fit [keV]	N_H [10^{20} cm^{-2}]	Photon Index	F(1 keV) [10^{-6} cm^{-2} s^{-1}]	kT [keV]	Norm[a]	χ^2/d.o.f.
0.3-3	$4.27^{+2.73}_{-2.02}$	$1.84^{+0.29}_{-0.32}$	7.14±1.14	—	—	2.32/6
0.6-3	4.28 (fixed)	$1.74^{+0.51}_{-0.42}$	6.96±1.32	—	—	2.12/6
0.3-3	$4.28^{+4.28}_{-2.61}$	—	—	$2.09^{+1.77}_{-0.69}$	3.60±0.57	2.43/6
0.6-3	4.28 (fixed)	—	—	$2.85^{+15.70}_{-1.18}$	3.20±0.60	1.55/6

[a] Normalization in units of $(10^{-20}(4\pi D^2)^{-1} \int n_e n_I \, dV)$, where D (in cm) is the distance to the source and n_e, n_I (both in cm^{-3}) are the electron and ion densities, respectively.

from the nucleus, at a location which does not show any excess radio emission, peculiar spectral index, variation in polarization, or excess rotation measure. To the south of hotspot B (upstream) we detect X-ray emission all along the 20 arcsec of the "jet neck" (CXO-N1 – CXO-N3) down to the location where the jet starts to deviate from its linear propagation from the core and starts to become radiative. We detect 3 distinct maxima in the X-ray emission with no significant extension of the emission perpendicular to the ridge line.

Similar as in the northern hotspot region, the data of the southern hotspot region shows a X-ray source (here called CXO-S1 which connects to CXO-S2) which lies along the straight extrapolation of the inner jet, but which has no obvious radio counterpart. Further to the east are two X-ray sources (CXO-S3 and CXO-S4) which are associated with two radio features, hotspot E and another feature previously not named. We do not detect X-ray emission from the 2 radio channels north of the hotspot region which seem to feed the hotspot region. Being 3.7 arcmin off-axis from the aimpoint, the spatial resolution of the X-ray image is degraded in comparison to the northern hotspot region and we do not detect a spatial extension perpendicular to the ridge line connecting the X-ray features.

III DISCUSSION

Harris et al. (1998) already noted that the Synchrotron Self-Compton mechanism as well as Inverse Compton processes off (unbeamed) photons from the Cosmic Microwave Background are not expected to contribute noticeably to the observed X-ray emission of the X-ray luminous feature hotspot B. The only single mechanism which is able to consistently explain the X-ray emission of all observed X-ray features is synchrotron radiation from in-situ accelerated electrons. The highly non-stationary and non-uniform character of the plasma flow within the hotspot regions predicts that both the production rate and high energy cutoff of accelerated particles strongly depend on the characteristic time scales of plasma instabilities and plasma reorganization [10]. The different radiative cooling times of X-ray electrons (electron Lorenz factor $\gamma_X = 5 \cdot 10^7$, cooling time $t_X = 150$ yr for an equipartition magnetic field of 60μG (hotspot B)) compared to the ones of radio electrons ($\gamma_R = 5 \cdot 10^3$, $t_R = 1.5$ Myr) can therefore result in substantial spatial variation of the radio to X-ray energy spectra and account for the difference in radio and X-ray morphologies.

The most likely flow patterns within multiple hotspots have intensively been discussed in the literature (see [11,12] and references therein). The X-ray emission now adds genuinely new information by tracing the sites of ongoing electron acceleration. The Chandra data strongly suggest that particles are accelerated all along the neck of the northern hotspot region and at hotspot B, and at various locations of the southern hotspot region. The non-detection of the outer radio features (feature A in the north, and feature D in the south) could either mean that these hotspots are in the process of dying and particle acceleration stopped recently, or that the physical conditions in these "terminal" hotspots do not allow electrons to be accelerated to sufficiently high energies to emit synchrotron X-rays.

Acknowledgments: We thank W. Alef for sharing with us the 5 GHz radio map of 3C 390.3. HK is grateful to P. Coppi and D. Harris for enlightening discussions on jets and acknowledges support by the NASA (NAS8-39073). SW acknowledges the support by the DFG (SFB 439) and DLR (DARA:500R96186).

REFERENCES

1. Chartas G., Worrall D.M., Birkinshaw M., et al., 2000, ApJ, 542, 655
2. Schwartz D.A., Marshall H.L., Lovell J.E.J., et al., 2000, ApJ, 540, 69
3. Marshall H.L., Harris D. E., Grimes J. P., et al., 2001, ApJ, 549, L167
4. Sambruna R.M., Urry C.M., Tavecchio F., et al., 2001, ApJ, 549, L161
5. Alef W., Wu S.Y., Preuss E., Kellermann K.I., Qiu Y.H., 1996, A&A, 308, 376
6. Prieto M.A., Kotilainen J.K., 1997, ApJ, 491, L77
7. Harris D.E., Leighly K.M., Leahy J.P., 1998, ApJ, 499, L149
8. Leahy J.P., Perley R.A., 1995, MNRAS, 277, 1097
9. Ebeling H., White D.A., Rangarajan F.V.N., 2001, submitted to MNRAS.
10. Jones T.W., Ryu D., Engel A., 1999, ApJ, 512, 105
11. Valtaoja E., 1984, A&A, 140, 148
12. Lonsdale C.J., Barthel P.D., ApJ, 1998, 115, 895

THE EXTRAORDINARY CAMPAIGNS ON MKN 421 OF 2000 (SAX) and 2001 (RXTE)

G. Fossati[*], M. Jordan[†] and J. Buckley[†]

[*]*UCSD/CASS, 9500 Gilman Dr., La Jolla, CA, 92093–0424*
[†]*Washington University, St. Louis, & Whipple Collaboration*

Abstract. We report on two recent long X–ray(/TeV) observations of the bright blazar Mkn 421, performed with *Beppo*SAX in April 2000 and *Rossi*XTE in March 2001, with simultaneous optical and TeV observations. The goal is to exploit the dramatic broadband spectral variability to constrain the physical conditions and processes responsible for particle acceleration in blazar jets.

INTRODUCTION

Blazars are radio–loud AGNs characterized by strong variability, high and variable polarization, and high luminosity. Their extreme properties are successfully interpreted in terms of radiation produced in relativistic jets [16]. They are the ideal laboratories for studying the physics of inner jets because relativistic beaming provides us with a "magnifying lens" to view of the region where the jets are formed and powered. Blazars spectral energy distributions νF_ν (SEDs) consist primarily of two components [3]. The first component, extending up to X–rays, is likely to be synchrotron radiation from relativistic electrons, while the second one (peaking in the γ-ray range), is thought to be produced via inverse Compton scattering (IC) of soft (IR-UV) photons [6]. Correlated variability suggests that they may be produced by the *same* electrons [15].

The importance of the "peaks" – The energy at which the SED peaks occur provide a powerful diagnostic tool to investigate the properties of emitting material [14]. In fact, for synchrotron radiation due to a single population of electrons with a spectral distribution with a break at γ_{peak}, the synchrotron peak is at $E_{sync,peak} \propto \gamma_{peak}^2 B \delta$, where B is the magnetic field, and δ is the Doppler "beaming" factor of the blob of emitting material. Irrespective of the nature of the seed photons, the peak synchrotron and IC power derive from the same electrons (except when Klein-Nishina effects are dominant), and this provides a further constraint. The value of γ_{peak} is likely determined by the balance between the particle acceleration and cooling mechanisms, and the variability observed at/above $E_{sync,peak}$ traces the evolution of the electron distribution, in the regime where acceleration and cooling are approximately balanced (e.g. [7, 10]).

HBLs and Mkn 421 – A sub-class of blazar, the HBLs (for High-energy-peak BLazar), provide the most interesting target because their synchrotron component peaks in the X–ray band. Hence, they allow us to *investigate directly* the behavior of the highest energy electrons, which are the most direct probe of the details of the acceleration and cooling mechanisms. Moreover, the IC peak occurs in the TeV band, that can be observed with

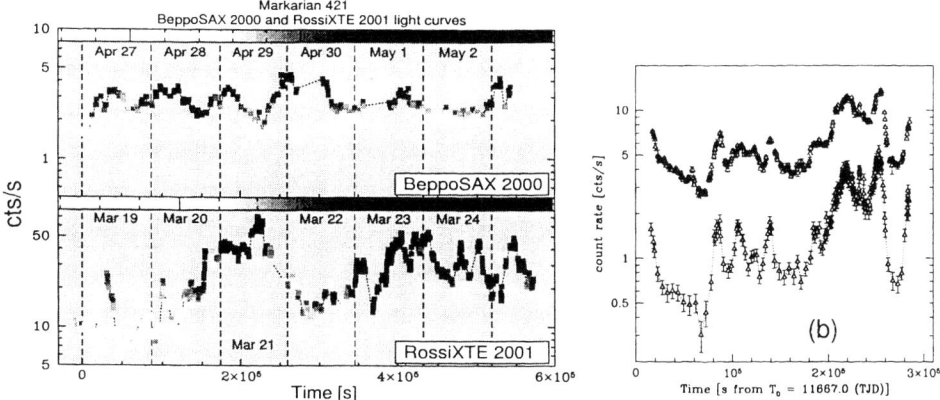

FIGURE 1. [a] 4–10 keV light curves, color coded according to hardness ratios (scales are different). Y-axes have the same dynamic range. [b] MECS 2–10 keV and PDS 12–60 keV (bottom) light curves for a section of the spring 2000 campaign. MECS binning is 500 s. The PDS data have a variable time binning to obtain S/N$\geq$10, but $\Delta T \leq$ 5ks.

the ground based atmospheric Cherenkov telescopes (ACT) [1].

Mkn 421 is the archetypical HBL, and the brightest at X–ray and UV wavelengths. It is also the only TeV source that is consistently detected across a large range in brightness (spanning more than a factor of 10, and reaching up to 5 times the flux of the Crab), and allows us to sample the main features of its variability with time resolution allowing a meaningful comparison with the X–ray data. Therefore, Mkn 421 has been target of several multi-wavelength campaigns (e.g. [12, 4, 5, 13, 11]).

In this paper we present the main features of the two latest X–ray/TeV campaigns on Mkn 421. Thorough analyses of the *Beppo*SAX and *Rossi*XTE data are in progress and they will be published in a series of separate papers. Preliminary accounts of these two campaigns are reported also by Fegan and Fossati [2] for *Beppo*SAX /Whipple 2000, and Jordan et al. [8] for *Rossi*XTE /Whipple 2001.

*BEPPO*SAX 2000

*Beppo*SAX observed Mkn 421 for two un-interrupted periods lasting 7 and 3 days respectively, and separated by about one week. Light curves are reported in Fig. 1a (top panel, the first 7 days, LECS data because of a 1.5 days gap in the MECS data), and Fig. 1b (second 3-day pointing, MECS and PDS). During this campaign Mkn 421 was consistently above its (then) historical maximum X–ray flux, reaching a flux of about 10^{-9} erg cm^{-2} s^{-1} in the 2–10 keV band. As it became clear with the ASCA long-look of 1998 [13] the source boasts a significant amount of intraday variability. The Whipple ACT observed a spectacular flare on April 30 (followed for about 6 hours) and with a factor of three increase in count rate over one hour [2]. Most unfortunately this flare coincides exactly with a gap also in the LECS data set.

$E_{sync,peak}$ *vs. Flux* – One of the most important findings of the analysis of the *Beppo*SAX 1998 data was that synchrotron peak energy and flux/luminosity, two quantities which

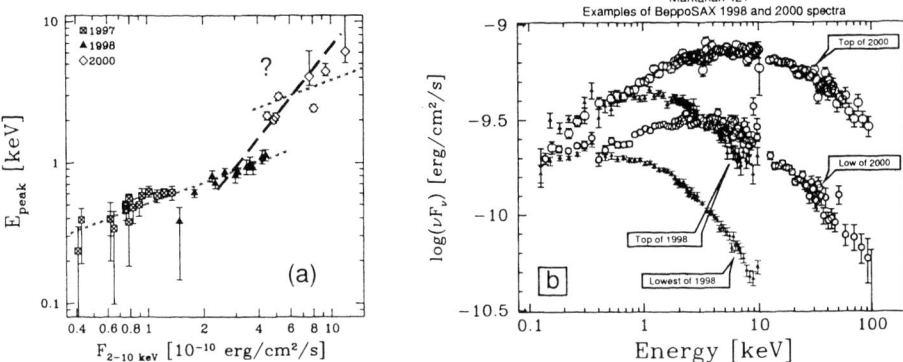

FIGURE 2. [a] The relationship between $E_{sync,peak}$ and flux, including preliminary results from 2000 (all data from *Beppo*SAX). The bottom short-dash line shows the best fit relationship to 1997–1998 data. The 2000 data do not obey the same straight relationship. They could follow a relationship with the same slope, but different zero-point (top short-dash line), or mark a new branch of the same trend, holding at higher luminosities (e.g. the long-dash line). [b] Examples of LECS+MECS+PDS *Beppo*SAX (inferred) spectra representing low/high state during the 2000 campaign. For comparison, we plotted also LECS+MECS data representing the lowest and highest states of *Beppo*SAX 1998.

are the indicators of the bulk of the radiative dissipation, track each other quite tightly (Fig. 2a). Fossati et al. [5] presented a tight relationship between peak energy and flux, with $E_{sync,peak} \propto F^{0.55\pm0.05}$ (based on 1997 and 1998 *Beppo*SAX data) (see also [9]). The diagram has been extended with *Beppo*SAX 2000 data towards the brighter states. The new data seem to depart from the straight relationship obeyed by 1997 and 1998 data. However, none of these past "experiments" has been able to track the development of spectral parameters over a series of flares, as we will be able to do with the 2000 dataset. The quantitative determination of such dependence can clearly provide a crucial clue on the mechanism(s) responsible for the variability and thus for the emission in jets. For example a different behavior is expected depending on whether an increase in dissipation corresponds to a variation in the acceleration process (e.g. in the energy of the emitting particles), in the intensity of the magnetic field or in the bulk Lorentz factor of the emitting plasma. In order to quantify the correlation it is therefore of chief importance to estimate the relative behavior during different events and especially covering a large range in the values of these two quantities.

Detection in the PDS – An extremely exciting result of the 2000 campaign is that the source is detected also with the higher energy detector, the PDS. In the past campaigns Mkn 421 was detected at few sigma significance by integrating over the whole dataset, and the PDS "spectrum" has always been consistent with being a hard –inverted– power law, most likely the onset of the IC component [5]. On the contrary, in the spring 2000 the synchrotron component was extending without significant steepening well above a few keV (Fig 1b). The PDS light curve is terrific (Fig. 2b), and for the first time we are able to extend the short timescale variability studies in the hard X–rays. Examples of broad band (3 full decades) spectra are shown in Fig. 2b, to demonstrate how well the synchrotron peak is defined when broad band bandpass is available.

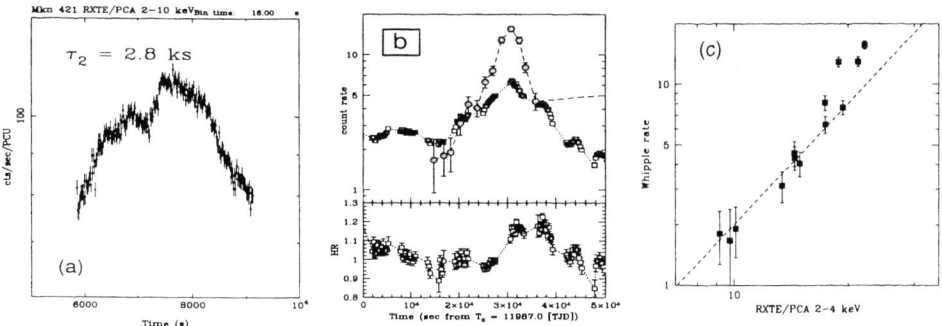

FIGURE 3. [a] Blow up of a section of the *Rossi*XTE 2001 dataset. It is the small –very hard– flare occurring in the middle of March 24. The τ_2 is the doubling time of the rise of the flare. [b] Detail of the simultaneous X–ray/TeV flare of March 19 2001. X–ray data are the empty squares, Whipple date the grey circles. In the bottom panel we plotted the X–ray hardness ratio. [c] Whipple (TeV) vs *Rossi*XTE /PCA for the flare shown in [b]. The dashed line represents the $F_{TeV} \sim F_X^2$ locus.

*ROSSI*XTE 2001

In this last section we illustrate by means of a few figures the unprecedented phenomenology and quality of the data gathered during this latest campaign. The 350 ks approved for *Rossi*XTE cycle 6 were successfully scheduled as a long un-interrupted observation during one of the time of the year with best TeV visibility (up to 7 hours/night). All major Atmospheric Cherenkov Telescopes (Whipple, HEGRA, CAT) observed Mkn 421. Moreover, we had 7 night at the 48" telescope of the Whipple Observatory, and we used the Antipodal Transient Observatory (ATO), a new robotic telescope deployed by the group of Washington University. Jordan et al. [8] give a more complete description of the Whipple and optical data reduction and analysis.

The first thing to note is that Mkn 421 once again broke its (observed) brightness record, reaching a 2–10 keV flux of about 8×10^{-9} erg cm^{-2} s^{-1}, and spent a good 50% of the campaign above the 2000 top flux. The (almost) full light curve is shown in Fig. 1a. The comparison with the *Beppo*SAX 2000 LECS light curve, that seemed already quite amazing, very effectively shows that the variability patterns and amplitude of the 2001 campaign are unprecedented. The total amplitude of variability is of about a factor of 12 (4–10 keV band). Frequent hour-scale flares were observed. Figure 3a shows the short timescale details that can be observed thanks to PCA throughput. For the whole duration of the observation the count rate is high enough to be able to work on 16 seconds, or shorter, time bins. This alone adds at least a factor of 20 in dynamic range in the variability studies, with respect to what is achievable with *Beppo*SAX.

The correlation of variation in the X–ray and TeV bands seems to be definitely confirmed, with (again) unprecedented detail, supporting the idea that the same electron distribution, in the same physical region, is responsible for the emission in both energy bands. This is illustrated with Figs. 3b and 3c. In the first one we plot a segment of the *Rossi*XTE and Whipple light curves, corresponding to the flare of March 19 (see Fig. 1a). In Figure 3c we plot the simultaneous X–ray and Whipple TeV rates for this flare. The dashed line represents a squared relationship between the two, that is closely followed by the data.

SUMMARY

The combination of energy coverage and duration of the 2000/2001 campaigns is unique, and they provided us with a great wealth of information, but at the same time revealed a richer than expected phenomenology. The advancement is due to *Beppo*SAX and *Rossi*XTE characteristics. *Beppo*SAX's unparalleled broad energy coverage allows a robust determination of the synchrotron peak energy. *Rossi*XTE's throughput extends the timescale domain that we can investigate, and also *Rossi*XTE can observe Mkn 421 at the epoch of best visibility by the ground based ACTs. The previous best observation was the 10 days pointing by ASCA in 1998 [13], but it missed the energy bandpass, the throughput, and the quality of TeV simultaneous coverage.

With these data we are going to address most of the interesting physical questions for which blazars are ideal laboratories. There is only one major piece of information that can not be learned with *Beppo*SAX and *Rossi*XTE, that is the characterization of the hard/soft inter-band X–ray lags, which is a necessary passage to pin-point the competitive interplay between acceleration and cooling of the emitting particles. *Beppo*SAX and *Rossi*XTE are not suitable to tackle this issue, because of the frequent gaps in the time series caused by the short orbital period, which unfortunately is also of the same order of magnitude of the expected timescale of the phenomena under investigation. To properly tackle this issue we will need XMM.

ACKNOWLEDGMENTS

We acknowledge the many co-investigators on the *Rossi*XTE proposal, J. Swank and E. Smith for implementing the best *Rossi*XTE scheduling possible, L. Maraschi and S. Wagner for interesting discussions, and L. Chiappetti and M. Capalbi for the invaluable help with the PDS data reduction. GF is partly supported by NASA grant NAG5-9122.

REFERENCES

1. Catanese M. & Weekes T. 2000, PASP, 111, 1193
2. Fegan D.J, & Fossati G. 2001, in Proc. of 27th ICRC
3. Fossati G., et al. 1998, MNRAS, 299, 433
4. Fossati, G., et al. 2000a, ApJ, 541, 153
5. Fossati, G., et al. 2000b, ApJ, 541, 166
6. Ghisellini G., et al. 1998, MNRAS, 301, 451
7. Inoue S. & Takahara F. 1996, ApJ, 463, 555
8. Jordan M., et al. 2001, in Proc. of 27th ICRC
9. Kataoka J. et al. 2001, astro–ph/0105029
10. Kirk J. et al. 1998, A&A, 333, 452
11. Krawczynski, H. et al. 2001, astro–ph/0105331
12. Maraschi L., et al. 1999, ApJ, 526, L81
13. Takahashi T., et al. 2000, ApJ, 542, L105
14. Tavecchio F., Maraschi L., Ghisellini G, 1998, ApJ, 509, 608
15. Ulrich M.-H., et al. 1997, ARA&A, 35, 445
16. Urry C.M, & Padovani P., 1995, PASP, 107, 83

COMPTEL Observations of the Virgo Blazars 3C 273 and 3C 279

W. Collmar[1], V. Schönfelder[1], S. Zhang[1,5], H. Bloemen[2], W. Hermsen[2], M. McConnell[3], K. Bennett[4], O.R. Williams[4]

[1] *Max-Planck-Institut für extraterrestrische Physik, Garching, Germany*
[2] *Space Research Organization Netherlands, Utrecht, The Netherlands*
[3] *Space Science Center, University of New Hampshire, Durham, USA*
[4] *Astrophysics Division, ESTEC, Noordwijk, The Netherlands*
[5] *High Energy Astrophysics Lab, IHEP, P.O.Box 918-3, Beijing, China*

Abstract. We report the main MeV properties (detections, light curves, spectra) of the Virgo blazars 3C 273 and 3C 279 which were derived from a consistent analysis of all COMPTEL Virgo observations between 1991 and 1997.

INTRODUCTION

The Virgo blazars 3C 273 and 3C 279 are well-known flat-spectrum radio quasars. Both sources have been detected at γ-ray energies by different instruments aboard the Compton Gamma-Ray Observatory (CGRO) from ~50 keV (OSSE) up to ~10-20 GeV (EGRET). In this paper we summarize their main MeV properties, which are derived from a consistent analysis of all COMPTEL observations between 1991 and 1997. A complete description of the analysis results, put into multifrequency perspective, is in preparation [1].

OBSERVATIONS AND DATA ANALYSIS

The imaging Compton Telescope COMPTEL - one of four experiments aboard CGRO - was sensitive to γ-rays in the energy range 0.75-30 MeV (for more details on COMPTEL see [2]). During 1991 and 1997 (i.e., CGRO observational phases I to VI) both sources were many times within the COMPTEL field-of-view (see Table 1). These data have been analysed – as a whole as well as individually – by using the standard COMPTEL maximum-likelihood analysis procedure including a filtering technique for background generation. Point spread functions assuming an E^{-2} power law shape for the source spectra were applied in our analyses.

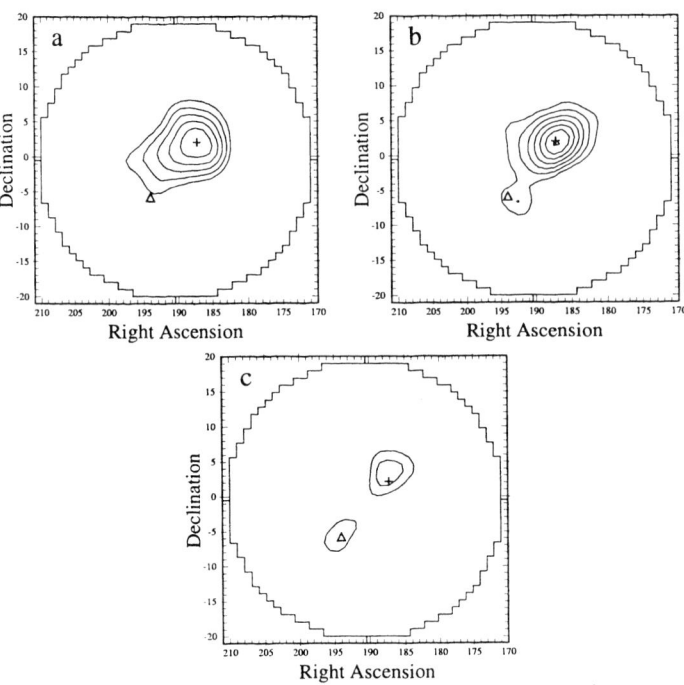

FIGURE 1. Time-averaged COMPTEL maximum-likelihood significance skymaps (1991 to 1997) of the Virgo region in 3 different bands (a: 1-3 MeV; b: 3-10 MeV; c: 10-30 MeV). The locations of the Virgo quasars 3C 273 (+) and 3C 279 (Δ) are given. The contour lines for all maps start at 4σ with a step of 1σ (χ_1^2-statistics for a known source).

TABLE 1. COMPTEL exposures on 3C 273 and 3C 279 during CGRO Phases I to VI (1991 - 1997). The time periods of the different phases, the total observation times (days within COMPTEL field-of-view), and the effective COMPTEL exposures of 3C 273 and 3C 279 are given.

CGRO Phase	Time Period mm/yy - mm/yy	Total Obs. Time [days]	3C 273 Eff. Exp. [days]	3C 279 Eff. Exp. [days]
I	05/91 - 11/92	27	8.95	10.12
II	11/92 - 08/93	21	5.58	5.51
III	08/93 - 10/94	64	10.85	13.90
IV	10/94 - 10/95	42	11.46	9.64
V	10/95 - 10/96	14	6.42	6.21
VI	10/96 - 11/97	49	15.08	12.07
Sum	05/91 - 11/97	217	58.3	57.5

RESULTS

3C 273

In the sum of the 1991 - 1997 data (Table 1), 3C 273 is significantly detected in the 3 standard COMPTEL bands above 1 MeV. The detection significances are $\sim 9\sigma$ in the 1-3 MeV, $\sim 9.5\sigma$ in the 3-10 MeV, $\sim 6\sigma$ in the 10-30 MeV band (Fig. 1), which make 3C 273 the most significant ($\sim 15\sigma$, 1 - 30 MeV) COMPTEL AGN overall. Evidence for the source is found in about 80% of the individual Virgo pointings (Fig. 2). Within uncertainties, no time variability is seen in each of the 3 bands on timescales of a few weeks.

The time-averaged COMPTEL MeV spectrum for the sum of the observations between 1991 and 1997 (Table 1) is shown in Fig. 3 in two representations. The left panel gives the spectrum in the 4 standard COMPTEL bands. The fluxes in these bands indicate a curved time-averaged MeV spectrum of 3C 273, which changes from a soft shape at upper COMPTEL energies to a harder one at lower energies. The emission maximum, and therefore the peak of the nonthermal inverse-Compton radiation of 3C 273 (for a multifrequency spectrum see e.g. [3]), is located in the 3-10 MeV band. A power-law fit to this summed spectrum results in an unacceptable χ^2-value of ~ 13 for 2 degrees of freedom, proving that a power law is not a viable model. The right panel shows a preliminiary 'high-resolution' (11 data points) spectrum of the same data which could be obtained because of the relatively high statistics for 3C 273. Again a spectral turnover at MeV energies is evident. This spectrum suggests that the spectral turnover of 3C 273 is on average broad and rather smooth. No spectral features (e.g., hints for a blueshifted annihilation line) are obvious.

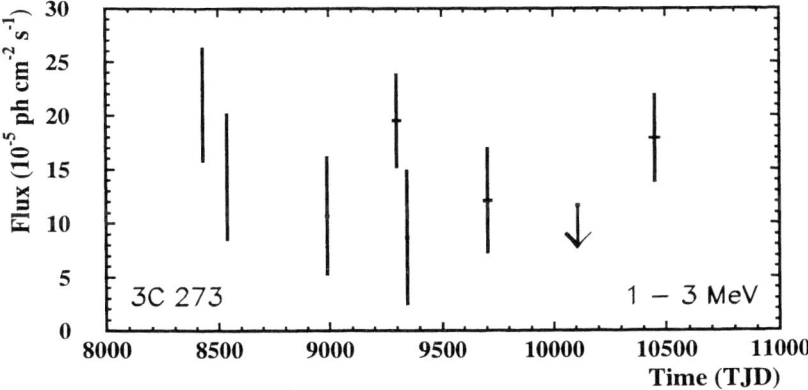

FIGURE 2. COMPTEL light-curves of 3C 273 in the 1-3 MeV band along the CGRO mission between 1991 and 1997. The blazar shows a rather stable γ-ray emission. The errors are 1σ and the upper limit is 2σ.

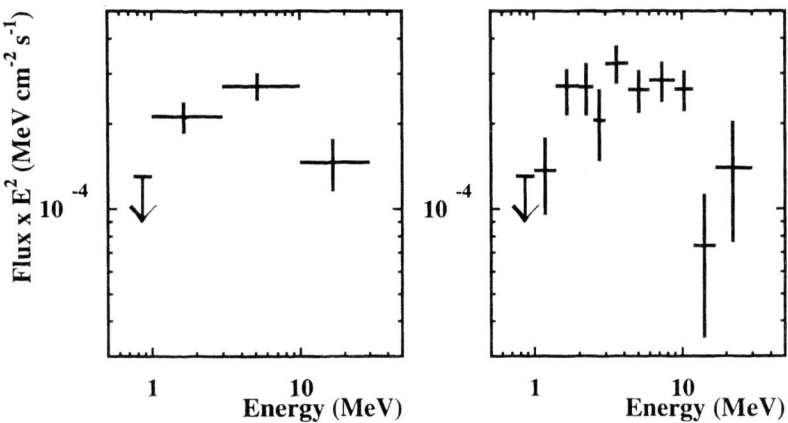

FIGURE 3. Time-averaged (i.e., for the sum of the observations in Table 1) COMPTEL MeV spectra of 3C 273. The left panel shows the spectrum in the 4 standard COMPTEL bands (0.75-1, 1-3, 3-10, 10-30 MeV), and the right one shows the same spectrum with increased spectral resolution (11 spectral points). The errors are 1σ and the upper limits are 2σ.

3C 279

In the sum of all 1991 - 1997 data there is evidence for 3C 279 at energies above $\sim$1 MeV, more significantly above $\sim$3 MeV (Fig. 1). The detection significances for this period are $\sim 3.5\sigma$, $\sim 4.5\sigma$, and $\sim 4.5\sigma$ in the 1-3 MeV, 3-10 MeV, and 10-30 MeV COMPTEL bands, respectively. The detections and non-detections along the course of the mission indicate time variability (Fig. 4). This is most obvious by the high flux in the 10-30 MeV band during CGRO VP 511.5 (February '96), which is about 3σ above the previous flux measurement and the upper limit in the next observational period, and which is simultaneous to the largest γ-ray flare ever observed from 3C 279 by EGRET. There is the trend, that 3C 279 becomes visible in the uppermost COMPTEL band, when EGRET observes a strong flaring event (Fig. 4).

The COMPTEL MeV spectra of 3C 279 can be well represented by power-law shapes. The 6-year time-averaged spectrum is consistent with a power-law shape ($E^{-\alpha}$) with a photon index α of 1.78$\pm$0.15 (Fig. 5). The spectra for the individual time periods are less well determined, but they indicate the trend of a spectral hardening with EGRET-measured flux. To investigate this trend, we subdivided the COMPTEL data according to the γ-ray state (high vs. low) above 100 MeV, and generated MeV spectra for the sum of both states (Fig. 5). These two spectra show different spectral shapes with a crossover at $\sim$2 MeV. This indicates that the 'EGRET flares' are a high-energy γ-ray ($>$3 MeV) phenomenon, and that the peak of the inverse-Compton emission changes with source flux in 3C 279.

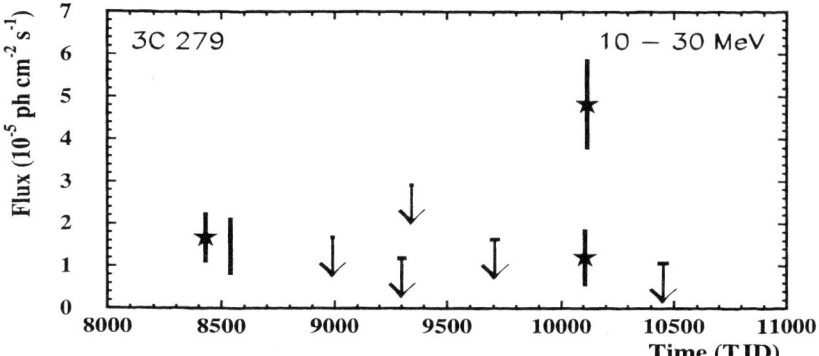

FIGURE 4. COMPTEL light curve of 3C 279 in the 10-30 MeV band from 1991 to 1997. The detections and non-detections indicate time variability. The blazar is always detected by COMPTEL when EGRET reports γ-ray flaring (i.e, high fluxes) at energies above 100 MeV. These periods are marked by a '★'.

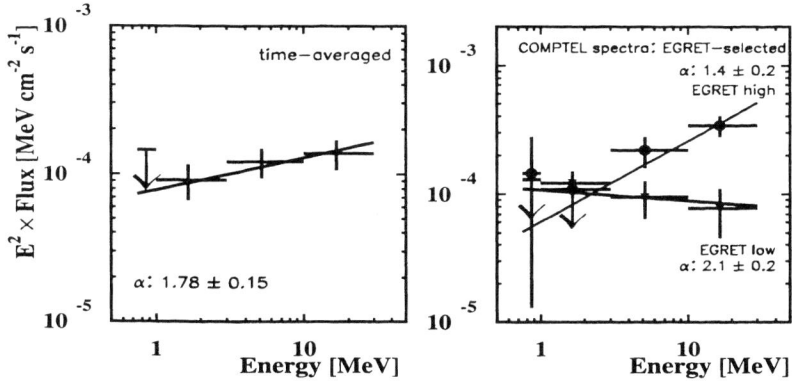

FIGURE 5. Time-averaged COMPTEL spectra of 3C 279. The left panel shows the spectrum averaged over the sum of the '91 to '97 data, while the right panel shows spectra averaged over times of γ-ray flaring ('EGRET high') and γ-ray quiescent ('EGRET low') periods. The spectral difference is obvious.

ACKNOWLEDGMENTS: The COMPTEL project is supported by the German government through DARA grant 50 QV 9096 8, by NASA under contract NAS5-26645, and by the Netherlands Organization for Scientific Research (NWO).

REFERENCES

1. Collmar, W., et al., in preparation (2002).
2. Schönfelder, V., et al., *ApJS* **86**, 657 (1993).
3. Lichti, G., et al., *A&A* **298**, 711 (1995).

Shock Structures in Relativistic Jets

Philip E. Hardee

University of Alabama, Tuscaloosa, AL 35487, USA

Abstract.
The dynamical structures associated with relativistic jets as revealed by numerical simulations and theoretical calculations, and as suggested by observations are presented. Mechanisms for the production of nearly stationary and rapidly moving shock structures are illustrated. These types of structures are expected to arise naturally on the relativistic jets that are thought responsible for galactic microquasars, Blazar activity and γ-ray bursts. [1]

INTRODUCTION

There is compelling evidence for the existence of relativistic jets in galactic and extragalactic objects. In the galactic microquasars GRS 1915+105 [1] and GRO J1655–40 [2][3] the observed motions indicate jet flow speeds that are 92% of lightspeed. Among the extragalactic superluminals, observed motions indicate jet flow speeds as large as 99.9% of lightspeed, e.g., 3C 345 [4]. Furthermore, it is suspected that ultrarelativistic jets with Lorentz factors in excess of 100 are responsible for the γ-ray bursts [5][6]. Resolved relativistic jets exhibit complex time-dependent structures. Here we present the types of time dependent structures associated with relativistic jet flows.

AXISYMMETRIC NUMERICAL SIMULATIONS

Production of Stationary Internal Shocks

Figure 1 is a Schlieren-type image that shows the gradient of the density from a hydrodynamical simulation of a propagating jet initialized with a Lorentz factor $\gamma_j = 5$ and an adiabatic index of 5/3. The jet was assigned a rest frame mass density $\rho_j = 0.1\rho_x$, i.e., 1/10 of the ambient density, and the jet and ambient pressures were taken to be equal [7][8]. Readily apparent in the image is the bow shock and contact discontinuity separating jet material from external shocked material. Jet material passes through the Mach shock disk to form a cocoon of shocked jet material around the jet.

---

[1] Various aspects of the work discussed here were performed in collaboration with I. Agudo (CSIC), A. Alberdi (CSIC), M-A. Aloy (Max-Planck), J. Biretta (STScI), C. Duncan (Bowling-Green S.U.), E. Gomez (U. Alabama), J-L. Gomez (U. Valencia), P. Hughes (U. Michigan), J-M. Ibanez (U. Valencia), A. Marscher (Boston U.), J-M. Marti (U. Valencia), & C. Walker (NRAO).

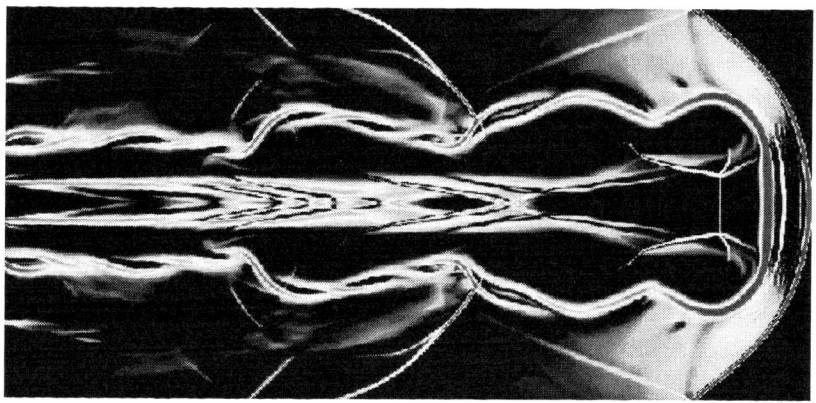

FIGURE 1. Density schlieren showing, from left to right, internal nearly stationary oblique shocks, internal Mach shock disk (narrow vertical line), contact discontinuity and external bow shock.

Behind the Mach shock disk, pressure waves are driven into the jet by vortices in the surrounding cocoon. Schlieren type pressure images show reflection of the cocoon driven pressure waves from the jet axis and diamond patterns within the jet [8]. The weak oblique conical shocks thus formed are nearly stationary and move with the axial speed of the cocoon vortices which can be backwards. Additional closely spaced oblique weak conical shock structures are produced as the initial perturbation couples to a normal pinch mode of the jet [8]. In this simulation the initial perturbation induces overpressures in the jet of about a factor of 5 and the excited normal mode induces overpressures of about a factor of 2.

Production of Differentially Moving Internal Shocks

In a different type of numerical simulation, shown in Figure 2, a steady constantly expanding jet with a small opening angle, $\sim 0.3°$, is established in equilibrium with a decreasing pressure isothermal atmosphere across a computational grid of length $400\,R_b$, where R_b is the beam radius at the injection position. At the inlet the jet density is $\rho_j = 10^{-3}\rho_x$, Lorentz factor is $\gamma_j = 4$ and the Mach number $M_j = 1.69$. The pressure gradient accelerates jet material and $\gamma_j \sim 12$ at $z = 400\,R_b$. This simulation represents a jet in equilibrium with a cocoon medium far behind the jet front. We concentrate our attention on the evolution of the flow after a square-wave increase of the beam flow Lorentz factor from the quiescent value $\gamma_b = 4$ to $\gamma_p = 10$ and an increase in pressure by a factor of two over a time interval $\tau_p = 0.75 R_b/c$, where c is the speed of light. The fluid piles up in front of the velocity perturbation, creating a fast shocked state with a mean speed of $0.995\,c$ ($\gamma_p \sim 10$). This shocked state is followed by a more slowly moving (mean speed of $0.973\,c$) rarefaction where the fast flow separates from the slower upstream flow [9].

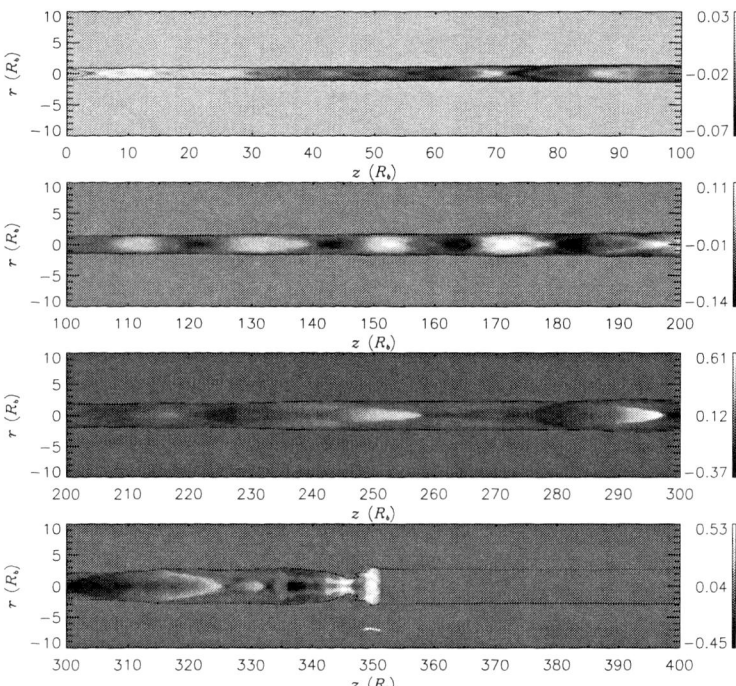

FIGURE 2. Lorentz factor variation associated with slowly moving conical shocks behind a rapidly moving shock perturbation located at $z \sim 350\, R_b$ in this image.

The passage of the main perturbation triggers a local pinch instability which propagates behind the main perturbation, leading to the formation of a series of conical "*trailing shocks*" following the main perturbation. The overpressure associated with these trailing shocks is about a factor of 2 and the Lorentz factor varies by $\pm 20\%$. These trailing components can be easily distinguished because they appear to be released from the primary rapidly moving component, instead of being ejected from the core. Those appearing closer to the core show small apparent motions. Those appearing farther downstream reach higher apparent motions. The spacing and velocity increase of the trailing components result as pinch modes with group velocity comparable to the speed of the rarefaction associated with the main perturbation are excited but slow to the phase speed of the excited wavelength. The excited wavelength is longer and the wave speed is higher at larger distance as an indirect consequence of the expanding jet's acceleration. The existence of these trailing components indicates that not all observed components (internal shocks) need be identified with velocity fluctuation at the central engine. Thus, multiple emission components (shocks) can be generated by a single velocity fluctuation at the central engine.

JET STRUCTURES IN 3D

Simulations and Theory

In the simulation shown here in Figure 3, a 'pre-existing' jet flow is established across the computational volume. For this simulation we take the ratio of rest frame densities to be $\rho_j = 10.0\,\rho_x$. The jet flow has $v_j = 0.9165c$ and the Lorentz factor is 2.5. The adiabatic index is 5/3 and the relevant sound speeds are $a_x = 0.6121c$ and $a_j = 0.2753c$. The jet is perturbed by a periodic precession at the inlet with $\omega R_j/u = 0.93$ but with a very small transverse velocity [10].

In the simulation, spiral shock waves are driven into the external medium as the helical pattern moves faster than the external sound speed (although slower than the jet speed). Complex pressure and velocity structure inside the jet is shown to be produced by a combination of the helical surface and first body modes predicted by a normal mode analysis of the relativistic hydrodynamic equations [11]. The surface and first body mode have different wave speed and wavelength, are launched in phase by the periodic precession, and exhibit a beat pattern in line-of-sight images (e.g., from $z = 15 - 20$). Wave speeds are a large fraction of the flow speed but the beat pattern remains stationary. Thus, in this simulation we find a mechanism that can produce differentially moving helically twisted and stationary features in the 3D jet and strong helical shocks in the external medium.

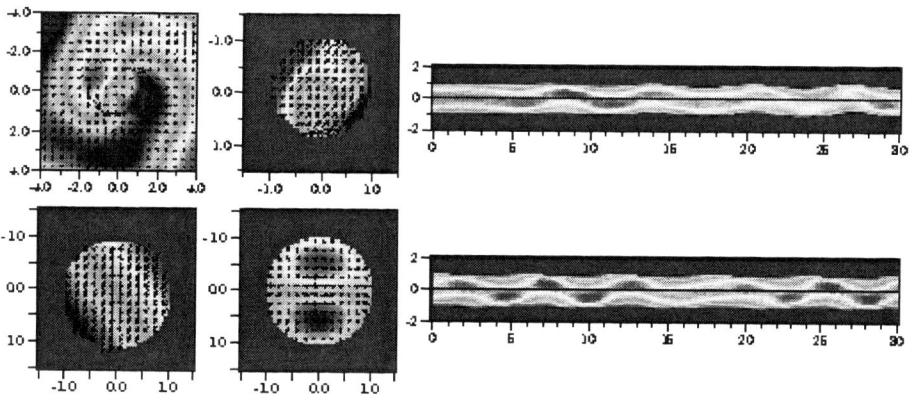

FIGURE 3. The top row shows from the numerical simulation: (left) a pressure cross section across the entire width of the grid, (middle) a pressure cross section showing only the rapidly moving jet material and (right) a line-of-sight integration through the data cube of the square of the pressure showing only the rapidly moving jet material. The bottom row shows from the theory: (left) a pressure cross section of the helical surface mode, (middle) a pressure cross section of the helical first body mode and (right) a line-of-sight integration through a theoretical data cube of the square of the pressure of combined surface and body modes showing only the rapidly moving jet material.

Observations and Theory

Resolved relativistic jets in extragalactic sources, exhibit time-dependent curved structures with both superluminally moving and much more slowly moving or stationary components, e.g., M 87 [12][13] & 3C 120 [14]. A combination of slowly and rapidly moving components can be the result of enhanced features moving with the jet flow through slower moving helical structures [14][15]. While the internal jet pressure structure in present 3D simulations only involves pressure fluctuations on the order of $\pm 15\%$, comparison between theory and observationally observed twisted structures in the M 87 and 3C 120 jets indicate the need for overpressures of about a factor of 2. Thus, we infer the existence of significant weak oblique twisted shock structures inside relativistic jets.

CONCLUSION

We conclude that relativistic jets will naturally contain a plethora of weak internal oblique shocks in addition to the strong shocks associated with the jet front, with velocity fluctuation, or with large amplitude sideways perturbation to the jet flow. By implication, the initial jet flow can be steadier than previously thought, while still producing highly variable emission associated with slowly or rapidly moving internal shock structures.

ACKNOWLEDGMENTS

Aspects of this research were supported by Spanish Dirección General de Investigación Científica y Técnica (grants PB97-1164,-1432), by NASA Astrophysical Theory grant NAG5-3839, by the Fulbright commission for collaboration between USA and Spain, by the guest program of the Max-Planck-Institut für Astrophysik, and by the National Science Foundation through grant AST-9802955 to the University of Alabama.

REFERENCES

1. Mirabel, I.F. & Rodríguez, L.F., *Nature* **371**, 46 (1994).
2. Hjellming, R.M., & Rupen, M.P., *Nature* **375**, 464 (1995).
3. Tingay, S.J. et al., *Nature* **374**, 141 (1995).
4. Zensus, J.A., Cohen, M.H., & Unwin, S.C., *ApJ* **443**, 35 (1995).
5. Woosley, S.E., "Gamma-Ray Bursts: The Central Engine", in *Gamma-Ray Bursts: 5th Huntsville Symposium*, eds. R.M. Kippen et al., AIP Conference Proceedings 526, New York, 2000, pp.555-564.
6. Müller et al., "2D Hydrodynamic Simulations of Relativistic Jets from Collapsars",in *Gamma-Ray Bursts: 5th Huntsville Symposium*, eds. R.M. Kippen et al., AIP Conference Proceedings 526, New York, 2000, pp.565-569.
7. Duncan, G.C., & Hughes, P.A., *ApJL* **436**, L119 (1994).
8. Hardee, P.E., Rosen, A., Hughes, P.A., & Duncan, G.C., *ApJ* **500**, 599 (1998).
9. Agudo, I. et al., *ApJL* **549**, L183 (2001).
10. Hardee, P.E., Hughes, P.A., Rosen, A., & Gomez, E.A., *ApJ* **555**, in press (2001).
11. Hardee, P.E., *ApJ* **533**, 176 (2000).
12. Biretta, J.A., Zhou, F., & Owen, F.N., *ApJ* **447**, 582 (1995).
13. Biretta, J.A., Sparks, W.B., & Macchetto, F., *ApJ* **520**, 621 (1999).
14. Walker, R.C. et al., *ApJ*, July 20, in press (2001).
15. Alberdi, A., Gómez, J.-L., Marcaide, J.M., Marscher, A.P., Pérez-Torres, M.A., *A&A* **361**, 529 (2000).

Recent Observations of 1ES2344+514 Using the Whipple Gamma-Ray Telescope

H. M. Badran

for the VERITAS Collaboration
Smithsonian Institution, Whipple Observatory, P.O.Box 97, Amado, AZ 85645
Department of Physics, Faculty of Science, Tanta University, Tanta, Egypt

Abstract. BL Lacertae objects are a class of AGN which are believed to be highly beamed. Very high energy gamma-rays emitted from such objects provide important information that improves our understanding of this class of objects. Only two AGNs, Mrk 421 and Mrk 501, are well established TeV emitters in the northern hemisphere. The Whipple collaboration previously reported a weak signal from the BL Lac object 1ES 2344+514. This object has been extensively monitored with atmospheric Cherenkov experiments from 1997 to 1999 with no other reported detection. The results of observations in 1999-2001 of 1ES 2344+514 taken with the new Whipple 490 pixel camera are presented. These observations resulted in a detection at the 3 σ level.

INTRODUCTION

Blazars are dominated by non-thermal continuum emission. They are characterized by emitting plasma in relativistic motion closely aligned with our line of sight (e.g. [1]). A subclass of these objects radiate in GeV-TeV part of the spectrum and can emit up to 90% of their output in gamma-rays during flares. Their spectra usually reveal two broad components. The low energy component peaks in the IR-X-ray range and the high energy component peaks in the MeV-TeV range. Analysis of flaring activity from these objects seems to provide some evidence for correlation between GeV and optical bands in high luminosity blazars [2] and [3] and TeV and X-ray in low luminosity blazars [4, 5, 6, 7]. In the synchrotron self-Compton model [8], synchrotron photons are scattered to higher energies by their parent electrons. This process is thought to be responsible for the production of X-rays. On the other hand, GeV-TeV emission from blazars is produced through the inverse Compton scattering of soft photons by ultra-relativistic electrons.

Table 1 summarizes the detection of TeV photons from AGNs. Unfortunately, only two of these detections are independently confirmed [9]. Increasing the number of confirmed TeV sources will increase the opportunities to test, in detail, the emission models for these objects, e.g. place constraints on the mechanism responsible for accelerating electrons to TeV energies. In addition, TeV gamma-rays from blazars can be used to probe the intergalactic infrared radiation field. The blazar 1ES 2344 is one of the best candidates for monitoring, particularly since it is one of the closest AGN (z=0.044). Very high energy gamma-rays from 1ES2344+514 were first detected by the Whipple Collaboration [10]. On Dec. 20th, 1995, the Whipple telescope detected a burst of TeV

TABLE 1. Reported detection of TeV photons from AGN. Some characteristics of these sources are also given.

Source	Discovery	Confirmed TeV source	Type	z	EGRET source
Markarian 421	Whipple [11]	Yes	HBL	0.031	Yes
Markarian 501	Whipple [12]	Yes	HBL	0.034	Yes
1ES 2344+514	Whipple [10]	Yes*	HBL	0.044	No
1ES 1959+650	TA [13]	No	HBL	0.048	No
BL Lac	Crimea [14]	No	HBL	0.069	Yes
PKS 2155-304	Durham [15]	No	HBL	0.116	Yes
1H 1426+428	Whipple [16]	No	HBL	0.129	No
3C66A	Crimea [17]	No	LBL	0.444	Yes

* this work

photons from the source corresponding to a flux (E>350 GeV)=$(6.6 \pm 1.9) \times 10^{-11}$ photons/cm^{2}s. The detected flux for non-flaring activity during Oct. 1995-Jan. 1996 was $(1.1 \pm 0.4) \times 10^{-11}$ photons/cm^{2}s. For the period 1996–97 an upper limit of 8.2×10^{-12} photons/cm^{2}s was reported. The HEGRA experiment found no evidence in their 1997-98 data for TeV emission from this object [18].

OBSERVATIONS AND DATA REDUCTION

The TeV activity of 1ES 2344 was monitored using the 10-m imaging atmospheric Cherenkov telescope at the Whipple Observatory. The observations were made with the present 490 pixel camera [19]. This camera was installed in Oct. 1999. The first phase of operation involved calibration of the electronics. A mirror re-coating program started about the same time [20]. As of Dec. 1999, the system was stable and a significant fraction of the mirrors were re-coated and mounted on the telescope. The data considered in this work was taken during the 1999-2000 and the 2000-2001 observing seasons. The discriminator thresholds were reduced on Oct. 25th, 2000 from 36 mV to 32 mV to reduce the energy threshold. At nearly this same time the program of mirror re-coating was completed.

As a result the peak energy, at which the rate of photons per unit energy of the Crab Nebula is highest, was reduced from 430 to 390 GeV. For these reasons the data were divided into two sets. Table 2 shows the selection criteria (Supercuts, [21]) for gamma-ray-like events for the two periods. The Crab rate for the first and second period was found to be 2.5 and 4.2 gammas/min, respectively.

Observations are carried out using two modes; On/Off and Tracking [22]. Each observation has a 28 minute duration. Only data taken in good weather and with elevation above 55° were included in this analysis. In addition, unstable raw data were excluded. Almost all of these data were taken in the Tracking mode. The tracking ratio was calculated using both zenith runs and Off runs from different sources with elevation above 55°. Fig. 1 shows the distribution of the alpha parameter for the On-source data that passed the selection criteria given in Table 2, for the two periods, and the sum of all

TABLE 2. Gamma-ray selection criteria for the two observing periods of the 10 m Whipple telescope with the current 490 pixels camera. Results of the 1ES2344+514 observations for 1999-2000 and 2000-2001 seasons.

Parameter	Jan. 1–Oct. 24, 2000 (36 mV, 3 fold) Cuts	Oct. 25-Dec. 28, 2000 (32 mV, 3 fold) Cuts	Total
picture/boundary	$\prec 2.25/4.25\sigma$	$\prec 5.0/4.5\sigma$	
Nbr3	on	on	
Max1/Max2	$\prec 30/30$ d.c.	$\prec 50/40$ d.c.	
Distance	0.4-$1.0°$	0.4-$1.0°$	
Width	0.05-$0.12°$	0.05-$0.13°$	
Length	0.13-$0.25°$	0.09-$0.26°$	
Length/Size	$\prec 0.0004°$/d.c.	$\prec 0.0004°$/d.c.	
Alpha	$\prec 15°$	$\prec 15°$	
Tracking ratio	3.20 ± 0.02	3.11 ± 0.02	
Hours	14.9	9.1	24.0
Significance	3.74	1.84	3.10
Peak Energy	430 GeV	390 GeV	
Flux (photons/cm^{2}s)	$1.21 \pm 0.34 \times 10^{-11}$	$0.91 \pm 0.51 \times 10^{-11}$	

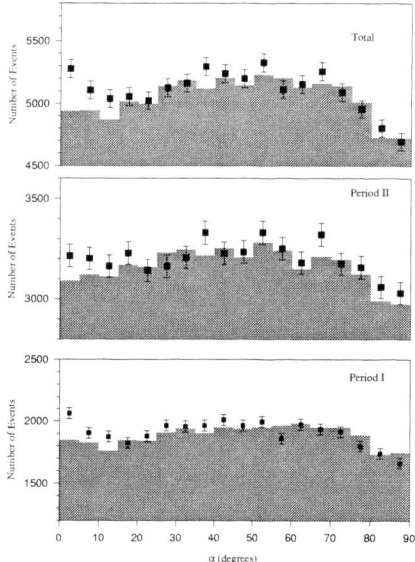

FIGURE 1. The α distribution of 1ES2344+514 data. These distributions are for all of the event that passed all cuts listed in Table 2 except for the α cut for period I (bottom), period II (middle) and the total of both periods (top). For comparison, the dark histogram is the expected distribution from a nonsource.

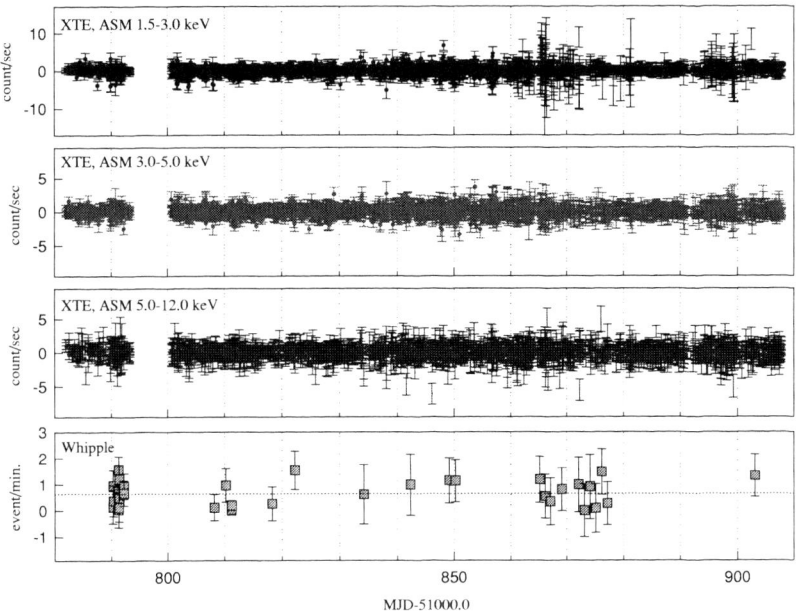

FIGURE 2. TeV gamma-ray light curve for all on-source data from 1ES2344+514 (bottom). Also shown are the light curves of ASM/RXTE data in three energy bands.

of the data. The gamma-ray signal is calculated for the $\alpha = 0° - 15°$ region and the background is estimated from the $\alpha = 20° - 65°$ region.

RESULTS

The observing time, the calculated significance of 1ES2344+514 data and the average flux for the two periods are given in Table2 . On the basis of these results we believe that we have confirmed 1ES2344+514 as a TeV source at the 3 σ level. The present result together with the previous Whipple detection [10] indicate that the object is a very high energy gamma-ray source. X-ray data from All-Sky Monitor (ASM) on board the Rossi X-Ray Timing Explorer satellite (RXTE) for the same time period were examined. X-ray light curves obtained from the data from the ASM/RXTE detector in three energy bands; 1.5-3.0, 3.0-5.0 and 5.0-12.0 keV are shown in Fig. 2. Also shown is the TeV gamma-ray rate from the source. Neither X-ray (especially the 5.0-12.0 keV range) nor TeV data exhibit any significant variation in the detected rate. This supports the idea that the source was not flaring during the period of the observations, hence these data do not provide a good opportunity for analysis of a possible correlation between X-rays and

TeV gamma-rays. The average flux in this work is consistent with the previous Whipple flux. The previous and current Whipple measurements together with the HEGRA upper limit may indicate a sharp decrease of the flux of 1ES2344+514 with energy. The absence of a simultaneous detection of the source by two different experiments does not allow any speculations about the spectral index of the source in the energy range 350 GeV - 1 TeV.

CONCLUSION

1ES2344+514 was observed by the Whipple 10-m gamma-ray telescope for a total of 24 hours. The TeV signal from 1ES2344+514 is consistent with the ASM/RXTE measurements. No obvious flaring activity was observed during 1999-2000 in either X-ray or TeV gamma-ray. The cumulative excess exceeds 3σ. While detection of 1ES2344+514 at TeV energies is not as yet confirmed by independent observations the consistently positive results are suggestive. Continued monitoring of the source with the Whipple Cherenkov telescope is planned.

ACKNOWLEDGMENTS

The VERITAS Collaboration is supported by the U.S. Dept. of Energy, N.S.F., the Smithsonian Institution, P.P.A. R.C. (U.K.) and Enterprise-Ireland.

REFERENCES

1. Urry, C.M. and Padovani, P., *PASP*, **107**, 803-845 (1995).
2. Wagner, S.J. et al., *ApJ*, **454**, L97-L100 (1995).
3. Wehrle, A.E., et al., *ApJ*, **497**, 178-187 (1998).
4. Macomb, D.J., et al., *ApJ*, **449**, L99-L103 (1995).
5. Takahashi, T., et al., *ApJ*, **470**, L89-L92 (1996).
6. Petry, D. et al., *ApJ*, **536**, 742-755 (2000).
7. Sambruna, R.M. et al., *ApJ*, **538**, 127-133 (2000).
8. Jones, T.W., et al., *ApJ*, **192**, 261-278 (1974).
9. Weekes, T.C., in *Proc. GeV-TeV Gamma Ray Astrophysics Workshop* (Snowbird, Utah), AIP 515, 3-15 (2000).
10. Catanese, M., et al., *ApJ*, **501**, 616-623 (1998).
11. Punch M., et al., *Nature*, **358**, 447-478 (1992).
12. Quinn, J., et al., *ApJ*, **456**, L83-L86 (1996).
13. Nishiyama, T., et al., *in Proc. 26th ICRC* (Salt Lake City, USA), **3**, 370-373 (1999).
14. Neshpor, Yu.I., et al., *Astronomy Reports*, **45**, 249-254 (2001).
15. Chadwick, P.M., et al., *Astropart. Phys.*, **11**, 145-151 (1999).
16. Horan, D., et al., *in Proc. 27th ICRC* (Hamburg, Germany), (2001).
17. Neshpor, Yu.I., et al., *Astron. Lett.*, **24**, 134-138 (1998).
18. Aharonian, F.A., et al., *A&A*, **353**, 847-852 (2000).
19. Finley, J.P. et al., *in Proc. 27th ICRC* (Hamburg, Germany), (2001).
20. Badran, H.M. and Weekes, T.C., *in Proc. 27th ICRC* (Hamburg, Germany), (2001).
21. Reynolds, P.T., et al., *ApJ*, **404**, 206-218 (1993).
22. Catanese, M., et al., *in Proc. 25th ICRC* (Durban, South Africa), **3**, 277-280 (1997).

Model for the X-ray Emission in the Jets and Hot Spots of Radio Galaxies

Charles D. Dermer[1] and Hui Li[2]

[1] Naval Research Laboratory, Code 7653, Washington, DC 20375-5352
[2] Theoretical Astrophysics, MS B288, Los Alamos National Laboratory, Los Alamos, NM 87545

Abstract. We propose a model for the origin of extended X-ray emission in the large-scale jets and hot spots of radio galaxies. If radio galaxies are misaligned blazars, then they are sources of collimated high-energy gamma radiation from the inner jet. Gamma rays with $\gtrsim 100$ TeV energies will be attenuated by the microwave background and surrounding radiation fields, causing high-energy electrons and pairs to be injected far from the central engine. The high-energy pairs will produce X-ray synchrotron radiation that could account for the emission detected from sources such as Pictor A and Cygnus A with the *Chandra X-ray Observatory*.

I INTRODUCTION

Blazars, which include flat spectrum radio quasars, optically violently variable quasars, highly polarized quasars, and BL Lac objects, are thought to be radio-loud AGN with relativistic jets that are aligned towards the observer. The EGRET instrument on the *Compton Gamma Ray Observatory* detected more than 60 blazars in the energy range between 100 MeV and 5 GeV [1]. Ground-based Cherenkov telescopes have detected six BL Lac objects at energies $\gtrsim 200$ GeV [2] with varying degrees of confidence. These latter sources are at low redshifts, consistent with the interpretation that TeV photons from large redshifts are attenuated by the diffuse intergalactic infrared and microwave radiation fields.

Observations of extragalactic jet sources with the *Chandra X-ray Observatory* have revealed X-ray emission from extended jets, hot spots, and lobes [3–8]. Although X-ray jets from sources such as M87 and 3C 273 were known prior to the launch of *Chandra*, the better imaging and sensitivity of *Chandra* has made it possible to resolve X-ray jets in many more sources than before [9]. Due to the absence of hot gas to confine the jet, thermal processes are generally discounted as the origin of the jet X-ray emission. Different nonthermal processes have been proposed to account for the X-ray emission, for example, synchrotron self-Compton emission in Cygnus A [5] and Compton-scattered cosmic microwave background (CMB) radiation in PKS 0637-752 [10]. Because different components appear to contribute to the radio-through-optical and to the X-ray spectra, nonthermal X-ray synchrotron radiation has been dismissed as a possible origin for the X-ray jet emission.

Here we propose a model for extended X-ray emission from extragalactic jet sources arising from nonthermal synchrotron X-ray emission emitted by pairs formed through $\gamma\gamma$ attenuation. If blazar γ-ray emission extends to $\gg 100$ TeV, these high-energy photons will be subject to pair-production attenuation by ambient photon fields, including internal synchrotron radiation, the direct and scattered accretion-disk radiation, and the diffuse extragalactic infrared and CMB radiation. A weak, randomly oriented magnetic field will isotropize the high-energy leptons. In addition to the high-energy γ-ray pair halos formed by Compton-scattered CMB radiation in the cascade around jet sources [11], nonthermal X-ray synchrotron radiation will be emitted. Deposition of the energy of the pair cascade in the extended jets, hot spots, and radio lobes could produce the X-ray emission observed with *Chandra*.

II PAIR CASCADE MODEL FOR JET EMISSION

In the standard model for blazars, collimated ejection of plasma at relativistic speeds produces beamed, high-energy γ radiation. The γ rays are formed within ≈ 1 pc from the central engine by leptonic or hadronic processes in the inner jets of blazars. The maximum γ-ray photon energy is unknown because of the various $\gamma\gamma$ attenuation processes that operate between the inner jet and the observer. In particular, the CMB and diffuse intergalactic infrared radiation (DIIRF) will attenuate $\gtrsim 100$ TeV and $\gtrsim 20\text{-}30$ TeV radiation, respectively, from extragalactic sources. In fact, the highest energy radiation detected from extragalactic sources is at ≈ 10 TeV from the low-redshift ($z \approx 0.03$) BL Lac objects Mrk 421 and Mrk 501 [12].

We therefore have no direct way of knowing the highest energy photons that are emitted by blazars. Here we assume that photons with energies $\gg 100$ TeV are emitted by blazar jets, and work out the consequences of this assumption.

A Attenuation in the Emitting Region

The model is viable only if very high-energy photons can escape from the jet plasma without significant attenuation due to the internal synchrotron emission. This places constraints upon the Doppler factor δ of the outflowing radiating plasma. The $\gamma\gamma$ transparency optical depth $\tau_{\gamma\gamma}(\epsilon'_1) \cong \sigma_T u(\epsilon')r'_b/(3m_e c^2)$, where $u'(\epsilon')$ is the spectral energy density, r'_b is the comoving blob radius, assumed spherical, and $\epsilon' = h\nu'/m_e c^2$ is the dimensionless photon energy [13]. Primes refer to quantities in the comoving blob frame. The threshold condition $\epsilon'_1 = 2/\epsilon'$ relates the energies of the interacting photons in the delta-function approximation for this process. The blob radius can be written in terms of the observed variability time scale t_{var} and δ through the relation $r'_b \cong c\delta t_{var}/(1+z)$.

Let the νF_ν flux at frequency $\nu = m_e c^2 \epsilon/h$ be denoted by $f(\epsilon)$. Because $\epsilon' u'(\epsilon') \cong 2d_L^2 f(\epsilon)/(c\delta^4 r_b'^2) \cong 2d_L^2 (1+z)^2 f(\epsilon)/(c^3 \delta^6 t_{var}^2)$, where d_L is the luminosity distance, one finds that the $\gamma\gamma$ transparency condition $\tau_{\gamma\gamma}(\epsilon_1') < 1$ implies

$$\delta \gtrsim [\frac{\sigma_T \epsilon_1}{3 m_e c^2} \frac{d_L^2 f(\epsilon)(1+z)^2}{c^2 t_{var}}]^{1/6}, \quad \epsilon = \frac{2\delta^2}{(1+z)^2 \epsilon_1}. \quad (1)$$

For γ rays with energy $100 E_{100 \text{ TeV}}$ TeV, $\epsilon_1 \cong 2 \times 10^8 E_{100 \text{ TeV}}$. Consider a low redshift radio galaxy at $d_L = 10^{27} d_{27}$ cm. If the measured νF_ν flux at frequency $\nu \cong 10^{12} \delta^2/E_{100 \text{ TeV}}$ Hz is $10^{-11} f_{-11}$ ergs cm^{-2} s^{-1} then, from equation (1), the blob is transparent to pair production attenuation if $\delta \gtrsim 14[d_{27}^2 E_{100 \text{ TeV}} f_{-11}/t_{var}(\text{d})]^{1/6}$, where $t_{var}(\text{d})$ is the variability time scale in days. The $\gtrsim 100$ TeV photons are therefore transparent to $\gamma\gamma$ attenuation if $\delta \gtrsim 15\text{-}20$. Such values of δ are consistent with measurements of superluminal motion in many jet sources.

B Attenuation by the Accretion-Disk Radiation Field

Very high-energy γ rays will be subject to $\gamma\gamma$ attenuation by photons from the direct and scattered radiation field. Considering only the scattered disk radiation field, we find that $\tau_{\gamma\gamma}(\epsilon_1) \cong \sigma_T \tau_{sc} L(\epsilon_s)/(12\pi m_e c^2 R_{sc} c)$ [13], where $\epsilon_s = 2/\epsilon_1$, and $\tau_{sc} = 0.01\tau_{-2}$ is the Thomson depth and R_{sc} is the radial extent of the broad-line region scattering clouds. A Shakura-Sunyaev accretion-disk spectrum with quasi-isotropic disk luminosity $10^{46} L_{46}$ ergs s^{-1} and maximum accretion-disk photon energy $\epsilon_{max} = 10^{-4} \epsilon_{-4}$ implies that $L(\epsilon) = (4/3) 10^{46} L_{46} \epsilon_s^{1/3}/\epsilon_{max}^{4/3}$. We therefore find that

$$\tau_{\gamma\gamma}^{ad}(E_{100 \text{ TeV}}) \cong 0.014 \frac{\tau_{-2} L_{46}}{R_{sc}(\text{pc}) \epsilon_{-4}^{4/3} E_{100 \text{ TeV}}^{1/3}} \quad (2)$$

for the $\gamma\gamma$ optical depth of the accretion-disk radiation field. We see that 100 TeV photons will pass through this scattered radiation field without significant attenuation, although ~ 10 GeV photons may be strongly attenuated by the disk radiation field. Thus the emission from a Shakura-Sunyaev disk field may imprint a signature on the high-energy radiation spectrum that can be observed with GLAST and high-energy γ-ray telescopes [14]. The high-energy jet radiation may also be subject to $\gamma\gamma$ attenuation by emission from the surrounding torus [15].

C Attenuation by the CMB Radiation Field

The highest energy photons will unavoidably be attenuated by the CMB radiation field to materialize as high-energy leptons, forming an electromagnetic cascade and pair halo [11]. These leptons will also produce synchrotron X-rays. Here we outline the radiation processes involved in the production of this emission. The *Compton Observatory* found many cases of jets with apparent isotropic γ-ray luminosity of

$10^{48}L_{48}$ ergs s^{-1}, with $L_{48} \cong 1$ [1]. We can therefore write the differential jet luminosity as $L_j(\epsilon) \cong 10^{45}k_{-3}L_{48}\epsilon^{-1}$ (ergs s^{-1} ϵ^{-1}), for a typical blazar spectrum extending from $\epsilon \lesssim 100$ to $\epsilon \gtrsim 10^9$ with a -2 photon index. The factor $10^{-3}k_{-3}$ takes into account the beaming factor of the radiation and the normalization of the total jet luminosity.

Let $\lambda(\epsilon)$ denote the attenuation mean-free-path due to $\gamma\gamma$ interactions with the CMB. An electron and positron pair will be formed with Lorentz factor $\gamma \cong \epsilon/2$. Thus the injection rate per differential distance dx and differential Lorentz factor $d\gamma$ is

$$\frac{dN_e}{d\gamma dx dt} \cong -\int_0^\infty d\epsilon \, \frac{2\delta(\gamma - \epsilon/2)}{m_e c^2 \epsilon} \frac{\partial [L_j(\epsilon) e^{-x/\lambda(\epsilon)}]}{\partial x} \cong \frac{4 \times 10^{45} k_{-3} L_{48}}{m_e c^2 (2\gamma)^2} \frac{e^{-x/\lambda(2\gamma)}}{\lambda(2\gamma)}. \quad (3)$$

The mean-free-path for $\gamma\gamma$ attenuation in a blackbody radiation field at temperature $\theta = k_B T/m_e c^2 = 4.55 \times 10^{-10}(1+z)$ is given by $\lambda^{-1}(\epsilon) = \alpha_f^2 \theta^3 f(\mu)/(\pi \lambda_C)$, where $\alpha_f = 1/137$, $\lambda_C = 3.86 \times 10^{-11}$ cm, and $\mu = (\epsilon\theta)^{-1}$ [16,17]. When $\mu \gg 1$ so that $\epsilon \ll 1/\theta$, $f(\mu) \to \sqrt{\pi\mu}\exp(-\mu)$. Thus $\lambda(E_{100\text{ TeV}}) \cong 1.3 E_{100\text{ TeV}}^{1/2} \exp(11.2/E_{100\text{ TeV}})$ kpc for low redshift sources with $E_\gamma \ll 1000$ TeV. One finds that $\lambda(E_{100\text{ TeV}}) \cong 500$, 84, and 21 kpc for $E_{100\text{ TeV}} = 2$, 3, and 6, respectively. Essentially all of the energy of photons with $E_\gamma \gtrsim 200$ TeV will be reprocessed into high-energy leptons with $\gamma \cong 10^8 E_{100\text{ TeV}}$ along the length of the jet. This represents one-half decade or more of the blazar emission spectrum. If the total jet power is $\sim 10^{46} L_{48}$ ergs s^{-1} for a 1% beaming factor, then $\approx 5 \times 10^{44}$ ergs s^{-1} could be injected along the jet for a -2 blazar jet spectrum that spans 10 decades in photon energy.

When the jet magnetic field $B_{\mu G} \gtrsim 3(1+z)\mu$G, most of the lepton energy will be lost through synchrotron emission rather than through Compton scattering, provided either that the leptons are locally isotropized in the comoving outflow, or that the magnetic field has a randomly oriented component. The peak synchrotron power is detected at $\cong 3 \times 10^{16} B_{\mu G} \delta_j E_{100\text{ TeV}}^2/(1+z)$ Hz, where δ_j is the local jet Doppler factor. This corresponds to the range of *Chandra* observing frequencies. Because of the relatively short energy-loss time scale $t_{syn} \cong 2.5 \times 10^5/[B_{\mu G}^2 E_{100\text{ TeV}}]$ yr, the synchrotron emission will display a -1.5 photon spectrum at photon energies below the peak emission, and a spectral softening and cutoff at higher energies.

III DISCUSSION

High-energy radiation from a blazar provides a collimated injection source of high-energy leptons far from the central nucleus due to $\gamma\gamma$ attenuation with the CMB radiation. The existence of high-energy radiation from extragalactic jet sources follows from the standard model of blazars where radio galaxies are the parent population of blazars [18]. X-ray synchrotron radiation and γ-ray pair halos [11] will provide indirect evidence for the existence of $\gg 100$ TeV photons from blazars, which is not directly available due to attenuation by the CMB and the DIIRF.

As much as $\approx 5 \times 10^{44} L_{48}$ ergs s^{-1} could be deposited in the form of high-energy leptons far from the central source. This is considerably more power than required for the X-ray luminosity measured from the extended jets of Pictor A and 3C 66B ($\sim 10^{41}$ ergs s^{-1}) [7,6], the western hot spot of Pictor A ($\sim 5 \times 10^{42}$ ergs s^{-1}) and knot WK7.8 in PKS 0637-752 ($\sim 4 \times 10^{43}$ ergs s^{-1}) [4]. This power is, however, injected uniformly with distance x according to expression (3), whereas resolved jet X-ray emission exhibits hot spots and irregularities. A spatially resolved model for the extended jets and hot spots in radio galaxies will require further consideration of energy transport through beam instabilities, magnetic field generation of the jet plasma, and a detailed treatment of radiation and plasma processes.

ACKNOWLEDGMENTS

The work of CD was supported by the Office of Naval Research. We thank Armen Atoyan and Reinhard Schlickeiser for many useful discussions.

REFERENCES

1. Hartman, R.C., et al., Astrophys. J. Suppl. **123**, 79 (1999).
2. Weekes, T.C., in High Energy Gamma Ray Astronomy, ed. F.A. Aharonian and H.J. Völk (New York: AIP), p. 15 (2001).
3. Schwartz, D.A., et al., Astrophys. J. **540**, 69 (2000).
4. Chartas, G., et al., Astrophys. J. **542**, 655 (2000).
5. Wilson, A.S., Young, A.J., and Shopbell, P.L., Astrophys. J. **542**, 655 (2000).
6. Hardcastle, M.J., Birkinshaw, M., and Worrall, D.M., MNRAS, in press (astro-ph/0106029) (2001).
7. Wilson, A.S., Young, A.J., and Shopbell, P.L., Astrophys. J. **547**, 740 (2001).
8. Sambruna, R.M., et al., Astrophys. J. **549**, L161 (2001).
9. Harris, D.E., in Particles and Fields in Radio Galaxies, ed. R.A. Laing and K.M. Blundell (astro-ph/0012374) (2000).
10. Tavecchio, F., Maraschi, L., Sambruna, R.M., Urry, C.M., Astrophys. J. **544**, L23 (2000).
11. Aharonian, F.A., Coppi, P.S., and Völk, H.A., Astrophys. J. **423**, L5 (1994).
12. Krennrich, F., et al., Astrophys. J. **511**, 149 (1999).
13. Dermer, C.D., and Schlickeiser, R., Astrophys. J. Suppl. **90**, 945 (1994).
14. Böttcher, M., and Dermer, C.D., Astron. Astrophys. **302**, 37 (1995).
15. Protheroe, R.J., and Biermann, P.L., Astropar. Phys. **6**, 293 (1997).
16. Gould, R.J., and Schréder, G.P, Phys. Rev. **155**, 1404 (1965).
17. Brown, R.W., Mikaelian, K.O., and Gould, R.J., Astrophys. Lett. **14**, 203 (1973).
18. Urry, C. M., and Padovani, P., Publ. Astron. Soc. Pacific **107**, 803 (1995).

The central black hole masses of Gamma-ray loud blazars

J.H. Fan* and K.S. Cheng†

*Center for Astrophysics, Guangzhou University, Guangzhou 510400, China
†Department of Physics, the University of Hong Kong, China

Abstract. In 1999, Cheng, Fan, Zheng proposed a method to determin the parameters: the central black hole mass, M, the Doppler factor, δ, the propagation angle of the γ-rays with respect to the symmetric axis of a two-temperature accretion disk, Φ, and the distance (i.e. the height above the accretion disk), d at which the γ-rays are created. In that method, available variability timescales in gamma-ray bands, the absorption effect of a γ-ray and the beaming effect have been taken into account. In this paper, we adopte this method to a larger sample. Our results indicate that, if we take the intrinsic γ-ray luminosity to be λ times the Eddington luminosity, $L_\gamma^{in} = \lambda L_{Edd.}$, the masses of the blazars are in the range of $(1.5 \sim 83) \times 10^7 M_\odot$ ($\lambda = 1.0$) or $(2.2 \sim 131) \times 10^7 M_\odot$ ($\lambda = 0.1$), the Doppler factors (δ) lie in the range of 0.17 to 3.72 ($\lambda = 1.0$) or 0.25 to 5.33 ($\lambda = 0.1$), the angle (Φ) is in the range of 10° to 68° ($\lambda = 1.0$) or 10° to 63° ($\lambda = 0.1$), and the distance (d) is in the range of $20.5 R_g$ to $750 R_g$ ($\lambda = 1.0$) or $18 R_g$ to $680 R_g$ ($\lambda = 0.1$).

INTRODUCTION

High energy gamma-rays have been detected from more than 60 AGNs (Hartman et al. 1999). Their high and rapidly variable gamma-ray emissions suggest that it is likely arisen from the jet of a blazar.

Various models for γ-ray emission from AGNs have been proposed. Generally, they are of two kinds: leptonic and hadronic models. But there is no consensus yet on the dominant emission process (see von Montigny et al. 1997 for 3C273, Ghisellini et al. 1996 for 3C279, Comastri et al. 1997 for 0836+710, Böttcher & Collmar 1998 for PKS 0528+134).

It is generally believed that the escape of high energy γ-rays from an AGN depends on $\gamma - \gamma$ pair production process because there are lots of soft photons around the central black hole. Becker & Kafatos (1995) have calculated the γ-ray optical depth in the X-ray field of an accretion disk. They found that the γ-rays should escape preferentially along the symmetric axis of the disk, due to the strong angular dependence of the pair production cross section. The phenomenon of $\gamma - \gamma$ "focusing" is related to the more general issue of $\gamma - \gamma$ transparency, which sets a minimum distance between the central black hole and the site of γ-ray production (Bednarek 1993, Dermer & Schlickeiser 1994, Becker & Kafatos 1995, Romero et al. 2000; Zhang & Cheng 1997). Therefore the γ-rays are focused in a small solid angle, $\Omega = 2\pi(1 - cos\Phi)$, suggesting that the apparent observed luminosity should be expressed as $L_\gamma = \Omega D^2 (1+z)^{\alpha_\gamma - 1} F_\gamma^{obs.} (> 100 MeV)$, where $F_\gamma^{obs.}$ is the observed energy flux of the γ-rays, D the distance to the AGN, and z

the redshift. The observed γ-rays from the AGN require that the jet almost points to us and the optical depth $\tau \leq 1.0$. In this sense, both the absorption and beaming (boosting) effects should be considered when the properties of a γ-ray loud blazar are discussed, which is the focus of the present paper. In section 2 we give the method and the results for 22 objects. $H_0 = 75$ km s^{-1} Mpc^{-1}, and $q_0 = 0.5$ are adopted throughout the paper.

METHOD AND RESULTS

Method

Now we describe our method of estimating the basic parameters (M, δ, Φ and d) of the blazars with short timescale variabilities in the γ-ray band (Cheng et al. 1999). As mentioned above, high energy γ-rays can escape only when the optical depth of γ-γ pair production is not larger than unity. Based on Becker & Kafatos (1995), we can obtain an approximate relation for the optical depth at an arbitrary angle, Φ,

$$\tau_{\gamma\gamma}(M_7,\Phi,d) = \frac{1}{3}(51-8\omega) \times \Phi^{2.5}(\frac{d}{R_g})^{-\frac{2\alpha_X+3}{2}} + kM_7^{-1}(\frac{d}{R_g})^{-2\alpha_X-3} , \quad (1)$$

where k is given by

$$k = 4.50 \times 10^9 \frac{\Psi(\alpha_X)(2-\omega)(1+z)^{3+\alpha_X}F_0'(1+z-\sqrt{1+z})^2}{(2\alpha_X+4-\omega)(2\alpha_X+3)} \times$$
$$[\frac{(\frac{R_0}{R_g})^{2\alpha_X+4-\omega} - (\frac{R_{ms}}{R_g})^{2\alpha_X+4-\omega}}{(\frac{R_0}{R_g})^{2-\omega} - (\frac{R_{ms}}{R_g})^{2-\omega}}](\frac{E_\gamma}{4m_ec^2})^{\alpha_X} , \quad (2)$$

Here, $\Psi(\alpha_X)$ is a function of the X-ray spectral index, α_X, F_0' the X-ray flux parameter in units of cm^{-2} s^{-1}, m_e the electron mass, c the speed of light, $R_g = \frac{GM}{c^2}$ the gravitational radius, E_γ the average energy of the γ-rays, and R_0 and R_{ms} are the outer and inner radii of the accretion disk respectively. ω is a free parameter, $\omega = 3$ is for a two-temperature disk while $\omega = 0$ is for a uniformly bright disk.

From Eq. (1), the optical depth depends on d, Φ and M. At first, d can be determined if the variability timescale (ΔT_D in days) for a blazar is observed; it is given by

$$\frac{d}{R_g} = 1.73 \times 10^3 \frac{\Delta T_D}{1+z} \delta M_7^{-1} \quad (3)$$

Furthermore, using the observed γ-ray flux, $F_\gamma^{obs}(>100MeV)$ in units of ergs cm^{-2} s^{-1}, the relationship among the intrinsic luminosity (L_{in}), the Doppler factor(δ), the mass of the central black hole (M), and the propagation angle (Φ), is given by $F_\gamma^{obs}(>100MeV) = (1+z)^{1-\alpha_\gamma}\delta^{\alpha_\gamma+4}L_{in}/\Omega D^2$. We can define an isotropic luminosity as $L_{iso} = 4\pi D^2(1+z)^{\alpha_\gamma-1}F_\gamma^{obs}(>100MeV)$ in units of 10^{48} ergs s^{-1}, which can be expressed as

$$L_{iso}^{48} = \frac{\lambda 2.52 \cdot 10^{-3}\delta^{\alpha_\gamma+4}}{1-\cos\Phi}M_7 , \quad (4)$$

where $L_{in} = \lambda L_{Edd} = \lambda 1.26 \times 10^{45} M_7$, and λ is a parameter depending on the specific γ-ray emission model.

Substituting Eqs. (3) and (4) into Eq. (1), we obtain a function of M and Φ. From this equation, a minimum value of $\tau_{\gamma\gamma}$ for a given mass, M, can be determined by $\frac{\partial \tau}{\partial \Phi}|_M = 0$, i.e. solving

$$\frac{2.5}{3}(51-8\omega)\Phi^{1.5}(1-\cos\Phi) - \frac{1}{3}(51-8\omega) \times \frac{2\alpha_X+3}{2\alpha_\gamma+8}\Phi^{2.5}\sin\Phi$$

$$-\frac{2\alpha_X+3}{\alpha_\gamma+4}kM_7^{-1}A^{-\frac{2\alpha_X+3}{2}}(1-\cos\Phi)^{-\frac{2\alpha_X+3}{2\alpha_\gamma+8}}\sin\Phi = 0 \qquad (5)$$

where

$$A = 1.73 \times 10^3 \frac{\Delta T_D}{1+z} M_7^{-\frac{\alpha_\gamma+5}{\alpha_\gamma+4}} \left(\frac{L_{iso}^{45}}{\lambda 2.52}\right)^{\frac{1}{4+\alpha_\gamma}}$$

Finally, letting the minimum of $\tau(M_7, \Phi)$ equal to 1.0, we have

$$\frac{1}{3}(51-8\omega) \times \Phi^{2.5}\left(\frac{d}{R_g}\right)^{-\frac{2\alpha_X+3}{2}} + kM_7^{-1}\left(\frac{d}{R_g}\right)^{-2\alpha_X-3} = 1 \qquad (6)$$

For a source with available data in the X-ray and γ-ray bands, the masses of the central black holes, M_7, the Doppler factor, δ, the distance (height), d, and the propagation angle with respect to the axis of the accretion disk, Φ, can be derived from Eqs. (3), (4), (5) and (6), where $R_{ms} = 6R_g$, $R_0 = 30R_g$, $E_\gamma = 1\text{GeV}$ and $\omega = 3$ (a two-temperature disk) are used.

Results

Since we are interested in the variability timescale, we consider here only those γ-ray loud blazars with short timescales of variation, detected in the γ-ray region or other lower energy bands. We use the doubling timescale, $\Delta T_D = (F_{min}/\Delta F)\Delta T$, as the variability timescale, where $\Delta F = F_{max} - F_{min}$ is the variation of the flux over the time ΔT. There are few simultaneous observations of the X-ray and γ-ray bands available, the data considered here are not simultaneous. The highest γ-ray flux given in the paper by Hartman et al (1999) are used to calculate the γ-ray luminosity. The X-ray data are from recent publications, particularly the paper by Fossati et al. (1998). The doubling time is from the paper by Dondi & Ghisellini (1995) except for two objects: 1226+023 (Courvoisier et al. 1988) and 2230+114 (Pica et al. 1988).

We do not know the intrinsic γ-ray luminosity, so we assume it is close to the Eddington luminosity, say $\lambda L_{Edd.}$. Using the available X-ray and γ-ray data, we estimate the four parameters (M_7, Φ, δ, d) and find that the derived values of the four parameters are not sensitive to the value of λ. The results are shown in Table 1. Col. 1 gives the name, Col. 2 the Doppler factor ($\lambda = 1.0$), Col. 3 the Doppler factor ($\lambda = 0.1$), Col. 4, the central black hole mass in units of $10^7 M_\odot$ ($\lambda = 1.0$), Col. 5, the central black hole mass ($\lambda = 0.1$), Col. 6, the propagation angle, Φ in the units of degree(°) ($\lambda = 1.0$), Col. 7, the propagation angle ($\lambda = 0.1$), Col. 8, the distance (height), $\frac{d}{R_g}$, where the γ-rays are

TABLE 1. Determined Results for the γ-ray loud Blazars

Name	δ	δ	M_7	M_7	Φ	Φ	$\frac{d}{R_g}$	$\frac{d}{R_g}$
(1)	(2)	(3)	(4)	(5)	(6)	(7)	(8)	(9)
0208-512	1.00	1.33	82.94	131.5	25.5	21.7	71.9	61.3
0219+428	0.62	0.88	19.81	29.92	41.0	38.0	47.0	44.0
0235+164	1.33	1.89	35.	54.	34.4	31.7	101.6	93.3
0420-014	1.02	1.35	10.3	15.6	33.8	28.0	126.	110.
0458-020	1.2	1.71	28.6	42.9	35.	31.5	140.	126.
0521-365	0.17	0.25	21.5	30.5	10.	10.	40.	40.
0528+134	3.72	5.33	5.09	8.21	43.	39.2	411	366
0537-441	1.83	2.51	12.56	19.02	38.1	35.3	86.7	80.3
0716+714	1.01	1.42	2.19	3.27	56.	53.	49.	46.
0735+178	0.50	0.66	19.4	28.4	19.5	19.0	35.8	34.
0829+046	0.20	0.41	10.9	16.5	11.0	10.	40.	37.0
0836+710	1.4	2.7	1.5	2.2	48.	45.	750.	680.
1101+384	0.39	0.58	1.5	2.3	61.	61.	35.	34.
1226+023	0.54	0.75	20.7	30.1	68.	63.	39.	37.
1253-055	1.43	2.03	6.67	10.62	21.3	19.2	121.	107.
1253-055	2.11	3.00	5.40	8.47	23.8	21.7	110.	99.6
1510-089	0.38	0.53	27.3	40.8	12.0	11.	43.	40.
1622-297	2.42	3.45	5.71	9.06	14.8	13.5	81.	73.
1633+382	3.22	4.60	3.81	6.15	36.8	33.5	347.	309.
2155-304	0.31	0.45	3.35	5.2	9.4	8.5	20.	18.
2200+420	0.57	0.79	4.45	6.63	14.1	13.0	27.7	25.8
2230+114	0.87	1.23	14.5	21.7	25.6	25.	103.	97
2251+158	0.55	0.85	2.12	2.89	38.5	33.6	285.	252

Notes to Table 1: Col. 1 gives the name, Col. 2 Doppler factor ($\lambda = 1.0$), Col. 3 Doppler factor ($\lambda = 0.1$), Col. 4, the central black hole mass in units of $10^7 M_\odot$ ($\lambda = 1.0$), Col. 5, the central black hole mass ($\lambda = 0.1$), Col. 6, propagation angle, Φ in the units of degree(°) ($\lambda = 1.0$), Col. 7, propagation angle ($\lambda = 0.1$), Col. 8, the distance (height), $\frac{d}{R_g}$, where the γ-rays are created ($\lambda = 1.0$), Col. 9, the distance (height) for $\lambda = 0.1$.

created ($\lambda = 1.0$), and finally Col. 9, provides the distance (height) when $\lambda = 0.1$. The mass found here ranges from $\sim 10^7 M_\odot$ to $\sim 10^9 M_\odot$.

ACKNOWLEDGEMENTS

This work is supported by the NSFC(19973001), the National 973 Project of China (NKBRAF G19990754) and the Outstanding Researcher Awards of the University of Hong Kong, a Croucher Foundation Senior Fellowship.

REFERENCES

1. Becker P., Kafatos, M. *ApJ* **453**, 83(1995)
2. Bednarek W. *A&A* **278**, 307(1993)
3. Böttcher M., Collmar W. *A&A* **327**, L57(1998)
4. Cheng, K.S. Fan J.H., Zhang, L., *A&A* **352**, 32(1999)
5. Comastri A. et al. *ApJ* **480**, 534(1997)
6. Courvoisier, et al. *Nat.* **335**, 683(1988)
7. Dermer C.D., Schlickeiser R., *ApJS* **90**, 945(1994)
8. Dondi, L., Ghisellini, G., *MNRAS* **273**, 583(1995)
9. Fan, J.H., Xie, G.Z., Bacon, R., *A&AS* **136**, 13(1999)
10. Fossati, G. et al. *MNRAS* **289**, 136(1998)
11. Ghisellini G., Maraschi L., Dondi L. *A&AS* **120**, 503(1996)
12. Hartman R.C., et al. *ApJS* **123**, 79(1999)
13. Pica, A.J. et al. *AJ* **96**, 1215(1988)
14. Romero G.E., Combi J.A., Cellone S.A., *Proceedings of the fifth Compton Symposium, Ed. M. McConnell and J.M. Ryan (eds), AIP, NY*, **333**(2000)
15. von Montigny C., Aller H, Aller M. et al. *ApJ***483**, 161(1997)
16. Zhang L., Cheng K.S. *ApJ* **475**, 534(1997)

VHE observations of unidentified EGRET sources

S.J. Fegan[*][†] and the VERITAS collaboration[*]

[*]*Fred Lawrence Whipple Observatory, PO Box 97, Amado, AZ 85645, USA*
[†]*Physics Department, University of Arizona, Tucson, AZ 85721, USA*

Abstract. Observations of unidentified EGRET sources were made with the Whipple 10m imaging atmospheric Čerenkov telescope between Fall 1999 and Spring 2001. During this period, a high resolution 490 pixel camera with 4° field of view was present on the telescope. Characterization of the off-axis response of this instrument was done using observations of the Crab Nebula. No significant emission was detected from the eight unidentified EGRET sources observed and upper limits are presented as a function of position.

INTRODUCTION

Very High Energy (VHE) γ-ray astronomy is the term used to describe observations in the energy range from 300GeV to 100TeV. Ground-based instruments operating in this energy domain typically have large collecting areas, good angular resolution and relatively large fields of view. The atmospheric Čerenkov imaging technique is described in detail elsewhere [8].

Twenty five years of gamma ray observations in the MeV to GeV range, have produced nearly 300 cataloged sources. During its lifetime, the EGRET experiment aboard CGRO made the most significant contribution to the list of detected sources, although its relative insensitivity to the arrival direction of 100 MeV photons means that the location of many sources are only known to within $\sim 0.5°$. The majority of sources are, as yet, not firmly associated with objects at other wavelengths. In many cases the EGRET error circle is populated by a number of prospective X-ray, optical and radio sources which are all candidate associations.

OBSERVATIONS

Observations of eight unidentified EGRET sources, listed in Table 1, were made with the Whipple 10m imaging atmospheric Čerenkov telescope in Arizona USA. The instrument and its characteristics are described in Finley et al. [2].

For off-axis and extended sources the telescope is operated in ON-OFF mode. Each 28 minute scan of the source region is followed by a 28 minute control run offset from the source by 30 minutes in right ascension and in time. Taking the control data in this manner compensates for differences in brightness that are a function of elevation and azimuth.

TABLE 1. Unidentified EGRET sources selected for observation.

Source	RA	Position Dec	l	b	Observation dates	Exposure (min)
3EG J0423+1707	04:23:00	17:06:60	178.48	-22.14	2000/12 - 2001/02	248
GeV J0433+2907	04:33:38	29:05:56	170.50	-12.58	1999/11 - 2000/01	500
3EG J0450+1105	04:50:00	11:05:00	187.86	-20.62	2000/11 - 2001/01	274
3EG J0634+0521	06:33:12	05:53:07	206.18	-1.41	2000/11 - 2001/03	275
3EG J1323+2200	13:23:03	21:59:41	359.33	81.15	2001/01 - 2001/02	83
GeV J1907+0556*	19:07:41	05:57:14	40.08	-0.88	2000/05 - 2000/06	277
GeV J2020+3658†	20:20:43	36:58:38	75.29	0.24	1999/10 - 1999/11	139
3EG J2227+6122	22:27:14	61:22:15	106.53	3.18	2000/09 - 2000/10	341

* Roberts et al. [10] note that this source is over 1° away from 3EG J1903+0550, with which it is associated in Hartman et al. [3]. They conclude that this association is likely to be incorrect.

† This source is incorrectly associated with 3EG J2016+3657 in the third EGRET catalog. Roberts et al. [10] note that 3EG J2021+3716 is consistent with the GeV source.

ANALYSIS

Before any analysis is performed, all data are subject to a number of standard operations. First, the data is flat-fielded, a process which compensates for any non-uniformities in the camera. Second, artificially generated noise is added to each image to remove any biases that exist between the on-source and control observations, a process referred to as *software padding*. These biases result from the control data being taken while pointing to a different part of the sky which has different background light characteristics. Finally, each image is cleaned by ignoring all channels which do not have sufficient signal in them. The details of these procedures are described elsewhere [9, 1].

For extended sources or sources where the source location is not well determined, it is essential to reconstruct the arrival direction of the primary. The arrival direction must be inferred from the "shape" of the observed image. There are a number of methods available, the approach taken here, described in detail in Lessard et al. [7], is to assume that the arrival direction of the primary lies along the major axis of the shower image and is displaced from the center of the shower image by a distance given by,

$$disp = \xi \left(1 - \frac{width}{length}\right)$$

where *width* and *length* describe the shape of the recorded image and ξ is a scaling parameter.

This method yields two possible arrival directions, each of which is on the major axis of the shower image, seperated from the centroid by the calculated parameter, *disp*. When creating a 2D map the origin of each event is assigned to both possible directions in the hope that one will have an excess as more event origins are superimposed.

A sky map is then produced by building up a 2-dimensional histogram of the reconstructed arrival direction with respect to the center of the camera. Errors in reconstructing both the image axis and *disp* are accounted for by convolving the final 2D map with a Gaussian function $g(\vec{r};r_0) = \exp(-r^2/2r_0^2)$, where r_0 is a scaling parameter chosen to maximise the significance of an excess.

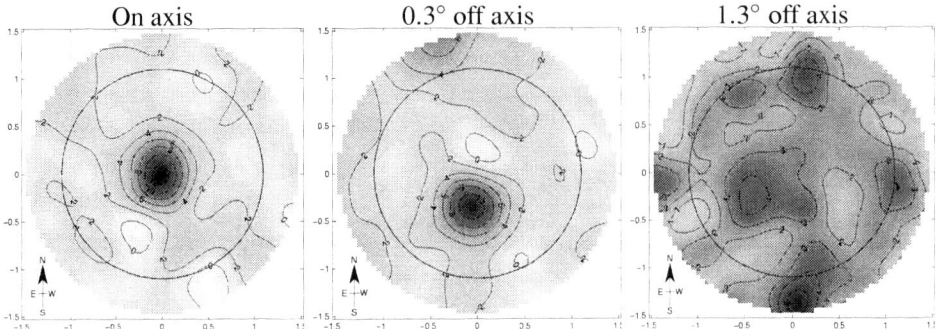

FIGURE 1. Observations of the Crab Nebula, offset by varying amounts from the center of the field of view. The contours show detection significance. Positions are in degrees from the center of the field of view with RA and Dec increasing toward the left and top respectively. A circle of radius $1.1°$ denotes the geometrical extent of the camera used. The observations at an offset of $1.3°$ place the Crab outside of this.

Calculation of excess signal, significance and upper-limit maps ($S(\vec{r})$, $\sigma(\vec{r})$ and $UL(\vec{r})$ respectively) is then done by convolving the ON and OFF counts with the smoothing function $g(\vec{r})$ in the appropriate manner, $S(\vec{r}) = \sum_{\vec{r'}} [ON(\vec{r'}) - OFF(\vec{r'})] g(\vec{r'} - \vec{r})$ and $\Delta S(\vec{r})^2 = \sum_{\vec{r'}} [ON(\vec{r'}) + OFF(\vec{r'})] g^2(\vec{r'} - \vec{r})$. Then $\sigma(\vec{r}) = S(\vec{r})/\Delta S(\vec{r})$ and $UL(\vec{r})$ is calculated from $S(\vec{r})$ and $\Delta S(\vec{r})$ by the method of Helene [4].

CALIBRATION

Calibration of the two dimensional analysis method was done using sets of observations of the Crab Nebula. Taking observations with the source location offset from the center of the field of view by various degrees and calculating the relative γ-ray rate allows a model of the detector response for off-axis and extended sources to be made.

Figure 1 shows significance maps for the Crab Nebula offset by three different amounts. In each of them the Crab is clearly visible. At an offset of $0.3°$ the γ-ray collection efficiency is 84% of what it is on axis. At an offset of $1.3°$, with the source outside of the geometrical extent of the camera, the efficiency is 30%. The significance map for this data shows appreciable background contamination over the field due to the simple reconstruction approach of assigning the arrival direction of each photon to two points on the shower axis. More sophisticated approaches can reduce such false sources [7].

Figure 2 shows the relative collecting efficiency for offset sources. This curve is used to normalize detected emission rates or upper limits to the Crab flux.

RESULTS

In one case, GeV J1907+0556, the analysis indicated significant emission throughout the 7 square degree field, the result of large brightness differences between ON and OFF

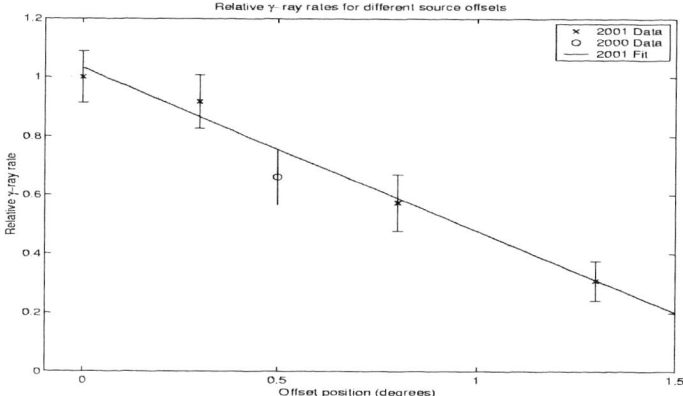

FIGURE 2. Relative Crab rate as a function of source offset. The off-axis response can be fit by a straight line.

TABLE 2. Upper limits derived from Figure 3.

Source	Positional Error* (degrees)	Upper Limit † (E >430GeV)
3EG J0423+1707	0.77	3.6
GeV J0433+2907	0.35	2.1
3EG J0450+1105	0.64	3.8
3EG J0634+0521	0.67	2.4
3EG J1323+2200	0.47	5.9
GeV J1907+0556	0.38	2.7
GeV J2020+3658	0.28	4.1
3EG J2227+6122	0.48	2.6

* 95% confidence circle from Lamb and Macomb [6] or Hartman et al. [3] as appropriate.
† Fluxes in units of $10^{-11} cm^{-2} s^{-1}$ calculated from measured Crab flux of Hillas et al. [5].

observations that was not fully compensated for in padding. For this source alone, the ON source counts were scaled by a value calculated by examining the number of counts in the region of $1.4° < r < 1.8°$ from the center of the field of view.

No significant emission was detected from any source. Figure 3 shows upper limits on emission from the sources observed. Table 2 summarizes these results for the error circle of each object. In each case, the highest limit found in each region is quoted.

REFERENCES

1. Cawley, M.F., et al., Exper. Astr., 1, 173, 1990
2. Finley, J.P., et al., Proc. 27th International Cosmic Ray Conference (Hamburg), in press, 2001
3. Hartman, R.C., et al., ApJS, 123, 79–202, 1999
4. Helene, O., NIM, 212, 319, 1983

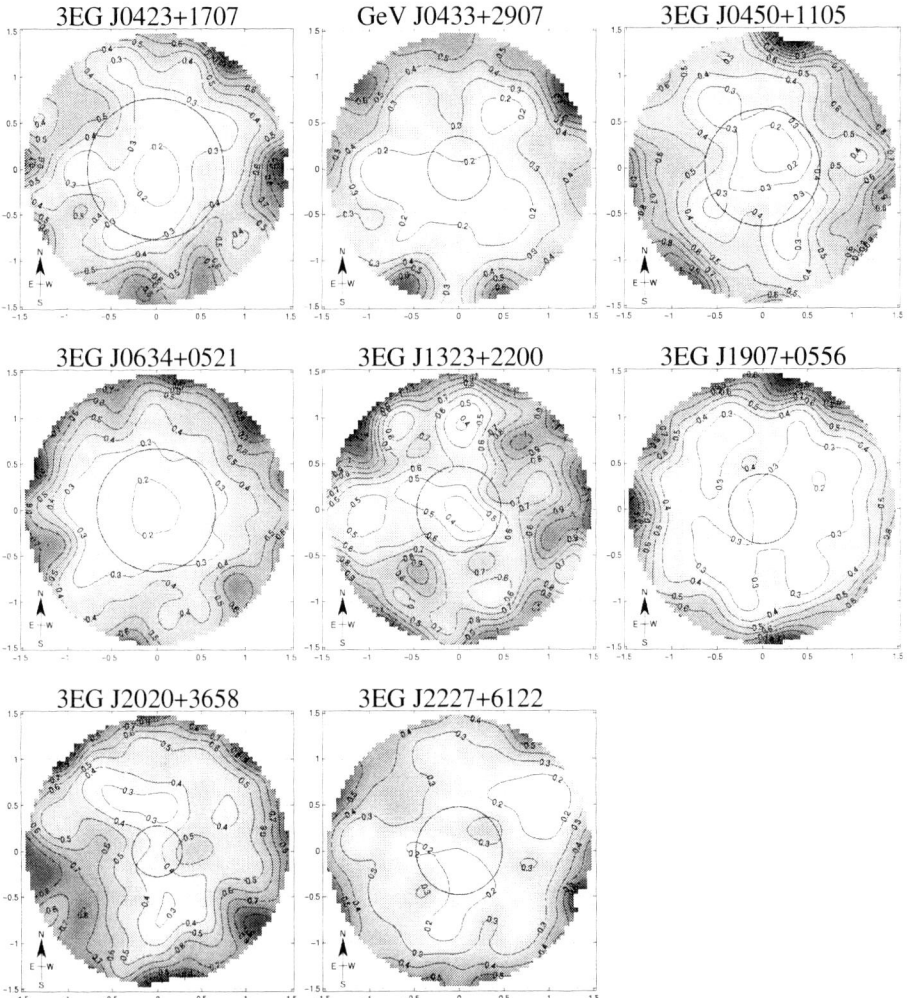

FIGURE 3. VHE upper limits on emission. Upper limits are given in units of the Crab Nebula flux. The axes are in degrees from the center of the field of view as given in Table 1. Increasing declination is toward the top of each plot, increasing RA to the left. The circle indicates the 95% confidence circle from Lamb and Macomb [6] or Hartman et al. [3] as appropriate. Where error ellipses have been given in in Lamb and Macomb [6], a circle with radius equal to the semi-major axis is displayed. The maximum upper limit in each 95% confidence region is given in Table 2.

5. Hillas, A.M., et al., ApJ, 503, 744, 1998
6. Lamb, R.C. and Macomb, D.J., ApJ, 488, 872-880, 1997
7. Lessard, R.W., et al., Astropart. Phys., 15, 1-18, 2001
8. Ong, R.A., Phys. Rep., 305, 93, 1998
9. Reynolds, P.T., Akerlof, C.W., Cawley, M.F., ApJ, 404, 206, 1993
10. Roberts, M.S.E., Romani, R.W., Kawai, N., ApJS, 133, 451, 2001

The effect of the SED shape on the gamma-ray vs. radio emission dependence in AGNs

A. Lähteenmäki[*] and E. Valtaoja[†]

[*]Metsähovi Radio Observatory, Metsähovintie 114, FIN–02540 Kylmälä, Finland
[†]Tuorla Observatory, FIN–21500 Piikkiö, Finland

Abstract. The gamma-ray emission in AGNs is produced at the beginning of radio outbursts and is correlated with the radio emission. The type of the correlation, however, seems to depend on the spectral behavior of the source (low-peaking SED, high-peaking SED). We compared the properties of AGNs with the location of the synchrotron peak and found that there is a sequence in the source properties (e.g., class) that depends on the synchrotron peak frequency. Another connection can be found between the gamma-ray vs. radio emission and source luminosity, and also the Doppler factor.

INTRODUCTION

The gamma-ray emission in AGNs is produced at the beginning of radio outbursts, and is clearly correlated with high frequency radio emission (e.g., Lähteenmäki et al. [6, 8], Lähteenmäki & Valtaoja [9]). Further studies of individual EGRET pointings have shown that more connections can be found if the properties of the sources are compared with synchrotron peak (ν_{peak}) data, source luminosity, or type.

We have used EGRET gamma-ray observations from Cycles 1 – 4 (Hartman et al. [4]) and high frequency (22 & 37 GHz) radio data from Metsähovi Radio Observatory. The entire EGRET/Metsähovi database consists of 85 sources (34 sources have been detected by EGRET, the rest are upper limits) for which there are 415 individual pointings. From the database we selected those sources which have ≥ 3 EGRET detections (21 sources). The synchrotron peak frequencies for 51 sources were obtained from Ghisellini et al. [3], the exact values kindly provided by G. Ghisellini (private communication). Thus we ended up with 18 sources for which we have synchrotron peak data, ≥ 3 EGRET detections, plus radio data.

SYNCHROTRON PEAK DEPENDENCE

The dependence of the (K-corrected) gamma-ray emission on high frequency radio emission applies also for individual EGRET pointings within one single source. However, the dependence varies from source to source according to the synchrotron peak frequency characteristic of each object. A correlation is found between the TYPE of the source (HPQ = high polarization quasar, LPQ = low polarization quasar, BLO = BL Lac object) and the gamma-ray emission. There is a gamma-ray vs. radio emission dependence sequence starting from sources with low synchrotron peak frequency (weak gamma-rays, strong radio emission, preferably LPQs) to sources with high synchrotron peak frequency (strong gamma-rays, weak radio emission, preferably HPQs and BLOs). It is interesting that there are no BLOs below $log\nu_{peak} = 14.6$ but the lowest peak frequencies are dominated solely by LPQs. Viewing angle θ seems to grow larger towards higher ν_{peak} (with correlation of 86 %), as found also by Georganopoulos [2], while Lorentz factor Γ and Doppler factor D decrease (with correlations of 98 % and 94 %, respectively). (θ, Γ and D are taken from Lähteenmäki & Valtaoja [7])

Examples of sources in the sequence

- *3C 273* (LPQ) has relatively weak gamma-ray emission but is very strong in the radio domain. $log\nu_{peak} = 12.2, \Gamma = 6.2, D = 5.7, \theta = 10.1$
- *1222+216* (LPQ) is an intermediate case between the two extremes, the gamma-ray and radio emissions being approximately equal. $log\nu_{peak} = 13.2, \Gamma = 4.3, D = 8.2, \theta = 2.4$
- *0528+134* (LPQ) is also an intermediate case. The correlation of gamma-ray and radio emission has an interesting structure, from high gamma-rays and low radio to intermediate gamma-rays and radio, implying that the correlation may vary according to the state of the source (quiescent or flaring). $log\nu_{peak} = 13.5, \Gamma = 8.1, D = 14.2, \theta = 2.6$
- *0954+55* (HPQ) has clearly strong gamma-rays but is extremely weak in the radio domain. $log\nu_{peak} = 14.9, \Gamma = ?, D = 4.6, \theta = ?$
- *ON 231* (BLO) has strong gamma-rays and very weak radio emission. $log\nu_{peak} = 15.1, \Gamma = 2.4, D = 1.6, \theta = 36.4$

LUMINOSITY

The correlation of radio luminosity with ν_{peak} is 90 %, the luminosity increasing with peak frequency. This is opposite to the correlation found by Fossati et al. [1] between ν_{peak} and radio luminosity at 5 GHz.

The gamma-ray to radio emission relation (i.e. the Compton dominance) is correlated with the luminosity (99.9 %), being identical with Ghisellini et al. [3].

UNIFICATION OF BLAZARS ?

The synchrotron peak frequency dependence sequence is identical with the unification scheme of Ghisellini et al. [3] and Fossati et al. [1]. They have found that there is a continuity between the different source types, from BLOs to HPQs to LPQs. Our relation of gamma-ray emission to average radio emission is correlated well with the synchrotron peak frequency (99.7 %) but this correlation we have found (Compton dominance increasing with ν_{peak}) is the opposite of Ghisellini et al. [3], where the

Compton dominance decreases towards higher ν_{peak}. However, the 18 sources we have looked at are almost all quasars while Ghisellini et al. have studied a sample of BLOs as well. The correlation they have found between the Compton dominance and $log\nu_{peak}$ *for quasars only* is not very clear if all BLOs are removed. The same applies for the correlation and unification sequence between ν_{peak} and radio luminosity at 5 GHz found by Fossati et al. [1]. Their correlations seem to be caused by the combination of quasar and BLO populations whereas we have mainly studied the properties within the class of quasars only (i.e. within the population of FR II counterparts). Another explanation arises naturally from the fact that when the SED is higher, the radio fluxes at, e.g., 22 GHz, are weaker. This introduces a selection effect since our sample consists of both intrinsically bright and weak low-SED sources but of only intrinsically bright high-SED sources.

There is also a correlation between the gamma-ray to radio emission relation and the Doppler factor (99.1 %). Huang et al. [5] have conducted a similar study and found a poor opposite correlation.

SUMMARY

We have found that for the population of quasars (HPQs and LPQs) when the jet intrinsic power increases, the SED also moves upwards, and the Lorentz factor (and consequently Doppler factor) decreases. This is opposite to the general trend found earlier when considering all radio sources (quasars plus BLOs), but not necessarily in conflict with the results found by Fossati et al. [1] and Ghisellini et al. [3]. There also seems to be a tendency for the more powerful (higher-SED) quasars to be stronger gamma-ray emitters (as measured by the gamma-ray to radio relation), but this may be a selection effect.

Our results also show that the gamma-ray to radio emission behavior of individual sources may depend on the source type. LPQs have weak gamma-ray emission but are rather strong radio emitters while (at least some) BLOs have very high gamma-ray fluxes but weak radio fluxes.

The complete results with figures will be published in Lähteenmäki & Valtaoja [10].

REFERENCES

1. Fossati et al. 1998, *MNRAS* 299, 433
2. Georganopoulos 2000, *ApJ* 543, L15
3. Ghisellini et al. 1998, *MNRAS* 201, 451
4. Hartman et al. 1999, *ApJS* 123, 79
5. Huang et al. 1999, *A&A* 341, 74
6. Lähteenmäki et al. 1997, *Proc. 4th Compton Symposium*, eds. C. D. Dermer, M. Strickman & J. D. Kurfess (New York, AIP), p.1452.
7. Lähteenmäki & Valtaoja 1999, *ApJ* 521, 493
8. Lähteenmäki et al. 2000, *Proc. 5th Compton Symposium*, eds. M. L. McConnell and J. M. Ryan (New York AIP), p. 372
9. Lähteenmäki & Valtaoja 2001a, *ApJ*, in preparation
10. Lähteenmäki & Valtaoja 2001b, *ApJ*, in preparation

The Nature of the EGRET Source 3EG J1621+8203

R. Mukherjee*, J. Halpern, N. Mirabal, D. Stern*, E. V. Gotthelf

Columbia Astrophysics Lab, Columbia University,
New York, NY 10027

* *Dept. of Physics & Astronomy, Barnard College, Columbia University,*
New York, NY 10027

Abstract. We present broad-band observations of 3EG J1621+8203 in an effort to understand the nature of this source. We have examined X-ray images of the field from the *ROSAT* PSPC, *ROSAT* HRI, and *ASCA* GIS to search for a possible counterpart to the EGRET source. We find several faint X-ray point sources in the gamma-ray error circle. Preliminary analysis indicates that most of the point sources correspond to stars or to faint radio sources. Of the nearly 40% identified sources in the 3EG Catalog, the vast majority are blazars, but there is no blazar candidate in the error circle of 3EG J1621+8203. Of the notable objects in the EGRET error circle, one is the bright FR I radio galaxy NGC 6251 at a redshift of 0.0249. If NGC 6251 is the counterpart to the EGRET source 3EG J1621+8203, then it would be the second radio galaxy to be detected by EGRET. The first was Centaurus A. Cen A provided the first clear evidence of the detection above 100 MeV of an AGN with a large-inclination jet. If the identification with NGC 6251 is correct, the apparent gamma-ray luminosity of 3EG J1621+8203 is lower than that of other EGRET blazars, just as in the case of Cen A.

INTRODUCTION

The Energetic Gamma Ray Experiment Telescope (EGRET) on the Compton Gamma Ray Observatory (CGRO) observed the γ-ray sky from 30MeV to 30GeV, from April 1991 to June 2000, and detected 271 point sources (Hartman et al. 1999). Interestingly, the majority of the EGRET point sources are unidentified due to lack of convincing counterparts at other wavelengths. Identification of the EGRET sources on the basis of position alone has been challenging because the sizes of the EGRET error contours are typically large, $\sim 0.5° - 1°$. Most of the identified EGRET sources are blazars at high Galactic latitudes or pulsars at low latitudes. Studying multiwavelength data on EGRET unidentified sources is often used as an approach to aid in the identification of these sources. Recently, several efforts have been made to identify the γ-ray sources using data from X-ray imaging studies (e.g. Roberts et al. 2001; Halpern et al. 2001). Here we present gamma-ray and X-ray observations of one particular unidentified EGRET source, namely, 3EG J1621+8203, in an effort to learn more about its nature.

I GAMMA RAY OBSERVATIONS

3EG J1621+8203 is a high latitude source located at l=115.53, b=31.77, with a 95% confidence error radius of $0°.85$ (Hartman et al. 1999). Although individual viewing periods yielded near-threshold detections by EGRET, 3EG J1621+8203 was clearly detected in the cumulative exposure from multiple EGRET viewings. The flux above 100 MeV was 10.7×10^{-8} photon cm^{-2} s^{-1} (Hartman et al. 1999). The measured spectral index of 3EG J1621+8203 in the EGRET energy range was 2.29±0.46.

II X-RAY OBSERVATIONS

The error circle of 3EG J1621+8203 is covered (although, not completely) by archival X-ray imaging observations acquired with the *ROSAT* (Roentgen Satellite) and *ASCA* (Advanced Satellite for Cosmology and Astrophysics) observatories. Data for the region were available for the *ROSAT* Position Sensitive Proportional Counter (PSPC) in the 0.2 – 2.0 keV range. There was also partial coverage of the region by the *ROSAT* High Resolution Imager (HRI) and the *ASCA* Gas Imaging Spectrometer (GIS).

Figure 1 shows the *ROSAT* PSPC image of the field of 3EG J1621+8203. The image was made by co-adding exposure corrected sky maps from observations made on 1991 March 12-15. The total exposure for the PSPC image was 14.7 ks. The circle in figure 1 corresponds to the 95% error contour of 3EG J1621+8203. Our study of the archival X-ray (*ASCA* and *ROSAT*) data yields several faint sources in the error circle of the EGRET source 3EG J1621+8203. The detected positions of the sources are indicated in figure 1 and Table 1. Preliminary analysis indicates that most of the X-ray point sources correspond to stars, or faint radio sources. The table shows the source correlations for the point sources in the X-ray image. Numbers 14, 15, and 16, correspond to three point sources in the field covered by the *ROSAT* HRI observations, not shown in the image. Of the notable sources, the brightest X-ray source in the *ROSAT* image of 3EG J1621+8203 is source # 8, which is a variable star of the RS CVn type. This source is unlikely to be the counterpart of 3EG J1621+8203. Source # 2, seen in both ROSAT and ASCA images, is the galaxy cluster RX J1641.2+8233, also unlikely to be the counterpart of the EGRET source. Most of the other sources have positional correspondence with stars in the USNO star catalog. We plan to observe several of these sources with the MDM 1.3 m telescope in the near future. We note that there is no potential spectrally flat, radio-loud blazar-like counterpart for 3EG J1621+8203.

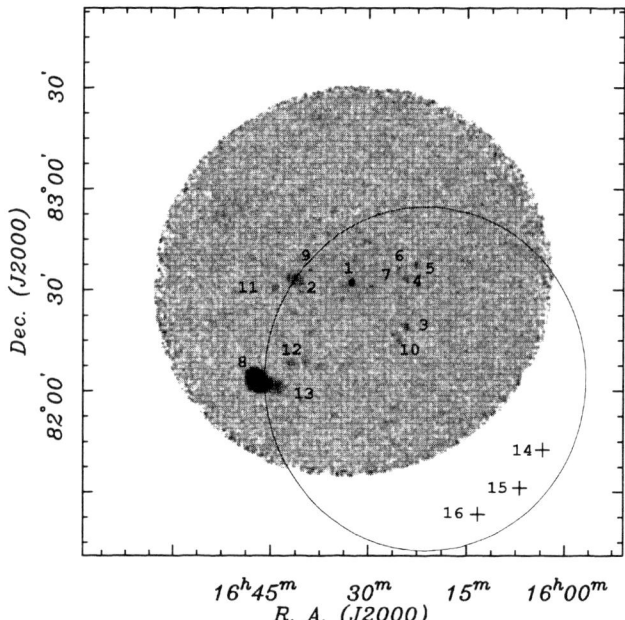

FIGURE 1. *ROSAT* PSPC image of 3EG J1621+8203. The circle indicates the 2σ error contour of the EGRET position for the source. The numbers correspond to X-ray point sources in the field of view, and are described in the text.

Of special interest is source # 1 seen in the *ROSAT* image, which is the bright FR I radio galaxy, NGC 6251, at a redshift of 0.023 (implied distance 91 Mpc for $H_0 = 75$ km s^{-1} Mpc^{-1}). This source is also seen in the *ASCA* image of the region. NGC 6251 has a radio jet from pc to Mpc scale which makes an angle of $\sim 45°$ to our line of sight (Sudou & Taniguchi 2000). If this source is the counterpart to the EGRET source, that would have interesting consequences. This is discussed more in the following section.

Several sources in Table 1 have no obvious counterparts. We intend to observe them optically, in order to learn more about their properties.

III DISCUSSION

Of the 67 AGN detected previously by EGRET (Hartman et al. 1999), only one was a radio galaxy, Cen A, which happened to be the brightest and closest radio galaxy. The majority of EGRET AGN are blazars, which are believed to have nearly-aligned jets along our line of sight. Cen A provides the first clear evidence of the detection of an AGN above 100 MeV with a large-inclination (70°) jet (Sreekumar et al. 1999).

TABLE 1. X-ray sources in the fields of 3EG J1621+8203

Number[a]	RA	Dec	Source Name/Comments
1	16 32 32.5	82 32 15	NGC 6251, Radio Galaxy
2	16 41 05.1	82 32 54	2E 1646.6+8238, Galaxy Cluster, $z = 0.26$
3	16 24 30.0	82 19 00	USNO Star, $R = 17.4$, $B = 16.9$
4	16 24 12.7	82 32 48	USNO Star, $R = 18.4$, $B = 19.4$
5	16 22 32.6	82 37 06	....
6	16 25 13.2	82 36 13	USNO Star, $R = 15.0$, $B = 16.2$
7	16 25 27.3	82 34 49	USNO Star, $R = 18.9$, $B = 19.1$
8	16 46 04.1	82 02 34	HD 153751, Star, Variable of RS CVn type
9	16 39 00.9	82 35 47	USNO Star, $R = 14.5$, $B = 15.6$
10	16 25 29.6	82 14 39	WNB1630.5+8221, radio source, 52mJy
11	16 44 12.6	82 30 07	USNO Star, $R = 18.3$, $B = 18.0$
12	16 41 18.9	82 07 56	USNO Star, $R = 13.1$, $B = 14.4$
13	16 42 58.1	82 01 02	....
14[b]	16 03 24.5	81 42 19	BD+82 477, $B = 10.5$, G0
15[b]	16 06 52.0	81 30 28	....
16[b]	16 13 17.3	81 23 33	USNO Star, $R = 13.6$, $B = 14.3$

(a) Identifying number in *ROSAT PSPC* image (Fig. 1).
(b) ROSAT HRI source, indicated in Fig. 1.

It is possible that the radio galaxy NGC 6251 (Source # 1) is the counterpart of the EGRET source 3EG J1621+8203. Based on the current evidence, of all the X-ray sources in the EGRET field-of-view, NGC 6251 appears to be the most likely candidate identification for the EGRET source. If this is indeed the case, NGC 6251 would be the second radio galaxy to be detected by EGRET at energies above 100 MeV. The high energy gamma-ray emission from 3EG J1621+8203 represents a lower luminosity (3×10^{43} ergs/s) than that of other EGRET blazars (typically $10^{45} - 10^{48}$ ergs/s), just as in the case of Cen A (Sreekumar et al. 1999).

Off-axis emission from blazars, whose jets are pointed away from our line-of-sight, have recently been discussed by Weferling & Schlickeiser (1999). Figure 2 shows the decrease in scattered energy for off-axis emission for different viewing angles, corresponding to two typical values of Lorentz factors (Γ) for blazars. The figure shows that a decrease in observer angle from 70° (e.g. Cen A) to 45° (e.g. NGC 6251) corresponds to an increase in the scattered energy by about a factor of 10. Cen A represents an example of a large inclination jet, which is still detected by EGRET due to its close proximity. NGC 6251 is much further away, but because of its smaller jet angle, its scattered energy does not decrease by as large a factor as Cen A's. It is therefore likely that the source is still detectable by EGRET.

FR I galaxies have been hypothesized to be the likely parent population of BL Lac objects. Using X-ray data, Padovani & Urry (1990) have found that the hypothesis that BL Lac objects are beamed FR I galaxies could explain their observed X-ray properties. Due to the limitations of EGRET's sensitivity and the intrinsic

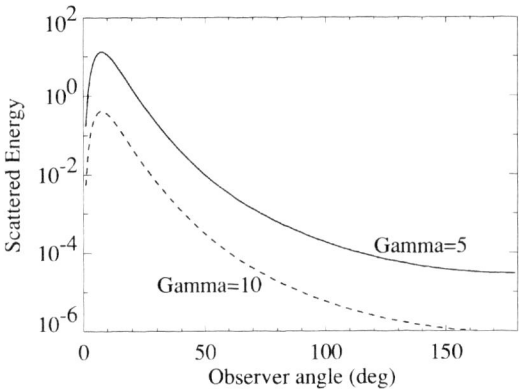

FIGURE 2. Decrease in the observed emission from a blazar as a function of jet orientation with respect to the observer (Weferling & Schlickeiser 1999; Sreekumar 1999).

low luminosity of these radio galaxies, it is not surprising that more such sources have not been detected above 100 MeV. Assuming a unification model for AGN (Dermer, Sturner & Schlickeiser 1997) and the fact that the high energy emission increases with decreasing jet angle, one can expect the detection of more distant radio-loud AGN with intermediate inclination angle jets in the future with more sensitive gamma-ray instruments. It has been suggested that since the number density of radio-loud FR I sources is nearly 1000 times larger than FSRQs and BL Lac objects, such "misaligned blazars" could contribute to the extragalactic gamma-ray background above 1 MeV (Sreekumar et al. 1999; Watanabe et al. 2001). It is likely that sensitive instruments like GLAST will detect a much larger number of radio galaxies with intermediate inclination angle jets at distances much further away than Cen A. NGC 6251 = 3EG J1621+8203 could be an example of such sources.

This research has made use of data obtained from HEASARC at Goddard Space Flight Center and the SIMBAD astronomical database. RM acknowledges support from NSF grant PHY-9983836.

REFERENCES

Dermer, C. D., Sturner, S. J., Schlickeiser, R. 1997, ApJS, 109, 103.
Halpern, J. P., et al. 2001, ApJL, 552, L125.
Hartman, R. C., et al. 1999, ApJS, 123, 79.
Padovani, P. & Urry, C. M. 1990, ApJ, 356, 75.
Roberts, M. S. E., Romani, R. W., & Noboyuki, K. 2001, ApJS, 133, 451.
Sudou, H. & Tanaguchi, Y. 2000, AJ, 120, 697.
Sreekumar, P., et al. 1999, APh, 221, 11.
Watanabe, K. & Hartman, D. H. 2001, these proceedings.
Weferling, B. & Schlickeiser, R. 1999, A&A, 344, 744.

High energy observations of Blazars- Archival analysis

Giridhar Nandikotkur*, Keith M. Jahoda*, Jean H. Swank*, P. Sreekumar**
and Rita M. Sambruna†

*Goddard Space Flight Center, Greenbelt, MD
†George Mason University, Fairfax, VA
**Indian Space Research Organization, Bangalore, India

Abstract.
We are conducting a systematic study of the archival data of 14 AGNs observed simultaneously by RXTE and EGRET. The sample includes eight flat spectrum radio quasars, three high-energy-peaked BL Lacs (HBLs), and three low-energy peaked BL Lacs (LBLs). Four sources have been observed for at least two consecutive EGRET observation periods (14 days). We have extracted spectral indices using the power-law models in both XTE and EGRET energy ranges and we are studying the X-ray-to-gamma-ray spectral energy distributions and their variability as a function of classification. Preliminary results show that FSRQs and LBLs have X-ray photon indices in the energy band 3.0-20 keV in the range 1.3-2.0, while HBLs have indices in the range 1.8-2.9. Several blazars were repeatedly observed in X-rays with XTE, enabling us to study flux and spectral variation on the time scales of half a day. Correlated variations between the photon index and flux are observed in HBLs with a trend of flatter slopes for increasing fluxes. In LBLs and FSRQs, both the flux and slope vary, but with no clear trends. The results are overall consistent with the current unification schemes for blazars where HBLs are dominated in X-rays by the variable tail of synchrotron emission, while in FSRQs and LBLs the XTE spectra are dominated by the inverse Compton emission extending to higher energies.

INTRODUCTION

The EGRET instrument aboard the Compton Gamma Ray Observatory has reported the detection of more than 65 blazars since its launch in 1991. The VLBI structure of these sources reveal compact cores with jet-like structures which often show evidence of superluminal motion (Vermeulen and Cohen [1]). The broad band spectrum of blazars (in the νF_ν space) shows two peaks: The first one is at infra-red/optical frequencies for *red blazars* or the "low-frequency-peaked" blazars (LBLs), and at UV/X-rays for *blue blazars* or the "high-frequency-peaked" blazars (HBLs) while the second peak is in the gamma-ray range (MeV-GeV) for LBLs and TeV range for HBLs.

It has been widely accepted that the radio to X-ray emission is due to synchrotron emission of relativistic electrons moving along the jet away from the core of the active galactic nucleus. The high energy emission is often attributed to inverse compton scattering of relativistic electrons either from the synchrotron photons themselves (SSC models Maraschi et al. [2] 1992; Bloom and Marscher [3] 1996) or from external soft photons from the accretion disk or broad-line region (ERC models) (ERC– Dermer and Schlickeiser [4] 1994; Sikora et al. [5] 1994, Blandford and Levinson [6] 1995; Ghisellini and

Madau [7]1996).

Due to its location on the multi-frequency spectrum, "X-rays" is an interesting region to study blazars as a function of their classification. HBLs have their peak synchrotron emission in X-rays. Significant variations in flux and spectral index are observed in time scales as short as an hour. For LBLs, X-rays and gamma-rays form the low and high energy tail of inverse-compton emission bump. Hence, simultaneous observations at these two wave-length help resolve the relative importance of the SSC and the ERC processes. For some of the blazars the transition from synchrotron to inverse-compton emission occurs in X-ray energy range leading to a complicated dependence of spectral index on flux. Time lags between hard and soft x-rays are also observed. Three such sources: BL Lac Tanihata et al. [8], S4 0954+645 Raiteri et al. [9], ON231 Tagliaferri et al. [10] have been reported so far.

SOURCE SELECTION AND ANALYSIS

We have identified 14 blazars that have been observed simultaneously by EGRET and RXTE. The sources have been listed in Table 1. Some of them have been a part of planned multiwavelength campaigns while some underwent a flare at one wavelength, thus initiating a target-of-oppotunity at the other.

These sources have been divided into three categories of LBL, HBL, and Flat-spectrum-radio Quasars (FSRQ) based on literature. BL LAC has shown clear indications of what could be termed as an *Intermediate Blazar (IBL)* due to a spectral break in X-rays. But since its emission lies on the low-energy tail of the high-frequency bump for most part of XTE's energy range, it has been classified as an LBL.

A homogenous analysis has been carried out for all the sources. For RXTE (PCA), data from PCUs 0, 1, and 2 have been chosen with the energy range restricted to 3.0 keV-20.0 keV. HEXTE data from cluster 0 and cluster 1 has been included in the energy-range 13.0 keV-200 keV. An absorbed powerlaw model has been used to characterize the spectra with absorption frozen at the galactic nH value. For a few sources, analysis has also been carried out with the absorption value reported in literature that differed from the galactic nH value.

All observations by EGRET that were used were after the end of phase 4 since RXTE was launched at the beginning of Phase 5. Due to degradation of gas chamber on board EGRET, it was frequently operated in a narrow field of view mode. Hence, some care was required while fitting power-law model to data. A four point spectrum was extracted for all sources which comprised of energy ranges 30MeV-100MeV, 100MeV-300MeV, 300MeV-1000MeV and 1000MeV-10000MeV. (These ranges were a natural choice with EGRET data products.) In cases where a source underwent a flare, each individual energy range was divided further if it had enough counts to justify such a step.

For three of the sources- 3C273, BLLac, PKS 1156+295, a flare was captured by both the datasets. For PKS 2255-282, XTE data was available after the gamma ray flare. PKS 1510-089 also underwent a minor flare in X-rays but EGRET was switch to a narrow field of view mode and did not capture the source.

Most of the LBLs and FSRQs were too weak to be detected by HEXTE with the

Spectral Index Distribution of Blazars

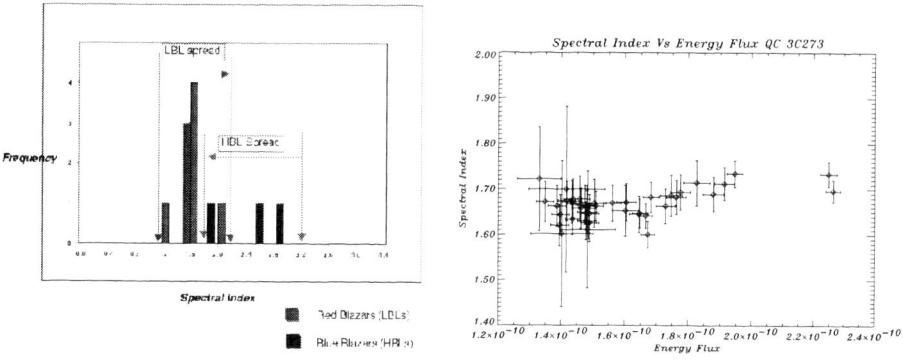

FIGURE 1. right:Spectral Index Vs Flux - 3C273, FSRQ

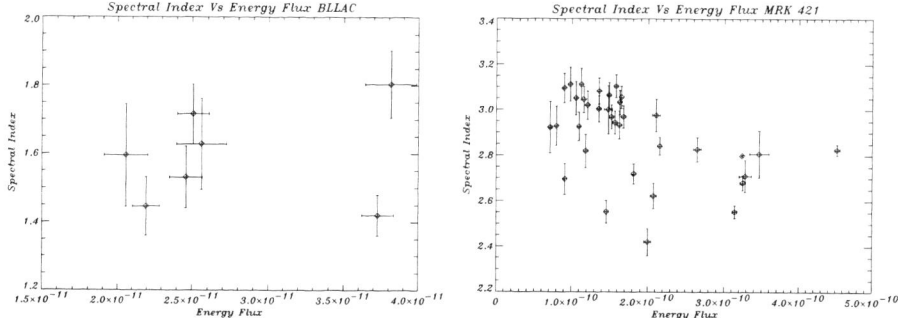

FIGURE 2. Spectral Index Vs Flux: left- BL LAC; right- Mrk 421

exception of 3C273. The spectral indices in 13 keV-200 keV range for sources that were detected were as following. 3C273:1.6±0.14; PKS2155-304:2.15±0.47; Mrk 501: 1.78±0.08. Table 2 shows the flux ($ergs/cm^2/s$) and photon spectral indices for PCA and EGRET. Spectral indices were not calculated for sources where the EGRET detection was less than 3σ.

DISCUSSION

Figure 1 shows a difference in distribution of spectral indices between HBLs and LBLs. HBLs have softer spectral indices (1.8-2.9) while LBLs and FSRQs have a harder spectral indices (1.3-2.0). This is consistent with two different processes doiminating the X-ray regime.

For HBLs, a significant variation in spectral index is observed. LBLs and FSRQs were too faint for a precise determination of spectral index to be made for a single RXTE

TABLE 1. Source Summary with XTE and EGRET observation details

Source	RA	DEC	Class	XTE Details			EGRET Details		
				ObsID	Start mm/dd/yy	Time in ksec	ObsId	Start mm/dd/yy	Time in Days
PKS 1156+295	179.89	29.24 till	FSRQ	30408-01-01-01	03/20/98	20.7	VP715.5	03/20/98	7
				30408-01-12-00	03/31/98				
QC354.3	343.49	16.15 till	FSRQ	20346-01-25-00	05/30/97	7.84	VP623.5	07/15/97	7
				20346-01-40-00	09/18/97				
PKS 0528+134	82.74	13.53 till	FSRQ	10359-02-01-00	08/16/96	12.5	VP527.0	08/13/96	14
				10359-02-07-00	09/03/96		VP5280	08/20/96	
3C273	187.28	2.05 till	FSRQ	20349-02-02-00	11/13/96	24.8	VP606.0	10/12/96	42
				20349-02-33-00	01/16/97		VP611.1	01/21/97	
3C279	194.04	-5.79 till	FSRQ	20349-01-14-10	12/18/96	16.8	VP606.0	10/12/96	42
				20349-01-57-10	01/10/97		VP610.0	01/21/97	
PKS 1510-089	228.21	-9.10 till	FSRQ	20345-02-31-00	07/13/97	10.9			
				20345-02-45-00	11/12/97				
PKS 1622-297	246.52	-29.85 till	FSRQ	20348-01-01-00	08/18/97	42	VP615.1	08/19/97	7
				20348-01-30-00	08/27/97				
PKS 2255-282	344.52	-27.97 till	FSRQ	30403-01-01-00	01/15/98	5.6	VP708.0	12/30/97	14
							VP709.1	01/06/98	7
PKS 0235+164	39.66	16.62 till	LBL	20429-01-01-00	11/03/97	12.3	VP631.1	11/03/97	
				20429-01-08-00	11/06/97				
BL LAC	330.68	42.28 till	LBL	10359-03-01-00	07/16/97	10.8	VP623.5	07/15/97	7
				20423-01-03-00	07/21/97	9.8			
S0716+714	110.48	71.34 till	LBL	10343-01-01-00	04/06/96	8.6	VP518.5	04/03/96	7
				10343-01-24-00	04/22/96				
Mrk 421	166.11	38.21 till	HBL	30261-01-02-00	03/24/98	26.5	VP716.5	03/27/98	7
				30261-01-42-00	04/13/98				
Mrk 501	253.47	39.76 till	HBL	20421-01-01-00	03/18/97	10.6	VP617.8	04/09/97	7
				20340-01-01-00	05/15/97				
PKS 2155-304	329.72	-30.23 till	HBL	10356-01-23-00	05/21/96	9.8	VP520.4	05/21/96	7
				10356-01-43-00	05/23/96				

pointing (approximately 1000 seconds).

HBLs have a clear trend of harder spectral indices (flatter slopes) with increasing flux. The case for Mrk421 is shown in Figure 2. This trend was observed in all three HBLs. The LBLs and FSRQs seem to have softer spectral indices with increasing fluxes when they were in a relatively active state (see Figure 1 for 3C273). But in most of the cases this correlation is not observable, if present due to the errors in the parameters. Figure 2

TABLE 2. Source Summary with XTE and EGRET observation details

Source	PCA Results		EGRET results		
	Flux in $ergs/cm^2/s$	Phot. Spec. Index	Flux > $100 MeV$ in 10^{-8} $photons/cm^2/s$	Phot. Spec. Index	Sig of Det.
PKS 1156+295	1.0e-11±4.0e-13	1.63±0.08	76.3±22.8	2.45±0.42	5.2σ
QC354.3	1.9e-11±6.7e-13	1.72±0.07	Not	detected	
PKS0528+134	1.0e-11±2.9e-13	1.67±0.10	74.1±19.6	2.33±0.38	4.6σ
3C273	1.6e-10±1.2e-13	1.66±0.01	74.1±9.5	2.37±0.13	11.0σ
3C279	5.6e-12±4.1e-13	1.49±0.15	12.4±4.1	2.12±0.32	3.5σ
PKS 1510-089	1.2e-11±4.9e-13	1.38±0.09	Narrow	FOV mode	
PKS1622-297	1.0e-11±6e-13	2.0±0.07	< 35	-	< 3σ
PKS2255-282	8.9e-12±8.7e-13	1.73±0.12	235.9±42.9	1.55±0.19	9.1σ
PKS 0235+164	3.25e-12±4.9e-13	1.665±0.31	15±17	-	< 3σ
BL LAC	3.2e-11±9.1e-13	1.55±0.06	149±21	1.83±0.15	10.7σ
S0716+714	Low σ		24±5	1.7±0.2	6.2σ
Mrk 421	1.73e-10±1e-12	2.84±0.01	Not	detected	
Mrk 501	3.9e-10±2.3e-12	1.84±0.01	Not	detected	
PKS 2155-304	1.9e-10±1.1e-12	2.55±0.01	16±8	-	< 3σ

(left) shows the plot for BL Lac during the summer of 1997 when it underwent a flare at optical, X-ray and gamma-ray wave lengths. Inspite of statistically significant values of spectral index and fluxes, there is no clear trend. The *intermediate characteristics* due to presence of both synchrotron and inverse-compton processes and a possible shift in the *break-energy* during the flare opens up large parameter space that cannot be studied with one data set. Repeated observations in a flaring, intermediate and quiescent state will help address the problem better.

REFERENCES

1. Vermeulen, R. C., and Cohen, M., *ApJ*, **430**, 467 (1994).
2. Maraschi, L., Ghisellini, G., and Celloti, A., *ApJ*, **397**, L5 (1992).
3. Bloom, S., and Marscher, A., *ApJ*, **467**, 657 (1996).
4. Dermer, C., and Schlickeiser, R., *ApJS*, **90**, 945 (1994).
5. Sikora, M., Begelman, M. C., and Rees, M. J., *ApJ*, **421**, 153 (1994).
6. Blandford, R., and Levinson, A., *ApJ*, **441**, 79 (1995).
7. Ghisellini, G., and Madau, P., *MNRAS*, **280**, 67G (1996).
8. Tanihata, C., et al., *ApJ*, **543**, 124T (2000).
9. Raiteri, C. M., et al., *A & A*, **352**, 19 (1999).
10. Tagliaferri, G., et al., *A & A*, **354**, 431T (2000).

γ-ray spectral changes during blazar outbursts

P. Sreekumar*, R.C. Hartman[†], R. Mukherjee** and M. Pohl[‡]

*ISRO Satellite Centre, Bangalore, India
[†]Code 661, NASA/Goddard Space Flight Center, MD, USA
**Barnard College and Columbia University, NY, USA
[‡]Ruhr-Universitaet Bochum, Germany

Abstract.
We examined the archival EGRET data on time-variable γ-ray blazars for systematic variations in the spectra during different intensity states (flare/quiet). Although the changes in the individual source spectra can be well measured in only a few cases, the collective results from the blazar sample clearly indicate a correlation between γ-ray intensity and spectrum, with a systematic trend towards a harder spectrum ($\sim$ 0.3 shift in the spectral index) during high γ-ray state. We discuss implications on models addressing the blazar-origin of the extragalactic γ-ray background.

INTRODUCTION

The Energetic Gamma-Ray Experiment Telescope (EGRET) on board the Compton Gamma-Ray Observatory (CGRO) covers the high-energy γ-ray range from approximately 30 MeV to 30 GeV. The latest EGRET source catalog (Hartman et al. 1999) reports 271 γ-ray sources, almost 2/3 unidentified. The identified sources include 66 sources firmly classified as being associated with the blazar-class of active galactic nuclei (AGN). The spectral energy distribution (SED) for most γ-ray blazars show significant power output at γ-ray energies especially during outbursts (von Montigny et al. 1995). EGRET observations have also shown that these blazars exhibit significant time variability at γ-ray energies, with variability time scales as short as 1 day or less (Mattox et al. 1997, Wehrle et al. 1998).

Sikora & Madejski (2001) reviewed blazar phenomena and addressed theoretical models used to explain the high-energy emission from blazars. Most models require a central compact source, maybe an accreting supermassive blackhole producing relativistic plasma outflow in the form of a jet, believed to be nearly aligned to our line-of-sight. γ-ray observations constrain intrinsic source luminosities and emission geometry (Maraschi, Ghisellini, & Celotti 1992; Schlickeiser 1996), providing strong evidence for beaming. For flat spectrum radio quasars (FSRQs) and radio-selected BL Lac objects (RBLs or LBLs), the radio to optical/UV emission is believed to be synchrotron emission. In the case of X-ray-selected BL Lac objects (XBLs or HBLs), the synchrotron peak extends to X-ray energies. At γ-ray energies, the emission is believed to arise from either soft photons boosted up in energy via inverse-Compton interactions with the energetic electrons or from energetic hadrons producing pions which decay into secondaries producing high-energy γ rays (Mannheim 1993). Hybrid models with contributions from both electrons and protons have also been proposed (Mannheim et al. 1996).

EGRET observations show that most of the γ-ray bright AGNs display significant time variability that is often correlated with changes at other wavelengths (Wagner et al. 1995; Mukherjee et al. 1997; Bloom et al. 1997, Hartman et al. 2001), though not always consistent with regards to the scale of the outburst and/or phase at lower frequencies. Dedicated multiwavelength campaigns that cover the range from radio to X-rays or γ-rays have been carried out on a limited number of sources and have yielded mixed results. Studies suggest the existence of intrinsically complex processes associated with outbursts in blazars; the exact temporal sequence remains to be fully understood.

In this paper we examine changes in the γ-ray spectrum during high and low intensity states for a subset of γ-ray blazars. Previous studies showed marginal evidence for spectral hardening during flares (Mücke et al. 1994, Mukherjee et al. 1995, Sreekumar et al. 1996, von Montigny et al. 1996, Bloom et al. 1997). We present here a more systematic analysis which includes all necessary energy-dependent instrument response corrections that otherwise could adversely influence the spectral analysis.

OBSERVATIONS AND ANALYSIS

The analysis used archival data from the start of the mission (April 1991) to the end of Dec 1996. Also included are data from viewing period 623.5 (15–22 July 1997), which was a Target of Opportunity observation of BL Lacertae in outburst. Since observations on any given source were often spaced apart in time by months to years, significant efforts were made to minimize systematic effects in the data. EGRET uses a spark chamber to image pair-production events above 30 MeV. The spark chamber gas in EGRET degrades with use and in the course of the period covered by the data, the spark chamber was replenished with fresh gas on five different occasions. This resulted in varying instrument sensitivity with time. We have included in our analysis appropriate energy-dependent sensitivity corrections (Esposito et al. 1999) prior to determining source spectra to ensure an unbiased dataset.

Of the 66 strongly detected blazars in the EGRET data, 10 were selected to carry out this analysis. The choice of blazar candidates was based on requirements such as strong time-variability and multiple high-significance detections.

RESULTS AND DISCUSSION

Figure 1 shows a plot of the γ-ray spectral index variation in two sources during *high* and *low* γ-ray states. Data on PKS 0208-512 shows good correlation of source intensity with spectral index, the spectrum consistently hardening with increasing intensity throughout the 5.5 years of observations. Similarly, blazars 3C273, 0528+138, 1633+382 and 1622-297 show a common tendency to display a harder spectrum as the source flux increases. At the other extreme, for the well observed source 3C279, the correlation is very weak (Kniffen et al. 1993). During the largest increase in γ-ray luminosity observed in 3C279 during Feb '96 (VP511.5) (Wehrele et al. 1998), the spectral slope remained unchanged from that derived during lower intensity observations. However, if one examines the

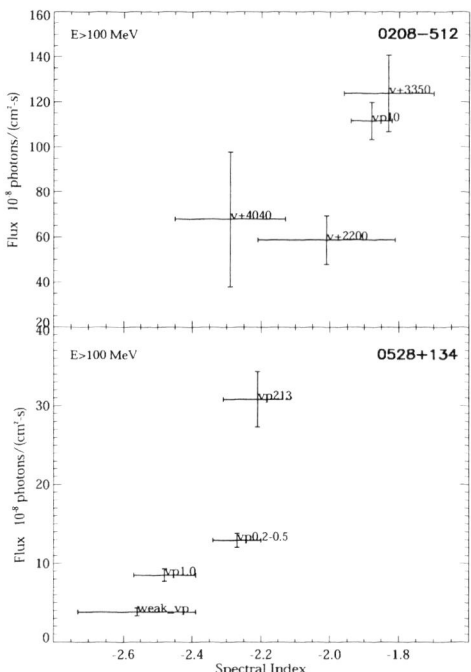

FIGURE 1. Spectral index variations in 0208-512 and 0528+134

pre-flare spectrum in Jan '96 (VP 511.0), there is marginal indication that the spectrum hardened during the peak of the 3C279 outburst. Figure 2 shows the cumulative evidence for changes in the spectrum for fractional change in the *low* state emission in all 10 blazars. There is a clear tendency for the single power law γ-ray photon spectra to harden (∼ 0.3 shift in the spectral index) when sources transition from a low state to a high state. If the observed changes in the spectral indices arise purely from statistical fluctuations, the observed data points would lie scattered about the solid line. It is well understood that the rather high detection threshold of EGRET permits examination of spectral changes in only the more intense, strongly-detected γ-ray sources. However the use of almost 5.5 years of data and extended sky coverage during these years, together with clear source selection criteria, prevents introduction of any other selection bias in this study. This analysis provides the clearest evidence yet for a systematic spectral evolution low-frequency peaked γ-ray blazars, viz., a spectral hardening when the γ-ray flux increases. No conclusions are derived on XBL/HBL sources since none of three known γ-ray XBLs satisfy the selection criteria.

Since it is generally believed that the bulk of the high-energy γ-ray emission in blazars arise from inverse Compton scattering of low-energy photons off energetic electrons in the jet, changes in the γ-ray spectrum could arise from either changes in the charged particle spectrum or the soft-photon distribution. Studying the low-energy photon distribution has been a significant observational challenge. Furthermore, it is still unclear if the bulk of these photons arise from the accretion disk (direct or reprocessed) or from the jet itself. We discuss below results from multiwavelength studies towards understanding blazar emission.

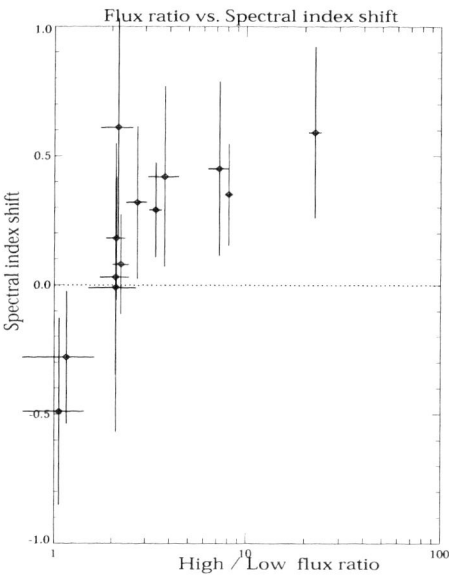

FIGURE 2. Changes in the spectral index for different intensity states

Results from multiwavelength studies

Mukherjee et al. (1999) studied the spectral energy distribution of a FSRQ (PKS 0528+134), and concluded that the synchrotron peak shifts towards lower energies during γ-ray outbursts if energetic electrons lose energy primarily via Compton scattering off external photons. Böttcher (1999) concludes that such a trend maybe typical of all FSRQ that are γ-ray bright, and predicts a resulting γ-ray spectrum that is harder during outbursts. A similar analysis of multi-year observations of 3C279 by Hartman et al. (2001) found no clear evidence for a shift in the synchrotron peak (peak is poorly defined due to lack of adequate FIR observations) However the spectral slope beyond the peak became harder at higher source intensity. Analysis of the large γ-ray flare in BL Lacertae during July 1997 (Bloom et al. 1997) indicated a harder γ-ray spectrum compared to previous quiet phase data. As reported by Webb et al. (1998), the optical observations indicate a harder spectrum when the object was brighter, though statistically the effect is not compelling. Considering the nearly identical spectral variations at γ-ray energies, this is more suggestive of a harder spectrum for the parent electrons that gives rise to the flatter synchrotron emission at optical energies with a corresponding harder inverse Compton γ-ray emission. Analysis by Cappi et al. (1998) of ASCA observations on 3C273, showed a weak correlation between the 2-10 keV spectral index and the corresponding X-ray flux; the spectrum being harder when the source got brighter. This is similar to that seen in X-ray selected BL Lac objects (Urry et al. 1997) where the X-ray emission is assumed to represent the tail of the synchrotron component and the spectral variations arise due to changes in synchrotron losses. In 3C273 however, it is believed that the 2-10 keV X-ray emission is primarily from upscattered inverse Compton photons (von Montigny et al. 1997), similar in origin to the high-energy γ-rays seen by EGRET.

Implications on diffuse extragalactic emission models

A significant fraction of the diffuse extragalactic emission above 10 MeV is believed to be made up of emission from unresolved blazars (Stecker & Salamon 1996; Sreekumar, Stecker & Kappadath 1997 and references therein). An important evidence in this regard is provided by the near consistency between the average spectral index of the observed γ-ray blazars (-2.14±0.04 Mukherjee et al. 1997) and the spectral index of the diffuse background emission (-2.1±0.03 Sreekumar et al. 1998). As stated earlier, it is a fair conclusion that the observations are more characteristic of high-luminosity phases of each blazar. However, as discussed by many authors (Kazanas & Perlman 1997), the small duty cycle of γ-ray flares in blazars implies that the bulk of the blazar contributions arises from the quiet-phase emission. In this regard, it is important to address spectral changes during blazar intensity variations. The results presented here suggest that for blazar-origin models the mean spectrum of unresolved blazars that contribute primarily to the observed diffuse extragalactic background is ∼-2.4 which is steeper than the average value derived from EGRET observations. Hence this finding suggests a reduced blazar contribution to the extragalactic γ-ray background and creates the need for increased contributions from other source classes.

REFERENCES

1. Bloom, S. D. et al. 1997, ApJ, 490, L145
2. Böttcher, M. 1999, ApJL, 515, 21
3. Cappi, M., Matsuoka, M., Otani, C., & Leighly, K.M., 1998, PASJ, 50, 213
4. Esposito, J. A. et al. 1999, ApJS, 123, 203
5. Hartman, R. C. et al. 1999, ApJS, 123, 79
6. Hartman, R. C. et al. 2001, ApJ, 553,
7. Kazanas, D. & Perlman, E. 1997, ApJ, 476, 7
8. Kniffen, D.A. et al. 1993, ApJ, 411, 133
9. Mannheim, K. 1993, A&A, 269, 67
10. Mannheim, K., Westerhoff, S., Meyer, H., & Fink, H.-H. 1996, A&A, 315, 77
11. Maraschi, L., Ghisellini, G., & Celotti, A. 1992, ApJ, 397, L5
12. Mattox, J.R. et al. 1996, ApJ, 461, 396
13. Mattox, J.R. et al. 1997, ApJ, 481, 95
14. Mücke, A. & Pohl, M. et al. 1997, AIP Conf. Proc. 410, C. Dermer, M. Strickman, J. Kurfess, 1233
15. Mukherjee, R. et al. 1995, ApJ, 445, 189
16. Mukherjee, R. et al. 1997, ApJ, 490, 116
17. Mukherjee, R. et al. 1999, ApJ, 527, 132
18. Schlickeiser, R. 1996, A&AS, 120, 481
19. Sikora, M., & Madejski, G. 2001, astro-ph/0101382
20. Sreekumar, P. et al. 1996, ApJ, 464, 628
21. Sreekumar, P., Stecker, F., & Kappadath, C. 1997, AIP Conf. Proc. 410, C. Dermer, M. Strickman, J. Kurfess, 344
22. Sreekumar, P. et al. 1998, ApJ, 494, 523
23. Urry, C.M. et al. 1997, ApJ, 486, 799
24. von Montigny, C. et al. 1995, ApJ, 440, 525
25. von Montigny, C. et al. 1997, ApJ,
26. Wagner, S. et al. 1995, ApJ, 454, 97
27. Webb, J. et al. 1998, AJ, 15, 2244
28. Wehrle, A.E. et al. 1998, ApJ, 497, 178

Multiwavelength Observations of 3EG J2006-2321 and 3EG J0433+2908

P. M. Wallace[1], M. Eracleous[2], J. V. Foreman[1], J. P. Halpern[3], O. Reimer[4], and D. J. Thompson[4]

[1]*Department of Physics and Astronomy, Berry College, Mt. Berry, GA 30149*
[2]*Dept. of Astronomy & Astrophysics, Pennsylvania State University, University Park, PA 16802*
[3]*Columbia Astrophysics Laboratory, Columbia University, New York, NY 10027*
[4]*Code 661 NASA Goddard Space Flight Center, Greenbelt, MD 20771*

Abstract. We present multiwavelength data for the variable EGRET sources 3EG J2006-2321 and 3EG J0433+2908. The former is listed as unidentified in the 3rd EGRET Catalog and the latter is listed as an AGN; however, as yet there has been no formal presentation of its broadband characteristics. The most likely radio counterpart for 3EG J2006-2321 is PMN J2005-2310 (S_5=260 mJy); optical observations indicate a V=18.7 point-like counterpart with z=0.83. No X-ray counterpart has been detected, but an upper limit on the X-ray flux is derived from ROSAT data. EGRET data indicate that 3EG J2006-2321 displays strong short-term variability above 100 MeV. These data indicate that 3EG J2006-2321 is probably a blazar. For 3EG J0433+2908, the most likely radio counterpart is 87GB 0430+2859 (S_5=481 mJy). This flat-spectrum source was monitored at 2.25 and 8.3 GHz at GBI; the light curves indicate that the source is variable at these frequencies. The optical spectrum of the V=17.8 counterpart is featureless; no redshift is known. A ROSAT source, RX J0433.6+2906, is coincident with the radio position. Based on these data and the evidence that 3EG J0433+2908 is variable above 100 MeV, we confirm that this source is a member of the BL Lac subclass of AGN.

3EG J2006-2321

Gamma-ray Observations

Between 22 April 1991 and 27 September 1995, the time span covered by the 3rd EGRET catalog (Hartman et al. 1999), 3EG J2006-2321 was within 30° of the instrument axis during 9 viewing periods (VP's). In only two was it detected. During VP 5.0 (1991 July 12-26) the aspect was 28°3 and the source was detected at the 4.4σ level. Three weeks later during VP 7.2 the source was 13°6 from the instrument axis and was not detected. During VP 13.1 (1991 October 31-November 7), 3EG J2006-2321 was again detected and it exhibited transient behavior on time scales of 12 hours (Wallace et al. 2000). During this VP the source was 13°6 from the instrument axis and the overall significance of the detection was 4.8σ. There was no detection of 3EG J2006-2321 in six VP's after 13.1, although the aspect was often much less than 20°. Combining the data from the two positive detections gives the source a 95% confidence radius of 0°80. The peak flux during VP 13.1 was (1.75±0.53)×10^{-6}

photons/cm^2/s and the ratio of peak to average flux is 5:1. Applying a χ^2 test to the light curve yields a variability index (McLaughlin et al. 1996) of 3.2, corresponding to a probability of 0.0006 that these data are consistent with an intrinsically nonvariable source.

Radio Observations

The NRAO/VLA Survey at 1.4 GHz (Condon et al. 1983) lists only four sources above 100 mJy within the error circle of 3EG J2006-2321. The source with the smallest separation (10.9') from the EGRET position is NVSS J200556-231028, with a 1.4-GHz flux of 302 mJy. The flux of the strongest source, NVSS J200711-233435, is slightly higher than this, at 319 mJy; this source is 23.0' from the gamma-ray position. The other two sources have weaker fluxes and are further from the gamma-ray source. Only one of five 5-GHz sources within the error circle is coincident with any of the NVSS sources: PMN J2005-2310 (S_5=260 mJy). This source is coincident with NVSS J200556-231028. This radio source is also the only one to feature a flat spectrum; its flux at 365 MHz is 260 mJy; α_r at this frequency is 0.7±02 (Douglas et al. 1996). In light of these data, we associate PMN J2005-2310 with the gamma-ray source.

Optical Observations

A CCD image in the V band centered on PMN J2005-2310 was taken at the 2.1-m telescope at Kitt Peak on 2000 June 2. A point-like optical counterpart with intrinsic V=18.7 was found within 2" of the position of PMN J2005-2310. An optical spectrum was also taken at Kitt Peak on 2000 June 2 and features a single broad Mg II 2897 Å emission line at 5129 Å, corresponding to z=0.83. This large redshift is typical of AGN.

X-ray Observations

The region surrounding PMN J2005-2310 has had little exposure to X-ray instruments. The sole data come from the ROSAT All-sky Survey; in both the Bright and Faint Source Catalogs there is no source within 30' of the radio position. This places an upper limit on the X-ray flux of the source. A typical dim source in the FSC has a PSPC count rate of ~8×10^{-13} counts/s. Assuming a power law spectrum with α=2.0, this corresponds to a flux between 0.1 and 2.0 keV of ~2.4×10^{-13} erg/cm^2/s.

Spectral Energy Distribution and Discussion

A rough SED of 3EG J2006-2321 is shown below in Figure 1. From the two-peaked profile it is evident that the distribution is consistent with a blazar identification. The conclusion of the present analysis is that 3EG J3006-3432 is probably a blazar, but in some respects it is unusual among EGRET AGN. In particular, its 5-GHz flux is much weaker than any other blazar with comparable peak gamma-ray flux. With 99.998% confidence, Mattox et al. (1997) find that the peak gamma-ray flux of EGRET blazars is linearly correlated with their 5-GHz flux densities. They determine that the probability of EGRET detecting a radio source with $S_5=260$ mJy is only 0.015. This probability is a factor of 12 less than that of any other EGRET blazar (B2115-304, p=0.18). Gamma-ray sources like 3EG J2006-2321 must be uncommon. If the same fraction of blazars with $S_5<1.0$ Jy displayed bright (100 MeV flux > 10^{-6} photons/cm^2/s) gamma-ray flares as those with $S_5>1.0$ Jy, EGRET would have detected more such flares than it did. Quantitative analysis of this conclusion is underway.

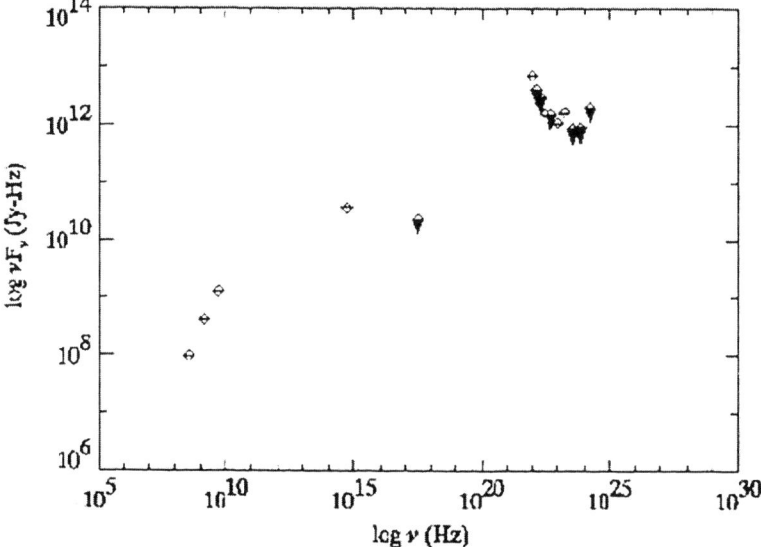

FIGURE 1. Rough spectral energy distribution for 3EG J2006-2321. Arrows indicate upper limits. The X-ray upper limit is derived from a negative detection in the ROSAT catalogs, and the gamma-ray upper limits are statistical. No error bars are shown.

3EG J20433+2908

Gamma-ray Observations

This source is listed in the 3EG catalog as an AGN; however, no formal presentation of the data to support this identification has yet been made. Over the span covered by the catalog, this source (l=170°48, b=-12°58) displays a mean high-energy gamma-ray flux of $(2.20\times0.28)\,10^{-7}$ photons/cm^2/s and a power-law spectrum with $\alpha = 1.90\pm0.10$. This is a harder spectrum than is displayed by most EGRET AGN; this source is listed in the GEV catalog. The source is well-exposed to EGRET, having been in the FOV almost 20 times from Phase 1 through Cycle 4; it has a 95% error radius of 0°18. From 1995 August 8-22 it was in a high state; its average flux during this time was $(7.57 \times 2.21) \, 10^{-7}$ photons/cm^2/s, a factor of ~3.5 times higher than its mean flux. This high state was reported to the IAU by Lundgren et al. (1995).

Radio Observations

Dingus et al. (1995) state that the strongest flat-spectrum source within the 95% error contour, 87GB 0430+2859, is the most likely counterpart to the EGRET source. Our results from analysis of the NED database are consistent with this conclusion; there are no other flat-spectrum radio sources within the contour with $S_5 > 100$ mJy. 87GB 0430+2859 is located 7' from the gamma-ray position; the probability of this being a chance coincidence is calculated by Dingus et al. (1995) to be 0.6%. The radio source has $S_5 = 481$ mJy and $\alpha = 0.2$ at this frequency. In response to the high state reported by Lundgren et al. (1995), 87GB 0430+2859 was monitored at GBI at 2.25 and 8.3 GHz. Preliminary light curves display evidence of variability at these frequencies.

Infrared and Optical Observations

There is an IR source in the 2MASS database that is coincident with 87GB 0430+2859. The magnitudes in the IR bands are J=14.28, H=13.34, and K=12.50.

An optical counterpart appears in the Palomar Sky Survey within 2" of the VLA position of the radio source; its magnitude is near the POSS limit. Condon et al. (1983) optically identified this source with an extended (5") galaxy with estimated magnitude ~19; work is underway to confirm or reject this association. Several other optical photometric observations have been made of this source. On 1997 February 6, it was observed with the KPNO 2.1-m telescope and was found to have V=17.8 (Halpern et al. 1997), about 2 magnitudes brighter than in the POSS field. Another photometric point taken in 2000 December indicates R = 17.4. Optical monitoring may help to confirm the variability suggested by these numbers.

Optical/NIR spectroscopy has also been performed for this source. Two spectra have been obtained: one from 4000-7000 on 2000 February 5 at KPNO and one from 4200-10000 on 2000 November 27 at the Hobby-Eberly Telescope. Both spectra show no emission or absorption features and they rise steeply toward the red; this redness is probably due to considerable extinction at the low Galactic latitude of the source. The spectrum is typical of the BL Lac subclass of AGN.

X-ray Observations

Voges et al. (1995) report that they have found a hard X-ray source in the ROSAT All-sky Survey within 22" of the position of 87GB 0430+2859. The PSPC count rate in the 0.1-2.0 keV ROSAT band is 0.05 counts/s during 1990 August 21-23, corresponding to an energy flux of 2×10^{-12} erg/cm^2/s. This source, RX J0433.6+2906, is the only X-ray source within 1° of the radio position.

Spectral Energy Distribution and Discussion

The data presented here support the AGN identification of 3EG J0433+2908 found in the 3EG catalog. It is a variable gamma-ray source. It is a strong, flat-spectrum, variable radio source; its featureless optical spectrum is evidence that it is a member of the BL Lac subclass of blazars. Optical monitoring and polarization measurements would be useful as final pieces of evidence for this identification. A rough SED of 3EG J0433+2908 is shown below in Figure 2. From the bimodal profile it is evident that the distribution is consistent with a blazar identification.

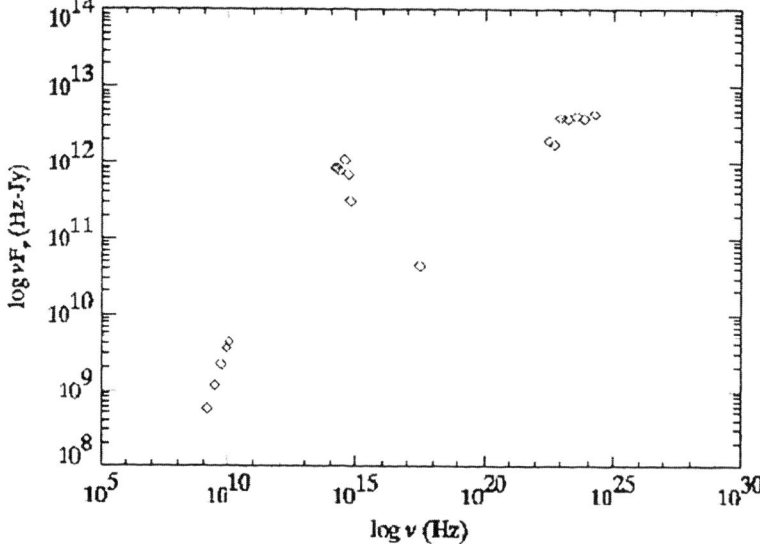

FIGURE 2. Rough spectral energy distribution for 3EG J0433+2908. No error bars are shown.

REFERENCES

1. Condon, J. J. et al. 1983, AJ, 115, 1693
2. Dingus, B. L. et al. 1996, ApJ, 467, 589
3. Douglas, J. N. et al. 1996, AJ, 111, 1945
4. Hartman, R. C. et al. 1999, ApJS, 123, 79
5. Halpern, J. P. et al. 1997, IAU Circular 6639
6. Lundgren, S. C. et al. 1995, IAU Circular 6258
7. Mattox, J. R. et al. 1997, ApJ, 481, 95
8. McLaughlin, M. A. et al. 1996, ApJ, 473, 763
9. Voges, W. & Kanbach, G. 1995, IAU Circular 6264
10. Wallace, P. M. et al. 2000, ApJ, 540, 183

Observations of the BL Lac Object, 1H1426+428 at TeV Gamma-ray Energies

D. Horan* and the VERITAS Collaboration*

Fred Lawrence Whipple Observatory, Harvard-Smithsonian CfA, P.O. Box 97, Amado, AZ 85645, USA

Abstract. TeV gamma-ray observations are reported of 1H1426+428 (1H1426), an X-ray selected BL Lacertae object at a redshift of 0.129. The X-ray spectrum appears to peak near 100 keV; if this is the peak of the synchrotron emission, then this AGN is a prime candidate for TeV gamma-ray emission assuming a Compton-synchrotron model. During the 1999, 2000 and 2001 observing seasons, the source has been intensively studied with the Whipple 10m imaging atmospheric Čerenkov telescope; these observations, together with the indications from earlier observations confirm this hypothesis. The average signal was at the 4.2σ level during the 2000 observing season, while the average signal recorded during 2001 was at the 4.9σ level.

INTRODUCTION

1H1426 is an X-ray selected BL Lacerate object at a redshift of 0.129. Throughout the past few years it has been observed extensively with the Whipple 10m Telescope. Since 1992, the Whipple Collaboration, using the imaging atmospheric Čerenkov telescope on Mt. Hopkins, has been searching for TeV γ-ray emission from AGN. In this paper we describe the properties of 1H1426 which make it particularly interesting for TeV γ-ray observations. We summarize the observations made in 1999, 2000 and 2001 and compare the TeV observations taken during this period with X-ray observations by RXTE. Finally we discuss the implications of the detection of 1H1426 at TeV energies.

1H1426: AN X-RAY PREDICTION

In 1998-1999, the *Beppo*SAX collaboration undertook an observing campaign with the aim of finding and studying other sources as "extreme" as Mrk 501 is in its flaring state [2]. The candidates chosen for the *Beppo*SAX survey were selected from the Einstein Slew Survey and the RASSBSC catalogs.

These *Beppo*SAX observations [2; 3; 4] revealed four new "extreme" High frequency peaked BL Lacs (HBLs), selected to have high synchrotron peak frequencies. These four candidates for TeV emission are: 1ES 0120+340, PKS 0548-322, 1ES 1426+428 (i.e. 1H1426) and H2356-309. The spectra for three of these objects are well fitted by convex broken power laws, with the break and hence the peak of the synchrotron emission, occurring at about 1.4keV for 1ES0120+340, 4.4keV for PKS0548-322 and 1.8keV for H2356-309. For 1H1426 however, no evidence for a spectral break up to 100keV was found. Instead, its spectrum is well fitted by a single power-law, with a flat spectral index of 0.92 up to 100keV, thus constraining the peak of the synchrotron emission to lie near or above this value. The best fit of a pure homogeneous synchrotron self compton (SSC) model for 1H1426 [3] predicted detectable γ-ray emission at TeV energies.

At the time of the BeppoSAX observation, the observed x-ray flux from 1H1426 was not very high, indicating that 1H1426, unlike Mrk501 mentioned earlier, was not in a flaring state at the time when the spectrum was derived. This implies that the synchrotron peak could reach values even higher than 100keV in the event of a flare, indicating the presence of highly relativistic electrons. These observations make 1H1426 a prime candidate for TeV emission.

OBSERVATIONS & RESULTS

The observations reported in this paper were taken with the 10m reflector at the Whipple Observatory in southern Arizona with a variety of camera configurations [5]. The analysis techniques used are described in Reynolds et al. [11] and in Catanese et al. [1]. The tracking ratio was calculated using all of the OFF source data available for each observing year, assuming there is no source contribution. In addition, the tracking ratio was checked using the OFF data from 1H1426+428 pairs; within statistical errors these tracking ratios were consistent with the standard tracking ratio derived from the full yearly database.

1999 Observations

A total of 24.35 hours of observations of 1H1426+428 were carried out in 1999, the main results of which are summarized in Table 1. Part of the rate curve for 1H1426+428 during 1999 is shown in Figure 1 (left panel). It can be seen to rise above its quiescent level for a few nights in mid-March (MJD 51248-51251), perhaps as the result of a flare. Each point corresponds to the total rate calculated for that night so in some cases that is just 28 minutes while at other times it is up to 3 hours. The significance reached a peak value of 3.1σ on March 14 (MJD 51251) after 2.3 hours. The significance stayed close to this peak value until March 15, 1999 (MJD 51252) before dropping off. Hence almost all the positive effect seen in 1999 can be accounted for by an apparent flare lasting for a few nights. The net excess for the observing season was not significant (0.9σ).

2000 Observations

1H1426 was observed quite extensively during 2000 with a total of 26.17 hours being spent on source. The main results are summarized in Table 1. These observations once again showed evidence for flaring activity, and indicated that some low-level continuous emission seems to be present. The cumulative significance for the 26.17 hours of TRACKING data taken during 2000 was at the 4.2σ level. Out of this, 14.2

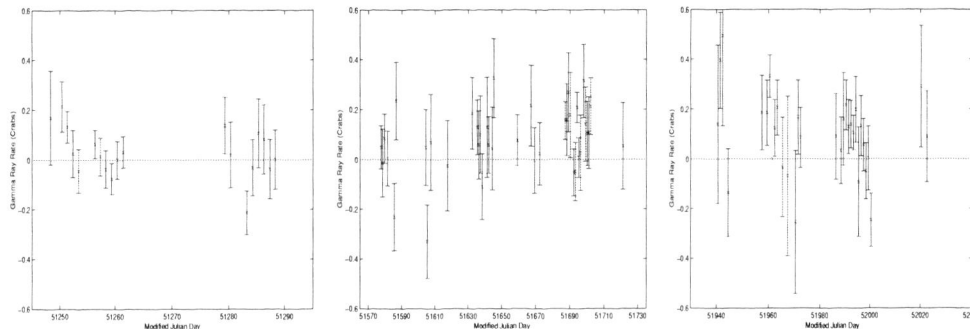

FIGURE 1. The light curves for 1H1426+428 for 1999, 2000 and 2001; *left* March 11, 1999 (MJD 51248) to April 20, 1999 (MJD 51288); *middle:* February 3, 2000 (MJD 51577) to June 26, 2000 (MJD 51721); *right:* January 31, 2001 (MJD 51940) to March 24, 2001 (MJD 52022).

TABLE 1. 1H1426 results from 1999 - 2001

Period	Exp. (hrs)	Total σ	Max. σ Month	Max. σ Night	Flux x 10^{-12}* (cm^{-2} s^{-1} TeV^{-1})	Peak Response Energy (GeV)
1999:03-06	24.35	0.88	1.57	2.13	<6.7 x 10^{-11}	500
2000:02-06	26.17	4.19	3.49	3.33	2.9 ± 1.1 x 10^{-11}	430
2001:01-04	30.55	4.94	4.68	3.87	4.9 ± 1.5 x 10^{-11}	390

* The integral fluxes are quoted above the peak energy response for the observation period, as given in column 7.

hours of data were taken in the ON/OFF mode, which amounted to a significance of 0.9σ.

There is a clear rise in the significance around May 30, 2000 (MJD 51694), possibly as the result of a one-day flare. From the 2.6 hours of data taken that night, a signal at the 3.3σ level was recorded. The rate curve for the year can be seen in Figure 1 (middle panel).

2001 Observations

During 2001, there has once again been evidence for both continuous and flaring emission from 1H1426+428. After 30.55 hours of observation, a signal at the 4.94σ level has been detected. There were 11.4 hours of data taken in the ON/OFF mode which gave a total significance of 1.1σ. The combined alpha plot for the 30.55 hours of data taken on source is shown in Figure 2, where a clear excess is visible in the 3 leftmost bins, as would be expected for a source of γ-rays. There was evidence for a one-day flare on February 20 (MJD 519601) with a significance of 3.87σ detected on that night. The rate curve for the 2001 data is shown in Figure 1 (right panel).

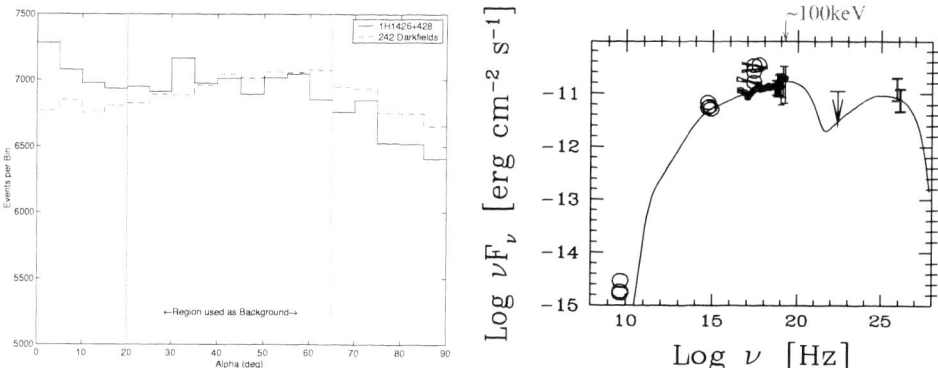

FIGURE 2. *left:* The alpha plot for the 30.55 hours of data taken on 1H1426 during 2001; *right:* The SED of 1H1426 from Costamante et al.(2000a) with a pure homogeneous SSC model fitted to the data. The points derived from the Whipple 2000 & 2001 data are shown.

Spectral Energy Distribution

Using all of the 2000 1H1426 data, a differential energy point was calculated. For this calculation, a peak energy response of 430GeV was used. A differential flux point was also calculated using the 2001 data, assuming a peak energy response of 390GeV. These points are added to the spectral energy distribution [2] and can be seen in Figure 2. No account is made for any absorption of the TeV photons by the infrared background.

Comparing the γ-ray & X-ray Flux from 1H1426+428

The γ-ray rates for 1999, 2000 and 2001 were compared with the X-ray flux from the All Sky Monitor (ASM) instrument on board the Rossi X-ray Timing Explorer (RXTE) [9]. No correlation was found between the Whipple daily averages and the ASM "one-day average" data points for the full 1999, 2000 or 2001 datasets.

DISCUSSION

Observations of 1H1426 with the Whipple telescope have been consistently positive since 1995 [7], although the statistical significance has not always been high. Since 1H1426 was on the very short list of predicted TeV emitters published by the *Beppo*SAX collaboration [2] and was also singled out as the most probable TeV emitter amongst these, we conclude that the detection is significant.

After 18.1 hours of observations on 1H1426, the CAT collaboration derived an average flux above 250 GeV, of $9.5 \pm 8.6 \times 10^{-11}$ cm^{-2} s^{-1} [10].

CONCLUSION

If independently verified, the results reported here represent an important addition to the catalog of TeV-emitting BL Lacs. That the source was predicted to be a TeV γ-ray source based on its X-ray spectrum is important in that it signifies the maturity of the observational techniques and the theoretical understanding of BL Lacs. It is remarkable also in that this source, like the other extreme blazars, Mrk 501 and 1E2344+514, is not included in the 3rd EGRET Catalog, demonstrating that while the AGN seen by EGRET and ground-based telescopes have many similarities, they also have significant differences.

Although the properties of 1H1426 reported here are relatively poorly defined in comparison with the better studied TeV BL Lacs, they agree in principle with the characteristics now well established for Mrk 421 and Mrk 501 (and to a lesser extent, 1E2344). Time variability on time-scales from days to years is evident in 1H1426. There is no clear evidence for a correlation of X-ray and TeV γ-ray fluxes but the sensitivity of observations to date is limited and even in the stronger sources this correlation has been shown to be complicated.

1H1426+428 is the most distant of the TeV-detected BL Lacs classified as HBL, and hence it has promising implications for the detection of more BL Lacs at $z > 0.1$. Deeper detections of such sources, which allow for more accurate measurements of the TeV energy spectrum, may significantly limit the density of the intergalactic background light.

ACKNOWLEDGMENTS

We acknowledge the technical assistance of K. Harris and E. Roache. The VERITAS Collaboration is supported by the U.S. Dept. of Energy, N.S.F., the Smithsonian Institution, P.P.A.R.C. (U.K.) and Enterprise Ireland.

REFERENCES

1. Catanese, M., et al., in Proc. 25th ICRC (Durban, South Africa), 3, 277, 1997
2. Costamante, L., et al., in Proc. "X-ray Astronomy '99"(Bologna, Italy) (Astro-Ph/0001410), 2000
3. Costamante, L., et al., in Proc. "Fourth Italian Conference on AGNs" (Astro-Ph/0007020), 2000
4. Cosatmante, L., et al., to appear in A&A (Astro-Ph/0103343), 2001
5. Finley, J.P., et al., in GeV-TeV Astrophysics: Toward a Major Atmospheric Čerenkov Telescope - VI," (Snowbird, Utah), (1999)
6. Finley, J.P., et al., To appear in the proceedings of the 27th ICRC (Hamburg, Germany), 2001
7. Horan D., et al., In preparation, 2001
8. Laurent-Muehleisen, Kollgaard, R. I., Feigelson, E. D., Brinkmann, W., Siebert, J., 1999, ApJ, 525, 127, 1999
9. Levine, A.M., et al., ApJL, 469, L33, 1996
10. Piron, F., Laboratoire de Physique Nucleaire de Haute Energie (CNRS/IN2P3 - Ecole Polytechnique),(unpublished), 2000
11. Reynolds, P.T., et al., ApJ, 404, 206, 1993

Optical Identification and Monitoring of High Energy Gamma-Ray Sources

Steven D. Bloom* and Daniel A. Dale[¶]

Department of Physics and Astronomy, Hampden-Sydney College, Box 821, Hampden-Sydney, VA, 23901

[¶] *Infrared Processing and Analysis Center, JPL/Caltech, Pasadena, CA, 91125*

Abstract. We will be presenting updated results to our optical survey of the "error boxes" of EGRET unidentified sources at high Galactic latitude. It is our intention to search for potential blazars that may have been missed in the original identification process. We have first searched the error boxes of unidentified sources at $|b| > 20°$ for flat spectrum radio sources using NASA Extragalactic Database (NED). For each such radio source found we conducted optical searches for counterparts using the Palomar 60-inch telescope. Many of the radio sources have plausible optical counterparts, and spectroscopy will be conducted at a later date to determine which sources are quasars or active galaxies (AGN's). Though it is plausible that several of these sources are extragalactic, no blazar-like activity has yet been observed.

INTRODUCTION

Though the EGRET instrument detected 271 sources, most of these sources remain unidentified with counterparts at another wavelength [7]. However, the distribution of sources on the sky does suggest at least two major components: Those near the Galactic plane, where 90 % of the sources are unidentified, and those above the plane, where 50% are unidentified [2]. Grenier [5] further suggests that those sources above the plane may have three subcomponents: extragalactic, Gould Belt, and Galactic halo. Though nearly all of the firmly identified extragalactic sources are blazars, there have been recent studies suggesting that some sources may be related to galaxy clusters [3]. The task of investigating all of these sources individually seems daunting. Therefore, several investigators have gone the route of examining several individual sources in depth ([6], [11], [13]). The advantage of this technique is that such searches are much more likely to be fruitful. This has particularly been true for finding new blazar-like objects at high Galactic latitude. We continue along these lines by searching within the error contours of high Galactic latitude EGRET unidentified sources for new blazars. We will then ascertain whether any such objects found are likely to be the source of the gamma rays.

SOURCE SELECTION

Using the NASA Extragalactic Database (NED), we have searched for flat spectrum radio sources (spectral index, $-0.5<\alpha<0.5$) within the 95% error contours of the EGRET unidentified sources with $|b|>20°$. Aside from a handful of pulsars, blazars with flat radio spectra are the only objects that have been detected by EGRET as a class. We have chosen this range of latitudes so that there will be minimal overlap with Galactic objects. We further restricted ourselves to the ranges of right ascension and declination that would be visible from Mount Palomar during the nights in November 1999. These radio data referenced in NED were usually from the Parkes or Green Bank surveys, which had positional accuracies of about 1-3 arc minutes (depending on observing frequency and brightness of the source), which is far too coarse for optical searches. We have therefore additionally searched for these radio sources in the NRAO-VLA Sky Survey (NVSS) in order to obtain positions accurate to approximately 1" [4]. In addition to the flat spectrum radio sources, we have compiled a secondary observing list consisting of radio sources of unknown spectral index, and/or X-ray and optical sources with accurate positions.

An extensive study of the identification of radio sources with EGRET sources was conducted by Mattox et al. [10]. Essentially they conclude that it is difficult to positively identify a gamma-ray source with any radio source with flux density less than about 500 mJy at 5 GHz mostly because the sky density of such sources is so high, and thus the probability for chance correlations is correspondingly high. They assume that the radio spectra would be near their peak fluxes at these frequencies, and thus 5 GHz flux densities would be representative of the synchrotron output of the source. This assumption was made out of necessity, since there are no complete high frequency surveys with sensitivity less than one Jansky. However, other investigators have shown that several sources might be fairly dim at these low frequencies, and have a brighter (and flatter) spectrum extending to at least 200 GHz (eg., [1]). Thus, some dim 5 GHz radio sources, which by the method of [10] would not be considered to be counterparts to EGRET sources, may indeed be counterparts to gamma-ray blazars. We intend to make up for this possible deficiency by searching for the optical counterparts to sources that may be fairly dim (< 100 mJy) at 5 GHz. If a counterpart is found we will search for blazar-like activity, and go on to investigate the optical and radio spectra of these objects.

OBSERVATIONS

We have conducted the search for optical counterparts using the 60-inch telescope at Mount Palomar with a Tek 2048x2048 CCD. The pixel size is 0.37 arc seconds, however we have 2x2 binned the pixels to a size of 0.74 arc seconds. Since the seeing for those nights (and most at Palomar) was worse than 1 arc second, no crucial information on source structure is lost. In addition, we would expect most of our sources to be point-like. We have used the Johnson B, V and Kron R filters for determining magnitudes in those bandpasses. The procedure was to first observe a candidate source in R for 10 minutes (and then adjust to a shorter or longer exposure depending on results). If no obvious counterpart was seen after 20 minutes, we did not go to the other filters. If a potential source was seen, we observed in V and B as well. In addition to the program sources, we have observed Landolt [8] standard stars throughout the evening to determine the photometric scale and atmospheric absorption in the usual manner (more details on procedure and reductions will appear in a forthcoming journal article by the authors). Galactic extinction corrections have been applied using the values from NED. We have repeated our observations on a second night to test for variability. We have used the Guide Star Catalogue to determine the astrometric plate solutions for our CCD fields. Typical rms uncertainties in our plate solutions were less than that of the individual stellar positions from the catalogue (about 0.3 arc seconds).

RESULTS

We have observed 10 radio sources (within four different EGRET error boxes) and found optical counterparts for seven, breaking into magnitude in the following manner. Of the seven sources seen visually, five are in the range of V=16-19, one is in the range of V=19-21 and one is dimmer than 21 (near the magnitude limit). A summary of the results for the visually detected sources is in Table 1. A more complete summary with multicolor photometry will be presented in a forthcoming paper. Column (1) gives the name of the EGRET source, Column (2) gives the name of the radio source that is a candidate counterpart, Column (3) gives the V magnitude attained by us as described in the previous section, Column (4) give the radio flux density at 5 GHz (in milliJanskys) as found in NED and Column (5) gives the spectral index between 1.4 and 5 GHz and is provided by NED if the information is known. We note that no source has shown significant variability across the two nights. Of particular interest is that two candidate counterparts (B0214+108 and NVSS J032850+21282) are in the catalogue of ROSAT/Green Bank ("RGB") sources [9]. Punsly [12] suggested that "RGB" sources would be likely gamma-ray blazars and that the counterparts are BL Lac objects with their spectral energy distributions (SED) peaked at intermediate to high frequencies (that is, optical to X-ray). These particular objects in our study, however, do not have well determined SED's. The brighter objects have been slated for spectroscopic and high frequency radio observations within the next several months (to be conducted by other groups, in collaboration with us in this study).

TABLE 1. Possible Counterparts to Gamma-Ray Sources

EGRET Source	Radio Source	V	Radio	S.I.
(1)	(2)	(3)	(4)	(5)
3EG J0038-0949	NVSS J003906-094247	21.5	214	-0.1
3EG J0215+1123	NVSS J021527+112318	17.9	56	-0.8
3EG J0215+1123	B0214+108	16.3	440	-0.8
3EG J0245+1758	NVSS J024437+172221	19.8	33	-0.1
3EG J0245+1758	NVSS J024640+180144	18.1	53	...
3EG J0245+1758	NVSS J024611+182330	18.8	214	0.1
3EG J0329+2149	NVSS J032850+212825	18.1	57	...

CONCLUSIONS

After examining the error boxes of four EGRET sources for coincident radio/optical objects, we have found four plausible counterparts to the gamma-ray sources. A summary of the results is shown in Table 1. For 3EG J0038-0949 there is only one plausible counterpart. For 3EG J0215+1123, we consider B0214+108 to be the most plausible counterpart, though even this source has difficulties. The extended radio source, 4C +10.06 is ambiguously identified with this catalogued QSO. The two nearest compact NVSS radios source are over 40 arc seconds from the optical object. Search results from NED indicate that there are many galaxies in this field, at least one of which has a red shift similar to the QSO. For 3EG J0245+1758, the second NVSS source listed in the table is the more likely counterpart (brighter, flat spectrum). This source is also a known QSO. At first, it may not seem that the suggested counterpart for 3EG J0329+2149 is plausible, due to its relatively low radio flux density; however, we note that this source is a coincident optical, IR, radio, and X-ray source. The compact core is known to have a flat spectrum [9]. We emphasize that we do not yet consider any of these suggestions to be firm identifications. We suggest that optical monitoring of these objects continue, and that spectroscopy conducted for any of the objects for which the optical counterpart has not yet been classified spectrally. The Hampden-Sydney College 16 inch telescope could be used to monitor the brightest 5 or so of these sources when they next become visible in northern Autumn 2001.

ACKNOWLEDGMENTS

This work has made use of the NASA Extragalactic Database and the Astrophysical Data System web abstract server.

REFERENCES

1. Bloom, S. D. et al., *Ap. J.*, **488**, L23 (1997)
2. Caraveo, P., these proceedings (2001)
3. Colafrancesco, S., these proceedings (2001)
4. Condon, J. J. et al., *A. J.*, **115**, 1693 (1998)
5. Grenier, I., these proceedings (2001)
6. Halpern, J. et al., *Ap. J*, **551**, 1016 (2001)
7. Hartman, R. C. et al., *Ap. J. Suppl.* **123,** 79 (1999)
8. Landolt, A. U., *A. J.*, **104**, 340 (1992)
9. Laurent-Muelheisen, S. A. et al., *A&A Suppl.*, **122**, 235 (1997)
10. Mattox, J. R. et al., *Ap. J.*, **481**, 95 (1997)
11. Mukherjee, R. et al., *Ap J*, **542**, 740 (2000)
12. Punsly, B., *Ap. J.*, **516**, 141(1999)
13. Wallace, P. M., these proceedings (2001)

CCD Photometry of Blazars at Abastumani:Progress Report

Omar M. Kurtanidze[1,2,3] and Maria G. Nikolashvili[1]

[1] *Abastumani Observatory, 383762 Abastumani, Georgia*
[2] *Astrophysikalisches Institut Potsdam, 14482 Potsdam, Germany*
[3] *Landessternwarte Heidelberg-Königstuhl, D-69117 Heidelberg, Germany*

Abstract. We give a brief summary of the ongoing Abastumani Blazar Monitoring Program started in the May of 1997. More than 45000 frames are obtained during 507 nights of observation for about 50 target objects, among them X-ray, γ-ray and Optical Blazars. All observations were done in the BVRI bands using ST-6 CCD based Photometer attached to the Newtonian focus of 70-cm meniscus telescope. Image reductions have been done using different software packages of image reduction systems such as IRAF, MIDAS and STARLINK. Most objects under study show light variations in optical band over one magnitude. Largest one was observed for AO 0235+116 - 4.0 mag in R band.

INTRODUCTION

One of the distinguishing characteristics of the blazar class of AGN which includes BL Lacertae type objects, high polarization quasars (HPQ) and optical violently variable (OVV) quasars is that their flux densities are highly variable at all wavelength from Radio to γ-Rays. Therefore the optical multiband monitoring along with other ones gives unique clues into the size and structure of the radiating region. Variability time scales have been derived for many blazars from monitoring programs which attain a time resolution of days and to years [18]. The best example of the international cooperation is the multiwavelength study of a selected blazars in OJ-94 project carried out during the last decade [13].

The observations by CGRO show that strongly variable and radio-loud quasars emit a significant fraction of their energy in γ-ray band. Variability of the γ-ray flux on a time scale as short as four hours has also been observed [7]. Unfortunately, existing multiwavelength data are not adequate yet to permit definite conclusions to be drawn about the nature of blazars since the optical coverage in the previous campaigns has been much too sparse. Therefore, Mattox [8] suggests that a coherent use of existing ground-based optical facilities will dramatically improve the quality of the optical data obtained during multiband blazar campaigns.

We started systematic multiband optical monitoring of blazars at Abastumani Observatory in May of 1997. In late October 1997 we joined the Whole Earth Blazar Telescope (WEBT, http://astro.fmarion.edu/WEBT). The aim of the programme is to study short-term and long-term variability of blazars and their correlations with that in radio, X-ray and γ-ray bands.

OBSERVATION AND DATA REDUCTION

Abastumani Observatory is located in the South-Western part of Georgia at a latitude of 41°.8 and a longitude of 42°.8 on the top of the Mt. Kanobili at 1700 m above mean sea level. The weather and seeing are very good in Abastumani (150 nights per year, 30%<1 arcsec). The mean values of the night sky brightness are B=22.0, V=21.2, R=20.6 and I=19.8.

Blazar Monitoring Program at Abastumani Observatory was started in the May 1997 and is carried out with Peltier cooled ST-6 CCD Imaging Camera attached to the Newtonian focus of the 70-cm meniscus telescope (1/3). The pointing accuracy of the meniscus telescope is one-two arcminutes and it is good enough to locate target object inside of the full frame field of view 14.9x10.7 sq. arcminute. The ST-6 Imaging Camera uses the TC241 CCD chip (375x242, 23x27 sq.micron) with a maximum quantum efficiency 0.7 at 675 nm. The readout, digitizing, downloading time of a full frame is 37 seconds.

All observations are performed using combined filters of glasses which match the standard B, V (Johnson) and Rc, Ic (Cousins) bands well. Reference sequences in the blazar fields are calibrated using the Landolt's equatorial standard stars [6]. In photometric nights at least one equatorial field is observed with different exposures. Because of the scale of CCD and resolution of the meniscus telescope are equal to 2.3x2.7 sq. arcsec per pixel and 1.5 arcsec respectively, the images are strongly undersampled, therefore to improve sampling it is needed to defocus frames. Unfortunately, dark current limits the exposure time to 900 sec.

The primary data analysis software systems used are IRAF, MIDAS and STAR-LINK installed on Pentium PC (RH Linux 6.2, 250Mhz MMX, 64Mb, 20.4 Gbts). The highest differential photometric accuracy reached is 0.007 (r.m.s.) magnitude in R band (180 sec) in the field of OJ 287 (comparison stars no. 9, 10).

RESULTS

List of target objects was compiled using three Catalogues [11,12,16]. In the period from May 1997 to November 2000, during 507 observing nights, more than 3.0Gbt (45000 frames) data were collected. In the Table 1 the list of the target objects along with the number of nights observed and frames obtained in every of the BVRI bands in excess of twenty are given. Last column shows the number of frames obtained to study the intraday (IDV) and intrahour (IHV) variability of selected blazars. Several campaigns have been conducted during great outburst and

TABLE 1. The list of target objects

RA (2000)	DEC (2000)	Name	Nights	N_B	N_V	N_R	N_I	$N_{IDV,Nights}$
01 12 05.7	22 44 38	S2 0109+22	72	72	72	92	48	402, 4
02 19 42.8	02 19 42	3C 66A	78	74	76	95	50	782, 6
02 38 38.8	16 36 59	AO 0235+164	74	49	77	118	48	1102, 15
03 26 13.9	02 25 15	1ES0323+022	24	18	24	34	21	
04 24 46.8	00 36 06	PKS 0422+004	38	38	41	51	22	268, 8
05 07 56.2	67 37 24	1ES0502+675	37	28	35	45	26	
07 21 53.4	71 20 36	S5 0716+714	105	155	151	188	151	4179, 52
07 57 06.6	09 56 35	OI 09.4	29	26	28	32	25	
08 09 49.0	52 18 56	1ES0806+524	34	34	36	38	32	
08 31 48.9	04 29 39	OJ 049	23	21	21	23	21	
08 54 48.9	20 06 31	OJ 287	50	66	68	81	58	
09 58 47.2	65 33 55	S4 0954+658	67	43	68	80	67	249, 14
10 15 04.2	49 26 00	1ES1011+496	26	25	25	28	21	
10 31 18.6	50 53 34	1ES1028+511	28	25	27	28	24	
11 04 27.3	38 12 32	MKR421	44	77	77	78	74	1906, 47
11 59 32.1	29 14 42	4C 29.45	48	57	55	68	56	150, 5
12 21 31.7	28 13 58	ON 231, W Com	78	75	74	99	76	
12 29 06.7	02 03 09	3C 273	28	31	29	32	30	
12 56 11.2	-05 47 22	3C 279	47	46	45	53	42	127, 4
14 19 46.6	54 23 15	OQ 530	38	34	36	45	33	
14 27 00.4	23 48 00	OQ 240	32	30	28	32	29	
15 01 01.9	22 38 06	MS 1458	21	15	21	20	21	
16 35 15.5	38 08 04	4C 38.41	20	12	16	26	17	
16 42 58.8	39 48 37	3C 345	54	50	55	64	52	208, 8
16 53 52.1	39 45 36	MRK 501	54	59	69	72	49	
17 25 04.3	11 52 15	H 1722+119	38	37	34	45	29	
17 28 18.5	50 13 11	I ZW 187	56	53	53	67	63	
17 48 30.1	70 05 54	S4 1749	51	45	45	60	37	
18 00 45.7	78 28 04	S5 1803+784	49	50	44	71	35	
18 06 50.7	69 49 28	3C 371	57	60	54	42	76	
19 59 59.9	65 08 55	1ES1959+650	45	45	43	57	41	594, 17
20 05 30.9	77 52 43	S5 2007+777	21	20	18	26	4	
20 35 22.4	10 56 07	OW 154.9	44	30	41	31	29	
22 02 43.2	42 16 40	BL Lacertae	183	138	144	142	88	12810, 172
22 32 36.2	11 43 51	CTA 102	20	20	20	31	20	
22 53 57.7	16 08 53	3C 454.3	34	34	34	54	37	59, 3
22 57 17.3	07 43 12	S2 2254+074	48	45	51	64	33	
23 47 04.8	51 42 18	1ES2344+514	37	19	36	51	35	

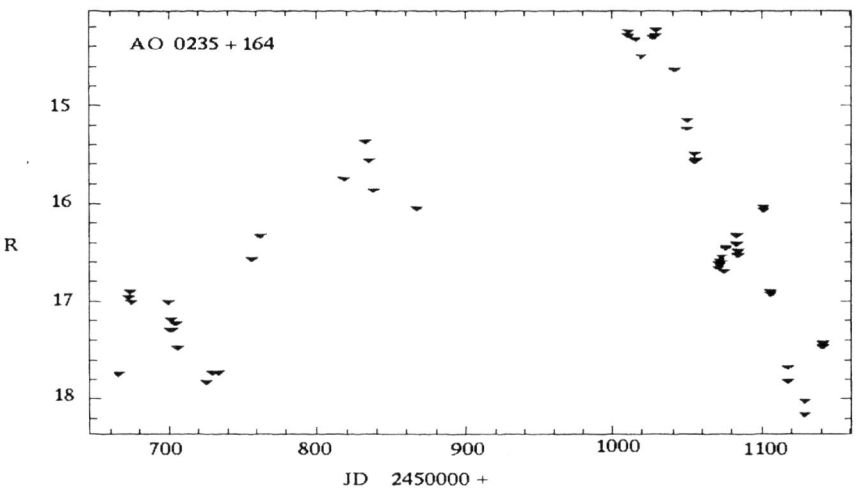

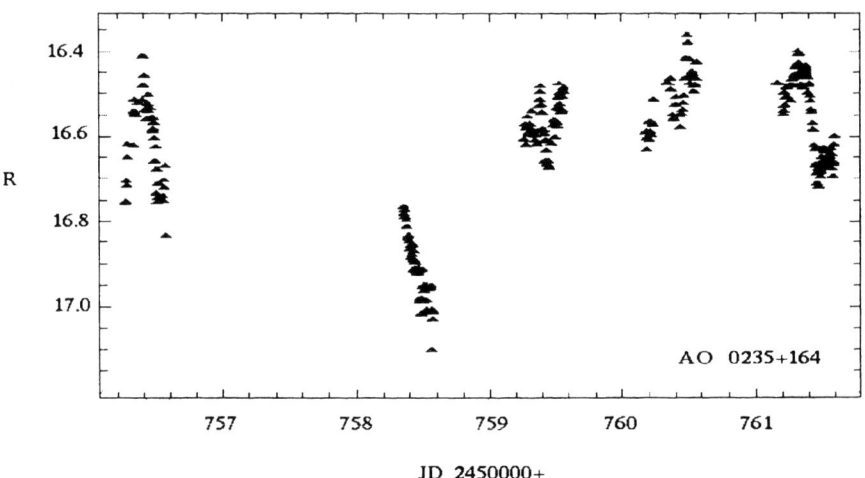

FIGURE 1. The LTV (1997-98) and IDV (Nov 3-8, 1997) of AO 0235+16.

post-outburst era of BL Lacertae, AO 0235+164, 3C 279, Mrk 241, 4C 29.45, S4 0954+65 and others. Part of the results were published [3,4,14,15], while others are submitted for publication [1,2,5]. Most objects under study show light variations over one magnitude in the optical band. Largest one was observed for AO 0235+164 and equals to 4 mag in R band. The lightcurves illustrating the activity of AO 0235+164 are shown in Figure 1. A few faint variable stars (B∼16-17 mag) with amplitude 0.3-0.4 and periods 3-5 hours were also identified. One of the surprising example is the comparison star number 5 in the field of TeV BL Lacertae object ES1959+650 [17].

ACKNOWLEDGMENTS

We thank the SOC and LOC for the kind invitation and financial support. O.M.K. gratefully acknowledges the hospitality and invaluable financial support of the Astrophisikalisches Institute Potsdam and thanks Dr G.M. Richter for the kind collaboration of many years, without which this Program would never have been conducted.

REFERENCES

1. Hartmann R. C., Bottcher M., et al., *ApJ*, **553**, 683, (2001).
2. Hartmann R. C., Villata M., et al., *ApJ*, accepted, (2001).
3. Kurtanidze O. M., et al., *Pubbl. Osserv. Astron. Univ. de Perugia*, **3**, 193, (1998).
4. Kurtanidze O. M., et. al., in Blazar Monitoring Towards the Third Millenium, eds. Raiteri C., et al., *Osservatorio Astronomico di Torino*, 29, (1999).
5. Kurtanidze O. M., et. al., IAU Symp. 205, in press, (2001).
6. Landolt A.U., *A.J.*, **104**, 340, (1992).
7. Mattox J.R., et al., *ApJ*, **476**, 692 (1997).
8. Mattox J.R., in *CCD Precision Photometry*, ASP Conf. Ser. **189**, 95, (2000).
9. Nikolashvili M. G., et al., in Blazar Monitoring Towards the Third Millenium, eds. Raiteri C., et al., *Osservatorio Astronomico di Torino*, 33, (1999).
10. Nikolashvili M. G., et al., in Blazar Monitoring Towards the Third Millenium, eds. Raiteri C., et al., *Osservatorio Astronomico di Torino*, 36, (1999).
11. Padovani P., Giommi P., *MNRAS*, **277**, 1477, (1996).
12. Perlman E., et al., *ApJS*, **104**, 251, (1996).
13. Takalo L., et al., in *BL Lac Phenomenon*, eds. Takalo L., et al., ASP Conf. Ser. **159**, 253, (1999).
14. Tosti G., et al., in *BL Lac Phenomenon*, eds. Takalo L., et al., ASP Conf., Ser., **159**, 145, (1999).
15. Tosti G., et al., *Blazar Data*, **2**, 1, (1999).
16. Véron-Cetty M., et al., *ESO Sci. Rep.*, **13**, (1993).
17. Villata M., et. al., *A&AS*, **130**, 305, (1999).
18. Wagner S.J, Witzel A., *ARA&A*, **35**, 607 (1995).

Colour Variation in BL Lacertae

Omar M. Kurtanidze[1,2,3] and Maria G. Nikolashvili[1]

[1] *Abastumani Observatory, 383762 Abastumani, Georgia*
[2] *Astrophysikalisches Institute Potsdam, 14482 Potsdam, Germany*
[3] *Landessternwarte Heidelberg-Königstuhl, D-69117 Heidelberg, Germany*

Abstract. We present preliminary results of the optical observations of BL Lacertae during the period from Aug 1997 to Aug 1998 carried out with ST-6 CCD Camera attached to the Newtonian focus of the 70-cm meniscus telescope of the Abastumani Observatory. The aim of these observations is to study the long-term, intranight and intrahour variability of BL Lacertae. They were studied on the bases of 172 nights of observations, while the long-term variability during 183 nights. The maximum variation is observed in B band and equals to 3.00 (rms=0.03). The variations in the V and R bands are within 2.71 (0.02) and 2.53 (0.01), respectively. This means that variations are larger at shorter wavelength or the object become bluer in the active phase. It were also demonstrated that BL Lacertae shows the short time scale variability in R band within 0.30 (0.02) and 0.10 (0.01).

INTRODUCTION

The BL Lacertae was discovered in 1929 by Guno Hoffmeister, who found it to vary by more than a factor two in one week and classified it as a short period variable star [2]. It's identification with unusual radio source VRO42.2201, rapid variation in the optical and radio bands, high degree of linear polarization and continuos spectrum with neither emission nor absorption lines were the indications of very peculiar character of this object. Its true extragalactic nature was determined when Miller [6] found faint emission lines at a redshift of 0.069. It exhibits a core-jet morphology and has ejected several superluminal components, effectively modeled as planar shocks embedded within a relativistic flow [3]. The core of the elliptical galaxy it is one of the most extreme examples of Active Galactic Nuclei. Historically, BL Lac is known to show $\sim$5 mag variation in optical band with episodic outbursts [1]. During the summer 1997 outburst it it showed a very strong activity including intranight ones [9]. The strong activity was also detected in the radio, X-ray and γ-ray bands.

OBSERVATIONS

We are intensively monitoring BL Lacertae at Abastumani Observatory since August 4, 1997, when it remained in a high state for more than two months. Rapid and large-amplitude flux variations characterised the source during this period. All observations presented here were carried out with 70-cm meniscus telescope and CCD Camera ST-6 attached to the Newtonian focus, using combined filters of glasses which match the standard B, V (Johnson) and Rc, Ic (Cousins) bands well [5]. To study the long-term variability (LTV) we observed BL Lac during 183 nights and collected more than 1100 frames in BVRI bands. More than 12000 frames were obtained in R band during 172 nights to study intranight (IDV) and intrahour variability (IHV). The duration of observational runs varied from two hours to six hours. The exposure times varied from 60 to 180 sec depending on the brightness of the object and the filter used. Instrumental magnitudes were obtained using DAOPHOT II routines [12]. Magnitudes are calculated relative comparison stars C and H, which have nearly the same colours as the object under study [10].

RESULTS

The preliminary results of observations of BL Lac during great summer 1997 outburst are presented by Blazar Monitoring Groups at the Annual OJ-94 Meeting [4,13], with detailed ones in [7,8,11,14–16]. We present here observations carried out during the great outburst and after the great outburst, when BL Lac possessed the second outburst. The light curves in B, V, R and I bands are presented in Figure 2. In both cases the maximum variation is observed in B band and equals to 3.00 (rms=0.03) and 1.25 mag (0.04). The variations in the V, R bands are within 2.71 (0.02), 2.53 (0.01) and 1.11 (0.03), 1.02 (0.01) magnitudes respectively. This means that variations are always larger at shorter wavelength or the object become bluer in the active phase. Evidence of IHV and IDV was practically found during many nights of observations. The typical amplitudes in R band were within 0.30 mag (rms=0.02) and 0.10 (rms=0.01) for intraday and intrahour variability respectively These variations are clearly demonstrated in the Figure. 1.

ACKNOWLEDGMENTS

We thank the SOC and LOC for the kind invitation and financial support. O.M.K. gratefully acknowledges the hospitality and invaluable financial support of the Astrophisikalisches Institute Potsdam and thanks Dr G.M. Richter for the kind collaboration of many years, without which this Program would never have been conducted.

REFERENCES

1. Fan J. H., et al., *ApJ*, **507**, 173, (1998).
2. Hoffmeister G., et al., Veranderlichte Sterne, Leipzig, (1990).
3. Hughes P.A., et al., *ApJ*, **341**, 68, (1989).
4. Kurtanidze O. M., et al., in Multifrequency Monitoring of Blazars, eds. Tosti J., et al., *Pubblicazioni Osservatorio Astronomico Universit de Perugia*, **3**, 193, (1998).
5. Kurtanidze O. M., et. al., in Blazar Monitoring Towards the Third Millenium, eds. Raiteri C., et al., *Osservatorio Astronomico di Torino*, 29, (1999).
6. Miller J.S., et al., *ApJL*, **219**, L85, (1978).
7. Nikolashvili M. G., et al., in Blazar Monitoring Towards the Third Millenium, eds. Raiteri C., et al., *Osservatorio Astronomico di Torino*, 33, (1999).
8. Nikolashvili M. G., et al., in Blazar Monitoring Towards the Third Millenium, eds. Raiteri C., et al., *Osservatorio Astronomico di Torino*, 36, (1999).
9. Nesci R., et al., *Pubbl. Osserv. Astron. Univ. de Perugia*, **3**, 189, (1998).
10. Smith P.S., et al., *AJ*, **90**, 1184, (1985).
11. Sobrito G., et al., *Blazar Data*, **1**, no.5, (1999).
12. Stetson P.B., *PASP*, **99**, 191, (1987).
13. Takalo L.O., et al., *Pubbl. Osserv. Astron. Univ. de Perugia*, **3**, 200, (1998).
14. Tosti G., et al., in BL LAC Phenomenon, eds. Takalo L.O., et al., 145, (1999).
15. Tosti G., et al., *Blazar Data*, **2**, no.1, (1999).
16. Webb J., et al., *AJ*, **115**, 2244, (1998).

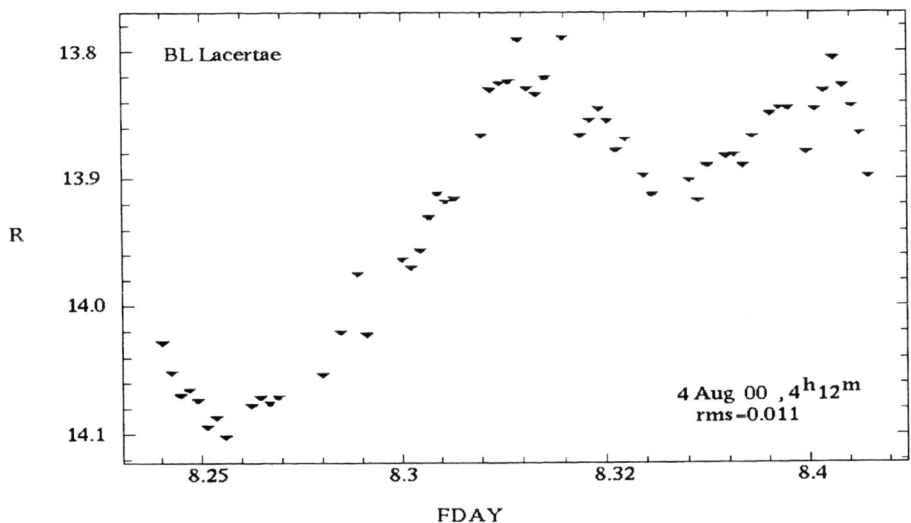

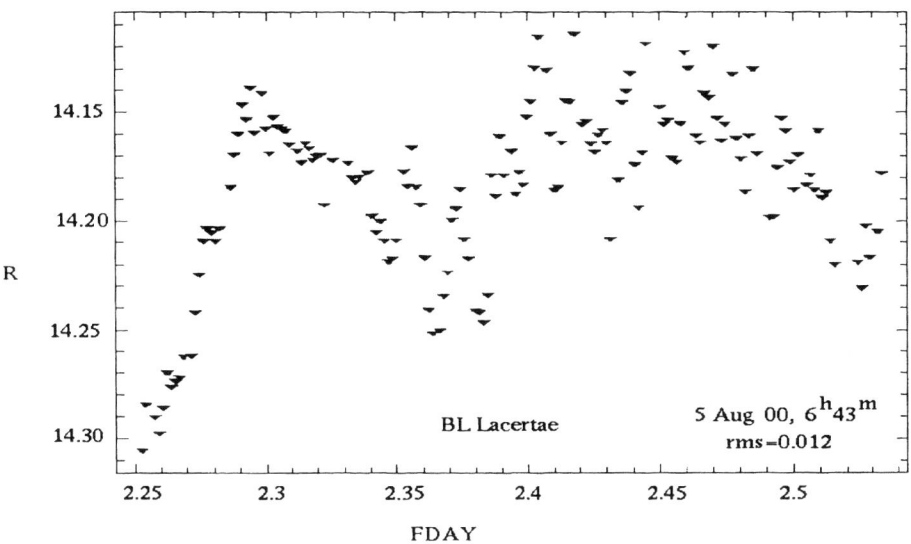

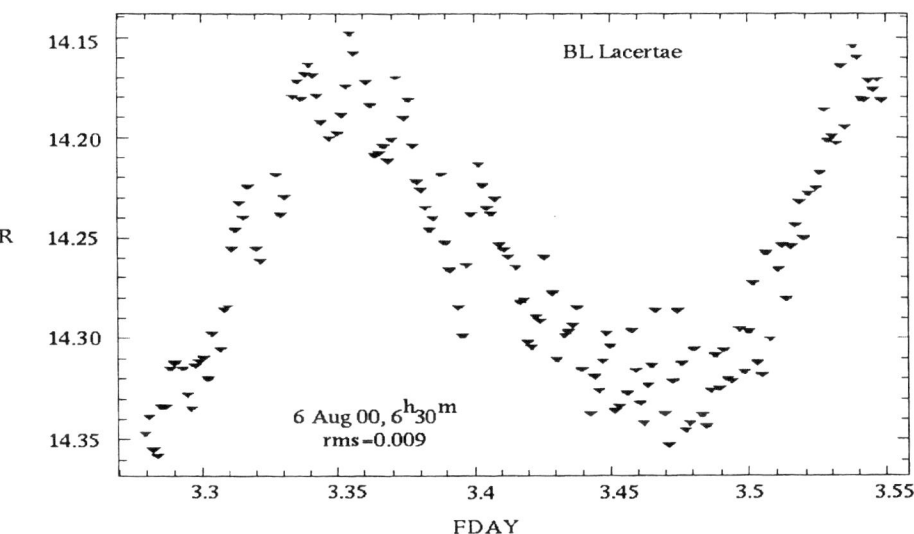

FIGURE 1. BL Lacertae's scales of variability in R band (4-6 Aug, 2000).

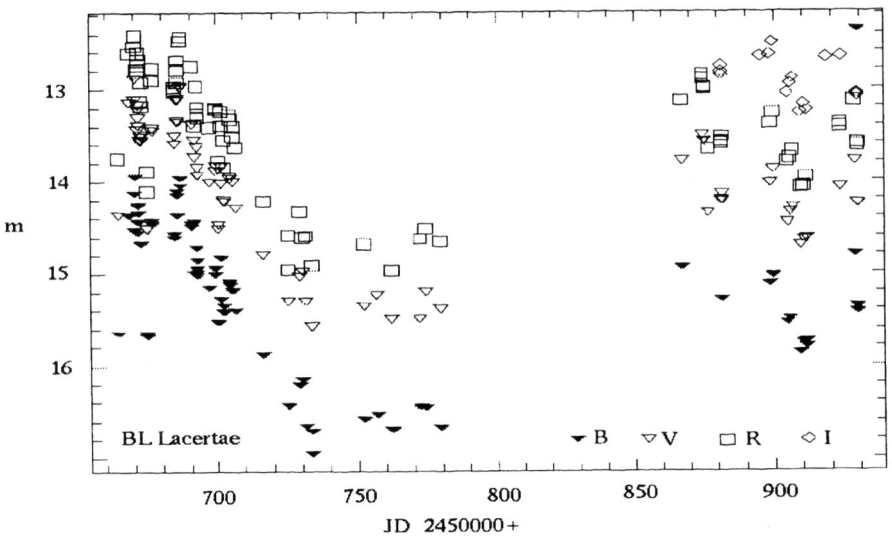

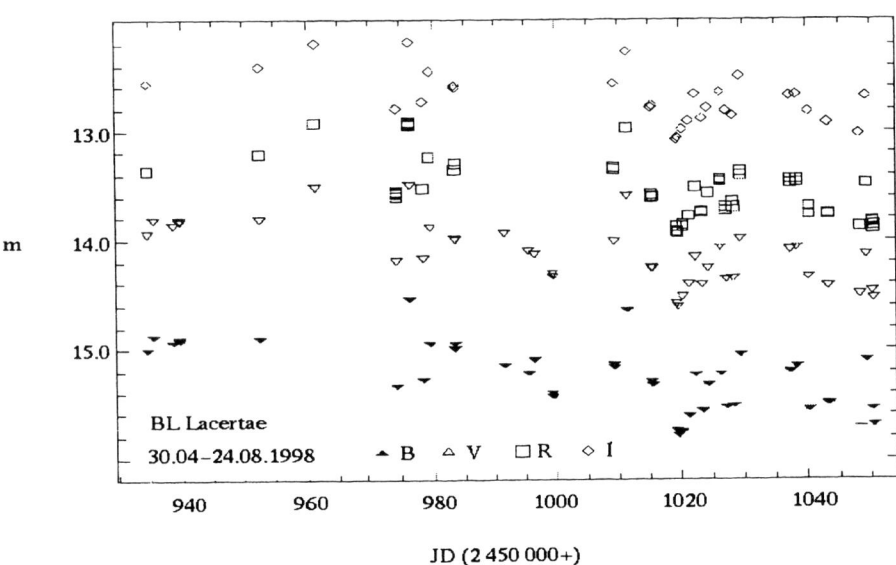

FIGURE 2. The lightcurves of BL Lacertae in BVRI bands (1997-1998).

COMPTEL Observations of the Blazars 3C 454.3 and CTA 102

S. Zhang[1,5], W. Collmar[1], V. Schönfelder[1], H. Bloemen[2],
W. Hermsen[2], M. McConnell[3], K. Bennett[4], O.R. Williams[4]

[1] *Max-Planck-Institut für extraterrestrische Physik, Garching, Germany*
[2] *Space Research Organization Netherlands, Utrecht, The Netherlands*
[3] *Space Science Center, University of New Hampshire, Durham, USA*
[4] *Astrophysics Division, ESTEC, Noordwijk, The Netherlands*
[5] *High Energy Astrophysics Lab, IHEP, P.O.Box 918-3, Beijing, China*

Abstract. We have analyzed the two blazars of 3C 454.3 and CTA 102 using all available COMPTEL data from 1991 to 1999. In the 10-30 MeV band, emission from the general direction of the sources is found at the 4σ-level, being consistent with contributions from both sources. Below 10 MeV only 3C 454.3 is significantly detected, with the strongest evidence (5.6 σ) in the 3-10 MeV band. Significant flux variability is not observed for both sources, while a low emission is seen most of the years in the 3-10 MeV light curve for 3C 454.3. Its time-averaged MeV spectrum suggests a power maximum between 3 to 10 MeV.

INTRODUCTION

Gamma-ray emission from the two blazars 3C 454.3 and CTA 102, located $\sim 7°$ apart on the sky, was discovered by EGRET during the early Compton Gamma-Ray Observatory (CGRO) mission [1,2]. COMPTEL measurements at MeV energies of the same time periods, reported earlier by Blom et al. [3], show indications that both sources are weakly detected in the COMPTEL uppermost (10-30 MeV) band. If combined with EGRET, the trend for a spectral flattening at MeV energies becomes visible. At even lower energies (hard X-rays) OSSE detected variability of both sources in 1994 [4].

In this paper, we present the results of the COMPTEL observations on both blazars during the whole CGRO mission. The data have been consistently analyzed for both blazars and compared to the results obtained early in the mission, published by Blom et al. [3]. A more detailed presentation of the analysis results will be given by Zhang et al. [5].

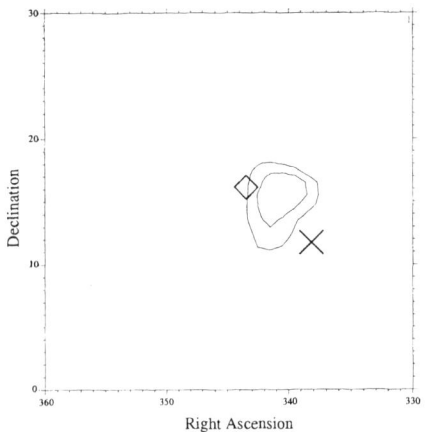 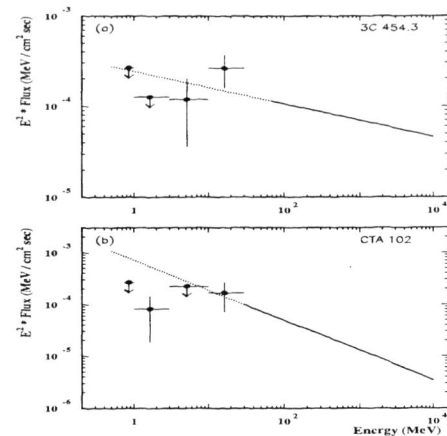

FIGURE 1. Left: COMPTEL 10-30 MeV map for 3C 454.3 (◊) and CTA 102 (×) from Phase 1 observations in 1992. The contour lines start at a detection significance level of 3 σ with steps of 0.5 σ. **Right:** Combined COMPTEL/EGRET spectra for 3C 454.3 (a) and CTA 102 (b). The filled circles represent the COMPTEL spectra of Phase 1 (VPs 19.0, 26.0, 28.0 and 37.0), the solid lines the EGRET spectra of VP 19 [1, 2] and the dotted lines their extrapolations toward lower energies. The error bars are 1 σ and the upper limits 2 σ.

DATA AND ANALYSIS METHOD

The imaging Compton Telescope COMPTEL - one of four experiments aboard CGRO - was sensitive to γ-rays in the energy range 0.75-30 MeV [6]. During its whole mission (1991 - 2000), the sources were in 22 CGRO viewing period (VPs) within 40° of the COMPTEL pointing direction. These VPs were selected in our analyses.

The analyses were carried out by using the standard COMPTEL maximum-likelihood analysis procedure including a filtering technique for background generation. Point spread functions of the instrument which assume an E^{-2} power law shape for the input spectrum were applied in our analyses.

RESULTS

Phase 1

For CGRO Phase 1 (April '91 to November '92) our results are similar to those reported by Blom et al. [3]. There is evidence for emission from the general direction of 3C 454.3 and CTA 102 in the 10-30 MeV band with a detection significance of $\sim$ 3.9 σ, which is consistent with contributions from both sources (Figure 1). At lower

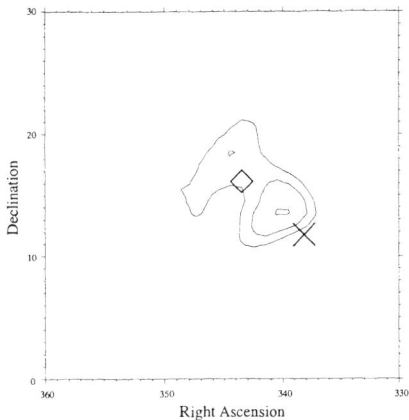 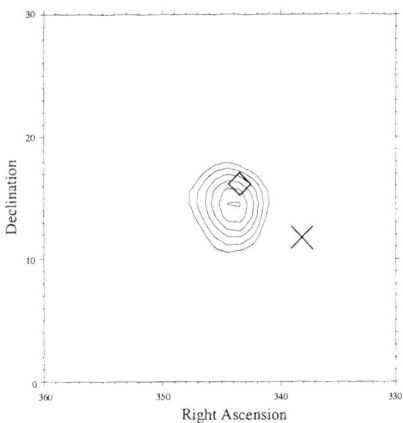

FIGURE 2. COMPTEL 10-30 MeV map (left) and 3-10 MeV map (right) for 3C 454.3 (◇) and CTA 102 (×) from 8 years of observations. The contour lines start at a detection significance level of 3 σ with steps of 0.5 σ.

energies, only marginal detections and upper limits are obtained. The combined COMPTEL/EGRET spectra indicate spectral turnovers at MeV energies for both sources (Figure 1).

Phases 1-8

No significant ($>3\sigma$) detection is found for either 3C 454.3 or CTA 102 in any individual COMPTEL observation during the whole CGRO mission in the four standard energy ranges. However, combining all the COMPTEL data provides evidence for 3C 454.3 at energies above 1 MeV and for CTA 102 at energies above 10 MeV. The 10-30 MeV skymap (Figure 2) shows that a 4σ-excess is located just between 3C 454.3 and CTA 102, again consistent with emission from both. No signal can be derived from CTA 102 in the 3-10 MeV band, while an obvious excess is detected near 3C 454.3 at a significance level of $\sim 5.6~\sigma$ (see Figure 2). 3C 454.3 is located at the 4.5σ-detection contour and is within the 3 σ error location contour. Since 3C 454.3 is the only known γ-ray source in the COMPTEL error box, we attribute this significant excess to emission from 3C 454.3.

A search for flux variability has been carried out in all individual observations on 3C 454.3 and CTA 102 in the four standard COMPTEL energy bands. All light curves are consistent with a constant flux for both sources over a period of about 8 years. While only marginal hints or upper limits are found along the CGRO mission for CTA 102, 3C 454.3 seem to be always emitting at somehow a low level

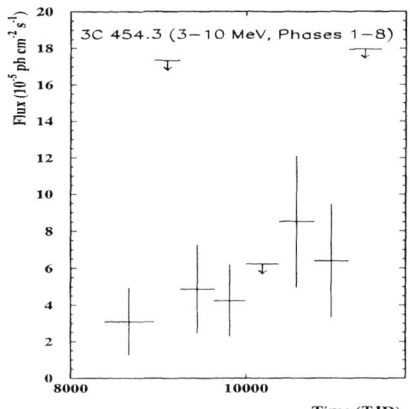
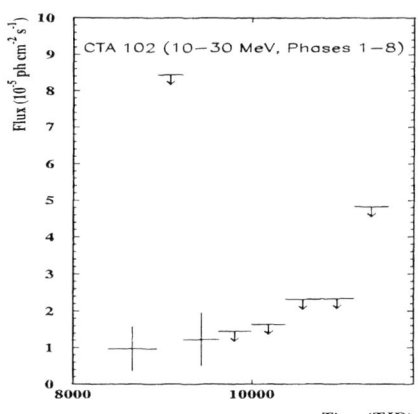

FIGURE 3. The 3-10 MeV light curves for 3C 454.3 (left) and the 10-30 MeV light curve for CTA 102 (right) with each bin averaged over one CGRO Phase which typically covers a time period of 1 year. The high upper limits show times of low source exposures. The error bars are 1 σ and the upper limits 2 σ.

in the 3-10 MeV band (Figure 3). This is consistent with the significant detection of 3C 454.3 in the sum of the 3-10 MeV data. The 10-30 MeV light curve of CTA 102 (Figure 3) shows mostly upper limits, indicating the source was weak along the CGRO mission.

Figure 4 shows the COMPTEL spectra averaged over a time period between April 1991 and November 1999 for both sources. The spectrum of 3C 454.3 (Figure 4) suggests a power maximum (at least with respect to MeV energies) in the 3-10 MeV band. No conclusion can be derived from the spectrum of CTA 102 because only in the 10-30 MeV band a weak detection is obtained (Figure 4).

DISCUSSION AND SUMMARY

Our analyses of the data of the early COMPTEL observations of the γ-loud flat-spectrum radio quasars 3C 454.3 and CTA 102 provide similar results with respect to the weak source detections and spectra as reported by Blom et al. [3]. The consistent analysis of all available COMPTEL data, covering 8 years, reveals some new features for these two blazars. The most important one is the significant detection of 3C 454.3 in the 3-10 MeV band. Furthermore, 3C 454.3 seems to be a emitter always detectable in the 3-10 MeV band, where it appears to reach its power maximum in the time-averaged spectrum. It resembles 3C 273 which is a rather stable MeV emitter with a power maximum in the 3-10 MeV band [10], if the upper limits in the 3-10 MeV light curve are detections of 3C 454.3 near the threshold of COMPTEL. Therefore, 3C 454.3 might internally be similar to

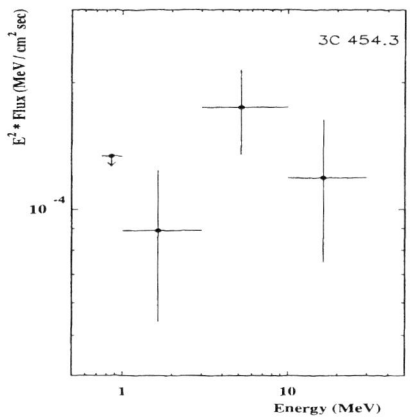 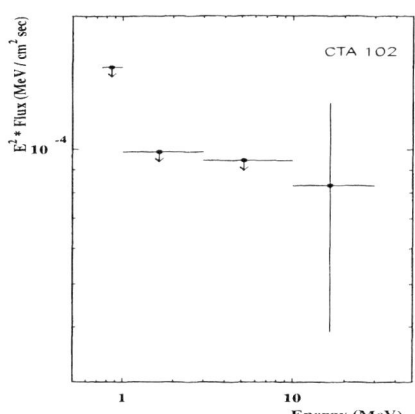

FIGURE 4. COMPTEL spectra for 3C 454.3 (left) and CTA 102 (right) averaged over observations in Phases 1-8. The error bars are 1 σ and the upper limits 2 σ.

3C 273 however at a lower flux level since we just see the 'tip of the iceberg' with COMPTEL. Compared to 3C 454.3, the MeV emission from CTA 102 is weak and the detection significance marginal. Indications for MeV emission from the source is found only in some early Phases. Its time-averaged MeV spectrum does not allow to draw any conclusions.

The COMPTEL results have to be compared to those in additional energy bands, in particular to the neighboring γ-ray bands covered by EGRET and OSSE. This work is in progress and will be given by Zhang et al. [5].

ACKNOWLEDGMENTS: The COMPTEL project is supported by the German government through DARA grant 50 QV 9096 8, by NASA under contract NAS5-26645, and by the Netherlands Organization for Scientific Research (NWO).

REFERENCES

1. Hartman, R.C., et al., *ApJ* **407**, L4 (1993).
2. Nolan, P.L., et al., *ApJ* **414**, 82 (1993).
3. Blom, J.J., et al., *A&A* **295**, 330 (1995).
4. McNaron-Brown, K., et al., *ApJ* **451**, 575 (1995).
5. Zhang, S., et al., *A&A* in preparation (2001).
6. Schönfelder, V., et al., *ApJS* **86**, 657 (1993).
7. Hartman, R.C., et al., *ApJS* **123**, 79 (1999).
8. Lichti, G.G., et al., *A&A* **298**, 711 (1995).
9. Dermer, C.D., et al., *ApJ* **416**, 458 (1993).
10. Collmar, W., et al., this proceedings (2001).

Optical photometric observations of γ-ray loud Blazars in the October of 2000

J.H. Fan* and O.M. Kurtanidze, M.G. Nikolashvili[†]

*Center for Astrophysics, Guangzhou University, Guangzhou 510400, China
[†]Abastumani Observatory, 383762 Abastumani, Georgia

Abstract. In the Oct. of 2000, we observed 10 gamma-ray loud blazars, PKS 0219+428 (3C 66A), PKS 0420-014 (OA129), S5 0716+714, 0754+100 (OT 090.4), 0827+243 (OJ 248), 1652+398 (Mrk 501), 2200+420 (BL Lacertae), 2230+114 (CTA 102), 2251+158 (3C 454.3), and 2344+514, with the 70 cm optical telescope at Abstumani Observatory, Georgia. We found intraday variations in PKS 0420-014, S5 0716+714, BL Lacertae, and CTA 102. A variation of 0.3 magnitude over a time scale of about 3 hours was observed in R band in BL Lacertae on JD 2451827. We did not detected variation from 3C 66A, Mrk 501, 3C 454.3 or 2344+514 during our observational period. For the TeV gamma-ray source 2344+514 no clear variation was detected in an observational period of two weeks.

INTRODUCTION

The nature of active galactic nuclei (AGNs) is still an open problem. Photometric observations of AGNs are important for constructing their light curves and to study their variation behavior on different time scales. Blazars are an extreme subclass of AGNs and often show large and violent variations through the whole spectrum (Wills & Wills 1981; Carini et al. 1992; Sillanpaa et al. 1996; Romero et al. 2000a,b; Takalo 1999; Miller & Noble 1996; Fan et al. 1997, 1998, 2001; Fan & Lin 1999, 2000; Villata et al. 1999; Raiteri et al. 1999; Terasranta et al. 1998; Webb et al. 1998; Hartman et al. 1999; Kraus et al. 1999; Massaro & Nesci 1999; Nikolashvili et al. 1999a,b; Tosti & Nucciarelli 1999; Urry 1999; Qian, Tao, & Fan 2000; Xie et al. 2001)

With the EGRET instrument on the GRO, more than 60 AGNs were detected. They share some common observational properties: flat radio spectrum, violent variations, two-component structure in the spectral energy distribution (with the lower frequency component being from the synchrotron emission and the higher frequency one from the inverse Compton process, although the origin seed photons for the inverse Compton radiation are unclear, e.g. Ghisellini et al., 1996; Comastri et al., 1997; Bottcher & Collmar 1998).

Gamma-ray flares are also found correlated with those in the lower energy bands. The short time scales in the gamma-ray bands are in the same range of those in the optical and X-ray bands. So, the short time scales detected in the lower energy bands can be used for the constraints of basic parameters of the gamma-ray loud sources (Dondi & Ghisellini, 1995, Cheng et al. 1999, Fan et al., 1999). Therefore, the optical monitoring is important for understanding the physical processes in the sources, not only in the optical band itself but also in other bands.

Since 1997, blazars have been monitored at the Abstumani observatory, Gorgia (see Kurtanidze & Nikolashvili 1999). In the September and October 2000, we monitored some gamma-ray blazars with the 70-cm telescope in Abstumani Observatory, Georgia. In this paper, we present the observational results. In section 2, we will describe the observations and data reduction, wheras in section 3, we will provide a brief conclusion.

OBSERVATIONS AND RESULTS

Data Reduction

Observations were carried out with Peltier cooled ST-6 CCD imaging Camera attached to the Newtonian focus of the 70-cm meniscus telescope (1/3) located at Abastumani Observatory, Georgia. All observations are performed with filters combined of glass, which match the standard B,V (Johnson) and R_C, I_C (Cousins) bands well (e.g. Kurtanidze & Nikolashvili 1999). The primary data analysis software system used are IRAF, MIDAS and STARLINK installed on Pentium PC (Linux 4.2, 200 MHz MMX, 64 MB, 1.3 and 6.4 GB)

To determine the magnitudes, we used stars in the field of the target sources. We determined the differential magnitudes of the target source and the comparison stars and those between the comparison stars, namely, O-S1 and S1-S2, from the instrumental magnitudes of the target object (O), the Star 1 (S1), and star 2 (S2). When there are more than two comparison stars in the field, we have calculated their mutual differential magnitudes and then determined their deviations, and then we choose the two stars which show the smallest deviation as our comparison stars, S1 and S2.

The curves S1-S2 indicate observational uncertainties in variability of the stars. The variability of the target object is investigated by means of the variability parameter, C (see Cellone et al. 2000). The rms errors are calculated as we did in our previous paper (Fan et al. 2001). During our observations, some objects displayed variations. We will discuss them detail below.

Results

PKS0420-014, OA 129

This is a strongly variable quasar in the optical band. The light curve from April 1969 to January 1986 was shown in the paper by Webb et al. (1988). Variations over time scales of a few days to about 20 days were detected in the source. Wagner (1995) suggested the fast flux variations with time scales of the order of 1-10 days. A 2.64 mag fall over 40 days from September 15 to October 15, 1995 as well as a short time scale variability of 0.12 mag in 40 minutes were observed by Villata (1997). Variation over similar short time scale was also noticed by Xie et al. (2001), who detected a fall of 0.62 mag within 49 minutes.

In our observation period JD 2451820 to JD 2451825, it showed intraday variation. A fall of 0.42 mag in R band from R = 15.52 mag to R = 15.94 mag was detected between JD 2451823.5 and JD 2451824. (see Fig. 1, where the upper panel stands for the difference between the target object and the comparison star, the lower panel for the difference between the two comparison stars), which suggests a doubling time of ΔT_D = 2.85 days.

S5 0716+714

S5 0716+714 is classified as a BL Lacertae object. A large amplitude variation of 3.3 mag was noticed by Biermann et al. (1981). A similar variation of 2.5 mag was detected by Heidt & Wagner (1996). On 1995 January 8 a burst with a duration of 6 hours was detected (Qian, Tao, and Fan, 2000). Intraday variations were also found in radio band (Quirrenbach et al. 1992).

During period of JD 2451825 to JD 2451834, this source showed brightness fluctuations. A fall of 0.7 magnitudes over a time scale of 3.1 days followed by a brightness increase of 0.73 mag over a time scale of 3.0 days is clearly shown in Fig. 2, suggesting that the doubling time scales are ΔT_D = 4.79 days and ΔT_D = 4.35 days, respectively. The falling and increasing doubling time scales are quite similar for this object.

PKS 2200+420, VRO42.22.01, BL Lacertae

The prototypical object of its class lies in a giant elliptical galaxy at a redshift of 0.07. It has been observed for about 27 years in the infrared band (see Fan et al. 1998a) and about 100 years in the optical (Fan et al. 1998b) band. A 14-year period and a maximum optical variation of ΔB = 5.31 are found from the B light curve (Fan et al. 1998b). It is a well studied object in the optical band (see Bloom et al. 1997; Miller et al. 1999; Nikilashvili et al. 1999a,b; Tosti et al. 1999; Lainela et al. 1999; Ghosh et al. 2000; Fan et al. 2001 and references therein).

In Oct 2000, it showed intraday variations as shown in Fig. 3. A rapid change of 0.3 mag over a time scale of 3 hours was detected on Oct. 9, which corresponds to a doubling time scale of ΔT_D = 11 hours.

1ES 2344+514

1ES 2344+514 (z = 0.044) was only recently identified as a BL Lac object (Perlman, 1996). It was discovered in TeV gamma-rays(> 350 GeV) with Whipple Observatory telescope and identified as a TeV BL Lac object (Catanese et al. 1998). But there is sparse data in other bands for this source. Miller et al. (1999) observed the object in the optical band and found a positive detection of microvariability.

In Oct 2000, we observed this source for two weeks, no clear variation was detected with C = 1.2 (see Fig. 4).

DISCUSSION AND CONCLUSION

We observed several gamma ray loud blazars at Abstumani Observatory in Oct 2000, and found variability in PKS 0420-014, S5 0716+714, BL Lacertae and CTA 102. But we did not find variations from other objects, which is partially caused by the fact that we did not get enough data from them. It is possible that those objects were in their quiescent state in our observation period or that they are relatively stable as it is the case for 3C 454.3 (Zhang et al 2001). Regarding to the TeV gamma-ray loud blazars, they are not so variable as GeV gamma-ray loud blazars.

ACKNOWLEDGEMENTS

This work is supported by the NSFC(19973001), the National 973 Project of China (NKBRAF G19990754) and the Outstanding Researcher Awards of the University of Hong Kong, a Croucher Foundation Senior Fellowship.

REFERENCES

1. Biermann, P., et al., *ApJ* **247**, L53(1981)
2. Bloom, S.D. et al. 1997,*ApJ* **490**, L145(1997)
3. Bottcher, M., Collmar, W., *A&A* **327**, L57(1998)
4. Carini M. T., Miller H. R., Noble J. C., Goodrich B. D. *AJ* **104**, 15(1992)
5. Catanese M., et al. *ApJ* **501**, 616(1998)
6. Cellone et al. *AJ* **119**, 1534(2000)
7. Cheng K.S., Fan J.H., Zhang L., *A&A* **352**, 32(1999)
8. Comastri, A. et al. *ApJ* **480**, 534(1997)
9. Dondi L., Ghisellini G., *MNRAS***273**, 583(1995)
10. Fan J.H., et al. *A&AS* **125**, 525(1997)
11. Fan J.H., Xie G.Z., Pecontal E., et al. *ApJ* **507**, 173(1998)
12. Fan J.H., Xie G.Z., Bacon R., it A&AS **136**, 13(1999)
13. Fan J.H., Lin R.G., *ApJS* **121**, 131 (1999),
14. Fan J.H., Lin R.G., *ApJ* **537**, 101(2000)
15. Fan J.H., Qian, B.C., Tao, J. *A&A*, (in press)(2001)
16. Ghisellini, G. et al. *A&AS* **120**, 503(1996)
17. Ghosh K.K., Ramsey B. D., Sadun A. C. et al. *ApJ* **537**, 638(2000)
18. Hartman, R.C., Bertsch, D.L., Bloom, S.D. et al. *ApJS* **123**, 79(1999)
19. Heidt J., Wagner S. *A&A* **305**, 42(1995)
20. Kurtanidze, O., Nikolashvili M.G, in *C.M. Raiter et al.(eds.) OJ-94 Ann. Meeting 1999*, p25(1999)
21. Lainela M., Takalo L.O., Sillanpaa A. et al. *ApJ* **521**, 561(1999)
22. Massaro E., Nesci R., *Blazar Data News* **24** (1999)
23. Miller H.R., et al. *ASP Conf. Ser.* **159**, 75(1999)
24. Miller H.R., Noble J.C. *APS Conf. Ser.* **110** (1999)
25. Nikolashvili M.G., et al. 1999a, in *C.M. Raiter et al.(eds.) OJ-94 Ann. Meeting 1999*, p33 (1999a)
26. Nikolashvili M.G., et al., in *C.M. Raiter et al.(eds.) OJ-94 Ann. Meeting 1999*, p36(1999b)
27. Perlman, S.M., et al. *ApJS* **104**, 251(1996)
28. Qian B.C., Tao J., Fan J.H., *PASJ* **52**(2000)
29. Quirrenbach, A. et al. *A&A* **258**, 279(1992)
30. Raiteri C.M., et al.,in *C.M. Raiter et al.(eds.) OJ-94 Ann. Meeting 1999*, p76(1999)
31. Romero, G.E. et al. *AJ* **120**, 1192(2000a),

32. Romero, G.E. et al. *A&AL* **360**, L47(2000b)
33. Sillanpaa, A., Takalo, L., Pursimo, T. et al. *A&AL* **315**, L13(1996)
34. Takalo, L.O. *ASP Conf. Ser* **159**, 253(1999)
35. Teraesranta H., Tornikoski M., Mujunen A. et al. *A&AS* **132**, 305(1998)
36. Tosti G. & Nucciarelli, G. *Blazar Data News* **22** (1999)
37. Urry, C.M. 1999, *ASP Conf. Ser.* bf159, 3(1999)
38. Villata M., Raiteri C.M., Tosti G., et al., it bmtm.proc, 73(1999)
39. Villata M., Raiteri C.M., Ghisellini G., et al., it A&AS **121**, 119(1997)
40. Wills D., Wills B.J. 1981, Nat 289, 384(1981)
41. Webb, J.R., Smith, A.G., Leacock, R.J. et al. *AJ* **95**, 374(1988)
42. Webb J.R., Freedman I., Howard E., et al. *AJ* **115**, 2244(1998)
43. Xie G.Z., et al., *ApJ*, (in press) (2001)
44. Zhang S., Collmar W., Schonfelder V. et al., (in this proceeding)(2001)

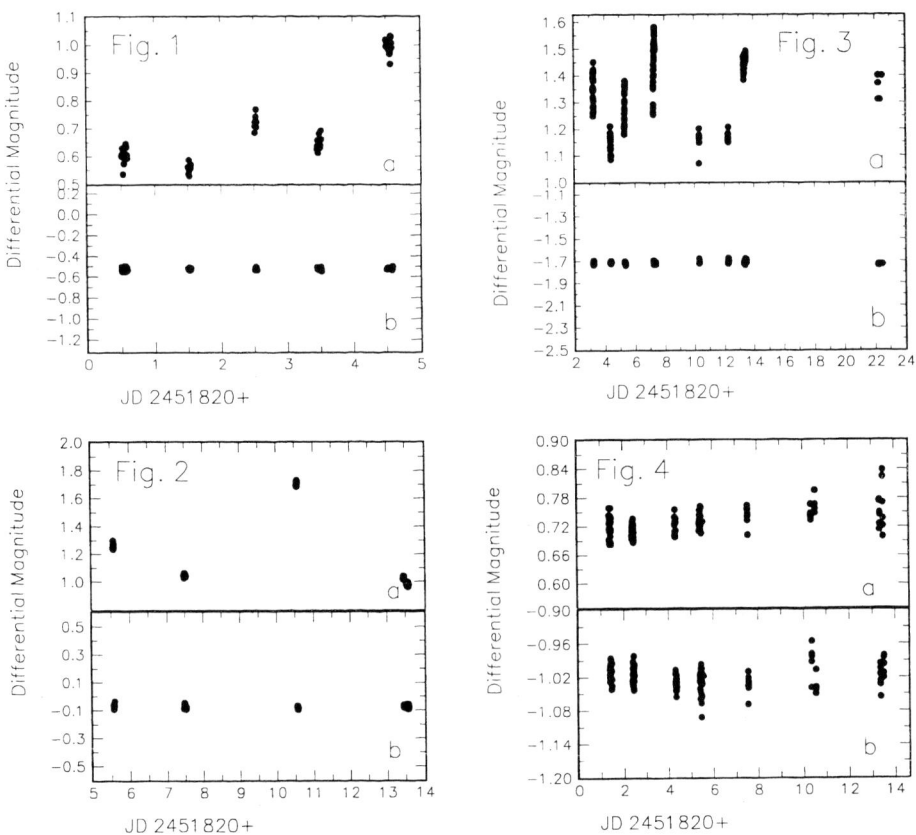

FIGURE 1. Light curves of 4 blazars

The spectral energy distribution of Centaurus A (NGC 5128)
– A summary of all observations including all CGRO results –

Helmut Steinle

Max-Planck-Institut für extraterrestrische Physik
Postfach 1312, 85748 Garching, Germany

Abstract. Due to its proximity and its importance for the understanding of active galaxies and their active nuclei (AGN), Centaurus A has been observed frequently within the last 150 years in all accessible wavelength bands. Thus a wealth of data exists which has been compiled into the "NASA Extragalactic Database" (NED). Missing from this compilation are to date almost completely the results of recent high energy observations as e.g. obtained by the Compton Gamma Ray Observatory (CGRO). A combination of those recent high energy results with all other observations in the NED enables us for the first time to establish the important spectral energy distributions (SEDs) of this closest AGN in different emission states. The combined data have been analyzed to produce SEDs which are from contemporaneous data and in addition an attempt was made to derive SEDs which are spatially resolved, i.e. SEDs from observations which can resolve nucleus and jet in Cen A are treated separately.

INTRODUCTION

The elliptical galaxy NCG 5128 is the stellar body of the giant double radio source Centaurus A (Cen A). With a distance of only 3 – 4 Mpc [3], Cen A is the nearest active galaxy. It contains an active nucleus (AGN) and a jet with a large inclination ($\sim 70°$) to the line-of-sight which is detected in all wavelength bands where the spatial resolution is sufficient. Cen A belongs to the Fanaroff-Riley type I galaxies and is often also classified as a Seyfert 2.
Its proximity makes Cen A uniquely observable among such objects and it is a very well studied and frequently observed galaxy in all wavelength bands. Its emission is detected from radio to high-energy gamma-rays [7,6,4] making it the only radio galaxy detected in MeV gamma-rays. All other AGN detected in MeV gamma-rays (and identified) are blazars [5] where the jet is aligned almost parallel to our line-of-sight. Because Cen A is seen under a much larger angle, it may be a representataive of the many other "normal" active galaxies which are just too far

away to be detected with present day instruments sensitive in gamma rays.
To study the global spectral energy distribution (SED) of Cen A over all available frequencies (energies) gives insight into the emission processes in AGN and may even provide hints to the source of the cosmic diffuse background at gamma-ray energies.

DATA

All available data have been combined into Fig. 1. About 40 % of the data (122 data points as of March 2001) are from the NASA Extragalactic Database (NED) [8]. This data base is rather complete up to about 10^{18} Hz (4 keV (EINSTEIN data)), but lacks all high energy observations. Thus all available data from the Compton Gamma-Ray Observatory (CGRO) taken during its more than 9 years of operation and very-high-energy (VHE) observations summarized by [4] have been added to the data set so that almost 300 data points are now available.

SPATIAL RESOLUTION OF THE OBSERVATIONS

Because Cen A is so close, the galaxy can be resolved into the nucleus and the outer regions, including the jet, with many of the instruments used. However, especially the instruments observing in the gamma-ray regime lack this resolution (OSSE several degrees, COMPTEL few degrees, EGRET half degree; all on board CGRO). Many authors, however, assume that the high energy emission observed can only originate in the nucleus and that emission from the jet is not visible if the object is viewed far from the jet axis (as is the case in Cen A with a viewing angle of $\sim 70°$). Therefore, other than in cases where the spatial resolution of the observations was unknown, the CGRO data are included in the plots of the nuclear data, but they are marked differenly.

TEMPORAL RESOLUTION OF THE OBSERVATIONS

Centaurus A is known to be a highly variable object in all wavelength bands and to show distict emission states [1,2,11,9] Therefore it is very important to measure complete SEDs simultaneous at a given time and in the different emission states. This is mandatory to avoid confusion in the interpretation of the data and difficulties when models are fitted to the data. However, simultaneous multiwavelength observations covering a large interval in frequencies have so far only been organized once in 1995 [10] when Cen A was observed in a low emission state. All other data have been taken at random times.
A problem related to the variability is the fact, that especially observations with low sensitivity instruments require very often long integration times, which are much longer than the typical time scales for the Cen A variability. Gamma-ray

measurements by the instruments on board CGRO lasted typically several weeks, whereas Cen A is known to be variable in the adjacent hard X-ray band on time scales of less than a day.

CLASSIFICATION OF DATA

To help to draw conclusions from this large collection of data (Fig. 1) in a reasonable manner, the data have been separated into the following groups:

- according to spatial resolution:
  - spatial resolution unknown
  - spatial resolution not sufficient to resolve Cen A
  - spatial resolution sufficient to observe the nucleus alone or if it is reasonably assumed that the emission is from the nucleus only (Fig. 2)
- simultaneous observations:
  - observations without exact date or averages of many observations
  - simultaneous observations (including "long" observations of low sensitivity instruments as e.g. the gamma-ray instruments on board CGRO) (Fig. 2)

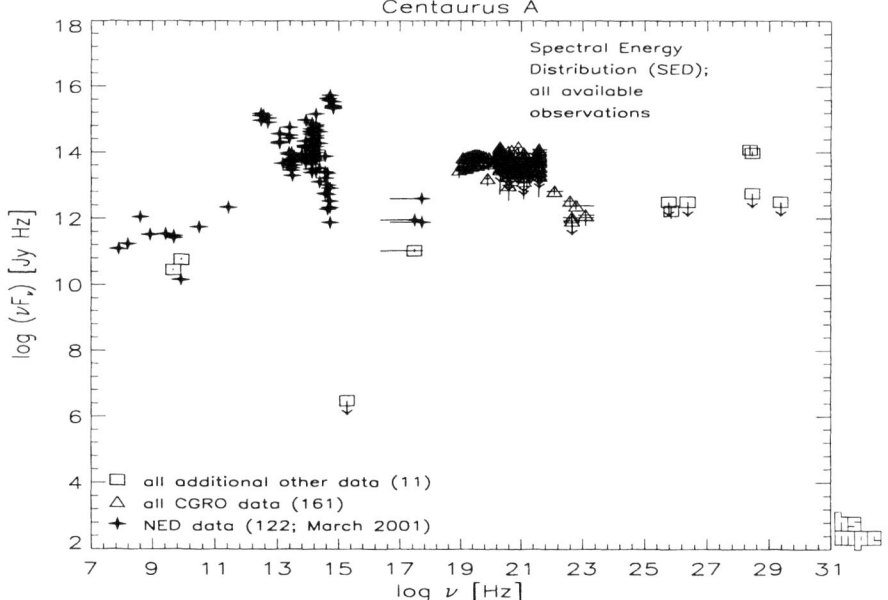

FIGURE 1. All available data from NED, CGRO and other observations

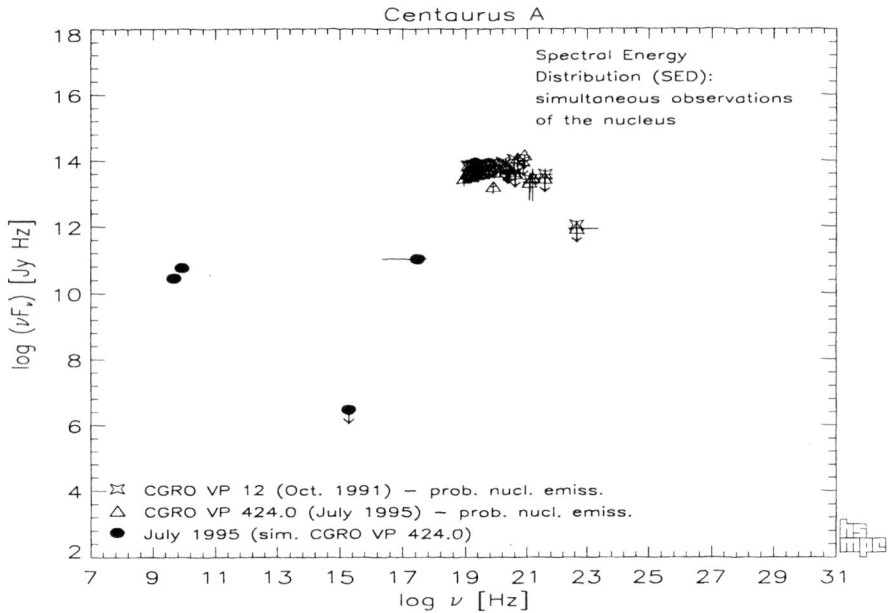

FIGURE 2. All simultaneous observation of the nucleus (see text).

Due to the limited information available on some of the data, an analysis of the original papers reporting the observations is necessary and will be conducted in the future.

RESULTS

In Fig. 1, the global structure of the spectral energy distribution of Cen A shows the typical two "bumps" which are usually (for Blazars see e.g. [12]) attributed to synchrotron emission and Compton-scattering for the low frequency (here: $\sim 10^{14}$ Hz) and high frequency peak (here: $\sim 10^{20}$ Hz) respectively. Possibly there are two more "bumps" at very low and very high frequencies. However at low frequencies, this impression may be caused by the scatter of the data points. The simultaneous observations in 1995 (see Fig. 2) do not support the presence of a "bump" at lower frequencies. On the high frequency side of the SED, the two detections of Cen A at $\sim 10^{28.5}$ Hz have been questioned and await confirmation from instruments available soon with much more sensitivity.

As one can see from the figures, despite the huge amount of data, only few data sets end up in the most interesting Figure 2. This shows dramatically the lack of coordinated simultaneus observations, an omission which hardly can be corrected, as the Compton Gamma-Ray Observatory, which covered a very important spectral region with its four instruments and which contributed many important measurements, was eliminated before further scheduled coordinated observations had taken place. No near-term future gamma-ray instrument will be able to close this gap in energy and in observational data.

ACKNOWLEDGEMENTS

I thank R.C. Hartman and F.W. Stecker for pointing out an error which I made in the plots of the EGRET data in the first version of the paper. This research has made use of the NASA/IPAC Extragalactic Database (NED) which is operated by the Jet Propulsion Laboratory, California Institute of Technology, under contract with the National Aeronautics and Space Administration.

REFERENCES

1. Bond I. A., Ballet J., Denis M., et al., *Astron. Astrophys* **307**, 708 (1996).
2. Baity W. A., Rothschild R. E., Lingenfelter R. E., et al., *Astrophys. Journ.* **244**, 429 (1981).
3. Hui X., Ford H. C., Ciardullo R., Jacobi G. H., *Astrophys. Journ.* **414**, 463 (1993).
4. Clay R.W., Dawson B.R., Meyhandan R., *Astropart. Phys* **2**, 347-352 (1994).
5. Collmar W., Bennett K., Bloemen H, et al., *Astrophys. Lett. Commun.* **39**, 57/(525) (1999).
6. Israel F.P., *Astron. Astrophys. Review* **8**, 237 (1998).
7. Johnson W.N., Zdziarski A.A., Madejski G.M., et al., "Seyferts and Radio Galaxies", in *4th Compton Symposium*, edited by C.D. Dermer, M.S. Strickman, J.D. Kurfess, 1997, AIP Conf. Proc. **410**, 283
8. NASA Extragalactic Database (NED), URL http://nedwww.ipac.caltech.edu/
9. Steinle H., Bennett K., Bloemen H., et al., *Astron. Astrophys.* **330**, 97-107 (1998).
10. Steinle H., Bonnell J., Kinzer R.L., et al., *Adv. Space Res.* **23**, 911 (1999).
11. Turner T.J., George I.M., Mushotzky R.F., Nandra K., *Astrophys. Journ.* **475**, 118 (1997).
12. Urry C.M., *Adv. Space Res.* **21**, 89 (1998).

The Gamma-Ray Horizon

Tanja M. Kneiske*, Karl Mannheim* and Dieter H. Hartmann[†]

*Universitäts-Sternwarte Göttingen
Geismarlandstrasse 11, DE-37083 Göttingen, Germany
[†]Clemson University, Departement of Astrophysics and Astronomy, Clemson, SC 29634-0978, USA

Abstract. We discuss several effects due to absorption of high energy gamma-rays by pair production with ambient background photons. This absorption process leads to the concept of the gamma-ray horizon, which is the energy dependent distance beyond which the optical depth due to this process exceeds unity. At low redshift we investigate the cut-off in the spectrum of Mkn501 (z=0.034), which is observed to be in the vicinity of $E_\gamma = 10$ TeV. To date there are no GeV-TeV sources with well established spectra and redshifts larger than that of Mkn501. Therefore we can only consider absorption for hypothetical sources as a function of redshift, to predict the locii of the spectral cut-offs for various redshifts. Our study shows that absorption of photons above 50 GeV is severe for redshifts beyond ≈ 3. Thus to sample high energy sources to even larger redshifts will require observations well below 50 GeV. In this energy regime the GLAST experiment, to be launched in 2005, will provide a significant improvement in sensitivity over previous missions. Observations from the ground with the next generation of Cherenkov telescopes (MAGIC, VERITAS, HESS...) will also provide increased sensitivity and, more importantly, lower thresholds. The combined effort of space- and ground-based high energy observatories is expected to provide data on a large sample of GeV/TeV sources, which will shed light on the central engines of active galaxies and probe the metagalactic radiation field through the pair creation opacity. High energy radiation from a variety of sources throughout the universe accumulates to the diffuse extragalactic gamma-ray background, as measured by EGRET. The spectrum of this background is also effected by absorption due to pair production. Our study shows that this effect occurs above ≈ 100 GeV, just beyond the reach of the EGRET experiment.

INTRODUCTION

The EGRET experiment aboard the Compton Gamma-Ray Observatory detected many blazars at GeV energies [1]. Only a few of these blazars were detected at TeV energies with existing Cherenkov telescopes. This lack of detection is probably not the result of reduced intrinsic emission, but most likely due to intergalactic absorption. Pair production modifies the spectra of distant sources radiating very high gamma-ray photons. Blazars, which are a subset of AGNs, exhibit strong vari-

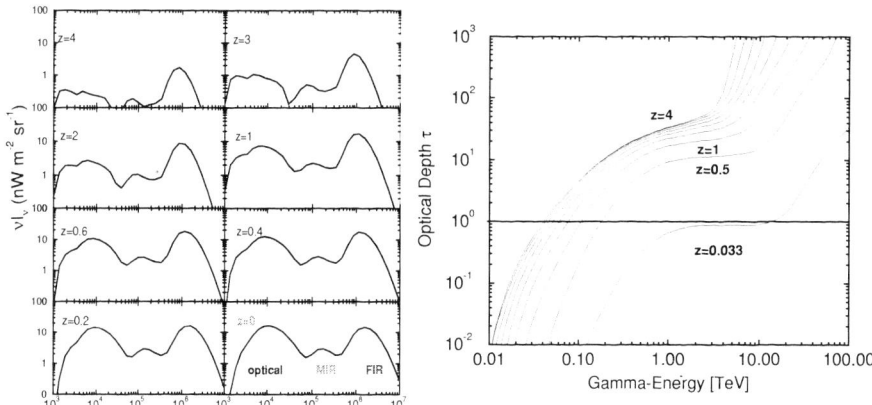

FIGURE 1. *Left panel*: Evolution of the Metagalactic Radiation Field from the IR to optical wavelengths [λ in Å](see [5]). The present day EBL, which is the MRF at $z = 0$, is reproduced with reasonable accuracy. *Left panel*:Optical depth to pair creation as a function of energy for sources at various redshifts. The solid horizontal line marks an optical depth of $\tau = 1$, which for Mkn501 (z=0.034) is reached at photon energies above 10TeV.

ablity and weak spectral features. For Mkn501, a very nearby blazar (z=0.034), high quality spectral data were obtained with HEGRA (e.g. [2], [3]). Based on these observations [4] estimated the intrinsic spectrum of Mkn501 by correcting for extinction by pair creation. These authors found that the intrinsic spectral shape is consistent with a power law of slope $\alpha = 2.00 \pm 0.03$. This approach requires a detailed model of the evolving metagalactic radiation field (MRF). We recently developed such a model [5] based on optical emission from evolving galaxies and infrared(IR) emission due to reprocessing of star light in dusty interstellar medium. For high energy sources at low redshift photon interaction with the IR background dominates, whereas gamma-rays from sources at high redshift preferentially pair create off the UV and optical field. Therefore it is important to simulate the MRF over a large frequency domain, from the optical to the UV band. There are several models in the literature that describe the present day extragalactic background light (EBL) [6] [7] [8]. Many of these models do not provide details of how this background evolves in time, which is essential for the study at hand. The optical depth for gamma-rays in the universe has been calculated by [9] using a semiempirical model, or by [10] based on structure formation scenarios. Our model utilizes direct observations of the luminosity density as a function of redshift [5]. Below we describe the effect of pair creation off photons from the MRF(z) for Mkn501 and also consider the cumulative gamma-ray background.

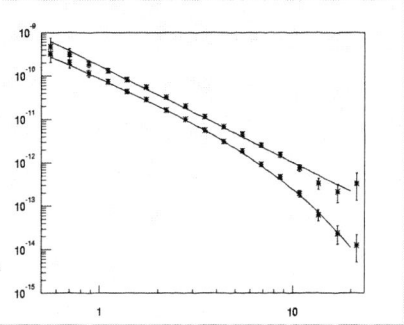

FIGURE 2. Recent observations of Mkn501 at TeV energies. Data (lower points) taken from [3]. The upper points follow from the data after the absorption correction is applied for $z = 0.034$. The solid lines are fits through the points.

GAMMA-RAY ABSORPTION

The kinematics of photon-photon pair production requires

$$E_\gamma E_{MRF} \geq 4(m_e c^2)^2 \approx 1 \text{MeV}^2. \quad (1)$$

where $\sqrt{E_\gamma \cdot E_{MRF}}$ is the center of mass energy, which must be larger than the rest mass energy of the electron-positron pair. Equation 1 implies that 10 TeV photons will preferentially be absorbed by NIR photons. On the other hand, 10 GeV photons interact mostly with 100 eV photons (UV). The probability for this process is proportional to the distance of the source, and the evolution of the MRF photons.

$$\tau_{\gamma\gamma}(E_\gamma, z_s) = c \int_0^{z_q} \int_0^2 \int_{\epsilon_{gr}}^\infty \frac{dl}{dz'} \frac{\mu}{2} n(z, \epsilon) \\ \sigma_{\gamma\gamma}(E_\gamma, \epsilon, \mu, z') \, d\epsilon \, d\mu \, dz' \quad (2)$$

Let us take a closer look at $n(z, \epsilon)$, in particular in the NIR and UV part of the sprectrum. Most of the low energy radiation is emitted by galaxies, while only a few percent is due to AGNs and other objects, such as supernovae. We determine the MRF using the global star formation rate history in conjunction with population synthesis models from [11]. To determine the emissivity in the IR we include, in a simle way, the absorption of star light by dust in the ISM and treat the reemission in the IR with three independent blackbody components. We also consider the effects of metallicity, which is essential for obtaining a good fit of the present day background. The evolution of the MRF is shown in Figure 1 (left panel) and the corresponding optical depth functions are shown along side (right panel).

Figure 2 shows the spectrum at TeV energies for the blazar Mkn501. The data clearly show a turnover around 10 TeV. To estimate the intrinsic spectrum of

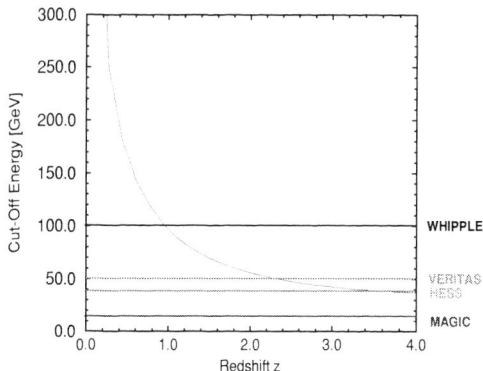

FIGURE 3. Cut-off energy as a function of redshift for an optical depth of 1. Also shown are the lower threshold energies of the Whipple telescope and three next generation Cherenkov telescopes.

Mkn501 each data point is multiplied by the factor $exp(-\tau)$. This extinction correction thus reveals that the intrinsic high energy spectrum of this blazar is consistent with a single power law, which has been noted before. Such a simple power law spectrum would be consistent with theoretical predictions beased on the self-synchrotron compton model. However uncertainties in the MRF model are still substantial (factor two), especially at high redshifts where do not yet know enough about the evolution of galaxies and the intergalactic medium. In particular, the evolution of metallicity with redshift is highly uncertain and could significantly modify the MRF and thereby the extinction correction. We therefore believe that power law gamma-ray spectra of blazars are not yet firmly established.

THE GAMMA-RAY HORIZON

We do not observe blazars at high redshift with existing Cherenkov telescopes. Our estimate of the extinction suggests essentialy zero flux above 50 GeV from sources past redshift of $z \approx 3$. Threshold energies of existing Cherenkov telescopes are near 100 GeV (Whipple) or larger still. New telescopes with lower threshold energies (HESS, MAGIC, VERITAS) are under developement, and expected to detect many new sources at medium and even large redshifts (see Fig 3).

THE GAMMA-RAY BACKGROUND

The pair creation absorption effect should also be visible in the spectrum of the diffuse gamma-ray background, which is believed to be the superposition of unresolved blazars [13]. We calculate the gamma-ray background with the following

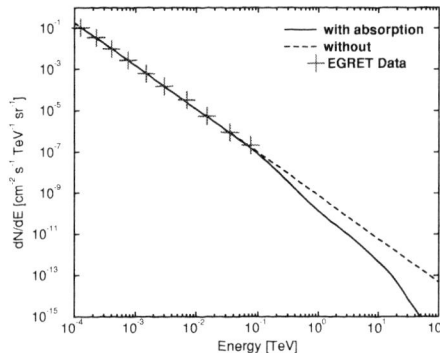

FIGURE 4. The observed, diffuse gamma-ray background extrapolated into the TeV regime with absorption evident above 0.1 TeV. Data taken from [12].

assumptions: 100% of the background is due to blazars, without density or luminosity evolution, and the cascade radiation from the absorbed energy was not included. We normalize the background to the EGRET data above 100 MeV. Although this model is rather crude the effect of pair opacity is noticeable above 0.1 TeV. The exact location of this turn-off is model dependent, but it is clear that future experiments should observe a significant decrease of the flux above 1 TeV. To utilize this effect as a probe of cosmic chemical evolution it is essential that experiments in space and on the ground operate simultaneously and ensure a significant spectral overlap.

REFERENCES

1. Hartman, R. C. et al. *ApJ* **123**, 79(1999)
2. Aharonian et al., *A&A* **349**, 11(1999)
3. Aharonian et al., *A&A* **366**, 62(2001)
4. Konopelko et al., *ApJ* **518**, 13(1999)
5. Kneiske, T. M., Mannheim, K. & Hartmann, D. H.(2001), submitted
6. Dwek,E. et al.*ApJ* **508**, 106(1998)
7. Madau, P., Pozzetti, L., & Dickinson, M., *ApJ* **498**, 106(1998)
8. Salamon, M.H., & Stecker, F.W., *ApJ* **439**, 547(1998)
9. Malkan, M. & Stecker, F.W., *ApJ* **496**, 13(1998)
10. Somerville, R. & Primack, J.R., *MNRAS* **310**, 1087(1999)
11. Bruzual, A.G., & Charlot,S., *ApJ* **405**, 538(1993)
12. Sreekumar et al. *ApJ* **494**, 532(1998)
13. Stecker, F.W. & Salamon M.H., astro-ph/0104368(2001)

Neutrino-Emission from Active Galactic Nuclei as a diagnostic tool

Claudia Schuster*, Martin Pohl* and Reinhard Schlickeiser*

Institut für theoretische Physik IV, Ruhr-Universität Bochum, 44780 Bochum, Germany

Abstract. AGN, active galactic nuclei, are luminous objects at cosmological distances, which have been reported as sources of high energy γ-rays. The emission is probably nonthermal radiation from relativistic jets belonging to the AGN. Earlier investigations of these processes have suggested that neutrinos are among the radiation products of the jets of AGN. Our calculation of the high energetic neutrino emission from the jets of AGN is based on a recently published model for γ-ray production by a collimated, relativistic blast wave [1]. In this scenario a strong electron-proton beam, the jet of the AGN, is assumed to move with bulk Lorenz factor Γ and to collide with ambient matter. In that process the beam sweeps up interstellar matter. It is important to note that the swept-up interstellar particles retain their relative velocities with respect to the jet plasma, but get isotropised in the jet rest frame by self-excited Alfénic turbulence, which leads to a deceleration of the beam because of momentum conservation. The spectral evolution of the energetic particles is determined by the interplay between the injection rate, i.e. the density of the interstellar medium, the energy losses from electromagnetic radiation, and diffusive escape. The neutrino production resulting from the proton-proton collisions in the highly relativistic plasma of the jet is calculated via pion and muon decay. γ-ray emission produced via π^0-decay and leptonic emission of secondary electrons has been discussed in [1]. Here we show that the resulting neutrino emission is strongly correlated with the simultaneously emitted γ-radiation.

CALCULATION OF NEUTRINO EMISSION

Here we calculate the neutrino emission resulting from the decay of charged pions[1] $\pi^\pm \to \mu^\pm + \nu_\mu(\overline{\nu}_\mu)$ and the emission arising through the subsequent decay of charged muons $\mu^\pm \to e^\pm + \overline{\nu}_\mu(\nu_\mu) + \nu_e(\overline{\nu}_e)$. We also take into account the neutrino emission resulting from the neutron β-decay $n \to p + e + \overline{\nu}_e$ of secondary neutrons produced in the proton-proton collisions in the blast wave. The resulting emission rates for neutrino production are calculated by:

$$Q(E_\nu, t) = \int_1^\infty N(\gamma_p, t)\, q(E_\nu, \gamma_p)\, d\gamma_p \qquad (1)$$

where $N(\gamma_p, t)$ is the proton spectrum as determined by the model of Pohl and Schlickeiser [1] and $q(E_\nu, \gamma_p)$ are the source functions, which describe the respective decays analogous to Marscher et al. [2].

---

[1] The quantities in parenthesis refer to the negatively charged particles

The temporal evolution of the proton spectra $N(\gamma,t)$ can be described by the continuity equation

$$\frac{\partial N(\gamma,t)}{\partial t} + \frac{\partial(\dot{\gamma} N(\gamma,t))}{\partial \gamma} + \frac{N(\gamma,t)}{T_E} + \frac{N(\gamma,t)}{T_N} = N^*(\gamma,t) \quad (2)$$

which includes the injection rate $N^*(\gamma,t)$, the energy losses from electromagnetic radiation $\dot{\gamma}$ and particle losses arising from diffusive escape and $p \rightarrow n$ reactions, described by the timescales for losses T_E and T_N, respectively. The injection rate is $N^*(\gamma,t) = N_0\, \delta(\gamma - \Gamma)$ in the case of particle isotropisation by pure electromagnetic turbulence [1].

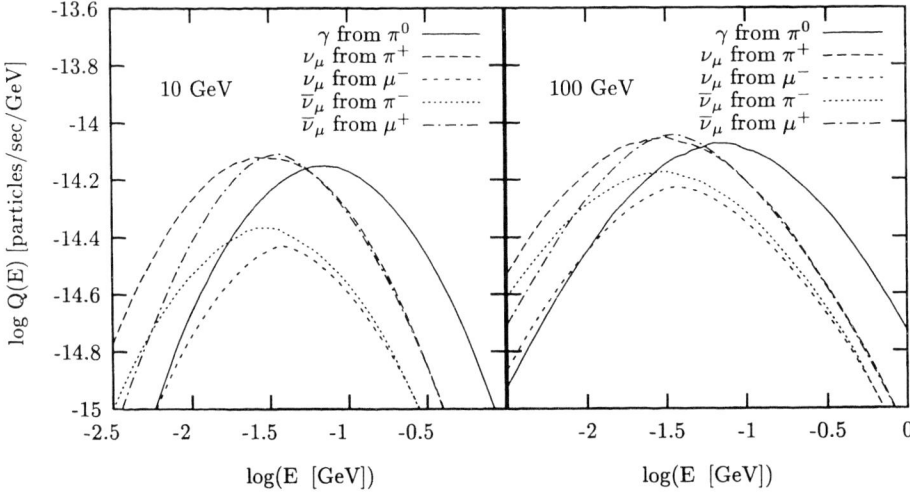

FIGURE 1. The production rate of muon neutrinos resulting from the various decay modes calculated for one proton of kinetic energy 10 GeV and 100 GeV, respectively. Additionally we show the emission of γ-rays resulting from the decay $\pi^0 \rightarrow 2\gamma$ displayed by the solid line. The emission from the positively charged pions and muons is slightly higher than the emission from the negative ones, for the respective pion production cross section is larger.

We focus our analysis on the high energy muon neutrino emission resulting from the decay mode of pions $\pi^\pm \rightarrow \mu^\pm + \nu_\mu(\bar{\nu}_\mu)$ and subsequently $\mu^\pm \rightarrow e^\pm + \bar{\nu}_\mu(\nu_\mu) + \nu_e(\bar{\nu}_e)$. Fig. 1 and 2 show the resulting production rates of neutrinos resulting from the various decay modes calculated for one proton of kinetic energy 10 GeV and 100 GeV, respectively. Here we see, that the emission rates do not change in the energy regime of a proton up to a range of 100 GeV. In Fig. 3 and Fig. 4 we show two examples of the spectral evolution of the total muon neutrino emission calculated with the blast wave model for the proton spectra. We assume a constant density of the background plasma. The production rate of γ-rays resulting from the decay $\pi^0 \rightarrow 2\gamma$ is depicted as well for reference. The bulk of muon neutrino emission occurs in the range between 100 GeV and 1 TeV and strictly follows the evolution of γ-ray production. The strong correlation between neutrino production and γ-ray production allows us to specifically search for neutrino emission from γ-ray bright AGN.

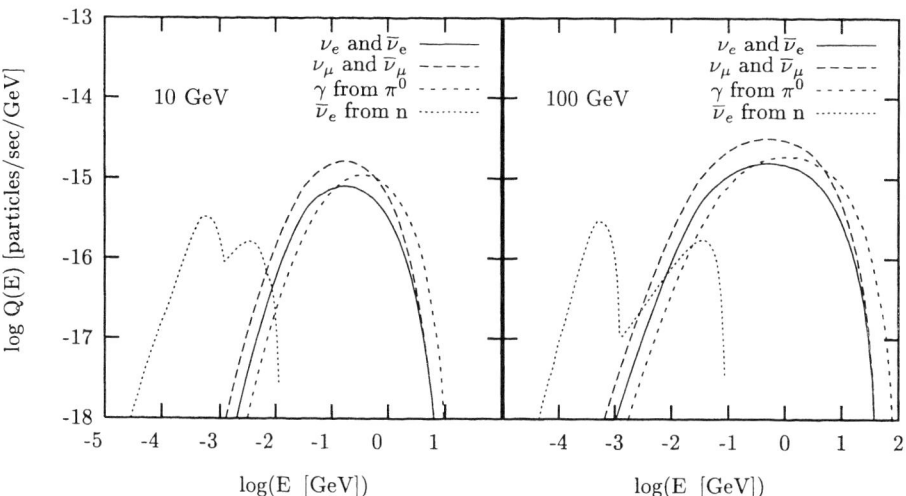

FIGURE 2. Here we show the electron neutrino emission for a single proton with the same kinetic energy as in Fig. 1. The production of $\bar{\nu}_e$ and ν_e resulting from the muon decay is displayed by the solid line. The energy of anti-electron neutrinos produced by β-decay is about 2 orders of magnitude smaller, as shown by the dotted line. For the description of the neutron source function $q_n(\gamma_n,t)$ we sum the intensities of two monoenergetic neutrons, one for the thermal proton and one for the relativistic proton. The complete rates of muon neutrinos and γ-rays are also depicted for reference.

DISCUSSION

We have calculated the neutrino emission resulting from jets of AGN, based on the assumption of the channeled blast wave model [1], for which we have investigated the decay modes of pion and subsequent muon decay. Neutrino emission resulting from the decay of secondary neutrons has been studied as well, but may be neglectable in most cases. We have shown that neutrinos resulting from β-decay of neutrons possess an energy about two orders of magnitude lower than that of the other decay channels. The bulk of the neutrino emission is expected in the energy range between 100 GeV and 1 TeV for TeV γ-ray sources. The emission rate resulting from protons in the energy regime between 1 GeV and 100 GeV does not vary much. Therefore the time dependence of the emission spectra is completely determined by the changing spectra of incoming protons. We additionally discuss some spectra resulting from the channeled blast wave model of Pohl and Schlickeiser [1] and observe that their shape is determined by the ratio between the cooling rate of the particles in the blast wave and the deceleration rate of the blast wave itself. Neutrino emission should be correlated with the emission of γ-rays. This allows us to distinctly look for neutrino emission from the jets of AGN by using the TeV γ-ray light curves to drastically reduce the temporal and spatial parameter space in the search for neutrino outbursts. Given the observed TeV photon fluxes from nearby BL Lacs the neutrino flux can exceed the atmospheric background flux and therefore be detectable with future neutrino observatories. We have estimated the neutrino detector

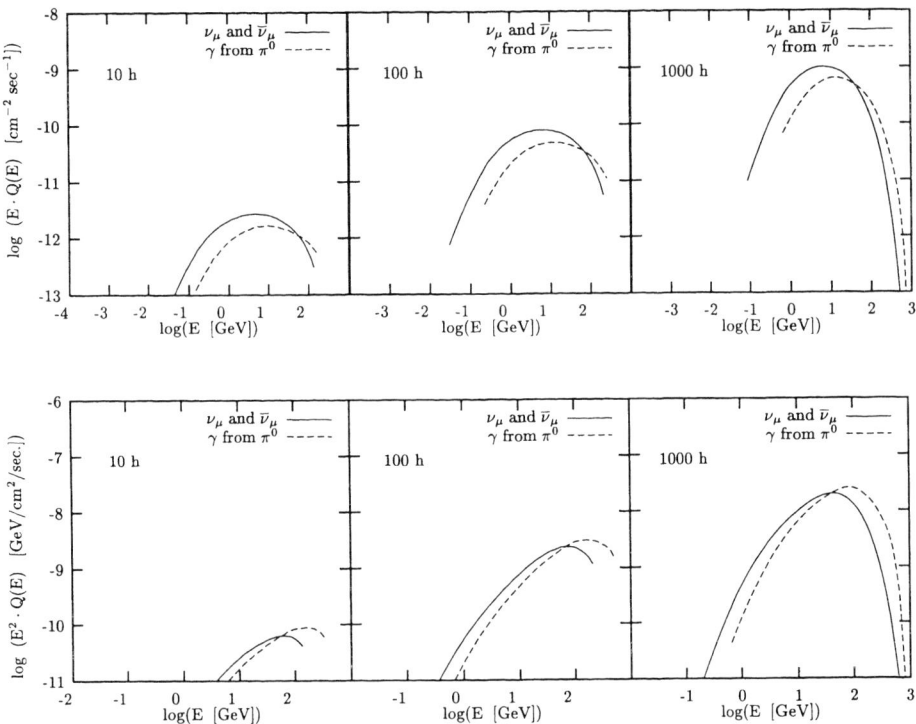

FIGURE 3. The evolution of the muon neutrino emission resulting from the protons in the blast wave in comparison with the γ-ray spectra. In this example the following parameters have been used: The radius of the plasma disk is $R = 10^{14}$ cm, the thickness of the disk is $d = 3 \cdot 10^{13}$ cm, and the initial Lorentz factor $\Gamma_0 = 300$. The constant densities inside and outside the jet are $n_b = 5 \cdot 10^8$ cm^{-3} and $n_i^* = 0.2$ cm^{-3}, respectively. Note that the mentioned time refers to the observer frame and therefore depends on the viewing angle θ, which we choose to be $0.1°$ in this example. The emission is calculated for a redshift of the AGN of $z = 0.5$. In the top row we depict the F_ν spectra and in the bottom row the νF_ν spectra. Obviously the spectral evolution of the neutrino spectra follows strictly the γ-ray production.

event rate for a typical TeV γ-ray source to be about one per month [3] for the planned ICECUBE experiment, which is the same as the event rate of atmospheric neutrinos per angular resolution element.

ACKNOWLEDGMENTS

Partial support by the Bundesministerium für Bildung und Forschung through the DESY, grant *05 CH1PCA 6*, is gratefully acknowledged.

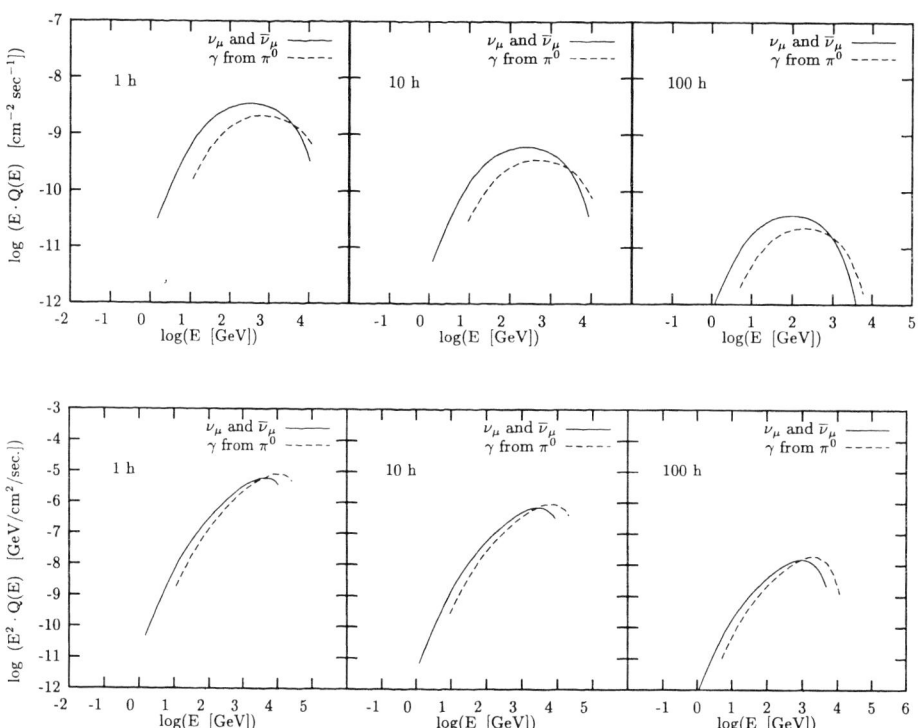

FIGURE 4. The emission spectra calculated with another set of parameters, the most important of which is the viewing angle $\theta = 2°$. In this case we use for the radius of the plasma disk $R = 2 \cdot 10^{15}$ cm and $d = 10^{14}$ cm for the thickness of the disk. The density in the jet have changed to $n_b = 10^8$ cm^{-3} and for the density outside the blast wave $n_i^* = 1.5$ cm^{-3} is assumed. Again we depict the F_ν spectra in the top row and the νF_ν spectra in the bottom row. The strong dependence of the observed emission on the angle θ is due to the high Lorentz factors in our model. This is responsible for the strong rise after 1000 h.

REFERENCES

1. Pohl, M., Schlickeiser, R., *A&A* **354**, 395 (2000)
2. Marscher, A. P., Vestrand, W. T., Scott, J. S., *Astrophys. J.*, **241**, 1166 (1980)
3. Schuster, C., Pohl, M., Schlickeiser, R., *A&A*, submitted, (2001)

Space VLBI: Past, Present and Future, and Connections to Gamma-Ray Astronomy

P.G. Edwards* and H. Hirabayashi*

*Institute of Space and Astronautical Science, 3-1-1 Sagamihara, Kanagawa 229-8510, Japan

Abstract. The rapid time-variability observed at gamma-ray energies places strong constraints on the size of the emitting regions in active galactic nuclei (AGN), but the best imaging resolution remains at the other end of the electro-magnetic spectrum. Very Long Baseline Interferometry (VLBI) enables sub–parsec-scale resolution imaging of AGN, and the addition of an orbiting telescope to existing ground arrays — space VLBI — is likely before the end of the decade to have achieved 25 micro-arcsecond imaging capability. Results from the on-going VLBI Space Observatory Programme (VSOP) mission are presented, and plans are described for a next generation mission which, with a possible launch in 2008, will have good overlap with GLAST.

INTRODUCTION

In the early days of radio astronomy, the angular resolutions of a degree or so achievable with single-telescope observations resulted in difficulties in finding counterparts in other parts of the spectrum (the same situation that exists for weaker EGRET sources!). Connected-element interferometry was developed in the 1950s and early 1960s, with a central frequency standard being used to supply a reference signal to all telescopes, and with the telescope signals being combined, or correlated, in real time. The separate developments of magnetic tape recording and stability in frequency standards led to 'independent Local Oscillator' observations in the mid-1960s, the technique being known today as Very Long Baseline Interferometry (VLBI). At 5 GHz, the limit to the angular resolution that can be achieved by ground-based VLBI is ~ 1 milli-arcsecond (mas). Many radio sources contain components that remain unresolved at this resolution, particularly in the cores of flat-spectrum active galactic nuclei (AGN). The first extension of VLBI beyond an Earth diameter was made in the late 1980s in a series of experiments with a TDRSS satellite [1, 2]. The first dedicated space VLBI satellite to be placed in orbit was HALCA, launched by the Institute of Space and Astronautical Science in 1997.

THE HALCA SATELLITE AND VSOP

HALCA is the main element of the VLBI Space Observatory Programme, VSOP [3, 4]. Imaging VLBI observations are made at 1.6 GHz (λ18 cm) and 5 GHz (λ6 cm) with the HALCA satellite and arrays of ground radio telescopes to make sub–milli-arcsecond resolution images of celestial radio sources.

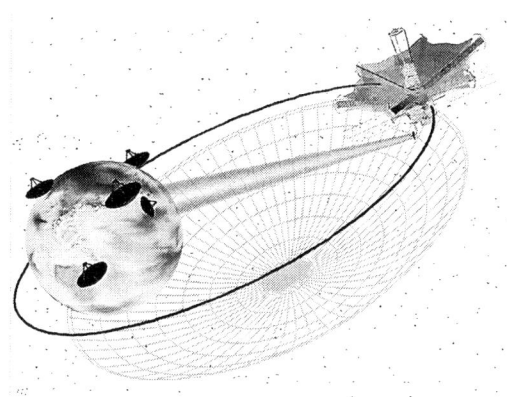

FIGURE 1. Schematic illustration of a VSOP observation: an array of ground radio telescopes and the HALCA satellite observe a source to synthesize a telescope bigger than the Earth. A real-time two-way link is used between a ground tracking station and the satellite to uplink a reference tone and down-link the scientific data. (©ISAS, used with permission)

HALCA carries an 8 metre diameter telescope in an elliptical orbit between 560 km and 21,400 km above the Earth's surface, with an orbital period of 6.3 hours. The elliptical orbit results in a wide range of baseline lengths and orientations being sampled, enabling imaging VLBI observations on baselines over three times longer than those achievable on Earth [3, 4]. Correlated flux densities of at least ∼100 mJy are required in order for sources to be detected on baselines to HALCA.

Approximately three-quarters of HALCA's observations are projects which are selected by international peer-review from open proposals submitted by the astronomical community in response to Announcements of Opportunity. These observations last typically ∼10 hours and use an array of ∼8 radio telescopes. The remaining 25% is devoted to the VSOP Survey Program, a mission-led survey of AGN at 5 GHz. Survey observations are in general of shorter duration and made with fewer telescopes. Nevertheless, the Survey is providing a large sample of data on sub–milli-arcsecond radio structures for studying the statistical properties of AGN and planning future VLBI (and space VLBI) observations [5]. Over 600 VSOP observations have been made to date, predominantly of compact extra-galactic radio sources. Of sources in the third EGRET catalog [6], VSOP observations have been made of about half of the 66 "high confidence" AGN and a third of the lower-confidence AGN identifications. Details about the VSOP project and VSOP observations are available from http://www.vsop.isas.ac.jp.

RECENT RESULTS FROM VSOP OBSERVATIONS

The X-Ray Jet of PKS 0637−752

Chandra observations revealed a prominent jet extending up to 11.5 arcseconds from the core of the $z = 0.651$ quasar PKS 0637−752 [7]. VSOP observations of

PKS 0637−752 were made in November 1997 and September 1999. Combining the VSOP observations with ground-only VLBI observations revealed that components in the jet have apparent speeds of $17.8 \pm 1.0\,c$ (for $H_0 = 50\,\mathrm{km\,s^{-1}\,Mpc^{-1}}$). This enables a lower limit of 17.8 to be placed on the Lorentz factor of the jet, and an upper limit of 6.4° for the angle between the jet and our line of sight. These are important parameters for constraining possible emission mechanisms within the jet [8].

Brightness Temperatures

VSOP observations of well-defined samples of AGN have been used to derive the brightness temperature distribution, revealing that a significant proportion have brightness temperatures greater than the theoretical limits of 10^{11}–10^{12} K, implying significant Doppler boosting [5, 9]. The lower limits on jet Lorentz factors estimated from these VSOP observations are starting to challenge numerical simulations that predict low Lorentz factor jets. An apparent relationship between intraday variability and VSOP-measured brightness temperature confirms that space VLBI resolutions are approaching the micro-arcsecond–scale structures inferred from intraday variability studies [10].

Magnetic Field Structure

Dual polarization VSOP imaging is possible for strong radio sources. Ground-only polarization observations of the blazar 1803+784 shows electric vectors that appear to be at an oblique angle to the radio jet, which is difficult to explain in standard models. However, a full-resolution VSOP image shows that the jet actually bends to the north and then back to the south, with the electric vectors following this curvature perfectly. This indicates that the magnetic field in the jet is, in fact, perpendicular to the jet, and retains this orientation as the jet bends. This may indicate that the magnetic field lines are compressed in a series of shock components, or possibly that we are seeing the dominant toroidal component of a helical magnetic field wrapped around the jet [11]. Such field configurations are expected in many theoretical models, but direct evidence for them had never before been obtained on such small scales.

LINKS WITH GAMMA-RAY ASTRONOMY

One of the biggest surprises from the CGRO mission was the predominance of the blazar class of sources at EGRET energies. Connections between the gamma-ray activity and radio flux density measurements were soon noted [12, 13]. These relationships have been studied for a number of individual sources, but the most extensive study to date has been that of Jorstad et al. [14], who monitored 42 EGRET sources at 22 and 43 GHz with the Very Long Baseline Array (VLBA) between November 1993 and July 1997. A total of 23 gamma-ray "flares", or high-states of gamma-ray emission, occurred in periods where there was sufficient VLBI data to search for correlations. Ten of these flares coincided, within 1 σ, with the extrapolated epoch of emergence of a new radio jet component from the core. A further six flares were marginal coincidences, with the gamma-ray flare within 3 σ of the radio ejection epoch. For seven cases no new component was seen to be ejected around the time of the gamma-ray flare, and although there were extenuating circumstances in some cases, it seems clear that not all gamma-ray high states are associated with a new VLBI component.

An inspired addition to this VLBI monitoring was a study of University of Michigan (single-dish) radio polarized flux density variations for these sources. As superluminal radio knots are relatively highly polarized, the emergence of a new component would be expected to be associated with a change in the polarized flux density. In fact, it was found that new components emerged near a minimum in polarized flux density, suggesting that the ejected components are cross-polarized with respect to the core [14].

The utility of the polarized flux density monitoring can be seen in the case of 3C 279. At the time of the very bright gamma-ray high-state of mid-1991 there is no evidence for the ejection of a new component in the data of Jorstad et al., however, this time does coincide with a pronounced minimum in the regular polarized flux density monitoring. (No new components were observed to emerge in the Jorstad et al. data set, however from the analysis of a much more extensive 3C 279 VLBI monitoring program it appears that new components emerged ∼6 months before and after the observed gamma-ray high state [15].)

From these studies, it was concluded that the gamma-ray flares are caused by inverse Compton scattering in the parsec-scale regions of the jet rather than closer to the core. This is consistent with the observation that gamma-ray flares occur after the onset of a mm-wave flare [13]. A secondary result, with implications for future space VLBI monitoring observations, is that the average visible "lifetime" of VLBI jet components at 22 and 43 GHz was 1.5±0.5 years.

Much still remains to be learned. Is each gamma-ray flare accompanied by a new superluminal component? Is each new superluminal component accompanied by a gamma-ray flare? Why have some sources that eject superluminal components and display variations in polarized flux density not been detected as gamma ray sources? Clearly, many of the answers will be revealed by more frequent, more sensitive, higher resolution observations of more sources!

THE FUTURE OF SPACE VLBI

The nominal lifetime of the VSOP project is five years, though, barring any catastrophic on-board failure, it is likely the the spacecraft will be able to continue for several years beyond this time. Planning for a follow-up mission to HALCA, currently dubbed VSOP-2, is underway at ISAS, in discussion with both domestic and international collaborators. Observing frequencies up to 43 GHz, cryogenically cooled receivers, an increased bandwidth and a larger telescope diameter will result in gains in resolution and sensitivity by factors of ∼10 over the VSOP mission [16]. Formal submission of the VSOP-2 proposal will take place within the next year, and launch on an ISAS M-V rocket could be as early as 2008, which would allow good overlap with the GLAST mission. (The ARISE mission, http://arise.jpl.nasa.gov, favorably reviewed in the McKee–Taylor report "Astronomy and Astrophysics in the New Millennium", is a candidate mission for the next decade.)

VSOP-2 science goals include: the study of emission mechanisms in conjunction with the next generation of X-ray and gamma-ray satellites; full polarization studies of magnetic field orientation and evolution in jets, and measurements of Faraday rotation towards AGN cores; high linear resolution observations of nearby AGN to probe the formation and collimation of jets and the environment around supermassive black holes; and the highest resolution studies of spectral line masers and mega-masers and circumnuclear disks. By the end of this decade, ground VLBI observing at 86 GHz is likely to be routine, however the highest resolution imaging possible will be VSOP-2 imaging at 43 GHz. An angular resolution of $\sim 25 \mu$as at 43 GHz will be achievable, corresponding to ~ 10 Schwarzschild radii at the distance of M87. Multi-epoch space VLBI observations of flaring gamma ray sources, backed up by regular ground-based millimetric and polarimetric monitoring, will be an important key to unlocking the secrets of these powerful objects.

ACKNOWLEDGMENTS

We gratefully acknowledge the VSOP Project, which is led by the Japanese Institute of Space and Astronautical Science in cooperation with many organizations and radio telescopes around the world.

REFERENCES

1. Levy, G.S., et al., *Science*, **234**, 187 (1986).
2. Linfield, R.P. et al., *ApJ*, **358**, 350 (1990).
3. Hirabayashi, H., et al., *Science*, **281**, 1825 (1998).
4. Hirabayashi, H., et al., *PASJ*, **52**, 955 (2000).
5. Hirabayashi, H., et al., *PASJ*, **52**, 997 (2000).
6. Hartman, R.C., et al., *ApJSupp*, **123**, 79 (1999).
7. Schwartz, D.A. et al., *ApJ*, **540**, L69 (2000).
8. Chartas, G. et al., *ApJ*, **542**, 655 (2000).
9. Tingay, S.J. et al., *ApJ*, **549**, L55 (2001).
10. Lister, M.L. et al., *ApJ in press (astro-ph/0102276)* (2001).
11. Gabuzda, D.C., *New Astr. Rev.*, **43**, 691 (1999).
12. Reich, W. et al., *A&A*, **273**, 65 (1993).
13. Valtaoja, E. and Teräsranta, H., *A&A*, **297**, L13 (1995).
14. Jorstad, S.G. et al., *astro-ph/0102012* (2001).
15. Wehrle, A.E. et al., *ApJ*, **133**, 297 (2001).
16. Hirabayashi, H. et al., "The VSOP-2 Mission", in *Astrophysical Phenomena Revealed by Space VLBI*, edited by H. Hirabayashi, P.G. Edwards and D.W. Murphy, ISAS, Sagamihara, 2000, p. 277.

Seyfert Galaxies

Modelling the R-Γ correlation in compact sources

Julien Malzac

Osservatorio Astronomico di Brera, via Brera, 28, 20121 Milano, Italy

Abstract. The observed correlation between slope Γ and the amplitude of the reflection component R in accreting black hole sources suggests that both the soft seeds photons of the Comptonisation process, and the reflection component originate from the same cold medium. Based on this assumption, several models have been proposed, that may account for the correlation. Accurate radiative transfer calculations and energy balance arguments, together with good quality data, can put some constraints and help in discriminating between them. We emphasize the problems encountered by the standard inner hot disk model to reproduce the correlation. We discuss possible alternative models.

THE HARD X-RAY SPECTRA OF BLACK HOLE SOURCES

The hard X-ray spectral properties of accreting galactic black holes sources (GBHs) in the low/hard state and Seyfert galaxies (AGN) are very similar. They both have power law spectra cutting-off at few hundred keV. These spectra are thought to form through Comptonisation process (Sunyaev & Titarchuk 1980). In addition the spectra exhibit reflection features, in particular, the Compton bump and the fluorescent iron line which are the signatures for the presence of cold matter reflecting the primary hard radiation (George and Fabian 1991; Nandra and Pound 1994). There are two parameters that are commonly used by to quantify the shape of the hard X-ray spectra of AGNs and GBHs: the photon index Γ and the reflection amplitude R. Γ is the slope of the primary Comptonisation spectrum in the 2-10 keV range. It depends both on the Thomson optical depth of the hot plasma τ, and its temperature T. The spectra are generally harder (i.e. Γ lower) when the product $T\tau$ is larger. Spectral fits indicate $\tau \sim 1$, $kT \sim 100$ keV. If the system is in radiative equilibrium, the temperature T, and thus Γ, is controlled by the amplification factor A, defined as: $A = L_h/L_s + 1$, where L_h is the power injected as heating in the hot plasma, and L_s is the incoming soft cooling luminosity. The larger A, the harder is the spectrum. The second important spectral parameter, R, is the normalization of the reflection component relative to the primary emission. R is defined so that $R = 1$ for an isotropic point source above a disc. If the hard X-ray source is isotropic, and Ω is the solid angle subtended by the reflector as seen from the hot plasma, at first order, $R \sim \Omega/2\pi$.

Observations of samples of sources as well as the time evolution of individual sources have shown a correlation between R and Γ (Zdziarski et al. 1999; Gilfanov et al. 2000). Sources with harder spectra tend to have lower reflection. Such a correlation suggests that the "amount" of reflector is linked together with the amount of cooling soft radiation entering the hot plasma source. Indeed, a larger soft luminosity means a lower

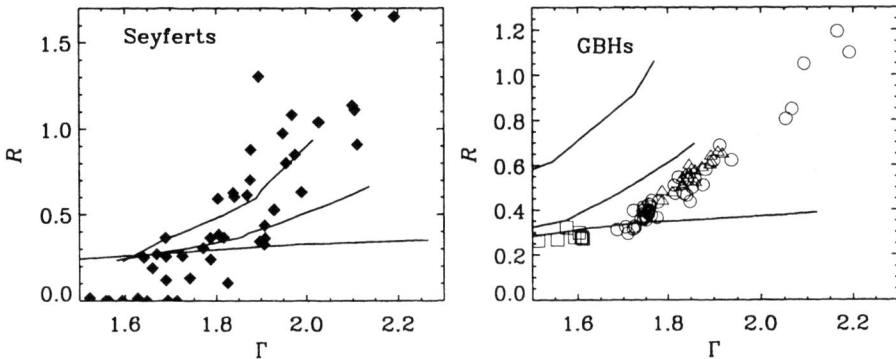

FIGURE 1. The R-Γ correlation predicted by the hot inner disc model compared with the data. Left panel: data from Seyfert galaxies (Zdziarski et al. 1999). Right panel: data from galactic black holes (Gilfanov et al. 2000). The solid curves represent the correlation obtained when the cold disk penetrates in the hot spherical cloud. The inner radius varying from $r_{in}=r_h$ (the radius of the hot cloud) down to 0. In each panel, the three curves stand for three different fixed Thomson optical depths in the hot medium; τ=0.3,1 and 3, respectively from top to bottom. The blackbody temperature of the soft photons are 5 eV and 100 eV respectively in the Seyfert and GBHs panels.

amplification and thus a softer spectrum. The simplest and most natural explanation is that the main fraction of the soft cooling photons are emitted by the reflector itself. Then, for a wide variety of configurations, a larger reflection amplitude corresponds to a larger soft luminosity entering the hot plasma. The aim of this paper is to discuss such models.

THE HOT INNER DISC MODEL

Currently, the most popular model for the hard X-ray emission of both GBHs and AGNs consists in a hot quasi-spherical Comptonising plasma, surrounded by a cold accretion disc which can overlap with the hot inner region (e.g. Poutanen Krolik and Ryde 1997; Esin et al. 1998; Gilfanov, Churasov and Revnivtsev 2000). Both the soft cooling luminosity and the reflection component originate from the disc. The main parameter determining the spectrum is the cold disc inner radius. It is allowed to move between the last stable orbit and the radius of the spherical hot plasma r_h. When $r_{in} = r_h$, the soft photons are emitted outside the hot phase and a small fraction of them enters the hot phase. When the disc inner radius decreases, i.e. the disc penetrates inside the hot phase, the cooling luminosity gradually increases and the spectrum softens. Simultaneously, R increases due to the enlarged reflection area. This model thus qualitatively accounts for the R-Γ correlation. We use the Monte-Carlo code of Malzac and Jourdain (2000) which is based on the non-linear method proposed by Stern et al. (1995a). The assumed geometry is very similar to the 'sphere+disc' geometry used in the simulations of Dove et al. (1997b). The main difference is that we allow the inner disc radius to change from zero, up to the radius of the hot phase. We assume that the disc luminosity entering the hot phase is due *only to reprocessing* of the hard X-ray radiation – i.e. in the inner parts of the cold disc, the intrinsic viscous dissipation is assumed to be negligible compare to the impinging high energy flux from the hot phase. We also assume that the disc is dense enough, so that the ionization parameter is low ($< 10^3$) and the matter can be considered as neutral. The R and Γ parameters are estimated from the simulated spectra, as described in Malzac, Beloborodov and Poutanen (2001, hereafter MBP). Our results, are presented in Fig. 1, together with R-Γ data from Seyfert galaxies and GBH sources (see caption of Fig. 1).

The model correlation appears to depend on the optical depth. This is due to the Compton smearing effect. Indeed, the reflection component originates from two contributions: (i) The *internal reflection* produced on the cold disc but *inside* the hot phase. This component is very sensitive to τ. Indeed, a fraction of the reflected photons may be Compton scattered in the hot plasma before escaping to the observer. The net effect is to decrease the apparent amount of reflection in the observed spectrum (see MBP). R is decreased roughly according to $\exp-\tau$. At large optical depths, reflection can be almost completely destroyed. (ii) The *external reflection* produced on the outer parts of the disc. Only a negligible fraction of this component has to travel through the hot phase. It is thus independent of τ. It represents an irreducible contribution of ~ 0.3, to the total reflection amplitude.

At large inner radii, the internal reflection is negligible and $R \sim 0.3$, independently of τ. When the inner radius decreases, the internal reflection becomes potentially important, and the total reflection amplitude then depends on τ. For large τ, there is almost no correlation (see curves for $\tau = 3$), Then, $R \sim 0.3$ independently of r_{in}. R values approaching unity can be obtained only for optically thin plasma ($\tau = 0.3$) at $r_{in} \sim 0$. For intermediate optical depth the maximum possible R is lower ~ 0.7. One can see from Fig. 1 that the observed range of reflection amplitudes is roughly 0–2 for AGNs and 0.2–1.2 in GBHs. In both case these are wider range than what can be produced by our model which is hardly 0.3–1.

Concerning Seyferts galaxies, even in the range of spectral parameters produced by the model, the correlation slope is not reproduced. The model correlation seems always flatter than the observed one, whatever the optical depth may be. On the other hand, for GBHs, it seems that for optical depth in the range 1–2, it is possible to qualitatively reproduce the observed correlation slope. This range of τ is moreover consistent with optical depth estimates derived from spectral fits. Taking into account the dissipation in the disc would certainly be more realistic. It requires, however, to make additional assumptions (e.g. on the fraction of energy dissipated in the Compton cloud and its size), and to relie on a specific disc model for the description of the disc intrinsic emissivity as a function of the radius. Recently, Beloborodov estimated the R and Γ dependence on r_{in}, for a standard disc model and a fixed total luminosity (see Beloborodov 2001). He found that the additional cooling increases dramatically when the inner radius decreases. This leads to a completely flat R(Γ) relation, clearly ruling out the model, for both GBHs and AGNs. We thus conclude that in GBHs the hot inner disk model may reproduce the correlation only if the viscous dissipation is negligible i.e. most of the accretion power has to be dissipated in the hot phase. In Seyferts galaxies, as long as we can be confident in the R-Γ data, this model does not account for the correlation.

DISCUSSION

It cannot be fully excluded that the problems encountered by the hot inner disk model could be due to measurement errors since the derived value of the spectral parameters are quite dependent on the spectral model as well as to eventual additional components in the observed spectra (such as absorbing column density or soft excess). Future more accurate broad band spectra measurements will allow to test this possibility.

On the other hand, if one believes the present data, it seems difficult to improve the model so that it over-passes its problems. For example the lower limit of $R \sim 0.3$ imposed by the model could become lower if the disc is strongly ionized (e.g. Ross and Fabian 1999). However, if the external disc is ionized, the internal disc should be ionized as well, and the upper limit should be decreased, making the range of predicted R even narrower. The highest predicted R could be increased by assuming that a large fraction of the reflection is produced far away, on an additional cold structure (like the torus presumably present in Seyfert galaxies, see Haardt, Ghisellini, Matt 1994), or on a flared disc. However producing a reflection as high as the highest value observed in Seyferts $R \sim 1.4 - 2$ would require the remote material to subtends a very large solid angle. A face-on torus does not produce enough reflection, even if it has a large covering angle. The cold material should have a specific geometry to produce enough reflection. Even so, the feedback from this remote cold matter would be very low and then we do not expect any R-Γ correlation. It is worth noting however that, in such case, in Seyferts the time delay between the fluctuation of the primary spectrum and the response of an extended reflector can explain most of the correlation (Nandra et al. 2000; Malzac & Petrucci in prep.). If this is really the case, the R–Γ correlation does not bring information on the physics of the central engine in AGN, and thus cannot help to constrain any model.

The alternative model of a MHD corona above an accretion disc (e.g. Galeev et al. 1979; Haard et al. 1994) disagrees with the observations of some black-hole sources in the hard state, for instance, Cyg X-1 (Gierliński et al. 1997). Moreover, MBP showed that, in the framework of the static corona model, simple changes in geometry produce an anti-correlation between R and Γ, opposite to what is observed. However, Beloborodov (1999) argued that the flaring plasma atop a cold disc is likely to acquire a mildly relativistic bulk velocity, $\beta = v/c \lesssim 0.5$, as a result of magnetic dissipation and/or radiative pressure in the corona. The bulk velocity may be directed away or towards the disc (and it may change) with the preferential direction away from the disc. The spectral characterics are then very sensitive to β. The observed $R - \Gamma$ correlation is well explained by varying β (B99,MBP).

Other possible models include the cloudlet models. In these scenarii the reflection and soft photons originate from cold dense clouds inside the hot phase (e.g. Guilbert & Rees 1988; Kunzic et al. 1997; Krolik 1998). Then, variations in the amount of cold matter mixed inside the Comptonising plasma may, in principle, produce a correlation. However, in the case of a uniform distribution of cold clouds inside the hot plasma, the radiative cooling is very efficient. The amount of cold matter is constrained to be low in order to reproduce the observed Γs (Malzac, these proceedings). Then the reflection arising from the cold clouds is negligible. An external reflector is required to reproduce the data and, again, the R-Γ correlation is difficult to explain. The cold material is then more likely to be located outside the hot phase, for example the clouds could be spherically distributed around it (Collin-Souffrin et al. 1996). the R-Γ correlation can be explained by changes in the cold clouds covering fraction (Malzac 2001).

CONCLUSIONS

The observation of a correlation between spectral slope and reflection amplitude may be of prime interest when trying to distinguish between models. The commonly accepted inner hot disk model meet serious difficulties in producing the range of observed parameters. As far as the data are believable, it is ruled out in the case of AGNs. The case of GBHs is still debatable, as long as one do not consider the internal dissipation in the disc. Alernative models matching better the data include the quasi-spherical accretion and the dynamic corona scenarii.

ACKNOWLEDGMENTS

This work was supported by the Italian MURST grant COFIN98-02-15-41.

REFERENCES

. Beloborodov, A. M., 1999, ApJ, 510, L123 (B99)
. Beloborodov, A.M., "Accretion disk models for luminous black holes", 33rd COSPAR Assembly, Adv. Space Research in press, astro-ph/0103320
. Collin-Souffrin, S., Czerny, B., Dumont, A.-M., Zycki, P. T., 1996, A&A, 314, 393
. Dove, B.D., Wilms, J., Maisack, M., Begelman, M.C., 1997b, ApJ, 487, 759
. Esin, A.A., Narayan, R., Cui, W., Grove, J.E., Zhang, S.-N., 1998, ApJ, 505, 854
. Galeev, A.A., Rosner, R., Vaiana, G.S., 1979, ApJ, 229, 318
. George, I.M., Fabian, A.C., 1991, MNRAS, 249, 352
. Ghisellini G., Haardt F., Matt H., 1994, MNRAS, 432, L95
. Gierliński, M. et al., 1997, MNRAS, 288, 958
. Gilfanov, M., Churazov, E., Revnivtsev, M., 2000, in Proc. 5th CAS/MPG Workshop on High Energy Astrophysics, in press, astro-ph/0002415 (GCR00)
. Guilbert, P.W., Rees, M.J., 1988, MNRAS, 233, 475
. Haardt, F., Maraschi, L., 1993, ApJ, 413, 507
. Haardt, F., Maraschi, L., Ghisellini, G., 1994, ApJ, 432, L95
. Kuncic, Z., Celotti, A., Rees, M.J., 1997, MNRAS, 284, 717
. Krolik, J.H., 1998, ApJ, 498, L13
. Malzac, J., 2001, MNRAS in press, astro-ph/0104273
. Malzac, J., Jourdain, E., 2000, A&A, 359, 843
. Malzac, J., Beloborodov, A., Poutanen, J., 2001, MNRAS in press, astro-ph/0102490
. Nandra, K., Pounds, K.A., 1994, MNRAS, 268, 405
. Nandra, K. et al., 2000, ApJ, 544, 734
. Poutanen, J., Krolik, J.H., Ryde, F., 1997, MNRAS, 292, L21
. Stern, B.E., Begelman, M.C., Sikora, M., Svensson, R., 1995a, MNRAS, 272, 291
. Sunyaev, R.A., Titarchuk, L.G., 1980, A&A, 86, 121
. Zdziarski, A.A., Lubiński, P., Smith, D.A., 1999, MNRAS, 303, L11 (ZLS99)

BeppoSAX observations of the radio-quiet QSO MR 2251-178

A. Orr*, P. Barr*, M. Guainazzi†, A. Parmar* and A. Young**

*Astrophys. Division of ESA, ESTEC, Postbox 299, NL-2200 AG Noordwijk, The Netherlands
†XMM SOC, VILSPA, Villafranca del Castillo, Spain
**Dep. of Astronomy, Univ. of Maryland, College Park, MD 20742, U.S.A.

Abstract.
We present the 0.1–200 keV *BeppoSAX* spectrum of MR 2251-178 observed at two epochs in 1998 separated by 5 months. Both epochs show identical spectral shape and X-ray flux. Analysis of the combined spectra allow us to confirm the presence of the ionized Fe Kα line detected by ASCA and to test the presence of reflection from ionized material. A good spectral fit is obtained when including a contribution from a mildly ionized reflector ($\xi_{0.01-100\text{keV}} \sim 1625$ erg cm s^{-1}) with a reflection normalization $R_{ion} \sim 0.11$. An exponential cut-off to the direct power-law continuum is then required at $E \sim 100$ keV.

Finally, we briefly discuss the spectral and variability properties of a second, much weaker X-ray source in the same field of view. This source, EXO 2251.1-1737, most likely is an AGN, although it has so far been classified as a star.

ARE RADIO QUIET QSOS JUST HIGH LUMINOSITY SY 1?

Some identical characteristics are observed among radio quiet QSOs and Sy 1, see e.g. the detailed study of George et al. [4]. For example, the 2–10 keV power-law spectral

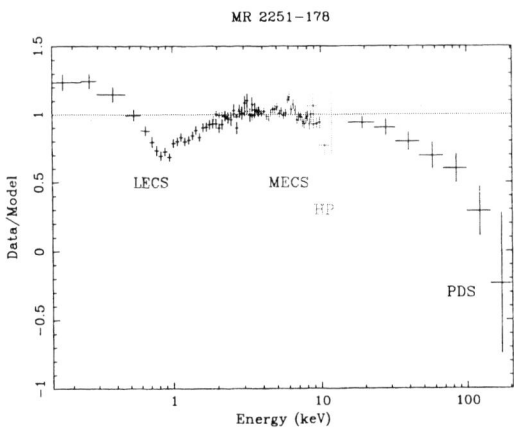

FIGURE 1. June+November 1998 *BeppoSAX* spectrum of MR 2251-178. Data-to-model ratio for a power-law fit with galactic nH

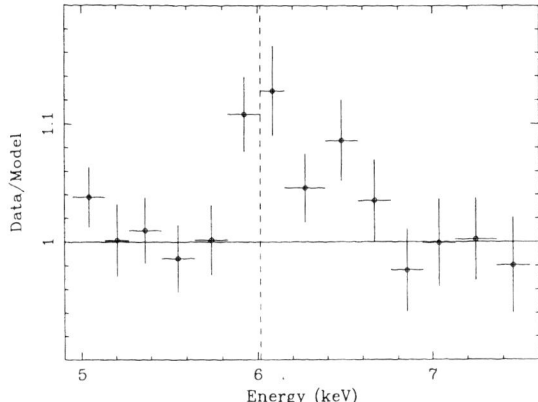

FIGURE 2. Fe Kα line profile: MECS data to model ratio in observer's reference frame. The rest frame value 6.4 keV is indicated by the dashed line. The model does not include any Fe K α line

slopes of low-z RQ QSOs are similar to those of low luminosity Sy 1s ($\Gamma \sim 1.9$–2). Furthermore, variability and warm absorber properties appear to be comparable in RQ QSOs and Sy 1s.

However, there are also some striking differences. The strong Compton reflection "hump" which is common in Sy 1s has not been clearly detected in luminous RQ QSOs, see e.g. Williams et al. [19], Lawson & Turner [6], Reeves et al. [14], George et al. [4], Reeves & Turner [15]. An analysis of ASCA data for high z RQ QSOs by Vignali et al. [18] shows that the 2–30 keV (source frame) spectra are well described by a single power-law. Nandra et al. [9] showed that the Fe Kα line emission in RQ QSOs is weaker than in Sy 1s and that the line profile and centroid energy both seem to depend on the AGN luminosity. This trend has recently been confirmed by George et al. [4].

A possible scenario is that the X-ray luminosity of the central source regulates the ionization state of the Compton "reflector". In this scenario one can make the following predictions:

1. At high ionization states the Fe Kα line is emitted at higher energies and its flux may decrease, see e.g. Nandra et al. [9] and references therein; Matt [7]; Ross et al. [17].

2. At high L_X, $\tau_{photo-electric}$ can become so low that the continuum flux < 30 keV is no longer absorbed but instead is reflected, see e.g. Basko et al. [2]; Ross et al. [17].

3. No "hard tails" are detected in RQ QSOs because the reflection component is confused with the underlying continuum emission. However, at high X-ray energies the cut-off due to Compton recoil and Klein-Nishina effects may be detectable.

A VERY STEADY SOURCE...

MR 2251-178 is a nearby (z = 0.064) X-ray bright RQ QSO ($L_X \sim 10^{45}$ erg s^{-1}). It is well known as being the first AGN in which a "warm absorber" was detected. *BeppoSAX* (Boella et al. [3]) obtained *the first* simultaneous 0.1–200 keV spectra of MR 2251-178 at two epochs in 1998, see Orr et al. (submitted) and Orr et al. [12]. The net exposure times with the MECS detector on-board *BeppoSAX* are $\sim$83 ks and $\sim$61 ks. MR 2251-178 shows no evidence of any change of flux (Δ Flux < 17.9%) or spectral shape between the 2 epochs which are separated by 5 months. For this reason, the two data sets for the LECS, MECS HPGSPC and PDS detectors have been added to obtain a high S/N spectrum.

A HIGH ENERGY CUT-OFF BUT NO COMPTON HUMP...

Fits of the resulting 0.1–200 keV range show that the X-ray spectrum is complex. In Fig. 1 one can see the large residuals to a power-law fit. Here is a summary of the fit results:

- A *high energy cut-off* is clearly present at $E_{cut-off} >$ 70 keV. A good fit is obtained with an exponential cut-off at $E_{cut-off} = 133^{+64}_{-35}$ keV, assuming a power-law continuum, plus warm absorber, plus Fe Kα line.
- There is *no* evidence for a Compton reflection "hump": the neutral reflection scaling factor is weakly constrained with an upper limit $R_{neut.refl.} <$ 0.38.
- The Fe Kα line is *ionized* and its width is unresolved by *BeppoSAX*: E_{line}(rest) = $6.53^{+0.14}_{-0.12}$ keV, EW = 62^{+12}_{-25} eV. These values fit in well with the findings of Reeves et al. [14]. The line "profile" is shown in Fig. 2.
- A soft excess is *not* required by the data...
- ...but a warm absorber is, see Orr et al. [11] and Orr et al. (submitted).

AN IONIZED REFLECTOR IS POSSIBLE, WITH SOME CAVEATS...

We tested the self-consistent ionized reflection model published by Ross et al. [17] which includes the Fe Kα line and the Auger effects. We assume a solar Fe abundance, include a warm absorber and an exponential high-energy cut-off of the incident power-law.

This model gives a good fit to the data but, most importantly, it leaves *no* systematic fit residuals. By itself, it can account for the ionized Fe line, part of the high energy cut-off and the lack of a reflection hump. (We note that the XSPEC `pexriv` model gives a slightly better fit, however, because it does not include Fe K emission it is not self-consistent). The fit results are listed in Table 1.

On the other hand, the model assumes a reflector with constant density. The upper layers of the reflector, which are subject to strong external illumination probably have lower densities and higher effective ionization parameters than the inner layers. This would tend to suppress Fe Kα emission in the outer layers as shown by Ross et al.

TABLE 1. Fit with self-consistent ionized reflection

Γ	E_{cut}	R
1.58 ± 0.03	102^{+39}_{-26}	$0.11^{+0.06}_{-0.05}$

ξ [erg s^{-1} cm]	χ^2_ν (dof)	
1625^{+1422}_{-930}	0.94 (155)	

[17]. Nayakshin et al. [10] and Ballantyne et al. [1] have recently calculated models of reflection on an ionized atmosphere in hydrostatic equilibrium. Ballantyne et al. [1] show that the constant density models of Ross & Fabian [16] actually give accurate fits to their "hydrostatic" reflection spectra.

AN AGN IN THE SAME FIELD OF VIEW?

EXO 2251.1-1737 is a weak ($m_V \gtrsim 16$) ($F_{0.1-10} = 1.43 \times 10^{-12}$ ergs cm^{-2} s^{-1}) X-ray source located $\sim 13.6'$ away from MR 2251-178, within the field of view of *BeppoSAX* (diameter of LECS FOV = $37'$). At this angular distance the flux from EXO 2251.1-1737 does not contribute to the spectrum extracted for MR 2251-178, whose 0.1–10 keV flux is over 50 times higher.

This object was first detected as a weak X-ray source by *Einstein* (Moran et al. [8]: 2E 2251.2-1737) and EXOSAT observatories (Giommi et al. [5]: EXO 2251.1-1737). The angular distance between the positions measured by *Einstein* and EXOSAT is only $0.66'$. The EXOSAT source was classified as a star by Giommi et al. [5] on the basis of its steep optical to X-ray slope ($\alpha_{ox} = 1.8$, $m_v = 16$).

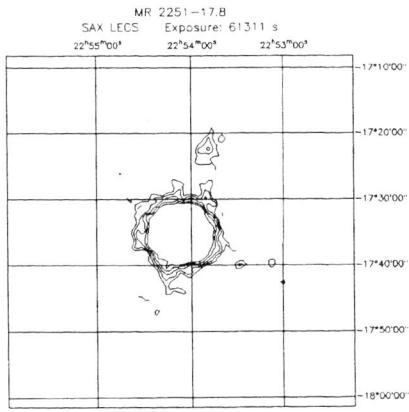

FIGURE 3. Logarithmic contour plot of 0.1–2 keV LECS image for the June 1998 observation. The bright central source is QSO MR 2251-178. EXO 2251.1-1737 is located northwards of it

The 2 *BeppoSAX* observations which include EXO 2251.1-1737 are separated by 5 months.

In the first epoch, the source has a soft, unabsorbed X-ray spectrum, best detected with the LECS instrument on-board *BeppoSAX*. It is well fit with thermal Bremsstrahlung (kT= 0.76, [0.55–1.11] keV) or a power-law ($\Gamma = 2.60[2.26 - 2.93]$)and shows evidence for a factor 3.6 decrease of 0.1–10 keV X-ray flux between the two observations separated by 5 months, however no decay-like, intra-observation variability is seen, as would be expected for a stellar flare.

During the second epoch the source was too faint to derive any spectral information. The *EXOSAT* source was classified as a star on the basis of its value of α_{ox}. However, because of its properties described above and because $F_X/F_V \sim 1$ we suggest it is an AGN, possibly a NLSy 1.

REFERENCES

1. Ballantyne D., Ross R., Fabian A., *Mon. Not. Roy. Astron. Soc.* accepted
2. Basko M., Sunyaev R., Titarchuk L., *Astron. Astrophys.* **31**, 249 (1974).
3. Boella G., Butler R., Perola G., et al., *Astron. Astrophys. Suppl. Ser.* **122**, 299 (1997).
4. George I., Turner T., Yaqoob T., et al., *Astrophys. Journal* **531**, 52 (2000).
5. Giommi P., et al., *Astrophys. Journal* **378**, 77 (1991).
6. Lawson A., Turner M., *Mon. Not. Royal Astron. Soc.* **288**, 920 (1997).
7. Matt G., *Nuclear Physics B (Proc. Suppl.)* **69/1–3**, 467 (1998).
8. Moran E.C., Helfand D., Becker R., White R., *Astrophys. Journal* **461**, 127 (1996).
9. Nandra K., George I., Mushotzky R., Turner T., Yaqoob T., *Astrophys. Journal* **488**, L91 (1997).
10. Nayakshin S., Kazanas D., Kallman T., *Astrophys. Journal* **537**, 833 (2000).
11. Orr A., Parmar A., Barr P., Guainazzi M., *Adv. Space Res.* **Vol. 25, No. 3/4**, 459 (2000).
12. Orr A. et al., 4rth Integral Workshop, ESA-SP Series (2001).
13. Orr A. et al., submitted to *Astron. Astrophys.*
14. Reeves J., Turner M., Ohashi T., Kii T., *Mon. Not. Royal Astron. Soc.* **292**, 468 (1997).
15. Reeves J., Turner M., *Mon. Not. Royal Astron. Soc.* **315**, 234 (2000).
16. Ross R., Fabian A., *Mon. Not. Royal Astron. Soc.* **261**, 74 (1993).
17. Ross R., Fabian A., Young A., *Mon. Not. Royal Astron. Soc.* **306**, 461 (1999).
18. Vignali C., Comastri A., Cappi M., et al., *Astrophys. Journal* **516**, 582 (1999).
19. Williams O., Turner M., Stewart G., et al., *Astrophys. Journal* **389**, 157 (1992)

X-ray spectra emitted by a hot plasma containing cold clouds

Julien Malzac

Osservatorio Astronomico di Brera, via Brera, 28, 20121 Milano, Italy

Abstract. We compute the hard X-ray spectra produced by Comptonisation process in a hot plasma which volume is pervaded by small cold dense clouds. Hard X-rays impinging on the clouds are partly reprocessed and thermally reemited, contributing in feeding the hot phase with soft photons, partly reflected, contributing to the formation of a hump in the high energy spectrum. The main cooling mechanism of the hot plasma is Compton cooling by the soft thermal emission from the cold clouds. Using a non-linear Monte-Carlo code, we compute the equilibrium temperature together with the escaping spectrum. The spectrum depends mainly on the amount of cold matter filling the hot phase. It is constrained to be very low in order to produce spectra similar to those observed in Seyfert galaxies and X-ray binaries. An additional external reflector is required in order to reproduce the full range of observed reflection amplitudes.

INTRODUCTION

The physical conditions in the inner parts of the accretion flow surrounding a black hole are likely to be very chaotic. A situation that has been often considered is that of the so-called 'cauldron', where a soup formed by a very hot plasma contains small grains constituted by small dense clouds of much colder matter. The first papers underlining the possible effects of cold matter in the centre of compact sources (reflections features, reprocessing) referred to such situations where cold clouds are distributed *inside* the hot Comptonising plasma (e.g. Guilbert and Rees 1988). Several later works were devoted to explain how such a configuration could be physically realized (e.g. Celotti, Fabian and Rees 1992; Kuncic, Blackman and Rees 1996; Kunzic, Celotti and Rees 1997; Krolik 1998) and compute the spectrum emitted by the clouds. These works focussed on the physics and radiative processes in the clouds. Here, we study the radiative coupling between the clouds and the hot plasma. Indeed the soft radiation emitted by the cold material constitutes the main cooling mechanism for the hot phase. It affects the temperature of the hot Comptonising plasma, and thus the emitted X and γ-ray spectrum. In addition, the presence of cold matter inside the hot plasma may contribute to the formation of a reflection component, forming a bump in the hard X-ray domain.

MODELING THE INHOMOGENEOUS PLASMA EMISSION

The geometry could be close to spherical, or disc-like. Here we will consider that the two-phase medium constitute an infinite slab. We checked that the geometry does not

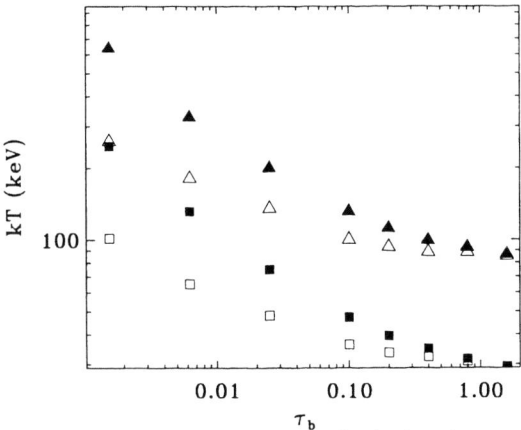

FIGURE 1. The volume average electron temperature in the hot phase versus τ_b. Filled and open symbols correspond to $kT_{bb} = 150$ eV (GBHs) and $kT_{bb} = 5$ eV (AGN), respectively. Squares and triangles show the cases $\tau_T = 0.6$ and $\tau_T = 2$.

affect qualitatively the results. The cold matter is assumed to be homogeneously distributed inside the plasma volume. The ambient high energy radiation which is intercepted by the cold clouds, is partly reprocessed as low energy (UV, EUV) radiation, and partly reflected in the X and γ-ray domains. The system is assumed to be in radiative equilibrium. The three main parameters, controlling the spectral shape are the following:

1. The amount of cold matter pervading the hot plasma. We assume that the individual cold clouds are much smaller than the characteristic size of the hot plasma. Then, we can quantify the amount of cold matter by using a "cloud" optical depth τ_b, such that a photon crossing the slab vertically has a probability $\exp(-\tau_b)$ of escaping without intercepting a cloud. τ_b may be understood/defined using a cross section formalism:

$$\tau_b = \frac{N}{V} <A> H \qquad (1)$$

where V is the plasma volume, N is the total number of cold clouds, $<A>$ is the average geometric cross section of the clouds, H is is the geometrical thickness of the slab.

2. The Thomson optical depth τ_T of the plasma defined along the height H of the slab.
3. The characteristic energy of the soft photons emitted by the clouds. We assume a blackbody emission with fixed temperature kT_{bb}.

We use the Monte-Carlo code of Malzac & Jourdain (2000) which is based on the non-linear Monte-Carlo method proposed by Stern (Stern 1985; Stern et al. 1995).

The slab is divided in 10 layers with equal volume. It is assumed to have a uniform density. The equilibrium temperature is computed in each zone according to the local balance heating=Compton cooling.

The interaction between a photon and a cloud is dealt as a regular Monte-Carlo interaction between 2 large particles (LPs; cf. Stern et al. 1995a). The only difference

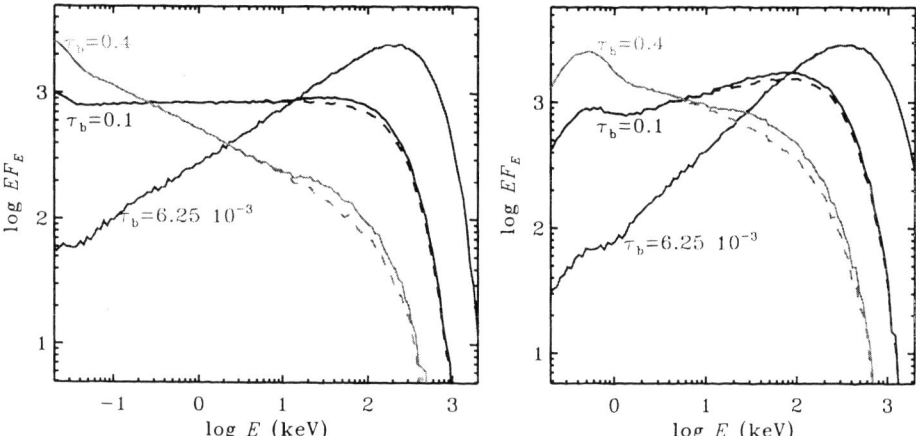

FIGURE 2. Spectra produced by the lumpy model. Dashed lines show the spectra without the reflection component. Left panel corresponds to the case of AGN ($T_{bb} = 5$ eV). Right panel corresponds to GBHs ($T_{bb} = 150$ eV). In both panel spectra with the indicated values of τ_b. The spectra are angle-averaged and normalized to the same total luminosity. The plasma Thomson depth is $\tau_T = 0.6$.

is that the cold medium forms a fixed background. The cold phase structure is assumed to be unaffected by the radiation. Then τ_b is simply the reaction rate for 1 photon, with time expressed in units of the light crossing time H/c of the medium.

When a LP photon interacts with a cloud, it is assumed to enter a semi-infinite slab medium with standard abundances and neutral matter (Morrisson & Mc Gammon 1983). The linear Monte-Carlo code of Malzac et al. (1998), is then used to track the path and the energy changes of the LP photon in this medium, until it is either absorbed in cold matter either escapes from the cold matter slab. The energy deposited in the cold cloud is reinjected in the hot plasma in the form of thermal soft photons LPs.

At fixed τ_T and T_{bb}, increasing τ_b increases the cooling of the plasma, due to the enhancement of reprocessing, and the subsequent increase of the soft radiation field. Then, as shown in Fig. 1, the temperature decreases with τ_b. This affects the shape of the emitted spectrum which becomes softer and cuts off at lower energy, as shown in Fig. 2 Fig. 3 shows the evolution of the 2-10 keV spectral slope when τ_b increases. The spectrum soften very quickly with larger τ_b,

When the amount off cold matter filling the cold plasma increases the amplitude of the reflection component R increases as well. The results of the numerical simulations shown in Fig. 3, are well represented by the simple formula:

$$R = (1 - e^{-\tau_b})e^{-\tau_T/2} \qquad (2)$$

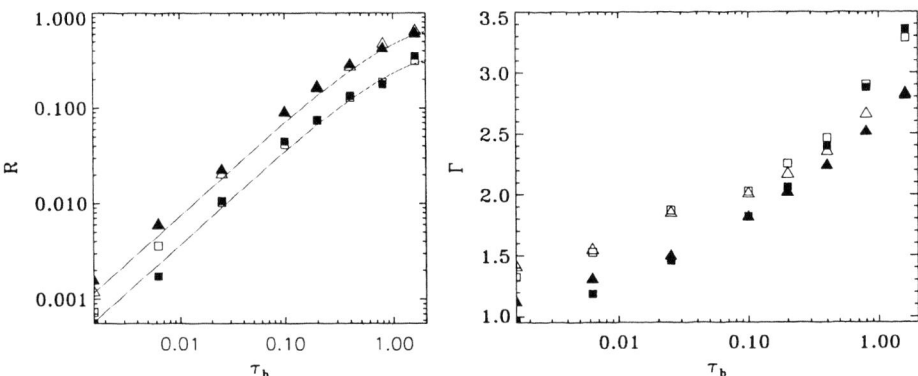

FIGURE 3. Angle averaged reflection (left) and spectral index (right) versus τ_b Filled and open symbols correspond to $kT_{bb} = 150$ eV (GBHs) and $kT_{bb} = 5$ eV (AGN), respectively. Squares and triangles show the cases $\tau_T = 0.6$ and $\tau_T = 2$. On the left panel, the solid lines show the approximation to the numerical results given by: $R = (1 - e^{-\tau_b})e^{-\tau_T/2}$

COMPARISON WITH THE DATA – CONCLUSION

From Fig. 3 one can see that the range of spectral slopes observed in Seyfert galaxies and galactic black holes in the hard state (1.4–2.2), is achieved for τ_b in the approximative range 10^{-3}–0.4. A larger τ_b leads to a too soft spectrum.

Equation 2, together with the observational constraint $\tau_b < 0.4$, forbids to produce a reasonable spectral slope together with a reflection amplitude larger than ~ 0.3. The situation is even more critical since $R \sim 0.3$ is reached only for the softest sources for which the observed reflection amplitudes ($R \gtrsim 1$) are the largest, as shown in Fig. 4. In order to reproduce the spectral characteristics of the numerous objects (both Seyfert and galactic sources) with significant reflection the model requires an additional reflecting medium: e.g. an external distribution of cold matter (Malzac 2001), a cold outer accretion disc, reflection on a torus in the case of Seyfert galaxies or on the companion star in the case of X-ray binaries.

The observed correlation between R and Γ shown in Fig. 4, is generally interpreted as being due to plasma cooling by soft photons emitted by the same medium that produce the reflection component (see Zdziarski et al. 1999; Gilfanov et al. 2000; Malzac in these proceedings). In this framework, the thermal emission from the additional reflector should thus contribute to, and even dominate, the radiative cooling of the hot plasma. If the above mentioned interpretation of the R-Γ correlation is correct, the amount of cold matter mixed together with the hot phase is constrained to be very low, so that its effects are negligible. Thus, in this scheme, the main source of soft photon/reflection is thus unlikely to be mixed inside the hot Comptonising plasma. The hot and cold phase have to be clearly distinct.

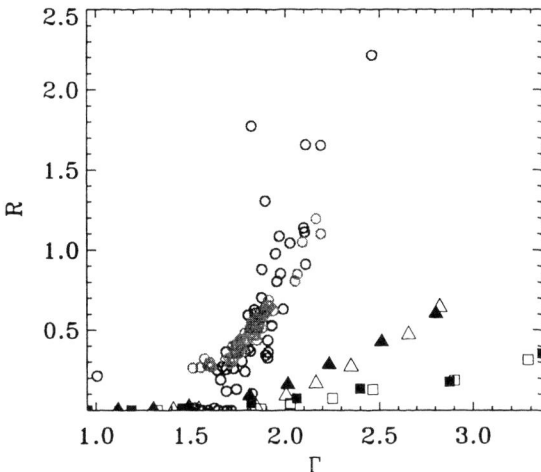

FIGURE 4. The circles show the data in the R–Γ plane, for the set of Seyfert galaxies (red circles) of Zdziarski et al. (1999) and for the set of galactic black holes (green circles) of Gilfanov et al. (2000). Filled and open black symbols correspond to $kT_{bb} = 150$ eV (GBHs) and $kT_{bb} = 5$ eV (AGN), respectively. Squares and triangles show the cases $\tau_T = 0.6$ and $\tau_T = 2$.

ACKNOWLEDGMENTS

This work was supported by the Italian MURST grant COFIN98-02-15-41.

REFERENCES

. Gilfanov M., Churazov E., Revnivtsev M., 2000, in Proc. 5th CAS/MPG Workshop on High Energy Astrophysics, in press (astro-ph/0002415)
. Celotti A., Fabian A.C., Rees M.J., 1992, MNRAS, 255, 419
. Guilbert P.W., Rees M.J., 1988, MNRAS, 233, 475
. Krolik J., 1998, ApJ, 498, L13
. Kuncic Z., Celotti A., Rees M.J., 1997, MNRAS, 284, 717
. Kuncic Z., Blackman E.G., Rees M.J., 1996, MNRAS, 283, 1322
. Kuncic Z., Celotti A., Rees M.J., 1996, MNRAS, 284, 717
. Malzac J., Jourdain E., Petrucci P.O., Henri G., 1998, A&A, 336, 807
. Malzac J., Jourdain E., 2000, A&A, 359, 843
. Malzac J., 2001, MNRAS in press, astro-ph/0104273
. Morrison R., McGammon D., 1983, ApJ, 270, 119
. Stern B.E., 1985, SvA, 29, 306
. Stern B.E., Begelman M.C., Sikora M., Svensson R., 1995a, MNRAS, 272, 291
. Zdziarski A. A., Lubiński P., Smith D. A., 1999, MNRAS, 303, L11

Thermal Comptonization and Disk Reprocessing in Type 1 Seyfert Galaxies

James Chiang[*] and Omer Blaes[†]

[*]*NASA/GSFC, Code 661, Greenbelt MD 20771 and Dept. of Physics, University of Maryland, Baltimore County, Baltimore MD 21250*
[†]*Department of Physics, University of California, Santa Barbara CA 93106-9530*

Abstract. Using a geometry consisting of a spherical Comptonizing region surrounded and penetrated by a thin accretion disk, we model the optical through γ-ray spectral energy distribution of the type 1 Seyfert galaxy NGC 5548. We take into account thermal reprocessing by the disk and how the resulting disk emission affects the energy balance of the Comptonizing region. These calculations allow us to infer the properties of the inner disk region including the size, temperature, and optical depth of the Comptonizing plasma. We discuss how *INTEGRAL* observations of AGNs such as NGC 5548 will provide stringent tests of this and other thermal Comptonization models.

The spectral energy distributions (SEDs) of type 1 Seyferts (Sy1s) are believed to be due to a combination of thermal emission from a multitemperature thin disk at optical and UV wavelengths and, at X-ray energies, thermal Comptonization from a hot plasma of order unit Thomson depth. The observations of relativistically broadened iron Kα fluorescence lines and the so-called Compton reflection humps in the hard X-rays imply that a substantial fraction of the X-ray continuum is intercepted by the accretion disk in these systems and reprocessed [22, 17, 13]. Since the typical albedos of the disk material are expected to be fairly low, $\lesssim 0.2$ [14], most of the reprocessed radiation must emerge from the disk as thermal emission. It has been argued that the highly correlated variability universally seen in the optical and UV continuum spectra of Sy1s provide evidence of thermal reprocessing in these objects [12, 7, 6, 19].

Detailed thermal Comptonization modeling has shown that the soft photons from thermal reprocessing can have a profound effect on the equilibrium state of the Comptonizing plasma [21, 9]. For models in which the hot plasma takes the form of a corona lying atop an accretion disk, because of the unity covering fraction by the disk of the X-ray flux, all of the soft photons due to thermal reprocessing re-enter the corona and cool the plasma significantly. As a result, since the temperature of the plasma determines (in part) the spectral index of the X-ray emission, the spectra produced by plane-parallel disk-corona models are too soft to account for the spectra observed from type 1 Seyfert galaxies [21]. In general, a geometry in which a smaller fraction of the thermally reprocessed emission re-enters the hot plasma is required.

Zdziarski, Lubiński, & Smith [23] (hereafter ZLS) showed that the apparent correlation between the X-ray spectral index, Γ, and the relative magnitude of the Compton reflection continuum, R, can be explained by considering global energy balance in geometries in which the disk covering fraction varies. In one such geometry, the Comptonizing region consists of a sphere of hot plasma, and a thin accretion disk penetrates the sphere at its equator by varying amounts. For larger overlaps, the reflection fraction is also larger, more of the disk photons re-enter the hot plasma to provide cooling via Compton losses, and softer X-ray spectra result. ZLS have calculated the expected R-Γ relationship in this geometry and found that it is very similar to the correlation which has been claimed for Sy1s and Galactic X-ray binaries [23].

Clearly, the soft photon flux from thermal reprocessing should also be related to the X-ray spectral index in these models. In fact, recent analyses of the UV and X-ray data from the 1997 IUE/RXTE observations of the Sy1 NGC 7469 show that a correlation between UV flux and X-ray spectral index does exist for this object [18]. In order to probe the relationship between the thermal Comptonization emission and disk thermal emission further, we have developed a Monte Carlo code to compute the contribution of these components, in energy balance, to the SEDs of type 1 Seyfert galaxies. We apply our calculations to multiwavelength data from the Sy1 NGC 5548 [3, 15].

We adopt the modified "sphere+disk" geometry proposed by ZLS [23]. Our Monte Carlo calculation follows the trajectories of individual photons from production as thermal photons in the accretion disk, to subsequent Compton up-scattering in the hot plasma, and possible reprocessing in the disk itself. The full Klein-Nishina cross-section is used, and the electrons in the Comptonizing plasma have a relativistic Maxwellian distribution. At each interaction point within the plasma, the momenta of the scattered photons are determined using the scalar Compton redistribution function [20, 11]. For each fixed geometry, specified by the disk inner radius r_{min} and sphere radius r_s, the temperature of the plasma is adjusted so that the reprocessed thermal emission from the disk matches the observed optical flux (at 5100Å) for the specific epoch we are trying to fit. The overall calculation is iterated until global energy balance is achieved. For X-ray photons from the plasma which strike the disk, we use a separate Monte Carlo code similar to that used by Magdziarz & Zdziarski [14] to compute the "exact" disk albedo and Compton reflection spectrum at each disk radius. For the disk material, we assume solar abundances and absorption cross-sections given by Bałucińska-Church & McCammon [1]. The atoms H and He are fully ionized, but all other elements are neutral. The disk flux includes internal viscous dissipation, but in the spectra we show, thermal reprocessing dominates the disk emission. The more notable simplifications we assume are a uniform temperature Comptonizing plasma, no ionization of elements heavier than He in the disk, and flat space-time trajectories for the photons.

Previous authors have fit an empirical relationship between the X-ray spectral index and the amplification factor for Comptonization: $\Gamma = \Gamma_0 (A - 1)^{-1/\delta}$ [2, 16]. Here Γ is the photon spectral index, and we define the amplification factor, A, to be the ratio of the thermal Comptonization luminosity to the luminosity of the accretion disk photons which are scattered in the hot plasma. In the limit where local viscous dissipation is negligible, the value of A is determined by the geometrical factors r_s and r_{min} and the radial Thomson depth of the sphere τ. We find $\Gamma_0 = 2.21$ and $\delta = 14$ similar to results found by other authors [2, 16]. Not surprisingly, we also see a slight dependence on Thomson depth [5]. In contrast to the constant albedo approximation assumed by ZLS [23], we find the disk albedo varies significantly as a function of disk radius and as well as for differing spectral indices of the incident spectrum. Furthermore, we find substantial anisotropy of the thermal Comptonization emission from the sphere with a factor of ~ 2 difference between emission emerging at inclinations $i \sim \pi/2$ versus $i \sim 0$, whereas ZLS assume that the emission from the sphere is uniform and isotropic. Despite these differences in detail with the assumptions of ZLS, we find a very similar relation between the X-ray spectral index and the ratio r_s/r_{min} [5].

In Fig. 1, we show the effect of varying the geometry of the sphere+disk model, using values for the Comptonizing sphere radius of $r_s = 2, 2.5, 3, 3.5, 4 \times 10^{14}$ cm, disk inner radius $r_{min} = 2 \times 10^{14}$ cm, and a Thomson depth $\tau = 1$. These SEDs and the ones which follow have been fit to the measured optical flux at 5100Å for each epoch in question. In Fig. 2, we plot the spectral energy distributions and data for epochs 2.1 and 2.2 from our 1998 EUVE/ASCA/RXTE campaign and epochs 5 and 9 from the 1989–1990 IUE/Ginga observations [15]. These plots illustrate the dependence of the model SEDs on various parameters. For the 1998 epochs, we fixed the disk inner radius at $r_{min} = 2 \times 10^{14}$ cm. This value was motivated by our estimate of the length scale of the Comptonizing region based on interband X-ray lags seen in the EUVE, ASCA, and RXTE/PCA light curves [3] and the fact that the observed values of the X-ray spectral index, $\Gamma \sim 1.9$ require $r_{min} \sim r_s$ in the sphere+disk model [23, 5]. In Fig. 2a, we set $r_s = 3 \times 10^{14}$ cm and $\tau = 1$; in Fig. 2b, the same value of r_s was used but it was found that a Thomson depth of $\tau = 3$ provided a better fit to the X-ray component. Since the spectral index depends predominantly on the ratio r_s/r_{min}, the major difference between the $\tau = 1$ (solid histogram) and $\tau = 3$ (dashed) spectra can be seen in the lower energy thermal roll-over for the latter case.

Since UV fluxes exist for the IUE/Ginga data, the disk inner radius can be better constrained for these epochs. Fig. 2c shows the SED for IUE/Ginga epoch 5 with $r_{min} = 1.5 \times 10^{14}$ cm, $r_s = 2 \times 10^{14}$ cm, and $\tau = 1$; and Fig. 2d shows the SEDs for IUE/Ginga epoch 9 in which optical/UV continuum was substantially redder than it was for epoch 5. By setting $r_{min} = 2 \times 10^{14}$ cm, while keeping $r_s = 2 \times 10^{14}$ cm and $\tau = 1$, we obtain a far better fit to the UV data, and reassuringly, we confirm the size estimates of r_s and r_{min} based on the interband X-ray lags. The dependence of the optical/UV spectrum on r_s and r_{min} shown in Figs. 1 and 2 illustrates a general feature of the sphere+disk model: For a given optical flux, the shape of disk spectrum is largely insensitive to changes in the sphere radius r_s (Fig. 1), but is very sensitive to changes in the disk inner radius r_{min} (Fig. 2d). A simple theoretical interpretation of this is given by Chiang & Blaes [4].

The existing simultaneous optical, UV, and X-ray data provide just enough information, through the optical and UV fluxes, the X-ray spectral index and ~ 1–30 keV fluxes, to infer values for the model parameters r_{min}, r_s, τ, and T_e. In the limit that thermal reprocessing dominates the disk emission, these parameters are sufficient to describe the entire spectral energy distribution. Therefore, a crucial test of this general scheme for linking thermal Comptonization and thermal reprocessing will be provided by simultaneous optical, UV, and *broad band* X-ray monitoring observations which include reliable measurements of the thermal roll-over at energies $\gtrsim 50$ keV. These latter measurements are necessary for making estimates of the X-ray luminosity L_x which can then be compared with the luminosities required by the model. Forthcoming *INTEGRAL* observations of Sy1s will provide some of the necessary spectral coverage to address this issue.

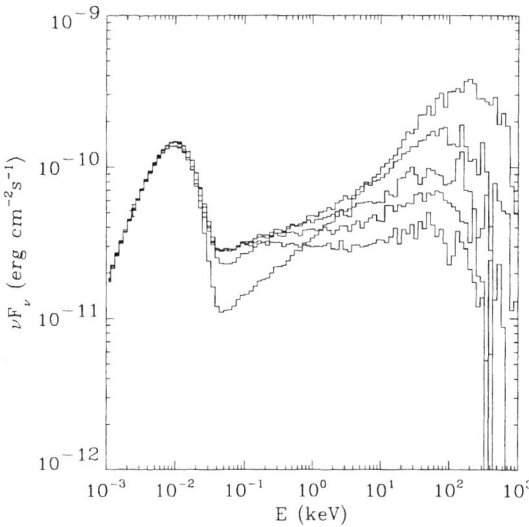

FIGURE 1. Model SEDs illustrating the dependence of the X-ray spectrum on the radius of the Comptonizing sphere. Both the disk inner radius, $r_{\min} = 2 \times 10^{14}$ cm, and radial Thomson depth of the sphere, $\tau = 1$, have been fixed. The sphere radii are $r_s = 2, 2.5, 3, 3.5,$ and 4×10^{14} cm. The softer X-ray spectra correspond to larger sphere/disk overlaps and therefore to the larger values of r_s. The near constancy of the optical/UV continuum despite large changes in the X-ray luminosity is explained by Chiang & Blaes [4].

We thank M. Dietrich [8] and M. Eracleous for providing optical fluxes for the 1998 EUVE/ASCA/RXTE observations. JC was partially supported by NASA ATP grant NAG 5-7723 during the course of this work. OB acknowledges support from NASA grant NAG 5-7075.

REFERENCES

1. Bałucińska-Church, M., & McCammon, D. 1992, ApJ, 400, 699
2. Beloborodov, A. M. 1999a, in High Energy Processes in Accreting Black Holes, eds. J. Poutanen and R. Svensson, (San Francisco: Astronomical Society of the Pacific), 295
3. Chiang, J., et al. 2000, ApJ, 528, 292
4. Chiang, J., & Blaes, O. 2001a, ApJL, submitted
5. Chiang, J., & Blaes, O. 2001b, in preparation
6. Collin-Souffrin, S. 1991, A&A, 249, 344
7. Courvoisier, T. J.-L., & Clavel, J. 1991, A&A, 248, 389
8. Dietrich, M., et al. 2001, A&A, 371, 79
9. Dove, J., et al. 1997, ApJ, 487, 747
10. Edelson, R. E., et al. 2000, ApJ, 534, 180
11. Jones, F. C. 1968, Phys. Rev., 167, 1159
12. Krolik, J. H. et al. 1991, ApJ, 371, 541
13. Madejski, G. M., et al. 1995, ApJ, 438, 672

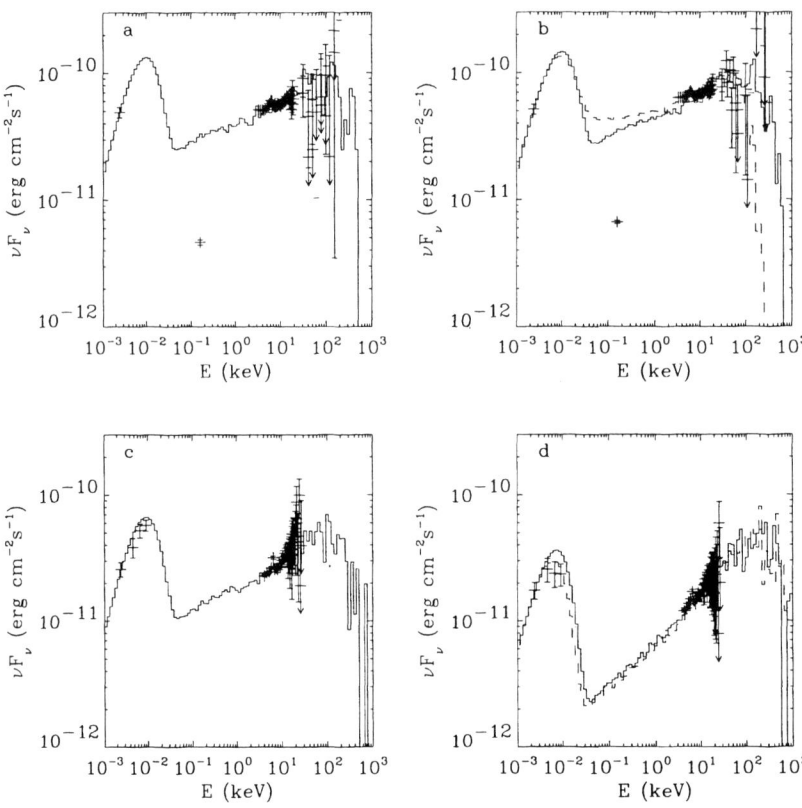

FIGURE 2. (a) Model SED fit to the data from RXTE epoch 2.1 [3]. (b) Fit to data from RXTE epoch 2.2. The solid histogram corresponds to $\tau = 1$ and the dashed histogram to $\tau = 3$. The data points at ~ 0.15 keV are fluxes measured by EUVE, uncorrected for Galactic absorption. (c) Fit to IUE/Ginga epoch 5. (d) IUE/Ginga epoch 9. The solid histogram has $r_{min} = 1.5 \times 10^{14}$ cm and the dashed histogram has $r_{min} = 2 \times 10^{14}$ cm. See the text for details.

14. Magdziarz, P., & Zdziarski, A.A. 1995, MNRAS, 273, 837
15. Magdziarz, et al. 1998, MNRAS, 301, 179
16. Malzac, J., Beloborodov, A. M., & Poutanen, J. 2001 (astro-ph/0102490)
17. Mushotzky, R., et al. 1995, MNRAS, 272, L9
18. Nandra, K., et al. 2000, ApJ, 544, 734
19. Peterson, B. M., et al. 1998, PASP, 110, 660
20. Poutanen, J., & Svensson, R. 1996, ApJ, 470, 259
21. Stern, B., et al. 1995, MNRAS, 272, 291
22. Tanaka, Y, et al. 1995, Nature, 375, 659
23. Zdziarski, A. A., Lubiński, P., & Smith, D. A., 1999, MNRAS, 303, L11 (ZLS)

Constraints on the high energy source of Seyfert 1 from *Beppo*SAX observations

P.O. Petrucci*, L. Maraschi*, F. Haardt[†], P. Grandi**, J. Malzac*, G. Matt[‡], F. Nicastro[§], L. Piro**, G.C. Perola[‡] and A. De Rosa**

Osservatorio Astronomico di Brera, Milano, Italy
[†]*Universitá dell'Insubria, Como, Italy*
**IAS/CNR, Roma, Italy*
[‡]*Universitá degli Studi "Roma 3", Roma, Italy*
[§]*IAS/CNR, Roma, Italy; CfA, Cambridge Ma., USA*

Abstract. We used high quality BeppoSAX data of 6 Seyfert galaxies to test realistic thermal Comptonization models. Our main effort was to adopt a Comptonization model taking into account the anisotropy of the soft photon field. The best fit parameter values of the temperature and optical depth of the corona and of the reflection normalization obtained fitting this class of models to the data are substantially different from those derived fitting the same data with the power law + cut–off model commonly used. The two models also provide different trends and correlation between the physical parameters, which has major consequences for the physical interpretation of the data

INTRODUCTION

The broad band X-ray spectra (2-300 keV) of Seyfert 1 galaxies are generally well fitted by a cut-off power law continuum + reflection. This model depends on three parameters: the spectral index Γ, the high energy cut-off E_c and the reflection normalization R. The physical interpretation of these parameters is generally done in the framework of thermal Comptonization mechanism, which is commonly believed to be the origin of the X-ray emission of Seyfert galaxies. Using approximate relations (cf. below), it is then possible to derive, from values of Γ and E_c, values of the temperature kT_e and optical depth τ of the "comptonizing" hot plasma (the so-called corona).
If such approximations are sufficient in isotropic geometries, where Comptonization process produces roughly cut–off power law spectra, strong discrepancies may appear in anisotropic ones, especially for small optical depth and large temperature. Consequently the physical parameter obtained fitting this class of models to broad band X-ray spectra are substantially different from those derived fitting the same data with the power law + cut–off model. In two recent papers (Petrucci et al., 2000, 2001, hereafter P01), we have applied accurate Comptonization models to fit high quality data of BeppoSAX observations of Seyfert 1 galaxies. We report in the following the main results of this study.

MODEL FITTING RESULTS

From the complete sample of Seyfert 1 observed by BeppoSAX we have selected six objects (NGC 5548, IC 4329A, NGC 4151, ESO 141-G55, Mkn 509 and NGC 3783) whose observations have high signal-to-noise ratios and hard spectra yielding a good detection in the PDS instrument.

We have fitted the data using two different models for the primary continuum: (1) an exponentially cut–off power law plus a reflection component from neutral material (PEXRAV model of XSPEC, Magdziarz & Zdziardski, 1995) and (2) a thermal Comptonization spectrum from a disk+corona configuration in slab geometry (Anisotropic Comptonization Code of Haardt 1994, hereafter AC2). The fit parameters of AC2 are the temperature of the corona kT_e, its optical depth τ, the temperature of the disk kT_{bb} (assuming a black body soft emission) and the reflection normalization R. On the other hand, the PEXRAV continuum depends only on 3 parameters: the e–folding energy of the cut–off power law E_c, the photon index Γ and the reflection normalization R. The temperature kT_e inferred from the PEXRAV fits is simply computed as $kT_e \equiv E_c/2$, keeping in mind that such approximation roughly holds for $\tau \lesssim 1$. For $\tau \gg 1$, $kT_e \equiv E_c/3$ would be more appropriate. Knowing the temperature, the spectral index derived from the PEXRAV fit can be used to determine the optical depth using the following relation

$$\Gamma - 1 \simeq \left[\frac{9}{4} + \frac{m_e c^2}{kT_e \tau(1+\tau/3)}\right]^{1/2} - \frac{3}{2}$$ (Shapiro, Lightman & Eardley, 1976). This equation is valid for $\tau > 1$, and we have checked *a posteriori* that such condition is roughly matched in all cases.

For all objects, the estimated corona temperatures (respectively optical depths) from PEXRAV are substantially smaller (respectively larger) than those inferred with an anisotropic Comptonization model (cf. Fig. 1a of P01). Large differences (up to a factor 8) are found between the two temperature estimates. The reflection normalizations obtained with AC2 are, in all cases but one (ES0 141-G55), larger than those found using the simple cut–off power-law model (cf. Fig. 1b of P01). In some cases we obtain differences of factors 4–5.

The reason of these differences is relatively simple. Within PEXRAV type models, the slope of the power law, determined with small errors by the LECS and MECS data, cannot change, by hypothesis, at higher energies. A cut–off around 100 keV is then required to fit the PDS data. In Comptonization models, the LECS and MECS data determine the slope below the anisotropy break. Above this break the intrinsic spectrum is steeper. It can thus fit the PDS data without an additional steepening beyond 100 keV, allowing for a larger temperature (and consequently a smaller optical depth to keep, roughly, the same power law slope) and a larger value of the reflection component normalization.

CONSTRAINTS ON THE X–RAY SOURCE GEOMETRY

In Fig. 1a and 1b the values of τ vs. kT_e, obtained for the 6 sources using the spectral models AC2 and PEXRAV, are compared with the theoretical relations expected for a plane and hemispherical Comptonizing region in energy balance (the temperature and optical depth then satisfy a definite and univocal relation which corresponds to roughly

constant Compton parameters $y \simeq 4 \left(\dfrac{kT_e}{m_e c^2} \right) \left[1 + 4 \left(\dfrac{kT_e}{m_e c^2} \right) \right] \tau(1+\tau).$)
The best fit results obtained with AC2 (Fig. 1a) are relatively close to the theoretical

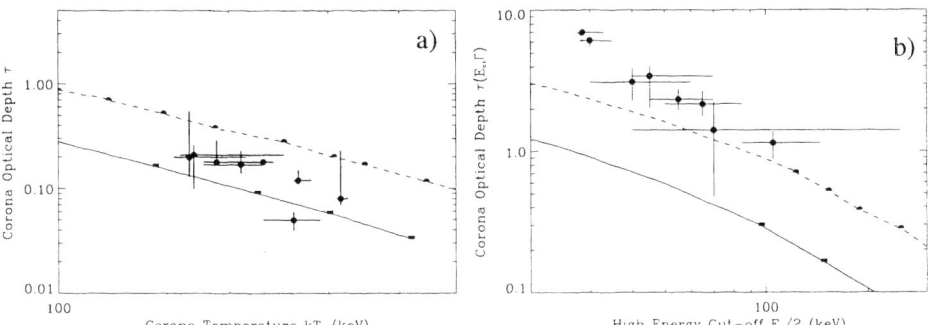

FIGURE 1. Optical depth τ versus temperature kT_e from (a) AC2 and (b) PEXRAV fits. The theoretical relations between τ and kT_e for a plane and hemispherical Comptonizing region in energy balance, in solid and dashed line respectively (from Stern et al., 1995), are shown for comparison.

expectations for the slab case even if they tend to fall preferentially above the solid line. The data therefore indicate a Comptonization parameter larger than for a pure slab geometry, that is a more "photon-starved" configuration. The values of kT_e and τ obtained with PEXRAV (cf. Fig. 1b) show the same trend as found with AC2 that is larger values of τ for smaller values of kT_e. They are also above the theoretical expectation for a slab and even above the theoretical expectation for a hemisphere. Therefore this set of parameters suggests, for each source of our sample, a configuration more "photon-starved" than a hemisphere.

An important difference with respect to the slab model is that the optical depths derived from PEXRAV fits are generally larger than 1. The corona should then be optically thick reducing or canceling the effects of anisotropy. Therefore also the PEXRAV model has an internal consistency. However a corona with large optical depth may wash out discrete features from the underlying disk (e.g. lines and reflection itself) more than would be desirable. This problem could be alleviated if the corona was "patchy".

We conclude that both a hot, optically thin corona with significant anisotropic effects and a less hot, optically thick, patchy corona with negligible anisotropy are consistent with the available data.

CONSTRAINTS ON THE X–RAY SOURCE NATURE

A correlation between the reflection normalization R and the photon index Γ has been claimed by Zdziarski et al. (1999, hereafter Z99) from the study of a large number of GINGA observations of Seyfert and galactic black hole objects.

We have plotted in Fig. 2a the reflection normalization R versus the photon index Γ obtained we obtained with PEXRAV. Our data don't show a clear correlation between the two parameters. However our sample is biased in favor of objects with hard spectra, to ensure a good detection in the PDS instrument. When the 13 objects actually observed by BeppoSAX are taken into account a stronger correlation is observed (Matt, 2000), in agreement with Z99.

Z99 interpret the correlation between R and Γ in the framework of thermal reprocessing

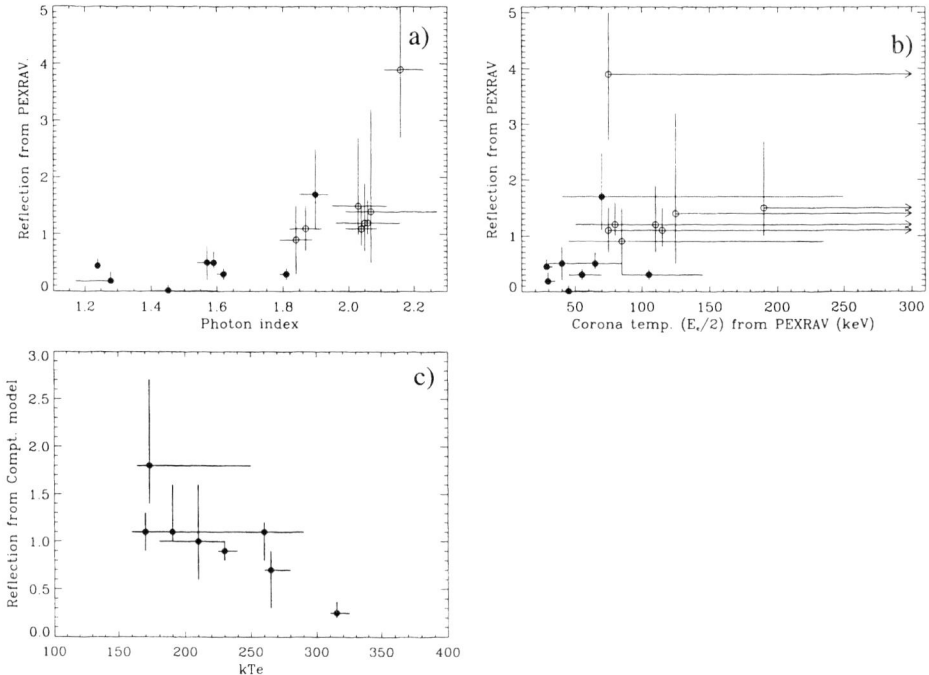

FIGURE 2. a) Reflection R_{PEXRAV} versus PEXRAV photon index Γ. b) Reflection component R_{PEXRAV} versus PEXRAV corona temperature $E_c/2$ from PEXRAV. The open circles are from Matt (2000) while the solid ones are the results of this work. c) R_{slab} versus slab corona temperature kT_e

models where R is directly proportional to the solid angle subtended by the cold matter surrounding the corona. In this case, the larger R and the larger the cooling of the corona resulting in softer spectra (larger Γ). It is interesting that the BeppoSAX data can provide the temperatures associated with sources with different values of R or Γ, which were not available from the GINGA data used by Z99. The result is shown in Fig. 2b where we have plotted R_{PEXRAV} vs. the corona temperature $E_C/2$. This plot suggests a *positive* correlation between the two parameters, that is, the temperature of the corona is *larger* for larger values of the reflection normalization. We note that the slab model analysis suggests different trends. If we plot R vs. kT_e for the latter model we find that sources with larger R tend to have lower kT_e (cf. Fig. 2c).

It is interesting to note that, in the interpretation of Z99, the $R_{PEXRAV}-E_c/2$ correlation of Fig. 2b suggests that the corona temperature is larger for larger cooling. Such behavior is expected in pair dominated thermal plasma in pair equilibrium (i.e. where pair creation is balanced by pair annihilation). Indeed, in this case, an increase of the cooling corresponds to a decrease of the number of particles in the hard tail of the thermal particle distribution, and thus to a decrease of the pair production rate. The number of particles in the thermal bath thus decreases, i.e. τ decreases. If the heating is kept constant, the available energy is now shared among less particles, resulting in an increase of the temperature.

On the contrary, the anticorrelation between R and kT_e obtained with the slab corona model (Fig. 2c) is naturally expected in the framework of reprocessing models but for low pair density corona, in complete opposition with the conclusions deduce from our PEXRAV fits.

CONCLUSION

We have testing thermal Comptonization models over the high signal to noise BeppoSAX observations of a sample of six Seyfert 1. We use two types of model: a detailed Comptonization code in slab geometry, which treats carefully the anisotropy effects in Compton processes, and a simple cut–off power law plus reflection model (PEXRAV model of XSPEC). The latter is a relatively good approximation to Comptonization spectra in isotropic geometry and/or optically thick corona. The main results of this work are the following:

- The data are well fitted by both models and there is no statistical evidence for a model to be better than the other. Both models give results in agreement with a X-ray source geometry more "photon starved" than the slab case.

- There are strong differences between the best fit values of the temperature and optical depth of the corona and the reflection normalization obtained with the two models. We generally obtained larger corona temperature, smaller optical depth and larger reflection normalization with the slab geometry in comparison to *PEXRAV*

- The two models leads to strong different relationships between physical parameters. For instance, we obtain a correlation between the reflection normalization R and the corona temperature with PEXRAV, and an anticorrelation with the slab corona model. This has major consequences for the physical interpretation of the data.

Forthcoming observations with CHANDRA, XMM-Newton and INTEGRAL are expected to bring substantial progress to discriminate between these two types of models for the high energy continuum of Seyfert galaxies.

ACKNOWLEDGMENTS

POP acknowledges a grant of the European Commission under contract number ERBFMRX-CT98-0195 (TMR network "Accretion onto black holes, compact stars and protostars").

REFERENCES

Haardt, F. 1994, PhD dissertation, SISSA, Trieste
Magdziarz, P. & Zdziarski, A. A. 1995, MNRAS,273, 837
Matt, G., "X-Ray Astronomy '99" proceedings, 1999, astro-ph/0007105
Petrucci, P. O. et al. 2000, ApJ, 540, 131
Petrucci, P. O. et al. 2001, in press (P01)
Shapiro, S. L., Lightman, A. P. & Eardley, D. M. 1976, ApJ, 204, 187
Stern, B. E., Poutanen, J. , Svensson, R. , Sikora, M. & Begelman, M. C. 1995, ApJL, 449, L13
Zdziarski, A. A., Lubinski, P. and Smith, D. A. 1999, MNRAS, 303, L11 (Z99)

Mrk 335: a NLSy1 with an unusual warm absorber?

Astrid Orr

Astrophysics Division of ESA, ESTEC, Postbox 299, NL-2200 AG Noordwijk, The Netherlands

Abstract. Mrk 335 is a NLSY1 well known for its rapid and erratic variability on time scales as short as a couple hours. We present the BeppoSAX data for this source. The BeppoSAX light curves shows that the highest X-ray variability occurs in the soft X-rays (0.1–2 keV). The variability possibly extends, at a lower level, into to the 15–70 keV range. Spectral analysis shows that, contrary to previous studies of Mrk 335, the data can be very well fit without a soft excess. Our best fit combines a power-law, a warm absorber and a small amount of neutral reflection. Remarkably, Mrk 335 seems to stand out among other NLSY1s by having an unusually strong and highly ionized warm absorber.

VARIABILITY FROM SOFT TO HARD X-RAYS

Mrk 335 is a NLSY1 well known for its rapid and erratic variability on time scales as short as a couple hours, but also over longer periods, as seen in Fig. 1.

Mrk 335 was observed by *BeppoSAX* from December 10 to 12, 1998, with net exposure time in the LECS and MECS instruments of ~38 and 89 ks. Figures 2 and 3 show lightcurves in the different BeppoSAX energy ranges corresponding to the LECS, MECS and PDS instruments. The most erratic variability occurs in the soft X-rays (LECS:0.1-2 keV), with flux changes up to a factor of 2.6 in ~11 hours. In the 2-10 keV "medium" range (MECS), besides a little of the "flickering" seen at lower energies, the light curve also shows a gradual but significant decline of the flux by ~25–30% over

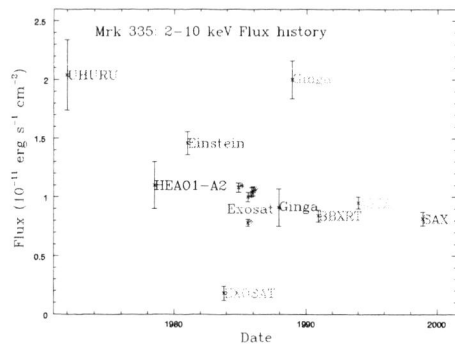

FIGURE 1. Flux history of Mrk 335. The BeppoSAX exposure was made during "medium" flux state

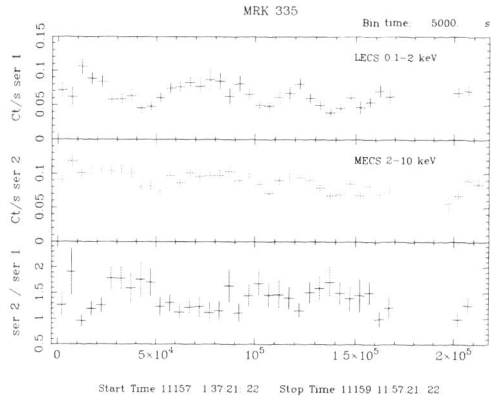

FIGURE 2. LECS and MECS light curves with hardness ratio

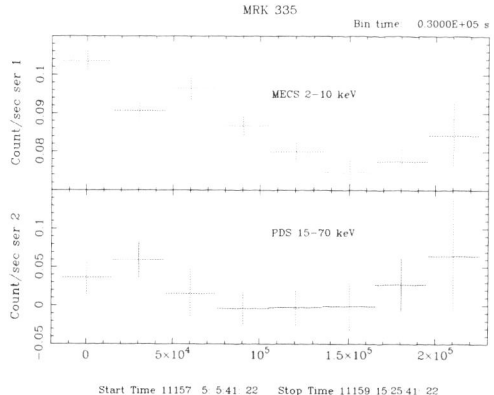

FIGURE 3. MECS and PDS light curves. A larger time bin size was chosen so as to allow comparison of the two curves

the ~3 day exposure. In both the soft and medium ranges constant fluxes can be clearly ruled out.

At the highest energies, from 15–70 keV, there is a hint of variability, as suggested in Fig. 3. The flux seems to be highest during the first 50000 s. This agrees with the trends observed at lower energies; however, it is a marginal effect because a constant flux also gives an acceptable fit to the PDS light curve.

CONTINUUM SPECTRUM: SOFT EXCESS VERSUS REFLECTION?

Mrk 335 has an overall steep and complex X-ray spectrum to which several components contribute in emission and absorption (see Fig. 4). Previous spectral studies of Mrk 335

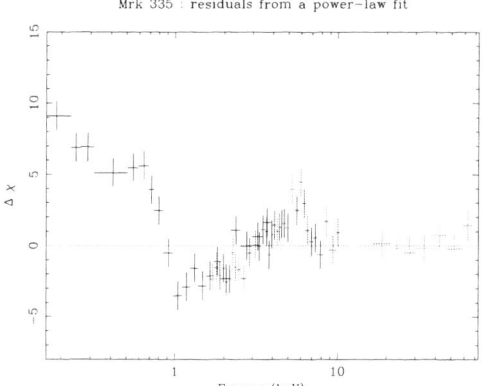

FIGURE 4. Residuals to a simple power-law fit including galactic neutral absorption. The residuals are plotted in terms of sigmas with error bars of size one

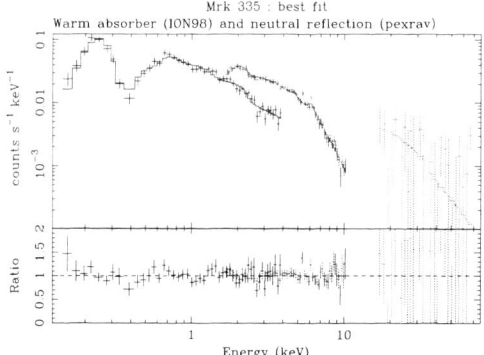

FIGURE 5. Best fit and residuals to the *BeppoSAX* spectrum. The model is composed of a power-law, a neutral reflection hump (XSPEC pexrav) and a warm absorber (ION98 code). No soft excess is present

based on data from Einstein, EXOSAT or ASCA have all shown the need to include a strong (& variable) soft excess component. The ASCA-based NLSy1 survey by Leighly (1999) indicates that soft excesses appear even more frequently in NLSy1 than in broad-line Seyfert galaxies (e.g. 17 of 19 NLSy1s).

However, the BeppoSAX spectrum can be very well explained without a soft excess (see Fig.5). Our best fit is obtained using a power-law continuum with a small amount of neutral reflection (XSPEC pexrav) combined with a warm absorber based on the ION98 photo-ionization code (Netzer 1996). The parameters for the continuum and neutral reflection are $\Gamma = 2.6 \pm 0.2$ and $R = 6.5 \pm 5$. The statistics are $\chi^2_\nu = 116.5/119$. A nearly as good fit is obtained by replacing the pexrav component by a double power-law ($\Gamma_1 = 3.1$, $\Gamma_2 = 2.1$). On the other hand, a power-law plus black-body soft-excess does not give an acceptable fit.

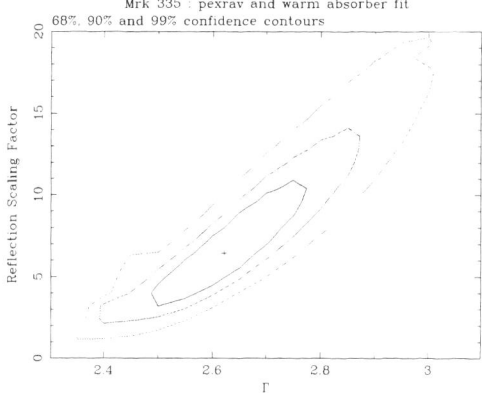

FIGURE 6. Confidence contours for power-law slope and reflection scaling factor. Same model as in Fig. 5

AN UNUSAL WARM ABSORBER FOR A NLSY1?

Systematic studies of warm absorbers in NLSy1 (*BeppoSAX*: Orr 1999, ASCA: Leighly 1999) show that, when modeled by two oxygen K-shell absorption edges, warm absorbers are less frequent in NLSy1s than in broad line Sy1s ($\sim$20 vs $\sim$50%).

In our best fit model the warm absorber (ION98 code) is described by $N_W = (3.39 \pm 0.68) \times 10^{22}$ cm^{-2} and $U_{oxy} = 0.073 \pm 0.012$ (where U_{oxy} is unitless and defined over the oxygen K-shell continuum, 0.538–10 keV).

The warm absorber can also be parametrized by fitting simple absorption edges. The fit is however not as good as with the ION98 code. With two oxygen edges (OVII & OVIII) plus a third edge above 1 keV one obtains: $E_{O7} = 0.74$, $E_{O8} = 0.87$, $E_3 = 1.45 \pm 0.09$; and $\tau_{O7} \leq 0.14$, $\tau_{O8} = 1.16 \pm 0.26$, $\tau_3 = 0.59 \pm 0.22$, respectively. The third edge may originate from ionized Mg (K-edge), ionized Fe or Ni (L-edges).

τ_{O8} is clearly large and much higher than τ_{O7}. On the other hand, using a big sample of NLSy1s (23 sources), Leighly (1999) obtains an average value of $\tau_{O7} = 0.19 \pm 0.04$, $\tau_{O8} = 0.053 \pm 0.020$. As a NLSy1 Mrk 335 therefore seems to have an unually strong and highly ionized warm absorber. We are currently investigating a larger sample of *BeppoSAX* archive data for NLSy1s to verify this conclusion.

REFERENCES

1. Leighly K., *Astrophys. Journal Suppl. Ser.* **125**, 317-348 (1999).
2. Netzer H., *Astrophys. Journal* **473**, 781 (1996).
3. Orr A., *New Astronomy Review* **44**, Issue 7-9, 487-490 (2000).

A quasi-spherical inner accretion flow in Seyfert galaxies

Julien Malzac

Osservatorio Astronomico di Brera, via Brera, 28, 20121 Milano, Italy

Abstract. We study a phenomenological model for the continuum emission of Seyfert galaxies. In this quasi-spherical accretion scenario, the central X-ray source is constituted by a hot spherical plasma region surrounded by spherically distributed cold dense clouds. The cold material is radiatively coupled with the hot thermal plasma. Assuming energy balance, we compute the hard X-ray spectral slope Γ and reflection amplitude R. This simple model enables us to reproduce both the range of observed hard X-ray spectral slopes, and reflection amplitude R. It also predicts a correlation between R and Γ which is very close to what is observed. Wide changes in the cloud covering fraction from source to source would be reponsible for most of the observed spectral variations. Moreover, if some internal dissipation process is active in the cold clouds, obscuration effects may provide a simple explanation for the observed distributions of reflection amplitudes, spectral slopes, and UV to X-ray flux ratios.

INTRODUCTION

The possibility for the existence of cold dense clouds in the innermost part of active galactic nuclei was suggested by Guilbert & Rees 1988. It has been further studied in details and shown to be physically realisable (Celotti, Fabian & Rees 1992; Kuncic, Blackman & Rees 1996; Kuncic, Celotti & Rees 1997). Collin-Souffrin et al. (1996) proposed a particular geometry where the cold material is spherically distributed around the central hard X-ray source. This cold material could be the residuals of a disc disrupted in its internal parts, or be part of an outflow or a wind, as well as spherically accreting material. Different variations on this model have been further studied, in the context of multi wavelength variability and line formation (e.g. Czerny & Dumont 1998; Abrassart & Czerny 2000; Collin et al. 2000). In all these studies the authors focussed on the cold blobs themselves and not on the problem of energy balance in the hot phase. The influence of the cold matter distribution on the primary emission was not considered. Indeed, a fraction of the cold clouds thermal radiation enters the central hot region. The heating of the hot plasma is thus balanced by the Compton cooling due to the incoming soft photon flux. The resulting equilibrium temperature T, (and thus the emitted the spectrum) then depends mainly on the cloud distribution. Here we derive analytical estimates for the hard X-ray spectral slope Γ and the amplitude of the reflection component R, and the UV to X luminosity ratios F_{uv}/F_x that are compared to the corresponding observed quantities.

SET UP

The black hole is at the center of a spherical hot plasma cloud, with radius r, forming the hard X-ray and soft γ-ray source trough Comptonisation process. Cold clouds are spherically distributed at some distance d from the center. They provide the seed photons for comptonisation in the hot phase. They intercept a fraction of the primary X-ray radiation. The main part of the intercepted high energy flux is reprocessed as low energy (UV) radiation, the rest is reflected in the X and γ-ray domain. The system is assumed to be in radiative equilibrium. The two main parameters, controlling the hard X-ray spectral slope Γ, and the reflection amplitude R are the following: (i) The relative (average) distance of the cold material from the centre d/r, (ii) the fraction of the luminosity which is intercepted by the clouds, i.e, the covering factor C. Some intrinsic soft emission from the cold material (due e.g. to shock dissipation) cannot be excluded. Then the additional soft luminosity entering the hot phase may strongly affect the radiative equilibrium. The spectral characteristics thus also depends on ratio of the power dissipated in the cold phase to that dissipated in the hot phase L_d/L_h. We will consider two simplifying cases: (i) $L_d/L_h = 0$: no dissipation in the cold phase, (ii) $L_d/L_h = \eta C$: dissipation in the cold phase growing linearly with C.

THE SPECTRAL PARAMETERS R AND Γ

With the idealized geometry described above one can derive an estimate for the reflection amplitude (see Malzac 2001 for details):

$$R \sim \frac{L_R}{L_x}\frac{2}{af(\cos i)} = \frac{2CK}{1-a_R CK}f^{-1}(\cos i), \qquad (1)$$

where a is the energy and angle integrated hard X-ray albedo of the cold matter for the incident comptonised spectrum ($a \sim 0.1$ for neutral matter), a_R the energy and angle integrated hard X-ray albedo of the cold matter for a reflection like incident spectrum (typically $a_R \sim 0.4$). The reflection amplitude is generally measured by fitting the spectrum with a slab reflection model. The measured values of R depend on the slab inclination angle i assumed when fitting the data. The f function in equation 1 represents the angular dependence of reflection amplitude in the slab case. We use the approximation given by Ghisellini et al. (1994):

$$f(\mu) = \frac{3\mu}{4}[(3-2\mu^2+3\mu^4)\ln\left(1+\frac{1}{\mu}\right)+(3\mu^2-1)\left(\frac{1}{2}-\mu\right)]. \qquad (2)$$

An estimate for the amplification factor A can be written as follow:

$$A = L/L_s = \frac{(1-KCa_R)[(1-C+C\xi)L_h+\xi L_d/2]}{\xi C(\omega-a)L_h+\xi(\omega-aC)L_d/2}, \qquad (3)$$

where: $\xi = 1-\sqrt{1-\left(\frac{r}{d}\right)^2}$, $K = (1-\xi+\xi\exp-\tau)$, $\omega = 1-KC(a_R-a)$, and τ is the hot plasma optical depth (observed to be ~ 1). Then Γ is directly related to A through the

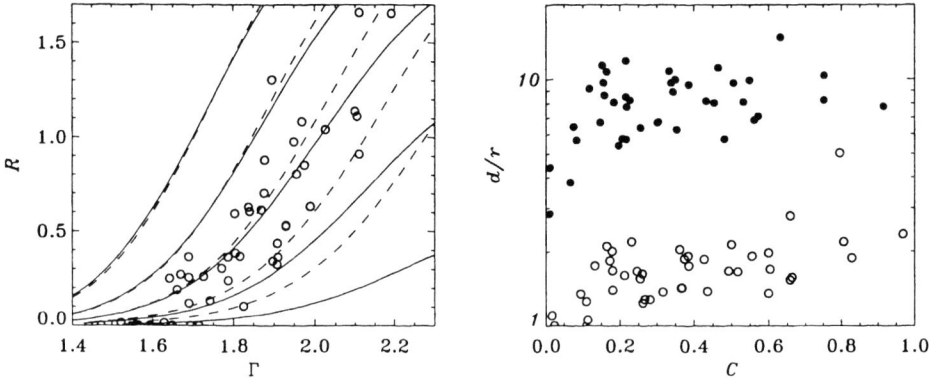

FIGURE 1. Left panel: The predicted R-Γ correlation compared with observations of Seyfert galaxies. The circles are the best fits parameters to the *Ginga* data (Zdziarski et al. 1999) in the R vs Γ plane. The lines represent the correlation obtained for cold blobs at a fixed distance from the hot plasma and covering fraction varying from 0 to 1. The solid lines stand for models with negligible dissipation in the blobs ($L_d/L_h=0$), from left to right for $d/r=4$, 2.5, 1.7, 1.2, and 1, respectively. The dashed lines stand for models with dissipation in the cold medium proportional to the covering factor $L_d/L_h = 60C$. The assumed distances from the centre are $d/r =22.1$, 13.6, 9.0, 5.9, 4.0 respectively from left to the right hand side. Right panel: The same R-Γ data transposed into the d/r vs C plane. The position of the open circles was determined assuming no dissipation in the cold phase, while $L_d/L_h = 60C$ was assumed for the filled circles. Data with $R=0$ were excluded (then d/r cannot be determined univocally). In both panel, the Thomson optical depth is fixed to unity, the reflector albedo is $a=0.1$, the reference inclination angle for reflection is $i=30°$ (as fixed by ZLS when determining R from spectral fitting).

phenomenological relation proposed by Beloborodov (1999):

$$\Gamma = \gamma(A-1)^{-\delta} \quad (4)$$

We found that for a spherical geometry with external soft photon injection, $\gamma = 2.26$ and $\delta = 1/12$ provides a good representation of the non-linear Monte-Carlo simulations (Malzac 2001).

It can be seen from equations 1,3,4 that both R and Γ increase with C. Indeed, a larger C means a larger effective reflecting surface as well as an extra amount of soft photons cooling the hot phase. It is obvious that changes in the covering factor will produce a correlation between these two spectral parameters. The left panel of Fig. 1 shows the correlations obtained when varying C, at different fixed d/r values, and compared to the correlation found by Zdziarski et al. (1999, hereafter ZLS) fitting a sample of ginga data. The observed correlation is qualitativelly reproduced for d/r in the range 1–3 and 4–10 respectively for $\eta = 0$ and $\eta = 60$. Equations 3, 4 and 1 are invertible. For any couple of observed R ($\neq 0$) and Γ, one can derive the corresponding C and d/r. The plot of the data in the C vs d/r plane is also shown in Fig. 1. The distance ratio d/r appears to be comparable in all sources, while the covering factor C can change widely. Without dissipation in he cold phase ($L_d/L_h = 0$) the average distance in our data sample is $<d/r>=1.7$, the rms spread is about 35 per cent. With dissipation we found $<d/r>=8.5$ for $\eta = 60$. For η larger than a few, the average distance increases with η like $(d/r) \sim \sqrt{\eta}$.

DISTRIBUTION OF SPECTRAL PARAMETERS

The observed distribution of covering factors can be determined directly from the data by determining C for each couple of $R\,\Gamma$ values in the ZLS sample. It is reasonably represented by the following analytic form:

$$p_C(C) = (1+\alpha)(1-C)^\alpha. \qquad (5)$$

The α parameter was determined by fitting the observed distribution and found to be in the range 1–1.5, depending slightly on η. The largest values being obtained in the limit of an infinite η. Then, in order to test the effects of the spread in d/r, we computed the R and Γ distribution expected if all the sources had the same distance ratio d/r (fixed to the observed average value), and the C distribution given by Eq. 5. We found the observed distributions to be qualitatively reproduced. This suggests that the source to source spread in d/r is actually negligible, the observed spectral difference are mainly controlled by the covering factor. The distribution of covering factors has to be a decreasing function of C, i.e. we observe much more sources with a low covering fraction than largely covered ones. Such a distribution may be due to selection effect against the the highly covered sources. Such sources are indeed more likely to be completely hidden by the highly absorbing cold material. Simple geometrical considerations show that such an effect can be effective only if the clouds are at a distance wich is large enough ($d/r > 3$), i.e. if there is some dissipation process in the cold clouds.

The ratio of the observed UV luminosity (coming from the cold clouds) to the hard X/γ one may be estimated as follow (Malzac 2001):

$$\frac{F_{uv}}{F_x} = \frac{[\omega - a]KCL_h + [1 + K\omega - KC(a_R + a)]L_d/2}{\omega[(1-KC)L_h + (1-K)L_d/2]}. \qquad (6)$$

Using this expression, the fitting C distribution functions 5 and setting d/r at the average value determined from the ZLS data, one can derive the expected distribution of F_{uv}/F_x. We compared it to the observed distribution in the sample of Walter and Fink (1994). The results are the following: Without dissipation in the clouds: the model does not match the data. For F_{uv}/F_x larger than a few, the model probability density is very close to 0. In contrast, the observed distribution extends up to quite large UV to X flux ratios ($F_{uv}/F_x \sim 50$). Clearly, the inner clouds do not emit enough UV radiation to reach the highest observed UV to X luminosity ratios. Actually, the data just imply that some additional source of UV radiation, not directly related to the central cloud system is present. For example, an external disc could be responsible for a significant (if not the largest) fraction of the observed UV emission. Such an outer source of UV radiation is unlikely to contribute sensibly to the cooling of the central plasma neither to the reflected component.

With dissipation in the clouds: the UV emission can be understood as being totally emitted by the cold clouds, in a way which is also consistent with the high energy data and model. A qualitative agreement is found for η in the range 40–80. In any case the η coefficient can not be much larger than the maximum observed $F_{uv}/F_x \sim 100$, otherwise too large F_{uv}/F_x ratios would be produced. This set an absolute upper limit on the distance $d/r \lesssim 13$

CONCLUSIONS

The R-Γ data are consistent with the spherical cloud model. It successfully reproduces the range of observed spectral slopes and reflection amplitudes, the R-Γ correlation and the individual R and Γ distributions. The observations put the following constraints on the model: the distance d/r of the cold material should be comparable in all objects. On the other hand, wide changes in the covering factor C from source to source would be responsible for the observed spectral differences. Without dissipation in the cold phase: the clouds are constrained to be in the immediate vicinity of the central hot plasma with an average relative distance $<d/r>=1.7$ and a small spread in distance from source to source. If there is some internal dissipation in the cold phase: the cloud distance may be somewhat larger, depending on the amount of dissipation in the cold phase. However, the distance is constrained to be lower than $d \sim 13r$, by the maximum observed UV to X flux ratios. Fixing $d \sim 8.5r$ (and $L_d/L_h \sim 60C$) enables us to understand the bulk of the UV emission as being emitted by the spherical reflector itself. If the cloud distance is lower than $8.5r$, the F_{uv}/F_x data suggest that part of the observed UV flux is emitted by some additional source, external to the cloud system. The observed distributions of spectral parameters requires the distribution of covering factors to decrease with C. If d is large ($d > 3r$) this kind of distribution may be explained by obscuration effects. Otherwise, if $d < 3r$ the observed C-distribution is more likely to reflects the intrinsic one.

ACKNOWLEDGMENTS

This work was supported by the Italian MURST grant COFIN98-02-15-41.

REFERENCES

- Abrassart A., Czerny B., 2000, A&A, 356, 475 (AC00)
- Beloborodov A. M., 1999b, in Poutanen J., Svensson R., eds, ASP Conf. Series Vol. 161, High Energy Processes in Accreting Black Holes. Astron. Soc. Pac., San Francisco, p. 295 (B99b)
- Celotti A., Fabian A.C., Rees M.J., 1992, MNRAS, 255, 419
- Collin-Souffrin S., Czerny B., Dumont A.-M., Zycki P. T., 1996, A&A, 314, 393
- Collin S., Abrassart A., Dumont D., Mouchet M., in proc. "AGN in their Cosmic Environment", Eds. B. Rocca-Volmerange & H. Sol, EDPS Conf. Series in Astron. & Astrophysics, in press, astro-ph/0003108
- Czerny B., Dumont A.M., 1998, A&A, 338, 386
- Ghisellini G., Haardt F., Matt G., 1994, MNRAS, 267, 743
- Haardt F., Maraschi L., 1993, ApJ, 413, 507
- Kuncic Z., Blackman E.G., Rees M.J., 1996, MNRAS, 283, 1322
- Kuncic Z., Celotti A., Rees M.J., 1997, MNRAS, 284, 717
- Malzac J., 2001, MNRAS in press, astro-ph/0104273
- Walter R., Fink H. H., 1993, A&A, 274, 105
- Zdziarski A. A., Lubiński P., Smith D. A., 1999, MNRAS, 303, L11 (ZLS)

Galaxy Clusters and Diffuse Gamma-Ray Background

Accelerated Particles from Shocks Formed in Merging Clusters of Galaxies

Robert C. Berrington[*], Charles D. Dermer[†] and S. J. Sturner[**]

[*]*ASEE Postdoctoral Fellow, Naval Research Laboratory, Code 7653, Washington, DC 20375-5352*
[†]*Naval Research Laboratory, Code 7653, Washington, DC 20375-5352*
[**]*NASA's GSFC and USRA, 7501 Forbes Blvd. #206, Seabrook, MD 20706-2253*

Abstract. Subcluster interactions within clusters of galaxies produce shocks that accelerate nonthermal particles. We treat Fermi acceleration of nonthermal electrons and protons by injecting power-law distributions of particles during the merger event, subject to constraints on maximum particle energies. The broadband nonthermal spectrum emitted by accelerated electrons and protons is calculated during and following the subcluster interaction for a standard parameter set. The intensity of γ-ray emission from primary and secondary processes is calculated and discussed in light of detection capabilities at radio and γ-ray energies.

INTRODUCTION

Rich clusters contain thousands of galaxies and are the largest gravitationally bound systems in nature. Masses for rich clusters are $\sim 10^{15} M_\odot$, with ~ 5–10% of the mass found in a hot intergalactic gas at temperatures of 2–12 keV. Rich clusters emit thermal bremsstrahlung with luminosities $L_x \sim 10^{44} - 10^{45}$ ergs s^{-1}[1]. Poor clusters contain hundreds of galaxies, have total masses $\sim 10^{14} M_\odot$, have a hot intergalactic gas of temperatures 1–5 keV and X-ray luminosities $L_x \sim 10^{42} - 10^{43}$ ergs s^{-1}[2]. Approximately 90% of the total mass of clusters is in the form of nonluminous dark matter.

In the hierarchical merging cluster scenario, poor clusters merge together to form richer clusters. Approximately 30–40% of galaxy clusters show evidence of substructure in both the optical [3] and X-ray wavelengths [4]. Velocity differences between the observed structures is ≈ 1000–2000 km s^{-1}. With gravitational forces driving the interaction between the two systems, cluster mergers are consistent with highly-parabolic orbits. Typical sound speeds within the intergalactic medium (IGM) are ≈ 800 km s^{-1}, so shocks will form at the interaction boundary of the two systems. Computer models of merging clusters support the development of shocks in the IGM [5, 6, 7]. Dimensional arguments show that a cluster merger releases $10^{63} - 10^{64}$ ergs of gravitational potential energy when initial separations are of the order $\sim$Mpc.

Only the most massive and X-ray luminous galaxy clusters have extended diffuse radio sources. With projected linear sizes ~ 1 Mpc, these diffuse sources have no known optical counterparts. The diffuse radio emissions have two distinct characteristics. The extended diffuse emission found in the central region of a galaxy cluster with a regular, azimuthally symmetric shape is known as a *radio halo*. The diffuse emission found on the cluster periphery are the cluster *radio relics*. These features often have irregular shapes with signs of filamentary structure. Radio relics are associated only with clusters that show evidence of a recent or ongoing merger event.

Shock fronts that form in the IGM as a result of a cluster merger event are thought to be associated with the cluster radio relics. The shock compression will orient any existing cluster magnetic field into the plane of the shock. Radio relics are characterized by highly organized magnetic fields with field strengths in the ~ 1 μG range with linearly polarized field lines in the vicinity of the shock [8]. The shock front will accelerate a fraction of the thermal particles within the IGM by first-order Fermi acceleration.

Recent work has highlighted the importance of nonthermal radiation from particles accelerated by shocks formed in merging clusters. Loeb and Waxman [9] have proposed that cluster mergers are the dominant contributor to the diffuse γ-ray background, provided the efficiency to convert the available gravitational energy into nonthermal electron energy is $\sim 5\%$. Some unidentified EGRET sources are claimed to be associated with γ-ray emission from galaxy clusters [10]. Excess EUV emission from Coma can be explained by nonthermal electrons accelerated at merger shocks [11]. Variations in radio surface brightness will result from superposition of cluster emissions [12].

In order to examine this question in more detail, we have modified a supernova remnant code [13] to calculate nonthermal radiation spectra from primary and secondary particles in merger shocks. We go beyond previous treatments by considering both primary electron and proton acceleration, a time-dependent treatment of radiation losses, and radiation signatures of secondaries from proton-nuclear interactions.

MODEL

We have adapted a supernova remnant code [13] to treat the cluster merger scenario. The code is designed to calculate time-dependent particle distribution functions evolving through adiabatic and radiative losses for electrons and protons accelerated by the first-order Fermi process at the cluster merger shock.

The electron and proton distribution functions originate from a momentum power-law injection spectrum. In terms of total particle energy $E = m\gamma c^2$, the injection function is

$$Q_{e,p}(E,t) = Q_{e,p}^0 \left[\frac{(pc)^{-s}}{\beta} \right] \exp\left[-\frac{E}{E_{max}(t)} \right], \quad (1)$$

where $p = \beta\gamma$ is the dimensionless momentum and s is the injection index. The maximum particle energy E_{max} is determined by the maximum energy associated with the available time since the beginning of the merger event, by a comparison of the Larmor radius with the size scale of the system, and by a comparison of the energy-gain rate through first-order Fermi acceleration with the energy-loss rate due to adiabatic, synchrotron and Compton processes. Particle injection ceases after the age of the shock front exceeds $t_{acc} = 10^9$ yrs. The constant $Q_{e,p}^0$ normalizes the injected particle spectrum over the entire volume $V(t)$ swept out by the shock front, and is determined by

$$E_{e,p}^{tot} = \int_0^{t_{acc}} dt \int_0^{E_{max}} dE\, E\, Q_{e,p}(E,t)\, V(t). \quad (2)$$

We assume a total available energy $E_{e,p}^{tot} = \eta_{e,p} 10^{63}$ ergs, and an efficiency factor $\eta_{e,p} = 5\%$ for both protons and electrons. Although the injection index s depends upon the Mach number of the shock, here we present calculations for a fixed index $s = 2$.

The time evolving particle spectrum is determined by solving the Fokker-Planck equation in energy space for a spatially homogeneous IGM, given by

$$\frac{\partial n(E,t)}{\partial t} = -\frac{\partial}{\partial E}[\dot{E}_{tot}(E,t)n(E,t)] + \frac{1}{2}\frac{\partial^2}{\partial E^2}[D(E,t)n(E,t)] + Q(E,t) - \frac{n(E,t)}{\tau_{pion}(E,t)}. \quad (3)$$

The quantity $\dot{E}_{tot}(E,t)$ represents the total synchrotron, Compton, Coulomb, and adiabatic energy-loss rate for electrons, and the sum of the Coulomb and adiabatic energy-loss rates for protons. Both protons and electrons are subject to diffusion in energy space by Coulomb interactions. The protons experience catastrophic losses due to proton-proton collisions on the time scale τ_{pion}. The spectra of secondary electrons and positrons are calculated from pion-decay products, and are subject to the same physical processes as the primary electrons.

The synchrotron, Compton, bremsstrahlung, and pion-decay γ-ray spectral components are calculated from the particle spectra following the methods described by Sturner, et al. [13]. We use a standard parameter set with a mean IGM number density $n_{IGM} = 10^{-3}$ cm^{-3}, a uniform cluster magnetic field of $B = 0.1$ μG, a constant shock speed $v_s = 1000$ km s^{-1}, and an acceleration period of 1 Gyr. Thus, $V(t) = A(t)v_s t$. We assume a constant surface area with a 1 Mpc radius. Particles loose energy due to adiabatic expansion according to the relation $-\dot{\gamma}/\gamma = \dot{V}(t)/V(t) = 1/t$.

RESULTS AND DISCUSSION

Figure 1 shows nonthermal photon spectra calculated at 1 Gyr and 5 Gyr for our standard parameters. The system is very luminous at radio frequencies during the particle acceleration phase because of intense Compton losses of nonthermal electrons on the CMB. After the acceleration period ends, the radiation from primary electrons is dominated by emission from secondaries. The π^0 bump at 70 MeV is hidden beneath the Compton-scattered CMB radiation from primary electrons during the acceleration period, but dominates the γ-ray spectrum after the acceleration period is over. The intensity of the π^0 bump depends sensitively upon the relative efficiencies $\eta_{p,e}$ for proton and electron acceleration. If no pion signature is found in nonthermal γ rays from merging clusters, then $\eta_p \lesssim \eta_e$ for our standard parameters.

From the total photon spectra in Figure 1, we have calculated light curves at various photon energies in Figure 2 for a merging cluster at a distance of 100 Mpc. The maximum peak 1.4 GHz radio flux density of ~ 12 Jy occurs at the end of the acceleration period. The radio emission from the merger event is, however, distributed over an angular region $\sim 0.5°$ of the merger shock, making it difficult to detect.

Figure 2 presents calculations of the energy fluxes from a merger shock at 100 Mpc, for photon energies >100 MeV, >1 GeV, and >300 GeV. The maximum luminosity occurs at the end of the acceleration period due to the accumulation of nonthermal protons. Cooling times due to synchrotron and Compton energy losses are very short for

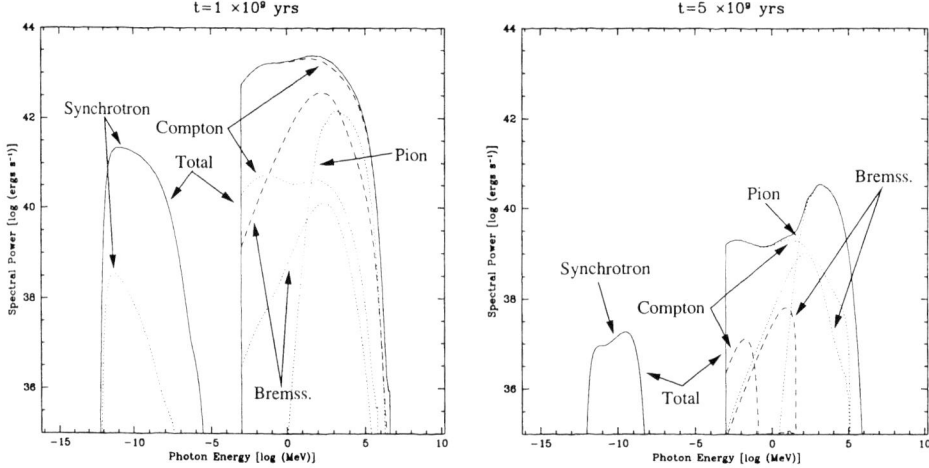

FIGURE 1. Total nonthermal photon energy spectra from a cluster merger shock. The solid curves are the total photon spectra summed from all processes. The dashed curves are the photons produced from the primary electrons, and the dotted curves are the secondary emission components. The left and right panels show the total nonthermal photon spectra at $t = 1$ Gyr and $t = 5$ Gyr, respectively, after the onset of the cluster merger event.

high-energy electrons; electrons that Compton-scatter microwave photons to energies $E = 100E_{100}$ MeV will cool on a time scale of $t_{cool} \cong 6 \times 10^6/\sqrt{E_{100}}$ yrs. Consequently, the γ-ray emission declines sharply after particle acceleration ceases, and approaches a level where all the γ-ray production originates from proton interactions. The maximum flux for secondary electrons is ≈ 2 orders of magnitude less than the maximum flux for the primary electrons, as can be seen by comparing the catastrophic proton loss time scale of ~ 30 Gyr with the 1 Gyr injection time scale. Because of adiabatic losses and the long catastrophic pion production cooling times, protons inject a slowly declining rate of secondary electrons in the 1-4 Gyrs after the initial acceleration event.

SUMMARY

We have studied the acceleration of protons and electrons in shocks formed by merging clusters of galaxies, and calculated the expected synchrotron, Compton, bremsstrahlung, and pion emission from the accelerated particles. Both the radio flux density at 1.4 GHz and the γ-ray flux at photon energies >100 MeV, >1 GeV and >300 GeV for a source at 100 Mpc as a function of time are presented. The flux is greatest at the end of the particle acceleration phase, and merging clusters of galaxies at a characteristic distance of 100 Mpc should be detectable with radio telescopes and GLAST. Radio and γ-ray detectability of merger events must contend with the angular extent of the emission.

Because the merger event lasts for only ~ 1 Gyr, a large fraction of clusters will no longer be experiencing ongoing nonthermal particle acceleration from cluster mergers.

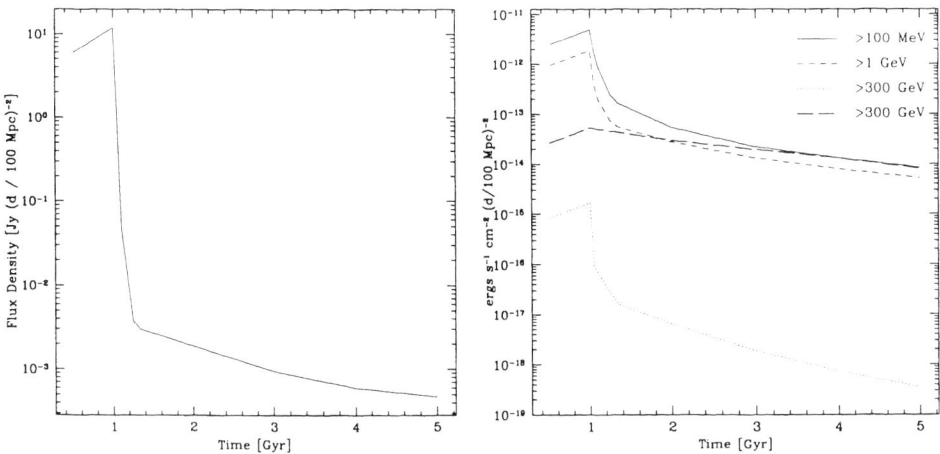

FIGURE 2. Calculated 1.4 GHz radio flux density (left panel) and integrated flux of high energy photons (right panel) from a cluster merger shock. Both panels are normalized for a system at a distance of 100 Mpc. The radio flux density is in the 1.4 GHz band, and the integrated energy fluxes for the high energy photons are shown in the right panel. Also shown is the >300 GeV emission for a merger where $v_s = 2000$ km s^{-1} and $B = 1$ μG as the boldfaced, long-dashed line.

GLAST, with a limiting sensitivity of 4×10^{-13} ergs cm^{-2} s^{-1} for a one year all-sky survey, will still be able to detect such systems if they are at optimal distances. Depending on the shock speed and magnetic field, cluster merger events can also be detectable sources of TeV radiation (see Fig. 2). We therefore expect detectable γ-ray emission from clusters at distances ~100 Mpc. Within this distance a number of radio relics have been detected [14, 15].

REFERENCES

1. Allen, S. W. and Fabian, A. C., *MNRAS*, **269**, 409 (1994).
2. Dahlem, M. and Thiering, I., *Publ. Astr. Soc. Pacif.*, **112**, 148 (2000).
3. Beers, T. C., Geller, M. J., & Huchra, J. P., *ApJ*, **257**, 23 (1982).
4. Forman, W., *et al.*, *ApJ*, **243**, L133 (1981).
5. Roettiger, K., Burns, J., and Loken, C., *ApJL*, **407**, 53 (1993).
6. Ricker, P. M., *ApJ*, **496**, 670 (1998).
7. Takizawa, M., *ApJ*, **520**, 514 (1999); *ApJ*, **532**, 183 (2000).
8. Enßlin, T. A., *et al.*, *Astron. & Astrophys.*, **332**, 395.
9. Loeb, A. and Waxman, E., *Nature*, **405**, 156 (2000).
10. Colafrancesco, S. and Blasi, P., *Astroparticle Physics*, **9**, 227 (1998).
11. Atoyan, A. M. and Völk, H. J., *ApJ*, **535**, 45 (2000).
12. Waxman, E. and Loeb, A., *ApJ*, **545**, L11 (2000).
13. Sturner, S. J., *et al.*, *ApJ*, **490**, 619 (1997).
14. Giovammini, G. and Feretti, L., *New Astronomy*, **5**, 335 (2000).
15. Kempner, J. C. and Sarazin, C. L., *ApJ*, **548**, 639 (2001).

The Soft γ-ray Spectrum of Galaxy Clusters

Mark Henriksen* and James Chiang*

Joint Center for Astrophysics, Physics Department, University of Maryland, 1000 Hilltop Circle, Baltimore, MD 21250

Abstract. Galaxy clusters appear to have a hard X-ray component based on recent SAX observations. The emission is best fit by a powerlaw and is likely non-thermal in origin. For the Coma cluster, the non-thermal emission is 10% of the thermal component in the 2 - 10 keV range and dominates above 25 keV. Thus it should be the dominant component in the soft gamma-ray regime. The spectral index is not well constrained by the current X-ray data but appears to be flatter than predicted by the standard Synchrotron-inverse Compton model for both Coma and A2256. INTEGRAL will be able to measure the spectral index of Coma, for example, to 20% and provide an important constraint on emission models. INTEGRAL observations should also reduce possible systematic errors associated with these early detections. The SAX observations challenge the standard model for non-thermal emission since some NT clusters have radio halos (Coma, A2256), others do not (A2199). Alternative emission processes have been suggested to explain the lack of radio halo emission in these clusters. Future missions such as GLAST will provide an important constraint on models of cluster formation and evolution as traced by cosmic-rays.

A BRIEF HISTORY OF CLUSTER NON-THERMAL EMISSION

The radio halo in the Coma cluster (A1656) has been known for some 3 decades now (Willson 1970). The Coma spectrum is well described by a synchrotron spectrum over the approximate range, 10 MHz–3 GHz (Giovanni et al. 1993). Other clusters of galaxies have since been found to have diffuse radio halos, including: A401, A754, A2255, A2256, A2319 and several more to have relics: A1367, A85. Non-thermal X-, γ-ray emission is expected from inverse Compton scattering of the cosmic microwave background photons by the same relativistic electrons which give rise to the synchrotron emission. The Coma cluster has been known for over 3 decades as an X-ray emitter. However, detection of the predicted non-thermal component is complicated since the X-ray spectrum of galaxy clusters appears to be dominated by thermal emission from the hot intergalactic medium over the energy range of 0.5–20 keV.

Over the years, studies of broad band X-ray spectra have placed increasingly strict limits on the amount of non-thermal emission in the hard X-ray/soft γ-ray regime from clusters known to have diffuse radio halos (Rephaeli & Gruber 1988 using HEAO1-A4; Henriksen 1998 using ASCA and HEAO 1-A2). Further upper limits for the Coma cluster came from the OSSE experiment on the Compton Gamma-ray Observatory (Rephaeli, Ulmer, & Gruber 1994) and EGRET (Sreekumar et al. 1996). The EGRET observation provided the best lower limit on the average cluster field (0.4μG) of these observations with a flux upper limit of 4×10^{-8} ph cm^{-2}s^{-1} at >100 MeV.

RECENT PROGRESS

The next generation of hard X-ray observatories have given new and more sensitive limits on non-thermal emission. Hard X-ray emission is detected above 20 keV with SAX for A2256 (and also for A2199, and Coma). The non-thermal X-ray spectrum detected from A2256 with the RXTE/SAX is marginally consistent(see Figure 1) with the flat spectrum, $\alpha_r = 0.8$, extended radio source in A2256 (Fusco-Femiano et al. 2000).

For A2256, RXTE and SAX are only weakly sensitive to the non-thermal component because of its flatness. The IBIS instrument aboard INTEGRAL is needed to characterize the spectral index in the soft γ-ray where the non-thermal component is uncontaminated by thermal emission. This also appears to be the case for the non-thermal component detected from the Coma cluster. JEM-X will be able to constrain the thermal component and provide a second method of constraining the non-thermal emission: fitting both data sets jointly.

The detection of non-thermal emission from A2199 (Kaastra et al. 1999) is puzzling since it does not appear to have an extended radio source. The detected flux level implies a very small average magnetic field ($0.05 \mu G$) if due to inverse Compton (Kempner and Sarazin 1999). Such small fields are inconsistent with the typical cluster fields derived from Faraday rotation studies (Kim, Tribble,& Kronberg 1991; Clarke, Kronberg, & Bohringer 1999). In contrast, RXTE derived upper limits found for A1367, and A754 imply average magnetic fields of 0.4–0.8 μG. MHD simulations by Dolag, Bartelmann, & Lesch (1999) show that a small, primordial field of $10^{-3} \mu G$ can be amplified to an average field of several μG during a cluster merger. Alternative models such as non-thermal bremsstrahlung (Blasi 2000; Sarazin & Kempner 2000) have also been suggested which alleviate some of these problems (e.g., lack of observable radio emission and very small magnetic fields) associated with an inverse-Compton interpretation of the X-ray emission.

NEW SCIENCE WITH INTEGRAL: MEASURING THE SOFT γ-RAY SPECTRUM

INTEGRAL should be able to measure the γ-ray spectrum for galaxy clusters that have X-ray emission detected: the Coma cluster, A2256, and A2199. The SAX limits on the spectral index for A2256 and Coma are poor but favor a flatter spectral index than the radio. The standard synchrotron-inverse Compton model predicts that the spectral index should be the same from the radio through the gamma-ray. Brunetti (1999) has proposed a model for the Coma cluster which can account for a flatter spectral index at higher energy. This model is characterized by 3 phases of cosmic ray evolution: initial injection, subsequent losses, and continued injection. INTEGRAL will provide the first accurate measurements of the cluster spectral index in the low energy γ-ray which will test this model and undoubtedly provide a critical constraint on all other evolutionary models for the cosmic ray production in galaxy clusters. It is notable that the spectral index derived from INTEGRAL observations will have better accuracy than that of the radio.

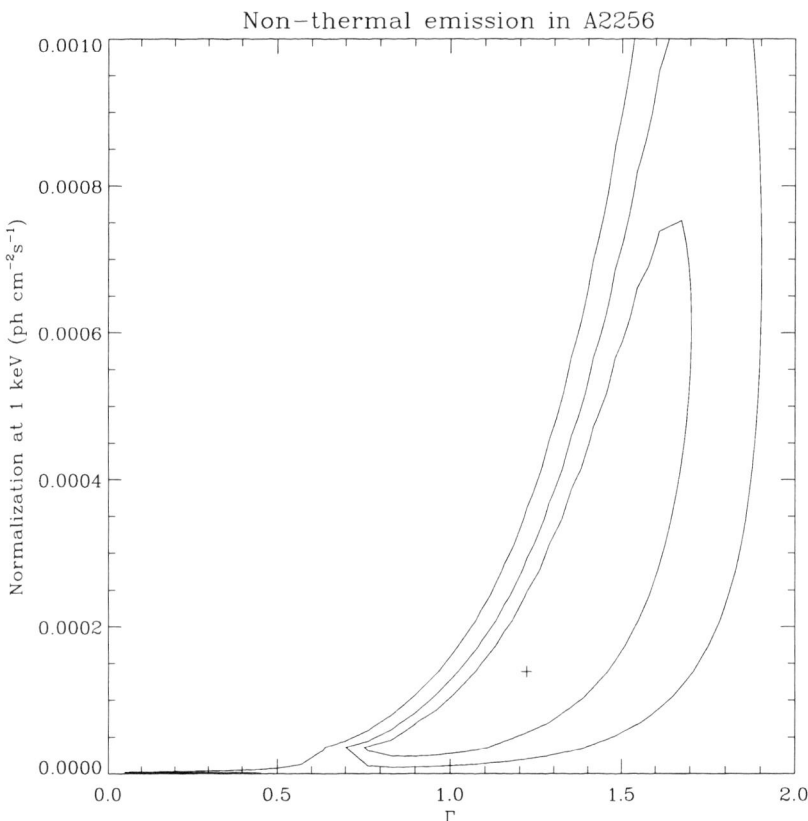

FIGURE 1. 68%, 90%, and 95% confidence contours for γ derived from fitting RXTE and SAX spectra.

FEASIBILITY

Results from modeling the simulated INTEGRAL spectrum from the Coma cluster were carried out for observations of $\sim 5 \times 10^5$ seconds. INTEGRAL should detect strong non-thermal emission from 20–160 keV (at >3σ) and weak emission out to 500 keV. Over the energy range 200–500 keV, the SPI has about the same S/N as the ISGRI. The IBIS simulator was used to make ISGRI pha files for Coma, A2256, and A2199. The input γ-ray spectral indices ($\alpha_x = \alpha_r + 1$) are based on the radio observations (Bridle 1981; Giovaninni et al. 1993) for Coma and A2199. For A2256, we have fit the archival BeppoSAX data directly (cf. Figure 1) and used the γ-ray index obtained from that analysis. We note that the value we obtain is consistent with the 90% confidence range found by Fusco-Femiano et al. (2000). Table 1 shows the results of fitting the simulated INTEGRAL spectra and Figure 2 shows the fit to the data for the Coma cluster.

While the search for non-thermal emission is best carried out at energies beyond the

Cluster	Flux, 20 - 80 keV 10^{-11} erg cm^{-2} sec^{-1}	α (Model)	α (Fit)	Exposure (10^5) sec
Coma	2.2	2.34	$2.4^{+0.4}_{-0.3}$	5.4
A2256	1.2	1.3	1.4 ± 0.3	4.5
A2199	1.2	1.8	$1.8^{+0.7}_{-0.5}$	4.5

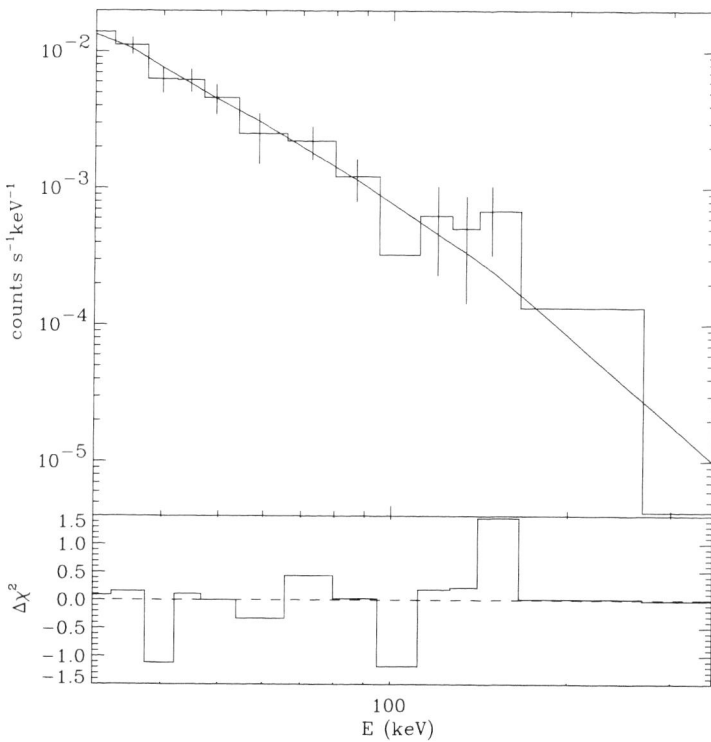

FIGURE 2. Fit to the Simulated INTEGRAL Observation for Coma

exponential decay of the thermal component, joint fitting of the JEM-X and IBIS data may provide the best constraint on a steep or multi-component non-thermal spectrum. For example, in the case of A2256 which has both a steep spectrum radio component $\alpha = 1.8$, and a flat spectrum radio component, $\alpha = 0.8$. Figure 3 shows the limits on the γ-ray spectrum for both components along with the thermal component. The steep component (dotted) is fixed at the upper limit derived from ASCA/RXTE (Henriksen 1999) (4% of the thermal component in the 2–10 keV band) while the flat spectrum

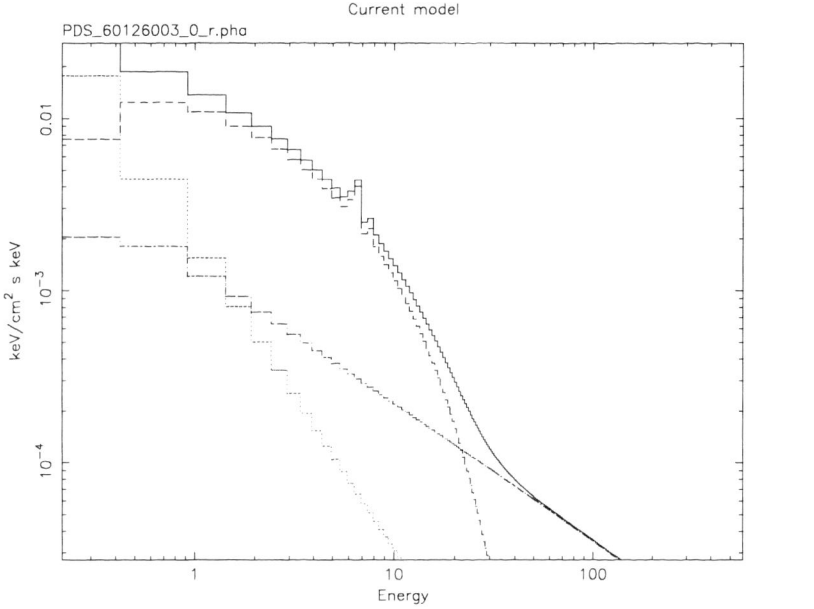

FIGURE 3. Spectrum for A2256 with non-thermal limits from SAX (flat spectrum) and RXTE (steep spectrum).

(dash-dot) is fixed at the PDS detection (~10% of the thermal component in the 2–10 keV band). The thermal component is shown by a dashed line. It is apparent that: (1) the steep component dominates the flat component at <2 keV, (2) the flat component is comparable to the thermal component at ~25 keV and is the dominant component at >30 keV, and (3) the non-thermal emission distorts the thermal continuum in the JEM-X energy band. Thus, a joint fit of the JEM-X and IBIS spectra will provide a second method of analysis (modeling the IBIS at >30 keV being the first). Both INTEGRAL data sets can therefore be used to model the complex emission in A2256.

The IBIS instrument provides a large field of view which is important for galaxy clusters since their emission may extend 6 degrees or more in diameter and radio relics are often fairly distant from the cluster center, as in the case of A1367 (~25 arc min). In addition, the A2199 non-thermal emission appears to increase with radius. A large field of view will thus be assured of getting all of the emission. On the other hand, there is the liability of contamination by point sources.

Contamination of the cluster spectrum by active galaxies is a possibility for the hard X-ray detections found using the SAX PDS. The non-thermal component is only in the PDS, which has a FWHM of 1 degree. The Coma cluster has several strong γ-ray emitters; most notably the Seyfert 1 galaxy, X Comae, and the Seyfert 2, Circinus. All large field-of-view collimators are subject to contamination by AGN and imaging such as that provided by the IBIS will allow direct measurement of contamination. The problem of contamination is complex since the AGN may be variable in the X-ray and γ-ray. Thus, the IBIS will give a much more accurate characterization of the AGN contribution than the alternative of using soft X-ray fields taken by the ROSAT PSPC or the ASCA GIS to examine the PDS field-of-view. Unlike these images, the IBIS will show the AGN brightness as it was during the cluster soft γ-ray observation. This capability makes the IBIS a uniquely powerful tool with which to study the cosmic-ray population and magnetic field structure in galaxy clusters.

REFERENCES

Blasi, P., 2000, ApJ, 532, 9.
Bridle, A., et al., 1979, AA, 80, 201.
Brunetti, G., et al., 1999, in MPE Report 271: "Diffuse Thermal and relativistic Plasma in Galaxy Clusters", 263.
Clarke, T., Kronberg, P., & Bohringer, H., 1999, in MPE Report 271: "Diffuse Thermal and Relativistic Plasma in Galaxy Clusters", 82.
Dolag, k., Bartelmann, M., & Lesch, H.,1999 , in MPE Report 271: "Diffuse Thermal and Relativistic Plasma in Galaxy Clusters", 257.
Henriksen, M., 1999, ApJ, 511, 666.
Henriksen, M., 1998, PASJ, 50, 389.
Henriksen, M., & Mushotzky, R., 2001, ApJ, 553, 84.
Fusco-Femiano, R., et al., 2000, ApJ, 534, L7.
Giovannini, G., 1993, ApJ, 406, 399.
Kaastra, J., et al., 1999, ApJ, 519, 119.
Kempner, J., & Sarazin, C., 2000, ApJ, 533, 73.
Kim, Tribble, & Kronberg 1991
Rephaeli, Y., & Gruber, D., 1988, ApJ, 333, 133.
Rephaeli, Y., Ulmer, M., & Gruber, D., 1994, ApJ, 429, 554.
Sarazin, C., & Kempner, J., 2000, 533, 73.
Sreekumar, P., et al., 1996, ApJ, 464, 628
Willson, M., 1970, MNRAS, 151, 1.

Clusters Of Galaxies – The EGRET Observations between 1991 and 2000

O. Reimer[1], P. Sreekumar[2]

[1]*NAS/NRC Research Associate, NASA GSFC, Greenbelt, MD 20770, USA*
[2]*ISRO Satellite Center, Bangalore, India*

Abstract. Various emission mechanism suggest clusters of galaxies to exhibit high-energy gamma-ray radiation. Galaxy clusters are predicted to be at the edge of the instrumental sensitivity currently accessible with gamma-ray telescopes. It is suggested that galaxy clusters contribute to the extragalactic diffuse background and, in few individual cases, they might be already detectable as individual sources. On the assumption that a flux limited sample of X-ray bright clusters will suit as a reasonable selection, gamma-ray fluxes (E > 100 MeV) are determined using EGRET data throughout the entire CGRO mission. In order to investigate beyond the case of the individual X-ray bright cluster, the gamma-ray data of individual clusters are cumulative stacked in a cluster-centered coordinate system and the resulting images have been analyzed. The results from EGRET are given and discussed in the light of predictions already found in the literature as well as in perspective of upcoming gamma-ray mission like INTEGRAL and, primarily, GLAST

INTRODUCTION

Clusters of galaxies are excellent representatives for formation and evolution of structures in the universe and extensively studied at radio, optical and X-ray wavelengths. Within the last decade, radio, EUV and X-ray observations revealed emission features, which also gave rise to predict galaxy clusters as emitters of high-energy gamma-rays: the discovery of diffuse radio halos, the controversially discussed EUV excess emission, and the evidence for a distinct nonthermal emission component in X-rays. Various scenarios are suggested: pp interactions of high-energy cosmic rays with intracluster gas (Berezinsky et al. 1997), relativistic electrons, which scatter background photons to higher energies (Enßlin & Biermann, 1998), electrons generated as secondaries in cosmic ray interactions in the intracluster medium (Blasi & Colafrancesco 1999), heating of cluster gas with injected cosmic-ray protons and magnetic field densities (Enßlin et al. 1997, Völk & Atoyan 1999, Atoyan & Völk 2000), and bow shocks of supersonically moving galactic halos (Bykov et al. 2000).

Generally, all of these models predict galaxy clusters to contribute to the extragalactic diffuse gamma-ray background. Rather different estimates of the quantity of such contribution could be found in the literature, comparable to the situation of the contribution of AGN to the extragalactic gamma-ray background (Dar & Shaviv 1996). In contrast to the well-observable population of Active Galactic Nuclei by EGRET, so far no galaxy cluster has been discovered in high-energy gamma-rays. Nevertheless, for several individual clusters flux predictions at EGRET energies (30 MeV - 10 GeV) exist (i.e. Dar & Shaviv 1995, Enßlin et al. 1997, Colafrancesco & Blasi 1999). However, the predicted gamma-ray fluxes are close to or even below the sensitivity of the EGRET telescope. The only existing study of galaxy clusters in high energy gamma-rays by McGlynn et al. 1994 was performed on earliest EGRET data only, and one upper limit has been determined for the Coma-cluster only (Sreekumar et al. 1996). Here we have expanded the analysis of Reimer 1999 into the relevant data throughout the entire CGRO mission from 1991 to 2000 in order to determine the currently best observational estimates on the high-energy emission of clusters of galaxies.

THE SELECTED SAMPLE OF GALAXY CLUSTERS

For analyzing the emission characteristics of galaxy clusters in the high-energy gamma-rays a sample of X-ray emitting clusters of galaxies has been compiled. This sample consists of the X-ray flux limited cluster detections from EINSTEIN (Edge et al., 1990), EXOSAT (Edge and Steward, 1991) and ROSAT surveys (XBACs: Ebeling et al. 1996, BCS north: Ebeling et al. 1998, BCS south: De Grandi et al. 1999). Cluster selections in X-rays currently provide the best way to obtain complete samples without introducing biases (i.e. projection effects). Appearing as extended sources with radii (r_{VTP}) of several arcminutes in X-rays, the limited angular resolution of existing gamma-ray telescopes justify the attempt to analyze clusters of galaxies as point-like excesses in gamma-rays. For 58 individual X-ray bright galaxy clusters within $z < 0.14$ gamma-ray data from the Compton GRO high-energy telescope EGRET were analyzed (see table in Reimer 1999). Although additional cluster surveys are on the way or have been recently completed (NORAS, REFLEX, HIFLUGCS, MACS) the chosen selection adequately represents the more energetic side of the log N- log S distribution from galaxy clusters in X-rays. Almost all clusters extensively discussed in individual papers due to evidence of nonthermal X-ray emission, EUV-excess features and/or characteristic radio halos are among this sample chosen to analyze for high-energy gamma-ray emission.

THE GAMMA-RAY ANALYSIS OF GALAXY CLUSTERS

So far, no galaxy cluster has been reported being positional coincident with gamma-ray point sources in existing EGRET source catalogues.[1] Only for the Coma cluster the result of an EGRET analysis has been published, based on observations from CGRO cycle 1 and 2 (Sreekumar et al. 1996). In the analysis described here, EGRET data of individual viewing periods from CGRO observation cycles 1- 9 were used for

---

[1] At this conference, S.Colafrancesco suggested a positional correlation between unidentified EGRET sources and Abell clusters. Despite the fact that there are 2712 clusters listed in the catalog (Abell 1958), and therefore *several* chance coincidences expected, the fact has not been explained why Coma, Virgo, and Perseus, the most obvious candidates for detectable gamma-ray emission still avoid their detection, and therefore do not coincide with any known yet unidentified gamma-ray source.

the analysis of 58 individual clusters. The latest and probably final improvements in the efficiency correction of EGRET have been fully implemented. Each galaxy cluster has been individually analyzed by means of standard EGRET data reduction procedures (likelihood source finding algorithm and subsequent flux determination at the X-ray position of the cluster center). This goes beyond the work presented by Reimer 1999, where four years of EGRET observations have been analyzed in strict congruence with the 3EG catalog of gamma-ray point sources (Hartman et al. 1999). The cumulative stacked maps are searched for residual sources after modeling and subtracting cataloged (and therefore well-known) gamma-ray point sources by using the maximum-likelihood technique. At the positions of the cluster center the gamma-ray flux has been determined.

Applying the same detection criteria as used and described in the EGRET source catalogs, none of the 58 galaxy clusters could be detected in the EGRET data. Special attention has been taken when existing EGRET sources are close to a considered cluster position. Four of these sources are identified blazar-class AGN. Only one of the remaining two cases shows considerable interference at the position of the analyzed cluster (A85 with the unidentified source 3EG J0038-09). Keeping in mind that the EGRETs instrumental point spread function has a width of 5.85° at 100 MeV, that the 3EG catalog contains about 170 unidentified sources, and also that our cluster sample consists of 58 candidates, such occurrence is perfectly in agreement with pure chance coincidence. The strongest gamma-ray excess for any cluster in the sample is a 1.6σ excess in the case of A3532, but this is well below the threshold of seriously being considered as a detection. Therefore for all galaxy clusters upper limits have been determined.

Triggered from this negative result on individual clusters, an approach has been made to study whether or not galaxy clusters radiate in gamma-rays as a population. For this purpose EGRET counts, exposure, and intensity maps from CGRO observation cycles 1- 9 were used, whenever an EGRET pointing has been within 30° of the considered cluster position for standard field-of-view observations or 19° for narrow field-of view observations, respectively. Before co-adding of individual maps a coordinate transformation into a cluster-centered system has been performed. The subsequent step in co-adding of individual maps in cluster-centered coordinates into the final stacked image required the exclusion of 8 galaxy clusters due to poor angular separation from the Galactic disk or dominant EGRET sources within the center region of the 40° by 40° map for each individual cluster. This assures that the central region of the final stacked image is not dominated from already identified gamma-ray sources or diffuse emission from the Galactic bulge. The central bin in the exposure map is $3.4 \cdot 10^{10}$ cm^{2}s (E > 100 MeV), the lowest values in the map about $1.4 \cdot 10^{10}$ cm^{2}s.

No excess at all is indicated in the central bin or even the central region of the constructed intensity image. Only after the individual gamma-ray intensities at the center position of the X-ray bright galaxy clusters have been weighted corresponding to their individual exposure and are analyzed on a adapted diffuse emission model, a final quantitative result on galaxy clusters as population will be given.

DISCUSSION

The negative results from an analysis of the final gamma-ray data from EGRET at positions of 58 individual galaxy clusters as well as from a superposition of 50 galaxy clusters might provoke some suspicion. Categorically, the question of an appropriate selected sample of galaxy clusters might arise. The assumption has been made that the brightest and closest clusters detected at X-ray wavelengths should be the best candidates to radiate in the gamma-rays, supported from various models of multifrequency emission properties. Because almost all clusters which show unusual multifrequency emission characteristics (EUV-excess emission, nonthermal X-ray emission and/or a radio halo) are naturally included here, the above assumption is certainly not artificially. To determine a quantitative results from the cumulative image is far from being trivial, it will incorporate detailed and precise modeling of the Galactic diffuse emission, should reflect the different cluster sizes and distances as well as the individual exposure histories. Despite that a recent modeling of the gamma-ray emission from galaxy clusters predicts values below the sensitivity of EGRET (Colafrancesco and Blasi 1998), some upper limits from individual galaxy clusters are already sensitive enough to restrict earlier model predictions found in the literature, for instance on Abell 426 (Enßlin et al. 1997, Dar & Shaviv 1995) or Virgo (Dar & Shaviv 1995). At this point, considering the combined exposure of $3.4 \cdot 10^{10}$ cm^{2}s by EGRET for 50 galaxy clusters without any hint of gamma-ray emission from the population as well, chances for INTEGRAL to detect clusters in the soft gamma rays are rather moderate due to the modest continuum sensitivity of SPI and IBIS-ISGRI, their limited field-of-view (i.e. limited long-term accumulation of required exposure time) and an observation strategy dominated by Galactic Plane observations. However, the nonthermal emission of some clusters might be observable by JEM-X beyond energies currently accessible. Speaking of gamma-rays, the (predicted) detection of galaxy clusters has to await the GLAST mission, which will combine the resolution and sky coverage required to go well below theoretical predictions on the high-energy emission of galaxy clusters.

ACKNOWLEDGMENTS

O.R. acknowledges a National Academy of Science/National Research Council Associateship at NASA/Goddard Space Flight Center.

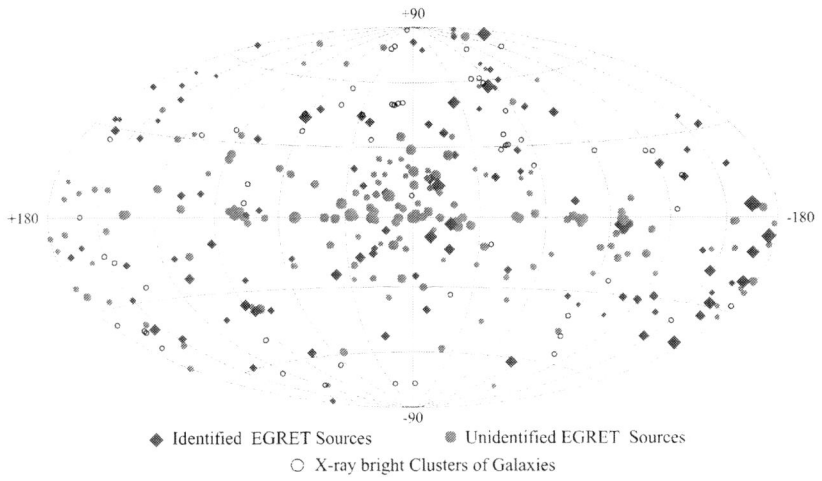

FIGURE 1. Positional arrangement of the studied 58 X-ray brightest clusters and gamma-ray sources detected by EGRET (E > 100 MeV)

REFERENCES

1. Abell, G.O., *Astrophys. J. Suppl.Series* **3**, 211 (1958).
2. Atoyan, A.M. and Völk, H.J., *Astrophys. J.* **535**, 45 (2000).
3. Berezinsky, V.S., Blasi, P. and Ptuskin, V.S., *Astrophys. J.* **487**, 529 (1997).
4. Blasi, P. and Colafrancesco, S., *Astroparticle Phys.* **12**, 169 (1999).
5. Bykov, A.M., Bloemen, H. and Uvarov, Y.A., *Astron. Astrophys.* **362**, 886 (2000).
6. Dar, A. and Shaviv, N.J., *Phys. Rev. Letters* **75**, 3052 (1995).
7. Dar, A. and Shaviv, N.J., *Astroparticle Phys.* **4**, 343 (1996).
8. Colafrancesco, S. and Blasi, P., *Astroparticle Phys.* **9**, 227 (1998).
9. DeGrandi, S. et al. *Astrophys. J.* **514**, 148 (1999).
10. Ebeling, H. et al., *Monthly Not. Royal Astron. Soc.* **281**, 799 (1996).
11. Ebeling, H. et al., *Monthly Not. Royal Astron. Soc.* **301**, 881 (1998).
12. Edge, A.C. et al., *Monthly Not. Royal Astron. Soc.* **245**, 559 (1990).
13. Edge, A.C. and Steward, G.C., *Monthly Not. Royal Astron. Soc.* **252**, 414 (1991).
14. Enßlin, T.A. et al. *Astrophys. J.* **477**, 560 (1997)
15. Enßlin, T.A. and Biermann, P.L. *Astron. Astrophys* **330**, 90 (1998)
16. Hartman, R.C. et al., *Astrophys. J. Suppl.Series* **123**, 79 (1999)
17. Hunter, S.D. et al., *Astrophys. J.* **481**, 205 (1997)
18. McGlynn, T.A., Vestrand, W.T. and Jennings, D., "A High Energy Gamma Ray Survey of Clusters of Galaxies" in *The 2nd Compton Symposium*, edited by C.E.Fichtel, N.Gehrels, J.P.Norris, AIP Conference Proceedings **304**, New York: American Institute of Physics, 1994, pp.669-673
19. Reimer, O., "EGRET Observations of Clusters of Galaxies" in *Proc. 26th International Cosmic Ray Conference*, edited by D.Kieda, M.Salamon, B.Dingus, Salt Lake City, 1999, Vol. **4**, pp.89-93
20. Sreekumar, P. et al., *Astrophys. J.* **464**, 628 (1996).
21. Völk, H.J. and Atoyan, A.M., *Astroparticle Phys.* **11**, 73 (1999).

Gamma-Rays from Galaxy Clusters: Preliminary Evidences and Future Expectations

S. Colafrancesco

Osservatorio Astronomico di Roma, Via Frascati 33, I-00040, Monteporzio, ITALY.

Abstract. We report here the preliminary evidence of a probable association between galaxy clusters and EGRET unidentified gamma-ray sources at high galactic latitude. Most of the clusters likely associated with EGRET sources show evidence of radio emission either because they host radio galaxies/sources in their environments or because they have a radio halo or relic inhabiting their intracluster medium. The cluster radio emission suggests that the relativistic particles (electrons, protons, ...), which are diffusing in the intracluster medium, might be also responsible for their gamma-ray emission. Beyond the spatial associations of clusters with unidentified gamma-ray sources, we further found a correlation between the X-ray luminosity of galaxy clusters and the gamma-ray luminosity of the associated gamma-ray source under the hypothesis that the EGRET sources have the same cluster redshifts. Such an intrinsic correlation strengthens the probability of a true, physical association between galaxy clusters and EGRET gamma-ray sources.

I INTRODUCTION

The large part of the gamma-ray sources detected with the EGRET instrument [1] on board the CGRO satellite have not yet been identified with secure counterparts at other wavelenghts because of the poor spatial resolution of the EGRET instrument ($\gtrsim 1$ degree). In facts, 170 gamma ray sources out of the 271 found in the Third EGRET catalogue [2] are not yet identified with firmly established counterpart. Most of the unidentified gamma-ray sources are found at low galactic latitudes, $|b| \lesssim 20°$, and are likely to belong to our Galaxy [3]. Fifty of these sources are found at high galactic latitudes, $|b| \gtrsim 20°$, and there are several hints that they are of extra-galactic nature [4]. Among the probably identified extra-galactic EGRET sources, most of them are AGNs [2] but there are no firm evidences that the remaining unidentified EGRET sources can be associated with another population of active galaxies.

Galaxy clusters are bright sources of X-rays produced through bremsstrahlung emission from a hot ($T \sim 10^7 - 10^8$ K), optically thin ($n \sim 10^{-3}$ cm^{-3}), highly ionized intracluster gas (mainly consisting of a population of thermal electrons and protons) in nearly hydrostatic equilibrium with the overall gravitational potential of the structure. Many galaxy clusters also show non-thermal emission phenomena like extended radio halos [5], likely produced by synchrotron emission of relativistic electrons either accelerated in the intracluster medium (hereafter ICM) by merging shocks or produced in the decay of dark matter annihilation products [6]. The presence of relativistic particles in the ICM has been also suggested to explain the emission excesses observed – in several clusters – in the EUV and in the hard X-rays (see [6] for a review). No evidence has been found yet for gamma-ray emission in the direction of a few selected clusters like Coma [7] and Virgo. There are, nonetheless, several theoretical motivations to expect that galaxy clusters can indeed be extended sources of gamma-rays emitted in the decay of neutral pions, produced either in the interaction of cosmic ray protons with the ICM protons [9,10] ($pp \rightarrow X + \pi^0 \rightarrow \gamma + \gamma$) or in the annihilation of dark matter particles [11] ($\chi\chi \rightarrow X + \pi^0 \rightarrow \gamma + \gamma$). Also primary cosmic ray electrons can produce a diffuse flux of gamma-rays due to non-thermal bremsstrahlung [6,7]. On top of such diffuse emission, the gamma-ray emission emerging from individual galaxies [12,13] living into the cluster is also expected.

II SPATIAL CORRELATION

We analyzed the available data for the gamma-ray sources in the Third EGRET catalogue [2] looking for a correlation between the position of gamma-ray sources and the positions of galaxy clusters in the Abell catalogue, in the ROSAT all sky survey and pointed observations and in the BeppoSAX cluster catalogue. We also looked for radio sources associated with galaxy clusters in the NVSS radio survey as well as in the available literature.

We found that 50 EGRET sources at high galactic latitude, $|b| > 20°$, are spatially correlated (within 1 degree from the center of the EGRET source) with the position of 70 galaxy clusters in the Abell catalogue. We performed Monte Carlo simulations to check if such a spatial association can be understood as a simple random projection effect. We found that, on average, 33 EGRET sources can be randomly associated with simulated cluster positions. Based on a Kolmogorov-Smirnov test, the probability that the remaining 17 EGRET unidentified sources are still randomly associated with galaxy clusters is $\lesssim 0.5\%$. We found that 18 of the original 50 EGRET sources associated with galaxy clusters have also an AGN whose position falls within the 95% confidence level probability contours of the gamma-ray source. We also found a Gamma Ray Burst in association with the EGRET source 3EG J2255-5012 and the clusters A1073 – A1074. Also a SN remnant is found in the field of the

Abell 1758

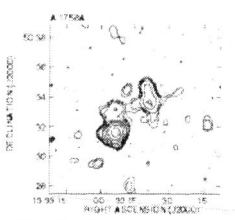

Radio halo-relic map
(VLA – 1.4 GHz)

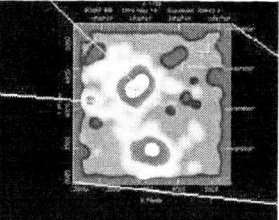

X-ray source
(ROSAT HRI)

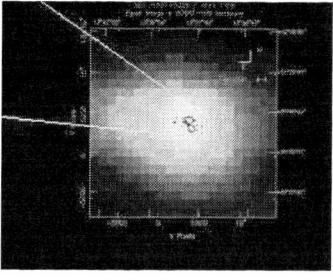

Fig. 1.a

EGRET source + X-ray contours

FIGURE 1. One of the most probable associations between galaxy clusters and EGRET unidentified gamma-ray sources: A1758. Shown are the EGRET image with the cluster X-ray brightness contours (right), the ROSAT X-ray image of the cluster (center) with the NVSS radio sources in the field (red circles) and the radio halo/relic image (left).

source 3EG J1235+0233 associated to the cluster A1564. We then excluded these EGRET sources and the associated clusters from the list of probable physical associations. Finally, we found that 24 galaxy clusters are associated to 18 unidentified EGRET sources for which there is no firmly established counterpart at other wavelenghts (see [14] for details). Many of these clusters (18 out of 24 clusters) have also bright NVSS radio sources within their Abell radius ($\approx 3h_{50}^{-1}$ Mpc), six clusters have identified bright radio galaxies and three clusters (e.g., A1758, A1914 and A85) have also a radio halo or radio relic inhabiting their ICM (see Fig.1). The presence of such relativistic particles strongly suggests that themselves and/or their parent population (e.g., relativistic protons, dark matter particles) can be responsible for a substantial gamma-ray flux at energies $E > 100$ MeV through different mechanisms, like non-thermal bremsstrahlung [6,8] and pion decay produced in pp collisions [9,10] or in dark matter particle annihilation [11]. The flux variability for the probable cluster-EGRET source associations is $\lesssim 20\%$, on average, and only in three cases (3EG J0616-3310, 3EG J2034-3110 and 3EG J1234-1318) it is $\gtrsim 30\%$. We also consider these three cases as suspiciously due to projection effects. The gamma-ray spectral indexes for the probable associations lie in the range $\approx 2-2.8$, values which are consistent with those expected from the viable mechanisms for gamma-ray emission in clusters [9].

III THE $L_\gamma - L_X$ CORRELATION

To strengthen our argument, we look for further, more intrinsic correlations among galaxy clusters and unidentified gamma-ray sources. We correlated the cluster X-ray luminosity, L_X, with the gamma-ray luminosity, L_γ, of the associated EGRET source under the assumption that it is physically associated to the cluster. Because different gamma-ray fluxes are given [2] for different EGRET viewing periods, we consider, in our analysis, both *i)* the fluxes given only in the viewing period P1234 and *ii)* the fluxes, selected among the different source viewing periods, which minimize the dispersion of the data in the $L_\gamma - L_X$ plane. The $L_\gamma - L_X$ correlation shown by the data in this last case (see Fig. 2) is fitted by $L_\gamma = AL_X^b$ with best fit values $A = 0.046 \pm 0.006$ and $b = 0.77 \pm 0.08$ (1 sigma errors). A similar result, however, is found also for viewing periods P1234 which are not optimized for both S/N ratio and data dispersion: specifically, we found best fit parameters $A = 0.048 \pm 0.036$ and $b = 0.57 \pm 0.08$. Such a $L_\gamma - L_X$ correlation is indeed expected in the viable model for the gamma-ray emission of galaxy clusters. In fact, both the diffuse emission arising from the interaction of relativistic particles with the cluster ICM and the one arising from a superposition of the gamma-ray emission associated with individual cluster galaxies predict a relation $L_\gamma \sim L_X^a$ with $a \approx 0.45 - 0.85$ [9,6,11,14], in agreement with our results shown in Fig.2.

The diffuse gamma-ray fluxes predicted [14] for the galaxy clusters associated to the EGRET sources are usually a factor 2-5 below the observed fluxes of the EGRET sources. So, to recover the gamma-ray fluxes of the EGRET sources, at least a comparable fraction of the cluster gamma-ray flux should be also contributed by the cluster galaxies. Thus, the EGRET data require that the gamma-ray fluxes associated to the galaxy clusters are likely due to a superposition of diffuse and concentrated gamma-ray emission.

IV FUTURE PERSPECTIVES

In this paper we report the preliminary evidence for an association of galaxy clusters with unidentified, high-latitude gamma-ray sources in the Third EGRET source catalogue [2]. While at the moment we have the first, preliminary evidence for the first gamma-rays coming from galaxy clusters, their detailed study will have a full bloom with the next generation space-borne (AGILE, GLAST, MEGA) and ground-based (VERITAS, ARGO, MAGIC) gamma-ray instruments. The next generation gamma-ray telescopes, and especially the GLAST mission, will have the spatial and spectral capabilities to confirm our preliminary results and to disentangle between the diffuse and concentrated nature of the cluster gamma-ray emission. Gamma-ray observations of galaxy clusters in the range $\sim 0.01 - 10^4$ GeV can probe directly the existence of different populations of relativistic particles (e.g., electrons,

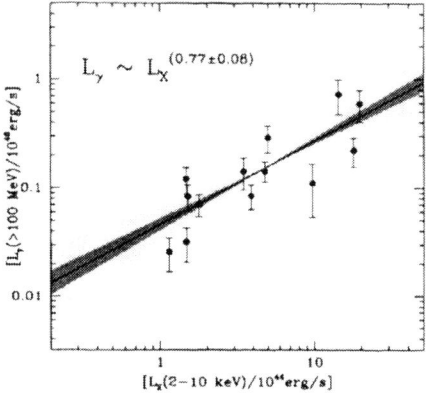

FIGURE 2. The $L_\gamma - L_X$ correlation shown by the clusters likely associated with EGRET sources. The best fit curve (solid blue) is shown together with the 68.3% (green area) and 99.7% (yellow area) confidence level regions for the fitting parameters.

protons, dark matter particles) in the ICM opening a new window on the astrophysical studies of large scale structures in the universe. Moreover, they can disentangle the production mechanism of high energy radiation and constrain at the same time the origin of the variety of puzzling non-thermal phenomena (radio halos/relics, EUV and hard X-ray excesses) which are observed in many galaxy clusters.

REFERENCES

1. Kanbach, G., *Space Sci. Rev.*, **49**, 69, (1988)
2. Hartman, R.C. et al., *ApJS*, **123**, 79 (1999)
3. Gehrels, N. et al., *Nature*, **404**, 363 (2000)
4. Grenier, I.A., these Proceedings (2001)
5. Giovannini, G. & Feretti, L., *New Astronomy*, **5**, 535 (2000)
6. Colafrancesco S., in *Constructing the universe with clusters of galaxies*, Eds. F. Durret & D. Gerbal, (2001)
7. Sreekumar, P. et al., *ApJ*, **464**, 628, (1996)
8. Longair, M. *High Energy Astrophysics*, Cambridge University Press, (1993)
9. Colafrancesco S. & Blasi, P. *Astroparticle Physics*, **9**, 227, (1998)
10. Völk H. & Atoyan, A., Astroparticle Physics, (astro-ph/9812458) (1999)
11. Colafranesco S. & Mele B., *ApJ* in press, (astro-ph/0008127) (2001)
12. Berezinsky V. et al. *Astrophysics of Cosmic Rays*, Nort-Holland (1990)
13. Dar, A. & DeRujula, A., *CERN-TH/2000-216* (astro-ph/0007306) (2000)
14. Colafrancesco, S., submitted to A&A, (2001)

The Extragalactic Gamma-Ray Background

F.W. Stecker

Laboratory for High Energy Astrophysics, Code 661, NASA/Goddard Space Flight Center, Greenbelt, MD 20771, USA.

M.H. Salamon

Physics Department, University of Utah, Salt Lake City, UT 84112, USA

Abstract.
  The *COMPTEL* and *EGRET* detectors aboard the Compton Gamma-Ray Observatory measured an extragalactic γ-ray background (EGRB) extending from ~ 1 MeV to ~ 100 GeV. Calculations performed making reasonable assumptions indicate that blazars can account for the background between ~ 10 MeV and ~ 10 GeV. Below 30 MeV, the background flux and spectrum are not very well determined and a dedicated satellite detector will be required to remedy this situation. Below 10 MeV, supernovae and possibly AGN may contribute to the extragalactic background flux. Above 10 GeV, the role of blazars in contributing to the background is unclear because we do not have data on their spectra at these energies and because theoretical models predict that many of them will have spectra which should cut off in this energy range. At these higher energies, a new component, perhaps from topological defects, may contribute to the background, as well as X-ray selected BL Lac objects. *GLAST* should provide important data on the emission of extragalactic sources above 10 GeV and help resolve this issue. *GLAST* may also be able to detect the signature of intergalactic absorption by pair production interactions of background γ-rays of energy above ~ 20 GeV with starlight photons, this signature being a steepening of the background spectrum.

INTRODUCTION

The EGRB measured by *EGRET* can be represented as of the power-law form

$$\frac{dN_\gamma}{dE} = (7.32 \pm 0.34) \times 10^{-6} \left(\frac{E}{0.451 \text{GeV}}\right)^{-2.10 \pm 0.03} \text{cm}^{-2}\text{s}^{-1}\text{sr}^{-1}\text{GeV}^{-1} \quad (1)$$

between 0.1 and ~ 50 GeV (statistics limited) [1]. At energies below 30 MeV, the EGRB spectrum appears to be steeper, as determined from an analysis of *COMPTEL* data [2].

Figure 1, taken from Ref. [3], shows a comparison between the diffuse innergalactic and extragalactic spectra measured by *EGRET*. It shows that these diffuse spectra have fundamentally different origins. The galactic spectrum shows evidence of the predicted "bump" from neutral pion decay [4], [5] whereas the extragalactic spectrum shows no such feature as would be expected from cosmic ray $p-p$ interactions. This type of direct spectral information eliminates purely diffuse extragalactic cosmic-ray interaction origin models, such as have been proposed [6] as explanations for the EGRB.

THE EGRB FROM 0.03 TO 10 GEV

The most promising model proposed for the origin of the GeV range extragalactic γ-ray background (EGRB), first detected by *SAS-2* and later confirmed by *EGRET* [1], is that it is the collective emission of an isotropic distribution of faint, unresolved blazars (See Ref. [7] and references therein.). Such unresolved blazars are a natural candidate for explaining the EGRB since, they are the only significant non-burst sources of high energy extragalactic γ-rays detected by *EGRET*.

The Unresolved Blazar Model:

To determine the collective output of all γ-ray blazars, one can use the observed *EGRET* distribution of γ-ray luminosities and extrapolate to obtain a "direct" γ-ray luminosity function (LF) per comoving volume, $f_\gamma(l_\gamma, z)$ [8]. Alternatively, one can make use of much larger catalogs at other wavelengths and assume a relationship between the source luminosities at the catalog wavelength and the GeV region [9], [10]. Both methods have uncertainties.

With regard to the former method, only the "tip of the iceberg" of the γ-ray LF has been observed by *EGRET*. Lower luminosity γ-ray sources whose fluxes at Earth would fall below *EGRET*'s minimum detectable flux, *i.e.* *EGRET*'s point source sensitivity (PSS), are not detected. Extrapolating the γ-ray LF to fainter source luminosities must then involve some extra assumption or assumptions.

We have chosen to use the latter method and have assumed a linear relation between the luminosities of a source at radio and γ-ray wavelengths in an attempt to estimate a LF which would hold at fainter luminosities. The extent of such a correlation is by no means well established [10] – [12]. However, since most theoretical models invoke the same high energy electrons as the source of both the radio and γ-ray emission, a quasi-linear relation between radio and γ-ray luminosities is a logical assumption. In fact, recent observations support this supposition [13].

We used this latter method to estimate the contribution of unresolved blazars to the EGRB, and found that up to 100% of the EGRB measured by can be accounted for [7]. Our model assumes a linear relationship between the differential γ-ray luminosity l_γ at $E_f = 0.1$ GeV and the differential radio luminosity l_r at 2.7 GHz for all sources, $l_\gamma \equiv \kappa l_r$ with κ determined by the observational data. One can

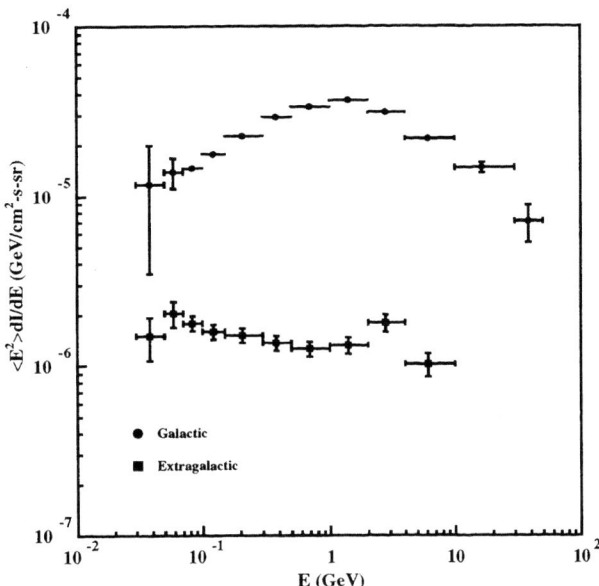

FIGURE 1. Comparison of Galactic and extragalactic diffuse spectra as determined by *EGRET*.

then used the measured radio LF $f_r(l_r, z)$ for blazars (flat spectrum radio sources) [14] to calculate the collective γ-ray output of all blazars. This LF is shown in Figure 2.

The simplified elements of our calculations are as follows: We assume that blazars spend 97% of their time in a quiescent state and the remaining 3% of their time in a flaring state We assume that the γ-ray and radio LFs in their quiescent state are related by $f_\gamma(l_\gamma, z) = \kappa^{-1} f_r(\kappa^{-1} l_\gamma, z)$. This relation changes by an average γ-ray "amplification factor", $\langle A \rangle = 5$, when the blazars are flaring. We assume that γ-ray spectra for all sources are of the power-law form $l(E) = l_\gamma (E/E_f)^{-\alpha}$, where α is assumed to be independent of redshift. We have taken the distribution of such spectral indeces, α, from appropriately related *EGRET* data. We also assume a slight hardening of the blazar spectra when they are in the flaring state which is supported by the *EGRET* data. For further details, see Ref. [7].

The number of sources $\mathcal{N}$ detected is a function of the detector's PSS at the fiducial energy E_f, $[F(E_f)]_{\min}$, where the integral γ-ray photon flux F is related to l_γ by

$$F(E) = l_\gamma (E/E_f)^{-\alpha} / 4\pi \alpha (1+\alpha)^{\alpha+1} R_0^2 r^2, \qquad (2)$$

where $R_0 r(1+z)$ is the luminosity distance to the source. The number of sources at redshift z seen at Earth with an integral flux $F(E_f)$ is given by

$$\frac{d\mathcal{N}}{dF(E_f)} \Delta F(E_f) = \int 4\pi R_0^3 r^2 \, dr \, f_\gamma(l_\gamma, z(r)) \Delta l_\gamma, \qquad (3)$$

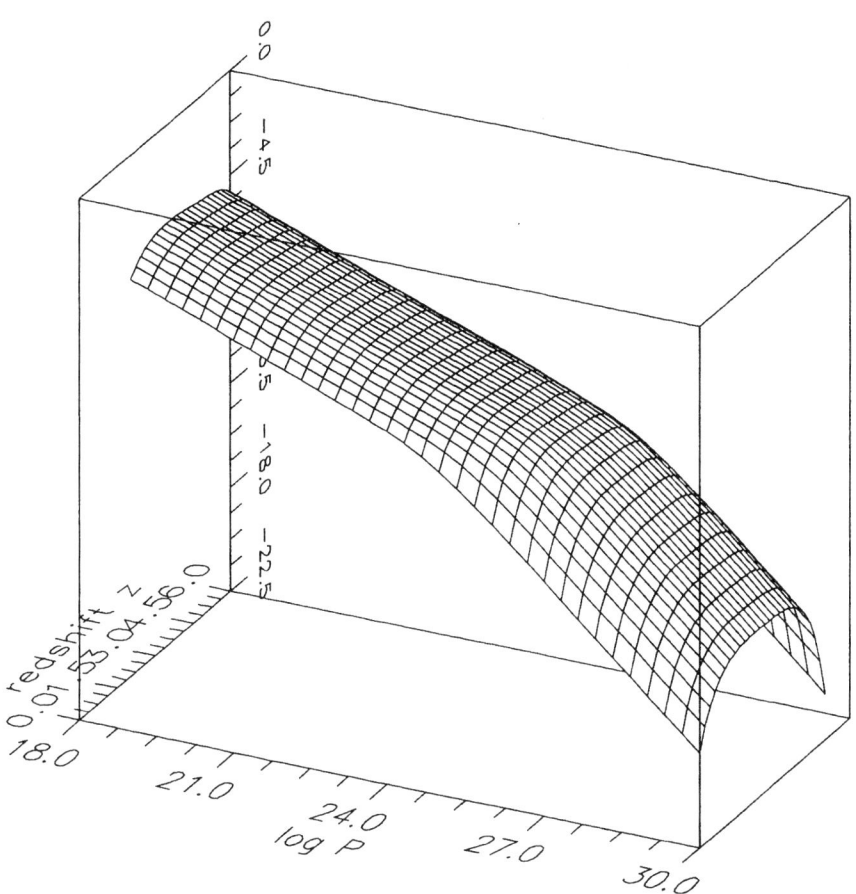

FIGURE 2. Radio luminosity (power) function at 2.7 GHz after Dunlop and Peacock [14].

where l_γ in the integrand depends on $z(r)$ and $F(E_f)$ from eq.2. The LF, f_γ, includes both quiescent and flaring terms. Figure 3 shows the results of our calculation of the number of sources versus flux above 0.1 GeV, *i.e.*, our predicted source count curve, compared to the *EGRET* detections [7]. The cutoff at $\sim 10^{-7}\mathrm{cm}^{-2}\mathrm{s}^{-1}$ for $E_f = 0.1$ GeV, their quoted PSS, is evident by the dropoff in the detected source count below this flux level.

To calculate the EGRB, we integrate over all sources *not* detectable by the telescope to obtain the differential number flux of EGRB photons at an *observed* energy E_0:

$$\frac{dN_\gamma}{dE}(E_0) = \int 4\pi R_0^3 r^2\, dr \int d\alpha\, p(\alpha) \int_{l_{min}}^{l_{max}} \frac{dF}{dE}(E_0(1+z)) f_\gamma(l_\gamma, z) e^{-\tau(E_0, z)}\, dl_\gamma. \quad (4)$$

This expression includes an integration over the probability distribution of spectral indices α based on the second *EGRET* Catalog [15].

There is also an important attenuation factor in this expression; the attenuation occurring as the γ-rays produced by blazars propagate through intergalactic space and interact with cosmic UV, optical, and IR background photons to produce $e^\pm$ pairs. If a substantial fraction of the EGRB is from high-z sources, a steepening in the spectrum should be seen at energies above ~ 20 GeV caused by the attenuation effect [16]. Figure 4, from Ref. [16], shows the calculated EGRB spectrum (based on the *EGRET* PSS) compared to *EGRET* data. The slight curvature in the spectrum below 10 GeV is caused by the distribution of unresolved blazar spectral indeces; the harder sources dominate the higher energy EGRB and the softer sources dominate the lower energy EGRB. The steepened spectra above ~ 20 GeV in Figure 4 show the attenuation effect and its uncertainty.

Critique of the Assumption of Independence of Blazar Gamma-Ray and Radio Luminosities

Chiang and Mukherjee [17] have attempted to calculate the EGRB from unresolved blazars assuming complete independence between blazar γ-ray and radio luminosities. They then used the intersection between the sets of flat spectrum radio sources (FSRSs) of fluxes above 1 Jy found in the Kühr catalogue and the blazars observed by *EGRET* as their sample, optimizing to the redshift distribution of that intersection set to obtain a LF and source redshift evolution. Using this procedure, they derived a LF which had a low-end cutoff at 10^{46} erg s^{-1}. Then, with no fainter sources included in their analysis, they concluded that only $\sim 1/4$ of the 0.1 to 10 GeV EGRB could be accounted for as unresolved blazars and that another origin must be found for the EGRB in this energy range.

We have argued above that it is reasonable to expect that the radio and γ-ray luminosities of blazars are correlated. Any such correlation will destroy the assumption of statisitical independence made by Chiang and Mukherjee and introduce a

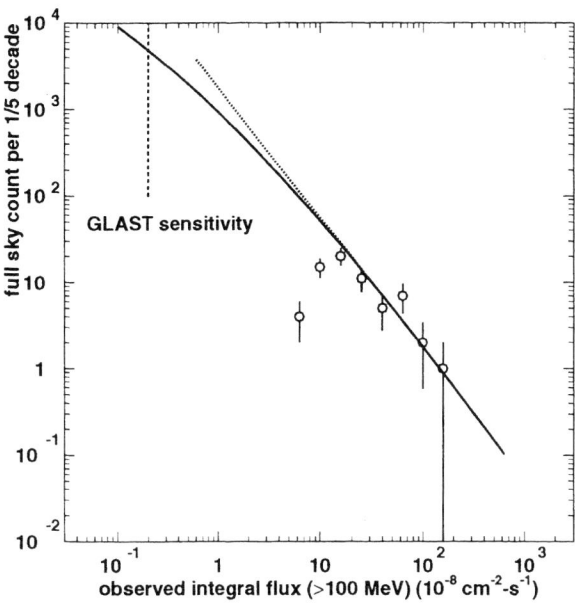

FIGURE 3. Source number count per one-fifth decade of integral flux at Earth The straight dotted line is the Euclidean relation $\mathcal{N}(>F) \propto F^{-3/2}$ for homogeneous distribution of sources. The open circles represent the *EGRET* blazar detections and the solid line is the model prediction.

bias in their analysis. In fact, their analysis leads to many inconsistencies. Among them are the following:

A. The LF derived by Chiang and Mukherjee [17] allows for no sources with luminosities below 10^{46} erg s^{-1}. In fact, *all* of the six sources found by *EGRET* at redshifts below ~ 0.2 have luminosites between $\sim 10^{45}$ erg s^{-1} and $\sim 10^{46}$ erg s^{-1} [18]. Elimination of fainter sources from the analysis can only lead to a lower limit on the EGRB from unresolved blazars. The fainter sources contribute significantly in acounting for unresolved blazars being the dominant component of the EGRB. (In this regard, see also, Ref. [19].)

B. Chiang and Mukherjee limit the *EGRET* sources in their analysis only to the FSRSs in the Kühr catalogue. However, if there is truly no correlation between blazar radio and γ-ray luminosities, then any of the millions of FSRSs given by the Dunlop and Peacock radio LF [14] are equally likely to be *EGRET* sources. In that case, of the 50 odd sources in the 2nd *EGRET* catalogue, virtually *none*, *i.e.* $\sim 10^{-6}$, should be Kühr sources.

The above discussion indicates that the assumption of non-correlation between the radio and γ-ray fluxes of blazars made by Chiang and Mukherjee in their analysis is not a good one and that this assumption invalidates their conclusions.

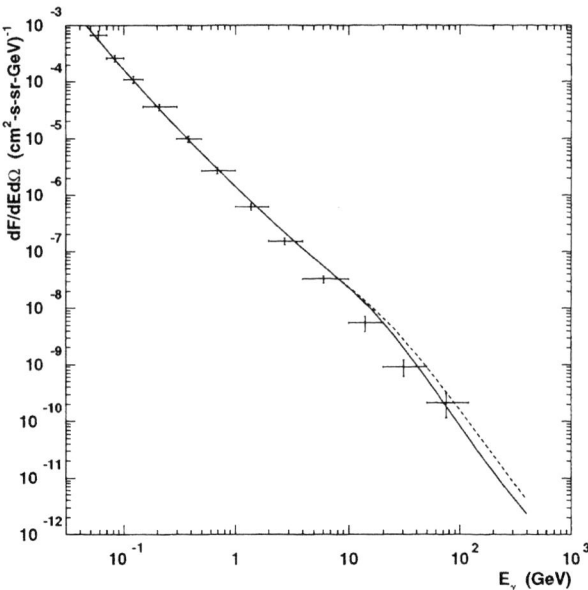

FIGURE 4. The predicted *EGRET* EGRB from unresolved blazars compared with the *EGRET* data. *GLAST* should see an EGRB about a factor of 2 lower at energies above 1 GeV (see text).

GLAST and the EGRB:

With an estimated point source sensitivity (PSS) nearly two orders of magnitude lower than *EGRET*'s, *GLAST* will be able to detect $\mathcal{O}(10^2)$ times more blazars than *EGRET*, and measure the EGRB spectrum to > 1 TeV (assuming the *EGRET* power law spectrum). These two capabilities will enable *GLAST* to either strongly support or reject the unresolved-blazar hypothesis for the origin of the EGRB.

Figure 3 shows that $\mathcal{O}(10^3)$ blazars should be detectable by *GLAST*, assuming it achieves a PSS of $\sim 2 \times 10^{-9}$ cm^{-2}s^{-1}. Using this PSS and our derived source count curve as shown in Figure 3, we have estimated that the remaining "diffuse" EGRB seen by *GLAST* should be a factor of ~ 2 lower for $E > 1$ GeV. Below 1 GeV, this factor of 2 will not apply because source confusion owing to the poorer angular resolution of *GLAST* at these lower energies will reduce the number of blazars resolved out of the background.

We conclude that *GLAST* can test the unresolved blazar background model in three ways:

A. *GLAST* should see roughly 2 orders of magnitude more blazars than *EGRET* because of its ability to detect the fainter blazars which contribute to the EGRB in our model. It can thus make a much deeper determination of the source count curve. *GLAST* can also determine the redshift distribution of many more identified γ-ray blazars, using its better point source angular resolution to make identifications with optical sources having measured redshifts. With its larger dynamic range, *GLAST* can then test the assumption of an average linear relation between the γ-ray and radio fluxes of identified blazars. All of these determinations will test the basic assumptions and results of our model.

B. With its better PSS, *GLAST* will resolve out more blazars from the background. Thus, fewer unresolved blazars will be left to contribute to the EGRB. reducing the level of the measured EGRB compared to *EGRET*'s by a factor of ∼2 if our predictions are correct.

C. The much greater aperture of *GLAST* at 100 GeV will allow a determination of whether or not a steepening exists in the EGRB, since the number of EGRB γ-rays recorded by *GLAST* above 100 GeV will be of order 10^3 to 10^4, assuming a continuation of the *EGRET* power-law spectrum. Such a steepening can be caused by both absorption and intrinsic turnovers in blazar spectra. Given enough sub-TeV spectra of individual blazars with known redshifts, these two effects can be separated.

THE EGRB BETWEEN 0.5 AND 30 MEV

The explanation for the origin of the EGRB at energies in the range of several MeV must be a non-blazar explanation. The reason for this is that while the EGRB spectrum in this energy range appears to be *softer* than that at higher energies [2], the data from *OSSE* and *COMPTEL* on individual blazars in this energy range indicate a *harder* spectrum than that at higher energies. The measured blazar spectra appear to break below ∼ 10 MeV to spectra with a typical power-law index of ∼ 1.7 [20]. Thus, even if unresolved blazars account for almost all of the EGRB in the 0.1 to 10 GeV range, this cannot be the case at lower energies.

Calculations have shown that a superposition of redshifted lines from Type Ia and Type II supernovae should reasonably provide a significant component of the EGRB at energies ∼ 1 MeV. The important line emission is from the decay chain ^{56}Ni → ^{56}Co → ^{56}Fe and also from the decay of ^{26}Al, ^{44}Ti and ^{60}Co [21], [22]. However, supernovae cannot account for the entire EGRB in this energy range, since they produce no line emission above 3.5 MeV.

Another serious possibility as a significant contributor to the multi-MeV EGRB is non-thermal tails in the energy spectra of the AGN [23]. These would be the same AGN which have recently been resolved out by the *Chandra* telescope and found to be the dominant component of the once unresolved X-ray background [24].

A recent discussion of AGN models fitting the X-ray background has been given in Ref. [25]. While there are no data on individual AGN in the multi-MeV energy range at the present time, Stecker, Salamon and Done [23] have pointed to the galactic black hole candidate Cyg. X-1 as an example of a black hole source which has been shown from *COMPTEL* data to have a non-thermal tail extending to multi-MeV energies [26]. If the extragalactic black hole sources which make up the X-ray background have such non-thermal tails, they may account for most of the

EGRB in the multi-MeV range.

It should be noted that the extraction of the $\sim$ MeV EGRB from the raw *COMPTEL* data is a difficult process, in part owing to the fact that this double Compton scattering telescope was not designed to measure this background. In our opinion, a dedicated low-mass, free flyer satellite, specifically designed to measure the EGRB at low γ-ray energies will be required in order to accurately determine its characteristics.

THE EGRB ABOVE 10 GEV

It has already been pointed out that the EGRB should break above $\sim$ 20 GeV energy owing to absorption of high energy γ-rays by pair-production interactions with lower energy starlight photons [16]. There is also another potential cause for a steepening in the EGRB from blazars. The *EGRET* detector obtained rough power-law spectral indeces for blazars in the 0.1 to 10 GeV energy decade, however, we presently have no data for these objects in the 10 to 100 GeV decade. Presently popular theoretical models predict that the spectra of highly luminous blazars will exhibit a cutoff at energies in the 10 to 100 GeV range, whereas the less luminous X-ray selected BL Lac objects can have spectra extending into the TeV energy range [27], [28].

Indeed, there have now been ground based detections of at least 5 X-ray selected BL Lac objects (Weekes, these proceedings), some of whose spectra extend to multi-TeV energies. While no other types of blazars have been seen at TeV energies, this may be an result of intergalactic γ-ray absorption [16], [29], [30] so that we do not really know if their intrinsic spectra turn down at energies in the 10 to 100 GeV decade. The *GLAST* telescope should provide this knowledge in the not-too-distant future.

If the spectra of most blazars possess intrinsic cutoffs above 10 GeV, then the EGRB from unresolved blazars would be expected to turn over as well. This effect should be more dramatic than the steepening in the EGRB predicted from the effect of intergalactic absorption [16]. In that case, if the *EGRET* results on the EGRB up to 100 GeV are correct, a new component may be present in this higher energy range. Such a component has been predicted to be produced by the decay of $\sim$ TeV mass higgs bosons from cosmic string processes in flat-potential supersymmetric models [31]. Of course, there may be other unknown possibilities as well.

CONCLUSIONS

We have a workable and testable hypothesis for the origin of the extragalactic γ-ray background measured by *EGRET*, *viz.*, that it is made up primarily of unresolved blazars. The *GLAST* γ-ray telescope, to be flown in the near future, will be able to test this hypothesis in three ways, *i.e.*, (a) by potentially resolving out and detecting thousands of more sources, (b) by measuring the remaining background flux, and (c) by determining the shape of the EGRB up to TeV energies. The many new ground-based detectors now under construction will supplement this information by discovering new extragalactic sources of γ-rays of energies above 50 GeV.

On the other hand, the mystery of the origin of the EGRB in the MeV energy range must await a better determination of this background by a future dedicated satellite detector.

REFERENCES

1. Sreekumar, P. et al., Ap.J. 494, 523 (1998).
2. Kappadath, S.C. et al. Astron. and Astrophys. Suppl. 120, 619 (1996).
3. Stecker, F.W. and Salamon, M.H., Phys. Rev. Letters 76, 3878 (1996).
4. Stecker, F.W., ApJ 212, 60 (1977).
5. Hunter, S.D. et al. ApJ 481, 205 (1997).
6. Dar, A. and Shaviv, N.J., Phys. Rev. Letters 75, 3052 (1995).
7. Stecker, F.W. and Salamon, M.H., ApJ 464, 600 (1996).
8. Chiang, J. et al., ApJ 452, 156 (1995).
9. Stecker, F.W., Salamon, M.H., and Malkan, M.A., ApJ 410, L71 (1993).
10. Padovani, P. et al., MNRAS 260, L21 (1993).
11. Mücke, A., et al., Astron. and Astrophys. Suppl. 120, 541 (1996).
12. Mattox, J.R., et al., ApJ 481, 95 (1997).
13. Jorstad, S.G. et al., ApJ Suppl. 134, 181 (2001).
14. Dunlop, J.S. and Peacock, J.A., MNRAS 247, 19 (1990).
15. Thompson, D.J. et al., ApJ Suppl. 101, 259 (1995).
16. Salamon, M.H. and Stecker, F.W., ApJ 493, 547 (1998).
17. Chiang, J. and Mukherjee, R., ApJ 496, 752 (1998).
18. Mukherjee, R., in Proc. Intl. Symp. on High Energy Gamma-Ray Astronomy, Heidelberg, in press, e-print astro-ph/0101301.
19. Mücke, A. and Pohl, M., Astron. and Astrophys. 312, 177 (2000).
20. McNaron-Brown, K. et al., ApJ 451, 575 (1995).
21. The, L.S., Leising, M.D. and Clayton, D.D., ApJ 403, 32 (1993).
22. Watanabe, K., et al., ApJ 516, 285 (1999).
23. Stecker, F.W., Salamon, M.H. and Done, C., e-print astro-ph/9912106.
24. Mushotzky, R.F. et al., Nature 404, 459 (2000).
25. Gilli, R., Salvati, M. and Hasinger, G., e-print astro-ph/0011134.
26. McConnell, M.L. et al., ApJ 543, 928 (2000).
27. Stecker, F.W., De Jager, O.C. and Salamon, M.H., ApJ Letters 473, L75 (1996).
28. Fossati, G., et al., MNRAS 299, 433 (1998).
29. Stecker, F.W., De Jager, O.C. and Salamon, M.H., ApJ Letters 390, L49 (1992).
30. Stecker, F.W. and De Jager, O.C., Astron. and Astrophys. 334, L85 (1998).
31. Bhattacharjee, P., Shafi, Q. and Stecker, F.W., Phys. Rev. Letters 80, 3698 (1998).

Contributions of GRBs and Cen A-like Radio Galaxies to the Cosmic Gamma-ray Background

K. Watanabe[1] & D.H. Hartmann[2]

1, EIT/LHEA, NASA/GSFC, Code 660, Greenbelt, MD 20771, USA
2, Clemson University, Clemson, SC 29634-0978, USA

Abstract. The contribution to the cosmic diffuse gamma-ray background (CGB) from Gamma Ray Bursts (GRBs) is studied in the 40 keV - 2 MeV regime. We use High Energy Resolution (HER) data from the Burst And Transient Source Experiment (BATSE) aboard the Compton Gamma-Ray Observatory (CGRO) to generate a GRB template spectrum. Although the GRB contribution to the CGB is generally small, in comparison to the dominant flux from Type Ia supernovae, the integrated GRB flux is in fact comparable to that from SNIa in the narrow 10-40 keV range. GRBs contribute to the CGB at the same level as Type II supernovae do. Although BATSE data are not available below ~40 keV, extrapolation of the template spectrum suggests that bursts can fill a significant part of the existing gap between Seyfert galaxies (dominating the CGB below ~ 100 keV) and SNIa (dominating at ~1 MeV). We estimate contributions from Cen A-like (FR I) radio galaxies in this energy regime, where INTEGRAL data is expected to provide major advances.

INTRODUCTION

We have studied the CGB in the MeV region both observationally and theoretically (see e.g., Watanabe et al. 1999a,b). Recent observations with COMPTEL and SMM show no evidence for a "MeV bump" (Weidenspointner 1999, Watanabe et al. 1999b) which was indicated by previous experiments. Suggested sources for the CGB include unresolved Seyfert galaxies (e.g., Zdziarski 1996) in the lower energy band (up to 300, keV) and Blazars (e.g., Sreekumar et al. 1998) in the high energy band (> 100 MeV). Although cosmological SNIa contribute a significant flux to the CGB around 1 MeV (The et al. 1993, Watanabe et al. 1998 & 1999a, Ruiz-Lapuente et al. 2001), there remains a gap between Seyfert and SNIa contributions where a continuous power-law like CGB spectrum has been observed. There is no generally accepted explanation for the origin of the CGB flux in this gap. Here we consider the contribution from GRBs and Cen A- like (FR I) radio galaxies, and show that neither of these sources can adequately fill the gap.

DATA ANALYSIS

Gamma-ray Bursts (GRBs)

We use the BATSE GRB trigger data from the 4B catalog, which is available online from http://cossc.gsfc.nasa.gov/cossc/batse/4Bcatalog/. For 1234 bursts in the catalog duration (T90) information is available. Using the FTOOLS CGRO sub-package and XSPEC, GRB energy spectra with background subtraction are constructed. High Energy Resolution (HER) data obtained from the detector which received the brightest gamma-ray flux among BATSE's eight detectors were selected for spectral analysis. In some cases HER data are not available or the burst is too faint to allow high quality spectral analysis. In the end, 781 spectra out of 1234 triggered events were generated. We sum all spectra (multiplied by T90) to obtain an average fluence template in units of $photons/cm^2/keV$. In order to obtain the GRB contribution to the CGB, the template was divided by 4π and multiplied by an average all sky GRB rate of one per day.

Cen A-like Radio Galaxies

Centaurus A (Cen A) is a radio-loud Seyfert galaxy, belonging to a class intermediate between radio-quiet (normal) Seyferts and Blazars, also classified as FR I radio galaxy. Cen A is the brightest radio galaxy (with a Seyfert 2 nucleus) detected with OSSE, the Oriented Scintillation Spectrometer Experiment on CGRO. The spectrum of Cen A appears to extend well into the GeV regime (Kinzer et al. 1995).

We use the BATSE occultation data for Cen A available online from http://cossc.gsfc.nasa.gov/cossc/batse/hilev/CEN_A/cen_a.html/. An average Cen A spectrum was obtained from 385 individual spectra in the energy range of 20 keV $<$ E $<$ 1 MeV by using the FTOOLS CGRO sub-package and XSPEC. We use the average spectrum as template for all the Cen A-like (FR I) radio galaxies. We fit the spectrum with a power law with index of 1.7, and extrapolate to 10 MeV. We adopt a present-day FRI galaxy density of $n_o = 2.0 \times 10^{-6}$ galaxies/Mpc^3 (e.g., Canosa, et al. 1999, Colina, et al., 1995, & Colla, et al., 1975). The contribution of the Cen A-like (FR I) radio galaxies to the CGB in units of (photons cm^2 s^{-1} keV^{-1} sr^{-1}) is given by

$$F_e = L_H D^2 \int dz\, n(z)\, \dot{N}(E_\gamma \times (1+z)) E(z)$$

where n(z) accounts for density evolution: $n(z)=n_0(1+z)^m$ with m = 3, 2, 1, 0, -1, -2, -3, $L_H=c/H_0$ is the Hubble length, D is the distance to Cen A (3.4 Mpc), the dotted N is the observed Cen A spectrum (photons cm^2 s^{-1} keV^{-1}), and where E(z) represents the evolution of the Hubble constant (see eq. [13.3] in Peebles 1993). The integration over red shift was carried out to z = 3. FIGURE 1 shows the resulting CGB contributions.

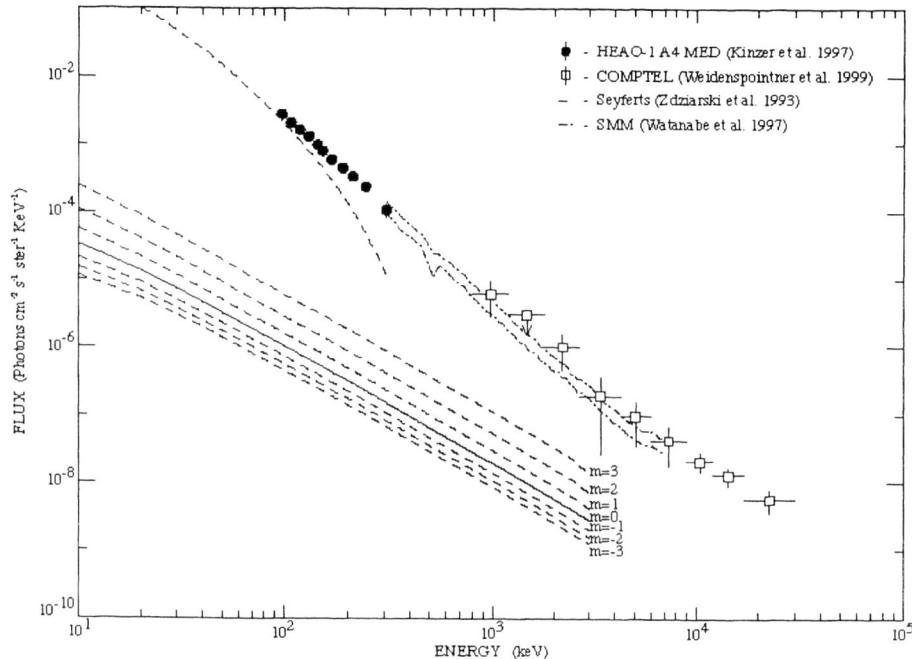

FIGURE 1. Estimated contribution of Cen A-like radio galaxies to the CGB for various assumptions about density evolution: $n(z) = n_0 (1+z)^m$ with $n_0 = 2.0 \times 10^{-6}$ galaxies Mpc^{-3} and m = (3, 2, 1, 0, -1, -2, -3).

Results

FIGURE 2 shows the contributions of GRBs and Cen A-like (FRI) radio galaxies to the CGB in comparison to contributions from SNIa and SNII (Watanabe et al. 1999a). The observed data are HEAO-1 (Kinzer et al.1997), COMPTEL (Weidenspointner 1999), and SMM (Watanabe et al. 1997). At low energies the contribution from Seyferts dominates (e.g., Zdziarski et al. 1993), while blazars dominate above a few MeV. Supernovae (mostly SNIa) dominate the MeV regime, where GRBs and Cen A-like radio galaxies (without density evolution) account for ~1% of the observed flux.

Conclusion

While Seyfert galaxies and SNIa are the major contributors to the CGB in the 0.1-1 MeV band, GRBs and Cen A-like radio galaxies provide a non-negligible portion of the flux near 300 keV. However, the apparent flux deficit is not fully accounted for by the contributions from bursts and radio galaxies. The fact that a gap between Seyferts and SNIa remains suggests that there are contributing sources that have not yet been recognized. INTEGRAL might discover a new class of sources, but will also improve our understanding of radio galaxies and thereby improve existing models of the CGB.

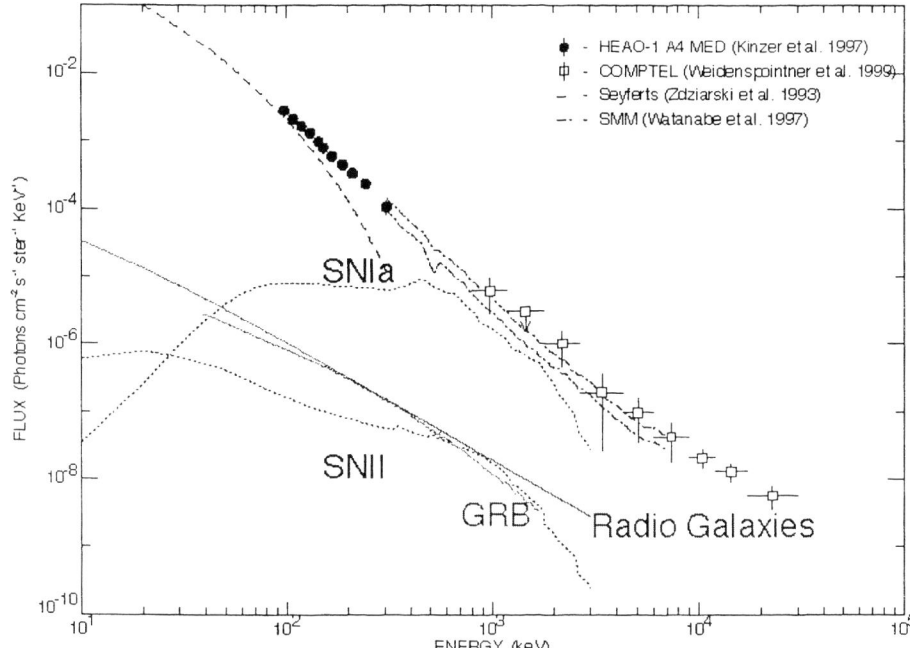

FIGURE 2. The observed CGB and estimated contributions from supernovae, gamma-ray bursts, and radio galaxies (without density evolution). While SNIa clearly account for the bulk of the observed CGB, up to 10% of the flux in the MeV regime could be due to SNII, GRBs, and radio galaxies.

REFERENCES

1. Canosa, C.M., et al. 1999, MNRAS 310:30
2. Colina, L. et al., 1995, ApJ, 448:548
3. Colla,,G. et al.,1975, A&A, 38:209
4. Sreekumar, P., et al.1998, ApJ, 494, 523
5. Kinzer,R.L., et al. 1995, ApJ,449:105
6. Kinzer, R. L., et al. 1997, ApJ , 475,361
7. Ruiz-Lapuente, P. et al., 2001, ApJ, 549:483
8. Peebles, P.J.E. 1993, Principles of Physical Cosmology (Princeton Univ. Press)
9. The, L.-S.,et al. 1993, ApJ ,403,32
10. Watanabe, K., et al. 1997, in the Proceedings of the 4th CGRO Symposium, pp.1223
11. Watanabe, K., et al. 1998, Astron. Nachr. 319, 67
12. Watanabe, K., et al. 1999a, ApJ, 516, 285
13. Watanabe, K., et al. 1999b, AIP 510, eds. M. L. McConnell and J. M. Ryan, pp. 471
14. Weidenspointner, G. 1999, Ph. D. Thesis, Technische Universität München
15. Zdziarski, A. A., et al. 1993, ApJ, 414, L81
16. Zdziarski, A. A. 1996, MNRAS, 281, L9

Particles and Cosmic Rays

The Origin of Cosmic Rays and the Diffuse Galactic Gamma-Ray Emission

Seth W. Digel[1]*, Stanley D. Hunter*, Igor V. Moskalenko[2]*, Jonathan F. Ormes* and Martin Pohl[†]

NASA/GSFC, Code 660, Greenbelt, MD 20771 USA
[†]*Ruhr-Universität Bochum, 44780 Bochum, Germany*

Abstract. Cosmic-ray interactions with interstellar gas and photons produce diffuse gamma-ray emission. In this talk we will review the current understanding of this diffuse emission and its relationship to the problem of the origin of cosmic rays. We will discuss the open issues and what progress might be possible with GLAST, which is planned for launch in 2006.

1. INTRODUCTION

The problem of the origin of cosmic rays (CRs) has existed for almost a century. In 1912, Victor Hess carried an electroscope aloft in a hot air balloon and discovered radiation increasing with altitude and exhibiting no day-night variation. It had to be some kind of highly-penetrating radiation coming from beyond the solar system. Then in 1948, Freier et al. [1] discovered that this radiation included nuclei of heavy elements with relative abundances similar to the solar system.

Much has been learned since about the important dynamical consequences of CRs for the Galaxy [2]. CRs are tied to magnetic fields, magnetic fields are anchored in the gas phase of the Galaxy, and the gas is held in place by gravitational forces. From radio mapping of external spiral galaxies like our own Milky Way, CRs are known to be common to galaxies. Evidence for the shock acceleration of electrons in supernova remnants (SNR) has recently been found (see Sec. 2). However, no convincing evidence has been found for the commonly held view that CR nuclei are also accelerated in SNR. A few percent of the energy released in SNR must find its way into energetic CRs in order to keep the Milky Way galaxy supplied with CRs.

Interactions of CRs with the interstellar medium (ISM) and photons produce diffuse high-energy gamma radiation that is diagnostic of the CRs. The γ-ray fluxes are low, but γ-rays do have the advantages of not being deflected by magnetic fields and low optical depth for attenuation by intervening matter. Results from observations of diffuse γ-ray emission by the Energetic Gamma Ray Experiment Telescope (EGRET) on the

---

[1] Universities Space Research Association

[2] NRC Senior Research Associate; on leave from Institute of Nuclear Physics, M. V. Lomonosov Moscow State University, 119 899 Moscow, Russia

Compton Gamma Ray Observatory have shown the potential of γ-ray measurements to contribute to solving the problem of the origin of CR nuclei ([3] and references therein).

In this paper we review the current understanding of CR origin and sources, CR diffusion throughout the interstellar medium, and the production of γ-rays by both CR nucleons and electrons. We also review the discoveries of EGRET and explore the problems relating to the interpretation of those results. We document some open questions and discuss how the Gamma Ray Large Area Telescope (GLAST) can contribute.

2. THE ORIGIN OF COSMIC RAYS

Observations of the Magellanic clouds with EGRET have shown that CRs in the GeV range are almost certainly Galactic [4]. However, only a few classes of objects in the Galaxy provide sufficient energy and power to replenish the CRs, one of which is SNR. In fact, particle acceleration at SNR shock waves is regarded as the most probable mechanism for providing CRs at energies below 10^{15} eV.

The distribution of SNR in the Galaxy is so poorly known and the propagation range of CRs, which is related to the halo size, is so weakly constrained, that a comparison of the CR source distribution, which may be inferred from the γ-ray gradient (Sec. 3), with the distribution of SNR must be inconclusive. A direct search for γ-ray emission from SNR is an alternative strategy. In fact, a few unidentified EGRET sources are positionally coincident with radio-bright SNR [5], but similar correlations exist with OB associations and SNR-OB associations (SNOBs) [6], not to mention the possibility of these sources actually being radio-quiet or highly dispersed pulsars. The EGRET data alone do not permit a firm identification of γ-ray sources with SNR, for the angular resolution is too coarse.

TeV observations using the atmospheric Čerenkov technique can be performed with much higher angular resolution and sensitivity than is possible with EGRET. However, the much higher threshold energy implies that CRs with energies around 100 TeV are probed. The most simple calculations of shock acceleration [7] indicate that particle spectra with number index ~ 2 may be produced, but also that acceleration cut-offs have to be expected at 0.1–1 PeV [8]. Corresponding models of hadronic γ-ray emission from SNR have suggested that a number of sources should be detectable with present-day telescopes [9], in particular those embedded in dense gas [10]. Following these expectations, very deep surveys of TeV emission from SNR were performed, but no SNR has been unambiguously detected as a source of hadronic TeV γ-rays to date [11].

More careful models of shock acceleration, which include non-linear effects arising, e.g., from the influence of the accelerated CRs on the shocked plasma, predict particle spectra that deviate from pure power laws [12]. Also, the energy and momentum carried by electromagnetic turbulence and the motion of the CR scattering centers relative to the plasma modify the process of shock acceleration, such that an E^{-2} spectrum is not the canonical result it was thought to be [13]. In fact a distribution of spectral indices should exist, and this is actually observed in the radio spectra of shell-type SNR [14]. Therefore, the non-detection of hadronic TeV emission from SNR may be not incompatible with CR acceleration in these sources; nevertheless, the results seem to conflict with the notion that SNR accelerate CR hadrons up to the "knee" (Sec. 3).

The recent detections of non-thermal X-ray synchrotron radiation from the SNR SN1006 [15], RX J1713.7-3946 [16, 17], IC443 [18], Cas A [19], and RCW86 [20], and the subsequent detections of SN1006 [21] and RX J1713.7-3946 [22] at TeV energies support the hypothesis that at least Galactic CR electrons are accelerated predominantly in SNR. The relative intensities of the keV synchrotron emission and the TeV inverse Compton (ICS) radiation are independent of the acceleration process and the model thereof [23]. Care has to be exercised in separating thermal from non-thermal X-ray emission [24], though, and the substructure of the SNR shock as well as propagation effects have to be taken into account when interpreting the TeV data [25, 26]. Nevertheless SN1006 and RX J1713.7-3946 are obvious examples of leptonic TeV γ-ray emission. The most recently detected SNR, Cas A [27], is a less clear case, but presumably the emission is also of leptonic origin [28].

It is interesting to note that for all SNR the X-ray flux, synchrotron or not, is less than the extrapolated radio synchrotron spectrum. Since many of the sources, in particular the five historical remnants, are too young for their electron spectra to be limited by energy losses, acceleration cut-offs must occur at electron energies of 100 TeV or less [29], which would be intrinsic to the actual acceleration process.

The production of CR electrons in SNR has important consequences. At energies higher than about 100 GeV the lifetime of electrons, and thus their range, is rather short; only a few SNR would contribute to the locally observable CR electron spectrum. Therefore the locally measured spectrum would not be representative of the spectrum elsewhere in the Galaxy. This conclusion is the basis for the ICS models of the GeV excess [30] (Sec. 3).

3. PROPAGATION OF COSMIC RAYS

The spectrum of CRs can be approximately described by a single power law with index −3 from 10 GeV to the highest energies ever detected, $\sim 10^{20}$ eV. The only feature observed is a "knee" around 10^{15} eV. Because of this featureless spectrum, CR production and propagation are believed to be governed by the same mechanism over decades of energy; a single mechanism works below the knee and the same or another one works above the knee, although the origin of the CR spectrum is not still understood.

Energetic CR interactions with gas or magnetic and radiation fields in the interstellar medium produce γ-rays that carry information about these interactions such as the spectrum and flux of CR and physical conditions such as the gas density and radiation field. Some portion of γ-rays is produced near the CR sources, like SNR and pulsars, that generally appear as point sources to γ-ray telescopes. The rest are produced in CR interactions in the ISM and are therefore diffuse.

In the ISM particles diffuse and lose or gain energy, and so their spectra change from their initial forms. Freshly-accelerated particles (Sec. 2) propagate in the ISM where they produce secondary particles and γ-rays. Electrons produce synchrotron photons and γ-rays via ICS and bremsstrahlung. Nucleons spallate and produce secondary nucleons, antiprotons, and charged pions that give rise to secondary positrons and electrons.

Decays of secondary π^0's also produce γ-rays.

CRs can be measured directly only in the solar system, in the outskirts of our Galaxy. Those observed in the solar system are a complicated mixture of primary and secondary particles which are diffusing through the ISM from their sources. Only photons or γ-rays are able to deliver the information directly from other parts of the Milky Way, but this information is integrated over the line of sight. To extract information about CRs from γ-rays models of CR propagation are needed.

CRs propagate throughout the Galaxy guided by magnetic fields. They are held in the Galaxy by scattering on magnetic irregularities, the process described mathematically as diffusive propagation. The models can be complex and distinguish between the diffusion rates in the thin galactic disk and extensive halo [31]. The diffusion coefficient may depend on the local plasma conditions and is probably different in the disk and the halo. It is often assumed to depend on particle rigidity with index 0.33–0.6, with the former value reflecting a "Kolmogorov" spectrum of magnetic scattering centers and the latter obtained phenomenologically from fitting the measured B/C ratio. Propagation may be influenced by a galactic wind (convection of particles away from the Galaxy) and/or diffusive re-acceleration (a second order Fermi process).

The diffusive model can be shown to be equivalent to, and because of the complexities mentioned above is often replaced by, a simpler empirical formalism known as the "leaky-box model." In this model the principal parameter is an effective escape length or grammage (column density of matter traversed, in g cm^{-2}) and the sources and particles are uniformly distributed in space and time throughout the Galaxy. The grammage parameter is determined by fitting to the data and found to be a broken power law in rigidity, increasing at low energies and decreasing with index –0.6 at relativistic energies.

Detailed CR diffusion models have been developed over the last 30 years. The GAL-PROP model [32], for example, is a recent numerical model which combines the Galactic structure with diffusive reacceleration and convection in the ISM. This model incorporates all CR species with $Z < 29$ including leptons and antiprotons, together with γ-rays and synchrotron emission.

Measurements of primary CR abundances or stable secondary/primary ratios do not permit distinguishing between the leaky box and diffusion models. However, the propagation of radioactive secondary isotopes depends on the lengths of time that particles spend in the disk (production) and the halo (decay), therefore making them sensitive probes of the propagation model.

Propagation parameters are usually derived using B/C and (Sc+Ti+V)/Fe ratios, while radioactive secondary isotopes ^{10}Be, ^{26}Al, ^{36}Cl, ^{54}Mn (all with $T_{1/2} \sim 0.3$–2 Myr) allow the size of the halo to be constrained. Figure 1 shows examples of B/C calculations and halo size constraints derived from radioactive isotopes [33]. Once defined from B/C and Be measurements, the propagation parameters determine other isotopic ratios.

Typical values of the diffusion coefficients depend on the propagation model and values of cross sections employed, but all of them are of the order of few times 10^{28} cm^2 s^{-1}. Larger values make the propagation more rapid and produce smaller amounts of secondaries, smaller values increase the secondary/primary ratios.

The spectrum of diffuse γ-rays from the inner Galaxy has excesses at high and low energies compared to calculations based on the assumption that the nucleon and electron spectra do not change shape throughout the Galaxy [39, 40] (see Sec. 4 and Fig. 3). This

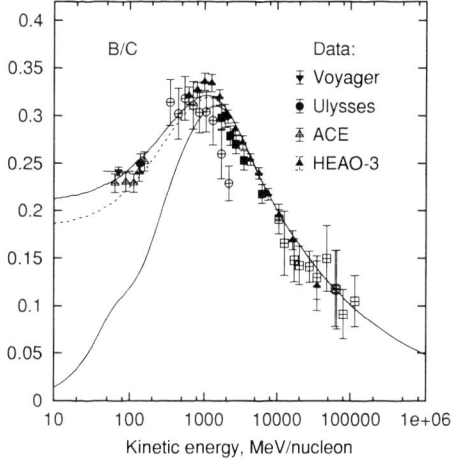 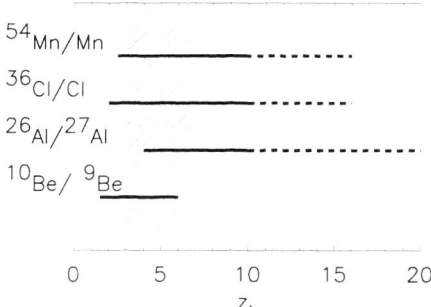

FIGURE 1. Left: B/C ratio calculated for CR halo scale height $z_h = 4$ kpc [33]. Lower curve local interstellar spectrum, upper modulated: solid curve – $\Phi = 500$ MV, dotted curve – 400 MV. Data below 200 MeV/nucleon: ACE [34], Ulysses [35], Voyager [36]; high energy data: HEAO-3 [37], for other references see [38]. **Right:** Halo size limits [33] derived from the abundances of the 4 radioactive isotopes and ACE data. The ranges reflect errors in ratio measurements, source abundances, and production cross sections. Dashed lines indicate uncertain upper limits. The shaded area indicates the range consistent with all ratios.

implies that the local spectra may be not representative for the Galaxy as a whole. At low energies, the excess may be due to unresolved point sources that dominate in the MeV range and below [41]. At high energies, where most photons come from ICS and π^0 decay, the excess can be explained by spectra of nucleonic [42] and/or leptonic [30] components that are harder than those observed locally.

Secondary antiprotons and positrons are produced in the same nucleonic interactions in the ISM as π^0's and thus provide information complimentary to that of γ-rays (without direction information), but in this case the agreement is good [40]. The leptonic hypothesis of the origin of the excess at high energies therefore may be more plausible, especially because of large energy losses of high energy electrons.

A new test for the origin of the excess will be feasible with the new generation telescope GLAST (Sec. 5), which will be able to measure γ-rays up to 300 GeV for the first time. While improved angular resolution will allow better discrimination of the point source contribution, the capability for spectral measurements at sub-TeV energies is essential to distinguish between diffuse nucleonic and ICS spectral components. The shape of the γ-ray spectrum from π^0-decay resembles the spectrum of the nucleonic component of CRs; the ICS spectrum is flatter and its cut off energy is determined by the maximum energy to which SNR can accelerate electrons. Because this energy is in the 1–100 TeV range, the spectrum of γ-rays in the 10–100 GeV range is therefore crucial [43].

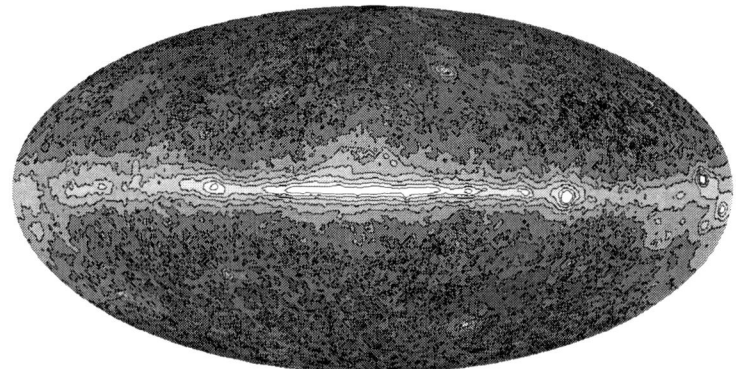

FIGURE 2. Intensity of gamma rays (> 100 MeV) observed by EGRET. The broad, intense band near the equator is interstellar emission from the Milky Way. The intensity scale ranges from 1×10^{-5} cm^{-2} s^{-1} sr^{-1} to 5×10^{-4} cm^{-2} s^{-1} sr^{-1} in ten logarithmic steps. The data have been smoothed slightly by convolution with a gaussian of FWHM 1.5°.

4. RESULTS AND QUESTIONS FROM EGRET

EGRET completed the first all-sky survey at high-energies (> 30 MeV, Fig. 2). More than 60% of the γ-rays that EGRET detected are from CR interactions in the Milky Way. The sensitivity as well as the excellent background rejection of EGRET enabled great progress in the study of interstellar γ-ray emission.

EGRET observations of the Large and Small Magellanic clouds confirmed that CRs are galactic in origin, rather than metagalactic or universal. The γ-ray flux of the LMC measured by EGRET is consistent with a CR density similar to that in the Milky Way [44]. However, the upper limit measured by EGRET for the SMC implies a CR density several times less [4].

The EGRET γ-ray spectrum of the inner Milky Way (Fig. 3) provided the first clear evidence for both electron and proton CRs across the Galaxy, with the spectrum following the π^0 "shoulder" for energies greater than the proton emissivity peak at half the π^0 rest mass [39]. Above ~ 100 MeV, π^0 decay γ-rays from proton-nucleon interactions are the dominant spectral component. At lower energies the spectrum is dominated by electron interactions via bremsstrahlung and ICS.

Models of the interstellar gamma-ray emission of the Milky Way based on the known γ-ray production mechanisms, together with inferred distributions of interstellar gas, low-energy photons, and CRs, predict γ-ray intensities consistent with the observations on scales of degrees [45, 39, 30, 40]. Owing to the limited statistics and angular resolution of the data, such models are essential for determining accurate positions and fluxes of γ-ray point sources at low latitudes. They are also the means to discover the distribution of CRs in the Milky Way. Broadly speaking, the CR distributions derived from the models are consistent with each other. However, differences in approach, including techniques used to resolve the non-unique inversion of radio and millimeter spectral line surveys into the 3-dimensional distribution of gas, presently limit the usefulness of detailed comparisons between models.

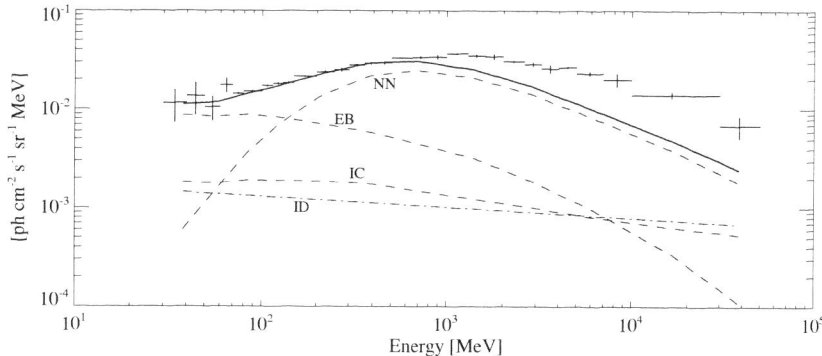

FIGURE 3. Spectrum of the inner Milky Way ($|l| < 60°$, $|b| < 10°$) with calculated components from bremsstrahlung (EB), inverse Compton (IC), π^0 decay (NN), and extragalactic isotropic emission (ID) [39].

For the nearest interstellar clouds, EGRET provided linear resolutions of ~ 10 pc. Studies of these, e.g. [46], permit the tightest direct constraints on confusion of unresolved gamma-ray point sources with diffuse emission, and also on the assumptions that CR densities and the molecular mass calibration, i.e., the relation of 2.6 mm CO line intensity W_{CO} to $N(H_2)$, are uniform on the scale of interstellar clouds.

The GeV excess (see Sec. 3) is evident in Fig. 3, as well as in spectra for smaller angular scales. Several explanations of the excess emission are possible: EGRET calibration error, uncertainty in the π^0 production, unresolved hard point sources, and a Galactic average proton and/or electron spectrum that may be harder than observed locally. An explanation involving ICS from electrons is perhaps the most plausible (see Sec. 3). Calibration error is unlikely, as the observed spectra of EGRET point sources, which are generally well described by single power laws, do not harden above 1 GeV (e.g., [47]). The π^0 production function has recently been re-evaluated using nuclear interaction Monte Carlo codes, and the hardening of emissivity predicted at GeV energies is not sufficient to explain the GeV excess [48]. Hunter et al. [39] concluded from the shape of the spectrum of the inner Galaxy that the contribution from unresolved sources with power-law spectra is $< 10\%$. This estimate is, however, rather uncertain because a large contribution from unresolved sources distributed closely like the molecular gas – although not evident in studies of local clouds – could be accounted for by reducing the $N(H_2)/W_{CO}$ ratio. Pulsars not detected individually could contribute significantly above 1 GeV, although apparently not with the correct latitude distribution [49, 50].

A halo of high-energy (> 100 MeV) γ-rays about the Milky Way remains when the EGRET team's interstellar emission model is subtracted from the observations [51, 52]. The simplest interpretation for the halo is incomplete accounting of the ICS emission at high latitudes (e.g., [53]). Other interpretations based on CR interactions in the halo with very cold H_2 in dense clumps, a dark matter candidate, have been proposed (e.g., [54]). The ICS interpretation alone appears sufficient. EGRET data may in fact offer little insights about baryonic dark matter in the halo. Even if the dark matter is in dense clumps, they could be dense enough to attenuate any γ-rays produced in them [55].

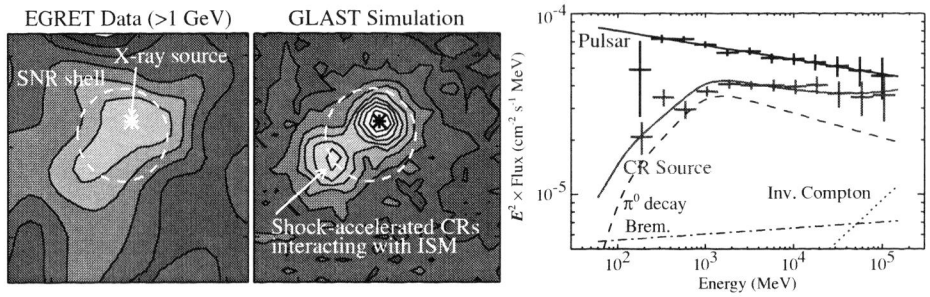

FIGURE 4. EGRET observation (summed Phase 1–4) and GLAST simulation (1-year sky survey) of the γ-Cygni SNR. The dashed circle is the location of the shell [58]. The spacing of the tick marks is 1°. Also shown are simulated measurements with GLAST of the spectra of the pulsar and CR source. See text for discussion of the γ-Cygni model.

5. ADVANCES ANTICIPATED WITH GLAST

The general capabilities of the Large Area Telescope on the Gamma-ray Large Area Space Telescope (GLAST), which is planned for launch in 2006, are described elsewhere in this volume [56]. The advantages of the large effective area, high observing efficiency, and narrow point-spread function of GLAST for the study of CR production and diffuse γ-ray emission are considered here.

Supernova remnants have been established as acceleration sites of leptons, but the acceleration of hadrons has not yet been detected (Sec. 2). The observational signature of proton acceleration would be expected to be π^0-decay γ-ray emission from proton collisions with nucleons in interstellar clouds being overtaken by SNR shocks. The γ-ray spectrum would appropriately reflect the hard spectrum of the CRs.

Although several EGRET γ-ray sources are spatially coincident with SNR [5, 57], the large positional uncertainties for the EGRET sources and the limited photon statistics do not permit π^0-decay emission from the shell to be distinguished from leptonic emission processes, or even from γ-ray emission from associated pulsars.

The potential for GLAST to clarify the nature of the associations is illustrated in Figure 4, which shows EGRET data and a GLAST simulation for the γ-Cygni SNR. This region was detected as a point source by EGRET (and designated as 3EG J2020+4017 [59]). An X-ray source at the position indicated in the figure has the spectral characteristics of a pulsar and has been proposed as the counterpart to the EGRET source [60]. For the simulation, it was assumed that the flux could be divided 60%/40% between the prospective pulsar and a source at the shell of the SNR where the EGRET data suggest an extension. Spectra for the two sources were selected to be consistent with the overall spectrum of 3EG J2020+4017 and the spectral components of the shell source were chosen to be consistent with models of the γ-ray emissivity of SNR [61] (for details see [62]). The GLAST intensity map shows that the CR source will be resolved from the pulsar at high energies. In addition, the spectra of the sources will be separately measurable at energies above 150 MeV, and the π^0-decay component of the shell source will be strongly detected.

For the study CRs in the Milky Way, the superior effective area and angular resolution of GLAST will permit important advances. GLAST will reveal whether unresolved (by EGRET) point sources are only a minor component of the celestial γ-ray flux, as is currently suspected (Sec. 4; [39]). The angular resolution of GLAST also will be important for studying the coupling of CRs to the spiral arms in the inner Milky Way. The diffuse interstellar emission is intense and quite structured near the Galactic equator, and evidence for coupling will require careful separation of components at different distances, most likely by study of the emission near the tangent directions of the arms.

GLAST observations of interstellar clouds at GeV energies may permit a new test of the origin of the GeV excess. CR electrons fully penetrate molecular clouds and produce γ-rays via bremsstrahlung and ICS with low-energy (microwave to ultraviolet) photons within the cloud. The radiation field inside a cloud varies with visual extinction to its center: the ultraviolet and optical radiation intensities decrease rapidly with increasing depth whereas the far-infrared radiation is ~ 10 times more intense throughout the cloud than the Galactic interstellar radiation field [63]. The resulting variation of the gamma-ray spectrum across the face of a cloud could be used to determine the ICS component.

GLAST will map the interstellar γ-ray emission of the LMC, SMC, and possibly M31, and additionally will likely detect several other local group galaxies and starburst galaxies as point sources, greatly expanding the potential for high-energy γ-ray studies of CRs.

ACKNOWLEDGEMENTS

IVM acknowledges support from the NRC/NAS Research Associateship Program. MP acknowledges partial support by the Bundesministerium für Bildung und Forschung, grant DLR 50 QV 0002.

REFERENCES

1. Freier, P., et al., *Phys. Rev.* **74**, 1818–1827 (1948).
2. Parker, E.N., *Space Sci. Rev.* **9**, 654–712 (1969).
3. Bertsch, D. L., et al., *Astrophys. J.* **416**, 587–600 (1993).
4. Sreekumar, P., et al., *Phys. Rev. Lett.* **70**, 127–129 (1993).
5. Esposito, J. A., et al., *Astrophys. J.* **461**, 820–827 (1996).
6. Kaaret P., and Cottam, J., *Astrophys. J.* **462**, L35–L38 (1996).
7. Bell, A., R., *Mon. Not. Royal Astron. Soc.* **182**, 147–156 (1978).
8. Lagage P. O., and Cesarsky C. J., *Astron. Astrophys.* **125**, 249–257 (1983).
9. Drury L. O'C., Aharonian, F. A., and Völk, H. J., *Astron. Astrophys.* **287**, 959–971 (1994).
10. Aharonian F. A., Drury, L. O'C., and Völk, H. J., *Astron. Astrophys.* **285**, 645–647 (1994).
11. Buckley, J. H., et al., *Astron. Astrophys.* **329**, 639–658 (1998).
12. Baring M. G., et al., *Astrophys. J.* **513**, 311–338 (1999).
13. Lerche, I., Pohl, M., and Schlickeiser, R., *J. Plas. Phys.*, in press (2001).
14. Green, D. A., in *High Energy Gamma-Ray Astronomy: International Symposium*, eds. Aharonian, F. A. and Völk, H. J., AIP Conf. Proc. 558, New York, 2001, in press
15. Koyama, K., et al., *Nature* **378**, 255–258 (1995).
16. Koyama, K., et al., *Pub. Astron. Soc. Japan* **49**, L7–L11 (1997).
17. Slane, P., et al., *Astrophys. J.* **525**, 357–367 (1999).

18. Keohane, J. W., et al., *Astrophys. J.* **484**, 350–359 (1997).
19. Allen, G. E., et al., *Astrophys. J.* **487**, L97–L100 (1997).
20. Borkowski, K. J., et al., *Astrophys. J.* **550**, 334–345 (2001).
21. Tanimori, T., et al., *Astrophys. J.* **497**, L25–L28 (1998).
22. Muraishi, H., et al., *Astron. Astrophys.* **354**, L57–L61 (2000).
23. Pohl, M., *Astron. Astrophys.* **307**, L57–L59 (1996).
24. Dyer, K. K., et al., *Astrophys. J.* **551**, 439–453 (2001).
25. Aharonian, F. A., and Atoyan, A. M., *Astron. Astrophys.* **351**, 330–340 (1999).
26. Atoyan, A. M., et al., *Astron. Astrophys.* **354**, 915–930 (2000).
27. Aharonian, F., et al., *Astron. Astrophys.* **370**, 112–120 (2001).
28. Atoyan, A. M., et al., *Astron. Astrophys.* **355**, 211–220 (2000).
29. Reynolds, S. P., and Keohane, J. W., *Astrophys. J.* **525**, 368–374 (1999).
30. Pohl, M., and Esposito, J. A., *Astrophys. J.* **507**, 327–338 (1998).
31. Jones, F. C., et al., *Astrophys. J.* **547**, 264–271 (2001).
32. Strong, A. W., and Moskalenko, I. V., *Astrophys. J.* **509**, 212–228 (1998).
33. Strong, A. W., and Moskalenko, I. V., to appear in *Adv. Space Res.* (astro-ph/0101068) (2001).
34. Davis, A. J., et al., in *Proc. ACE-2000 Symp.*, eds. Mewaldt, R. A., et al., AIP Conf. Proc. 528, New York, 2000, pp. 421–424.
35. DuVernois, M. A., Simpson, J. A., and Thayer, M. R., *Astron. Astrophys.* **316**, 555–563 (1996).
36. Lukasiak, A., McDonald, F. B., and Webber, W. R., Voyager Measurements of the Charge and Isotopic Composition of Cosmic Ray Li, Be and B Nuclei and Implications for Their Production in the Galaxy," in *Proc. 26th ICRC*, edited by D. Kieda et al., 1999, 3, 41–45.
37. Engelmann, J. J., et al., *Astron. Astrophys.* **233**, 96–111 (1990).
38. Stephens, S. A., and Streitmatter, R. A., *Astrophys. J.* **505**, 266–277 (1998).
39. Hunter, S. D., et al., *Astrophys. J.* **481**, 205–240 (1997).
40. Strong, A. W., Moskalenko, I. V., and Reimer, O., *Astrophys. J.* **537**, 763–784 (2000); Erratum: ibid., **541**, 1109 (2000).
41. Valinia, A., Kinzer, R. L., and Marshall, F. E., *Astrophys. J.* **534**, 277–282 (2000).
42. Mori, M., *Astrophys. J.* **478**, 225–232 (1997).
43. Strong, A. W., and Moskalenko, I. V., these Proceedings.
44. Sreekumar, P., et al., *Astrophys. J.* **400**, L67–L70 (1992).
45. Strong, A. W., and Mattox, J. R., *Astron. Astrophys.* **308**, L21–L24 (1996).
46. Digel, S. W., et al., *Astrophys. J.* **520**, 196–203 (1999).
47. Ulmer, M. P., et al., *Astrophys. J.* **448**, 356–364 (1995).
48. Chang, J., et al., *Astron. Astrophys.*, submitted (2001).
49. Pohl, M., et al., *Astrophys. J.* **491**, 159–164 (1997).
50. Zhang, L., and Cheng, K. S., *Mon. Not. R. Astron. Soc.* **301**, 841–848 (1998).
51. Dixon, D. D., et al., *New Astron.* **7**, 539–561 (1998).
52. Sreekumar, P., et al., *Astrophys. J.* **494**, 523–534 (1998).
53. Moskalenko, I. V., and Strong, A. W., *Astrophys. J.* **528**, 357–367 (2000).
54. De Paolis, F., et al., *Astrophys. J.* **510**, L103–L106 (1999).
55. Kalberla, P. M. W., Shchekinov, Yu. A., and Dettmar, R.-J., *Astron. Astrophys.* **350**, L9–L12 (1999).
56. Michelson, P. F., these Proceedings.
57. Sturner, S. J., and Dermer, C. D., *Astron. Astrophys.* **332**, L17–L20 (1995).
58. Higgs, L. A., Landecker, T. L., and Roger, R. S., *Astron. J.*, **82**, 718–724 (1977).
59. Hartman, R. C., et al., *Astrophys. J. Suppl.* **123**, 79–202 (1999).
60. Brazier, K. T. S., et al., *Mon. Not. R. Astron. Soc.* **281**, 1033–1037 (1996).
61. Gaisser, T. K., Protheroe, R. J., and Stanev, T., *Astrophys. J.* **492**, 219–227 (1998).
62. Allen, G. E., Digel, S. W., and Ormes, J. F., "What Can Be Learned About Cosmic Rays with GLAST?" in *Proc. 26th ICRC*, edited by D. Kieda et al., 1999, 3, pp. 515–518.
63. Mathis, J. S., Mezger, P. G., and Panagia, N., *Astron. Astrophys.* **128**, 212–229 (1983)

Gould's Belt and the local cosmic-ray electron spectrum

Martin Pohl[*], Christophe Perrot[†] and Isabelle Grenier[†]

[*]*Institut für theoretische Physik IV, Ruhr-Universität Bochum, 44780 Bochum, Germany*
[†]*Université de Paris VII & CEA Saclay, Service d'Astrophysique, 91191 Gif-sur-Yvette, France*

Abstract. In a recent paper Pohl & Esposito [1] demonstrated that if the sources of cosmic-rays are discrete, as are Supernova Remnants (SNR), then the spectra of cosmic-ray electrons largely vary with location and time and the locally measured electron spectrum may not be representative of the electron spectra elsewhere in the Galaxy, which could be substantially harder than the local one. They have shown that the observed excess of γ-ray emission above 1 GeV can in fact be partially explained as a correspondingly hard inverse Compton component, provided the bulk of cosmic-ray electrons is produced in SNR.

As part of a GLAST IDS program to model the Galactic γ-ray foreground we have continued the earlier studies by investigating the impact of the star forming region Gould's Belt on the local electron spectrum. If the electron sources in Gould's Belt were continous, the local electron spectrum would be slightly hardened. If the electron sources are discrete, which is the more probable case, the variation in the local electron spectrum found by Pohl & Esposito persists.

THE LOCAL COSMIC-RAY ELECTRON SPECTRUM

The recent detections of non-thermal X-ray synchrotron radiation from the supernova remnants SN1006 [2], RX J1713.7-3946 [3], IC443 [4, 5], Cas A [6], and RCW86 [7] and the subsequent detections of SN1006 [8], RX J1713.7-3946 [9], and Cas A [10] at TeV energies support the hypothesis that at least Galactic cosmic-ray electrons are accelerated predominantly in SNR.

The Galactic distribution and spectrum of cosmic-ray electrons are intimately linked to the distribution and nature of their sources. Supernova remnants are transient features, which happen stochastically in space and time. Because effects of the discreteness of sources show up only at higher particle energy, we may describe the propagation of cosmic-ray electrons at energies higher than a few GeV by a simplified, time-dependent transport equation

$$\frac{\partial N}{\partial t} - \frac{\partial}{\partial E}(bE^2 N) - DE^a \nabla^2 N = Q \qquad (1)$$

where we consider continous energy losses by synchrotron radiation and inverse Compton scattering, an energy-dependent diffusion coefficient DE^a, and a source term Q. The Green's function for this problem can be found in the literature [11].

$$G = \frac{\delta\left(t - t' + \frac{E - E'}{bEE'}\right) \exp\left(-\frac{(r-r')^2}{4\lambda}\right)}{bE^2 (4\pi\lambda)^{3/2}} \qquad \text{with} \qquad \lambda = \frac{D\left(E^{a-1} - E'^{a-1}\right)}{b(1-a)} \qquad (2)$$

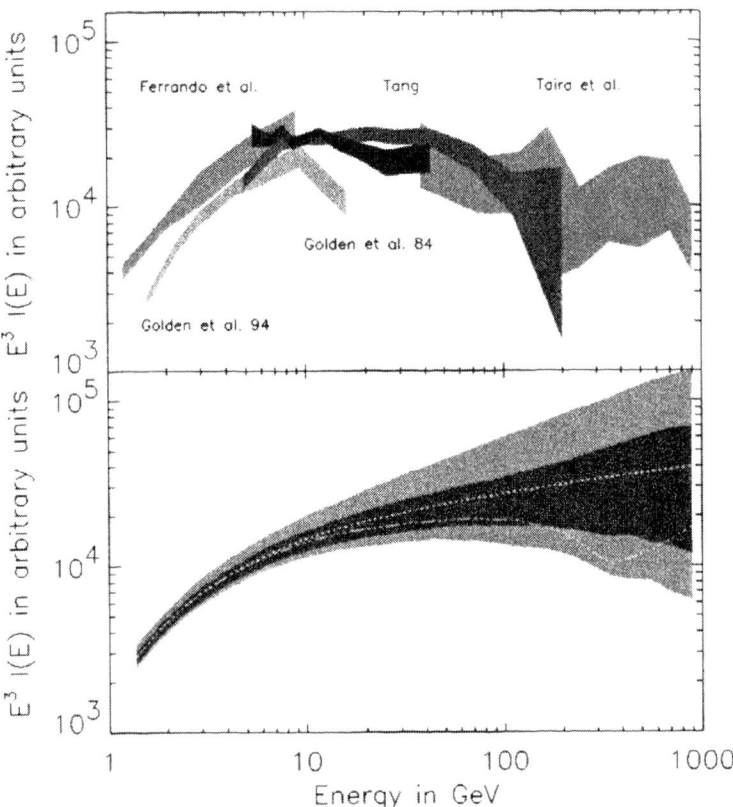

FIGURE 1. The locally observed electron spectrum in the upper panel compared with the range of possible spectra in the calculation of Pohl & Esposito [1] for a homogeneous population of sources in the Galactic disc with an initial spectral index of $s = 2.0$ at the sources. For each experiment the 1 σ uncertainty range is indicated by a grey-shaded band which connects the data points at the mean energies of the corresponding energy bins. The range of possible spectra in our model is given by the grey-shaded bands in the lower panel. At each energy the locally observed spectra will be in the dark grey shaded region during 68% of the time, and during 95% of the time they will be within the light grey shaded region. The white dash-dotted line shows one of 400 random spectra as a particular example of what may be observed. The white dotted line indicates the time-averaged spectrum.

The spatial distribution of sources modifies the local electron spectrum, while the randomness in time induces a time variability in the electron flux at higher energies, that stems from the fluctuations in the number of SNR within a certain distance and time interval. Thus the discreteness of sources does not only cause a cutoff in the electron spectrum, but makes it variable with time and thus unpredictable beyond a certain energy [1]. In Fig.1 we show the range of variability in comparison with the measured local electron spectrum for a homogeneous distribution of SNR in the Galaxy. While in steady-state models the observed electron spectrum requires an electron source spectral index of $s = 2.4$, it is in the range of possible local spectra with $s = 2.0$ in the time-

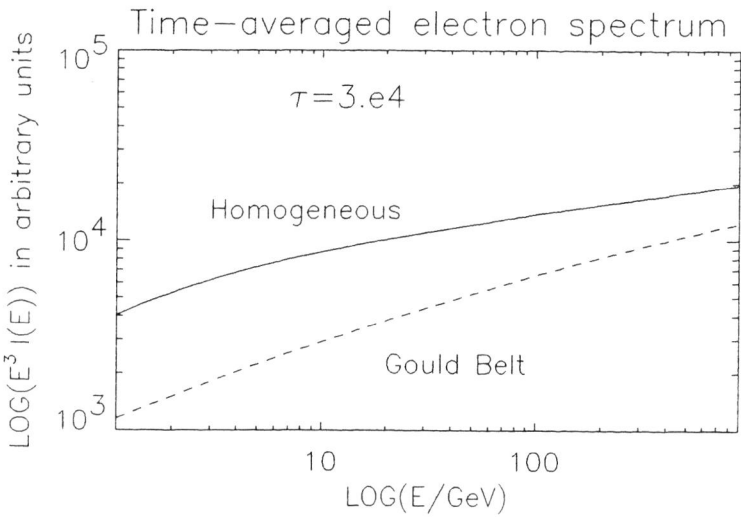

FIGURE 2. The steady-state local electron spectrum for a homogeneous distribution of SNR in comparison with that for SNR in Gould's Belt, both with the same SN rate per area. Here the life-time of SNR was assumed $\tau = 30000$ yrs.

dependent calculation. This implies that the average electron spectrum in the Galaxy, e.g. probed by line-of-sight integrals of leptonic emission through the Galactic plane, can be much harder than would be deduced in steady-state models. Pohl & Esposito have shown that the observed excess of gamma ray emission above 1 GeV [12] can in fact be explained as a correspondingly hard inverse Compton component.

THE IMPRINT OF GOULD'S BELT ON THE ELECTRON SPECTRUM

In the earlier calculations the supernova rate per area was assumed uniform throughout the Galactic plane. In this paper we investigate the impact of a local, non-uniform SNR distribution on the local cosmic ray electron spectrum. The most prominent local star-forming region is Gould's Belt [13], an expanding disc-like region 600 pc in diameter, in which the supernova rate per area is three to five times higher than the Galactic rate at the solar circle [14]. Based on its stellar content the age of this structure can be estimated to be 30 to 40 Myr. Assuming the evolution of Gould's Belt to be entirely determined by kinematical effects following an initial explosive event, its expansion history can be modelled [15]. This model can serve as a probability distribution of supernovae in space and time to be used in Eq. 1 for the source term Q.

In Fig. 2 we show the results for the average (steady-state) local electron spectrum. At higher electron energies the radiative energy losses permit only local SNRs to contribute to the local electron flux. Thus a locally enhanced SNR rate is increasingly important

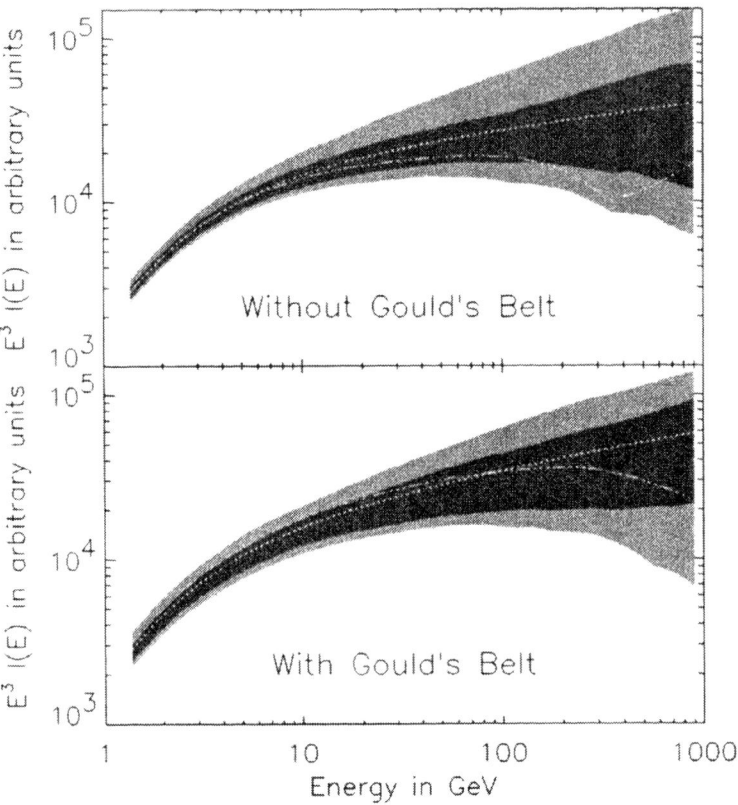

FIGURE 3. The range of possible spectra for a homogeneous Galactic distribution of SNR (top, the same as in Fig.1) compared with the range obtained for a Galactic distribution plus Gould's Belt (bottom). The level of variability is marginally smaller in the Gould's Belt case.

with increasing electron energy. Consequently, the contribution of SNR in Gould's Belt has a slightly harder spectrum than that of Galactic disc distribution of SNR, as has the summed electron spectrum. The hard spectrum of electrons from the Gould's Belt SNR is entirely a geometrical effect. Here we have assumed that enhanced SN activity commences without delay in the regions which are over-run by the expanding front of Gould's Belt. In reality this may not be true, in which case the effect of Gould's Belt would be less than calculated here. We have also assumed that the SN rate per area is constant within the Belt disc. We have then calculated the range of possible local electron spectra for the time-dependent case, shown in Fig.3 in comparison with that for a homogeneous Galactic disc distribution of SNR. The variability amplitude is marginally smaller for the Galactic disc distribution plus Gould's Belt, because then we expect more local SNR than in the earlier calculation, and therefore the relative width of the Poissonian distribution, which governs the variability amplitude, is smaller. At the same time the cosmic-ray electron source power required to reproduce the observed local electron flux is slightly reduced to ~ 60% of the SNR source power in the old calculation.

DISCUSSION

In this paper we have investigated the effect of Gould's Belt as a local system of enhanced star-forming and hence supernova activity on the local cosmic-ray electron spectrum under the assumption that the electron are solely produced in SNR. On average the local electron spectra would be slightly harder than in the case of a homogeneous Galactic disc distribution of SNR. At the same time the variability induced by the discrete nature of SNR in space and time would be only marginally reduced compared with the case of homogeneously distributed SNR [1].

Consequently, the electron source spectra in SNR must be slightly softer by $\Delta s \simeq 0.07$ than previously thought. Hard electron source spectra and the correspondingly hard inverse Compton gamma-ray spectra have been discussed as a possible explanation of the excess of Galactic GeV γ-rays observed with EGRET. Our results do not rule out this hypothesis, but certainly reduce the available parameter space for such models.

The conclusions presented here rely on the assumption that cosmic-ray electron are solely produced in SNR, as a consequence of which the local electron spectrum above 50 GeV may show deviations from power-law behaviour. New electron measurements in the energy range above 100 GeV are urgently required to possibly detect such spectral structures and thereby confirm an electron origin in discrete sources.

ACKNOWLEDGEMENTS

Partial support by the Bundesministerium für Bildung und Forschung, grant DLR 50QV0002, is acknowledged.

REFERENCES

1. Pohl, M., Esposito, J. A., *Ap.J.* **507**, 327 (1998)
2. Koyama K. et al., *Nature* **378**, 255 (1995)
3. Koyama K. et al., *PASJ* **49**, L7 (1997)
4. Keohane J.W. et al., *Ap.J.* **484**, 350 (1997)
5. Slane, P., et al., *Ap.J.* **525**, 357 (1999)
6. Allen G.E. et al., *Ap.J.* **487**, L97 (1997)
7. Borkowski, K.J., Rho, J., Reynolds, S.P., Dyer, K.K., *Ap.J.* **550**, 334 (2001)
8. Tanimori, T., et al., *Ap.J.* **497**, L25 (1998)
9. Muraishi, H., Tanimori, T., Yanagita, S. et al., *A&A* **354**, L57 (2000)
10. Aharonian, F., et al., *A&A* **370**, 112 (2001)
11. Ginzburg V.L., Syrovatskii S.I., *The origin of cosmic-rays*, Pergamon Press (1964)
12. Hunter, S.D., et al., *Ap.J.* **481**, 205 (1997)
13. Pöppel, W., *Fund. Cosm. Phys.* **18**, 1 (1997)
14. Grenier, I., *A&A* **364**, L93 (2000)
15. Grenier, I., Perrot, C., this volume

A Study of Very High Energy Gamma-Rays and Electrons with GLAST

R. Terrier*, A. Djannati-Ataï*, A. Chehktman[†], J.E. Grove[†] and W.N. Johnson[†]

*PCC Collège de France, Paris
[†]Naval Research Laboratory, Washington DC

Abstract. The GLAST instrument is dedicated to gamma-ray astronomy in a wide energy range of 20 MeV to 300 GeV, but could be used to study gamma-rays or cosmic electrons up to a few TeV. In this paper we study the high energy response of GLAST from 10 GeV up to one TeV. Reconstruction algorithms are presented both for energy and direction measurements using the GLAST calorimeter. Performance estimates are made based on the application of these algorithms to detailed simulated data. Using these estimates we study the sensitivity of GLAST to very high energy cosmic electrons.

INTRODUCTION

The GLAST calorimeter consists of 16 modules of 8 layers of 12 CsI(Tl) crystals in an hodoscoping arrangement, this is to say alternatively oriented in X and Y directions, to provide an image of the shower [4]. It is designed to measure energies from 30 MeV to 300 GeV and even 1 TeV.

However, the GLAST calorimeter is only $8.5X_0$ thick and therefore cannot provide good shower containment for high energy events, though these events are very precious for several astrophysics topics. Indeed, the mean fraction of the shower contained can be as low as 30% at 300 GeV normal incidence. In this case, the energy observed becomes very different from the incident energy, the shower development fluctuations become larger and the resolution decreases quickly.

We present here the results of a study on the reconstruction of energy of high energy gamma-rays and electrons in the GLAST calorimeter. We will first show different algorithms of shower leakage correction, and give the performances obtained. Finally we will apply these results to the observation of high energy electrons with GLAST, giving a strategy for background rejection at high energies.

These results are based on Monte-Carlo simulations using GLASTSIM, GISMO based simulation program for GLAST. The whole instrument response was tested with photons, electrons and protons beams, with energies ranging from 5 GeV to 1 TeV, and uniform angle exposure.

ENERGY RECONSTRUCTION

Electromagnetic longitudinal shower profile model

The electromagnetic shower model proposed by [3] has been used in this work. It will allow us to have an idea of the shower behaviour in general, and to fit the mean profile on the observed distribution.

We can describe a shower's longitudinal energy density as:

$$f_L(z) = E_0 \frac{(z/\lambda)^{\alpha-1} e^{-z/\lambda}}{\Gamma(\alpha)}$$

The parameters α and λ are random variables, approximately log-normally distributed. Here the shower fluctuations come from the fluctuations of the parameters, this is to say from the fluctuations of the shower maximum position and of the shower length. The distributions of these parameters in CsI were determined for various energies using a simple GEANT3.21 simulation, and were interpolated, giving a parametrization of the mean shower profile for a given energy.

Profile fitting

The first solution to correct for the energy loss is to fit the mean shower profile described above to the observed longitudinal profile. There are 2 free parameters, E_0 and the starting point of the shower to take into account early fluctuations. The shower shape parameters λ and α are fixed to their mean value for the energy E_0.

We minimize the following expression: $\chi^2 = \sum \frac{(E_i - \bar{E}_i)^2}{\sigma_i^2}$ where $\bar{E}_i$ is the amount of energy in the ith layer given by the model, and E_i is the measured energy. The errors σ_i are $\sqrt{\bar{E}_i}$.

One can see on figure 1 the performances obtained at several energies. The profile fitting method proves to be an efficient way to correct for shower leakage, specially at low incidence angles when the shower maximum is not contained. The resolution is 18 % for on axis 1 TeV photons, which is a 50 % improvement compared to the raw sum of the energies recorded in the crystals.

Last layer correction

The second method uses the correlation between the escaping energy and the energy deposited in the last layer of the calorimeter. The last layer carries the most important information concerning the leaking energy: the total number of particles escaping through the back should be nearly proportional to the energy deposited in the last layer. The measured signal in that layer can therefore be modified to account for the leaking energy.

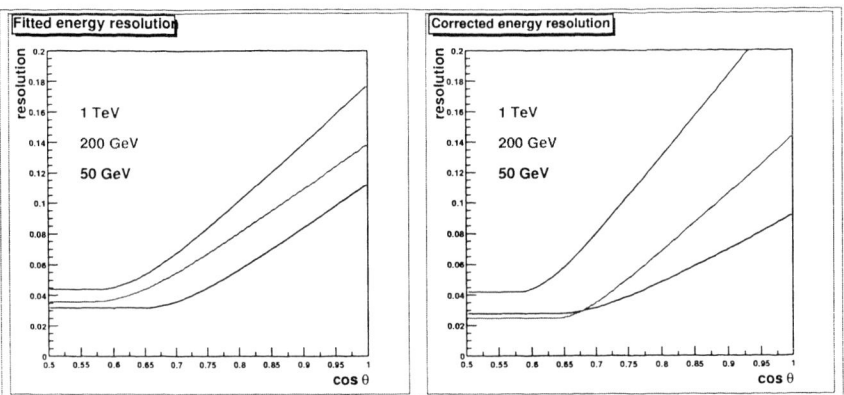

FIGURE 1. Energy resolution obtained using mean profile fitting on the left plot, and last layer correction on the right plot.

We determined this correlation at several energies and angles. For one particular incident energy, the bidimensionnal distribution of the energy escaping and the energy deposited in the last layer can be fitted by a simple linear function:

$$E_{leak} = \alpha E_{last} + \beta$$

where the α parameter is proportional to the logarithm of the incident energy. Since the only information we have, initially, on the incident energy is the total measured energy E_m, we have to use it as the estimator of E_0. The reconstructed energy is then:

$$E_{rec} = E_m + \alpha(E_m)E_{last} + \beta(E_m)$$

To improve the result, one can iterate using the new estimator to determine better values of α and β. The obtained resolutions are shown on figure 1.

Discussion

The methods presented improve significantly the resolution. Up to 1 TeV, the resolution on axis is better than 20 %, and for large incident angles (more than 60 degrees) it is around and even less than 4 %. It should be noted that the best layer correction is more robust since it doesn't rely on a fit, but its validity is limited to relatively well contained showers, making it difficult to use at more than 70 GeV for low incidence events. There is still some room for improvements, especially by correcting for losses between the different calorimeter modules and through the sides. This is particularly relevant for low incidence particles and explains some of the differences from the resolution studies reported for the SLAC test beam in [2]

APPLICATION : COSMIC ELECTRON SPECTRUM MEASUREMENT WITH GLAST

Up to now, the observations poorly constrain the electron spectrum at energies over 100 GeV [1]. It is fitted by a power-law of index 3.3 and a flux of $1.5 10^{-8} \text{cm}^{-2}\text{s}^{-1}\text{sr}^{-1}\text{GeV}^{-1}$ at 100 GeV.

If electrons are accelerated in SNRs, the local spectrum above 30 GeV should strongly be time variable, because of the Poisson fluctuations of SNR number [5]. Changing the electron injection spectrum index in the galactic diffuse emission model could therefore explain the GeV excess, and still be consistent with the local spectrum. Indeed, above 100 GeV, it becomes unpredictable. Using simulations, these authors provide the distribution and spread of possible spectra.

GLAST should be able to help to constrain this spectrum. We will here give an example of a possible observation.

Background rejection

The main source of background comes from cosmic-ray hadrons. The protons flux is a few hundred times larger than electrons at 100 GeV. Nevertheless, high energy hadrons produce very recognizable signatures in the calorimeter allowing us to easily reject them. Since the interaction length in the CsI is larger than the vertical dimension of the calorimeter, only a limited fraction will interact and deposit enough energy to mimic a high energy lepton. For he remaining evens, we use cuts based on:

- the ratio of crystals above a given threshold
- the shower shape, in particular the lateral spread of hadronic showers is much larger

Using these cuts we can reduce the background to a significantly lower level than the electron flux.

The other source of noise come from gamma-rays themselves. At high energies, the distinction is much more difficult: because of the backsplash in the ACD tiles, they cannot be tagged as charged particles. However, at high latitudes, the galactic diffuse emission is less intense than on the galactic plane. The intensity should be around ten times lower than electron intensity at 300 GeV if we extrapolate the extragalactic background flux observed by EGRET. Even if the IC emission from the galactic background were much larger than previously thought its flux at high latitudes would be of the same order than the extragalactic flux [6].

Limiting the observations to high latitudes ($|b| > 70^o$) in order to keep the gamma-ray contamination as low as possible, we obtain, after the cuts, an effective area for electrons of 8000 cm^2 averaged over the whole field of view.

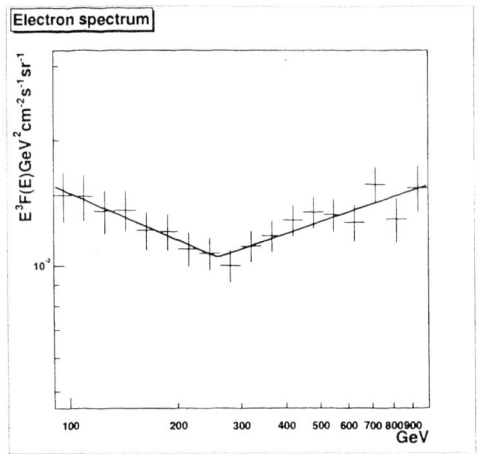

FIGURE 2. Broken power-law electron spectrum from a one year, high-latitude ($|b| > 70°$) GLAST observation

Electron spectrum

A simple broken power-law spectrum which fits in the range of possible spectra has been assumed here. The chosen spectrum flux is $F_0 = 1.5 10^{-8} \text{cm}^{-2}\text{s}^{-1}\text{sr}^{-1}\text{GeV}^{-1}$ at 100 GeV and the spectral index is $\alpha_0 = 3.3$ for $E < 300 \text{GeV}$ and $\alpha_1 = 2.8$ for $E > 300 \text{GeV}$.

15% energy bins were used. This is very conservative choice, since the energy resolution is much lower on most of the angle range. We supposed a one year observation of the high latitude emission ($|b| > 70°$)in order to keep the gamma-ray contamination as low as possible.

The figure 2 gives the results of the simulation and the reconstructed spectrum. The spectral break is easily seen, and the two spectral indexes are fitted with a precision of about ±0.05, but the spectral break position is not well constrained. Some further analysis will take into account the dependence of energy resolution on incident angle to improve the results and in particular the spectral break position determination.

REFERENCES

1. Barwick, S.W., et al., 1998, Astrophysical Journal, 298, 779
2. Do Couto e Silva, E., et al. SLAC-PUB-8682, Nov 2000
3. Grindhammer,G., Peters, S., 1993, Int. Conf. on Monte Carlo Simulation in High Energy and Nuclear Physics, Tallahassee, Florida
4. Johnson, W.N. et al. Fifth Compton Symposium, Portsmouth, NH, September 1999.
5. Pohl, M., Esposito, J.A. 1998, ApJ 507
6. Strong, A.W., Moskalenko, I.V. Proc. 26th ICRC (Salt Lake City), 1999, v.4, p.52

Excess GeV radiation and cosmic ray origin

Ingo Büsching*, Martin Pohl* and Reinhard Schlickeiser*

Institut für theoretische Physik IV, Ruhr-Universität Bochum, 44780 Bochum, Germany

Abstract. Particle acceleration at supernova remnant (SNR) shock waves is regarded as the most probable mechanism for providing Galactic cosmic rays at energies below 10^{15} eV. The Galactic cosmic ray hadron component would in this picture result from the injection of relativistic particles from many SNRs. It is well known that the superposition of individual power law source spectra with dispersion in the spectral index value, which behaviour is observed in the synchrotron radio spectra of shell SNR, displays a positive curvature in the total spectrum and in particular shows a hardening at higher energies.

Recent observations made with the EGRET instrument on the Compton Gamma-Ray Observatory of the diffuse Galactic γ-ray emission reveal a spectrum which is incompatible with the assumption that the cosmic ray spectra measured locally hold throughout the Galaxy: the spectrum above 1 GeV, where the emission is supposedly dominated by π^0-decay, is harder than that derived from the local cosmic ray proton spectrum. We demonstrate that in case of a SNR origin of cosmic ray nucleons part of this γ-ray excess may be attributed to the dispersion of the spectral indices in these objects.

THE DISPERSION OF COSMIC RAY SPECTRA IN SNR

The recent detections of non-thermal X-ray synchrotron radiation from the supernova remnants (SNRs) SN1006 [1], RX J1713.7-3946 [2], IC443 [3, 4], Cas A [5], and RCW86 [6] and the subsequent detections of SN1006 [7] and RX J1713.7-3946 [8] at TeV energies support the hypothesis that at least Galactic cosmic ray electrons are accelerated predominantly in SNR. To date, there is still no unambiguous proof that cosmic ray (CR) nucleons are similarly produced in SNR.

The spectrum of the diffuse Galactic γ-ray radiation above 1 GeV where the emission is supposedly dominated by π^0-decay, is harder than that derived from the local CR proton spectrum [9]. In the following we will demonstrate that in case of a SNR origin of CR nucleons part of the γ-ray excess may be attributed to the dispersion of the spectral indices in these objects. In global averages, as are γ-ray line-of-sight integrals, this dispersion leads to a positive curvature in the composite spectrum, and hence to modified π^0-decay γ-ray spectra. The synchrotron spectra of shell SNRs show power law behaviour $I_\nu \propto \nu^{-\alpha}$ [10, 11, 12] with $<\alpha> \simeq 0.5$ and a significant dispersion σ_α in the spectral index. We therefore assume a power law energy distribution for CR hadrons leaving the SNR, which is modified by the propagation through the Galaxy due to a moment-dependent diffusion coefficient $D \propto p^b$

$$N(p) = N_0 \left(\frac{p}{mc}\right)^{-s} \tag{1}$$

with a spectral index having a mean value $<s> = 1 + b + 2 <\alpha> \simeq 2.7$ and dispersion

$\sigma = 2\sigma_\alpha$. This is justified, because the age of the SNRs is much smaller than the characteristic radiative loss times of both CR nucleons and GeV electrons in the remnant, and because both the particle acceleration processes and the spatial propagation scale with rigidity. We represent the distribution of hadron spectral indices by the Gaussian

$$n(s) = \frac{1}{\sqrt{2\pi}\sigma} \exp\left[-\frac{(s-<s>)^2}{2\sigma^2}\right] \quad (2)$$

and obtain from Eq. (1) for the averaged hadron spectrum in the Galaxy

$$<N(p)> = \int_{-\infty}^{\infty} ds\, N(p,s) n(s) = N_0 \left(\frac{p}{mc}\right)^{-<s>+\frac{\sigma^2}{2}\ln\left(\frac{p}{mc}\right)} \quad (3)$$

PION DECAY GAMMA RAYS

To compare our model with recent observations of the diffuse Galactic γ-emission, the pion source spectrum is calculated using the cosmic hadron distribution function in Eq. (3) as input for the Monte-Carlo code DTUNUC (V2.2) [13, 14, 15, 16], which is based on a dual parton model [17].

We compare the γ-ray spectra thus derived with the observed EGRET spectra of diffuse γ-rays from the inner Galaxy. As can be seen in Fig. 1, a dispersion in the index

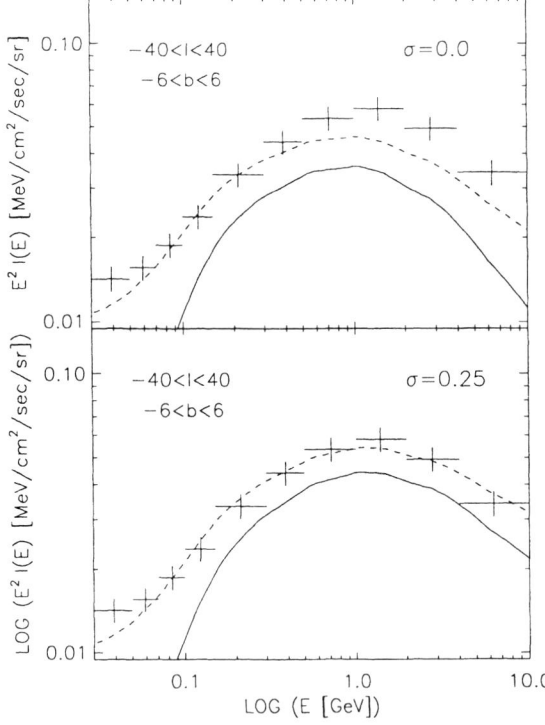

FIGURE 1. The observed intensity spectrum from the inner Galaxy shown in comparison with π^0-decay spectra with (bottom panel) and without (top panel) dispersion in the cosmic ray spectrum. The solid lines display the π^0-decay spectra, and the dashed lines are the total γ-ray spectra including the leptonic contribution, which here is simply given as a power law $\propto E^{-2}$ with an intensity determined by the data for $E \leq 100$ MeV. Whereas in the top panel, i.e. without dispersion, the GeV excess can be clearly seen, the bottom panel proves that a dispersion with $\sigma = 0.25$ can in fact explain the data and remove the GeV excess.

of the CR source spectra of $\sigma \simeq 0.25$ is sufficient to explain the observed intensity spectrum and remove the GeV excess. We can therefore conclude that, if dispersion is to be made responsible for the GeV excess, it must be at a level of $\sigma \simeq 0.25$.

COMPATIBILITY WITH UPPER LIMITS AT HIGHER GAMMA RAY ENERGIES

A dispersion in the CR source spectra – and thus in the CR flux throughout the Galaxy – has impact also on the resultant γ-ray spectra at higher energies, both for the diffuse emission and for individual SNRs. To date, observations of TeV γ-ray emission from individual SNR have yielded only a few detections, one of which (Cas A, [18]) is presumably [19] and the others are clearly caused by leptonic emission. Also, upper limits for the diffuse galactic γ-radiations at 500 TeV have been recently derived by the Whipple-team [20]. To investigate the compatibility of our model to these observational facts, we use an analytical approximation for high CR proton energies. We approximate the differential cross-section by the total cross section $\sigma_{pp}^{\pi} \simeq \sigma_{p,inel} \simeq 3 \cdot 10^{-26}$ cm^2, the multiplicity $\xi \simeq E_p^{1/4}$ (GeV), and a δ-function in energy, centered at the mean pion Lorentz factor $\bar{\gamma}_\pi \simeq \gamma_p^{3/4}$ [21].

$$\sigma^{\pi^0}(\gamma_\pi, \gamma_p) \simeq 3 \cdot 10^{-26} ((\gamma_p - \gamma_{\text{thr}}) 0.938)^{1/4} \delta(\gamma_\pi - \gamma_p^{3/4}) \frac{\text{cm}^2}{\text{GeV}} \quad (4)$$

As can be seen in Fig. 2, a dispersion of $\sigma = 0.25$ clearly violates the upper limit at 500 GeV derived by the Whipple team, if hadrons are accelerated in the sources up to $\gamma_{\text{max}} = 10^6$, but the upper limit is clearly satisfied, if $\gamma_{\text{max}} = 10^5$. Therefore, if a spectral dispersion exists at the level required to explain the GeV excess, there must be a high energy cut-off in the CR source spectra somewhere between $\gamma_{\text{max}} = 10^5$ and $\gamma_{\text{max}} = 10^6$.

DISCUSSION

We have calculated the high energy diffuse Galactic γ-ray emission produced by the hadronic component of CR under the assumption that the interstellar CR spectrum is a superposition of individual power-law spectra with a dispersion in the spectral index as a result of CR production in SNR. We have shown that a diffuse γ-ray spectrum thus derived can in fact explain the GeV excess, provided the dispersion in the individual SNR production spectral indices is $\sigma \simeq 0.25$. To comply with data in the TeV energy range both for individual SNR and for the diffuse emission, the existence of a high energy cut-off in the CR source spectra at proton energies not higher than somewhere between Lorentz factors $\gamma_{\text{max}} = 10^5$ and $\gamma_{\text{max}} = 10^6$ is required.

The hypothesis presented here has a number of consequences for observable quantities which can be used to test the viability of the scenario.

Is this model compatible with the SNR radio synchrotron spectra? Taking the radio spectral index data for Galactic shell SNR from [12], the actual distribution of observed

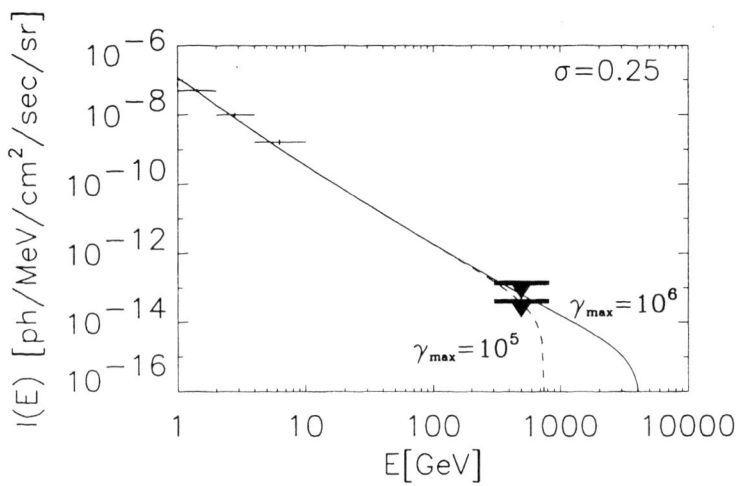

FIGURE 2. Composite spectrum between 1 GeV and 1 TeV of the diffuse Galactic γ-ray emission from the Galactic plane at $l \approx 40°$ based on EGRET and Whipple data. The upper limit at 500 GeV, which is at the 99.9% confidence level, depends on the γ-ray spectrum. The lower mark applies for a γ-ray spectral index $s = 2.0$ whereas the upper mark is appropriate for a spectral index $s = 2.6$. The data are compared with π^0-decay spectra calculated for a cosmic ray spectrum produced by individual sources with mean spectral index $<s> = 2.7$ and dispersion $\sigma = 0.25$. The solid line refers to a high energy cut-off in the cosmic ray proton spectrum at $\gamma_{max} = 10^6$. It should be compared with the lower mark for the upper limit at 500 GeV, which it clearly and significantly exceeds. For $\gamma_{max} = 10^5$ the dashed line shows no contradiction with the Whipple data, when compared to the upper mark for the upper limit at 500 GeV.

spectral indices is statistical compatible with a dispersion $\sigma_\alpha = 0.10$ and $<\alpha> = 0.53 \pm 0.02$ [22]. This analysis indicates that a dispersion in the radio spectral indices of shell SNR exists a a level corresponding to $\sigma_s = 0.2$ which falls marginally short of $\sigma = 0.25$ required to explain the GeV excess. Therefore, given the uncertainties, we conclude that $\sigma = 0.25$ is compatible with the radio synchrotron spectra of SNR.

Why don't we observe a dispersion in the local CR spectrum? Taking the local supernova rate of about $S = 30$ Myr^{-1}kpc^{-2} and the life time identical to the escape time $\tau \simeq 20\gamma^{-0.6}$ Myr, the number of SNR contributing to the local CR (proton) flux is then $N = \pi H^2 \tau S \simeq 17000\gamma^{-0.6}$. This means while at a proton energy of 50 GeV some 1600 SNR would contribute, above 100 TeV the number would be down to 17 SNR. In this case the local CR spectrum could strongly deviate from the average spectrum in Eq. 3.

What about deviations from power law behaviour in CR sources? We have found that there must be a cut-off at proton energies not higher than a few hundred TeV. This can be established as a sharp cut-off, or alternatively in the form of a spectral steepening at somewhat smaller energies, which behaviour is predicted in models of non-linear shock acceleration [23]. As a result the average spectrum at higher energies would be softer. This would change the expected the local CR proton spectrum above 10 TeV, which is observed to be slightly softer than that obtained by direct measurements at lower energies [24, 25].

Is the high energy limit in the CR proton source spectra compatible with the observed all-particle spectra near the "Knee"? We have not investigated the composition and in particular not studied the possibility of different source spectra for different species. This issue remains for further investigations.

Since there is observational evidence for a dispersion in the spectral indices of the CR spectra in SNR, the composite CR spectrum must be curved, if the CR are predominantly produced in SNR. We have shown that if the dispersion is as strong as $\sigma = 0.25$, its effect on the interstellar CR spectrum would explain the GeV excess in the diffuse Galactic γ-ray spectrum. If the actual dispersion is weaker than this, it would still contribute to the GeV excess and therefore should not be neglected.

ACKNOWLEDGEMENTS

Partial support by the Bundesministerium für Bildung und Forschung through the DLR, grant *50 OR 0006*, is gratefully acknowledged.

REFERENCES

1. Koyama K. et al., *Nature* **378**, 255 (1995)
2. Koyama K. et al., *PASJ* **49**, L7 (1997)
3. Keohane J.W. et al., *ApJ* **484**, 350 (1997)
4. Slane, P., et al., *ApJ* **525**, 357 (1999)
5. Allen G.E. et al., *ApJ* **487**, L97 (1997)
6. Borkowski, K.J., Rho, J., Reynolds, S.P., Dyer, K.K., *ApJ* **550**, 334 (2001)
7. Tanimori, T., et al., *ApJ* **497**, L25 (1998)
8. Muraishi, H., Tanimori, T., Yanagita, S. et al., *A&A* **354**, L57 (2000)
9. Hunter, S.D., et al., *ApJ* **481**, 205 (1997)
10. Clark, D. H., Caswell, J. L., *MNRAS* **174**, 267 (1976)
11. Milne, D. K., *Austr. J. Phys.* **32**, 83 (1979)
12. Green, D.A., in *High energy gamma-ray astronomy*, eds. F.A. Aharonian and H.J. Völk, AIP Conference Proceeding **558**, 59 (2001)
13. Möhring H.-J., Ranft J., *Z. Phys. C* **52**, 643 (1991)
14. Ranft, J., Capella A., Trân Thanh Vân J., *Phys. Lett. B* **320**, 346 (1994)
15. Ferrari A., Sala P.R., Ranft J., Roesler S., *Z. Phys. C* **70**, 413 (1996)
16. Engel R., Ranft J., Roesler S., *Phys. Rev. D* **55**, 6957 (1997)
17. Capella A., Sukhatme U., Tan C.-I., Trân Thanh Vân J., *Phys. Rep.* **236**, 227 (1994)
18. Aharonian, F., Akhperjanian, A., Barrio, J. et al., *A&A* **370**, 112 (2001)
19. Atoyan, A.M., Aharonian, F.A., Tuffs, R.J., Völk, H.J., *A&A* **355**, 211 (2000)
20. LeBohec, S, et al., *ApJ* **539**, 209 (2000)
21. Mannheim, K., Schlickeiser, R., *A&A* **286**, 983 (1994)
22. Büsching, I.,Pohl, M., Schlickeiser, R., submitted to *A&A* (2001)
23. Baring, M.G., Ellison, D.C., Reynolds, S.J., Grenier, I.A., Goret, P., *ApJ* **513**, 311 (1999)
24. Asakimori, K., et al., *ApJ* **502**, 278 (1998)
25. Amenomori, M., et al., *Phys. Rev. D* **62**, 112002 (2000)

Channeled blast wave behaviour based on longitudinal, electrostatic instabilities

Martin Pohl*, Ian Lerche[†] and Reinhard Schlickeiser*

*Institut für Theoretische Physik IV, Ruhr-Universität Bochum, 44780 Bochum, Germany
[†]Department of Geological Sciences, University of South Carolina, Columbia, SC 29208, USA

Abstract. To address the important issue of how kinetic energy of collimated blast waves is converted into radiation, Pohl and Schlickeiser [1] have recently investigated the relativistic two-stream instability of electromagnetic turbulence. They have shown that swept-up matter is quickly isotropized in the blast wave, which provides relativistic particles and, as a result, radiation. Here we present new calculations for the electrostatic instability in such systems. It is shown that the electrostatic instability is faster than the electromagnetic instability for highly relativistic beams. However, even after relaxation of the beam via the faster electrostatic turbulence, the beam is still unstable with respect to the electromagnetic waves, thus providing the isotropization required for efficient production of radiation. While the emission spectra in the model of Pohl and Schlickeiser have to be modified, the basic characteristics persist.

THE DISPERSION RELATION

We consider the following situation: a blast wave, idealized as a dense cloud shaped like a thick disk, moves relativistically through the ambient medium. Viewed in the blast-wave frame, the interstellar medium is a relativistic beam of electrons and protons which enters the blast wave. This situation is unstable and waves will be excited which backreact on the incoming beam. In the earlier analysis [1] the stability of this beam was examined under the assumption that the background magnetic field is uniform and directed parallel to the direction of motion. It was shown that the beam very quickly excites low-frequency electromagnetic waves, which quasi-linearly isotropize the incoming interstellar electrons and protons in the blast wave plasma, thus providing relativistic particles. We now expand on the previous treatment by calculating the two-stream instability for longitudinal, electrostatic waves. In contrast to electromagnetic waves, which scatter the particles in pitch angle but preserve their kinetic energy until the distribution is isotropized, the electrostatic waves change the particle's energy until a plateau distribution is established.

The initial particle distribution function for protons and electrons is

$$f(\vec{p}, t=0) = \frac{n_i \delta(p_\perp) \delta(p_\parallel + P)}{2\pi p_\perp} + \frac{n_b \delta(p_\perp) \delta(p_\parallel)}{2\pi p_\perp} \quad (1)$$

with $P = mV\Gamma$. Then, for charge e, mass m particles under the action of an electric field $\vec{E} = \vec{e}_\parallel E_\parallel \exp(\imath k(x - at))$, in the direction parallel to an ambient magnetic field the

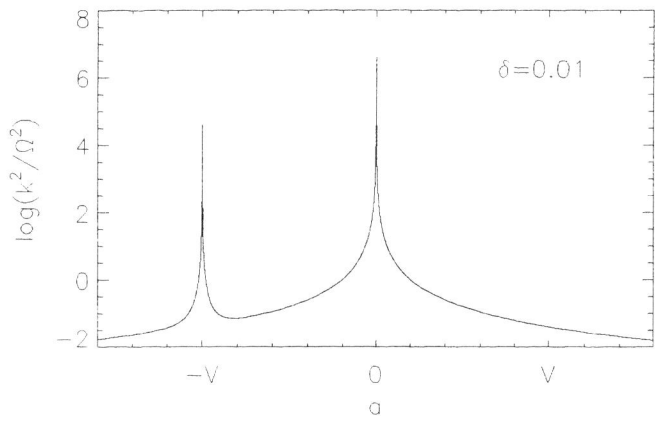

FIGURE 1. The dispersion relation for longitudinal, electrostatic waves. There is always a real solution in the range $a \leq -V$, and another real solution in $a \geq 0$. In the remaining range $-V \leq a \leq 0$ the dispersion relation returns two real values of a only if k^2 exceeds k_{min}^2.

perturbations, $\delta f \exp(\imath k(x-at))$, to the original distribution function satisfy

$$\frac{\partial \delta f}{\partial t} + \vec{v}\frac{\partial \delta f}{\partial \vec{x}} + e\vec{E}\frac{\partial f}{\partial \vec{p}} = 0 \qquad (2)$$

which, to first order in $\delta f/f$, yields the dispersion relation

$$k^2 = 4\pi \sum e^2 \int \frac{\frac{\partial f}{\partial p_\|}}{v_\| - a} d^3p = \Omega^2 \left[\frac{1}{a^2} + \frac{\delta}{(a+V)^2}\right] \qquad (3)$$

where $\Omega^2 = \omega_{p,p}^2 + \omega_{p,e}^2$ and $\delta = n_i/(n_b \Gamma^3)$. A plot of k^2 versus a is given in Fig.1. The waves are unstable for negative k with

$$k^2 \leq k_{min}^2 = \frac{\Omega^2}{V^2}(1+\delta^{1/3})^3 \quad \text{and} \quad a_{min}(k_{min}) = -V(1+\delta^{1/3})^{-1} > -V, \qquad (4)$$

in which range the dispersion relation is solved by $aV^{-1} = -\sin^2\phi + \imath y_I$ with

$$\frac{k^2}{k_{min}^2} = \left(\frac{\tan^2\phi+1}{\tan^2\phi-1}\right)^2 \frac{(\tan\phi-\sqrt{\delta})(1-\tan\phi\sqrt{\delta})}{\tan\phi(1+\delta^{1/3})^3} \qquad (5)$$

$$\text{and} \qquad y_I^2 = \frac{\tan\phi(\sqrt{\delta}\tan^3\phi-1)}{(1+\tan^2\phi)^2(\tan\phi-\sqrt{\delta})} \qquad (6)$$

The range of ϕ is restricted to $\delta^{-1/6} \leq \tan\phi \leq \delta^{-1/2}$. On $\tan\phi = \delta^{-1/6}$ note that $k^2 = k_{min}^2$ and $a_R = a_{min}$, and on $\tan\phi = \delta^{-1/2}$, $k = 0$ and $a_R = a_{max} = -V(1+\delta)^{-1}$. Thus $\tan\phi$ covers the complete wavenumber spectrum where instability can occur.

In Fig.2 we show the unstable region of the dispersion relation in phase velocity; k^2 is basically independent of the phase velocity a_R.

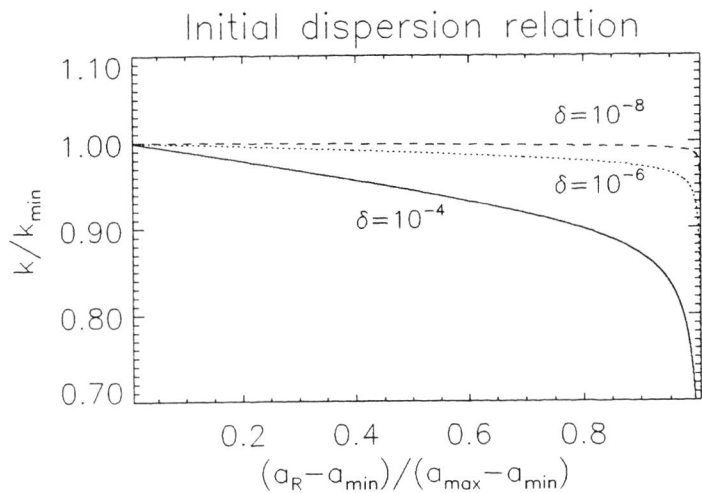

FIGURE 2. The dispersion relation for longitudinal waves in the unstable phase velocity range. For small perturbation parameters, δ, k/k_{min} is essentially flat, unless a_R is very close to a_{max}.

THE BACKREACTION ON THE PROTON-ELECTRON BEAM

The time-dependent behavior of the intensities $I(k,t)$ of the excited waves is given by $\frac{\partial}{\partial t} I(k,t) = 2\gamma I(k,t)$ [2, 3]. To describe the long-term influence of the excited waves on the beam particles, one uses the quasi-linear Fokker-Planck equation. Because the longitudinal waves act with an electric vector only, and because that vector parallels the magnetic field, the phase space density for the resonant particles then has only its momentum parallel to the ambient field influenced by the longitudinal turbulence. Hence, the corresponding Fokker-Planck equation reads

$$\frac{\partial f}{\partial t} = \frac{\partial}{\partial p_\parallel} \left(D \frac{\partial f}{\partial p_\parallel} \right) \quad (7)$$

where the diffusion coefficient, D, is given by

$$D = \frac{\langle \Delta p_\parallel^2 \rangle}{\Delta t} = 16\pi^2 e^2 \int_{k_{min}}^{0} dk\, I(k)\, \delta[k(v_\parallel - a_R)] = 16\pi^2 e^2 \int_{k_{min}}^{0} dk\, \frac{I(k)}{|k|}\, \delta(v_\parallel - a_R) \quad (8)$$

Because of the unknown nature of the initial wave spectrum, and because other processes ignored in the development will also influence the generation of waves, one of the standard devices is to ignore the wave intensity spectrum generation and to use *models* of how one believes the waves spectrum has evolved to its current state.

For the growth rate as a function of phase velocity, Fig.3 shows that the growth rate is essentially independent of a_R. Neglecting the possible effects of damping and cascading, the intensity spectrum in phase velocity, $I(a_R) = I(k)\frac{dk}{da_R}$ is then flat between

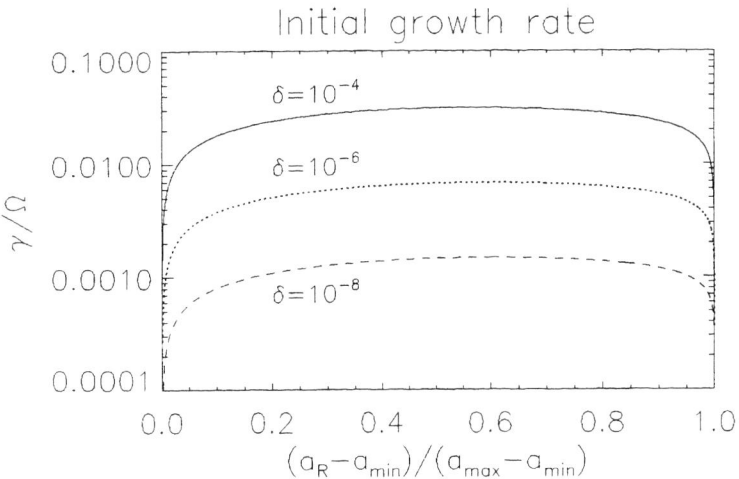

FIGURE 3. The growth rate γ as a function of the phase velocity a_R for three values of the perturbation parameter δ. The perturbation parameter controls only the speed of the instability, but not its spectral form.

a_{min} and a_{max}, and zero outside this range. Upon interaction with this wave spectrum the beam will relax to a plateau distribution between the velocities $-V$ and $v_\| = a_{min}$. The energy available for the build-up of the waves comes from the beam. Because the wave growth is much faster than the relaxation of the beam we may estimate the final wave intensity spectrum as the ratio of the energy lost by the beam during relaxation to a plateau between the velocities $-V$ and $v_\| = a_{min}$ and the phase velocity range in which effective growth occurs, namely $\delta^{1/3} V$. Therefore

$$I(a_R) \simeq \frac{\Delta E}{\Delta a_R} \simeq \frac{n_i m_p c^2 \Gamma^3}{2V} = I_0 \qquad (9)$$

The diffusion coefficient is then

$$D(p_\|) = 16\pi^2 e^2 I_0 \int_{a_{max}}^{a_{min}} da_R \frac{1}{|k|} \delta(v_\| - a_R) \simeq 2\pi m_e m_p c^2 \Gamma^3 \Omega \frac{n_i}{n_b} \qquad (10)$$

where we have again used that $k \simeq k_{min} \simeq -V/\Omega$ in the resonant range. The time scale for the relaxation of the beam by the electrostatic instability can then be estimated as

$$\tau \simeq \frac{p^2}{D} \simeq \frac{m_p n_b}{2\pi m_e n_i \Omega \Gamma} \simeq 5 \cdot 10^{-3} \frac{\sqrt{n_b}}{\Gamma n_i} \sec = 5 \cdot 10^{-3} \frac{\sqrt{n_8}}{\Gamma_2^2 n_i^*} \sec \qquad (11)$$

where n_i^* is the interstellar matter density in the laboratory frame, $n_8 = 10^{-8} n_b$ and $\Gamma_2 = 10^{-2} \Gamma$. The relaxation time scale is two orders of magnitude shorter than the time scale for the electromagnetic instability derived in [1].

477

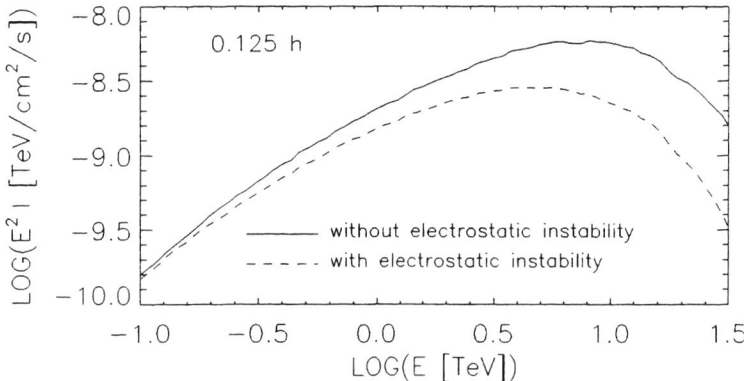

FIGURE 4. π^0-decay spectra calculated with and without the electrostatic instability. The spectra refer to an observed time of 0.125 hours after the blast wave over-runs an isolated cloud. At this time the spectrum of swept-up particles has evolved very little. At the peak, about 50% of the beam particle energy is lost due to the effect of the electrostatic instability. There is little effect at smaller photon energies.

CONSEQUENCES

The electrostatic instability provides a plateau-distribution in $p_\parallel$, which is still unstable with respect to electromagnetic waves, implying that isotropization in the blast wave still occurs on roughly the same time scale as calculated in [1]. Therefore, the effect of the electrostatic instability is to provide a change in the spectrum of swept-up particles:

$$\dot{N}(\gamma) = N_0\, \delta(\gamma - \Gamma) \qquad \text{without electrostatic instability} \qquad (12)$$

$$\dot{N}(\gamma) = \frac{N_0}{\Gamma - 1}\, \Theta(\Gamma - \gamma) \qquad \text{with electrostatic instability} \qquad (13)$$

In the radiation spectra this change corresponds to a reduction of efficiency by roughly 50% at the highest energies and little change at smaller energies, as shown in Fig.4. Therefore the isotropization and radiation conclusions drawn by Pohl & Schlickeiser [1] remain valid.

ACKNOWLEDGEMENTS

Partial support by the Bundesministerium für Bildung und Forschung, grant DESY Verbundforschung 05AG9PCA, is acknowledged.

REFERENCES

1. Pohl, M., Schlickeiser, R., *A&A* **354**, 395 (2000)
2. Lerche I., *Ap.J.* **147**, 689 (1967)
3. Lee M.A., Ip W.-H., *Jour. Geoph. Res.* **92**, 11041 (1987)

Gamma-ray detection of particle dark matter

L. Bergström

Department of Physics, Stockholm University, Box 6730, S-113 85 Stockholm, Sweden

Abstract. Recent results from N-body simulations of structure formation indicate that galactic dark matter halos may have a clumpy structure, with the density profiles of halos and subhalos showing steep cusps near the center. These features imply that the detection of particle dark matter through gamma-rays from annihilations in the Galactic halo, or from the integrated effects of smaller subhalos out to moderate readshifts, may be a promising method to detect supersymmetric dark matter.

INTRODUCTION

A variety of independent estimates of the matter density in the universe point to a value larger than the maximal value provided by baryons alone according to nucleosynthesis. The need for nonbaryonic dark matter is therefore striking. A satisfactory description of most cosmological observations is obtained by a "ΛCDM model" with $\Omega_B \sim 0.05$, $\Omega_{CDM} \sim 0.25$, $\Omega_\Lambda \sim 0.7$, The nature of the dark matter is one of the basic fundamental research problems in present-day cosmology and astrophysics.

It may not be unexpected that massive, electrically neutral, weakly interacting and long-lived particles make up a substantial fraction of the average cosmic mass density. If such a massive particle species has roughly the same type of gauge couplings as the known quarks and leptons, it must have been produced in large abundances in the earliest universe when the thermal energies were high enough to produce it in collisions between ordinary particles.

WEAKLY INTERACTING MASSIVE PARTICLES - SUPERSYMMETRIC PARTICLES

One of the prime candidates for the non-baryonic component is provided by the lightest supersymmetric particle, plausibly the lightest neutralino χ, which is a mixture of the supersymmetric partners of the photon, the Z and the two neutral CP-even Higgs bosons present in the minimal extension of the supersymmetric standard model (for reviews, see [1, 2]). The attractiveness of this candidate, besides its particle physics virtues, stems from the fact that it is electrically neutral and thus neither absorbs nor emits light, and stable so that it can have survived since the big bang. Furthermore, it has gauge couplings and a mass which for a large range of parameters in the supersymmetric sector imply a relic density in the required range to explain the observed $\Omega_M \sim 0.3$. Its couplings to ordinary matter also means that its existence as dark matter in our Galaxy's halo may be experimentally tested.

The rates of the processes $\chi\chi \to \gamma\gamma$ [3, 4] and $\chi\chi \to Z\gamma$ [5] have the property of giving very distinct, "smoking gun" signals of monoenergetic photons with energy $E_\gamma = m_\chi$ (for $\chi\chi \to \gamma\gamma$) or $E_\gamma = m_\chi(1 - m_Z^2/4m_\chi^2)$ (for $\chi\chi \to Z\gamma$) emanating from annihilations in the halo.

The detection probability of a gamma ray signal, either continuous or line, will of course depend sensitively on the density profile of the dark matter halo. The integral which determines the gamma-ray flux $J(\hat{n})$ is given by

$$J(\hat{n}) = \int_{\text{line-of-sight}} \rho^2(\ell)d\ell(\hat{n}), \qquad (1)$$

and is evidently very sensitive to local density variations along the line-of-sight path of integration. In the case of a smooth halo, its value varies by three orders of magnitude from high galactic latitudes to a small-angle average towards the galactic center in the cusped models based on N-body simulations of halo structure [6]. Since the neutralino velocities in the halo are of the order of 10^{-3} of the velocity of light, the annihilation can be considered to be at rest. The resulting gamma ray spectrum from the 2γ process is a line at $E_\gamma = m_\chi$ of relative linewidth 10^{-3} which in favourable cases will stand out against the background.

To compute $J(\hat{n})$ in Eq. (1), a model of the dark matter halo has to be chosen. As shown by detailed N-body simulations (see, e.g., [7, 8] and references therein), in the current picture of structure formation large structures forms by successive merging of small substructures, with smaller objects generally being denser. The N-body simulations also show that the dark matter density profile in clusters of galaxies and in single galaxies develops a steep cusp near the center, $\rho_{CDM}(r) \sim r^{-\alpha}$ with α ranging from 1[9] to 1.5 [10]. If applicable to the Milky Way, this would lead to a much enhanced annihilation rate towards the galactic centre, and also to a very characteristic angular dependence of the line signal.

Space-borne gamma ray detectors, like the projected GLAST satellite [11], have a relatively small area (on the order of 1 m^2 compared to $10^4 - 10^5$ m^2 for groundbased Air Cherenkov Telescopes), but a correspondingly larger angular acceptance so that the integrated sensitivity is in fact similar. This is at least true if the Galactic center does not have a very large dark matter density enhancement which would favour ACTs. A line signal can be searched for with higher precision using GLAST, since its energy resolution will be at the few percent level. Also, GLAST with its wide field of view has the potential of dicovering local substructures within the halo, which may appear as "hot spots" on the gamma-ray sky.

Recently, a new signature has been proposed [12], which alleviates the need for exceptional energy resolution while still being sensitive to the striking gamma-ray line signal. The idea is that the integrated effect of all structure in the Universe, out to a redshift of order unity (where absorption of gamma-rays in the 100 GeV range due to pair production on near-infrared and optical photons becomes large) may give a visible signal.

The differential rate of change of the comoving density of neutralinos is given by

$$\frac{dn_c(z)}{dz} = \kappa \frac{(1+z)^2}{h(z)} n_c(z)^2, \qquad (2)$$

where $\kappa = \langle \sigma v \rangle / H_0$, $h(z) = \sqrt{\Omega_M(1+z)^3 + \Omega_K(1+z)^2 + \Omega_\Lambda}$, with Ω_M, Ω_Λ and $\Omega_K = 1 - \Omega_M - \Omega_\Lambda$ being the present fractions of the critical density given by matter, vacuum energy and curvature. The differential spectrum of the number density n_γ of photons generated by annihilations is then:

$$\frac{dn_\gamma}{dz} = N_\gamma \frac{dn_c}{dz} = \int_0^{m_\chi} \frac{dN_\gamma(E)}{dE} \frac{dn_c}{dz} dE. \qquad (3)$$

Here, dn_c/dz can be computed directly from (2), replacing to an excellent approxiamtion the exact solution $n_c(z)$ by the present number density of neutralinos n_0 on the right hand side.

An optical depth to gamma-ray absorption on the extragalactic background light of order unity is reached for a redshift which can be approximated by $z_{max}(E_0) \sim 3.3(E_0/10 \text{ GeV})^{-0.8}$ [13, 14].

Approximating $\Omega_\chi \sim \Omega_M$ (since the baryonic contribution is constrained by nucleosynthesis to be much smaller), $n_0 = \rho_\chi/m_\chi = \rho_{crit}\Omega_M/m_\chi$, where $\rho_{crit} = 1.06 \cdot 10^{-5} h^2$ GeV/cm^3, the gamma-ray flux is given by [12]:

$$\phi_\gamma = \frac{c}{4\pi} \frac{dn_\gamma}{dE_0} = 8.3 \cdot 10^{-14} \frac{\Gamma_{26} \Omega_M^2 h^3}{m_{100}^2} I_c \text{ cm}^{-2} \text{s}^{-1} \text{sr}^{-1} \text{GeV}^{-1}, \qquad (4)$$

where $\Gamma_{26} = \langle \sigma v \rangle / (10^{-26} \text{ cm}^3 \text{s}^{-1})$, m_{100} the mass in units of 100 GeV, and

$$I_c = \int_0^{z_{max}} dz \frac{(1+z)^3 \Delta^2(z)}{h(z)} \frac{dN_\gamma(E_0(1+z))}{dE}. \qquad (5)$$

The presence of structure is taken into account by including an enhancement factor $\Delta^2(z)$, which is unity in the (unrealistic) structureless case. The gamma line contribution to (5) is particularly simple, just picking out the integrand at $z+1 = m_\chi/E_0$; it has the very distinctive and potentially observable signature of being asymmetrically smeared to lower energies (due to the redshift) and of suddenly dropping just above m_χ. The continuum emission will produce a less conspicuous feature, a smooth "bump" below one tenth of the neutralino mass which may be more difficult to detect.

The important effect of non-linear structure formation is to increase the average of the square of the dark matter density by a large factor compared to the smooth case, and thus enhance the annihilation rate.

The increase of average squared overdensity per halo of radial extent R_M is given by [12]:

$$\Delta^2 \equiv \langle \left(\frac{\rho_{DM}}{\rho_0}\right)^2 \rangle_{r<R_M} = \left(\frac{\rho'_{DM}}{\rho_0}\right) \frac{I_2}{I_1}, \qquad (6)$$

where
$$I_n = \int_0^{R_M/a} y^2 dy (f(y))^n. \tag{7}$$

A value of Δ^2 of $2.3 \cdot 10^5$ for a Milky-Way sized halo is obtained for the Moore profile, $1.5 \cdot 10^4$ for the Navarro-Frenk-White profile[9], and $7 \cdot 10^3$ for a cored, modified isothermal profile (modified such that the density falls as $1/r^3$ at large radii [15]). The flux ratios, $30:2:1$ for these three models should be compared with the ratios $1000:100:1$ obtained within a 5-degree cone encompassing the galactic center [15].

The number density of halos is scaling like $\sim 1/M^2$, and small-mass halos are denser. To a good approximation [12], $\Delta^2 \sim 2 \cdot 10^5 M_{12}^{-0.22}$, where M_{12} is the halo mass in units of 10^{12} solar masses. This means that the total flux from a halo of mass M scales as $M^{0.78}$. Since the number density of halos goes as M^{-2}, the fraction of flux coming from halos of mass M scales as $M^{-1.22}$.

Thus the gamma-ray flux will dominantly come from the smallest CDM halos. In simulations, substructure has been found on all scales (being limited only by numerical resolution). Setting $10^5 - 10^6 M_\odot$ as the minimal scale, we find that the flux from small halo structure is enhanced by roughly a factor $2 \cdot 10^6$ compared to the smooth case, giving a possibly observable signal. (For very small dark matter clumps, there will be no gain in overdensity, since once the matter power spectrum enters the k^{-4} region a constant density is predicted [16].)

We use the results obtained with the DarkSUSY package [17]. Models with large $\gamma\gamma$ rates ($(\sigma v)_{2\gamma} \gtrsim 10^{-29}$ cm^{3}s^{-1}) exist in all the mass range from $m_\chi = 70$ GeV to several TeV. Consider a high-rate model with $m_\chi = 86$ GeV, $\Gamma_{26} \sim 6$, $b_{\gamma\gamma} \sim 3 \cdot 10^{-3}$, in the "concordance" cosmology $\Omega_M = 0.3$, $\Omega_\Lambda = 0.7$, $h = 0.65$. The continuous gamma-ray rest frame energy distribution per annihilating particle is conveniently parametrized as

$$\frac{dN_{\text{cont}}}{dE}(E) = \frac{0.42}{m_\chi} \frac{e^{-8x}}{x^{1.5} + 0.00014} \tag{8}$$

where $x = E/m_\chi$. In Fig. 1, we show the results for the 86 GeV DarkSUSY model, and a model of 166 GeV mass, $\Gamma_{26} = 59$, $b_{\gamma\gamma} = 1.2 \cdot 10^{-4}$.

As can be seen, a substantial fraction of the diffuse extragalactic gamma-ray background measured by EGRET could in fact be attributed to dark matter annihilation, and with GLAST there is a hope to detect the conspicuous feature near $E_\gamma = m_\chi$ predicted in these models, which are based on the minimal supersymmetric standard model and the results of the most accurate N-body simulations to date.

ACKNOWLEDGMENTS

I would like to thank my collaborators, Joakim Edsjö and Piero Ullio, for numerous discussions. This work was supported by the Swedish Research Council.

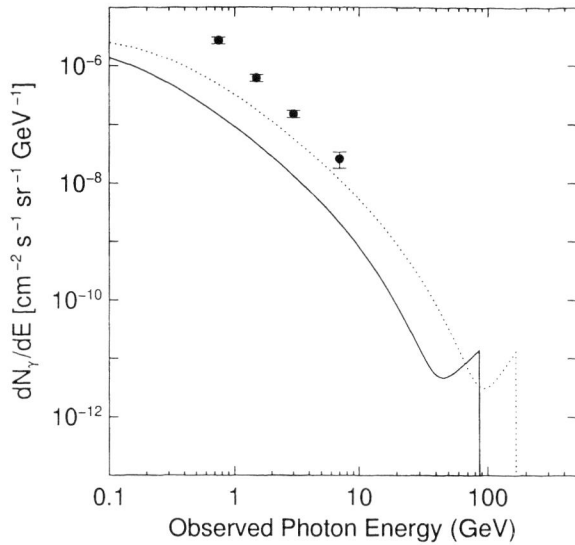

FIGURE 1. The predicted diffuse γ-ray flux [12], both from cosmic annihilations into continuum gamma-rays (giving low-energy gammas), and from annihilations into a gamma-ray line with rest-frame energy given by the neutralino mass. The redshifted line gives the conspicuous feature at the highest energies. Shown are cosmic annihilation of 86 GeV (solid line) and 166 GeV (dotted line) neutralinos with parameters as described in the text. The EGRET data [18] on the extragalactic flux are the data points with error bars shown. We only include those data points where direct energy measurements were made.

REFERENCES

1. G. Jungman, M. Kamionkowski and K. Griest, Phys. Rep. **267** (1996) 195.
2. L. Bergström, Rept. Prog. Phys. **63**, 793 (2000).
3. L. Bergström and H. Snellman, Phys. Rev. D **37**, 3737 (1988).
4. A. Bouquet, P. Salati and J. Silk, Phys. Rev. **D40**, (1989) 3168; G. Jungman and M. Kamionkowski, Phys. Rev. **D51** (1995) 3121.
5. M. Urban et al., Phys. Lett. **B293** (1992) 149; L. Bergström and J. Kaplan, Astropart, Phys. **2** (1994) 261; P. Ullio and L. Bergström, Phys. Rev. **D57** (1998) 1962.
6. L. Bergström, P. Ullio and J.H. Buckley, Astropart. Phys. **9** (1998) 137.
7. A. Jenkins et al. (The Virgo Consortium), Astrophys. J. **499**, 20 (1998).
8. B. Moore et al., Astrophys. J. **499**, L5 (1998).
9. J.F. Navarro, C.S. Frenk and S.D.M. White, Astrophys. J. **462**, 563 (1966).
10. S. Ghigna et al., Astrophys. J. **544**, 616 (2000).
11. Gamma-ray Large Area Space Telescope (GLAST), homepage http://www-glast.stanford.edu.
12. L. Bergström, J. Edsjö and P. Ullio, eprint astro-ph/0105048 (2001).
13. M.H. Salamon and F.W. Stecker, Astrophys. J. **493**, 547 (1998).
14. J.R. Primack, R.S. Somerville, J.S. Bullock and J.E.G. Devriendt, eprint astro-ph/0011475 (2000).
15. C. Calcáneo-Roldan and B. Moore, Phys. Rev. D **62**, 123005 (2000).
16. E. A. Baltz, C. Briot, P. Salati, R. Taillet and J. Silk, Phys. Rev. D **61**, 023514 (2000).
17. DarkSUSY package, homepage http://www.physto.se/~edsjo/darksusy. See P. Gondolo, J. Edsjö, L. Bergström, P. Ullio and E. A. Baltz, eprint astro-ph/0012234 (2000).
18. P. Sreekumar et al., Astrophys. J. **494**, 523 (1998).

Nucleosynthesis and Galactic Diffuse Emission

Gamma-Ray Signatures of Supernovae and Hypernovae

Ken'ichi Nomoto*, Keiichi Maeda*, Yuko Mochizuki†, Shiomi Kumagai**,
Hideyuki Umeda*, Takayoshi Nakamura* and Isao Tanihata†

*Department of Astronomy and Research Center for the Early Universe, School of Science,
University of Tokyo, Bunkyo-ku, Tokyo 113-0033, JAPAN
†RIKEN (The Institute of Physical and Chemical Research), Hirosawa 2-1, Wako, Saitama
351-0198, JAPAN
**Department of Physics, College of Science and Technology, Nihon University, Kanda-Surugadai
1-8, Chiyoda-ku, Tokyo 101, JAPAN

Abstract.
We review the characteristics of nucleosynthesis and radioactivities in 'hypernovae', i.e., supernovae with very large explosion energies ($\gtrsim 10^{52}$ ergs) and their γ-ray line signatures. We also discuss the ^{44}Ti line γ-rays from SN1987A and the detectability with INTEGRAL. Signatures of hypernova nucleosynthesis are seen in the large [(Ti, Zn)/Fe] ratios in very metal poor stars. Radioactivities in hypernovae compared to those of ordinary core-collapse supernovae show the following characteristics: 1) The complete Si burning region is more extended, so that the ejected mass of ^{56}Ni can be much higher. 2) Si-burning takes place in higher entropy and more α-rich environment. Thus the ^{44}Ti abundance relative to ^{56}Ni is much larger. In aspherical explosions, ^{44}Ti is even more abundant and ejected with velocities as high as $\sim 15,000$ km s^{-1}, which could be observed in γ-ray line profiles. 3) The abundance of ^{26}Al is not so sensitive to the explosion energy, while the ^{60}Fe abundance is enhanced by a factor of ~ 3.

INTRODUCTION

Massive stars in the range of 8 to $\sim 100 M_\odot$ undergo core-collapse at the end of their evolution and become Type II and Ib/c supernovae (SNe II and SNe Ib/c). These SNe II and SNe Ib/c release large explosion energies and eject explosive nucleosynthesis products, thus being major sources of radioactive species. Until recently, we have considered supernovae with the explosion energies of $E = 1 - 1.5 \times 10^{51}$ ergs. These energies have been estimated from the observations of nearby supernovae, such as SNe 1987A, 1993J, and 1994I. Also the progenitors of these SNe are estimated to be 13 - 20 $M_\odot$ stars (e.g., [1]).

Recently, there have been a number of candidates for the gamma-ray burst (GRB)/supernova (SN) connection (see [1] for references), including GRB980425/SN Ic 1998bw, GRB971115/SN Ic 1997ef, GRB970514/SN IIn 1997cy, GRB980910/SN IIn 1999E, GRB980326, and GRB970228. Among the SNe with a possible GRB counterpart, SNe Ic 1998bw[2, 3] and 1997ef[4, 5] are characterized by a very large kinetic explosion energy, $E \gtrsim 10^{52}$ erg. This is more than one order of magnitude larger than in typical SNe, so that these objects may be called "Hypernovae". These SNe produced more ^{56}Ni than the average core collapse SN. The masses of these hypernova progenitors are estimated to be $M \gtrsim 25 M_\odot$. These massive stars are likely to form black holes, while less massive stars form neutron stars (see, however, [6]).

Regarding γ-ray signatures of such hypernovae, whether and how the hypernovae actually induce gamma-ray bursts needs further study of aspherical explosions (e.g., [7, 8]). Another γ-ray signatures, we discuss here, are the line γ-ray emissions. We review the characteristics of nucleosynthesis and radioactivity in hypernovae and their γ-ray line signatures. We also discuss the ^{44}Ti line γ-rays from SN1987A and the detectability with INTEGRAL. For line γ-rays from Type Ia supernovae, see [9] for a review.

Before discussing γ-rays, we first point out that signatures of hypernova nucleosynthesis are seen in the large [(Ti, Zn)/Fe] ratios in very metal poor stars.

NUCLEOSYNTHESIS IN HYPERNOVA EXPLOSIONS

In core-collapse supernovae/hypernovae, stellar material undergoes shock heating and subsequent explosive nucleosynthesis. Iron-peak elements are produced in two distinct regions, which are characterized by the peak temperature, T_{peak}, of the shocked material. For $T_{\text{peak}} > 5 \times 10^9$K, material undergoes complete Si burning whose products include Co, Zn, V, and some Cr after radioactive decays. For 4×10^9K $< T_{\text{peak}} < 5 \times 10^9$K, incomplete Si burning takes place and its after decay products include Cr and Mn (e.g., [10, 11, 12]).

We note the following characteristics of nucleosynthesis with very large explosion energies[13]:

1) Both complete and incomplete Si-burning regions shift outward in mass compared with normal supernovae, so that the mass ratio between the complete and incomplete Si-burning regions is larger. As a result, higher energy explosions tend to produce larger [(Zn, Co)/Fe], small [(Mn, Cr)/Fe], and larger [Fe/O]. The elements synthesized in this region such as ^{56}Ni, ^{59}Cu, ^{63}Zn, and ^{64}Ge (which decay into ^{56}Co, ^{59}Co, ^{63}Cu, and ^{64}Zn, respectively) are ejected more abundantly than in normal supernovae.

2) In the complete Si-burning region of hypernovae, elements produced by α-rich freezeout are enhanced because nucleosynthesis proceeds at lower densities (i.e., higher entropy) and thus a larger amount of ^{4}He is left. Hence, elements synthesized through capturing of α-particles, such as ^{44}Ti, ^{48}Cr, and ^{64}Ge (decaying into ^{44}Ca, ^{48}Ti, and ^{64}Zn, respectively) are more abundant.

3) Oxygen burning takes place in more extended, lower density regions for the larger explosion energy. Therefore, O, C, Al are burned more efficiently and their abundances in the ejecta are smaller, while a larger amount of burning products such as Si, S, and Ar are synthesized. Therefore, hypernova nucleosynthesis is characterized by large abundance ratios of [Si/O], [S/O], [Ti/O], and [Ca/O].

ASPHERICAL EXPLOSIONS

[14] and [15] have identified some signatures of asymetric explosion in the late light

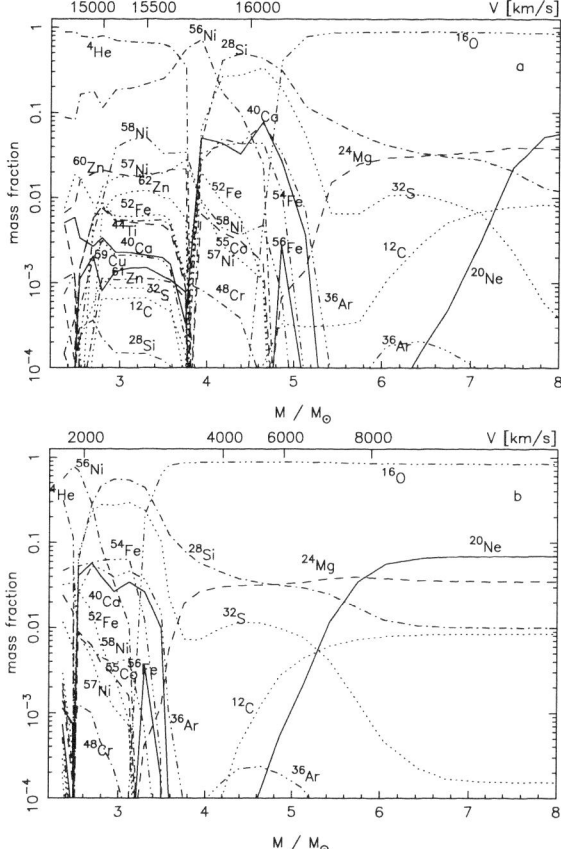

FIGURE 1. The isotopic composition of the ejecta in the direction of the jet (upper panel) and perpendicular to it (lower panel). The ordinate indicates the initial spherical Lagrangian coordinate (M_r) of the test particles (lower scale), and the final expansion velocities (V) of those particles (upper scale) [16].

curve and spectra of SN 1998bw. [16] have examined the effect of aspherical (jet-like) explosions on nucleosynthesis in hypernovae. The progenitor model is the 16 $M_\odot$ He core of the 40 $M_\odot$ star and the explosion energy is $E = 1 \times 10^{52}$ ergs.

Figure 1 shows the isotopic composition of the ejecta of asymmetric explosion model in the direction of the jet (upper panel) and perpendicular to it (lower panel).

In the z-direction, where the ejecta carry more kinetic energy, the shock is stronger and post-shock temperatures are higher. Therefore, larger amounts of α-rich freeze-out elements, such as ^{4}He, ^{44}Ti, and ^{56}Ni are produced in the z-direction than in the r-direction.

On the other hand, along the r-direction ^{56}Ni is produced only in the deepest layers, and the elements ejected in this direction are mostly the products of hydrostatic nuclear burning stages (O) with some explosive oxygen-burning products (Si, S, etc).

In the spherical case, Zn is produced only in the deepest layer, while in the aspherical model, the complete silicon burning region is elongated to the z (jet) direction, so that [Zn/Fe] is enhanced irrespective of the mass cut. On the other hand, ^{55}Mn, which is produced by incomplete silicon burning, surrounds ^{56}Fe and located preferentially in the r-direction.

In this way, larger asphericity in the explosion leads to larger [Zn/Fe] and [Co/Fe], but to smaller [Mn/Fe] and [Cr/Fe]. Then, if the degree of the asphericity tends to be larger for lower [Fe/H], the trends of [Zn, Co, Mn, Cr/Fe] follow the ones observed in metal-poor stars, as discussed later.

SIGNATURES OF HYPERNOVA NUCLEOSYNTHESIS IN GALACTIC CHEMICAL EVOLUTION

Several observational signatures of hypernova nucleosynthesis have been noticed in several objects [13]. The abundance pattern of metal-poor stars with [Fe/H] < -2 provides us with very important information on the formation, evolution, and explosions of massive stars in the early evolution of the galaxy.

In the early galactic epoch when the galaxy is not yet chemically well-mixed, [Fe/H] may well be determined by mostly a single SN event [18]. The formation of metal-poor stars is supposed to be driven by a supernova shock, so that [Fe/H] is determined by the ejected Fe mass and the amount of circumstellar hydrogen swept-up by the shock wave[19]. Then, hypernovae with larger E are likely to induce the formation of stars with smaller [Fe/H], because the mass of interstellar hydrogen swept up by a hypernova is roughly proportional to E [19, 20] and the ratio of the ejected iron mass to E is smaller for hypernovae than for canonical supernovae.

The observed abundances of metal-poor halo stars show quite interesting pattern. There are significant differences between the abundance patterns in the iron-peak elements below and above [Fe/H]~ -2.5 - -3, which cannot be explained with the conventional chemical evolution model that uses previous nucleosynthesis yields.

1) For [Fe/H]$\lesssim -2.5$, the mean values of [Cr/Fe] and [Mn/Fe] decrease toward smaller metallicity, while [Co/Fe] increases [21, 19].

2) [Zn/Fe]~ 0 for [Fe/H] $\simeq -3$ to 0[22], while at [Fe/H] < -3.3, [Zn/Fe] increases toward smaller metallicity ([23, 24]).

The larger [(Zn, Co)/Fe] and smaller [(Mn, Cr)/Fe] in the supernova ejecta can be realized if the mass ratio between the complete Si burning region and the incomplete Si burning region is larger, or equivalently if deep material from complete Si-burning region is ejected by mixing or aspherical effects. This can be realized if (1) the mass cut between the ejecta and the collapsed star is located at smaller M_r[26], (2) E is larger to move the outer edge of the complete Si burning region to larger M_r[27], or (3) asphericity in the explosion is larger.

Also a large explosion energy E results in the enhancement of the local mass fractions of Zn and Co, while Cr and Mn are not enhanced [25]. Models with $E_{51} = E/10^{51}$ergs do not produce sufficiently large [Zn/Fe]. To be compatible with the observations of [Zn/Fe] ~ 0.5, the explosion energy must be much larger, i.e., $E_{51} \gtrsim 20$ for $M \gtrsim 20 M_\odot$.

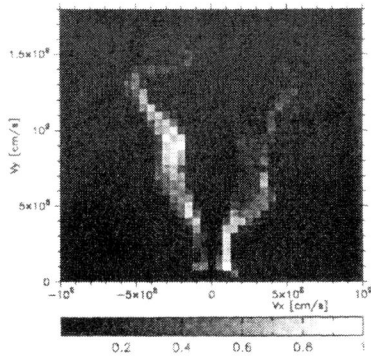

FIGURE 2. The density distributions of ^{44}Ti (left half) and ^{56}Ni (right half). The densities of each element are represented with linear scale, from 0 to the max density of each element[28].

Therefore, if hypernovae made significant contributions to the early Galactic chemical evolution, it could explain the large Zn and Co abundances and the small Mn and Cr abundances observed in very metal-poor stars.

NUCLEAR RADIOACTIVITY IN HYPERNOVAE

Hypernovae produce radioactive species in larger amount than normal SNe II, thus being important sources of line γ-rays and positrons.

^{56}Ni and Positrons

A large amount of ^{56}Ni can be produced as actually observed: $\sim 0.5\ M_\odot$ from SN1998bw, $\sim 0.15 M_\odot$ from SN1997ef, and even much larger from SN1999as. Thus hypernovae are the important source of positrons as well. In particular, positrons in the galactic center region might be significantly contributed by hypernovae, because as observed in M82, hypernovae would make an important contribution in nucleosynthesis associated with starburst events.

^{44}Ti

^{44}Ti is synthesized in the α-rich freezeout, in complete Si burning. In the very high entropy environment in hypernovae, the photons dissociate all the preexisting nuclei down to essentially α-particles and neutrons. The abundances then shift to those in nuclear statistical equilibrium but freezes out with excess α-particles. The ^{44}Ti yield depends strongly on the location of the mass cut, electron fraction, and entropy condition

TABLE 1. The mass ratio ^{44}Ti/^{56}Ni for $40M_\odot$ models. Values of spherical models are taken from [27].

Spherical ($E_{51} = 1$)	Spherical ($E_{51} = 10$)	Aspherical ($E_{51} = 10$)
6.51E-4	2.79E-3	1.01E-2

TABLE 2. Mass of ^{26}Al and ^{60}Fe ($M_\odot$) (z=0.02).

E_{51}	1	10	1	10
M	^{26}Al		^{60}Fe	
$20M_\odot$	8.87e-5	5.73e-5	1.37E-5	1.31E-4
$25M_\odot$	1.17e-4	1.22e-4	1.86E-4	3.38E-4
$40M_\odot$	8.61e-5	1.20e-4	1.82E-5	2.47E-5

in the α-rich freezeout. Therefore the initial yield of ^{44}Ti provides unique information to constrain the explosion models.

Iron-peak elements are produced in the deep region near the mass cut, so that their production is strongly affected by asphericity of an explosion. Figure 2 shows the 2D density distribution of ^{44}Ti and ^{56}Ni, which is distributed preferentially in the z-direction. ^{44}Ti, which is produced by the strong α-rich freezeout, is distributed preferentially in the z-direction. Moreover, ^{44}Ti production is strongly enhanced compared with a spherical model[29], since the post-shock temperature along the z-direction is much higher than that of a spherical model.

Table 1 summarize the mass ratio ^{44}Ti/^{56}Ni produced in the spherical and the aspherical explosion of the 40 $M_\odot$ models. We can see that ^{44}Ti is strongly enhanced in hypernova and aspherical models. The γ-rays from the decays of ^{44}Ti are possible tools to investigate the Galactic hypernova remnants.

^{26}Al and ^{60}Fe

Preliminary results of ^{26}Al and ^{60}Fe are summarized in Table 2.

1) The ^{26}Al abundance does not depend much on E. It is a little smaller in hypernovae than supernovae, because more ^{26}Al is consumed in oxygen burning as O, Mg, Ne are.

2) ^{60}Fe abundance is larger by a factor of ~ 3.

RX J0852-4622/GRO J0854-4622

COMPTEL has detected the ^{44}Ti 1157 keV γ-ray line from the supernova remnant (SNR) RX J0852-4622 [30, 31, 32], though the evidence is found at the 2σ to 4σ significance level[33]. ASCA has observed RX J0852-4622 and detected the 4.1 keV X-ray emission line from Ca[34]. [34] has provided the following analysis and interpretation. The abundance of Ca is oversolar by a factor of 8 ± 5, while other elements are subsolar.

The mass of Ca is $\sim 1.1 \times 10^{-3} M_\odot$. The excess Ca is very likely ^{44}Ca, the decay product of ^{44}Ti. This feature is seen only in the north-west shell, which suggests that the supernova ejecta has just collided with the interstellar material there.

[35] has assumed that the width of the ^{44}Ti line is due to Doppler broadening and thus ^{44}Ti is expanding at $\sim 15{,}000$ km s^{-1}[30]. Then they suggested that such high velocity ^{44}Ti is ejected from a Type Ia supernova of sub-Chandrasekhar mass[36, 37], because the model produces $\sim 10^{-3} M_\odot$ ^{44}Ti in the outer He detonation zone which expands at $\sim 15{,}000$ km s^{-1}.

Here we suggest an alternative model for the high velocity ^{44}Ti. As seen in Figure 2, the asymmetric hypernova explosion ejects ^{44}Ti at $\sim 15{,}000$ km s^{-1} in the jet direction. The amount of ^{44}Ti is as large as $1 \times 10^{-3} M_\odot$ (Table 1), being consistent with the observation.

INTEGRAL observations of the ^{44}Ti lines and their line profiles from RX J0852-4622 are important to discriminates the models and clarify the energetics of the explosion.

SN 1987A

SN 1987A in the LMC has shown for the first time that the energy source of supernova ejecta directly comes from the decays of radioactive nuclei (e.g., [38, 39] for reviews). In this section, we investigate the initial abundance of ^{44}Ti to discuss the detection possibility of the line gamma-rays from SN 1987A, by comparing theoretical light curves with the observed bolometric luminosity of SN 1987A.

In SN 1987A, it is established that the observed light curve in early time is first governed by ^{56}Ni [$t_{1/2}$ (half-life) = 6.1 d] and then its daughter ^{56}Co ($t_{1/2}$ = 77.3 d). The synthesized ^{56}Co nuclide decays mainly by positron emission to stable ^{56}Fe. As we shall see later, the observed light curve in the wavelength ranging from ultraviolet (UV) to infrared (IR) has been successfully modeled with the energy supply from the decay of ^{56}Co until ~ 800 days (e.g., [38]). This has been directly confirmed by the detection of the hard X-rays and the line γ-rays from the decay sequence of ^{56}Co.

Afterwards, the decline of the observed light curve apparently slowed down, due to the decay of ^{57}Co ($t_{1/2}$ = 272 d). The slowness of the decline of observed light curve becomes distinguished in particular after ~ 1500 days from the explosion. The dominant energy source at this moment can be attributed to ^{44}Ti decay. ^{44}Ti decays by orbital electron capture to ^{44}Sc, emitting 67.9 keV (100 %) and 78.4 keV (98 %) lines. ^{44}Sc then decays mainly by positron emission into ^{44}Ca, which emits a 1157 keV (100 %) de-excitation line.

Obviously, a luminosity observation in SN 1987A of late years is crucial to study the property of the extra energy source, namely, the ^{44}Ti production. Recently, this crucial luminosity at 3600 days, i.e., 10 years after the explosion was reported by [40]. The observed luminosity in the UV-IR range is $L = (1.9 \pm 0.6) \times 10^{36}$ erg sec^{-1}. The result is obtained by the collaboration of CTIO with HST, and hereafter we refer to this as the CTIO+HST luminosity. In the following, we investigate whether the ^{44}Ti decay provides enough nuclear energy that can account for the observed CTIO+HST luminosity under the possible range of the half-life value of ^{44}Ti [41].

The Half-Life of ^{44}Ti

The energy release from the ^{44}Ti decay depends strongly on its half-life. It is known that the published half-life values of ^{44}Ti that were measured in laboratories display a large spread, ranging from ~ 35 to ~ 68 years. However, experimental efforts have been concentrated especially in the last a few years, and at this moment the half-life appears to be settled in 60 ± 3 years, including the errors up to 3 σ (see, [42] and references therein).

We note here that the half-life values obtained in laboratories are for neutral atoms. Since ^{44}Ti undergoes pure orbital electron capture decay, its decay rate becomes smaller than the experimental value if ^{44}Ti is highly ionized under the condition of a supernova remnant. For example, the decay rates of hydrogen-like and helium-like ions are, respectively, ~ 44 % and ~ 88 % of that of neutral atoms. Details are found in [43, 44].

It is expected that ^{44}Ti were neutral when the CTIO+HST observation was carried out. We thus adopt 60 ± 3 years to calculate theoretical light curves to compare the observation.

Theoretical Light Curve of SN 1987A

Our calculation of the light curves is based on [38] and the adopted nuclear decay property data are updated. We perform Monte Carlo simulations of the Compton degradation of the line γ-rays emitted from the decays of ^{57}Co (14 keV, 122 keV, 136 keV, etc.) and ^{44}Ti (68 keV, 78 keV, 1157 keV) to obtain the UV-IR light curves. The UV, optical, and IR photons originate from the energy loss of the emitted γ-rays during the radiative transfer in the ejecta. The calculated UV-IR luminosity is the result of subtracting the energy of the X-ray and γ-ray photons which have managed to get out of the remnant.

For the velocity distribution of particles, we adopt the explosion model 14E1 proposed by [45] whose main-sequence mass, ejecta mass, and explosion energy are $20\,M_\odot$, $14.6\,M_\odot$ ($4.4\,M_\odot$ core material plus $10.2\,M_\odot$ hydrogen-rich envelope), and 1×10^{51} erg, respectively. This model was derived from a detailed analysis of the plateau shape of the light curve of SN 1987A which is observed until 120 days after the explosion, and well accounts for the earlier optical, X-ray, and γ-ray light curves of SN 1987A[46]. Note that the ^{56}Ni mass in SN 1987A has been constrained as $0.07 M_\odot$ from the intensity during the observed exponential decline. The synthesized mass of the radioactive ^{57}Ni is adopted from [10], and the input value of the initial ^{44}Ti abundance is varied within the uncertainty studied in [10] to compare the CTIO+HST luminosity.

In Figure 3, we show the calculated UV-IR light curve of SN 1987A and the observed luminosities. The solid line denotes the calculated evolution of the total luminosity. For this, the ^{44}Ti half-life of 60 yrs and $<^{44}\text{Ti}/^{56}\text{Ni}> = 1$ have been adopted. Here, $<^{44}\text{Ti}/^{56}\text{Ni}>$ is defined as the ratio of $^{44}\text{Ti}/^{56}\text{Ni}$ in amounts in the supernova remnant to $^{44}\text{Ca}/^{56}\text{Fe}$ in the solar neighborhood, i.e., $<^{44}\text{Ti}/^{56}\text{Ni}> \equiv [X(^{44}\text{Ti})/X(^{56}\text{Ni})]/[X(^{44}\text{Ca})/X(^{56}\text{Fe})]_\odot$. The three decay sequences of ^{56}Ni, ^{57}Ni and ^{44}Ti are used as the energy source, and contributions from each decay sequence are shown in Figure 3, respectively, with the labeled dotted lines and the

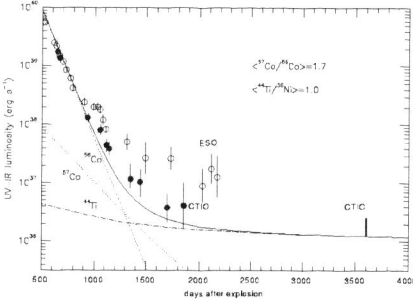

FIGURE 3. The calculated light curve and the observed bolometric luminosity of SN 1987A, including the latest observed luminosity (CTIO/HST).

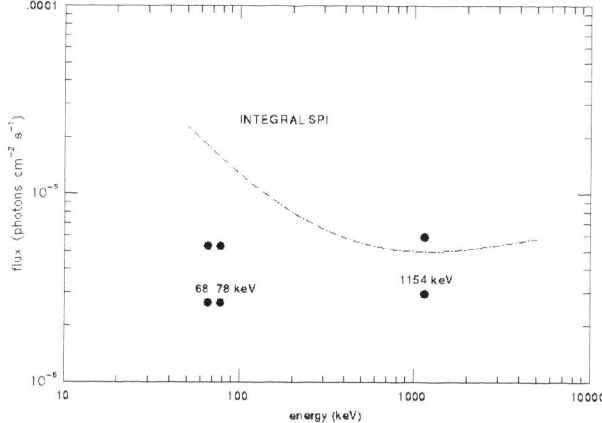

FIGURE 4. Predictions of the line gamma-ray fluxes from the ^{44}Ti decay sequence (filled circles) for the initial mass of ^{44}Ti of $1 \times 10^{-4} M_\odot$ and $2 \times 10^{-4} M_\odot$, respectively. The sensitivity limit of INTEGRAL SPI (10^6 sec) is also shown by the dash-dotted curve.

dash-dotted lines. We found that the observed CTIO+HST luminosity is reasonably explained by the energy release from the ^{44}Ti decay for its half-life of 60 ± 3 yrs.

Production of ^{44}Ti in SN 1987A

The $<^{44}$Ti/^{56}Ni$>$ values that are required to explain the CTIO+HST luminosity are found to be roughly in between 1 and 2. Since the synthesized mass of ^{56}Ni has been determined for SN 1987A, the obtained $<^{44}$Ti/^{56}Ni$>$ values can directly be translated to the initial mass of ^{44}Ti. Note that the amount of ^{44}Ti is the total value that is responsible for the overall energy in γ-rays, X-rays, and in the UV-IR range.

We thus obtained that the initial ^{44}Ti mass is $(1.1 - 2.5) \times 10^{-4} \, M_\odot$ for the half-life value of 60 ± 3 yrs. We remark that the ^{44}Ti yield calculated from recent explosive nucleosynthesis models overlap with our estimate (e.g., [47, 48, 49]). However, it should be also mentioned that these model predictions are subject to nuclear reaction cross sections which have not yet measured so far (see, [50, 51]).

Note that [52] estimated that positrons from $(1\text{-}2) \times 10^{-4} M_\odot$ of ^{44}Ti provides the overall luminosity of the FeII emission lines, and [53, 54] obtained the upper limit of the ^{44}Ti mass, $1 - 1.5 \times 10^{-4} \, M_\odot$, based on ISO SWS/LWS observations.

^{44}Ti in SN 1987A with INTEGRAL

Detection of ^{44}Ti line γ-rays from SN 1987A should be an important target for coming INTEGRAL mission. In Figure 4, we show the expected fluxes from SN 1987A on Earth for 68, 78, and 1157 keV lines, respectively. For each γ-ray line in Figure 4, the upper filled circle is calculated for the ^{44}Ti yield of $2 \times 10^{-4} \, M_\odot$, and the lower filled circle for $1 \times 10^{-4} \, M_\odot$. In Figure 4, the sensitivity limit of INTEGRAL SPI (10^6 sec) is also shown, which strongly depends on energy. We see from Figure 4 that it is possible to detect the 1157 keV line marginally if the produced amount of ^{44}Ti is as large as $2 \times 10^{-4} \, M_\odot$, but for other cases the expected fluxes lie below the sensitivity limit of SPI.

In the above discussion of the expected fluxes, no ionization effect on the ^{44}Ti decay is taken into account. Recently, helium-like and hydrogen-like ions of O, Ne, Mg, and Si have been observed with Chandra X-ray observatory[55]. The ionization of the ejected material was caused because the supernova blast wave struck against the ring of matter that was ejected by the progenitor star. As discussed previously, the decay rate of highly ionized ^{44}Ti can become considerably small compared with that of neutral ^{44}Ti. This means, as claimed by [44], all the ^{44}Ti γ-ray line fluxes from SN 1987A possibly lie below the detection limit of INTEGRAL SPI.

Finally, we should note that INTEGRAL has capabilities both in the detection of nuclear lines and pulsed emission. It has not been clear yet whether SN1987A has formed a neutron star or a black hole. It is certainly worth trying fo INTEGRAL and Astro E-II to search for pulsation.

REFERENCES

1. Nomoto, K., et al. 2000, in "Supernovae and Gamma Ray Bursts" eds. M. Livio, et al. (Cambridge Univ. Press) (astro-ph/0003077)
2. Iwamoto, K., Mazzali, P.A., Nomoto, K., et al. 1998, Nature, 395, 672
3. Woosley, S.E., Eastman, R.G., & Schmidt, B.P. 1999, ApJ, 516, 788
4. Iwamoto, K., Nakamura, T., Nomoto, K., et al. 2000, ApJ, 534, 660
5. Mazzali, P.A., Iwamoto, K., & Nomoto, K. 2000, ApJ, 545, 407
6. Wheeler, J. C., Yi, I., Höflich, P. A., & Wang, L. 2000, ApJ, 537, 810
7. MacFadyen, A.I. & Woosley, S.E. 1999, ApJ 524, 262
8. Khokhlov, A.M., Höflich, P.A., Oran, E.S., Wheeler, J.C., Wang, L., & Chtchelkanova, A.Yu. 1999, ApJ, 524, L107
9. Kumagai, S., & Nomoto, K. 1997, in Thermonuclear Supernovae, ed. P. Ruiz-Lapuente, et al. (Kluwer)

10. Hashimoto, M., Nomoto, K., & Shigeyama, T. 1989, A&A, 210, L5
11. Woosley, S. E., & Weaver, T. A. 1995, ApJS, 101, 181
12. Thielemann, F.-K., Nomoto, K., & Hashimoto, M. 1996, ApJ, 460, 408
13. Nomoto, K., Maeda, K., Umeda, H., & Nakamura, T. 2001, in The Influence of Binaries on Stellar Populations Studies, ed. D. Vanbeveren (Kluwer), in press (astro-ph/0105127)
14. Nakamura, T., Mazzali, P. A., Nomoto, K., & Iwamoto, K. 2001a, ApJ, 550, 991
15. Mazzali, P.A., Nomoto, K., Patat, F., & Maeda, K. 2001, ApJ, in press (astro-ph/0106095)
16. Maeda, K., Nakamura, T., Nomoto, K., Mazzali, P.A., & Hachisu, I. 2000, ApJ, submitted (astro-ph/0011003)
17. Patat, F., et al. 2001, ApJ, 555, in press (astro-ph/0103111)
18. Audouze, J., & Silk, J. 1995, ApJ, 451, L49
19. Ryan, S.G., Norris, J.E. & Beers, T.C. 1996, ApJ, 471, 254
20. Shigeyama. T., & Tsujimoto, T. 1998, ApJ, 507, L135
21. McWilliam, A., Preston, G.W., Sneden, C., & Searle, L. 1995, AJ, 109, 2757
22. Sneden, C., Gratton, R.G., & Crocker, D.A. 1991, A&A, 246, 354
23. Primas, F., Reimers, D., Wisotzki, L., Reetz, J., Gehren, T., & Beers, T.C. 2000, in The First Stars, ed. A. Weiss, et al. (Springer), 51
24. Blake, L.A.J., Ryan, S.G., Norris, J.E., & Beers, T.C. 2001, Nucl.Phys.A.
25. Umeda, H., & Nomoto, K. 2001, ApJ, submitted (astro-ph/0103241)
26. Nakamura, T., Umeda, H., Nomoto, K., Thielemann, F.-K., & Burrows, A. 1999, ApJ, 517, 193
27. Nakamura, T., Umeda, H., Iwamoto, K., Nomoto, K., Hashimoto, M., Hix, R.W., Thielemann, F.-K. 2001b, ApJ, 555, in press (astro-ph/0011184)
28. Maeda, K. 2001, Master Thesis, University of Tokyo
29. Nagataki, S., Hashimoto, M., Sato, K., Yamada, S. 1997, ApJ, 486, 1026
30. Iyudin, A.F., Schönfelder, V., Bennett, K., et al. 1998, Nature, 396, 142
31. Aschenbach, B. 1998, Nature, 396, 141
32. Aschenbach, B., Iyudin, A.F., & Schönfelder, V. 1999, A&A, 350, 997
33. Schönfelder, V., Bloemen, H., Collmar, W., Diehl, R., et al. 2000, 5th Compton Symp., AIP, 510, 54
34. Tsunemi, H., Miyata, E., Aschenbach, B., Hiraga, J, & Akutsu, D. 2000, PASJ, 52, 887
35. Iyudin, A.F., & Aschenbach, B. 2001, in New Century of X-ray Astronomy, ed. Kunieda (AIP)
36. Livne, E., & Arnett, D. 1995, ApJ, 452, L62
37. Arnett, W. D., Supernovae and Nucleosynthesis (Princeton Univ. Press)
38. Kumagai, S., Nomoto, K., Shigeyama, T., Hashimoto, M., & Itoh, M. 1993, A&A, 273, 153
39. Nomoto, K., et al. 1994, in Supernovae (Les Houches, Session LIV) ed. S. Bludman et al. (Elsevier Science Pub.), 199
40. Suntzeff, N.B. 1997, in SN1987A: Ten Years After, eds. M.M. Phillips and N.B. Suntzeff (PASP).
41. Mochizuki, Y.S., Kumagai, S., Tanihata, I. 1998, in Origin of Matter and Evolution of Galaxies, eds. S. Kubono et al., World Scientific, in press
42. Hashimoto, T., Nakai, K., Wakasaya, Y., et al. 2001, Nucl. Phys. A686, 591
43. Mochizuki, Y., Takahashi, K., Janka, H.-Th., Hillebrandt, W., & Diehl, R. 1999, A&A, 346, 831
44. Mochizuki, Y. 2001, Nucl. Phys. A688, 58c
45. Shigeyama, T. & Nomoto, K. 1990, ApJ, 360, 242
46. Nomoto, K., Shigeyama, T., Kumagai, S., Yamaoka, H. 1991, in Supernovae, ed. S.E. Woosley (Springer), 176
47. Thielemann, F.-K., Hashimoto, M., & Nomoto, K. 1990, ApJ, 349, 222
48. Timmes, F.X., Woosley, S.E., Hartmann, D.H., & Hoffman, R.D. 1996, ApJ, 464, 332
49. Woosley, S.E. & Hoffman, R.D. 1991, ApJ, 368, L31
50. The, L.-S., Clayton, D.D., Jin, L., & Meyer, B.S. 1998, ApJ, 504, 500
51. Sonzogni, A.A., Rehm, K.E., Ahmad, I., et al. 2000, Phys. Rev. Lett. 84, 1651
52. Chugai, N.N. et al. 1997, ApJ, 483, 925
53. Lundqvist, P., Sollerman, J., Kozma, C. et al. 1999, A&A, 347, 500
54. Lundqvist, P., Kozma, C., Sollerman, J., & Fransson, C. 2001, A&A, in press (astro-ph/0105402)
55. Burrows, D.N., Michael, E., Hwang, U., et al. 2000, ApJ, 543, L149

Gamma-ray signatures of classical novae

Margarita Hernanz*, Jordi Gómez-Gomar† and Jordi José**

*Institute for Space Studies of Catalonia (IEEC) and Instituto de Ciencias del Espacio (CSIC),
Edifici Nexus, C/Gran Capità, 2-4, E-08034 Barcelona, Spain
†Institute for Space Studies of Catalonia (IEEC)
**Institute for Space Studies of Catalonia (IEEC) and Departament de Física i Enginyeria Nuclear
(UPC), Avda. Víctor Balaguer, s/n, E-08800 Vilanova i la Geltrú (Barcelona), Spain

Abstract. The role of classical novae as potential gamma-ray emitters is reviewed, on the basis of theoretical models of the gamma-ray emission from different nova types. The interpretation of the up to now negative results of the gamma-ray observations of novae, as well as the prospects for detectability with future instruments (specially onboard INTEGRAL) are also discussed.

INTRODUCTION

Classical novae are explosive phenomena occurring in close binary systems of the cataclysmic variable type. In these binaries, a normal main sequence star overflows its Roche lobe, transferring H-rich matter to its companion white dwarf star through an accretion disk. Matter accumulates on top of the degenerate white dwarf star, where it is gradually compressed and heated, until hydrogen reaches conditions for ignition. This ignition happens in a degenerate regime, thus leading to a thermonuclear runaway, because of the inability of matter to thermally readjust itself through expansion. During explosive hydrogen burning, radioactive nuclei (with lifetimes ranging from ~ 100 s to $\sim 10^6$ s) are synthesized. The radioactive isotopes with lifetimes around 100 s, like ^{14}O (τ=102 s), ^{15}O (τ=176 s) and ^{17}F (τ=93 s), are responsible for the explosion itself, because they can be transported by convection to the outer envelope, during the thermonuclear runaway (since $\tau_{conv} < \tau$). These nuclei are prevented from destruction in the outer cooler shells, and their subsequent decay releases energy which is largely responsible for the expansion and large increase in luminosity of the nova.

Other radioactive isotopes synthesized in novae, with longer lifetimes, are responsible for the gamma-ray emission of these objects. Two types of emission are expected: prompt emission, related with e^--e^+ annihilation (with e^+ coming from the decay of the short-lived ^{13}N, τ=862 s, and ^{18}F, τ=158 min) and long-lasting emission, caused by the decay of ^{7}Be (τ=77 days) and ^{22}Na (τ=3.75 yr). The prompt emission appears very early (before optical maximum, i.e., usually before nova discovery), has short duration (a couple of days) and consists of a 511 keV line plus a continuum below it (see below for details). The long-lasting emission consists of lines (either 478 keV from ^{7}Be decay or 1275 keV from ^{22}Na decay), lasting around 2 months and 3 years, respectively.

The potential role of classical novae as sources of gamma-ray emission was pointed out long ago [2, 1, 20], but detailed models combining both the explosion modeling and the production and propagation of gamma-rays are more recent [10, 11, 5, 12]. Up to now, there have been unsuccessful attempts to detect gamma-ray emission from novae. Efforts have been made mainly to detect the ^{22}Na line, at 1275 keV, with the COMPTEL instrument onboard the Compton Gamma-Ray Observatory, CGRO [14, 15]. Previous attempts to detect the ^{7}Be line, at 478 keV, and the 1275 keV line were made with the GRS instrument onboard the Solar Maximum Mission, SMM, satellite [6] All these efforts have only provided upper limits, fully compatible with our theoretical predictions [18, 19]

Other attempts have concentrated on the annihilation emission (511 keV line plus continuum below it), with large field of view instruments, like WIND/TGRS [8] and CGRO/BATSE [13], without success and, again, with upper limits compatible with theoretical predictions. The possible detection of this type of emission from novae with the CGRO/BATSE instrument had been pointed out by Fishman et al. (1991) prior to CGRO launch. The sensitivities of the instruments were too low to detect the emission, which is more intense than that in the 478 and 1275 keV lines but has much shorter duration. In addition to search for gamma-ray emission in particular objects, there have been attempts to look for the Galactic accumulated emission at 478 and 1275 keV, both with CGRO/OSSE and SMM/GRS [21, 6, 7]. In this case, more flux is accumulated since more sources are contributing, because the typical period between two succesive nova explosions in the Galaxy is shorter than the lifetimes of ^{7}Be and ^{22}Na. But again not enough sensitivity was available. We have recently made predictions about the detectability of this accumulated emission with INTEGRAL/SPI [17]; the cumulative emission around the Galactic center has some chance of being detected with SPI, during the deep survey of the central radian of the Galaxy (or, at least, better upper limits than those of SMM or COMPTEL are expected).

GAMMA-RAY EMISSION: LINES AND CONTINUUM

A summary of the main radioactive nuclei synthesized in novae is shown in table 1. It is important to stress that these nuclei are not produced in the same amounts in all the nova types, since their synthesis is closely related to the nuclear paths followed by the nova during its evolution. These paths depend on the initial chemical composition of the accreted envelope, which is related to that of the underlying white dwarf core, because some mixing between the core and the envelope should be invoked in order to explain the observed abundances of novae. It turns out that CO novae are the main producers of ^{7}Be, whereas ONe novae are responsible for ^{22}Na synthesis. In table 2 we show some examples of nova models, with their relevant yields of radioactive isotopes. The specific kinetic energy of the ejecta is also shown for completeness. These results have been obtained by means of a hydrodynamic code, which computes the nova evolution from the accretion phase up to the explosion and ejection of the envelope (see José & Hernanz 1998 for details). The ^{18}F yields still suffer from some uncertainty, mainly because of the not well known ^{18}F(p,α) reaction (see [3] for a recent analysis).

The gamma-ray output of a particular nova model at different epochs after the outburst (defined as the epoch of peak temperature), has been computed with a Monte Carlo code, which handles gamma-ray production and transfer in the expanding envelope (see [5] for details), with properties derived from the hydro code models. In figure 1 we show the spectral evolution of a CO and an ONe nova (M_{wd}=1.15 and 1.25 $M_\odot$, respectively), at distance 1 kpc. For all models there is a continuum between (20-30) and 511 keV, and a line at 511 keV ($\sim$ 8 keV full width half-maximum, FWHM), with intensities decreasing very fast [12]. The 511 keV line comes from the direct annihilation of positrons and from the positronium (in singlet state) emission, whereas the continuum originates in both the positronium continuum (triplet state positronium) and the Comptonization of photons emitted in the 511 keV line. There is a cutoff of the continuum at low energies (20-30 keV, depending on the chemical composition), related to photoelectric absorption, which acts as a sink of the Comptonized photons. In addition to this prompt and short-duration emission, there is a longer duration gamma-ray output, consisting of a line at 478 keV ($\sim$(3-8) keV FWHM), in CO novae, or at 1275 keV ($\sim$20 keV FWHM), in ONe novae. The general trends for other CO and ONe models are similar to those shown here. It is worth noticing that models with lower masses are more opaque (i.e., the 0.8 $M_\odot$ CO nova), because of the smaller expansion velocities (see table 2).

The light curves for the different types of emission are shown in figures 2, 3, 4 and 5. Figure 2 shows the light curve of the 511 keV line (FWHM between 3 and 8 keV) for all models, and those of different energy bands in the continuum, for an ONe nova. The continuum emission at energies lower than 511 keV dominates, being the band between 20 and 250 keV the one with the highest flux (but also the one which decreases faster, as can also be seen in figure 1). This prompt emission gives a direct insight of the dynamics of the expanding envelope, as well as information about its content on the radioactive nuclei ^{13}N and ^{18}F. In the case of ONe novae, there is also the contribution of positrons from ^{22}Na decay, which produces smaller fluxes but lasts a longer time (up to complete transparency of the envelope, which occurs at around 1 week after peak temperature, the exact value depending on the expansion velocities of the envelope).

We have analyzed the influence of the mass and the velocity of the ejecta on the prompt emission, by means of some extra models, in which we scale either the mass of the ejecta or its terminal velocity. These are in some way not self-consistent models, because they are not the result of evolutionary calculations, but they are good for illustrative purposes. Figure 3 shows the 511 keV line light curves for a CO and an ONe nova (both of 1.15 $M_\odot$), for a range of parametrized ejected masses (the value obtained in the evolutionary model is shown in table 2). The effect of increasing the ejected mass is twofold, depending on the epoch. At early times, the larger the mass the lower the flux, because of the increasing opacity. On the contrary, later on the opacity doesn't play an important role, and the larger the mass the larger the flux, because of the larger amount of radioactive isotopes. It is worth reminding that in ONe novae the emission lasts longer than in CO ones (see figure 3, right), because of the contribution of the e^+ from ^{22}Na decay.

The influence of the velocity of the ejecta is shown in figure 4. At early times, larger velocities imply larger transparency and thus larger fluxes (both for CO and ONe novae). At later times (after $\sim$ 2days), only the case of ONe novae is relevant, since there are still e^+ from ^{22}Na decay; then, the larger the velocity the earlier the flux disappears, because the envelope becomes transparent before, thus allowing e^+ to freely escape (see figure 4, right). This facts demonstrates again that the analysis of the prompt gamma-ray emission

TABLE 1. Radioactive isotopes ejected by novae relevant for gamma-ray emission

Isotope	Lifetime	Main disintegration process	Type of γ-ray emission	Nova type
^{13}N	862 s	β^+–decay	511 keV line & continuum	CO and ONe
^{18}F	158 min	β^+–decay	511 keV line & continuum	CO and ONe
^{7}Be	77 days	e^-–capture	478 keV line	CO
^{22}Na	3.75 years	β^+–decay	1275 keV & 511 keV lines	ONe
^{26}Al	10^6 years	β^+–decay	1809 keV & 511 keV lines	ONe

TABLE 2. Radioactivities in novae ejecta (^{13}N and ^{18}F at 1h after T_{peak})

Nova	$M_{wd}(M_\odot)$	$M_{ejec}(M_\odot)$	KE (erg/g)	^{13}N ($M_\odot$)	^{18}F ($M_\odot$)	^{7}Be ($M_\odot$)	^{22}Na ($M_\odot$)
CO	0.8	6.2×10^{-5}	8×10^{15}	1.5×10^{-7}	1.8×10^{-9}	6.0×10^{-11}	7.4×10^{-11}
CO	1.15	1.3×10^{-5}	4×10^{16}	2.3×10^{-8}	2.6×10^{-9}	1.1×10^{-10}	1.1×10^{-11}
ONe	1.15	2.6×10^{-5}	3×10^{16}	2.9×10^{-8}	5.9×10^{-9}	1.6×10^{-11}	6.4×10^{-9}
ONe	1.25	1.8×10^{-5}	4×10^{16}	3.8×10^{-8}	4.5×10^{-9}	1.2×10^{-11}	5.9×10^{-9}

TABLE 3. SPI 3σ detectability distances (in kpc) for lines and continuum (see text for details about T_{obs}).

Nova type	$M_{wd}(M_\odot)$	511 keV line	478 keV line	1275 keV line	(170-470) keV
CO	0.8	0.7	0.4	-	0.4
CO	1.15	2.4	0.5	-	2.0
ONe	1.15	3.7	-	1.1	3.0
ONe	1.25	4.3	-	1.1	3.0

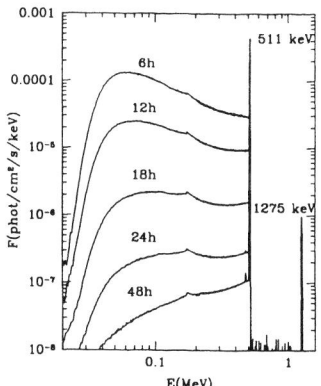 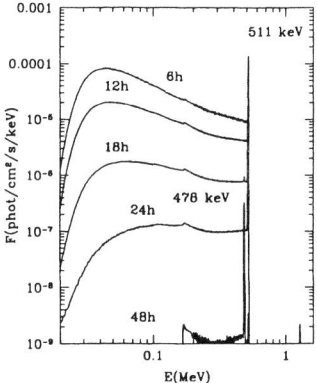

FIGURE 1. (Left) Gamma-ray spectra for an ONe nova of $1.25M_\odot$, at different epochs after the outburst (defined as the peak temperature time) and at distance 1 kpc. (Right) Same for a CO nova of $1.15M_\odot$

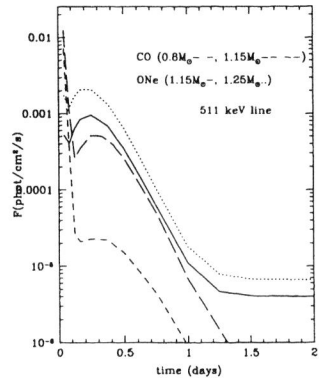 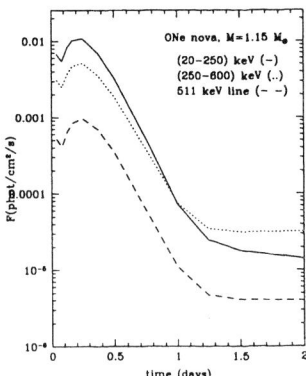

FIGURE 2. (Left) Light curves for the 511 keV line of the 4 nova models shown in table 2, placed at a distance of 1 kpc. (Right) Continuum light curves for the ONe nova of 1.15 $M_\odot$ at the same distance.

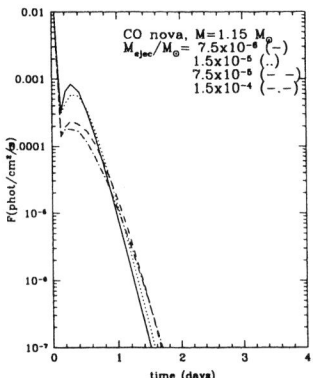 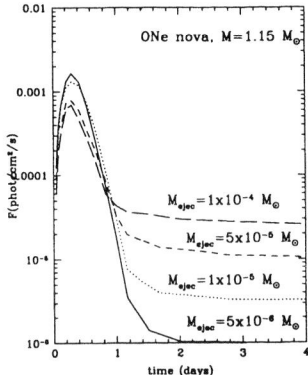

FIGURE 3. (Left) Light curves for the 511 keV line for a CO nova of 1.15 $M_\odot$, for a range of ejected masses. (Right) Same for an ONe nova of 1.15 $M_\odot$. Distance is 1 kpc.

of classical novae would provide a great deal of information about the dynamics of the expanding envelope, as well as about the ratio between its ^{18}F and ^{22}Na contents.

In figure 5 we display the light curves of the 478 keV line, for the two CO novae from table 2, and those of the 1275 keV line, for the ONe novae in table 2. These light curves show a first phase of increasing flux, related to the increasing transparency of the envelope, followed by the characteristic exponential decay phase, when the envelope is already transparent. The light curve of the 478 keV line shows in addition an intense peak at early times, which comes from the Comptonization of the 511 keV photons (see above). The fluxes of the 478 and 1275 keV lines during the exponential decay phase, directly reflect the amount of ^{22}Na and ^{7}Be in the envelope.

PROSPECTS FOR DETECTABILITY OF INDIVIDUAL NOVAE

In order to predict detectability distances of the gamma-ray emission from novae, the abovementioned light curves for the different types of emission have been used. The fluxes are quite small, leading to detectability distances with INTEGRAL/SPI of around 1 kpc, for the 1275 keV line, and 0.5 kpc, for the 478 keV line, for the nominal observation time of 10^6s (see table 3 for exact values). Concerning the 511 keV line and the continuum, detectability distances with SPI are around 3 kpc (see table 3), adopting 10 h of observation time, starting 5 h after peak temperature. For the continuum we have adopted the range (170-470) keV, which is optimal for SPI, since it avoids the 478 keV line and the low energies, where the background is too high. The width of the lines has been fully taken into account to derive all the detectability distances. As it is known, the instrument INTEGRAL/SPI will have a very good spectral resolution, which means that its nominal sensitivity for narrow lines is degraded when they are broad.

Our time origin in the figures is at peak temperature, which happens before the maximum in visual luminosity. The time interval between peak temperature and maximum in visual luminosity depends on the particular nova model, mainly on its speed class (rate of decline of the visual luminosity). It ranges from some days to some weeks, but its exact value is difficult to establish, because novae are usually discovered at or after visual maximum. Therefore, the epoch of peak temperature is close to peak gamma-ray luminosity (corresponding to the e^--e^+ annihilation emission), but it is not reachable from visual observations. The early appearance, before optical detection, of the prompt gamma-ray emission from novae, makes its detection with SPI problematic. It will be only possible if a close enough nova falls in the field of view of the instrument when it is doing another observation (i.e., during the Galactic plane survey -GPS- or during the Galactic center deep exposure -GCDE). We have also considered alternative ways to detect this intense emission, by means of the SPI shield, which provides a large detection area with a wide field of view, but without spectroscopic capability [16]. In summary, the prompt gamma-ray e^--e^+ annihilation emission can almost only be detected with wide field of view intruments scanning all the sky very often (like the future EXIST, MEGA, Advanced Compton Telescope). Up to now, "a posteriori" analyses (provided that there was some observation of the right field at the right moment) of the CGRO/BATSE [13] and WIND/TGRS [8] data have been performed; the negative results are fully compatible with our theoretical predictions and are related to the not enough sensitivity of these instruments.

DISCUSSION

The main factor affecting detectability of novae is distance (see table 3), but the distances of novae are not easy to determine accurately. The visual luminosity (i.e., the absolute visual magnitude) of a classical nova at maximum is not directly correlated with its amount of the radioactive nucleus ^{22}Na, or any other radioactive nucleus (in contrast with SNIa, where ^{56}Ni is responsible for both the visual and the gamma-ray luminosities at early times). Therefore, some other characteristic, such as apparent visual magnitude, should be used as distance indicator. But, as with any cosmic object, novae which are apparently bright visually can be farther away than novae which are dim, if the visual extinction (intrinsic plus interstellar) of the apparently bright object is much smaller than that of the apparently dim object. Once the preliminary visual light curve and visual extinction are obtained, a distance determination is possible through indirect methods, which suffer from large uncertainties. They depend on various not well known nova properties. First, the empirical relationship between absolute magnitude at maximum, M_V^{max}, and speed class of the nova (MMRD relation); the speed class is measured by the time of decline of the visual magnitude by 2 or 3 magnitudes (t_2 or t_3). Second, the visual extinction of the nova, A_V, which has intrinsic plus interstellar contributions; the latter varies a lot depending on the location of the nova in the Galaxy.

Once M_V^{max} and A_V are known, the derivation of the distance from the apparent magnitude at maximum, m_V^{max}, is straightforward. Therefore, the main uncertainties affecting distance determinations are: general validity of the empirical M_V^{max}-t_2 (or t_3) relationship, determination of A_V, in addition to the determination of t_2 (or t_3) and of m_V^{max} (often it is not known if the nova has been caught at the maximum or after it) from the observations. In figure 6, we show a m_V^{max}-distance diagram, for novae discovered in the last century (up to 1995). The data shown are taken from the samples of Shafter [22]. We have superimposed two curves indicating the apparent magnitudes at maximum, m_V^{max}, one could expect, provided that novae are standard candles with absolute magnitude at maximum M_V^{max}=-7.5, and that visual extinction, A_V, ranges from 0 to 3 magnitudes. For distances up to 1 kpc, m_V^{max} should be smaller (brighter) than 5.5 (for 3 kpc, m_V^{max} ranges from 8 to 5, or brighter if M_V^{max} is < -7.5). If we include novae after 1995, two outstanding points at m_V^{max}=2.8 and 4, and d$\sim$ 2 and 4 kpc (Nova Vel 1999 and Nova Aql 1999b, respectively) would appear (with $M_V^{max} < -7.5$; Nova Vel 1999 probably had $M_V^{max} \sim -8.7$ (IAUC 7193)), in addition to more "normal" points with distances larger than 5 kpc and m_V^{max} larger than 8. The number of novae discovered during the period 1991-1995 versus m_V^{max} is also shown in figure 6.

In order to estimate the probability of having a nova at a particular distance, it is instructive to look at figure 7, which shows an histogram of the novae distances for the same nova set mentioned above [22], as well as for the subset of novae in the 1991-1995 period. The sample of years 1991-1995 suffers from small number statistics, but it is more representative of recent more accurate observations. Although the distances have

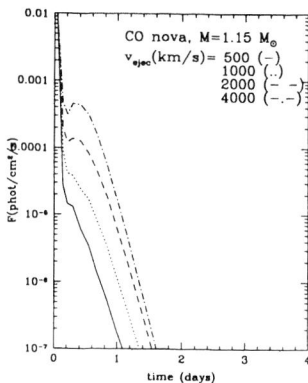
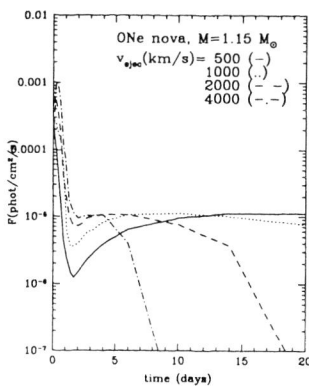

FIGURE 4. (Left) Light curves for the 511 keV line for a CO nova of 1.15 $M_\odot$, for a range of parametrized velocities of the ejecta. The value indicated corresponds to the outermost shell. (Right) Same for an ONe nova of 1.15 $M_\odot$. Distance is 1 kpc.

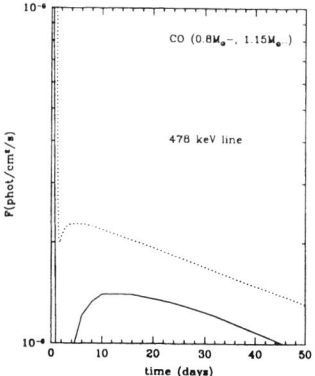
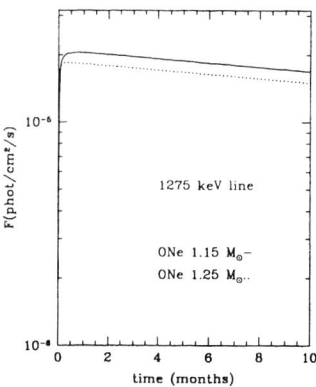

FIGURE 5. (Left) Light curves for the ^{7}Be line (478 keV) for two CO nova models. (Right) Light curves for the ^{22}Na (1275 keV) for two ONe models. Distance is 1 kpc.

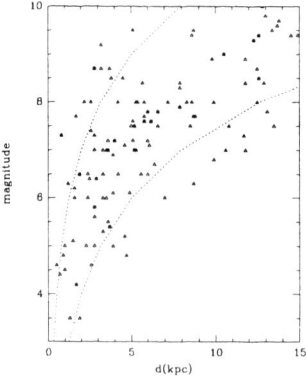 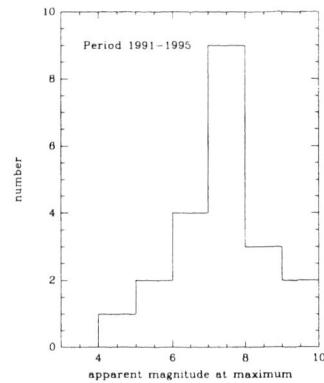

FIGURE 6. (Left) Apparent visual magnitudes at maximum, m_V^{max}, versus distances. Filled squares correspond to the 1991-1995 period and open triangles to the 1901-1990 period. The dashed curves represent the m_V^{max} vs. distance relationship obtained for an absolute $M_V^{max} = -7.5$ (typical for novae) and a range of visual extinctions (from right to left $A_V = 0$ and $A_V = 3$ magnitudes). (Right) Histogram of novae apparent magnitudes at maximum, for the novae in the period 1991-1995.

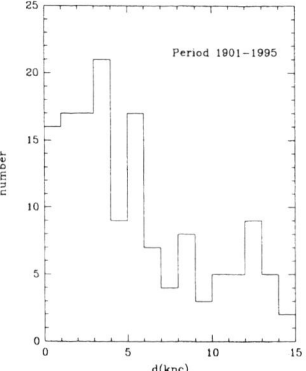 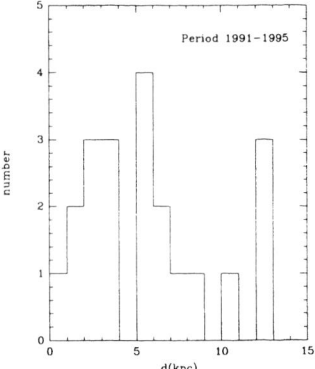

FIGURE 7. (Left) Histogram of novae distances for the novae discovered in the last century (until 1995). (Right) Same for the subset of the recent novae in the period 1991-1995

a large uncertainty, some general trends can be extracted: the observed nova rate for novae at distances shorter than 1 kpc is 1/5=0.20 yr^{-1} (1991-1995 set), or 16/95=0.17 yr^{-1} (complete set 1901-1995), which is not very large. If we relax the distances of detectability of novae by INTEGRAL by a factor of 3 (i.e., we adopt 3 kpc instead of 1 kpc, invoking the effect of the uncertain ejected masses -for some observed novae- by a factor of 10), the observed nova rate increases to 6/5=1.20 yr^{-1} (1991-1995 nova set), or 50/95=0.53 yr^{-1} (complete set 1901-1995). Therefore, there is some chance to have a close nova during INTEGRAL's lifetime (2 to 5 years).

Concerning future instrumentation, if an increase of sensitivity by a factor of 10 (for broad lines) is achieved, the detection of novae would be a routine instead of a chance. If, in addition, these instruments have wide fields of view and are designed to perform frequent surveys of the sky in the hard X-ray domain (E>100 keV), then the prompt e^--e^+ annihilation gamma-ray emission of novae could be detected for many Galactic novae. This fact would be crucial not only for the understanding of the nova explosion mechanism itself, but also for the knowledge of the nova distribution in the Galaxy ([9]. This distribution is not at all known, since only 3-5 of the 35±11 Galactic novae exploding every year are discovered optically nowadays.

REFERENCES

1. Clayton D.D., 1981, ApJ, 244, L97
2. Clayton D.D., Hoyle F., 1974, ApJ, 187, L101
3. Coc A., Hernanz M., José J., Thibaud J.P., 2000, A&A, 357, 561
4. Fishman G.J. et al., 1991, in Durouchoux P., Prantzos N., eds., Gamma-Ray Line Astrophysics. AIP, New York, p. 190
5. Gómez-Gomar J., Hernanz M., José J., Isern J., 1998, MNRAS, 296, 913
6. Harris M.J., Leising M.D., Share G.H., 1991, ApJ, 375, 216
7. Harris M.J., et al. 1996, A&AS, 120, 343
8. Harris M.J., et al. 1999, ApJ, 522, 424
9. Harris M.J., et al. 2000, ApJ, 542, 1057
10. Hernanz M., Gómez-Gomar J., José J., Isern J., 1997a, in 2nd INTEGRAL Workshop "The transparent Universe". ESA SP-382, Noordwijk, p. 47
11. Hernanz M., Gómez-Gomar J., José J., Isern J., 1997b, in 4th COMPTON Symposium. AIP, New York, 1125
12. Hernanz M., José J., Coc A., Gómez–Gomar J., Isern J., 1999, ApJ, 526, L97
13. Hernanz M., Smith D.M., Fishman J., Harmon A., Gómez-Gomar J.,José J., Isern J., Jean P., 2000, in 5th COMPTON Symposium. AIP, New York, 82
14. Iyudin A.F., et al. 1995, A&A, 300, 422
15. Iyudin A.F., et al. 1999, Astrophys. Lett. & Comm., 38, 371
16. Jean P., Gómez-Gomar J., Hernanz M., José J., Isern J., Vedrenne G., Mandrou P., Schönfelder V., Lichti G., Georgii R., 1999, Astrophys. Lett. & Comm., 38, 421
17. Jean P., Hernanz M., Gómez-Gomar J., José J., 2000, MNRAS, 319, 350
18. José J., Hernanz M., 1998, ApJ, 494, 680
19. José J., Coc A. & Hernanz M. 1999, ApJ, 520, 347
20. Leising M.D., Clayton D., 1987, ApJ, 323, 159
21. Leising M.D., Share G.H., Chupp E.L., Kanbach G., 1988, ApJ, 328, 755
22. Shafter A.W. 1997, ApJ, 487, 226
23. Starrfield S., Truran J.W., Wiescher M.C., Sparks W.M. 1998, MNRAS, 296, 502

Study of the Galactic Distribution of Nova-Produced ^{22}Na with COMPTEL

A.F. Iyudin*, V. Schönfelder*, A.W. Strong*, K. Bennett[†], R. Diehl*, W. Hermsen**, G.G. Lichti* and J. Ryan[‡]

*Max-Planck-Institut für extraterrestrische Physik, Postfach 1312, 85741 Garching, Germany
[†]Astrophysics Division, ESTEC, 2200 AG Noordwijk, The Netherlands
**SRON-Utrecht, Sorbonnelaan 2, 3584 CA Utrecht, The Netherlands
[‡]UniversityofNewHampshire,InstituteforStudiesofEarth,OceansandSpace, DurhamNH03284,USA

Abstract.
The COMPTEL telescope on board the Compton Gamma-Ray Observatory (CGRO) is capable of imaging γ-ray line sources, like classical novae, in the MeV regime at a level of sensitivity up to a few 10^{-5} photons cm^{-2}s^{-1}. At this level of sensitivity quite high expectations can be placed on the detection of the predicted ^{22}Na γ-ray line at 1.275 MeV from nearby novae. Unfortunately, no positive detection of the nova-produced ^{22}Na was reported until recently.

We have used COMPTEL data collected up to the 2nd CGRO reboost to assess (1) - sodium production by individual novae, and (2) - the Galactic distribution of the nova-produced ^{22}Na.

Introduction

The classical nova outburst has been modelled as a thermonuclear runaway in the accreted hydrogen-rich envelope of the white dwarf companion of a close binary system, e.g. [29, 30]. In general, observations of novae support such models [10, 11].

It is currently believed that novae may be an important source of Galactic ^{22}Na [3]. ^{22}Na decays with a 3.75 yrs life-time to a short lived excited state of ^{22}Ne at 1.275 MeV. In addition to the ^{22}Na line novae are prolific producers of 511 keV γ-ray line emission, which accompanies the decay of the β$^+$-unstable products of nucleosynthesis in novae, as well as of 478 keV emission originating from the ^{7}Be decay to ^{7}Li. Until now, there have been no positive detections of any of these lines, at 1.275 MeV, 511 keV and 478 keV [12, 16, 17, 18, 24, 25].

Model calculations of the novae outbursts provide theoretical estimates for X_{22}, M_{ej} and v_{exp} which allows a comparison between predicted and measured fluxes from the ^{22}Na decay in shells ejected by novae. It has been common belief that detectable amounts of ^{22}Na are synthesized almost exclusively in the high-mass, fast ONe novae, which have enough Ne seed nuclei. Unfortunately, attempts to measure the fluxes $F_{1.275}$ from a number of recent ONe novae seems to counter this common belief. These results can be summarized as follows: *(i)* only upper limits were derived with COMPTEL at the level of ($\sim 2 \times 10^{-5}$ photons/(cm^{2}s)) for the ^{22}Na γ-ray line from the studied ONe novae (V693 CrA, V1370 Aql, QU Vul, V838 Her, NSgr 1991, NSct 1991, NPup 1991, V1974 Cyg,

NCir 1995, NAql 1995, NCen 1995, NCru 1996 and V382 Vel) [16, 17]; *(ii)* upper limits from other γ-ray telescopes are usually less constraining [12, 16, 23, 24]; *(iii)* COMPTEL derived upper limits on the 1.275 MeV fluxes appear to be in general agreement with theoretical predictions [20, 21, 32], provided the adopted estimates of distances to these novae are correct.

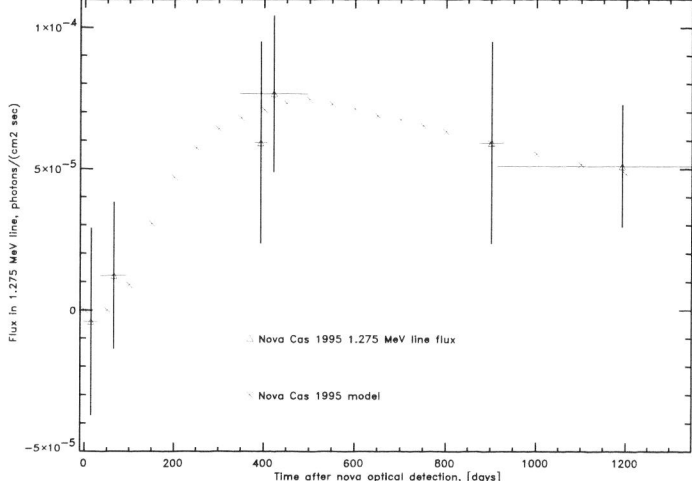

Fig. 1. Light curves of NCas 1995 in the γ-ray line emission at 1.275 MeV. The 1.275 MeV line flux values were derived from the image analysis. The curve marked by asterisks represents a model for the decaying ^{22}Na embedded in the NCas 1995 ejecta that expands with the velocity of ~150 km/s.

Thus, the preliminary attempts to detect the 1.275 MeV γ-ray line emission from ^{22}Na produced in the *fast* ONe novae have led to a negative result. As will be discussed below the only likely positive detection of such emission was found to be related to the *very slow* CO-type Nova Cas 1995 (V723 Cas).

The full model of the global Galactic distribution of 1.275 MeV line emission includes, apart from the volume distribution of the classical novae, also the distribution of different types of novae, e.g., CO-, and ONe- type novae, or, equivalently, slow novae and fast novae. We note, that the disk and bulge novae components might be represented by different types, and/or speed classes of novae! Global Galactic distributions of novae were proposed in papers [15, 31, 22, 4, 13, 28]. As discussed below, preliminary COMPTEL results point toward a very strong bulge component of the Galactic 1.275 MeV line emission.

Instrument and Analysis Methods

COMPTEL, due to its combination of imaging and spectroscopic capabilities [27], provides a unique opportunity to measure line emission from point-like sources or from extended regions (e.g. the Galactic bulge).

Generally, different viewing periods covering the position of the relevant nova were combined to achieve the best possible sensitivity. Imaging and flux evaluation were done in a ±2 σ energy window around the 1.275 MeV line, where σ is the instrumental energy resolution for this line. The maximum-likelihood method was extensively used to derive individual novae fluxes (upper limits), while the maximum-entropy maps were mostly used for the ^{22}Na global Galactic distribution study. For the detection of weak sources (small fluxes) it is essential to optimise the signal-to-noise (S/N) ratio of COMPTEL, which is at the one-percent level only. A powerful tool to optimize the S/N ratio is event selection. The used selections are described by a time-of-flight (ToF) window of 115-130, the minimum Earth-horizon event angle $\zeta \geq 10^o$, and avoid the use of the "faulty" D2-modules, e.g. three D2-modules which have one out of seven photomultipliers (PMs) switched off (see [27] for a detailed instrument description).

The other important requirement for the safe detection of a source is the correct handling of the background underlying the source signal. For detection of the ^{22}Na line-emitting sources one possibility is to model the background in the line interval from the data at adjacent energy intervals (method B1). The background model derived in this way still contains systematic uncertainties due to the underlying continuum emission and small differences in the event distributions in the (χ, ψ) space. A second method derives the background model (B2) from the line energy interval itself [1, 2]. Both methods have been extensively used in the studies of the ^{22}Na line emission from novae.

To distinguish between a possible transient source with a 1.275 MeV line emission and a nova with a line due to the decaying ^{22}Na, we have used the method introduced in [17] and described in more detail in [18]. Namely, we have measured the decay curve of the nova-produced ^{22}Na in the 1.275 MeV line emission, and compared it with a model calculation (e.g. see Fig. 1). The time delays between the nova maximum brightness and the time of the COMPTEL measurements have been taken into account.

This method of following the nova light-curve in the 1.275 MeV line emission for more than 4 years (Fig. 3 in [18]) was crucial in establishing of what we believe now to be a tentative detection of the ^{22}Na 1.275 MeV line emission from the slow nova NCas 1995. The relative distribution of the $F_{1.275}$ data points in Fig. 1 is interpreted now as evidence for how the shell ejected by the N Cas 1995 first became transparent to γ-rays (an initial increase of $F_{1.275}$ at the early dates) and, after all of the ^{22}Na synthesized during the nova explosion became observable, how it decayed at later times. An imaging analysis performed in the 1.275 MeV line using all the data yielded a total significance of the NCas 1995 detection of $\sim 4\sigma$ (based on statistical uncertainties only). From results of the fitting one may state that the most probable evolution of the NCas 1995 flux in the 1.275 MeV line was consistent indeed with that of the slowly expanding shell with a total mass of ^{22}Na of

$$X_{22} \sim 5.4 \times 10^{-8} \left(\frac{D}{2.4 kpc}\right)^2.$$

More discussion of this result is given in a forthcoming publication [19].

Global Galactic Distribution of the ^{22}Na

COMPTEL results on the global galactic distribution of the ^{22}Na were derived by constructing longitude and/or latitude profiles of the 1.275 MeV line emissivity extracted from the maximum-entropy map. The map itself was made for the combination of all COMPTEL observations till the 2nd CGRO reboost. The latitude profile is shown in

Fig. 2.

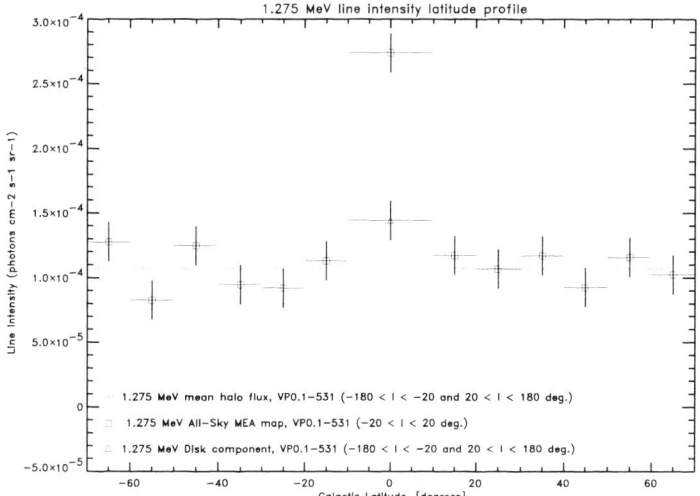

Fig. 2. Latitude profile of the galactic bulge region in the light of the 1.275 MeV emission is shown by squares for the combination of all observations. The disk component outside the bulge is shown also by a triangle. A dotted line shows a mean halo intensity in the 1.275 MeV line.

The bulge component clearly sticks out from the distributions shown. The error bars plotted in the latitude profile were derived from the scatter of the intensity values outside the bulge region.

The main results on the global galactic distribution of the ^{22}Na line emission can be summarized as follows:

• COMPTEL's 1.275 MeV line intensity profiles point toward a bulge shape with the ratio of the major-to-minor bulge axis of ~ 2. This bulge shape is consistent with the bulge model [8, 9] derived from the COBE measurements of the Galactic IR-emissivity;

• Assuming a yield of the ^{22}Na in nova is between $3 \times 10^{-9}\ M_\odot$ and $1.2 \times 10^{-8}\ M_\odot$ as was modelled in [14, 21] we derive the range for the nova rate in the Galactic bulge as,
$20.5\ \mathrm{yr}^{-1} \leq R_{CN} \leq 82\ \mathrm{yr}^{-1}$,
where the upper bound value is higher than normally quoted [6, 13].

• Further assuming that Galactic bulge novae comprises $\sim 75\ \%$ of all Galactic novae we evaluate the Galactic disk nova rate at $\sim 7\ \mathrm{yr}^{-1}$, compared with the value of $5 \pm 2.5\ \mathrm{yr}^{-1}$, which was evaluated in [5].

• If one assumes that all bulge 1.275 MeV emission is due to the ONe-type novae, then the space density of active classical novae systems in the bulge derived under this assumption is very high, namely
$1.4 \times 10^{-5}\ \mathrm{pc}^{-3} \leq D_{CN} \leq 5.8 \times 10^{-5}\ \mathrm{pc}^{-3}$,
i.e. more than order of magnitude higher than the value favoured in [26].

• From the above it follows that the bulge novae have to be represented by mostly slow, CO-type novae. This conclusion is consistent with that of [7].

Acknowledgments

The COMPTEL project is supported by the German "Ministerium fuer Bildung und Forschung" through DLR grant 50 QV 9096 8. AFI acknowledges financial support from the German "Ministerium für Bildung und Forschung" through the DLR grant 50 OR 0002.

REFERENCES

1. Bloemen, H. et al., *ApJSS* **92**, 419 (1994).
2. Bloemen, H. et al., *Proc. 5th Compton Symposium*, (1999).
3. Clayton, D.D., & Hoyle, F., *ApJ* **187**, L101 (1974).
4. Dawson, P.C., & Johnson, R.G., *J.R.Astron.Soc.Can.* **88**, 369 (1994).
5. Della Valle, M., & Duerbeck, M., *A&A* **271**, 175 (1993).
6. Della Valle, M., & Livio, M., *A&A* **286**, 786 (1994).
7. Della Valle, M., & Livio, M., *ApJ* **506**, 818 (1998).
8. Dwek, E., et al., *ApJ* **445**, 716 (1995).
9. Freudenreich, H.T., *ApJ* **492**, 495 (1998).
10. Gallagher, J.S. & Starrfield, S., *ARAA* **16**, 171 (1978).
11. Gehrz, R.D., Truran, J.W., Williams, R.E. and Starrfield, S., *PASP* **100**, 3 (1998).
12. Harris, M.D., et al., astro-ph/0004164, (2000).
13. Hatano, K., Branch, D., Fisher, A., Starrfield, S., *MNRAS* **290**, 113 (1997).
14. Hernanz, M., Gomez-Gomar, J., Jose, J., Coc, A., Isern, J., *Astrophysical Letters and Communications* **38**, 407 (1999).
15. Higdon, J.C., Fowler, W.A., *ApJ* **317**, 710 (1987).
16. Iyudin, A.F., Bennett, K., Bloemen, H., et al., *A&A* **300**, 422 (1995).
17. Iyudin, A.F., Proceedings of the 10th Workshop on "Nuclear Astrophysics", Ringberg Castle, Tegernsee, Germany, March 20-25, 2000, *MPA/P12* 118 (2000).
18. Iyudin, A.F., Bennett, K., Bloemen, H., et al., in Proc. of the 4th INTEGRAL Workshop, Alicante, Spain, September 04-08, 2000, *ESA-SP* in press (2001a).
19. Iyudin, A.F., Bennett, K., Bloemen, H., et al., *A&A* , submitted (2001b).
20. Jose, J., Hernanz, M., *ApJ* **494**, 680 (1998).
21. Jose, J., Coc, A., Hernanz, M., *ApJ* **520**, 347 (1999).
22. Kent, S.M., Dame, T.M., Fazio, G., *ApJS* **127**, 131 (1991).
23. Leising, M.D., Share, G.H., Chupp, E.L. & Kanbach, G., *ApJ* **328**, 755 (1988).
24. Leising M.D., Clayton D.D., The L.-S., et al., in AIP Conf. Proc., 280, 137 (1993).
25. Mahoney, W.A., et al., *ApJ* **262**, 742 (1982).
26. Patterson, J., *ApJS* **54**, 443 (1984).
27. Schönfelder, V., Aarts, H., Bennett, K., et al., *ApJS* **86**, 657 (1993).
28. Shafter, A.W., *ApJ* **487**, 226 (1997).
29. Starrfield, S., Sparks, W.M. & Truran, J.W., *ApJS* **28**, 247 (1974).
30. Truran, J.W., in Essays in Nuclear Astrophysics, ed. C.A. Barnes, D.D. Clayton, & D.N. Schramm (Cambridge: Cambridge Univ. Press), 467 (1982).
31. Van der Kruit, P., Buser, R., King, I.R., *The Milky Way as a Galaxy*, University Science Books, Mill Valley, 331, (1990).
32. Wanajo, S., Hashimoto, M.-A., & Nomoto, K., *ApJ* **523**, 409 (1999).

TGRS and the 478 keV Line from ^{7}Be in Novae

M. J. Harris, D. M. Palmer, G. Weidenspointner, H. Seifert

Universities Space Research Association

B. J. Teegarden, T. L. Cline, N. Gehrels, R. Ramaty

Code 661, NASA/Goddard Space Flight Center, Greenbelt, MD 20771, USA

Abstract.
We obtain limits on γ-ray line emission at 478 keV from both known and undiscovered novae. The instrument has good spectral resolution and a complete view of the southern ecliptic hemisphere. Although most of this field of view is unmodulated and acquires only background spectra, an occulter, taking advantage of the 3 s rotation period of the *Wind* satellite, modulates signals from the Galactic center (GC). The backgrounds are low and extremely stable, allowing us to extract cosmic signals from the background spectrum. The instrumental background contribution to the 478 keV line is described by a simple model; no departures from it (i.e. nova signals) were detected during 1995–1997. In a separate analysis the quasi-steady emission at 478 keV from integrated novae towards the GC is measured in the occulted data set.

Novae are expected to produce ^{7}Be from the reaction ^{3}He$(\alpha,\gamma)^7$Be, since the accreted material which undergoes explosive H burning will be enriched in ^{3}He and ^{4}He from pp-chain nucleosynthesis. The decay of ^{7}Be gives rise to a γ-ray line at 478 keV in 10.4% of cases, on a time-scale of 77 d. The nucleosynthesis depends on whether the underlying white dwarf is a degenerate CO remnant or an ONe remnant, with ^{7}Be produced mainly in CO events [2,6].

Spacecraft and Instrument. The TGRS detector is attached to the southern ($+z$) surface of the *Wind* spacecraft, oriented to the S. ecliptic pole. A concentric Pb occulter fixed to the same surface subtends 90° at the detector — *Wind* rotates every 3 s, so that signals from the ecliptic plane are modulated. The detector is surrounded by a Be radiative cooler. It is a 35 cm^2 Ge crystal, whose resolution prior to fall 1997 was 3–4 keV FWHM at 511 keV [4]. The effective area at 478 keV is $\simeq 15$ cm^2 for 0° zenith angle ϕ. The TGRS detector is unshielded and open to the entire $+z$ hemisphere. The projected area and thickness do not vary much with ϕ [9]. Background spectra were accumulated continuously except for brief

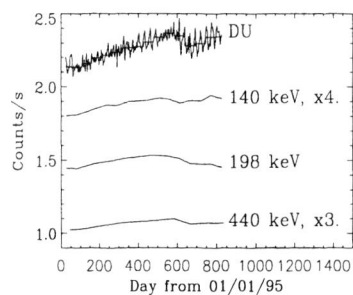

FIGURE 1. Variability of monitors of GCR intensity through 1996 solar minimum. (Top, full line) Count rate at energies > 8 MeV on 1 d time-scale. (Top, histogram) Count rate at energies > 8 MeV on 53 d time-scale. (Lower full lines) Intensities of prompt instrumental de-excitation lines.

(~few hr) perigee passes, γ-ray bursts, and solar flares. The detector performance deteriorated over time due to accumulated cosmic-ray-induced damage. We restrict our analyses to high-quality data obtained during 1995–1997, in which live time coverage is about 90% [5].

The occulter modulates an area $\simeq 16° \times 90°$, along the ecliptic[1] every 3 s. The data are summed into 128 angle bins around the ecliptic plane. A dip in the signal in any angle bin indicates the presence of a source in that direction [12].

Background Analysis. *Wind* has spent long periods in interplanetary space near the Earth-Sun L1 point, where background count rates are low, due to the avoidance of Earth's trapped radiation belts and to the absence of Earth albedo radiation. The most important background comes from radioactivities excited by Galactic cosmic-ray (GCR) impacts.

As well as being low, the background has also been extremely stable, tracking the variation of the GCR flux, which peaked during the 1996 solar minimum. Time series for two proxies for the GCR flux are shown in Fig. 1, energy losses > 8 MeV in the detector and the count rate measured in known prompt de-excitation lines whose lifetimes are $\leq \mu s$. There is a weak ~ 14 d modulation due to crossings of the heliospheric neutral sheet by the *Wind* orbit, but the variability is small if the rate is averaged over time-scales ~ 53 d.

The stability of the background makes it possible to search for sudden increments in the strength of an interesting line even if there is a background line at the same energy. This is the case at 478 keV, where the ^{9}Be in the cooler undergoes spallation by neutrons and protons to ^{7}Be. This analysis cannot detect a DC level of cosmic 478 keV line emission, given the strong background line. It can detect sudden increases in 478 keV line strength, both at the times of known events, and in a search of the entire 1995–1997 data.

---

[1] It is slightly offset from the ecliptic plane in order to maximise modulation of the GC.

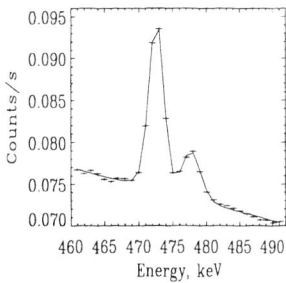

FIGURE 2. TGRS background spectrum during 1995 July 31–1995 September 22, fitted by power law and lines at 472 keV (^{24m}Na) and 478 keV (^{7}Li).

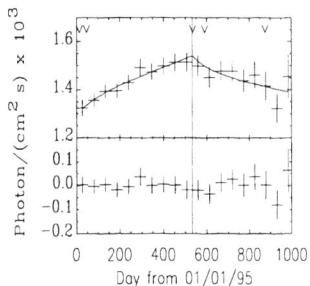

FIGURE 3. (*Top*) Time series of 478 keV line strengths in 53 d spectra, fitted to a model of the behavior of the instrumental background line (full line). Arrows — epochs of known Southern novae: in sequence, BY Cir, V888 Cen, V4361 Sgr, CP Cru, and N Sco 1997. (*Bottom*) Residual flux after subtracting the above model.

Spectra summed in 53 d intervals were fitted using a power law and two lines, one at 472.2 keV due to ^{24}Na, and the ^{7}Be line from the radiative cooler. The former was slightly distorted due to cosmic-ray damage even before fall 1997, and was fitted by an asymmetric function [10]. The latter is somewhat broader than the instrument resolution, being slightly blended, and we fitted it with a single Gaussian (Fig. 2). We then fitted the line amplitude time series (Fig. 3) with a simple model where ^{7}Be is created in the instrument by a time-dependent GCR flux and decays with a 53.28 d half-life. Fig. 1 suggests that a simple model for the GCR flux would be a linear increase between 1995 January and 1996 June, followed by a sharp change of slope. Let the ^{7}Be production rate be $R+Ct$ (C changes sign at solar minimum). In equilibrium the ^{7}Be abundance is $\left(\frac{R}{\lambda} - \frac{C}{\lambda^2}\right)[1-exp(-\lambda t)] + \frac{Ct}{\lambda} + N(t=0)\,exp(-\lambda t)$ where λ is the decay constant. We fitted the time series to this function with 5 free parameters and searched for significant sudden flux increases over and above it.

No flux increases appeared either at the times of known S. hemisphere novae, or at any other time (Fig. 3). Clearly no hitherto undiscovered novae were discovered using this method. The limits for this general case, and for the known times of five novae, are given in Table 1.

Occulted data analysis. Leising pointed out [7] that at any one time, there will be a number of novae emitting γ-ray lines in the Galaxy, which for the 478 keV line is of the order $R_N \tau(^7Be)$, averaging ~ 10 for a typical nova rate ~ 50 yr^{-1}. This can be searched for as a quasi-constant diffuse source. Most of the novae will lie in the GC direction, and the TGRS occulted data can be searched in exactly the same way as for the diffuse 511 keV line [4].

The principle behind the analysis was to extract a spectrum of the GC by channels from the amplitude of the drop in count rate in each channel in the TGRS angle bins from the GC direction, assuming the same distribution along the ecliptic as the diffuse 511 keV line [4]. This spectrum was fitted by a power law with a line at 478 keV superimposed.

Results. Our results for the increase in flux due to individual novae are shown in Table 1. The count rate was divided by photopeak effective area for the appropriate ϕ.[2] Table 1 also gives the absolute flux from the Galactic center direction integrated over all novae. The earlier value from *SMM* is shown for comparison [3].[3]

From the flux upper limits in Table 1 we calculate the ^{7}Be abundances in the sources, given the (very uncertain) distances. Theoretical values are also given. Our upper limits improve upon the *SMM* values by an order of magnitude for individual events, and by a factor 2 for the GC integrated flux.

Conclusions. Our best upper limit for a CO nova, BY Cir, exceeds the theoretical prediction by a factor at least 30.[4] This result constrains some earlier suggestions that the models underestimate the ^{7}Be yield for various reasons. For example, models generally predict the ejection of a mass $\sim 10^{-5}$ $M_\odot$, whereas several different spectroscopic methods suggest ejection of $\sim 10^{-4}$ $M_\odot$ [13]. In addition, the initial abundance of ^{3}He which burns to ^{7}Be in the explosion is usually assumed to be solar, but it is known that the ^{3}He abundance should be enhanced in the donor due to pp-chain nucleosynthesis [11]. Older main sequence star models showed ^{3}He enhancements ~ 10 times solar, while more recent models suggest a value about 50% of this [8]. However the effect turns out to be weak, the nova yield of ^{7}Be being proportional not to the ^{3}He abundance but to its logarithm [1]. Yields of ^{7}Be are thus probably enhanced by a factor < 2.5 by ^{3}He enrichment. The combination of this factor with a factor $\times 10$ from more massive ejecta raises the ^{7}Be yield to a value close to our measurement. We conclude that our results do not permit enhancements of the ^{7}Be yield by factors more optimistic than realistic

---

[2] In the general case this is the Galactic nova distribution average, 60.0° [5].
[3] Note that the integrated flux measurements are not exactly comparable, since the *SMM* value comes from a very broad region $\sim 130°$ across around the GC, whereas the TGRS value reflects the signal from the $16° \times 90°$ occulted region only.
[4] A tighter limit could be implied by the integrated GC flux measurement for a high enough Galactic nova rate R_N (Table 1).

Source	Sub-class	Date Estimated	Zenith Angle	Flux^a γ cm^{-2} s^{-1}	Distance pc	^{7}Be mass^a per nova, $M_\odot$
Individual Novae						
General case		1995–1997	60°	1.0×10^{-4}		
BY Cir	CO	1995 Jan 24	45°	6.8×10^{-5}	3575	3.9×10^{-8}
V888 Cen		1995 Feb 23	42°	6.3×10^{-5}	7740	1.7×10^{-7}
V4361 Sgr		1996 Jul 1	95°	1.1×10^{-4}	5655	1.6×10^{-7}
CP Cru	ONe	1996 Aug 24	37.°0	8.8×10^{-5}	9790	3.8×10^{-7}
Nova Sco		1997 Jun 4	97°	1.6×10^{-4}	12200	1.1×10^{-6}
GC Integrated						
TGRS			84.5°	7.7×10^{-5}	8000	$3.4 \times 10^{-6}/R_N$ ^{b}
SMM				1.5×10^{-4}	8000	$3.5 \times 10^{-6}/R_N$ ^{b}
Theory						$0.8\text{--}1.1 \times 10^{-10}$ ^{c}

^{a} 3 σ upper limit.
^{b} The effective apertures of TGRS and SMM will contain respectively 0.06 R_N and 0.12 R_N novae at any one time, where R_N is the total Galactic nova rate per year. The total ^{7}Be masses inferred from the flux upper limits have been divided by these factors to get the yield per nova.
^{c} Model predictions for 0.8 and 1.15$M_\odot$ CO white dwarfs [2].

TABLE 1. Results for 478 keV line fluxes and ^{7}Be yields.

factors currently accepted for ejected mass and ^{3}He initial abundance.

REFERENCES

1. Boffin, H. M. J., Paulus, G., Arnould, M., & Mowlavi, N., *A&A*, **275**, 96 (1993).
2. Gomez-Gomar, J., Hernanz, M., Jose, J., & Isern, J., *MNRAS*, **296**, 913 (1998).
3. Harris, M. J., Leising, M. D., & Share, G. H., *ApJ*, **375**, 216 (1991).
4. Harris, M. J., et al., *ApJ*, **501**, L55 (1998).
5. Harris, M. J., et al., *ApJ*, **542**, 1057 (2000).
6. Hernanz, M., & Jose, J., in *Cosmic Explosions*, ed. S. Holt & W. W. Zhang, AIP Conf. Proc. 522, New York, 2000, p. 339.
7. Leising, M. D., in AIP Conf. Proc. 170, *Nuclear Spectroscopy of Astrophysical Sources*, ed. N. Gehrels & G. H. Share, 1988, 130.
8. Morel, P., Pichon, B., Provost, J., & Berthomieu, G., *A&A*, **350**, 275 (1999).
9. Owens, A., et al., *Space Sci. Rev.*, **71**, 273 (1995).
10. Phillips, G. W., & Marlow, K. W., *Nucl. Instr. Methods*, **137**, 525 (1976).
11. Starrfield, S., Truran, J. W., Sparks, W. M., & Arnould, M., *ApJ*, **222**, 600 (1978).
12. Teegarden, B. J., et al., *ApJ*, **463**, L75 (1996).
13. Warner, B., *Cataclysmic Variable Stars*, CUP, Cambridge, 1995, 260–269.

Gamma-Ray Line Emission from Superbubbles in the Interstellar Medium: The Cygnus Region

S. Plüschke[1], R. Diehl[1], K. Kretschmer[1], D.H. Hartmann[2], U. Oberlack[3]

[1] MPE Garching, Giessenbachstrasse, 85748 Garching, Germany
[2] Dep. of Physics & Astronomy, Clemson Univ., Clemson, SC 29634, USA
[3] Astrophys. Lab. Columbia Univ., New York, NY 10027, USA

Abstract. The star forming process in the Milky Way is non-uniform in time and space. The scale of star forming regions ranges from groups within a few pc to large segments of spiral arms with linear dimension of order kpc. When many stars form in a relatively small volume over a short duration, a localized starburst ensues. The energetic impact of such a burst of star formation can severely affect the dynamic structure of the gaseous disk. Stellar winds and supernova explosions drive an expanding superbubble, whose size eventually exceeds the scale height of the disk and thus drives a disk-wind blowing metal enriched gas into the halo. We discuss the basic scenario of superbubble evolution, emphasizing the associated gamma-ray line signatures. In particular, we discuss nuclear line emission from ^{26}Al and ^{60}Fe in the Cygnus region.

INTRODUCTION

The COMPTEL experiment aboard the Compton Observatory has produced a nine year all sky survey in the MeV regime [18]. Here we emphasize the all sky map in a narrow band centered on 1.809 MeV which selects the radioactive decay line from ^{26}Al (Figure 1). With a mean lifetime of roughly 1 Myr there are many individually unresolved sources which contribute to this diffuse glow of the Galaxy (for a recent review see [17]). The bulk of the observed flux is due to ^{26}Al emerging in the winds of massive stars and the ejecta of core-collapse supernovae with relative contributions of 60% and 40%, respectively [8]. Nucleosynthesis of ^{26}Al in Novae and AGB stars also contributes to the glow, but probably to a lesser extent. Population synthesis studies suggest that the galactic disk presently contains about 2 $M_\odot$ of radioactive ^{26}Al. In steady-state a global galactic star formation rate of a few solar masses per year corresponds to a supernova rate of a few events per century, which implies a typical yield of 10^{-4} $M_\odot$ of ^{26}Al per supernova. This observationally determined yield is consistent with the theoretical values derived from hydrostatic and explosive nucleosynthesis simulations (e.g. [11,23]).

Multi-wavelength observations of the Milky Way and external galaxies clearly show that star formation exhibits a hierarchical pattern. Most stars form in groups, associations, clusters, and still larger conglomerates [4]. Thus, star formation is is a strongly correlated process. The transformation of 10^4 to 10^5 solar masses of gas into stars (with a conversion efficiency of roughly 10%) leads to a subsequent dynamic

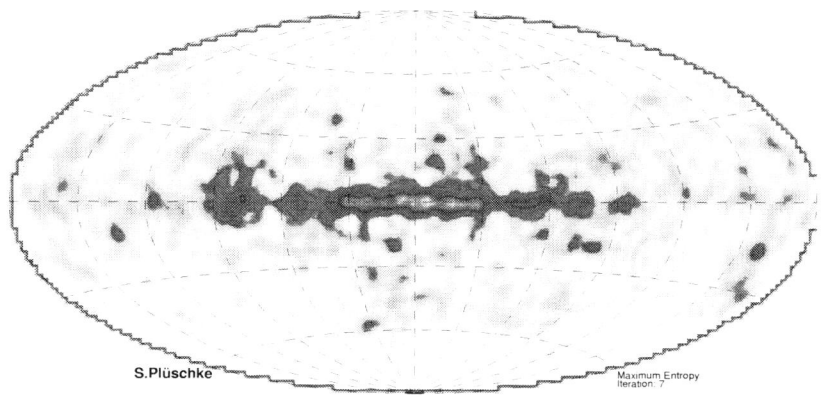

Figure 1. Galactic map at 1.8 MeV obtained with COMPTEL [15]. The prominent extended emission feature between $l = 70°$ and $l = 90°$ coincides with the Cygnus star forming region.

epoch due to the winds and explosions from several hundred stars in the 8 to 100 $M_\odot$ range. The kinetic energy input into the surrounding interstellar medium from these stars drives the expansion of a "super-remnant" which can even blow a "hole" into the gaseous disk. There is much observational evidence for such superbubbles in the Milky Way and other nearby galaxies. Superbubbles provide the means by which the disk and the halo are chemo-dynamically coupled. The interior of a superbubble, similar to a supernova remnant, contains a tenuous, high temperature gas. If the ^{26}Al ejected by massive stars attains high velocities in this environment we may expect significant broadening of the observed 1.8 MeV line. In fact, the GRIS experiment has measured a FWHM of 5.4 ± 1.2 keV, which corresponds to temperatures somewhat higher than 10^8 K [12]. Temperatures in this regime are actually very hard to accomplish in the ISM and also hard to maintain on timescales comparable to the mean life of ^{26}Al. Alternatively, the radioactive aluminum nuclei could be embedded in high velocity dust grains (perhaps accelerated in superbubble interiors), so that the observed line broadening is not thermal in nature but a purely kinematic effect. In that case typical velocities should be of order 500 km/s. Assuming free expansion these velocities would lead to an angular scale height that exceeds the COMPTEL observations. It is currently unclear if a broad 1.8 MeV line is a global feature, or if only localized regions on the sky exhibit this effect while others show less or perhaps no broadening. This issue is likely to be resolved with data from the INTEGRAL mission [22].

Here we focus on one particular star forming region in the Galaxy, the Cygnus region. We model the observed 1.8 MeV flux with a variety of OB associations whose properties are discussed in the literature. We also provide a flux estimate at 1.173 and 1.332 MeV from ^{60}Fe.

MODELLING THE CYGNUS REGION

The Cygnus region (defined by the extended 1.8 MeV emission feature at $l = 80\pm 10$ deg, see Fig. 1) contains numerous massive stars. The galactic O star catalogue [5] lists 96 O stars in this field. In addition, there are 23 Wolf-Rayet stars of which 14 are of WN-type, 8 of WC-type, and 1 is classified as WO [21]. Between 5 and 10 of these stars are believed to be members of OB associations. In addition, the Galactic SNR Catalogue lists 19 remnants in this region. For 9 of these remnants reliable age and distance estimates are available [7]. In our model WR stars and SNRs that do not belong to one of the Cygnus OB associations are treated as isolated sources whose estimated 1.8 MeV flux is subtracted from the integrated COMPTEL measurement. The resulting residual flux is shown in Fig. 3.

In the Cygnus region one finds numerous open clusters and nine OB associations [1,6,10]. Two of those (OB 5 & 6) could be unrelated due to projection and selection effects [6]. Cygnus OB4 has poorly determined properties. Therefore, the remaining six OB associations (1,2,3,7,8,9) were chosen as a basis set for our model.

Optical studies of the Cygnus region are hampered by a giant molecular complex along the line of sight [3]. From the 2MASS NIR survey of a field centered on Cygnus OB2 the estimated number of O stars has tripled [9]. Extinction corrections in this region are severe. Knödlseder [9] inferred 120 ± 20 O stars and 2500 ± 500 B stars. Similar corrections are expected for the remaining OB associations in Cygnus. We used the CO map [3] to estimate the extinction correction factors. The procedure was calibrated with observations in the optical and NIR of Cyg OB2.

Based on the observed numbers of O stars, corrected for extinction, we generate synthetic stellar populations for each association by means of Monte Carlo sampling of an appropriate initial mass function. MC sampling allows us to derive statistical uncertainties due to the relative small size of the associations. Our population synthesis model (e.g. [13,14]) provides the time evolution of the interstellar masses of ^{26}Al and ^{60}Fe, the kinetic power due to stellar winds and supernova explosions, and the emission of ionizing photons. We also determine the number of WR stars as function of time. Adding these WR stars from each association the expected total number of WR stars potentially observable is of order 10 ± 3. This estimate is consistent with the reported number of associated WR stars in the Galactic WR Star Catalogue [21].

Gamma-ray lines from the decay of ^{26}Al and ^{60}Fe follow directly from the predicted mass budget. Uncertainties in flux predictions are due to a combination of statistical errors (small population size), systematic uncertainties (e.g. stellar yields), and the often poorly known distances to the associations [14].

The release of freshly synthesized nuclei is directly coupled to the injection of kinetic energy into the interstellar medium. Typical stellar wind power is $\sim10^{37}$ erg/s and supernovae deliver $\sim10^{51}$ erg of kinetic energy to the surrounding medium (but see [20]). Thus OB associations drive large cavities into the interstellar medium. The interior of these cavities (superbubbles) is filled with a hot, tenuous plasma, and their size can reach several 100 pc. Because of the high ejection velocities (several 1000 km/s) and the low density in the cavity the freshly ejected nuclei travel large distances before they decay. Assuming a mean velocity of 1000 km/s ^{26}Al nuclei could traverse a distance of ~200 pc within one fifth of their mean lifetime. Therefore the diffuse nature of the observed 1.809 MeV emission has two contributions, the spatial distribution of the many contributing sources and the dynamic expansion.

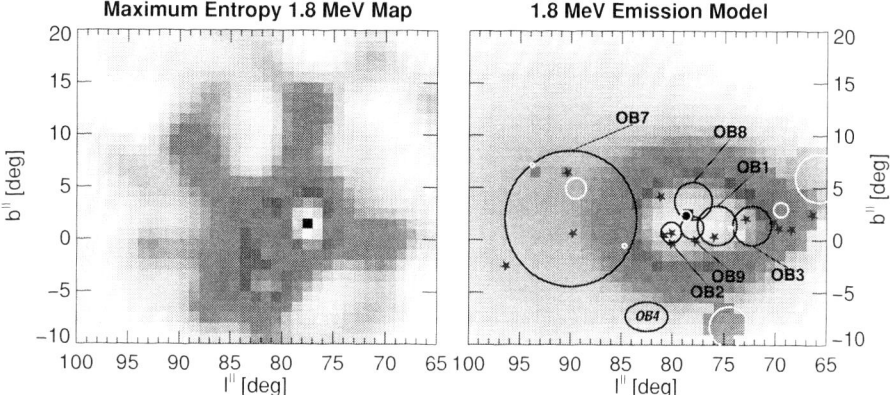

FIGURE 2. (left side) Detailed COMPTEL 1.8 MeV map of the Cygnus region. (right side) angular distribution of known OB associations, WR stars and SNRs overlaid on the ^{26}Al distribution model.

We developed a 1D analytical model of expanding superbubbles [16], similar to the model of [19], to simulate the dynamic evolution of the ^{26}Al distribution in the ISM. The model assumes a uniform ambient density and drives an expanding thin shell by a central, variable energy source. The time dependent luminosity of the central source is obtained from the population synthesis model. Energy losses due to radiation from the bubble interior are taken into account. We assume that ^{26}Al fills the bubble interior such that the surface brightness in the γ-ray line is constant. We compare the model (Fig. 2 right) with the data (Fig. 2 left) by means of Maximum Likelihood fitting. The model fits suggest a mean ambient density between 20 and 30 cm^{-3}, consistent with values inferred by other methods (e.g. [2]).

In addition, the fits imply the presence of a low-intensity component across the whole Cygnus region, which could be the result of a small amount of ^{26}Al in the nearby foreground. It is also possible that several unidentified novae and/or AGB stars explain this low-intensity component. Yet another alternative is a small contribution from the very extended associations Cygnus OB5 & 6 [1], which were not included in this study for reasons given above.

CONCLUSIONS AND PREDICTIONS FOR ^{60}FE

Fig. 3 shows that the multiple OB association model for Cygnus successfully describes the observed 1.8 MeV flux. Roughly 70% of the observed flux is due to the OB associations, while about 20% of the flux is attributed to isolated WR stars and SNRs in this region. The origin of the remaining 10% is not yet clear.

In addition, from this model of the massive star population in the Cygnus region we can predict the γ-ray line fluxes at 1.173 and 1.332 MeV due to radioactive decay of

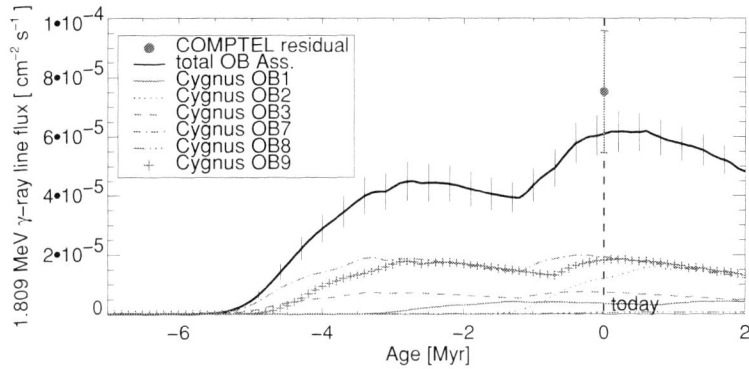

FIGURE 3. Flux at 1.8 MeV from ^{26}Al as a function of time predicted for the Cygnus star forming region. The present day value (Age = 0) is reproduced by the superposition of ^{26}Al produced by six well known OB associations in this region. The data point shown here represents the integrated flux minus an estimated contribution from isolated, non-associated WR stars and SNRs.

^{60}Fe, which is also abundantly produced by massive stars. Our best fit model predicts a total flux of $(9 \pm 5)\ 10^{-6}$ photons cm^{-2} s^{-1} for each line. Even if these lines are combined, a detection with COMPTEL is rather unlikely. The INTEGRAL mission [22] is significantly more sensitive to narrow lines in this energy regime, but a detection of ^{60}Fe specifically in Cygnus would require a substantial fraction of the available observation time. However, a detection of ^{60}Fe from the inner Galaxy is indeed expected within the first year of the INTEGRAL mission.

REFERENCES

1. Alter, J. et al., *Catalogue of star clusters and associations*, Budapest: Akademiai Kiado, 1970.
2. Comeron, F. & Torra., J., *ApJ* **423**:652+, 1994.
3. Dame, T.M. et al., *ApJ* **547**:792-813, 2001.
4. Elmegreen, B.G. & Efremov, Y.N., *ApJ* **466**:802+, 1996.
5. Garmany, C.D. et al., *ApJ* **263**:777+, 1982.
6. Garmany, C.D. & Stencel, R.E., *A&AS* **94**:211-244, 1992.
7. Green, D.A., *Catalogue of gal. SNRs (August 2000)*, URL: http://www.mrao.cam.ac.uk/surveyz/snrs/, 2000.
8. Knödlseder, J., *ApJ* **510**:915-929, 1999.
9. Knödlseder, J., *A&A* **360**:539-548, 2000.
10. Massey, P. et al., *ApJ* **454**:151+, 1995.
11. Meynet, G. et al., *A&A* **320**:460-468, 1997.
12. Naya, J. et al., *Nature* **384**:44-46, 1996.
13. Plüschke, S. et al., in *Astronomy with Radioactivities*, eds. R. Diehl & D. Hartmann, MPE Rep. 274, 1999.
14. Plüschke, S. et al., in *The Influence of Binaries on Population Studies*, Brussels, Aug. 2000, eds. W. van Rensbergen, D. Vanbeveren & B. de Loore, KAP (in press), 2000.
15. Plüschke, S. et al., *Proc. of the 4th INTEGRAL Workshop*, Alicante/Spain, ESA-SP series (in press), 2001.
16. Plüschke, S., PhD thesis, TU München, 2001.
17. Prantzos, N. & Diehl, R., *Phys. Rep.* **267(1)**:1-70, 1996.
18. Schönfelder, V. et al., *A&AS* **143**:145-179, 2000.
19. Shull, M.J. & Saken, *ApJ* **444**:663-671, 1995.
20. Thornton, K. et al., *ApJ* **500**:95+ ,1998.
21. van der Hucht K. et al., *New Astr. Rev.* **45(3)**:135-232, 2001.
22. Winkler, C., in *The 5th Compton Symposium*, Portsmouth, Sep. 1999, AIP Conf. Proc. Vol. 510, 2000.
23. Woosley, S.E. & Weaver, *ApJS* **101**:181+, 1995.

Unveiling the true age of the "Vela Junior" Supernova Remnant

S. Mereghetti* and A. Pellizzoni*

Istituto di Fisica Cosmica CNR, Milano

Abstract.
The supernova remnant G266.1–1.2 has attracted considerable interest owing to its possible detection in the 1.156 MeV line of ^{44}Ti by COMPTEL. If this controversial result is confirmed, G266.1–1.2 well deserves the "Vela Junior" nickname, since the observed gamma-ray line flux implies a very small age ($\sim$700 years) and distance ($\sim$200 pc).

We discuss the implications of recent X-ray observations on the SNR distance and age. Two sources were detected with BeppoSAX close to the geometrical center of G266.1–1.2. Independent of which one is the neutron star associated to G266.1–1.2, their properties imply an age of a few 10,000 years and a distance greater than one kpc.

We also present a preliminary analysis of the brightest portion of the SNR shell, in which significant upper limits on the presence of X–ray emission lines are derived.

INTRODUCTION

The supernova remnant G266.1–1.2 consists of a circular radio/X–ray shell, with $\sim 2°$ diameter, seen in projection against the much more extended emission of the Vela supernova remnant [1]. G266.1–1.2 (also known as RX J0852.0–4622) has attracted considerable interest since it has possibly been detected with COMPTEL [2] in the γ–ray line of ^{44}Ti (1.156 MeV). This isotope, with a lifetime of only $\sim$90 years, is considered a good tracer for young supernova remnants (SNRs). The observed γ–ray line flux of $\sim 4 \times 10^{-5}$ photons cm^{-2} s^{-1} and the relatively small angular dimensions of the remnant imply an age of only $\sim$680 years and a small distance of $\sim$200 pc [3]. Thus G266.1–1.2 could be the remnant of the closest supernova event to have occurred in recent historical times.

Recent ASCA observations of G266.1–1.2 show that the X–rays from the SNR shell have a non-thermal spectrum and the X–ray spectral fits require a high absorption value [4], favoring a distance of the order of $\sim$1-2 kpc. This would place G266.1–1.2 well beyond the Vela SNR. The ASCA data revealed also a central point source, surrounded by diffuse X–ray emission, that was interpreted [4] as the neutron star associated to G266.1–1.2. Here we report on BeppoSAX observations that give new information relevant to estimate the distance and age of this interesting SNR.

THE CENTRAL POINT SOURCES

A BeppoSAX observation performed in May 1999 revealed the presence of *two* X–ray sources close to the geometrical center of G266.1–1.2 [5]. As visible in soft and hard X–ray images shown in Fig. 1, the two sources have clearly different spectra. The softer one, only visible below $\sim$5 keV, coincides with the source detected with ASCA, AX J0851.9–4617.4 [4]. The hard X–ray image shows instead the presence (at more than 7σ) of a second, weaker source, SAX J0852.0–4615 [5]. The angular separation between the two sources is more than 3$'$, significantly greater than the statistical error associated to the source positions. Probably SAX J0852.0–4615 was not detected as a separate source by ASCA due to the limited angular resolution and short exposure time.

The BeppoSAX spectrum of AX J0851.9–4617.4 is equally well fitted by a power law (α_{ph} = 3.6$\pm$0.6), a thermal bremsstrahlung (kT$_{Br}$ = 1.3$\pm$0.4 keV), or a blackbody (kT$_{BB}$ = 0.5 $\pm$ 0.1 keV). The absorption value is not well constrained: only upper limits of a few $\times 10^{22}$ cm^{-2} could be derived. The unabsorbed 2-10 keV flux for the best fit power law parameters is 6.6$\times 10^{-13}$ erg cm^{-2} s^{-1}. The corresponding values for the bremsstrahlung and blackbody models are 6.0 and 5.6 $\times 10^{-13}$ erg cm^{-2} s^{-1}, respectively. No pulsations were found and the flux measured with BeppoSAX is consistent with that obtained with ASCA five months earlier.

SAX J0852.0–4615 , with only $\sim$80 net counts detected above 5 keV, is too weak for a detailed spectral and timing analysis. It is clearly harder, and possibly more absorbed than AX J0851.9–4617.4 . Assuming a power law spectrum with α_{ph} between 1 and 3 and N$_H$=4$\times 10^{21}$ cm^{-2} (i.e. the absorption derived from the fits to the G266.1–1.2 shell), the observed 2-10 keV flux lies between $\sim$ 3 and 4 $\times 10^{-13}$ erg cm^{-2} s^{-1}.

Both AX J0851.9–4617.4 and SAX J0852.0–4615 , located close to the geometrical center of the SNR, are in principle good candidates to be the neutron star associated to G266.1–1.2 . Of course, the possible nature of the two sources must be first discussed by looking at their counterparts in the optical band. While the brightest object in the error circle (1$'$ radius) of SAX J0852.0–4615 has a magnitude $\sim$15, implying an X–ray to optical flux ratio $\gtrsim$ 0.1 that makes a stellar identification unlikely, two bright early type stars, HD76060 and Wray 16-30, are compatible with the position of AX J0851.9–4617.4 .

HD76060, is a bright (V=7.9) B8 type star, with a parallax distance of $\sim$270 pc [6], probably belonging to the OB association Trumpler 10. For the AX J0851.9–4617.4 thermal bremmstrahlung best fit, its 2-10 keV luminosity would be L$_x \sim 5 \times 10^{30}$ (d/270 pc)^2 erg s^{-1}, which is not unreasonable for an early type star.

Wray 16-30 is a B[e] star with V=13.8 [7] also detected with IRAS. Its apparent magnitude suggests a distance in excess of 1 kpc; it is therefore unlikely that Wray 16-30 contributes significantly to the X–ray emission from AX J0851.9–4617.4 .

We thus believe that AX J0851.9–4617.4 is probably associated to HD76060 and the harder source SAX J0852.0–4615 is the one most likely for being the neutron star produced in the G266.1–1.2 supernova. Its hard X–ray spectrum is not consistent with thermal emission from a hot neutron star surface. A more plausible interpretation is in terms of non-thermal processes of magnetospheric origin. However, if SAX J0852.0–4615 were at the very small distance of $\sim$200 pc proposed for G266.1–

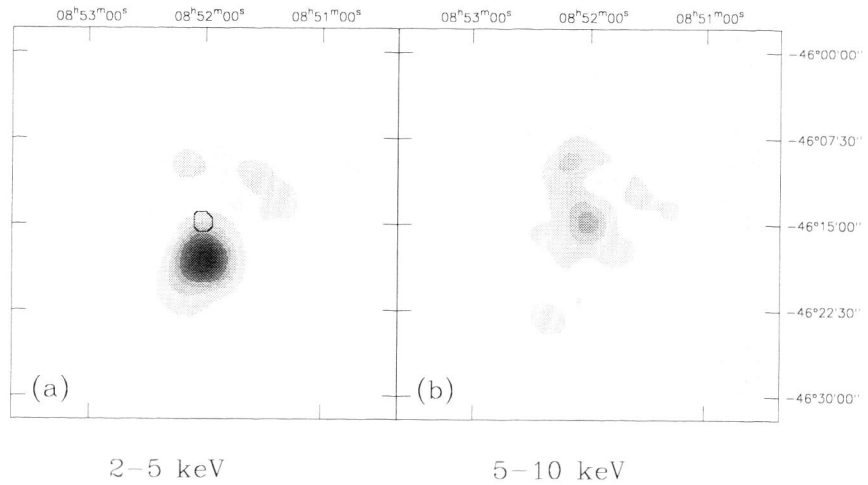

FIGURE 1. BeppoSAX MECS images of the sources at the center of G266.1–1.2 in the energy ranges 2-5 keV (left) and 5-10 keV (right). Both images have been smoothed with a Gaussian with σ=1′. The highest contour level of the 5-10 keV image is drawn on the soft X–ray image to show the different positions of the two detected sources. Coordinates are J2000.

1.2 [3], its X–ray luminosity would be extremely low compared to other known young neutron stars (see, e.g. ref.[8]). On the other hand a distance of the order of ∼1-2 kpc would give a more plausible luminosity of $\sim 10^{32}\,d_{kpc}^{2}$ erg s^{-1}.

Note that, as discussed in detail in ref.[5], the above considerations would also apply to the alternative (though less likely) possibility that AX J0851.9−4617.4 is the neutron stas associated with G266.1–1.2 and SAX J0852.0−4615 an unrelated source, e.g. an AGN falling by chance within ∼7′ from the geometrical center of G266.1–1.2.

X–RAY EMISSION FROM THE NW SHELL

The northern part of the G266.1–1.2 shell was pointed in a 120 ks long BeppoSAX observation performed in May 1999. The 2-10 keV image of the brightest part of the NW rim is shown in Fig. 2. A spectral analysis of these data gave results very similar to those obtained in the ASCA observation [4]. In particular, a power law spectrum gives the best fit with $\alpha_{ph} = 2.8 \pm 0.1$ and $N_H = 8 \times 10^{21}$ cm^{-2}. We note that although also a thermal bremsstrahlung gives a formally acceptable fit, the temperature is rather high (> 3 keV), thus suggesting a non-thermal nature of the X–ray emission. In any case, the absorption value is well constrained above the typical column density of the Vela SNR ($\sim 10^{20}$cm^{-2}), in agreement with the hypothesis that G266.1–1.2 is beyond Vela.

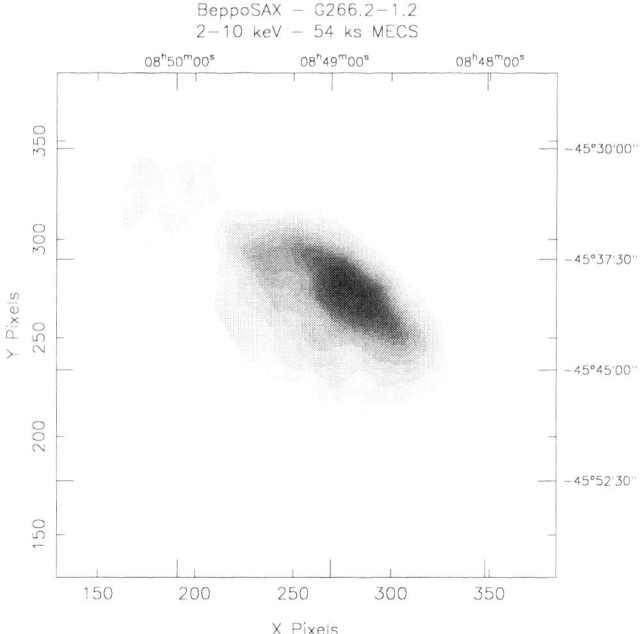

FIGURE 2. BeppoSAX MECS images in the 2-10 keV of the Northern part of the shell of G266.1–1.2.

We searched for X–ray emission lines that might be expected in relation to the detection of ^{44}Ti in gamma-rays. The possible presence of a line at 4.1 ± 0.2 keV from a small region of the G266.1–1.2 NW shell was inferred from the ASCA data [9]. This feature, with a flux of $(5\pm2)\times10^{-5}$ photons cm^{-2} s^{-1}, was interpreted as an helium-like Ca-K emission line. This would require a significant overabundance of Ca, which is a product of the ^{44}Ti decay. An alternative interpretation of such a feature can be given in terms of fluorescence from ^{44}Sc. However, in this case an expected flux of only $\sim6.6\times10^{-6}$ photons cm^{-2} s^{-1} was estimated [4] based on the reported flux in the ^{44}Ti gamma-ray line [2].

We do not find any evidence for such a line and can place an upper limit of $\sim5\times10^{-6}$ photons cm^{-2} s^{-1}.

CONCLUSIONS

A recent BeppoSAX observation of the central region of the supernova remnant G266.1–1.2 has led to the detection of two X-ray sources that can be considered good neutron star candidates [5]. The softer source, AX J0851.9−4617.4 was already observed with ROSAT and ASCA. The newly discovered X–ray source, SAX J0852.0−4615, does not have bright stellar counterparts. It has a harder spectrum and properties consistent with a neutron star with an age of a few 10^4 years at a distance of $\sim$1-2 kpc.

Unless very unusual properties are invoked, both sources are difficult to explain as the neutron star associated to G266.1–1.2, if the latter is really as young and nearby (∼680 years, d∼200 pc) as implied by the claimed detection of a ^{44}Ti γ–ray line [2].

We confirm the likely non-thermal nature of the emission from the brightest part of the G266.1–1.2 shell. The relatively high absorption also suggests a location farther than the Vela SNR. A preliminary analysis does not confirm the strong emission line at 4.1 keV observed with ASCA [9].

In conclusion, the BeppoSAX X–ray data do not support the interpretation of the previous ROSAT and COMPTEL observations in terms of a very young and nearby SNR. Also considering the recent reassesment [10] of the statistical significance of the ^{44}Ti emission from the direction of G266.1–1.2, we believe that there are no compelling reasons to require a particularly small distance and age.

REFERENCES

1. Aschenbach, B. 1998, Nature, 396, 141
2. Iyudin A.F. et al. 1998, Nature, 396, 142
3. Aschenbach, B., Iyudin, A.F., & Schönfelder V. 1999, A&A, 350, 997
4. Slane P. et al. 2001, ApJ 548, 814
5. Mereghetti S. 2001, ApJ 548, L213
6. de Zeeuw P.T. et al. 1999, AJ, 117, 354
7. Thé P.S., de Winter D. & Pérez M.R. 1994, A&ASS, 104, 315
8. Becker W. & Trümper J. 1997, A&A, 326, 682
9. Tsunemi H. et al. 2000, PASJ 52, 887
10. Schönfelder V. et al. 2000, Proc. 5th Compton Symposium, M.L.McConnell & J.M.Ryan eds., AIP Conf. Proc. 510, 54

On the Nature of the Nonthermal Emission from the Supernova Remnant IC 443

S. J. Sturner[a,b], O. Reimer[a,c], J. W. Keohane[d], C. M. Olbert[d],
R. Petre[a], & C. D. Dermer[e]

[a]*NASA/GSFC, Greenbelt, MD 20771-0001*
[b]*USRA, 7501 Forbes Blvd., #206, Seabrook, MD 20706-2256*
[c]*National Academy of Science/National Research Council*
[d]*North Carolina School of Science & Mathematics, P.O. Box 2418, Durham, NC 27715*
[e]*U.S. Naval Research Laboratory, Code 7653, Washington DC 20375-5352*

Abstract. The supernova remnant (SNR) IC 443 is a nearby (~ 1.5 kpc) remnant of intermediate age which appears to be interacting with a molecular cloud and is spatially coincident with the unidentified EGRET source 3EG J0617+2238. We present new spectra of IC 443 obtained using the PCA on *RXTE*. The spectrum is well fit by a two-component model consisting of a non-equilibrium ionization collisional plasma model with kT~0.6 keV plus a power-law with index ~2.2. We compare our results with the earlier results of *HEAO 1* A-2, *Ginga*, *ASCA*, and *BeppoSAX*, and find generally good agreement. We also discuss the possible association of 3EG J0617+2238 with IC 443 given that recent *Chandra* results indicate that much of its nonthermal X-ray emission originates from a pulsar wind nebula. We find that unless the calculation of the position of 3EG J0617+2238 is affected by nearby strong sources such as Geminga, the *Chandra* source is excluded from being associated with the EGRET source.

INTRODUCTION

IC 443 is a canonical mixed-morphology SNR that has a complex morphological structure due to interactions with local molecular clouds. In particular, Braun & Strom [1] have characterized the morphology of IC 443 as a series of three interconnected shells resulting from these interactions. The distance to IC 443 is somewhat uncertain but is generally thought to be 1.5-2.0 kpc [1,2,3,4].

The age of IC 443 has been controversial until very recently. Using X-ray observations made with the *Einstein Observatory* and the *HEAO 1* A-2 experiment, Petre et al. [5] determined the age of IC 443 to be between 2800 and 3400 years. More recently, Chevalier [6] has modeled IC 443 as an explosion inside a molecular cloud and suggests an age of about 30,000 years, which was confirmed by Olbert et al. [7].

IC 443 is particularly interesting because it has been observed to emit a non-thermal X-ray spectrum up to ~15 keV [8,9,10] and thus it belongs to the select subclass of SNRs that exhibit non-thermal X-ray emission which includes Cas A [11], SN 1006 [12], and G347.3-0.5 [13]. It is also spatially-coincident with an unidentified high-energy gamma-ray source seen by the EGRET experiment on the *CGRO* [14,15,16].

RXTE PCA OBSERVATION OF IC 443

IC 443 was observed using the Proportional Counter Array (PCA) on *RXTE* between August 13 and 16, 1996 with the 1° field-of-view (FOV) centered at $\alpha = 6^h 17^m 1.5^s$, $\delta = +22° 34' 52.3"$ (J2000) and a net exposure time of 35,488 seconds. Analysis of the data was performed using the standard suite of FTOOLS and the latest background model developed for faint sources, L7-240. A detailed description of the analysis method can be found on the HEASARC website.

The spectrum we derived is for the spatially integrated X-ray emission from the SNR. We fit the combined data from all observation segments in XSPEC using a two-component model that included a thermal and non-thermal power-law component. We found that the generalized nonequilibrium ionization model (gnei) [17] best characterized the low-energy thermal component of the spectrum. We show the results of the fitting procedure in Figure 1 and list the best-fit parameters in Table 1. We find that the high-energy X-ray spectrum of IC 443 can be well characterized by a power-law to energies beyond 20 keV. A two-temperature thermal (gnei) model can also adequately fit the data ($\chi_v^2 = 1.66$) with temperatures of 0.63 keV and 12.4 keV but this requires an unacceptably large absorbing column of 5.2×10^{22} cm^{-2} which we feel invalidates this model.

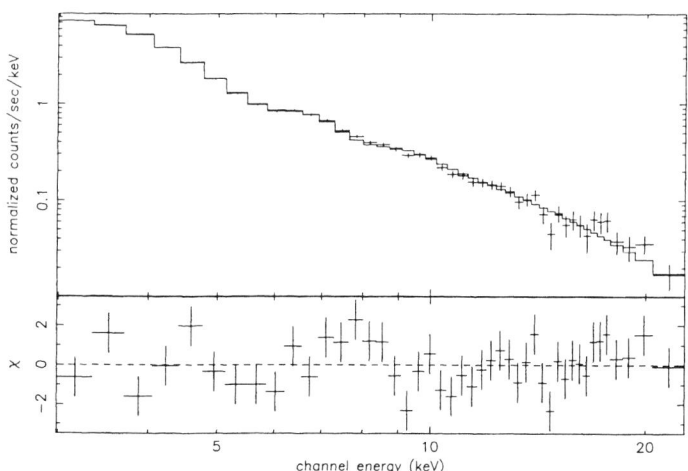

FIGURE 1. *RXTE* PCA data for IC 443 with the best-fit folded gnei+pow model. Also plotted are the residuals in units of χ^2.

TABLE 1. Best-Fit Model Parameters.

Parameter	Best-Fit Value	1σ Error
n_H (x10^{22} cm^{-2})	0.89	+0.44/-0.41
kT (keV)	0.60	+0.009/-0.018
τ (kyr cm^{-3})	18.97	+1.07/-7.46
<kT> (keV)	1.89	+0.26/-0.10
Photon Spectral Index	2.21	+0.07/-0.07
χ_v^2	1.32	

In Table 2 we compare the *RXTE* PCA spectral results with those of previous missions. Since this is a heterogeneous set of observations, it is difficult to precisely compare the results. We find that our *RXTE* PCA results closely match those from the *Ginga* LAC, except that Wang et al. [10] report a significantly higher integrated flux in the 2-20 keV band. Keohane et al. [9] previously suggested the possibility of a calibration discrepancy between the *Ginga* LAC and *ASCA* GIS because of similar discrepancies in the observed fluxes.

Keohane et al. [9] found that ~60% of the remnant-integrated 7 keV emission could be attributed to 2 unresolved sources in the south-central region (Source A) and eastern rim (Source B) of IC 443. Source A was found to contribute ~40% of the total 7 keV emission and its spectrum can be well characterized by a thermal+nonthermal two-component model with $kT = 0.89$ keV and spectral index = 2.3. Our integrated 2-10 keV flux is consistent with that observed by Keohane et al. [9]. The spectral index we derive for the nonthermal component is consistent with the index they derived for Source A as would be expected if Source A is indeed the largest source of >7 keV photons in the field-of-view.

We find that our spectral index is marginally consistent with the index of Source A determined by Bocchino & Bykov [8] using the *BeppoSAX* MECS. Their best-fit spectrum, a single power law with index 1.96, is significantly harder than any of the previously derived spectra for either IC 443 in its entirety or for Source A in particular. The 2-10 keV integrated flux from Source A + Source B is also significantly higher than both that observed by *RXTE* PCA for the entire remnant as well as that reported by Keohane et al. [9] using the *ASCA* GIS. On the other hand it appears consistent with the flux reported by Wang et al. [10].

TABLE 2. Comparison of *RXTE* PCA Fit Parameters with Other Mission Results.

Instrument	Source	n_H (10^{22} cm^{-2})	kT (keV)	Spectral Index	Integrated Flux (10^{-11} ergs/cm^2/s)
HEAO 1 A-2 MED	Entire SNR	-----	1.05 (+0.29,-0.25)	-----	7±1 (2-10 keV)
Ginga LAC	Entire SNR	0.79	0.68	2.2	9 (2-20 keV)
ASCA GIS	Source A	0.18±0.03	0.89*	2.3±0.2	-----
	Source B	<0.03	0.89*	0.1±1.3	-----
	~SNR†				5±1 (2-10 keV)
BeppoSAX MECS	Source A	-----	-----	1.96 (+0.21,-0.12)	7.5±1.2 (2-10 keV)
	Source B	-----	1.07 (+0.23,-0.31)	1.5±0.9	1.8±0.2 (2-10 keV)
RXTE PCA	Entire SNR	0.89 (+0.44,-0.41)	0.60 (+0.009,-0.018)	2.21±0.07	6.22 (2-10 keV) 6.62 (2-20 keV)

*Fixed at off source value
†Summed over 2 GIS fields-of-view, estimated to be ~90% of total SNR emission

IS 3EG J0617+2238 ASSOCIATED WITH IC 443?

The EGRET experiment on the *Compton Gamma Ray Observatory* has found a 17.4σ source, 3EG J0617+2238, that is spatially consistent with IC 443 [14,15,16]. The spectrum of this source is well fit by a power-law with index 1.98, a broken power-law with index 1.79 below 1 GeV and 2.65 above 1 GeV, and a 1.68 power-law with exponential cut-off at 2.2 GeV [18]. The source shows very marginal variability [19], so that there is no evidence that it is a background blazar.

This lack of variability and its positional coincidence with IC 443 have led to proposals that this EGRET source is due either to locally accelerated cosmic rays [20,21,22,23] or a radio-quiet pulsar [24]. Both models have their merits. SNRs have long been thought to be the sources of the Galactic cosmic rays [25] and there is very strong evidence that IC 443 is interacting with dense interstellar clouds [26,27]. The discovery that Geminga was a radio-quiet, gamma-ray emitting pulsar [28] has also led to speculation about the Galactic population of such objects and their contribution to the population of unidentified EGRET sources. Here we examine these possibilities in light of the recent X-ray observations of the region.

Recently, Olbert et al. [7] have shown, using the *VLA* and *Chandra*, that Source A is a comet-shaped nebula of hard emission which contains a softer point source at its apex. They have designated this source CXOU J061705.3+222127, and have argued that this structure is a pulsar and an associated wind nebula. Not much is known about Source B. As shown in Table 2, its spectrum is harder than Source A/CXOU J061705.3+222127.

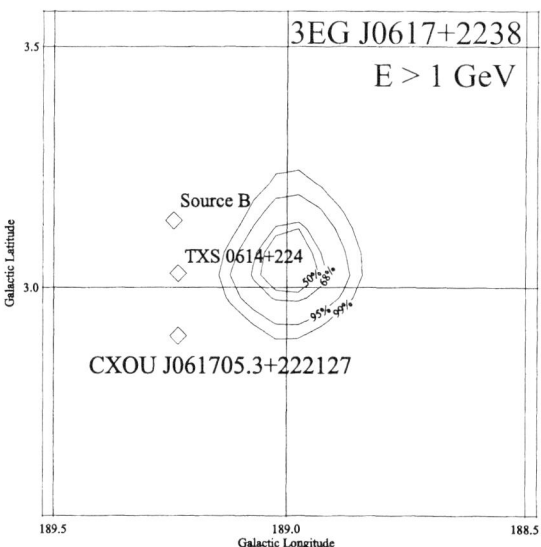

FIGURE 2. This plot shows the 50%, 68%, 95%, and 99% confidence contours for the EGRET source 3EG J0617+2238 for photons >1 GeV [14]. Also shown are the positions of the two hard X-ray sources seen by *ASCA*, *BeppoSAX*, and *Chandra*, as well as the radio source TXS 0614+224 which is thought to be extragalactic. Note that all of the sources are positionally inconsistent at the 99% confidence level.

In addition to these sources, there is a radio source in the same region which is thought to be extragalactic in origin [29]. This radio source was seen at flux densities of 2.1 Jy at 151 MHz and < 0.2 Jy at 1419 MHz. We identify the radio source with the source TXS 0614+224 [30].

We plot the EGRET location for 3EG J0617+2238 for photons energies >1 GeV as well as the locations of the X-ray and radio sources in Figure 2. Note that all of these sources are outside the 99% confidence contour and are offset from the most likely gamma-ray source position by ~15'. Even though the spectrum of the EGRET source is pulsar-like [31], this offset suggests that the EGRET source and the pulsar nebula are unrelated unless the EGRET source position has been affected by the presence of strong nearby sources beyond the currently understood systematics [32]. Deep observations of the region containing the GeV source with new, powerful x-ray instruments such as *Chandra* and *XMM* are necessary to probe for possible x-ray counterparts to the high-energy gamma-ray source.

REFERENCES

1. Braun, R., and Strom, R. G., *A&A* **164**, 193 (1986).
2. DeNoyer, L.K., *ApJ* **212**, 416 (1977).
3. Fesen, R.A., *ApJ* **281**, 658 (1984).
4. Asaoka, I., and Aschenbach, B., *A&A* **284**, 573 (1994).
5. Petre, R., Szymkowiak, A.E., Seward, F.D., and Willingale, R., *ApJ* **335**, 215 (1988).
6. Chevalier, R.A., *ApJ* **511**, 798 (1999).
7. Olbert, C.M., Clearfield, C.R., Williams, N.E., Keohane, J.W., and Frail, D.A., *ApJ* (2001) in press.
8. Bocchino, F., and Bykov, A.M., *A&A* **362**, L29 (2000).
9. Keohane, J.W., et al., *ApJ* **484**, 350 (1997).
10. Wang, Z.R., Asaoka, I., Hayakawa, S., & Koyama, K., *PASJ* **44**, 303 (1992).
11. Allen, G.E., et al., *ApJ* **487**, L97 (1997).
12. Koyama, K. et al., *Nature* **378**, 255 (1995).
13. Slane, P. et al., *ApJ* **525**, 357 (1999).
14. Hartman, R.C. et al., *ApJS* **123**, 79 (1999).
15. Sturner, S.J., & Dermer, C.D. *A&A* **293**, L17 (1995).
16. Esposito, J.A., Hunter, S.D., Kanbach, G., & Sreekumar, P. *ApJ* **461**, 820 (1996).
17. Borkowski, K.J., Sarazin, C.L., and Blondin, J.M., *ApJ* **429**, 710 (1994).
18. Bertsch, D.L. et al., "Spectral Modeling of the EGRET 3EG Gamma Ray Sources Near the Galactic Plane," in *The Fifth Compton Symposium*, edited by M. L. McConnell and J. M. Ryan, AIP Conference Proceedings 510, New York, 2000, pp. 504-508.
19. Tomkins, W., Ph.D. Thesis, Stanford University (1999).
20. Sturner, S.J., Skibo, J.G., Dermer, C.D., and Mattox, J.R., *ApJ* **490**, 619 (1997).
21. Gaisser, T.K., Protheroe, R.J., and Stanev, T., *ApJ* **492**, 219 (1998).
22. Baring, M.G., et al., *ApJ* **513**, 311 (1999).
23. Bykov, A.M., Chevalier, R.A., Ellison, D.C., and Uvarov, Y.A., *ApJ* **538**, 203 (2000).
24. Yadigaroglu, I.-A., and Romani, R.W., *ApJ* **449**, 211 (1995).
25. Ginzburg, V.L., Syrovatskii, S.I., *Origin of Cosmic Rays*, Macmillan, New York, 1964.
26. Cesarsky, D., et al., *A&A* **348**, 945 (1999).
27. Wang, Z., and Scoville, N.Z., *ApJ* **386**, 158 (1992).
28. Halpern, J.P., and Holt, S.S., *Nature* **357**, 222 (1992).
29. Green, D.A., *MNRAS* **221**, 473 (1986).
30. Douglas, J.N., et al., *AJ* **111**, 1945 (1996).
31. Cheng, K.S., & Zhang, L., *ApJ* **498**, 327 (1998).
32. Esposito, J.A., et al., *ApJS* **123**, 203 (1999).

SNR and fluctuations in the diffuse Galactic γ-ray continuum

A.W. Strong* and I.V. Moskalenko[1][†]

*Max-Planck-Institut für extraterrestrische Physik, Postfach 1312, 85741 Garching, Germany
[†]NASA Goddard Space Flight Center, Code 660, Greenbelt, MD 20771, U.S.A.

Abstract.
In studies of cosmic-ray propagation and diffuse continuum γ-ray emission from the Galaxy it has usually been assumed that the source function can be taken as smooth and time-independent, an approximation justified by the long residence time of cosmic-rays in the Galaxy. However, especially for electrons at high energies where energy losses are rapid, the effect of the stochastic nature of the sources becomes apparent and indeed has been invoked to explain the GeV excess in the diffuse emission observed by EGRET. In order to address this problem in detail a model with explicit time-dependence and a stochastic SNR population has been developed, which follows the propagation in three dimensions. The results indicate that although the inhomogeneities are large they are insufficient to easily explain the GeV excess. However the fluctuations should show up in the gamma-ray distribution at high energies and this should be observable with GLAST. We also show that estimates of the TeV continuum emission from the plane are consistent with the Whipple upper limit.

INTRODUCTION

The diffuse continuum γ-ray emission from the Galaxy measured by EGRET and COMPTEL has been the subject of many studies relating to cosmic-ray origin and propagation. Usually it has been assumed that the source function can be taken as smooth and time-independent, an approximation justified by the long residence time ($> 10^7$ years) of cosmic-rays in the Galaxy. We have previously described a numerical model for the Galaxy encompassing primary and secondary cosmic rays, γ-rays and synchrotron radiation in a common framework. Up to now our GALPROP code handled 2 spatial dimensions, (R, z), together with particle momentum p. This was used as the basis for studies of cosmic-ray (CR) reacceleration, the size of the halo, positrons, antiprotons, dark matter and the interpretation of diffuse continuum γ-rays. Some aspects cannot be addressed in such a cylindrically symmetric model: for example the stochastic nature of the cosmic-ray sources in space and time, which is important for high-energy electrons with short cooling times, and local inhomogeneities in the gas density which can affect radioactive secondary/primary ratios. The motivation for studying the high-energy electrons is the observation of the >1 GeV excess in the EGRET spectrum of the Galactic

---

[1] NRC Senior Research Associate; on leave from Institute of Nuclear Physics, M. V. Lomonosov Moscow State University, 119 899 Moscow, Russia

emission, which has been proposed to originate in inverse-Compton emission from a hard electron spectrum; this hypothesis can only be reconciled with the local directly-observed steep electron spectrum if there are large spatial variations which make our local region unrepresentative of the large-scale average spectrum. Strong et al. [1] presented a study of diffuse γ-rays based on the 2D model. First results from an extension of the model to 3D, which can cover these issues, are presented here.

MODEL

The GALPROP code, which solves the cosmic-ray propagation equations on a grid, has been entirely rewritten (in C++) using the experience gained from the original version and including both 2D and 3D spatial grid options. Cosmic-ray nuclear reaction networks are included with a comprehensive new cross-section database. This allows the models to be tuned on stable and radioactive CR primary/secondary ratios, in particular B/C and ^{10}Be/^{9}Be. The 2D mode essentially duplicates the original version. In 3D (x,y,z,p) the propagation is solved as before using a Crank-Nicolson scheme. The additional dimension increases the computer resources considerably but a 200 pc grid cell is still practicable. The main enhancement is the inclusion of stochastic SNR events as sources of cosmic rays. The SNR are characterized by the mean time t_{SNR} between events in a 1 kpc^3 unit volume, and the time t_{CR} during which an SNR actively produces CR. The propagation is first carried out for a smooth distribution of sources to obtain the long timescale solution; then the stochastic sources are started and propagation followed on a fine time scale for the last 10^7 years or so. For high-energy electrons (TeV) which lose energy on timescales of 10^5 years the effect is a very inhomogeneous distribution. The amplitude of the fluctuations depends on the two parameters t_{SNR} and t_{CR} which are both poorly known. t_{SNR} is adjusted to be consistent with the observed present number of SNR in the Galaxy and estimates of the SNR rate [2]; models for shock acceleration in SNR indicate $10^4 < t_{CR} < 10^5$ yr, the sources switching off at the adiabatic/radiative transition [3].

SNR AND EFFECT ON THE γ-RAY SKY

We consider a model with a hard electron injection spectrum (index 1.8) as used in Ref. [1] following the suggestion [4] that fluctuations could allow such a spectrum to be consistent both with the GeV γ-ray excess and the locally observed electron spectrum. For $t_{SNR} = 10^4$ years (corresponding to a "standard" Galactic SN rate 3/century) the TeV electron distribution is inhomogeneous, but still none of the spectra around $R = R_\odot$ resemble even remotely that observed locally (Fig. 1). For $t_{SNR} = 10^5$ years (Galactic SN rate 0.3/century) the distribution above 100 GeV is even more inhomogeneous and the spectrum fluctuates greatly (Fig. 2). Some of the spectra resemble that observed locally within a factor of a few, but still none is fully compatible with the local spectrum.

We conclude that the "hard electron spectrum" hypothesis for the EGRET γ-ray excess would require an SN rate much lower than standard, with correspondingly large power

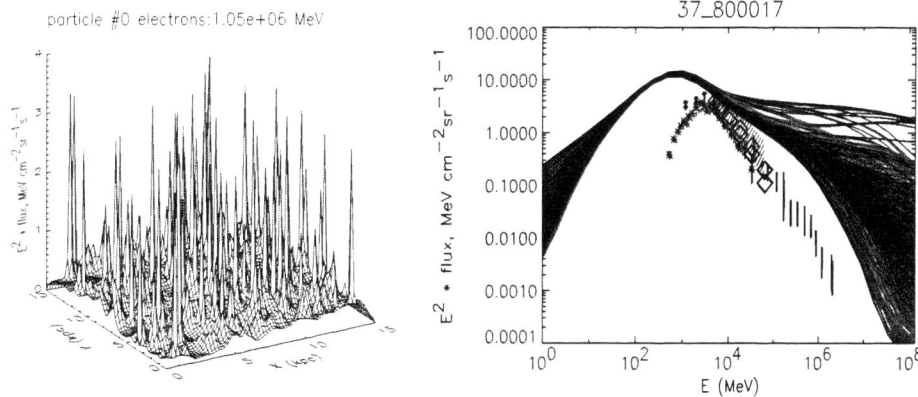

FIGURE 1. Distribution of 1 TeV electrons at $z = 0$ (left) and spectral variations in $4 < R < 12$ kpc (right) for $t_{SNR} = 10^4$ yr. Data points: locally measured electron spectra; references are given in Ref. [1], with additional data from Ref. [5].

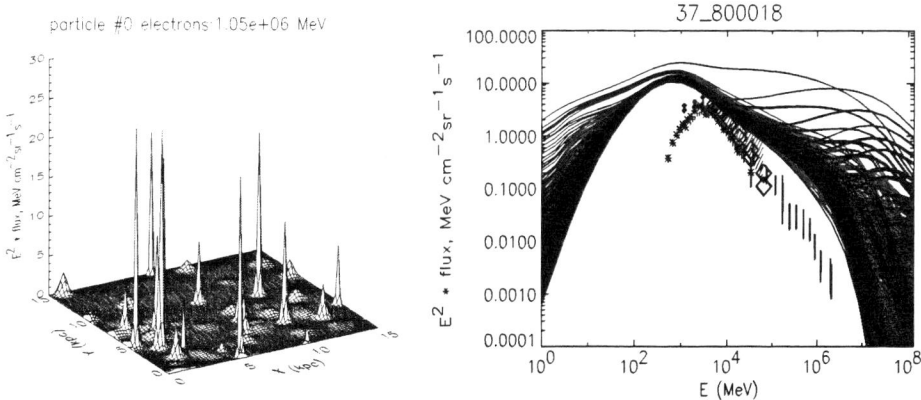

FIGURE 2. Distribution of 1 TeV electrons at $z = 0$ (left) and spectral variations in $4 < R < 12$ kpc (right) for $t_{SNR} = 10^5$ yr. Data as Fig. 1.

requirements for acceleration of electrons per SNR. It is possible that the rate of *CR-producing* SNR could be lower than that of all SNR, so that a sufficiently low rate could be possible, but this is unlikely in view of the power requirements. The CR luminosity in this model is 2×10^{41} erg s^{-1}; for 3 SNR/century we require 2×10^{50} erg/SNR in CR which is plausible (for 10^{51} erg and 20% acceleration efficiency: see Ref. [6] for discussion) but for 0.3 SNR/century we need 2×10^{51} erg/SNR in CR which would require quite another class of objects.

Our conclusion differs from that of Ref. [4], who found that a hard electron spectrum model is consistent with observations considering the fluctuations, but they included

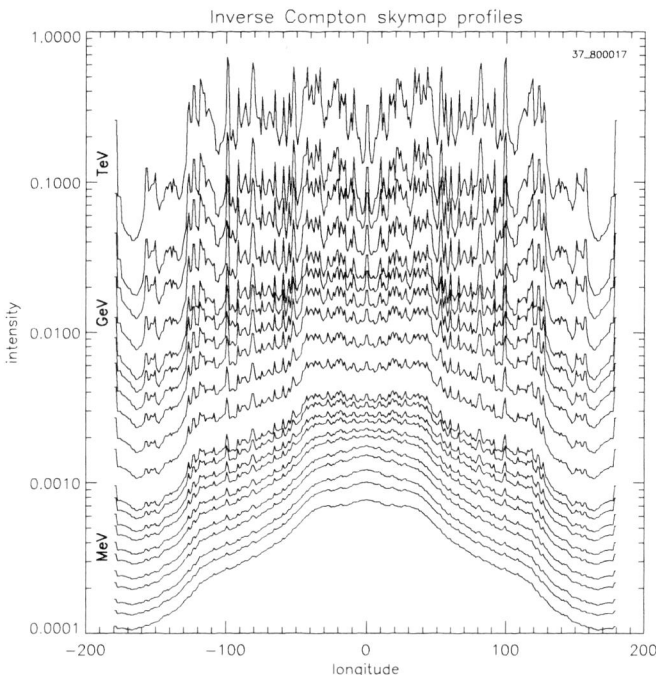

FIGURE 3. Inverse Compton γ-ray longitude distributions at $b = 0$ for γ-ray energies from 1 MeV (bottom) to 1 TeV (top). Note: the intensity scale is arbitrary and the curves have been shifted vertically for clarity.

however a dispersion in the electron injection spectral index which increases the variations further, although our tests of such a dispersion do not suggest that it eases the problem since a large number of SNR still contribute to the local electron spectrum at 1 TeV (see Fig. 1).

The inverse-Compton emission becomes increasingly clumpy at high energies due to the effect of SNR (Fig. 3). The effect is already visible at 1 GeV and will be an important signature at GLAST energies up to 100 GeV.

GALACTIC DIFFUSE TEV γ-RAYS

Recently observations of the Galactic plane (around $l = 40°$) have been reported by the Whipple Observatory [12], which place limits on the > 500 GeV intensity. We have extended our predicted spectrum for a hard electron injection spectrum to the TeV range (Fig. 4). Since the maximum energy of accelerated electrons is unknown we consider the extreme case of 100 TeV. Even for this case the predicted spectrum from inverse-Compton emission is compatible with the Whipple upper limit, and a lower cutoff energy will be also consistent with Whipple; it is clear that an improved limit would quickly

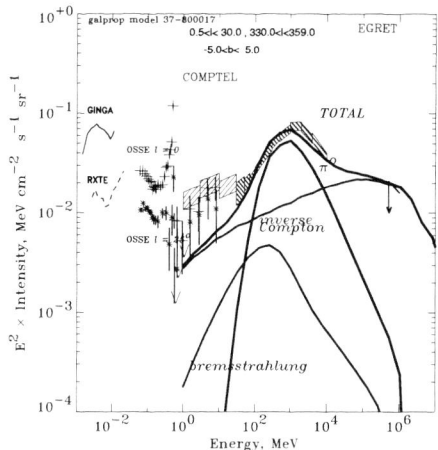

FIGURE 4. Spectrum of the inner Galaxy ($330° < l < 30°$, $|b| < 5°$ for model with $t_{SNR} = 10^4$ yr and hard electron injection spectrum (index 1.8) and cutoff at 100 TeV. Data: EGRET [7], COMPTEL [8], OSSE [9], GINGA [10], RXTE [11], Whipple [12]. Note that the Whipple upper limit is for $l = 40°$.

provide a critical test for models with hard electron injection spectra. The SNR shock-acceleration models [13] suggest a cutoff around 1 TeV which imply a cutoff in the γ-rays around 10 GeV in which case the predicted intensities are well below the Whipple limit and detection will be difficult.

ACKNOWLEDGMENTS

IVM acknowledges support from the NRC/NAS Research Associateship Program.

REFERENCES

1. Strong, A. W., Moskalenko, I. V., and Reimer, O., *Astrophys. J.* **537**, 763–784 (2000).
2. Dragicevich, P. M., Blair, D. G., and Burmann, R. R., *Mon. Not. Roy. Astr. Soc.* **302**, 693–699 (1999).
3. Sturner, S. J., et al., *Astrophys. J.* **490**, 619–632 (1997).
4. Pohl, M., and Esposito, J. A., *Astron. Astrophys.* **507**, 327–338 (1998).
5. Kobayashi, T., et al., in *Proc. 26th ICRC* **3**, 61–64 (1999).
6. Fields, B. D., et al., *Astron. Astrophys.* **370**, 623–634 (2001).
7. Strong, A. W., and Mattox, J. R., *Astron. Astrophys.* **308**, L21–L24 (1996).
8. Strong, A. W., et al., *Astrophys. Lett. Comm.* **39**, 209–212 (1999).
9. Kinzer, R. L., et al., *Astrophys. J.* **515**, 215–225 (1999).
10. Yamasaki, N. Y., et al., *Astrophys. J.* **481**, 821–831 (1997).
11. Valinia, A., Marshall, F.E., *Astrophys. J.* **505**, 134–147 (1998).
12. LeBohec, S., et al., *Astrophys. J.* **539**, 209–215 (2000).
13. Baring, M. G., et al., *Astrophys. J.* **513**, 311–338 (1999).

EGRET Observations of Diffuse Gamma-Ray Emission in Taurus and Perseus

Seth W. Digel[*] and Isabelle A. Grenier[†]

[*]*USRA/NASA GSFC, Code 661, Greenbelt, MD 20771 USA*
[†]*Université Paris 7 and Service d'Astrophysique, CE Saclay, 91191 Gif/Yvette, France*

Abstract. We present an analysis of the interstellar gamma-ray emission observed toward the extensive molecular cloud complexes in Taurus and Perseus by the Energetic Gamma-Ray Experiment Telescope (EGRET). The region's large size (more than 300 square degrees) and location below the plane in the anticenter are advantageous for straightforward interpretation of the interstellar emission. The complex of clouds in Taurus has a distance of ~ 140 pc and is near the center of the Gould Belt. The complex in Perseus, adjacent to Taurus on the sky, is near the rim of the Belt at a distance of ~ 300 pc. The findings for the cosmic-ray density and the molecular mass-calibrating ratio $N(H_2)/W_{CO}$ in Taurus and Perseus are compared with results for other nearby cloud complexes resolved by EGRET. The local clouds that now have been studied in gamma rays can be used to trace the distribution of high-energy cosmic rays within 1 kpc of the sun.

INTRODUCTION

The gamma-ray emissions from the interactions of high-energy cosmic rays (CRs) with interstellar nucleons have long been used to study the densities of CRs and the column densities of molecular gas in the Milky Way (e.g., [1]). The low instrumental background, good point-spread function (PSF), and long life of the Energetic Gamma-Ray Experiment Telescope (EGRET) made possible study of individual interstellar clouds, e.g., to search for variations of CR density and molecular mass calibration within and among the clouds in the solar vicinity.

We report a study of Taurus and Perseus ($l = 150°$ to $185°$, $b = -30°$ to $-5°$), which contain two of the last large, local (distances ~ 140 pc and ~ 300 pc, respectively [2]) cloud complexes to be analyzed with EGRET data. This region poses some unique challenges to the interpretation of the diffuse gamma-ray emission, owing to its closeness to the Galactic anticenter (where velocity is not a good distance indicator for interstellar gas) and to the potential for contamination from the bright emission of the Crab pulsar. We describe our findings in the context of studies of other local clouds with EGRET.

DATA

We use composite photon and exposure maps for all data from Phases 1–5 with the standard sensitivity corrections. For viewing periods when EGRET was operated in full field-of-view mode, only directions within $30°$ of the instrument axis are included. The

PSF, sensitive area, and energy resolution rapidly degrade at larger inclination angles. By cutting the data at 30°, little data are lost and the simplifying assumption can be made that a single set of PSFs applies to the composite dataset. The exposure for 100–10,000 MeV in Taurus/Perseus is $(6.2–29) \times 10^8$ cm^2 s, with the large gradient due to the many observations of the Crab.

The integrated intensity W_{CO} of the 2.6-mm line of CO is commonly used as a tracer for the column density of H_2, which although the dominant constituent of the dense interstellar medium is rarely directly observable at interstellar conditions. Empirically, W_{CO} is proportional to $N(H_2)$ (e.g., [3]), with the ratio $N(H_2)/W_{CO}$ conventionally denoted X. We use the new composite CO survey of Dame, Hartmann, & Thaddeus [4] for the analysis presented here.

We derive $N(H\,I)$ from the 21-cm line Leiden-Dwingeloo survey [5]. A spin temperature of 125 K was assumed for the minor correction for optical depth.

ANALYSIS

Because high-energy CRs uniformly penetrate interstellar gas and the gas is optically thin to gamma rays, in principle the diffuse gamma-ray intensity can be well modelled as a linear combination of the $N(H\,I)$ and W_{CO} maps plus the isotropic extragalactic background. However, the closeness of the Taurus/Perseus region to the anticenter complicates the analysis; lines of sight close to the plane include gas at great distances in the outer Galaxy (where the CR density is lower) that cannot be distinguished by line-of-sight velocity. A simple approach that led to the quantitatively best fitting models was to assume that the gamma-ray emissivity of the atomic gas is uniform for $b < -15°$ (where most of the interstellar gas is nearby) but varies linearly with latitude for $b > -15°$ (to accommodate variation of CR density across the outer Galaxy, to first order). The constant and slope are free parameters.

Taking into account the exposure map ε and the effective point-spread function PSF, our model for the gamma-ray photons detected in Taurus/Perseus may be written as

$$\theta(l,b) = [A + \alpha\Theta(b+15°) \times (b+15°)]N(H\,I)_c + BW_{CO,c} + C\varepsilon_c$$
$$+ \sum_{i=1}^{N} F_i PSF(l-l_i, b-b_i)\varepsilon(l_i, b_i), \quad (1)$$

for any given energy range. Θ is the Heaviside step function; $\Theta(x) = 0$ for $x < 0$ and $\Theta(x) = 1$ for $x \geq 0$. A is the gamma-ray emissivity for $b < -15°$, and α is the slope for $b > -15°$. $B = 2AX$ is the effective emissivity of the CO, and C is the intensity of the isotropic background emission plus foreground inverse Compton (IC) emission. The F_i are the fluxes of the point sources. The subscript c indicates multiplication with the exposure map and convolution with $PSF(x,y)$. ε_c is the exposure map itself convolved with PSF.

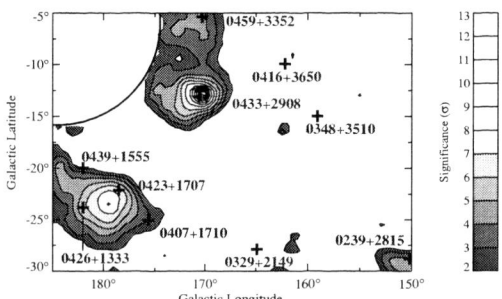

FIGURE 1. Map of likelihood test statistic with contours in units of source significance. Maps for five energy ranges (100–150, 150–300, 300–500, 500–1000, and 1000–10,000 MeV) were summed to increase the sensitivity of the analysis. Contours are in units of source significance. Crosses mark the positions of 3EG catalog sources [7]. The arc delineates the 10° radius around the Crab that was excluded.

RESULTS

The results of a maximum likelihood search for point sources [6] are shown in Figure 1, a composite map of likelihood test statistic. Several of the 3EG sources are not detected in this analysis, most likely because they are variable and best detected in maps for individual viewing periods. In the present analysis, we include four point sources: 3EG J0239+2815, 3EG J0433+2908, 3EG J0459+3352, and a newly-detected (7.8σ) source G179.3-23.7. The latter may correspond to one or both of the unidentified sources 3EG J0423+1707 and 3EG J0426+1333. The interstellar emission model used for the 3EG catalog has a defect along $l = 180°$ and the two sources may be artifacts from the splitting of G179.3-23.7. We do not investigate possible counterparts here, but note that MRC 0418+148 (179.80°, $-24.03°$), a flat-spectrum radio source with flux 0.31 Jy at 4.85 GHz [8], lies just outside the 99% confidence contour (diameter 1.2°).

Figure 2 compares the gamma-ray intensity observed by EGRET with the maximum likelihood model (including the four point sources) for the representative energy range 100–10,000 MeV. No large-scale deviations are evident; the greatest difference between the maps, near (163°, $-11°$), is likely associated with 3EG J0416+3650, which was only marginally significant in this composite dataset and not included in our model.

The model was fit to the EGRET data for several energy ranges to derive the spectrum of the gamma-ray emissivities (Table 1). The region within 10° of the Crab pulsar was excluded. The Crab is so bright that its emission must otherwise be carefully modelled in the tail of the PSF. We do not investigate the variations of the gradient term ('α') here; the general decrease of emissivity with increasing b that the negative values found here imply is consistent with a decreasing CR density across the outer Galaxy. The differential gamma-ray emissivity in Taurus/Perseus is compared with other local cloud complexes in Figure 3. The only significant difference is at high energies: the well-known 'GeV excess' [9] is not evident in Taurus. The origin of the GeV excess is uncertain. On large angular scales, Strong, Moskalenko, & Reimer [10] explain it in terms of a harder spectrum of CR electrons and a slightly modified spectrum of protons (their 'HEMN' model). In their model, most of the GeV excess is in fact IC emission.

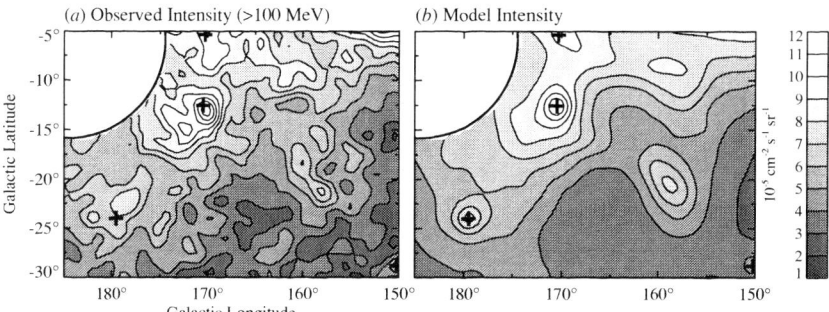

FIGURE 2. Observed and model gamma-ray intensity for the representative energy range > 100 MeV. The model was calculated using the parameter values in Table 1 and the maximum likelihood fluxes of the 4 point sources (positions indicated with crosses). All maps were smoothed with a gaussian of FWHM 1.5° for display. The arc indicates the region around the Crab that was excluded from the analysis.

TABLE 1. Diffuse Model Parameters by Energy Range

Band (MeV)	$A(10^{-26}$ s^{-1} $sr^{-1})$	$\alpha(10^{-28}$ s^{-1} sr^{-1} $deg^{-1})$	$B(10^{-6}$ cm^{-2} s^{-1} sr^{-1} $(K\,km\,s^{-1})^{-1})$	$C(10^{-6}$ cm^{-2} s^{-1} $sr^{-1})$	$X(10^{20}$ cm^{-2} $(K\,km\,s^{-1})^{-1})$
100–10,000	2.02 ± 0.14	-7.2 ± 1.4	4.33 ± 0.25	6.3 ± 2.1	1.08 ± 0.10
300–10,000	0.82 ± 0.07	-3.0 ± 0.6	1.72 ± 0.11	< 1.1	1.06 ± 0.16
30–100	2.79 ± 0.38	-14 ± 3	< 4.0	< 22	–
100–150	0.41 ± 0.06	-1.7 ± 0.6	0.66 ± 0.15	< 1.2	0.82 ± 0.22
150–300	0.50 ± 0.06	-1.2 ± 0.5	1.32 ± 0.11	< 2.1	1.34 ± 0.19
300–500	0.25 ± 0.03	-0.8 ± 0.3	0.46 ± 0.06	< 0.3	0.93 ± 0.16
500–1000	0.17 ± 0.03	-0.6 ± 0.2	0.43 ± 0.05	< 0.1	1.30 ± 0.27
1000–10,000	0.09 ± 0.02	< -0.3	0.33 ± 0.03	< 0.2	1.92 ± 0.44

We cannot conclude based on the analyses of local clouds that the excess is correlated with the interstellar gas, which would argue against an IC interpretation. In fact, the absence of a GeV excess in the Taurus/Perseus emissivity may be an artifact of our analysis, if IC emission that might otherwise be assigned to the 'A' term is being accounted for in the gradient term described above. Indeed, the latter may also subsume the extragalactic isotropic emission, which is detected here only below 300 MeV, and with an intensity lower than expected (parameter C in Table 1).

The emissivity above 300 MeV is mostly due to CR protons with GeV energies (see Fig. 3), and no significant variation among the local clouds is seen. Below 300 MeV, electrons have a much more significant contribution, and for this energy range variations are suggested in Figure 3. Density variations of CR electrons are expected on even ~ 100 pc intercloud distance scales owing to their rapid energy losses [11].

The value of X found here, $(1.08 \pm 0.10) \times 10^{20}$ cm^{-2} $(K\,km\,s^{-1})^{-1}$ (> 100 MeV), is consistent with the values for other local clouds studied with EGRET, which range from 0.92 ± 0.14 (Ophiuchus [12]) to 1.64 ± 0.31 (Monoceros [13]) in the same units. No variation of X within or between Taurus and Perseus clouds is indicated by our results, although the statistical uncertainties are large.

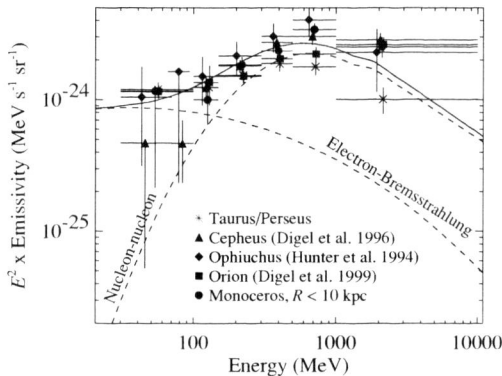

FIGURE 3. Differential gamma-ray emissivities for Taurus/Perseus and other local interstellar cloud complexes. The dashed lines show the predicted bremsstrahlung and pion-decay contributions to the emissivity based on measurements of CR electrons and protons within the solar system [9].

CONCLUSIONS

The interstellar emission observed by EGRET in Taurus/Perseus can be described satisfactorily by a simple model that accounts for the decrease of CR density across the outer Galaxy. Within the region analyzed, four point sources were detected in the Phase 1–5 composite data set. One of them, G179.3-23.7, is not in the 3EG catalog and may be an improved characterization for two closely separated unidentified sources in the catalog. The integral emissivity above 300 MeV, which primarily traces the density of GeV CR protons, does not vary significantly among any of the local clouds studied with EGRET. The variations of emissivity below 300 MeV can be understood in terms of expected density variations of CR electrons on even 100 pc scales. The molecular mass calibrating ratio X in Taurus/Perseus is consistent with that found for other local clouds, and no significant variation within the region was found.

REFERENCES

1. Lebrun, F., et al., *Astrophys. J.* **274**, 231–236 (1983).
2. Ungerechts, H., and Thaddeus, P., *Astrophys. J. Suppl.* **63**, 645–660 (1987).
3. Dickman, R. L., *Astrophys. J. Suppl.* **37**, 407–427 (1978).
4. Dame, T. M., Hartmann, Dap, and Thaddeus, P., *Astrophys. J.* **547**, 792–813 (2001).
5. Hartmann, Dap, and Burton, W. B., *Atlas of Galactic Neutral Hydrogen*, Cambridge University Press, Cambridge, 1997.
6. Mattox, J. R., et al., *Astrophys. J.* **461**, 396–407 (1996).
7. Hartman, R. C., et al., *Astrophys. J. Suppl.* **123**, 79–202 (1999).
8. Gregory, P. C., and Condon, J. J., *Astrophys. J. Suppl.* **75**, 1011–1291 (1991).
9. Hunter, S. D., et al., *Astrophys. J.* **481**, 205–240 (1997).
10. Strong, A. W., Moskalenko, I. V., & Reimer, O., *Astrophys. J.* **537**, 763–784 (2000).
11. Pohl, M. & Esposito, J. A., *Astrophys. J.*, **507**, 327–338 (1998).
12. Hunter, S. D., et al., *Astrophys. J.* **436**, 216–228 (1994).
13. Digel, S. W., et al., *Astrophys. J.* **555**, in press (2001).

Pulsars

Gamma-Ray Pulsars: At the Tip of the Iceberg

Alice K. Harding

NASA Goddard Space Flight Center, Greenbelt MD 20771

Abstract.
Although the theory of pulsar acceleration and high energy emission has been studied for over 25 years, the origin of the pulsed γ-rays is a question that remains unanswered. Characteristics of the seven γ-ray pulsars detected by CGRO could not clearly distinguish between an emission site at the magnetic poles (polar cap models) and emission from the outer magnetosphere (outer gap models). The two types of models make contrasting predictions for the numbers of radio-loud and radio-quiet γ-ray pulsars and for their spectral characteristics. GLAST will probably detect at least 50 radio-selected pulsars and possibly more radio-quiet pulsars. With this large sample, it will be possible to fully test the model predictions and finally resolve this longstanding question.

INTRODUCTION

Although the Compton Gamma-Ray Observatory (CGRO) made significant advances in the study of high-energy pulsar emission, we are still in the discovery phase. The crop of seven clearly detected γ-ray pulsars, as well as several more candidates, provided important model constraints and directions for study. But the γ-ray pulsars are still a tiny fraction of the known radio pulsars, of which there are currently over 1000 [6]. To cross the threshold into the definition phase we will need to detect γ-ray emission from a much larger fraction of pulsars. The next-generation of γ-ray telescopes, both in space and on the ground, will not only breach the unexplored territory between 20 and 200 GeV, but are expected to make an unprecented increase in the γ-ray pulsar population. The Gamma-Ray Large Area Space Telescope (GLAST) will probably detect one hundred or more radio-selected pulsars, with the predicted number being very model dependent. However, the number of radio-quiet γ-ray pulsars could dwarf the number of radio-selected γ-ray pulsars and even approach the total radio pulsar population.

We are thus poised on the tip of the γ-ray pulsar iceberg, ready to plunge below the present sensitivity level, into the bulk of the known (and probably also the unknown) pulsar population. With vastly increased numbers of γ-ray pulsars, one can begin to study trends and model predictions. I will review the current high energy emission models and discuss the key tests which will best be able to distinguish between competing models.

MAJOR RESULTS FROM THE COMPTON OBSERVATORY

During its nine-year mission, observations by CGRO increased the number of known γ-ray pulsars from two to seven. In addition, EGRET made three "candidate" detections with somewhat lower significance, including the millisecond pulsar PSR J0218+4232 [28]. A pattern which emerged from the γ-ray light curves was that, with the one exception of PSR1509-58, all have double-peaked pulses with interpeak emission. This strongly suggests that we are observing emission associated with just one of the magnetic poles and that the emission pattern is a hollow cone or wide fan beam.

Another strong pattern identified in the CGRO pulsars was a correlation, shown in Figure 1, between high-energy luminosity and the quantity $P^{-3/2}\dot{P}^{1/2}$, where P and $\dot{P}$ are the pulsar rotation period and period derivative, respectively. For a pure dipole field, this quantity is proportional to the polar cap current, the voltage across open field lines and $\dot{E}^{1/2}$, where $\dot{E}$ is the spin-down luminosity. Thus the γ-ray luminosity seems to be closely tied to the primary particle acceleration of the pulsar. Such a correlation was actually predicted in polar cap models [17], but it remains to be seen whether this relation is expected in outer gap models. One interesting feature to be seen in Figure 1 is that the luminosity of pulsars having low voltage is approaching the spin-down luminosity, so that the γ-ray efficiency is reaching high levels. Obviously the efficiency must saturate before it reaches 100% and how this happens will be intriguing to investigate with more sensitive detectors.

The spectra of all the CGRO pulsars have clear high-energy turnovers since none have detected pulsed emission at TeV energies. Three pulsars, Vela, Crab and Geminga, have spectral turnovers in the EGRET range, around 5 GeV and one, PSR1509-58, has a sharp turnover in the COMPTEL range around 10 MeV. Below the turnovers, there is a trend of increasing spectral hardness with dipole spin-down age, $\tau = P/\dot{P}$ [38].

Among the most important discoveries of CGRO was the identification of the mysterious γ-ray source Geminga as a γ-ray pulsar [4]. Geminga is the first detected γ-ray pulsar which was not a previously known radio pulsar. Although very weak radio pulses have since been reported (e.g. [30]), it opened speculation that many more pulsars exist that are not picked up in radio surveys. This has turned out to be true, as recent X-ray telescopes such as RXTE, ASCA and Chandra have detected a dozen or so new X-ray pulsars that are radio-quiet, many of them in young supernova remnants [19].

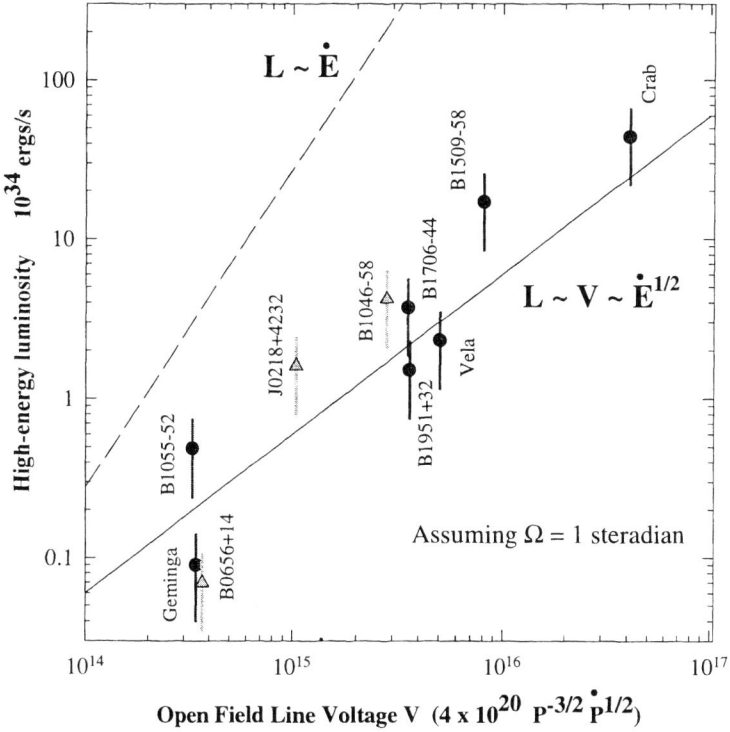

FIGURE 1. High-energy observed luminosity vs. open field line voltage. Circles: high confidence pulsars, triangles: lower confidence pulsars. From Thompson [39].

EMISSION MODELS

Because it is not yet clear how and where in the pulsar magnetosphere the nonthermal high-energy radiation originates, two competing γ-ray emission models have developed. These models generally agree that the particle Lorentz factors must be in the range of at least $10^5 - 10^7$ and that these energies are the result of acceleration by large-scale electric fields. Polar cap models [11] [40] advocate that particle acceleration occurs near the neutron star surface and that γ-rays result from a curvature radiation or inverse Compton induced pair cascade in a strong magnetic field. There is a large variation among polar cap models, with the primary division being whether or not there is free emission of particles from the neutron star surface. The subclass of polar cap models based on free emission of particles of either sign, called space-charge limited flow models, assumes that the surface temperature of the neutron star (many of which have now been measured in the range $T \sim 10^5 - 10^6$ K) exceeds the ion and electron thermal emission temperatures. Although $E_\parallel = 0$ at the neutron star surface in these models, the space charge along open field lines above the surface falls short of the corotation charge, due to the curvature of the field [1] or to general relativistic inertial frame dragging [31]. The $E_\parallel$ generated by the charge deficit accelerates particles, which radiate inverse Compton (IC) photons (at particle Lorentz factors $\gamma \sim 10^2 - 10^6$) and curvature (CR) photons (Lorentz factors $\gamma \gtrsim 10^6$). It has been realized for some time that only younger pulsars are capable of producing pairs (and pair cascades) through CR emission in dipole magnetic fields [1]. Recent studies [25] [23] have found that virtually *all* pulsars are capable of producing pairs by either resonant scattering (in the case of high-field pulsars) or non-resonant scattering (in the case of older and weaker field pulsars), at particle energies much lower than required for CR-produced pairs. However, it is found [23] that in most pulsars, the IC pair formation fronts do not produce sufficient pairs to screen the $E_\parallel$ completely, thus allowing acceleration to the Lorentz factors $\sim 10^7$ sufficient to produce CR. Some gamma-ray pulsars may thus have dual cascades initiated by both IC and CR photons.

Outer-gap models [8] [34] assume that acceleration occurs in vacuum gaps along null charge surfaces in the outer magnetosphere and that γ-rays result from photon-photon pair production-induced cascades. The gaps arise because charges escaping through the light cylinder along open field lines above the null charge surface cannot be replenished from below. Pairs from the polar cap cascades, which flow out along all the open field lines, will undoubtedly pollute the outer gaps to some extent, but this effect has yet to be investigated. The electron-positron pairs needed to provide the current in the outer gaps are produced by photon-photon pair production. In young Crab-like pulsars, the pairs are produced by CR photons from the primary particles interacting with non-thermal synchrotron X-rays from the same pairs. In older Vela-like pulsars, where non-thermal X-ray emission is much lower, the pairs were assumed to come from interaction of primary particle inverse Compton photons with infra-red photons. However, this original Vela-type model [8] predicted large fluxes of TeV emission, from inverse Compton scattering of the infra-red photons by primary electrons, which violates the observed upper limits [32] by several orders of magnitude. Cheng [7] revised the outer gap model for Vela-type pulsars by proposing another self-sustaining gap mechanism where thermal X-rays from the neutron star surface interact with primary particle radiation to produce pairs, replacing the infra-red radiation (which has also never been observed). Some of the accelerated pairs flow downward to heat the surface and maintain the required thermal X-ray emission. The modern outer gap Vela-type models [34] [44] all adopt this picture.

PREDICTIONS FOR HIGH ENERGY EMISSION

Polar cap and outer gap models and their variations make contrasting predictions for the numbers of radio-quiet and radio-loud γ-ray pulsars and of their spectral characteristics. CGRO did not have the spectral resolution at high energies or the sensitivity to discriminate between polar cap and outer gap models. Improved energy resolution and sensitivity around 1 GeV and in the unexplored region above 20 GeV is needed to test the theoretical predictions.

Spectral shape and cutoffs

Polar cap models predict that the γ-ray spectra cutoff very sharply (as a "super-exponential") due to one-photon pair production attenuation, at a field-dependent energy, while outer gap model spectra cutoff more slowly (as a simple exponential) due to a particle acceleration limit. The highly relativistic particles emit photons at very small ($\theta \sim 1/\gamma$) angles to the open magnetic field lines. The photons of energy ϵ (in units of mc^2), emitted near the neutron star surface, are initially below the threshold for one-photon pair production ($\epsilon_{th} = 2/\sin\theta$), but may reach threshold by increasing θ in the course of propagating across curved field lines. The γ-ray spectrum will exhibit a cutoff at the pair escape energy (cf. [20]), i.e. the highest energy at which photons emitted at a given location can escape the magnetosphere without pair producing. A rough estimate of this cutoff energy, assuming emission along the polar cap outer rim, $\theta_{PC} \simeq (2\pi r/cP)^{1/2}$, at radius r, is

$$E_c \sim 2 \text{ GeV } P^{1/2} \left(\frac{r}{R_0}\right)^{1/2} \max\left\{0.1, B_{0,12}^{-1}\left(\frac{r}{R_0}\right)^3\right\} \quad (1)$$

where P, R_0 and B_0 are the neutron star period, radius and surface magnetic field in units of 10^{12} G. At all but the highest fields there is a prediction that the spectral cutoff energy should be inversely proportional to surface field strength, or $B_0 = 6.4 \times 10^7 P\dot{P}^{1/2}$ G for a dipole field. In fields above $\sim 2 \times 10^{13}$ G, photon splitting, in which a single photon splits into two lower energy photons, becomes the dominant attenuation process and lowers the photon escape energy [20] [3]. The observed cutoff energies of eight γ-ray pulsars seem to increase with decreasing surface field [18], although a larger number of measurements are needed to confirm this trend, with a dependence even stronger than predicted by the polar cap model for a constant emission radius. However, some increase in emission radius is expected due to the trend for larger acceleration zones in older pulsars, but work to quantify this trend is still in progress. It appears from Eqn (1) that pulsars with long periods and/or low magnetic fields ("old" pulsars) will be the best candidates for detection above 20 GeV. Millisecond pulsars, having surface fields in the range $B_0 \sim 10^8 - 10^{10}$ G, should allow high-energy photons with energies above 100 GeV to escape from near the neutron star surface if such radiation is actually produced. Polar cap cascade model calculations [5] predict a weak IC radiation component at TeV energies from ms pulsars, which is possibly detectable with air Cherenkov telescopes like MAGIC.

Polar cap models predict that the number of pair generations in cascades will decrease as the pulsar period increases or the magnetic field decreases. For a particular pulsar, the cascade terminates when the photon energy drops below the pair escape energy, E_c, so that the maximum number of generations (which can be a non-integer) is [43] (see also [2])

$$n = \frac{\log(E_c/E_0)}{\log(\kappa)} + 1, \qquad (2)$$

where E_0 is the primary photon energy and κ is an energy degradation factor for each generation which depends on the radiation process. For synchrotron cascade branches, $\kappa_{SR} = \frac{3}{4}\chi$, where $\chi = 0.5\max[(B/B_{cr})\sin\theta, 0.1]$, $B_{cr} = 4.4 \times 10^{13}$ G and θ is the angle between the photon wavevector and the local magnetic field. The cascade photon spectrum steepens at each generation due to energy losses, such than $\alpha_{i+1} = (\alpha_i + 1)/2$, giving the final cascade spectral index as [21]

$$\alpha_n = 2 - \frac{2 - \alpha_1}{2^{n-1}}. \qquad (3)$$

where α_1 is the photon index of the primary electrons. In the case of curvature radiation by particles undergoing radiative energy loss, $\alpha_1 = 5/3$ [10]. In the limit of $n \to \infty$, $\alpha \to 2$. Thus, the larger the cascade generation number n, the larger the photon spectral index, and since $\kappa_{SR} \propto B_0/P \propto \tau^{1/2}$, these models predict an increase in spectral hardening with age, as observed.

The picture is quite different in outer gap models. When the high-energy photons are emitted in the outer magnetosphere, where the local magnetic field is orders of magnitude lower than the surface field, one-photon pair production plays no role in either the pair cascade or the spectral attenuation. In this case the high-energy cutoffs in the photon spectrum come from the upper limit of the accelerated particle spectrum, due to radiation reaction. The shape of the cutoff is thus a simple exponential, more gradual than in polar cap model spectra. Due to the large errors of the EGRET data points above 1 GeV, the measurements at present do not definitely discriminate between model spectra. GLAST should have the energy resolution and dynamic range to measure the shape of the cutoffs seen by EGRET and should be able to rule out either the simple exponential or super-exponential shape. In addition, GLAST will detect enough γ-ray pulsars with different field strengths to look for a correlation between surface field strength and cutoff energy.

Outer gap models predict an emission component at TeV energies due to inverse Compton scattering by gap-accelerated particles. TeV photons should escape from the outer gaps of pulsars with even strong surface fields and this component is thus expected to be observable in many pulsars. The original predictions of Cheng et al. [8] were not verified by observations of ground-based detectors [32] [29], requiring a revision of the Vela-like model [7]. However, even later models which predicted lower TeV fluxes [34] are above CANGAROO upper limits on pulsed emission from Vela [42]. The most recent outer gap models [26] [27], have predicted TeV inverse-Compton fluxes which are below the present observational upper limits, but which should be detectable with the next generation of TeV detectors. Unfortunately, while a TeV emission component is an essential prediction of all outer gap models, the inverse Compton flux level depends on the pulsed emission spectrum in the infra-red (IR) band which is notoriously difficult to measure in most pulsars.

Unlike in polar cap models, pair production in outer gap models plays a critical role in production of the high energy emission: it allows the current to flow and particle acccleration to take place in the gap. Beyond a death line in period-magnetic field space, and well before the traditional radio-pulsar death line, pairs cannot close the outer gap and the pulsar cannot emit high energy radiation. This outer gap death line for γ-ray pulsars falls around $P = 0.3$ s for $B \sim 10^{12}$ G [9]. Geminga is very close to the outer gap death line and recent self-consistent models [27] have difficulty accounting for GeV γ-ray from pulsars of this age. Polar cap models, on the other hand, predict that all pulsars are capable of γ-ray emission at some level. Which pulsars are detected as γ-ray pulsars is thus a matter of sensitivity. Detection of pulsars with periods much exceeding that of Geminga would thus be strong evidence in favor of polar cap models.

Population statistics and radio-quiet pulsars

Polar cap and outer gap models predict different ratios of radio-loud to radio-quiet γ-ray pulsars, primarily due to the different geometry of the high-energy emission regions and its location relative to the radio emission region. Numerous studies of radio emission morphology of many pulsars (e.g. [33]) argue in favor of an origin in the polar regions, within tens of stellar radii fo the neutron star surface. Thus, polar cap γ-ray emission is expected to have a much higher correlation with radio emission. In fact, the radio emission is physically linked to the γ-ray emission in polar cap models if pairs from the high-energy cascades are a necessary requirement for coherent radio emission. On the other hand, the high energy emission in the outer gap is generally radiated in a different direction from the radio emission, which allows these models to account for the observed phase offsets of the radio and γ-ray pulses. At the same time, there will be fewer radio-γ-ray coincidences and thus a larger number of radio-quiet γ-ray pulsars. In Romani & Yadigaroglu's geometrical outer gap model [35], the observed radio emission originates from the magnetic pole opposite to the one connected to the visible outer gap. Many observer lines-of-sight miss the radio beam but intersect the outer-gap γ-ray beam, having a much larger solid angle. When the line-of-sight does intersect both, the radio pulse leads the γ-ray pulse, as is observed in most γ-ray pulsars.

Simulations of the radio and γ-ray pulsar populations in both models confirm these intuitive ideas. Table 1 summarizes a number of γ-ray pulsar population studies and their predictions for both EGRET and GLAST. Although there are significant variations in the numbers of predicted γ-ray pulsars due to different model assumptions in the various studies, outer gap models clearly predict a much larger ratio of radio-quiet to radio-loud γ-ray pulsars. Polar cap models predict that EGRET detected only a few radio-quiet pulsars, which would imply that not many of the unidentified sources in the plane (see [16] for review) are pulsars. On the other hand, outer gap model simulations would predict that many EGRET unidentified sources at low latitudes are pulsars. However, all of these studies have treated the γ-ray and radio emission geometry in an oversimplified way. The population studies of polar cap γ-ray pulsars have assumed that both γ-ray and radio emission are beamed with the same direction and have the same solid angle. A more realistic treated of the emission geometry will probably result in a larger number of radio-quiet pulsars. The outer gap simulations assume only that the γ-ray beam solid angle is larger than the radio beam solid angle. Several studies have predicted the number of radio-loud and radio-quiet pulsars which GLAST should detect. The polar cap model simulation of Gonthier et al. [13] [14] finds that GLAST should detect about equal numbers of radio-loud and radio-quiet pulsars as point sources. Although GLAST will have the capability to detect pulsed γ-ray emission, the required sensitivity is much higher, about equal to the EGRET point source sensitivity. Thus, only about 10% of the radio-quiet sources will have detected γ-ray pulsations. The outer gap simulations [45] predict that GLAST may detect 13 times as many radio-quiet as radio-loud pulsars as point sources, with the detected number of radio-quiet pulsars equaling the present radio pulsar population! This would have profound consequences for neutron star evolution and supernova rates in the galaxy.

TABLE 1 - Predicted Pulsar Populations

	EGRET		GLAST	
	Radio Loud	Radio Quiet	Radio Loud	Radio Quiet (Pulsed)
POLAR CAP				
Sturner & Dermer [37]	4	1		
Gonthier et al. [13]:				
no B decay	7	1	76	74 (7)
B decay[a]	9	2	90	101 (9)
OUTER GAP				
Yadigaroglu & Romani [41]	5	17		
Zhang et al. [45]	10	22	80	1100

[a] Assuming a magnetic field decay with timescale 5×10^6 yr.

Recently, another possible population of radio-quiet γ-ray pulsars has been suggested [24]. According to the polar cap model (e.g. [11]), γ-ray emission occurs throughout the entire pulse phase. Primary electrons that initiate pair cascades at low altitude continue to radiate curvature emission on open field lines to high altitudes beyond the cascade region, producing a lower level of softer off-beam emission. Due to the flaring of the dipole field lines, this emission may be seen over a large solid angle, far exceeding that of the main beams. Since the radio emission is expected to originate within ten stellar radii of the neutron star surface, it is quite probable to see off-beam γ-ray emission and miss the radio beam. Harding

& Zhang [24] estimate that the probability of detecting such off-beam emission is a factor of $\sim 4-5$ times higher than that of the on-beam emission. At least some of the radio-quiet Gould Belt sources detected by EGRET [12] [15] could therefore be such off-beam gamma-ray pulsars.

TABLE 2 - Gamma-ray Pulsar Model Predictions

Observation	Polar cap	Outer gap
HE Cutoff Correlated with B_0	⇑	⇓
Pulsed Emission Above 50 GeV		
Normal PSRs	⇓	⇑
ms PSRs	⇑	⇑
Long-period γ-ray PSRs	⇑	⇓
N(radio-quiet) >> N(radio-loud)	⇓	⇑

SUMMARY

Table 2 summarizes some key predictions of high-energy pulsar emission models which can potentially be tested by future instruments having both higher sensitivity and larger energy range. Probably the most discriminating tests will be measurement of pulsar spectra at energies from 1 GeV to 10 TeV. In this range, polar cap models predict steep spectral cutoffs due to magnetic pair production attenuation and essentially no detectable emission above about 50 GeV from normal-period pulsars, although both models predict an IC emission component at TeV energies from ms pulsars. The emission from long-period pulsars predicted by polar cap models should also be more easily detectable above 1 GeV, due to the hardness of their spectra. The number of radio-quiet pulsars detected by GLAST will be a very important diagnostic. Although both polar cap and outer cap models now expect significant numbers of radio-quiet γ-ray pulsars detectable with GLAST sensitivity, the outer gap models predict a much larger ratio of radio-quiet to radio-loud pulsars.

REFERENCES

1. Arons, J., *ApJ*, **266**, 215 (1983).
2. Baring, M. G., in proc. of "Unidentified γ-ray Sources", in press (2001)
3. Baring, M. G. & Harding, A. K., *ApJ*, **547**, 929 (2001).
4. Bertsch, D. L. et al., *Nature*, **357**, 306 (1992).
5. Bulik, T. et al., MNRAS, **317**, 97 (2000).
6. Camilo, F., et al., in Pulsar Astronomy: 2000 and Beyond - IAU Coll. **177**, 3 (2000).
7. Cheng, K. S., *Proc. Toward a Major Atmospheric Cherenkov Detector*, ed. T. Kifune (Tokyo: Universal Academy), 25 (1994).
8. Cheng, K. S., Ho, C. & Ruderman, M. A., *ApJ*, **300**, 500 (1986).

9. Chen, K. & Ruderman, M. A., *ApJ*, **402**, 264 (1993).
10. Daugherty, J. K. & Harding, A. K., *ApJ*, **252**, 337 (1982).
11. Daugherty, J. K., & Harding A. K.: *ApJ*, **458**, 278 (1996).
12. Gehrels, N., et al., *Nature*, **404**, 363 (2000).
13. Gonthier, P. G. et al., *ApJ*, submitted (2001).
14. Gonthier, P. G. et al., these proceedings (2001).
15. Grenier, I. A. et al., *A & A*, in press (2000)
16. Grenier, I. A., these proceedings (2001).
17. Harding, A. K., *ApJ*, **245**, 267 (1981).
18. Harding, A. K., in "High Energy Gamma-Ray Astronomy", (AIP Conf. 558), ed. F.A. Aharonian & H.J. Volk. p. 115 (2001).
19. Harding, A. K., in "Young Supernova Remnants", (AIP Conf. 565), ed. S.S.Holt & U. Hwang, p. 351 (2001).
20. Harding, A. K., Baring, M. G. & Gonthier, P. L. , *ApJ*, **476**, 246 (1997).
21. Harding, A. K. & Daugherty, J. K., Adv. Space Res. *21(1/2)*, 251 (1998).
22. Harding, A. K. & Daugherty, J. K. , *Lett. Comm.*, **38**, 25 (1999).
23. Harding, A. K. & Muslimov, A. M., in prep (2001).
24. Harding, A. K. & Zhang, B., *ApJ*, **548**, L37 (2001).
25. Hibschman, J. A. & Arons, J., astro-ph/0102175 (2001).
26. Hirotani, K. , *ApJ*, **549**, 495 (2001).
27. Hirotani, K. & Shibata, S., MNRAS, in press (2001).
28. Kuiper, L. et al., *A & A*, **359**, 615 (2000).
29. Lessard, R. W. et al., *ApJ*, **531**, 942 (2000).
30. Malofeev, V. M. & Malov, O. I., *Nature*, **389**, 697 (1997).
31. Muslimov, A. G. & Tsygan, A. I., *MNRAS*, **255**, 61 (1992).
32. Nel, H. I. et al., *ApJ*, **418**, 836 (1993).
33. Rankin, J. M. , *ApJ*, **405**, 285 (1993).
34. Romani, R. W., *ApJ*, **470**, 469 (1996).
35. Romani, R. W. & Yadigaroglu, I.-A., *ApJ*, **438**, 314 (1994).
36. Ruderman, M. A. & Halpern, J., *ApJ*, **415**, 286 (1993).
37. Sturner, S. J. & Dermer, C. D. , A&AS, **120**, 99 (1996).
38. Thompson, D. J. et al., *ApJ*, **436** 229 (1994).
39. Thompson, D. J., in High Energy Gamma-Ray Astronomy, ed. F. Aharonian & H. Volk (AIP: New York), p. 103.
40. Usov, V. V. & Melrose, D. B., *Aust. J. Phys.*, **48**, 571 (1995).
41. Yadigaroglu, I.-A. & Romani, R. W., *ApJ*, **449**, 211 (1995).
42. Yoshikoshi, T. et al. *ApJ*, **487**, 65 (1997).
43. Zhang, B. & Harding, A. K. , *ApJ*, **535**, L51 (2000).
44. Zhang, L. & Cheng, K. S. , *ApJ*, **487**, 370 (1997).
45. Zhang, L., Zhang, Y. J. & Cheng, K. S., A & A, 357, 957 (2000).

The Parkes Multibeam Pulsar Survey and the Discovery of New Energetic Radio Pulsars

N. D'Amico[*], V. M. Kaspi[+†], R. N. Manchester[**], F. Camilo[‡], A. G. Lyne[§], A. Possenti[*], I. H. Stairs[¶], M. Kramer[§], G. Hobbs[§] and J. F Bell[**]

[*]*Osservatorio Astronomico di Bologna, via Ranzani 1, 40127 Bologna, Italy*
[†]*McGill University, Physics Department, Rutherford Physics Building, 3600 University Street, Montreal, Quebec, H3A 2T8 Canada*
[+]*Department of Physics and Center for Space Research, Massachusetts Institute of Technology, Cambridge, MA 02139*
[**]*Australia Telescope National Facility, CSIRO, P.O. Box 76, Epping, NSW 1710, Australia*
[‡]*Columbia Astrophysics Laboratory, Columbia University, 550 W 120th Street, New York, NY 10027*
[§]*University of Manchester, Jodrell Bank Observatory, Macclesfield, Cheshire, SK11 9DL, UK*
[¶]*NRAO, P.O. Box 2, Green Bank, WV 24944*

Abstract.
The Parkes multibeam pulsar survey is a deep search of the Galactic plane for pulsars. It uses a 13-beam receiver system operating at 1.4 GHz on the 64-m Parkes radio telescope. It has much higher sensitivity than any previous similar survey and is finding large numbers of previously unknown pulsars, many of which are relatively young and energetic. On the basis of an empirical comparison of their properties with other young radio pulsars, some of the new discoveries are expected to be observable as pulsed γ-ray sources. We describe the survey motivation, the experiment characteristics and the results achieved so far.

INTRODUCTION

Since the pioneering γ-ray observations conducted early in the 70's with balloon-borne experiments [1], radio pulsars were privileged as targets for γ-ray emission searches. One of the reasons was partly observational. The angular resolution of γ-ray telescopes is rather poor, and the timing analysis is the only effective technique to identify point sources in γ-ray astronomy. The first systematic γ-ray observations conducted by the SAS-II satellite [2], showed that the Crab and Vela pulsars were strong sources of pulsed γ-ray emission [3, 4]. One of the other major contribution of the SAS-II satellite to γ-ray astronomy was the observation of significant γ-ray emission from the Galactic disk. In principle, on the basis of the SAS-II observations, most of Galactic γ-rays could be interpreted as due to diffuse emission only [5]. On the other hand, in the following years, the better statistics achieved by the γ-ray satellite COS-B [6], proved the existence of several localised sources of γ-ray emission along the Galactic disk [7]. Since then, the identification of these sources has represented a challenge for γ-ray astronomy. The hypothesis that they could be young energetic pulsars similar to the Crab and Vela pulsars is interesting [8]. However, the low counting statistics available in γ-ray astronomy and the long integration times (several weeks) required to obtain a significant detection prevent an unbiased periodicity search of the γ-ray data. The discovery of potential radio pulsar counterparts, and the possibility of phasing the γ-ray counts using the radio timing parameters, represents the unique tool for the identification.

The first searches of the COS-B γ-ray error boxes for radio pulsars conducted at Arecibo and Parkes in the early 80's [9, 10], were unsuccessful. But in 1992, in a deep survey of the Galactic disk conducted at Parkes, Johnston et al (1992) [11], discovered a new sample of radio pulsars, including a few relatively young and energetic objects. One of them, PSR B1706-44, was coincident with the COS-B source 2CG342-02. Precise radio timing parameters obtained at the epoch of the EGRET observation of this source led to the detection of pulsed γ-ray [12]. Indeed, a long term radio timing program of a large sample of known pulsars conducted during the CGRO mission led to the detection of pulsed γ-ray emission from a few other radio pulsars [13, 14]. On the other hand, EGRET has also discovered a large number of new localised γ-ray sources [15], making their identification a serious challenge. If some of the Galactic EGRET sources are pulsars, they are likely to be relatively young and energetic, similar to the Crab and Vela pulsars. This hypothesis is strongly supported by the discovery that "Geminga", the third strongest γ-ray source in the sky, is indeed a pulsar [26].

Although young pulsars evolve relatively fast, so that they tend to be intrinsically rare in the population, the paucity of young pulsars in the observed sample is believed to be further enhanced by observational selection effects. Pulsars, being the remnant of massive stars, are believed to be born as a disk population. So young pulsars, which are expected to be found close to their birth place, tend to be located at low Galactic latitudes. The observation at radio wavelengths of pulsars (and in particular short-period pulsars) at low Galactic latitudes is limited by several factors, including pulse smearing due to dispersion and scattering in the interstellar medium, and Galactic background radiation.

Beside their potential role as counterparts of the Galactic gamma-ray sources, young pulsars are interesting for several reasons: they are likely to be associated with supernova remnants [16], they generally exhibit rotational instabilities including glitches [17, 18, 19], and they sometimes emit detectable pulsed radiation at optical [20] and X-ray frequencies [21, 22].

In this paper we report on a new survey of the Galactic plane for pulsars which is in progress at Parkes, including the discovery of a substantial number of new young and energetic radio pulsars.

THE PARKES MULTIBEAM PULSAR SURVEY

The new deep survey of the Galactic plane for pulsars is being carried out using a 13-beam receiver on the Parkes 64-m radio telescope [23, 24]. Observations are made using dual-polarization feeds with a bandwidth of 288 MHz centered on 1374 MHz. A large filterbank system gives 96 3-MHz channels for each polarization of each of the 13 beams. Signals from individual frequency channels are detected, added in polarization pairs, high-pass filtered, integrated, 1-bit digitised every 250 μs and output to digital linear tapes (DLTs). The excellent system noise temperature ($\sim$ 21 K), large bandwidth

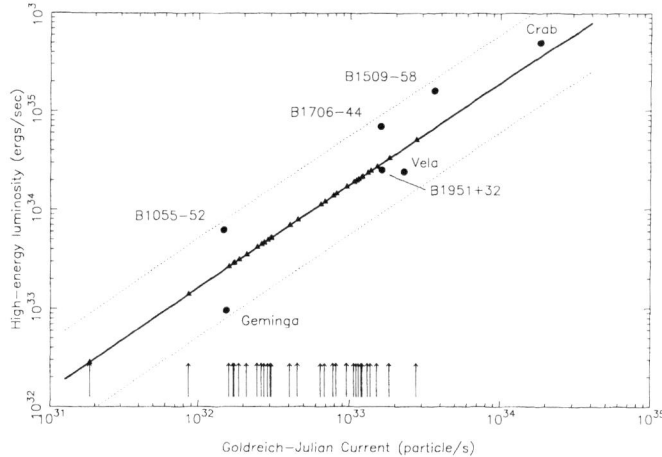

FIGURE 1. High energy luminosity of pulsars as a function of the so-called Goldreich-Julian current parameter (thick line)(e.g., Thompson et al (1999) [30]). Dots correspond to known γ-ray pulsars. The dotted lines represent a region containing all the known γ-ray pulsars. For each of the new 30 young radio pulsars, the value of the Goldreich-Julian current is indicated with a vertical arrow. Dark triangles show where on the line the new young pulsars would be located

and relatively long integration time of 35 min per pointing give a sensitivity limit for long-period pulsars of about 0.2 mJy.

Offline processing is performed on networked workstations at Jodrell Bank Observatory, ATNF, McGill and Bologna. Data are searched for periodic signals over a range of dispersion delays, using an analysis procedure similar to that employed in the Parkes Southern pulsar survey [25]. Candidates are re-observed for confirmation using the centre beam of the multibeam receiver. Confirmed candidates are then regularly observed at Parkes or Jodrell Bank for at least a year, in order to measure precise positions, pulse periods P and period derivatives $\dot{P}$, and orbital parameters if binary motion is present.

SURVEY STATUS AND RESULTS

At the time of writing, the survey is about 90% complete, and has resulted in the discovery of more than 600 new pulsars. Among them are a number of interesting objects including: a double-neutron-star system, PSR J1811-1736 [23]; two young pulsars, PSRs J1119-6127 and J1814-1744, which have the highest dipole magnetic field strengths among radio pulsars [27]; a young binary pulsar in an eccentric 5-hour orbit, PSR J1141-6545, for which the measurement of the relativistic precession of periastron was already obtained [28]; a high-mass binary system, PSR J1740-3052, for which the minimum mass for the companion is 11 $M_\odot$ [29].

Timing observations have also shown that at least 30 of the new discoveries are young pulsars, with characteristic age $\tau_c = P/2\dot{P} < 100,000$ years. These pulsars are prime

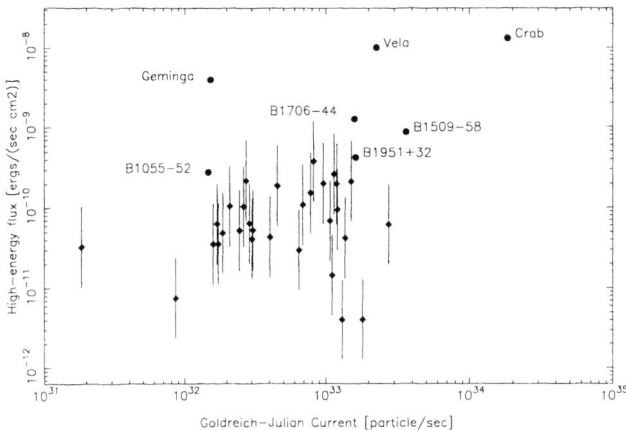

FIGURE 2. High energy flux of pulsars as a function of the Goldreich-Julian current. Full dots indicated known γ-ray pulsars. Predicted values for the new young radio pulsars are indicated. Their uncertainty is derived from the dotted line region of Fig. 1. The distance adopted for each pulsar is the nominal value derived from the observed dispersion measure

candidates for pulsed γ-ray emission searches. There is a strong correlation of high-energy luminosity and the spin-down luminosity $\dot{E}$. In particular, Thompson et al (1999) [30] have shown that the high-energy luminosity, integrated over photon energies above 1 eV, is proportional to the Goldreich-Julian current $\dot{N} \simeq 1.7 \times 10^{38} \dot{P}^{1/2} P^{-3/2}$. Fig. 1 shows the predicted high-energy luminosity for the 30 new young pulsars as a function of the Goldreich-Julian current parameter, computed from the observed spin parameters.

Adopting for each pulsar the distance derived from the observed dispersion measure, we can convert the predicted high-energy luminosity into a predicted flux. Fig. 2 shows the observed and predicted high-energy flux from pulsars. Some of the new young radio pulsars have predicted flux values similar to those already observed from other known radio pulsars. Indeed, some of the new pulsars are coincident with EGRET sources [31]. Future γ-ray experiments like AGILE and GLAST should search these new young radio pulsars for pulsed γ-ray emission.

REFERENCES

1. Leray, J.P., Vasseur, J., Paul, J., Parlier, B., Forichon, M., Agrigner, B., Boella, G., Maraschi, L., Treves, A., Buccheri, R., Cuccia, A., Scarsi, L. 1972, A&A, 16, 443
2. Fichtel, C.E., Hartman, R.C., Kniffen, D.A., Thompson, D.J., Ogelman, H., Ozel, M.E., Tumer, T., Bignami, G.F. 1975, ApJ, 198, 163
3. Kniffen, D.A., Hartman, R.C., Thompson, D.J., Bignami, G.F., Fichtel, C.E., Ogelman, H., Turner, T. 1974, Nature, 251, 397
4. Thompson, D.J., Bignami, G.F., Fichtel, C.E., Kniffen, D.A. 1974, ApJL, 190, L15
5. Kniffen, D.A., Hartman, R.C., Thompson, D.J., Fichtel, C.E., 1973, ApJL, 186, L105
6. Bignami, G.F. at al, 1975, Space Sci. Instr., 1, 245

7. Swanenburg, B. N. Bennett, K., Bignami, G. F., Buccheri, R., Caraveo, P. Hermsen, W., Kanbach, G., Lichti, G. G., Masnou, J. L., Mayer-Hasselwander, H. A., Paul J. A., Sacco, B., Scarsi, L., Wills, R. D. 1981, ApJL, 243, L69
8. D'Amico N. & Scarsi L., 1980, in: Gravitational radiation, collapsed objects and exact solutions; Proceedings of the Einstein Centenary Summer School, Perth, Australia, January 1979. (A81-33409 14-90) Berlin, Springer-Verlag, 1980, p. 67-87.
9. D'Amico N., 1983, Space Sci. Rev., 36, 195
10. Manchester, R. N., Damico, N., Tuohy, I. R. 1985, MNRAS, 212, 975
11. Johnston, S., Lyne, A. G., Manchester, R. N., Kniffen, D. A., D'Amico, N., Lim, J., Ashworth, M. 1992, MNRAS, 255, 401
12. Thompson, D. J., Arzoumanian, Z., Bertsch, D. L., Brazier, K. T. S., D'Amico, N., Fichtel, C. E., Fierro, J. M., Hartman, R. C., Hunter, S. D., Johnston, S. 1992, Nature, 359, 515
13. Nolan, P. L., Fierro, J. M., Lin, Y. C., Michelson, P. F., Bertsch, D. L., Dingus, B. L., Esposito, J. A., Fichtel, C. E., Hartman, R. C., Hunter, S. D., von Montigny, C., Mukherjee, R., Ramanamurthy, P. V., Thompson, D. J., Kniffen, D. A., Schneid, E., Kanbach, G., Mayer-Hasselwander, H. A., Merck, M. 1996, A&AS, 120, 61
14. Kaspi, V. M., Lackey, J. R., Mattox, J., Manchester, R. N., Bailes, M., Pace, R. 2000, ApJ, 528, 445
15. Hartman, R. C., Bertsch, D. L., Bloom, S. D., Chen, A. W., Deines-Jones, P., Esposito, J. A., Fichtel, C. E., Friedlander, D. P., Hunter, S. D., McDonald, L. M., Sreekumar, P., Thompson, D. J., Jones, B. B., Lin, Y. C., Michelson, P. F., Nolan, P. L., Tompkins, W. F., Kanbach, G., Mayer-Hasselwander, H. A., Mucke, A., Pohl, M., Reimer, O., Kniffen, D. A., Schneid, E. J., von Montigny, C., Mukherjee, R., Dingus, B. L. 1999, ApJS, 123, 79
16. Kaspi, V. M. 2000, in Kramer M., Wex N., & Wielebinski R., eds, Pulsar Astronomy - 2000 and Beyond, IAU Colloquium 177, Astronomical Society of the Pacific, San Francisco, 485
17. Arzoumanian, Z., Nice, D. J., Taylor, J. H., & Thorsett, S. E. 1994, ApJ, 422, 671
18. Wang, N., Manchester, R. N., Pace, R., Bailes, M., Kaspi, V. M., Stappers, B. W., & Lyne, A. G. 2000, MNRAS, 317, 843
19. Lyne, A. G., Shemar, S. L., & Graham-Smith, F. 2000, MNRAS, 315, 534
20. Middleditch, J., Pennypacker, C. R., & Burns, M. S. 1987, ApJ, 315, 142
21. Becker, W. & Trumper, J. 1997, A&A, 326, 682
22. Marshall, F. E., Gotthelf, E. V., Zhang, W., Middleditch, J., & Wang, Q. D. 1998, ApJ, 499, L179
23. Lyne, A. G. et al 2000, MNRAS, 312, 698
24. Manchester, R. N. et al. 2001, MNRAS, Submitted
25. Manchester, R. N., Lyne, A. G., D'Amico, N., Bailes, M., Johnston, S., Lorimer, D. R., Harrison, P. A., Nicastro, L., Bell, J. F. 1996, MNRAS, 279, 1235
26. Bertsch, D. L., et al. 1992, Nature, 357, 306
27. Camilo, F., Kaspi, V. M., Lyne, A. G., Manchester, R. N., Bell, J. F., D'Amico, N., McKay, N. P. F., Crawford, F. 2000, ApJ, 541, 367
28. Kaspi, V. M., Lyne, A. G., Manchester, R. N., Crawford, F., Camilo, F., Bell, J. F., D'Amico, N., Stairs, I. H., McKay, N. P. F., Morris, D. J., Possenti, A. 2000, ApJ, 543, 321
29. Stairs, I. H., Manchester, R. N., Lyne, A. G., Kaspi, V. M., Camilo, F., Bell, J. F., D'Amico, N., Kramer, M., Crawford, F., Morris, D. J., Possenti, A., McKay, N. P. F., Lumsden, S. L., Tacconi-Garman, L. E., Cannon, R. D., Hambly, N., Wood, P. W. 2001, MNRAS, in press (astro-ph/0012414)
30. Thompson, D. J., Bailes, M., Bertsch, D. L., Cordes, J., D'Amico, N., Esposito, J. A., Finley, J., Hartman, R. C., Hermsen, W., Kanbach, G., Kaspi, V. M., Kniffen, D. A., Kuiper, L., Lin, Y. C., Lyne, A., Manchester, R., Matz, S. M., Mayer-Hasselwander, H. A., Michelson, P. F., Nolan, P. L., Ogelman, H., Pohl, M., Ramanamurthy, P. V., Sreekumar, P., Reimer, O., Taylor, J. H., Ulmer, M. 1999, ApJ, 516, 297
31. D'Amico, N., Kaspi, V. M., Manchester, R. N., Camilo, F., Lyne, A. G., Possenti, A., Stairs, I. H., Kramer, M., Crawford, F., Bell, J. F., McKay, N. P. F., Gaensler, B. M., Roberts, M. S. E. 2001, ApJL, 552, L45

Rotationally-induced asymmetry in the double-peak lightcurves of the bright EGRET pulsars?

J. Dyks* and B. Rudak[†]

Nicolaus Copernicus Astronomical Center, Toruń, Poland
[†]*Nicolaus Copernicus Astronomical Center, also TCfA NCU, Toruń, Poland*

Abstract. Pulsed emission from the bright EGRET pulsars - Vela, Crab, and Geminga - extends up to $\lesssim 10$ GeV. The generic gamma lightcurve features two peaks separated by 0.4 to 0.5 in phase. According to Thompson (2001) the lightcurve becomes asymmetrical above ~ 5 GeV in such a way that the trailing peak dominates over the leading peak above ~ 5 GeV. We attempt to interpret this asymmetry within a single-polar-cap scenario. We investigate the role of rotational effects on the magnetic one-photon absorption rate in inducing such asymmetry. Our Monte Carlo simulations of pulsar gamma-ray beams reveal that in the case of oblique rotators with rotation periods of a few millisecond the rotational effects lead to the asymmetry of the requested magnitude. However, the rotators relevant for the bright EGRET pulsars must not have their inclination angles too large in order to keep the two peaks at a separation of ~ 0.4 in phase. With such a condition imposed on the model rotators the resulting effects are rather minute and can hardly be reconciled with the magnitude of the observed asymmetry.

INTRODUCTION

High quality gamma-ray data for three pulsars - Vela, Crab, and Geminga - provided by EGRET aboard the CGRO enable an analysis of properties of pulsar high-energy radiation as a function of both, photon energy and phase of rotation. The spectra of pulsed radiation from these sources (as well as from three other EGRET pulsars: B1706-44, B1951+32, and B1055-52) extend up to $\lesssim 10$ GeV. All three pulsars feature gamma lightcurves characterised by two strong peaks separated by 0.4 to 0.5 in rotational phase. The pulses are asymmetrical in the sense that their leading peaks (LP) exhibit lower energy cutoffs (about ~ 5 GeV) than their trailing peaks (TP). In other words, the trailing peaks dominate over the leading peaks above ~ 5 GeV [6].

The high-energy cutoffs in pulsar spectra are interpreted within polar cap models as due to one-photon absorption of gamma-rays in strong magnetic field with subsequent $e^{\pm}$-pair creation. A piece of observational support for such an interpretation comes from a strong correlation between the inferred 'spin-down' magnetic field strength and the position of the high-energy cutoff [6]. This, in turn, opens a possibility that the observed asymmetry between LP and TP, i.e. the dominance of LP over TP above ~ 5 GeV, is a direct consequence of propagation effects (which eventually lead to stronger one-photon absorption for photons forming LP than TP) rather than due to some inherent property of the gamma-ray emission region itself. The aim of this research note is to investigate

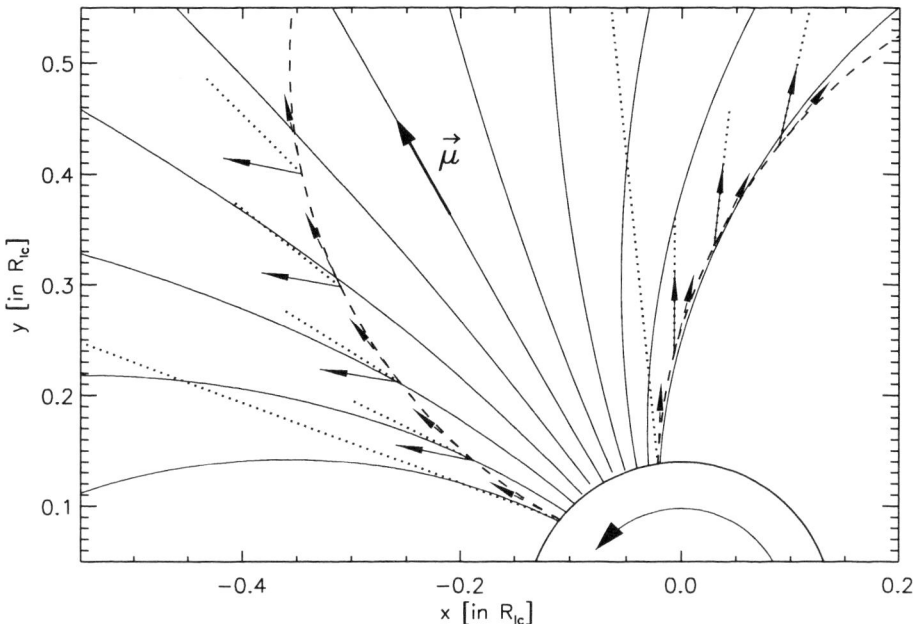

FIGURE 1. The view of the orthogonally rotating pulsar with the period $P = 1.5$ ms. The magnetic field lines correspond to the static-like dipole (see text for other details).

the role of rotation in the built-up of such asymmetry.

We consider the purely rotational effects: due to the presence of the rotation-induced electric field $\vec{E}$, aberration of photon direction and slippage of magnetosphere under the photon's path. All these effects are of the same order of magnitude: $\propto \beta$, where β is a local corotation velocity v expressed in units of the speed of light c. We assume that the magnetic field within the radius of a few pulsar radii has a shape of a rigidly rotating static-like dipole. In reality the rotation distorts this shape because of both retardation effects as well as toroidal currents due to plasma dragging. Fortunately, these are higher order effects ($\propto \beta^2$) and will be ignored below. Another expected disturbance of the magnetic field structure comes from longitudinal currents suspected to flow within the region of open field lines. No self-consistent solution of this problem has been found so far (see [1] and references therein). Nevertheless, the longitudinal currents are expected to modify the dipole magnetic field by a factor of $\propto \beta^{3/2}$, and therefore they will be neglected too.

IS THE LEADING PEAK ABSORBED MORE EFFICIENTLY?

Let us consider the photon propagation in the equatorial plane of an orthogonal rotator with rotation period $P = 1.5$ ms (this case is ideal for instructive purposes, and it is shown in Fig 1). For definitness, we constrain to its northern magnetic hemisphere. Then, at any point in the plane, photons propagating upwards cross the magnetic field $\vec{B}$ at angles $\theta_B < 90°$ and the rotation-induced electric field $\vec{E} = -\vec{\beta} \times \vec{B}$ is directed towards the reader, at right angle to the page.

One of the striking features of one-photon magnetic pair production is that its rate R is not axisymmetric around a local $\vec{B}$ direction if a weak electric field $\vec{E} \perp \vec{B}$ is present [3]. Within the accesible range of angles θ_B the rate vanishes in the unique direction which lies in the plane perpendicular to $\vec{E}$ and deviates from $\vec{B}$ by angle $\sim E/B$ towards the rotation direction. In local coordinate frame with $\hat{z} \parallel \vec{B}$, $\hat{y} \parallel \vec{E}$, and $\hat{x} \parallel \vec{E} \times \vec{B}$ this "free propagation" direction is given by $\hat{\eta}^{FP} = [\eta_x^{FP}, \eta_y^{FP}, \eta_z^{FP}] = [E/B, 0, (1-E^2/B^2)^{1/2}]$. If $E \ll B$ the pair production rate R is approximately symmetric around the free propagation direction $\hat{\eta}^{FP}$ instead of around $\vec{B}$. Moreover, R increases monotonically for angles which depart from $\hat{\eta}^{FP}$. Accordingly, the projection of photon momentum on $\hat{\eta}^{FP}$ is a better measure of R than its projection on $\vec{B}$. The directions of $\hat{\eta}^{FP}$ at various points within the magnetosphere are shown in Fig. 1 as solid arrows. Note that the free propagation direction deviates from the local $\vec{B}$ by an angle $\theta_B = \arcsin(E/B) \simeq E/B$ which increases with altitude.

High-energy photons are emitted from the outer rim of standard polar cap tangentially to the magnetic field in the corotating frame (dashed arrows at the star surface in Fig. 1). In the observer frame (OF) the photons propagate at the aberrated direction (dotted lines) which at the emission point is just the free propagation direction (at this point the angle between the photon direction and the magnetic field line equals $\theta_B \simeq E/B$, and therefore $R = 0$ [5], [7]). Initially, therefore, the rate R is symmetric for photons in the leading and in the trailing peak both in the corotating and in the observer frame. As the photons propagate outward, however, the free propagation direction starts to deviate from the photon direction $\hat{\eta}$ and this occurs in a different way for photons of the leading and the trailing peak. The reasons for which the local $\hat{\eta}^{FP}$ diverges from $\hat{\eta}$ include: (1) the magnetic field line curvature, (2) the increase in E/B ratio with altitude, and (3) the slippage of magnetic field lines under a photon's path. For photons in the leading peak the effects (1) and (2) cumulate whereas for the trailing peak they effectively tend to cancel out each other. In consequence, the photons in the leading peak suffer stronger absorption than the photons of the same energy in the trailing peak. This is why the high energy cutoff in the LP spectrum occurs at a slightly lower energy than the cutoff for the TP. The difference becomes more pronounced for smaller curvature of magnetic field lines. The slippage (3) does not change this picture.

Photon propagation direction $\hat{\eta}$ as seen in the observer frame (dotted lines) and the local free propagation direction $\hat{\eta}^{FP}$ in the OF (solid arrows) are shown in Fig. 1 for a few positions along the photon trajectory in the corotating frame. The stronger absorption of the leading peak is evident (for the TP, $\hat{\eta}$ nearly coincides with $\hat{\eta}^{FP}$).

Another way to understand the asymmetry in the pair production rate is to follow photon trajectories in a reference frame where $\vec{E}' = 0$ is fulfilled[1] and the pair pro-

---

[1] This filled and corotating magnetosphere assumption is an approximation within the polar gap where accelerating electric field is expected. Its strength is lower than the corotational electric field by $\beta^{1/2}$.

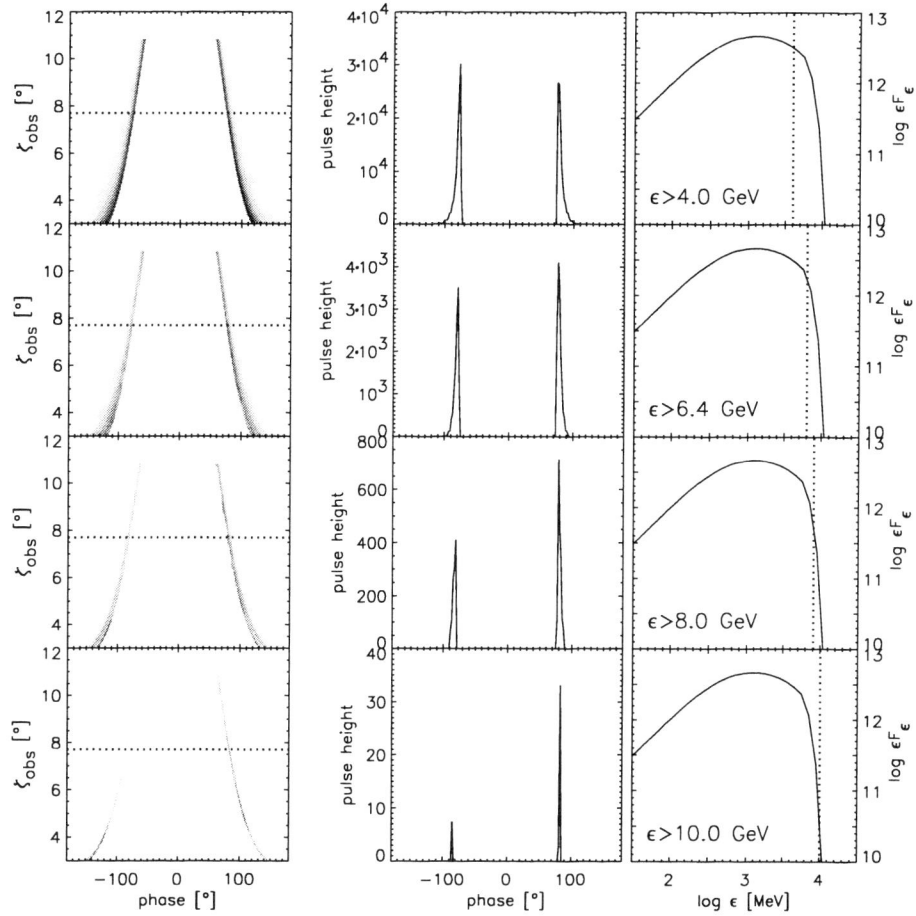

FIGURE 2. Evolution of double peak pulse profile with photon energy. For details see text.

duction rate R' can be approximated by the well-known formula for the pure-$\vec{B}$ case: $R'(\varepsilon', \vec{B}', \sin\theta'_B) = c_1 \sin\theta'_B B' \exp[-c_2/(\varepsilon' \sin\theta'_B B')]$ where $\theta'_B = \angle(\hat{n}', \vec{B}')$ and ε' is photon energy. If a frame of local $\vec{E} \times \vec{B}$ drift is chosen, with $\vec{\beta}_D = \vec{E} \times \vec{B}/B^2$, the rate R in the observer frame can be expressed as $R = c_1 \gamma_D (1 - \eta_x \beta_D) R'(\varepsilon', \vec{B}', \sin\theta'_B)$ where $\varepsilon' = \varepsilon \gamma_D (1 - \eta_x \beta_D)$, $\vec{B}' = \vec{B}/\gamma_D$, and $\sin\theta'_B = [(\eta_x - \beta_D)^2 + \eta_y^2(1 - \beta_D^2)]^{1/2}/(1 - \eta_x \beta_D)$. In the equatorial plane of the orthogonal rotator $\eta_y = 0$ so that $\eta_x = \mp \sin\theta_B$ where the signs 'minus' and 'plus' correspond to the leading and the trailing peak, respectively. Typically $\eta_x \ll 1$ which implies that the difference between the pair production rates in the locally drifting frame and in the observer frame results primarily from the aberration of photon direction whereas the change in ε or $\vec{B}$ is a second order effect. Obviously, the aberration is asymmetric for the leading and the trailing peak ($\eta_x < 0$ and $\eta_x > 0$ respectively). Fig. 1 presents this "aberration effect" in the rigidly corotating frame where

$\vec{E}' = 0$ is assumed. Photon trajectories in this frame are marked with dashed lines which indicate clearly that photons of the leading peak encounter larger $B'_\perp$ than photons of the trailing peak of the same energy.

We have performed Monte Carlo simulations of radiative processes above polar cap, including the emission of curvature radiation with subsequent one-photon pair production (for description of the model see [2]). We find that a difference by a factor of ~ 2 in the position of high-energy cutoffs for the leading and the trailing peak can be generated for Vela-like objects with emission regions placed a few stellar radii above the surface provided the inclination angle α of the magnetic dipole exceedes ~ 45 degrees. However, for such large α the observed peak-to-peak separation of about 0.4 cannot be reproduced [4]. Therefore, we now turn to the case of nearly aligned rotators. As an example we choose the model with parameters of the Vela pulsar ($B_{\text{dip}} = 3.4$ TG, $P = 0.0893$ s) and we take the inclination angle $\alpha = 7.6°$. We then place the polar accelerator at the altitude of $\sim 4\,R_{\text{NS}}$ to ensure that the magnetosphere is not entirely opaque to curvature photons of energy $\lesssim 10$ GeV. The numerical results are presented in Fig. 2. The three columns of Fig. 2 present: 1) the photon distribution in the observer's colatitude-phase space $(\zeta_{\text{obs}}, \phi)$ (left), 2) pulse profiles for $\zeta_{\text{obs}} = 7.7°$ integrated above a photon energy ε (middle), and 3) the phase-integrated spectrum with ε marked with the dotted vertical line (right). Four rows correspond to increasing energy of $\varepsilon = 4.0, 6.4, 8.0,$ and 10 GeV (top to bottom). The peak-to-peak asymmetry (middle column) does agree qualitatively with the observed data even in this nearly aligned case. However, the really strong fading of the leading peak LP occurs only at the very end of the curvature spectrum, where its level drops below the detectability level of EGRET.

Therefore, the rotational effects alone probably cannot account for the asymmetries observed in the bright EGRET pulsars. However, the photon statistics at the highest energy bins are too low to treat this conclusion as being firm. Alternatively, the asymmetries may be generated by some properties inherent in the region of the gamma-ray emission.

ACKNOWLEDGMENTS

We are grateful to V.S. Beskin for useful comments on the issue of magnetospheric distortions. This work was supported by KBN grant 2P03D02117 and NCU grant 405A.

REFERENCES

1. Beskin, V. S., *Physics-Uspekhi*, **42**, 1071–1098 (1999)
2. Daugherty, J. K., and Harding, A. K., *ApJ*, **252**, 337–347 (1982)
3. Daugherty, J. K., and Lerche, I., *ApSS*, **38**, 437–445 (1975)
4. Dyks, J., and Rudak, B., *MNRAS*, **319**, 477–483 (2000)
5. Harding, A. K., Tademaru, E., Esposito, L. W., *ApJ*, **225**, 226–236 (1978)
6. Thompson, D. J., astro-ph/0101039 (2001)
7. Zheng, Z., Zhang, B., and Qiao, G. J. *A&A*, **334**, L49–L52 (1998)

Resolving the Crab σ–Problem

D. Kazanas[1] and J. Contopoulos[2]

[1] *NASA/Goddard Space Flight Center, Code 661, Greenbelt, MD 20771*
[2] *200 Akti Themistokleous Str., Piraeus 18539, Greece*

Abstract. Using the exact solution of the axisymmetric pulsar magnetosphere derived in a previous publication and the conservation laws of the associated MHD flow, we show that the Lorentz factor of the outflowing plasma increases linearly with distance from the light cylinder. Therefore, the ratio of the Poynting to particle energy flux, generically referred to as σ, decreases inversely proportional to distance, from a large value (typically $\gtrsim 10^4$) near the light cylinder to $\sigma \simeq 1$ at a transistion distance R_{trans}. Beyond this distance the inertial effects of the outflowing plasma become important and the magnetic field geometry must deviate from the almost monopolar form it attains between R_{lc} and R_{trans}. We anticipate that this is achieved by collimation of the poloidal field lines toward the rotation axis, ensuring that the magnetic field pressure in the equatorial region will fall-off faster than $1/R^2$ (R being the cylindrical radius). This leads both to a value $\sigma = \sigma_s \ll 1$ at the nebular reverse shock at distance R_s ($R_s \gg R_{\text{trans}}$) and to a component of the flow perpendicular to the equatorial component, as required by observation. The presence of the strong shock at $R = R_s$ allows for the efficient conversion of kinetic energy into radiation. We speculate that the Crab pulsar is unique in requiring $\sigma_s \simeq 3 \times 10^{-3}$ because of its small translational velocity, which allowed for the shock distance R_s to grow to values $\gg R_{\text{trans}}$.

I INTRODUCTION

The most crucial element in the efficient conversion of the Crab (and other plerion) pulsar power into radiation is thought to be the presence of a (reverse) strong shock at an angular distance of 10″ (corresponding to a distance $\simeq 3 \times 10^{17}$ cm) from the location of the pulsar. This shock randomises the highly relativistic upstream MHD wind which is produced by the pulsar, thereby causing the wind to radiate away a major fraction of its available energy. The presence of this strong shock is predicated on the dominance of the relativistic MHD wind emanating from the pulsar by particles rather than magnetic field, i.e. that the magnetization parameter (defined below) at the shock distance has a value $\sigma_s \ll 1$.

The value of σ_s has been estimated e.g. by [8] from the detailed structure of the MHD flow downstream from the shock and concluded that matching the nebular expansion velocity at the nebular edge requires $\sigma_s \simeq 3 \times 10^{-3}$. Thia value is consistent with that obtained by [6], who used fits to the surface brightness of the nebula, under the assumption that the emission above 10 GeV is due to inverse Compton scattering of lower frequency photons, which presumably represent the synchrotron emission from the same electron distribution.

However, as shown in [15], matching the expansion velocity v_{ex} of the nebula at its edge at $R = R_N$ is just a statement of conservation of the momentum flux injected by the pulsar wind through an MHD shock at $R = R_s$, leading to $R_s/R_N \sim (v_{\text{ex}}/c)^{1/2}$, independent of the value of σ_s. In [8] it was shown that, if in addition $\sigma_s \ll 1$, one obtains $v_{\text{ex}}/c \sim \sigma_s$; however, the latter is not a condition necessary for matching the nebular expansion velocity to that of the MHD wind at R_s.

The values of σ_s inferred for the Crab (and also other plerions, e.g. Vela) MHD winds raise the following problem: the value of this parameter near the pulsar light cylinder is estimated to be quite high $\sigma \sim 10^{4-5}$ [4]. Given that in a MHD wind $B_\phi \propto 1/R$, it is thought that both the magnetic and ram pressures should decrease like $1/R^2$, with their ratio thus remaining roughly constant at the value it attains near the light cylinder. Therefore, values $\sigma_s \sim 1$ (let alone $\sigma_s \sim 10^{-3}$) are hard to understand and yet overwhelmingly favored by observation.

This problem led to the suggestion that annihilation of magnetic field energy and conversion of the resulting energy into that of the outflowing particles could indeed provide for the required reduction in σ_s with distance [4,14]. Such a solution is in principle possible (see though [9]), however, this process would work only on the magnetic dipole field component perpendicular to the direction of the pulsar angular velocity Ω. The component of the magnetic dipole field which is parallel to Ω is simply advected away with no possibility of such an annihilation. Since the observations [1] seem to suggest that, at least for the Crab, the magnetic dipole is closely aligned with the pulsar rotation axis, it appears unlikely that a large fraction of the available magnetic energy could in fact annhilate.

Motivated by our recent exact solution of the axisymmetric pulsar magnetosphere [3], we have decided to take a closer look at the problem of the entire MHD wind and its impact on the nebular morphology and dynamics. The exact solution of [3] provides the complete, global, magnetic field and associated electric current structure for an aligned rotator in the force free (i.e. with negligible inertia) MHD approximation, including their distribution across the crucial light cylinder surface.

II THE σ-PROBLEM AND ITS RESOLUTION

We provide below a summary of our knowledge of axisymmetric pulsar magnetospheres (the most likely configuration of the Crab [1]) based on the solution given in [3] and using the Crab pulsar values as fiducial figures.

The field lines that cross the light cylinder emanate from a region near the pole, the polar cap, and are necessarily open. We calculate the polar cap radius to be equal to

$$R_{pc} = \sqrt{1.36}\, r_* \left(\frac{r_*}{R_{lc}}\right)^{1/2} = 0.9 \left(\frac{P}{33\text{ ms}}\right)^{-1/2} \text{ km} \qquad (1)$$

Here, $r_* = 10$ km is the canonical radius of a neutron star, and

$$R_{lc} = \frac{cP}{2\pi} = 1576 \left(\frac{P}{33 \text{ ms}}\right) \text{ km} \tag{2}$$

is the light cylinder radius (P the period of the neutron star rotation)[1]. At the footpoints of the magnetic field lines on the polar cap, the magnitude of the magnetic field B_* is of the order of 10^{12} G, the number density n_* of electrons/positrons in the outflowing wind is equal to

$$n_* = \kappa n_{GJ} \equiv \kappa \frac{B_*}{ePc} = 2 \times 10^{16} \left(\frac{\kappa}{10^4}\right) \left(\frac{B_*}{10^{12} \text{ G}}\right) \left(\frac{P}{33 \text{ ms}}\right)^{-1} \text{ cm}^{-3}, \tag{3}$$

where $\gamma_* \sim 200$, $\kappa \sim 10^{3-4}$ are respectively their Lorentz factor and multiplicity, adopted from the cascade models of [5] (i.e. from physics outside the context of ideal MHD) and e is the electron charge.

The open field lines contain an amount of magnetic flux

$$\Psi_{op} = \pi R_{pc}^2 B_* = 2.7 \times 10^{12} \left(\frac{P}{33 \text{ ms}}\right)^{-1} \left(\frac{B_*}{10^{12} \text{ G}}\right) \text{ G km}^2 \tag{4}$$

and an electron/positron wind with mass loss rate

$$\dot{M} = \pi R_{pc}^2 n_* m_e = 7.8 \times 10^{-31} \kappa \left(\frac{P}{33 \text{ ms}}\right)^{-2} \left(\frac{B_*}{10^{12} \text{ G}}\right) M_\odot \text{ yr}^{-1} \tag{5}$$

from each polar cap (m_e is the electron rest mass). The wind carries a kinetic energy flux

$$W_{\text{Kinetic}} = \gamma \dot{M} c^3 = 7 \times 10^{-5} \kappa \left(\frac{\gamma}{200}\right) \left(\frac{P}{33 \text{ ms}}\right)^{-2} \left(\frac{B_*}{10^{12} \text{ G}}\right) L_\odot, \tag{6}$$

and the magnetic field carries a Poynting flux

$$W_{\text{P}} = \frac{\Omega}{2\pi c} \int_0^{\Psi_{op}} I(\Psi) d\Psi = 10^4 f \left(\frac{I}{I_*}\right) \left(\frac{P}{33 \text{ ms}}\right)^{-4} \left(\frac{B_*}{10^{12} \text{ G}}\right)^2 L_\odot \tag{7}$$

from each polar cap. Here, $I_* \equiv \Omega \Psi_{op}/2 \equiv e n_{GJ} c \cdot \pi R_{pc}^2/4$ and I are the total amount of electric current flowing through the polar cap and the magnetosphere respectively. f is a factor of order unity which depends on the distribution of the electric current $I(\Psi)$ across open field lines ($f = 0.67$ for the numerical solution of [3]). The energy reservoir is obviously the neutron star spindown energy loss rate

$$W_{\text{Spindown}} = W_{\text{Kinetic}*} + W_{\text{Poynting}*} = W_{\text{Kinetic}} + W_{\text{Poynting}} \tag{8}$$

at all distances. The magnetization parameter σ is thus defined as

---

[1] Henceforth, we will denote cylindrical radii with capital R, and spherical radii with small r.

$$\sigma \equiv \frac{W_{\text{Poynting}}}{W_{\text{Kinetic}}} = \frac{\Psi_{\text{op}} I}{Pc\gamma \dot{M}c^3} = \left(\frac{I}{I_*}\right)\left(\frac{\gamma_*}{\gamma}\right)\sigma_* . \tag{9}$$

Here, σ_* is the value of the magnetization parameter near the surface of the neutron star, or equivalently near the light cylinder.

Using eq. (8) divided through with $W_{\text{Kinetic}*}$ one obtains

$$\sigma_* + 1 = \frac{\gamma}{\gamma_*}(\sigma + 1) , \tag{10}$$

It is apparent from eq.(10), that a decrease in σ must be accompanied by a concomitant increase in γ. Following the work of [3], we know that acceleration in dissipation zones near the light cylinder are not convincing anymore, while models showing MHD acceleration at large scales [16] define a-priori the field geometry. To answer this question we turn to the study of the basic equations of the problem, focusing our analysis on the energy flux conservation equation along open field lines.

Energy flux conservation implies that

$$\gamma\left(1 - \frac{R}{R_{lc}}\frac{v_\phi}{c}\right) = \gamma_* \tag{11}$$

along any open field line (e.g. [10,2]), with γ_* the initial value of the electron Lorentz factor. This is just the differential form of the energy flux conservation equation (8). The induction equation further gives that

$$\frac{v_\phi}{c} = \frac{R}{R_{lc}} + \frac{v_p}{c}\frac{B_\phi}{B_p} . \tag{12}$$

In order to simplify the notation, we will concentrate our discussion on the last open field line along the equator. For force–free conditions one can show that

$$B_\phi = -\frac{R}{R_{lc}} B_p \tag{13}$$

when $R \gg R_{lc}$. This is identically valid in the analytic monopole solution [13] and in the asymptotically monopole-like part of the dipole solution of [3]. Eqs. (11) and (12) then yield

$$\gamma\left[1 - \left(\frac{R}{R_{lc}}\right)^2\left(1 - \frac{v_p}{c}\right)\right] = \gamma_* ,$$

which further yields

$$\gamma = \left[\gamma_*^2 + \left(\frac{R}{R_{lc}}\right)^2\right]^{1/2} \to \frac{R}{R_{lc}} \tag{14}$$

for $R \gg R_{lc}$. This is a very important result which indicates that, under an aligned dipole geometry, the wind's Lorentz factor increases linealy with the (cylindrical) distance.

The linear growth, however, cannot continue beyond a distance

$$\gamma_* \sigma_* R_{lc} = 2 R_{\text{trans}} = 3 \times 10^6 R_{lc} \ll r_s ,\quad (15)$$

(r_s is the spherical shock radius) at which the Lorentz factor reaches the asymptotic value implied by mass conservation and the observed spin-down luminosity. The problem we are presented with has arisen from our neglect of matter in our assumption of negligible inertia (i.e. force–free) conditions. Note that eq. (13) is not valid at the distance where inertial and magnetic forces are comparable.

In this case one can write for the ratio of σ's

$$\frac{\sigma_s}{\sigma_{\text{trans}}} = \sigma_s = \frac{I_s}{I_{\text{trans}}} \frac{\gamma_{\text{trans}}}{\gamma_s} = \frac{1}{2} \frac{I_s}{I_{\text{trans}}} = \frac{1}{2} \frac{(RB_\phi)|_s}{(RB_\phi)|_{\text{trans}}} = 3 \times 10^{-3} \quad (16)$$

It can be shown that, when $\sigma \ll 1$, $B_\phi \to -(R/R_{lc}) B_p$, and thus

$$\frac{(R^2 B_p)|_s}{(R^2 B_p)|_{\text{trans}}} \sim 6 \times 10^{-3} .\quad (17)$$

In other words, at $R > R_{\text{trans}} \sim 5 \times 10^5 \, R_{lc}$, $B_p R^2$ decreases with distance, and consequently, *the field/flowlines diverge away from monopolar towards the axis of symmetry.* Therefore, the field energy density decreases faster than that of the particles with (the cylindrical) R, yielding the observed value for σ.

REFERENCES

1. Aschenbach, B. & Brinkmann, W., *A& A*, **41**, 147 (1975)
2. Contopoulos, J., *ApJ*, **446**, 67 (1995)
3. Contopoulos, J., Kazanas, D. & Fendt, C., *ApJ*, **511**, 351-358 (1999)
4. Coroniti, F. V., *ApJ*, **349**, 538 (1990)
5. Daugherty, J. K. & Harding, A. K., *ApJ*, **252**, 337 (1982)
6. De Jager, O. C. & Harding, A., *ApJ*, **396**, 161 (1992)
7. Harding, A. K., *ApJ*, **245**, 267-273 (1981)
8. Kennel, C. F. & Coroniti, F. V., *ApJ*, **283**, 694 (1984)
9. Lyubarsky, Y. & Kirk, J. G, *ApJ*, **547**, 437 (2001)
10. Mestel, L & Shibata, S., *MNRAS*, **271**, 621-638 (1994)
11. Michel, F. C., *ApJ*, **180**, L133-L137 (1973)
12. Michel, F. C., *ApJ*, **180**, 207-225 (1973)
13. Michel, F. C., 1991, Theory of Neutron Star Magnetospheres (University of Chicago Press: Chicago)
14. Michel, F. C., *ApJ*, **431**, 397 (1994)
15. Rees, M. J., & Gunn, J., *MNRAS*, **167**, 1 (1974)
16. Takahashi, M. & Shibata, S., *PASJ*, **50**, 271 (1998)

RXTE Observations of the Vela Pulsar: The Pulsar Rosetta Stone

M.S. Strickman

Naval Research Laboratory, Washington DC

A.K. Harding

NASA Goddard Space Flight Cente,Greenbelt, MD

C. Gwinn

University of California Santa Barbara, Santa Barbara, CA

P. McCulloch, D. Moffett

University of Tasmania, Tasmania, Australia

Abstract. We report on our analysis of a 274 ks observation of the Vela Pulsar with the Rossi X-Ray Timing Explorer (RXTE). The double-peaked, pulsed emission at 2-30 keV, which we had previously detected during a 93 ks observation, is confirmed with much improved statistics. There is now clear evidence, both in the spectrum and the light curve, that the emission in the RXTE band is a blend of two separate components. The spectrum of the harder component connects smoothly with the OSSE, COMPTEL and EGRET spectra and the peaks in the light curve are in phase coincidence with those of the high-energy light curve. The spectrum of the softer component is consistent with an extrapolation to the pulsed optical flux and the soft component of the second RXTE peak is in phase coincidence with the second optical peak. In addition, we see a peak in the 2-8 keV RXTE light curve at the radio peak phase.

PSR B0833-45: THE VELA PULSAR

The Vela Pulsar is the brightest celestial gamma-ray source in the sky, but has traditionally been a difficult object to detect in X-rays. It is intrinsically faint at these energies and is embedded in a bright synchrotron nebula that represents additional background for the pulsar itself. The first firm X-ray detection of pulsed emission from the Vela Pulsar was by ROSAT in the 0.1-2 keV band [1]. The pulsed spectrum was described by a black body model with a pulsed fraction of about 11% for energies less than 1.2 keV.

The first convincing hard X-ray/soft gamma-ray observations were by the OSSE [2] and COMPTEL [3] instruments on CGRO. The total pulsed spectra measured by these instruments in the 50 keV – 30 MeV band is harder than the benchmark CGRO/EGRET spectrum at higher energies [4]. Hence, the general appearance of the spectrum in the combined CGRO energy range is a very hard shape in the OSSE range ($\Gamma=1.3$) rolling off to a rather softer power law in the EGRET range ($\Gamma=1.7$). Neither OSSE nor COMPTEL had adequate sensitivity to measure phase-resolved spectra, but EGRET found that best fit power law models had photon indices ranging from 1.4 to 2.2.

The behavior of the pulsar light curve measured previous to RXTE seemed to have three "regimes" consisting of the single-peaked radio light curve, the thermal X-ray light curve, and the double-peaked optical and gamma-ray light curves. The peak separation in the optical case is smaller than in the gamma-ray and the optical peak phases differ from the gamma-ray phases. The ROSAT light curve is rather different, with a broad emission region roughly spanning the entire gamma-ray "on-pulse" region.

These observations lead to some obvious questions: What happens to the pulsed spectrum between optical and hard X-ray bands? Where does the switch between "optical" light curve behavior and "gamma-ray" light curve behavior occur? Whither the radio pulse?

In order to address these and other concerns, we have performed two long observations of the Vela Pulsar with the PCA instrument on RXTE. The first observation of 93 ks, performed in January 1997, was reported by Strickman, et al [5]. Using epoch-folding analysis with the standard RXTE ftools and deriving a spectrum based on "on-pulse" minus "off-pulse" rates, we found light curves grossly similar to gamma-ray light curves and a hard total pulsed spectrum with substantial variations from phase to phase. We speculated that "peak 2", the peak corresponding to the second gamma-ray peak, may have multiple components and that some of these components might be "optical-like" while others might be "gamma-ray-like". However, statistics were inadequate to prove either claim.

Our second observation of 274 ks, during April/May and July/August 1998, had sufficient statistics to confirm our speculations. We describe this observation in the next section.

RESULTS

For examination of light curve behavior, we summed our epoch-folded 1998 data into three broad energy bands. The light curves for each of these bands, together with those at lower and higher energies are shown in Fig. 1. The first X-ray peak clearly aligns with the first gamma-ray peak in all three broad bands (as indicated by the left-most dotted line). The second X-ray peak is clearly resolved into two peaks with phase separation 0.09 ± 0.01. As shown by the third and fourth dotted lines, the lower phase, softer component aligns with the optical second peak while the higher phase, harder component aligns with the second gamma-ray peak. Note also that there are no features in the RXTE light curves corresponding to the first optical peak (second

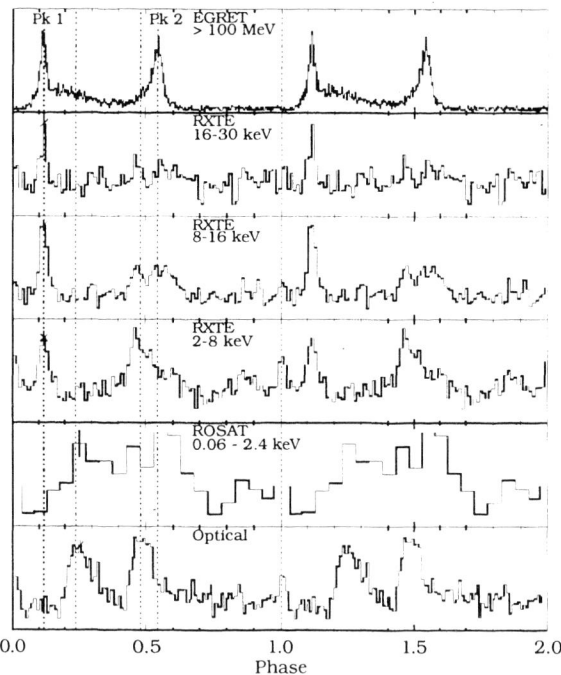

FIGURE 1. Broad band light curves from EGRET [4], the 1998 RXTE observation, ROSAT [1] and optical [6]. In each case, the radio peak is centered at phase 0.0.

dotted line). Unexpectedly, we also see a definite feature in the 2-8 keV band light curve at the radio phase. This feature is relatively soft and does not appear at higher energies. Note that it is also present in the optical light curve.

Phase resolved spectra for the 1998 observation were derived in a different fashion from the earlier RXTE result. Since we wanted to separate the spectra from the two components of the second peak, we fit sinusoid functions to each component for each native RXTE channel. Since the per-channel statistics are rather limited, we constrained sinusoid center phase and width by functions derived from similar fits to the broad-band light curves, fitting amplitude only for individual channels. Spectra were then derived from the sinusoids by integration.

Phase resolved spectra are shown in Fig. 2, together with optical, ROSAT and OSSE spectra for context. Statistically, the firmest conclusion to be drawn is that the peak one spectrum is considerably harder than the rest. The peak 2 soft component appears to be softer than the hard component, but it is not well represented by a simple power law model. The difference in hardness between the two is only marginally significant. The radio peak spectrum is similar in shape to the peak 2 soft spectrum. Note also, that these two components extrapolate approximately to the optical points.

The total pulsed spectrum from RXTE and a number of other measurements is shown in Fig. 3. The RXTE result (whose complex shape is due to the superposition of the varying phase-resolved spectra), smoothly fills the gap between thermal and hard X-ray spectra. Comparison to the Chandra "point source" (i.e. pulsed plus

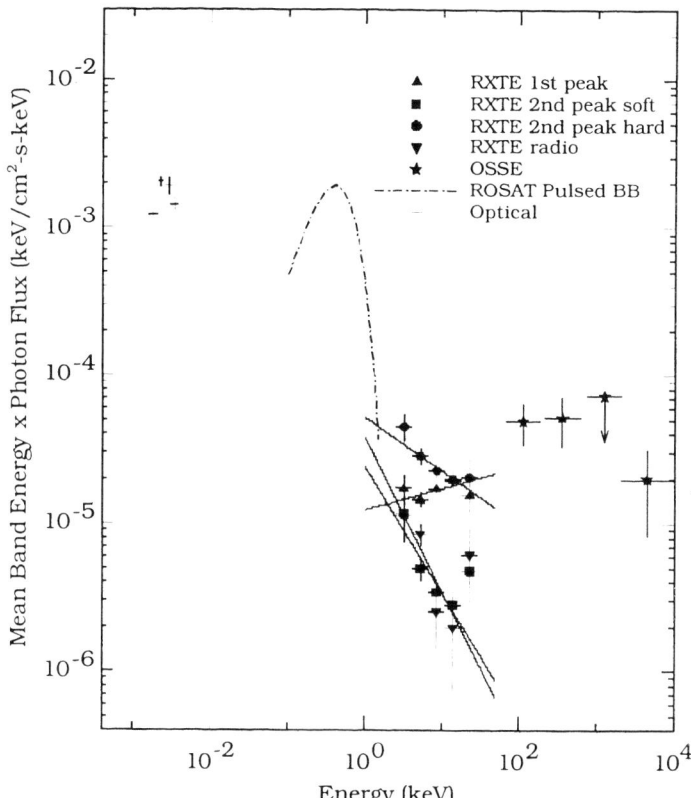

FIGURE 2. RXTE phase-resolved spectra shown with optical [6], ROSAT [1] and OSSE [2] total pulsed spectra.

unpulsed minus nebular and other background) spectrum [7] indicates rough agreement in spectrum shape and a much higher pulsed fraction than in the thermal component measured by ROSAT.

DISCUSSION

Our 1998 observations of the Vela Pulsar confirm our speculations from our previous observation concerning the multicomponent nature of the second light curve peak. The 2-30 keV band exhibits features of emission from radio through high energy gamma rays and is clearly a transitional region between lower and higher energy emission processes. Behavior of the observed peaks 1 and 2 components and the feature at the radio phase are correlated in phase and spectrum shape. Soft components are phase-aligned with features at lower energies, hard components with features at higher energies.

The complex behavior of nonthermal pulsed X-ray emission from Vela presents challenges to modeling the global spectrum (e.g. [8]). This band, when combined with

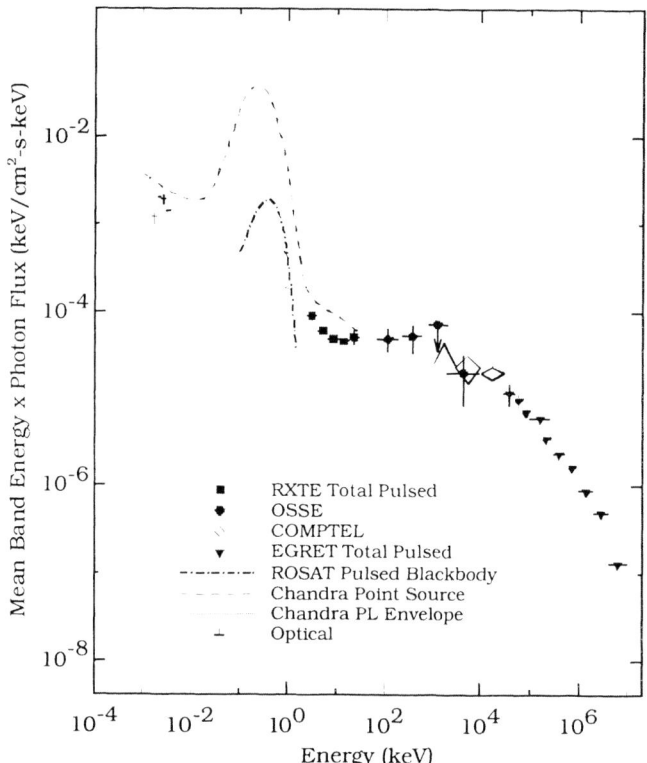

FIGURE 3. Total pulsed spectrum of the Vela Pulsar.

other spectral features will pose an important constraint on physical parameters such as emission altitude. None of the existing models are yet sufficiently refined to make predictions concerning phase-resolved spectrum behavior in this band, but these observations will supply an impetus to move in that direction.

REFERENCES

1. Ogelman, H., in *Lives of Neutron Stars*, edited by M.A. Alpar et al., Dordrecht: Kluwer, 1993, p. 101.
2. Strickman, M.S. et al.., *Ap. J.* **460**, 735 (1996).
3. Bennett, K. et al., *Ap J. Suppl.* **90**, 823 (1994).
4. Kanbach, G. et al., *A.&A.* **289**, 855 (1994).
5. Strickman, M. et al., *Ap. J.* **574**, 373 (1999).
6. Nasuti, F.P. et al., *A.&A.* **323**, 839 (1997).
7. Pavlov, G. et al., *Ap. J. in press* (2001).

BeppoSAX observations of the γ-ray pulsars PSR B0656+14 and PSR B1706−44

E. Massaro*, T. Mineo†, G. Cusumano†, B. Sacco† and W. Becker**

*IAS-CNR and Univ. La Sapienza, Roma, Italy
†IFCAI-CNR, Palermo, Italy
**MPI-IEP, Garching, Germany

Abstract. The Italian-Dutch satelliet for X-ray astronomy BeppoSAX observed the two γ-ray pulsars PSR B0656+14 and PSR B1706-44 in March 1999. Both sources were detected in the LECS and MECS images. No evidence of modulation with the pulsars' period was found. The X-ray spectrum of PSR B0656+14 is complex and requires a three component model, while that of PSR B1706-44 can be fitted by a single power law.

INTRODUCTION

The study of the X-ray emission of γ-ray pulsars is important for the understanding the physical processes occurring in the magnetosphere of these sources. Several models predict the production of a large number of secondary electron-positron pairs originated by the interaction of primary high energy curvature photons with the intense magnetic field. These pairs are expected to radiate X rays via the synchrotron mechanism. Other important emission processes of X-ray photons are the thermal (blackbody) radiation from the polar caps, heated by the impinging of high energy particles, and the inverse Compton effect. Finally, in the case of young pulsars, X rays can be emitted in a compact synchrotron nebula surrounding the neutron star. In order to discriminate among these possibilities it is important to perform very detailed and precise observations of several sources of this class. In this contribution we present the preliminary results of the spectral analysis of the X-ray emission from two EGRET γ-ray pulsars observed by the Italian-Dutch satellite BeppoSAX.

PSR B0656+14 was discovered by Manchester et al. (1978) and, in the X-ray band, a pulsed signal at the radio period was observed by Cordova et al. (1989). γ-ray emission in the EGRET band has been reported by Ramanamurthy et al. (1996). ROSAT PSPC observations (Finley et al. 1992) showed that the 0.1–2.4 keV spectral distribution can be well fitted by two blackbodies or by a blackbody plus a power law. Greiveldinger et al. (1996), on the basis of ASCA and ROSAT data, proposed a three component model (two blackbodies plus a power law), while Wang et al. (1998) found that only two blackbodies without a power law are sufficient to have a satisfactory spectral fit.

PSR B1706−44, a young pulsar with a period of 0.102 s was discovered in the radio band by Johnston et al. (1992) and an unpulsed X-ray was first detected by Becker et al. (1995) with ROSAT PSPC. In the more recent X-ray observations performed with ASCA

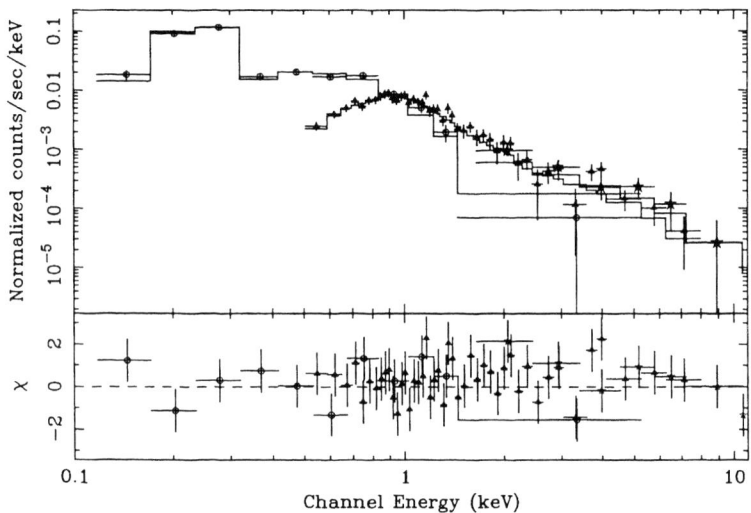

FIGURE 1. Spectral fit of the LECS (open circles), MECS (stars) and ASCA GIS (triangles) spectra of PSR B0656+14 with two blackbody plus power-law model

TABLE 1. Best fit values for the three component spectrum of PSR B0656+14

Parameter (unit)	Value
N_H (10^{20} cm^{-2})	3.4±1.1
kT_1 (keV)	(5.89±0.48)×10^{-2}
Norm$_1$ (erg s^{-1} kpc^{-2})	(3.64±0.11)×10^{-4}
kT_2 (keV)	0.12 ±0.01
Norm$_2$ (erg s^{-1} kpc^{-2})	(4.26±0.13)×10^{-5}
Photon Index	2.10 ±0.23
PL Norm.	(6.61±0.86)×10^{-5}

(SIS+GIS) (Finley et al. 1998) and RossiXTE (Ray et al. 1999) no pulsed emission was detected. SIS and GIS spectra in the (0.5–5) keV range were fitted by a power law with a photon index ranging from 1.6 to 1.9 and column densities of (1.3–2.2) 10^{21} cm^{-2}, but these values are poorly constrained because of their quite large (1 standard deviation) uncertainties of about 0.3 and 1.3 10^{21} cm^{-2} (and even more), respectively. A spectral fit with the higher column density fixed at 5 10^{21} cm^{-2} gave the steeper photon index of 2.3 ± 0.3. A γ-ray source was detected by COS B (Swanenburg et al. 1981) and the pulsation at energies greater than about 50 MeV was found by EGRET-CGRO (Thompson et al. 1992). An unpulsed source at TeV energies has been detected by the CANGAROO collaboration (Kifune et al. 1995). McAdam et al. (1993) proposed a possible association of PSR 1706−44 with SNR G343.1-2.3, but the VLA images by

Frail et al. (1994) indicated that it may be located inside a plerionic nebula. Evidence for a X-ray compact nebula (with a radius of about 27") was also found by Finley et al. (1998) in a ROSAT-HRI image.

OBSERVATIONS AND DATA REDUCTION

BeppoSAX observed PSR B0656+14 and PSR B1706−44 in 1999: the former from March 9 to 11 and the latter from March 29 to 31. The LECS and MECS images show sources at positions fully compatible with the radio coordinates. In the case of PSR B0656+14 the events for the time and spectral analysis were selected within circular regions, centred at the radio position, with radii of 4' and 3' for the LECS and MECS, respectively. The background was estimated from annular regions in the same fields and from a collection of blank field images. The the local background evaluation for PSR B1706−44 was more difficult because of the presence of the near bright LMXRB 4U 1705-44. In order to get a reliable evaluation of the local background in the MECS image, we computed the count level in a series of adjacent small circular regions, radially located with respect to the binary source in the pulsar direction. Then we fitted these values with an analytical formula excluding the pulsar region, and we assumed the value interpolated at its position as the background estimate. The same procedure applied to the LECS data, which have a much poorer statistics and a wider PSF, gives a detectable signal only in the energy channels up to about 1.5 keV. We therefore considered only the low energy photons included in a single bin from 0.1 to 1.5 keV.

We also searched for pulsed emission from both sources. We used the folding technique with extrapolated radio ephemeris to the observations epochs and also a period search using the Z^2 statistics with one and two harmonics, but no significant signal was detected, even considering various energy ranges.

Finally, because the spectrum of PSR B0656+14 resulted quite complex, to obtain a more accurate estimate of the various components we considered another observation of this pulsar, performed by ASCA on 1998 October 11 and available from the archive, and joined it to our BeppoSAX data.

SPECTRAL ANALYSIS

PSR B0656+14

On the basis of literature results, we used multicomponent spectral models to fit the LECS and MECS data. The fits with an absorbed blackbody + powerlaw and with two blackbodies gave the not acceptable values of the reduced χ^2 of 2.98 (13 d.o.f.) and 4.10 (12 d.o.f.), respectively, while a power law plus two blackbodies gave the better value of 1.21 (11 d.o.f.). The absorbing column density resulted equal to $(3.1 \pm 0.8)\, 10^{20}$ cm^2 and the photon index of the power law equal to 2.08 ± 0.41, while the blackbodies' parameters were poorly constrained because of the limited statistics. We, therefore, added to the BeppoSAX data set a 153 ks long ASCA (GIS) observation and performed a joint spectral analysis. Any attempt to obtain an acceptable fit with only two components

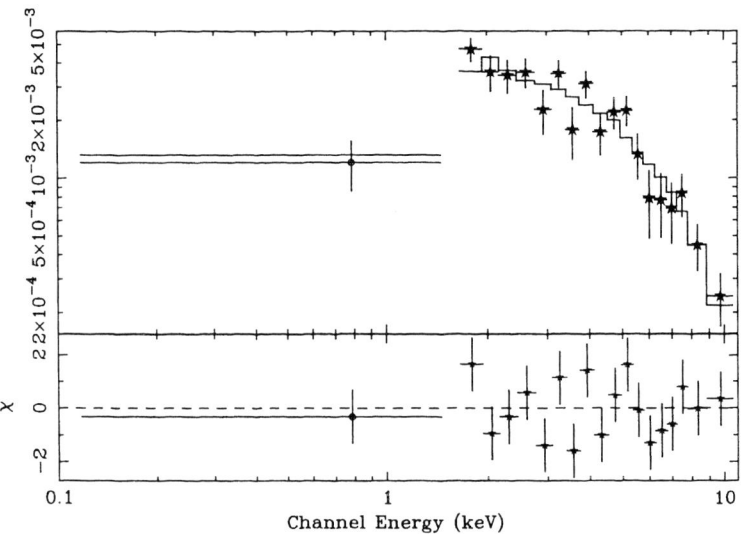

FIGURE 2. Spectral fit of the LECS and MECS spectrum of PSR B1706-44 with a power-law model

failed and a good fit was reached again with two blackbody distributions plus a power law ($\chi_r^2 = 0.97$, 55 d.o.f.), confirming that the X-ray spectrum of this source is complex (see Fig. 1). Our spectral results, reported in Table 1, are generally in agreement with the previous ones: typical differences in the blackbody temperatures are of the order of 20 % while a difference of a factor of two is between our estimate of N_H is and that of Greiveldinger et al. (1996). Similar results have been recently obtained by Zavlin, Pavlov and Halpern (2001) using the same ASCA data added to a ROSAT observation.

PSR B1706-44

The spectral analysis of the LECS and MECS data confirmed that the spectrum of this source can be well described by a single power law. The fit in the energy range 1.6–9 keV, for which the evaluation of the local background contribution due to the near LMXRB is easier, gave a photon index of 1.66 ± 0.13 and a column density $N_H = (2.6 \pm 1.5)\, 10^{21}$ cm^{-2}, with a reduced $\chi^2 = 1.28$ (15 d.o.f.) (Fig. 2). The residuals show an irregular scatter with respect to the power law continuum, but this may be an effect of the model adopted for the local background evaluation. Using two times wider energy bins, this scatter disappears and the reduced χ^2 lowers to a value smaller than unity but the spectral parameters remain unchanged. The power law best fit with the N_H value fixed at the ROSAT-PSPC result of $5\, 10^{21}$ cm^{-2} (Becker et al. 1995) gave a photon index of 1.72 ± 0.12 (reduced $\chi^2 = 1.25$, 16 d.o.f.), practically coincident with the previous value. Furthermore, to be more confident that this result was independent of the local background, we used different estimates of its intensity and spectrum derived from the modified models for the LMXRB contribution. The changes of the best fit spectral parameters' values were always smaller than the statistical uncertainties. We consider this result the best available information on the X-ray spectrum PSR B1706−44: in particular, the our statistical error of the photon index is about a factor of 3 smaller than that of Finley et al. (1998). Our result then confirms that this radiation is originated in the compact synchrotron nebula around the pulsar.

REFERENCES

Becker W., et al. 1995, A&A 298, 528
Cordova F.A., et al. 1989, ApJ 345, 451)
Finley J.P., et al. 1992, ApJ 394, L21
Finley J.P., et al. 1998, ApJ 493, 884
Frail D.A., et al. 1994, ApJ 437, 781
Greiveldinger C., et al. 1996, ApJ 465, L35
Johnston M., et al. 1992, MNRAS 225, 401
Kifune T., et al. 1995, ApJ 438, 91
McAdam D., et al. 1993, Nature 361, 516
Manchester R.N., et al. 1978, MNRAS 185, 409
Ramanamurthy P.V., et al. 1996, ApJ 458, 755
Ray R.D., et al. 1999, ApJ 513, 919
Swanenburg B.N., et al. 1981 ApJ 450, 784
Thompson D.J., et al. 1992, Nature 359, 615
Wang F.J., et al. 1998, ApJ 498, 373
Zavlin V.E., Pavlov G.G., Halpern J.P., 2001, preprint

Galactic Population of Radio and Gamma-Ray Pulsars

Peter L. Gonthier*, Michelle S. Ouellette[†], Shawn O'Brien[‡], Joel Berrier* and Alice K. Harding[||]

Hope College, Department of Physics, 27 Graves Place, Holland, MI 49422-9000
[†]*Michigan State University, Physics and Astronomy Department, East Lansing, MI 48824-1116*
[‡]*University of Notre Dame, Department of Physics, Notre Dame, IN 46556*
[||]*NASA - Goddard Space Flight Center, LHEA, Greenbelt, MD 20771*

Abstract. We simulate the characteristics of the Galactic population of radio and gamma-ray pulsars using Monte Carlo techniques. At birth, neutron stars are spatially distributed with supernova-kick velocities in the Galactic disk and randomly dispersed in age over the past 10^9 years. From their birth location, the neutron stars are evolved in the Galactic gravitational potential to the present time. With a radio luminosity model, we estimate the radio flux and filter each pulsar through a selected set of radio-survey parameters determining a flux threshold. Using the features of recent polar cap acceleration models invoking space-charge-limited flow, a pulsar death region further attenuates the population of radio-loud pulsars, and gamma-ray luminosities are assigned. Assuming a featureless emission geometry of 1 steradian and with an alignment of the radio and gamma-ray beams, and for the case of no field decay and a death valley, our model predicts that EGRET should have seen 10 radio loud and 2 radio quiet, gamma-ray pulsars. GLAST, on the other hand, is expected to observe 93 radio-loud and 87 radio-quiet, gamma-ray pulsars of which 8 are expected to be identified as pulsed sources.

With the advent of the *Compton Gamma Ray Observatory* (CGRO), the number of γ-ray pulsars has grown to eight, with several additional candidates. We anticipate that many more pulsed sources will be added to the list with the future telescope, *Gamma-Ray Large Area Space Telescope* (GLAST) scheduled for launch in late 2005. Among the known γ-ray pulsars, only Geminga is radio-quiet or at least radio weak [1,2]. Of the 271 sources listed in the Third EGRET Catalog [3], about 170 of these γ-ray point sources have not been identified with sources at other wavelengths. Recently Grenier & Perrot [4] and Gehrels et al. [5] suggested that some of these unidentified sources are correlated with the Gould Belt of massive stars from a nearby Galactic structure consisting of an expanding disk of gas with young stars (≤ 30 million years) inclined about $20°$ to the Galactic plane. Harding & Zhang [6] suggest that some of the sources associated with the Gould Belt are indeed radio-quiet, off-beam γ-ray pulsars seen at large angles to the magnetic pole.

We develop a model to simulate the production of neutron stars within the Galaxy, evolving their trajectories, periods and period derivatives from their birth forward in time to the present. Assuming that the radio and γ-ray beams are aligned, we supply the radio and γ-ray characteristics to each neutron star and filter its properties through a set of radio surveys and γ-ray thresholds (in and out of plane) associated with EGRET and expected for GLAST. These γ-ray thresholds correspond to the flux required for the instrument to identify the object as a point source, whereas higher thresholds apply to identify the object independently as a pulsed source. The birth rate of neutron stars is assumed to be constant during the history of the Galaxy, and the age of the pulsar is randomly selected from the present to 10^9 years in the past. We have taken the modified expressions describing the γ-ray luminosity from the work of Zhang & Harding [7] where a polar-cap model simulates the pair cascade region with curvature radiation or inverse Compton scattering of the primary particles and synchrotron radiation and inverse Compton scattering of subsequent higher generation pairs. This model uses the self-consistent acceleration model of Harding & Muslimov [8] to produce the primary particles. For the case of no field decay, we find the need to introduce a death valley between the dipole and multipole death lines, suggested by the recent work of Zhang, Harding & Muslimov [9], in order to reduce an excess of pulsars that would otherwise accumulate near the multipole death line. In addition, we consider the case in which the magnetic field is allowed to decay as suggested by several studies by Tauris & Konar [10] and Goldreich & Reisenegger [11]. We find that good agreement is achieved with a decay constant of 5×10^6 years. With field decay, the death valley was no longer required to obtain fairly good agreement with the observed pulsar population.

In Figure 1, we compare in a $(\dot{P}P)$ plot the select group of 496 observed pulsars (1a) with distributions of simulated pulsars for the case of no field decay with a pulsar death valley (1b) and for the case of field decay with no death valley (1c). The dotted lines are shown for the locus of constant magnetic field with the indicated strengths. The dashed lines represent the indicated ages of pulsars assuming a dipole spin-down of a constant field. The solid lines show the pulsar death lines for dipole and multipole magnetic field distributions in the space-charge-limited-flow model (SCLF) [9]. The radio pulsars observed and those simulated are filtered through the select group of surveys and displayed in the figure with solid dots. Figure 1b and 1c represent the simulated pulsar distributions assuming no field decay with a death valley and field decay without a death valley, respectively. The clear absence of high-field, high-period observed pulsars in Figure 1a as compared to those simulated in Figure 1b, might be suggestive of the decay of the magnetic field, indicated by the improved agreement in Figure 1c. In fact, the whole pear-shaped distribution of observed pulsars in Figure 1a may be explained by field decay as seen in Figure 1c. The excess of simulated high-field, high-period pulsars in Figure 1b have ages of the order of 10^7 years. Therefore, a decay constant of this order is required for these high-field pulsars to have their fields decay by an order

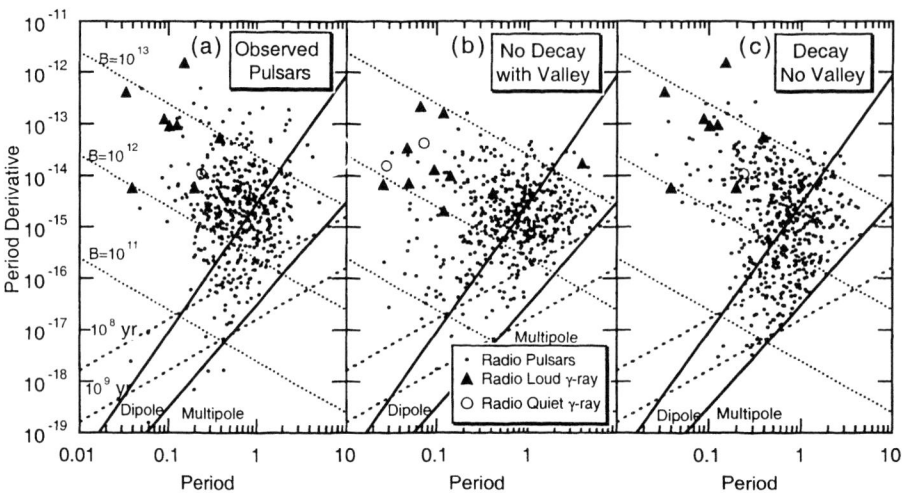

FIGURE 1. Distributions of observed pulsars (a), simulated pulsars assuming no field decay and death valley (b) and with field decay and no death valley (c) as a function of the period derivative and period of the pulsars. Solid dots indicate radio pulsars, solid triangles represent radio-loud, γ-ray pulsars and open circles symbolize radio-quiet, γ-ray pulsars observed (a) and predicted (b and c) for EGRET.

of magnitude. We have used a decay constant of 5×10^6 years in Figure 1c. While this decay constant is short, we anticipate exploring in a future study the effects of the decay of the magnetic field.

In Figure 2, we compare various distributions of the indicated features of observed pulsars (shaded) and of simulated pulsars. Smooth simulated distributions have been obtained from a group of 10,000 calculated pulsars and then normalized to a total of 496 pulsars, which is equivalent to the number of observed pulsars of the selected group. The dark histograms represent the case in which there is no field decay with a pulsar death valley is assumed between the dipole and multipole death lines indicated in Figure 1. The light histograms result from the simulation of the field decay case with no death valley. Under the assumption of field decay, the pulsar age depends on the decay constant of the magnetic field (5 Myr). As a result, we show separate figures for the pulsar age distributions. As noted for the case of no field decay, too many older pulsars with long periods and with small period derivatives are produced in our simulation. However, the simulated distributions of radio flux and distance from Earth along with the magnetic field for the case of no field decay (1b) agree very well with those observed. The distributions for the case of field decay overall appear to better describe the observed distributions without the necessity of introducing a pulsar death valley between the dipole and multipole death lines.

For the case of no field decay with a death valley, the model predicts that GLAST

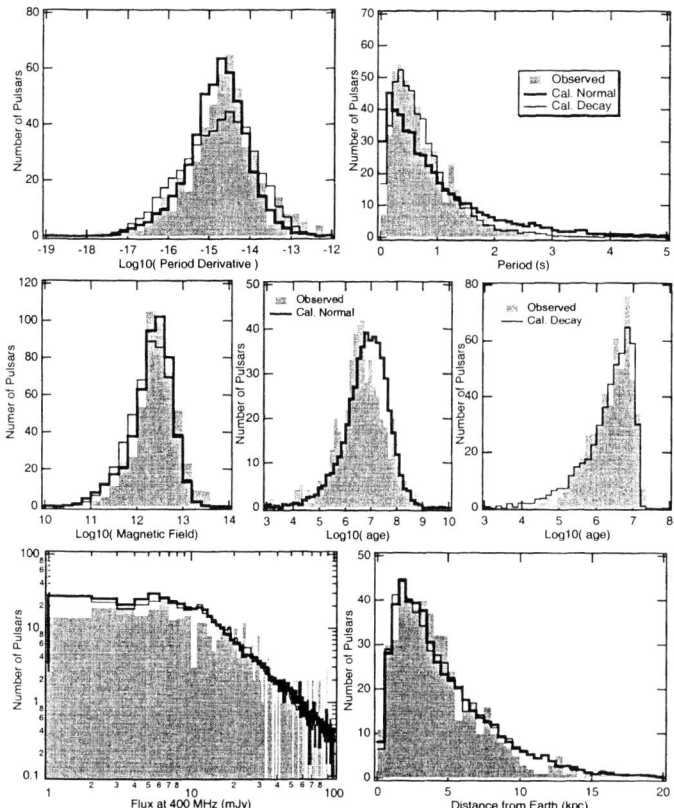

FIGURE 2. Distributions of various pulsar characteristics indicated as shaded histograms (observed pulsars) and plain histograms (simulated pulsars). Thick histograms represent the distributions of the simulated pulsars assuming no field decay and pulsar death valley, while the thin histograms result from the case assuming field decay and no death valley.

should observe 93 radio-loud, γ-ray pulsars compared to 10 predicted for EGRET detected as point sources. The model predicts that GLAST should observe 87 radio-quiet, γ-ray pulsars compared to 2 predicted for EGRET. The GLAST sensitivity for blind period searches is expected to be about the same as the EGRET point-source detection sensitivity (S. Ritz, priv. comm.). GLAST will therefore be expected to identify 8 of these 87 objects as pulsed sources. The results predicted for GLAST are interesting in that about as many radio-quiet, γ-ray pulsars are expected to be observed as radio-loud, γ-ray pulsars. Given the flux thresholds we have used, these are objects observed as point sources. Thus the simulations suggest that due to the increased sensitivity, GLAST will be able to detect more pulsars that are further away than even those detected by the radio surveys used in this study.

In Table 1, we summarize the number of radio-quiet and radio-loud, γ-ray pulsars simulated for the two main cases explored in this study in which we have assumed no magnetic field decay requiring a pulsar death valley and field decay without a death valley. The results for the case of no field decay without a valley are also included for comparison. In addition, we have indicated the number of pulsed sources that GLAST would observe from the radio-quiet, γ-ray pulsar group.

TABLE 1. Simulated Pulsar Statistics

Case	EGRET		GLAST		Neutron Star Birth
	Radio Quiet	Radio Loud	Radio Quiet (Pulsed)	Radio Loud	Rate (per century)
No decay with valley	2	10	87 (8)	93	2.6
No decay no valley	1	4	42 (4)	51	0.4
Decay no valley	2	14	113 (11)	113	1.5

From these simulations, we can estimate a neutron star birth rate in the Galaxy. As mentioned previously, there are 4π more pulsars whose beams do not point in the direction of the Earth than those used in the simulation. We indicate in Table 1 the estimated neutron star birth rates per century for each of the indicated cases. The birth rates represent lower limits as we have assumed 100% detection within the 1 steradian, solid angle of the radio beam. A Gaussian pulse shape and intrinsic duty cycle may raise the birth rates by a factor of three.

REFERENCES

1. Kuzmin, A.D. & Losovsky, B. Ya., *IAU Circular*, 6559 (1997).
2. Malofeev, V.M. & Malov, O.I., *Nature*, **389**, 697 (1997).
3. Hartman, R. et al., *ApJ Supp.*, **123**, 79 (1999).
4. Grenier, I.A. & Perrot, C., *Proc. XXVI Int. Cosmic Ray Conf. Salt Lake City*, **3**, 476 (1999).
5. Gehrels, N. et al., *Nature*, **404**, 363 (2000).
6. Harding, A.K., & Zhang, B., *ApJ Letters*, in press (2001).
7. Zhang, B., & Harding, A. K., *ApJ*, **532**, 1150 (2000).
8. Harding, A. K., & Muslimov, A. G., *ApJ*, **508**, 328 (1998).
9. Zhang, B., Harding, A. K., & Muslimov, A. G., *ApJ*, **531**, L135 (2000).
10. Tauris, T. M. & Konar, S. 2001, astro-ph/0101531.
11. Goldreich, P. & Reisenegger, A. 1992, ApJ, 395, 250.

Gamma-ray emission in relativistic pulsar magnetospheric plasmas

Q. Luo[*], D.B. Melrose[*] and G.Z. Machabeli[†]

[*]*RCfTA, School of Physics, University of Sydney, NSW 2006, Australia*
[†]*Abastumani Astrophysical Observatory, Tbilisi 380060, Georgia*

Abstract. We consider γ-ray emission as the result of plasma instabilities. The pulsar magnetosphere is populated with an outflowing relativistic pair plasma, which contains a fast energetic particle beam. Such a system is well known to be unstable with certain types of plasma instabilities being able to develop, leading to quasilinear diffusion. Due to such diffusion, particles initially in the ground Landau state can acquire pitch angles so that they can emit cyclotron/synchrotron radiation. This process should contribute to the observed high energy spectrum in addition to that from the widely discussed synchrotron-cascade process. Application of this mechanism to high energy emission from both normal young pulsars and millisecond pulsars is discussed.

1. Introduction

A small fraction of the over 1000 known radio pulsars radiate pulsed X- and γ-rays with high efficiency [17, 6]. Despite the small population of this type of pulsar, study of their emission mechanism can provide insight to how pulsars operate in general.

There are two types of emission model: the polar cap (PC) model [1, 4, 14], which assumes that the emission is produced above the PC where acceleration occurs, and the outer gap (OG) model [3], which assumes the emission is due to acceleration in the outer magnetospheric region. In most current models for pulsar high energy emission, γ-ray production is due to cyclotron, synchrotron, curvature radiation or inverse Compton scattering, with cyclotron/synchrotron radiation being dominant in some cases. Due to the strong pulsar magnetic field, the synchrotron life time of radiating particles is extremely short, implying that synchrotron emission is possible only for secondary pairs produced with nonzero pitch angles from cascades near or in the acceleration region.

A pulsar magnetosphere is populated with a dense electron-positron pair plasma that is produced above the PC through cascades by accelerated primary particles [15, 1, 4]. Such a plasma system, which includes a bulk secondary pair plasma and an energetic primary particle beam, is unstable, subject to various instabilities [12].

Plasma instabilities can cause particles to diffuse in momentum space, leading to increase in a particle's pitch angle that otherwise decreases due to synchrotron radiation. Through such processes electrons that initially move one-dimensionally along the field lines can acquire nonzero pitch angles [9] radiating cyclotron/synchrotron emission to contribute to the high energy spectrum. In this paper we propose that this mechanism can produce observable X- and γ-ray emission for millisecond pulsars (MSPs) as well as normal young pulsars.

2. Conventional synchrotron-cascade models

In the usual synchrotron-cascade models, e.g. ref [4], the dominant process is synchrotron radiation by secondary pairs produced from the cascade. In the strong pulsar magnetic field, electrons rapidly radiate away their perpendicular energy so that they are in the ground Landau state, moving one-dimensionally along the field lines. The time-scale for particles to transit to the ground state is $\tau_c = (3/4)(c\gamma/r_e\Omega_e^2) = 10^{-5}(5 \times 10^8 G/B)^2(\gamma/10^4)$ s, where γ is the Lorentz factor of electrons, $\Omega_e = eB/m_e c$ is the cyclotron frequency, r_e is the classical electron radius, and B is the magnetic field. Electrons lose their relativistic perpendicular energy (with the perpendicular momentum $p_\perp/m_e c > 1$) through synchrotron radiation on a much faster time scale, given by $(m_e c/p_\perp)^2 \tau_c$.

Because of the very short synchrotron lifetime, the emission can occur only for secondary pairs which are created with nonzero pitch angles. Once the particles relax to the ground state, there is no further synchrotron radiation. Therefore, in the conventional models, the emission region must be located near or in the acceleration region where the cascade occurs.

Particles in the ground Landau state can be excited to higher Landau levels through (a) Compton scattering [5], or (b) plasma processes such as cyclotron absorption [2, 8] and nonresonant diffusion in pitch angles [9]. If the relevant process of increasing the pitch angle is efficient, synchrotron radiation can occur well beyond the acceleration region, which should have major observational consequences.

3. Scattering in pitch angles

We assume that a relevant (plasma) wave can grow, leading to magnetic field fluctuations [9, 13]. Scattering in pitch angles is considered as quasilinear diffusion [11], in which case the wave growth is linear with the backreaction included in diffusion in momentum space. There are resonant diffusion (RQD), involving resonant interactions between waves and particles, and nonresonant diffusion (NQD) due to nonresonant interactions. In both cases, one may write the distribution into the form $f(p_\perp, p_\parallel) = f_0(p_\perp, p_\parallel) + f_1(p_\perp, p_\parallel)$ where $p_\perp$, $p_\parallel$ are, respectively, the perpendicular and parallel momenta, f_1 is the linear response to the growing wave, and f_0 varies on a slower time scale through either RQD or NQD. The particular effect of diffusion here involves redistribution of the particle's perpendicular and parallel energy, leading to overall an increase in the perpendicular energy (increasing pitch angles) for particles with small pitch angles.

Electrons can acquire a nonzero pitch angle through RQD such as diffusion due to cyclotron absorption, given by

$$\psi = \langle p_\perp \rangle / \langle p_\parallel \rangle = (2D_{\psi\psi}/\alpha\gamma^2)^{1/2}, \tag{1}$$

where $\alpha = 2e^2\Omega_e^2/3c^2$, $D_{\psi\psi} \approx (r_e/2m_e c\gamma^3)\hbar\Omega_e n_k$ is the resonant pitch angle diffusion coefficient [11], n_k is the occupation number of the unstable wave, γ is the Lorentz factor of the radiating particles. Since $\psi \propto 1/\gamma^{5/2}$, the process is effective for electrons with a relatively small Lorentz factor.

In the NQD processes, electrons obtain a pitch angle through scattering by fluctuations in magnetic fields [9, 10]

$$\psi \approx B_\perp/B, \qquad (2)$$

where B is the local static magnetic field and $B_\perp$ is the magnetic field fluctuations due to unstable waves. The characteristic pitch angle derived through this process depends only on the wave amplitudes, allowing diffusion for energetic electrons with a large Lorentz factor.

4. Cyclotron/synchrotron radiation

As an electron acquires pitch angle, it emits cyclotron radiation ($p_\perp/m_e c < 1$) or synchrotron radiation ($p_\perp/m_e c > 1$). For synchrotron radiation, the characteristic photon energy (in $m_e c^2$) is given by $\varepsilon = (3/2)\varepsilon_B \gamma^2 \psi$ where $\varepsilon_B = B/B_c$ and $B_c \approx 4.4 \times 10^{13}$ G is the critical field (corresponding to the cyclotron energy being equal to the electron rest energy $\hbar\Omega_e = m_e c^2$). Consider the NQD case only and assume that the ratio of the wave to beam particle energy density in the plasma rest frame is $\xi < 1$, i.e. $\bar{B}_\perp^2/4\pi = \xi m_e c^2 \bar{n}_{GJ} \bar{\gamma}_b$, where $\bar{n}_{GJ}$ and $\bar{\gamma}_b$ are respectively the Goldreich-Julian density and the Lorentz factor of the beam in the plasma rest frame. Then, the photon energy is

$$\varepsilon \approx 3.6 \times 10^2 \left(\frac{B_*}{10^{12}\,\text{G}}\right)^{1/2} \left(\frac{0.1\,\text{s}}{P}\right)^{1/2} \left(\frac{20}{x}\right)^{3/2} \left(\frac{\gamma}{10^6 \Gamma_s}\right)^2, \qquad (3)$$

where we assume $\xi = 0.1$, $\gamma_b = 10^7$, x is the radial distance in the star's radius (R_*), B_* is the magnetic field on the PC, Γ_s is the bulk Lorentz factor. All quantities are in the pulsar frame. For the beam particles with $\gamma = \gamma_b$, this effect leads to synchrotron radiation above MeV energy, and for the bulk plasma with $\gamma \ll \gamma_b$ the radiation is in the X-ray range.

For synchrotron emission to occur, one requires $\bar{\gamma}\bar{\psi} \gg 1$, which gives the minimum radial distance

$$x \gg 5 \left(\frac{B_*}{10^{12}\,\text{G}}\right)^{1/3} \left(\frac{P}{0.1\,\text{s}}\right)^{1/3} \left(\frac{\gamma}{10^7}\right)^{-2/3}. \qquad (4)$$

For the nominated parameters, synchrotron radiation occurs only for $x \gg 5$. As we consider regions inside the light cylinder with $x < x_{LC} \equiv R_{LC}/R_*$, the particle's Lorentz factor must satisfy $\gamma > 3 \times 10^4 (B_*/10^{12}\,\text{G})(0.1\,\text{s}/P)$.

The luminosity can be estimated from $L_{\text{syn}} = P_{\text{syn}} n_{GJ} M \Delta V$ where P_{syn} is the single particle synchrotron power, ΔV is the volume of the emission region, M is the multiplicity of the secondary pairs. Then, we have

$$L_{\text{syn}} \approx 1.2 \times 10^{33}\,\text{erg s}^{-1} \left(\frac{0.1\,\text{s}}{P}\right)^2 \left(\frac{B_*}{10^{12}\,\text{G}}\right)^2 \left(\frac{20}{x}\right)^6 \left(\frac{\gamma}{10^7}\right)^2 \left[\frac{\Delta V}{(xR_*)^3}\right] M, \qquad (5)$$

where $\xi = 0.1$, $\Gamma_s = 10$, and x satisfies (4). If the radiating particles are the primary beam particles we have $M = 1$.

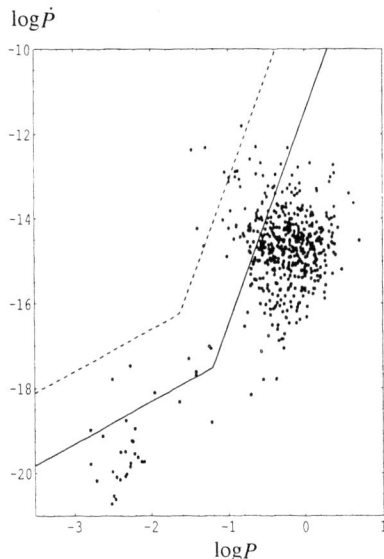

FIGURE 1. Pulsars with synchrotron emission as the result of diffusion. The solid, dashed curves correspond respectively to $L_{\rm syn} = 10^{33}\,{\rm erg\,s^{-1}}$, $5\times 10^{34}\,{\rm erg\,s^{-1}}$. Pulsars with the lominosity $\geq L_{\rm syn}$ are located on the left side of each curve.

Electrons can be scattered to the cyclotron regime with $\bar{\psi}\bar{\gamma} < 1$. Photons radiated from the fundamental states have the characteristic energy $2\varepsilon_B\gamma$ (the blue-shifted cyclotron energy). The cyclotron power is $L_{\rm cyc} = P_{\rm cyc} n_{\rm GJ} M \Delta V$ with $P_{\rm cyc} = \sigma_T c (B_\perp^2/4\pi)\gamma^4$ the single particle power [11]. Then, the luminosity is

$$L_{\rm cyc} \approx 3 \times 10^{31}\,{\rm erg\,s^{-1}} \left(\frac{M}{10^3}\right)\left(\frac{0.1\,{\rm s}}{P}\right)^2 \left(\frac{B_*}{10^{12}\,{\rm G}}\right)^2 \left(\frac{20}{x}\right)^6 \left(\frac{\gamma}{10^3}\right)^4 \left[\frac{\Delta V}{(xR_*)^3}\right]. \quad (6)$$

Since the secondary pairs have a spread in γ, one expects that cyclotron lines are smeared out to form a continuous spectrum.

5. Application to pulsar high-energy emission

Synchrotron radiation can be sustained provided that the perpendicular diffusion time $t_\perp \approx \psi^2 t_\parallel$ is much shorter than the synchrotron cooling time (where $t_\parallel$ is the parallel diffusion time, typically shorter than the plasma flow time). Since $\psi \ll 1$, this condition is satisfied for the region that is not too close to the PC.

Figure 1 shows a distribution of pulsars with synchrotron radiation due to plasma instabilities. The plots with $\xi = 0.8$ correspond respectively to the synchrotron luminosity 10^{33} and $5 \times 10^{34}\,{\rm erg\,s^{-1}}$. The emission region is assumed to have a volume $\Delta V = 0.1(xR_*)^3$ at $x = 4\times$ the lower limit to Eq. (4). For MSPs, the lower limit is replaced by the radial distance of the pair production front, which is assumed to be $2.5R_*$.

For young pulsars, consider the Crab pulsar as an example. For emission from the beam particles with $\gamma_b = 10^7$ the typical photon energy is about 10^2 MeV with the emission region being at $x \approx 50$ stellar radii (using $P = 33$ ms, $B_* = 6.6 \times 10^{12}$ G). Using Eq. (4) and (5) and assuming $\gamma = \gamma_b$ and $M = 1$, the luminosity can be as high as $L_{\rm syn} \approx 8 \times 10^{34}$ erg s^{-1} for $x = 50$, $\Delta V = 0.1(xR_*)^3$, $\xi = 0.5$. If the acceleration occurs near the polar cap, the beam Lorentz factor is about 4×10^6, limited by energy loss due to curvature radiation. Then, the process should contribute to the spectrum below 100 MeV. GLAST will have better sensitivity and resolution (than EGRET) and should be able to test the model.

Radiation from the bulk plasma is in the cyclotron regime. For $x = 20$, photons have the blue-shifted cyclotron energy $\varepsilon \approx 4 \times 10^{-2}(B_*/6.6 \times 10^{12}\,{\rm G})(20/x)^3(\gamma/10^3)$, in the keV range. The luminosity is $L_{\rm cyc} \approx 10^{34}$ erg s^{-1}, which should produce an observable keV spectrum.

Figure 1 shows that the mechanism can be effective for MSPs with relatively strong magnetic fields such as PSR 1937+21. The condition (4) does not constrain MSPs as diffusion is allowed near the PC. For PSR 1937+21 with $B_* = 7.5 \times 10^8$ G and $P = 1.6$ ms, the maximum photon energy radiated by the primary beam can be in the GeV range, which can initiate a pair cascade producing more pairs. The luminosity is $L_{\rm syn} \approx 4 \times 10^{33}$ erg s^{-1} for $x \approx 2.5$ stellar radii and $\Delta V = 0.5(xR_*)^3$, which should be observable with GLAST. As in normal young pulsars, the mechanism produces significant cyclotron radiation in the keV range that forms part of the observed X-ray spectrum [16].

In summary, pulsar plasmas are intrinsically unstable with various instabilities occurring and these can cause diffusion in pitch angle. Electrons initially in the ground Landau state can acquire pitch angle such that they radiate cyclotron or synchrotron photons with energy up to about 100 MeV for normal young pulsars and a few GeV for MSPs. This mechanism should contribute to the overall spectrum in addition to the usual synchrotron-cascade process. To calculate the spectrum, a detailed model for plasma instabilities is required and will be investigated further.

REFERENCES

1. Arons, J & Scharlemann, E.T. 1979, ApJ, 231, 854.
2. Blandford, R. & Scharlemann, E.T. 1976, MNRAS, 174, 59.
3. Cheng, K.S., Ho, C. & Ruderman, M. 1986, ApJ, 300, 522.
4. Daugherty, J.K. & Harding, A.K. 1982, ApJ, 252, 309.
5. Daugherty, J.K. & Harding, A.K. 1986, ApJ, 337, 362.
6. Fierro, J.M., Michelson, P.F. & Nolan, P.L. 1998, ApJ, 494, 734.
7. Gedalin, M.E., Melrose, D.B. & Gruman, E. 1998, Phys. Rev. E, 57, 3399.
8. Lyubarskii, Y.E. & Petrova, S.A. 1998, A&A, 337, 433.
9. Machabeli, G.Z., Luo, Q., Melrose, D.B. & Vladimirov, S.V. 2000, MNRAS, 312, 51.
10. Machabeli, G.Z., Vladimirov, S.V., Melrose, D.B. & Luo, Q. 2000, Phys. Plasmas, 7, 1280.
11. Melrose, D.B. 1980, Plasma Astrophysics, Vol. 1 (Gordon and Breach: New York).
12. Melrose, D.B. 2000, in Pulsar Astronomy–2000 and Beyond, ASP Conf. Series, Vol. 202, P. 721.
13. Melrose, D.B. & Gedalin, M.E. 1999, ApJ, 521, 351.
14. Sturner, S.J. & Dermer, C.D. 1994, ApJ, 420, L79.
15. Sturrock, P. 1971, ApJ, 164, 529.
16. Takahashi, M. et al. 1999, Astron. Nachr. 320, 340.
17. Thompson, D.J. 1996, in Problems and Progress, ASP Conf. Series, Vol. 105, p. 307.

Observations of the Crab Nebula with CAT and CELESTE

Conor Masterson for the CAT and CELESTE collaborations

PCC - Collège de France,11 pl. Marcelin Berthelot, Paris 75231, France

Abstract. The Crab Nebula is the *standard candle* source of TeV γ-rays, due to its stability and high flux. It has been observed by the CAT collaboration from October 1996 to March 2000, above an energy threshold of 250 GeV. CAT first detected the Crab in 1996.

The combined flux and energy spectrum are reconstructed from these observations using the standard CAT analysis method and are presented here. Observations of the Crab with the CELESTE detector are also presented here. Steady emission from the Crab has been detected by this experiment at an energy of 60 GeV and an upper limit on the pulsed emission from this source is also given.

An independent method for reconstruction of the flux and energy spectrum has also been developed and is described. The results of the application of this method to the CAT Crab database are presented and compared with those of the standard method.

IMAGING ČERENKOV TELESCOPES AT THÉMIS

The CAT (Čerenkov Array at Thémis) imaging telescope, equipped with a very-high-definition camera, started operation in September 1996 on the site of the former solar plant at Thémis (France), at an altitude 1600 Metres a.s.l. It is described in [1]. The camera consists of 546 photomultiplier tubes (PMTs) with an angular size of $0.12°$. The high resolution allows accurate determination of shower parameters at energies above 250 GeV.

CAT shares the Thémis site with the CELESTE experiment, which uses the former solar furnace to collect Čerenkov light. It currently consists of 40 heliostats, each with an area of 54 m^2, each focusing light onto a separate PMT via a system of secondary optics mounted on a 100 m tower. The experiment is described in detail in [2]. Thanks to the very large light collection area and the ability to reject the local μ-meson background a threshold of 60 GeV may be achieved.

OBSERVATIONS OF THE CRAB NEBULA

The Crab Nebula is the *standard candle* source of TeV γ-rays, due to its stability and high flux. It has been observed by the CAT collaboration from October 1996 to March 2000, above an energy threshold of 250 GeV [3]. CAT first detected the Crab in 1996. Since then a total of 100 hours of observations have been made on the source, giving a steady rate of 1.8 $\gamma\,min^{-1}$ after selection cuts, using the standard CAT analysis [4]. Data taken at zenith angles up to $60°$ are included in this analysis [see 5].

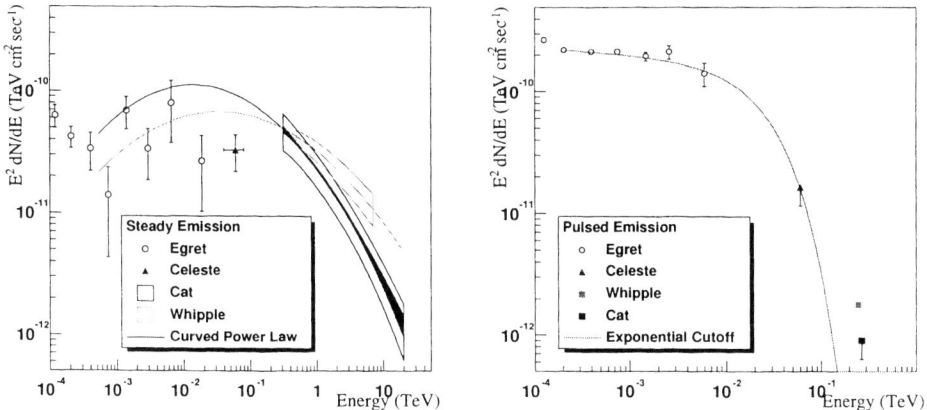

FIGURE 1. Power Spectrum of steady and pulsed γ-ray emission from the Crab Nebula

This data set has been used to derive the spectral characteristics of the emission from the Crab Nebula, the resulting power spectral distribution is shown in figure 1. A Maximum Likelihood technique is used as described in the next section. In order to compare the measured spectrum with fluxes measured at lower energies, a curved spectral hypothesis (H^{cb}) was used in the fit. The resulting characterisation is as follows:

$$\frac{dN}{dE} = 2.33 \pm 0.07 \pm 0.60 \times 10^{-11} E^{-2.74 \pm 0.04 \pm 0.08 + (0.20 \pm 0.07 \log(E))} \text{cm}^{-2}\text{s}^{-1}\text{TeV}^{-1}$$

The probability of this result being consistent with a straight power law (H^{pl}) can be estimated from the ratio of the likelihood values between the two hypotheses:

$$\lambda = 2 \log \left[\frac{L^{cb}}{L^{pl}} \right] \quad (1)$$

This value behaves (asymptotically) like a χ^2 with one degree of freedom. Here the resulting value is: $P(\lambda) = 0.0074$.

The statistical error in the estimation of the spectrum is indicated by the filled region, while the unfilled box indicates the systematic error. The CAT spectral curve is extended to lower energies for comparison with the steady flux measured by the EGRET experiment (taken from [8]). The spectrum as measured by the Whipple collaboration [9] is also shown for comparison, along with the systematic error in this measurement.

The CELESTE experiment has observed the Crab nebula between 1999 and 2001. The above plot shows the flux measured by CELESTE at an energy of 60 GeV. The measured Celeste flux is:

$$\frac{dN}{dE} = 3.3 \pm 0.5_{\text{stat}} \pm 1.1_{\text{syst}} \times 10^{-8}/E^2 (\text{GeV}) \text{cm}^{-2}\text{s}^{-1}\text{GeV}^{-1}$$

It can be seen that the CELESTE measurement bridges the gap between the *High Energy* and *Very High Energy* ranges and thus provides useful information on the shape of the emission spectrum in this region.

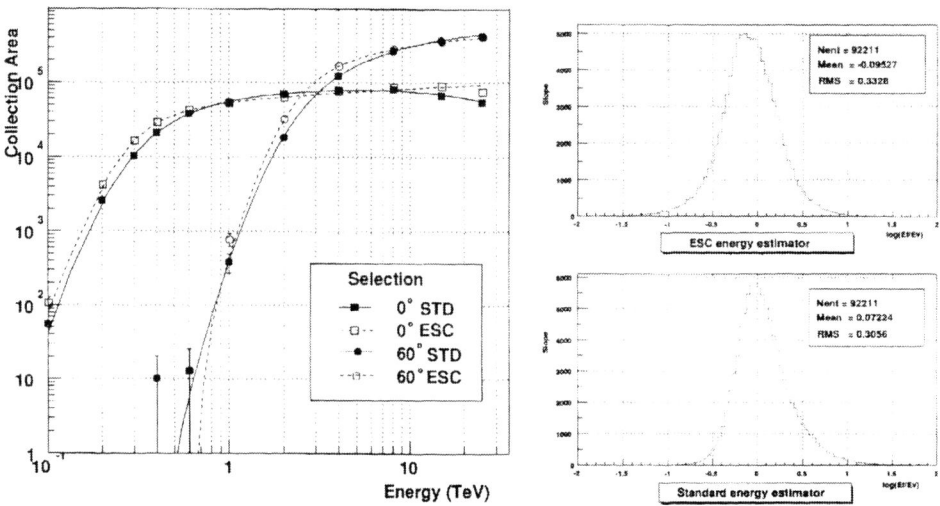

FIGURE 2. Effective area and energy resolution comparisons between the standard CAT spectral analysis and the independent (ESC) method.

Pulsed emission from the Crab

Figure 1 also shows the upper limits for pulsed γ-ray emission from the Crab nebula at VHE energies. Included are limits from the CAT and CELESTE experiments [10, 11], as well as that from the Whipple collaboration [12].

The CELESTE limit is derived by fitting a power law model ($\phi \propto E^{-2.0}$) with an exponential cutoff component ($\phi \propto e^{-E/E_0}$), such that the cutoff energy (E_0) is consistent with the EGRET observations [13] and the signal seen by CELESTE at an energy of 60 GeV. The maximum cutoff energy compatible with this result is 30 GeV.

ENERGY SPECTRUM MEASUREMENTS WITH CAT

The energy spectrum of γ-ray emission from the Crab has been estimated using the standard CAT Crab database. The technique uses a Maximum likelihood method to predict the shape of an *a priori* model of the energy distribution, based on an estimated energy for each event. A forward-folding technique is used to take account of the detection efficiency and energy resolution of the camera as a function of the zenith angle and energy of the primary particle [14, 15].

In order to confirm the accuracy of the CAT spectral analysis it was decided to develop an independent method of estimation, to eliminate the possibility of bias in the event selection or energy estimation routines used in the CAT technique. A subset of the Crab database was selected at small zenith angles in order to minimise possible systematic problems. Due to the smaller range of sensitive energies inherent in this, the spectral distribution for this dataset is best fitted by a straight power law distribution as follows:

$$\frac{dN}{dE} = 2.21 \pm 0.051 \pm 0.60 \times 10^{-11} E^{-2.78 \pm 0.03 \pm 0.08} \text{cm}^{-2}\text{s}^{-1}\text{TeV}^{-1}$$

A selection system based on the *Extended Supercuts* (ESC) [16] technique was chosen to ensure an unbiased response over the maximum energy range. This selection is not highly optimised for signal to noise ratio, thus the number of events passing the selection cuts is higher than for the standard CAT analysis.

Figure 2 shows the estimated Collection Area of the detector, based on Monte-Carlo simulated events over a range of energies from 100 GeV to 25 TeV. The curves are shown for two zenith angles of 0° and 60°. The solid line indicates the standard CAT analysis, while the dashed line indicates the new ESC analysis. The greater collection area due to the looser selection cuts can be seen.

The energy of each event was estimated using a fitted function based on the Hillas parameters. This function uses less information than the standard CAT energy estimator, which uses a model of the particle distribution in each shower to fit the observed Čerenkov light image. The energy resolution functions ($\log(E_{\text{fitted}}/E_{\text{real}})$) are compared in figure 2 with the standard function, for events at 0° zenith angle, between 100 GeV and 10 TeV.

It can be seen that the standard analysis method produces a better estimate of the energy of each event, as the error per event is on average smaller. However the difference is not large, on the order of 10%, despite the extra information used in the standard energy estimator.

The functions for collection area and energy resolution are used to fit the spectral model using the standard forward-folding technique. Figure 3 shows the resulting differential energy spectrum, it can be seen that it is well fitted by a straight power law, as follows:

$$\frac{dN}{dE} = 2.29 \pm 0.119 \times 10^{-11} E^{-2.93 \pm 0.081} \text{cm}^{-2}\text{s}^{-1}\text{TeV}^{-1}$$

Comparison of Spectral parameters

Figure 3 also shows the likelihood distribution for differential power law spectral parameters. The standard CAT analysis and the independent ESC analysis of the same data set are shown. It can be seen that the errors for the standard method are smaller due to the greater efficiency of the event selection and energy estimation procedures.

The two methods produce results that agree within the statistical errors for the data set, thus it may be said that it is unlikely that the standard CAT event reconstruction procedure introduces systematic a bias into the spectral analysis. Both methods do, however, depend on the same set of Monte Carlo simulations; this is the subject of further study to investigate possible systematic biases in the simulation of the CAT detector.

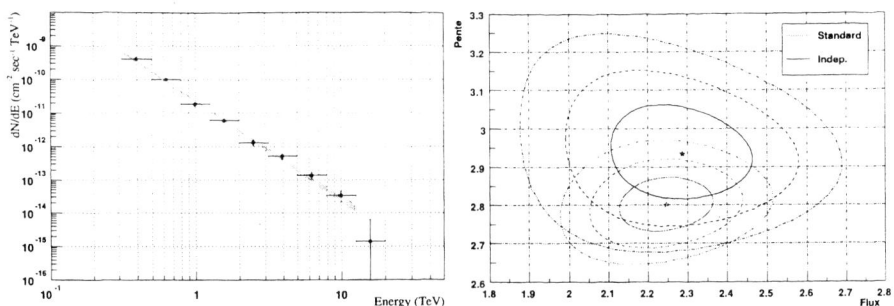

FIGURE 3. (a) Power law fit to Crab data using independent method. (b) Comparison of fitted energy spectral parameters.

REFERENCES

1. Barrau, A., et al., *Nuclear Instruments and Methods*, **A416**, 278 (1998).
2. Dumora, D., et al., Celeste experimental proposal (1996), available at http://polywww.in2p3.fr/celeste/public/cxp.ps.gz.
3. Tavernet, J. P., et al., "Measurement of the Gamma-ray Spectrum of the Crab Nebula Above 250 GEV with the CAT Cherenkov Telescope", in *Texas Symposium on Relativistic Astrophysics*, 1998, p. E645.
4. Le Bohec, S., et al., *Nuclear Instruments and Methods*, **A416**, 425 (1998).
5. Masterson, C., et al., "Observations of the Crab Nebula with the CAT imaging Atmospheric Cherenkov Telescope", in *Proc. Int. Symp. High Energy Gamma-Ray Astronomy*, 2001, p. 753.
6. Harding, A. (2000), private communication.
7. Hirotani, K., and Shibata, S., astro-ph 0011488 (2000).
8. de Jager, O., *Astrophysical Journal*, **457**, 253 (1996).
9. Hillas, A. M., et al., ApJ, **503**, 744 (1998).
10. Musquere, A., et al., "Search for VHE pulsed emission from the Crab with the CAT telescope", in *Proceedings of the 26th International Cosmic Ray Conference (Salt Lake City)*, 2000, vol. 3, p. 460.
11. de Naurois, M., et al. (2001), ap. J. (In preparation).
12. Lessard, R. W., et al., ApJ, **531**, 942 (2000).
13. Fierro, J., et al., *Astrophysical Journal*, **494**, 734 (1998).
14. Djannati-Ataï, A., et al., A&A, **350**, 17–24 (1999).
15. Piron, F., Ph.D. thesis, Universite de Paris-Sud (2000).
16. Mohanty, G., Biller, S., Carter-Lewis, D. A., Fegan, D. J., Hillas, A. M., Lamb, R. C., Weekes, T. C., West, M., and Zweerink, J., *Astroparticle Physics*, **9**, 15–43 (1998).

Variable High-Energy γ–Ray Emission From Pulsar Wind Nebulae

Mallory S.E. Roberts[*], Bryan M. Gaensler[†] and Roger W. Romani[**]

[*]*Dept. of Physics, McGill University*
[†]*Center for Space Research MIT*
[**]*Dept. of Physics, Stanford University*

Abstract. There is growing evidence that some pulsar wind nebulae (PWN) are significant sources of variable emission above 100 MeV. We review the observational evidence associating five PWN and candidate PWN with the four sources showing the most > 100 MeV variablility among Galactic sources that are bright above 1 GeV. We discuss potential physical mechanisms for such variability, examine the apparent X-ray and radio eficiencies, and place limits on the magnetic field required for variable synchrotron emission at such high energies.

INTRODUCTION

The nature of the majority of γ–ray sources in the Galaxy is one of the oldest mysteries of high energy astrophysics. The only firmly identified class of objects emitting at $E > 100$ MeV is isolated young pulsars, whose phase-averaged emission appears constant. However, the *EGRET* data clearly show that there is at least one class of variable emitters in the Galaxy [1, 2] (Figure 1). Several suggestions as to the nature of these objects have been put forward (eg. isolated accreting black holes, [3]), but no strong observational evidence has previously supported a particular class.

In a recent flux-limited survey of the brightest sources of emission above 1 GeV, Roberts, Romani and Kawai (2001) noted that the four Galactic sources in their survey showing the most variability above 100 MeV (Figure 1) all contained extended sources of hard X-ray emission. Although relatively rare, these sources are likely to be pulsar wind nebulae (PWN). Indeed, two of the sources are coincident with known radio pulsars having high spin-down energy, and non-thermal radio sources presumed to be PWN. Here we review the evidence showing that these extended X-ray sources are PWN, including new radio and X-ray observations that show the source in GeV J1809-2327 is a probable PWN. We then examine the efficiency of conversion of spin-down energy into radio and X-ray emission, and discuss the implications of variable γ–ray emission above 100 MeV coming from PWN.

γ–RAY PULSARS WITH WIND NEBULAE

The number of known PWN around detected pulsars are around 9 in the radio [5] and 8 (two in the LMC) in X-rays [6] with 5 belonging to both groups. A similar number

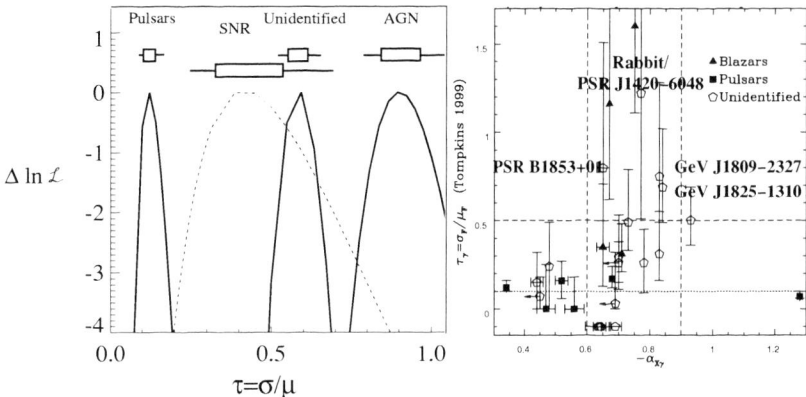

FIGURE 1. (left) liklihood distributions (curves) and the corresponding 68% and 90% confidence regions (bars and lines) of the average variability index τ (defined as the standard deviation of source flux over the mean source flux) for the > 100 MeV emission of γ–ray pulsars, AGN, sources coincident with supernova remnants, and unidentified sources in the 3rd *EGRET* Catalog (from Tompkins, 1999). (right) Variability index versus broadband energy "spectral index" $-\alpha_{X\gamma}$ between 1 keV and 1 GeV for the brightest sources above 1 GeV (see Roberts, Romani, and Kawai 2001 for details).

of good PWN candidates of each group exist wwhere a pulsar hasn't been detected. Remarkably, the above groups contain 5 and 4 (respectively) of the known γ–ray pulsars. Only the oldest γ–ray pulsars, such as Geminga and PSR B1055-52 show no sign of having PWN. Therefore, it might be expected that other γ–ray pulsars might reveal themselves through PWN.

None of the known γ–ray pulsars total flux above 100 MeV is significantly variable as measured using the τ statistic (the standard deviation of the flux meaurements over mean flux) [2] (Figure 1), which analysis is sensitive to time scales of a few months to a few years. However, in the 70-150 MeV range, the unpulsed flux of the Crab was observed to vary [7], with the suggestion that the synchrotron component is due to the inner nebula, perhaps associated with the variable optical wisps [8], and has a variable exponential cutoff at around 25 MeV. de Jager et al. (1996) argue that this cutoff is a general upper limit to the synchrotron emission by balancing the acceleration timescale equal to the electron gyroperiod across a strong shock and the synchrotron loss timescale:

$$E_{max} = 25(D\alpha/sin\theta)\text{MeV}$$

Where D is the Doppler boost factor, θ is the electron pitch angle, and α is a parameter expected to be less than 1. If this is the true limit on emission from shock generated electrons, then to reach energies of several hundred MeV, either the Doppler factor is large or the pitch angle is small.

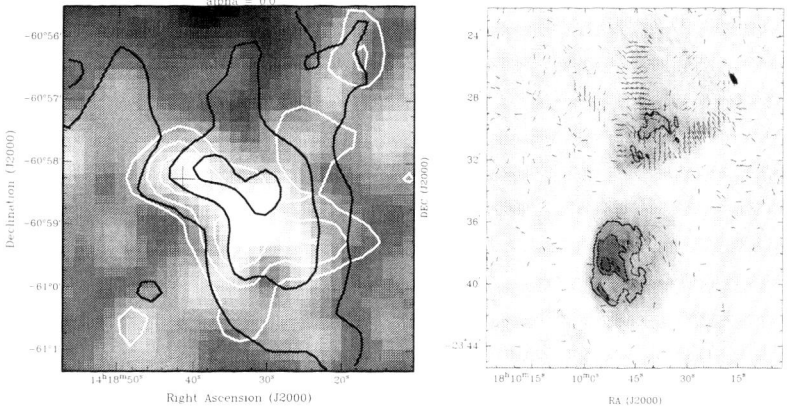

FIGURE 2. (left) Spectral Tomography map of Rabbit for spectral index 0.0. The brighter regions are steeper than 0. The black contours are the continuum radio emission, the white contours are the linearly polarized emission, and the cross marks the X-ray peak seen with the ASCA SIS. (right) Polarization vector map of GeV J1809-2327 at 6 cm, on a greyscale map of the 6 cm continuum.

THE CANDIDATE VARIABLE γ–RAY PWN

PSR B1853+01 in the supernova remnant W44 is coincident with GeV J1856+0115 and is known to have an X-ray [9] and radio [10] PWN. The morphology of the radio PWN suggests a plume or wake of material coming from the pulsar, which is at the southern tip of the nebula. The X-ray nebula is not well resolved in the *ASCA* image, but also appears to be primarily to the north of PSR B1853+01.

GeV J1417-6100 is coincident with the Kookaburra complex of radio sources [11]. There are two extended, hard X-ray sources in the complex, which seem to have associated radio emission. One has recently been found to contain PSR J1420-6038, a young pulsar with high spin-down energy at a dispersion measure distance of 8 kpc [12, 13]. The brighter X-ray source, however, is coincident with the Rabbit nebula, whose polarization and spectral index strongly suggest a PWN identification (Figure 2).

GeV J1825-1310 contains an extended hard X-ray source or cluster of sources [4]. Sensitive radio mapping of the region is complicated by the presence of the Sharpless 53 cluster of HII regions, and has not yet been done. Extensive VLA observations of this source are being planned.

GeV J1809-2327 contains an extended hard X-ray source north of an OB association with numerous X-ray emitting stars. It is embedded in the Lynds 227 dark nebula which has a bubble in the molecular gas near the X-ray source [14]. We have observed this region with the VLA in D configuration at 20 and 6cm. We find two nebula, one associated with the stellar cluster and the other with the hard X-ray source. The one around the stellar cluster is clearly thermal. The nebula around the hard X-ray source is significantly polarized (Figure 2) and has a spectral index of $\alpha \sim -0.25 - -0.4$, typical for a PWN. A short Chandra ACIS observation shows an X-ray point source at the southern tip of the nebula, with a small tail or jet pointing towards the center of the radio nebula.

X-ray Emission

An apparent correlation between spin-down luminosity and X-ray luminosity for the known X-ray PWN has been noticed by several authors. We can study this by examining the X-ray efficiency $\varepsilon_X = L_X/\dot{E}$. Chevalier(2000) lists values of ε_X for 8 PWN ranging from 0.0003 for Vela to 0.03 for the crab. Below we list 1-10 keV efficiencies for the PWN candidates. If there is no measured pulse, we assume the efficiency to estimate $\dot{E}$. Distances are determined from either the pulsar dispersion measure [15] or inferred from nearby young objects [16] and Γ is the photon index used to determine the unabsorbed flux.

Name	$\dot{E}$	ε_X	d(kpc)	Γ
PSR B1853+01	4.3×10^{35}	0.005	3.0	2.3
PSR J1420-6048	1.0×10^{37}	0.005	8	1.4
Rabbit	1.2×10^{36}	0.003	1.5	1.9
GeV J1809-2327	1.1×10^{36}	0.003	1.9	2.1
GeV J1825-1310	6.8×10^{36}	0.004	4.1	2.2

Radio Efficiency

The efficency ε_R of converting spin-down energy to radio emission is defined by:

$$L_R = \varepsilon_R \dot{E}$$

Typical values of $\varepsilon_R = (1-5) \times 10^{-4}$ [17, 5], although there are a few that go as low as $\varepsilon_R = 2 \times 10^{-6}$. For a particular PWN spectral index, the 20cm flux can be related to the spin down energy and radio efficiency by integrating across the 10 MHz to 100 GHz band.

Name	$\dot{E}$	ε_R	d(kpc)	α
PSR B1853+01	4.3×10^{35}	4.4×10^{-4}	3.0	-0.12
PSR J1420-6048 (excess, see [13])	1.0×10^{37}	6×10^{-6}	8	-0.3
PSR J1420-6048 (entire wing)	1.0×10^{37}	3×10^{-4}	8	-0. 3
Rabbit (assumed ε_R, d)	8.8×10^{34}	2.5×10^{-4}	1.5	-0.25
Rabbit (assumed $\dot{E}, d$)	1.2×10^{36}	1.9×10^{-5}	1.5	-0.25
GeV J1809-2327 (assumed ε, d)	2.6×10^{35}	2.5×10^{-4}	1.9	-0.4
GeV J1809-2327 (assumed $\dot{E}, d$)	1.1×10^{36}	6.2×10^{-5}	1.9	-0.4

MAGNETIC FIELD LIMITS FROM γ–RAY VARIABILITY

Assuming that the variable γ–ray emission is synchrotron radiation, we can immediately put a lower limit on the magnetic field in the emitting region. It is reasonable to assume that the synchrotron cooling timescale must be shorter than the variability time scale.

$$t_{cool} \leq t_{var}$$

Therefore, a lower limit on the magnetic field is given by:

$$B \geq 10^{-4} \left(\frac{10 \text{MeV}}{E} \right)^{1/3} t_{var}^{-2/3}$$

Variability analysis of *EGRET* data is generally sensitive to time scales of a few months to a few years [2]. For $E = 100$ MeV and $t_{var} = 0.2 - 2$ yr., the magnetic field is $B > 0.3 - 1.4 \times 10^{-4}$ G. Compare this to the inferred magnetic fields in the Crab of 3×10^{-4} and in Vela, whose spectrum cuts off somewhere below 40 MeV [18], of 6×10^{-5}.

ACKNOWLEDGMENTS

The National Radio Astronomy Observatory is a facility of the National Science Foundation operated under cooperative agreement by Associated Universities, Inc. MSER is a Quebec Merit Fellow, BMG is a Hubble Fellow.

REFERENCES

1. McLaughlin, M. A., Mattox, J. R., Cordes, J. M., and Thompson, D. J., *ApJ*, **473**, 763 (1996).
2. Tompkins, W. F., *Applications of Likelihood Analysis in Gamma-Ray Astrophysics*, Ph.D. thesis, Stanford University, Stanford, CA 94305 (1999).
3. Punsly, B., *ApJ*, **516**, 141 (1999).
4. Roberts, M. S. E., Romani, R. W., and Kawai, N., *ApJS*, **133**, 451 (2001).
5. Gaensler, B. M., Stappers, B. W., Frail, D. A., Moffett, D. A., Johnston, S., and Chatterjee, S., *MNRAS*, **318**, 58 (2000).
6. Chevalier, R., *ApJ*, **539**, L47 (2000).
7. de Jager, O. C., Harding, A. K., Michelson, P. F., Nel, H. I., Nolan, P. L., Sreekumar, P., and Thompson, D. J., *ApJ*, **457**, 253 (1996).
8. Scargle, J. D., *ApJ*, **156**, 401+ (1969).
9. Harrus, I. M., Hughes, J. P., and Helfand, D. J., *ApJ*, **464**, L161 (1996).
10. Frail, D. A., Giacani, E. B., Goss, W. M., and Dubner, G., *ApJL*, **464**, L165 (1996).
11. Roberts, M. S. E., Romani, R. W., Johnston, S., and Green, A. J., *ApJ*, **515**, 712 (1999).
12. D'Amico, N., Kaspi, V. M., Manchester, R. N., Camilo, F., Lyne, A. G., Possenti, A., Stairs, I. H., Kramer, M., Crawford, F., Bell, J. F., McKay, N. P. F., Gaensler, B. M., and Roberts, M. S. E., *ApJ*, **552**, L45 (2001).
13. Roberts, M. S. E., Romani, R. W., and Johnston, S., *ApJLsubmitted* (2001).
14. Oka, T., Kawai, N., Naito, T., Horiuchi, T., Namiki, M., Saito, Y., Romani, R. W., and Kifune, T., *ApJ*, **526**, 764 (1999).
15. Taylor, J. H., and Cordes, J. M., *ApJ*, **411**, 674–684 (1993).
16. Yadigaroglu, I.-A., and Romani, R. W., *ApJ*, **476**, 347 (1997).
17. Frail, D., and Scharringhausen, .., *ApJ*, **480**, 364 (1997).
18. de Jager, O. C., Harding, A. K., Sreekumar, P., and Strickman, M., *A&AS*, **120**, 441 (1996).

Solar and Stellar Flares

Solar Gamma-Ray Physics Comes of Age

Gerald H. Share and Ronald J. Murphy

E.O. Hulburt Center for Space Research
Naval Research Laboratory, Washington D.C., 20375

Abstract. The launch of NASA's *HESSI* satellite will provide solar scientists with ≥ 2arcsec imaging/≥ 2 keV spectroscopy of solar flares from ~10 keV to 10 MeV for the first time. We summarize recent developments in our understanding of solar flares based primarily on gamma-ray line measurements made over the past twenty years with instruments having moderate spectral resolution. These measurements have provided information on solar ambient abundance, density and temperature; accelerated particle composition, spectra, and transport; and flare energetics.

RELATIONSHIP TO CELESTIAL GAMMA RADIATION

This conference primarily focuses on gamma rays from galactic and extra-galactic sources. These sources produce radiation that has been measured up to TeV energies [1] and is primarily in the form of continuum emission from relativistic electrons. Gamma-ray lines have been detected only from de-excitations of freshly produced radioactive nuclei in supernovae [2] and in relic radiation from earlier explosions [3,4]. To date there have been no observations of celestial gamma-ray lines from nuclei excited by collisions with energetic ions [5]. The most likely source of these lines is cosmic-ray proton interactions with the interstellar medium [6]. Interactions of higher-energy protons create pions that decay and produce a characteristic broad feature near 70 MeV that has been detected from the Galactic plane [7]. High-energy bremsstrahlung from relativistic electrons can mask the presence of this feature, however [8].

GAMMA-RAY PRODUCTION DURING SOLAR FLARES

The sun is a prolific source of high-energy electrons, ions, and neutrons during solar flares. These flares can produce particles with energies up to tens of GeV that are revealed by ground-based measurements [9] and gamma-ray observations in space [10]. The primary source for particle acceleration is the energy contained in the sun's magnetic field. How this conversion of magnetic energy to kinetic energy takes place is still not known, however, reconnection is a likely mechanism [11].

We show a simple representation of this process in Figure 1. The image in the upper left corner was taken during a limb flare by the Yohkoh satellite [12]. The contours show the hard X-ray sources and the gray scale shows the image in soft X rays. The full magnetic loop filled with hot plasma is revealed in soft X-rays. In contrast the hard X rays reveal the footpoints of the loops where the accelerated particles interact. A hard X ray source also appears above the soft X-ray loop. This region may represent the location where the energy for the flare originates although particle acceleration may occur all along the loop structure [11]. The hard X-ray intensities from the foot points are often different from one another and have an inverse relationship with microwave emissions. This suggests that the electrons have a higher probability of impacting the solar atmosphere where the magnetic field is weaker.

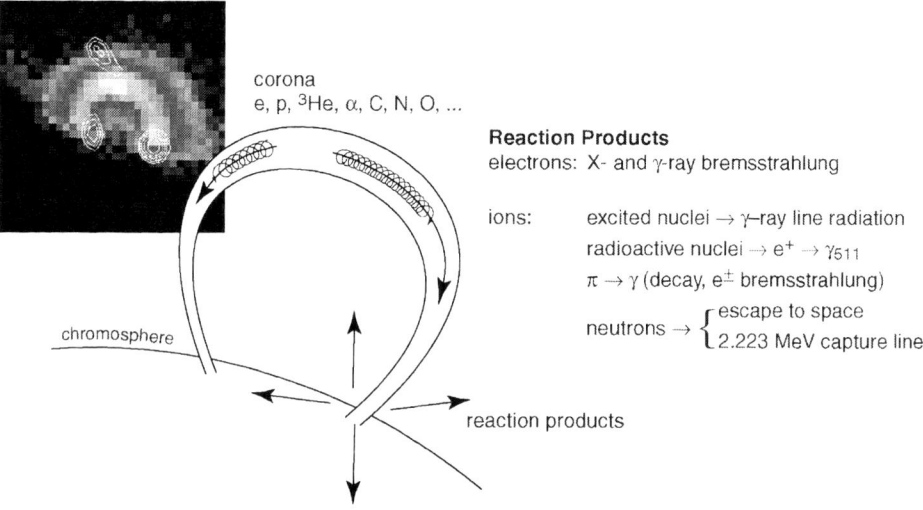

FIGURE 1. Schematic showing particle acceleration and interaction in magnetic loops during solar flares.

With the upcoming launch of the *HESSI* satellite that offers the promise of high-resolution spectroscopy of solar flares [13], we shall focus our remarks primarily on what has been learned from γ-ray line spectroscopy. Figure 2 shows the γ-ray spectrum of the 1991 June 4 X12+ solar flare (N30E70) observed by the *Compton Gamma Ray Observatory (CGRO)* OSSE experiment [14]. The captions describe how γ-ray line and continuum studies reveal the physics of flares [6,15-19]. We use this figure to illustrate what has been learned about ion acceleration, transport, and interaction in flares. The spectrum covers the range from 100 keV to 10 MeV and exhibits an electron bremsstrahlung continuum that has a power-law shape extending to high energies. The continuum doesn't always follow a single power law. It often steepens or hardens above a few hundred keV, and sometimes further hardens ≥ 1 MeV [20]. Nuclear lines are superposed above the power law continuum. These include lines from de-excitation of directly excited nuclei and spallation products in the solar atmosphere after impact by flare-accelerated protons and α particles. De-excitation lines from Mg, Si, Fe, Ne, C and O referred to in the figure are broadened by ~1-2% by nuclear recoil. The spectrum also contains very broad lines produced when flare-accelerated heavy nuclei impact atmospheric H and He. Two other prominent line features appear in the spectrum at 0.511 MeV and 2.223 MeV. These lines originate from positron-electron annihilation and neutron capture on H, respectively. Below we discuss our current understanding of the characteristics of nuclear lines and their relationship to the physics of solar flares.

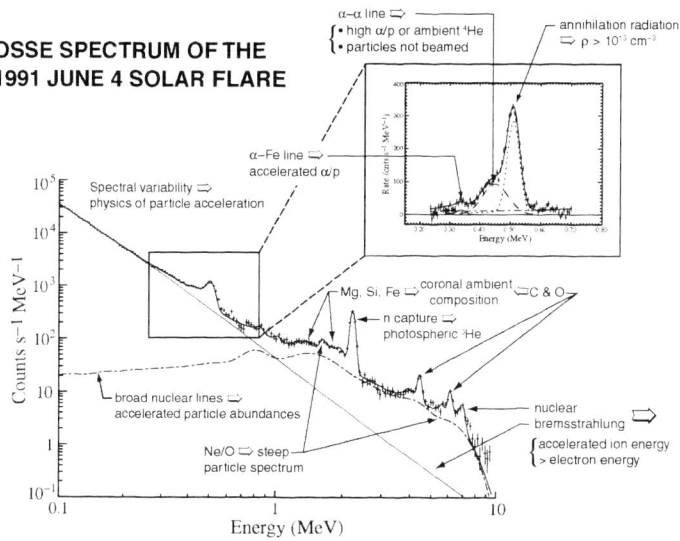

FIGURE 2. OSSE spectrum of the 1991 June 4 solar flare summarizing the physics revealed by γ-ray spectroscopy.

Narrow Gamma-Ray Line Studies

About 30% of all flares with emission >0.3 MeV exhibit clear evidence for nuclear lines. This evidence is usually provided by detection of the neutron capture line at 2.223 MeV or other strong narrow lines. The presence of a nuclear component may also be revealed by general characteristics of the spectrum such as hardening above ~1 MeV and steepening >7 MeV, where few nuclear lines are produced. Additional studies based on these characteristics indicate that an even higher percentage of high-energy flares have a nuclear component [21] (see also Young, et al. in these proceedings). Narrow γ-ray line observations are key to understanding the characteristics of accelerated protons and α particles at the Sun. They also provide information on ambient composition, temperature, and density of the flare plasma [6,15,19,20].

Hydrogen Neutron-Capture Line

The 2.223 MeV neutron-capture line is the most prominent line observed in spectra of flares that are not too close to the limb (see Figure 2). Because it is produced by capture of neutrons after they slow down deep in the solar atmosphere, the line is narrow, delayed in time, and significantly attenuated for flares near the limb. The line's narrow width and strength makes it an excellent indicator of the presence of ions in flares. It therefore has been used to search for a threshold in ion acceleration using combined data from *SMM*/GRS and *CGRO*/OSSE [22] and for continuous acceleration of ions during solar quiet times [23,24]. Measurement of its intensity and temporal variation relative to prompt de-excitation lines has provided information on the spectra of ions above ~10 MeV and on the concentration of ^{3}He in the photosphere [25]. The latter measurements are possible because ^{3}He nuclei capture neutrons in competition with photospheric hydrogen and therefore affect the decay time of the 2.223 MeV capture line. Observations with *SMM*, *GRANAT*, *CGRO* and *Yohkoh* suggest photospheric ^{3}He/H ratios of ~ 2 - 4 x 10^{-5} [26]. These ratios are dependent on assumptions concerning the depth of interaction and solar atmospheric model, however.

Positron-Electron Annihilation Line

Positrons from decay of radioactive nuclei and high-energy pions annihilate with electrons to produce the 0.511 MeV line. Comparison of its intensity profile with those from de-excitation lines provides information on both the species of radioactive nuclei and the density of the medium where the positrons annihilate [15]. It is also an excellent diagnostic for flares that exhibit a distinct second stage during which the proton spectrum is hard enough to produce pions. A classic example is the 1991 July 11 flare observed by the EGRET, COMPTEL, and OSSE instruments on *CGRO* [10]. The annihilation line peaks in coincidence with the high-energy emission suggesting a π^+ origin [27].

Measurements of the 0.511 MeV line width and positronium continuum provide information about the temperature and density of the ambient material. The inset of Figure 1 shows the region of the annihilation line observed by OSSE. Detailed measurements of the line and continuum have been made in seven flares using the moderate resolution spectrometer on *SMM* [28]. We plot the measured 3γ/2γ ratio vs. line width in Figure 3; the curve shows the results of a calculation. The positronium continuum/line ratio in the flares are all significantly lower than that measured by *SMM* for the Galactic plane emission; these low ratios indicate temperatures >10^5 K in the flare plasma. The width of the annihilation line becomes broad enough to be resolved by *SMM* for temperatures >10^6 K; the line in flare '14' appears to have been formed at such high temperatures. There is also one flare, '9', for which the 3γ/2γ ratio falls significantly below the curve; this suggests a local density >10^{14} cm^{-3}.

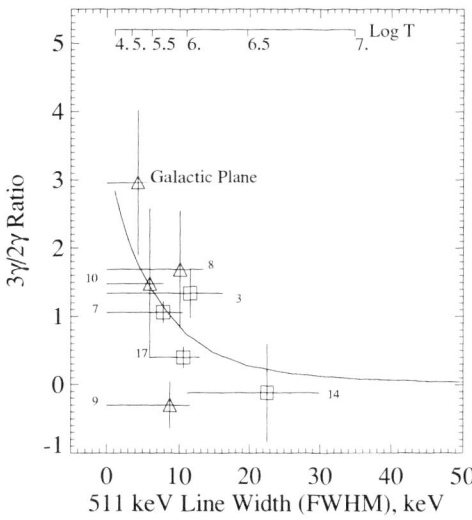

FIGURE 3. *SMM* measurements of the 3γ/2γ positron annihilation ratio compared with the 511 keV line width for 7 flares.

Narrow De-excitation Lines

To date about 18 relatively narrow de-excitation lines have been identified in solar flares: ^{59}Ni (0.339 MeV), ^{7}Be (0.429), ^{7}Li (0.478), ^{56}Fe (0.847), ^{18}F (0.937), ^{18}F/^{58}Co/^{58}Ni/^{59}Ni (1.00-1.08), ^{56}Fe (1.238), ^{55}Fe (1.317), ^{24}Mg (1.369), ^{20}Ne (1.633), ^{28}Si (1.779), ^{32}S /^{14}N (2.230/2.313), ^{20}Ne (3.334), ^{12}C (4.439), ^{15}N/^{15}N/^{15}O (~5.3), ^{16}O (6.130), ^{11}C (6.337, 6.476), ^{14}N/^{16}O (7.028/6.919). These lines were identified from the summed spectrum of 19 intense flares observed by the *SMM* spectrometer [28, 29] that we plot in Figure 4.

Our fits to some of the prominent de-excitation lines, e.g. ^{12}C, in this summed spectrum indicated widths that were about a factor of two larger than expected [6, 20]. We therefore studied whether this increase could be due to changes in the line energies from flare to flare. Not only was this true, but we also found that the fitted energy was dependent on the flare's heliocentric angle. We show this variation in Figure 5 where we plot the fitted energies of the ^{12}C, ^{16}O, ^{20}Ne de-excitation lines in flares grouped according to the cosine of heliocentric angle. The lines are Doppler-shifted by ~1% for flares near the center of the solar disk. Note that there is no significant redshift for the 2.223 MeV neutron capture line, which is formed at rest (other higher energy lines contribute near the limb where the line is weak). This is the first evidence for such a redshift in nuclear lines during flares and indicates that accelerated ions preferentially interact in a downward direction.

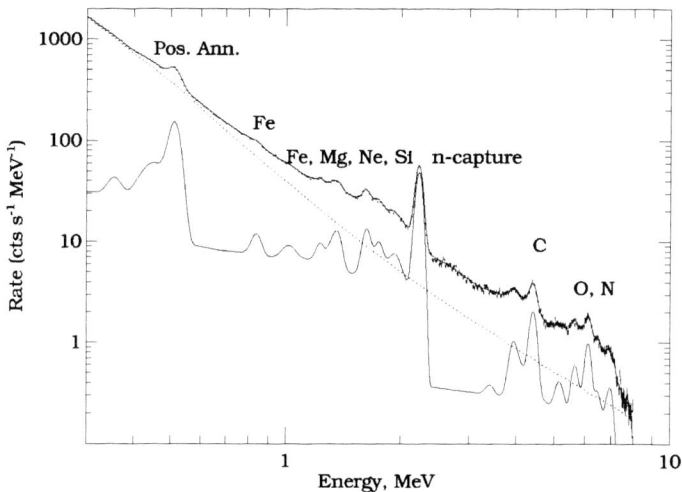

FIGURE 4. Count spectrum derived from the sum of 19 flares observed by *SMM*. Solid curve drawn through the points is best fit. The thin curve shows the best fit to the narrow lines.

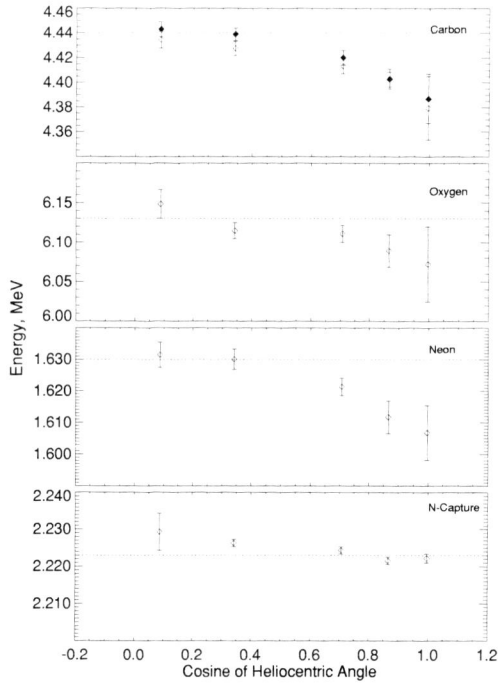

FIGURE 5. Variation of narrow de-excitation line energies with respect to heliocentric angle of the flare.

We found flare-to-flare variations in relative line fluences in the 19 solar flares [29], suggesting that the abundances of elements in the plasma group with respect to first ionization potential (FIP). Using these published line fluences, measured cross sections, and kinematical calculations, Ramaty, *et al.* [30] showed that the composition of the flare plasma is, on average, close to coronal. This is illustrated in Figure 6 where they compare the observed ^{24}Mg (low FIP)/^{16}O (high FIP) line ratios. A photospheric ambient abundance would require extremely steep accelerated particle spectra that are inconsistent with observations for most of the flares. However, flare '9' appears depleted in low FIP emission lines; this suggests a composition closer to that of the photosphere. As discussed in the section on positron annihilation above, there is other evidence for believing that the γ rays from this flare may have been produced at depths where the ambient composition was closer to photospheric. This suggests that flare particles may interact in plasmas with compositions ranging from those found in the upper photosphere to those in the corona. Recently reported spectroscopic measurements of flares with OSSE [14] suggest that the ambient composition may also change within flares. This implies that ions accelerated in different flares and at different times in flares may interact at significantly different depths in the solar atmosphere. This could happen, for example, if the height of magnetic mirroring increases with time.

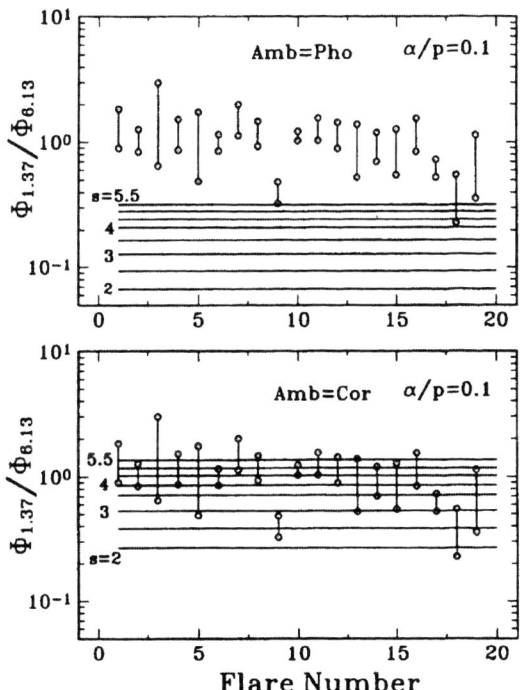

FIGURE 6. Measured Mg/O line ratios for 19 flares compared with calculations for photospheric and coronal ambient composition [30].

Another narrow line ratio, ^{20}Ne/^{16}O, can be used to estimate the spectrum of accelerated ions above ~5 MeV [29]. The measurements are consistent with power laws [30]. Assuming that these power laws extend down to 1 MeV, Ramaty and Mandzhavidze [19] compared the energies contained in accelerated ions and electrons (>20 keV). We show this comparison in Figure 7 for the 19 SMM flares and also for the 1991 June 4 flare observed by OSSE [14]. From this comparison we see that there is variability in the relative energies imparted to electrons and ions during acceleration in flares, but on average they are comparable. It is important to note, however, that this sample of flares is biased as it was specifically selected because of the strong nuclear contribution. HESSI has the potential for detecting a proton capture line near 2.37 MeV that can measure the energy contained in protons below 1 MeV [31]. These low-energy protons can have other measurable effects during flares if with they are energetically important.

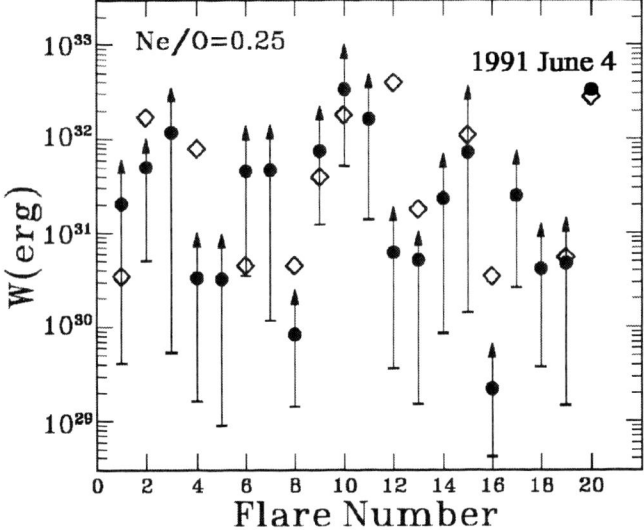

FIGURE 7. Calculated energy content in flare-accelerated ions (filled circles) and electrons (diamonds) [19].

Accelerated Helium in Flares

The inset of Figure 1 shows a detail of the region containing α-He fusion lines. We have found high fluxes in these lines relative to the de-excitation lines in the 1991 June 4 flare [14] and in the 19 *SMM* flares [32]. This led us to conclude that the accelerated α/p ratio typically had to be large, ~0.5, for an assumed ambient ^{4}He/H abundance ratio of 0.1. Mandzhavidze, *et al.* [33] suggested that the ambient ratio might also be higher in some flares and described a way in which γ-ray spectroscopy could distinguish between the two explanations. This required the measurement of other lines that only result from interactions of α-particles on ^{56}Fe. There is evidence for a weak line at 0.339 MeV from such interactions (inset of Figure 1) in the spectra. Based on this we concluded that, on average, the ambient ^{4}He abundance is consistent with accepted photospheric values and a high accelerated α/p ratio is needed [34]. Mandzhavidze et al. performed studies of individual flares and concluded that there is evidence for both a higher accelerated α/p ratio and enhanced ambient ^{4}He [35].

These same spectral studies have provided information on the accelerated ^{3}He/^{4}He ratio in flares. We studied the summed spectrum of 19 *SMM* flares in the 0.7 to 1.5 MeV region (see Figure 4). This region contains the relative strong 0.847 and 1.238 MeV lines from ^{56}Fe, the weak line from ^{55}Fe at 1.317 MeV, and the strong ^{24}Mg line at 1.369 MeV. The key line features for understanding the ^{3}He abundance appear near 0.937 MeV and ~1.03 MeV. The relative strength of the ~1.03 MeV feature suggests a high accelerated α/p ratio from interactions on ^{56}Fe and/or a high ^{3}He flux producing lines after interaction with ^{16}O. There is evidence for ^{3}He in the 4 σ detection of the 0.937 MeV line and in the shift of the ~1.03 MeV feature to higher energies in comparison with a model for ^{3}He/^{4}He=0. This suggests a flare-averaged ^{3}He/^{4}He ratio of ~0.1, about 10^3 times the photospheric ratio [34,35].

Accelerated Heavy Ions

We have revealed the broad γ-ray lines from interactions of accelerated ions with ambient H and ^{4}He in spectral data from OSSE and *SMM* [36]. Lines attributable to accelerated ^{56}Fe and ^{12}C appear reasonably well resolved in the summed 19-flare *SMM* spectrum shown in Figure 8 after bremsstrahlung and narrow lines have been removed. The broad lines have widths of ~30% FWHM and appear to be redshifted. Broad lines from ^{24}Mg, ^{20}Ne, and ^{28}Si cannot be resolved from each other and the contribution from unresolved lines. The higher energy N and O lines are also blended. Comparison of broad-line fluxes from accelerated nuclei with the respective fluxes in narrow lines from the ambient material measure relative enhancements in the accelerated particles. We find that the accelerated ^{56}Fe abundance is enhanced over its ambient concentration by about a factor of 5 - 10, consistent with that measured in impulsive solar-energetic particles in space [37].

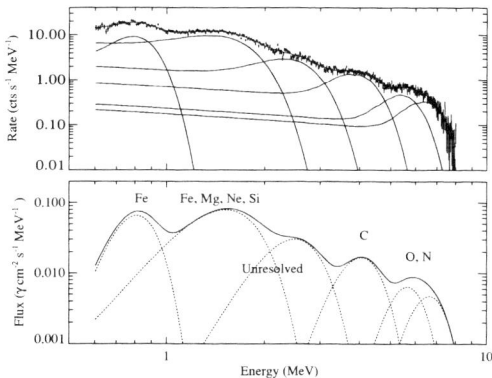

FIGURE 8. Gamma-ray spectrum revealing broad lines from accelerated heavy ions. a) count spectrum after subtracting narrow lines and bremsstrahlung; b) inferred photon spectrum with lines identified.

ACKNOWLEDGMENTS

The "coming of age of solar γ-ray physics" was in large part due to the theoretical and interpretative work of Reuven Ramaty and Natalie Mandzhavidze. Regretfully, they both recently succumbed to chronic illnesses. This work was supported under NASA DPR W-18995.

REFERENCES

1. Weekes, T. C., *Physica Scripta* **T85**, 195-209 (2000).
2. Leising, M. D., and Share, G. H., *Astrophys. J.* **357**, 638-648 (2000).
3. Diehl, R., Iyudin, A., Oberlack, U., *et al., Nucl. Phys. A* **62**, C79-C82 (1997).
4. Purcell, W. R., Grabelsky D. A., Ulmer M. P., *et al., Astrophys. J.* **413**, L85-L88 (1993).
5. Harris, M. J., Share, G. H., Messina, D.C., *Astrophys. J.* **448**, 157-163 (1995).
6. Ramaty, R., Kozlovsky, B., and Lingenfelter, R.E, *Astrophys. J. Supp.* **40**, 487-526 (1979).
7. Hunter S. D., Bertsch D. L., Catelli J. R., *et al.*, Astrophys. J. **481**, 205-240 (1997).
8. Strong A.W., Moskalenko I. V., and Reimer O., *Astrophys. J.* **537**, 763-784 (2000).
9. Falcone, A. D., *et al.*, in *AIP Conf. Proc. 528,* edited by R. A. Mewaldt, *et al.*, 2000, pp 193-196.
10. Kanbach G., Bertsch D. L., Fichtel C. E., *et al., Astron. Astrophys. Sup.* **97**, 349-353 (1993).
11. Miller, J. A., Cargill P. J., Emslie A. G., *et al.*, J. Geophys. Res. **102: (A7)**, 14631-14659 (1997).
12. Masuda, S., Kosugi, T., Hara H., *et al., Nature* **371**, 495-497 (1994).
13. Lin, R.P., *et al.*, in *High Energy Solar Physics-Anticipating HESSI,* edited by R. Ramaty and N. Mandzhavidze, San Francisco: Astron. Soc. Pacific, 2000, pp. 1-12.
14. Murphy, R. J., Share, G. H., Grove, J. E., *et al., Astrophys. J.* **490**, 883-900 (1997).
15. Ramaty, R. and Murphy, R J., *Space. Sci. Rev.* **45**, 213-268 (1987).
16. Chupp, E. L., *Science* **250**, 229-236 (1990).
17. Mandzhavidze N., and Ramaty R., *Nucl. Phys. B* **33**, 141-160 (1993).
18. Hudson H., and Ryan J., *Ann. Rev. Astron. Astr*. **33**, 239-282 (1995).
19. Ramaty, R., and Mandzhavidze, N., "Gamma Rays from Solar Flares," in *Highly Energetic Physical Processes and Mechanisms for Emission from Astrophysical Plasmas,* edited by P.C.H. Martens, S. Tsuruta, and M. A. Weber, San Francisco: Astron. Soc. Pacific, 2000, pp. 123-131.
20. Share, G.H., and Murphy, R. J., in *High Energy Solar Physics-Anticipating HESSI,* edited by R. Ramaty and N. Mandzhavidze, San Francisco: Astron. Soc. Pacific, 2000, pp. 377-386.
21. Share, G. H., and Murphy, R. J., in *AIP Conf. Proc. 528,* R. Mewaldt, *et al.*, ed., 2000, pp.181-184.
22. Murphy, R. J., *et al.*, in *AIP Conf. Proc. 280,* M. Friedlander, *et al.*, eds., 1993, pp. 619-630.
23. Harris, M. J., Share, G. H., Beall, J. H., Murphy, R. J., *Solar Physics* **142**, 171-185 (1992).
24. Feffer, P. T., *et al., Solar Physics* **171**, 419-445 (1997).
25. Hua, X.-M., and Lingenfelter, R. E., *Astrophys. J.* **319**, 555-566 (1987).
26. Yoshimori, M., Shiozawa, A., and Suga, K., *Proc. 26th Int. Cosmic Ray Conf.* **6**, 5-8 (1999).
27. Murphy, R. J., and Share, G. H., in *AIP Conf. Proc. 510,* M. McConnell and J. Ryan, eds., 2000, pp. 559-563.
28. Share, G. H., Murphy, R. J., and Skibo, J. G., in *AIP Conf. Proc. 374,* R. Ramaty, N. Mandzhavidze, X.-M. Hua, eds., 1996, pp. 162-170.
29. Share, G. H., and Murphy, R. J., *Astrophys. J.* **452**, 933-943 (1995).
30. Ramaty, R., Mandzhavidze, N., Kozlovsky, B., *et al.*, *Astrophys. J .* **455**, L193-L196 (1995).
31. Share, G. H., Murphy, R. J., and Newton E. K., *Solar Physics*, in press.
32. Share, G. H., and Murphy, R. J., *Astrophys. J.* **485**, 409-418 (1997).
33. Mandzhavidze, N., Ramaty, R., and Kozlovksy, B. *Astrophys J.* **489**, L99-L102 (1997).
34. Share, G. H., and Murphy, R. J., *Astrophys. J.* **508**, 876-884, (1998).
35. Mandzhavidze, N., Ramaty, R., and Kozlovksy, B. *Astrophys J.* **518**, 918-925 **(**1999).
36. Share, G. H., and Murphy, R. J., *Proc. 26th Int. Cosmic Ray Conf.* **6**, 13-16 (1999).
37. Reames, D. V., *Space Sci. Rev.* **90**, 413-491 (1999).

COMPTEL Gamma-Ray Observations of the C4 Solar Flare on 20 January 2000

C. A. Young[a], M. B. Arndt[b], K. Bennett[c], A. Connors[d], H. Debrunner[e], R. Diehl[f], M. McConnell[g], R. S. Miller[g], G. Rank[f], J. M. Ryan[g], V. Schoenfelder[f], C. Winkler[c]

[a]*Emergent-IT, Inc., Goddard Space Flight Center*
[b]*Bridgewater State College*
[c]*European Space Research and Technology Center*
[d]*Eureka Scientific*
[e]*University of Bern*
[f]*Max Planck Institute for Extraterrestrial Physics*
[g]*University of New Hampshire*

Abstract. The "Pre-SMM" (Vestrand and Miller 1998) picture of gamma-ray line (GRL) flares was that they are relatively rare events. This picture was quickly put in question with the launch of the Solar Maximum Mission (SMM). Over 100 GRL flares were seen with sizes ranging from very large GOES class events (X12) down to moderately small events (M2). It was argued by some (Bai 1986) that this was still consistent with the idea that GRL events are rare. Others, however, argued the opposite (Vestrand 1988; Cliver, Crosby and Dennis 1994), stating that the lower end of this distribution was just a function of SMM's sensitivity. They stated that the launch of the Compton Gamma-ray Observatory (CGRO) would in fact continue this distribution to show even smaller GRL flares. In response to a BACODINE cosmic gamma-ray burst alert, COMPtonTELescope on the CGRO recorded gamma rays above 1 MeV from the C4 flare at 0221 UT 20 January 2000. This event, though at the limits of COMPTEL's sensitivity, clearly shows a nuclear line excess above the continuum. Using new spectroscopy techniques we were able to resolve individual lines. This has allowed us to make a basic comparison of this event with the GRL flare distribution from SMM and also compare this flare with a well-observed large GRL flare seen by OSSE.

On January 20, 2000, a GOES C4.1 class solar flare occurred. The soft X-ray flux began at 8640 s (02:21 UT) peaked at 8760 s (02:26 UT) and ended at 9000 s (02:30 UT). There was no Hα identification but the Nobeyama radio telescope observed a radio burst from the flare at N15W33 corresponding to NOAA active region number 8829. COMPtonTELescope's rapid gamma ray burst response system was triggered by BATSE at 8642 s. The trigger from BATSE also alerted OSSE, which subsequently slew its four detectors to the direction of the Sun.

The automated system of COMPTEL imaged the Sun at a significance of 7.2 σ using 94 events recorded from annuli within 1° of the Sun. Significant emission was detected from approximately 8640 s to 8740 s (Figure 1). There is evidence in the energy loss spectrum for nuclear line emission from ~1 – 10 MeV (Figure 2). The flare was only observed in the telescope mode, no emission was detected in the raw or processed burst data (see Schönfelder et al. 1993 for a discussion of COMPTEL).

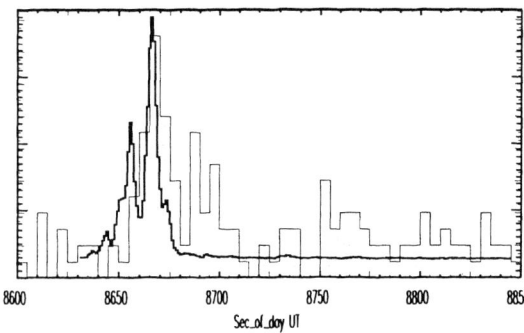

FIGURE 1. Light curves of the 20 January 2000 events observed in BATSE 1024 ms data (solid line) and the COMPTEL telescope (histogram).

OSSE observed emission above one MeV during the time from 8663 s to 8696 s (Murphy 2000). A single power law with photon spectral index -2.85±0.3 (Murphy 2000) provided an adequate fit to the OSSE data with no evidence of nuclear line emission. The total fluence above 50 keV and 1 MeV was 86±1 and 0.8±0.6 photons cm^{-2}, respectively (Murphy 2000). The OSSE data place a 2-σ upper limit for the 2.223 MeV neutron capture line of 0.4 photon cm^{-2}.

We selected the analysis interval for the COMPTEL telescope data to be 8640 s to 8740 s. Normally, the background is modeled by using the average of 15 and 16 orbits before and after the event so that the geomagnetic conditions during the flare are similar. However, due to large data gaps only data 15 orbits before the flare are available. This limits the ability to produce the most accurate background model. Consequently, a different approach to background estimation was necessary. The next reasonable choice was to choose intervals just before and just after the flare as the background estimate.

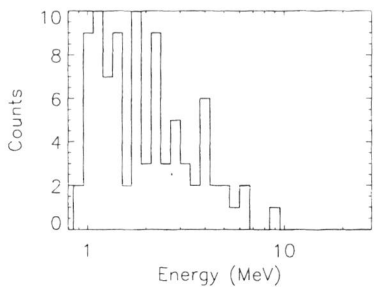
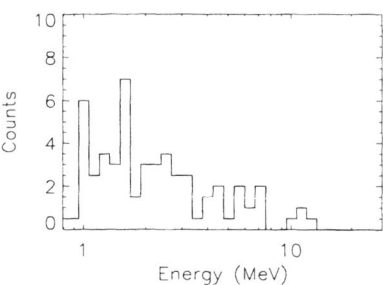

FIGURE 2. COMPTEL energy loss spectrum and background energy loss spectrum of the 20 January 2000 solar flare.

After a suitable background was selected, the energy loss spectrum (Figure 2a) and the selected background (Figure 2b) were deconvolved (using a new deconvolution technique, see Young et al. 2001) with the instrument response to obtain an estimate of the flux spectrum. In order to deal effectively with few counts in the energy loss spectrum at energies greater than ~8 MeV, two energy binning were used. The flux spectrum was computed for a binning of 32 energy bins and 8 energy bins. These two were then combined, using the finer binning up to 8 MeV and the coarser binning greater than 8 MeV.

Ideally, one would fit a gamma-ray spectrum with the individual components of the spectrum varying all parameters. Even for a large, intense event such as the June 4, 1991 event, COMPTEL does not have the statistics to resolve all the components such as the broad lines. A first approach even for a large event is to fit the expected strong lines and a composite spectrum for the remaining lines and continuum, varying only the amplitude of the lines and the composite spectrum. However, for the case of the January 20 flare, the fitting process is more difficult due the small number of counts.

Our approach to obtaining a reasonable model fit was to use data from the BATSE instrument to estimate the intensity and shape of the bremsstrahlung continuum. Using the most solar-facing BATSE detector, the data from 30 keV to 1000 MeV was fit with a broken power law with a first index of -3.13±0.5, a break energy of 86.7 keV and a second index of -2.85±0.02 yielding a continuum flux above 1 MeV of $(3.2\pm0.2)\times10^{-3}$ γ cm^{-2} s^{-1}. The higher energy power law for the BATSE data is consistent with the fit obtained by OSSE.

In addition to the BATSE based power law continuum, different combinations of several nuclear components were tested. Eight different combinations were used. The two standard components of all the models were the power law from BATSE and a narrow line of unknown strength at 2.223 MeV. The best fit model (Figure 3) contained the addition of 3 narrow lines at 1.1, 1.8, and 4.4 MeV to account for the 3 strongest features in addition to the 2.2 MeV line (These four lines were determined significant with a P-value of 3×10^{-3} (3 σ)) by calculating the probability, under the assumption the mean rate of the source is zero, of obtaining as many events as observed or more given the background (Cowan 1998)). The model also included a composite spectrum of previously identified broad lines (Share and Murphy 1995).

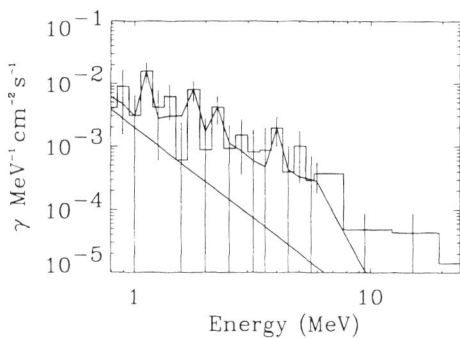

FIGURE 3. The best-fit model for the January 20, 2000 event. The model consists of a power law based on the BATSE data fit, a broad line template, and lines at 1.1,1.8,2.2, and 4.4 MeV. The straight solid line is the power law fit and the other solid line is the complete model. (Error bars extending to the bottom of the plots are 1 σ upper limits.)

Using the results from the OSSE solar flare web page (Murphy 2000) and those of Murphy et al. 97, we can compare this small flare (model 2) to the X12+ class flare from 4 June 1991. These values are shown in Table 1.

TABLE 1. A comparison of the COMPTEL measurement of the January 20 2000 C4 GOES event and the OSSE measurement of the June 4 1991 X12+ GOES event.

	2000 January 20	1991 June 4
2.223 MeV Fluence	0.23±0.1	1050±19
4.4 MeV fluence	1.28±0.1	189±9
> 1 MeV fluence	0.813±0.67	~3500±63
2.223/4.4 fluence ratio	0.18±0.14	5.56±0.28
Proton spectral index (Ramaty 1996)	> 5.5 ($\alpha/p = 0.1$)	3.37±0.1 ($\alpha/p = 0.1$)
2.223/>1 MeV fluence ratio	0.24	~0.3
# > 30 MeV protons	1×10^{31}	$(6.7 \pm 1.2) \times 10^{32}$

One final important comparison for this flare is with the set of gamma-ray line events observed by SMM. SMM was pivotal in dispelling the idea that gamma-ray line emission was rare and only occurred in the largest of flares. However, due to its sensitivity SMM never had a positive detection of nuclear lines from any of the C class flare that it observed. COMPTEL, which is roughly an order of magnitude more sensitive than SMM has extended the SMM distribution of gamma ray flares by roughly an order of magnitude smaller. Figure 4 shows the SMM distribution of flare narrow nuclear line fluence as a function of the continuum power law fluence and GOES soft X-ray class (Vestrand et al. 1999).

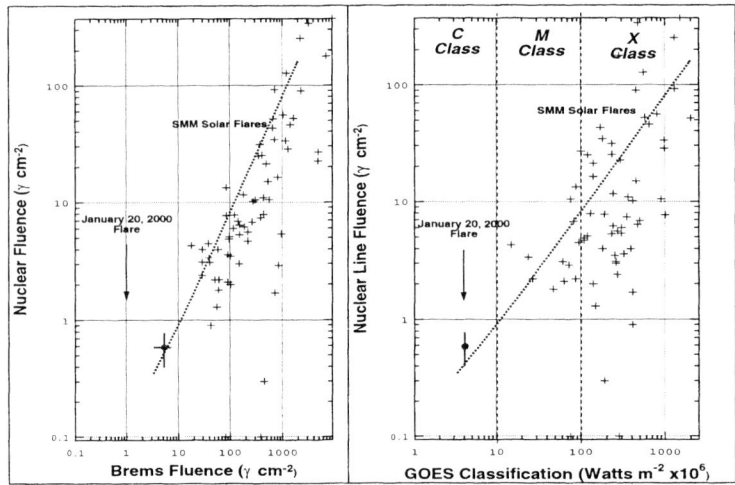

FIGURE 4. Scatter plots of narrow nuclear line fluence vs. continuum fluence and GOES classification respectively for the SMM GRS catalog and the 20 January 2000 event observed with COMPTEL.

All of these comparisons indicate that as compared to the June 4, 1991 flare and the SMM observed gamma-ray flares, the solar flare on January 20, 2000 is a normal or ordinary gamma-ray line solar flare. Actually, this comparison has shown that the 4

June 1991 flares is the odd flare of the bunch. It shows an unusually large electron bremsstrahlung component.

In response to a BACODINE cosmic gamma-ray burst alert, COMPTEL on the CGRO recorded gamma rays above 1 MeV from the C4 flare at 0221 UT 20 January 2000. This event, though at the limits of COMPTEL's sensitivity, clearly shows a nuclear line excess above the continuum. Using the new spectroscopy techniques we are able to resolve individual lines. This has allows us to make a basic comparison of this event with the GRL flare distribution from SMM and also compare this flare with a well-observed large GRL flare seen by OSSE. We show that this flare is normal, i.e., it is a natural extension of the SMM distribution of flares. The analysis of this flare means there is no evidence for a lower flare size for proton acceleration. Protons even in small flares contain a large part of the accelerated particle energy.

ACKNOWLEDGMENTS

This work was supported through NASA contract NAS5-26645 and NASA's Supporting Research and Technology program.

REFERENCES

1. Bai, T., *Ap. J.* **308**, 912 (1986).
2. Cliver, E. W. et al., *Ap. J.* **426**, 767-773 (1994).
3. Cowen, G., *Statistical Data Analysis*, Oxford: Claredon Press, 1998, p. 59.
4. Murphy R. J. et al., *Ap. J.* **490**, 883-900 (1997).
5. Murphy, R. J., http://osse-www.nrl.navy.mil/solarflare/flarelib.htm,2000.
6. Ramaty, R. et al., "Solar Atmospheric Abundances from Gamma Ray Spectroscopy" in *High Energy Solar Physics Workshop*, edited by R. Ramaty et al., AIP Conference Proceedings 374, New York: American Institute of Physics, 1996, pp. 172-183.
7. Schönfelder, V. et al., *Ap. J. Suppl. Series* **86**, 657-692 (1993).
8. Share, G.H. and Murphy, R. J., *Ap. J.* **452**, 933 (1995).
9. Vestrand, W. T., *Solar Phys* **118**, 95 (1988).
10. Vestrand, W. T., and Miller, J. A., "Particle Acceleration in Flares," in *The Many Faces of the Sun: A Summary of the Results from NASA's Solar Maximum Mission*, edited by K. T. Strong et al., New York City: Springer-Verlag, 1998, pp. 231-272.
11. Vestrand, W. T. et al., *Ap. J. Suppl. Series* **120**, 409-467 (1999).
12. Young, C. A. et al., "Bayesian Multiscale Deconvolution Applied to Gamma-ray Spectroscopy" in *These Proceedings*, 2001.

X- and Gamma-Ray Observations of the 15 November 1991 Solar Flare

M. B. Arndt[1], A. Connors[1], J. Lockwood[1], M. McConnell[1], R. Suleiman[1], J. Ryan[1], C.A. Young[1], G. Rank[2], V. Schönfelder[2], H. Debrunner[3], K. Bennett[4], O. Williams[4], and C. Winkler[4]

[1]*University of New Hampshire,* [2]*Max Plank Inst. für Extraterrestrische Physik,* [3]*Univ. Bern*
[4]*SSD/ESTEC*

Abstract. This work expands the current understanding of the 15 November 1991 Solar Flare. The flare was a well observed event in radio to gamma-rays and is the first flare to be extensively studied with the benefit of detailed soft and hard X-ray images. In this work, we add data from all four instruments on the Compton Gamma Ray Observatory. Using these data we determined that the accelerated electron spectrum above 170 keV is best fit with a power law with a spectral index of -4.6, while the accelerated proton spectrum above 0.6 MeV is fit with a power law of spectral index -4.5. From this we computed lower limits for the energy content of these particles of ~10^{23} ergs (electrons) and ~10^{27} ergs (ions above 0.6 MeV). These particles do not have enough energy to produce the white-light emission observed from this event. We computed a time constant of 26^{+20}_{-15} s for the 2.223 MeV neutron capture line, which is consistent at the 2σ level with the lowest values of ~70 s found for other flares. The mechanism for this short capture time may be better understood after analyses of high energy EGRET data that show potential evidence for pion emission near ~100 MeV.

INTRODUCTION

The goals of this dissertation work were to add to the extant body of knowledge of the 15 November 1991 solar flare in two ways: by analyzing high-energy data from the Compton Gamma Ray Observatory (CGRO) that have been underutilized in previous studies and by applying another flare model to explain the most intense high-energy emission from the event. In these proceedings we only have space to discuss the CGRO data and analyses.

This X1.5 event was a well observed flare in a broad range of wavelengths [1-26] and is the first to be extensively studied with the benefit of detailed X-ray images. This flare was located near disk center and lasted on the order of minutes in X-rays.

DATA ANALYSIS

COMPTEL data analysis was done using the Maximum Entropy Method (MEM). In this method, a test photon spectrum is folded through an instrument response and compared to the measured count spectrum using a χ^2 test. EGRET photon spectra were also generated by a MEM approach.

BATSE data were fit with a double power law where E_B is the break energy:

$$I(E) \propto \begin{cases} E^{-\gamma_1} \text{ for } E < E_B \\ E^{-\gamma_2} \text{ for } E > E_B \end{cases} \quad (1)$$

RESULTS

Figure 1 is a composite spectrum from the impulsive phase (22:36:36 – 22:38:14 UT) of the flare. All data are background subtracted. The BATSE curve (0.2 – 10 MeV) is an extrapolated fit where $\gamma_2 = -3.6$. The discrepancy between COMPTEL (0.6 – 10 MeV) and EGRET (1 – 300 MeV) spectra near 10 MeV and the emission near 60 MeV are most likely due to background subtraction issues with EGRET data.

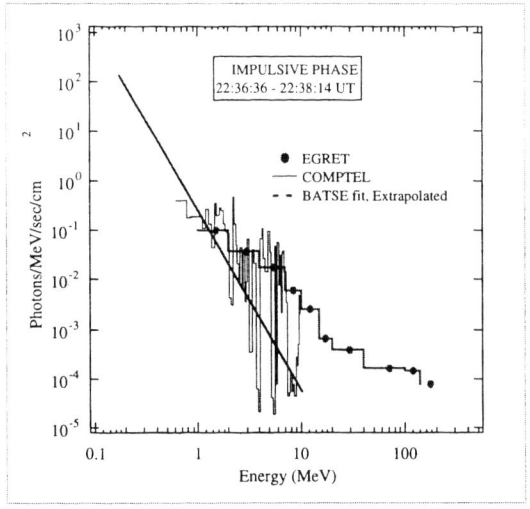

FIGURE 1. Composite Spectrum of the impulsive phase of the 15 November 1991 solar flare.

Accelerated Protons

Using fluences derived from COMPTEL spectra we are able to deduce the shape of the accelerated proton spectrum. Two accelerated proton spectra shapes are considered: the Bessel function, parameterized by αT and the power law, parameterized by s. Our values (bold) and previously published values of αT and s [27-29] are listed in Table 1 along with fluence ratios derived from several high-energy emission lines.

The proton spectrum above 0.6 MeV is best fit with a power law of s = −4.5. The energy content of these ions is ~10^{27} ergs. These particles do not have enough energy to produce the observed white light emission, which has an energy content on the order of 10^{30} ergs[6].

TABLE 1. Shape of accelerated proton spectra.

Fluence Line Ratio	Value	αT	s
$\phi_{4.4}/\phi_{0.42}$	0.035 – 0.065[13]	--	--
$\phi_{4.4}/\phi_{2.223}$	0.52 ± 0.14	0.009±0.002[11-12]	4.5-5
$\phi_{4.7}/\phi_{2.223}$	1.6 ± 0.34	0.008 -0.015	4-5
$\phi_{4.7}/\phi_{2.223}$	--	0.010 ± 0.002[25]	4-5
$\phi_{4.4}/\phi_{6.13}$	2.11 ± 0.47	--	--
$\phi_{1.63}/\phi_{6.13}$	4.95 ± 1.0	--	4.5-5.5

2.223 MeV Emission

Using emission between 3.956 and 7.055 MeV as a template for the neutron production rate S(t′), we compute the 2.223 MeV time constant τ using the expression

$$F_{2.223MeV}(t) \propto \int_{-\infty}^{t} S(t')e^{-(t-t')/\tau} dt'. \quad (2)$$

This time constant τ is a function of the neutron decay time as well as proton and ^{3}He neutron capture times. Subsequently, τ can provide information about proton and ^{3}He densities. We found the best fit to be 26^{+20}_{-15} s, which is consistent at the 2σ level with values of ~70 s found for other flares.

A low value of τ suggests the presence of either an unusually high ^{3}He abundance or that neutron capture is occurring in a dense environment where neutrons thermalize and are quickly captured on ^{1}H.

Given our τ and a typical chromospheric density, we find a ^{3}He/H abundance ratio nearly an order of magnitude higher than values computed for other flares, but in agreement with recent results [30].

The white light emission from this flare suggests that very high energy protons are penetrating into the photosphere. The neutrons created in this dense layer would be captured quickly, also resulting in a low τ.

Accelerated Electrons

Table 2 is a summary of spectral indices from various instruments. Included are satellite viewing angles, power law spectral indices below (γ_1) and above (γ_2) break energies. If error bars are not included, they were not present in the literature. The data do not all agree within error bars, however the discrepancies may be explained by the different viewing angles of each instrument. We use the indices derived from BATSE data in our work. The energy content (lower limit) from electrons above 170 keV is ~10^{23} ergs.

TABLE 2. Spectral Indices of Accelerated Electrons During Impulsive Phase.

Satellite (Viewing Angle)	γ_1	E_B (KeV)	γ_2
BATSE (~20°)	2.66 ± 0.27	168 ± 51	3.61 ± 0.23
OSSE (~20°)	3.0 ± 0.7	100	--
Yohkoh (~20°)	2 – 4.5[19] 3.7 ± 0.3[31]	93	--
PVO (~52°)	--	150	3.37 ± 0.05[15]
Ulysses (~80°)	3.08[8]	166(γ_1), 150 (γ_2)	2.72 ± 0.07[15]
Yohkoh (HXS)	3.20[8]	87	3.82[2] 3.70 ± 0.03[15]
Yohkoh (HXT)	3.39[8]	93	--

CONCLUSIONS

Our goals with this work were to add to the extant body of knowledge of the 15 November 1991 solar flare by introducing new high-energy data from the CGRO.

These data allowed us to confirm previous results and to compute the 2.223 MeV time constant, which is consistent (but only at the 2 σ level) with the lowest values computed for other flares. We computed the accelerated particle spectra and subsequent energy content of these particles. We also found that the accelerated protons do not have enough energy to produce the observed white light emission.

Future Work

Once we have had the opportunity to analyze EGRET data in more depth we will further improve our understanding of the high-energy particle dynamics within this flare. If EGRET does observe extended pion emission we have further evidence that high-energy protons are reaching deep into the chromosphere or photosphere. This extended pion emission would also allow us to reclassify this event as a long duration gamma-ray flare.

The 15 November 1991 solar flare was unique because it was observed in a broad energy range with detailed X-ray images. We look forward to HESSI providing us with similarly well observed events.

ACKNOWLEDGMENTS

I would like to thank my advisor Dr. J. M. Ryan and my other committee members (Drs. M. McConnell, J. Hollweg, S. Habbal, and D. Meredith) for their help and support. I would also like to thank Drs. T. Sakao, M. Lee, T. Forbes, R. Murphy, H. Debrunner, and M. Yoshimori for our enlightening conversations. I would also like to acknowledge R. Murphy (OSSE), M. Yoshimori (YOHKOH), D. Bertsch (EGRET), and R. Schwartz (BATSE) for their help and access to data. This work was funded by the University of New Hampshire, a NASA Space Grant and NASA grants NAG5-2350, NAG5-7179, NAG5-3802, and NAS5-26645.

REFERENCES

1. Aschwanden, M.J., Kosugi, T., Hudson, H.S., Wills, M.J., and Schwartz, R.A., *Astrophys.J*, **470**, 1198, (1996).
2. Aschwanden, M.J., Wills, M.J., Hudson, H.S., Kosugi, T., and Schwartz, R.A., *Astrophys.J*, **468**,398,(1996).
3. Canfield, R.C., Hudson, H., Leka, K.D., Mickey, D.L., Metcalf, T.R., Wuelser, J.-P., Acton, L.W., Strong, K.K.T., Kosugi, T., Sakao, T., Tsuneta, S., Culhane, J.L., Phillips, A., and Fludra, A., *Publ. Astron. Soc. Japan.*, **44**, No.5, L111, (1992).
4. Canfield, R.C. and Reardon, K.P., *Sol. Phys.*, **182**, 145, (1998).
5. Culhane, J.L., et al., *Adv. Space Res.*, **13**, No. 9, 303, (1993).
6. Hudson, H.S., Acton, L.W., Hirayama, T., and Uchida, Y., *Publ. Astron. Soc. Japan*, **44**, No. 5, L77, (1992).
7. Inda-Koide, M., Makishima, K., Kosugi, T., and Kaneda, H., *Publ. Astron. Soc. Japan*, **47**, 661, (1995).
8. Kane, S.R., Hurley, K., McTiernan, J.M., Boer, M., Niel, M., Kosugi, T., and Yoshimori, M., *Astrophys.J*, **500**,1003, (1998).
9. Kane, S.R., Hurley, K., McTiernan, and J.M, Sommer, M., *Adv. Space Res.* **13**, No. 9, 241, (1993).
10. Kane, S.R., McTiernan, J.M., Loran, J., Lemen, J., Yoshimori, M., Ohki, K., and Kosugi, T., *Adv. Space Res.*, **13**, No. 9, 237, (1993).
11. Kawabata, K., Yohimori, M., Suga, K., Morimoto, K., Hiraoka, T., Sato, J., and Ohki, K., *Astrophys.J Suppl. Ser.*, **90**, 701, (1994).
12. Kawabata, K., Yohimori, M., Suga, K., Morimoto, K., Hiraoka, T., Sato, J., and Ohki, K., in *Proceedings of Kofu Symposium*, NRO Report No. 360, 123, (1994).
13. Kotov, YU.D., Bogovalov, S.V., Endalova, O.V., and Yoshimori, M., *Astrophys.J*, **473**, 514, (1996).
14. Matthews, S.A., Brown, J.C., and van Driel-Gesztelyi, L., *Astron. Astrophys.*, **340**, 277, (1998).
15. McTiernan, J.M., Kane, S.R., Hurley, K., Laros, J.G., Fenimore, E.E., Klebsadel, R.W., Sommer, M., and Yoshimori, M., in *Proceedings of Kofu Symposium*, NRO Report No. 360, 389, (1994).
16. Sakao, T., Ph.D. Thesis, National Astronomical Observatory, Japan, (1994).
17. Sakao, T., Kosugi, T., and Masuda, S., in *Proceedings of Observational Plasma Astrophysics, Five Years of Yohkoh and Beyond*, Kluwer Academic Publishers, (1998).
18. Sakao, T., et al., *Publ. Astron. Soc. Japan*, **44**, No. 5, L83, (1992).
19. Sakao, T., Kosugi, T., Masuda, S., Yaji, K., Inda-Koide, M., and Makishima, K., in *Proceedings of Kofu Symposium*, NRO Report No. 360, 169, (1994).
20. Sylwester, B. and Sylwester J., Evolution of Yohkoh Observed White Light Flares, Kluwer Academic Publishers, 2000 (in press).
21. Takakura, T., Kosugi, T., Sakao, T., Makishima, K., Inda-Koide, M., and Masuda, S., *Publ. Astron. Soc. Japan*, **47**, 355, (1995).
22. Wülser, J.-P., Canfield, R.C., Acton, L.W., Culhane, J.L., Phillips, A., Fludra, A.J., Sakao, T., Masuda, S., Kosugi, T., and Tsuneta, S., *Astrophys.J*, **424**, No.1, 459, (1994).
23. Yoshimori, M., Kawabata, K., and Ohki, K., in *Proceedings of the XXIII International Cosmic Ray Conference*, University of Calgary, (1993).
24. Yoshimori, M., Morimoto, K., Suga, K., and Matsuda, T., in *Magnetodynamic Phenomena in the Solar Atmosphere – Prototypes of Stellar Magnetic Activity*, Kluwer Academic Publishers, (1996).
25. Yoshimori, M., Suga, K., Morimoto, K., Hiroko, T., Sato, J., Kawabata, K., and Ohki, K., *Astrophys.J Suppl. Ser.*, **90**, 639, (1994).
26. Yoshimori, M., Takai, Y., Morimoto, K., Suga, K., and Ohki, K., *Publ. Astron. Soc. Japan*, **44**, No. 5, L107, (1992).
27. Ramaty, R., Mandzhavidze, N., and Kozlovsky, B., in *Proceedings of High Energy Sol. Phys.*, Greenbelt Maryland, AIP # 374, (1995).
28. Ramaty, R., Mandzhavidze, N., Kozlovsky, B., and Murphy, R.J., *Astrophys.J Lett.*, **455**, L193, (1995).
29. Ramaty, R., Mandzhavidze, N., Kozlovsky, B., and Skibo, J.G., *Adv. Space. Res*, **13**, 275, (1993).
30. Young, C.A., these proceedings.
31. Yoshimori, M. personal communication, (2000).

Energetic Proton Spectra in the 11 June 1991 Solar Flare

C. A. Young[a], K. Bennett[b], A. Connors[c], R. Diehl[d], M. McConnell[e],
G. Rank[d], J. M. Ryan[e], R. Suleiman[f], V. Schönfelder[d], C. Winkler[b]

[a]*Emergent-IT, Inc., Goddard Space Flight Center*
[b]*European Space Research and Technology Center*
[c]*Eureka Scientific*
[d]*Max Planck Institute for Extraterrestrial Physics*
[e]*University of New Hampshire*
[f]*Havard-Smithsonian Center for Astrophysics*

Abstract. The June 11,1991 gamma-ray flare seen by the Compton Gamma-ray Observatory (CGRO) displays several features that make it a dynamic and rich event. It is a member of a class of long duration gamma-ray events with both 2.223 MeV and greater than 8 MeV emission for hours after the impulsive phase. It also contains an inter-phase between the impulsive and extended phases that presents a challenge to the standard gamma-ray line (GRL) flare picture. This phase has strong 2.223 MeV emission and relatively weak 4.44 MeV emission indicative of a very hard parent proton spectrum. However, this would indicate emission greater than 8 MeV, which is absent from this period. We present the application of new spectroscopy techniques to this phase of the flare in order to present a reasonable explanation for this seemly inconsistent picture.

During the 22nd solar cycle NOAA active region 6659 crossed the solar disk from 1 June to 15 June 1991 and produced some of the largest flares of that cycle. Six X-class flares occurred on 1,4,6,9,11, and 15 June 1991. After the X12 flare on 4 June 1991, the Sun was declared a CGRO target-of-opportunity and CGRO was re-oriented toward the Sun on 9 June 1991. This placed the Sun into the FoV of all CGRO's instruments from 9 to 15 June 1991. On 11 June 1991 a X12/3B flare started at 0156 UT as measured by the 1-8 Å SXR channel of GOES-7. The flare sit was at a heliographic location of N31W17. COMPTEL measured γ-ray emission from 0.8 to 30 MeV and neutrons for several hours (Ryan et al. 1993; McConnell et al. 1994; Suleiman et al. 1994). This included nuclear line emission, 2.223 MeV emission lasting over 5 hours (Rank 1996) and 8-30 MeV π^0 decay emission. The EGRET spark chamber could not observe the impulsive phase due to dead-time effects but observed > 1 GeV emission for at least 8 hours after the peak. The EGRET spectrum showed no sign of a high-energy cut-off (Kanbach et al. 1993). EGRET/TASC impulsive phase measurements of 2.223 MeV emission and nuclear line emission were reported (Dunphy et al. 1999; Schneid et al. 1994) along with evidence for π^0 emission, neutrons and spectral evolution (Dunphy et al. 1999). OSSE reported prolonged 2.223 MeV emission (Murphy et al. 1993) and nuclear emission, 0.511 MeV positron-annihilation emission, >16 MeV γ-rays and neutrons. BATSE-LADs measured HXRs and γ rays for about one hour in the energy range of 20 keV to ~1.9 MeV. CGRO was not the only γ-ray experiment that observed the 11 June 1991 flare. GRANAT/PHEBUS also reported observations of Bremsstrahlung, 2.223 MeV, and nuclear emission during the impulsive phase of the flare (Trottet et al. 1993;1994).

Figure 1 shows a lightcurve of the 11 June 1991 phase with both the Telescope and Burst modes of COMPTEL. The flare was subdivided into the three phases as defined by Rank (1996). Also included are the similar phases defined for an OSSE analysis (Murphy and Share 1999) and an EGRET/TASC analysis (Dunphy et al. 1999).

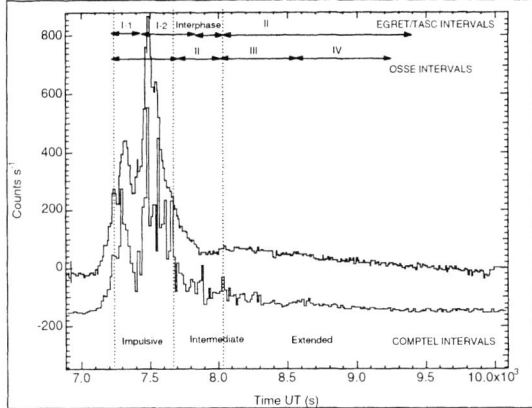

FIGURE 1. Light curves of the 11 June 1991 X-class solar flare as measured by the COMPTEL telescope (lower-blue) and burst modes (upper-green). Included are the observation intervals for COMPTEL, OSSE, and EGRET/TASC. The curves are slightly offset arbitrarily.

The COMPTEL (see Schönfelder et al. 1993 for a discussion of COMPTEL) event data (full source and 4 background sets) were binned in energy space. These spectra were deconvolved (using a new deconvolution technique, see Young et al. 2001) with the telescope response generated for the 11 June 1991 flare (based on its location within the COMPTEL FoV). Figure 2 is the photon flux spectrum for the full flare. Included in the plot is the best fit model composed of the 19 flare SMM broad line template (Share and Murphy 1995), a power law for the electron bremsstrahlung component (determined with BATSE data and PHEBUS (Trottet et al. 1993)), the 10 strongest narrow lines (based on flare modeling) and another power law to account for pion decay secondary Bremsstrahlung.

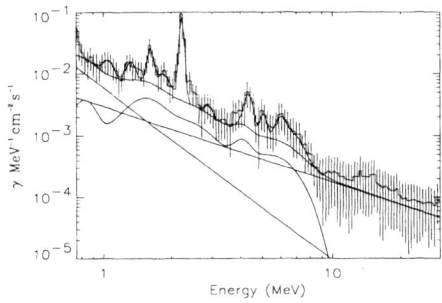

FIGURE 2. The COMPTEL flux spectrum for the entire 11 JUNE 91 flare. The 2 straight solid lines are from the bremsstrahlung model fit (steeper is primary component – see text). The solid line rolling over at 10 MeV is the broad line model fit. The solid line just above that is the sum of the 2 bremsstrahlung components and the broad line model. The solid line closely following the data is the sum of all the models including the narrow line component.

We have separately analyzed the data for each of the 3 phases (impulsive, intermediate, and extended) (See Figure 1) but we only discuss the intermediate phase here.

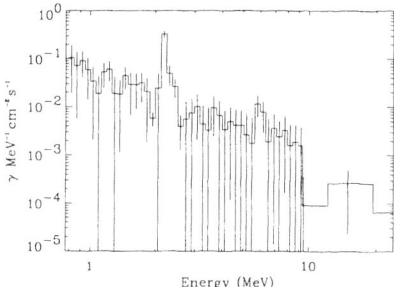

FIGURE 3. COMPTEL flux spectrum for the intermediate phase of the 11 JUNE 91 flare. (Error bars extending to the bottom of the plots are 1 σ upper limits.)

Figures 3 contain a plot of the deconvolved photon flux for the intermediate phase. The Intermediate (Rank 1996) or Interphase (Murphy and Share 1999, Dunphy et al. 1999) immediately following the peak of the impulsive phase is difficult to explain. All three analysis of this flare using COMPTEL, OSSE and EGRET data obtained a hard proton spectrum with an index around 2 using the 2.2 to 4-7 MeV fluence ratio. This hard a spectrum would suggests the possible presence of a high energy pion component above 8 MeV and an emission line at 5.3 due to spallation of C and O. However, none of the three instruments observed such a component. Murphy and Share (1999) argued that the inconsistencies in their measurements, in the interphase (II) and IV, indicate a two-component spectrum. Though this is plausible for the extended interval (IV), we do not agree with this conclusion for the interphase (II). We contend that this interval only contains a soft spectrum. If it did contain a hard spectrum, we would expect to see significant emission at 5.3 MeV and above 8 MeV. This is not seen in OSSE, EGRET/TASC, and not by the more sensitive instrument COMPTEL.

Determining the amount of expected charged pion component in the data with the COMPTEL data is difficult but we can estimate the expected amount of emission at 5.3 MeV. We were not able to fit a line at 5.3 MeV in the intermediate phase spectrum but we can estimate an upper limit for the emission. An upper limit for the measured emission at 5.3 MeV is $0.046 \, \gamma \, cm^{-2}$. If the parent proton spectrum was very hard with a spectra index of 2 (as indicated by 2.2-to-4.4 fluence ratio) then based on the measured 4.44 and 6.13 MeV flux we would expect a fluence of $0.28 \pm 0.07 \, \gamma \, cm^{-2}$ (Ramaty et al 1996; Mandzhavidze and Ramaty 2000). On the other hand, the softer proton spectrum of 5 (indicated by the 1.6/6.1 MeV fluence ratio) would produce a fluence of $0.06 \pm 0.02 \, \gamma \, cm^{-2}$. So the upper limit on the 5.3 MeV fluence is consistent with the softer proton spectrum.

We contend that the reason a hard spectrum is indicated by the 2.223-to-4.44 ratio is simply an indication of the inability of the ratio to give reliable results and its double valueness in this regime. The ratio turns up sharply because when the spectrum is soft the only neutron production channel is the p and α on CNO process. At these lower energies, the ratio turns up because the production of neutrons continues in this channel but the production of lines from CNO quickly turns off. So the ratio sharply turns up which is not accounted for in many of the published results

Another indicator of the soft spectrum comes from the study of the time decay of the 2.223 MeV line. The measurements from COMPTEL have been used to model

(Rank 1996) the 2.223 MeV emission with a decay constant of τ=230 s (long compared to the average 100 s). The time constant τ can be defined as (Prince et al. (1983)),

$$\tau = \left[\frac{1}{\tau_H} + \frac{1}{\tau_{He}} + \frac{1}{\tau_d}\right]^{-1}; \tau_H = \frac{1.4 \cdot 10^{19}}{n_H} s; \tau_{He} = \frac{8.5 \cdot 10^{14}}{r \cdot n_H} s \quad (1)$$

where τ_H is the capture time on hydrogen, τ_{He} is the capture time on ^{3}He, τ_d is the neutron decay time, n_H is the hydrogen number density, and r is the ^{3}He/H ratio. τ is maximum when r = 0 (no competing radiation less capture on ^{3}He). So if we set r = 0 and solve for n_H when τ = 230 s we find that n_H = 4.56·10^{16} cm^{-3}. This corresponds to a depth above the base of the photosphere of about 170 km (Fontenla et al. 1993). For an average r of 5·10^{-05} (Prince et al. 1983) the depth of would be about 50 km above the photosphere's base. These calculations suggest that the capture time of 230 s means neutrons are being captured higher up in the photosphere. For neutrons to be captured at a point of lower hydrogen density and higher height in the photosphere they would have to be of lower energy or traveling at a shallow angle in the atmosphere. The existence of low energy neutrons would indicate that the proton spectrum during this part of the flare was soft not hard. So given the indication of a soft spectrum in terms of unambiguous line ratios and the extremely long 2.223 MeV decay the most plausible explanation for the proton spectrum during the interphase is that it has a very soft spectrum.

Given a very soft proton spectrum we can estimate the ^{3}He content of the photosphere. What we will use is the neutron thermalization model we have developed (Young 2001). This will allow us to calculate the distribution of thermal neutrons in the solar atmosphere and then calculate the neutron absorption rate and decay time using a model for the density distribution for the lower chromosphere and the photosphere (Fontenla et al. 1993). Previous calculations of the ^{3}He content in the photosphere have not used information about the spatial distribution of the thermal neutrons. They only used a single average depth and so a single atmosphere density to calculate the ^{3}He/H ratio. We will now use our calculations of neutron transport to calculate a more realistic ratio using the spatial distribution of the thermal neutrons. Since the protons spectrum is soft, the neutron spectrum is dominated by low energy neutrons (<10 MeV). This allows the use of the analytical neutron transport model.

Using the distribution for thermal neutrons we calculated, we obtain a ^{3}He/H ratio of 8.7e-05 with a 1 σ range of 1.96e-04 to 1.75e-05. Previous values that have been reported are 5e-05 (no error reported) for the 7 June 1980 flare (Chupp et al. 1981), (2.3±1.2)e-05 for the 3 June 1982 flare (Hua and Lingenfelter 1987), 2.3e-05 (2 σ upper-limit) for the 4 June 1991 flare (Murphy et al. 1997) and (3.0±1.6)e-05 for this flare (Rank 1996).

The purpose of this work is to present an explanation of the puzzle presented by the spectral observations by CGRO of the intermediate phase of the 11 June 1991 solar flare. Analysis of observations by OSSE (Murphy and Share 1999), COMPTEL (Rank 1996) and EGRET (Dunphy et al. 1999) using standard spectroscopy methods indicates the presence of both a hard and a soft parent proton spectrum. We present an application of a new spectroscopy technique to the COMPTEL observations. We then show that the theoretically expected emission from a hard proton spectrum is not observed by COMPTEL. We conclude that the lack of this predicted emission and the longer than normal 2.223 MeV emission decay time (Rank 1996) can only be due to a soft parent proton spectrum. This means that the region of 2.223/4-7 MeV fluence space is largely unexplored for soft proton spectra. The use of this ratio must be reexamined for proton spectra with indices greater than 5 or 6. We then apply a model we developed for the transport of neutrons created from a soft proton spectrum to determine the photospheric ^{3}He abundance during this flare. We calculated a ^{3}He/H ratio of 8.7e-05 with a 1 σ range of 1.96e-04 to 1.75e-05 for this flare using this new model. This is larger than all previous values reported.

ACKNOWLEDGMENTS

This work was supported through NASA contract NAS5-26645 and NASA's Supporting Research and Technology program.

REFERENCES

1. Chupp, E. L., et al., *Ap. J. Letters* **244**, 171 (1981).
2. Dunphy et al., *Solar. Phys.* **65**, 2503-2504 (1999).
3. Fontenla, J. M. et al., *Ap. J.* **406**, 319 (1993).
4. Hua, X. M. and Lingenfelter, R. E., *Ap. J.* **319**, 555-566 (1987).
5. Kanbach, G. D., et al., *Astron. & Astrophy (Suppl. Series)* **97**, 349-353 (1993).
6. McConnell, M., "An Overview of Solar Flare Results from COMPTEL" in *High-Energy Solar Phenomena – A New Era of Spacecraft Measurements*, edited by J. M. Ryan and W. T. Vestrand, AIP Conference Proceedings 294, New York: American Institute of Physics, 1994, pp. 21-25.
7. Murphy R. J. et al., "OSSE Observations of Solar Flares" in *Compton Gamma-Ray Observatory Workshop*, edited by M. Friedlander et al., AIP Conference Proceedings 280, New York: American Institute of Physics, 1993, pp. 619-630.
8. Murphy R. J. et al., *Ap. J.* **490**, 883-900 (1997).
9. Murphy R. J. and Share G. H., "Accelerated-Particle Spectral Variability in the 1991 June 11 Solar Flare" in *The Fifth Compton Symposium*, edited by M. McConnell and J. M. Ryan, AIP Conference Proceedings 510, New York: American Institute of Physics, 1998, p. 559.
10. Prince, T. A. et al., "The Time History of 2.22 MeV Line Emission in solar Flares" in *18th International Cosmic Ray Conference*, Bangalore, India, 1983.
11. Ramaty, R. et al., "Solar Atmospheric Abundances from Gamma Ray Spectroscopy" in *High Energy Solar Physics Workshop*, edited by R. Ramaty et al., AIP Conference Proceedings 374, New York: American Institute of Physics, 1996, pp. 172-183.
12. Ramaty, R. and Mandzhavidze, N., "Particle Acceleration and Abundances from Gamma-Ray Line Spectroscopy" in *High Energy Solar Physics Workshop – Anticipating HESSI*, edited by R. Ramaty and N. Mandzhavidze., ASP Conference Series 206, San Francisco: Astronomical Society of the Pacific, 2000, pp. 64-70.
13. Rank, "Gamma Rays and Neutrons of the Solar Flares on 11 and 15 June 1991 Measured with COMPTEL" Ph.D. Thesis, Technical University of Munich, 1996.
14. Ryan, J. M. et al., "COMPTEL Gamma Ray and Neutron Measurements of Solar Flares" in *Compton Gamma-Ray Observatory Workshop*, edited by M. Friedlander et al., AIP Conference Proceedings 280, New York: American Institute of Physics, 1993, pp. 631-642.
15. Schneid, E. J. et al., "EGRET Observations of Extended High Energy Emissions from the Nuclear Line Flares of June 1991" in *High-Energy Solar Phenomena – A New Era of Spacecraft Measurements*, edited by J. M. Ryan and W. T. Vestrand, AIP Conference Proceedings 294, New York: American Institute of Physics, 1994, pp. 95-99.
16. Schönfelder, V. et al., *Ap. J. Suppl. Series* **86**, 657-692 (1993).
17. Share, G.H. and Murphy, R. J., "Gamma-Ray Measurement of Energetic Heavy Ions at the Sun." in *26th International Cosmic Ray Conference*, Salt Lake City, 1999.
18. Suleiman, R. et al., "COMPTEL's Solar Flare Catalog" in *High-Energy Solar Phenomena – A New Era of Spacecraft Measurements*, edited by J. M. Ryan and W. T. Vestrand, AIP Conference Proceedings 294, New York: American Institute of Physics, 1994, pp. 51-54.
19. Trottet, G., et al., *Astron. & Astrophy (Suppl. Series)* **97**, 337-339 (1993).
20. Trottet, G., "X-Ray and Gamma-Ray Observations of Solar Flares by GRANAT" in *High-Energy Solar Phenomena – A New Era of Spacecraft Measurements*, edited by J. M. Ryan and W. T. Vestrand, AIP Conference Proceedings 294, New York: American Institute of Physics, 1994, pp. 3-14.
21. Young, C. A. et al., "Bayesian Multiscale Deconvolution Applied to Gamma-ray Spectroscopy" in *These Proceedings*, 2001.
22. Young, C. A., "Solar Flare Gamma-Ray Spectroscopy with CGRO-COMPTEL" Ph.D. Thesis, University of New Hampshire, 2001.

Expected Gamma-Ray Fluxes from Interactions of Flare Energetic Particles with Solar Wind Matter

Lev I. Dorman[1,2]

[1] Israel Cosmic Ray Center and Emilio Segre' Observatory, affiliated to Tel Aviv University, Technion, and Israel Space Agency; P. O. Box 2217, Qazrin 12900, ISRAEL

[2] IZMIRAN, Troitsk, Moscow region, 142092, RUSSIA

Abstract. On the basis of data on solar flare energetic particle (FEP) generation and propagation in the Heliosphere we calculate the space-time-energy distribution of these particles in the Heliosphere in the periods of great FEP events. On the basis of observation data and investigations of cosmic ray nonlinear processes in the Heliosphere we determine the space-time distribution of solar wind matter. Then we calculate the generation of gamma-rays (GR) by decay of neutral pions generated by nuclear interactions of FEP with solar wind matter and determine the expected space-time distribution of GR emissivity. Then we calculate the expected time variation of the angle distribution and spectra of GR fluxes generated by interaction of FEP with solar wind matter for observations from the Earth or from space-probes on different distances from the Sun. For some simple diffusion model of solar FEP propagation we obtain analytical approximation described the time evolution of GR flux angle distribution as well as time evolution of GR spectrum. It is shown that by observations of GR generated by solar FEP interactions with wind matter can be obtain additional important information on FEP propagation and matter distribution in the inner Heliosphere in periods of great FEP events.

INTRODUCTION

The generation of gamma rays (GR) by interaction of flare energetic particles (FEP) with solar wind matter shortly was considered in [1,2]. Here we will give a development of this research with much more details. GR generation in the Heliosphere by solar FEP in periods of great events, determined mainly by 3 factors:
1st - by space-time distribution of solar FEP in the Heliosphere, their energetic spectrum and chemical composition (see review in [3-6]); for this distribution can be important nonlinear collective effects of FEP pressure and kinetic stream instability [7-10].
2nd - by the solar wind matter distribution in space; for this distribution are important pressure and kinetic stream instability of galactic cosmic rays and solar FEP [8-11].

3rd- by properties of solar FEP interaction with solar wind matter accompanied with GR generation through decay of neutral pions [12-14].

THE FIRST FACTOR: FEP SPACE-TIME DISTRIBUTION

The problem of solar FEP generation and propagation through the solar corona and in the interplanetary space as well as its energetic spectrum and chemical and isotopic composition was reviewed in [3-6]. In the first approximation according to numeral data of observations of many events for about 5 solar cycles the time change of solar FEP and energy spectrum change can be described by the solution of isotropic diffusion (characterized by the diffusion coefficient $D(E_k)$) from some pointing instantaneous source $Q(E_k,\mathbf{r},t) = N_o(E_k)\delta(\mathbf{r})\delta(t)$ of solar FEP by

$$N(E_k,\mathbf{r},t) = N_o(E_k)\left[2\pi^{1/2}(D(E_k)t)^{3/2}\right]^{-1} \times \exp\left(-\mathbf{r}^2/(4D(E_k)t)\right), \quad (1)$$

At the distance $r_1 = 1\,AU$ the maximum of solar FEP density will be reach according to Eq. (1) at the moment

$$t_m(r_1, E_k) = r_1^2/6D(E_k). \quad (2)$$

THE SECOND FACTOR: SPACE-TIME DISTRIBUTION OF SOLAR WIND MATTER

Let us suppose the model of Parker [15] of radial solar wind expanding into the interplanetary space. In this case solar wind matter density

$$n(r,\theta) = n_1(\theta)u_1(\theta)r_1^2/(r^2 u(r,\theta)), \quad (3)$$

where $n_1(\theta)$ and $u_1(\theta)$ are the matter density and solar wind speed at the latitude θ on the distance $r = r_1 = 1\,AU$ from the Sun. The dependence $u(r,\theta)$ is determined mainly by galactic cosmic ray nonlinear processes in the Heliosphere [10]. According to [11]

$$u(r) \approx u_1(1 - b(r/r_o)), \quad (4)$$

where r_o is the distance to the terminal shock wave and parameter $b \approx 0.13 \div 0.45$. We will use here $r_o = 100\,AU$ and $b=0.3$.

THE THIRD FACTOR: GR GENERATION BY FEP IN INTERACTIONS WITH SOLAR WIND

Generation of Neutral Pions

According to [12-14] the neutral pion generation caused by nuclear interactions of energetic protons with hydrogen atoms through reaction $p + p \to \pi^o + anything$ (characterized by inclusive cross section $\langle \varsigma\sigma_\pi(E_k)\rangle$) will be

$$F^{\pi}_{pH}(E_{\pi},r,\theta,t) = 4\pi n(r,\theta,t) \int_{E_{k\,min}(E_{\pi})}^{\infty} dE_k \, N_p(E_k,r,t) \varsigma\sigma_{\pi}(E_k) \left(\frac{dN(E_k,E_{\pi})}{dE_{\pi}}\right), \quad (5)$$

where $n(r,\theta,t)$ is determined by Eq. (4), $E_{k\,min}(E_{\pi})$ is the threshold energy for pion generation, $N_p(E_k,r,t)$ is determined by Eq. (3), and $\int_0^{\infty}(dN(E_k,E_{\pi})/dE_{\pi})dE_{\pi} = 1$.

Space-Time Distribution of Gamma Ray Emissivity

GR emissivity caused by nuclear interactions of FEP protons with solar wind matter will be determined according to [12-14] by (here $E_{\pi\,min}(E_{\gamma}) = E_{\gamma} + m_{\pi}^2 c^4 / 4E_{\gamma}$):

$$F^{\gamma}_{pH}(E_{\gamma},r,\theta,t) = 2 \int_{E_{\pi\,min}(E_{\gamma})}^{\infty} dE_{\pi} \left(E_{\pi}^2 - m_{\pi}^2 c^4\right)^{-1/2} F^{\pi}_{pH}(E_{\pi},r,\theta,t), \quad (6)$$

Let us introduce Eq. (2) in (5) and (6) by taking into account Eq. (5):

$$F^{\gamma}_{pH}(E_{\gamma},r,\theta,t) = B(r,\theta,t) \int_{E_{\pi\,min}(E_{\gamma})}^{\infty} \left(E_{\pi}^2 - m_{\pi}^2 c^4\right)^{-1/2} dE_{\pi} \times$$

$$\int_{E_{k\,min}(E_{\pi})}^{\infty} N_{op}(E_k) \varsigma\sigma_{\pi}(E_k)(t/t_m)^{-3/2} \exp\left(-3r^2 t_m / 2r_1^2 t\right) dE_k . \quad (7)$$

where

$$B(r,\theta,t) = 3^{3/2} 2^{7/2} \pi^{1/2} r_1^2 n_1(\theta,t) u_1(\theta,t) / r^2 u(r,\theta,t). \quad (8)$$

Expected Angle Distribution and Time Variations of Gamma Ray Fluxes for Observations Inside the Heliosphere

Let us assume that the observer is inside the Heliosphere, on the distance $r_{obs} \leq r_o$ from the Sun and situation in the Heliosphere can be considered as spherically-symmetrical. In this case the expected angle distribution and time variations of GR fluxes will be

$$\Psi^{\gamma}_{pH}(E_{\gamma},r_{obs},\varphi,t) \approx F^{\gamma}_{pH}(E_{\gamma},r = r_{obs}\sin\varphi,t)(\varphi_{max} - \varphi_{min})r_{obs}\sin\varphi, \quad (9)$$

where φ is the angle between direction on the star and direction of observation, $\varphi_{max} = \arccos(r_{obs}\sin\varphi/r_i)$; $\varphi_{min} = -\arccos(r_{obs}\sin\varphi/r_i)$ if $r_{obs} > r_i$ and $\varphi_{min} = \varphi - \pi/2$, if $r_{obs} \leq r_i$ (here $r_i = \eta(2t/3t_m)^{1/2}$). For the great solar event with the total energy in FEP 10^{32} ergs, $n_1 = 5 cm^{-3}$ for $r_{obs} = 1\,AU$, Eq. (9) gives

$$\Psi^{\gamma}_{pH}(E_{\gamma} > 0.1 GeV, r_{obs},\varphi,t) \approx \frac{6.7 \times 10^{-6}}{\sin\varphi} \left(\frac{t}{t_m}\right)^{-\frac{3}{2}} \exp\left(-\frac{3t_m \sin^2\varphi}{2t}\right) ph.cm^{-2} s^{-1}, \quad (10)$$

where t_m is determined by Eq. (2). Results for different φ are shown in Fig. 1 (for φ from 2° to 26°) and in Fig. 2 (for φ from 28° to 179°).

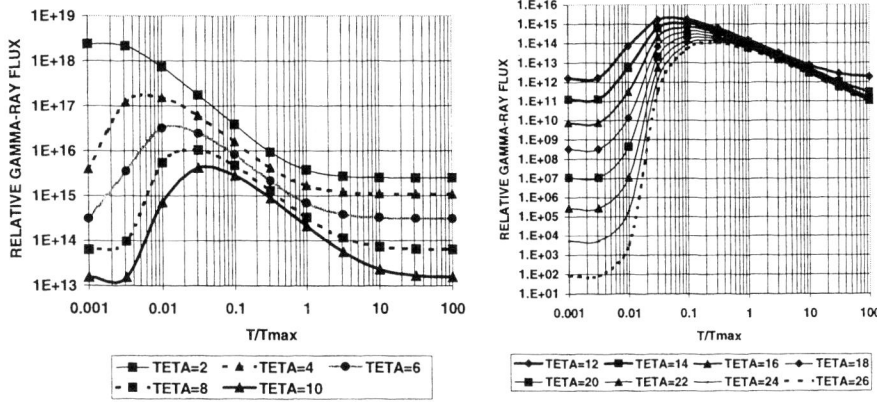

Figure 1. Expected GR fluxes for directions 2° to 26° from the Sun in dependence of T/Tmax, where Tmax is determined by Eq. (2). To obtain absolute fluxes of GR for event with energy 10^{32} *ergs* in FEP necessary to multiply relative fluxes on factor 10^{-19}; for FEP event with energy 10^{31} *ergs* this factor will be 10^{-20}.

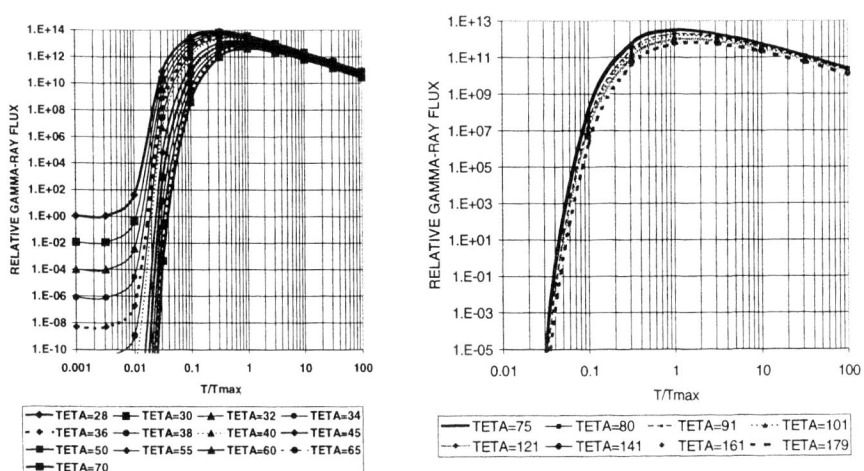

Figure 2. The same as in Figure 1, but for directions 28° to 179° from the Sun.

DISCUSSION AND CONCLUSIONS

Estimations according to Eq. (10) show that in periods of great solar FEP events with total energy $\approx 10^{32}$ ergs the expected flux of GR with energy > 100 MeV in direction 2° from the Sun at $t/t_m = 1/100$ reaches $\approx 10^{-1}\, ph.cm^{-2}.sr^{-1}.sec^{-1}$, at $t/t_m = 1/3$ reaches $\approx 10^{-3}\, ph.cm^{-2}.sr^{-1}.sec^{-1}$ (see Fig. 1). In direction 30° from the Sun (see Fig. 2) expected GR fluxes are much smaller: the maximum will be at $t/t_m = 1/3$ and reaches value only $\approx 10^{-5}\, ph.cm^{-2}.sr^{-1}.sec^{-1}$. Expected GR fluxes are characterized by great specific time variations, which depend from direction of observations relative to the Sun, total FEP flux from the source, parameters of FEP propagation (summarized in value of t_m), and properties of solar wind. It is important that present GR telescopes might measure expected GR fluxes in periods of great FEP events. These observations of gamma rays generated in interactions of FEP with solar wind matter can give important information on solar wind 3d-distribution as well as on properties of solar FEP and its propagation parameters. Moreover, the monitoring of GR observations in directions at few degrees from the Sun can give important possibility to predict expected radiation hazard from FEP on the Earth and in space.

REFERENCES

1. Dorman, L.I., Cosmic ray nonlinear processes in gamma-ray sources, *Astronomy and Astrophysics*, Suppl. Ser., **120**, No. 4, 427-435, 1996.
2. Dorman, L.I., Angle distribution and time variation of gamma ray flux from solar and stellar winds, 1. Generation by flare energetic particles, in C.D. Dermer, M.S. Strickman, and J.D. Kurfess (eds.), *Proc. 4th Compton Symposium.* Williamsburg, pp.1178-1182, 1997.
3. Dorman, L.I., *Cosmic Ray Variations*, Gostekhteorizdat, Moscow, 1957.
4. Dorman, L.I., and Miroshnichenko, L.I., *Solar Cosmic Rays*, Fizmatgiz, Moscow, 1968.
5. Dorman, L.I. and Venkatesan, D., Solar cosmic rays, *Space Sci. Rev.* **64**, 183-362, 1993.
6. Stoker, P.H., Relativistic solar cosmic rays, *Space Sci. Rev.*, **73**, 327, 1995.
7. Berezinskii, V.S., Bulanov, S.V., Ginzburg, V.L., Dogiel, V.A., and Ptuskin, V.S., *Cosmic Ray Astrophysics*, Fyzmatgiz, Moscow, 1990.
8. Dorman, L.I., Ptuskin, V.S. and Zirakashvili, V.N., Outer Heliosphere: pulsations, cosmic rays and stream kinetic instability, in S. Grzedzielski and D.E. Page (eds.) *Physics of the Outer Heliosphere*, Pergamon Press, pp. 205-209, 1990.
9. Zirakashvili, V.N., Dorman, L.I., Ptuskin, V.S. and Babayan, V.Kh., Cosmic ray nonlinear modulation in the outer Heliosphere. *Proc.22-th Intern. Cosmic Ray Conf.*, Dublin, Vol.3, pp 585-588, 1991.
10. Dorman, L.I., Cosmic ray nonlinear effects in space plasma, 2. Dynamic Heliosphere, in M.M. Shapiro, R. Silberberg and J.P.Wefel (eds.). *Currents in High Energy Astrophysics*, Kluwer Academic Publishers., Dordrecht/Boston /London, NATO ASI Serie, Vol. 458, pp. 193-208, 1995.
11. Le Roux, J.A. & Fichtner, H., *Ap. J*, **477**, L115, 1997.
12. Stecker, F.W., *Cosmic Gamma Rays*, Mono Book Co, Baltimore, 1971.
13. Dermer, C.D., *A&A*, **157**, 223, 1986.
14. Dermer, C.D., *Astrophys. J.*, **307**, 47, 1986.
15. Parker, E.N., *Interplanetary Dynamically Processes*, Intersci. Publ., New York-London. 1963.

Interactions of Flare Energetic Particles with Stellar Wind Matter: Expected Gamma Ray Fluxes from Local Stars

Lev I. Dorman [1, 2]

[1] Israel Cosmic Ray Center and Emilio Segre' Observatory, affiliated to Tel Aviv University, Technion, and Israel Space Agency; P. O. Box 2217, Qazrin 12900, ISRAEL

[2] IZMIRAN, Troitsk, Moscow region, 142092, RUSSIA

Abstract. For different types of local stars with flare activity we calculate expected gamma-ray fluxes in periods of flare energetic particle (FEP) generation. We suppose that main processes of FEP generation and propagation in the stellar-sphere are similar with processes in the Heliosphere but with much bigger energetic. We calculate the space-time-energy distribution of these particles in the stellar-sphere in the periods of FEP events. On the basis of investigations of cosmic ray nonlinear processes we determine the space-time distribution of stellar wind matter. Then we calculate the generation of gamma rays by decay of neutral pions generated in nuclear interactions of FEP with stellar wind matter and determine the expected space-time distribution of gamma-ray emissivity. Then we calculate the expected time variation of the angle distribution and spectra of gamma ray fluxes. For some simple diffusion models of stellar FEP propagation we obtain analytical approximation described the time evolution of gamma ray flux angle distribution as well as time evolution of gamma ray spectrum. It is shown that by observations from local stars of gamma rays generated by stellar FEP interactions with stellar wind matter can be obtain important information on stellar activity, on FEP spectrum, on mode of FEP propagation, and on matter distribution in the inner stellar-sphere.

INTRODUCTION

It is well known (see in [1-3]) that among 33 nearest to the Sun stars at least 13 are flare stars, and total number of known flare stars are about 100 in the vicinities of the Sun and about 1000 in nearest star associations Orion, Pleiads, Hyades, Praesepe and others. The most part of flare stars are red dwarfs of UV Cet-type with radius from 0.1 to 0.8 $R_\odot$, and mass from 0.06 to 0.6 $M_\odot$. They have spotness photosphere, good developed chromospheres and coronas. The time distribution of flares is accidental, and average interval between flares is from about one hour to about several tens of days. The dependence of total energy E_{UV} of stellar flare in UV radiation from frequency v of flare occurrence for 15 nearest flare stars and from some stars in the nearest star associations Orion and Pleiades have power form of type

$$E_{UV} \propto \nu^{-\beta}, \qquad (1)$$

where $0.5 \leq \beta \leq 2$. For example, for the star YYGem the frequency of flares with $E_{UV} \approx 10^{34}$ ergs is about one per 100 hours, and with $E_{UV} \approx 10^{32}$ ergs about one per 10 hours; for flare stars in Orion the frequency of flares with $E_{UV} \approx 10^{36}$ ergs is one per 10^4 hours, and with energy $E_{UV} \approx 10^{34}$ ergs the frequency was $\nu \approx 3 \times 10^{-3}$ $hour^{-1}$. It is important that for solar flares the connection between total energy in FEP and frequency of events is described by the same relation as Eq. (1) with $\beta \approx 0.8$ (see in [4]). Convective zones in flare stars are deeper than in the Sun and total energy in flare can be several orders higher, up to 10^{36} ergs. We assume that in FEP go about the same energy as in soft X-rays and UV radiation (let us note that in great solar flares in soft X-rays and UV radiation go energy $(3-5) \times 10^{31}$ ergs, about the same as in solar FEP). For gamma ray generation through decay of neutral pions are important also the properties of stellar winds from flare stars. According to [5, 6], the mass-loss rates of active late-type dwarf stars can be $10^{-13} \div 10^{-11}$ $M_\odot$ yr^{-1} (in comparison with about $2 \times 10^{-14} M_\odot$ yr^{-1} from the Sun); the stellar wind speed is expected about the same as for solar wind (300-600 km/s). Measurements of GR from stellar winds generated by stellar FEP will give additional information not only on FEP but also on stellar winds. Here we will develop research, started in [7, 8].

THE FIRST FACTOR: STELLAR FLARE ENERGETIC PARTICLE GENERATION AND PROPAGATION

The problem of stellar FEP generation and propagation through the stellar corona and in the stellar-sphere space as well as its energetic spectrum and chemical and isotopic composition we do not know exactly, but we assume that they are in the first approximation similar to what we know now on solar FEP generation and propagation (see review in [9-12]). In the first approximation according to numeral data of observations of many events for about 5 solar cycles the time change of stellar FEP and energy spectrum change can be described by the solution of isotropic diffusion (characterized by the diffusion coefficient $D(E_k)$) from some pointing instantaneous source $Q(E_k, \mathbf{r}, t) = N_o(E_k) \delta(\mathbf{r}) \delta(t)$ of stellar FEP by

$$N(E_k, \mathbf{r}, t) = N_o(E_k) \left[2\pi^{1/2} (D(E_k)t)^{3/2} \right]^{-1} \times \exp\left(-\mathbf{r}^2 / (4D(E_k)t)\right), \qquad (2)$$

At the distance $r_1 = 1\ AU$ the maximum of stellar FEP density will be reach according to Eq. (2) at the moment

$$t_m(r_1, E_k) = r_1^2 / 6D(E_k). \qquad (3)$$

THE SECOND FACTOR: SPACE-TIME DISTRIBUTION OF STELLAR WIND MATTER

If we assume for the first approximation the model of Parker [13] of radial stellar wind expanding into the space of stellar-sphere (what is in good agreement with all available data of direct measurements for solar wind), then the behavior of matter density of stellar wind will be described by the relation

$$n(r,\theta) = n_1(\theta) u_1(\theta) r_1^2 / (r^2 u(r,\theta)), \qquad (4)$$

where $n_1(\theta)$ and $u_1(\theta)$ are the matter density and stellar wind speed at the latitude θ on the distance $r = r_1$ from the star ($r_1 = 1 AU$).

THE THIRD FACTOR: GR GENERATION BY FEP IN INTERACTIONS WITH STELLAR WIND

Generation of Neutral Pions

According to [14-16] the neutral pion generation caused by nuclear interactions of energetic protons with hydrogen atoms through reaction $p + p \to \pi^o + anything$ (characterized by inclusive cross section $\langle \varsigma \sigma_\pi(E_k) \rangle$) will be

$$F_{pH}^\pi(E_\pi, r, \theta, t) = 4\pi n(r,\theta,t) \int_{E_{k\,min}(E_\pi)}^\infty dE_k\, N_p(E_k, r, t) \langle \varsigma \sigma_\pi(E_k) \rangle \left(\frac{dN(E_k, E_\pi)}{dE_\pi} \right), \qquad (5)$$

where $n(r,\theta,t)$ is determined by Eq. (4), $E_{k\,min}(E_\pi)$ is the threshold energy for pion generation, $N_p(E_k, r, t)$ is determined by Eq. (2), and $\int_0^\infty (dN(E_k, E_\pi)/dE_\pi) dE_\pi = 1$.

Space-Time Distribution of Gamma Ray Emissivity

GR emissivity caused by nuclear interactions of FEP protons with stellar wind matter will be determined according to [14-16] by (here $E_{\pi\,min}(E_\gamma) = E\gamma + m_\pi^2 c^4 / 4 E\gamma$):

$$F_{pH}^\gamma(E_\gamma, r, \theta, t) = 2 \int_{E_{\pi\,min}(E_\gamma)}^\infty dE_\pi \left(E_\pi^2 - m_\pi^2 c^4 \right)^{-1/2} F_{pH}^\pi(E_\pi, r, \theta, t), \qquad (6)$$

Let us introduce Eq. (2) in (5) and (6) by taking into account Eq. (5), and let us use

$$B(r,\theta,t) = 3^{3/2} 2^{7/2} \pi^{1/2} r_1^2 n_1(\theta,t) u_1(\theta,t) / r^2 u(r,\theta,t). \qquad (7)$$

As result we obtain for space-time distribution of GR emissivity (see also Fig. 1):

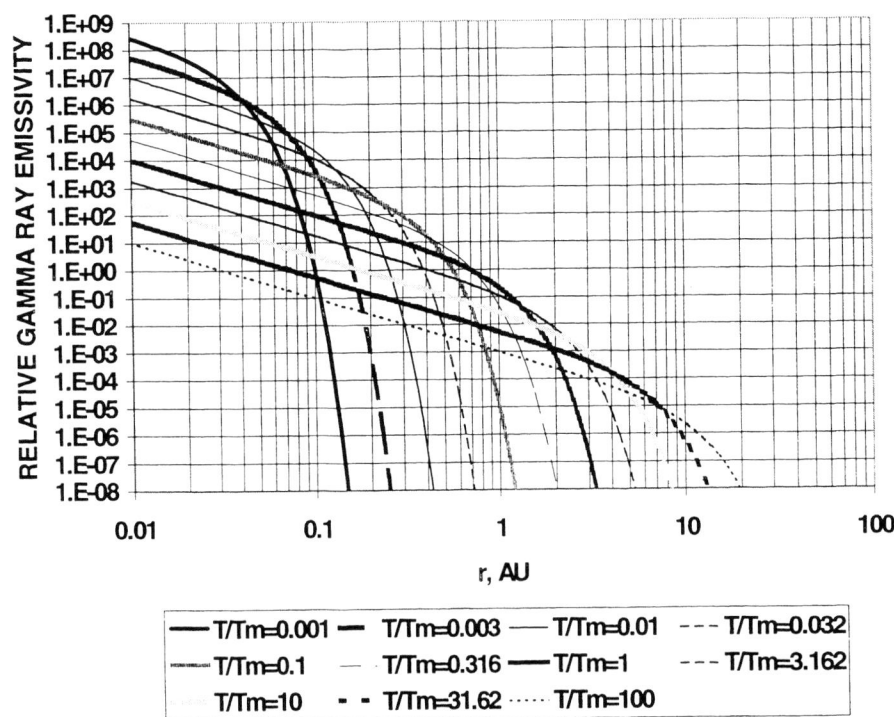

Figure 1. Expected space distribution of gamma ray emissivity in stellar wind for different time T after FEP generation in units of time maximum Tm on 1 AU, determined by Eq. (3). The curves are from T/Tm=0.001 up to T/Tm=100.

$$F^\gamma_{pH}(E_\gamma,r,\theta,t) = B(r,\theta,t) \int_{E_\pi \min(E_\gamma)}^{\infty} \left(E_\pi^2 - m_\pi^2 c^4\right)^{-1/2} dE_\pi \times$$

$$\int_{E_k \min(E_\pi)}^{\infty} N_{op}(E_k)\langle\varsigma\sigma_\pi(E_k)\rangle (t/t_m)^{-3/2} \exp\left(-3r^2 t_m / 2r_1^2 t\right) dE_k. \tag{8}$$

EXPECTED GR FLUXES FROM LOCAL FLARE STARS CAUSED BY FEP INTERACTIONS WITH STELLAR WIND MATTER

Expected GR flux Ψ from flare star on the distance $r_{obs} \gg r_o$, where r_o is radius of stellar-sphere will be:

$$\Psi^\gamma_{pH}(E_\gamma,r_{obs},t) = 2\pi r_{obs}^{-2} \int_{-\pi/2}^{\pi/2} \cos\theta d\theta \int_0^{r_o} r^2 dr F^\gamma_{pH}(E_\gamma,r,\theta,t). \tag{9}$$

For spherical-symmetrical modes of FEP propagation and stellar wind matter distribution with taking into account Eq. (8), we obtain:

$$\Psi_{pH}^{\gamma}(E_{\gamma},r_{obs},t) = 4\pi r_{obs}^{-2} F_{pH}^{\gamma}\left(E_{\gamma},\eta,t_m\right)\left(\frac{t_m}{t}\right)r_1^3 \Phi\left(\frac{r_o}{\eta}\left(\frac{3t_m}{t}\right)^{1/2}\right) ph.cm^{-2}s^{-1}, \quad (10)$$

where $\Phi(x)$ is the probability function.

DISCUSSION AND CONCLUSIONS

Let us consider Eq. (10). For a flare star with total energy in FEP event 10^{36} ergs and $n_1(\theta,t) \approx 500\ cm^{-3}$ the expected flux will be

$$\Psi_{pH}^{\gamma}(E_{\gamma} > 0.1 GeV, r_{obs}, t) = 2 \times 10^{28} r_{obs}^{-2}\left(\frac{t_m}{t}\right) \Phi\left(\frac{r_o}{\eta}\left(\frac{3t_m}{t}\right)^{1/2}\right) ph.cm^{-2}s^{-1}. \quad (11)$$

According to Eq. (11) for $t_m = 10^3 s$ at $t = 10\ s$ and $100\ s$ the value $(r_o/\eta)(3t_m/t)^{1/2} >> 1$ and $\Phi(x) \approx 1$, that at distance $r_{obs} = 3 \times 10^{18} cm$ expected flux $\Psi_{pH}^{\gamma}(E_{\gamma} > 0.1 GeV, r_{obs}, t)$ will be 2×10^{-7} and $2 \times 10^{-8}\ ph.cm^{-2}s^{-1}$. Eq. (11) shows that the total flux of gamma rays from stellar wind generated by FEP interaction with wind matter must fall inverse proportional with time what can be used for separation of GR generated in stellar wind from generated in stellar flare. Obtained results show that in future by GR monitoring of nearest flare stars with great FEP events can be determined the energy spectrum of FEP and parameters of FEP propagation, total energy in FEP, stellar wind and stellar activity properties.

REFERENCES

1. Gershberg, R.E., *Flares of Red Dwarf Stars*, Fizmatgiz, Moscow, 1970.
2. Gurzadyan, G.A., *Flare Stars*, Fizmatgiz, Moscow, 1973.
3. Gershberg, R.E., *Flare Stars of Small Masses*, Fizmatgiz, Moscow, 1978.
4. L.I. Dorman and L.A. Pustil'nik, *26-th ICRC*, **6**, 407-410, 1999.
5. Lim J. and White S.M., *Astrophys. J. Letters*, **462**, L91, 1996.
6. Wargelin B.J. and Drake J.J., *Astrophys. J.*, **546**, No. 1, L57-L60, 2001.
7. Dorman, L.I., Cosmic ray nonlinear processes in gamma-ray sources, *Astronomy and Astrophysics*, Suppl. Ser., **120**, No. 4, 427-435, 1996.
8. Dorman, L.I., Angle distribution and time variation of gamma ray flux from solar and stellar winds, 1. Generation by flare energetic particles, in C.D. Dermer, M.S. Strickman, and J.D. Kurfess (eds.), *Proc. 4th Compton Symposium*. Williamsburg, pp.1178-1182, 1997.
9. Dorman, L.I., *Cosmic Ray Variations*, Gostekhteorizdat, Moscow, 1957.
10. Dorman, L.I., and Miroshnichenko, L.I., *Solar Cosmic Rays*, Fizmatgiz, Moscow, 1968.
11. Dorman, L.I. and Venkatesan, D., Solar cosmic rays, *Space Sci. Rev.* **64**, 183-362, 1993.
12. Stoker, P.H., Relativistic solar cosmic rays, *Space Sci. Rev.*, **73**, 327, 1995.
13. Parker, E.N., *Interplanetary Dynamically Processes*, Intersci. Publ., New York-London. 1963.
14. Stecker, F.W., *Cosmic Gamma Rays*, Mono Book Co, Baltimore, 1971.
15. Dermer, C.D., *A&A*, **157**, 223, 1986.
16. Dermer, C.D., *Astrophys. J.*, **307**, 47, 1986.

Surveys and Population Studies

A Multiwavelength Strategy for Identifying Celestial γ-ray Sources

Patrizia A. Caraveo

Istituto di Fisica Cosmica "G. Occhialini"
Via Bassini, 15 –20133 Milano, Italy
pat@ifctr.mi.cnr.it

Abstract. The vast majority of the high-energy γ-ray sources discovered by EGRET are still unidentified. Percentages range from 50% at high galactic latitudes, where blazars are responsible for almost all identified sources, to more than 90% near the galactic plane, where isolated neutron stars appear to be the only certified class of sources of high energy γ-rays. In spite of all the efforts devoted to the identification problem, the only success story, so far, appears to be the chase for Geminga, where X-rays led the way to eventual optical identification. Similar searches are now starting to produce encouraging results, although none has reached, as yet, a certified identification.

HISTORICAL OVERVIEW

Unidentified sources are as old as γ-ray astronomy itself. As soon as the NASA SAS-2 satellite was able to discriminate point-like sources from the underlying diffuse emission, an unidentified source appeared in the γ-ray sky. The very first images of the galactic anticentre unveiled γ 195+5, later to become Geminga. When, in 1973, the SAS-2 mission ended prematurely, γ-ray sources, encompassing the Crab and Vela pulsar together with the unidentified one, were a reality (Fichtel et al, 1975).

COS-B continued on the same track. Its longer active life span allowed for a significant increase in the number of photons collected, yielding a grand total of 25 sources, almost a ten-fold increase with respect to SAS-2. The second COS-B catalogue is the legacy to high energy astronomy of a successful mission. It recognized, for the first time, the paramount importance of unidentified objects which, at the time, appeared to be responsible for 21 of the 25 detections. Of course, the degree-size positional uncertainty of each source, prevented straightforward identifications. Only objects with unambiguous timing signature (such as pulsars) could be reliably identified. Indeed, apart from the Crab and Vela pulsars, COS-B was able to pinpoint two prominent objects in its error boxes, namely the first extragalactic source, 3C273, and the molecular cloud ρ-Oph. (see Bignami & Hermsen, 1983, for a review). In spite of the COS-B uneven sky coverage, well visible in Figure 1, the concentration of sources along the galactic plane appeared to be real, pointing towards a galactic population with a rather small scale height and an average distance between 2 and 7 kpc, implying an average luminosity in the range $(0.4-5)10^{36}$ erg/sec (Swanenburg et al, 1981).

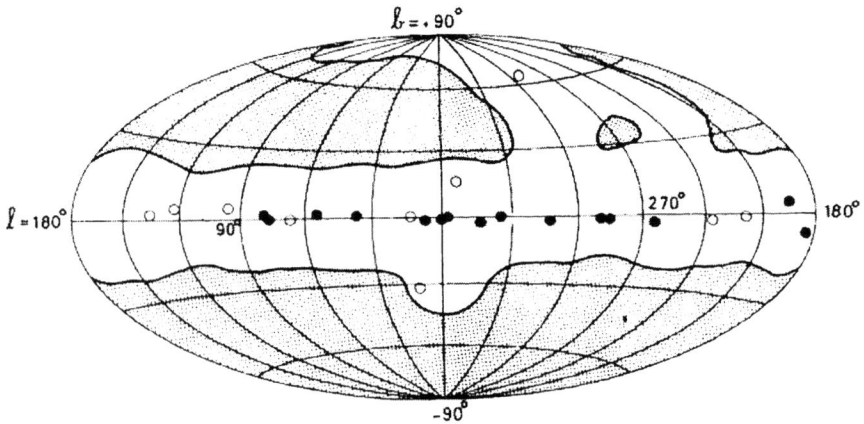

FIGURE 1. Second (and last) COS-B Catalogue of γ-ray sources (Swanenburg et al. 1981). Open circles denote faint sources, filled circles brighter ones. The dividing line is at $1.3\ 10^{-6}$ ph/cm^{2}sec. The shaded area has not been searched for sources since the coverage was not adequate.

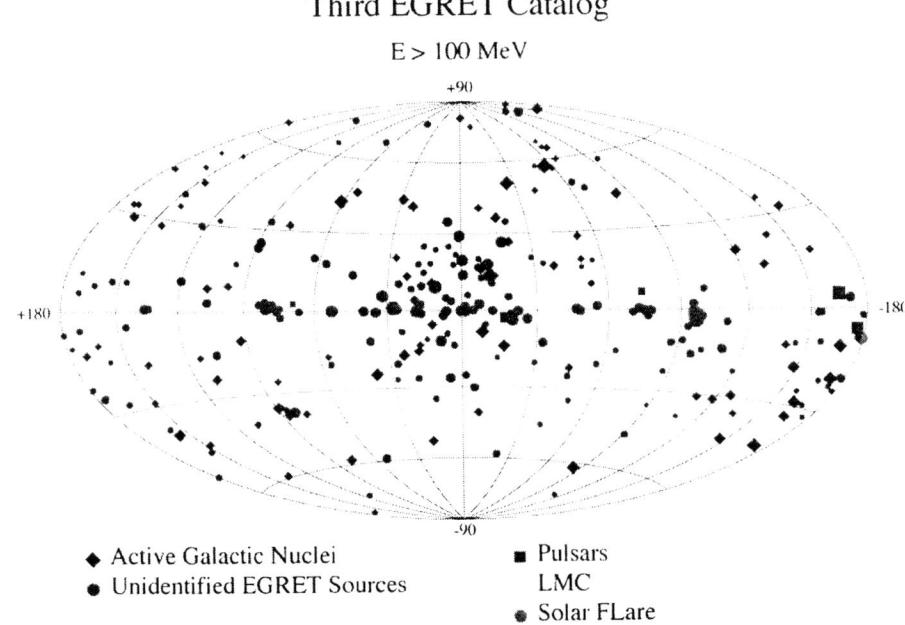

FIGURE 2. Third EGRET catalogue (Hartman et al., 1999). The size of the symbols, the meaning of which is spelled out in the legenda, is proportional to the measured source flux.

EGRET, onboard the Compton Gamma ray Observatory, provided another ten-fold increase in the number of sources (see Figure 2). Owing to a much bigger sensitive area, coupled with a better angular resolution, EGRET was able to lower significantly the COS-B detection limit. Moreover, by covering the entire sky, EGRET unveiled a new class of powerful γ-ray emitters: the Blazars. At variance with low latitude galactic sources, Blazars are best seen in high latitude, less crowded fields. Often, their γ-ray flux shows variability correlated to the radio and optical ones, making the identification job much more easy. Unfortunately, the identifications of the low latitude sources continue to be hampered by the error box dimensions. Although smaller than the COS-B ones, the EGRET error boxes are still too big to allow for direct identifications. Statistical studies on the low latitude sources, performed by Mukherjee et al (1995), confirmed the findings of Swanenburg et al, (1981). The unidentified low latitude EGRET sources are at distances between 1.2 and 6 kpc and their luminosities are in the range (0.7- 16.7) 10^{35} erg/sec.

It is worth mentioning that EGRET discovered short-term variability from GRO J1838-04 (Tavani et al ,1997), later listed in the EGRET catalogue as 3CGJ1837-0423. However, the lack of any real time notification of such a variability hampered the search of an equally varying counterpart, thus missing a precious chance for source identification.

Table 1 summarizes the source growth-rate from COS-B to EGRET. The low latitude (b<10°), presumably galactic, sources went from 22 to 80, while the high latitude (b>10°), presumably extragalactic, ones jumped from 3 to 181, with Blazars accounting for roughly half of them. If we forget this classical (and somewhat arbitrary) separation and look at the distribution of all the EGRET sources, we realize that there is an excess of sources at middle latitudes. Gehrels et al (2000) have proposed this to be a new classes of γ-ray sources, possibly linked to the Gould's Belt structure (see also Grenier, 2000). On average, middle latitude sources are fainter than the low latitude ones, closely aligned with the galactic plane, and their spectral shapes tend to be softer. Moreover, middle latitude sources should be closer to us than low latitude ones, thus their average luminosity should be lower.

Among the 80 low latitude sources, Table1 shows how pulsars numbered from 2 to 6, with three more possible entries (Thompson, 2001). However important, such an increase confirms pulsars (be they radio loud or radio quiet) as an established class of gamma-ray emitters, although, probably not the only one. Indeed, the comparison between the measured pulsar luminosities (e.g. Thompson et al., 1999) and the average values inferred from the statistical study shows clearly that only very young and energetic pulsars could fulfil the luminosity requirements of unidentified sources. This limits quite severely the total contribution of classical pulsars to the low latitude, galactic γ-ray source population

What about the remaining unidentified sources? Surprisingly enough, EGRET has done very little to clarify the nature of the galactic γ-ray emitters. While their number increased fourfold, successful identifications remain at a meagre <10% level. They encompass, anyway, only isolated neutron stars.

TABLE 1. From COS-B to EGRET.

Low Latitude (b<10°)	COS-B	EGRET
Total	22	80 (- 1 solar flare)
Identified	**pulsars**	**pulsars**
	Crab, Vela	Crab, Vela, PSR 1706-44,
		PSR1951+32, PSR1055-52
		Geminga (**radio quiet**)
% unidentified	90%	90%

High Latitude (b>10°)	COS-B	EGRET
Total	3	181
Identified	3C273, ρ Oph	66 *variable* **AGNs**
		28 probable **AGNs**
		LMC, CenA
% unidentified	-	50%

COUNTERPART SEARCHES

More of the same: Pulsars

In view of the overwhelming presence of pulsars among identified galactic γ-ray sources, it is natural to explore the possibility that at least a fraction of the remaining low latitude sources belong to the same class of compact objects.

Indeed, the search for new pulsars started as soon as COS-B discovered a population of unidentified sources, but no pulsars were unveiled. New searches were triggered by the EGRET findings, but the lack of results, experienced at the time of COS-B, appears to be unchanged. Dedicated radio searches (Nice and Sayer, 1997), aimed precisely at the search for radio pulsars inside the error boxes of 10 of the brightest EGRET sources, yielded null results. The implication is that straightforward radio pulsar identification is not the only possible solution to the enigma of the unidentified high-energy γ-ray sources. This is further strengthened by the work of Nel et al. (1996) who investigated 350 known pulsars. They found few positional coincidences but no significant γ-ray timing signature for any of the pulsars in the survey.

However, more powerful surveys, now underway, may be changing this negative trend. Kaspi et al.(2000) find evidence of an association between the 20,000 y old PSR B1046-58, one of the candidates listed by Thompson (2001), and 3EG J1048-5840. D'Amico et al. (2001), using the new Parkes data, discover two young, promising radio pulsars inside the error boxes of 3EG J1420-6038 and 3EG J1837-0606. In view of the time noise, usually present in young pulsars, it will be difficult to search for their time signature in the EGRET data. They will be certainly studied by the future gamma ray missions, as was the case for the unidentified COS-B source 2CG 342-02, identified by EGRET with the newly discovered PSR 1706-49.

An Elusive Template

Apart from classical radio pulsars, γ-ray astronomy does offer a remarkable example of an isolated neutron star (INS) which behaves as a pulsar as far as X-and-γ-astronomy are concerned but has little, if at all, radio emission. As an established representative of the non-radio-loud INSs (see Caraveo, Bignami and Trümper, 1996 for a review), Geminga offers an elusive template behaviour: prominent in high energy γ-rays, easily detectable in X-rays and downright faint in optical, with sporadic or no radio emission (see Bignami and Caraveo, 1996 for a review of the source multiwavelength phenomenology).

Although the energetic of Geminga ($L_\gamma = 3.3 \; 10^{34}$ erg/sec) is not adequate to account for the very low latitude (i.e. more distant) EGRET sources, it could satisfy the fainter middle latitude sources, presumably belonging to a more local galactic population (Gehrels et al. 2000). Their gamma yield is certainly compatible with the rotational energy loss of a middle-aged neutron star, like Geminga.

Thus, in spite of being the only confirmed identification, Geminga does not provide a viable template for the entire family of unidentified γ-ray sources. Some other object or class of objects is needed. However, the strategy devised for the chase of Geminga seems to still be the best one to bridge the positional accuracy gap intrinsic to γ-ray astronomy.

Ongoing Multiwavelength Efforts

The γ-to-X-to-optical multiwavelength approach has by now been applied to a number of COS-B and EGRET sources.

The Einstein coverage of the error box of 2CG 135+01 yielded the discovery of the X-ray emission of the periodically variable radio source GT61.303 (Bignami et al, 1981). In spite of ad hoc searches for γ-ray variability correlated with the radio one (e.g. Kniffen et al, 1997; Tavani et al., 1998), no conclusive proof has been brought forward for the identification of 2CG135+1, now 3EG 0241+6103, with this peculiar binary system. The association remains, however, tantalizing.

For 3EG J0634+0521, Kaaret et al. (1999) have suggested the identification with SAX J0635+0533, a binary system containing a compact object with a Be star companion. Recently, such a proposed identification has been strengthened by the discovery of a 34 msec pulsation (Kareet, Cusumano and Sacco, 2000), coupled with a high period derivative, pointing towards a young, energetic pulsar in a binary system. This is an absolute first in the panorama of known binary systems, making SAX J0635+0533 an interesting system "per se". For both 3EG J0241+6103 and 3EG J0634+0521, the physics behind the γ-ray production would be particle acceleration at the shock created by the pulsar interaction with the thick Be-star wind during periastron passage. Roberts, Romani and Kawai (2001) list a number of pulsar nebulae which could be associated, mostly on positional grounds, with their specially selected GeV sources.

Of special interest is the case of the Kookaburra Nebula, in the error box of 3EG J 1420-6038, where is located also one of the new pulsars of D'Amico et al. (2001). Oka et al (1999) invoke the interaction of a pulsar nebula with a dark cloud to account for 3EG J 1809-2328.

A newly discovered, energetic, young, isolated pulsar, showing 51.6 msec X-ray and radio pulsations, has been proposed by Halpern et al (2001a) as the counterpart of 3EG J2227+6122. As we have already remarked in the case of the young pulsars discovered by D'Amico et al (2001), only future missions will be able to confirm such a promising identification by detecting the pulsar time signature in γ-rays.

Radio quiet INS identifications have been proposed for 3EG J1835+5918, the brightest unidentified γ-ray source, (Mirabal and Halpern 2001; Reimer et al., 2001), 3EG J0010+7309 (Brazier et al, 1998) and 3EG J2020+4026 (Brazier et al., 1996). It is interesting to note that both 3EGJ 1835+5918 and 3EG J0010+7309 are middle latitude sources, thus their energetic requirements are easily compatible with a Geminga-like identification.

3EG J2016+3657, on the other hand, has been identified with a Blazar behind the Galactic plane (Mukherjee et al, 2000; Halpern et al, 2001b). This extragalactic "contamination", deep in the galactic plane, should not come as a surprise. The isotropic distribution of Blazars, coupled with the negligible absorption suffered by γ-ray photons through the galactic plane, should result in quite a few galactic Blazars.

Thus, years of multiwavelength world-wide efforts have yielded the following counterparts:

4 energetic young radio pulsars, one of which has been discovered in X-rays. So far, only PSR B1046-58 is considered a candidate EGRET pulsar;

3 Geminga-like, radio quiet INSs;

2 peculiar binary systems (one with a young pulsar);

1 galactic Blazar;

few pulsar nebulae.

All in all, about a dozen EGRET sources have a tentative, more-or-less likely, identification. If compared to the scores of sources awaiting an identification, the number is small, however, the panorama is a rapidly evolving one. Two years ago, a similar review would have resulted in not more than 4 tentative identifications. Indeed, the most recent (and sometimes most promising) entries in our summary list are the outcome of the searches prompted by the second EGRET catalogue (Thompson et al. 1995, 1996). This gives us an idea of the time needed to bring to completion a thorough search for counterparts requiring one, or more, cycles of X-rays observations followed by one, or more, cycles of optical (and radio) ones.

The multiwavelength approach, although promising, is a time consuming exercise.

THE NEW MILLENIUM

New instruments, now operational both in the optical and X-ray domains, promise to speed up significantly the X and optical coverage of γ-ray error boxes.

On the X-ray side, two great observatories such as Chandra and Newton-XMM can cover with few, relatively short, pointings each EGRET error box, pushing the source detection limit to unprecedented levels.

It is easy to predict that dozens of serendipitous sources (mainly stars and AGNs) will be detected in each EGRET error box by these powerful X-ray telescopes.

Thus, it will be critical to plan for a "massive" approach to the optical identification work, which is bound to become the bottleneck of this multiwavelength chain.

Optical wide-field-imagers, with typical field-of-views of a fraction of square degree, offer just such a new perspective since they can speed-up considerably the tedious X-to-optical comparison work, aimed at discarding candidates with obvious identifications. Moreover, taking advantage of field of views comparable to those of the X-ray telescope, the optical work can proceed independently from the X-ray one, bypassing the need to wait for the results of the X-ray observations to plan the optical follow-up exposures.

In spite of the new instruments, selecting promising targets will remain a difficult matter. So far, γ-ray astronomy has not been able to classify its sources in families characterized by different templates at various wavelengths. Source classification should proceed together with the identification work and should be based on a trial-and-error approach.

Our Program

For our Newton-XMM exploratory program, we shall apply the Geminga strategy to two middle latitude EGRET sources. First of all, one has to single out potential neutron star candidates taking advantage of the high F_x/F_v of these objects. All efforts must then converge towards the identification of the neutron star candidate.

Our two EGRET sources, namely 3EG J1249-8330 and 3EG J0616-3310, have been selected on the basis of their positional accuracy, spectral shape, galactic location and lack of candidate extragalactic counterparts. Each source error box, a circle of 30' radius, will be covered by EPIC with four 10,000 sec exposure pointings, yielding a homogeneous coverage of about 1 sq deg. On the basis of similar pointings, we expect to have 50-100 sources, as faint as 10^{-14} erg/cm^2 sec, in each EGRET error box.

The X-ray coverage will be complemented by the optical one, done in three colours by the European Southern Observatory WFI (Wide Field Imager), operating at the ESO 2.2m telescope in La Silla, Chile. Its 8 CCD detectors cover a 30' x 30' f.o.v., a value directly comparable to the EPIC one. In order to be able to discard non-neutron-star optical IDs, we ought to reach a limiting magnitude of m_v 25, a value typical for a 1 h dithered exposure. Thus, in order to do a multicolour optical coverage of a 10,000 sec EPIC exposure, we will need about the same observing time with a 2 m class optical telescope. Cross correlation of the optical and X-ray images will yield the F_x/F_v parameter for the X-ray sources discovered by EPIC. Different colours will provide a further handle to solve ambiguous cases, where more than one optical entry will be compatible with the X-ray position

Of course, should new templates arise, they could be immediately implemented taking advantage of our unbiased X and optical coverage.

CONCLUSIONS

Irrespective of the nature(s) of unidentified γ-ray sources, it is clear that we are still far from a general solution, if any. A lot of observing time, both in X-rays and in the optical, should be devoted to these mysterious objects, in order to identify as many as possible of them before future high energy missions will start their active life. With some a-priori knowledge of the nature of the sources they are going to observe, missions such as Agile or Glast could optimize their observing strategies

REFERENCES

1. Bignami,G.F et al ., *Ap.J.* **247** pp. L85-L88 (1981)
2. Bignami,G.F and Hermsen W., *A.R.A.A.*, **21** pp. 67-108 (1983)
3. Bignami,G.F. and Caraveo, P.A., *A.R.A.A.* , **34,** pp. 331-381 (1996)
4. Brazier, K.T.S. et al , *M.N.R.A:S.* **281**, pp.1033-1037 (1996)
5. Brazier, K.T.S. et al , *M.N.R.A:S.* **295**, pp.819-824 (1998)
6. Caraveo, P.A., Bignami, G.F. and Trumper J. *A.&A. Rev.* **7** pp. 209-216 (1996).
7. D'Amico, N. et al., *Ap.J. 552,* pp. L45-L48 (2001)
8. Fichtel C.E. et al, *Ap.J.* **198**   pp. 163-182 (1975)
9. Gehrels N. et al *Nature .* **404**   pp. 363-365 (2000)
10. Grenier, I., *A &A* , **364**, L93-L96 (2000)
11. Halpern J. P. et al , *Ap.J.,* **552**   pp.L125-L128   (2001a)
12. Halpern J. P. et al , *Ap.J.,* **551**   pp.1061-1023   (2001b)
13. Hartman, R.C., *Ap.J.S.,* **123,** pp. 79 (1999)
14. Kaaret,P. et al, . *Ap.J,* **523**   pp. 197-202 (1999)
15. Kaaret, P., Cusumano, G. and Sacco, B., *Ap.J,* **542,**   pp. L41-L43 (2000)
16. Kaspi, V.M. et al. *Ap.J,* **528,** pp. 445-453 (2000)
17. Kniffen, D.A. et al. *Ap.J,* **486** , pp. 126-131 (1997)
18. Mirabal, N. and Halpern J.P. *Ap.J* , **547,**   pp. L137-140 (2001)
19. Mukherjee,R. at al, *Ap.J,* **441,**   pp. L61-64 (1995)
20. Mukherjee,R. at al, *Ap.J,* **542,** pp. 740-749 (2000)
21. Nel, H.I. et al, *Ap.J,* **465,** pp. 898 (1996)
22. Nice, D.J., and Sayer R.W., *Ap.J,* **476,** pp. 261 (1997)
23. Oka, T., et al. *Ap.J,* **526,** pp. 764-771 (1999)
24. Reimer, O. et al. *M.N.R.A:S.* in press (2201)
25. Roberts, M.S.E., Romani ,R.W., Kawai N. *Ap.J.S.,* **133,** pp. 451-465 (2001)
26. Swanenburg, B.N. et al. **243,** pp. L69-L73 (1981)
27. Tavani, M. et al *Ap.J,* **479,** pp. L109-L112 (1997)
28. Tavani, M. et al *Ap.J,* **497,** pp. L89-L91 (1998)
29. Thompson, D.J. et al, *Ap.J,S.* **101,** pp. 259 (1995)
30. Thompson, D.J. et al, *Ap.J.S.,* **107,** pp. 227(1996)
31. Thompson, D.J. et al, *Ap.J,* **516,** pp. 297-306 (1999)
32. Thompson, D.J. Symposium on *High-Energy Gamma-Ray Astronomy, Heidelberg, Germany,* astro-ph 0101039 (2001)

Neutron star contribution to the Galactic unidentified EGRET sources

Isabelle A. Grenier & Christophe A. Perrot

Université Paris 7 & Service d'Astrophysique CEA Saclay
91191 Gif/Yvette, France

Abstract. The nature of the Galactic unidentified EGRET sources frustratingly remains an enigma three decades after their discovery. No clear picture has emerged yet as to the nature of the objects involved, but neutron star activity in various forms appears as a dominant component. Several candidates have recently emerged from X-ray observations that reinforce the importance of neutron stars as powerful γ-ray sources in binary systems, as isolated pulsars, or as pulsar wind nebulae. In the solar neighbourhood, the enhanced rate at which the starburst Gould Belt has produced supernovae for the past few million years and the lack of more convincing candidates strongly suggest pulsars as counterparts to the EGRET sources correlated with the Gould Belt. Off-beam pulsed emission appears as an interesting possibility to match the source counts as well as their flux and spectral distributions. In addition, it is shown here that the population of pulsars born in the Belt exhibits the characteristic Belt signature across the sky over at least 5 Myr despite the dynamical evolution of the Belt and the rapid migration of pulsars.

PULSAR ACTIVITY IN THE GALACTIC DISC

The spatial, spectral, and temporal characteristics of the unidentified sources at low latitude point to the existence of a population of rather hard and steady sources in the inner Galactic disc. These two characteristics are suggestive of a pulsar population. Their marked concentration near the Galactic plane, with a sharp Gaussian profile in $|b|$ with $\sigma_b = 1.6° \pm 0.3°$, and their gathering in the first and second quadrants clearly reveal their origin in the thin Galactic disc, 1 to 4 kpc away, in the inner regions [39] (see Figure 1). As candle sources, their luminosities would range between 0.6 and 4 10^{35} erg/s [29]. A majority of them, ~80%, appear to be stable with a low τ fractional variability [59]. The fractional variability, i.e. the ratio of the standard deviation to the mean of the set of recorded fluxes, is a reliable indicator that allows the separation of constant sources from those with poor flux measurements above a structured background and in the presence of variable nearby sources. Another fluctuation index, I, points to the same lack of variability and finds only 3 variable sources among the 38 ones along the Galactic plane [69]. One should, however, keep in mind that variability indicators are less sensitive at low latitude because of the intense Galactic background. The sharp break near 3° in the latitude profile corresponds to obvious changes in flux and spectrum as well. The sources near the plane are harder with average spectral indices of -2.18 ± 0.04 and -2.49 ± 0.04 at $|b|<5°$ and $5° <|b|< 30°$, respectively [12]. Different log(N)-log(>S) distributions have been found in the two latitude ranges ([12]). Faint sources cannot be resolved above the intense Galactic background, but their integrated flux should not exceed 10% of the interstellar emission [25]. The subsequent limit on their number shows that the low-latitude population has a distinctly flatter log(N)-log(>S) distribution than at mid-latitude. The lack of bright sources off the plane is also truly remarkable. So, this set of results indicates that a distinct population of stable, rather bright and hard sources dominates in the inner

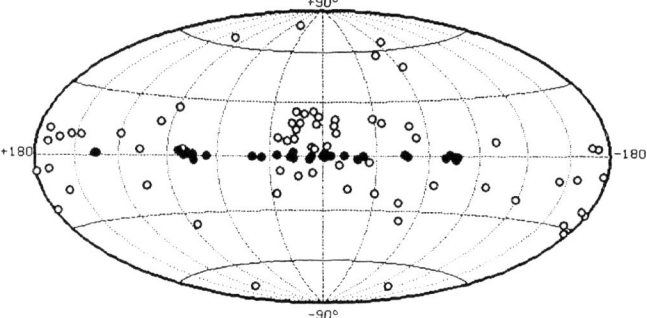

Figure 1. distribution, in Galactic coordinates, of the 99 persistent sources from the 3rd EGRET catalogue which are likely explained by pulsar activity. The 32 sources at |b| < 3° (in filled circles) are a few kpc away in the Galactic disc, most of the others (in open circles) belong to the Gould Belt.

Galaxy.

The daunting problem of obtaining firm identifications at low latitude results from source confusion in the large error box, in particular toward crowded regions in the spiral arms or in the Cygnus complex. There is growing evidence from positional correlations with tracers of star formation, such as HII regions, pulsars, supernova remnants, and OB associations, that a majority of the sources is related to active star-forming sites [54, 11, 64, 28]. The latest compilation [49] lists 26 sources coincident with an OB association and 22 with a supernova remnant. The combined chance probability is small in both cases ($<10^{-3}$ and $<10^{-5}$, respectively). Ten sources coincide with both an OB association and a remnant, i.e. a SNOB, in which the acceleration could proceed in 2 steps, the first using the supersonic wind of a massive O or WR star and the second stage being provided by the shock wave of a nearby supernova remnant [37]. This attractive theory has yet to be verified observationally. Adopting the stellar counterpart distance, the derived luminosities range from 0.1 to 2 10^{35} erg/s over 4π sr, i.e. in close agreement with the values inferred from the global spatial distribution. No clue can be gained from the large dispersion in spectral indices, ranging from 1.7 to 3.1, nor by comparing the luminosities and spectral indices of the sources toward an OB association or a SNOB complex. Supersonic winds of massive O stars are potentially active acceleration sites, at the terminal shock with the interstellar medium or along the turbulent wind [5, 6, 61, 62]. The resulting synchrotron radiation and inverse Compton emission on the stellar UV radiation field would produce γ rays. Yet, the lack of detection by EGRET of the numerous nearby O stars in the Gould Belt sets a stringent limit on their individual emission above 100 MeV. So, collective emission is required to account for the EGRET sources coincident with an OB association in the Galactic plane [13].

new pulsars candidates

Rotation-powered pulsars are obvious potential candidates for this population of hard and stable sources. Seven of the nine pulsars having the highest $\dot{E}/D^2$ value are now established sources of γ-ray pulsations, thus indicating a high probability of intercepting the γ-ray beam when the narrow radio beam sweeps across the Earth, and indicating a close relationship between the onset of high-energy showers and coherent bunching of radio emitting electrons. On the other hand, pulsed γ rays have been searched for without success for 350 radio pulsars ([40]). Seven γ-ray pulsars, with ages from 1 kyr to 0.5 Myr, reside at kpc distances in the Galactic disc and one (Geminga, 0.3 Myr) belongs to the Gould Belt. The latest detections are PSR B1046-58 [30] and probably PSR J2229+6114 [19]. PSR B1046-58 is a 20 kyr, $\dot{E}=2\ 10^{36}$ erg/s, 3 kpc distant, pulsar. As for Vela, its lightcurve exhibits two peaks above 400 MeV with a familiar phase separation of ~0.4 and an offset of ~0.15 from the radio pulse. It radiates ~1 % of its spin-down power into γ rays over 1 sr with a typical E^{-2} spectrum and steady flux. PSR J2229+6114 is quite an energetic, $\dot{E}=2.2\ 10^{37}$ erg/s, 10.5 kyr, 3 kpc distant, pulsar, second only to the Crab in spin-down power. If responsible for the flux from 3EG 2227+6122, it would radiate 0.1% of its power in γ rays over 1 sr [19]. Radio and X-ray pulsations have been recorded so far, but the extrapolation of the ephemerides is not reliable enough yet to search back for γ-ray pulsations in the EGRET data. Nonetheless, the lack of another likely candidate in the error box makes it a rather promising identification. The EGRET source is indeed very stable albeit slightly softer than expected for its young age (see Figure 3).

Another two tentative identifications have been recently proposed in rather complex regions [1]: a 13 kyr, $\dot{E}=10^{37}$ erg/s, 7.7 or 2 kpc distant pulsar, PSR J1420-6048, in the Kookaburra nebula, and a 34 kyr, $\dot{E}=2\ 10^{36}$ erg/s, 6.2 kpc distant pulsar, PSR J1837-0604 in the inner Galaxy. They rank 11 and 22, respectively, in the $\dot{E}/D^2$ list, but their distance estimate, based on their dispersion measure, is highly uncertain. The hard spectra recorded for the related 3EG sources, with respective indices of 2.02 ± 0.14 and 1.82 ± 0.14, are consistent with a pulsar hypothesis. Emission from 3EG J1837-0606 is very stable, but the variability of 3EG 1420-6038 is puzzling [59]. It should be noted that another hard X-ray source is found in the error box of 3EG J1837-0606, near the edge of an HII region.

Other interesting radio-quiet neutron star candidates have been proposed according to their Geminga-like characteristics, i.e. a faint soft X-ray source in the error box with no optical counterpart down to V magnitudes of 24 or 25. Large X-ray-to-optical flux ratios of a few hundred are indeed only seen among neutron stars. In this context, the X-ray source in the γ Cygni remnant toward 3EG J2020+4017 was proposed as a possible neutron star counterpart [4]. The EGRET source is stable, its spectral index of 2.08 ± 0.10 not too soft, but the association remains unclear because there is a 15th magnitude K0V star in the X-ray box and no compact keV nebula to support the pulsar hypothesis [47]. Two more promising candidates have been found at medium latitude in the CTA1 remnant toward 3EG J0010+7309 [2, 51], and toward the brightest unidentified source 3EG J1835+5918 [36]. The detailed spectrum from 3EG J0010+7309 exhibits a $E^{-1.6}$ spectrum cutting off at 2 GeV and a stable long-term behaviour which are strongly suggestive of a pulsar. The presence of compact keV nebula gives extra support to this hypothesis. It would be powered by a 20 kyr, $\dot{E} = 1.7 \; 10^{36}$ erg/s pulsar. The 3EG J1835+59 source is also very hard ($\gamma = -1.69 \pm 0.07$) and stable, but no keV nebula has been detected to reinforce the pulsar origin. The marginal variability seen in X rays, but not in γ rays, is puzzling and needs confirmation.

pulsar contribution to the Galactic disc sources

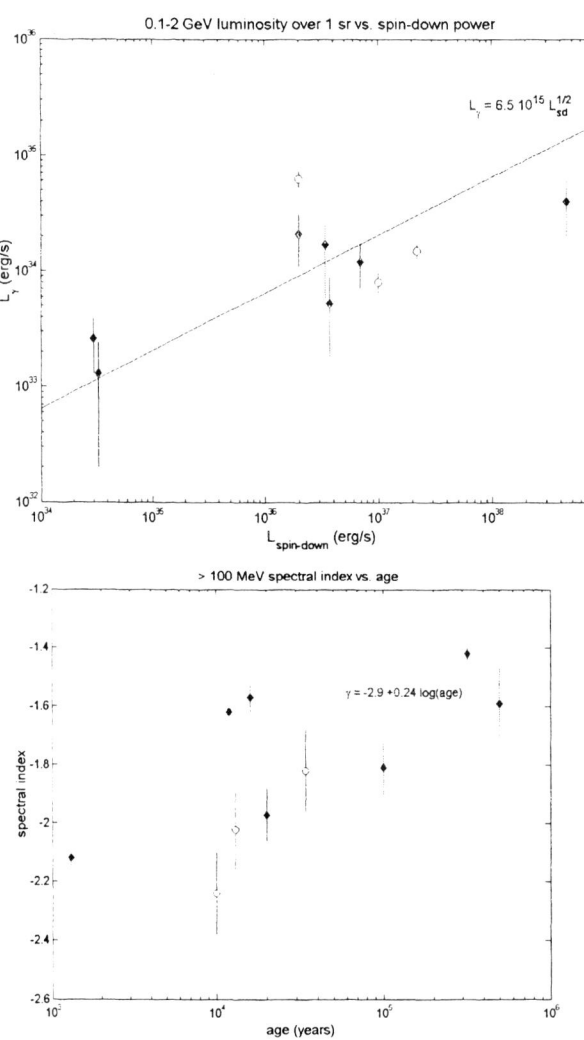

Figures 2 & 3: 0.1-2 GeV luminosity vs. spin-down power and spectral index vs. age for the 7 identified pulsars (Crab, Vela, Geminga, PSR1706-44, PSR1951+32, PSR1055-52, PSR1046-58, filled circles) and new candidates (PSR2229+61, PSR1420-60, PSR1837-06, open circles). The luminosity above 400 MeV has been used for PSR 1046-58. For PSR 1420-6048, the closer distance of 2 kpc suggested by the X-ray data has been adopted.

The present sample of γ-ray pulsars is strongly biased to high luminosity and youth, as confirmed by various pulsar models. The γ-ray efficiency, $L_\gamma/\dot{E}$, where L_γ notes the luminosity above 100 MeV and $\dot{E}$ the spin-down power, is seen to increase with pulsar age, provided that all beams have comparable solid angle [57, 66]. In fact, the luminosity L_γ scales as $\dot{E}^{1/2}$, as expected from the maximum Goldreich-Julian current that can be drawn from a polar cap when assuming that a constant energy loss per particle is converted into γ rays. This relationship holds over 4 decades [57, 68] and has been updated in Figure 2 using the newly claimed detections. Observations of pulsars older than $5 \; 10^5$ years are desperately needed to constrain the evolution of the γ-ray efficiency at low spin-down power and to search for a death line for pulsed γ radiation. Emission also tends to become harder with age [57] (see Figure 3). PSR J1046-58 as well as the PSR J2229+6114 and PSR J1420-6048 candidates nicely follow the general luminosity and spectral trends. PSR J1837-0604 presents a high efficiency, but one should remember the large uncertainty in its distance.

Beaming fractions as high as 10%, and decreasing with pulsar period as $P^{-1/2}$, are estimated for the polar-cap main beam [68]. Larger values will be discussed below for off-beam emission seen at large angle from the neutron star magnetic axis. In the outer-gap model, the beaming fraction evolves with the magnetic inclination and the thickness of the vacuum gap beyond the null charge surface. It drops from 80% to 25% over 1 or 2 Myr for the standard outer gap at 60° inclination. After that, the γ-ray emission is quenched because of the lack of thermal X rays from the stellar surface [48]. The thick outer gap uses the backflow of pairs towards the neutron star to heat up the polar cap surface. Slightly smaller beaming fractions, from 50% to 20%, are estimated in this case for the same 60° inclination over a timescale < 1 Myr. The beam opens up at lower magnetic inclination [67]. Conversely, the polar cap accelerator remains active much longer, at least for several million years.

These statistics have been used to show that pulsars likely dominate the unidentified source population at low latitude. Their average spectral index of -2.18 ± 0.04 weakly supports this idea. All models assume an initial magnetic field at birth of 2 or 3 10^{12} G, an initial period of 10 to 30 ms, a large random kick velocity from [35] and an average birth rate of 0.01 yr^{-1}. The outer-gap model predicts that young pulsars, produced in the spiral arms and massive-star forming regions up to 4 kpc from the Sun, can account for most of the unidentified sources at $|b|<2.5°$ and for a third of them at $2.5<|b|<10°$. Despite their rapid motion away from the plane, at most 1 or 2 pulsars are expected at higher latitude [64, 65]. This model can reasonably reproduce the spatial distribution of all sources at |b| < 5°. The thick outer-gap model reaches similar conclusions. Thanks to an average beam solid angle of 2.8 sr, young (<1 Myr-old) pulsars can account for most of the sources coincident with an OB association or a supernova remnant along the Galactic plane. As many as 32 are expected at |b| < 5° [69]. For the 'comptonized' version of the polar-cap model, using a beam solid angle 4 times larger than the nominal polar-cap one, the population of < 1 Myr-old pulsars can match the characteristics of the known γ-ray pulsars, but can only produce 2 to 4 radio-quiet γ-ray sources near the Galactic plane [53], unless the radio and γ-ray beams are misaligned (see Gonthier et al. in these proceedings for recent estimates from the polar cap). All models show that, because of the limited EGRET sensitivity at low latitude, the sample of visible candidates is severely flux-limited and age limited, as is the sample of the identified γ-ray pulsars. 90% and 95% of the detectable outer-gap and polar-cap pulsars, respectively, should be younger than 0.3 Myr and no older one would be visible beyond 2 or 3 kpc.

pulsar wind nebulae

Powered by a pulsar relativistic wind or confining relic high-energy electrons, pulsar wind nebulae have also emerged as interesting candidates. MeV to TeV emission from the Crab nebula is successfully interpreted as synchrotron-self-Compton emission in the nebular magnetic field from electrons energized at the pulsar wind terminal shock. For the compact synchrotron nebula of PSR1706-44 and the "bubble" of relic electrons trailing off the Vela pulsar, inverse-Compton scattering of the cosmic microwave background (CMB) radiation, or of the local IR field, is favoured. In these small nebulae, the necessary magnetic field falls 3 to 6 times below the equipartition value and raises confinement problems over long periods. Recent ASCA observations reveal that 6 or 7 EGRET sources could originate from pulsar wind nebulae [47, 16].

Towards 3EG 1809-2328 a non-thermal keV and radio nebula is possibly interacting with a 10^4 M$\odot$ CO cloud [42, 47]. Electrons accelerated up to at least 20 TeV may account for the very hard X-ray synchrotron radiation in a 20 µG field while bremsstrahlung radiation from GeV electrons in the surrounding cloud may explain the $E^{-2.06 \pm 0.08}$, mildly variable, EGRET source. The bremsstrahlung spectrum should interestingly extend to TeV energies, well above the current instrumental limits in sensitivity.

3EG 1420-6038 is another noteworthy example [47]. The error box overlaps much of the north-east wing of the Kookaburra nebula, a bird-like structure with a thermal body and two non-thermal wings. The north-east wing contains the PSR J1420-6048 pulsar discussed above. Its unpulsed keV flux and the coincident radio hot spot show that it powers an active wind nebula and the variability of the EGRET source [59] makes a nebular origin of the γ rays more tempting than a magnetospheric one. The 95% error box extends to the south-west and includes the Rabbit, a non-thermal, 10% polarized, center-filled radio and X-ray nebula, interpreted as a plerion ([47] and ref. therein). So, the EGRET source might result from the combination of two wind nebulae.

In the composite supernova remnant W44, γ rays can be produced by the hard shell electrons pervading a nearby cloud or up-scattering the remnant IR dust emission and ambient Galactic and CMB radiation fields. Bremsstrahlung radiation dominates in the cloud and inverse-Compton shell emission dominates away from it. Both can match the flux from 3EG 1856+0114 [26]. However, half of the shell lies outside the 99% 3EG contour and the latest 3EG and GeV error box are reasonably centered on the pulsar B1853+01 location and on the radio+keV plerion surrounding it [21, 16]. X-ray and radio emitting relic electrons are also present in a 2-pc-long cometary tail. The $E^{-1.93 \pm 0.10}$ source spectrum is typical of a pulsar, unlike the observed flux variability and no pulsation has been detected at the radio frequency [58]. So, an origin of the γ rays either in the compact wind nebula or in the cometary tail is quite plausible.

In the composite remnant CTA1, an interesting radio-silent pulsar candidate has been discussed above. Yet, plerionic emission from the energetic keV synchrotron nebula is also quite plausible to explain the source 3EG 0010+7309.

A surprisingly large fraction (40%) of the γ-ray emission seen from PSR 1046-58 above 400 MeV appears to be unpulsed. It therefore brings a second example of hard DC emission from a pulsar, probably from the compact (∅ <3') synchrotron nebula surrounding it [46].

Another example comes from the energetic radio and X-ray pulsar, PSR J2229+6114, found toward an EGRET source. It is also surrounded by a centrally peaked, non-thermal, X-ray nebula, inside an incomplete hard radio shell. The central nebula is evidently powered by the pulsar wind [18].

Another pulsar wind nebula association has been proposed between the GeV J1825-1310 source and PSR B1823-13 [47]. PSR B1900+05 also coincides with an EGRET source and no γ-ray pulsations were detected, but no evidence for a wind nebula has been reported. So, at least 7 EGRET sources are coincident with pulsar wind nebulae.

In the case of IC 443, a pulsar wind nebula [44] and/or cosmic-ray acceleration in the region where the remnant shock overtakes a dense cloud have been proposed as the origin of the 3EG 0617+2238 source [31]. The 3EG and GeV error boxes, however, exclude these possibilities. They point away from this region, to the center of the shell where no peculiar X-ray activity has been recorded to explain the stable $E^{-2.01 \pm 0.06}$ source.

Pulsars in binary systems

Binary systems where a pulsar closely orbits a massive star can produce γ rays at the shock between the pulsar relativistic wind and the outflow from the companion star [55]. Three EGRET sources have been tentatively associated with X-ray binary systems, such as 2CG 135+01 (alias 3EG 0241+6103) with the radio source LSI +61° 303. In this system, a neutron star, in an eccentric orbit around a B0Ve star, accretes matter from the stellar wind. Radio outbursts occur with a 26.5 day period and a 4-year modulation during supercritical accretion near periastron, or, alternatively, from pairs accelerated at the shock front between the pulsar and stellar winds. In the latter case, up-scattering of stellar optical photons by the radio emitting electrons can produce γ rays [10, 33]. The association is, however, not clear. The EGRET variability does not correlate with the radio phase, nor does the stable COMPTEL flux. The radio source lies outside the > 300 MeV 95% error box. On the other hand, the COMPTEL spectrum reasonably bridges the ROSAT and EGRET spectra, suggesting continuous emission from 10^2 eV to 10 GeV, and there are no other attractive counterparts in the field. So, the identification clearly needs confirmation.

Another very eccentric binary system, comprising a B2e star and a 48 ms pulsar, PSR B1259-63, in a 3.4 yr orbit, has been extensively studied near apastron [24] and periastron [56]. Unlike LSI +61° 303, it is not an accretion-powered system. Pairs are accelerated at the shock front between the pulsar and stellar winds. They radiate synchrotron radiation up to 10 keV near apastron and up to 200 keV near periastron, but no MeV to GeV flux has been recorded [17, 56]. The lack of γ-ray detection can be explained if the stellar photons are up-scattered to sub-TeV energies by the X-ray emitting TeV electrons [32]. If so, this system might be visible at a few 100 GeV a few days about periastron.

Another association has recently been proposed between the Be/X-ray binary system SAX J0635+0533 and the 3EG 0634+0521 source ([27] and ref. therein). The binary system contains a rather young (1.4 kyr) and energetic ($\dot{E} = 5 \ 10^{38}$ erg/s) pulsar with a 33.8 ms period. The X-ray source exhibits a hard $E^{-1.50\pm0.08}$ spectrum in the 1-40 keV band and a luminosity at a distance ≤ 5 kpc which is typical of such a young object. The variability of the γ-ray source has been revisited. Its steadiness and the 2.03 ± 0.26 spectral index above 100 MeV are both consistent with pulsar emission. If confirmed, this system would offer a unique opportunity to study the early phase of a binary system.

Reactivated neutron stars in the form of ms pulsars have become attractive counterparts since one, PSR J0218+4232, has been convincingly associated with an EGRET source (3EG J0222+4253) [34]. This system consists of a 460 Myr-old pulsar with a spin period of 2.3 ms and a white dwarf companion. The orbital period takes ~ 2 days. The identification is based on the alignment of the pulse profiles recorded in X rays, γ rays < 300 MeV, and in the radio ([34, Kuiper & Hermsen, in preparation]). The two peaks are separated by 0.45 in phase. At a distance of 5.7 kpc in the halo, but not in a globular cluster, the very soft $E^{-2.6}$ spectrum corresponds to a luminosity of 1.6 10^{34} erg/s over 1 sr above 100 MeV. The γ-ray emission is very stable [59]. The νF_ν spectrum of this pulsar remarkably peaks in the MeV range. Despite the low magnetic field of 10^{8-9} G, both polar-cap and outer-gap models predict that a large fraction of the spin-down power may be converted into γ rays [68, 66]. This fraction amounts to 7% for PSR J0218+4232 (over 1 sr, [34]). The weak B field is indeed compensated by a very compact magnetosphere: the field strength of PSR J0218+4232 at the light cylinder falls between that of Crab and Vela. Yet, both models fail to reproduce the spectra: the predicted emission is much too soft in the keV-0.1 MeV band compared with the data from PSR J0218+4232 and two other pulsars; it is much too hard in the EGRET band compared with the PSR J0218+4232 data [68, 60, 50, 3]. So, the confirmation of this ms pulsar activity and the discovery of others will unveil a whole new facet of pulsar γ-ray activity in their old age.

PULSAR ACTIVITY IN THE GOULD BELT

The sample of unidentified sources at |b|>3° is not homogeneous. Persistent and non-persistent sources are found in equivalent numbers and exhibit distinct spectral, temporal, and spatial properties. A source is labelled "persistent" (P) when significantly detected in the 4-yr (P1234) EGRET survey data, whereas a non-persistent ( ) one is seen in individual observations, but not in P1234. The persistent sources are significantly harder and more stable than the non-persistent ones. Average spectral indices $\gamma_P = 2.25 \pm 0.03$ and $\gamma = 2.52 \pm 0.06$ were obtained for the two samples, respectively, with a chance probability of $2 \ 10^{-7}$ of equal index. Similarly, averages $\tau_P = 0.38 \pm 0.06$ and $\tau = 0.95 \pm 0.18$ were obtained with a chance probability of $1.3 \ 10^{-4}$ of equal τ. The τ average is equivalent to that of 0.90 ± 0.07 found for the EGRET active galactic nuclei that are well known for their flaring activity. In contrast, most of the persistent sources show no or little variability [59]. The two groups also have distinct distributions across the sky [14, Grenier, in preparation] and the difference cannot be attributed to systematic biases due to the survey exposure and the intense Galactic background. Their excess at mid-latitude is quite significant and points to a Galactic origin for both populations. Whereas non-persistent sources appear to have a large scale-height above the Galactic plane, the persistent sources are correlated with the curved lane of the Gould Belt (see Fig. 1) and their distribution is significantly better correlated with the Gould Belt than with other Galactic structures [15]. As many as 45 ± 6 sources can be statistically associated with the Belt, out of which about 10 might be background sources originating from the Galactic disc. The Belt rim extends to 150 pc towards Ophiuchus and to 450 pc towards Orion, so the Belt sources have typical luminosities at 300 pc of 0.3 to 6 10^{33} erg/s over 4π sr for E^{-2} spectra above 100 MeV. An equivalent number of non-persistent sources is found at large scale height above the Galactic plane or in the halo, but the discussion on their origin is deferred to a forthcoming paper. At most 35 potentially extragalactic sources are present in the sample.

Gould Belt evolution

The Gould Belt is a local expanding structure which is well delineated both by the local group of massive stars and by a high concentration of HI and H_2 gas. Because of its starburst activity over the past ~30 Myr, it has recently produced supernovae at a larger rate than the local Galactic disc. A frequency of 20 to 27 SNe per Myr has been inferred from the stellar content of the Belt, which corresponds to a 75 to 95 Myr^{-1} kpc^{-2} rate that is 3 to 5 times larger than the local Galactic one [15]. The uncertainty results from the poor knowledge of the initial stellar mass function at high masses. This high rate is valid for the past few Myr. The Belt therefore provides candidate pulsars within a few hundred parsecs of the Sun, both in excess of Galactic disc pulsars and with its peculiar inclined geometry. For any of the pulsar simulations discussed above in the Galactic disc, the number of potential objects at high latitude cannot be increased by dramatically increasing the pulsars beam width, γ-ray efficiency or birth rate for fear of overproducing sources near the plane. For instance, an increase of the birth rate beyond 30% is not supported by the data [64]. Larger kick velocities or a larger scale height at birth are not supported by the radio data. As a first step, one should test whether the spatial distribution of the Belt pulsars still bears its characteristic spatial signature after a few Myr against the Belt expansion and their rapid migration.

The dynamical evolution of the Belt has been modelled in 3D and confronted to the spatial as well as velocity distributions of all HI and CO clouds within a few hundred parsecs from the Sun. The position and velocity centroid of the clouds were determined using the "clumpfind" tool [63] which allows to search for local peaks of emission within the longitude, latitude and velocity data cubes of the relevant CO and HI surveys [8, 23, 52]. Cuts in latitude and velocity were applied to eliminate obvious background clouds in the Galactic disc and in the Local Arm. A maximum-likelihood test was used to confront the modelled Belt and the clumps (l,b,v) coordinates. It included a distance probability for a subset of well-known clouds (Perrot & Grenier, in preparation).

The Gould Belt was modelled as an initially inclined and laterally expanding cylinder, split into 60 elementary contiguous surfaces sweeping momentum from the ambient interstellar gas. The centre was initially located in the Galactic plane. The local HI gas distribution was taken as the combination of two Gaussians and one exponential, with densities and dispersions or scale-height $N_{1HI} = 0.395$ cm^{-3}, $\sigma_1 = 90.03$ pc, $N_{2HI} = 0.107$ cm^{-3}, $\sigma_2 = 225.17$ pc, $N_{3HI} = 0.064$ cm^{-3}, $z_3 = 403.0$ pc respectively [9]. The local H$_2$ distribution was described by a gaussian with $N_{H2} = 0.2$ cm^{-3} and $\sigma_{H2} = 74.0$ pc [8]. While the Belt evolves into an elliptical section under the momentum flow and Galactic differential rotation, it tends to fall back onto the Galactic plane under the gravitational pull of the Galactic disc. The values recommended by the IAU were adopted for the Oort's constant A_c and for the solar galactocentric radius. The gravitational torque was calculated for the local stellar mass density distribution [7, 41]. We considered no internal pressure from supernovae or stellar winds, and no external pressure from the interstellar medium.

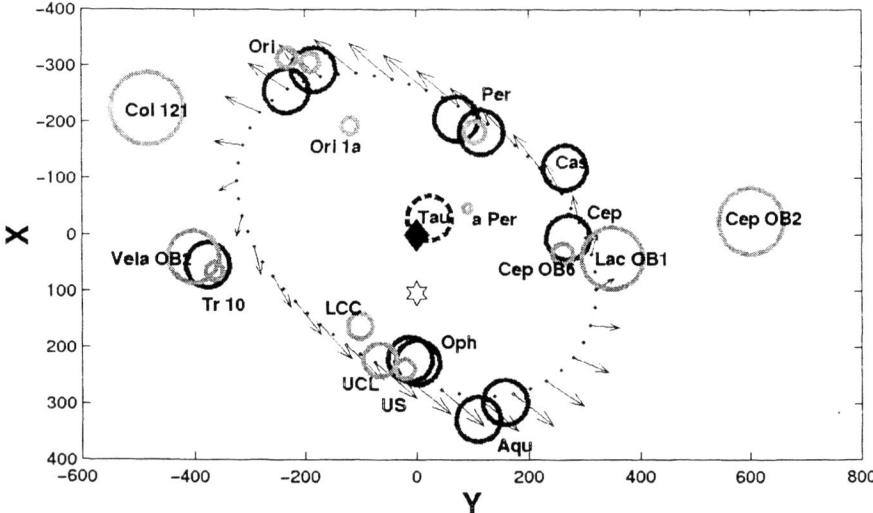

Figure 4. Present position of the Gould Belt projected onto the Galactic Plane in a non-inertial frame centered on the Belt center. The X-axis points to the Galactic center, while the Y-axis points in the direction of the Galactic rotation. The velocity field outlines the Belt expansion. The main cloud complexes are plotted as dark circles. The main nearby OB associations sizes and distances (light circles) were estimated from recent Hipparcos measurements.

The best fit yields characteristics for the Gould Belt in good agreement with the presence of massive clouds and active OB associations in the solar neighbourhood. As shown in Figure 4, the Belt rim reasonably coincides with Per OB2, Ori OB1b and 1c, LCC, UCL, US, and Cep OB6. The best fit yields a present inclination $\varphi = 17.2° \pm 0.3°$ to the Galactic plane, an age $\tau = 26.4 \pm 0.4$ Myr and an initial energy $E_i = (1.0 \pm 0.1)\,10^{52}$ erg. These values are comparable to those obtained in the modelling of the Belt versus the local stellar velocities [38] and to those obtained in the 2D modelling of the HI longitude-velocity distribution [43]. The centre of the Belt today is located

103 ± 4 pc away from the Sun in the direction $l_{centre} = 180° ± 2°$, and the current ellipse has a semi-major axis a = 354 ± 5 pc and a semi-minor axis b = 232 ± 5 pc. These results are independent of the presence or not in the fit of the Taurus clouds, which are located near the Belt centre.

recent pulsar production in the Gould Belt

To know if the Belt spatial signature is preserved over 5 Myr, we used this dynamical model to generate pulsars and we let them migrate in the Galactic potential [45]. They were generated at random, uniformly, in a thick torus (40 pc in height) covering the outer half of the Belt disc, as situated at the time of the pulsar birth. Their initial velocity resulted from the combination of the Galactic rotation velocity and of a random component following a Maxwellian distribution with dispersion $\sigma_v = 300 \sqrt{3}$ km/s, i.e. a mean of 480 km/s [35]. To evaluate the pulsar visibility, as seen by EGRET, standard parameters for γ-ray pulsed emission were assumed, i.e. a luminosity $L_{>100 MeV} = 1.5 \; 10^{16} \dot{E}^{1/2}$ erg/s over 1 sr [57]; a magnetic field B following a Gaussian distribution in log(B) with a mean of 12.4 and a dispersion of 0.3; an initial spin period $P_0 = 30$ ms evolving with time as $P(t) = [P_0^2 + \frac{16\pi^2 R_{NS}^6 B^2}{3Ic^3} t]^{1/2}$; a random magnetic inclination and aspect angle; a beam opening angle $\theta \approx 10 \; \theta_{polar\;cap}$; and birth rate of 25 Myr^{-1} in the Belt.

As illustrated on Figure 5, the Belt signature is well preserved over at least 5 Myr against pulsar migration and the Belt evolution. The asymmetry on both sides of the Galactic plane in the 1st+2nd or in the 3rd+4th quadrants corresponds to a 2:1 ratio in the total number of visible objects in these areas. The latitude profiles integrated over the 3rd+4th quadrants in Figure 6 show that this contrast is little reduced from 1 to 5 Myr. So, the Belt signature should be preserved beyond 5 Myr.

The stability of most of the Belt sources [59, 12] is consistent with a pulsar origin. Geminga is the first example of a Belt pulsar and notably the intrinsically faintest γ-ray pulsar observed so far. Given the Belt supernova rate, it should have produced a total of 7 to 9 pulsars of age < 0.34 Myr (the age of Geminga). With a luminosity equivalent to that of Geminga, they would be easily detectable anywhere inside or around the Belt. So, the beaming fraction is the limiting factor on their visibility and, for a polar-cap beaming fraction η ~7% [68], about 0.5 should be detectable by EGRET. So, the detection of Geminga is consistent with the Belt production rate and a beaming fraction of the bright "on" beam of the order of 10-20%. To detect as many as 30 Belt γ-ray sources requires the product of η by pulsar age T to be η×T ~ 1.2 over the entire Belt, or a longer timespan if the older pulsars are only visible to shorter distances. Using the $L\gamma \propto \dot{E}^{1/2}$, $\dot{E} \propto T^{-2}$, and $\eta \propto P^{-1/2}$ dependence, the on-beam flux from a 5 Myr-old pulsar scales as $F_{old-on} = F_{Gem} (T_{old}/T_{Gem})^{-3/4} (D_{old}/D_{Gem})^{-2}$ with respect to Geminga. It remains visible to 430 pc for a visibility threshold of ~6 10^{-8} γ cm^{-2} s^{-1} [22]. Over 5 Myr and with an old on-beam fraction η ~ 0.5×7%, the Belt would produce 3 to 5 visible sources with fluxes around several times 10^{-7} γ cm^{-2} s^{-1}. Five such persistent sources are indeed present in the sample.

Off-beam (or trailer) emission is observed outside the two main peaks in the recorded light curves. The off-beam to main peak contrast (per phase bin) decreases from ~30 to ~5 with age from the Crab to Geminga. A contrast of 5 to 10 and the assumption pulsar distribution for ages < 5 Myr

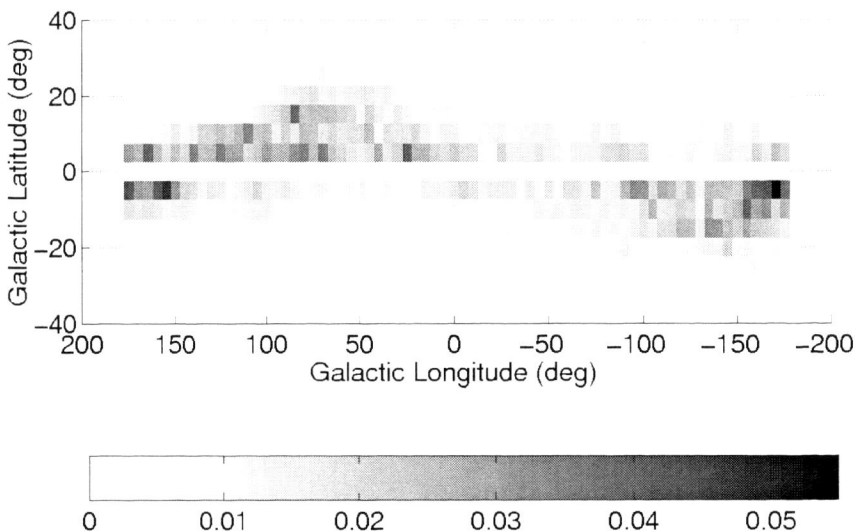

Figure 5. Number density map at |b|>2.5° for pulsars born in the Gould Belt with ages $0 < \tau < 5$ Myr. The visibility threshold is constructed from the interstellar γ-ray background model (Hunter et al. 1997), the extragalactic intensity and the EGRET 4-year exposure map, following the 4 and 5 σ levels adopted in the 3rd EGRET catalogue

Figure 6. Latitude profile of the number of visible pulsars integrated over negative longitudes for pulsar ages less than 1 and 5 Myr.

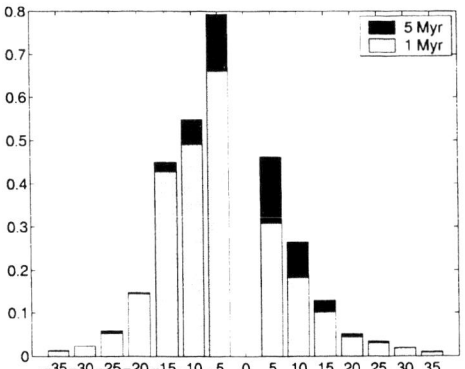

that off-beam photons are detected over half a period imply a flux ratio F_{on}/F_{off} between the on and off beams of order 5. A 5 Myr-old off-beam pulsar therefore remains visible to 200 pc (or 250 pc for $F_{on}/F_{off} = 3$). In fact, the bulk of the persistent Belt sources have fluxes 2 to 6 times fainter than the brightest ones mentioned above. In the polar-cap model, synchrotron cascades produce the hard on-beam at low altitude above the stellar surface while primary electrons at a higher altitude produce a wide and soft "off"-beam of γ rays by curvature radiation. The ratio of the off and on beam fractions is typically $D = \eta_{off}/\eta_{on} = 3$ for a Geminga-like pulsar and it rises to 5 for a 5 Myr-old object [20]. The softness of the off-beam emission (as observed and predicted [20]) may explain the soft average index $\gamma = 2.25 \pm 0.03$ of the Belt sources compared to that of the young on-beam Galactic pulsars. With the above parameters, the Belt would produce 11 to 15 detectable old off-beam pulsars. This number compares favourably with the number of Belt sources given the large uncertainties in the evolution of the pulsed emission to old ages and at large angle. Detailed simulations of the off-beam pulsar population in the Belt are underway. The outer-gap beam is naturally wide, but shorter lived. Because of its strong dependence on age and magnetic inclination, detailed simulations are required to quantify the number of visible Belt pulsars. They are underway.

So, within our present understanding of pulsar γ-ray emission, and in the absence of more compelling candidates from supersonic stellar winds, supernova remnants or accreting systems [15, 13], million year old pulsars born in the Gould Belt appear as the most promising counterparts for the Belt EGRET sources. They would provide a unique opportunity to constrain pulsar models to older ages and a variety of aspect angles, using GLAST to locate them and to detect their γ-ray periodicity. As supernova relics, they would bring valuable constraints on the initial stellar mass spectrum at large masses.

So, both locally and throughout the Galaxy, it appears that neutron star activity may explain numerous unidentified EGRET sources, most of them in fact, but perhaps not all. A few flaring sources near the Galactic plane and the halo sources may require other exciting possibilities, such as black holes.

REFERENCES

1. D'Amico N. et al., ApJ 552, L45, 2001
2. Brazier K. T. S., et al., 1998, MNRAS, 295, 819.
3. Bulik T. & Rudak B., 2000, MNRAS 317, 97
4. Carraminana A., et al., 1997, Proc. 4th Compton Symp., 1267.
5. Cassé M., Paul J.A., 1980, ApJ, 237, 236
6. Cesarsky C.J., Montmerle T., 1983, Space Science Rev. 36, 173
7. Crézé, M. et al., 1998, A&A, 329, 920-936
8. Dame, T. M. et al, 1987, ApJ, 322, 706
9. Dickey, J. M., Lockman, F. J., 1990, ARA&A, 28, 215
10. van Dijk R., et al., 1996, A&A, 315, 485.
11. Esposito J. A., et al., 1996, ApJ, 461, 820.
12. Gehrels N., et al., 2000, Nature, 404, 363
13. Grenier I.A., 2001, Proc. "Gamma Symp. 2000", Heidelberg, AIP 558, 191
14. Grenier I.A., 2001, Proc. "The nature of Galactic high-energy g-ray sources", Tonantzintla, AIP, in press
15. Grenier I.A., 2000, A&A, 364, L93
16. Grenier I.A., 2000, Proc. of 6th "Towards a major atm. Cherenkov Detector" Symposium, Snowbird, AIP 515, 261
17. Grove J.E. et al, ApJ, 447, L113
18. Halpern J.P. et al., ApJ 547, 323, 2001
19. Halpern J.P. et al., ApJ, 2001, in press, astro-ph/0104109
20. Harding A.K., Zhang B., ApJ 548, L37
21. Harrus I. M., Hughes J. P., Helfand D. J., 1996, ApJ, 464, L161.
22. Hartman R. C. et al., 1999, ApJS, 123, 79
23. Hartmann, D. et al., 1996, A&AS, 119, 115
24. Hirayama M., et al., 1999, ApJ, 521, 718.
25. Hunter S.D., Bertsch D.L., Catelli J.R., et al., 1997, ApJ 481, 205
26. de Jager O. C., Mastichiadis A., 1997, ApJ, 482, 874.
27. Kaaret P., Cusumano G., Sacco B., ApJ 542, L41, 2000

28. Kaaret P., Cottam J., 1996, ApJ, 462, L1.
29. Kanbach G., et al. 1996, A&AS, 120, 461
30. Kaspi V., et al., 2000, ApJ 528, 445
31. Keohane J. et al., ApJ 484, 350, 1997
32. Kirk J. G., Ball L., Skjaeraasen O., 1999, Aph, 10, 31.
33. Kniffen D. A., et al., 1997, ApJ, 486, 126.
34. Kuiper L., Hermsen W., et al., 2000, A&A 359, 615
35. Lorimer, D. R. et al., 1997, MNRAS, 289, 592
36. Mirabal N., Halpern J.P. ApJ 547, L137, 2001
37. Montmerle T., 1979, ApJ, 231, 95.
38. Moreno, E. et al., 1999, ApJ, 522, 276
39. Mukherjee R., Grenier I. A., Thompson D. J., 1997, Proc. 4th Compton Symp., AIP 410, 394.
40. Nel H. I., et al., 1996, ApJ, 465, 898.
41. Ojha, D. K. et al., 1996, A&A, 311, 456
42. Oka T., et al., 2000, ApJ, in press, astro-ph 9907261.
43. Olano, C. A., 1982, A&A, 112, 195
44. Olbert C.M. et al., ApJ Letters, 2001, in press, astro-ph/0103268
45. Paczynski, B., 1990, ApJ, 348, 485
46. Pivovaroff M. J., Kaspi V. M., Gotthelf E. V., 1999, astro-ph 9906374.
47. Roberts M.S.E. et al., ApJ, 2001, in press, astro-ph/0102471
48. Romani R. W., 1996, ApJ, 470, 469.
49. Romero G. E., Benaglia P., Torres D. F., 1999, A&A, 348, 868.
50. Rudak B. & Dyks J., MNRAS 303, 477, 1999
51. Slane P. et al., 1997, ApJ, 485, 221.
52. Strong, A. W. et al., 1982, MNRAS, 201, 495
53. Sturner S. J., Dermer C. D., 1996, ApJ, 461, 872.
54. Sturner S. J., Dermer C. D., 1995, A&A, 293, L17.
55. Tavani M; Arons J., 1997, ApJ, 477, 439.
56. Tavani M., et al., 1996, A&AS, 120, 221.
57. Thompson D. J. et al., 1997, Proc. 4th Compton Symp., AIP 410, 39.
58. Thompson D. J., et al., 1994, ApJ, 436, 229.
59. Tompkins W., 1999, PhD thesis, Stanford University.
60. Wei D.M., Cheng K.S., Lu T., 1996, ApJ 468, 207
61. White R.L., 1985, ApJ 289, 698
62. White R.L., Chen W., 1992, ApJ 387, L81
63. Williams, J. P. et al., ApJ, 428, 693, 1994.
64. Yadigaroglu I., Romani R. W., 1997, ApJ, 476, 347.
65. Yadigaroglu I., Romani R. W., 1995, ApJ, 449, 211
66. Zhang L. & Cheng K.S., MNRAS 294, 177, 1998
67. Zhang L. & Cheng K.S., ApJ 487, 370, 1997
68. Zhang B., Harding A.K., ApJ 532, 1150, 2000
69. Zhang L., Zhang Y.J., Cheng K.S., 2000, A&A 357, 957

Population studies of the Gamma-ray sources

A. W. Chen[*], S. Mereghetti[†], A. Pellizzoni[†], M. Tavani[†] and S. Vercellone[†]

[*]Consorzio Interuniversitario per la Fisica Spaziale, Viale Severo 63, Torino, I-10133 Italy
[†]Istituto di Fisica Cosmica G.Occhialini, CNR, via Bassini 15, Milano, I-20133 Italy

Abstract.
We present the status of ongoing modelling of the populations of the EGRET point sources. To analyze the two-dimenstional distribution, we wrote Monte Carlo simulations to produce model populations. We compared models containing an isotropic component and one or two galactic components to the EGRET catalog; preliminary results indicate that while one galactic component produces a spatial distribution consistent with the EGRET sources, the resulting flux distributions are inconsistent. Simultaneous comparison of spatial and flux distributions will allow physical models to be constrained.

INTRODUCTION

Two different approaches have been successfully adopted to understand the nature of gamma-ray sources, in the lack of obvious identifications as provided, e.g., by the detection of periodic variability.

The first one is based on observations at lower energies, to look for candidate counterparts in the gamma-ray error boxes. Typically such searches start from the X–ray and/or radio band, where it is easier to spot potentially unusual objects. Optical/IR follow-ups can then be targeted to study better a few selected objects.

The second method consists of "population studies" by which average flux, spectral and variability properties, spatial distribution, etc. , can be compared to those of other known populations of astrophysical objects (e.g., by Gehrels et al. (2000), Zhang, Zhang & Cheng (2000), Romero et al. (1999)).

We are undertaking a population study of the EGRET sources that, in several respects, differs from other analysis of this kind performed in the past. Here we discuss the motivations of our work, and present some preliminary results. Although for the moment the derived conclusions are not different from those reported by other authors, we believe that our approach is less affected by biases and uncertainties and will probably lead to more robust results.

OUR APPROACH TO POPULATION STUDIES

Our method is based on the following guidelines:

- Avoid data binning.

The use of binned distributions of small samples leads to histograms with only a few objects per bin. The most common binned technique, the χ^2 test, is not correct if the statistical errors are not normally distributed. An alternative representation of the data is possible by means of unbinned integral distributions, that can be compared to the expectations by means, e.g., of the Kolmogorov-Smirnov (KS) test.

- Avoid cuts in the sources population (i.e. model the whole 3EG Catalogue)

  Rather than concentrate on the unquantifiably biased unidentified source catalog, we instead model the spatial (and flux) distribution of the *whole sample of sources* of the 3EG Catalogue, independent of their identification.

- Take properly into account the sensitivity variation across the sky.

  The non uniform sky exposure and the presence of the diffuse galactic gamma-ray emission result in large spatial variations of the limiting flux. Due to the large point spread function, the sensitivity at any sky position is also affected by the presence of nearby (strong) sources. In our analysis we model the 3-D distribution of potential sources in the galaxy and select those that would be visible given the sensitivity of the instrument, which varies as a function of direction.

- Use extensive Monte Carlo simulations to obtain reliable estimates of significance.

  This is the most time consuming task in this kind of studies. On the other hand it is essential in order to assess correctly the confidence levels at which a given model can be accepted or rejected.

To illustrate the above guidelines with a simple example we considered the galactic latitude distribution of the sources in the 3EG catalogue (Mereghetti et al. 2001). We found a 2 σ upper limit to the number of isotropically distributed sources corresponding to $\sim 50\%$ of the total sample.

MODELING THE SKY DISTRIBUTION

To test the observed sky distribution of 3EG sources with different models, we use a two-dimensional generalization of the K-S test, the Peacock (1983) statistic as modified by Fasano and Franceschini (1987, hereafter FF87). While no formal proof of the robustness of the FF87 method exists as it does for the KS test, extensive Monte Carlo simulations, as well as use in other contexts, have demonstrated its robustness in a wide variety of physical situations, including (l,b) distributions.

Our first objective was to test the sky distributions against a very simple two-component model, consisting of an isotropic and a disk component. The expected (l,b) distributions were obtained with MC simulations in which we took into account the effect of sky exposure and galactic background.

As expected, the results of this analysis showed that single component models (i.e. disk or isotropic only) are clearly rejected. Surprisingly, however, we found that models based on only 2 components cannot be ruled out based only on the spatial distribution. For example, a two-component model in which 65% of the sources are from an exponential disk with scale height 2.5 kpc and radius 29 kpc, the rest from an isotropic component, was consistent with the spatial distribution of sources (Fig. 1).

One might speculate about the physical implications of such a model, but in the next section we see that the resulting flux distribution is inconsistent with that of the observed sources.

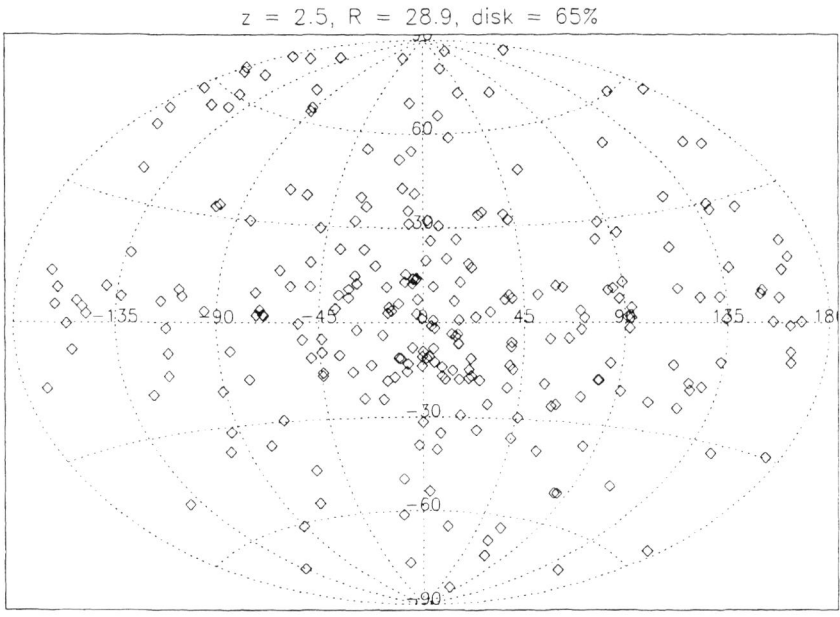

FIGURE 1. A sky distribution consistent with that of the EGRET sources, taken from a model parent distribution with an isotropic component and a single axisymmetric Galactic disk component with the parameters listed in the title. The distances are in kpc.

3-D MODELING: L, B, AND FLUX

If we wish to include the flux distributions in our comparisons, we must adopt a further generalization of the KS statistical test to higher dimensions. We use a three-dimensional generalization of the KS test developed by FF87 to compare simultaneously the spatial and flux distributions of EGRET sources to samples drawn from models with one or more components.

No exact analytical formula exists for the confidence level of the FF87 statistic. Monte Carlo simulations, not yet complete, are necessary to calibrate the statistic in the current context. Nonetheless, we can make qualitative estimates for comparison using the tables in FF87. We tested a single-disk + isotropic component model. The disk component, exponential in both z (scale height = 0.9 kpc) and R (scale length = 10 kpc), had a power-law luminosity function with index 2.67 between 10^{33} and 10^{36} erg/s. The isotropic component had fluxes distributed according to a power-law with the same index. This model produced a test statistic of 0.15, which according to the tables in FF87 would correspond to a significance level of 90-95%. The flux distributions of the model and catalog sources are shown in Figure 2. It is unclear whether we will be able to find a

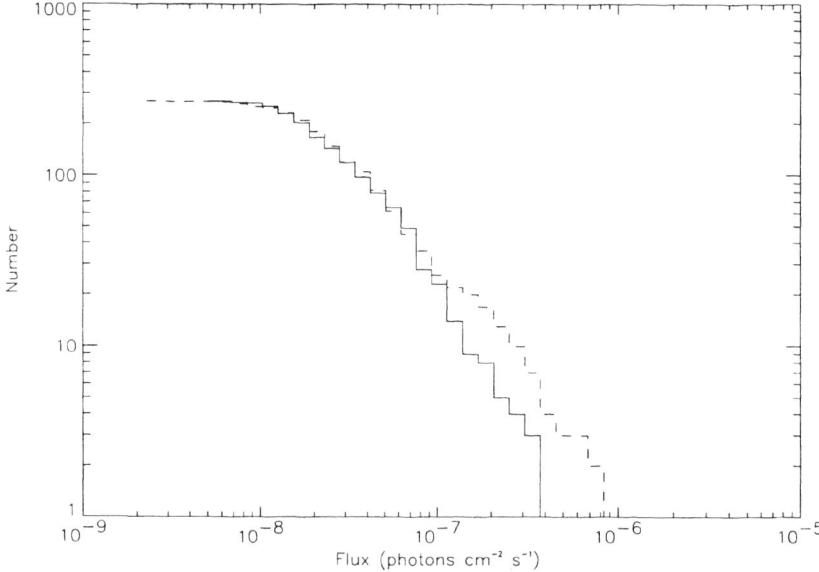

FIGURE 2. The flux distribution of the EGRET sources (solid), and that of a model distribution with an isotropic component and a single axisymmetric Galactic disk component. The model is inconsistent with the data. The fluxes are in units of photons cm^{-2} s^{-1}.

single-disk + isotropic model with different parameters that is more consistent with the observed source distribution.

Note that a high statistic, failing to rule out a theoretical model, does not indicate that the sources are actually derived from that model. It simply indicates that one cannot prefer other models on purely statistical grounds. For example, if it is found that a model with two disk components and an isotropic component is consistent with the data, and the flux distributions for each are physically plausible, then in order to require, for example, an association with star-forming regions or an asymmetrical warp, one would have to use other considerations, such as variability or spectra.

CALIBRATION OF THE TEST STATISTICS WITH MONTE CARLO SIMULATIONS

Extensive Monte Carlo simulations are required to properly characterize the distribution of the FF87 statistic. This is because preliminary calibrations indicate that hypothetical source distributions drawn from identical parent distributions frequently produce statistics far from the mean. Therefore, ruling out a given model requires a high degree of confidence, and, because exact analytic expressions do not exist for the distributions of the multi-dimensional KS-type statistics, quantifying that confidence level necessitates the Monte Carlo simulations.

We calculated the two-dimensional test statistics for an ensemble of 30 samples drawn from the same parent distribution. The mean test statistic was 0.10, corresponding formally to a significance level of $\approx 50\%$. However, two of the 30 samples had test statistics > 0.19, corresponding to significance levels $> 99\%$. Conversely, ensembles of test statistics comparing different parent distributions had tails extending below 0.19. The low-probability tails of the identical-parent cases and the overlap of the two probability distributions demonstrate the need for empirically calibrated ensembles.

CONCLUSIONS AND FUTURE WORK

A robust upper limit on the isotropic component (50% of the total sample) has been simply obtained by examining the latitude distribution of the EGRET sources (Mereghetti et al. 2001). If the only isotropic component were due to AGN's this would be a significant result. Unfortunately we cannot exclude the presence of a different classes of galactic sources, possibly extending also at medium and high galactic latitudes.

The spatial distribution of the sources alone is insufficient to rule out a model containing a single disk component and an isotropic component. Preliminary results suggest that it may not be possible to find a single-disk + isotropic component model consistent with the observed flux and angular distributions. If no such model exists, the next logical step is a two-disk + isotropic component model. Many physical models would be consistent with this framework, for example a thick disk of isolated pulsars and a thin disk containing both young pulsars and SNRs emitting cosmic rays interacting with high-density nearby material. The parameters of such a model would allow us to consider the possible physical origins of the two disk components. If such a model is also ruled out, we will be able to test more complex models, for example, including explicit dependence on spiral arm density or the structure of Gould's belt.

REFERENCES

1. Fasano, G. & Franceschini, A. (1987), MNRAS, 225, 155
2. Gehrels, N. et al. (2000), Nature, 404, 363
3. Gosset, E. (1987), A&A, 188, 258
4. Hartman, R.C. et al. (1999), ApJS, 123, 79.
5. Mereghetti, S. et al. (2001), "Population Studies of the Gamma-Ray Sources", in *The Nature of Unidentified Galactic High-Energy Gamma-ray Sources*, AIP, AIP conference proceedings, Tonantzintla, Mexico, October 9-11, 2000.
6. Peacock, J.A. (1983), MNRAS, 202, 615
7. Romero, G.E., Benaglia, P., & Torres, P.F. (1999), A&A 348, 868
8. Zhang. L., Zhang, Y.J., & Cheng, K.S. (2000), A&A 357, 957

Artifact Sources Near Bright EGRET Pulsars

D.J. Thompson*, D. L. Bertsch* and R.C. Hartman*

*Laboratory for High Energy Astrophysics, NASA/GSFC, Greenbelt, MD 20771 USA

Abstract. As noted in the Third EGRET Catalog, six sources near the bright Vela pulsar are thought to be artifacts: 3EG J0824−4610, 3EG J0827−4247, 3EG J0828−4954, 3EG J0841− 4356, 3EG J0848−4429, and 3EG J0859−4257. This conclusion is based on analysis of phase-resolved maps of the pulsar region. The artifact sources are statistically significant only in the on-pulse maps where Vela itself is bright. Details of this analysis show that there is at most one source in addition to Vela in this region. Additional analysis using the same phase-resolved-map technique suggests that one of the four sources near the Crab pulsar, 3EG J0521+2147, is also likely to be an artifact. Both of the sources closest to Geminga are detected at comparable levels in both on-pulse and off-pulse maps, supporting their validity. These results illustrate the difficulty of detecting weak sources near bright ones, even using modeling techniques such as maximum likelihood.

INTRODUCTION

In high-energy gamma rays, the Vela region is strongly dominated by one bright source - the Vela pulsar. Nevertheless, maximum likelihood analysis of the EGRET maps suggests that there are six additional sources in the vicinity of Vela, shown in the third EGRET catalog (Hartman et al. 1999). None of these sources is obvious in the intensity map of the Vela region. Because Vela is pulsed, we can greatly reduce its influence on the analysis by constructing phase-resolved maps, separating the data into on-pulse and off-pulse components. If these additional sources are real, they should appear consistently in both maps; if they are artifacts resulting from fluctuations in the distribution of photons from Vela or imperfect knowledge of the EGRET point spread function, then they would be seen mostly in the on-pulse map where Vela is bright. Similar analysis can be done for the other bright pulsars, Crab and Geminga.

VELA REGION

The light curve of the Vela gamma-ray pulsar (Fig. 1) shows two distinct regions: phase 0 - 0.6 (on-pulse) includes the vast majority of the photons, while phase 0.6 - 1.0 (off-pulse) is much weaker. In the off-pulse region, Vela itself is just a 5σ detection (Fierro et al 1998).

Maximum likelihood analysis was carried out for cataloged sources in the Vela region, using the E>100 MeV on-pulse and off-pulse maps. Because the on-pulse region occupies 60% of the phase, the "expected" counts in the unpulsed region should be 2/3 of those seen in the on-pulse map. As seen in Table 1, in all cases the off-pulse counts were well below the expected count rates. With the possible exception of 3EG J0848−4429 (which is about 2σ below the expected value), all these sources appear to be artifacts.

The likelihood map of the off-pulse portion of the data is shown in Fig. 2. The one source that exceeds the significance threshold is the Vela pulsar itself. Although there is

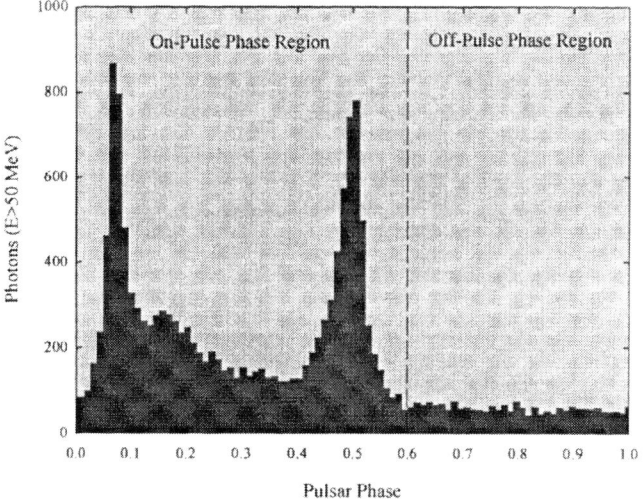

FIGURE 1. Light curve of the Vela pulsar for energies above 50 MeV.

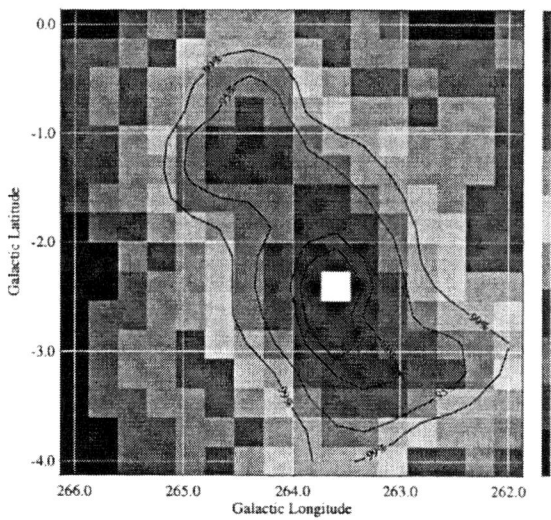

FIGURE 2. Likelihood map of the off-pulse portion of the Vela region exposure.

TABLE 1. Sources in the Vela Region

Source	Off-pulse σ	Off-pulse counts	Catalog σ	Expected counts
3EG J0824−4610	1.7	48 ± 30	9.3	235 ± 54
3EG J0827−4247	1.2	42 ± 36	6.1	205 ± 50
3EG J0828−4954	2.8	72 ± 28	5.9	153 ± 31
3EG J0841−4356	0.0	≤ 72	5.3	386 ± 67
3EG J0848−4429	3.8	147 ± 41	5.7	230 ± 56
3EG J0859−4257	1.6	51 ± 34	5.1	138 ± 36

some evidence of a second source consistent with the position of 3EG J0848−4429, its significance is below the threshold for acceptance.

CRAB REGION

The light curve of the Crab gamma-ray pulsar, Fig. 3, also shows two distinct regions: phase 0 - 0.6 (on-pulse) includes the majority of the photons, while phase 0.6 - 1.0 (off-pulse) is weaker. Unpulsed emission from the Crab, seen in the off-pulse phase region, is thought to originate from the nebula (DeJager et al. 1996)

Maximum likelihood analysis was carried out for cataloged sources in the Crab region, using the E>100 MeV on-pulse and off-pulse maps. Because the on-pulse region occupies 60% of the phase, the "expected" counts in the unpulsed region should be 2/3 of those seen in the on-pulse map, or 40% of those in the combined map that was used

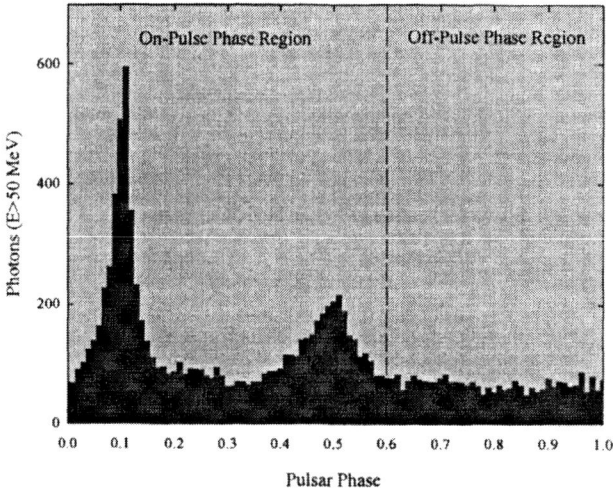

FIGURE 3. Light curve of the Crab pulsar for energies above 50 MeV.

TABLE 2. Sources in the Crab Region

Source	Off-pulse σ	Off-pulse counts	Catalog σ	Expected counts
3EG J0520+2556	4.4	138 ± 35	6.2	142 ± 24
3EG J0521+2147	2.3	71 ± 33	7.2	190 ± 28
3EG J0530+1323	22.0	869 ± 51	33.2	834 ± 80
3EG J0542+2610	2.2	85 ± 40	4.9	131 ± 29
3EG J0546+3948	4.0	93 ± 26	5.9	94 ± 18

to construct the EGRET catalog. As seen in Table 2, in most cases the off-pulse counts were consistent with the expected count rates. The exception is 3EG J0521+2147, which is about 3.6σ below the expected value. This source appears to be an artifact, while the rest are validated by the off-pulse analysis.

GEMINGA REGION

The light curve of the Geminga gamma-ray pulsar, Fig. 4, can be divided into regions: phase 0 - 0.75 (on-pulse) includes the majority of the photons, while phase 0.75 - 1.0 (off-pulse) is weaker. Unlike the Vela pulsar, Geminga itself shows strong emission at all phases (Fierro et al. 1998); unlike the Crab, Geminga has no prominent supernova remnant surrounding it.

Maximum likelihood analysis was carried out for cataloged sources in the Geminga region, using the E>100 MeV on-pulse and off-pulse maps. Because the on-pulse region

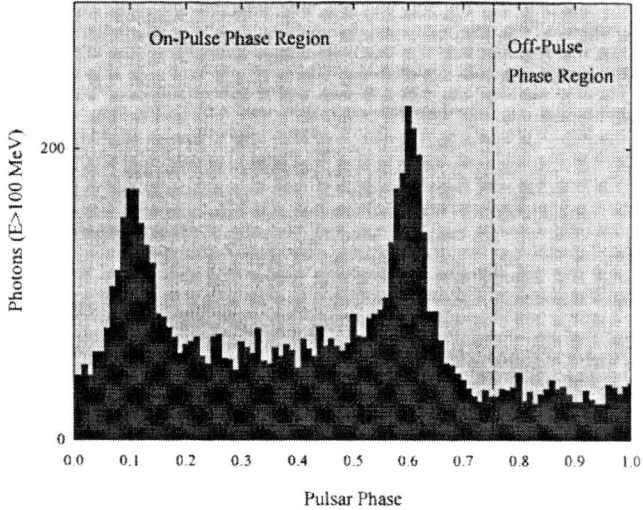

FIGURE 4. Light curve of the Geminga pulsar for energies above 100 MeV.

TABLE 3. Sources in the Geminga Region

Source	Off-pulse σ	Off-pulse counts	Catalog σ	Expected counts
3EG J0617+2238	6.8	192 ± 33	17.4	269 ± 18
3EG J0628+1847	4.2	123 ± 32	6.3	118 ± 20

occupies 75% of the phase, the "expected" counts in the unpulsed region should be 1/3 of those seen in the on-pulse map, or 25% of those in the combined map. As seen in Table 3, in both cases the off-pulse counts were consistent with the expected count rates. The 3EG J0617+2238 result is about 2.3σ below the expected value but remains a strong detection.

CONCLUSIONS

- Using phase-resolved maps in regions around bright pulsars provides an additional dimension to the study of nearby sources. Choosing off-pulse phases reduces the impact of the pulsars.
- Techniques such as maximum likelihood, although powerful, are limited by the quality of the modeled point spread function and fluctuations in the data. For EGRET, the long observations over its nine year life meant that the bright pulsar statistics were actually greater than the statistics used for calibrating the point spread function.
- At least six, and probably seven, of the sources in the third EGRET catalog are likely artifacts caused by the presence of bright sources nearby: one near the Crab, and five or six near Vela. Neither of the sources closest to Geminga appears to be in question.
- In regions around bright sources, there is no substitute for intrinsic angular resolution. The GLAST Large Area Telescope, with its combination of improved spatial resolution and much better sensitivity at the higher energies where the point spread function is much narrower, will improve greatly on the ability to resolve weak sources near bright ones.

REFERENCES

1. De Jager, O.C., Harding, A.K., Michelson, P.F., Nel, H.I., Nolan, P.L., Sreekumar, P., and Thompson, D.J., 1996 ApJ, 457, 253
2. Fierro, J.M. et al. 1998 ApJ, 494, 734
3. Hartman, R.C., et al. 1999, ApJS, 123, 79
4. Thompson, D.J. 2001, *High Energy Gamma-Ray Astronomy*, ed. F.A. Aharonhian, H.J. Volk, AIP Conf. Proc. 558, p. 103

Possible New Identifications for Southern EGRET Sources

M. Tornikoski*, A. Lähteenmäki*, M. Lainela† and E. Valtaoja†

*Metsähovi Radio Observatory, Metsähovintie 114, FIN–02540 Kylmälä, Finland
†Tuorla Observatory, FIN–21500 Piikkiö, Finland

Abstract. We have made total flux density observations at high radio frequencies (90 and 230 GHz) of 12 Southern sources that were classified as possible ERGET identifications in the Third EGRET Catalog. Our observations confirm the blazar-nature of 9 of them. We have also studied sources that we considered good candidates for AGN counterparts of previously unidentified EGRET sources and which had not been observed in the millimeter domain before. Four of them showed millimeter-domain activity which may be related to their gamma-ray activity, making them good candidates for the EGRET source identification.

INTRODUCTION

The third catalog of high-energy gamma-ray sources detected by EGRET [3], hereafter 3EG, includes 66 high-confidence identifications with blazar-type active galactic nuclei (AGNs) and 27 lower-confidence potential blazar identifications. 170 sources (96 of which have $|b| > 10°$) in 3EG were not identified firmly with known objects.

Our earlier work [15, 4, 13, 5, 14] has shown that there seems to be a connection between the high radio-frequency and EGRET gamma-ray activity, and that the most probable blazar to be detected with EGRET is a source with an ongoing and still rising high-frequency radio flare.

The case of PKS 2255−282 [13] and others have demonstrated that many AGNs remain undetected by EGRET until they go through an activity period which is observed both in the radio domain and at the gamma-ray energies. We believe that several of the unidentified EGRET sources in 3EG will turn out to be radio loud AGNs which probably would have shown millimeter-wavelength activity at the time of the EGRET detection.

The aim of this project was to make millimeter-domain observations of the potential EGRET counterpart AGNs. The observations were made with the 15-meter Swedish–ESO Submillimeter Telescope (SEST) on the European Southern Observatory site of the La Silla mountain in Chile. The observing frequencies were 90 GHz (3 mm) and 230 GHz (1.3 mm). (More information about the SEST continuum observations can be found in Tornikoski et al. [12].)

We had two major goals in this project:

- To make further studies of probable/potential candidates already suggested in the literature.
- To study the new unidentified sources listed in 3EG but not included in 2EG or

2EGS: to propose possible candidates for them and to study a sample of those candidates to see if our method is feasible for making new identifications.

SAMPLE 1: SOUTHERN LOWER-CONFIDENCE AGN INDENTIFICATIONS FROM 3EG

3EG lists twelve Southern sources which are potentially identified with blazars. The identifications are considered to be of low confidence either because the object has a low radio flux reported in the literature (historical low radio frequency data) or because it lies outside the 95% uncertainty contour. The low frequency radio data are from an observing epoch of ca. a decade ago, and because these sources are expected to be highly variable in the radio/mm domain, the historical low frequency fluxes do not necessarily correlate well with the mm-fluxes during the time of the EGRET observations.

Sample 1: Results

In order to study the 12 Southern low-confidence AGN identifications in 3EG we used millimeter-wave data from SEST observations (90 and 230 GHz), complemented with IRAM data from literature [8, 9, 7].

For two sources, B1716−771 and J1808−5011, we only had one 90 GHz data point, taken after the end of EGRET operation. For both of them the S/N of the 90 GHz observation was less than 4, giving us only the upper limit of the source flux density. Since both of these sources were faint at the time of the 90 GHz observation, they are neither very promising candidates of strong millimeter-wave activity nor good candidates for the EGRET-detection counterparts. For a third source, B0539−057, we have three 90 GHz data points collected between 1989.09 and 1991.16, and one 230 GHz upper limit observed in 1993.16. At least before the time of the EGRET operation the source was faint in the millimeter domain, and had a relatively steep ($\alpha_{5,90} = -0.40$) spectrum, which make it an unlikely blazar-counterpart for EGRET.

The remaining nine sources (B0130−171, B0506−612, B0521−365, B0537−286, B0805−077, B1127−145, B1313−333, B1504−166, B1514−241) were bright in the mm-domain. For some of these sources we have mm-wave monitoring data spanning a decade, giving us a very good estimate of the source behavior over a long period of time, and confirming their millimeter-domain variability. The maximum spectral indices $\alpha_{5,90}$ for these sources were very flat, the largest spectral index (for B1313−333) being +0.27, and all except one being $\alpha > -0.20$. The steepest spectral index was that of B1145−145, $\alpha = -0.43$, but also this source is bright and variable in the mm- domain, with the maximum flux densities at 90 GHz and 230 GHz being 2.15 Jy and 1.30 Jy, respectively.

SAMPLE 2: PROMISING CANDIDATES FOR NEW SOUTHERN EGRET IDENTIFICATIONS

Mattox et al. [6], hereafter MSM, have presented a method to make new identifications for the unidentified sources listed in the 2nd EGRET catalog [10], hereafter 2EG, and the Supplement to the 2nd EGRET catalog [11], hereafter 2EGS. MSM list 17 potential identifications for which they expect further investigation to be most useful.

When we exclude all Northern Hemisphere sources and all sources that have been either left out from 3EG (after the improved analysis), or identified in the 3EG (including the two sources identified using our group's observations [1]), this leaves us with two sources: J1249−8303 and J1650−5044.

MSM list radio sources as potential source identifications for the 2EG and the 2EGS with the probabilities of correct identification, based on the radio source position and flux density. The sources with the highest probabilities have been identified with the 2EG/2EGS sources either in the same paper (Table 3 of MSM, "EGRET identifications with a high probability of being correct") or later in 3EG. The probability listed in MSM is, however, based on low frequency radio fluxes of the Parkes-MIT-NRAO survey [2] obtained in 1990 at 4.8 GHz. We have composed a list of candidates for radio identifications of EGRET sources from MSM Tables 1 and 2, excluding all Northern Hemisphere sources, all sources that have been identified in MSM, or in Bloom et al. [1], or in 3EG, and all that have the identification probability $p < 0.001$. There were 20 sources left in our sample.

Sample 2: Results

The first of the two "promising candidates" from MSM, J1249−8303, was below the detection limit at 90 GHz. The second source, J1650−5044, has a millimeter flux and spectrum which would make it a very promising candidate for the AGN counterpart of the EGRET-detection, but in 3EG Hartman et al. comment that the corresponding 3EG source, 3EG J1638−5155, lies 2° away from the 2EG position, and that the possible identification by MSM of 2EG J1648−5042 with this AGN thus appears less convincing.

Out of the 20 "possible candidates" from MSM we observed 19 sources, 3 of them (J0726−4728, J1058−8003, J1703−6212) with a significant ($S/N > 4$) detection at 90 GHz and a very flat 5 to 90 GHz spectrum, which make them likely counterparts for the EGRET detections. In addition to these, there are two sources (J1454−1925 and J2005−2310) with a marginal 90 GHz detection and a flat spectrum. It is worth making further millimeter-wave observations of these sources at some other epochs to see whether their high radio-frequency variability corresponds to that of other flat-spectrum sources and thus makes them possible candidates for the EGRET identifications.

SAMPLE 3: POSSIBLE CANDIDATES FOR NEW SOUTHERN 3EG EGRET IDENTIFICATIONS

We chose potential new candidates for counterparts of unidentified 3EG sources according to the following criteria:

- Only Southern Hemisphere sources with $|b| > 10°$ were included.
- Only sources with no entry in 2EG or 2EGS were included, i.e., these are new sources not earlier discussed by MSM, and also these were more likely to be in an active radio state during our observations in 2000 if compared to "old" EGRET detections from 2EG.
- We searched for the candidates within a 1° radius of the 3EG position, which is the size of the EGRET error box.
- During the first round we selected candidates which are radio sources from the PMN survey [2] or the PKS survey [16]. There were several candidates for each 3EG source.
- From the candidates we excluded all of those that had a 5 GHz flux in the PMN survey [2] less than 200 mJy. Even though the PMN fluxes are one-epoch fluxes obtained almost 10 years ago of sources which are expected to be highly variable, it is reasonable to assume that very few if any of the extremely faint sources would be detected in the millimeter domain with the current sensitivity of the SEST instruments.

EGRET-detected blazars have extremely flat 5 to 90 GHz spectra. The median $\alpha_{5,90}$ during millimeter-domain activity, for the ones that we have data for at 90 GHz, is $\alpha = +0.08$. During this study our main goal was to search among a set of sources relatively faint at 5 GHz (catalog data), and thus routinely excluded from any high radio-frequency studies, for AGNs with unusually flat spectra. Our assumption was that if we find objects faint at 5 GHz being relatively bright ($\alpha_{5,90} = -0.20$ or larger) at 90 GHz, these are probably sources variable in the millimeter domain and thus good blazar candidates.

Sample 3: Results

We observed 19 of the EGRET source counterpart candidates at 90 GHz in April 2000 and/or August 2000. Among these sources there are two significant detections ($S/N > 4$) at 90 GHz. One of them, J1819−6345, is a known compact steep-spectrum source, and our 90 GHz data point confirms the steepness of its spectrum. It is an unlikely AGN counterpart for an EGRET-detection.

J1605−1139 has a flux density of 530 mJy at 90 GHz and a 5 to 90 GHz spectral index $\alpha = +0.12$, making it a good candidate for mm-domain activity. This source is located within 50.6′ of the EGRET position of 3EG J1607−1101, which makes it a very promising candidate for the source identification.

There were three other flat-spectrum ($\alpha > -0.20$) sources (J0710−3850, J1254−4425 and B2247−13) detected at a marginal level at 90 GHz. Since our study of Sample 3 was made using 1–2 observing epochs only, the mm-variability behavior of these sources is currently unknown, but the flatness of their spectra indicates that they are active also at the high radio-frequencies. We will continue the flux density monitoring of these sources in the millimeter domain.

CONCLUSIONS

We have made high radio-frequency observations of Southern AGNs that were considered to be possible counterparts of the unidentified EGRET sources. Nine AGNs classified as "possible AGN identifications" in 3EG were found to have flat spectra and to be bright and variable in the millimeter domain, thus confirming the identification. For some of the sources we have dense flux density monitoring data that show that the gamma-detection was made at the time of increased activity at high radio-frequencies.

We have also studied the millimeter flux densities and the 5 to 90 GHz spectra of sources that were considered possible identifications for the so far unidentified EGRET sources. We propose that the following four EGRET sources can possibly be identified with an AGN: 2EG 0720−4746 (id: J0726−4728), 2EGS 1050−7650 (id: J1058−8003), 2EGS 1703−6302 (id: J1703−6212), and 3EG J1607−1101 (id: J1605−1139). In addition to these, we found that five other AGNs, though faint at 90 GHz, seem to have flat 5 to 90 GHz spectra, which indicates mm-activity that may be related to the gamma-ray activity.

REFERENCES

1. Bloom et al. 1997, *ApJ* 488, L23
2. Griffith & Wright 1993, *AJ* 105, 1666
3. Hartman et al. 1999, *ApJS* 123, 79 (3EG)
4. Lähteenmäki et al. 1997, *Proc. 4th Compton Symposium*, eds. C. D. Dermer, M. Strickman & J. D. Kurfess (New York, AIP), p.1452.
5. Lähteenmäki et al. 2000, *Proc. 5th Compton Symposium*, eds. M. L. McConnell and J. M. Ryan (New York AIP), p. 372
6. Mattox et al. 1997, *ApJ* 481, 95 (MSM)
7. Reuter, H.-P. et al. 1997, *A&AS*, 122, 271
8. Steppe, H. et al. 1992, *A&AS*, 96, 441
9. Steppe, H. et al. 1993, *A&AS*, 102, 611
10. Thompson et al. 1995, *ApJS* 101, 259 (2EG)
11. Thompson et al. 1996, *ApJS* 107, 227 (2EGS)
12. Tornikoski et al. 1996, *A&AS* 116, 157
13. Tornikoski et al. 1999, *AJ* 118, 1161
14. Tornikoski et al. 2000, *Proc. 5th Compton Symposium*, eds. M. L. McConnell and J. M. Ryan (New York AIP), p. 372
15. Valtaoja et al. 1996, *A&AS* 120, 491
16. Wright & Otrupcek 1990, *Parkes Radio Sources Catalogue*, Version 1.01, Australia Telescope National Facility, Parkes

A Search for Supernova-Remnant Masers Toward Unidentified EGRET Sources

Z. Arzoumanian*, F. Yusef-Zadeh[†] and T. J. W. Lazio**

*LHEA, NASA-GSFC, Code 662, Greenbelt, MD 20771
[†]Dearborn Observatory, Northwestern University
**Remote Sensing Division, Naval Research Laboratory

Abstract. Supernova remnants expanding into adjacent molecular clouds are believed to be sites of cosmic ray acceleration and sources of energetic gamma-rays. Under certain environmental conditions, such interactions also give rise to unusual OH masers in which the 1720 MHz satellite line dominates over the more common 1665/7 MHz emission. Motivated by the apparent coincidence of a handful of EGRET sources with OH(1720 MHz) maser-producing supernova remnants, we have carried out a search using the Very Large Array for new OH(1720 MHz) masers within the error regions of 11 unidentified EGRET sources at low Galactic latitude. While a previously known maser associated with an HII region was serendipitously detected, initial results indicate that no new masers were found down to a limiting flux of, typically, 50 mJy. We discuss the implications of this result on the nature of the unidentified Galactic EGRET sources.

MOTIVATION

A search for OH(1720 MHz) maser emission toward unidentified EGRET sources is both theoretically and observationally well-motivated.

- Supernova remnant (SNR) shocks are believed to harbor sites of cosmic ray acceleration and production of high-energy γ rays: Fermi acceleration may produce relativistic protons that interact with ambient nuclei to create π^0, which decay into high-energy γ rays. A nearby molecular cloud can increase the density of target nuclei (e.g., Aharonian et al. 1994). OH maser emission is an unambiguous tracer of the type of interactions likely to produce high-energy γ rays (Claussen et al. 1997; Wardle 1999).

- The positions of γ-ray sources and SNRs on the sky are correlated (e.g., Sturner & Dermer 1995; Esposito et al. 1996), especially for nearby remnants. This correlation has been attributed to the presence of young, rotation-driven pulsars within the SNR, but the alternative hypothesis (that the remnants themselves are the γ-ray sources) has been largely unexplored observationally.

- OH(1720 MHz) maser emission is detected from nearly two dozen Galactic SNR adjacent to molecular clouds (e.g., Frail et al. 1996; Green et al. 1997; Yusef-Zadeh et al. 1999), including four remnants associated with EGRET sources. These sources have hard spectra and do not exhibit significant variability.

- Perhaps coincidentally, SNRs found to contain OH masers and hard low-latitude EGRET sources are both more prevalent in the inner Galaxy. Such a spectral disparity between inner- and outer-Galaxy EGRET sources would be difficult to explain in a pulsar-origin model.

For these reasons, we chose the satellite line of the hydroxyl radical (OH) at 1720.5 MHz as a tool to uncover remnants that may be obscured by nearby or surrounding molecular clouds.

OH(1720 MHZ) MASER EMISSION

OH(1720 MHz) masers have proven to be unique probes of C-type shocks, the magnetic fields of SNRs behind the shock front, and gas dynamics (Wardle 1999). 1720 MHz line emission is clearly evident in IC443, W28, W44 (Frail et al. 1996) and Sgr A East (Yusef-Zadeh et al. 1996), all of which are coincident with EGRET sources. Three other SNRs possibly associated with EGRET sources, CTA1, Monoceros, and γ-Cygni, apparently do not contain OH masers—the stringent criteria for producing such masers (an X-ray or cosmic ray flux to dissociate H_2O formed in the shock to OH, and molecular gas densities and temperatures appropriate for collisional pumping of the OH; Wardle, Yusef-Zadeh & Geballe 1998) are not often met, even if particle acceleration and γ-ray production are occuring. Indeed, Green et al. 1997 find that only 10% of the remnants they surveyed contain OH masers, most of which belong to the morphological class of radio shell, center-brightened thermal X-ray ("mixed morphology"; Rho & Petre 1998) SNRs. By contrast, four of seven putative EGRET-SNR associations exhibit OH maser emission.

TARGET SELECTION

Source selection was based on the Second EGRET Catalog, its Supplement, and Lamb & Macomb's (1997) catalog of GeV sources. Telescope pointings were based on source coordinates and error circles from the Third Catalog. We used available corollary information for source spectra and variability (Merck et al. 1996; McLaughlin et al. 1996). Consistent with the properties of the existing (postulated) SNR associations, two anti-center and nine inner-Galaxy EGRET sources were selected as targets according to the following criteria:

- visible from the VLA: $\delta \geq -35°$,
- low Galactic latitude: $|b| \leq 10°$,
- evidence of hard spectrum: $\alpha \leq -2.0$ (for $F \propto E^{-\alpha}$), or appearance in GeV source catalog.

Telescope scheduling constraints and the sizes of the EGRET error circles precluded a deep search of the full error region for each target of interest. Our chosen observing strategy (integration time vs. number of pointings) represents a compromise between sensitivity and coverage of the error regions with $\sim 25'$-diameter FOV imaging.

TABLE 1.

Source	l	b	V*	α/GeV†		Notes**
OH(1720 MHz) Maser Search Targets.						
3EG J0459+3352	170.30	−5.38	0.53	2.2	C	2EG J0506+3424
3EG J0634+0521	206.18	−1.41	0.13	1.9	C	2EG J0635+0521
3EG J1734−3232	355.64	0.15		GeV		GEV J1732−3130
3EG J1809−2328	7.47	−1.99	1.69	2.1/GeV	C	2EG J1811−2338
3EG J1812−1316	16.70	2.39	3.05	2.3/GeV	C	2EG J1813−1229
3EG J1823−1314	17.94	0.14	1.41	2.0/GeV	C	2EG J1825−1307
3EG J1837−0606	25.86	0.40		GeV		GEV J1837−0611
3EG J1903+0550	39.52	−0.05		GeV		GEV J1907+0556
3EG J2021+3716	74.76	0.98	1.40	1.9/GeV	C	2EG J2019+3719
3EG J2033+4118	75.58	0.33		GeV		GEV J2035+4210
3EG J2227+6122‡	106.53	3.18	0.34	2.1/GeV?		2EG J2227+6122
EGRET sources tentatively identified with SNRs.						
3EG J0010+7309	119.92	10.54	1.49	GeV		CTA1
3EG J0617+2238	189.00	3.05	1.52	2.0 GeV		IC443(OH)
GEV J0633+0645	204.83	−0.96		GeV		Monoceros
3EG J1746−2851	0.11	−0.04	1.88	GeV		Sgr A E(OH)
3EG J1800−2338	6.25	−0.18	0.05	1.9/GeV	C	W28(OH)
3EG J1856+0114	34.60	−0.54	1.14	GeV	C	W44(OH)
2EG J2020+4017	78.05	2.08	0.83	2.1/GeV	C	γ-cygni

* Variability index of McLaughlin et al. 1996. Values greater than 1 indicate $\sim 2\sigma$ flux variations on ~ 1 year timescales.
† Spectrum of γ-ray flux, $F \propto E^{-\alpha}$; "GeV" indicates $E_\gamma > 1$ GeV sources listed by Lamb & Macomb (1997; "?" for their low-significance sources).
** "C": confused—source flux and significance may be uncertain. "E": may be extended.
‡ In the time since our observations were made, X-ray and radio counterparts to 3EG J2227+6122 have been proposed (Halpern et al. 2001) that suggest a pulsar origin.

SUMMARY OF OBSERVATIONS

The 27 antennas of the NRAO Very Large Array were used in the CnD configuration. Two IF pairs measured the right and left circular polarizations simultaneously, with 128 channels for each IF spanning a bandwidth of 1.5625 MHz. The IF pairs were centered at $v = \pm 80$ km s^{-1}, yielding velocity coverage of ± 216 km s^{-1} when the IFs were combined. Dwell times on each position were typically 10–12 minutes.

The spectroscopic interferometer data were reduced using standard procedures from the AIPS software package (Greisen 2000). A continuum spectrum was fit and subtracted from each channel; the channels were then imaged, and statistics for pixel intensities were formed. We searched for excesses in the normally-distributed noise over and above a false-alarm probability of one pixel exceeding threshold, given the number of channels and pixels (roughly 6σ). In addition to searching for peaks in single channels, we used the SERCH prodecure within AIPS to search for line profiles spread across 2,

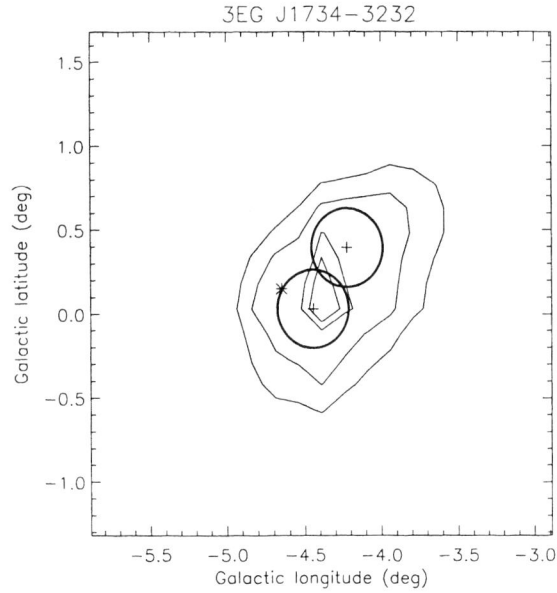

FIGURE 1. Contours of the 3EG likelihood test statistic (Hartman et al. 1999) with two VLA fields-of-view superposed. The asterisk indicates the position of a well-known HII region/OH maser.

3, and 4 channels, to increase sensitivity to weak line emission.

We observed the maser-producing remnant W28 as a test source: its two brightest maser spots were readily detected in a very short (30 sec) integration.

RESULTS

No new OH(1720 MHz) masers were detected in our survey, to a limiting flux of roughly 50 mJy for the inner-Galaxy pointings, and 35 mJy for the two anti-center pointings. Green et al. (1997) show (their Fig. 4) that just two of the 119 OH masers from 17 SNRs have measured flux below 40 mJy, which suggests that our search is sensitive to maser emission out to distances comparable to those of the ~ 160 remnants that have been surveyed for masers to date, i.e., better than half of all cataloged SNRs, and for distances of at least several kpc.

We detected, serendipitously, line emission from the well-known hydroxyl maser and HII region Maser 355.34+00.14 (in which the main OH lines at 1665 and 1667 MHz are not suppressed) as an unresolved source at position RAJ 17:33:28.9, DecJ −32:47:49, near the edge of one FOV (Fig. 1). The 1720 MHz line flux was 65 ± 18 mJy. Higher-resolution observations (Forster & Caswell 1999) distinguish a dozen maser spots with 1665 MHz line flux up to 17 Jy and LSR radial velocity ~ 20 km s^{-1}. The HII region is unlikely to be associated with the EGRET source.

CONCLUSIONS

The incompleteness of our survey, both in areal coverage of EGRET error regions and, to a lesser extent, in sensitivity, makes it impossible to draw firm conclusions about possible associations of unidentified EGRET sources with supernova remnants driving cosmic-ray acceleration and high-energy γ-ray emission. Nevertheless, and despite our present null result, continuing multiwavelength follow-up observations of the unidentified EGRET sources are bringing to light previously unknown supernova remnants that are apparently associated with the γ-ray sources—see, in particular, Combi, Romero & Benaglia (1998) for 2EGS J1703−6302 (beyond the VLA's southern limit), and Combi et al. (2001) for three EGRET sources near $(\ell, b) = (6°, -12°)$. The remnants they find are nearby, large, shell-like, and have low surface brightness, and they do appear to abut against molecular clouds. If such associations are borne out by future work and found to be common, the apparent absence of OH(1720 MHz) maser emission suggested by our survey may be attributable to the narrow range of environmental conditions under which such emission is expected to arise (Green et al. 1997; Wardle 1999).

ACKNOWLEDGMENTS

The National Radio Astronomy Observatory is a facility of the National Science Foundation operated under cooperative agreement by Associated Universities, Inc. This work was performed while ZA held a National Research Council Research Associateship at NASA/GSFC. Basic research in radio astronomy at the Naval Research Laboratory is supported by the Office of Naval Research.

REFERENCES

1. Aharonian, F. A., Drury, L. O'C., and Völk, H. J., *Astron. & Astrophys.* **285**, 645–647 (1994)
2. Claussen, M. J., et al. *Astrophys. J.* **489**, 143–159 (1997)
3. Combi, J. A., Romero, G. E., and Benaglia, P., *Astron. & Astrophys. Letters* **333**, 91–94 (1998)
4. Combi, J. A., et al., *Astron. & Astrophys.*, submitted `astro-ph/0103047` (2001)
5. Esposito, J. A., et al., *Astrophys. J.* **461**, 820–827 (1996)
6. Forster, J. R., and Caswell, J. L., *Astron. & Astrophys. Suppl. Ser.* **137**, 43–49
7. Frail, D. A., et al., *Astron. J.* **111**, 1651–1659 (1996)
8. Green, A. J., et al., *Astron. J.* **114**, 2058–2067 (1997)
9. Greisen, E. (ed.), *AIPS Cookbook*, NRAO, `www.cv.nrao.edu/aips/cook.html` (2000)
10. Halpern, J., et al., *Astrophys. J. Letters* **552**, 125–128, (2001)
11. Hartman, R. C., et al., *Astrophys. J. Suppl. Ser.* **123**, 79–202 (1999)
12. Lamb, R. C., and Macomb, D. J., *Astrophys. J.* **488**, 872–880 (1997)
13. McLaughlin, M. A., et al., *Astrophys. J.* **473**, 763–772 (1996)
14. Merck, M., et al., *Astron. & Astrophys. Suppl. Ser.* **120**, 465–469 (1996)
15. Rho, J., and Petre, R., *Astrophys. J. Letters* **503**, 167–170 (1998)
16. Sturner, S. J., and Dermer, C. D., *Astron. & Astrophys. Letters* **293**, 17–20 (1995)
17. Yusef-Zadeh, F., et al., *Astrophys. J. Letters* **466**, 25–29 (1996)
18. Yusef-Zadeh, F., et al., *Astrophys. J.* **527**, 172–179 (1999)
19. Wardle, M., Yusef-Zadeh, F. and Geballe, R. R., `astro-ph/9804146` (1998)
20. Wardle, M., *Astrophys. J. Letters* **525**, 101–104 (1999)

Galactic Plane EGRET Unidentified Source Distribution

D. Bhattacharya[*], A. Akyüz[†], T. Miyagi[*], J. Samimi[**] and A. Zych[*]

[*]*Institute of Geophysics and Planetary Physics, University of California, Riverside, CA 92521*
[†]*Dept. of Physics, Cukurova University, Adana, Turkey*
[**]*Dept. of Physics, Sharif University, Tehran, Iran*

Abstract. We compare different aspects of EGRET unidentified (EUI) source distribution in the Galactic plane with pulsar distribution. A EUI source LogN-LogS analysis is presented and compared with the Galactic radio pulsar LogN-LogS distribution. A number of systematic effects that could introduce errors to the EGRET LogN-LogS relation are discussed. The EUI source and pulsar LogN-LogS relations are similar to what is expected from a disk distribution of sources. A two-point angular correlation analysis of the EUI sources is presented where a clustering at $\sim 3^o$ is tentatively identified. A similar correlation analysis is carried out for the Galactic pulsars. We also compare the EUI source and pulsar densities along the Galactic plane, no apparent correlation is evident.

Introduction: The EGRET unidentified (EUI) sources located within a few degrees of the Galactic plane are possibly Galactic and it has been argued that counterparts for many of these sources could be off-beam pulsars (Halpern and Ruderman 1993, Helfand 1994, Mukherjee et al. 1995, Yadigaroglu and Romani 1995, Harding and Zhang 2001). In this paper, we present a preliminary analysis of the LogN-LogS, angular correlation and densities of the EUI sources within $\pm 10^o$ of the Galactic equator. These are compared with the same aspects of the known pulsar distribution. This analysis excludes the mid-latitude EUI sources which are suggested to be located in the Gould Belt (Grenier et al. 2000, Gehrels et al. 2000).

LogN-Logs Analysis: The LogN-LogS plot of the high latitude EUI sources was found to be consistent with an isotropic extragalactic distribution (Özel and Thompson 1996). The complete LogN-LogS curve for the Galactic plane EUI sources given in the Third EGRET Catalog (Hartman et al. 1999) should represent the disk distribution of these sources with respect to the Sun. From the center of a simple uniform disk distribution of sources with constant luminosity, one expects to detect N sources above the flux level S according to $N \sim S^{-\beta}$, where $\beta = 1$. However, when the observer moves from the center of the disk (as in the case of the Sun), the integral source counts are represented by two power laws:

$$N(S) \sim S^{-\beta_1}, \; S > S_c \; and \; N(S) \sim S^{-\beta_2}, \; S < S_c \qquad (1)$$

The break in the plot at a flux value of S_c is dictated by the position of the observer with respect to the center. For a disk with a radius of 16 unit length and the observer at 8 unit length from the center, $\beta_1 \sim 1$ and $\beta_2 \sim 0.6$ with the assumption of uniform source distribution, constant source luminosity and no extinction.

We follow other works to assume that the EUI sources are associated with star forming regions, OB associations, supernova remnants (SNR) or pulsars (Sturner and Dermer 1995, Kaaret and Cottam 1996, Romero, Benaglia and Torres 1999). In Figure 1a, we

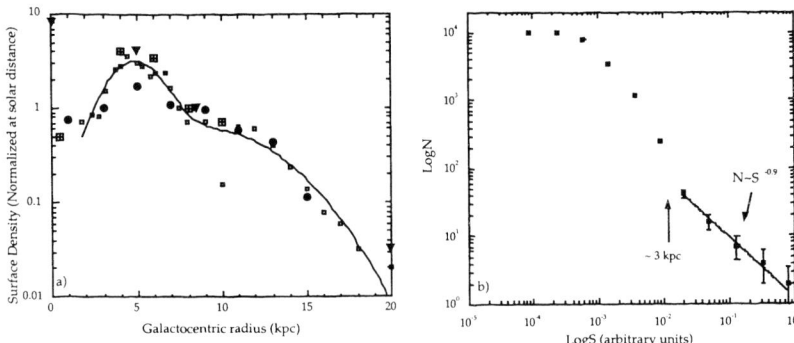

FIGURE 1. a) The distribution of molecular clouds (open square), OB associations (crossed square), supernova remnants (filled circle) and stars (inverted filled triangle) as functions of Galactic radius. A functional fit to the molecular gas distribution is also shown. b) A simulated LogN-LogS relation assuming a source distribution according to the molecular gas distribution and constant source luminosity.

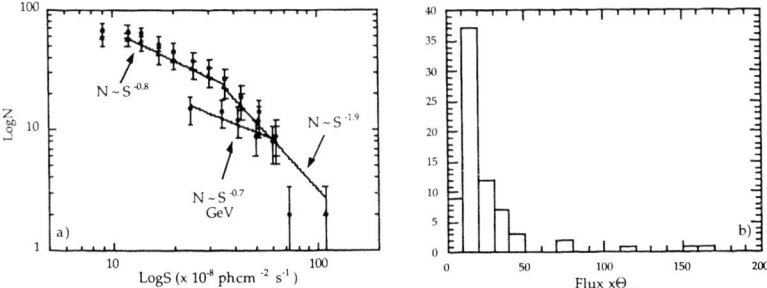

FIGURE 2. a) The LogN-LogS curve of EUI sources with $|b| < 10^o$ (open circles). A LogN-LogS curve for sources with $Flux * \theta_{95} > 50$ are also shown (filled triangles). LogN-LogS for EUI sources with GeV emission is given by filled circles. b) $Flux \times \Theta_{95}$ distribution for EUI sources.

show the distribution of molecular clouds (Wliams and McKee 1997), OB associations (McKee and Williams 1997), SNR (Case and Bhattacharya 1998) and stars (Lacey and Fall 1985) as functions of Galactocentric radius. We use a functional fit to the distribution of the molecular gas as our distribution function for sources in the Galactic plane. A simulated LogN-LogS for these sources, assuming a constant luminosity, is shown in Figure 1b. The relation can be approximated by $N(S) \sim S^{-1}$ within 3 kpc of the sun. We attribute the break at ~ 3 kpc to the presence of the peak surface density of the molecular clouds at ~ 5 kpc from the Galactic center.

In Figure 2a, we show the LogN-LogS of EUI sources with $b < 10^o$. A total of 68 sources were used. The cumulative number of sources was calculated per logarithmic flux intervals. A broken power law can be fit to the data:

$$N(S) \sim S^{-1.9}, \quad S > 3 \times 10^{-7} \text{ ph cm}^{-2} \text{ s}^{-1}$$
$$N(S) \sim S^{-0.8}, \quad S < 3 \times 10^{-7} \text{ ph cm}^{-2} \text{ s}^{-1} \qquad (2)$$

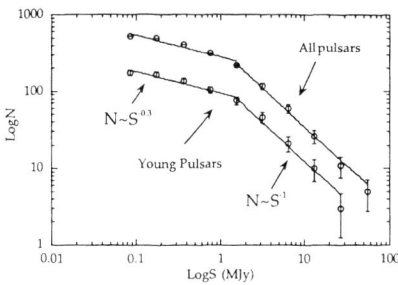

FIGURE 3. LogN-LogS of the Galactic pulsars. Pulsars from both the Princeton Catalog and Parkes Multibeam Pulsar Survey were used.

The power law index of the EUI source LogN-LogS curve below the flux level of 3×10^{-7} ph cm^{-2} s^{-1} is similar to a disk distribution (N $\sim$ S^{-1}) within errors. However, the steeper curve above 3×10^{-7} ph cm^{-2} s^{-1} indicates a source population that are quite nearby and are not part of the general population that make up the bulk of the Galactic plane EGRET sources.

Systematic Effects: The systematic effects that can introduce errors to the LogN-LogS curve (Hasinger et al. 1993) include i) the merging of two or more sub-threshold sources present in the same resolution PSF to produce a single detected source, ii) the merging of two or more above threshold sources present within the instrument resolution into a single bright source, and iii) the inability of the EGRET instrument to resolve discrete sources from the strong diffuse continuum. It can be argued that for the first two effects, either the flux or the PSF (Θ_{95}) value will be higher than the average so that a simple product of these two parameters ($Flux \times \Theta_{95}$) for a particular source would be high. We have plotted the $Flux \times \Theta_{95}$ for all Galactic plane EUI sources ($|b| < 10^\circ$) in Figure 2b. Here sources with high $Flux \times \Theta_{95}$ values would represent confused detections. Indeed, 8 of the 10 sources with $Flux \times \Theta_{95} > 50$ are termed either extended or confused in 3rd EGRET catalog. We have discarded the sources with $Flux \times \Theta_{95} > 50$ to create a new LogN-LogS curve that is shown in Figure 2a. However, $Flux \times \Theta_{95}$ corrected slope is almost identical to the uncorrected one. This seems to suggest that an elaborate fluctuation analysis is needed to derive a confusion free LogN-LogS plot.

EUI sources detected above 1 GeV are considered relatively confusion free due to its smaller PSF and can be used to construct the LogN-LogS plot. This curve is also shown in Figure 2a. The weighted fit of this curve can be described as N(S) $\sim$ S$^{-0.7}$ which is closer to a population distributed in a disk. The two GeV sources at very high fluxes are not included in the above fit.

In Figure 3, we show the LogN-LogS plot of pulsars within $|b| < 10^\circ$. The LogN-LogS plot of the young pulsars (with ages $< 10^6$ years), which are considered to be more probable gamma-ray sources, is also shown. The pulsar fluxes and ages are taken from the Princeton Pulsar Catalog (Taylor, Manchester and Lyne 2001) and the new Parkes Multibeam Pulsar Survey (2001). The plots are for observations at 1.4 GHz. In total, 527 pulsars of all ages and 175 pulsars of ages $< 10^6$ years were used. The plot suggests that pulsar distributions follow a Galactic disk distribution.

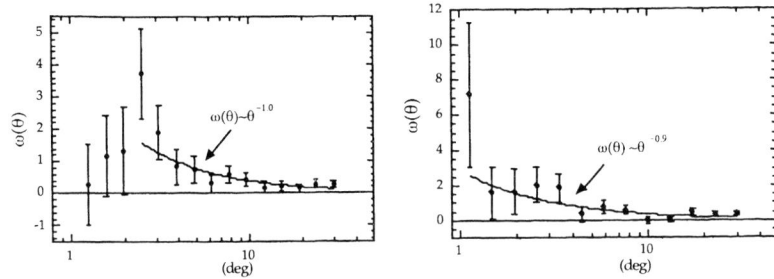

FIGURE 4. a) Angular correlation function of EUI sources within $|b| < 10°$. b) Angular correlation function of Galactic pulsars with ages less than 10^6 years.

Angular Correlation: The distribution of the EUI sources in the Galactic plane show signs of clustering in several regions. The clustered regions include the Vega, Cygnus and Carina OB associations. To quantify the scales of such clustering a two-point angular correlation function can be helpful. This function, $w(\theta)$, is needed to determine the probability $\delta P(\theta)$ of finding two objects separated by an angle θ and located within two small solid angles $\delta\Omega_1$ and $\delta\Omega_2$ (Peebles 1973). When $w(\theta) = 0$, the sources are not clustered. An estimation of the correlation function can be constructed by directly comparing the pairings of the EUI sources to that of a random sample. If N_d and N_r are the numbers of EUI sources and the random sources, respectively, then we can use the following definition of the angular correlation function:

$$w(\theta) = \left[\frac{2N_r}{(N_d-1)}\right]\frac{N_{dd}}{N_{dr}} - 1 \qquad (3)$$

where N_{dd} is the number of EUI source pairs and N_{dr} is the number of EUI source and random source pairs. The term within the square bracket is a normalization factor which is the ratio of the independent data source pairs to the number of data source-random source pairs. A uniform random sample 10 times larger than the number of EUI sources for $|b| < 10°$ is generated. The two point correlation function for EUI sources is presented in Figure 4a. We have used logarithmic bins between $1°$ and $30°$ with a bin of $\Delta log\theta = 1.25°$. The lower limit of $1°$ is assumed from EGRET angular resolution limits and the upper limit is constrained only by Galactic structures assumed to be less than $30°$. The function peaks at $\theta \sim 2.5°$ and then decays according to a power law of the form $\theta^{-\alpha}$ where α is 1.0. A general interpretation of the power law shape of the angular correlation function is that it indicates a hierarchical clustering of sources.

We also generated a correlation function for young pulsars (age $< 10^6$ years). This is presented in Figure 4b. The function is a power law with an index $\alpha \sim 0.9$. There can be clustering at levels of 1 degree and lower. At this point we cannot argue that clustering properties of pulsars and EUI sources are different. Future detections of more EUI sources possibly will resolve the issue.

EUI source density vs pulsar density: Finally, we tried to find out whether EUI source density vary with the pulsar density in the Galactic plane. Searches for pulsars and EUI sources were carried out within $6° \times 6°$ boxes along the entire Galactic

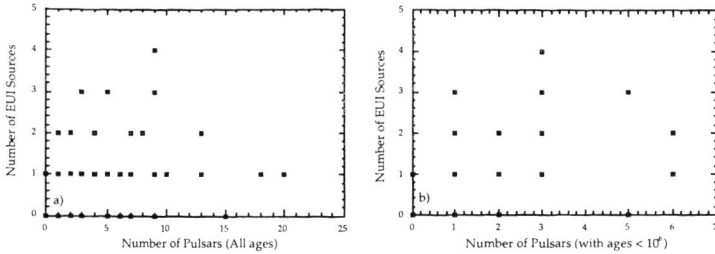

FIGURE 5. a) The number of EUI sources as a function of pulsars of all ages (Princeton Catalog) within a $6^o \times 6^o$ box. b) The same as in Figure 5a but for pulsars with ages $< 10^6$ years.

plane. We show the preliminary results in Fig.5. If EUI sources are off-beam pulsars, then, in principle, we might expect the EUI source population to correspond the pulsar population. However, no apparent correlation can be seen. Only Princeton Catalog was used and other criteria such as detection sensitivities were not taken into account. We also find pulsar densities no higher than the background in Vela, Carina and Cygnus regions, where the number of EUI sources are 5–6 within a 4° radius circle in each of these regions.

Conclusion: i) LogN-LogS distributions of Galactic plane EUI sources are similar to radio pulsars. ii) There might be EUI source clusterings at a $\sim 3^o$ level. Angular correlation functions of Galactic plane EUI and pulsars are not dissimilar. iii) However, no apparent correlation was found between pulsar number densities and EUI sources in the Galactic plane.

References
Case, G. and Bhattacharya, D. 1998, ApJ, 504, 761
Gehrels N. et al. 2000, Nature, 404, 3636
Grenier, I.A. et al. 2000, A&A, 364, L93
Hasinger G. et al. 1993, A&A, 275, 1
Halpern, J.P. and Ruderman, M.A. 1993, ApJ, 415, 286
Harding, A.K. and Zhang, B. 2001, ApJ, 548, L37
Hartman, R.C. et al. 1999, ApJS, 123, 1, 79
Hasinger, G. et al. 1993, A&A, 275, 1
Helfand, D.J. 1994, MNRAS, 267, 490
Kaaret, P. and Cottam J. 1996, ApJ, 462, L35
Lacey, C.G. and Fall, S.M. 1985, ApJ, 290, 154L
McKee, C.F. and Williams, J.P. 1997, ApJ, 476, 144
Mukherjee, R. et al. 1995, ApJ, 441, L61
Özel, M.E. and Thompson, D.J. 1996, ApJ, 463, 105
Parkes Multibeam Pulsar Survey 2001, http://www.atnf.csiro.au/research/pulsar/psr
Peebles, P.J.E. 1973, ApJ, 185, 413
Romero, G.E., Benaglia, P. and Torres, D.F. 1999, A&A, 348, 868
Sturner, S.J. and Dermer, C.D. 1995, A&A, 291, L17
Taylor, J.H., Manchester, R.N. and Lyne, A.G. 2001, htttp://pulsar.princeton.edu/pulsar
Williams, J.P. and McKee, C.F. 1997, ApJ, 476, 166
Yadigaroglu, I.A. and Romani, R.W. 1995, ApJ, 372, L99

Analysis Techniques

Effects of Background Counts in RMS Normalization

W. T. Bridgman

Scientific Visualization Studio, Goddard Space Flight Center, Code 935, Greenbelt, MD 20771

Abstract. Root-Mean-Square (RMS) normalization is a popular method of measuring the fast time variability in astronomical x-ray sources. The traditional definition contains the implicit assumption that the background contribution to the count rate is negligible. In this letter I illustrate that in cases where background counts are important, this can lead to erroneous results. However, I also illustrate a simple modification that can remove the effect of the background and makes RMS variability a more reliable measure of variability between observations from different instruments.

RMS NORMALIZATION

Root-Mean-Square (RMS) variability, as defined in van der Klis [7] and van der Klis [8], has become a popular measure of X-ray temporal variability in astronomical literature. It provides a single number for roughly parameterizing variability of a source. In addition, it can be applied to the power spectral density to generate a measure of variability over a specific frequency range. However, this measure is valuable only when it can provide some absolute measure on the variability of the source and is free of biases due to Poisson fluctuations in the background.

Some work has been done on this issue as far back as Groth [3] and more recently by Welsh [10]. However, both of these authors restricted their analyses to Gaussian distributed background noise. For high-energy astrophysics, high-time resolution and low count rates require this effect to be examined when the noise is Poisson distributed. The author became involved with this problem while working on time series analysis on CGRO/OSSE data[1].

NORMALIZATION WITHOUT BACKGROUND

Methodologies for doing these calculations are outlined in Papoulis [6]. The analysis and notations that follow are based on van der Klis [9] and are reproduced here for clarity and comparison.

Assume we have no photons from the background and x_k are the counts in bin k of a time series with N bins. We compute the (detrended) Fourier component in frequency bin j

$$a_j = \sum_{k=0}^{N-1} (x_k - \bar{x}) e^{2\pi i j k/N} \qquad (1)$$

where $\bar{x} = N/N_{ph}$ is the mean counts per bin and $N_{ph} = \sum_{k=0}^{N-1} x_k$ is the total number of photons or counts in the time series. Detrending removes power from the zero frequency bin due to any constant component in the flux. From this, we can compute a power estimate in the frequency bin with

$$P_j = \frac{2}{N_{ph}} |a_j|^2 \qquad (2)$$

The factor of two arises since we fold the power in the negative frequencies into the positive frequency bins keeping P_j defined over the frequency range of $0 \leq f \leq f_{Nyquist}$ (or equivalent binning range $0 \leq j \leq N/2 - 1$ where $f_{Nyquist} = 1/2\Delta T$). This normalization was originally suggested by Leahy et al. [5] and has the advantage that Poisson noise appears as $P_j = 2$ for all frequencies.

Next, we can define the RMS power in the frequency bin as

$$Q_j = \frac{P_j}{\lambda} \qquad (3)$$

where $\lambda = N_{ph}/T$ is the mean photon rate (per second) and T is duration of the time series in seconds. This gives Q_j units of $(rms/mean)^2/Hz$.

We define the quantity r^2, with units of $(rms/mean)^2$, as the integrated power from zero to the Nyquist frequency. Incorporating $\Delta f = 1/T$ and Equations 2 and 3, we find

$$r^2 = \int_0^{f_{Nyquist}} Q(f)\, df = \sum_{j=0}^{N/2-1} Q_j \Delta f = \sum_{j=0}^{N/2-1} \frac{1}{\lambda} \frac{2}{N_{ph}} |a_j|^2 \frac{1}{T} \qquad (4)$$

Using the discrete form of Parseval's Relation

$$\sum_{j=0}^{N-1} |a_j|^2 = N \sum_{k=0}^{N-1} (x_k - \bar{x})^2 \qquad (5)$$

and casting it into a more convenient form

$$2 \sum_{j=0}^{N/2-1} |a_j|^2 = N^2 \overline{(x - \bar{x})^2} \qquad (6)$$

we can reduce the definition of r^2 to

$$r^2 = \frac{1}{\lambda} \frac{1}{N_{ph}} N^2 \overline{(x-\bar{x})^2} \frac{1}{T} = \frac{N^2}{N_{ph}^2} \overline{(x-\bar{x})^2} = \frac{\overline{(x-\bar{x})^2}}{\bar{x}^2} \qquad (7)$$

or just

$$r = \frac{\sqrt{\overline{(x-\bar{x})^2}}}{\bar{x}} = \frac{\sigma_x}{\bar{x}} \qquad (8)$$

where σ_x is the standard deviation of x. This *rms/mean* value is sometimes reported as a percentage.

Not often mentioned in dealing with rms/mean is that for a Poisson-distributed signal, $\sigma_x = \sqrt{\bar{x}}$, and so $r = 1/\sqrt{\bar{x}}$. For count rates $< 1 count/bin$, rms/mean in fact exceeds unity. Even though an averaged power spectrum of such a signal shows constant power at all frequencies, the rms/mean for it can vary dramatically based on the average count rate alone. This raises some questions about just what rms/mean actually tells us and whether a better measure of variability is available. That is a topic for future work.

NORMALIZATION WITH BACKGROUND

Next we will examine the effect of adding background noise to the time series. Still denoting the source counts as x_k, we will add in a distribution of (uncorrelated) background counts, y_k. Regrettably, for a single pointing we can only detect their sum, $z_k = x_k + y_k$. We now compute the $(rms/mean)^2$ of the resulting time series and denote it by R^2

$$R^2 = \frac{\overline{(z - \bar{z})^2}}{\bar{z}^2} = \frac{\overline{((x+y) - \overline{x+y})^2}}{\overline{x+y}^2} \quad (9)$$

$$= \frac{\sigma_x^2 + \sigma_y^2 - 2\bar{x}\bar{y} + 2\overline{xy}}{\overline{x+y}^2} \quad (10)$$

$$= \frac{r^2 \bar{x}^2 + \sigma_y^2}{(\overline{x+y})^2} \quad (11)$$

where we've now defined $r = \sigma_x/\bar{x}$ to be the quantity computed if there is no background - the quantity which describes a characteristic of the source alone. Clearly, the quantities r and R are different. Due to the low count rates in gamma-ray astronomy, we cannot accurately perform a background subtraction on the raw time series. If we wish to remove the effects of the background, we have to make a little more effort.

First we solve for r^2:

$$r^2 = \frac{R^2 \bar{z}^2 - \sigma_y^2}{\bar{x}^2} \quad (12)$$

which gives us the *unbiased* RMS of the source from the RMS computed from the raw data.

Consider the OSSE observations of Cygnus X-1 in observing state 91159002, which had 35 counts/second in the 39-140 keV band. The total detector counts in the same band were 136.51 counts/second for a signal-to-noise value of ≈ 0.35. In 4 millisecond bins this gives a total count rate of 0.498 counts/bin and a source count rate of 0.173 counts/bin. Clearly the correction will be significant.

To apply this correction to power spectra, it is necessary redefine the power per unit frequency so that the definitions of R^2 and r^2 maintain their original form, i.e., one can define the source power, q_j, per unit frequency so that

$$R^2 = \sum_{j=0}^{N/2-1} Q_j \, \Delta f \tag{13}$$

and

$$r^2 = \sum_{j=0}^{N/2-1} q_j \, \Delta f \tag{14}$$

In this way, the definitions of r and q_j will be consistent with the definitions used in cases of no background.

$$r^2 = \frac{\bar{z}^2}{\bar{x}^2} R^2 - \frac{\sigma_y^2}{\bar{x}^2} \tag{15}$$

$$\sum_{j=0}^{N/2-1} q_j \, \Delta f = \frac{\bar{z}^2}{\bar{x}^2} \sum_{j=0}^{N/2-1} Q_j \, \Delta f - \frac{\sigma_y^2}{\bar{x}^2} \tag{16}$$

Now define a noise per unit frequency correction term, N_j, by

$$\sum_{j=0}^{N/2-1} N_j \, \Delta f = \frac{\sigma_y^2}{\bar{x}^2} \tag{17}$$

$$\sum_{j=0}^{N/2-1} q_j \, \Delta f = \frac{\bar{z}^2}{\bar{x}^2} \sum_{j=0}^{N/2-1} Q_j \, \Delta f - \sum_{j=0}^{N/2-1} N_j \, \Delta f \tag{18}$$

$$\sum_{j=0}^{N/2-1} q_j \, \Delta f = \sum_{j=0}^{N/2-1} \left[\frac{\bar{z}^2}{\bar{x}^2} Q_j - N_j \right] \Delta f \tag{19}$$

So the power per unit frequency can be written

$$q_j = \frac{\bar{z}^2}{\bar{x}^2} Q_j - N_j \tag{20}$$

At this point, we know Q_j and $\bar{z}$ from the original time series and $\bar{x}$ can be deduced from the background-subtracted energy spectra. This scaling term on Q_j can be shown to be the same as originally proposed in Grove et al. [4] to handle the effects of background counts in CGRO/OSSE high-time resolution data. However, to keep this consistent with the definition of the *integrated* RMS variability, we must include the additional term, N_j.

To proceed further, we must make some assumptions about the correction term. The simplest assumption supported by the data is that the background is Poisson-distributed white noise. In this case, we expect N_j to be a constant across all frequencies and we can define it as

$$N_j = \frac{1}{f_{Nyquist}} \frac{\sigma_y^2}{\bar{x}^2}. \tag{21}$$

We can obtain σ_y either from concurrent background pointings or, if such observations are unavailable, estimate it from the Poisson prescription

$$\sigma_y = \sqrt{\bar{y}} \tag{22}$$

yielding a corrected power per frequency bin of

$$q_j = \frac{\bar{z}^2}{\bar{x}^2} Q_j - \frac{1}{f_{Nyquist}} \frac{\bar{y}}{\bar{x}^2} \tag{23}$$

Naturally, the error bars for model fitting must be propagated through this process as well.

DISCUSSION

This renormalization of the RMS variability yields quantities where the differential and integral forms retain the same meaning as in the case of no background. It also has the virtue that it can be applied in cases where the background cannot be cleanly removed from the signal before performing time series analysis. In addition, one must include deadtime effects [11] and additional binning corrections [9].

One should caution that in cases of very low signal-to-noise ratios (less than ≈ 0.10), some Monte-Carlo simulations suggest that uncertainties in the measure of σ_y may dramatically alter the above analysis.

REFERENCES

1. Bridgman, W.T.; Leising, M.D., Clayton, D.D.; Strickman, M.S.; Kurfess, J.D.; Johnson, W.N.; Kinzer, R.L; Kroeger, R.A.; Grove, J.E.; Jung, J.V.; Grabelsky, D.A.; Purcell, W.R.; & Ulmer, M.P. 1993. Proceedings of the 2nd Compton Symposium, AIP 304. pp. 225–229.
2. Crary, D. J.; Kouveliotou, C.; van Paradijs, J.; van der Hooft, F.; Scott, D. M.; Paciesas, W. S.; van der Klis, M.; Finger, M. H.; Harmon, B. A.; & Lewin, W. H. G. 1996, ApJ 462, L71.
3. Groth, Edward J., 1975. ApJS 286, 29:285-302.
4. Grove, J.E.; Strickman, M.S.; Matz, S.M.; Hua, X.-M.; Kazanas, D.; & Titarchuk, L., 1998. ApJ Letters 502, p45.
5. Leahy, D. A.; Darbro, W.; Elsner, R. F.; Weisskopf, M. C,; Sutherland, P. G.; Kahn, S.; & Grindlay, J. E. 1983, ApJ 266, 160–170.
6. Papoulis, Athanasios. 1991, Probability, Random Variables, and Stochastic Processes (New York, NY: McGraw-Hill).
7. van der Klis, M. 1989, "Quasi-Periodic Oscillations and Noise in Low-Mass X-Ray Binaries," in Annual Reviews of Astronomy and Astrophysics 27, 517–553.
8. van der Klis, M. 1989, "Fourier Techniques in X-ray Timing," in Timing Neutron Stars, H. Ogelman & E.P.J. van den Heuvel(eds.): Kluwer, pp. 27–69.
9. van der Klis, M. 1997. "Quantifying Rapid Variability in Accreting Compact Objects" in Statistical Challenges in Modern Astronomy II, G. Jogesh Babu & Eric D. Feigelson(eds.): Springer-Verlag, pp 321-331.
10. Welsh, W.F., 1997. Astronomical Time Series, Kluwer Academic Publishers. pp. 171-174.
11. Zhang, W.; Jahoda, K.; Swank, J.H.; Morgan, E.H.; & Giles, A.B. 1995, ApJ 449, 930.

An observability study for the tentatively identified 3EG sources likely to be detected by the next-generation Cherenkov telescopes

Dirk Petry[*] and Olaf Reimer[†]

[*]*Dept. of Physics and Astronomy, Iowa State University, Ames, IA 50011, USA*
[†]*Laboratory for High Energy Astrophysics, NASA/GSFC, Greenbelt, MD 20771, USA*

Abstract.
We present a compilation of data on the 22 tentatively identified gamma-ray sources from the Third EGRET Catalog which may be detected by the next-generation imaging atmospheric Cherenkov telescopes.

INTRODUCTION

The Third EGRET Catalog (3EG, Hartman et al. 1999), comprises 271 objects. Among these, 197 are not identified with a counterpart at lower wavelengths (radio, optical or X-rays). Seven of these are now believed to be artefacts of the background model near bright sources. The remaining 190 are the unidentified EGRET objects (UNIDs). A sizable number of researchers is working on identifying the UNIDs and so far more than 38 have a published tentative ID which still needs to be confirmed by either more observations or improved analysis of archival data. We call these sources the tentatively identified EGRET objects (TIDs).

New contributors to the field will be the next-generation Cherenkov telescope (CT) observatories which are under construction in Australia (CANGAROO III, e.g. Mori et al. 1999), Namibia (HESS I, e.g. Hofmann et al. 1999), La Palma (MAGIC I, e.g. Lorenz et al. 1999) and Arizona (VERITAS, e.g. Krennrich et al. 1999).

These new instruments will reach thresholds below 100 GeV and source location accuracies of about 1'. All UNIDs are unidentified because their position is only known with insufficient accuracy, some of the position probability maps having 95% confidence level contour radii of more than 1°. With an order of magnitude increase in location accuracy, deep well-targeted observations in the radio, optical and X-ray range become possible and make an identification almost certain. In addition, the much improved photon statistics of CTs (collection areas $> 10^4$ m^2) result in a higher sensitivity for pulsed components and thus Pulsar identifications. However, CTs can only contribute for those sources which show emission above several 10 GeV.

In Petry (2001), a catalog was compiled which contains all UNIDs which may possibly be detectable by the next-generation Cherenkov telescopes under moderate assumptions about spectral steepening and taking into account the elevation-dependent sensitivity of the instruments. This catalog contains 78 objects. Among them are 22 TIDs.

These objects justify a closer examination since for their tentative counterparts various pieces of information exist which are not available for the other UNIDs: We have an exact source position which can be targeted. We know the source type and have therefore at least vague model predictions for the spectrum beyond the EGRET energy range. We have also model predictions for the variability characteristics of the source.

In this article we present first results of our data compilation and studies concerning the 22 TIDs which may exhibit significant emission beyond 10 GeV and for which the next-generation Cherenkov telescopes may provide the clue to their final identification.

THE DATA PRESENTED HERE

Table 1 gives a summary of the data presented at the conference. For each object we examine:

- What is the predicted emission of the TID in the energy regime near the threshold of the next-generation Cherenkov telescopes (CTs)?
- Which of the four observatories can observe the object?
- Is the emission variable?
- What is the angular size of the tentative counterpart?
- Are bright stars nearby which may influence the sensitivity of the CTs?
- Are there neighbouring EGRET objects which may lead to source confusion?
- What chances are there for a detection if the tentative identification turns out to be wrong? Will new pointings be necessary?

Due to the limited space in these proceedings, we refer for more detailed information to our poster which can be found in petry-reimer_poster.eps.gz at http://cossc.gsfc.nasa.gov/meetings/Gamma2001/session17/ and viewed using ghostview, and to Petry & Reimer (2001). We give here, however, the complete list of references.

SIMULATED CAMERA RESPONSES

We have examined the predicted response of the VERITAS photomultiplier camera to the starfield at each of the 22 source positions. The results were obtained from a simulation of the VERITAS optics and wavelength-dependent photomultiplier response and are shown here for the first time. The starfield information (star positions and spectra) was extracted from the SKY2000 master star catalog (Sande et al. 1998) which is reasonably complete up to magnitude 9. If not available from the catalog, the U band magnitude was calculated from the B and V magnitude assuming a main sequence star. Most important result of the simulation are the maps of the Poisson signal fluctuations in units of photoelectrons caused by the starlight and diffuse NSB in the field of view around each source. In order to not exceed the page limit of this publication, we only show two examples (figure 1), one for a very extended and one for a well constrained EGRET position probability map.

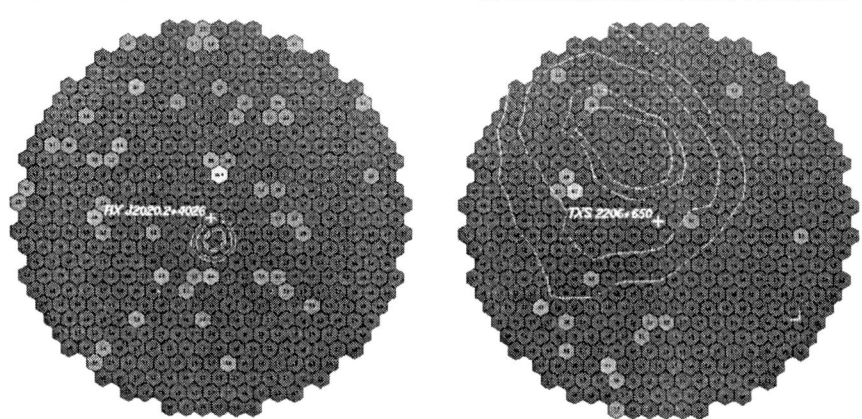

FIGURE 1. Examples of EGRET position probability maps for two of the tentatively identified sources discussed here: 3EG J2020+4017 (left) and 3EG J2206+6602 (right). The maps are superimposed on simulated representations of the field of view of the VERITAS Cherenkov telescope camera. The numbers in the individual photomultiplier pixels represent the expected signal fluctuations due to night sky background and star light in units of photoelectrons.

CONCLUSION

The next-generation Cherenkov telescopes (CTs) will be able to make an important contribution to the identification of some of the enigmatic unidentified sources of the third EGRET catalog. EGRET UNIDs for which a tentative identification exists are especially easy to target, and instruments on the northern hemisphere will be able to observe almost all such sources for which emission beyond 30 GeV can be expected. The short catalog of 22 such sources which we have compiled here, shows that a positive detection of any of these objects by CTs will be an interesting result in itself providing constraints for source models and, of course, leading to a clear identification of the 3EG source. Furthermore, the lessons learned form the observations of these objects will help in the examination of the remaining 57 EGRET UNIDs from the list compiled in Petry (2001) which have no identification whatsoever but which may have significant emission beyond 30 GeV.

For the beginning of an observation campaign, the most interesting object in our list is 3EG J1856+0114. This object has a flat spectrum with no obvious cut-off below 10 GeV. Due to its proximity to the SNR W44, it has been studied extensively (see e.g. the overview in Buckley et al. 1998). W44 is a radio shell-type SNR with an angular diameter of about 0.5° associated with PSR 1853+01. There is both evidence for a synchrotron nebula and interactions with molecular clouds. Extrapolating the thin outer gap model of Zhang & Cheng (1998) to 60 GeV yields a differential flux of 65 % of the Crab Nebula. The object is an ideal candidate for CT observations and is accidentally the only UNID which can be equally well observed both from the southern and the northern hemisphere. It could therefore be used to cross-calibrate the four CT observatories.

TABLE 1. The tentatively identified EGRET sources likely to be detected by the next-generation Cherenkov telescopes [Column titles: (1) name of the tentative counterpart, (2) equatorial coordinates of the counterpart, (3) galactic coordinates of the counterpart, (4) redshift for the extragalactic counterparts if known, (5) spectral index at 100 MeV from 3EG, (6) expected integral flux above 60 GeV (cm^{-2}s^{-1}), (7) source observability (C = CANGAROO, H = HESS, M = MAGIC, V = VERITAS), (8) variability index of the 3EG source as defined in Tompkins (1999), (9) problematic starfield?, (10) new pointings be necessary if the tentative identification turns out to be wrong?] [Abbreviations: ^{a} will probably require > 50 h of observation time (Petry 2001), ^{b} high energy cutoff visible in EGRET data, ^{c} soft low energy tail, ^{d} low statistics in EGRET data, ^{e} short-time variability observed, ^{f} predicted differential flux at 60 GeV (cm^{-2}s^{-1}MeV^{-1}), for comparison: predicted Crab Nebula flux at 60 GeV is 1.28×10^{-14}cm^{-2}s^{-1}MeV^{-1}.]

3EG name	tent. ID[1]	RA/DEC J2000[2]	l/b[3]	z[4]	α[5]	F(60 GeV)[6]	observ.[7]	var.[8]	s.[9]	p.[10]
0010+7309	RX J0007.0+7302 in SNR CTA1	00 07 02.2 +73 02 59	119.66 +10.46	-	-1.6^b	8.3E-16^f	M, V	0.31	no	no
0222+4253	3C 66A	02 22 39.6 +43 02 07.8	140.14 -16.77	0.444	-2.0	3.1E-10	M, V	^{e}	no	no
0241+6103	LSI+61°303	02 40 31.67 +61 13 45.6	135.68 +1.09	-	-2.2	1.3E-10	M, V	0.49	no	no
0617+2238	1WGA J0617.1+2221 in SNR IC443	06 17 06.1 +22 21 30	189.23 +2.90	-	-1.8^b	2.2E-10	H, M, V	0.26	yes	no
0808+4844^a	QSO B0809+483	08 13 36.09 +48 13 02.5	171.17 +33.24	0.87	-2.3	2.9E-11	M, V	n/a	no	yes
0812-0646^a	PKS 0805-077	08 08 15.54 -07 51 09.9	229.04 +13.16	1.84	-2.4	2.0E-11	C, H, M, V	n/a	no	yes
0917+4427	QSO B0917+449	09 20 58.46 +44 41 54.0	175.70 +44.82	2.18	-2.1	2.8E-11	M, V	n/a	no	yes
1009+4855^a	QSO B1011+496	10 15 04.23 +49 26 00.7	165.53 +52.71	0.20	-2.0^d	7.5E-11	M, V	n/a	no	yes
1323+2200	QSO B1324+224	13 27 00.86 +22 10 50.2	3.38 +80.53	1.40	-1.9	3.0E-10	M, V	2.69	no	no
1410-6147	RX J1420.1-6049 in SNR 312.4-00.4	14 20 07.8 -60 48 56	314.45 +1.38	-	-2.0^b	1.8E-10	C, H	0.33	yes	no
1800-2338	SNR W28	18 01.0 -23 11	6.71 -0.05	-	-2.0^c	2.2E-10	C, H, M, V	0.03	yes	no
1824-1514	LS5039	18 26 14.9 -14 50 51	16.88 -1.29	-	-2.4	7.5E-11	C, H, M, V	n/a	yes	yes
1835+5918	RX J1836.2+5925	18 36 13.82 +59 25 28.9	88.88 +25.00	-	-1.7^b	≈1.4E-15^f	M, V	0.15	no	no
1856+0114	PSR 1853+01 in SNR W44	18 56 10.89 +01 13 20.6	34.56 -0.50	-	-1.8	8.3E-15^f	C, H, M, V	0.80	yes	no
1903+0550	SNR 040.5-00.5	19 06.9 +06 33	40.52 -0.44	-	-2.5	3.7E-11	C, H, M, V	0.35	yes	yes
2016+3657	TXS 2013+370	20 15 28.89 +37 10 58.7	74.87 +1.22	?	-2.0^c	1.3E-10	M, V	0.37	yes!	no
2020+4017	RX J2020.2+4026 in SNR G78.2	20 20 17.1 +40 26 09	78.09 +2.27	-	-2.0^b	1E-17^f	M, V	0.07	yes!	no
2100+6012^a	B2101+6003	21 02 40.31 +60 15 09.8	97.96 +9.01	?	-2.1^d	3.5E-11	M, V	0.15	yes	no
2206+6602^a	TXS 2206+650	22 08 03.20 +65 19 38.8	106.94 +7.66	?	-2.3	2.6E-11	M, V	0.27	yes	yes
2227+6122	RX J2229.0+6114	22 29 04.97 +61 14 12.9	106.65 +2.95	-	-2.2	6.1E-11	M, V	0.10	no	no
2255+1943^a	QSO B2246+2051	22 53 07.36 +19 42 34.8	88.33 -35.09	0.284	-2.3^d	4.3E-11	C, H, M, V	0.41	no	yes
2352+3752^a	QSO 2346+385	23 49 20.91 +38 49 17.6	109.89 -22.47	1.03	-2.6^d	1.3E-11	M, V	24.92	no	yes

ACKNOWLEDGMENTS

D.P. would like to thank C. Duke, Grinnell College, Iowa, for contributing ray tracing code for the VERITAS optics simulation. This research has made use of the SIMBAD database, operated at CDS, Strasbourg, France.

REFERENCES

1. Bocchino, F. & Bykov, A.M., 2000, A&A, 362, L19
2. Bloom, S.D., et al., 1997, ApJ, 488, L23
3. Brazier, K.T.S., et al., 1996, MNRAS, 281, 1033
4. Brazier, K.T.S., et al., 1998, MNRAS, 295, 819
5. Buckley, J.H., et al., 1998, A&A, 329, 639
6. Cheng, K.S. & Zhang, L., 1998, ApJ, 498, 327
7. Fichtel, C.E., et al., 1994, ApJS, 94, 551
8. Halpern, J.P., et al., 2001, ApJ, 547, 323
9. Halpern, J.P., et al., 2001, ApJ, 551, 1016
10. Hartman, R.C., et al., 1999, ApJS, 123, 79
11. Hofmann, W., et al., 1999, in Dingus, B.L., et al. (eds.) Proc. "Towards a Major Atmospheric Cherenkov Detector VI", AIP Proc. 515, 500
12. Keohane, J.W., et al., 1997, ApJ, 484, 350
13. Kniffen, D.A., et al., 1997, ApJ, 486, 126
14. Krennrich, F., et al., 1999, in Dingus, B.L., et al. (eds.) Proc. "Towards a Major Atmospheric Cherenkov Detector VI", AIP Proc., 515, 515
15. Kuiper, L. et al., 2000, A&A, 359, 615
16. Lorenz, E., et al., 1999, in Dingus, B.L., et al. (eds.) Proc. "Towards a Major Atmospheric Cherenkov Detector VI", AIP Proc., 515, 510
17. Mattox, J.R., et al., 1997, ApJ, 481, 95
18. Mattox, J.R., Hartman, R.C. & Reimer, O., 2001, ApJS, 135
19. Merck, M., et al., 1996, A&ASS, 120, 465
20. Mirabal, N., et al., 2000, ApJ, 541, 180
21. Mirabal, N., et al., 2001, ApJL, 547, 137
22. Mori, M., et al., 1999, in Dingus, B.L., et al. (eds.) Proc. "Towards a Major Atmospheric Cherenkov Detector VI", AIP Proc., 515, 485
23. Mukherjee, R., et al., 1997, ApJ, 490, 116
24. Mukherjee, R., et al., 2000, ApJ, 542, 740
25. Neshpor, Yu. I., et al., 2000, Astronomy Report, 44, 641
26. Olbert, C.M., et al., 2001, ApJL, in press (astro-ph/0103268)
27. Paredes, J.M., et al., 2000, Science, 288, 2340
28. Petry, D., 2001, in Carramiñana, A., Reimer, O. & Thompson, D. (eds.), "The nature of unidentified high-energy γ-ray sources", IAU colloquia proceedings, in press (astro-ph/0101496)
29. Petry, D. & Reimer, O., 2001, Astropart. Phys., in preparation
30. Reimer, O., et al., 2000, Proc. 5th Compton Symp., AIP 510
31. Reimer, O., et al., 2001, MNRAS, 324, 772
32. Roberts, M.S.E., et al., 2000, in Carramiñana, A., Reimer, O. & Thompson, D. (eds.), "The nature of unidentified high-energy γ-ray sources", IAU colloquia proceedings, in press (astro-ph/0102471)
33. Romero, G.E., et al., 1999, A&A, 348, 868
34. Rowell, G.P., et al., 2000, A&A, 359, 337
35. Sande, C.B., et al., 1998, "SKY2000 - Master Star Catalog - Star Catalog Database, Version 2", Goddard Space Flight Center, Flight Dynamics Division
36. Strickman, M.S., et al., 1998, ApJ, 497, 419
37. Tavani, M. et al., 1997, ApJ, 497, L89
38. Tompkins, W., 1999, PhD Thesis, Stanford University
39. Yadigaroglu, I.A. & Romani, R.W., 1997, ApJ, 476, 347
40. Zhang, L. & Cheng, K.S., 1998, A&A, 335, 234
41. Zhang, L., Zhang, Y.J. & Cheng, K.S., 2000, A&A, 357, 957

Bayesian Multiscale Deconvolution Applied to Gamma-ray Spectroscopy

C. A. Young[a], A. Connors[b], E. Kolaczyk[c], M. McConnell[d], G. Rank[e], J. M. Ryan[d], V. Schönfelder[e]

[a] Emergent-IT, Inc., Goddard Space Flight Center
[b] Eureka Scientific
[c] Boston University
[d] University of New Hampshire
[e] Max Planck Institute for Extraterrestrial Physics

Abstract. A common task in gamma-ray astronomy is to extract spectral information, such as model constraints and incident photon spectrum estimates, given the measured energy deposited in a detector and the detector response. This is the classic problem of spectral "deconvolution" or spectral inversion. The methods of forward folding (i.e. parameter fitting) and maximum entropy "deconvolution" (i.e. estimating independent input photon rates for each individual energy bin) have been used successfully for gamma-ray solar flares (e.g. Rank, 1997; Share and Murphy, 1995). These methods have worked well under certain conditions but there are situations were they don't apply. These are: 1) when no reasonable model (e.g. fewer parameters than data bins) is yet known, for forward folding; 2) when one expects a mixture of broad and narrow features (e.g. solar flares), for the maximum entropy method; and 3) low count rates and low signal-to-noise, for both. Low count rates are a problem because these methods (as they have been implemented) assume Gaussian statistics but Poisson are applicable. Background subtraction techniques often lead to negative count rates. For Poisson data the Maximum Likelihood Estimator (MLE) with a Poisson likelihood is appropriate. Without a regularization term, trying to estimate the "true" individual input photon rates per bin can be an ill-posed problem, even without including both broad and narrow features in the spectrum (i.e. a *multiscale* approach). One way to implement this regularization is through the use of a suitable Bayesian prior. Nowak and Kolaczyk (1999) have developed a fast, robust, technique using a Bayesian multiscale framework that addresses these problems with added algorithmic advantages. We outline this new approach and demonstrate its use with time resolved solar flare gamma-ray spectroscopy.

Given a measured energy deposit spectrum we can extract spectral information such as constraints on model parameters and an estimate of the incident photon spectrum. This is the classic problem of spectral deconvolution or spectral inversion. In the ideal case this is equivalent to solving an integral equation for the photon spectrum f(E') given the measured energy loss spectrum C(E) and the energy response of the detector R(E',E),

$$C(E) = \int_0^{E'} R(E,E') f(E') dE' \tag{1}$$

The solution to this deceptively simple equation is part of a larger class of problems commonly referred to as inverse problems. If we had the exact continuous forms of C and R we could, in principle, solve the integral equation that is just a Fredholm equation of the first kind (Craig and Brown 1986; Hansen 1998). The count data C(E) are only known for a discrete set of energies, E_i, where i=0,...,n-1. We are then solving a set of n integral equations but many functions can satisfy a given discrete set of equations without satisfying our original equation. The instrument response is seldom described analytically so that we then must replace the set of integral equations with the matrix equation,

$$C_i = R_{ij} f_j \text{ for } i = 0,...,n-1 \text{ and } j = 0,...,m-1. \qquad (2)$$

The formal solution of this equation is $\mathbf{f}=\mathbf{R}^{-1}\mathbf{C}$ but usually $\mathbf{R}^{-1}$ is unbounded and its computation is sensitive and unstable to small perturbations in the data. The inverse problem is then termed "ill-posed".

There exist a number of "classical" methods for numerical inversion of matrix equations, which were discussed in detail by Craig and Brown (1986) and Hansen (1998). These methods are classical in the sense that they explicitly use prior information or make assumptions about the source function (Craig and Brown 1986). Due to many shortcomings we do not use any of these methods but instead use "non-classical" or regularization methods. The general idea of regularization is to introduce an extra term (or regularization function ϕ) to minimize irregular solutions, i.e.,

$$L(f) = \|C - Rf\|^2 + \phi(f,\alpha) \qquad (3)$$

The free parameter α is chosen to balance the minimization of the norm versus the suppression of noise. Below, we will discuss two commonly used techniques and we present the use of a novel new method applied to γ-ray spectroscopy.

The method of forward folding is commonly used in γ-ray and X-ray astronomy. A model μ is convolved (folded) with the detector response $\mathbf{R}$ yielding a model set of data $\mathbf{d}$. A maximum-likelihood fit of the real data $\mathbf{C}$ with the model data is then preformed maximizing the log of the probability (or in the case of Gaussian statistics minimizing χ^2). This procedure has the disadvantage that it is inherently restricted to the assumed model. One does not obtain an estimate of the real spectrum but rather a set of parameters associated with that model. However, when the model choice is appropriate and realistic and the response is dominated by diagonal elements the inferred spectrum is robust.

Maximum Entropy methods are also popular for the deconvolution of image data and have proven to be successful in spectral deconvolution (Gull and Skilling 1990). In addition to maximizing a likelihood, prior information in the form of information entropy S is maximized. Many versions of S have been defined in the literature.

The methods of forward folding and maximum entropy "deconvolution" have been used successfully for gamma-ray solar flares (e.g. Rank, 1996; Share and Murphy, 1995). These methods worked well under certain conditions but there are situations were they do not apply. These are: 1) when no reasonable model (e.g. fewer parameters than data bins) is yet known, for forward folding; 2) when one expects a mixture of broad and narrow features (e.g. solar flares), for the maximum entropy method; and 3) low count rates and low signal-to-noise, for both. For Poisson data the Maximum Likelihood Estimator (MLE) with a Poisson likelihood is appropriate. Without a regularization term, trying to estimate the "true" individual input photon rates per bin can be an ill-posed problem. One way to implement this regularization, though, is through the use of a suitable Bayesian prior. Nowak and Kolaczyk (1999) developed a fast, robust, technique using a Bayesian multiscale framework that addresses these problems with added algorithmic advantages. We outline this new approach so that we can apply it to solar flare gamma-ray spectroscopy.

Recent treatments of Poisson inverse problems have augmented the likelihood equations with a regularization or penalization term as discussed above. This regularization term stabilizes the otherwise ill posed ML problem. The regularization term can take the form of a Bayesian prior so that the MLE is replaced with the Maximum *a posteriori* (MAP) estimator. If we wish to use a MAP estimator, we first apply Bayes' theorem,

$$p(\lambda|y) = \frac{p(y|\lambda)p(\lambda)}{p(y)} \qquad (4)$$

This equation relates the likelihood to the posterior with the prior p(λ) and p(y) being a normalization based on the data. The prior can also be interpreted as a penalizing function giving the terminology "Penalized MLE". The MAP estimate is then the value of λ that maximizes the log of the posterior.

Multiscale analysis is the study of behavior or structure in data at various spatial and/or temporal scales. One way to address our ill-posed problem is through a multi-scale framework (Starck, Murtagh, and Bijaoui, 1998). The usual multi-scale model is formulated with a wavelet decomposition but wavelets and Poisson data are somewhat incompatible (Kolaczyk 1999). Nowak and Kolaczyk (1999) developed a deconvolution technique that uses a Bayesian multiscale framework that addresses these problems with other advantages. The technique uses an Estimator Maximization (EM) algorithm that has a closed-form step. Under reasonable choice of the multiscale priors, the EM algorithm converges to a unique, global MAP estimate.

The problem at hand is to estimate the photon flux, λ, from the observed count data y. The counts are related to the flux by the relation $y_n = P(\mu_n)$, n=0,...,N-1, where $P(\mu_n)$ is the Poisson distribution with mean counts μ_n. The mean counts μ_n are related to the flux by the relation $\mu = \mathbf{R} \bullet \lambda$, where R is an N × M matrix (the response) of transition probabilities.

It will be useful later in this discussion to introduce the idea of the "complete data" z(n,m). This is the total number of m to n (emission to detection) events, $z(n,m) = P(\lambda_m R_{n,m})$. The indirectly observed count data is then given by summing the complete data over m, $y_n = \sum_m z(n,m)$. Also, were we able to detect the photons directly without the detector we would have the direct data, $x_m \equiv \sum_n z(n,m)$, from which it follows that $x_m = P(\lambda_m)$.

To seek a solution of the general inverse problem we must first solve the direct-data Poisson estimation problem, $x_m = P(\lambda_m)$. The simplest multiscale data analysis is the unnormalized Haar analysis. The data $\{x_{j,k}\}$ are the unnormalized Haar scaling coefficients of $\mathbf{x}$. The index j refers to the resolution of the analysis, 2j, where j = J is the index for the highest or finest scale and j = 0 is the lowest or coarsest scale.

Using conditional probability relationships, the joint probability of the data in a multiscale representation can be expressed with the factorized form,

$$p(x) = P(x_{0,0}) \prod_{j=0}^{J-1} \prod_{m=0}^{2^j-1} \Pr(x_{j+1,2m} | x_{j,m}). \quad (5)$$

The parent $(x_{j,k})$, child $(x_{j+1,2k})$ relationship is expressed by the conditional likelihood, $\Pr(x_{j+1,2m}|x_{j,m})$. The MAP estimation of λ requires the likelihood function of $\mathbf{x}$, $p(\mathbf{x} | \lambda)$. The multiscale expansion of $p(\mathbf{x} | \lambda)$ requires that we define the multiscale analysis of the intensity λ, analogous to the analysis of $\mathbf{x}$.

The parameters $\{\lambda_{j,m}\}$ are the unnormalized Haar scaling coefficients of the intensity λ.

Using the definitions for the multiscale analysis of $\mathbf{x}$ and λ and the multiscale factorization of $\Pr(\mathbf{x})$ we can express the parent-child conditional likelihood as,

$$p(x_{j+1,2m} | x_{j,m}, \lambda_{j,m}) = B(x_{j+1,2m} | x_{j,m}, \rho_{j,m}), \quad (6)$$

where $B(x | n, \rho) = \binom{n}{x} \rho^x (1-\rho)^{n-x}$, is the binomial distribution with parameters n and ρ. The parameters $\rho_{j,m} = \frac{\lambda_{j+1,2m}}{\lambda_{j,m}}$ are the canonical multiscale parameters for the Poisson model and can be viewed as "splitting" factors, governing the multiscale refinement of the intensity. We can represent the multiscale analysis of x and λ as a binary tree where the splitting factors are multiplicative weights in the tree's links. The complete factorization of the likelihood is then

$$p(\mathbf{x} | \lambda) = P(x_{0,0} | \lambda_{0,0}) \times \prod_{j=0}^{J-1} \prod_{m=0}^{2^j-1} B(x_{j+1,2m} | x_{j,m}, \rho_{j,m}) \quad (7)$$

where $P(x_{0,0}|\lambda_{0,0})$ is just the Poisson probability function of $x_{0,0}$ with mean $\lambda_{0,0}$.

A maximum likelihood analysis of the binomial conditional likelihood leads to a MLE estimation of the splitting parameters (Nowak and Kolaczyk 1999) of $\hat{\rho}_{j,m} = \frac{\lambda_{j+1,2m}}{\lambda_{j,m}}$. There is a one-to-one mapping from (ρ,λ_{00}) to λ so using the multiscale synthesis equation (Nowak and Kolaczyk 1999) and the estimate $\hat{\rho}_{j,m}$ we find the MLE of each intensity element of the finest scale to be $\hat{\lambda}_{J,m} = x_{J,m} \equiv x_m$. The MLE returns the raw data as our MLE intensity estimate, an expected result (Nowak and Kolaczyk 1999). The next step is the MAP estimation.

The crucial ingredient in moving from a MLE estimation to a Bayesian estimation is the choice of a prior distribution $p(\lambda)$. A good choice of prior reflects known or assumed attributes of the intensity and matches the functional form of the likelihood (in our case Poisson and Poisson-binomial). Conjugate priors have the computational advantage that they are obtained by updating the parameters of the prior based on the measurements (Nowak and Kolaczyk 1999). The natural choice of the conjugate prior for the total intensity λ_{00} is the gamma probability density. The choice for modeling the splitting parameter is as an independent beta distributed random variable $Be(\rho|\alpha,\beta)$. We have no a priori knowledge of asymmetry therefore we use only a symmetric beta prior of mean 1/2 with $\alpha=\beta$. In our case the gamma prior has negligible effect so the important parameter is the beta prior in the splitting parameters (Nowak and Kolaczyk 1999). The beta priors $\{\alpha_j\}$ reflect our belief or prior knowledge of the intensities regularity. Combining the prior and the likelihood and using the conjugacy of the prior with the likelihood produces a posterior density (Nowak and Kolaczyk 1999). As with the MLE, the synthesis equations can be used to obtain a MAP estimate of λ.

Moving back to the more difficult Poisson inverse problem: As we showed for the analysis of the directly observed Poisson data, a multiscale factorization of the data likelihood played a key role. For the analysis of the indirectly observed data, the complete data likelihood plays a key role through the EM algorithm. Nowak and Kolaczyk (1999) show that the complete data likelihood is proportional to the direct data likelihood. The log complete data posterior is then just a combination of the log complete data likelihood and the log prior,

$$L(\lambda) \equiv \log p(\lambda_{0,0},\rho \mid z) \quad (8)$$
$$= \log P(x_{0,0} \mid \lambda_{0,0}) + \log G(\lambda_{0,0} \mid \gamma,\delta) +$$
$$\sum_{j=0}^{J-1}\sum_{m=0}^{2^j-1} \log B(x_{j+1,2m} \mid x_{j,m},\rho_{j,m}) + \log Be(\rho_{j,m} \mid \alpha_j,\alpha_j) + C.$$

The key to the EM algorithm for MLE (Nowak and Kolaczyk 1999) is the introduction of the complete data, $z(n,m)$. If we could observe these complete data, we have shown that a closed-form maximizer of the complete data posterior exists. The EM algorithm iteratively alternates between computing the expected complete data log-posterior and a maximizer of this function leading to a MAP estimate of the log-posterior. The problem with the MAP-EM algorithm is that now the M-step does not have a closed-form solution. Fortunately, Nowak and Kolaczyk (1999) solved this problem by taking a multiscale approach that does have a closed-form M-step. As an EM algorithm it has the standard property that the posterior probability does not decrease with subsequent iterations and the estimate is non-negative. The algorithm also has the feature that it is computationally simple and for a certain choice of prior parameters the MAP converges to a global solution.

Unfortunately, errors or confidence intervals in the traditional sense do not follow (Kolaczyk 1999). In order to produce spectra with which we can then calculate line fluxes and physicals parameters; we must be able to produce errors or uncertainties in our estimates. The most straightforward method for this is to use a parametric bootstrap (Connors 2000; Kolaczyk 2000).

Figure 1 shows the light curve for a gamma-ray solar flare divided into 3 time intervals and the deconvolved spectra for each of the time intervals.

We have shown this technique to be very useful for the analysis of solar flare gamma-ray spectra. This technique is fast, robust, and requires no knowledge of the underlying physics. This technique holds great promise for the deconvolution of gamma-ray spectra as well as spectra in other energy ranges.

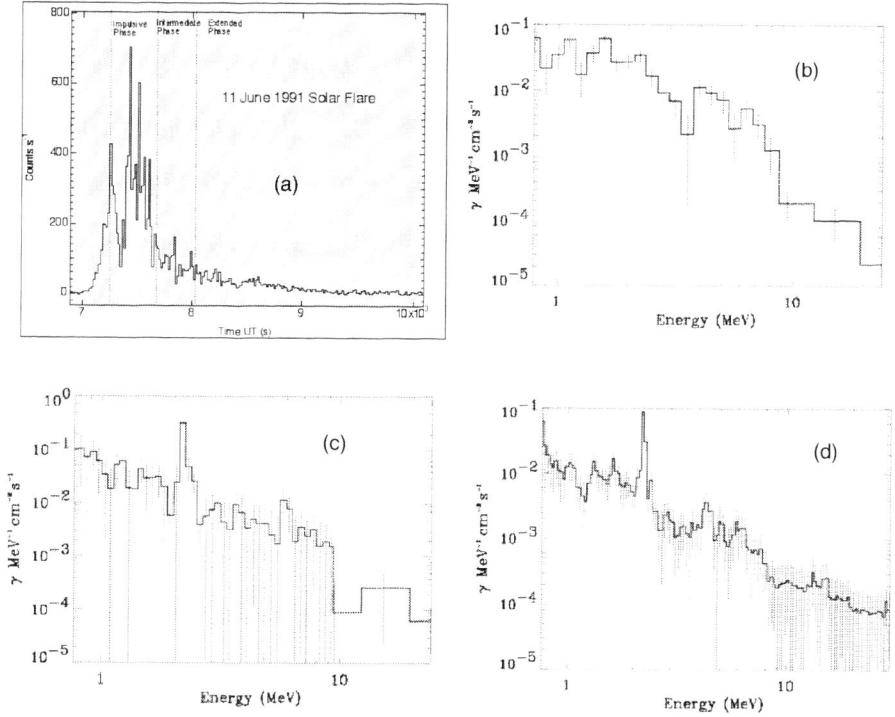

FIGURE 1. The light curve for a gamma-ray solar flare (a). The light curve is divided into 3 time intervals. Also shows the deconvolved spectra for the time intervals defined in the light curve (b-d). (Error bars extending to the bottom of the plots are 1 σ upper limits.)

ACKNOWLEDGMENTS

This work was supported through NASA contract NAS5-26645 and NASA's Supporting Research and Technology program.

REFERENCES

1. Connors, A., *Confidence Intervals for MAP estimation method*, Private Comm. (2000).
2. Craig, I. J. D., and Brown, J. C., *Inverse Problems in Astronomy*, Boston: Adam Hilger, Ltd., 1986.
3. Gull, S. F., and Skilling, J., *the MEMSYS5 User's Manual*, Royston: Max. Entropy Data Consultants Ltd., 1990.
4. Hansen, P. C., *Rank-Deficient and Discrete Ill-Posed Problems*, Philadelphia: Siam, 1998.
5. Kolaczyk, E. D., "Some Observations on the Tractability of Certain Multiscale models," in *Bayesian Inference in Wavelet-based Models*, edited by M. A. Vidkovic, New York City: Springer-Verlag, 1999.
6. Kolaczyk, E. D., *Bootstrapping Poisson Estimators*, Private Comm. (2000b).
7. Nowak, R. D., and Kolaczyk, E. D., "A Bayesian Multiscale Framework for Poisson Inverse Problems" in *Proceedings of the IEEE International Conference on Acoustics, Speech, and Signal Processing*, 1999.
8. Share, G.H. and Murphy, R. J., *Ap. J.* **452**, 933 (1995).
9. Starck, J.-L. et al., *Image Processing and Data Analysis: The Multiscale Approach*, New York City: Cambridge University Press, 1998.

EGRET's Detection Efficiency in the Later Phases of the Mission

D.L. Bertsch*, R.C. Hartman*, S.D. Hunter*, D.J. Thompson* and P. Sreekumar[†]

Laboratory for High Energy Astrophysics, NASA/GSFC, Greenbelt, MD 20771 USA
[†]ISRO Satellite Centre, Bangalore, India

Abstract. The detection efficiency of EGRET varied throughout the mission due to aging of the spark gas between gas refills, and later in the mission due to partial hardware failures. After the gas refill in 1995 September until the end of the mission in 2000 May, EGRET was operated for approximately 700 days – several times as long as for earlier gas fills. The efficiency degradation was severe during this time, and it was highly energy dependent affecting the low energies the most. This paper reports on an extensive effort to determine the efficiency factors that apply to data during this period.

INTRODUCTION

Between the time of the last gas exchange at the end of CGRO's viewing Cycle 4 and the end of the mission, EGRET's efficiency degradation was much more severe than for earlier fills, and it became strongly dependent on energy, affecting the low energies the most. The method of calibrating the efficiency that was used up to the end of Cycle 4 (Esposito et al. 1999) could not deal with the energy dependence adequately. A modified approach was developed and applied to all of the viewing periods since the end of Cycle 4 and the resulting efficiency measurements have been modeled to smooth out statistical effects. This model takes into account the failure of one-half of the spark chamber readout that occurred on 1997 Nov. 4 which caused a discrete drop in efficiency and it also accounts for a partial gas exchange done on 1998 Dec. 1 that improved performance somewhat. Otherwise, the assumption made in the modeling is that the efficiency degradation is a smoothly varying function.

The sections below briefly describe the method for determining the efficiency and model that was used to smooth the fluctuations. The resulting efficiencies for each viewing period since Cycle 4 are tabulated in a public ftp-site given below, and are they are also available in the CGRO archives. The flux and energy spectra of the Crab, Geminga, and Vela pulsars, using data since the last gas fill and with exposures corrected with the new efficiencies, compare favorably with values observed early in the mission.

VIEWING PERIOD ANALYSIS

A standard reference map for the whole sky was constructed using the EGRET likelihood analysis program (Mattox et al. 1996) to subtract the EGRET catalog sources (Hartman et al. 1999), and then locally match the residual to the diffuse gamma ray model (Hunter et al. 1997) with a scale factor plus an offset on a $0.5° \times 0.5°$ skybin basis. The model diffuse maps were then scaled with these factors to produce a so-called "ideal gas map" for each energy region. Then for each viewing period, a likelihood analysis was done for events above 100 MeV to determine the significant sources in the field of view. A second likelihood analysis using this same set of sources was performed for each of the EGRET standard 10 energy intervals (in MeV, 30-50, 50-70, 70-100, 100-150, 150-300, 300-500, 500-1000, 1000-2000, 2000-4000, 4000-10000) to subtract the source contributions. The remaining diffuse counts were then summed over a circle of $15°$ centered on the pointing direction. Corresponding sums for each energy regions were made on the "ideal gas maps" and the ratio was then used as measure of the instrument efficiency.

FUNCTIONAL FITS

The efficiencies obtained above have large statistical uncertainties, and since it is believed that the instrument characteristics change smoothly with time, a model was developed and fitted to the observed values. The approach assumes that the efficiencies depend on energy and time with parameters that differ between three zones: first for the time from last complete gas fill (1995 Sept. 9) until the failure of spark chamber B's readout (1997 Nov. 4), second from that time until the partial gas refill (1998 Dec. 1), and third for the time remaining until re-entry (2000 May 27). In terms of days of operation (EGRET was turned off for certain viewing periods), the first interval ended after 454 days, the second ended after 572 days, and the last interval ended at 700 days.

Plots of the efficiencies for each viewing period studied showed a reasonably linear fit with a logarithmic energy axis. Using this functional form, the efficiency model can be written as

$$eff(E,t) = eff(100,t) + v(t) \times log_{10}(\frac{E}{100}) \quad (1)$$

The coefficients, $eff(100,t)$ and $v(t)$, determined from fits of equation (1) for each viewing period were then plotted against time and modeled by the following functions

$$eff(100,t) = N \times exp\left(\frac{(t-t_o)}{\tau}\right) \quad (2)$$

$$v(t) = \alpha \times (t - t_o) + \beta \quad (3)$$

Except for the time constant, $\tau = 577.9 \pm 59.1\, days$, the constants in this model differ according to zone as defined above. Table 1 summarizes their values.

TABLE 1. Constants in the Efficiency Model

Zone	t_o, days	α, $days^{-1}$	β
1	0	$(1.336 \pm 0.615) \times 10^{-4}$	0.1040 ± 0.0181
2	454	$(5.87 \pm 4.35) \times 10^{-4}$	0.0870 ± 0.0307
3	572	$(9.49 \pm 35.4) \times 10^{-5}$	0.2105 ± 0.0289

Figure 1 compares the model fit to the observed efficiencies for two different energy regions. A conservative estimate of a systematic error of 10% has been added to all of the observed uncertainties. The reduced χ^2 values of 1.6 are reasonable, but suggest that the systematic errors might be even larger. Similar values of χ^2 were obtained for the other 8 standard energy intervals not shown here. The three zones are marked in these plots. The dashed lines show the envelope of uncertainty that results from the uncertainties of the parameters in Table 1.

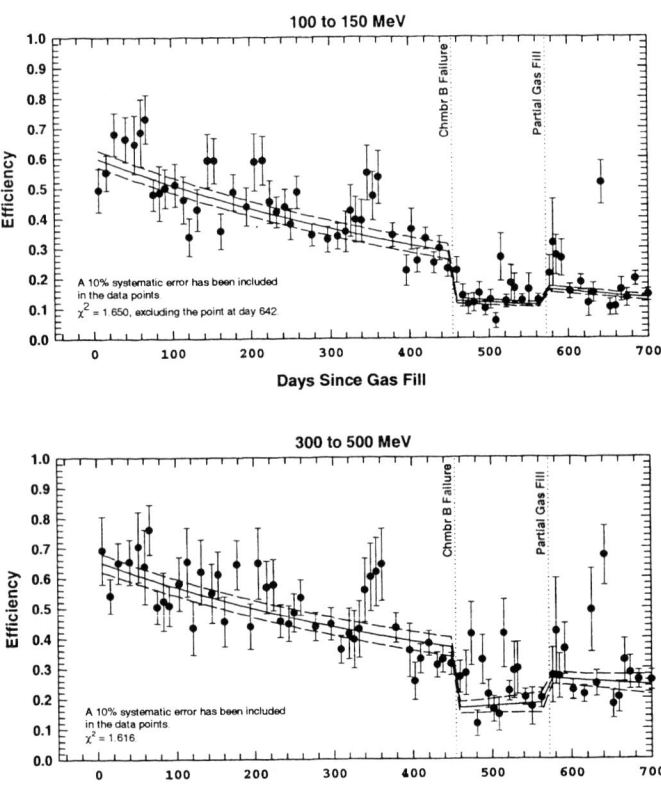

FIGURE 1. Observed efficiencies compared to the model values (shown as lines) for two selected energy intervals. The dashed lines are computed based on the uncertainties in the fit parameters given in Table 1.

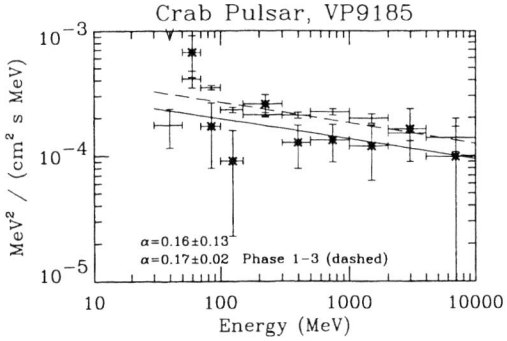

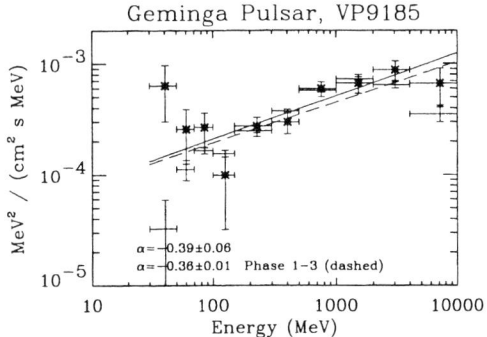

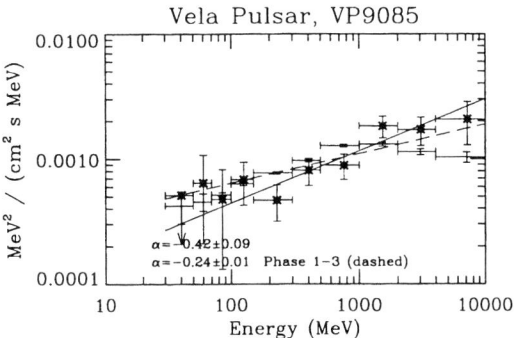

FIGURE 2. Comparison of spectra measured near the end of the mission with measurements observations summed over Cycles 1 through 3. The later data have been corrected by the efficiencies modeled in equations (1-3).

CHECKS USING THE PULSAR SPECTRA

Energy spectra of the Crab, Geminga, and Vela observations in all viewing periods that contain their exposures since the last gas fill and after corrections for efficiency were compared with spectra obtained from the summed Cycles 1 through 3 as a standard reference. The agreement is reasonably good as can be seen for the cases shown in Fig. 2 at time very near the end of the mission. The agreement for Geminga appears to be better than the Crab in this figure. However, the two sets of data were from the same viewing period so the differences are most likely statistical in nature.

CONCLUSIONS

A reasonable model for EGRET's efficiency degradation since the last gas fill has been developed and tested. All of the EGRET data have now been reprocessed and the entire summary database, efficiency correction factor table, and standard sets of maps have been installed in the CGRO Science Support Center archive. The updated "scale.factor" file used by EGRET software to calculate exposure can be downloaded from the anonymous ftp site, gamma@gsfc.nasa.gov from the directory, "/pub/Scalefactors".

REFERENCES

1. Esposito, J.A. et al. 1999, ApJS 123, 203
2. Hartman, R.C., et al. 1999, ApJS, 123, 79
3. Hunter, S.D. et al. 1997, ApJ, 481, 205
4. Mattox, J.R. et al. 1996, ApJ, 461, 396

Future Missions (GLAST/AGILE)

The Gamma-ray Large Area Space Telescope Mission: Science Opportunities

Peter F. Michelson
On behalf of the GLAST Science Team

Department of Physics and Stanford Linear Accelerator Center, Stanford University, Stanford, CA 94305, USA

Abstract. The Gamma-ray Large Area Space Telescope (GLAST), the next-generation high-energy gamma-ray mission, is scheduled for launch in 2006. The GLAST Observatory will have two scientific instruments: (1) The Large Area Telescope (LAT) is an imaging, wide field-of-view telescope sensitive to radiation over the energy range from 20 MeV to more than 300 GeV. The LAT will have, depending on energy, sensitivity more than 40 times better than that of EGRET. (2) The Gamma-ray Burst Monitor (GBM) operates from 10 keV to 25 MeV. The LAT sensitivity is 4×10^{-9} photons cm^{-2} s^{-1} (> 100 MeV) for a one year all-sky survey. The combination of the LAT and the GBM instruments will provide spectral observations of gamma-ray bursts over 6 orders of magnitude in energy. The upper end of the LAT's energy range overlaps with ground-based observatories to provide an unprecedented capability to observe sites of particle acceleration in the Universe. GLAST will operate for a minimum of 5 years, with the first year devoted to an all-sky survey, followed by observations determined by peer-reviewed Guest Investigator proposals.

INTRODUCTION

Our understanding of the high-energy Universe has experienced a revolution in the last decade. In particular, breakthrough observations by EGRET on the Compton Observatory of high-energy gamma-ray blazers, pulsars, unidentified sources (both Galactic and extragalactic), delayed emission afterglows from from gamma-ray bursts and solar flares, and diffuse radiation from our Galaxy and beyond, have all changed our view of the high-energy Universe and raised many new questions.

The Gamma-ray Large Area Space Telescope (GLAST) mission is designed to make the next major advance in our observational knowledge of the Universe above 20 MeV. GLAST, currently scheduled for launch in March 2006, is an international project with participation from the United States, France, Germany, Italy, Japan, and Sweden. The primary instrument on GLAST is the Large Area Telescope (LAT), a solid-state pair conversion telescope being developed by a team led by Stanford University (PI: P. Michelson). The Gamma-ray Burst Monitor (GBM) instrument is being developed by a team led by Marshall Space Flight Center (PI: C. Meegan).

Large Area Telescope

The LAT, shown in Figure 1, has the typical elements of a high-energy pair-conversion telescope. The major subsystems of the LAT are:

- *Tracker/Converter.* Incident photons convert in one of the layers of high-Z converter material, and the resulting e^+ - e^- particles are tracked by silicon-strip detectors through successive planes. The pair conversion signature is used to help reject the much larger background of charged cosmic rays.
- *Calorimeter.* CsI(Tl) bars, arranged in a segmented manner, give both longitudinal and transverse information about the energy deposition pattern. The calorimeter depth is 8.5 r.l., for a total instrument depth of 10 r.l. The shower imaging capability of the calorimeter enables the high-energy reach of the LAT by allowing correction for fluctuations in the leakage of the shower from the calorimeter. This capability also contributes significantly to background rejection.
- *Anticoincidence Detector (ACD).* The ACD array of plastic scintillator tiles provides most of the rejection of charged particle backgrounds. The ACD segmentation avoids the "backsplash" self-veto that affected EGRET's performance above a few GeV.
- *Trigger and Data Acquisition System.* This system collects data from the above subsystems, implements a multi-level event trigger, and provides an on-board capability to search for transients.

The LAT has a modular design, with 16 individual tracker and calorimeter modules set in a 4 × 4 array.

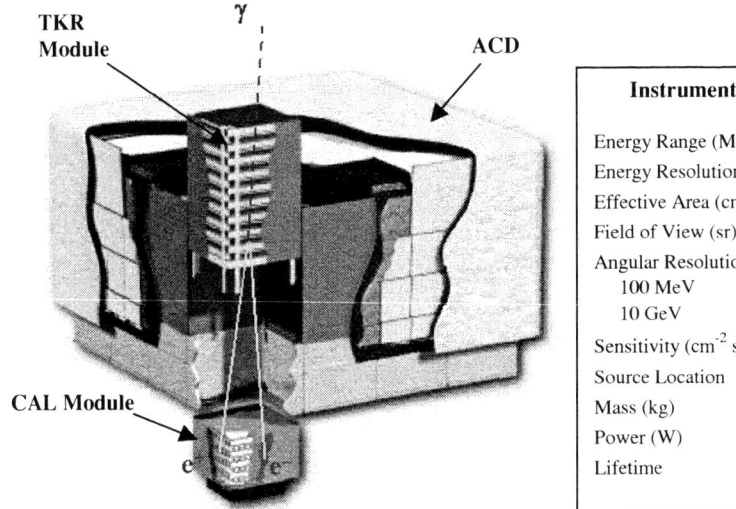

Instrument Parameters	
Energy Range (MeV)	20-300000
Energy Resolution	~0.1
Effective Area (cm^2)	12000
Field of View (sr)	2.5
Angular Resolution	
100 MeV	~3.5°
10 GeV	~0.1°
Sensitivity (cm^{-2} s^{-1})	~4 × 10^{-9}
Source Location	< 1 arcmin
Mass (kg)	3000
Power (W)	650
Lifetime	2006-2011

FIGURE 1. Schematic view of the Large Area Telescope showing the principal subsystems.

Gamma-ray Burst Monitor

The GBM complements the LAT instrument by providing spectral coverage of Gamma-Ray Burst (GRB) transients down to keV energies. The GBM also provides GRB trigger and position information to the LAT.

The GBM has twelve NaI detectors operating from 10 keV to 1 MeV and two BGO detectors operating from 150 keV to 25 MeV. The NaI detectors are oriented to provide a large (8 sr) field-of-view. This coverage, larger than that of the LAT, allows the GBM to detect bursts outside the view of the LAT and to direct repointing of the spacecraft to better target GRBs for the LAT.

GLAST Science Objectives

The science objectives of GLAST are largely motivated by the discoveries of EGRET and of ground-based atmospheric Cherenkov telescopes (ACT) above 300 GeV. The LAT will explore, with ~10% spectral resolution, the energy band beyond EGRET's reach and will overlap with ACTs up to 1 TeV, providing them with absolute calibration. Figure 2a shows the sensitivity of the LAT and various ground-based ACTs. The spectral overlap is illustrated in Figure 2b showing the LAT's capability to measure the unpulsed flux from the Crab from well below 100 MeV to 1 TeV. Sources below the EGRET detection threshold (~ 6×10^{-8} cm^{-2} s^{-1}, >100 MeV) will be localized within 1 arcminute.

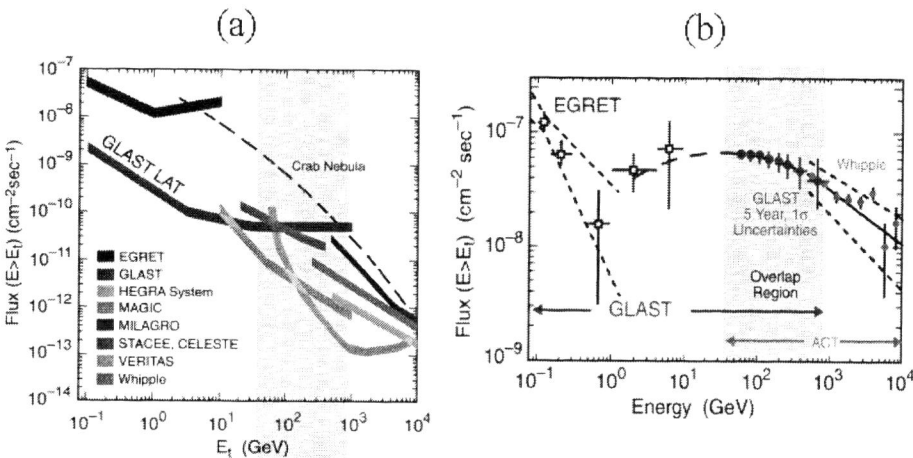

FIGURE 2. (a) Sensitivities of LAT and various ground-based atmospheric Cherenkov Telescopes (ACTs). (b) The LAT will provide measurement of the Crab unpulsed flux from below 100 MeV to ~ 1 TeV.

These capabilities enable a wealth of science investigations that can be summarized as follows:

Understand the mechanisms of particle acceleration in Active Galactic Nuclei, Pulsars, and Supernovae Remnants

Gamma-rays are a direct probe of particle acceleration mechanisms operating in astrophysical systems. We can anticipate how GLAST will advance our knowledge of these non-thermal processes by reference to discoveries with EGRET in three important source classes.

AGN Jets. With its detection of more than 60 AGN, almost all blazars, EGRET has strengthened the unified model of AGN as supermassive black holes with accretion disks and jets [1]. A simple extrapolation of the EGRET Log N-Log S curve indicates that the LAT may detect as many as 10,000 AGN in two years of observation. This is more than the number of currently identified blazars. Population studies with such a large sample will allow tests of the unified model, studies of jet formation and evolution with redshift, and studies of jet properties with AGN type and orientation. The likely EGRET detection of Cen A suggests that GLAST should detect many examples of other classes of AGN.

With the LAT's sensitivity and broad energy coverage, quiescent emission and spectral transitions to flaring states can be measured. The LAT's wide field-of-view is essential for AGN variability studies. Flares as bright as that observed from 3C 279 with EGRET will be measurable with a resolution of 2 minutes [2].

Pulsars. Electric fields generated by charge depletion along open magnetic field lines in pulsar magnetospheres are thought to accelerate particles to ~10 GeV and produce the pulsed gamma-ray emission observed by EGRET from at least six isolated neutron stars [3]. GLAST should increase this population database by at least an order of magnitude. Depending on whether the acceleration site resides near the polar cap of the neutron star or close to the light cylinder (outer gap), the number of pulsar detections expected with GLAST is very different [4]. Predictions based on the polar cap model suggest that GLAST will detect about 180 radio pulsars and ~10 radio-quiet (Geminga like) gamma-ray pulsars. The outer gap model predicts GLAST will detect significantly fewer radio pulsars (~50-80), but many more (~600-1100) radio-quiet pulsars. With GLAST, periodicity searches on sources as faint as $\sim 6\times10^{-9}$ cm^{-2} s^{-1} to find radio-quiet/gamma-ray loud pulsars will be feasible and will allow tests of these predictions.

Supernovae Remnants and Cosmic Rays. There is near consensus that cosmic-rays with energy less than 10^{15} eV are generated by shock-acceleration in supernovae remnants (SNRs). Recent X-ray and TeV observations have confirmed electron acceleration up to TeV energies by detecting non-thermal bremsstrahlung and inverse Compton emission from a few SNR shells, in particular from plerions [5, 6]. However, freshly accelerated protons have not yet been observed. GLAST could observe high-energy gamma rays from proton interactions that produce π^os that decay with a characteristic spectral signature.

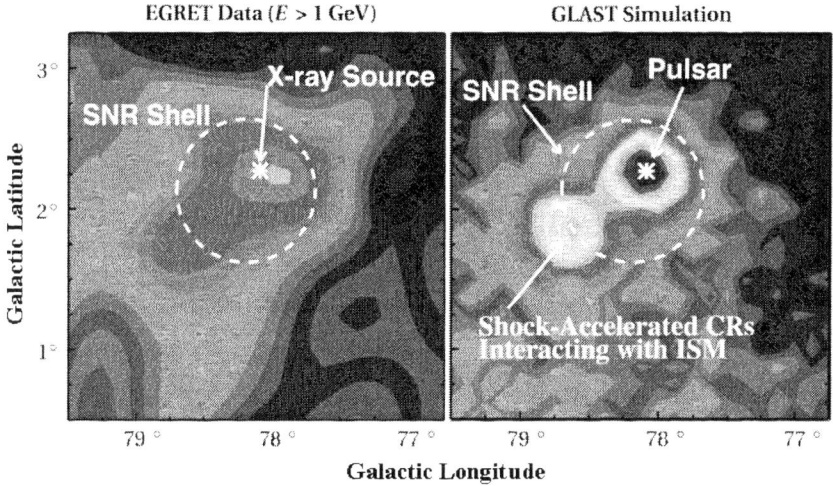

FIGURE 3. Simulation of GLAST likelihood map compared with a map derived from EGRET data for the region around γ-Cygni. The large circle in both maps is the EGRET location uncertainty. As shown on the right, GLAST will be able to localize both point sources and regions of enhanced diffuse emission significantly better than before. (simulation courtesy of Seth Digel)

The LAT's spatial resolution and energy resolution will, in some cases (of order 10 remnants), allow separation of the extended shell emission from an SNR from a compact source (e.g., pulsar) inside it. GLAST will also be able to search deeply for SNRs among the large population of unidentified EGRET sources and perhaps resolve some of those detected. In γ-Cygni for example, the central source, coincident with an X-ray source, is suspected to be a pulsar [7]. For the simulation shown in Figure 3, the EGRET flux detected from the direction of γ-Cygni was portioned between the pulsar and a shell segment. In this scenario, GLAST can localize both the point source and the shell.

Resolve the Gamma-Ray Sky: Unidentified Sources and Diffuse Emission

The interstellar emission of the Milky Way is an intense celestial background that must be modeled in detail in order to make a reliable source catalog and to determine the galactic gamma-ray background. With a reliable gamma-ray background model, GLAST will likely find several hundred or more new Galactic sources in addition to the $\sim 10^4$ expected extragalactic sources.

Unidentified Sources. Many EGRET unidentified sources are likely related to star-forming sites in the solar neighborhood or a few kiloparsecs away along the Galactic plane [8]. These sites harbor compact stellar remnants, SNRs and massive stars, i.e., many likely candidate gamma-ray sources. Pulsar populations may indeed explain a large fraction of unidentified sources close to the Galactic plane [9] and in the nearby starburst Gould Belt [10]. Other candidate objects for the unidentified sources include binary systems, systems with advection-dominated accretion flows onto a black hole, isolated accreting black holes, and Kerr black holes.

GLAST will help address the identification of these sources in three ways:
- Provide excellent source localization (95% confidence diameter) for 5σ (one-year survey) sources and EGRET sources: 14' and <0.3' respectively, for an E^{-2} source and 1° and 1.5' respectively, for a source with a spectral cut-off at ~3 GeV, as anticipated for pulsars.
- Provide good sensitivity up to ~300 GeV, to look for spectral signatures of inverse Compton emission from plerions. Recent studies have indeed found possible associations between X-ray synchrotron nebulae and EGRET sources near the Galactic plane [11].
- Look for periodicities on time scales from milliseconds to seconds. Extrapolating from EGRET analyses of Geminga, the LAT sensitivity will allow searches in sources as faint as ~5×10^{-8} without prior knowledge from radio data.

Interstellar Emission from the Milky Way, Nearby Galaxies, and Galaxy Clusters. Interstellar emission from the Milky Way is the most prominent feature of the gamma-ray sky. It is produced by the interaction of cosmic rays with nuclei and with low-energy photons.

Of particular importance to the study of extended interstellar emission is the LAT's excellent rejection of charged particle backgrounds while maintaining very large effective area for gamma rays.

The angular resolution and effective area of the LAT will allow the study of external galaxies in the light of their interstellar emission. For example, the LAT will resolve the LMC in detail, and, in particular, map the massive star-forming region of 30 Doradus (see Figure 4a). The LAT will also map M31, thus inaugurating the study of cosmic rays in spiral galaxies other than the Milky Way (see Figure 4b).

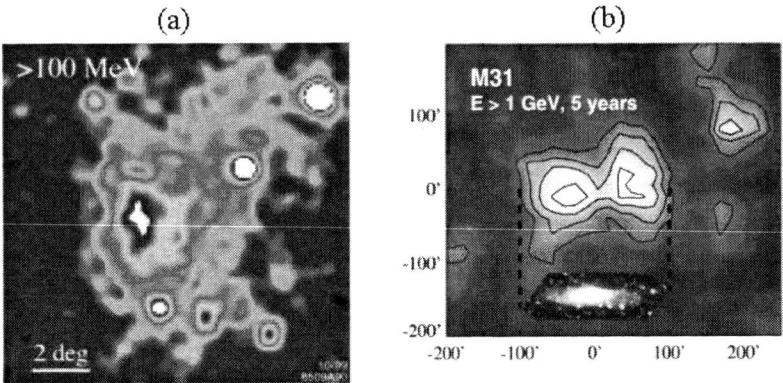

FIGURE 4. (a) Simulated GLAST map of the interstellar emission from the LMC observed after two years of exposure in sky-survey mode. (b) Simulated GLAST map of M31 above 1 GeV, illustrating the capability to map the cosmic-ray distribution in other galaxies.

Extragalactic Diffuse Emission. An isotropic, apparently extragalactic component of the high-energy gamma-ray flux was discovered by SAS-2 and observed by EGRET [11]. It is well fit by a power-law spectrum of index −2.1 over the range 30 MeV − 100 GeV.

Much of this emission is likely produced by unresolved blazers. However, this conclusion requires extrapolation of the relative contributions from flaring and quiescent blazar emission. EGRET detected most blazers only during their flaring states, so the quiescent emission is not well measured.

GLAST will observe the spectrum with better precision and over a broader energy range than EGRET. With reasonable assumptions, GLAST should detect more than 10 million diffuse photons above 100 MeV and more than 1000 above 1 TeV. GLAST will also directly measure the quiescent (and flaring) emission from thousands of blazars, allowing a detailed evaluation of the AGN contribution to the diffuse background. After the blazar component has been resolved, any truly diffuse cosmological flux remaining would be of great interest.

Determine the High-Energy Behavior of Gamma-Ray Bursts and Transients

EGRET detected two components of high-energy gamma-ray emission from GRBs: prompt emission, well defined at lower energies, and a delayed component extending to GeV energies that lasted more than an hour in the case of GRB940217 [12]. The initial pulsed component of this burst was poorly measured by EGRET because of the deadtime of the spark chamber (~100 ms/trigger). The LAT is designed with low deadtime (~20 microsec/trigger) so that even a high-flux burst like GRB940217 will be detected with very little (< few%) deadtime during the most intense part of the burst.

The delayed high-energy component of GRBs will also be much better measured because of the wide field-of-view of the LAT. Models of delayed high-energy emission, for example, involving production of gamma-rays from ultra-high-energy cosmic rays [13] and interaction with the intergalactic medium [14], can be tested.

Internal and external shock models are poorly constrained by spectral and temporal behavior observed at BATSE energies [15]. GLAST, by a combination of the LAT and the GBM, will force comparison of models with observations over an enormous dynamic range in energy $\sim 10^3 - 10^4$, instead of the factor of ~20 afforded by BATSE.

For bright bursts, the LAT can measure exponential high-energy spectral cutoffs expected from moderately high redshift GRBs caused by $\gamma\gamma$ absorption in the extragalactic background light (complementing AGN probes of this effect). This effect should be distinguishable from $\gamma\gamma$ attenuation internal to the source because internal absorption is expected to produce time-variable breaks in power-law energy spectra [16].

Finally, detailed simulations by J. Norris, et al. show that the LAT may detect ~200 GRBs per year, 40 times as many as EGRET detected during the life of the CGRO mission.

Potentially the most revolutionary discoveries that GLAST will make may come from searching for signatures of Galactic dark matter or from the use of AGN spectral cutoffs to probe the epoch of galaxy formation in the early universe.

Searching for Gamma-Ray Signatures of Dark Matter. Evidence of galactic dark matter is provided by several means: the rotation curves of galaxies, structure-formation arguments, and the dynamics and weak lensing of clusters of galaxies.

Exotic particle physics candidates have been suggested for the dark matter, including the LSP in R-parity conserving SUSY. If true, gamma-ray annihilation radiation generated by in a galactic halo may be detectable by the LAT. Although a null result from GLAST would not likely constrain SUSY parameter space, if SUSY is discovered at accelerators, GLAST may be able to determine its cosmological significance quickly.

Baryonic dark matter in the Milky Way may also exist in cold molecular clouds [17, 18]. Its signature would be a hardening of the interstellar gamma-ray spectrum above ~1 GeV. Such an excess should be measurable with GLAST with the LAT's background rejection and sensitivity. The LAT's angular resolution will allow precise measurements of the molecular emissivity at the periphery of the Milky Way to set limits on baryonic dark matter [19].

Probing the Era of Galaxy Formation. Photons above 10 GeV can probe the era of galaxy formation through absorption by near UV, optical, and near IR extragalactic background light (EBL). The latter depends sensitively on star formation rates and the presence of dust [20, 21, 22].

With the detection of $\sim 10^4$ AGN at redshifts up to ~4, LAT data will provide an important new probe of the EBL. Spectra to greater 50 GeV will be determined for several hundred sources. The large number of blazars that will be detected over a broad energy range will provide the data necessary to evaluate the gamma-ray optical depth as a function of redshift and energy, and discriminate against peculiar effects of individual sources. This program will require obtaining redshifts for these sources.

ACKNOWLEDGEMENTS

The author gratefully acknowledges assistance from members of the GLAST Science Team, many of whom contributed to the above summary. In particular, useful discussions with E. Bloom, S. Digel, N. Gehrels, I. Grenier, N. Johnson, T. Kamae, J. Norris, J. Ormes, S. Ritz, and D. Thompson are acknowledged.

REFERENCES

1. Hartman, R. C., et al., *Ap. J. Suppl.* **123**, 79 (1999).
2. Kniffen, D. A., et al., *Ap. J.* **411**, 133 (1993).
3. Thompson, D. J., et al., in *Proceedings of 4th Compton Symposium*, edited by C. D. Dermer, et al.., AIP Conference Proceedings **410**, New York, 1997, pp. 39-51.

4. Harding, A. K., in *High-Energy Gamma-Ray Astronomy*, edited by F A. Aharonian & H. J. Volk, AIP Conference Proceedings **558**, New York, 2001, pp. 115-126.
5. Tanimori, T., et al., *Ap. J. Lett.* **497**, L25 (1998).
6. Koyama, K., et al., *Nature* **378**, 255 (1995).
7. Brazier, K., et al., *MNRAS* **281**, 1033 (1996).
8. Gehrels, N. et al., *Nature* **404**, 363 (2000)
9. Yadigaroglu, Y., & Romani, R., *Ap. J.* **476**, 347 (1997).
10. Grenier, I. A., & Perrot, C., in *Proceedings of the 26th International Cosmic Ray Conference*, Salt Lake City, Utah OG 2.1.11 (2000).
11. Sreekumar, P., et al., *Ap. J.* **494**, 523 (1998).
12. Hurley, K. C., et al., *Nature* **372**, 652 (1994).
13. Mottcher, M., & Dermer, C. D., *Ap. J. Lett.* **499**, L131 (1998).
14. Plaga. R., *Nature* **374**, 430 (1995).
15. Fenimore, E. E., & Ramierez-Ruiz, E., Ap. J. 539, 712 (2000).
16. Baring, M. G., in *Proceedings GeV-TeV Gamma Ray Astrophysics Workshop VI*, Snowbird, Utah, edited by B. L. Dingus, M. H. Salamon, & D. B. Kieda, AIP Conference Proceedings **515**, New York, 2000, p. 238.
17. Sciama, D. , *MNRAS* **312**, 33 (2000).
18. De Paolis, F., et al., *Ap. J.* **510**, L103 (1999).
19. Digel, S. W., et al., *Ap. J.* **458**, 561 (1996).
20. Stecker, F., *Astroparticle Physics* **11**, 83 (1999).
21. Madau, P., & Phinney, E. S., *Ap. J.* **456**, 124 (1996).
22. MacMinn, D., & Primack, J., *Space Sci. Rev.* **75**, 413 (1996).

Gamma-Ray Large Area Space Telescope (GLAST) Project

Scott Lambros

*NASA/Goddard Space Flight Center
Greenbelt, Md. 20771*

Abstract--The Gamma-ray Large Area Space Telescope (GLAST) mission is a scientific spaceflight investigation that measures the direction, energy and arrival time of celestial gamma rays. It is planned for launch in March 2006. The scientific applications include determining the structure of high-energy astrophysical processes such as those found in active galactic nuclei, black holes, and supernovae. Several other natural phenomena in the universe will be investigated. One of the most exciting aspects of the GLAST mission is its potential for new discovery.

There are two instruments on the GLAST Observatory: the Large Area Telescope (LAT), a joint development of NASA and DOE, and the GLAST Burst Monitor (GBM). Both instruments have international collaborators.

The spacecraft will be a modified version of an existing design. The ground system elements and data flow are also described.

INTRODUCTION

The Gamma-ray Large Area Space Telescope (GLAST) is NASA's next major gamma-ray mission scheduled for launch in March 2006. It is a NASA managed mission with joint development with the Department of Energy for the Large Area Telescope (LAT) instrument, and with international participation for both the LAT and GLAST Burst Monitor (GBM) instruments. The GLAST Observatory will be launched into a 550 km. circular orbit at 28.5 degrees inclination. The design lifetime is 5 years, with a 10-year operational goal.

Science requirements for the mission were developed by an international group of scientists called the Facility Science Team. Since the two instruments were selected in March 2000, a Science Working Group (SWG) was formed which has refined the science requirements, encapsulated in a Science Requirements Document. Included in the SWG are four interdisciplinary scientists, also selected at the time of instrument selection, who perform science analysis to prepare for and utilize the GLAST data.

Since the Earth is protected from gamma rays by the atmosphere, a spaceflight mission offers prime observation. Because such high energy is required to produce gamma rays, they yield much information about high-energy engines in the universe. In addition, they are specifically suited for observation because they do not bend in a magnetic field; thus they point directly back to their source.

GLAST was rated the highest recommended space mission in its class by the National Research Council Decadel Review Committee.

The major elements comprising the GLAST mission are described in this paper. They include the two instruments, the spacecraft that provides the platform to service the instruments and provide communication to the ground, and the various ground system elements that provide the health and safety monitoring of the observatory and the processing of the science data obtained by the instruments.

SCIENCE

High-energy gamma rays probe the most energetic phenomena occurring in nature, providing insight into the mechanisms of particle acceleration in the universe. High-energy gamma rays are emitted from a diverse population of astrophysical sources. Stellar mass objects, in particular neutron stars and black holes, the nuclei of active galaxies that likely contain super massive black holes, interstellar gas in the galaxy that interacts with high-energy cosmic rays, the diffuse extragalactic background, supernovae that may be sites of cosmic-ray acceleration, and gamma-ray bursts are all gamma-ray emitters. The sun also produces high-energy gamma rays during active periods. Many transient sources, from the subsecond time scales of the fastest gamma-ray bursts to AGN flares lasting days or more, often radiate the bulk of their power at gamma ray energies.

High-energy gamma ray astronomy has undergone a blossoming of discovery in recent years. In particular, the Energetic Gamma Ray Experiment Telescope (EGRET) on the Compton Gamma Ray Observatory (CGRO) has opened the field to detailed studies of several classes of galactic and extragalactic objects. GLAST is an evolution of the instrumentation used in EGRET, with higher sensitivity, energy resolution, and location accuracy. GLAST also follows on the heels of the Burst and Transient Source Experiment (BATSE) instrument, also on CGRO, for observing gamma- ray bursts.

Some of the particular science objectives that GLAST will address are:

- Examine particle acceleration engines in active galactic nuclei jets, pulsars, supernova remnants, and solar flares
- Investigate the structure of super-massive black holes
- Use gamma rays to probe the optical-UV extragalactic background light, yielding valuable information from the era of galaxy formation
- Generate an extensive source catalog (where EGRET discovered 85 AGNs, GLAST expects on the order of 10,000) and resolve unidentified sources from EGRET
- Probe our understanding and theories regarding dark matter. This invisible unknown form of matter has been theorized to make up 80 to 90% of all matter in the universe
- Investigate whether the diffuse gamma-ray emission is isotropic or an integrated flux from many as yet unresolved sources
- Explore the mysteries of gamma ray bursts: an unknown localized phenomenon that generates energy brighter than the entire rest of the gamma-ray sky for typically on the order of a few seconds.
- Discovery! With the significant leaps in instrumentation on GLAST, a key objective is to uncover what we do not yet know.

INSTRUMENTS

Two instruments were selected to fly on the GLAST mission. The Large Area Telescope (LAT) utilizes the bulk of the mission resources. The GLAST Burst Monitor (GBM) is a smaller instrument for enhancing the science of the mission.

Large Area Telescope [1]

Organization

The LAT instrument (Figure 1) is a joint development with NASA and the Department of Energy (DOE). Participants include a large collaboration of developers, including several U.S. institutions and foreign participants. The Principal Investigator is Dr. Peter Michelson at Stanford University. Project management is located at the Stanford Linear Accelerator Center (SLAC), a DOE facility at Stanford University. Other domestic organizations contributing hardware include the University of California, Santa Cruz (UCSC); the Naval Research Laboratory (NRL); and the Goddard Space Flight Center (GSFC). Foreign contributors include France, Japan, Italy and Sweden.

The joint development aspect of this instrument is consistent with the Cosmic Connections Initiative, a planned collaboration among NASA, DOE, and NSF. Projects in this initiative are geared toward exploiting the synergy of combining particle physics with astrophysics.

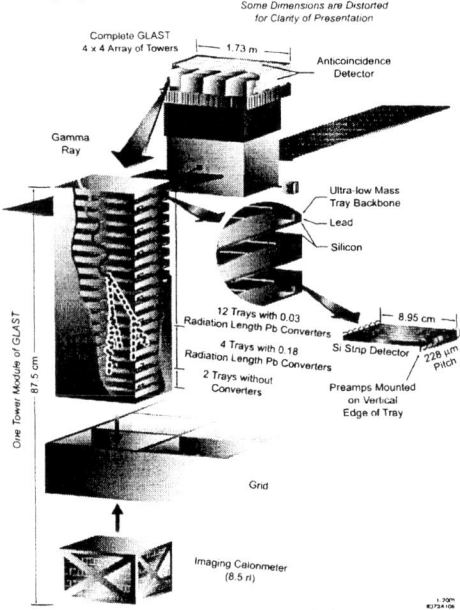

Figure 1 – LAT Instrument Subsystem

Technical Description

The LAT measures the direction, energy and arrival time of gamma rays in the energy range of 20 MeV to 300 GeV. This wide-field-of-view, high-energy pair conversion telescope includes the same essential elements as EGRET. Compared to EGRET, the LAT will have a greatly improved effective area ($>8000 cm^2$), sensitivity ($< 6 \times 10^{-9} cm^{-2} s^{-1}$ for energy $> 100 MeV$), angular resolution ($< 0.15°$ for energy $>10 GeV$; $< 3.5°$ for energy $= 100$ MeV), and energy range.

The LAT consists of four subsystems: a silicon-strip detector tracker for measuring the direction of an incident gamma ray; a hodoscopic cesium iodide (CsI) calorimeter for measuring the energy of an incident gamma ray; a segmented, plastic scintillator, anti-coincidence detector (ACD) for cosmic ray rejection; and a data acquisition subsystem for on-board triggering, cosmic-ray filtering, control, and interfacing with the spacecraft. The instrument consists of a modular 4x4 array of towers, each composed of a tracker and calorimeter, and their associated electronics. The ACD covers the entire array of towers.

The LAT instrument weighs 3000 kg. The required power is 650 watts, orbital average. The data rate averages 300 kbps. A few times during the mission, especially during initial checkout, the instrument will turn off its filtering mode (i.e., cosmic ray suppression) and telemeter all the data, so that the filtering algorithms can be verified on the ground. In this case, the peak data rate jumps to several megabits per second.

GLAST Burst Monitor [2]

Organization

The Principal Investigator for the GBM is Dr. Charles Meegan from the NASA/Marshall Space Flight Center (MSFC). Along with the MSFC, the University of Alabama in Huntsville (UAH) will provide design and hardware. A significant portion of the instrument (i.e., all the detector hardware) will be provided by Germany. Project management is centered at the MSFC.

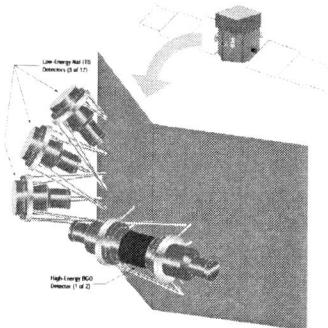

Figure 2 – GBM Instrument Concept.

Technical Description

The GBM (Figure 2) will produce spectral observations of gamma-ray bursts and provide approximate burst locations over a wide field of view. Its spectral range will span from 10 keV to 25 MeV, providing good overlap with the LAT yielding an unprecedented six decades of burst energy spectra with no gaps. On-board burst locations will be determined to within 15 degrees (refined to about 1.5 degrees on the ground), and used to repoint the observatory to allow the burst to enter the LAT field of view (FOV). The GBM effective FOV exceeds 8 steradians.

The GBM consists of two types of detectors and a data processing unit. Twelve sodium iodide (NaI) scintillation detectors directly coupled to photomultiplier tubes (PMTs) measure incident gamma rays in the range of 5 keV to 1 MeV. The detectors are positioned around the spacecraft, facing in different directions to achieve the large FOV and to determine burst location using relative rates. Two bismuth germinate (BGO) scintillation detectors provide spectral coverage in the range of 150 keV to 30 MeV. The detectors are directly coupled to two PMTs on opposite sides, whose outputs are summed. The Data Processing Unit (DPU) interprets the signals from the detectors and formats the data to output to the spacecraft recorder. It also generates alert messages for gamma-ray bursts, which includes a rough position.

The GBM weighs 70 kg and uses about 10 watts orbital average. The telemetry rate is about 10 kbps, orbital average and 10 Mbps peak.

SPACECRAFT

The spacecraft (Figure 3) utilized for the GLAST mission will be procured through

Figure 3 – Spacecraft Concept

the Rapid Spacecraft Development Office (RSDO) at NASA/GSFC. This contractual vehicle in place at GSFC allows a user to select and quickly procure a spacecraft from a catalog of options. To be in the catalogue, the spacecraft must have already been designed, built, qualified for flight, tested, passed through launch processing, and determined launch ready. This approach ensures that the spacecraft are a true catalog item in that they have been successfully built. When RSDO vendors propose a solicitation for a particular mission, they are allowed to propose their catalog spacecraft bus and mission-unique modifications. This procedure is particularly useful for missions that do not have many unique or complicated spacecraft interfaces.

For GLAST, the large mass is the most unique aspect of the spacecraft as compared to the capabilities of the current spacecraft buses in the RSDO catalog.

For proposing spacecraft, a unique structure will have to be defined to handle the mass of the two instruments.

The disadvantage of this approach is that no spacecraft vendor is on board early in the program to refine instrument interfaces and perform interface trade studies. To mitigate this situation, accommodation studies with potential RSDO vendors in the early years of mission development will be performed. These studies will help to refine the instrument interfaces so that the instrument teams can continue development. In turn, the required interfaces for the instruments will be incorporated into the Spacecraft Performance Specification for the next study, and eventual spacecraft procurement. This entire process of defining/refining, trading off, and specifying requirements is overseen by the GLAST Project engineering staff, which uses its technical judgement to monitor the process, as well as to perform technical trade studies to feed into the process.

GROUND SYSTEM AND OPERATIONS

Elements of the ground system include a Mission Operations Center (MOC); two Instrument Operations Centers (IOCs), one at each instrument developer's site; a Science Support Center (SSC); and ground stations and associated data routing networks.

The MOC, which operates the GLAST Observatory, sends and receives commands and telemetry. It monitors the health and safety of the spacecraft, and conducts limited health checks of the instruments. It routes the telemetry to other ground elements.

The IOCs perform health and safety monitoring of the instruments and data processing. The LAT IOC is located at Stanford/SLAC; the GBM IOC is located at MSFC. Science data products are sent from the IOCs to the SSC for archiving and distribution to the science community. The IOCs also support the instrument science teams in performing their investigations. The IOCs will provide instrument command structures, software updates, and table loads to the MOC to be uploaded to the instruments.

The SSC is used to process production data to higher levels of data products, and archive and distribute GLAST data products to the science community. It will also support selected guest investigators during their particular investigations. Tools for data use will be provided to data users. The SSC is responsible for generating the GLAST observing schedule, an activity led by the Project Scientist. The SSC is located at Goddard Space Flight Center.

The primary ground station is Malindi, provided by Italian partners, and allows for two nominal data downlinks per day.

The space network, i.e., the Tracking and Data Relay Satellite System (TDRSS), will be used for alert messages. These are messages generated by the instruments to signal special transient events such as gamma-ray bursts and will be sent to other experimenters via the Gamma-ray burst Coordinate Network, allowing them to observe the same events. This approach will enable multiwavelength studies of interesting phenomena.

There will also be an automatic repointing feature on board the spacecraft. If one of the instruments detects an interesting event (defined by a predetermined threshold), the spacecraft will automatically repoint to observe that event better.

CONCLUSIONS

The elements of the mission, as described in this paper are being developed by scientists and engineers from several different countries. Most of the engineering challenges have some basis in heritage, however, much of the development is new for this application (e.g., silicon strip detector technology in this space application; the RSDO spacecraft is a catalog item, but modified for this particular mission).

The GLAST mission will push forward the boundaries of gamma-ray science. Results of this mission will enhance our understanding of the structure of high energy phenomenon in the universe. The results will also increase our understanding of the early universe, and what interactions were occurring during the era of galaxy formation. Perhaps most importantly, it will lead us to new discoveries of things which cannot be predicted.

REFERENCES

[1] Peter F. Michelson, et. al., *GLAST Large Area Telescope Flight Investigation: An Astro-Particle Physics Partnership Exploring the High-Energy Universe,* proposal to NASA AO 99-OSS-03, Stanford University, November 1999.

[2] Charles Meegan, et. al., *GLAST Burst Monitor,* proposal to NASA AO 99-OSS-03, NASA/Marshall Space Flight Center, November 1999.

Science with AGILE

M. Tavani*, G. Barbiellini†, A. Argan**, N. Auricchio‡, P. Caraveo*,
A. Chen**, V. Cocco§, E. Costa¶, G. Di Cocco‡, G. Fedel†, M. Feroci¶,
M. Fiorini*, T. Froysland§, M. Galli‖, F. Gianotti‡, A. Giuliani*,
C. Labanti‡, I. Lapshov¶, P. Lipari††, F. Longo‡‡, E. Massaro††,
S. Mereghetti*, E. Morelli‡, A. Morselli§, A. Pellizzoni*, F. Perotti*,
P. Picozza§, C. Pittori§, C. Pontoni§§, M. Prest§§, M. Rapisarda¶¶, E. Rossi‡,
A. Rubini¶, P. Soffitta¶, M. Trifoglio‡, E. Vallazza†, S. Vercellone* and
D. Zanello††

*IFC-CNR, V. Bassini 15, 20133 Milano - Italy
†Univ. di Trieste and INFN, V. Padriciano 99, Trieste - Italy
**IFC-CNR Milano, and CIFS, Villa Gualino, Viale Settimio Severo 63, 10133 Torino - Italy
‡ITESRE-CNR, V. Gobetti 101, 40129 Bologna - Italy
§Univ. Roma-II and INFN, V. Ric. Scientifica 1, 00133 Roma - Italy
¶IAS-CNR, V. Fosso del Cavaliere, 00133 Roma - Italy
‖ENEA, Sez. Bologna, Italy
††Dip. Fisica Univ. Roma-I and INFN, P.le Aldo Moro 2, 00185 Roma - Italy
‡‡Univ. di Ferrara and INFN, V. del Paradiso 12, 44100 Ferrara - Italy
§§INFN Trieste, and CIFS, Villa Gualino, Viale Settimio Severo 63, 10133 Torino - Italy
¶¶ENEA, Sez. Roma, Italy

Abstract. AGILE is an ASI gamma-ray astrophysics space Mission which will operate in the 30 MeV – 30 GeV with imaging capabilities also in the 10–40 keV range. Primary scientific goals include the study of AGNs, gamma-ray bursts, Galactic sources, unidentified gamma-ray sources, diffuse Galactic and extragalactic gamma-ray emission, high-precision timing studies, and Quantum Gravity testing. AGILE will be the only Mission entirely dedicated to source detection above 30 MeV during the period 2003-2006.

THE AGILE MISSION

AGILE is a Small Scientific Mission dedicated to high-energy astrophysics supported by the Italian Space Agency and scientifically developed in CNR and INFN laboratories. AGILE is currently in Phase C, and planned to start operations during the year 2003. The AGILE instrument is highly innovative and designed to detect and image photons in the 30 MeV–50 GeV and 10–40 keV energy bands. AGILE is characterized by an excellent spatial resolution and timing capability, and by an unprecedently large field of view covering $\sim 1/5$ of the entire sky at energies above 30 MeV.

The AGILE spacecraft will be of the MITA class (total satellite weight of ~ 230 kg) to be launched in a low-background equatorial orbit of height near 550 km. AGILE scientific data will be of great relevance also for joint studies of high-energy sources with other scientific satellites and ground-based facilities for radio/optical/TeV observations.

THE INSTRUMENT

The AGILE scientific instrument is based on an innovative design allowing the simultaneous detection of hard X-rays and gamma-rays with unprecedented imaging and timing capabilities [23, 24]. The instrument consists of two imaging detectors: (1) Super-AGILE (SA) and, (2) the Gamma-Ray Imaging Detector (GRID) made of a Silicon Tracker and a Mini-Calorimeter. The Mini-Calorimeter is also capable of independently detect transient events. A description of the instrument can be found in Ref. [4]. Physical constraints on the instrument are tight in terms of absorbed power ($\lesssim 70$ W), volume and mass (~ 80 kg). We summarize here the main instrument characteristics.

A **Silicon Tracker (ST)**, consisting of 14 detection planes, is devoted to the detection and imaging of high-energy photons above ~ 20 MeV [4, 5, 6]. Particle track reconstruction is based on the floating strip readout of the **analog (deposited charge) signal** from Si microstrips of pitch of 121 μm (see Fig. 1). The total number of channels is $\sim 43,000$. The ST spatial resolution is excellent, reaching $\sim 40 \mu$m for a variety of incidence angles [6, 5]. Fast low-power electronics allows reaching very short gamma-ray detection deadtimes of order of 100 μs. The ST on-axis effective area near 100 MeV is ~ 500 cm^2.

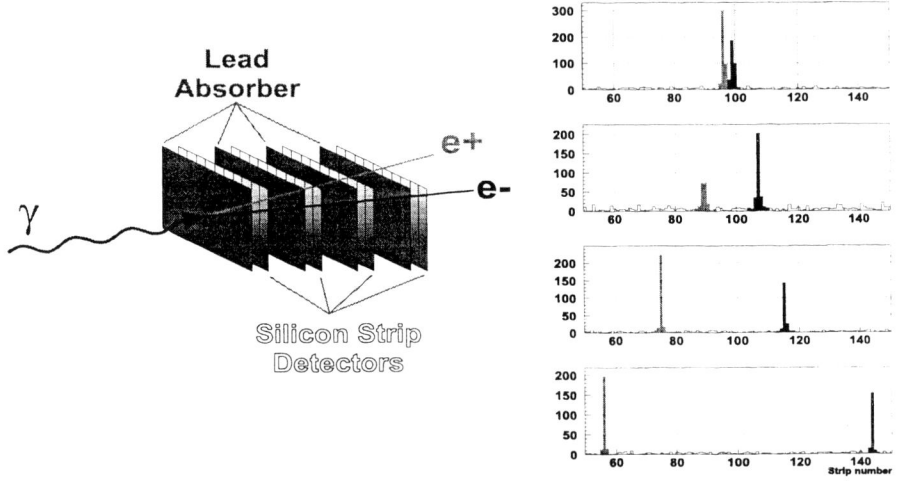

PHOTAG TESTBEAM - INFN TS/IFC MI

FIGURE 1. The AGILE Tracker will make a crucial use of the analog signal from the Si microstrips. An AGILE beamtest was carried out in August, 2000 at the CERN T11 beamline (East Hall, CERN PS). A photon beam (of energy range $\sim 30 - 500$ MeV) was produced by Bremsstrahlung of electrons of momentum ranging from 0.15 to 1 GeV/c hitting a thin lead target. The electron beam was deviated by a magnet spectrometer, and tagged by two delay wire chambers and a lead glass calorimeter. A typical gamma-ray photon event detected by 4 Silicon detectors spaced with lead converters (each of $\sim$0.07 radiation length) is shown on the right. The histograms represent the charge collected on the readout strips configured with the baseline AGILE tracker layout (Silicon microstrip pitch of 121μm, for a floating strip readout system of 242μm pitch). The spatial resolution achieved by this readout configuration is excellent (below 40 μm for a wide range of photon incidence angles). Figure and data from Ref. [4, 5, 6, 12].

Super-AGILE (SA) consists of an additional plane of Si detectors placed on top of the Si Tracker, and of an ultra-light coded mask system [9, 15]. SA is aimed at detecting hard X-rays in the energy range between 10 and 40 keV. Imaging capabilities are quite good (pixel size of ~ 6 arcmin) for an on-axis $(5-\sigma)$ sensitivity of ~ 5 mCrab (1-day integration time).

The Mini-Calorimeter (MCAL) is made of two layers of CsI(Tl) bars with independent fast readout [1, 11]. MCAL supports both the ST particle energy reconstruction (as part of GRID), and the independent detection of photons in the energy range $\sim 0.3-100$ MeV.

All AGILE active detectors are surrounded by an **Anticoincidence (AC) System** consisting of a top plastic scintillator plane and 12 lateral planes [17].

The **AGILE Data Handling System (DH)** provides the on-board data processing for the GRID, SA and MCAL events [25]. A DH essential task is the implementation of a GRB Search Procedure to be carried out for a large variety of trigger timescales (from $\lesssim$ 1ms to tens of seconds). Super-AGILE can also quickly image fast transients, and a fast communication channel for GRB alerts is currently envisioned.

Table 1 summarizes the scientific performance of the AGILE detectors.

SCIENTIFIC PERFORMANCE

Fig. 2 shows the AGILE-GRID effective area, and Fig. 4 shows a typical GRID error box for a weak off-axis AGN ($\sim 30°$). In both figures, we emphasize a comparison with EGRET capabilities.

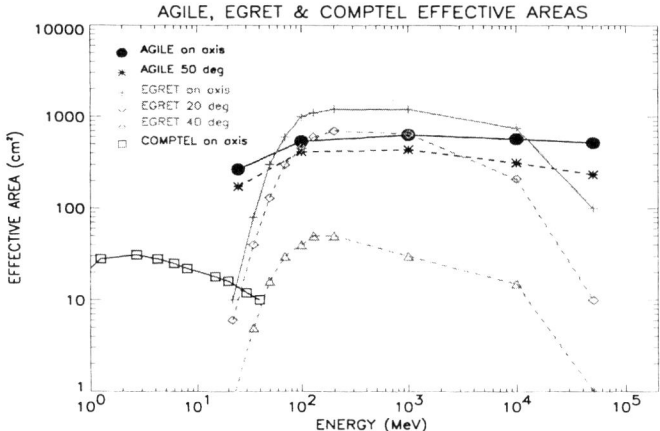

FIGURE 2. Effective area as a function of photon energy for the AGILE-GRID, EGRET [27], and COMPTEL [20]. GRID simulations results from Ref. [8].

TABLE 1. AGILE Detectors' Capabilities

Gamma-Ray Imaging Detector (GRID)	
Energy Range	30 MeV – 50 GeV
Field of view	~ 3 sr
Effective Area (on-axis, at 400 MeV)	$\sim 540\,cm^2$
Effective Area (50-60° off-axis, at 400 MeV)	$\sim 320\,cm^2$
Angular Resolution (68% cont. radius, 1 GeV)	36 arcmin
Source Location Accuracy (for S/N $\gtrsim 10$)	$\sim$5-20 arcmin
Energy Resolution (with MCAL, at 400 MeV)	$\Delta E/E \sim 1$
Deadtime	$\sim 100\,\mu s$
Absolute Timing Accuracy	$\sim 2\,\mu s$
Mini-Calorimeter (MCAL)	
Energy Range	250 keV – 200 MeV
Energy Resolution	~ 1 MeV
Effective Area (at 300-900 keV)	$\sim 100\,cm^2$
Effective Area (at 1-10 MeV)	$\sim 500\,cm^2$
Effective Area (at 10-100 MeV)	$\sim 1000\,cm^2$
Deadtime (single CsI bar)	$\lesssim 10 - 20\,\mu s$
Absolute Timing Accuracy	$\lesssim 5\,\mu s$
Super-AGILE (SA)	
Energy Range	10-40 keV
Field of view (Full Width at Zero Sens.)	$107° \times 68°$
Sensitivity (5σ in 1 day)	~ 5 mCrab
Angular Resolution (Pixel Size)	6 arcmin
Source Location Accuracy (for S/N$\sim$10)	$\sim$1-3 arcmin
Energy Resolution	$\Delta E < 4$ keV
Deadtime (single "daisy-chain" unit)	$\lesssim 5\,\mu s$
Absolute Timing Accuracy	$\lesssim 5\,\mu s$

Gamma-Ray Astrophysics with the GRID

The GRID has been designed to obtain:

- **excellent imaging capability in the energy range 100 MeV-50 GeV**, improving the EGRET angular resolution by a factor of 2;
- **a very large field-of-view**, allowing simultaneous coverage of $\sim 1/5$ of the entire sky per each pointing (FOV larger by a factor of $\sim$5 than that of EGRET);
- **excellent timing capability**, with absolute time tagging of uncertainty near $1\,\mu s$ and very small deadtimes ($\sim 100\,\mu s$ for the Si-Tracker and $\sim 20\,\mu s$ for each of the individual CsI bars, see Fig. 7);
- **a good sensitivity for point sources**, comparable to that of EGRET for *on-axis* sources, and substantially better for *off-axis* sources;
- **excellent sensitivity to photons in the energy range $\sim$30-100 MeV**, with an effective area above 200 cm^2 at 30 MeV;
- **a very rapid response to gamma-ray transients and gamma-ray bursts**, obtained by a special Quicklook Analysis program and coordinated ground-based and space observations.

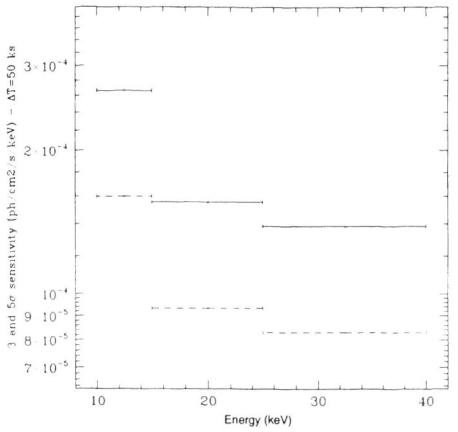 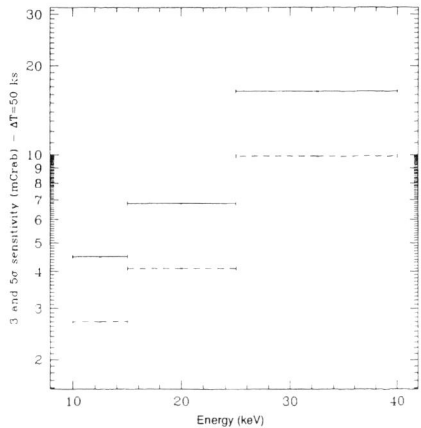

FIGURE 3. *Left Panel:* Super-AGILE simulated sensitivity (solid data points: 5σ, dashed data points: 3σ) for a 50 ksec integration and Crab-like spectrum for the combined four Si-detectors, in units of photons cm^{-2} s^{-1} keV^{-1}. *Right Panel:* simulated sensitivity (solid data points: 5–σ, dashed data points: 3σ) in mCrab units. Simulation results from Refs. [9, 15]

Super-AGILE

An imaging coded mask detector system (Super-AGILE) in addition to the GRID will provide a unique tool for the study of high-energy sources. The Super-AGILE FOV is planned to be ~ 0.8 sr. Super-AGILE can provide important information including:

- **source detection and spectral information in the energy range ~10-40 keV** to be obtained simultaneously with gamma-ray data (5 mCrab sensitivity at 15 keV (5σ) for a 50 ksec integration time);
- **accurate localization (~1-2 arcmins) of GRBs and other transient events** (for typical transient fluxes above ~ 1 Crab); the expected GRB detection rate is $\sim 1-2$ per month;
- **excellent timing**, with absolute time tagging uncertainty and deadtime near $5\mu s$ for each of the 16 independent readout units of the Super-AGILE Si-detector;
- **long-timescale monitoring (~2 weeks) of hard X-ray sources**;
- **hard X-ray response to gamma-ray transients detected by the GRID**, obtainable by slight repointings of the AGILE spacecraft (if necessary) to include the gamma-ray flaring source in the Super-AGILE FOV.

The combination of simultaneous hard X-ray and gamma-ray data will provide a formidable combination for the study of high-energy sources. Given the sensitivities of the GRID and Super-AGILE, simultaneous hard X-ray/gamma-ray information is anticipated to be obtainable for: (1) GRBs, (2) blazars with strong X-ray continuum

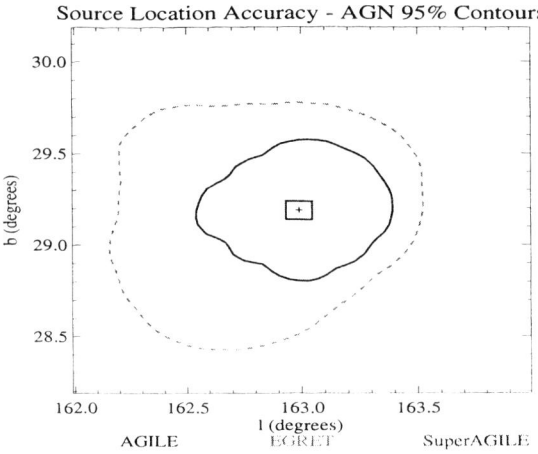

FIGURE 4. A comparison of simulated 99% contour levels for GRID (solid curve) and Super-AGILE (square) positioning of a relatively weak off-axis AGN, with what obtained by EGRET (dotted curve). We assumed a 1-week effective exposure time for a gamma-ray source of flux above 100 MeV equal to 30×10^{-8} ph.cm^{-2}s^{-1} positioned at ~ 28 degrees off-axis for AGILE and at ~ 17 degrees off-axis for EGRET. Simulations results from Refs. [15, 28].

emission such as 3C 273 and Mk 501, (3) Galactic jet-sources with favorable geometries, (4) unidentified gamma-ray sources.

IMAGING AND TIMING OF GAMMA-RAY SOURCES

Fig. 5 shows a typical AGILE pointing, and Fig. 6 shows a comparison of intensity maps obtained by a single 2-week viewing period of the anti-center region of the Galactic plane. Relatively bright AGNs and Galactic sources flaring in the gamma-ray energy range above a flux of 10^{-6} ph cm^{-2} s^{-1} can be detected within a few days by the AGILE Quicklook Analysis. We conservatively estimate that for a 3-year mission AGILE is potentially able to detect a number of gamma-ray flaring AGNs larger by a factor of several compared to that obtained by EGRET during its 6-year operative life (see Fig. 8). Furthermore, the large FOV will favor the detection of fast transients such as gamma-ray bursts. Taking into account the high-energy distribution of GRB emission above 30 MeV, we conservatively estimate that ~ 1 GRB/month can be detected and imaged in the gamma-ray range by the GRID and Super-AGILE.

The existence of a large number of variable gamma-ray sources (extragalactic and near the Galactic plane, e.g., [13]) makes necessary a reliable program for quick response to transients. Quicklook Analysis of gamma-ray data will be a crucial task to be carried out by the AGILE Team. Prompt communication of gamma-ray transients (typically requiring 1-3 days to be detected with high confidence for sources above 10^{-6} ph cm^{-2} s^{-1}) is planned. Detection of short timescale (seconds/minutes/hours) transients (GRBs, SGRs, solar flares and other bursting events) is possible in the gamma-

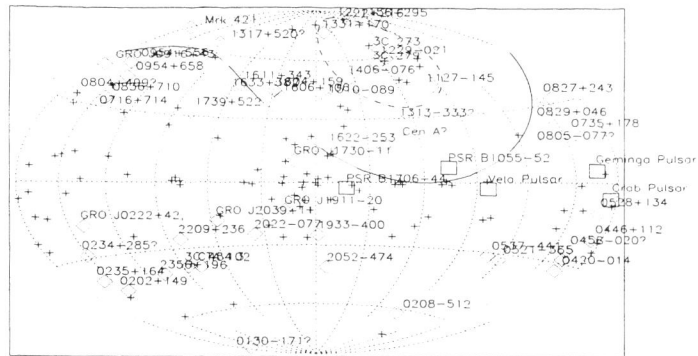

FIGURE 5. Comparison between a typical GRID pointing centered at the blazar 3C 279 region (area within the solid line circle of radius equal to 60°) and an EGRET pointing of the same source region (area within the dashed line of radius equal to 25°).

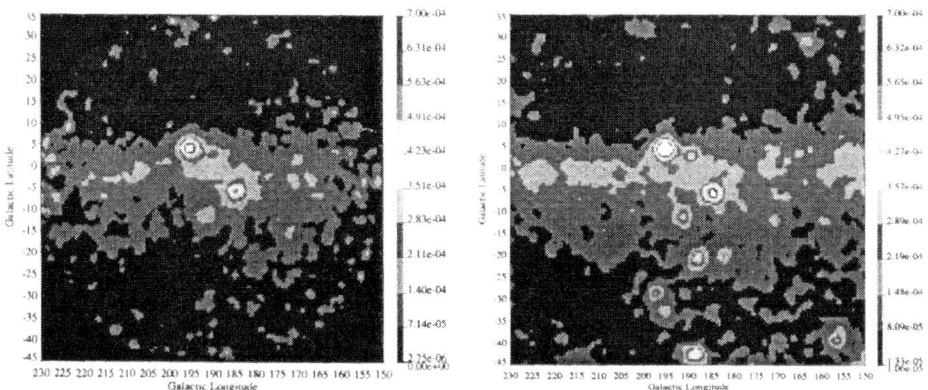

FIGURE 6. *Left Panel:* EGRET intensity map (photons energy above 100 MeV) of the GRO Viewing Period (VP) n. 1 of the field containing the Geminga and Crab pulsars (from the EGRET public database). *Right Panel:* AGILE simulated intensity map of the same sky region (above 100 MeV) assuming all sources at their average flux reported by the 3rd EGRET Catalog (from Ref. [28]). The variable nature of several of these sources is clear from their absence in the left panel giving the GRO VP 1 pointing. In both cases a 2-week total pointing duration was assumed. Intensity map scale in units of photons cm^{-2} s^{-1} sr^{-1}.

ray range. A primary responsibility of the AGILE Team will be to provide accurate positioning of transients, and to alert the community through dedicated channels.

After a 1-year all-sky pointing program, we expect the AGILE average exposure for a generic source to be larger by a factor of ~ 4 compared to what obtained by EGRET during the same time period. Therefore, AGILE average sensitivity for a generic gamma-ray source above the Galactic plane is expected to be better than EGRET by a factor ~ 2 (see Ref. [16]). Deep exposures for selected sky regions can be obtained by a program with repeated overlapping pointings. For selected regions, AGILE can then achieve a sensitivity larger than EGRET by a factor of $\sim 4-5$ at the completion of its program, reaching a minimum detactable flux near 5×10^{-8} ph cm^{-2} s^{-1}. This capability can be particularly important to study a selected list of persistent gamma-ray sources.

AGILE detectors will have optimal timing capabilities. The on-board GPS system can reach an absolute time tagging precision for individual photons near $2\,\mu$s. Depending on the detectors hardware and electronics, absolute time tagging can achieve values near $1-2\,\mu$s for the Si-Tracker, and $3-4\,\mu$s for the individual detecting units of the Mini-Calorimeter and Super-AGILE.

Furthermore, instrumental deadtimes will be unprecedently small for gamma-ray detection. The GRID deadtime will be $\lesssim 100\,\mu$s (improving by three orders of magnitude the performance of previous spark-chamber detectors such as EGRET). The deadtime of MCAL single CsI bars is near $20\,\mu$s, and that of single Super-AGILE readout units is $\sim 5\,\mu$s. Taking into account the segmentation of the electronic readout of MCAL and Super-AGILE detectors (32 MCAL elements and 16 Super-AGILE elements) the effective deadtimes will be much less than those for the individual units.

Fig. 7 shows the AGILE timing performance compared to other gamma-ray missions. Fast AGILE timing will, for the first time, allow investigations and searches for sub-millisecond transients in the gamma-ray energy range.

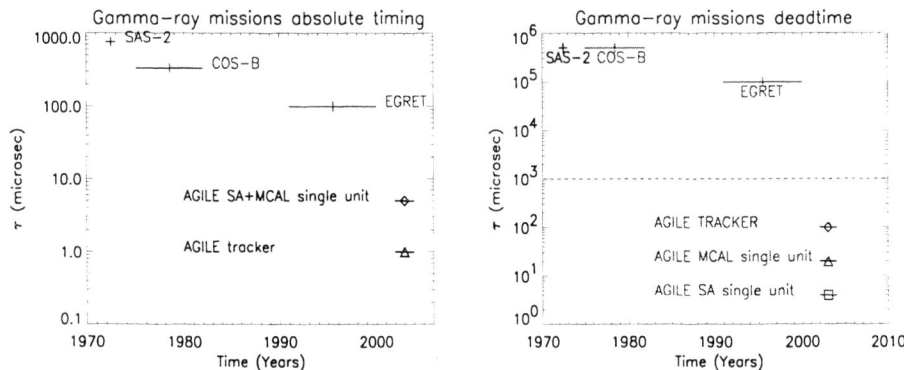

FIGURE 7. *Left Panel:* Absolute time-tagging uncertainty (τ) of AGILE and previous gamma-ray detectors. *Right Panel* Instumental deadtimes (τ) for the AGILE detectors and previous gamma-ray instruments.

SCIENTIFIC OBJECTIVES

We summarize here the main AGILE's scientific objectives.

• **Active Galactic Nuclei**. Simultaneous monitoring of a large number of AGNs per pointing will be possible. Several outstanding issues concerning the mechanism of AGN gamma-ray production and activity can be addressed by AGILE including: (1) the study of transient vs. low-level gamma-ray emission and duty-cycles; (2) the relationship between the gamma-ray variability and the radio-optical-X-ray-TeV emission; (3) the correlation between relativistic radio plasmoid ejections and gamma-ray flares; (4) hard

X-ray/gamma-ray correlations. A program for joint AGILE and ground-based monitoring observations is being planned. On the average, AGILE will achieve deep exposures of AGNs and substantially improve our knowledge on the low-level emission as well as detecting flares. We conservatively estimate that for a 3-year program AGILE will detect a number of AGNs ~ 3 times larger than that of EGRET (Fig. 8). Super-AGILE will monitor, for the first time, simultaneous AGN emission in the gamma-ray and hard X-ray ranges.

• **Gamma-Ray Bursts**. About ten GRBs were detected by the EGRET spark chamber during ~ 7 years of operations [19] (see also Fig. 8). This number was limited by the EGRET FOV and sensitivity and, from what we know today, not by the GRB emission mechanism normally producing gamma-rays above 100 MeV (Ref. [21]). The GRID detection rate of GRBs is expected to be a factor of ~ 5 larger than that of EGRET, i.e., ≥ 5–10 events/year). The small GRID deadtime ($\sim 10^3$ times smaller than that of EGRET) allows a better study of the initial phase of GRB pulses (for which EGRET response was in many cases inadequate). The remarkable discovery of 'delayed' gamma-ray emission up to ~ 20 GeV from GRB 940217 [14] is of great importance to model prompt and afterglow acceleration processes. AGILE is expected to be highly efficient in detecting photons above 10 GeV because of limited backsplashing. Super-AGILE will be able to locate GRBs within a few arcminutes, and will systematically study the interplay between hard X-ray and gamma-ray emissions. Special emphasis is given to fast timing allowing the detection of sub-millisecond GRB pulses independently detectable by the Si-Tracker, MCAL and Super-AGILE.

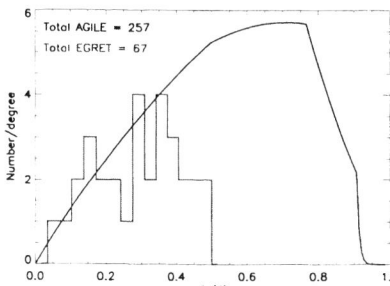 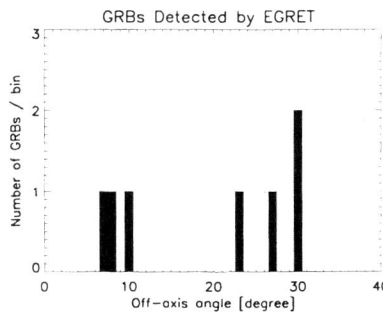

FIGURE 8. *Left Panel:* The histogram shows the distribution of $\sin(\alpha)$ (with α the off-axis angle) for the AGNs detected by EGRET and the solid curve gives the number of AGNs detectable by AGILE during a 3-year Mission lifetime [7]. *Right Panel*: Off-axis angle distribution for the EGRET detections of gamma-ray bursts.

• **Diffuse Galactic emission**. The AGILE good angular resolution and large average exposure will further improve our knowledge of cosmic ray origin, propagation, interaction and emission processes. We also note that a joint study of Galactic gamma-ray emission from MeV to TeV energies is possible by special programs involving AGILE and new-generation TeV observatories of improved angular resolution.

• **Gamma-ray pulsars**. AGILE will contribute to the study of gamma-ray pulsars in several ways: (1) searching for pulsed gamma-ray emission from the ~ 30 new young pulsars recently discovered in the Galactic plane (Ref. [10]); (2) improving photon statistics for gamma-ray period searches; (3) detecting possible secular fluctuations of the gamma-ray emission from neutron star magnetospheres; (4) studying unpulsed gamma-ray emission from plerions in supernova remnants and searching for time variability of pulsar wind/nebula interactions, e.g., as in the Crab nebula.

- **Search for non-blazar gamma-ray variable sources in the Galactic plane**, a new class of unidentified gamma-ray sources such as the mysterious GRO J1838-04 [22] and the variable 2CG 135+1.
- **Galactic sources, new transients**. A large number of gamma-ray sources near the Galactic plane are unidentified and can be monitored on timescales of months/years. Also Galactic X-ray sources (such as Cyg X-1, Cyg X-3, GRS 1915+10, GRO J1655-40 and others) can produce detectable gamma-ray emission for favorable source states and geometries, and a TOO program is planned to follow-up new discoveries of *microquasars*.
- **Fundamental Physics: Quantum Gravity**. AGILE detectors are suited for Quantum Gravity studies [26]. The existence of sub-millisecond GRB pulses lasting hundreds of microseconds [2] opens the way to study QG delay propagation effects with the AGILE detectors. If these ultra-short GRB pulses originate at cosmological distances, sensitivity to the Planck's mass can be reached [26].

REFERENCES

1. Auricchio N. et al., *these Proceedings* (2001)
2. Bhat C.L., et al., *Nature*, 359, 217 (1992)
3. Barbiellini G. et al., *Proceedings of the 5th Compton Symposium*, AIP Conf. Proceedings, ed. M. McConnell, 2001, Vol. 510, p. 750
4. Barbiellini G. et al., *these Proceedings* (2001a)
5. Barbiellini G. et al., *these Proceedings* (2001b)
6. Barbiellini G. et al., *NIM*, submitted (2001)
7. Chen A. et al., in preparation (2001)
8. Cocco V., Longo F. and Tavani M., *these Proceedings* (2001)
9. Costa E. et al., *NIM*, in preparation (2001)
10. D'Amico N., these Proceedings (2001)
11. Di Cocco G. et al., *NIM*, in preparation (2001)
12. Fedel G., Laurea Dissertation, University of Trieste (2000)
13. Hartman R.C. et al., *ApJS*, 123, 79 (1999)
14. Hurley K. et al., *Nature*, 372, 652 (1994)
15. Lapshov I. et al., *these Proceedings* (2001)
16. Pellizzoni A. et al., *these Proceedings* (2001)
17. Perotti F. et al., *NIM*, in preparation (2001)
18. Pittori C. et al., *these Proceedings* (2001)
19. Schneid E.J. et al.,in AIP Conf. Proc. no. 384, p.253 (1996)
20. Schoenfelder V. et al., *ApJS*, 86, 657 (1993)
21. Tavani M., *Phys. Rev. Letters*, 76, 3478 (1996)
22. Tavani M., et al., *ApJ*, 479, L109 (1997)
23. Tavani M. et al., *Proceedings of the 5th Compton Symposium*, AIP Conf. Proceedings, ed. M. McConnell, 2001, Vol. 510, p. 746
24. Tavani M. et al., *these Proceedings* (2001)
25. Tavani M. et al., in preparation (2001)
26. Tavani, M., in preparation (2001).
27. Thompson D.J. et al., *ApJS*, 86, 629 (1993)
28. Vercellone S. et al., *these Proceedings* (2001)

Gamma-Ray Imaging by Silicon Detectors in Space: Presentation of the AGILE Reconstruction Method and Kalman Filter Algorithms

Carlotta Pittori[*], Andrea Giuliani[†], Sandro Mereghetti[†] and Marco Tavani[†]

[*]*Dip. di Fisica, Univ. di Roma "Tor Vergata" and INFN, Sez. di Roma II, I-00133 Roma, Italy*
[†]*IFC-CNR, V. Bassini 15, 20133 Milano - Italy*

Abstract. We present the AGILE REconstruction Method (AREM) and the track finding optimization by Kalman filter algorithms. AREM is a method of γ-ray direction reconstruction to be applied to high-resolution Silicon Tracker detectors in space. It can be used in a "fast mode", independently on Kalman filters techniques, or in an "optimized mode", including Kalman filter algorithms for track identification. AREM correctly addresses three points of the analysis which become relevant for off-axis incidence angles: 1) intrinsic ambiguity in the identification of the 3-D e^+/e^- tracks and conversion plane; 2) proper identification of the 3-D reconstructed direction; 3) careful choice of an energy weighting scheme for the 3-D tracks. We present the preliminary results of the angular resolution obtained by analyzing simulated γ-rays in the AGILE detector. The excellent spatial resolution obtained by the AGILE Silicon Tracker allows to improve the angular resolution by a factor ~ 2 at energies $\gtrsim 400$ MeV with respect to previous spark chamber detectors (e.g. EGRET).

INTRODUCTION

AGILE (Astro-rivelatore Gamma a Immagini LEggero) is a Small Scientific Mission of ASI (Agenzia Spaziale Italiana) with a γ-ray imaging system based on state-of-the-art Silicon strip technology [1]. Thanks to the fast readout electronics and to the segmented anticoincidence system, AGILE will have, among other features, an unprecedently large field of view, ~ 3 sr (larger than previous γ-ray experiments by a factor ~ 5) and a very good intrinsic spatial resolution of the Si-Tracker (with a distance of 1.6 cm between contiguous planes). CERN testbeams show that, by using the analog information on the charge distribution released in Si-microstrips of pitch 121 μm, one can achieve a spatial resolution of order of ~ 40 μm [2]. For comparison, we recall here that the EGRET [3] spark chamber pitch was equal to 820 μm. The AGILE goal is to obtain the best sensitivity ever reached for off-axis events (up to $\sim 60°$), and an on-axis sensitivity comparable to that of EGRET despite the smaller dimensions and effective area. Therefore, the optimization of the angular resolution algorithms becomes a crucial point to fulfil the mission scientific objectives.

THE AREM METHOD

The γ-ray detection and direction reconstruction are based on the physical process of pair production, and are obtained from the identification and detailed analysis of the electron/positron tracks originating from a common vertex. Crucial to this task is a proper account of the effects of multiple Coulomb scattering and energy distribution between the e^+/e^- particles. The current customary simplification of analyzing separately the two tracks projections in the ZX and ZY Tracker views ("2-D projection method") induces two kinds of systematic error in the photon direction reconstruction:

A) the intrinsic ambiguity in the proper identification of the two 3-D tracks;

B) the problem of the identification of the true 3-D direction reconstruction.

Finally, we emphasize the importance of the:

C) choice of track weighting scheme.

In general, the photon energy is not evenly divided between the two particles. Since the direction of the most energetic particle is closer to that of the incident photon, an "energy-weighted" reconstructed direction should be computed[1]. As for the point A), we note that when the e^+/e^- pair hits simultaneously the active Tracker layers, the signal will correspond to two projected track points in each ZX and ZY view, but it could correspond to two possible couples of points in space, as shown in Fig. 1-A. This

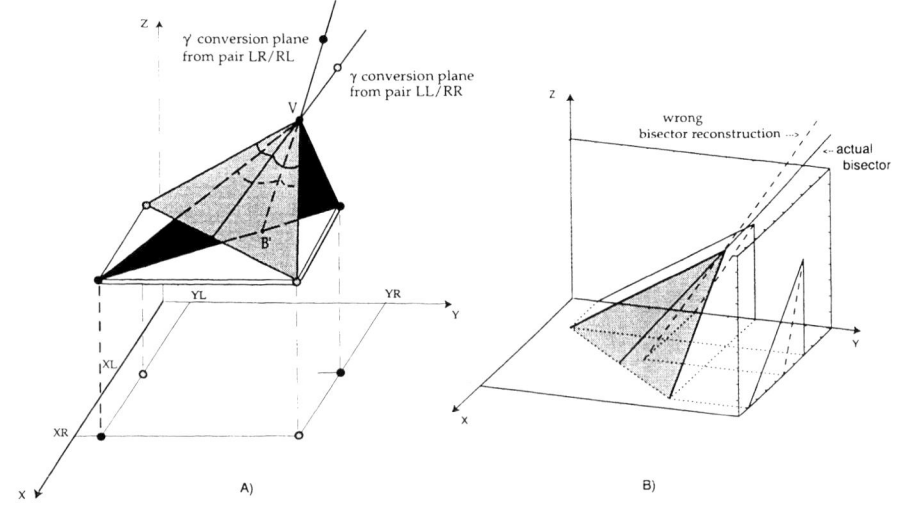

FIGURE 1. A) The "conversion plane problem": a correct event reconstruction in 3-D implies solving the coordinate track ambiguity and making the right choice for one out of two possible conversion planes. B) The "projection problem": the projections of the bisector on the ZX and ZY planes are different from the bisectors of the projections of the tracks.

---

[1] For simplicity, in the following we illustrate points A) and B) in case of an even energy sharing, i.e., when the photon direction coincides with the bisector.

gives rise to an ambiguity in the conversion plane identification. In terms of projected views the "conversion plane problem" can be phrased as: "Does the track to the left in the ZX view correspond to the left track in the ZY view - pair LL/RR? Or does it correspond to the track to the right - pair LR/RL?"

The "projection problem", point B), stems from the fact that in previous γ-ray experiments, after the identification of the two projected tracks in each view, the next step was to take their (eventually weighted) bisectors and compose them to obtain the reconstructed γ-ray direction. As illustrated in Fig. 1-B, this procedure is not correct since the true 3-D bisector (solid line in the shaded conversion plane) is different from the one obtained from the two bisecting lines in each projected view (dashed line from dashed projections). As shown in ref.[4], this systematic effect increases for increasing off-axis angles and large opening angles, up to values of $\sim 0.5°$. In the case of EGRET data[2], this effect is hidden by the relatively low spark chamber intrinsic resolution, but it would have a significant impact for AGILE. Furthermore, with high resolution Si-detectors it is possible to estimate the e^+/e^- energies from a few MeV to the GeV scale, by measuring deviations due to multiple scattering effects. This fact allows to properly define the weight of each track for the direction reconstruction (point C).

As described in detail in ref.[5], AREM is a 3-D reconstruction method, which takes into account these three points of the analysis. It provides a general baseline, to be optimized for each particular γ-ray instrument, for the photon direction reconstruction algorithms. Several approaches are under study: *(i)* first n-Planes Resolution, using only information from the first hit planes (2PR, 3PR, ...), *(ii)* algorithms based on the Kalman filter [6] for an optimal use of the information from all hit planes. In the following, we present some preliminary results of our analysis.

PRELIMINARY RESULTS

The Monte Carlo simulations of the AGILE-GRID imaging performance were done using the GEANT 3.21 code [7]. In Fig. 2 and in Fig. 3 we show, as an example, the 3-D Point Spread Function (PSF) obtained by using only information from the first 3 hit Tracker planes (3PR) for near-on-axis events at $E_\gamma = 1$ GeV and $E_\gamma = 200$ MeV. In Fig. 4 we show the 3-D PSF distribution profiles obtained from the AGILE Kalman filters algorithms (AKF) for several angles and energy values. The 3PR provides a satisfactory PSF for near on-axis events, compatible with AKF, even though with a lower reconstruction efficiency (85% vs. 97% at 1 GeV ; 76% vs. 94 % at 200 MeV). The AKF provides a good event reconstruction with very high efficiency (above 90% for $E_\gamma > 200$ MeV) for a variety of incidence angles. Finally, in Fig. 5 we compare the preliminary AGILE angular resolution, between 0° and 50° off-axis, with that of EGRET on-axis. The figure shows the 3-D 68% containment radius as a function of energy. The AGILE 3-D PSF is better than that of EGRET by a factor of ~ 2 above 400 MeV. We expect to further improve this performance, especially at low energies, by a more accurate study

---

[2] We warmly thank the EGRET team, in particular D.L. Bertsch and D. Thompson, for many discussions and for allowing us to perform a test of our reconstruction algorithms on EGRET calibration data.

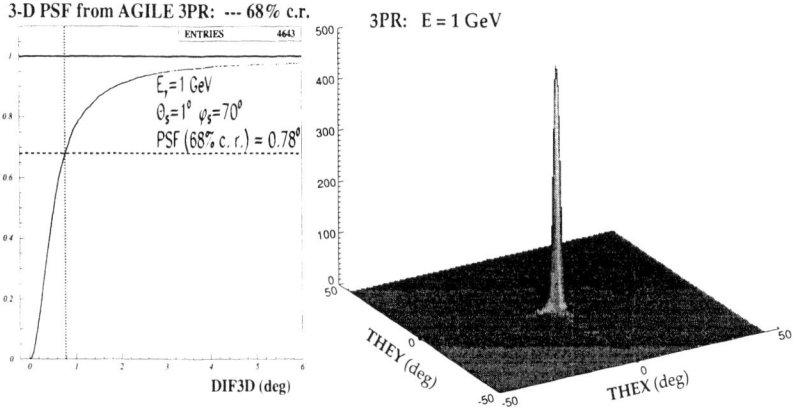

FIGURE 2. AGILE-GRID on-axis 3-D PSF from the 3PR reconstruction for 1 GeV photons. The left panel curve represents the integral distribution of the difference between true and reconstructed direction of each photon.

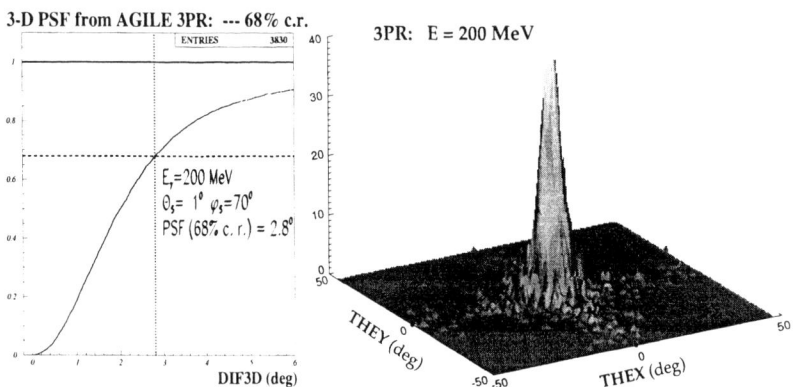

FIGURE 3. AGILE-GRID on-axis 3-D PSF from the 3PR reconstruction for 200 MeV photons.

of the charge deposition in the Si-microstrips, and by an optimization of the energy determination and weighting scheme based on extensive Monte Carlo simulations.

REFERENCES

1. Tavani, M. et al., ed.by M. McConnell, AIP Conf. Proc. 510, 2001, p. 746; and these Proceedings.

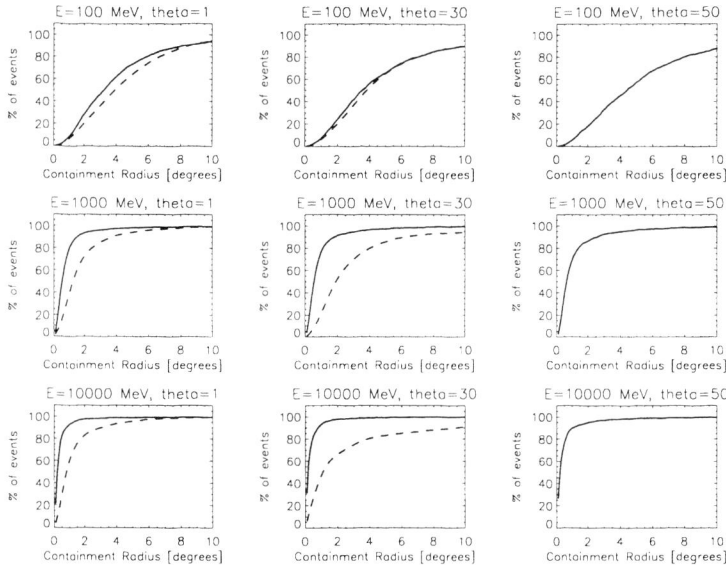

FIGURE 4. 3-D integral PSF profiles obtained with the AGILE-GRID Kalman filters algorithms (solid curve) compared to the corresponding EGRET values (dashed curve) when available (public data).

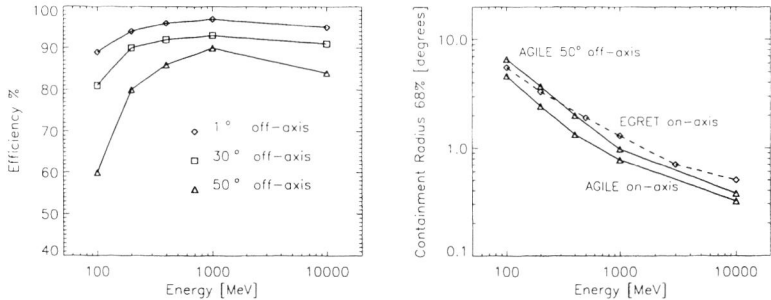

FIGURE 5. Left panel: Reconstruction efficiency of the AGILE Kalman filters algorithms for different off-axis directions. Right panel: Preliminary results for the AGILE 3-D containment radius (68%).

2. Barbiellini, G. et al., these Proceedings.
3. Thompson, D.J. et al., ApJSS, **86**, 629 (1993).
4. Giuliani, A., "Studio e Ottimizzazione della Risoluzione Angolare del Telescopio Spaziale per Astronomia Gamma AGILE", Laurea Dissertation, Università degli Studi di Pavia (2001).
5. Pittori, C., and Tavani, M., Gamma-Ray Imaging by Silicon Detectors in Space, Rome 2 preprint ROM-2F/2001/12, Rome, I-00133 (2001).
6. Frühwirth, R., NIM A, **262**, 444 (1987).
7. Cocco, V., Longo, F., and Tavani, M., AGILE Internal Tech. Note, AGILE-SIM-TN-001, Issue n.2 (2000), and preprint in preparation.

GEANT Simulation of the AGILE Gamma-Ray Imaging Detector

Veronica Cocco[*], Francesco Longo[†] and Marco Tavani[**]

[*]*Dip. di Fisica, Univ. di Roma "Tor Vergata" and INFN, sezione di Roma II, Italy*
[†]*Dip. di Fisica, Univ. di Ferrara and INFN, sezione di Ferrara, Italy*
[**]*Istituto Fisica Cosmica, CNR, Milano, Italy*

Abstract. We present results obtained with a GEANT-based Simulator of the Gamma-Ray Imaging Detector (GRID) developed for the AGILE space astrophysics mission. We describe the AGILE instrument geometry and the model assumed for the charged particle and the albedo-photon backgrounds. Using this simulator, we optimized the event trigger processing. In this paper, we present the main results on different levels of data processing and obtain the background rejection efficiency and the GRID effective area for photon detection in the energy range $\sim$ 30 MeV - 50 GeV.

INTRODUCTION

AGILE is an ASI Small Scientific Mission dedicated to high energy astrophysics [1]. The AGILE instrument is designed to detect and image photons in the 30 MeV - 50 GeV and 10 - 40 keV energy bands, with excellent spatial resolution and timing capability and an unprecedently large field of view covering $\sim$ 1/5 of the entire sky at energies above 30 MeV. Primary scientific goals include the study of AGNs, gamma-ray bursts, Galactic sources, unidentified gamma-ray sources, diffuse Galactic gamma-ray emission, and high precision timing studies. The AGILE gamma-ray mission requires a low-background orbit to maximize its scientific output.

The optimization of the AGILE design was obtained through a Montecarlo study of the detector performance. Simulations were done using the GEANT 3.21 code [2] which traces all possible interactions of particles with the apparatus, and reliably takes into account the deposited energy in the instrument detectors.

In this paper, we describe the AGILE instrument and the particle/albedo-photon background models assumed for the optimization of the on-board data processing. We outline the adopted trigger strategies and present the main results about the on-board background rejection and photon-detection efficiency (a first step to obtain the effective area).

THE AGILE INSTRUMENT MODEL

The AGILE scientific instrument is made of three integrated detectors with broad-band detection and imaging capabilities [1, 3]. The AGILE Gamma-Ray Imaging Detector (GRID) consists of a Silicon-Tungsten Tracker, a Cesium Iodide Mini-Calorimeter, an

Anticoincidence system made of segmented plastic scintillators, fast readout electronics and processing units. The Super-AGILE detector will provide detection and imaging capabilities in the hard X-ray range. It consists of an additional plane of four Silicon square units positioned on the top of the GRID Tracker plus an ultra-light coded mask structure with a top absorbing mask at a distance of 14 cm from the Silicon detectors. The CsI Mini-Calorimeter will also detect and collect events independently from the GRID in case of impulsive transients.

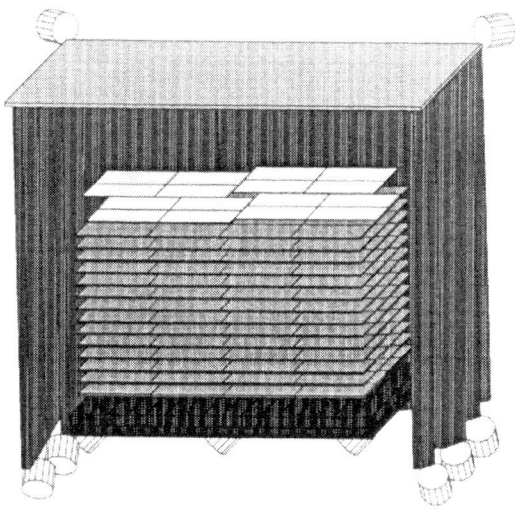

FIGURE 1. A simplified view of the AGILE instrument model. The Super-AGILE coded mask structure, the mechanical structure and the lateral electronics boards are not shown.

GRID. In the simulation code we modelled the GRID detector according to the AGILE design [3]. The Silicon Tracker is made of 14 detection planes with a distance between consecutive planes of 1.6 cm. Each plane is composed of two layers of 16 silicon tiles with a surface of 9.5×9.5 cm^2 each and with a thickness of 410 μm. The two layers consist of silicon microstrips detectors oriented in orthogonal directions with a readout pitch of 242 μm, but with the floating strip design. All planes (except the last two planes) contain a Tungsten layer of 245 μm (0.07 X_0). We modeled also the mechanical supports, the Aluminum honeycomb structure, the front-end electronics chips and other tray components. In our simulations, we make a crucial use of a parametrization of the capacitive coupling among contiguous Si-microstrips. Our parametrization reproduce the experimental data obtained at a CERN test beam in May 2000 [4, 5]. The Mini-Calorimeter is modeled by two planes, each containing 16 CsI bars oriented orthogonally. The bars are 1.4 cm thick and have a width of 2.4 cm. The Anticoincidence (AC) system is made of a top panel of plastic scintillator ($\sim 54 \times 54 \times 0.5$ cm^3) and 3 panels for each lateral side of the AGILE Tracker ($\sim 18.1 \times 44.4 \times 0.6$ cm^3). We also included a simplified description of the photomultipliers, GRID readout electronics, and mechanical structure.

Super-AGILE. The Super-AGILE detector layer is made of 16 Silicon detectors that are similar to those used for the Tracker. A gold mask ($\sim 90\,\mu$m thick) is placed at a distance of 14 cm from the active detector plane, and is supported by a light structure of Au-coated Carbon fibers acting as collimator for 10–40 keV X-rays. We studied in detail the background induced on the GRID by Super-AGILE.

Fig. 1 shows a simplified view of the AGILE instrument model used in our simulations.

BACKGROUND ASSUMPTIONS

Charged particle background

A quasi-equatorial orbit is preferred for the AGILE mission and will provide a relatively low-background environment. Taking into account data from SAS-2 and Beppo-SAX missions, we expect an average rate of charged particle background above ~ 1 MeV of ~ 0.3 particles cm^{-2} s^{-1} for a quasi-equatorial orbit near 550 km. The charged particle background for this orbit is known to be relatively stable, with an increase by a factor 10-100 near the South Atlantic Anomaly. The charged particle energy spectra assumed in our simulations are shown in Fig. 2. They are based on data from the 1998 AMS Shuttle flight [6, 7], and from the MARYA experiment on board of the MIR space station [8]. These data were selected for events detected near the geomagnetic equator, and their low-energy extrapolations are consistent with the total rates detected by SAS-2 and Beppo-SAX. We used the correct angle distributions for different particle components: an isotropic distribution for electrons, positrons and trapped protons, and an upper-hemispheric distribution for primary protons (for a zenithal AGILE pointing).

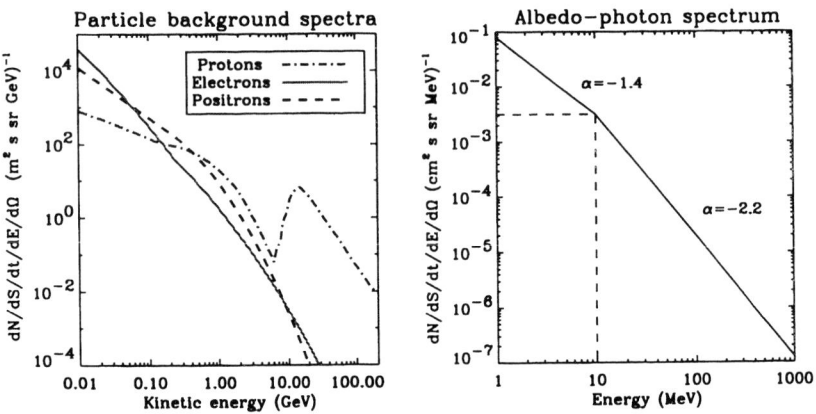

FIGURE 2. *Left Panel*: Charged particle background energetic spectra (from [6, 7, 8]) assumed in our simulations. *Right Panel*: Average albedo-photon energy spectrum (from [9, 10]).

Albedo-photon background

The interaction of the charged cosmic-rays with the upper atmosphere induces a relatively strong gamma-ray background peaking at the Earth horizon. This effect involves a localized increase of the gamma-ray emission that we properly took into account on the basis of SAS-2 [9] and other balloon data [10]. Fig. 2 shows the average flux of albedo photons over the solid angle of the subtended Earth surface at the height of 550 km.

TRIGGER STRATEGIES AND BACKGROUND REJECTION

We studied different GRID trigger configurations, and optimized their performance. The baseline GRID trigger logic consists of two different levels. A (hardware) Level-1 trigger logic uses the information from the Silicon detectors and AC panels and considers also a simplified view of the event topology obtained by the front-end chips. Level-1 trigger reduces the charged-particle background from a rate of ~ 2000 Hz to a rate of ~ 60 Hz. A (software) Level-2 on-board data processing makes a crucial use of the analog (charge) information in the Si-microstrips for a refined view of the event topology at the "cluster" level. Level-2 on-board processing also selects events based on a simplified photon direction reconstruction (necessary to reject Earth albedo photons). After the on-board Level-2 processing, we can reduce the total (charged particle and albedo-photon) background rate to $\sim 20-30$ Hz.

EFFECTIVE AREA

Using the trigger logic outlined in the previous section, we studied the AGILE-GRID efficiency to detect gamma-rays at different incidence angles and energies. The effective area is, by definition: $A_{eff} = \varepsilon A_{\perp}$, where $A_{\perp}$ is the detector "geometrical area" (equivalent area perpendicular to the incident flux direction) and ε is the detector efficiency, given by the product $\varepsilon = \varepsilon_i \cdot \varepsilon_t \cdot \varepsilon_r$, with ε_i the photon interaction probability, ε_t the trigger efficiency, and ε_r the photon event reconstruction efficiency.

The GRID effective area after Level-2 processing (without the photon-event reconstruction cut) is shown in Fig. 3. The GRID is characterized by an excellent performance off-axis, and by an effective area smaller by a factor of 2 than that of EGRET for on-axis events.

CONCLUSIONS

Our simulations show that the mechanical and electronic design of the GRID is appropriate for an efficient background rejection and optimized scientific performance. Despite its small volume and mass, the AGILE-GRID will reach a very good sensitivity and wide angle event acceptance. For each pointing, the GRID field of view will be unprecedently large, $\sim 1/5$ of the entire sky, for observations above 30 MeV.

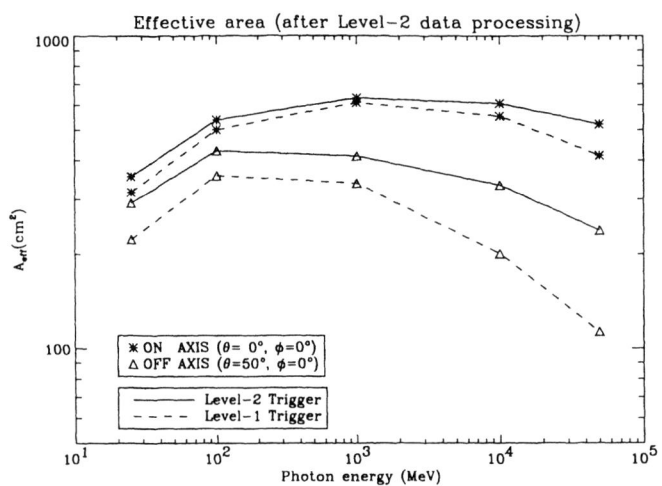

FIGURE 3. AGILE effective area after on-board Level-2 data processing (without photon event reconstruction efficiency to be obtained after an off-line data analysis. Simulations results from ref. [11, 12]).

We acknowledge the joint work and contributions to the material presented in this paper by G. Fedel, P. Lipari, and A. Pellizzoni.

REFERENCES

1. Tavani M. et al., these Proceedings
2. Giani S. et al., CERN Long Writeup W5013 (1994)
3. Barbiellini G. et al., these Proceedings
4. Fedel G. et al., *X-Ray and Gamma-Ray Instrumentation for Astronomy XI*, Flanagan K.A. and Siegmund O.H.W. eds., SPIE Conference Proceedings 4140, 274, (2000)
5. Barbiellini G. et al., NIM-A submitted, INFN/TC-01/006 (2001)
6. Alcaraz J. et al., Physics Letters B 472, 215 (2000)
7. Alcaraz J. et al., Physics Letters B 484,10 (2000)
8. Koldashov S.W. et al., 24th ICRC 4, 993, (1995)
9. Thompson D.J., Simpson G.A. and Özel M.E., Journal of Geophysical Research 86, 1265 (1981)
10. Costa E., Massaro E., Salvati M. and Appolloni A., Astrophysics and Space Science 100, 165 (1984)
11. Longo F., Cocco V. and Tavani M. (2001), in preparation
12. Cocco V., Longo F. and Tavani M. (2001), in preparation

Test Campaign of the Mini-Calorimeter for the AGILE Satellite

Natalia Auricchio[a], Enrico Celesti[a], Guido Di Cocco[a], Marcello Galli[b], Fulvio Gianotti[a], Claudio Labanti[a], Alessandro Mauri[b], Marco Malaspina[a], Elio Rossi[a], John B. Stephen[a], Alessandro Traci[a], Massimo Trifoglio[a]

[a]*Ist. TeSRE-CNR Bologna-Italy*
[b]*ENEA "E. Clementel" Bologna-Italy*

Abstract. The AGILE Mini-Calorimeter (MCAL, with total on-axis radiation length 1.5 X_0) comprises 2 orthogonal planes each consisting of 16 CsI(Tl) bars. Its primary purpose is the energy determination of gamma-rays detected by the Gamma-Ray Imaging Detector (GRID) Tracker (energy range 30 MeV - 50 GeV). In addition, the MCAL is capable of independently triggering and detecting gamma-ray bursts and other impulsive events in the energy range 0.25-250 MeV. A MCAL prototype, comprising 8 CsI(Tl) detector elements, was tested both with laboratory sources and with charged particles (p=1 GeV/c) during an AGILE Beam Test carried out in November 2000 at the CERN T11 beamline (East Hall, CERN PS). The test setup included a prototype of the electronic chain. A prototype of the digital data acquisition chain, which will be the baseline of the payload Electronic Ground Support Equipment, was also built and tested. We present the preliminary results of this test campaign dedicated to characterizing the detector unit and electronics.

1. MINICALORIMETER

The MCAL is an essential part of the AGILE scientific instrument [1]. It consists of two orthogonal planes each made of 16 bars of CsI(Tl) wrapped with white diffusive paper, having a cross section of 1.5x2.3 cm^2 and a length of 40 cm. The signal from each scintillator bar is collected by two photo-diodes (PDs) placed at each ends [2,3]. The MCAL operates in two modes:

1) **GRID mode:** acting as AGILE's calorimeter. The FEE processes the signals obtaining:
   - trigger pulses generated when a high energy deposit (greater than 10 MeV) is detected in the whole Calorimeter.
   - the two signals of each bar are Analog-to-Digital (AD) converted after receiving a command from the AGILE Data Handling system (DH).

2) **Burst Monitor mode:** to detect, independently of the GRID, gamma-ray transient events. In this mode, both ends of each bar (total 64) will act as individual gamma-ray detectors. The calorimeter FEE performs the AD conversion of signals from the triggered detectors and sends data to the AGILE Data Handling. The events detected

by the Burst electronics will be used to determine the address of counters in the AGILE DH to be incremented; these counters will be used in the AGILE burst search process.

2. MCAL PROTOTYPE

A MCAL prototype was tested with radioactive sources at the TeSRE Laboratory and at CERN with particles of 1 GeV/c. The prototype, shown in Fig. 1, is composed of 8 CsI(Tl) bars (6 SCIONIX and 2 CRISMATEC), having dimensions of 400x26x15 mm^3 and covered with the light diffusive paper (Millipore). On each bar end, 2 parallel PDs (Hamamatsu S3590-08) are attached by means of a transparent glue, and provide their signals to the LABEN Charge Sensitive Preamplifiers, having an electronic noise of ~ 900 e$^-_{rms}$. Two different Test Equipment (TE) were available to collect the output from the Preamplifiers: a prototype of the TE to be provided by LABEN for the characterization of the Proto Flight Model (PFM), and a TE provided by TeSRE.

The LABEN TE is composed by an electronic module including 16 channels for analogic signal shaping (shaping time 3 µsec) and stretching, representative of the scientific data acquisition chain in **Grid Mode**. They operate in Sample and Hold mode and convert the signal upon receiving an external trigger. The stretched signals are multiplexed towards an ADC DATEL ADS-942A low power, 14 bit, 2 MHz, input 0-10V. This TE provides also a channel which is representative of the Peak and Hold (auto-trigger) chain of the **Burst Mode**. The TE includes a WinNT PC Host Computer, which acquires the converted signals through a 32 bit parallel interface (NI PCI DIO32HS).

The TeSRE TE electronics is based on the refurbishment of an existing system which provides 16 auto-trigger channels. In this case, a QNX PC Host Computer acquires the analog signals through a DAQ I/F card (NI PCI-MIO16E-1), including an ADC and connected to a BNC rack-mount analogue breakout accessory (NI BNC-2090).

FIGURE 1. The MiniCalorimeter prototype.

As shown in Fig. 2, the two TEs are operated in parallel by the MCAL Science Console, a Linux PC which acquires, archive, and provides quick look access to the event data coming from the two data streams. A similar scheme applied at CERN.

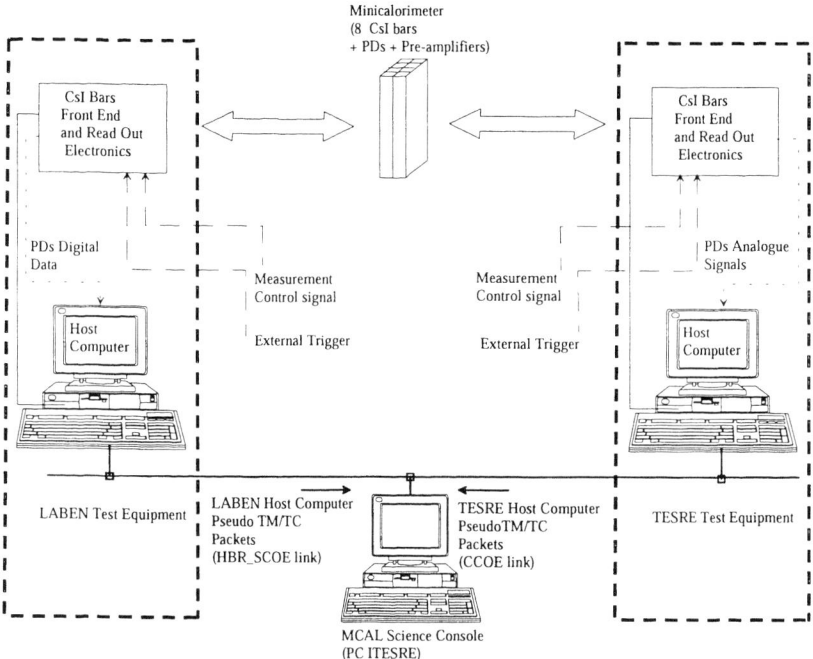

FIGURE 2. Experimental setup during CsI bars performances test.

2.1 THE MCAL SCIENCE CONSOLE QUICK-LOOK SOFTWARE

The MCAL Science Console provides a Quick Look Analysis (QLA) of the measurements and archives the data in both raw and FITS formats. Fig. 3 shows the beta-prototype graphical user interface (GUI) developed for the second run at CERN. In this run there were eight bars in the calorimeter arranged as two planes of 4 elements, one above the other. The GUI was designed to give a representation of the event data from either one of these two planes, but in subsequent versions the operator will be able to switch between any subset of the final arrangement of 32 bars as required.

If we take the outputs of the two photo-diodes from any one bar as OA and OB then the energy and position of the interaction producing these outputs can be reconstructed using:

$$P = \log \frac{OA}{OB}, \quad E = \sqrt{OA * OB} \tag{1}$$

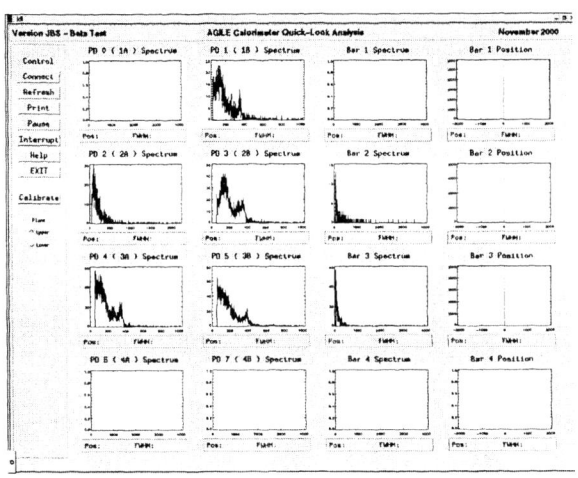

FIGURE 3. Graphical User Interface of the Quick Look Analysis.

In order to allow for variations in light roll-off between the bars, and to allow a calibration to be made, the operator can input 8 'calibration' parameters C1-8, comprising a 'gain' and 'offset' terms, so that what is actually displayed is:

$$OA' = C1 * OA + C2 , \quad OB' = C3 * OB + C4 \qquad (2)$$

$$P = C5 * \log\left(\frac{OA'}{OB'}\right) + C6 , \quad E = C7 * \sqrt{OA'*OB'} + C8 \qquad (3)$$

Finally the operator is also able to use mouse input on each of the displays to 'zoom' the display and also to compute the peak position and the full width at half maximum.

2.2 BARS' PERFORMANCES

The detectors were illuminated with a ^{22}Na collimated source at various positions from each couple of PDs in order to determine the light output as a function of interaction position along the bar and to calculate the light attenuation coefficient, reported in table 1 and shown in Fig. 4:

TABLE 1. Bars' Performances.

Bars Code Number	Light Output @ 1 cm by PDs (e⁻/keV)	Attenuation Coefficient (cm^{-1})	Bars Code Number	Light Output @ 1 cm by PDs (e⁻/keV)	Attenuation Coefficient (cm^{-1})
Bar 1a	12.49±0.05	0.041	Bar 1b	12.63±0.06	0.046
Bar 2a	12.45±0.04	0.023	Bar 2b	12.79±0.04	0.017
Bar 3a	12.30±0.05	0.026	Bar 3b	13.45±0.04	0.024
Bar 4a	16.31±0.05	0.034	Bar 4b	16.09±0.04	0.039
Bar 5a	13.79±0.04	0.040	Bar 5b	14.90±0.06	0.026
Bar 6a	17.36±0.07	0.042	Bar 6b	19.48±0.06	0.049
Bar 7a	15.18±0.06	0.032	Bar 7b	15.22±0.04	0.042
Bar 8a	15.87±0.05	0.036	Bar 8b	15.98±0.04	0.032

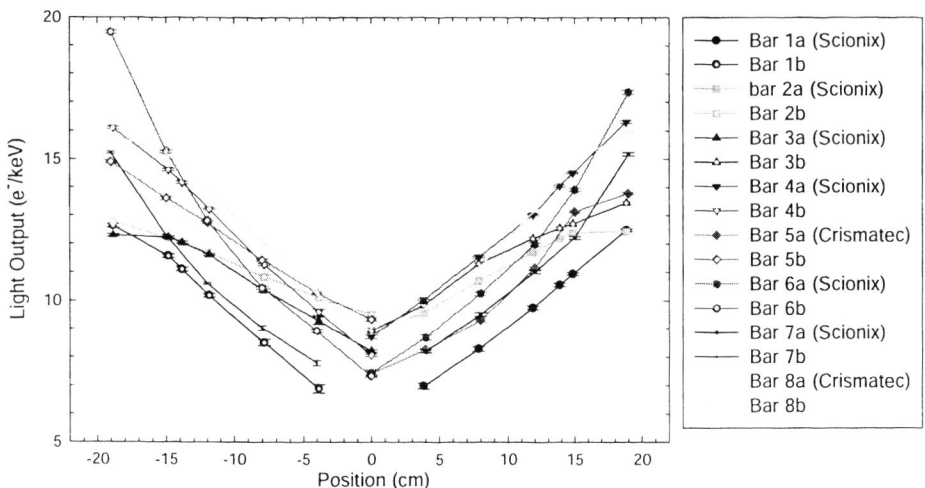

FIGURE 4. Light attenuation along the bars.

CONCLUSIONS

The light output at 1 cm from each couple of PDs is ~15 e$^-$/keV as expected, even if the bars show a high spread in the performance both in terms of light output and attenuation coefficient. This disuniformity is probably due to the insufficient accuracy of the bar preparation, such as polishing of the crystal surface, the PDs optical contact and bar wrapping. Clearly final production for the AGILE MCAL will require careful manufacturing and testing.

REFERENCES

1. Tavani, M., et al., these Proceedings, 2001.
2. Di Cocco, G., et al., *The Mini-Calorimeter for AGILE satellite*, Conf. Proceedings of the Italian Physical Society, 2000, Vol 68, pp. 227-230.
3. Rossi, E., et al., *Test campaign of the Mini-Calorimeter for the AGILE satellite*, SPIE Proceedings, 2000, Vol. 4140, pp. 486-492.

The next generation of high-energy gamma-ray detectors for satellites: the AGILE Silicon Tracker

Guido Barbiellini[*], Giuliano Bordignon[*], Giulio Fedel[*], Fernando Liello[*], Francesco Longo[†], Cristian Pontoni[**], Michela Prest[**] and Erik Vallazza[‡]

[*]*Department of Physics University of Trieste and INFN, sezione di Trieste*
[†]*Department of Physics University of Ferrara and INFN, sezione di Ferrara*
[**]*CIFS - Consorzio Interuniversitario di Fisica Spaziale and INFN, sezione di Trieste*
[‡]*INFN, sezione di Trieste*

Abstract.
AGILE (Light Imaging Detector for Gamma Astronomy) is a satellite for the detection of gamma-ray sources in the energy range 30 MeV - 50 GeV within a large field of view ($\sim 1/5$ of the sky) and it is planned to fly in the years 2003 - 2006, a period in which no other mission entirely dedicated to photon detection above 30 MeV is planned. AGILE is made of a Tungsten-Silicon Tracker, a CsI Minicalorimeter, an anticoincidence system and a X-ray detector sensitive in the 10 - 40 keV range. The Tracker consists of 14 planes, each one made of 2 layers of 16 single-sided, AC coupled, 410 μm thick, 9.5×9.5 cm^2 Silicon detectors with a readout pitch of 242 μm and a floating strip. The AGILE trigger is generated by the Silicon strips which are readout by the TAA1, a low noise, self triggering ASIC used in a very low power configuration (~ 400 μW/channel) with analog readout. The number of Tracker readout channels is 43000. We present a description of the Tracker and the performance of the detector (position resolution, cluster pulse height, readout and trigger logic) obtained during three testbeam periods at CERN.

INTRODUCTION

AGILE, the first ASI (Italian Space Agency) Small Scientific Mission, is dedicated to high energy astrophysics. The AGILE instrument is designed to detect and image photons in the 30 MeV - 50 GeV and 10 - 40 keV energy bands, with excellent spatial resolution and timing capability and an unprecedently large field of view covering $\sim 1/5$ of the entire sky at energies above 30 MeV. The AGILE scientific instrument is made of three integrated detectors. The AGILE Gamma-Ray Imaging Detector (GRID) consists of a Silicon-Tungsten Tracker, a Cesium Iodide Mini-Calorimeter[1], an Anti-coincidence system made of segmented plastic scintillators and fast readout electronics and processing units[2]. AGILE will have detection and imaging capabilities in the hard X-ray range provided by the Super-AGILE detector[3].

A detailed description of the Silicon Tracker is given in [6]. In the following, we describe briefly the detector and the prototype test results.

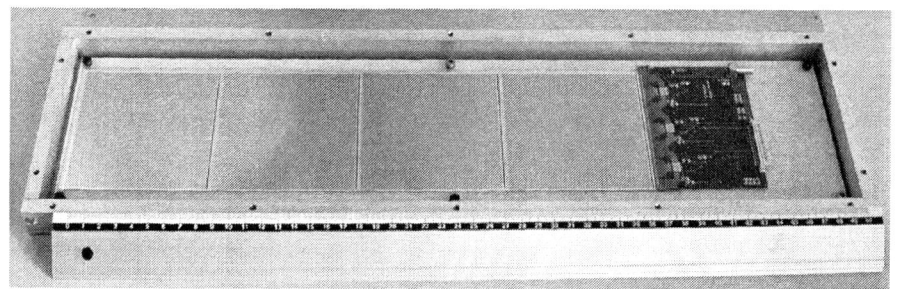

FIGURE 1. Prototype ladder tested during the May 2000 testbeam.

THE AGILE SILICON TRACKER

The AGILE Tracker [5, 6] is made of 14 detection planes, each consisting of

- 2 views of 16 single sided 9.5×9.5 cm^2 AC-coupled 410 μm thick Silicon microstrip tiles. The two views are positioned orthogonally one with respect to the other in order to obtain a x-y imaging system.
- for the first 12 planes, a Tungsten layer 245 μm thick, corresponding to 0.07 radiation lenghts.

The detector is used with a "floating strip" configuration, with a strip pitch of 121 μm, while the readout pitch is 242 μm. This configuration has been chosen in order to achieve an excellent spatial resolution while keeping under control the number of readout channels and hence the detector power consumptions.

The Silicon manufacturer is HAMAMATSU PK (Japan). Four Silicon detectors are connected together through wire bonding to obtain a "ladder" 38 cm long. Fig. 1 is a picture of the ladder built for the May 2000 testbeam at the CERN PS area.

The readout ASIC is the TAA1 (fig. 2), a 128 channel, low noise, low power, self triggering ASIC designed by IDE AS (Norway) and produced by AMS (Austria) with 0.8 μm N-well BiCMOS, double poly, double metal on epitaxial layer technology. To limit the power consumption of the Tracker, the ASIC is operated in a very low power configuration ($\sim$ 400 μW/channel).

The AGILE GRID trigger is mainly provided by the Silicon Tracker. The GRID trigger is divided into 3 levels, two hardware ones and a software one.

The Level 1 trigger is given by the coincidence of 3 out of 4 consecutive Silicon Tracker views. Veto signals from the top AC and a suitable combination of the lateral ACs are foreseen. The Level 1.5 philosophy is based on the use of the ratio between the number of triggered ASICs and the number of triggered views in order to reject charged particles, and on a coarse event topology at the ASIC level [6].

For the software level both the analog information given by the Tracker and the track reconstruction are essential.

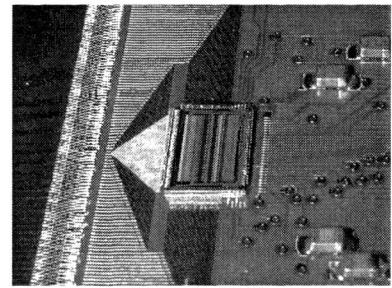

FIGURE 2. AGILE Silicon Tracker readout ASIC.

TRACKER TESTBEAMS

The final prototype of the Silicon detector has been tested during three testbeam periods in the year 2000 at the T11 beamline at the CERN PS[4].

May 2000

The purpose of the May testbeam (May 1-11, 2000) was to confirm the design of the detector in order to start the final production. An AGILE ladder has been tested. More than 2 millions of events have been collected to study the ladder behaviour in terms of cluster pulse height, signal to noise ratio and position resolution at different incidence angles of the beam with respect to the Silicon strips(see fig. 3). A dedicated set of runs has been acquired to study the trigger efficiency of the Silicon-frontend ASIC ensemble. The results of the testbeam and the comparison between data and simulation are presented in [6, 5].

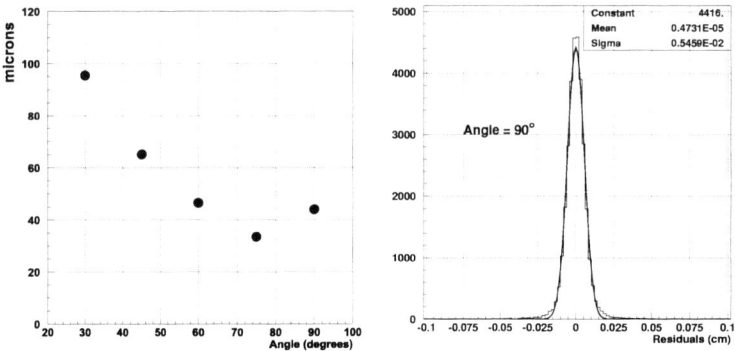

FIGURE 3. (Left Panel) Position resolution as a function of the incidence angle of the beam with respect to the strips direction. An angle of 90° corresponds to normal incidence on the Silicon detector (Right Panel) Position resolution at normal incidence (from [6]).

August 2000

The goal of the August testbeam (August 9-23) was to study a possible gamma-ray facility for the AGILE calibration (PHOTAG - photon tagged beam).
Gamma-rays in the energy range 60 MeV - 1 GeV have been produced from electron bremsstrahlung. The electron/positron pair from the photon conversion has been detected using four AGILE Silicon detectors with a lead converter layer of 0.05 radiation lenghts for each plane. Figure 4 shows an example of a gamma-ray detected by the AGILE detectors.

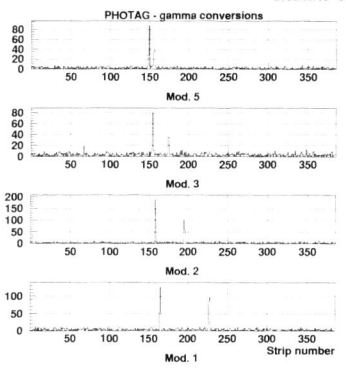

FIGURE 4. Conversion of a gamma-ray inside the prototype Silicon Tracker. The plot shows the two tracks reconstructed by the Silicon strips with their analog signal. The vertical scale is in ADC counts.

November 2000

During the November testbeam (November 22-29) the trigger and readout logic of the GRID have been studied in detail using a compact mini-Tracker with 4 detection planes (fig. 5) and with a prototype of the trigger electronics, with the following data acquisition flow chart:

(a) the trigger is generated by the coincidence of at least 3 out of 4 Silicon planes

(b) the trigger bits are latched and serialized to be transferred to the data acquisition board to compute the so called R-trigger (the ratio between the triggered ASICs and the triggered views) in order to reject single track charged particles

(c) the triggered ASICs addresses are decoded to operate the sparse readout, that is to read only the triggered ASICs in order to reduce the deadtime. Fig 6 shows two events: the event on the left is rejected due to the R-trigger, while the one on the right is accepted. In both cases, only the fired TA1s are read.

CONCLUSIONS

The Silicon Tracker is the heart of the AGILE instrument with its ~ 43000 Silicon channels and total Silicon detector area of 4 m^2. The Silicon detector described in this

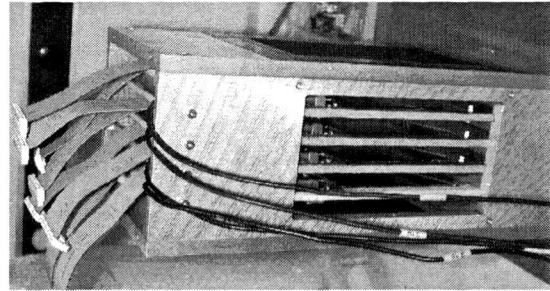

FIGURE 5. Compact mini-Tracker for the tests at the CERN T11 beamline. The distance between the planes is 1.6 cm, as in the final Tracker.

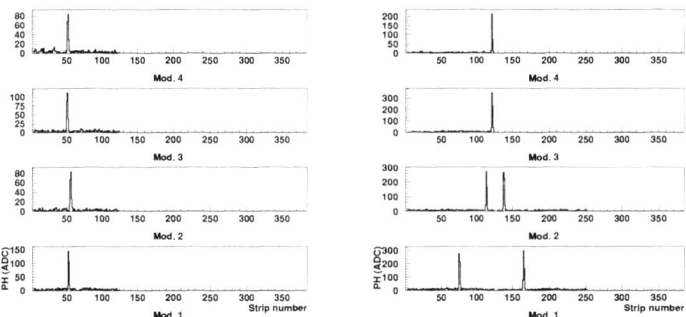

FIGURE 6. Rejected and accepted tracks.

paper is the largest ever developed. The several testbeam studies have demonstrated the validity of the analog readout and of the floating strip principle, the hardware based trigger logic, and the feasibility of a photon tagged beam at CERN. The August and November 2000 testbeam data are being analyzed and will be the subject of future publications.

REFERENCES

1. Auricchio N. et al. (2001), these Proceedings
2. Barbiellini G. et al. (2001a), these Proceedings
3. Lapshov I. et al. (2001), these Proceedings
4. http://psschedule.web.cern.ch/PSschedule/pindex.html
5. Fedel G. et al. (2000), The AGILE Silicon Tracker: an advanced gamma and X-ray detector for space, in *X-Ray and Gamma-Ray Instrumentation for Astronomy XI*, Flanagan K.A. and Siegmund O.H.W. editors, SPIE Conference Proceedings 4140, 274
6. Barbiellini G. et al. (2001), The AGILE Silicon Tracker: architectural design and prototype testbeam results, submitted to NIM, INFN/TC-01/006

AGILE Sky Exposure and Sensitivity Maps

A. Pellizzoni [a], A. Chen [b], A. Giuliani [a], S. Mereghetti [a], M. Tavani [a], S. Vercellone [a]

[a] *Istituto di Fisica Cosmica "G. Occhialini" (IFC/CNR), via Bassini 15, 20133, Milano, Italy*
[b] *Consorzio Interuniversitario Fisica Spaziale, Villa Gualino, 10133, Torino, Italy*

Abstract. The very large field of view (~1/5 of the sky) of the AGILE gamma-ray satellite allows us to monitor broad sky regions with a few pointings, increasing the chance of detecting variable sources and efficiently sampling their characteristics. We present the AGILE exposure and sensitivity maps for several examples of selected pointing programs taking into account Earth occultation effects and other satellite constraints. Considering instrument efficiencies, point spread function, particle background noise and a model for the diffuse gamma-ray emission, we estimate the global scientific performance of AGILE in observations simulating both single pointing and complete sky surveys (i.e. Galactic Plane scans and selected deep exposures). In this way we provide an estimate of the number of detectable sources giving some highlights bout the AGILE observation planning strategy.

INTRODUCTION

AGILE is a small scientific mission dedicated to observations of the gamma-ray sky in the energy range 50 MeV-10 GeV with simultaneous imaging also in hard X-rays (10-40 keV) [1].
The AGILE field of view (FOV) is very large (~3 steradians, ~6 times larger than the EGRET FOV) and flat (~70% of the on-axis sensitivity at ~40 degrees off-axis) allowing for (a) uniform and extensive exposures in only a few pointings, (b) very efficient discovery and monitoring of transients, (c) flexibility and compliance to spacecraft pointing constraints.
These capabilities can be quantified by calculating exposure and sensitivity maps for selected examples of observation programs, thus obtaining source detection levels and hints on possible observation planning strategies.
Exposure maps (in units of $cm^2 \cdot s$) are one of the input parameters for the likelihood data analysis envisaged for AGILE [2] and allow us to generate intensity maps starting from counts maps or (vice versa) to calculate expected counts from a given source flux. Sensitivity maps (in units of $ph/cm^2 \cdot s$) represent instead the minimum detectable fluxes from point sources for each direction of the sky and allow us to estimate the number of detectable point sources. We calculated exposure and sensitivity maps as a function of Galactic coordinates simulating either a single pointing or complete sky surveys (see examples below). We take into account Earth occultation effects and use instrument parameters (effective area, point spread function, background) obtained by

Monte Carlo simulations [3] [4] and gamma-ray diffuse emission maps from the EGRET public archive.

Earth occultation of the field of view strongly affects exposure and sensitivity. From any point of the AGILE nominal orbit (altitude ~550 km), the angular dimension of the Earth will be ~130 degrees, and source occultation ranges from 0% to ~40% of the orbit depending on the pointing direction. Note that there are two unocculted regions corresponding to directions perpendicular to the orbital plane (Figure 1).

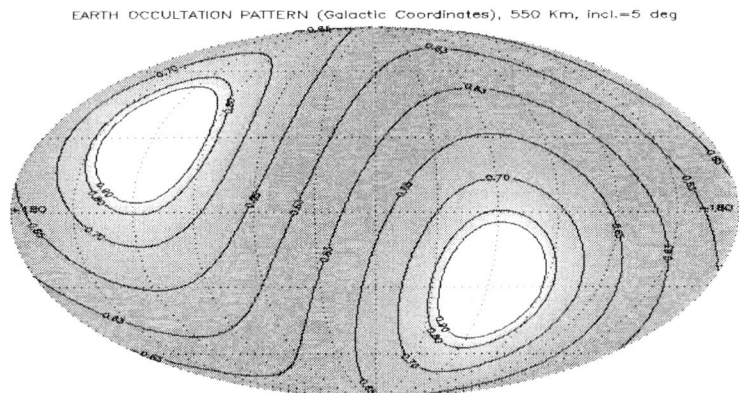

FIGURE 1. Sky visibility integrated over a nominal AGILE equatorial orbit of height 550 km and inclination 5°. The minimum value is 0.63 in the Galactic Plane near l=~30°. 100% visibility regions correspond to directions perpendicular to the orbital plane.

FEW DEEP POINTINGS OR MANY SHORT EXPOSURES?

The Galactic plane contains all the known gamma-ray pulsars and many interesting transient sources and represents a main target of the AGILE science program. The presence of strong diffuse emission affects the sensitivity but enriches the scientific content of Galactic plane observations. Furthermore, the large AGILE FOV allows us to efficiently monitor many AGNs with galactic latitudes up to 60-70 degrees even when the spacecraft is pointing along the plane.

Taking into account these AGILE features, is it better to observe the Galactic plane with a few long pointings or with many shorter observations?

A two-year survey of the Galactic plane (Figure 2) could be done with 24 pointings (~1 month each) separated by 15 degrees (|b|<60°-70°). About 250 EGRET sources including many AGNs could be monitored during this survey that is in practice an almost all sky survey.

Our simulations show that the Galactic plane could be observed by AGILE with high level of uniformity in a reasonable number of pointings, even when the observations are not optimized. The resulting flux limits are on the order of the faintest EGRET sources (detected with ~500 counts). In this example, covering >90% of the sky, more than 90% of the EGRET sources would be monitored, and due to the large and flat AGILE field of view, many new Galactic transients and blazars would be discovered.

An AGILE pointing at the Galactic center (GC), extending over 130 degrees in latitude and longitude, would also include part of the Gould Belt region in which unidentified Gamma-ray sources seem to be concentrated. In a GC AGILE pointing region, EGRET observed ~70 unidentified objects, ~25 AGNs, 1 pulsar; assuming source fluxes extrapolated from the 3rd EGRET Catalog, after one month of observation of the GC, we expect about ~780 counts from PSR B1706-44, ~70-350 counts from Unidentified sources and up to ~750-800 counts from flaring AGNs. In the GC region more than ~30 known sources are detectable above 5σ in only one month of observations including Earth occultation effects. This is just a rough estimate because the strong variability of many sources affects the detection levels.

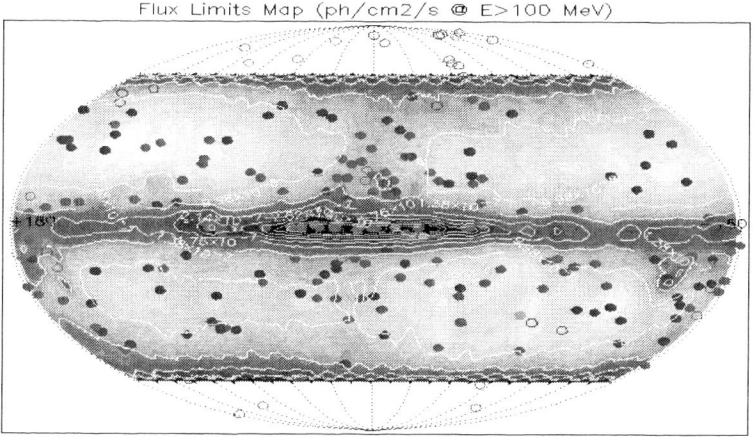

FIGURE 2. Sensitivity map (E>100 MeV) for a two-year Galactic Plane survey (see text). The averaged flux limit is ~10^{-7} ph/cm^2 s and the minimum flux achieved (~$6\cdot10^{-8}$ ph/cm^2 s) is of the order of the faintest EGRET source. Filled circles represent detected sources from the 3rd EGRET catalog.

Alternately, an AGILE deep observation of the Vela region would simultaneously include more than ~50 known Galactic and extragalactic sources. In particular, a single deep exposure (two-years) of this region (l=~264°, b=~-3°) would allow us to observe ~30 unidentified sources, 2 pulsars, ~17 AGNs and the Magellanic Clouds (MC) (Figure 3). Note that this pointing includes one of the two unoccluded windows, further improving the exposure level of the region around the MC. This extreme case of a very long observation of a single region should be compared with the survey described above (short pointings covering the entire Galactic plane).

CONCLUSIONS

The examples described here (a uniform Galactic plane survey and a deep pointing of the Vela region) represent two extreme opposite strategies on how to carry out an observation program with AGILE. The resulting exposure and sensitivity maps show that AGILE is optimally suited to monitor huge sky regions in a few pointings instead of carrying out a uniform survey of the Galactic plane. On the other hand, a deep exposure of a single unocculted sky region would improve the 2-year sensitivity only by a factor of ~2-3 compared to a typical survey (Figure 4). Therefore a good compromise could be the observation of ~5-6 Galactic and extragalactic regions monitoring the whole sky but concentrating at the same time on the study of relevant point sources and interesting diffuse emission regions.

AGILE exposure properties can be compared with those of EGRET. As an example, we obtained the EGRET exposure for the first cycle of observations (Apr.'91/Nov.'92) directly from the CGRO public archive and we compared it to an AGILE exposure map corresponding to 5 deep pointings of $\sim 10^7$ s each (~1.5 yrs observation time in total). Although the EGRET on-axis effective area is a factor of ~2 better, the averaged AGILE exposure ($\sim 2.5 \cdot 10^9$ cm$^2 \cdot$s) is a factor ~4 greater than that of EGRET (Figure 5) because of AGILE's larger and flatter field of view.

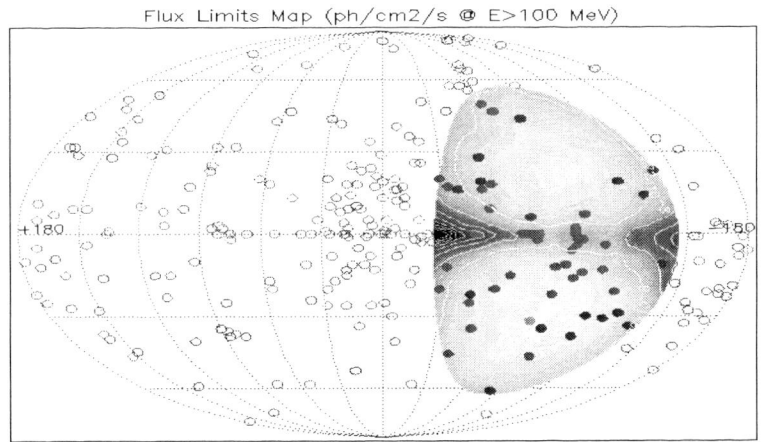

FIGURE 3. Sensitivity map for a deep pointing toward the Vela region. The minimum flux achieved is $\sim 2 \cdot 10^{-8}$ ph/cm^2 s). The lack of Earth occultation produces a very good exposure in the region around the LMC (l=~270°, b=~-30°).

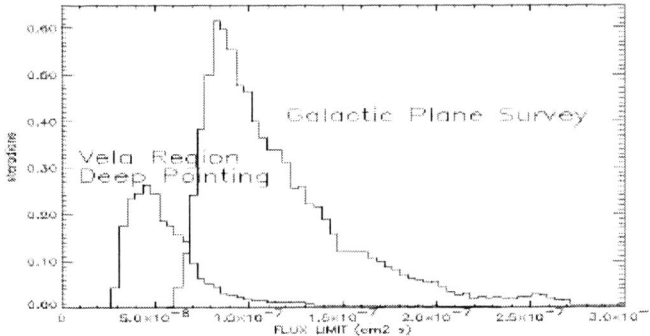

FIGURE 4. Sky distribution of sensitivities corresponding to the maps in Figures 2 and 3 (see text).

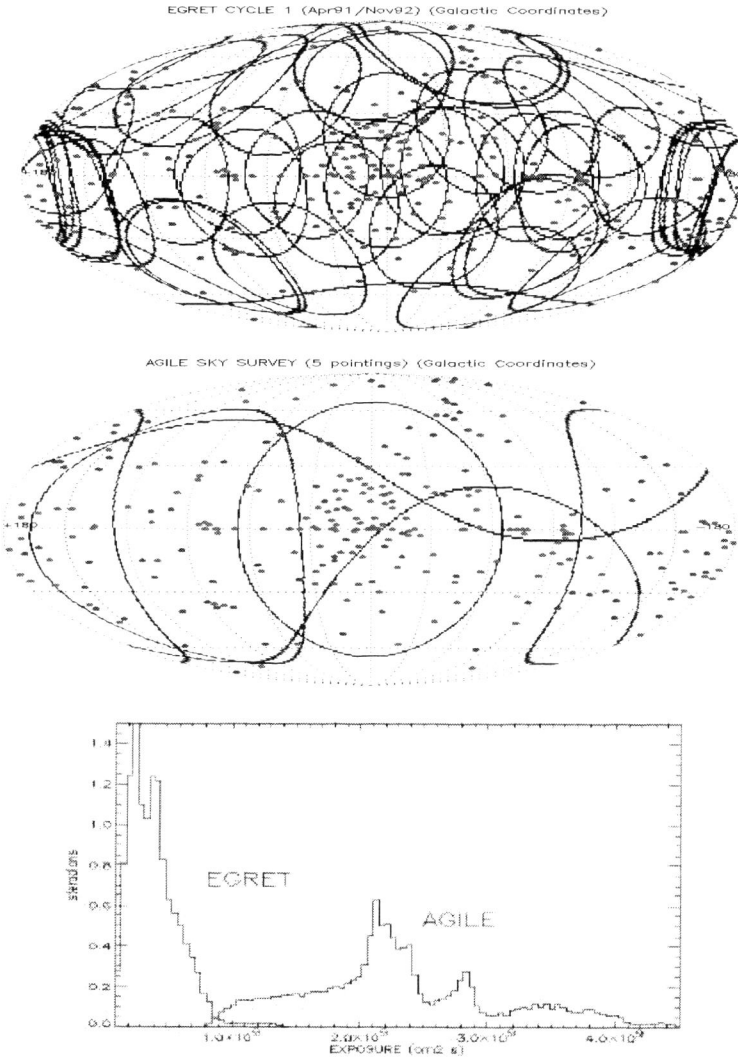

FIGURE 5. EGRET sky pointings distribution for the first cycle of observations (upper panel) compared with a possible AGILE sky pointing strategy of 5 deep exposures lasting $\sim 10^7$ s each (middle panel). Despite its smaller effective area, the AGILE total exposure will be significantly better than that of EGRET (lower panel).

REFERENCES

1. Tavani, M. et al., "*Science with AGILE*", these proceedings.
2. Vercellone, S. et al., "*Imaging of High Energy Sources with AGILE*", these proceedings.
3. Cocco, V. & Longo, F., "*GEANT Simulation of the AGILE Gamma-ray Imaging Detector*", these proceedings.
4. Pittori et al., "*Gamma-ray Imaging by silicon detector in space*", these proceedings.

Imaging of High Energy Sources with AGILE

S. Vercellone*, A.W. Chen[†], V. Cocco**, M. Feroci[‡], M. Galli[§],
A. Giuliani[†], I. Lapshov[‡], P. Lipari[¶], F. Longo[∥], S. Mereghetti*,
A. Pellizzoni*, C. Pittori**, P. Soffitta[‡] and D. Zanello[¶]

IFC-CNR, V. Bassini 15, 20133 Milano - Italy
[†]*IFC-CNR and CIFS, Villa Gualino, Viale Settimio Severo 63, 10133 Torino - Italy*
**Univ. Roma-II and INFN, V. Ric. Scientifica 1, 00133 Roma - Italy*
[‡]*IAS-CNR, V. Fosso del Cavaliere, 00133 Roma - Italy*
[§]*ENEA, Roma, Italy*
[¶]*Univ. Roma-I and INFN, P.le Aldo Moro 2, 00185 ROMA*
[∥]*Univ. di Ferrara and INFN, V. del Paradiso 12, 44100 Ferrara*

Abstract. We present the source imaging performance of the AGILE Satellite in the hard X-ray (10–40 keV) and gamma-ray (30 MeV–50 GeV) energy bands. The AGILE instrument will combine state-of-the-art Silicon detection technology for gamma-ray imaging (Gamma-Ray Imaging Detector – GRID) with a coded-mask imager sensitive in the hard X-ray band (Super-AGILE).
AGILE is designed to reduce the uncertainty in gamma-ray source localization to the level of a few arcmins, depending on the source intensity, spectrum and background properties.
The AGILE Event Simulator (AEV) is a tool to generate gamma-ray images taking into account the proper instrument characteristics. We present preliminary results obtained with AEV for several astrophysical sources.

INTRODUCTION

The *AGILE* scientific instrument [2, 3, 14, 15] is light (~ 80 kg) and effective in detecting and monitoring gamma-ray sources within a large field of view (FOV). We adopted the philosophy of one integrated instrument made of three detectors with broad-band detection and imaging capabilities: the Gamma-Ray Imaging Detector (GRID) [4], sensitive in the energy range 30 MeV–50 GeV, the hard X-ray monitor Super-AGILE [8] (SA), sensitive in the energy range 10–40 keV and a non-imaging Mini-Calorimeter [1] (MCAL), sensitive in the energy range 0.3–200 MeV.

The *AGILE Event Simulator* (AEV) has been developed to create maps of hard X-ray and gamma-ray sources, in order to test the AGILE astrophysical imaging performance in the energy ranges 10–40 keV and 30 MeV–50 GeV.

The current input consists of three main objects:

1. <u>Instrument & Observation Parameters</u>: they consist of the energy and position dependent effective area (A_{eff}) and Point Spread Function (PSF), and the direction and orientation of the instrument. The instrument parameters have been obtained by detailed Monte Carlo simulations, described in [5, 11].

2. <u>Sources of γ-ray emission</u>: containing physical parameters of gamma-ray point sources as well as Galactic and extragalactic diffuse emission.

3. Sources of hard X-ray emission.

The Simulator consists of a collection of *IDL* routines which: 1) create a source photon list, 2) display source positions, and 3) produce output data (counts map) in a format compatible with the EGRET analysis programs (EGRET *LIKE* Code, [9]).

For each coordinate bin (0.5×0.5 degrees), AEV computes the expected number of counts from gamma-ray sources and from the diffuse emission. We have used the 3rd EGRET Catalog [6] for source positions, fluxes (highest statistically significant detections) and spectra. The diffuse Galactic background contribution has been computed according to [7], while for the extragalactic component we refer to [13]. See also [10] for the AGILE exposure calculation.

THE SIMULATIONS

The Galactic Anticenter

We made a simulation of an AGILE observation similar to EGRET VP1.0, pointing at the galactic coordinates $l = 190.92$ and $b = -4.74$ for a net exposure time of 6.05×10^5 sec and integrating in the energy range 0.1–10 GeV.

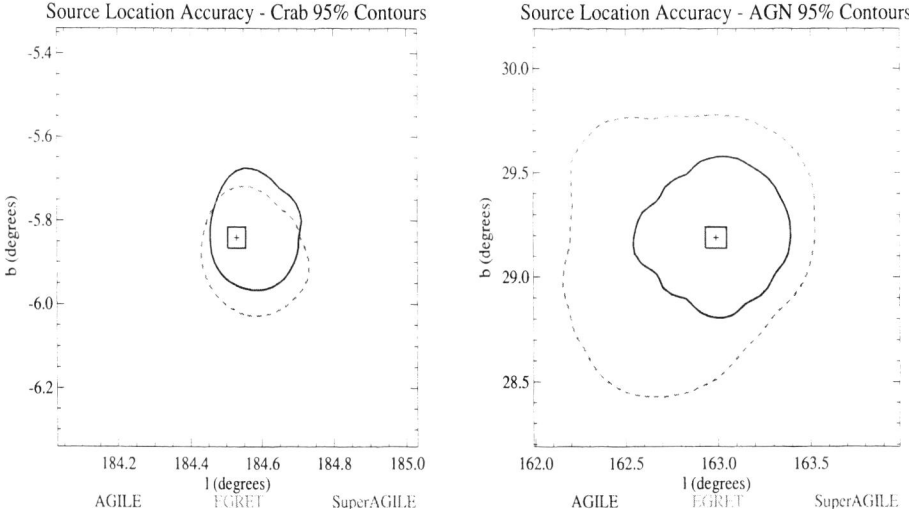

FIGURE 1. EGRET (dashed line), AGILE (solid line) source location accuracy (95% contour level) and Super-AGILE 99% error-box for the Crab pulsar (*Left Panel*) and for the AGN 3EG J0743+5447 (*Right Panel*).

We performed a likelihood analysis of the Crab pulsar using the EGRET *LIKE* program to evaluate the AGILE detection performance. Table 1 lists the source properties and the likelihood results. In Figure 1 (left panel) we show the source location accuracy 95% contour for EGRET, AGILE and Super-AGILE 99% error-box.

TABLE 1. Comparision between AGILE and EGRET likelihood parameters for the Crab.

	Off-axis angle (degrees)	Flux (E>100 MeV) (10^{-8} ph cm^{-2} s^{-1})	95% Elliptical fit semi-axis (arcmin)	Hard X-ray 99% error box
EGRET	6	~230	$a = 9.8, b = 8.8$	-
AGILE	6	~230	$a = 8.5, b = 7.7$	2'

From Figure 1 (left panel) and Table 1 we can derive the different on-axis and off-axis perfomance for AGILE and EGRET.

On-axis, the AGILE effective area is about one half of that of EGRET, but its PSF is much narrower (68% containment radius $\Theta_{68} \sim 0°.75$, $E \sim 1$ GeV). The total number of counts detected by AGILE for each source is smaller than for EGRET, but due to the better PSF it can still slightly improve the source localization.

Off-axis, ($\alpha \gtrsim 25°$ and E>400 MeV), the AGILE effective area is 2.5 times larger than the EGRET one and the AGILE Θ_{68} is 0.4 times the EGRET one. The off-axis performance of AGILE is almost the same as its on-axis one, while EGRET has no detection capability for off-axis angles larger than 40°.

Due to a large FOV (~ 3 sr), AGILE will be able to monitor effectively a wide portion of the sky, detecting and localizing highly off-axis gamma-ray sources, both transient and steady. Super-AGILE can reduce the uncertainty in gamma-ray source location accuracy to the level of a few arcmin for sources above 5–10 mCrab in the 10–40 keV range.

AGILE and mid-latitude AGNs

We simulated an AGN flaring at the edge of the fully-coded Super-AGILE FOV ($\sim 35° \times 35°$). We choose 3EG J0743+5447, observed by EGRET during VP227.0 at 17° off-axis angle. The AGN flux is 30×10^{-8} ph cm^{-2} s^{-1}, the spectral photon index is $\Gamma = 2.03$. We simulated this AGN at 28° off-axis angle ($l = 172$, $b = 17$) for a net exposure time of $T_{exp} = 6.05 \times 10^5$ sec (comparable to EGRET VP227.0).

Figure 1 (right panel) shows the source location accuracy 95% contour for EGRET, AGILE and Super-AGILE 99% error box and Table 2 lists the likelihood results.

TABLE 2. Comparison between AGILE and EGRET likelihood parameters for a mid-latitude AGN with flux $F_\gamma \simeq 30 \times 10^{-8}$ ph cm^{-2} s^{-1} (E>100 MeV).

	Off-axis angle (degrees)	95% Elliptical fit semi-axis (arcmin)	Hard X-ray 99% error box	(TS)$^{1/2}$ level
EGRET	17	$a = 41.5, b = 36.4$	-	8.3
AGILE	28	$a = 22.8, b = 20.9$	6'	10

This simulation shows how AGILE can significantly contribute in increasing the number of AGNs detected in a single Viewing Period, monitoring known sources and detecting flaring AGNs, and how Super-AGILE can permit an excellent localization of new blazars.

A 10σ detection of such an AGN in a one-week observation might allow to study, for the first time, the simultaneous AGN variability in the gamma-ray and in the hard X-ray energy bands.

AGILE and GRBs

We simulated the detection of a bright gamma-ray burst similar to the "Superbowl burst" GRB 930131. We placed the simulated GRB at $18°$ off-axis angle ($l = 184$, $b = 12$), with a gamma-ray flux $F_\gamma = 7.4 \times 10^{-6} \times (\frac{E}{147 MeV})^{-2.03}$ ph cm^{-2}s^{-1}MeV^{-1} (see [12]) for a net integration time of $T_\gamma = 100$ sec in the energy band 0.1–10 GeV, and a hard X-ray flux $F_{10-40keV} \sim 1$ Crab ($T_X = 10$ sec).

In Figure 2 we show the AGILE simulated counts maps (left panel) and the source location accuracy 95% contour (right panel).

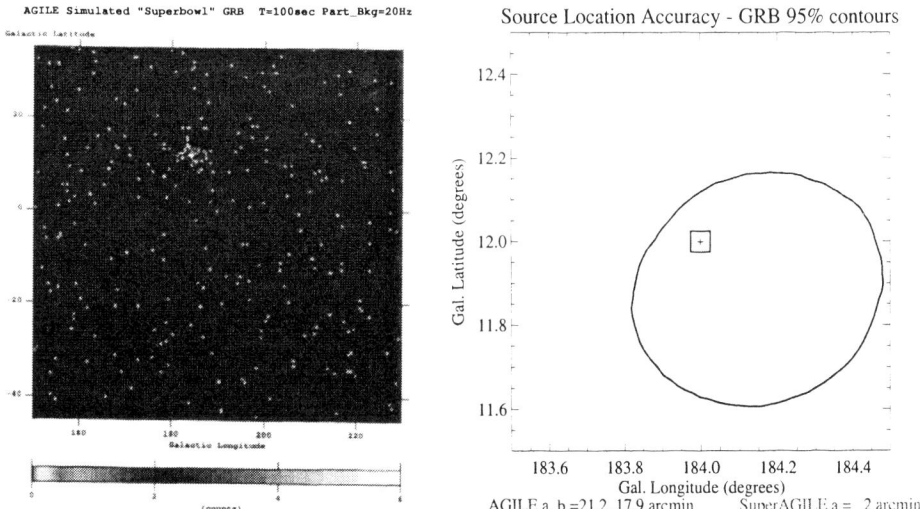

FIGURE 2. *Left Panel:* AGILE simulated counts map for a GRB similar to the "Superbowl burst" GRB930131. The effective observation time is 100 sec. *Right Panel:* GRID source location accuracy (95% contour level, solid line) and Super-AGILE 99% error-box.

For this simulation we have considered no *off-line* particle background rejection, keeping the particle background rate at the Level-2 trigger rate (20 Hz). This simulates the GRID *on-board* performance in imaging GRBs. AGILE can localize such a GRB within ~ 20 arcmin ($(TS)^{1/2} \sim 30$) at the Level-2 trigger background rate. Even at the edge of its field of view, Super-AGILE will be able to localize such a burst ($F_{10-40keV} \sim$ Crab) within 2'–3'.

CONCLUSIONS

We can summarize the expected AGILE scientific performance compared to EGRET as follows:

1. *On-Axis Sources*
   - Slightly improved source localization.

2. *Off-Axis Sources ($> 10°$)*
   - Significantly improved source localization.
   - Monitor of a large number of sources.

3. *AGNs*
   - Precise localization of medium intensity blazars.
   - Study of the gamma-ray flaring duty cycle.
   - Possible joint detection by the GRID and Super-AGILE, producing error boxes $\sim$ 6 arcminutes.

4. *GRBs*
   - Precise *on-board* localization of medium-high intensity burst.
   - *On-board* fast processing (5-10 sec) and the capability to transmit rapidly through a dedicated fast channel GRB coordinates and sky maps by Super-AGILE.

REFERENCES

1. Auricchio N. et al., "Test Campaign of the Mini-Calorimeter for the AGILE Satellite", *these Proceedings*, 2001
2. Barbiellini G. et al., *Proceedings of the 5th Compton Symposium*, AIP Conf. Proceedings, ed. M. McConnell, 2001, Vol. 510, p. 750
3. Barbiellini G. et al.,"The AGILE Scientific Instrument", *these Proceedings*, 2001a
4. Barbiellini G. et al., "The next generation of high-energy gamma-ray detectors for satellites: the AGILE silicon tracker", *these Proceedings*, 2001b
5. Cocco V., Longo F. and Tavani M., "GEANT Simulation of the AGILE Gamma-ray Imaging Detector" *these Proceedings*, 2001
6. Hartman R.C. et al., *ApJS*, 1999, 123, 79
7. Hunter S.D. et al., *ApJ*, 1997, 481, 205
8. Lapshov I. et al., "SuperAGILE: The X-ray Monitor on-board of AGILE", *these Proceedings*, 2001
9. Mattox J.R. et al., *ApJ*, 1996, 461, 396
10. Pellizzoni A. et al., "AGILE Sky Exposure and Senitivity Maps", *these Proceedings*, 2001
11. Pittori C. et al., "Gamma-Ray Imaging by Silicon Detectors in Space: the AGILE Reconstruction Method and Kalman Filter Algorithms", *these Proceedings*, 2001
12. Sommer E. et al., *ApJ*, 1994, 422, L63
13. Sreekumar P. et al., *ApJ*, 1998, 494, 623
14. Tavani M. et al., *Proceedings of the 5th Compton Symposium*, AIP Conf. Proceedings, ed. M. McConnell, 2001, Vol. 510, p. 746
15. Tavani M. et al., "Science with AGILE", *these Proceedings*, 2001

Super-AGILE: The X-ray Monitor on-board of AGILE

Igor Lapshov, Lidia Barbanera, Enrico Costa, Ettore Del Monte, Marco Feroci, Geiland Porrovecchio, Marcello Mastropietro, Luigi Pacciani, Alda Rubini, Paolo Soffitta*, Ennio Morelli[†], Massimo Rapisarda**, Guido Barbiellini, Francesco Longo, Michela Prest, Erik Vallazza[‡], Andrea Argan, Sandro Mereghetti, Marco Tavani, Stefano Vercellone[§] and Aldo Morselli[¶]

*IAS-CNR, Rome, Italy
[†]ITESRE-CNR, Bologna, Italy
**ENEA, Frascati (Rome), Italy
[‡]INFN, Trieste, Italy
[§]IFC-CNR, Milan, Italy
[¶]INFN, Rome, Italy

Abstract. Super-AGILE is the hard X-ray imaging detector of the AGILE gamma-ray mission. It is devoted to monitor X-ray (10-40 keV) sources with an on-axis sensitivity near 5 mCrab in one observing day and to detect X-ray transients in a field of view of 107 deg x 68 deg. Super-AGILE is well matched with the Gamma-Ray Imaging Detector (GRID), and potentially provides source arc-minute positioning depending on intensity, spectrum and background conditions.

Super-AGILE detects hard X-rays with one additional layer of four Silicon micro-strip detectors, for 1444 cm^2 total geometrical area. It will be placed on top of the AGILE Tracker and equipped with a system of four mutually orthogonal one-dimentional coded masks to encode the X-ray sky. Low-noise electronics based on ASICs technology is the front-end read out. We present here the instrumental and astrophysical performance of Super-AGILE as derived by Monte Carlo simulations and experimental tests.

DESCRIPTION OF SUPERAGILE

Super-AGILE is basically composed by a Detection Plane (DP), a Collimator equipped with a Coded Mask, Front-End Electronics and an Interface Electronics (SAIE). Table 1 summarizes the main instrument characteristics and Figure 1 shows the Detector layout (see also [5] for an extensive description of Super-AGILE). The DP is composed of 4 detection units (DUs), placed on the same Al honeycomb plane support, so that two of them sample the X-direction and the other two are devoted to the Y-direction. Each DU is composed by 4 Si microstrip tiles, bonded in pairs so that the effective length of each strip is approximately 19 cm. They are read-out through a set of IDE AS-XAA1 chips, based on ASIC technology, 12 for each of the DUs. The collimator is mounted on the same tray supporting the DP, and in turn supports the 4 orthogonal, one-dimensional coded masks. The coded masks have a 50% covering factor. They will be manufactured either of Gold or Tungsten. The SAIE is in charge of interfacing Super-AGILE with the

TABLE 1. The Basic Characteristics of Super-AGILE

Detector Type	Silicon Strip
Basic Detection Unit	4 Si Tiles, 19cm x 19cm
Total Geometric Area	1444 cm^2
On-Axis Effective Area	320 cm^2 (13keV)
Detector Strip Size	121 μm
Detector Thickness	400 μm
Energy Resolution (FWHM)	$\sim$ 3-4 keV
Timing Accuracy	$\sim$ 5 μs
Collimator Materials	75 μm Tungsten-Coated Carbon Fiber
Mask Size	1444 cm^2
Mask-Detector Distance	14 cm
Mask Transparency	50%
Mask Material	Tungsten
Mask Thickness	100 μm
Mask Element Size	242 μm
Field of View (FWZR)	107° x 68°
On-Axis Angular Resolution	5.9 arcmin
Source Location Accuracy	$\sim$ 2 arcmin for bright sources
Point Source Sensitivity	5 mCrab on axis

AGILE Data Handling System, allowing an event-by-event transmission with better than 5 μs timing resolution. The energy information will be provided in the extended energy range between 1 and 64 keV, with 64 channels, to allow a finer threshold calibration at low energies, and exploit for calibration purposes the Tungsten fluorescences at $\sim$58 keV. The combined capabilities of the SAIE and the AGILE Data Handling (see also [4]) allows the transmission to the ground of a relatively large set of scientific housekeeping data, including ratemeters and detector images. In particular, the AGILE Data Handling will be able to perform a continuous automatic search for transient events (e.g. gamma-ray bursts) on timescales from 1 ms to 100 s. Once a transient event is triggered onboard, the Data Handling will be able to provide attitude-corrected sky images for it, determining the location of the transient source on the sky (see Figure 2 for illustration). The possibility to distribute in almost real time the coordinates of the transient event through a fast link (e.g., TDRSS or similar) is currently under study.

SENSITIVITY AND EFFECTIVE AREA

We studied the sensitivity and expected astrophysical performances by means of analytical calculations and Monte Carlo simulations. In Figure 3 we show the combined effective area of four Super-AGILE DUs over the field of view (FOV) in the 10-50 keV energy band. The central 60°x60° area contains the overlapping FOVs of orthogonal detectors, providing an effective bi-dimensional source location capability. The outer regions of the FOV enable one-dimensional localization of sources. This effective area is sufficient to detect on-axis sources as weak as 5 mCrab for an integration time of 50

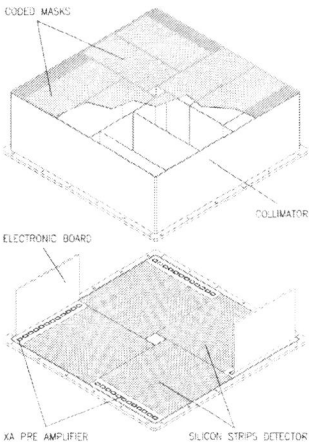

FIGURE 1. Schematic view of the Super-AGILE structure.

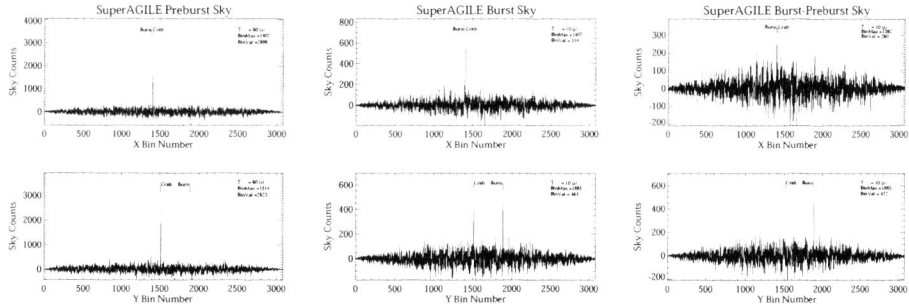

FIGURE 2. Illustration of possible on-board burst localizations by the Data Handling system. *Left:* X and Y sky images, integrated during the last 60 seconds before the burst trigger. The Crab pulsar is visible near the center of the field of view. *Center:* X and Y sky images obtained during 10 seconds of the burst. *Right*: The Data Handling system will be capable of normalizing the pre-burst data and subtracting it from the burst images, with consecutive determination of the burst position.

ks (with four detectors). The right plot in Figure 3 shows the sensitivity of one Super-AGILE detector over the field of view in two orthogonal directions. The plot with Y=0 corresponds to a coding direction.

SOURCE LOCALIZATION

From the Super-AGILE mask/collimator/detector geometry one can derive the angular resolution of the detector to be 5.9 arcmin. However, its source location accuracy should normally be better than that. We anticipate, that Super-AGILE will be able to locate bright sources within a 2 arcmin error box thanks to the fact that the detector bin size

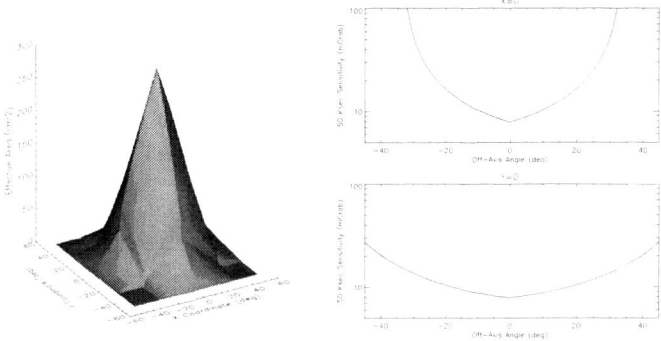

FIGURE 3. *Left Panel:* Effective area of 4 Super-AGILE detector units over the FOV. *Right Panel:* 5σ sensitivity of one Super-AGILE detector over the FOV for a 50 ksec observation

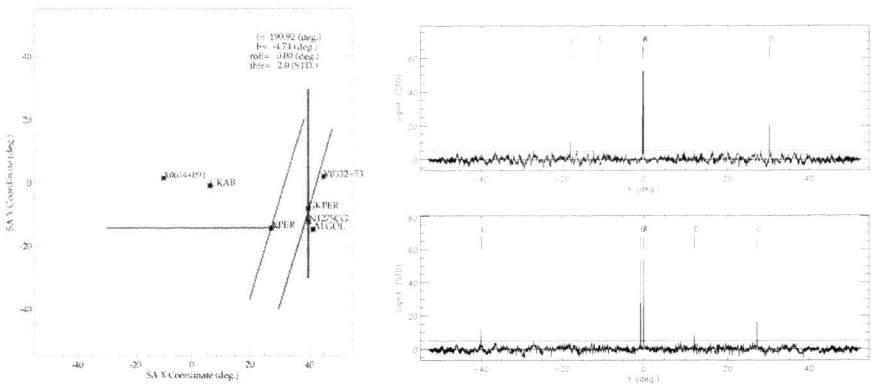

FIGURE 4. *Left Panel:* Simulated observation of a field near the Crab source. See the main text for the explanations. *Right Panel:* Super-AGILE sky images of a sample field with 5 sources. Each image, which is a combination of two detectors was attitude corrected for spacecraft pointing errors.

being half of the mask element size.

For the central 60°x60° region of the FOV we will normally have two one-dimensional positions of the source with an error 2-6 arcmin, depending on the source brightness. The outer regions of the FOV will enable detecting and locating sources in either of 2 orthogonal directions, thus making the source location error box to be 2 arcmin x 60°. For this case a better localization of a new discovered source can be derived using Earth occultation techniques, data from other operating spacecraft or a different AGILE satellite pointing. Figure 4 (Left Panel) presents two simulated Super-AGILE observations of a field around the Crab pulsar. The pointing in both observations remains the same, while the roll angle being different by 16 degrees. Both sources in the central part of the FOV were detected in both orthogonal directions, thus the location error box being within 6 arcmin. Some sources on the outer part on the FOV were detected by only one of two orthogonal detector units, hence their error boxes are extended in one direction. However, the second observation with a different roll angle can locate even these sources by intersecting error boxes. The AGILE spacecraft pointing stability is expected to be within a circle of 1 degree radius. This leads to the necessity of correcting the coordinates of events based on the data provided by star sensors. Without such correction images will be blurred and the sensitivity of the instrument will decrease. Figure 4 (Right Panel) shows the result of correcting of Super-AGILE data. The attitude correction has a positive effect on the quality of Super-AGILE images, although it cannot improve the experiment's sensitivity compared to the "ideal" pointing case. The software attitude correction procedure, designed for Super-AGILE, enables to suppress to some extent the coding noise from bright sources in the field of view, and to make more prominent fainter sources. This is possible by the large Super-AGILE FOV and by the fact that the pixel size expressed in angular units is strongly variable over the FOV. As a result, while reconstructing some part of the sky, we still have shadows cast by other sources blurred strongly enough not to produce hight coding noise peaks. The positive result of such procedure can be clearly seen in Figure 4 (Right Panel).

EXPECTED SCIENTIFIC PERFORMANCE

Super-AGILE will study a variety of X-ray sources of different types including Galactic and some bright extragalactic sources. We note that as a result of the long AGILE pointings (~ 2 weeks), Super-AGILE will monitor the same sky region for a long time. This pointing strategy will provide accurate energy spectra and flux monitoring, looking for short timescale variations. Super-AGILE is therefore well suited for studying the activity of fast X-ray transients. For example, it will be able to provide energy spectra of events similar to short recurring bursts from Soft Gamma-ray Repeaters (e.g., [1]). Furthermore, the giant flares from these sources (e.g., [3]) are very good candidates for emission of rapid and intense flashes of gamma-rays, and Super-AGILE can provide to the GRID accurate positions of possible new soft gamma-ray repeater events. A more detailed description of Super-AGILE and its scientific capabilities can be found elsewhere (see also [2, 5, 6]).

REFERENCES

1. Aptekar, R., et al. 2000, *astro-ph/0004402*
2. Costa, E., et al., NIM, in preparation
3. Feroci, M., et al., *ApJ, 549 (2001) 1021-1038*
4. Morselli, A., et al. 2000, *Proc. SPIE conference 4140, pp.493-499*
5. Soffitta, P., et al. 2000, *Proc. SPIE conference 4140, pp.283-292*
6. Tavani, M., et al. 2001, *these Proceedings*

The AGILE Scientific Instrument

G. Barbiellini*, M. Tavani[†], A. Argan**, N. Auricchio[‡], P. Caraveo[†],
A. Chen**, V. Cocco[§], E. Costa[¶], G. Di Cocco[‡], G. Fedel*, M. Feroci[¶],
M. Fiorini[†], T. Froysland[§], M. Galli[||], F. Gianotti[‡], A. Giuliani[†],
C. Labanti[‡], I. Lapshov[¶], P. Lipari[††], F. Longo[‡‡], E. Massaro[††],
S. Mereghetti[†], E. Morelli[‡], A. Morselli[§], A. Pellizzoni[†], F. Perotti[†],
P. Picozza[§], C. Pittori[§], C. Pontoni[§§], M. Prest[§§], M. Rapisarda[¶¶], E. Rossi[‡],
A. Rubini[¶], P. Soffitta[¶], M. Trifoglio[‡], E. Vallazza*, S. Vercellone[†] and
D. Zanello[††]

Univ. di Trieste and INFN, V. Padriciano 99, Trieste - Italy
[†]*IFC-CNR, V. Bassini 15, 20133 Milano - Italy*
**IFC-CNR Milano, and CIFS, Villa Gualino, Viale Settimio Severo 63, 10133 Torino - Italy*
[‡]*ITESRE-CNR, V. Gobetti 101, 40129 Bologna - Italy*
[§]*Univ. Roma-II and INFN, V. Ric. Scientifica 1, 00133 Roma - Italy*
[¶]*IAS-CNR, V. Fosso del Cavaliere, 00133 Roma - Italy*
[||]*ENEA, Sez. Bologna, Italy*
[††]*Dip. Fisica Univ. Roma-I and INFN, P.le Aldo Moro 2, 00185 Roma - Italy*
[‡‡]*Univ. di Ferrara and INFN, V. del Paradiso 12, 44100 Ferrara - Italy*
[§§]*INFN Trieste, and CIFS, Villa Gualino, Viale Settimio Severo 63, 10133 Torino - Italy*
[¶¶]*ENEA, Sez. Roma, Italy*

Abstract.
The AGILE scientific instrument is based on an innovative design based on three detecting systems: (1) a Silicon Tracker, (2) a Mini-Calorimeter, and (3) an ultralight coded mask system with Si-detectors (Super-AGILE). AGILE is designed to provide: (1) excellent imaging in the energy bands 30 MeV–50 GeV (5–10 arcmin for intense sources) and 10-40 keV (1–3 arcmin);, (2) optimal timing capabilities, with independent readout systems and minimal deadtimes for the Silicon tracker, Super-AGILE and Mini-Calorimeter; (3) large fields of view for the gamma-ray imaging detector ($\sim$3 sr) and Super-AGILE ($\sim$1 sr).

INTRODUCTION

The AGILE Mission is the first of the Italian Space Agency Small Scientific Missions [10]. It is devoted to high-energy astrophysics and is currently planned to be operational in 2003. The AGILE scientific instrument [2, 11, 12] is based on the state-of-the-art technology of solid state Silicon detectors developed by our group in INFN and CNR laboratories. The instrument is light ($\sim$ 80 kg) and very effective in detecting and monitoring hard X-ray/gamma-ray sources within a large field of view (FOV).

We adopted the philosophy of one integrated instrument made of three detectors with broad-band detection and imaging capabilities: the Gamma-Ray Imaging Detector (GRID) [4, 9] sensitive in the energy range 30 MeV–50 GeV, the hard X-ray imager named Super-AGILE (SA) [7] sensitive in the energy range 10–40 keV, and a non-imaging CsI(Tl) Mini-Calorimeter (MC) [1] sensitive in the energy range 0.3–200 MeV. We briefly describe here the main instrument's characteristics.

THE INSTRUMENT

FIGURE 1. Schematic view of the AGILE instrument (AC system partially displayed). The gamma-ray imager is made of a Tracker (14 Tungsten-Silicon planes) and a Mini-Calorimeter (two layers of 16 CsI(Tl) bars each). Super-AGILE has its detection plane with 4 independent Si-detectors on top of the first GRID tray, and an ultra-light coded mask system (CMS) positioned above it (the figure shows the CMS partition configuration). The instrument size is $63 \times 63 \times 58.5 \, cm^3$, including Super-AGILE and the AC system.

Fig. 1 shows the AGILE instrument configuration including the Si-Tracker, Super-AGILE, Mini-Calorimeter, the Anticoincidence system and electronics. The baseline AGILE instrument is made of the following elements.

- **Silicon-Tracker**, a gamma-ray pair-converter and imager made of 14 planes, with two Si-layers per plane providing the X and Y coordinates of interacting charged particles. The fundamental Silicon detector unit is a tile of area $9.5 \times 9.5 \, cm^2$, microstrip pitch equal to 121 μm, and thickness 410 μm. The adopted "floating readout strip" system has a total of 384 readout channels (readout pitch equal to 242 μm) and three readout TA1 chips per Si-tile. Each Si-Tracker layer is made of 4×4 tiles, for a total geometric area of $38 \times 38 \, cm^2$ and 1,536 readout channels. The first 12 planes are made of three elements: a first layer of Tungsten ($0.07 \, X_0$) for gamma-ray conversion, and two Si-layers (views) with microstrips orthogonally positioned. For each plane there are then $2 \times 1,536$ readout microstrips. Since the GRID trigger requires at least three Si-planes to be activated, two more Si-planes are inserted at the bottom of the Tracker without Tungsten layers. The total readout channel number of for the GRID Tracker is $\sim 43,000$. Both digital and analog information (charge deposition in Si-microstrip) is read by TA1 chips. The distance between mid-planes equals 1.6 cm (optimized by Montecarlo simulations). The GRID has an *on-axis* total radiation length near $\sim 1\ X_0$. Special algorithms applied off-line to telemetered data will allow optimal background subtraction and reconstruction of the photon incidence angle. Both digital and analog information are crucial for this task. The positional resolution obtained by these detectors in recent beam tests at CERN is excellent, being below 40 μm for a large range of photon incidence angles [3]. More information on the Silicon Tracker can be found in Refs. [4, 9].

- **Super-AGILE**, made of four square Silicon detectors ($19 \times 19\ \text{cm}^2$ each) and associated FEE placed on the first GRID tray plus an ultra-light coded mask system supporting a Tungsten mask placed at a distance of 14 cm from the Silicon detectors. Super-AGILE tasks are: *(i)* photon-by-photon detection and imaging of sources in the energy range 10-40 keV, with a field-of-view (FOV) of $\gtrsim 0.8$ sr, good angular resolution (1-3 arcmins, depending on source intensity and geometry), and good sensitivity (~ 5 mCrab for 50 ksec integration, and $\lesssim 1$ Crab for a few seconds integration); *(ii)* simultaneous X-ray and gamma-ray spectral studies of high-energy sources; *(iii)* excellent timing ($\lesssim 4\mu s$); *(iv)* burst trigger for the GRID and MC; *(v)* GRB alert and quick on-board positioning capability. Refs. [6, 7] describe the Super-AGILE structure and scientific capabilities.

- **Mini-Calorimeter** (MC), made of two planes of Cesium Iodide (CsI) bars, for a total (on-axis) radiation length of $1.5\ X_0$. The signal from each CsI bar is collected by two photodiodes placed at both ends. The MC tasks are: *(i)* obtaining additional information on the energy of particles produced in the Si-Tracker; *(ii)* detecting GRBs and other impulsive events with spectral and intensity information in the energy band $\sim 0.3 - 100$ MeV. We note that the problem of "particle backsplash" for AGILE is much less severe than in the case of EGRET. AGILE allows a relatively efficient detection of (inclined) photons near 10 GeV and above also because the AC-veto can be disabled for events with more than ~ 100 MeV total energy collected in the MC. Ref. [1] describes the MC characteristics.

- **Anticoincidence System**, aimed at both charged particle background rejection and preliminary direction reconstruction for triggered photon events. The AC system surrounds all AGILE detectors (Super-AGILE, Si-Tracker and MC). Each lateral face is segmented with three plastic scintillator layers (0.6 cm thick) connected to photomultipliers placed at their bottom. A single square plastic scintillator layer (0.5 cm thick) constitutes the top-AC layer whose signal is read by four photomultipliers placed at the four corners. Ref. [8] describes the AC system characteristics.

- **Data Handling System**, for fast processing of the GRID, Mini-Calorimeter and Super-AGILE events. The GRID trigger logic for the acquisition of gamma-ray photon data and background rejection is structured in two main levels: Level-1 and Level-2 trigger stages. The Level-1 trigger is fast ($\lesssim 5\mu s$) and requires a signal in at least three out of four contiguous tracker planes, and a proper combination of fired TA1 chip number signals and AC signals. An intermediate Level-1.5 stage is also envisioned (lasting $\sim 20~\mu s$), with the acquisition of the event topology based on the identification of fired TA1 chips. Both Level-1 and Level-1.5 have a hardware-oriented veto logic providing a first cut of background events. Level-2 data processing includes a GRID readout and pre-processing, "cluster data acquisition" (analog and digital information), and processing by a dedicated CPU. The Level-2 processing is asynchronous (estimated duration $\sim$ a few ms) with the actual GRID event processing. The GRID deadtime turns out to be $\sim 100~\mu s$ and is dominated by the Tracker readout.

The charged particle and albedo-photon background passing the Level-1+1.5 trigger level of processing is simulated to be $\lesssim 100$ events/sec for the nominal equatorial orbit of AGILE [5]. The on-board Level-2 processing has the task of reducing this background by a factor between 3 and 5. Off-line processing of the GRID data with both digital and analog information is being developed with the goal to reduce the particle and albedo-photon background rate above 100 MeV to ~ 0.01 events/sec.

In order to maximize the GRID FOV and detection efficiency for large-angle incident gamma-rays (and minimize the effects of particle backsplash from the MC and of "Earth albedo" background photons), the data acquisition logic uses proper combinations of top and lateral AC signals and a coarse on-line direction reconstruction in the Si-Tracker. For events depositing more than ~ 100 MeV in the MC, the AC veto can be disabled to allow the acquisition of gamma-ray photon events with energies larger than 1 GeV.

Appropriate data buffers and burst search algorithms are envisioned to maximize data acquisition for transient gamma-ray events (e.g., GRBs) in the Si-Tracker, Super-AGILE and Mini-Calorimeter, respectively.

The Super-AGILE event acquisition is conceptually simple. After a first "filtering" based on AC-veto signals and pulse-height discrimination in the dedicated FEE (XAA1 chips), the events are buffered and transmitted to the CPU for burst searching and final data formatting. The 4 Si-detectors of Super-AGILE are organized in 16 independent readout units, of $\sim 5~\mu s$ deadtime each.

Given the relatively large number of readable channels in the Si-Tracker and Super-AGILE ($\sim 50,000$ channels), the instrument requires a very efficient readout system. In order to maximize the detecting area and minimize the instrument weight and absorbed power, the GRID and Super-AGILE front-end-electronics is partly accommodated in special boards placed externally on the Tracker lateral faces. Electronic boxes, P/L memory (and buffer) units will be accommodated at the bottom of the instrument. Ref. [13] describes the AGILE Data Handling System.

Table 1 summarizes the main characteristics of the AGILE gamma-ray instrument and its performance compared to that of EGRET. We assumed a typical 2-week pointing duration and a $\sim 50\%$ exposure efficiency.

TABLE 1. A COMPARISON BETWEEN EGRET AND AGILE

	EGRET	AGILE
Mass	1830 kg	80 kg
Gamma-ray energy band	30 MeV – 30 GeV	30 MeV–50 GeV
Field of View	~ 0.5 sr	~ 3 sr
PSF (68% containment radius)	5.5° 1.3° 0.5°	4.7° (@ 0.1 GeV) 0.6° (@ 1 GeV) 0.2° (@ 10 GeV)
Deadtime for γ-ray detection	$\gtrsim 100$ ms	$\lesssim 100 \mu s$
Sensitivity for pointlike sources[†] (ph cm^{-2} s^{-1} MeV^{-1})	8×10^{-9} 1×10^{-10} 1×10^{-11}	6×10^{-9} (@ 0.1 GeV) 4×10^{-11} (@ 1 GeV) 3×10^{-12} (@ 10 GeV)
Required pointing reconstruction	~ 10 arcmin	~ 1 arcmin

CONCLUSIONS

The AGILE scientific instrument is innovative in many ways, and is designed to obtain an optimal gamma-ray detection performance despite its relatively small mass and absorbed power. The refined readout of the Silicon Tracker allows to reach an excellent spatial resolution ($\sim 40 \mu m$) that is crucial for gamma-ray imaging. The combination of hard X-ray (Super-AGILE) and gamma-ray imaging capabilities in a single integrated instrument is unique to AGILE. We anticipate a crucial role of Super-AGILE for studies of AGNs, GRBs, and Galactic sources. Positioning better than ~ 6 arcmin can be obtained for sources detectable in the hard X-ray range. Instrumental deadtimes for the different detectors are unprecedently small for gamma-ray instruments, and microsecond photon timing can be achieved. An optimal Burst Search Procedure is implemented in the on-board Data Handling System allowing a GRB search for a broad dynamic range of durations from milliseconds to hundreds of seconds.

REFERENCES

1. Auricchio N. et al., *these Proceedings*, 2001
2. Barbiellini G. et al., *Proceedings of the 5th Compton Symposium*, AIP Conf. Proceedings, ed. M. McConnell, 2001, Vol. 510, p. 750
3. Barbiellini G., Prest M. et al., *NIM*, 2001, submitted
4. Barbiellini G. et al., *these Proceedings*, 2001b
5. Cocco V., Longo F. & Tavani M., *these Proceedings* (2001)
6. Costa E. et al., 2001, in preparation
7. Lapshov I. et al., *these Proceedings*, 2001
8. Perotti F. et al., 2001, in preparation
9. Prest M. et al., 2001, in preparation
10. Tavani M. et al.,*The AGILE Mission*, 2001 htpp://www.ifctr.mi.cnr.it/Agile
11. Tavani M. et al., *Proceedings of the 5th Compton Symposium*, AIP Conf. Proceedings, ed. M. McConnell, 2001, Vol. 510, p. 746
12. Tavani M. et al., *these Proceedings*, 2001
13. Tavani M. et al., *NIM*, 2001, in preparation

Future Missions (Gamma Ray Bursts)

Swift: A Gamma Ray Burst MIDEX

Scott Barthelmy

NASA Goddard Space Flight Center
Greenbelt, Maryland, 20771

on behalf of the Swift Team

Abstract. Swift is a first of its kind multiwavelength transient observatory for gamma-ray burst astronomy. It has the optimum capabilities for the next breakthroughs in determining the origin of gamma-ray bursts and their afterglows as well as using bursts to probe the early Universe. Swift will also perform the first sensitive hard X-ray survey of the sky. The mission is being developed by an international collaboration and consists of three instruments, the Burst Alert Telescope (BAT), the X-ray Telescope (XRT), and the Ultraviolet and Optical Telescope (UVOT). The BAT, a wide-field gamma-ray detector, will detect ~1 gamma-ray burst per day with a sensitivity 5 times that of BATSE. The sensitive narrow-field XRT and UVOT will be autonomously slewed to the burst location in 20 to 70 seconds to determine 0.3-5.0 arcsec positions and perform optical, UV, and X-ray spectrophotometry. On-board measurements of redshift will also be done for hundreds of bursts. Swift will incorporate superb, low-cost instruments using existing flight-spare hardware and designs. Strong education/public outreach and follow-up programs will help to engage the public and astronomical community. Swift has been selected by NASA for development and launch in late 2003.

1. INTRODUCTION

The discovery by BeppoSAX and ground observers of afterglow [1-3] from gamma-ray bursts (GRBs) has revolutionized our understanding of these enigmatic events. We now know that they are cosmological with z in the range 0.3 to 4.5 and involve the most powerful explosions known. These explosions are thought to create super-relativistic blast waves resulting in afterglow that fades from gamma-rays to radio. A panchromatic approach to observations of GRBs is now essential for the next phase of discovery.

Much of the information on GRB afterglows is being lost due to the fact that it currently takes 5 to 12 hours to point X-ray and optical telescopes to look at burst sources. By this time the fluxes have dropped by a factor of thousands (Figure 1).

In addition, current instruments may not be sensitive to the full diversity of the GRB population. BATSE showed that the distribution of burst durations is bimodal [4,5], with the weaker bursts having, on average, longer durations than the stronger bursts [6]. Short bursts (less than a few sec) have not been observed by BeppoSAX, so we do not know the nature of this distinct class of GRB.

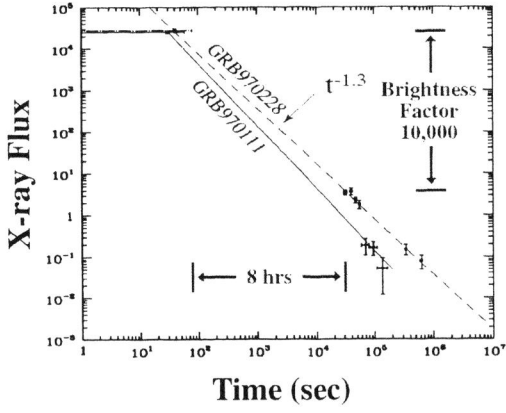

FIGURE 1. The Gap

Swift is a multiwavelength observatory that exploits the newly discovered afterglow characteristics to make a comprehensive study of ~1000 bursts over the course of its 3-year mission. It will determine the origin of GRBs, tell us how the blast wave interacts with its surroundings, and identify classes of bursts and their associated physical processes. In addition, Swift will investigate how GRBs can be used to study the early Universe.

2. SWIFT INSTRUMENTS

The Swift instrumentation was carefully chosen for GRB discovery. It incorporates a wide-field GRB detector, plus two sensitive narrow-field telescopes for identifying and observing the X-ray, UV, and optical afterglow (Figure 2).

The *Burst Alert Telescope (BAT)* covers the 15 to 150 keV energy band and uses a coded aperture mask to provide positions of 1-4 arcmin accuracy depending on burst brightness. The GRB detector has a CdZnTe (CZT) detector array with an area of 5200 cm^2 and a coded aperture mask, which covers 2 sr of the sky. The mask to detector array separation is 1 meter. The BAT will detect ~300 bursts per year. The large detector area and a sophisticated triggering system that takes advantage of the imaging capability of the instrument allows the BAT to detect bursts of all durations to a sensitivity 5 times better than BATSE (BAT threshold ~10^{-8} erg cm^{-2} for a 1 sec GRB).

The BAT instrument will also perform a hard x-ray survey while "staring" for GRBs and as the spacecraft slews from burst to burst for the NFI follow-ups. It will be ~30 times more sensitive than the HEAO A3 survey [30]. The BAT parameters are summarized in Table 1. The instrument is being developed at NASA's Goddard Space Flight Center. Further details of the BAT instrument are in Barthelmy [26] and its survey capabilities in Krimm [29].

FIGURE 2. Swift Instruments: BAT, XRT, and UVOT.

TABLE 1. BAT Parameters

Energy Range	15-150 keV
Aperture	Coded mask
Detecting Area	5200 cm2
Detector	CdZnTe
Detector Operation	Photon counting
Field of View	1,4 sr (half-coded)
Detection Elements	256 modules of 128 elements
Detector Size	4 x 4 x 2 mm^3
Mask cell size	5x5x1 mm^3 lead
Telescope PSF	17 arcmin
Burst sensitivity	0.2 photons cm^{-2} s^{-1}

The *X-Ray Telescope (XRT)* utilizes flight-spare X-ray optics from the JET-X instrument on the Spectrum X-Γ mission. This mirror has a 15 arcsec half-power diameter at 1.5 keV. The XRT will locate bursts to 5.0 arcsec accuracy. The detector is a 600x600 pixel CCD from the XMM program, giving a FOV of 23.6 arcmin square in an energy range of 0.2-10.0 keV. Compared to the BeppoSAX X-ray telescope, the XRT has twice the effective area (~110 cm^2 @ 1.5 keV) and four times better angular resolution. The XRT parameters are summarized in Table 2. The instrument is being developed at University of Leicester, Osservatorio Astronomico di Brera, and Penn State University. Further details of the XRT can be found in D. Burrows [25] and M. Zugger [28].

TABLE 2. XRT Parameters

Energy Range	0.2-10 keV
Telescope	JET-X Wolter Type 1
Detector	EPIC CCD
Effective Area	110 cm^3 @ 1.5 keV
Detector Operation	Photon counting, Integrated imaging, & Timing
Field of View	23.6 x 23.6 arcmin
Detection Elements	600 x 600 pixels
Pixel Scale	2.36 arcsec
Telescope PSF	15 arcsec HPD @ 1.5 keV
Sensitivity	2×10^{-14} erg cm^{-2} s^{-1} in 10^4 seconds

The *UV/Optical Telescope (UVOT)* is a 30-cm diameter modified Ritchey-Chrétien equipped with an image intensified CCD covering 170 to 650 nm. It has a FOV 17 arcmin square and is based closely on the design of the XMM Optical Monitor (OM). The UVOT is capable of reaching m_B = 24 in 1000 s (open filter). A filter wheel provides 6 broadband colors plus two grisms and a 4x magnifier. The grisms will obtain spectra with resolving power of $\lambda/\Delta\lambda$ = 200-400 for sources brighter than m_B = 17. The optical point spread function of the telescope is 0.3 arcsec allowing for excellent astrometry. By registering the field against foreground stars, the UVOT will provide < 0.3 arcsec positions. The UVOT parameters are summarized in Table 3. The instrument is being developed at Mullard Space Science Laboratory and Penn State University. Further details of the UVOT can be found in P. Roming [27].

TABLE 3. UVOT Parameters

Wavelength Range	170 nm - 650nm
Telescope	Modified Ritchey-Chrétien
Aperture	30 cm diameter
F-number	12.7
Detector	Intensified CCD
Detector Operation	Photon counting
Field of View	17 x 17 arcmin
Detection Elements	2048 x 2048 pixels
Telescope PSF	0.9 arcsec @ 350nm
Colors	6 broad-band plus 2 grisms
Sensitivity	B=24 in white light in 1000 s
Pixel Scale	0.5 arcsec

3. SWIFT MISSION

The strategy of the Swift mission is to slew to each new GRB position as soon as possible and to follow the GRB afterglows as long as they are visible. To observe the earliest phase of the afterglow, new BAT positions will trigger an autonomous slew of the spacecraft followed by a programmed sequence of observations with the XRT and UVOT.

The initial GRB position is normally determined by the BAT, but positions can also be uploaded from other satellites through a real-time TDRSS uplink (<10 min). Either case will trigger the spacecraft software to calculate and execute an autonomous slew. All calculations of slew path and pointing constraints will be done on-board. Figure of Merit (FoM) software will determine when to slew to a new position. The FoM has a flexible design that can accommodate more focused studies of specific GRB questions as the mission progresses.

Each of the three Swift instruments rapidly produces alert messages after a GRB is detected. To ensure prompt delivery, these messages are sent through a real-time TDRSS downlink to the ground, and routed immediately to the GRB Coordinates Network (GCN) [7] for delivery to the community. Based on the similar process already in place for BATSE, we estimate the entire acquisition and delivery process will take 20 seconds for BAT positions and 96 seconds for XRT positions.

The bulk of the data will be downloaded to the Malindi ground station (5-8 passes per day at 2.25 Mbps) and then sent to the Penn State Mission Operations Center (MOC). The Malindi service is contributed by the Italian Space Agency (ASI). A commercial ground station provides backup. The command uplink at 2 kbps is provided by Malindi and TDRSS.

When Swift is not engaged in prompt observations of the most recent bursts, it will follow a one week schedule uploaded from the ground each working day and as needed. This schedule will provide for long term follow-up of GRB afterglows and other science. The PSU Mission Operational Center (MOC) will be capable of generating a new schedule in < 2 hours.

TABLE 4. Mission Characteristics

Autonomous slew decision capability
Fast Slew - $50°$ in < 75 s
Low Earth Orbit at $22°$ Inclination
Launch Vehicle: Delta 2420 with 3 meter fairing
Mass: 1500 kg
Power: 1040 W
Launch Date: September 2003

4. SWIFT SCIENCE

Recent GRB discoveries have shown that X-ray, optical, and radio afterglows exist, continuing for days after the bursts, but fading quickly t^{-1} to t^{-2} is typical). Better data on faster time scales for many more bursts is needed. (See ref. [8] for discussion of requirements for future GRB missions.) The Swift mission provides the needed capability to answer the following four key science questions: What are the progenitors of GRBs? How does the blast-wave evolve and interact with its surroundings? Are there different classes of bursts with unique physical processes at work? What can GRBs tell us about the early Universe?

4.1. GRB Progenitors

To determine the origin of GRBs three parameters are needed, the total energy released, the nature of the host galaxy (if one exists), and the location within the host galaxy. The Swift mission is optimized to measure all three of these for many hundreds of bursts.

Obtaining the energetics requires a reliable redshift measurement. Ideally this should be done independently for both the afterglow and any host galaxy to check that there has not been a chance coincidence [9]. The UV grisms and filters of Swift can make redshift determinations by searching for the Ly-α cutoff in the UV and eliminate the $1.3 < z < 2.5$ deadband of current observations [9] during the early phase of the afterglow. In addition, time varying optical, UV and X-ray lines and edges are expected within the first hour following a burst from the illumination of the immediate (100 pc) environment by the initial event [10, 11]. The rapid response of Swift will enable a search for predicted X-ray lines (see next section) and again provide a direct redshift measure from the afterglow.

The UVOT will obtain < 0.3 arcsec positions by using background stars to register the field. This will provide a unique host galaxy ID and also allow later comparison with HST fields to determine the position within the galaxy.

There will probably be events where no optical afterglow is detected because of dust extinction surrounding the site of the GRB. The position from the XRT will then be crucial. By obtaining 5.0 arcsec positions, the XRT will enable unique identification of candidate host galaxies down to m_R ~26. Follow-up observations with Chandra made within a couple of days for a selection of these events will give sub-arcsec positions within the Swift 5.0 arcsec error circle.

4.2. Blast-wave Interactions

Afterglow is thought to be produced by the interaction of an ultra-relativistic blast-wave with the interstellar or intergalactic medium. The blast-wave model [12] predicts a series of stages as the wave slows. A key prediction is a break in the spectrum that moves from the gamma to optical band, and is responsible for the power law decay of the source flux. This break moves through the X-ray band in a few seconds, but takes up to 1000 s to reach the optical. Thus observations within the first 1000 s in the optical and UV are crucial to see this early phase. While it now seems likely that all the GRBs have X-ray afterglow, not all have bright optical afterglow (at least after several hours). This may be due to optical extinction, but it is also possible that in some cases the optical (and X-ray) afterglow is present but decays much more rapidly [13] and is a function of the density of the local environment [14]. Prompt high-quality X-ray, UV and optical observations over the first minutes to hours of the afterglow (inaccessible without Swift) are crucial to resolve this question. Continuous monitoring is important since model-constraining flares can occur in the decaying emission.

Star forming regions are embedded in large columns of neutral gas and dust. The presence of extinction can be readily determined by multi-band photometry in the optical and IR. The simultaneous detection of high X-ray absorption, coupled with photometric E (B-V) measurements with Swift, will determine whether dust and gas are present. Continuous monitoring over the first few hours to days will indicate whether dust is building up (due to condensation out of an expanding hot wind) or disappearing (due to ablation and evaporation).

4.3. Classes of GRB

Swift will determine whether sub-classes of GRBs exist and what fundamental differences in the source physics cause the classes. While some evidence of sub-classes has been obtained (e.g. bimodal duration distribution, possible correlation of hardness and logN-logP shape, short-bursts having V/V_{max} consistent with a Euclidean distribution) it is not clear if these are real differences in physical phenomena or simply represent the distribution function of GRB properties such as beaming angle, density of the local medium, or initial energy injection. Swift data will determine locations, redshifts, and afterglow properties of the different classes and thus allow physical understanding of their nature. Central to the confusion regarding potential classes is that we do not have a reliable standard candle. Swift remedies that by directly measuring the distance through redshift. This will give an exact determination of the GRB luminosity function.

Since BeppoSAX does not presently detect bursts shorter than a few seconds, we have no idea of the nature or even existence of optical, X-ray, or radio afterglows for these objects. Swift will be sensitive to the shortest events, and will provide far better coverage of these events than has been possible.

In the interesting scenario that Swift discovers some GRBs that do not have X-ray or UV/optical afterglow, the BAT will still provide positions of 1-4 arcmin, which is sufficient to look for radio or IR counterparts. Only the rapid response of Swift will be able to identify such a new and elusive sub-class of GRB event.

If there are classes of GRBs that are the signal of conventional supernova explosions (e.g., refs. [17, 18]), the UVOT will provide unique and unprecedented coverage of the optical and UV light curve during the early stage.

4.4. Burst as Astrophysical Tools

Since the lifetime of NS-NS or massive star progenitors is short compared to the Hubble time at z < 5, the cosmic GRB rate should be proportional to the star formation rate. The cosmic rate of massive star formation is at present controversial. Estimates that star formation peaks at z ~1-2 and declines sharply at high redshifts have been reported [19]. However, recent IR [20,21] and X-ray cluster [22] results show a considerably higher rate in dust enshrouded galaxies at higher redshifts. Swift, by obtaining a large sample of GRBs over a wide range of fluences and redshifts, will determine whether their evolution follows that of star formation in the Universe and, because the X-ray flux does not depend greatly on the line of sight column, these results will be independent of absorption.

Since GRBs are the most luminous objects in the Universe, they provide a unique opportunity to probe the intergalactic medium (IGM) and the ISM of the host galaxies via measurement of absorption along the line of sight [23]. Depending on evolution, GRBs might originate from redshifts up to ~15 and have a median redshift > 2, larger than that of any other observable population. By rapidly providing both accurate positions and optical brightness, Swift will enable the immediate follow-up of those GRBs bright enough for high resolution optical absorption line spectroscopy at redshifts large enough to study the reionization of the IGM [24]. This information on the high-z Ly-α forest will be unique because there are currently no known bright (m < 17) galaxies or quasars at z > 4.8 [23].

5. GROUND SYSTEM AND DATA ANALYSIS

A layered data analysis approach will be used to achieve rapid dissemination of Swift results and data to the community. The most urgently needed results, namely the GRB positions (gamma & x-ray) and the optical finding charts, are produced on the spacecraft. Quicklook results, including the multi-wavelength light curves, are produced in the MOC in near real-time and distributed using the GCN. Definitive standard products, including spectra, multi-band light curves, and images, will be made into production FITS files. This data base of Swift results will be augmented with contributed results from other observers.

The BAT survey data will be pipe-line processed and be available to the public in a couple days. It will contain an on-going catalog of known sources with lightcurves and spectra plus and newly detected hard x-ray transient sources. The transients sources will be distributed through the GCN and a web site.

All the Swift data will be processed at the Swift data center at Goddard and will be made available to the general public through the HEASARC in the US, a data center in the UK, and a data center in Italy. The end result will be easy access for the entire community to a broad range of timely information on GRBs.

6. SCIENCE TEAM

The Swift science team is made up of world experts in GRB astronomy, space instrumentation, data analysis, theory, GRB follow-up observation, and outreach.

An essential Swift capability is providing the world-wide community of ground and space observers with rapid arcsec positions for hundreds of GRBs. Until now such work has been based on ~10 GRBs/yr, with no information on the optical appearance of the field, and lagging at least 4-8 hours after the burst. We have assembled a Follow-up Team under the leadership of Kevin Hurley, to use their expertise on large facilities to guarantee systematic study of Swift GRBs.

We will actively encourage follow-up observations of all kinds, regardless of membership in the Swift team, by providing precise positions and other data in real-time to the GCN. The GCN founder, Scott Barthelmy, is a member of the Swift team. Observers will be free to use the data with no restrictions. We will encourage all observers to make their data public by maintaining a web database.

ACKNOWLEDGMENTS

I thank John D. Myers for assistance in writing this paper. I gratefully acknowledge Neil Gehrels (PI) and the Swift Science Team for preparing the Swift Proposal and Phase A study on which this paper is based, and the Swift Management and Engineering Team led by Tim Gehringer for the technical development of the mission.

REFERENCES

1. Costa, E., et al., *Nature*, **387**, 783, (1997).
2. Van Paradijs, J., et al., *Nature*, **386**, 686, (1997).
3. Frail, D.A., et al., *Nature*, **389**, 361, (1997).
4. Kouveliotou, C., et al., *ApJ*, **413**, L101, (1993).
5. Klebesadel, R.W., *Proc. Los Alamos Workshop on GRBs*, 1990 Taos, eds. C. Ho, R. Epstein, and E. Fenimore, Cambridge Univ. Press, Cambridge, 161, (1992).
6. Norris, J.P., et al., *ApJ*, **439**, 542, (1995).
7. Barthelmy, S., et al., *Proc. Fourth Huntsville GRB Workshop*, ed. C. Meegan, R. Preece, and T. Koshut, AIP, New York, 1998, 99.
8. Gehrels, N., *Cosmic Explosions*, ed. S. Holt and W. Zhang, AIP, New York, 2000.
9. Hogg, D.W. and Fruchter, A.S., *ApJ*, **520**, 54, (1998).
10. Perna, R. and Loeb, A., *ApJ*, **501**, 467, (1998).
11. Mészarós, P. and Rees, M., *ApJL*, **502**, L105, (1998).
12. Mészarós, P. and Rees, M., *ApJL*, **418**, L59, (1993).
13. Groot, P.J., et al., *ApJL*, **502**, L123, (1998).

14. Piran, T., astro-ph/9807253, (1998).
15. Fenimore, E., Madras, C., and Nayakshin, S., *ApJ*, **473**, 998, (1996).
16. Fenimore, E., et al., *ApJ* **512**, (1999).
17. Bloom, J.S., et al., *Nature*, **401**, 453, (1999).
18. Woosley, S.E., et al., *ApJ*, **516**, 788, (1999).
19. Madau, P., et al., *MNRAS*, **283**, 1388, (1996).
20. Blain, A.W., et al., *MNRAS*, **302** (1999).
21. Rowain-Robinson, M., et al., *MNRAS*, **289**, 490, (1998).
22. Mushotsky, R.F., and Lowenstein, M., *ApJL*, **481**, L63, (1997).
23. Lamb, D.Q. and Reichart, D.E., *ApJ* **536**, (2000).
24. Miralda-Escide, J., *ApJ*, **501**, 15, (1998).
25. Burrows, D. et al., *SPIE Proceedings, X-Ray and Gamma-Ray Instrumentation for Astronomy XI, The Swift Mission*, Swift X-Ray Telescope, [4140-07]
26. Barthelmy, S. D., *SPIE Proceedings, X-Ray and Gamma-Ray Instrumentation for Astronomy XI, The Swift Mission*, Burst Alert Telescope (BAT) on the Swift MIDEX mission, [4140-06]
27. Roming, P. W., et al., *SPIE Proceedings, X-Ray and Gamma-Ray Instrumentation for Astronomy XI, The Swift Mission*, Ultra-Violet/Optical Telescope of the Swift MIDEX Mission [4140-08]
28. Zugger, M. E., D., N. Burrows, and J. Shoemaker, *SPIE Proceedings, X-Ray and Gamma-Ray Instrumentation for Astronomy XI, The Swift Mission*, Laboratory x-ray camera electronics: a testbed for the Swift XRT, [4140-09]
29. Krimm, H., et al., these proceedings.
30. Levine, A.M., et al., *ApJS*, **54**, 581, (1984).

The Swift Ultra-Violet/Optical Telescope

Peter Roming[a], S. D. Hunsberger[a], John Nousek[a], & Keith Mason[b]

[a]*Department of Astronomy & Astrophysics, Pennsylvania State University,
525 Davey Lab, University Park, PA 16802, USA*
[b]*Mullard Space Sciences Laboratory, University College London,
Holmbury St. Mary, Dorking, Surrey RH5 6NT, UK*

Abstract. The Ultra-Violet/Optical Telescope (UVOT) provides the Swift Gamma-Ray Burst Explorer with the capability of quickly detecting and characterizing the optical and ultraviolet properties of gamma ray burst counterparts. The UVOT design is based on the design of the Optical Monitor on XMM-Newton. It is a Ritchey-Chrétien telescope with microchannel plate intensified charged-coupled devices (MICs) that deliver sub-arcsecond imaging. These MICs are photon-counting devices, capable of detecting low intensity signal levels. When flown above the atmosphere, the UVOT will have the sensitivity of a 4m ground based telescope, attaining a limiting magnitude of 24 for a 1000 second observation in the white light filter. A rotating filter wheel allows sensitive photometry in six bands spanning the UV and visible, which will provide photometric redshifts of objects in the 1-3.5z range. For bright counterparts, such as the 9th magnitude GRB990123, or for fainter objects down to 17th magnitude, two grisms provide low-resolution spectroscopy.

INTRODUCTION

Every day a bright flash of gamma rays appear from a random location on the sky and then quickly fades. These gamma ray bursts (GRBs) are the most energetic events since the Big Bang. Since their discovery in 1973 [1], progress in understanding GRBs has been slow due to the difficulty in identifying optical counterparts. From observations of GRB970111, Feroci et al. [2] determined a decay in optical brightness of $\sim t^{-1.3}$. If such rapid decay is typical, then prompt multi-wavelength observations of GRBs is essential.

The Swift Explorer mission is a unique multi-wavelength observatory designed to rapidly slew to GRBs and autonomously begin performing observations. The observatory consists of three co-aligned telescopes: the Burst Alert Telescope (BAT) [3], the X-Ray Telescope (XRT) [4], and the Ultra-Violet/Optical Telescope (UVOT). Collectively, these telescopes provide wavelength coverages of 0.2-150keV and 170-650nm. The primary scientific objectives are to explore the sites and nature of GRB progenitors, pioneer the use of GRBs as probes of the early Universe, and provide a sensitive all-sky hard x-ray survey. An overview of the Swift mission is provided elsewhere [5].

UVOT DESCRIPTION

The UVOT supplies UV/optical coverage within a 17′ x 17′ field. It is a significant supplement to the other Swift instruments despite its small aperture. Because it will be flown above atmospheric extinction, diffraction, and background, the sensitivity of the UVOT is equal to a 4 m telescope on the ground. In addition, its UV capacity permits observations not achievable from the ground. It has photon-counting detectors (able to preserve individual photon positions and timing information), therefore it functions in a mode more analogous to x-ray telescopes than to conventional optical telescopes.

The UVOT is co-aligned with the XRT and supported on an optical bench. An 11-position filter wheel permits low-resolution grism spectra of bright GRBs, magnification, and broadband UV/visible photometry. Photons register on a microchannel plate intensified charged-coupled device. A summary of the UVOT's characteristics is in Table 1 and a summary of the broadband filters is in Table 2.

The UVOT possesses a strong heritage from Optical Monitor (OM) on ESA's XMM-Newton mission. To minimize risk, the UVOT design deviates from the OM hardware design only when required by new interfaces to the spacecraft, OM part unavailability, or the inability of OM to accommodate a Swift project requirement.

The modifications from OM to the UVOT design comprise: a new Instrument Control Unit (ICU) safing software for bright object avoidance, a new Digital Processing Unit (DPU), new DPU processing software, ICU/DPU-to-spacecraft bus interface software, an increase in filter wheel revolutions, and a new telescope door.

The UVOT ICU software controls instrument safing and autonomous processes. As the UVOT DPU design has a dedicated 1553 spacecraft interface, the ICU will not be tied up with interprocessor communications, reducing CPU cycles for safing activities. The ICU can calculate whether the field-of-view is safe for UVOT to observe by accessing an onboard bright source catalogue.

The UVOT DPU is built around the RAD6000 microprocessor. New DPU software is being created to process the science data and new ICU/DPU-to-spacecraft bus interface software is being written to telemeter the data.

TABLE 1. UVOT Characteristics

Telescope	Modified Ritchey-Chrétien
Aperture	30 cm diameter
F-number	12.7
Detector	Intensified CCD
Detector Operation	Photon Counting
Field of View	17′ × 17′
Detection Element	256 × 256
Resolution	2048 × 2048 after centroiding
Telescope PSF	0.9″ FWHM @ 350nm
Wavelength Range	170-650nm
Filters	11 (One blocked)
Sensitivity	B=24.0 in white light in 1000s
Pixel Scale	0.5″

TABLE 2. Broadband Filters

Name	λ_C (nm)	FWHM (nm)
UVW2	180	80
UVM2	220	70
UVW1	260	100
U	360	70
B	420	100
V	550	90

Despite the increase in filter wheel rotations required by the UVOT as compared to OM, the OM filter wheel is rated to double the number of rotations required by the UVOT. A redesign of the filter wheel assembly will not be needed. A new UVOT door is being built.

The UVOT is broken down into 5 basic units: a Telescope Module enclosing a UV/optical telescope, two redundant photon counting detectors, two filter wheel mechanisms, a steering mirror mechanism, power supplies, and electronics; two redundant Digital Electronics Modules (DEMs), each one housing a DPU, an ICU, and power supplies for the DPU and ICU; and two Interconnecting Harness Units to connect the TM to the two DEMs. A more comprehensive description of the UVOT can be found elsewhere [6].

OBSERVING STRATEGY

A library of automated observing sequences will be included onboard as part of the UVOT flight software. The UVOT will execute a "standard" GRB observing sequence such as that summarized in Table 3. The science software generates two types of data: image and event. Images are produced to allow the telescope to function as a conventional optical telescope. The instrument can also record time-stamped photon events over the entire detector frame. Both types of data can be generated simultaneously. The software collects data within defined "windows" that are regions on the detector containing the GRB.

When the BAT detects a GRB, the spacecraft slews to position all instruments on target. During the slew, no data is collected. When the spacecraft is within 10' of the target position, the "settle" phase begins. As the spacecraft is maneuvering into its final position, the target is moving within the field-of-view of the UVOT detector. The collection of the earliest UV photons then begins. The time and location of each photon event is recorded. Messages from the Attitude Control System on the spacecraft are also saved and used to reconstruct the GRB data on the ground later.

TABLE 3. Standard GRB Observing Sequence

Phase	Data Type	Data Windows	Filter Position(s)	Observation Time
Slew	none	none	Blocked	~50sec
Settle	event	17' × 17'	UVW2	~35sec
Finding Chart	image	8' × 8'	V	100sec
	event	8' × 8'		
10sec exposures	event	2' × 2'	UVW2, UVM2, UVW1, U, B, V	10sec per filter, repeat 10 times for a total observation of 600sec
100sec exposures	image	4' × 4'	UVW2, UVM2, UVW1, U, B, V	100sec per filter, repeat 5 times for a total observation of 3000sec
	event	2' × 2'		
1000sec exposures	image	4' × 4'	UVW2, UVM2, UVW1, U, B, V	1000sec per filter for a total observation of 6000sec
	event	2' × 2'		

One of the mission requirements is to rapidly provide a finding chart to the GRB Coordinate Network (GCN). An 8' × 8' window is centered on the GRB position provided by the BAT and both image and event data are collected. In order to use the real-time downlink capability of TDRSS, the data buffer is limited to only 2000 bytes. This is insufficient to send the entire finding chart image. Instead, the UVOT software uses the image to create a list of the 50 brightest stellar sources. Each source is parameterized by position and a 5×5 pixel image. The source list is sent to the ground immediately via the TDRSS network and a reconstructed image is posted to the GCN website. The full finding chart image and event data will be relayed during routine ground station contact.

Next, a series of 10-second exposures is executed, followed by a series of 100-second and then 1000-second exposures. The observing sequence continually rotates through the six broadband filters. The idea is to capture data in each filter quickly enough to provide meaningful light curves. This is critical immediately after the burst when the decay in brightness is greatest. It is also important to reduce the data window sizes as much as possible. The image window size is constrained by the telemetry rate and by CPU resources, i.e., processing time and memory allocation. The window sizes listed in Table 3 assume that a more accurate position has been provided by the XRT.

DATA SIMULATION

Theoretical models can be used in a data simulation program. Results presented here focus on three variations of the dissipative fireball model [7]. There are two basic types of models. In type "a" models, the initial energy input is impulsive. As relativistic ejecta interact with the surrounding medium, much of the energy is radiated as external and reverse shocks. In type "b" models, the energy input is continuous over some period of time, e.g., a relativistic wind. The resultant radiation is due to internal shocks within the wind. As shown in Figure 1, there are significant differences in peak optical flux and decay rates. In the type a1 model, a forward blast wave is the dominant factor. In the type a2 model, a reverse shock is considered. Type b2 is a wind model. Specific model parameters are detailed in Mészáros and Rees [7].

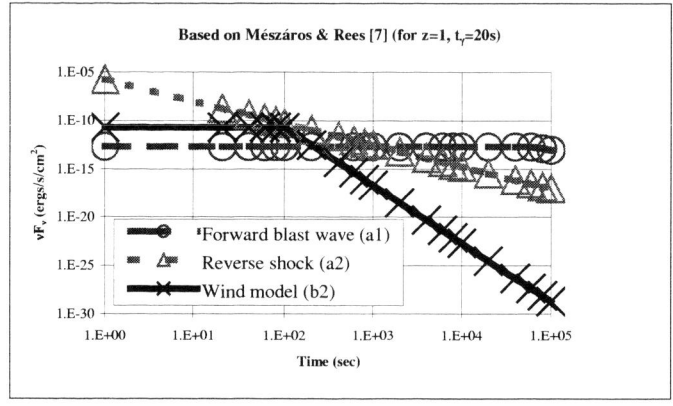

FIGURE 1. GRB Models

One goal of the simulation is to determine if differences between models can be seen in data collected by the UVOT. Figure 2 shows resultant V-band light curves for each model assuming that the standard UVOT GRB observing sequence is executed. Based on this example, the critical period for distinguishing between models is the first 1000 seconds (during the finding chart and 10sec exposures).

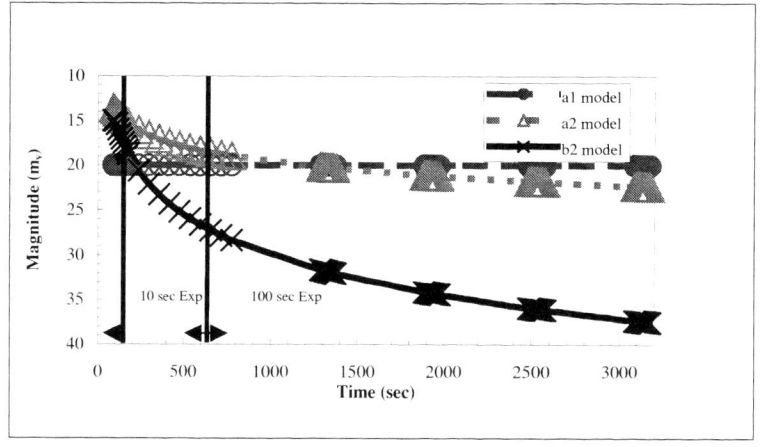

FIGURE 2. V-band Light Curves

ACKNOWLEDGMENTS

We would like to acknowledge the members of the UVOT team at Penn State University, Mullard Space Sciences Lab, and Southwest Research Institute for all their efforts. This work is sponsored at Penn State by NASA's Office of Space Science through contract NAS5-00136, and at MSSL by funding from PPARC.

REFERENCES

1. Klebesdael, R. W., Strong, I. B., and Olson, R. A., Astrophysical Journal **182**, L85-L88 (1973).
2. Feroci, M., et al., Astronomy and Astrophysics **332**, L29-L33 (1998).
3. Barthelmy, S. D., et al., "The Burst Alert Telescope (BAT) on the Swift MIDEX mission," in *X-Ray & Gamma-Ray Instrumentation for Astronomy* XI, edited by K. A. Flanagan & O. H. W. Siegmund, SPIE Proceedings Series 4140, 2000, pp. 50-63.
4. Burrows, D. N., et al., "The Swift X-ray Telescope," in *X-Ray & Gamma-Ray Instrumentation for Astronomy* XI, edited by K. A. Flanagan & O. H. W. Siegmund, SPIE Proceedings Series 4140, 2000, pp. 64-75.
5. Gehrels, N. A., "The Swift Gamma Ray Burst MIDEX," in *X-Ray & Gamma-Ray Instrumentation for Astronomy* XI, edited by K. A. Flanagan & O. H. W. Siegmund, SPIE Proceedings Series 4140, 2000, pp. 42-49.
6. Roming, P. W. A., et al., "The Ultra-Violet/Optical Telescope of the Swift MIDEX Mission," in *X-Ray and Gamma-Ray Instrumentation for Astronomy* XI, edited by K. A. Flanagan and O. H. W. Siegmund, SPIE Proceedings Series 4140, 2000, pp. 76-86.
7. Mészáros, P. and Rees, M. J., Astrophysical Journal **476**, 232-237 (1997).

Swift Burst Alert Telescope Hard X-Ray Monitor and Survey

Hans A. Krimm[*,¶], Louis M. Barbier[¶], Scott D. Barthelmy[¶], Anthony J. Dean[†], Ardeshir Eftekharzadeh[§,¶], Edward E. Fenimore[+], Neil Gehrels[¶], Derek D. Hullinger[§,¶], Hideki Ozawa[*,¶], David M. Palmer[+], Ann M. Parsons[¶], Tadayuki Takahasi[ψ], Makoto Tashiro[☆], Jack Tueller[¶], and Georg Weidenspointner[*,¶]

[*]*Universities Space Research Association, 7501 Forbes Blvd., Suite 206, Seabrook, MD 20706 USA*
[¶]*Code 661, NASA Goddard Space Flight Center, Greenbelt, MD 20771, USA*
[†]*University of Southampton, Highfield, Southampton, UK SO17 1BJ*
[§]*Department of Physics, University of Maryland, College Park, MD 20742, USA*
[+]*Los Alamos National Laboratory, PO Box 1663 Los Alamos, NM 87545 USA*
[ψ]*Institute of Space and Astronautical Science, 3-1-1 Yoshinodai, Sagamihara, Kanagawa 229-8510, JAPAN*
[☆]*University of Tokyo, 7-3-1 Hongo, Bunkyo-ku, Tokyo 113-8654 JAPAN*

Abstract. The Burst Alert Telescope (BAT) on the Swift gamma-ray burst mission will perform the first new all sky hard x-ray survey since 1977. BAT is a coded aperture instrument with 17 arcminute pixels and a 2 ster partially coded field of view. The imaging area has 32768 CdZnTe detectors, each 4X4X2 mm, with a total area of 5243 cm^2. Swift will perform pointings covering >64% of the sky each day and achieve an integrated sensitivity in three years of 0.6 milliCrabs for sources well off the Galactic plane. This survey is expected to identify hundreds of new highly obscured AGN.

THE BURST ALERT TELESCOPE (BAT)

The Burst Alert Telescope on the Swift gamma-ray burst mission is a coded aperture telescope which provides the gamma-ray burst triggers for the mission. BAT will locate several hundred bursts per year to an accuracy of better than 4 arc minutes. Upon receipt of a valid BAT trigger the Swift satellite will rapidly slew to point the narrow field instruments, the X-Ray Telescope (XRT) and the Ultra-Violet / Optical Telescope (UVOT), toward the gamma-ray burst. The BAT will simultaneously provide follow-up hard x-ray and gamma-ray observations of burst afterglows. In

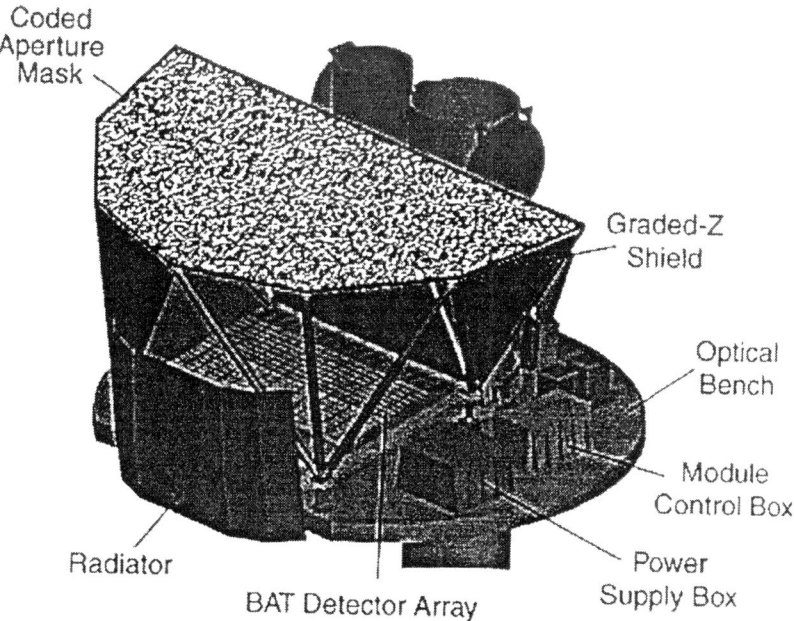

FIGURE 1. The BAT instrument. The coded aperture mask is built of 5 mm X 5 mm x 1 mm tungsten plates and is located one meter from the detector array. The graded-Z shield (partially cut away in the drawing) reduces the background from gamma rays coming from outside the aperture.

addition to detecting and observing gamma-ray bursts, the BAT will provide a hard x-ray survey of unparalleled sensitivity.

The BAT instrument is shown in Figure 1 and its parameters are listed in Table 1.

SURVEY SCIENCE

The BAT will produce the most sensitive hard x-ray survey ever made. The survey will be carried out during GRB pointings and daily sweeps of most of the sky.

TABLE 1. BAT Parameters

Aperture	Coded Mask
Detecting Area	5200 cm^2
Detector	CdZnTe
Detector Operation	Photon counting
Field of View	1.4 sr (half coded)
Detection Elements	256 modules of 128 elements
Detector Size	4 mm X 4 mm X 2 mm
Telescope PSF	17 arcmin
Energy Range	15-150 keV

Assuming uniform coverage to estimate sensitivities, the BAT instrument will provide an exposure of 6.5 X 10^{10} cm^2-s for each sky pixel yielding a 5σ statistical sensitivity of 200 mCrab in the 15 – 100 keV band. We estimate the BAT detection sensitivity to be systematics limited at ~0.6 mCrab at high galactic latitude (> 45°) and ~2 mCrabs when strong galactic sources are in the field of view (FOV). These levels correspond to 15σ and 50σ detections in a statistical sense. This survey will be 30 times more sensitive than the best complete hard x-ray survey (HEAO A-4 [3], complete to 17 mCrab) for 1/3 of the sky and 10 times more sensitive for the rest. For known sources >2 mCrab, we will be able to measure a 3σ flux in <1 day providing detailed hard x-ray light curves for hundreds of sources. Swift's large FOV ensures good coverage for the Constellation-X and GLAST missions.

More than 400 AGN will be detected in the survey. Recent studies with ASCA, Ginga, and BeppoSAX have shown the existence of a large population of highly absorbed Seyfert 2 galaxies with line-of-sight column densities > 10^{23} cm^{-2}. The large column makes the nuclei of these objects essentially invisible at optical and soft x-ray wavelengths. Detailed models [2,4] show that such a population of highly absorbed AGN is needed to produce the observed 30 keV bump in the hard x-ray background and that it comprises close to half of all AGN. The only known method of detecting such objects is an unbiased sky survey in the E > 15 keV band of sufficient sensitivity to detect a large population. We expect Swift to detect > 300 AGN brighter than 0.6 mCrab within 45 degrees of the galactic poles and >100 AGN brighter than 2 mCrab in the rest of the sky. At this time only ~20 AGN have high quality detections at energies > 30 keV [1].

BAT will be sensitive to Soft Gamma Repeaters (SGR) because of its excellent short burst trigger. While the SGR bursts are shorter than the Swift slew time, the XRT and UVOT will perform sensitive searches immediately after the burst for x-ray and optical counterparts and will likely be on-target when following bursts occur.

The Swift laboratory has the capability for rapid reaction science, using the TDRSS uplink. This will provide a unique ability to respond in a matter of minutes with sensitive gamma-ray, x-ray, UV, and optical observations to most events on the sky. This includes targets of opportunity for x-ray transients, pulsar glitches, outbursts from dwarf novae, and stellar flares. The highly variable black hole binary GX 339-4 is a good example of the importance of rapid multiwavelength observations for understanding accretion physics [5]. The BAT is > 10 times more sensitive as a monitor than BATSE and will initiate many of the targets of opportunity. As with any new observational capability, the potential for serendipitous science return is high.

SURVEY ANALYSIS

Transient Detection

Hard x-ray transients are dynamic on time scales of minutes to hours. On-board processing will allow rapid response to transients and transient alerts will be promptly relayed to the ground. Ground trasient analysis provides a more sensitive search for weak transients, but at a substantial time delay.

Description of Survey Procedure

The basic survey approach will be to record five minute detector plane images continuously as well as rates with as fast as 4 millisecond time resolution. The images for each pointing in each orbit will be converted to sky images without background subtraction. A peak search will be performed and all bright sources will be tested against the on-board source list. If there is an unidentified source or interesting variability of a known source, a TDRSS alert will be generated after screening by the BAT team.

On the ground, the BAT analysis program will do a full iterative clean deconvolution for the multiorbit integrated image of each pointing position. This procedure is described graphically in Figure 2. A peak search and error analysis will

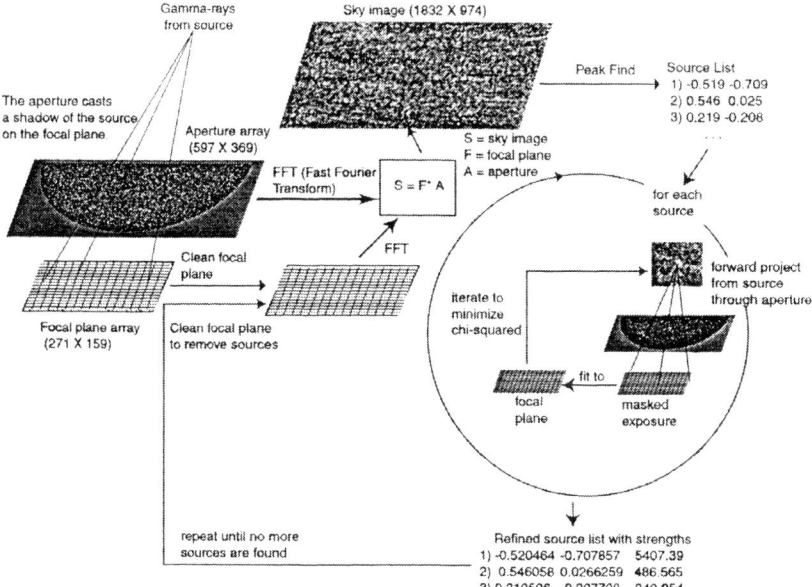

FIGURE 2. Flow Chart of Basic Image Analysis. The first steps in the analysis of the image data are shown. These include a deconvolution for each pointing and a clean algorithm for the background and for each source.

be performed for all significant source detections and for the positions of known hard x-ray emitters to extract refined positions, light curves, and energy spectra. Each source will be identified and interesting transients will be reported through the GRB Coordinates Network and other appropriate channels as soon as possible.

A deep survey will be achieved by adding together the individual sky images for each pointing location to make an all sky map. The BAT team anticipates that most of the sources will be readily identified using existing x-ray surveys and other archival information, but the expected BAT position uncertainty (~4 arcmin radius) is not small enough to ensure the identification of all sources. In the BAT error circle we expect to find an average of one to two extragalactic sources, but clustering will result in a significant number of fields with many galaxies. It will be critical to make followup observations of these positions with the Swift narrow field instruments to help identify these sources.

Simulations

As a tool to help us better understand the BAT survey sensitivity and to provide simulated data for use in developing the survey software, the BAT team is making extensive use of a photon tracking monte carlo program, *grmcflight*, developed by Ed Fenimore at Los Alamos National Laboratory. This program allows us to accurately model the BAT instrument and its environment and to simulated the spectra of AGN and other astrophysical sources including the diffuse photon and particle background. Additional modules allow us to simulate detector noise and efficiency and data pipeline processing.

REFERENCES

1. Dermer, C. and Gehrels, N., *ApJ* **447**, 103 (1995).
2. Haginger, G., *Astronomische Nachrichten*, **319**, 37 (1998).
3. Levine, A.M., *et al.*, *ApJS* **54**, 581 (1984).
4. Madau, P., *et al.*, *MNRAS* **270**, 117 (1994).
5. Smith, I. A.; Filippenko, A. V.; Leonard, D. C., *ApJ* **519**, 779 (1999).

The GLAST Burst Monitor (GBM)

R. M. Kippen*, M. S. Briggs*, R. Diehl[†], G. J. Fishman**, R. H. Georgii[†],
C. Kouveliotou**, G. G. Lichti[†], C. A. Meegan**, W. S. Paciesas*,
R. D. Preece*, V. Schönfelder[†] and A. von Kienlin[†]

*University of Alabama in Huntsville, Huntsville, AL 35899, USA
[†]Max-Planck-Institut für extraterrestrische Physik, 85748 Garching, Germany
**NASA/Marshall Space Flight Center, Huntsville, AL 35812, USA

Abstract. The study of gamma-ray bursts (GRBs) is one of the primary scientific objectives of the Gamma-ray Large Area Space Telescope (GLAST) mission. With its high sensitivity to prompt and extended 20 MeV to 300 GeV burst emission, GLAST's Large Area Telescope (LAT) is expected to yield significant progress in the understanding of GRB physics. To tie these breakthrough high-energy measurements to the known properties of GRBs at lower energies, the GLAST Burst Monitor (GBM) will provide spectra and timing in the 10 keV to 25 MeV energy range. The GBM will also have the capability to quickly localize burst sources to $\sim 15°$ over more than half the sky, allowing the LAT to re-point at particularly interesting bursts which occur outside its field of view. With combined LAT/GBM measurements GLAST will be able to characterize the spectral behavior of many bursts over nearly six decades in energy. This will allow the unknown aspects of high-energy burst emission to be explored in the context of well-known low-energy properties. In this paper, we present an overview of the GBM instrument, including its technical design, scientific goals, and expected performance.

GAMMA-RAY BURSTS AND GLAST

Recent breakthrough discoveries, along with decades of observations and theoretical speculations, have brought gamma-ray bursts (GRBs) into the forefront of astrophysics research. The enigmatic GRB phenomenon — the most powerful in the Universe yet discovered — has captured the wonder and imagination of a broad audience, such that any future high-energy observatories are bound to have GRBs as a major scientific objective. The Gamma-ray Large Area Space Telescope (GLAST), planned for launch in 2006, is no exception.

The GLAST Large Area Telescope [LAT; see 1] will provide particularly important insight into the physics of gamma-ray bursts as it will probe — with high sensitivity over a wide field of view — the relatively unknown aspects of GRB emission above 20 MeV, where the effects of high-energy particle acceleration, relativistic beaming, and intergalactic attenuation are most clearly observed. Measurements obtained with the GLAST-LAT will build on the handful of bursts detected by CGRO-EGRET in this energy range [2, 3], which have supplied unique and important constraints on the energy release and photon production mechanisms [see e.g., 4, 5]. In addition to providing high-quality spectral and temporal measurements, the GLAST-LAT will also be able to localize more than 100 GRB sources per year with $\sim 10'$ precision [6], and provide these locations to follow-up observers within minutes of burst onset. Thus, GLAST will be an important component in the new science of tracking GRB afterglow emission at different wavelengths over extended periods of time. This approach promises to yield fruitful scientific return, as it probes the GRB phenomenon in vastly different physical regimes.

Although the GLAST-LAT holds great promise for future GRB research, there are important limitations to its effectiveness as a burst detector. Foremost among these is the problem of continuity with the current knowledge of GRBs, which is based mainly on measurements of low-energy gamma rays below 1 MeV. With the LAT alone it will be difficult to evaluate how the ground-breaking high-energy observations fit into the known low-energy characteristics of GRB behavior. This problem is most evident in terms of GRB energy spectra, where the most characteristic known feature — a break typically in the energy range 100–500 keV [7, 8] — occurs at energies well below the LAT threshold. The EGRET observations of extended or delayed high-energy flux [2, 3] suggest different or evolving emission processes that will be difficult to evaluate in a limited energy range. Other significant concerns for the LAT as a burst detector are the technical problems associated with autonomous triggering and rapid source localization given the large on-orbit background rate and small number of detected source photons for weak bursts. In order to overcome or mitigate these problems, GLAST will include a secondary instrument: the GLAST Burst Monitor (GBM[1]). This paper [see also 9] provides a brief introductory description of the GBM instrument, and how it will perform to enhance the GLAST mission in the study of GRBs.

GBM ROLE AND REQUIREMENTS

The primary role of the GBM is to enhance the scientific return of GLAST GRB observations by providing simultaneous low-energy spectral and temporal measurements for all GRBs that occur within the LAT field of view. This requires an effective energy range extending low enough to measure well-below the typical GRB spectral break, and high enough to overlap with LAT measurements for inter-instrument calibration. Furthermore, the GBM sensitivity and field of view (FoV) must be commensurate with the LAT capabilities to ensure that many bursts will have simultaneous low-energy and high-energy measurements with similar statistical significance. The GBM will also assist the LAT in its ability to rapidly detect and localize bursts by providing prompt burst trigger notification information. The secondary GBM objective is to provide coarse burst locations over a wide FoV that can be used to re-point the LAT at particularly interesting bursts for performing afterglow observations, or to notify external follow-up observers.

The GBM instrument performance requirements to achieve its scientific objectives are listed in Table 1. In this context, a requirement is a key capability that the instrument will be designed to achieve, while a goal represents a capability to strive for that enhances the scientific measurement performance. Also listed in the table is the key item that drives each requirement. It is important to note that a fundamental requirement not included in

---

[1] http://gammaray.msfc.nasa.gov/GBM

TABLE 1. Summary of GBM Scientific Performance Requirements

Parameter	Requirement	Goal	Main Driver
Low Energy Limit	10 keV	5 keV	Characterize spectra below break
High Energy Limit	25 MeV	30 MeV	Overlap LAT energy range
Energy Resolution[a]	<25%	<18%	Continuum spectroscopy
Field of View[b]	>8 sr	>10 sr	Match, exceed LAT FoV
Time Accuracy[c]	<10 μs	<2 μs	Measure rapid variability
Average Dead Time	<10 μs/count	<3 μs/count	Measure intense pulses
Burst Sensitivity[d]	<0.5	<0.3	Consistent with LAT GRB sensitivity
Burst Alert Locations[e]	—	<15°	Sufficient to re-point LAT
Burst Alert Time Delay[f]	<2 s	<1 s	Less than typical GRB duration

[a] FWHM, 0.1–1 MeV
[b] Co-aligned with LAT field of view
[c] Relative to spacecraft time
[d] Peak flux for 5σ detection in ph$\cdot$cm$^{-2}\cdot$s^{-1} (50–300 keV)
[e] 1σ systematic error radius
[f] Time from burst trigger to spacecraft notification. Used to notify ground or LAT.

the table is that the GBM is constrained to consume only a small fraction ($\lesssim$5%) of the overall GLAST mission resources (e.g., mass, power, cost), and shall in no way detract from the LAT's operation or performance.

GBM INSTRUMENT DESIGN

Given the stringent resource limits, the GBM is forced to be a relatively modest instrument. To fit the required performance within the limitations, the design and technology borrow heavily from previous GRB instruments, particularly from CGRO-BATSE. Like BATSE, the GBM design is based on the use of two types of cylindrical crystal scintillation detectors, whose light is read out by photomultiplier tubes (PMTs).

An array of 12 sodium iodide (NaI) detectors (0.5-in thick, 5-in diameter) are employed to cover the lower end of the energy range up to $\sim$1 MeV. Each NaI detector consists of the crystal, an aluminum housing, a thin beryllium entrance window on one face, and a 5-in diameter PMT assembly (including a pre-amplifier) on the other. These detectors will be distributed around the GLAST spacecraft (see Figure 1) with different orientations so as to provide the required sensitivity and FoV. The thin NaI detectors produce a cosine-like off-axis response that will be used to localize burst sources by comparing the counting rates from detectors with different viewing angles. To cover higher energies, the GBM will also include two 5-in thick, 5-in diameter bismuth germanate (BGO) detectors. The BGO detectors have a combination of high-density (7.1 g cm^{-3}) and large effective $Z \approx 63$ that results in good stopping power up to the start of the LAT energy range $\sim$20 MeV. They will be placed on opposite sides of the GLAST spacecraft to provide high-energy spectral capability over approximately the same FoV as the NaI detectors. For redundancy, each BGO detector will have two PMTs, located at opposite ends of the crystal.

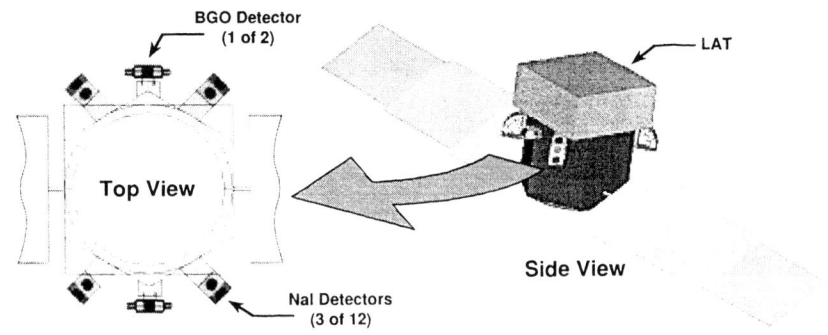

FIGURE 1. Preliminary concept for placement of the GBM detectors on the GLAST spacecraft.

The signals from all 14 GBM detectors will be collected by a central Data Processing Unit (DPU). This unit will digitize and time-tag the detector pulse height signals, package the resulting data into several different types for transmission to the ground (via the GLAST spacecraft), and perform various data processing tasks such as autonomous burst triggering. In addition, the DPU will be the sole means of controlling and monitoring the instrument. Included in this function is the ability to control adjustable detector power supplies to enable automatic control of the gain of each PMT.

There will be three basic types of science data: (1) continuous data will consist of the counting rates from each detector with various (selectable) energy and time integration bins; (2) burst trigger data will contain lists of individually time-tagged pulse height events from selected detectors for periods before and after each on-board burst trigger; (3) burst alert data will contain computed information from a burst trigger, such as intensity, location, and classification. The burst alert data will receive priority telemetry that will allow transmission to the ground at any time in less than 7 s. Alerts will also be made available to the LAT and to the spacecraft to aide in LAT burst detection and for making re-pointing decisions. The remaining data types will be transmitted via discrete ground contacts with a typical latency of $\lesssim 12$ hours.

EXPECTED PERFORMANCE

We have performed simulations to assess the GRB measurement performance of the GBM instrument. These, include Monte Carlo simulations of the physical detector response, measured detector performance properties, and background rates scaled appropriately from BATSE measurements. With these assumptions, and using detection criteria similar to those of BATSE ($>4.5\sigma$, 50–300 keV, in at least two detectors in 1.024 s), the predicted GBM burst detection rate ranges from 150–225 per year, depending on the pointing schedule of GLAST (i.e., the fraction of time Earth blocks the GBM FoV). This rate is commensurate with that expected for the LAT [6]. In practice, a higher GBM burst detection rate will be achieved with a more flexible trigger algorithm that provides improved background estimates, and uses several different energy ranges and time-scales.

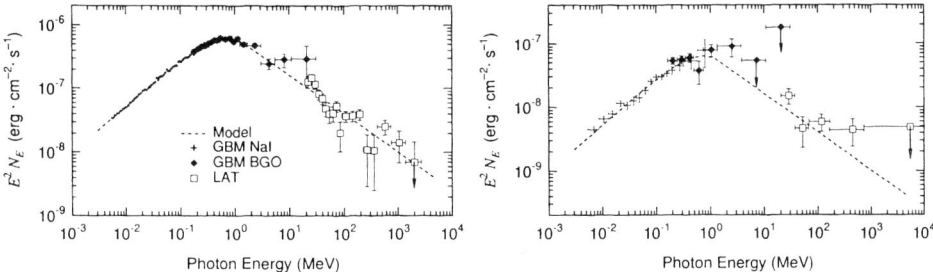

FIGURE 2. Simulated GBM and LAT spectral measurements of the bright GRB 940217 (*left*), and the same burst dimmed by a factor of ten (*right*).

The average GBM statistical location uncertainty for all triggered bursts is estimated to be $\sim 15°$ (1σ radius), and improves to $\sim 9°$ ($\sim 1.5°$) for the brightest 40% (5%) of the bursts. The systematic location error is estimated to be $\sim 1-2°$ for final ground processed data, and $\sim 5-10°$ for on-board processing.

To evaluate the spectroscopic performance of the GBM, detailed simulations were performed for a variety of different spectral characteristics, burst intensities, and observing conditions, including joint observations with the LAT. Figure 2 shows examples of simulated observations of the bright GRB 940217 (as characterized by CGRO measurements [3]). For this burst the joint GBM+LAT measurements constrain the time-averaged burst spectrum over more than five decades in energy with typical statistical uncertainties in the spectral parameters of less than 1% ($\sim 2-10\%$ when the burst is dimmed by a factor of ten). In addition to measuring the low-energy spectral regime below the LAT threshold, the GBM significantly improves the constraints on high-energy spectral behavior compared to those of the LAT alone. For instance, in the simulation of GRB 940217 the uncertainty in the high-energy spectral index improves by a factor of four compared to a fit of the LAT data alone. The combination of GBM and LAT will thereby provide a powerful tool to study GRB spectra and their underlying physics.

REFERENCES

1. Michelson, P., these proceedings.
2. Dingus, B. L., et al., in *Gamma-Ray Bursts: Second Huntsville Workshop*, eds. G. J. Fishman, J. J. Brainerd, & K. Hurley, AIP Conf. Proc. **307**, New York, 1994, p. 22.
3. Hurley, K., et al., *Nature* **372**, 652 (1994).
4. Baring, M. G., and Harding, A. K., *ApJ* **491**, 663 (1997).
5. Mészáros, P., and Rees, M. J., *MNRAS* **269**, L41 (1994).
6. Bonnell, J. T., et al., in *Gamma-Ray Bursts: Fourth Huntsville Symp.*, eds. C. Meegan, R. Preece, & T. Koshut, AIP Conf. Proc. **428**, New York, 1998, p. 884.
7. Band, D., et al., *ApJ* **413**, 281 (1993).
8. Preece, R. D., et al., *ApJSS* **126**, 19 (2000).
9. von Kienlin, A., et al., in *Proc. of the 4th INTEGRAL Workshop*, Alicante, Spain (Sept. 2000).

Future Missions (INTEGRAL)

Science with INTEGRAL in Perspective

V. Schönfelder

Max-Planck-Institut für extraterrestrische Physik, P.O. Box 1312, D-85741 Garching, Germany

Abstract. The ESA-Mission INTEGRAL (International Gamma-Ray Astrophysics Laboratory) is the next step in low-energy gamma-ray astronomy (up to 10 MeV) and will be launched in 2002. Its two main instruments, the spectrometer SPI and the Imager IBIS complement each other: SPI is dedicated to high-resolution line spectroscopy ($\Delta E = 2.5$ keV FWHM at 1.3 MeV) and IBIS to fine imaging (angular resolution : 12' FWHM). These two co-aligned main instruments are complemented by two monitors in the X-ray (3 - 35 keV) and optical (V, 550 nm) range. With this combination of instruments unprecedented sensitivities will be achieved for the study of astrophysical continuum and line emitting processes in compact and extended source regions. INTEGRAL will be an Observatory-type mission. Most of the observing time will be awarded to the scientific community (65 % during the first year of the mission). The remaining time is reserved for the INTEGRAL Science Working Team.

INTRODUCTION

The International Gamma-Ray Astrophysics Laboratory (INTEGRAL) of ESA will be the next step in space-borne gamma-ray astronomy after the successful missions of SIGMA on GRANAT and the Compton Gamma-Ray Observatory. INTEGRAL will be devoted to high-resolution spectroscopy and fine source imaging in the energy range 15 keV to 10 MeV, and in addition, will allow parallel monitoring of the observed gamma-ray sources in the neighbouring X-ray band and at optical wavelengths (Winkler and Hermsen, 1999; and Winkler 1999).

INTEGRAL was selected in 1993 as a Medium Size Mission of the ESA programme. It is worth mentioning that the present concept of INTEGRAL differs somewhat from that proposed to ESA in 1993. Most of the changes were made after it became clear that a major involvement of the US could not be realized and when the UK decided not to participate in the building of the scientific payload.

INTEGRAL is expected to be launched in 2002 by a Russian PROTON rocket, which will bring the laboratory into a 72 hour elliptical orbit (inital perigee 10 000 km, apogee: 153 000 km) of 51.6° inclination. The nominal mission life time will be 2 years. An extension by three additional years is possible. INTEGRAL will be operated as an observatory: most of its observation time will be made available to the scientific community.

THE SCIENTIFIC PAYLOAD

INTEGRAL will carry four instruments: the two main instruments SPI (Spectrometer INTEGRAL) and IBIS (Imager on-Board the INTEGRAL Satellite), and the two monitors

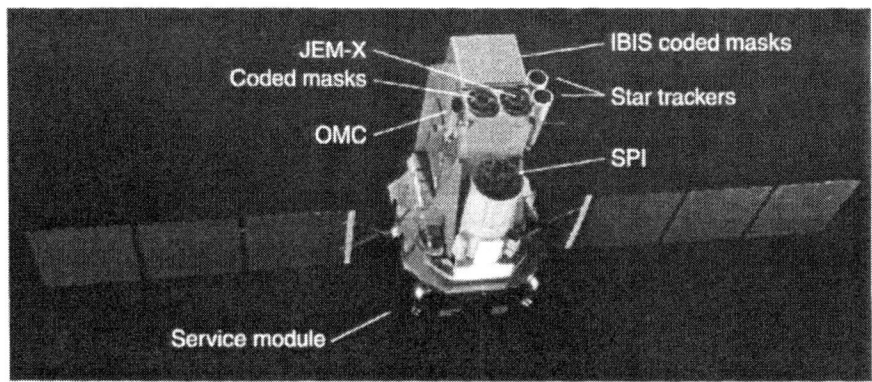

FIGURE 1. Schematic View of INTEGRAL with its four instruments.

JEM-X and OMC.

SPI (20 keV to 8 MeV) will be devoted to high-resolution spectroscopy with modest imaging. Within its fully coded field-of-view of 16° FWHM, its angular resolution is about 2.5° FWHM (Vedrenne et al., 1999). SPI will build upon the results which OSSE and COMPTEL obtained in the field of gamma-ray line spectroscopy. It will be able to measure the profiles of gamma-ray lines to an accuracy of about 2 keV. This is its most important aspect. The knowledge of the line profile allows a detailed study of the physical conditions of the sky regions in which the lines are produced, like temperature, expansion velocity, ionization degree, and emission geometries. A similar experiment (called GRSE) was also originally planned for the Compton Gamma-Ray Observatory, but was never realized.

The imager IBIS (15 keV to 10 MeV) will have an excellent angular resolution of 12 arc min within its fully coded field-of-view of 9° FWHM (Ubertini et al., 1999). IBIS may be considered as the follow-on step after SIGMA on GRANAT, which had practically the same angular resolution, but which was less sensitive by about a factor of 10.

The two monitor instruments will allow parallel monitoring of the gamma-ray sources in the adjacent X-ray band and at optical wavelengths. JEM-X (3 keV to 35 keV) has an excellent angular resolution of 3 arc min in its 4.8° FWHM field-of-view (Lund et al., 1999). The optical monitor camera OMC (500 to 600 nm) has a pixel resolution of 16.6 arc sec within its 5.0° x 5.0° field-of-view (Giménez et al., 1999).

SPI, IBIS and JEM-X are all based on the coded aperture imaging technique. SPI uses germanium detectors, IBIS CdTe and CsI detectors, and JEM-X microstrip Xenon gas detectors. OMC consists of a passively cooled CCD in the focal plane of a 50 mm lens.

THE INTEGRAL SENSITIVITIES

The scientific return of the INTEGRAL mission will mainly depend on the sensitivities of its instruments. The actual sensitivities will only be known after launch, when the in-orbit background is known. Much effort has been invested to predict the background from

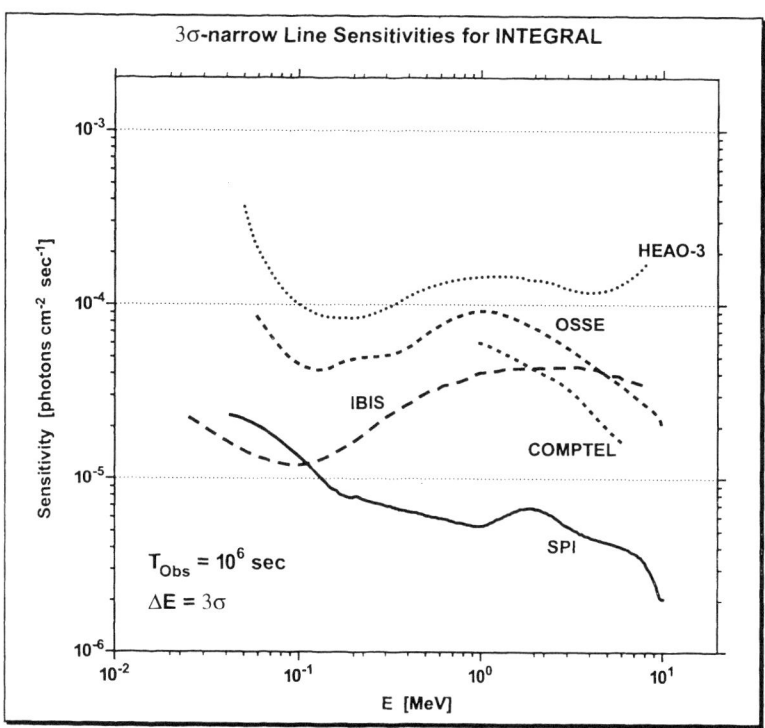

FIGURE 2. Point source sensitivity (3σ) for narrow line detections with SPI and IBIS for 10^6 sec observation time in comparison with the sensitivities of HEAO-3, OSSE, and COMPTEL.

previous missions and to derive realistic sensitivity estimates for the instruments.

Figure 2 shows the expected 3σ narrow line sensitivity of SPI and IBIS for an observation time of 10^6 sec in comparison with that of HEAO-3 and COMPTEL and OSSE on the Compton Gamma-Ray Observatory. The term "narrow" is defined to be identical to the energy resolution of the instruments (2.5 keV FWHM for SPI at 1.3 MeV, about 6 to 7 % FWHM for IBIS, 3 keV FWHM for HEAO-3, 10 % FWHM for COMPTEL, and 5 to 6 % FWHM for OSSE). SPI will be by far the most sensitive narrow gamma-ray line instrument ever flown. For broadened gamma-ray lines, the SPI sensitivity is somewhat degraded. The dependence of the degradation factor on the width of the line is illustrated in Fig. 3.

The IBIS, SPI and JEM-X 3σ continuum sensitivities (again for an observation time of 10^6 sec and for an energy interval $\triangle E = E/2$) are shown in Fig. 4, and are compared with the SIGMA sensitivity. Below 1 MeV, the point source detection sensitivity level is in the range between 0.5 to 20 m Crab strengths. Above 1 MeV, SPI and IBIS have comparable sensitivities (at the 100 m Crab level).

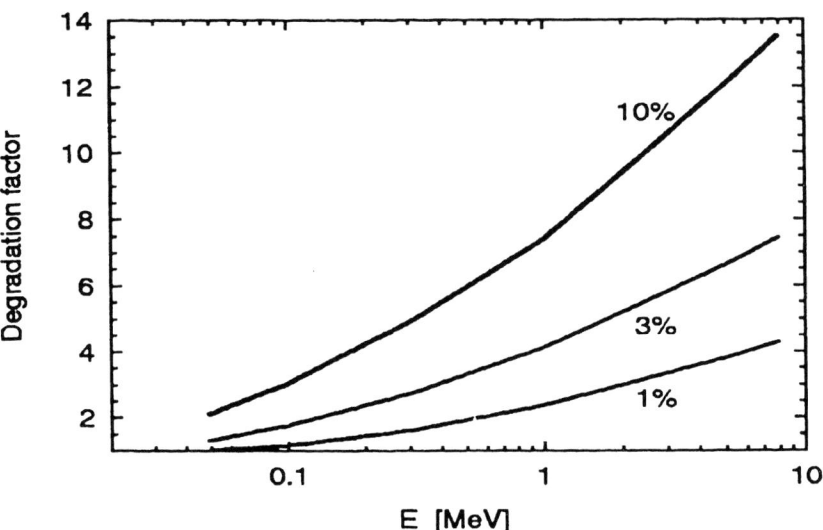

FIGURE 3. Degradation of SPI sensitivity for broadened lines. The degradation factor is plotted for line widths of 1 %, 3 %, and 10 % FWHM of the line energy.

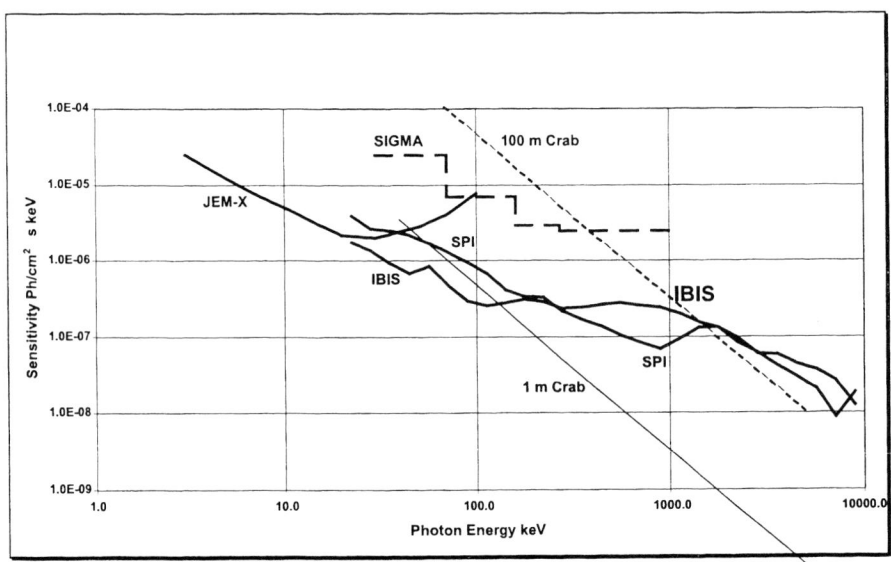

FIGURE 4. INTEGRAL continuum sensitivity (3σ) for an observation time of 10^6 sec and energy bins $\triangle E = E/2$ in comparison with the SIGMA sensitivity.

SCIENCE WITH INTEGRAL IN PERSPECTIVE

Considering the improvements in sensitivity - as illustrated in Fig. 2 and 4 - the most important and spectacular results from INTEGRAL can be expected in the field of gamma-ray line spectroscopy and in the study of continuum gamma-ray sources in the sub-MeV range. The key objects to be studied by INTEGRAL will be

- **compact objects**, like neutron stars, stellar black hole candidates, high-energy transients (X-ray novae), galactic micro-quasars, and gamma-ray bursters
- **stellar nucleosynthesis sites**, especially novae, supernovae, supernova remnants, but also hydrostatic nucleosynthesis sites like AGB and Wolf-Rayet stars
- **the interstellar medium** (mapping in continuum and gamma-ray line emission)
- **the Galactic center region**
- **external galaxies**, especially Seyfert galaxies and gamma-ray blazars
- **the cosmic gamma-ray background** (its spectral and spatial structure)
- **the unidentified gamma-ray sources** (especially from the EGRET source catalogue).

SPI's main research field is that of gamma-ray line spectroscopy. Here the key topics are twofold: first, the study of the broad-scale interstellar emission of the electron-positron annihilation line, the 1.809 MeV ^{26}Al line, the 1.17 and 1.33 MeV ^{60}Fe lines, and the 1.275 MeV ^{22}Na line. Of prime interest are the profiles of these lines. Second, the study of individual gamma-ray line sources: Candidates are the two supernova remnants Cas-A and RX 0852-46, from which the signature of the 1.157 MeV ^{44}Ti line was found, and other still to be discovered ^{44}Ti line sources, whose (broad) line profiles are expected to be measured by SPI for the first time (the latter of these two sources may also be identified as a 1.809 MeV line source), any nearby ($<$ 15 Mpc) supernova that may occur after the launch of INTEGRAL to search for ^{56}Co and ^{57}Co gamma-ray line emission (1 or 2 such supernovae are expected over the life time of the mission), any nearby ($\leq$ 1 kpc) nova to search for the 478 keV and 1.275 MeV lines from ^{7}Be and ^{22}Na decay, and possible 511 keV annihilation line sources, especially in the Galactic center region.

The search for nuclear interaction lines from interstellar space is another interesting topic for SPI. However, according to presently existing predictions, observation times much in excess of 10^6 sec will possibly be needed to detect such lines.

For IBIS Galactic binaries with compact objects like neutron stars, black-hole candidates, and transients are of special interest. Their observations will allow us to compare the spectral behaviour of these objects, to perform timing analysis and to search for transient spectral features like those observed near 511 keV from Nova Musca and the 1E 1740-2942 source. IBIS is expected to largely enhance our knowledge about transient source types discovered and investigated by SIGMA. Especially the Galactic center region is rich in such objects.

In the field of gamma-ray bursts 10 to 20 events will be detected within the fields-of-view of SPI and IBIS each year. Strong bursts will be located to an accuracy of better than 1 arc min, and high-resolution spectroscopy measurements of the burst will be possible. The location of the bursts will be made available to the world-wide community in less than 5 minutes. Follow-up observations of a burst as target-of-opportunity will be

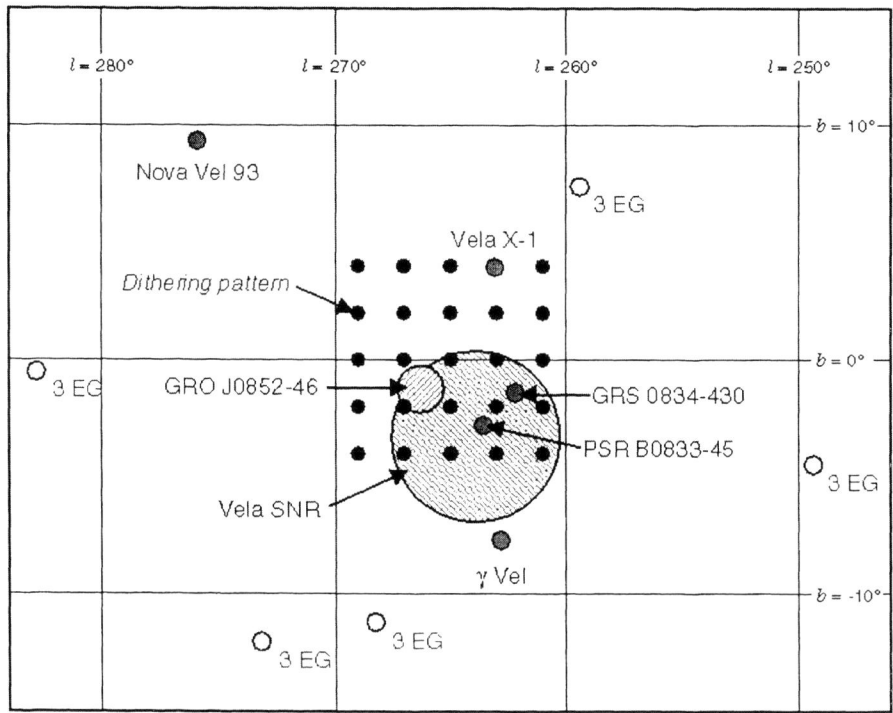

FIGURE 5. The Vela region will be used by INTEGRAL as a showcase for Galactic gamma-ray sources.

possible 36 hours after the initial event. The anticoincidence system of SPI will be used as part of the interplanetary network to locate bursts to about 10 arc sec.

THE OBSERVATION PROGRAM OF INTEGRAL

In the first year of the mission 65 % of the INTEGRAL observation time will be awarded to the scientific community in general (this is called the "open time"). The remaining fraction will be reserved for the INTEGRAL Science Working Team (called the "Core Programme"). In later years of the mission the fraction of the open time will further increase.

The core programme is well defined and consists of three elements (Winkler et al., 1999): a deep exposure of the Galactic central radian, weekly scans of the Galactic plane, and pointed observations.

The purpose of the deep exposure of the Galactic central radian is to map the Galactic ridge in continuum and line emission, in order to perform deep imaging and spectroscopic studies in that region. For each point along the plane in the central radian, the 10^6 sec sensitivity level (shown in Fig. 2 and 4) will be achieved after the first year.

The purpose of the Galactic plane scans is to discover transient sources (like e.g. Nova

Musca). For continuum sources around 100 keV the sensitivity limit of the plane scan is at the 20 m Crab intensity level for one single scan. The sum of all weekly scans over 1 year results on average in a 1-day exposure for each point along the Galactic plane, and can thus be used to build maps of the plane in continuum and line emission.

The pointed observations are of three different kinds: first, follow-up observations of transient events detected during the Galactic central radian deep exposure and the weekly Galactic plane scan; second, follow-up observations of externally triggered transient events; and third, observation of persistent sources. Candidates for the follow-up observations are X-ray novae, superluminal jet sources, supernovae, classical novae, and AGN outbursts. The time left in the Core Programme for the observation of persistent sources is only 10^6 sec. The INTEGRAL Science Team has decided to devote this time to the observation of the Vela region, which contains several exciting and spectacular objects (see Fig. 5). Among these are RX 0852-46, the closest historical supernova remnant in the Galaxy; γ^2-Velorum; the closest Wolf-Rayet star; the Vela pulsar; the Vela supernova remnant; Vela X-1; GRS 0834-430: an unusual Be-star binary accreting pulsar; Nova Vel 93: a black-hole X-ray nova in quiescence, and a few unidentified EGRET sources.

The INTEGRAL data first will arrive at the INTEGRAL Data Centre, which is located at Versoix, near Geneva. Its main task will be the data processing, quick-look analysis, instrument performance monitoring, target-of-opportunity recognition and gamma-ray burst alerts, standard science analysis, and archiving of the science data. All INTEGRAL data will be released for public use one year after they have been given to the individual observers.

REFERENCES

Giménez, A. et al., 1999, Proc. of 5th Compton Symposium, ed.: M.L. McConnell, and J.M. Ryan, AIP **510**, 732 (2000)

Lund, N. et al., 1999, Proc. of 5th Compton Symposium, ed.: M.L. McConnell, and J.M. Ryan, AIP **510**, 727 (2000)

Winkler, C. and Hermsen, W., 1999, Proc. of 5th Compton Symposium, ed.: M.L. McConnell, and J.M. Ryan, AIP **510**, 676 (2000)

Winkler, C., 1999, Proc. 3rd INTEGRAL Workshop, Astrophys. Letters & Communications **39**, 309

Winkler, C. et al., 1999, Proc. 3rd INTEGRAL Workshop, Astrophys. Letters & Communications **39**, 361

Ubertini, P., 1999, Proc. of 5th Compton Symposium, ed.: M.L. McConnell, and J.M. Ryan, AIP **510**, 684 (2000)

Vedrenne, G. et al., 1999, Proc. 3rd INTEGRAL Workshop, Astrophys. Letters & Communications **39**, 325

Gamma-Ray Polarisation Measurements with INTEGRAL/IBIS

J.B. Stephen[†], E. Caroli[†], R.C. da Silva[§] and L. Foschini[†]

[†]*Istituto TeSRE/CNR, Via P. Gobetti 101, Bologna, ITALY*
[§]*Laboratoire PHASE/CNRS, Strasbourg, FRANCE*

Abstract. The IBIS telescope on board the INTEGRAL satellite, while designed to produce high resolution images of the gamma-ray sky, will also be able to measure the polarisation of strong high energy sources. We present an estimate of the polarization sensitivities of both the high energy detector (PICsIT) operating in polarization mode and of the IBIS Compton events which scatter from the low energy to the high energy detector, for possible observations of the CRAB pulsar and for the case of a strong gamma-ray burst in the fully coded field of view.

INTRODUCTION

The IBIS instrument on board the INTEGRAL satellite [1] will create images of the gamma-ray sky over a wide energy range of ~20 keV - ~10 MeV. This is achieved by the use of two discrete detection planes, one (ISGRI, [2]) composed of 16384 CdTe micro-detectors for use between ~20 keV and ~1 MeV and the second (PICsIT, [3]) comprising 4096 CsI scintillator crystals functioning above ~150 keV. While the data from ISGRI are always transmitted in photon-by-photon mode, those from PICsIT are usually (the 'standard' mode) transmitted in the form of accumulated detector 'images' and as high resolution timing (without spatial information). However, due to the importance of potential polarisation measurements at high energy, a dedicated polarisation mode of operation has been incorporated in PICsIT. Furthermore, the data from Compton events between ISGRI and PICsIT are continuously transmitted in photon-by-photon mode, allowing possible polarisation measurements to be performed even when the high energy detector is in the 'standard' mode of operation. The IBIS instrument on board INTEGRAL may be used to investigate source polarisation in two modes of operation:
- The dedicated (PICsIT) Polarimetry mode
- The IBIS (ISGRI-PICsIT) Compton Mode

While the second mode is always present and so is particularly useful to study impulsive events such as gamma-ray bursts (GRBs) which may occur in the field of view, the former is a dedicated mode and would be useful to study persistent sources such as the CRAB pulsar. In the following sections we estimate the polarisation sensitivity in both modes by means of Monte-Carlo simulations.

THE MONTE-CARLO MODEL

The Monte-Carlo model employed uses both a very simplified IBIS design as well as a reduced data simulation. The ISGRI plane was considered to be a unique plane of 128 x 128 CdTe detectors i.e. with the correct number of detectors but ignoring the fact that in reality the plane is divided into 8 modules with a gap between each. In an analogous manner the PICsIT plane consisted of 64 x 64 CsI detectors but again without considering the modularity. The data for the polarimetry mode simulation were generated by irradiating only one pixel of CsI in the centre of the plane, while those for the Compton mode were generated by irradiating 4 (2x2) pixels of CdTe near the centre of the plane and constructing a map of Compton events in PICsIT. Obviously in neither case do we take into account edge effects or off-axis distortion. Three simulations were performed – at 250 keV, 511 keV and 1 MeV. It is important to note that not all Compton events are useful for the polarisation measurement. In particular, the only events which carry useful information are those which undergo a Compton scattering in ISGRI followed by ONE interaction in PICsIT. This is because during the on-board computations the position information for events which interact more than once in PICsIT is lost (a statistically reconstructed position is all that is recorded).

POLARISATION DEFINITIONS

The minimum detectable polarisation (MDP) in the absence of background noise can be described [4] by the formula:

$$MDP = \frac{n_\sigma}{Q_{100}} \sqrt{\frac{1}{C_C}} \quad ; \quad Q_{100} = \frac{N_\perp - N_\parallel}{N_\perp + N_\parallel} \qquad (1)$$

where n_s is the significance of the measurement, Q_{100} is the Q-factor (or sensitivity to 100% polarised photons) of the detector, $N_\perp$ and $N_\parallel$ are the number of Compton counts in areas perpendicular and parallel to the polarisation vector of the incoming photons and C_C refers to the number of Compton events *used in the calculation of* Q_{100}. Clearly $C_C = N_\perp + N_\parallel$.

In order to make explicit the inter-relationship between Q_{100} and the Compton efficiency, we can rewrite (1) using the total number of Compton events (C_T) and the *effective* Q_{100}:

$$MDP = \frac{n_\sigma}{Q_{100}^{eff}} \sqrt{\frac{1}{C_T}} \quad ; \quad Q_{100}^{eff} = Q_{100} \sqrt{C_C / C_T} \qquad (2)$$

This effective factor automatically takes into account the fact that Compton events with different linear separation may have both widely different Q-factors, but also widely different probabilities. We shall use the formulation of equation (2) for both modes of operation, assuming that for the IBIS Compton mode the data used can be selected for time around the GRB such that the source dominates over the background rate.

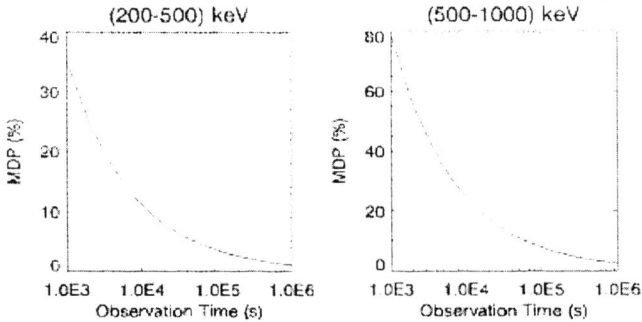

FIGURE 1. The MDP(3s) for observations of the CRAB pulsar in two energy bands PICsIT Polarimetry Mode Sensitivity

The PICsIT polarimetry mode takes into account those events which produce energy deposits in a CsI detection element and then Compton scatter and interact again *in an adjacent pixel*. The Q_{100} factor in this case is large, due to the almost 90° scattering angle, however the efficiency is low leading to an effective Q_{100} of less than 0.2 at 200 keV decreasing in an almost linear manner to a value of around 0.07 at 1 MeV. The total Compton efficiency over the same range may be approximated by two linear fits breaking at 511 keV. Then we can integrate equation (3) using these relationships to provide the minimum detectable polarisation as a function of observation time for an observation of the CRAB pulsar in two energy bands as shown in Figure 1. It must be remembered that these are upper limits as we have neglected the background term.

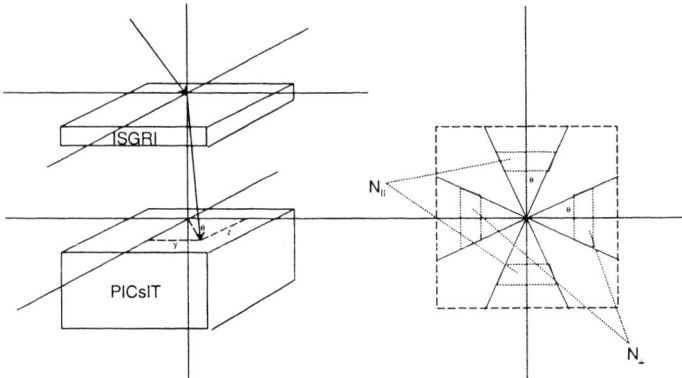

FIGURE 2. The linear separation and angle parameters for double events to be included in the calculation of the effective Q_{100}. IBIS Compton Mode Polarisation Sensitivity

It is the use of the Compton mode between ISGRI and PICsIT for polarisation measurements where the utility of defining the effective Q_{100} becomes apparent. This is because there is no restriction on the linear distance between the point of interaction in the two planes for valid events (within the overall dimensions of the planes), thus allowing optimisation of the effective Q_{100} by defining the acceptance areas of events to be included in the sums of $N_\perp$ and $N_\parallel$ depending on (see Figure 2):
- Minimum linear distance (y/z) from point of interaction in ISGRI to that in PICsIT
- Maximum linear distance (y/z) from point of interaction in ISGRI to that in PICsIT
- Angle of the second point of interaction with respect to first (?).

This can be useful because the Q_{100} increases with linear separation, while the efficiency decreases, leading to a maximum of the *effective* Q_{100} at some distance from the original interaction point, depending on the energy of the event as shown in figure 3. Once again we can integrate equation (3) and use the relationships of Q_{100} and Compton efficiency with Energy to produce the minimum detectable polarisation. In order to apply this to the number of gamma-ray bursts INTEGRAL is expected to see within the field of view we can work forward from the CGRO catalogue and assume an average burst spectrum which consists of a broken power law with spectral index of -1 and -2.25 below and above the break energy, which itself is in the range 100 keV - 1MeV. Furthermore we can use the distribution in fluence given in [5].

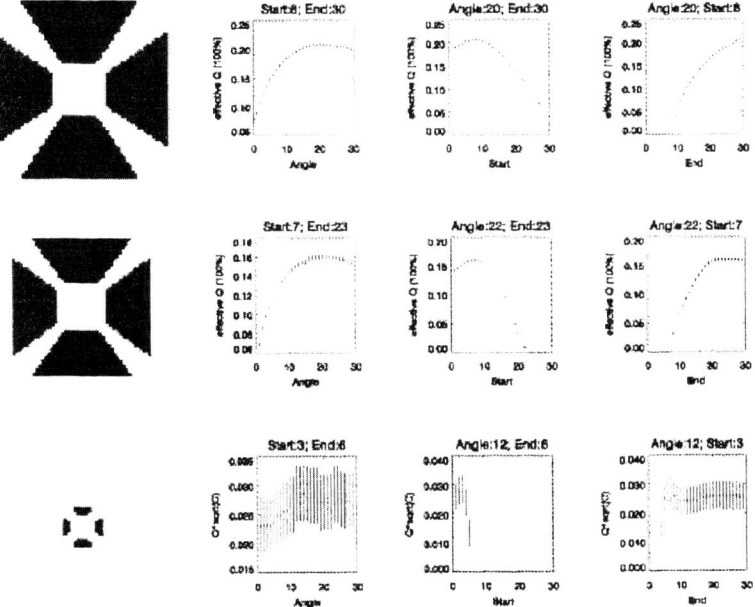

FIGURE 3. The optimum parameters y,z_{min} (start); y,z_{max} (end) (both in pixels) and acceptance angle for the calculation of the effective Q_{100} for 250 keV (*top*); 511 keV and 1 MeV (*bottom*) are depicted in the left column. In the other columns the variation in the effective Q_{100} with each parameter individually is shown. It can be seen that as energy increases, the maximum in the effective Q_{100} comes from the integration of events closer to the original interaction point as the decrease in efficiency more than cancels the effect of the increase in Q_{100} with distance

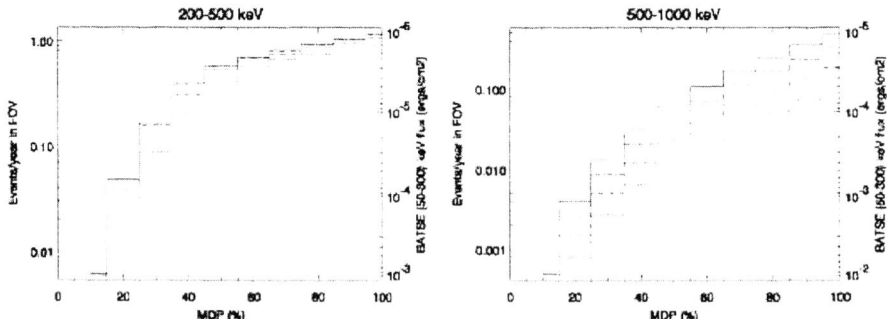

FIGURE 4. The number of events/year in the field of view for two energy bands – a low energy regime from 200 – 500 keV and a higher energy range from 500 – 1000 keV as a function of minimum detectable polarisation and for 4 GRB break points (200, 300 , 400 and 500 keV from the bottom). Also shown for comparison purposes is the equivalent energy fluence (in ergs/cm^2) for the BATSE 50-300 keV band.

The definition of the FOV of IBIS for Compton mode depends on the energy – at low energies (say < 500 keV) the hopper is opaque and so we can say that, roughly, the FOV is around 400 square degrees (corresponding to the half coded area), or one hundredth of the full sky. At higher energies (say > 500 keV) the hopper is increasingly more transparent, and so the FOV may be multiplied by up to a factor 5 (equivalent to about a 45° x 45° aperture). Using all this information we can estimate the number of GRB events per year which IBIS will be able to detect at a given MDP as a function of energy as shown in Figure 4.

CONCLUSIONS

We have shown, using a very simplified Monte-Carlo simulation that the IBIS telescope on board INTEGRAL will have some possibility of making polarimetric observations of gamma-ray sources. Detailed modelling of the instrument and more sophisticated methods of polarisation analysis will allow a more accurate evaluation of the polarimetric sensitivity to be performed.

REFERENCES

1. Gehrels, N. A.and Winkler, C., Proc. SPIE **2806**, 210-216 (1996).
2. Lebrun, F., Blondel, C., Fondeur, I., Goldwurm, A., Laurent, P. and Leray, J., Proc. SPIE **2806**, 258-268 (1996).
3. Labanti, C., Di Cocco, G., Malaguti, G., Mauri, A., Rossi, E., Schiavone, F., and Traci, A. Proc. SPIE **2806**, 269-279 (1996).
4. F. Lei, A.J. Dean and G.L. Hills, Space Science Reviews 82, 309-388 (1997).
5. Petrosian, V. & Lee, T. Astrophys. J. **467**, L29, (1996).

Evaluation of the INTEGRAL/IBIS photons detectors efficiencies by Monte Carlo simulation

G. De Cesare*, C. Ciocca[†], M. Del Santo*, G. Di Cocco[†], P. Laurent**, F. Lebrun**, G. Malaguti[†], L. Natalucci*, V. Reglero[‡] and P. Ubertini*

*IAS/CNR (Roma, Italy)
[†]TESRE/CNR (Bologna, Italy)
**CEA-SACLAY (France)
[‡]GAGE (University of Valencia, Spain)

Abstract. IBIS, the imager on board the INTEGRAL satellite, is a coded mask telescope for X and gamma-ray astronomy with imaging and spectroscopy capabilities from 15 keV to 10 MeV. In order to cover this energy range, IBIS uses two position sensitive detectors: the low energy detector layer ISGRI and the high energy detector layer PICsIT. The sensitivity of a coded mask instrument depends on the detectors efficiencies, imaging efficiency, and on the diffuse photons and particles background count-rate. In the IBIS energy range, also the opacity at low energies of the open mask elements and the transparency of the closed elements at high energies give a significant effect on the sensitivities curves. In this work we present a Monte Carlo evaluation of the IBIS detectors efficiencies. The mask transparency data from the Flight Model (FM) are also presented.

INTRODUCTION

The IBIS telescope uses a coded aperture mask imaging system to obtain the sky images and spectra of sources in the X and gamma ray band. The image in a given energy band is obtained by decoding the count maps produced by the Position Sensitive Detector (PSD), which consists of two detector planes:

- ISGRI, covering the low energy range from 15 keV to 800 keV;
- PICsIT, covering the high energy range from 150 keV to 10 MeV.

So, IBIS is sensitive to X and gamma-ray sources in the energy band from 15 keV to 10 MeV. At these energies, the photons interact with the instrument with photoelectric interaction, Compton scattering and electron-positron pairs production. Therefore the IBIS spectral response and the imaging performances depend in different way on these effects, according with their importance in different energy ranges.

To obtain a realistic simulation of the instrument we have developed a GEANT-based Monte Carlo (MC) code [4] in order to simulate the electromagnetic showers inside the instrument masses. Using the events list produced by the MC, images and spectra can be obtained and therefore the detectors efficiencies evaluated.

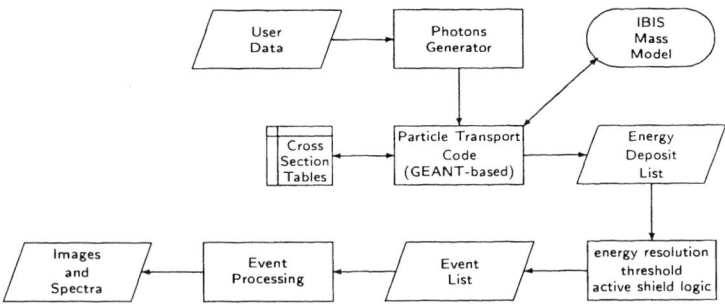

FIGURE 1. IBIS Monte Carlo environment flow chart. In order to obtain the final output (spectra and images) the following pipeline have to be maked: user input ⇒ the energy deposit list file ⇒ the event list files ⇒ spectra and images.

THE MONTE CARLO CODE

The detectors efficiencies curves presented in this paper have been obtained using a dedicated Monte Carlo code [4], based on GEANT package [5]. We use this library for the simulation of the photon and photon-induced radiation transport inside the IBIS masses, modelled by the so called Mass Model (MM).

The code gives the possibility to display the used instrument MM and the photons tracks. The transport simulation includes also the secondary particles, in particular the electrons scattered by the physical interactions and the 511 keV photons produced by the annihilation processes [5]. In the second stage the energy deposits in the IBIS detectors (ISGRI, PICsIT and Veto Units) are convolved with the energy spread distribution and filtered by the thresholds and the active shield [1] (anti-coincidence) logic in order to obtain the events list data. So these files are processed to obtain spectra and images.

The accuracy of the simulation results depends mainly on the instrument MM and on the detector response model (energy resolution, thresholds and on-board selection).

The simulation environment

Basically the software consists in three layers (fig. 1):

1. basic Monte Carlo code, which produces one energy deposit list file;
2. a tool to convert the previous file into three fits event list files (photon by photon): one for ISGRI, one for PICsIT and one for Compton events;
3. a collection of FORTRAN programs to extract shadowgrams, images and spectra from the photon list files.

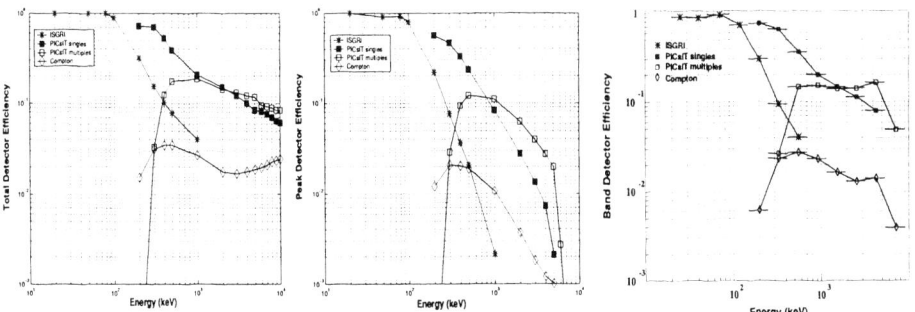

FIGURE 2. IBIS on-axis total, peak and band efficiency vs energy, evaluated by the Monte Carlo. For the total and peak efficiency, each point corresponds to a mono-energetic simulated beam incoming to the detector plane. The band efficiency has been obtained using a power law spectrum.

The energy deposit is the physical energy deposited inside the IBIS detection units, evaluated photon by photon using the Monte Carlo simulation. The event list is obtained convolving the energy deposits with the expected energy spread (due to the resolution of the involved detector unit) and taking into account the thresholds and the anti-coincidence (active shield) system.

The IBIS Mass Model

We have implemented in the geometry the following subsystems:

- ISGRI, the upper Position Sensitive Detector based on 16384 CdTe solid state units;
- PICsIT [2], the bottom Position Sensitive Detector based on 4096 CsI scintillator units;
- the active shield, based on BGO scintillator detectors modules;
- the Hopper, the tungsten passive shield, a truncated piramid on the top ISGRI detector plane;
- the Tube, the shield enclosing the volume from the Hopper to the Mask;
- the coded Mask, made up of a 95 × 95 pattern of open elements (holes) and closed elements (16 mm thick W alloy blocks).

The Mass Model (MM) can be updated without changing the other parts of the code. It is also possible to run the same Monte Carlo code with differents MM.

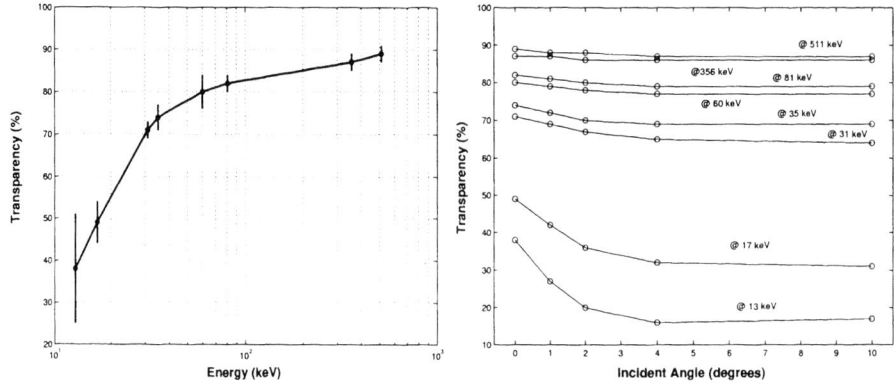

FIGURE 3. Transparency of the open mask elements, versus energy for a normal beam and versus the incident angle. Data from the Mask Flight Model tests.

DETECTORS EFFICIENCIES EVALUATION

The detector efficiency, at a given energy, is defined as the ratio of the "detected" to the incident photons. We define here three kind of efficiency (fig. 2):

a) <u>Total Detector Efficiency</u>. The source is a parallel and mono-energetic beam and all the photons in the detector spectrum are included in the evaluation of the "detected" photons.
b) <u>Peak Detector Efficiency</u>. The source is a parallel and mono-energetic beam and the photons in a band large 1 FWHM (centered on the peak) in the spectrum are included in the evaluation of the "detected" photons.
c) <u>Band Detector Efficiency</u>. The source is a power law parallel beam, with the Crab photon index. The "detected" photons at a given energy E are the photons in the energy band $[E - E/4, E + E/4]$, i.e. in a band large $E/2$ centered in E.

Note that the band efficiencies values at high energy are lower then the total efficiencies values. This fact can be explained by the large energy spread at high energies.

The detectors efficiencies have been evaluated for the IBIS event modes:

- ISGRI, the events detected by this detector only;
- PICsIT singles, the events detected by only one PICsIT CsI unit;
- PICsIT multiples, the coincident events in two or three PICsIT CsI units;
- Compton, the ISGRI-PICsIT coincidence events (both singles and multiples).

MASK TRANSPARENCY

The number of photons detected by the IBIS telescope depends on the detectors efficiencies and on the coded mask transparency.

The IBIS mask [3] is formed by three subsystems: code (MURA), support structure (SS) and interface frame. SS provides required mechanical stiffness and code positioning during launch and operations. The mask design driver was to minimise the amount of passive materials obstructing the FOV. Only 16 kg of carbon fibre support the total weight of the mask of 197 kg. The SS Transparency has been measured in 40 open pixels spread over the mask at 9 energies and 6 incident angles using a HP Ge Oxford-CN detector in a dedicated facility built at the University of Valencia (collimation angle 1.27°). Main conclusions can be summarised as follows:

1. No inhomogenities have been found within uncertainties. Standard deviations for 40 points measured are of the same order as their internal errors. A single matrix could be defined for the full code area as function of energy and incident angle.
2. Transparency variation with energy for a 0° pitch angle is given in Figure 3. Error bars are the standard deviations.
3. Transparency variation with incident angle is displayed in Figure 3. Support Structure absorption is nearly constant for higher energies and becomes more critical for lower energies and small angle variations. For incidence angles larger than 4° transparency remains invariable. We would like to point out that all data displayed have not been yet corrected by the effect of the pitch angle 1.27°. They represent an underestimation of the operational mask transparencies.

CONCLUSIONS

The IBIS detectors efficiencies vs energy have been evaluated using a Monte Carlo code, starting from a specific Mass Model. This code may be used also to obtain a fast evaluation of the IBIS spectral and imaging capabilities, with a suitable trade-off between accuracy and computation speed. Since for the present we have not yet implemented in the Mass Model described in this paper the mask details, to evaluate the scientific capabilities we use the experimental data to account for the mask induced effects.

The open mask elements transparencies for the IBIS Flight Model have been evaluated in a dedicated facility built at the University of Valencia.

REFERENCES

1. G. De Cesare et al., 2000, *Simulation of the BGO anticoincidence shield for IBIS on INTEGRAL*. To be published in proc. of the 4th INTEGRAL Workshop (Alicante, Spain, 2000), ESA-SP series
2. G. Malaguti et al., 2000, *PICsIT: Monte Carlo simulation results*. To be published in proc. of the 4th INTEGRAL Workshop (Alicante, Spain, 2000), ESA-SP series
3. V. Reglero et al., 2000, *IBIS mask pre-calibration matrix*. To be published in proc. of the 4th INTEGRAL Workshop (Alicante, Spain, 2000), ESA-SP series
4. G. De Cesare, 1999, *IBIS Monte Carlo Simulation*, IAS/CNR report, Ref.: IN.IM.IAS.RP.007/99
5. *GEANT, Detector Description and Simulation Tool*, CERN program Library Long Writeup W5013, CERN Geneva, 1994

In-flight calibration sources simulations for the IBIS telescope

M. Del Santo*, A. Bazzano*, A. J. Bird†, G. De Cesare*, P. Laurent**, G. Malaguti‡ and L. Natalucci*

*IAS/CNR, Roma (Italy)
†University of Southampton (UK)
**CEA, Saclay (France)
‡TESRE/CNR, Bologna (Italy)

Abstract. The early performance verification phase of the INTEGRAL satellite foresees basically several different observations as part of commissioning of the on-board instruments.
   The Cygnus region, two empty fields and the Crab have been selected as targets. These observations are necessary in order to obtain imaging performances for point sources and for crowded fields, spectral cross-calibration between the on-board instruments, a background variation study, absolute flux and timing calibration.
   We report simulation results for the IBIS coded mask telescope of planned calibrations concerning Crab and Cygnus X-1, performed in both staring and dithering observing modes. A Monte Carlo code including a full geometrical and physical model of the instrument has been used.

INTRODUCTION

The IBIS in-flight calibration [6] will be achieved by use of: on-board calibration radioactive sources; naturally occurring radioactive signatures due to spacecraft irradiation; selected observations of characteristic sources or source regions. Each of the above calibration methods is necessary to provide a full calibration of the instrument response; moreover these tests will be used in conjunction with instrument simulation to provide regular updates to the full IBIS calibrated model. At launch, *spectral response* of IBIS will be well-defined at certain energies, whereas it will be interpolated at all energies by the calibrated model. The *flux calibration* will be well-defined relative to the other instruments on board using PayLoad Ground Calibration, PLGC. The *imaging performance* will be determined by the PLGC tests and calibrated model of the instrument. The *timing performance* will be functionally checked by dedicated tests on ground.
   Optimisations and configurations carried out during the Performance Verification (PV) phase [2, 8] are aimed at: checking for immediate post-launch changes; optimising instrument configuration, whenever possible (veto logic, etc.); determining absolute flux Calibration in each mode; establishing imaging performance and PSF in all modes; establishing spectral performance on the overall energy range; obtaining primary measurements of in-orbit background in all modes.

TABLE 1. Crab and Cyg X–1 significances (1 σ) for a staring on–axis observation of 10^5 s with IBIS in standard mode (hs = hard state, ss = soft state).

Energy Range keV	Crab ISGRI	Cyg X–1/hs ISGRI	Cyg X–1/ss ISGRI	Crab PICsIT	Cyg X–1/hs PICsIT	Cyg X–1/ss PICsIT
20–35	1593.2	1717.4	1003.4	—	—	—
35–70	1255.1	1591.9	660.0	—	—	—
70–150	776.3	1068.8	257.7	—	—	—
150–250	177.7	200.8	33.5	149.4	185.0	27.9
250–400	34.8	28.6	5.7	56.3	39.6	9.6
400–650	10.0	4.0	—	18.2	7.2	3.1
650–1000	3.5	—	—	10.2	—	—
1000–2500	—	—	—	7.7	—	—

SPECIFIC CALIBRATION OBSERVATIONS

The total **Crab** observing time is 1×10^6 s, of which 7×10^5 s are in staring mode. The Crab observations in the Fully Coded Field Of View (FCFOV) will determine the telescope point spread function, update the response matrix in terms of absolute flux calibration, verify the timing, by recovering the Crab 33 *ms* periodicity and absolute phase. A single on–axis observation of 1×10^5 s will allow to fulfil goals as above and to verify the IBIS [10] performances, against the on–ground calibration, up to 2 MeV. Different observations at off–set angles (in the Partially Coded Field Of View, PCFOV) are necessary to map the imaging PSF as a function of the position within the FOV and to map the off–axis efficiency. They will be used also to check the shield leakage measurements obtained during the IBIS instrument-level calibrations and the PLGC activities. The following angles with respect to the telescope axis are been chosen: 9.6°, 13.2° at 0° azimuth angle, in order to get 50% and 15% illumination respectively; other orientations needed will be determined by the shield leakage measurements. The duration of each observation should be 10^5 s per pointing. Concerning the dithering strategy, 3×10^5 s are necessary to allow the same results above ~ 1 MeV.

The **Cygnus Region** will be observed for a total of 1.9×10^6 s, of which 9×10^5 s are staring observations (pointings as the Crab). In case of launch of INTEGRAL on April 2002, this source will provide the first in orbit calibration that will allow verification of the imaging performance for point source, for crowded field and the first spectral–cross calibration with SPI.

Empty Fields will be used to determine the instrument background and study the background variations. Two empty fields have been selected corresponding to $l_{II} = 240°$ $b_{II} = 40°$ and to $l_{II} = 60°$ $b_{II} = -45°$. They will be observed in staring mode (10^5 s each) and in dithering mode (2×10^5 s each), in order to verify whether we can image the sum of dither observations using matrices from the staring ones.

Early data from the **Galactic Plane Scan** will be used to verify the imaging capability in the presence of many sources with large dynamic range of intensity and timing characteristics. Additionally, field distortions (in excess of the Point Source Location Accuracy) can be measured by comparing the reconstructed source positions to the previously known X–ray positions.

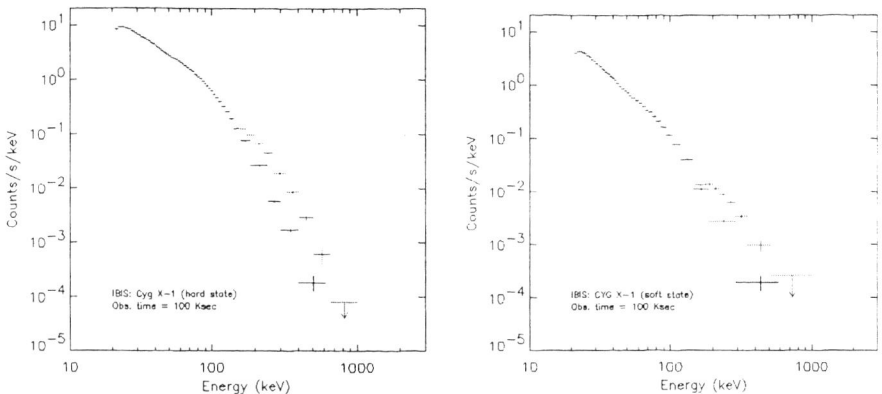

FIGURE 1. ISGRI (black) and PICsIT (light) count rate spectra for the Cygnus X−1 hard state (left) and soft state (right). The observing time is 10^5 s and the shown upper limits are 2σ.

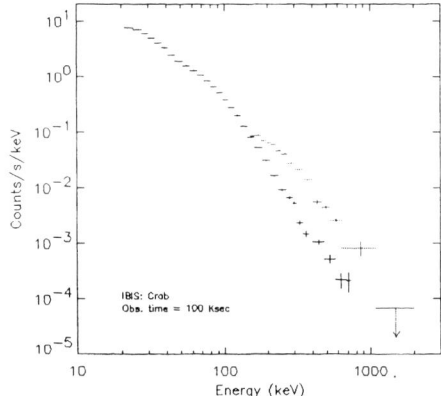

FIGURE 2. ISGRI (black) and PICsIT (light) count rate spectra for the Crab. The observing time is 10^5 s and the upper limit is 2σ.

DEVELOPED TOOLS

The simulations are obtained by injecting the input source spectra in an IBIS mass model (based on a Monte Carlo code which uses the GEANT library) developed at IAS [3]. This code comprises an implementation of the geometry and mass composition (with the active detectors, passive elements and coded mask) and a model of the instrument resolution and electronics logic. Starting by the events list we extract the detector spectra and the data maps (shadowgrams) in different energy ranges.

In order to obtain the sky images, the shadowgrams are then deconvolved. We developed several IDL procedures which use the linear cross–correlation as deconvolution

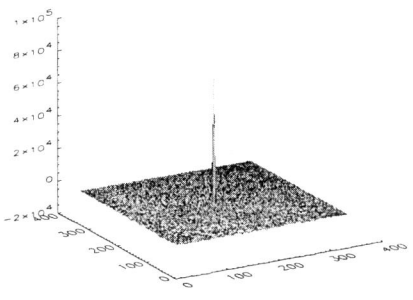
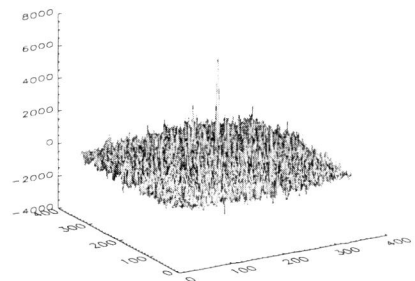

FIGURE 3. Crab dithering sky maps (in 3×10^4 s) with ISGRI in 150–250 keV (left) and with PICsIT in 650–1000 keV (right).

algorithm and others to estimate the signal to noise ratio from the images [4].

For our simulations, we used the IBIS background spectra and count rates estimated by Integral Mass Model Team in Southampton; they are available at http://www.integral.soton.ac.uk/~integral/results/v3.4.

We chose the following spectral models for the simulated sources:

- Crab: $46.0 \times 10^{-5} (E/100 \, keV)^{-2.12}$ ph cm^{-2}s^{-1}keV^{-1} [1]
- Cygnus X–1 (hard state): $2.34 \, E^{-1.64} \, e^{-E/188 \, keV}$ ph cm^{-2}s^{-1}keV^{-1} [9]
- Cygnus X–1 (soft state): $17.8 \, E^{-2.57}$ ph cm^{-2}s^{-1}keV^{-1} [5]

SIMULATION RESULTS

Staring on–Axis Observations

We report the count rate spectra (Figures 1, 2) and the signal to noise ratios (Tables 1) concerning three staring on axis observations of 100 ksec each: Crab, Cygnus X-1 (hard state), Cygnus X-1 (soft state). The spectra have been scaled by an imaging efficiency factor, in order to reproduce the expected source count spectra. They are shown only for the single events. In fact during the calibration phase every mode will be selected separately. As a further development, we will also consider the double and Compton events. The signal to noise ratios (SNR) are estimated by deconvolved images in several energy ranges for the two IBIS detectors. Note that for SNR evaluation the double events for PICsIT are also considered.

Dithering Observation Strategy

The raster dither foresees a "5 by 5" pointing pattern with the target nominally at the centre. The dwell time on each pointing is 1200 sec and the nominal angular separation between two contiguous pointings is 2 degrees [7]. One dithering observation will be composed by several cycles of this 5×5 pattern. We show two Crab images (Figure 3) in dither observing strategy for 3×10^4 seconds, corresponding to one single pattern. Every

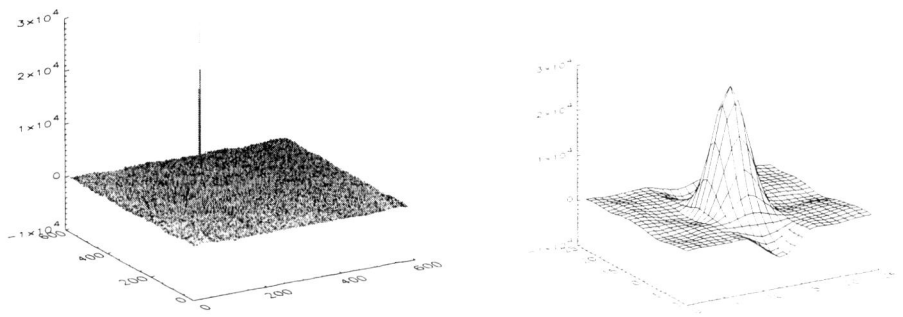

FIGURE 4. Crab in the PCFOV: ISGRI image in 35–70 keV energy range and 1 ksec observing time (left) and its PSF zoom (right)

image is obtained by adding up the 25 sky maps in which the target position is fully coded: the "dithering field of view" results then $16.6° \times 16.6°$. The cross-correlation is carried out using a decoding matrix with elements equals to 1/4 of the mask elements corresponding to a decoding step of $3'$.

Crab Observation in the PCFOV

The IBIS $9° \times 9°$ field of view is fully coded: every source in this field is coded with the full instrument sensitivity. The total field of view is $29° \times 29°$ full width at zero response. We simulated a Crab observation with source at $9.6°$ off–set and $0°$ azimuthal angle with respect to the telescope reference coordinate system; it gives a 50% transmission. We show the $35 \div 70$ keV image in 1 ksec (Figure 4, on the left): the signal to noise ratio estimated is 108 σ. We applied a filter to minimize the ghosts and to reduce systematic structures tipical of the images in the PCFOV. A zoom of the PSF (Figure 4, on the right) using a fine deconvolution is shown; note that, despite the off–set position of the source in the PCFOV, the PSF doesn't lose its symmetrical shape significantly.

REFERENCES

1. Bartlett L. M., 1994, PhD Dissertation, LHEA NASA
2. Commissioning phase science proposal, 04/07/00
3. De Cesare G., IN–IAS–RP–007/99
4. Del Santo M., IN–IB–IAS–RP–0014/00
5. Gierliński M.,et al.; 1999, M.N.R.A.S., 309, 496
6. IBIS Calibration Plan, IN–IB–IAS–RP–0016/99
7. INT–SOC–DOC–001, 28/07/98
8. INT–TN 26079, 1998
9. McConnell M. L., et al.; 2000, astro–ph/001484
10. Ubertini P. et al., Astro Lett. and Communication, Vol. 39, 331

Results From The SPI Imaging Test Setup

Cornelia B. Wunderer[1], R. Diehl[1], R. Georgii[1], A. v. Kienlin[1],
G. G. Lichti[1], V. Schönfelder[1], A. Strong[1],
P. Connell[2], F. Sanchez[3], G. Vedrenne[4]

[1]*Max-Planck-Institute for extraterrestrial Physics, Garching, Germany*
[2]*University of Birmingham, Birmingham, UK*
[3]*University of Valencia, Valencia, Spain*
[4]*Centre d'etudes spatiales des rayonnements, Toulouse, France*

Abstract. The SPI Imaging Test Setup (SPITS) was built at MPE to allow experimental verification of the imaging properties of the SPectrometer onboard INTEGRAL (SPI). Of special importance is the possibility to validate simulations – which are needed for SPI image reconstruction – with laboratory measurements. SPITS consists of a coded mask and two Germanium detectors. The coded mask has the same tungsten-alloy HURA mask coding as SPI and is made of SPI flight model materials. The two hexagonal Ge-detectors are from the SPI flight detector manufacturing line. Housed in a common Al vacuum cap and mounted on an XY-table, they can be moved to cover the 19 Ge detector positions of the SPI camera. We have measured the response of SPITS to radioactive sources (60 keV to 1.8 MeV) at a distance of 9 m from the detector plane. We use both image deconvolution algorithms foreseen for SPI data analysis (*spiros* and *spiskymax*) for our analysis. We present our findings for the angular resolution and the point-source-location capability of SPITS for several photon energies and for several source geometries relative to the mask coding.

INTRODUCTION

The INTEGRAL observatory [1] consists of 4 instruments: the imager IBIS, the spectrometer SPI which will deliver high resolution spectra (2.2 keV FWHM at 1.33 MeV) with an angular resolution of 2.5° FWHM in the energy range 20 keV – 8 MeV [2,3], an X-ray monitor (JEM-X) and an optical monitor (OMC). The latter two provide multi-wavelength information. All instruments except OMC use tungsten-alloy coded masks to obtain directional information on the incoming photons.

The coded mask of SPI consists of 63 hexagonal opaque tungsten-alloy mask elements. Each mask element is 30 mm thick, enough to absorb gamma-rays up to several MeV. The position-sensitive detector of SPI is a Germanium 'camera' with 19 hexagonal 'pixels'. Each Ge crystal is 7 cm long with a front area of 27 cm^2 (56 mm side-to-side).

Although the technique of using coded masks is well known in the X-ray domain (e.g. [4] and references therein), INTEGRAL's instruments will be the first space borne detector systems using coded masks up to 10 MeV. Studies of the imaging capabilities of SPI were done using GEANT and other simulation tools [5,6,7]. The SPI Imaging Test Setup (SPITS) was built to complement the theoretical studies of

SPI imaging performance and to provide the chance to test SPI data-analysis methods on experimental data long before launch. The SPITS results will complement findings from the SPI flight model calibrations which took place in April / May 2001.

THE SPI IMAGING TEST SETUP

SPITS has a flight instrument equivalent coded-aperture mask. Its coding pattern is identical to that of SPI, and it was constructed on the basis of the SPI mask development model. Instead of SPI's 19 Ge detectors, SPITS is equipped with only two. The individual Ge crystals in their hexagonal Al housings are from the SPI FM manufacturing line. Mounted in a common Al vacuum cap on an Al cold plate and cooled with liquid nitrogen, they are moved on an XY-table to cover the positions of all 19 SPI Ge detectors. Thus a "19-pixel-image" is recorded by combining several measurements. This is possible on the ground since neither source activity nor laboratory background change significantly during the 'image' measurement.

SPI is equipped with a BGO anticoincidence shield. On the ground in a laboratory environment, such protection is not needed. In addition, SPI has a 5 mm plastic scintillator mounted between mask and Ge detectors to further veto background. This is not needed for SPITS either, but since the 5 mm plastic absorbs some of the (lower energy) gamma-rays as well, it is emulated in the SPITS setup by an equivalent plexiglass sheet. A more detailed description of SPITS can be found in [8].

MEASUREMENTS

SPITS has been exposed to several laboratory point sources (active source diameter 1 mm) at 9 m distance from the detector plane. The energy range covered is 60 keV (^{241}Am) to 1.8 MeV (^{88}Y). The sources are mounted in a holder situated 9 m from the detectors which covers a 9° horizontal and 5° vertical field of view and allows easy reproducibility of source locations. The sources can be moved on lines with 0°, 15°, 30°, 45°, 60°, and 90° tilt towards the horizontal, allowing several different sampling directions of the mask pattern (which itself has a 120° rotational symmetry).

We used measurements with one single point source each to explore the point source location capability of SPITS at different energies and source locations. To determine the angular resolution (which we understand to be the minimum angular separation of two (equally strong) point sources that are properly separated), we added the measurement data taken with one source at two consecutive positions and then analyzed the combined data.

IMAGE RECONSTRUCTION ALGORITHMS

SPITS data analysis is performed using the image reconstruction algorithms developed for SPI. For proper analysis SPITS instrument response functions (IRFs) are needed that reflect both the finite source distance of 9 m (SPI IRFs are calculated for sources at infinity) and the differences in the mass model. While SPI IRFs are generated using a combination of full-fledged GEANT simulations and ray tracing, SPITS IRFs are generated using only ray-tracing methods.

Two methods are implemented for SPI image analysis: *spiros* [9], and *spiskymax* [10]. While *spiros* reconstructs the image by performing a cross-correlation and then iteratively removing the strongest remaining (point) source from the image, *spiskymax* reconstructs the whole image at once and is based on Maximum Entropy algorithms. The entropy of an image is a measure of structure in the image, and maximizing the image entropy is equivalent to determining the 'smoothest' image consistent with the data. Both algorithms are used to reconstruct SPITS images.

RESULTS

The raw SPITS data are count spectra at each of the 19 Ge detector positions. SPI analysis software requires binned counts in an energy band and the corresponding background estimate as input. The background estimate for the SPITS data is obtained by fitting line and continuum, then using the continuum counts in the binning region as background estimate. We bin a $\pm 2\,\sigma$ region around the peak.

SPITS' capability to locate a single point source was tested with laboratory point sources at different angles from the line of sight. The data were analyzed using both the *spiskymax* and the *spiros* algorithms. *spiros* returns source location coordinates, while *spiskymax* results are returned as a sky image only. We have determined source locations from the *spiskymax* images by two different methods: (1) the brightest pixel determines the source location and (2) the peak location of a 2-D gaussian fit to the image is used.

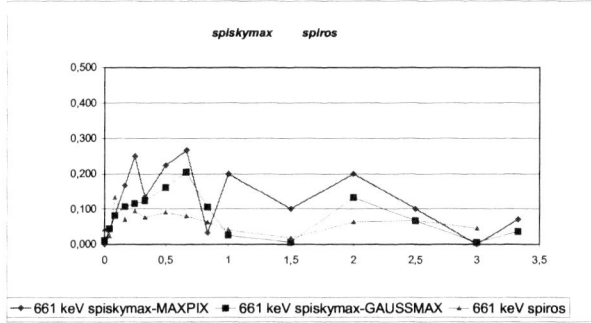

FIGURE 1. Localization accuracy for a single ^{137}Cs point source. The image is reconstructed using *spiros* and *spiskymax* algorithms. Deviations in source position reconstruction do not exceed 0.3°.

Figure 1 shows the discrepancies between true and reconstructed source locations for a ^{137}Cs source (661 keV) for several angular distances from the SPITS axis (source is moved horizontally). The signal-to-background ratio is ~ 60. *spiros* results are within 0.1° or the true source location, while *spiskymax* results are correct to 0.25°. Since the *spiros* algorithm is designed primarily for point source reconstruction and its first step is to determine the best location of the strongest source in the image, it stands to reason that spiros outperforms the *spiskymax* algorithm for the task of correctly locating a single strong point source. Because the source location capability of a coded aperture system depends strongly on measurement statistics, the above result reflects the inherent capabilities of such a mask-detector-configuration. In-flight SPI point source location accuracies may differ since for typical SPI observations signal-to-background ratios will be much lower.

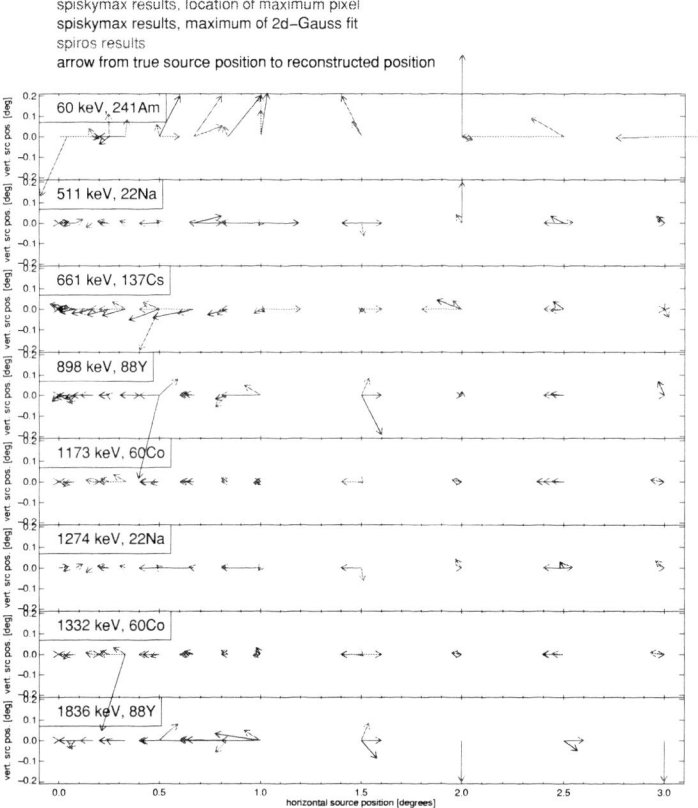

FIGURE 2. Localization accuracy for a single point source at several energies. Here, deviations in position are represented with arrows pointing from the true source position to the reconstructed one. In the results presented here, the influence of the s/b ratio dominates; further analysis is needed.

In Figure 2, we show the reconstruction of sources at several photon energies. Owing to the different continuum background levels, the signal-to-background (s/b) ratios vary from 1.6 for ^{241}Am (60 keV) to 60 for ^{137}Cs. The s/b-ratio for the 511 keV line is lower as for the other ^{22}Na line because it is also part of the background spectrum. Since the ^{88}Y source we used is about a factor of 5 weaker than the other sources, here the signal-to-background ratio is lower as well. Note that these lower s/b-ratios are reflected in the location accuracies. Further analysis is needed to compensate for this effect.

To determine SPITS' angular resolution, we have analyzed measurements with a ^{60}Co source using *spiskymax*, combining measurements at different positions to obtain several different instances of a given source separation. For high s/b ratios, in some instances proper separation of sources as close as 1° was possible. With ~ 10 samples per separation distance, we achieved reliable source separation with good positioning accuracy for separations of 2.5° and higher for both 1173 keV and 1332 keV ^{60}Co lines. For comparison, we also analyzed the ^{241}Am data in this way, but - as for the point source localization - results were considerably worse owing to the higher background levels in the spectra. Figure 3 shows two combinations of ^{60}Co sources separated by only 1°, showing that source separation for such small distances is possible in some cases.

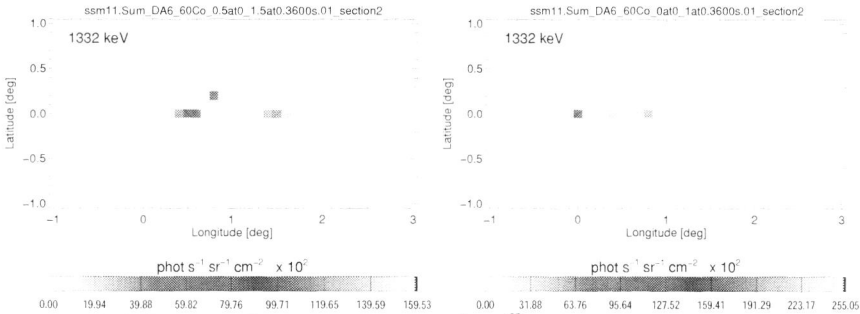

FIGURE 3. *spiskymax* reconstruction of two ^{60}Co sources separated by 1°.

SUMMARY

The point source location capability for the mask-detector geometry of SPITS (and SPI) is 0.15° or better for sufficiently high s/b ratios. Sources as close as 1° are separated well in some cases – provided statistics are good enough. Reliable source separation is achieved for source separations above 2°. Some aspects of the analysis are currently dominated by statistical effects – this is to be improved in further analysis. SPITS future plans include measurements up to 9 MeV at the Stuttgart Dynamitron and imaging of extended sources in the laboratory.

REFERENCES

1. Winkler, C., *Proc. 3rd INTEGRAL Workshop* (1999).
2. Mandrou, P., et al., *Proc. 2nd INTEGRAL Workshop*, 591-598 (1996).
3. Trams, N. R., "SPI Observer's Manual", *ISOC*, INT-SOC-DOC-022 (2000).
4. Badiali, M., et al., *Astron. Astrophys.* **151**, 259-263 (1985).
5. Skinner, G. K., et al., *Proc. 4th COMPTON Symposium* **2**, 1544-1548 (1997).
6. Connell, P. H., et al., *Proc. 3rd INTEGRAL Workshop* **2**, 397-400 (1999).
7. Strong, A.W., et al., *Proc. 3rd INTEGRAL Workshop* **2**, 221-224 (1999).
8. Wunderer, C. B., et al., *IEEE Trans. Nucl. Sci.*, accepted for publication (August 2001).
9. Connell, P. H., et al., *Proc. 4th INTEGRAL Workshop*, in press (2001).
10. Strong, A.W., et al., *Exp. Astronomy* **6**, 97-102 (1995).

First Results On SPI/INTEGRAL Flight-Model Gamma-Camera Calibration

P.PAUL, L.BOUCHET, JP.ROQUES

Centre d'Etude Spatiale des Rayonnements, 9 avenue du Colonel Roche, BP 4346, 31029 Toulouse, France

Abstract. SPI is a high resolution γ-ray telescope that will fly onboard the INTEGRAL observatory in 2002. The telescope uses cooled germanium detectors and operates over the energy range 20 keV to 8 MeV with an energy resolution of 1-5 keV (2 keV @ 1 MeV). A tungsten coded mask located 1.7 m above the detector array provides imaging capabilities over a 15 ° fully coded field of view with an angular resolution of ~3°. From calibrations of the gamma camera, we have measured various parameters such as the energy resolution, detection efficiency of the 19 individual Germanium detectors of the detector array. Furthermore, these calibrations allow us to check the homogeneity of the detector array.

INTRODUCTION

Our main objective is the test of the SPI Flight-Model (FM) camera. Calibrations were done between September 7, 2001 and October 3, 2001, with 9 radioactive sources whose energy is in the range 20 – 2000 keV (low energy range). A complete description of SPI can be found in Mandrou et al. (1997).

The camera consists of 19 n-type germanium detectors, with a total geometric detection area of 500 cm^2. One detector has an hexagonal shape, with a flat-to-flat distance of 6 cm and is 7 cm height. For these calibrations we have an experimental setup that could acquire through 19 channels (one channel per detector) the energy of the photon, the hinted detector and the dating of the interaction with a time resolution of 50 nanoseconds. The time resolution helps us to discriminate between two types of events : the "single event" and the "multiple event". We call "single event": a photon that interacts only with one detector. We call "Multiple event": a photon that interacts with several detectors. The timing system is unable to discriminate in which detector the primary interaction happens. In this case, to recover the energy of the incident photon we add the deposit energy in each of the hit detectors in a time window of ~250 nanoseconds.

EXPERIMENTAL SETUP

The detection array is placed in a cooling box (cooled down to 82 K). The preamplified signal of each detector is fed into 2 channels, one for the spectroscopy and the other for the timing analysis. The analog spectroscopy signal is converted into a channel number (0 to 16383) for the low-energy range (20 keV – 2 MeV). A radioactive source is placed 2 m above the detector array. A 3-cm slab of tungsten with an hexagonal hole simulates an element of the mask transparent to the radiation (open mask element). In flight, for on-axis source, the light projected through an open mask element onto the detector array is a 6-cm flat-to-flat hexagon. The slab is placed between the source and the detection array in order to fulfill this requirement. This slab can also move horizontally to simulate the illumination of each detector. It is then possible to use this system to derive the characteristics of each detector.

FIGURE 1. Calibration system for the detection array.

DATA ANALYSIS

During the calibration, we scanned the low range of SPI, we used 9 sources : ^{22}Na, ^{137}Cs, ^{57}Co, ^{85}Sr, ^{241}Am, ^{203}Hg, ^{109}Cd, ^{54}Mn and ^{88}Y. Background measurements are acquired regularly in order to subtract properly the background from source flux measurement. We have also compared some of the measurements with the simulations (Kandel, 1998). Furthermore, with this experimental setup we have been able to simulate an image of a source in the field-of-view of the telescope.

Measurements

As SPI is a spectrometer with a high spectral resolution, the measurements of its performances are essential. The "single event" energy resolution (figure 2) has been measured for different energy and detectors (for the ground electronics). We fulfill the required specifications and the energy resolution for the camera is homogeneous within 10 %. The simulation is also fit also for these measurements. Figure 3 shows the measured energy response for the single events for several energies. We defined the efficiency, at a given energy as the ratio of the measured counts to the number of events falling on the detection array.

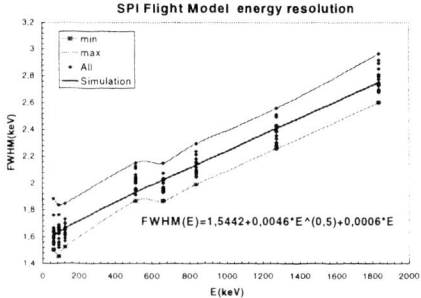
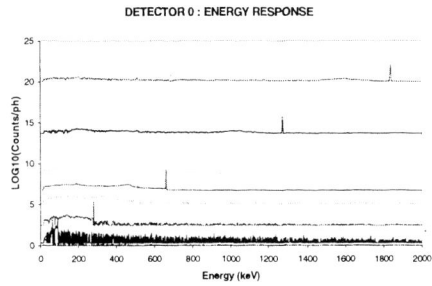

FIGURE 2. Energy resolution for the camera flight model with the ground electronics.

FIGURE 3. "Single event" spectra for the following line energy in keV from bottom to top: 59.59 keV, 88.05 keV; 279.19 keV, 513.99 keV, 661.65 keV, 834.81 keV, 1274.54 keV and 1836.08 keV.

Figures 4 and 5 show the efficiencies of SPI respectively for the "single events" and for the "multiples events" for all the detection array. The mean "single events" detection efficiency for the 19 Ge detectors is also shown. We have superimposed the Monte-Carlo (Geant code) simulations results. Note that we have some problems with the ^{22}Na source (1275 and 511 keV). Nevertheless, these measurements show that the detector array is uniform within 7% for the single events. The uncertainties in these measurements come essentially from the source activities.

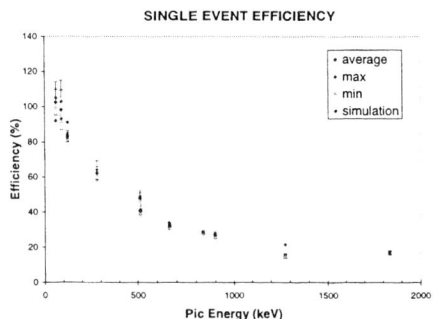
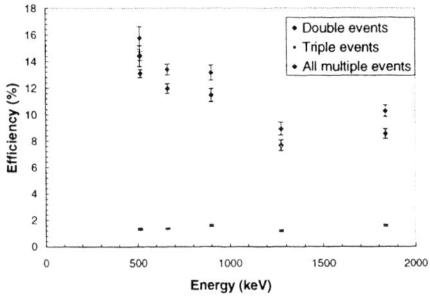

FIGURE 4. "Single event" detection efficiency for the 19 Ge detectors (black), minimum and maximum value for the single events. We have superimposed the Monte-Carlo (Geant code) results

FIGURE 5. "Multiple events" efficiency for all the detection array.

For the "multiple events" we add the deposited energy in each of the hit detector in a time window of ~250 nanoseconds. So we recover the energy of the incident photon which experiences multiple Compton diffusion or pair production in several detectors.

We classified these events according to the number of detectors involved for the detection of one incident photon. We have double events when two detectors are involved, triple events when three detectors are involved and so on. On figure 5 we plot the "multiple events" efficiency for the "double events", the triple events and for all events. In this energy range (20 keV 2 MeV) we notice that the detection array efficiency of "triple events" and for extension for the higher level are low compared to "double events" efficiency.

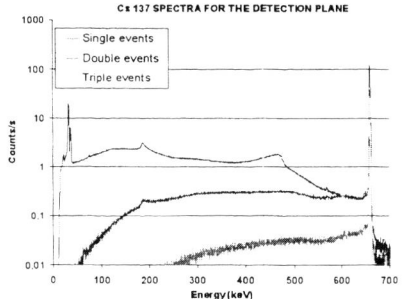

FIGURE 6. Single and Multiple events for the camera flight model..
Radioactive Source : ^{137}Cs

Comparison with simulations

All measurements and particularly "single events" spectra have been simulated using the standard GEANT code. The geometry used in this simulation included the main elements of the experimental setup, like the slab of tungsten, the cooling box and the detection array with its 19 germanium crystals. The main goal of this simulation is to validate the geometry used in the simulation for the γ camera. On figure 7 we compare the measured "single event" and simulated spectra for the ^{85}Sr radioactive source.

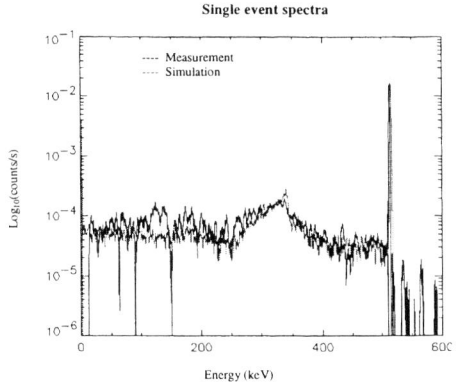

FIGURE 7. Single event spectra simulation and measurement for the detector central.
Radioactive source : ^{85}Sr

Imaging

In order to build the shadowgram of a on-axis source, we add the raw data of all the lighted detectors. Images can then be reconstructed using a kind of cross-correlation between the mask pattern and the data recorded on the detector or using the Iterative Removal Source (IROS) method (Bouchet, 2000)

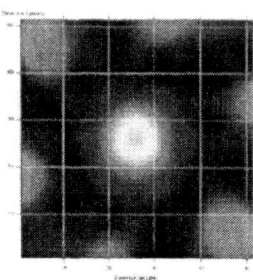

FIGURE 8. Image of a source in the field-of-view of the telescope (^{137}Cs, 661 keV).

CONCLUSIONS

These calibrations show that the performances of SPI flight-model camera are in agreement with the required specifications. Furthermore, recent calibrations have been done with the complete flight-model of the telescope. These measurements will allow to check and to complete the characteristics of SPI spectrometer.

REFERENCES

Bouchet, L., 2000, "Habilitation à diriger les Recherches', Université Paul Sabatier, Toulouse III
Kandel, B., 1998, Ph.D, Université Paul Sabatier, Toulouse III
Mandrou et al, 1997, "The INTEGRAL Spectrometer SPI", Proc. 2nd INTEGRAL Workshop

Mission Concepts: High-Energy Gamma Rays

Feasibility studies of coded masks for high-energy gamma-ray telescopes in space

Y. C. Lin*, P. F. Michelson*, P. L. Nolan* and D. J. Thompson[†]

W. W. Hansen Experimental Physics Laboratory, Stanford University, Stanford, CA 94305
[†]*Code 661, NASA/Goddard Space Flight Center, Greenbelt, MD 20771*

Abstract. It is a well-known fact that the high-energy gamma-ray telescopes in conventional designs are inherently limited in their spatial resolving powers. EGRET can locate a point source to 5 to 10 arcmin accuracy, while GLAST can achieve source location capabilities to better than 0.5 arcmin. Although there is still room for improvement in the angular resolution of the conventional high-energy gamma-ray telescopes, other means can also be explored at this time. Coded masks have been successfully used for X-ray and low-energy gamma-ray telescopes. Such devices have also been considered for high-energy gamma-ray telescopes from time to time. In this paper we investigate the feasibility of using coded masks to achieve much higher source location capabilities for a high-energy gamma-ray space telescope. We discuss the merits and the difficulties of using coded masks to achieve this goal. In particular, we look into the issues of size, weight, field of view, and the trade-off between the angular resolution and the field of view for such coded masks.

INTRODUCTION

The use of coded masks for source location is a well-established technique in X-ray and low-energy gamma-ray astronomy. Examples of such detectors can be found in [1, 2]. In the field of high-energy gamma-ray astronomy, the Soviet-French GAMMA-I [3] was the only telescope to adopt this technique for photon energies > 50 MeV. For a review of this subject, see [4]. Johansson *et al.* [5] made a laboratory study of coded masks for the energy range 50 - 500 MeV, with rather encouraging results. Thompson [6] has carried out an in-depth investigation in the use of coded masks for source location in high-energy gamma rays. He has come to the conclusion that coded masks do not possess a clear advantage over the conventional designs. But now with the experience of EGRET and GLAST telescopes in hand, the question of coded masks for high-energy gamma rays may merit another critical look.

BASIC PARAMETERS OF THE MASK

For high-energy gamma rays, we have to use active masks. Consider a mask with elements of square cross sections. There are three basic parameters for a mask:

$$a = element\ width,\ l = element\ length,$$

$$d = distance\ from\ mask\ to\ detector.$$

Then the two fundamental properties of a telescope, the angular resolution and the field of view, are given as:
$$angular\ resolution = a/d,$$
$$field\ of\ view = a/l.$$
(This is the half cone angle extended by the telescope.) To achieve good angular resolution, we need to make the elements very thin. On the other hand, because the element length cannot be too short in order to avoid too many unconverted photons going through the mask, thin elements will make the field of view small.

THE FLUX LIMIT: A SIMPLIFIED ANALYSIS

Let:
$$f = photon\ flux,\ in\ no.\ of\ photons\ cm^{-2}\ s^{-1},$$
$$A = detector\ face\ area,\ in\ cm^2,$$
$$\varepsilon = detector\ efficiency,\quad T = observation\ time,\ in\ s.$$

Put:
$$A_{eff} = effective\ area\ of\ the\ detector = \varepsilon \cdot A,$$
$$\tilde{N}_\gamma = no.\ of\ collected\ photons\ if\ there\ were\ no\ mask,$$
$$\tilde{N}_\gamma = f \cdot A_{eff} \cdot T = f \cdot \varepsilon \cdot A \cdot T.$$

Consider a mask with:
$$a^2 = element\ cross\ section,$$
$$\Pi_t = no.\ of\ transparent\ elements,$$
$$\Pi_d = no.\ of\ dark\ (opaque)\ elements,$$
$$\Pi = total\ no.\ of\ elements\ in\ the\ basic\ mask\ pattern = \Pi_t + \Pi_d.$$

Assume that one basic mask pattern is in the right size just to cover the detector face. Then:
$$A = a^2 \cdot \Pi,$$
$$\tilde{N}_\gamma = f \cdot \varepsilon \cdot a^2 \cdot \Pi \cdot T = f \cdot \varepsilon \cdot a^2 \cdot (\Pi_t + \Pi_d) \cdot T.$$

Let:
$$p_t = transmission\ probability\ of\ the\ transparent\ elements,$$
$$p_d = transmission\ probability\ of\ the\ dark\ (opaque)\ elements.$$

Then, with the mask in place, the no. of collected photons are:
$$N_\gamma = f \cdot \varepsilon \cdot a^2 \cdot (\Pi_t p_t + \Pi_d p_d) \cdot T.$$

In the central signal, from the transparent elements alone:
$$N_{\gamma c} = f \cdot \varepsilon \cdot a^2 \cdot \Pi_t p_t \cdot T.$$

In the side lobe, taken to be flat,

$$N_{\gamma l} = f \cdot \varepsilon \cdot a^2 \cdot \Pi_t \left(\frac{\Pi_t}{\Pi_t + \Pi_d} p_t + \frac{\Pi_d}{\Pi_t + \Pi_d} p_d \right) \cdot T,$$

where $\frac{\Pi_t}{\Pi_t+\Pi_d}$ is the probability that a transparent position in the basic mask pattern will find a match with a transparent position on the detector face. Similarly, $\frac{\Pi_d}{\Pi_t+\Pi_d}$ is the probability that a transparent position in the basic mask pattern will find a match with a dark (opaque) position on the detector face. The signal above the side lobe is:

$$N_{\gamma s} = N_{\gamma c} - N_{\gamma l}.$$

The detection significance is thus:

$$\eta = N_{\gamma s}/\Delta N_{\gamma s} = (N_{\gamma c} - N_{\gamma l})/\sqrt{N_{\gamma c} + N_{\gamma l}}.$$

In practice, people usually use:

$$\Pi_t = \Pi_d = \frac{1}{2}\Pi.$$

The Ideal Case

In the ideal case, the transparent elements are 100% transparent, while the dark (opaque) elements are 100% opaque. Then:

$$p_t = 1, \quad p_d = 0, \quad N_{\gamma c} = \frac{1}{2}\tilde{N}_\gamma, \quad N_{\gamma l} = \frac{1}{4}\tilde{N}_\gamma.$$

$$\eta = \left(\frac{1}{2}\tilde{N}_\gamma - \frac{1}{4}\tilde{N}_\gamma\right)/\sqrt{\frac{1}{2}\tilde{N}_\gamma + \frac{1}{4}\tilde{N}_\gamma} = \frac{1}{2\sqrt{3}}\sqrt{\tilde{N}_\gamma}.$$

If we require $\eta = 3$, then:

$$\sqrt{\tilde{N}_\gamma} = 6\sqrt{3}, \quad N_{\gamma c} = \frac{1}{2}\tilde{N}_\gamma = 54.$$

Thus we need to collect only 54 photons to obtain a 3-σ signal relative to the side lobe.

More Realistic Cases

Now Consider:

$$p_t = 1, \quad p_d = 0.1,$$

for 100 MeV photons on 4-X_0 BGO. Then:

$$N_{\gamma c} = \frac{1}{2}\tilde{N}_\gamma,$$

$$N_{\gamma l} = \frac{1}{2} \tilde{N}_\gamma \left(\frac{1}{2} + \frac{1}{2} \times 0.11\right) = \tilde{N}_\gamma \times \frac{1}{4} \times 1.11.$$

$$\eta = \frac{0.89}{2\sqrt{3.11}} \sqrt{\tilde{N}_\gamma}.$$

Again require $\eta = 3$. Then:

$$N_{\gamma c} = 71.$$

If:

$$p_t = 1, \quad p_d = 0.15,$$

for 50 MeV photons on 4-X_0 BGO. Then:

$$N_{\gamma c} = 79.$$

The required numbers of photons for moderately significant detections are not very large.

THE ANGULAR RESOLUTION AND THE FIELD OF VIEW

Consider a case of 4-X_0 BGO:

$$a = 4\ mm, \quad l = 40\ mm, \quad d = 5\ m.$$

Then:

$$angular\ resolution = a/d = 3.4',$$
$$field\ of\ view = a/l = 7.2°.$$

The angular resolution is comparable to the best cases of EGRET, but the field of view is much smaller. To improve the angular resolution, we reduce the element width to a = 2 mm, and increase the distance to d = 20 m. Then:

$$angular\ resolution = 0.34',$$

an improvement by a factor of 10. This is what GLAST can achieve under favorable conditions. But the field of view becomes:

$$field\ of\ view = 2.9°.$$

The field of view of coded masks are always much smaller than the conventional designs, although the angular resolution is not much better than what the conventional high-energy gamma-ray telescopes can achieve. In order to substantially improve the source location capability, we need to push the coded mask to the extreme, say to a = 1 mm, d = 50 m. Then we can achieve an angular resolution of $4''$.

SIZE AND WEIGHT

Consider a mask with a = 2 mm, l = 40 mm, d = 20 m. This mask will produce an angular resolution of 0.34' as indicated in the previous section. Assume the detector face is 160 cm × 160 cm. In order to cover the detector face within the field of view, the mask needs to be of a size of 360 cm × 360 cm × 4 cm. If we use BGO for the dark (opaque) elements and lucite for the transparent elements, then the weight of the mask is:

$$2,100\ Kg.$$

We have to mount this piece of mass 20 m away from the detector, in a mechanically rigid way such that the entire unit, mask plus detector, can be steered frequently for target pointing. The mechanical structure of such coded masks may pose quite a challenge in the actual constructions.

CONCLUSIONS

(1) It is technically feasible to use an active coded mask for high-energy gamma-ray telescopes in space, if the desired angular resolution is comparable to the telescopes in conventional designs. But the prices that one has to pay for reduced field of view, increased size and weight, and added mechanical complexity are not necessarily small.

(2) To achieve much improved angular resolution with coded masks, one has to push the design parameters to the extreme. Then the added size and weight, and the resulting mechanical complexity, can be prohibitive.

(3) In the field of high-energy gamma-ray astrophysics, the photon fluxes that one must be dealing with are inherently small. This means that one has to stay on the target for a long time in order to collect enough photons for the study. Thus a telescope with wide field of view, which can then be used to observe many sources at one time, is distinctly more advantageous over those with narrow field of view. This is the reason why the success of coded masks in the X-ray and low-energy gamma-ray telescopes, which are used to observe much higher fluxes, may not be easily transferred to the field of high-energy gamma rays.

(4) But if the research subjects demand a high-energy gamma-ray telescope with very high spatial resolving power, then coded masks may be the only kind of device available at present.

REFERENCES

1. Bouchet, L. e. a., *ApJ*, **548**, 990–1009 (2001).
2. Winkler, C., *ApJS*, **92**, 327–332 (1994).
3. Akimov, V. V. e. a., , in *Proc. 19th Int. Cosmic Ray Conf.*, 1985, p. 330.
4. Caroli, E. e. a., *Space Science Reviews*, **45**, 349–403 (1987).
5. Johansson, A. e. a., "The use of an active coded aperature", in *IEEE Transactions on Nuclear Science*, NS-27 1, IEEE, 1980, pp. 375–380.
6. Thompson, D. J., *Nucl. Instr. and Methods in Physics Research*, pp. 390–401 (1986).

Design of a Next Generation High-Energy Gamma-Ray Telescope

S.D. Hunter [1], D.L. Bertsch [1], P. Deines-Jones [1,2]

[1] NASA/Goddard Space Flight Center, Greenbelt, MD 20771 USA
[2] Universities Space Research Association, Seabrook, MD 20706 USA

Abstract. The Next Generation High-Energy Gamma-ray (NGHEG) mission is a recommended priority for a new mission after GLAST. One of the science goals of the NGHEG mission, mapping our Galaxy, clouds, supernova remnants, and nearby galaxies for cosmic ray sources at arcminute resolution, places stringent requirements on the design of the track imager. We examine the design requirements of a track imager for a gamma-ray telescope capable of arc-minute imaging.

NEXT-GENERATION HIGH-ENERGY GAMMA-RAY MISSION

The Next Generation High Energy Gamma Ray (NGHEG) mission is a GRAPWG recommended priority for a new mission after GLAST. The science goals of NGHEG are expected to include

- Unambiguously resolving the galactic plane into compact sources and true diffuse emission from cosmic-ray interactions,
- Mapping of nearby normal galaxies, and providing a complete catalog of fluxes from galaxies more than a few arc-minutes in extent,
- Imaging many supernovae remnants,
- Using distant sources (blazars, GRBs) as cosmological probes,
- Locating sources to about one arc-second, and
- Providing high-sensitivity polarization measurements.

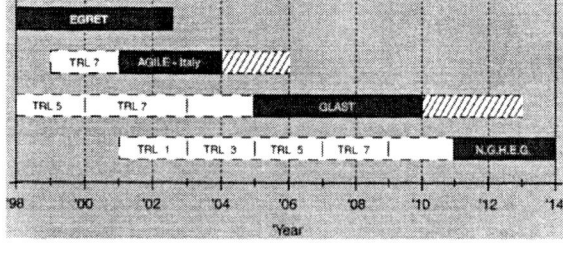

FIGURE 1. High-energy gamma-ray Astronomy Roadmap as suggested by the Gamma-Ray Astronomy Working Group (GRAPWG, 2000).

These science goals are ambitious. An order-of-magnitude improvement in angular resolution, i.e. arc-minute imaging, would allow mapping of a host of objects, and also yield an enormous increase in signal-to-background for compact objects whose size, or structure, is on the order of the GLAST resolution or smaller. (GLAST's best angular resolution, 0.1-0.2°, allows arc-minute positioning but not arc-minute imaging.). Polarization sensitivity would, in addition, provide an excellent new tool for probing pulsar geometry and dynamics, and separating electromagnetic and nuclear processes.

Designing a track imager with polarization sensitivity is consistent with the demands for arc-minute imaging. Both require that the electron and positron momenta be measured before Coulomb scattering dominates the initial electron directions. We examine the impact on the track imager design imposed by this constraint.

Meeting the Arc-Minute Challenge - Instrument Considerations

The angular resolution of conventional pair telescopes is the sum in quadrature of the
- angular error due to scattering,
- instrumental angular resolution imposed by the detector's finite spatial resolution, and
- unmeasured nuclear recoil momentum, ~1 MeV/c (Jost et al. 1950).

It follows that single-photon arc-minute imaging (0.3 mrad) must be done above 1 MeV/3×10^{-4} = 3 GeV. Above 3 GeV the angular resolution improves slightly whereas, below 3 GeV the angular resolution is dominated by the unmeasured nuclear recoil. We therefore set a design goal of an instrument point-spread function (PSF) which is within a factor of ~2 of the recoil limit at 3 GeV. In order to detect enough photons to form a meaningful image in a few week observation will require a large effective area, ~10 m^2. A moderately wide field of view, >60° cone angle, is also desirable. We view this target as achievable only if the conventional heavy calorimeter is either replaced, or supplemented, with a different energy measurement method. One approach is to supplement on-axis calorimetry by measuring the multiple scattering of the electrons produced by the gamma ray (Bertsch 1984).

Angular Resolution Near the Kinematic Limit

The direction of the electron and positron must be determined before scattering confuses their direction. One measure of the usable range can be obtained by equating the most probable opening angle of the pair (Borsellino, 1953; Hintermann 1954).

$$\omega_0 = E_\gamma \cdot mc^2 / E_+ E_- = 4 \cdot mc^2 / E_\gamma \quad \text{(for equipartition of energy)}, \qquad (1)$$

with the scattering angle [5]

$$\theta_0 = \frac{13.6\,\text{MeV}}{\beta c p} \sqrt{x/X_0}\left[1 + 0.038\ln(x/X_0)\right]. \qquad (2)$$

The most probable opening angle of the pair is only weakly dependent on the energy split if the split is more equal than ~20/80. The opening angle increases by a factor of ~2 for a 10/90 split [3].

We assume scattering does not confuse the electron and positron tracks if the track directions are determined within a distance where θ_0/ω_0 is ≤ 0.3, figure 2. This distance is a few milli-RL (mRL). To determine the electron/positron directions

Figure 2. The ratio of the RMS scattering angle to the RMS opening angle as a function of the range of the electrons in the tracker. Scattering is assumed to not confuse the electron and positron tracks if the track directions are determined within a distance where this ratio is ≤ 0.3.

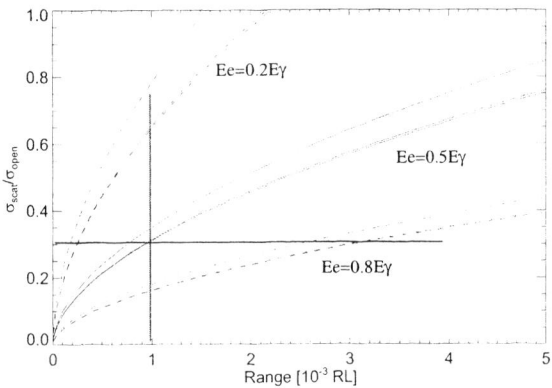

requires 10-1000 samples, depending on the gamma-ray energy, within a 1 mRL depth of the track imager. This corresponds to a density of 10^{-4} to 10^{-6} RL/sample. On the assumptions that the electron and positron directions are determined in 2 mRL, the spatial resolution is ¼ of the measurement distance, and equal energy split between the electron and positron, the angular resolution can be calculated. The angular resolution, expressed in terms of the track imager density, is shown in figure 3. One arc-minute single photon angular resolution can be achieved above ~ 3 GeV and then only if the track imager density less than ~2×10^{-5} RL/sample.

Figure 3. Angular resolution (68% containment angle) as a function of gamma-ray energy and track imager density (R.L. per sample distance along track). Arc-minute imaging can only be obtained above ~3 GeV and then only if the tracker density is less ~2×10^{-5} R.L. per sample distance. This plot was calculated

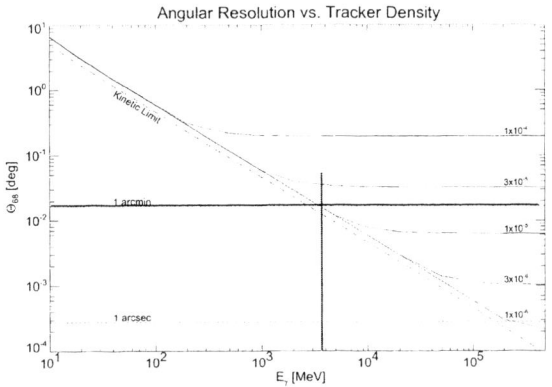

Momentum Determination by Multiple Coulomb Scattering

The RMS angle of multiple scattering is inversely proportional to $p\beta$, where p is the particle momentum and β is its velocity, eq. (2). The arithmetic mean of the projected scattering angle can be written as:

$$\langle|\vartheta|\rangle = \frac{k'zt^{1/2}}{p\beta} \tag{3}$$

where: z, p, and β denote the charge, momentum, and velocity of the scattered particle; t is the measurement cell length and k' is a constant depending on the material, cell length, and charge and energy of the particle.

The 'coordinate method' [2] of momentum determination is well suited to a low-density track-imaging detector. The distance between the electron (positron) track and a straight 'reference' line is measure at successive points along the track, figure 4. S_i and D_i, see figure 3, are the so-called first and second differences of the coordinates.

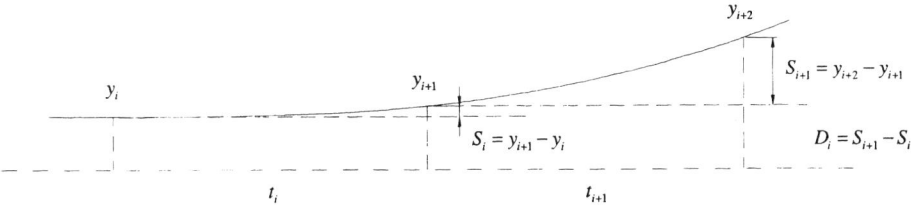

Figure 4. Diagram of the 'coordinate' method of momentum determination from Coulomb scattering.

Denoting the cell length by t gives

$$\vartheta_k^{(i)} = \frac{D_i}{t} = \frac{y_{i+2} - 2y_{i+1} + y_i}{t} \qquad (4)$$

On the assumption that the angles between successive intersections have a Gaussian distribution, it can be shown (Bertsch, 1984) that the arithmetic mean of multiple Coulomb scattering determined by the coordinate method can be written as

$$\langle |\vartheta_K| \rangle = \frac{K z t^{1/2}}{p\beta} = \left(\frac{2}{3}\right)^{1/2} \langle |\vartheta| \rangle \qquad (5)$$

This technique can be readily used for determination of the electron and positron momenta in a track imager [2].

Effective area

The effective area of a gamma-ray telescope is determined by the geometric area (launch vehicle fairing) and the mass to orbit (i.e. the amount of converter material). The effective area as a function of geometric area and converter mass is shown in figure 5. A track imager with 5×10^4 cm^2 effective area and depth of 0.5 RL would require a geometric area of $\sim 12\times10^4$ cm^2 and a converter mass of ~ 800 kg.

Conclusion:

A large area, low-density tracker design is required if arc-minute imaging and polarization sensitivity are to be achieved. These requirements present several unique challenges. The 5×10^4 cm^2 example above, with a ½ RL deep track imager, a density of 2×10^{-5} RL/sample, and geometric area of 12×10^4 cm^2, would require a track imager

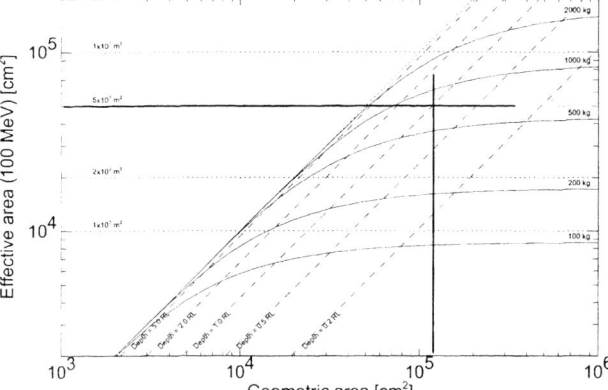

Figure 5. Effective area of a gamma-ray telescope is determined by the geometric area and the mass to orbit. The effective area is plotted as a function of the geometric area for different masses of converter material and 55% conversion probability. The dashed lines give the depth of the tracker in RL.

height of 10 m. We note that a 12×10^4 cm^2 CsI calorimeter even 4 RL thick (half of the EGRET thickness) would have a mass of over 4000 kg, almost *twice* the mass of the GLAST LAT. A detailed instrument design, incorporating these parameters, and Monte Carlo simulations are in progress to optimize the track imager design to achieve the science goals of the NGHEG Mission.

REFERENCES

1. Jost, R., Luttinger, J.M., & Slotnick, M. 1950, Phys. Rev., 80, 189-196.
2. Bertsch, D.L. 1984, Nucl. Inst. Meth. 220, 489-511.
3. Borsellino, A. 1953, Phys. Rev., 89, 1023.
4. Hintermann, K. 1954, Phys. Rev., 93, 898-899.
5. See e.g. Particle Physics Booklet, Particle Data Group, Springer.
6. Depaola, G.O., Kozameh, C.N., & Tiglio, M.H. 1999, Astroparticle Physics, 10, 175.

Mission Concepts: Gamma-Ray Line Spectroscopy

Diffractive/Refractive Lenses
- A Revolution In Gamma-Ray Astronomy?

G. K. Skinner

CESR, Toulouse, France
& University of Birmingham, U.K.

Abstract. The use of diffractive/refractive gamma-ray optics in the form of variants of the Fresnel lens offers the prospect of enormous advances in both sensitivity and angular resolution. Such lenses can concentrate gamma-ray flux from many square metres collecting area onto a small, low background, detector and can have imaging capability close to the diffraction limit, i.e. better than 1 micro arc second. In addition, they are simple in construction and robust. The drawback is that the focal lengths are extremely long ($\sim 10^6$ km). However the mission requirements are more modest than those of other projects currently being planned. It is argued that there are no technological hurdles that would prevent the construction and use of telescopes based on these principles. The scientific objectives which could be achieved with such an instrument include direct imaging of the surroundings of supermassive black holes in AGN.

INTRODUCTION

Despite the major advances made with Compton-GRO and expected with INTEGRAL, it can be argued that the current status of instrumentation for gamma-ray astronomy corresponds to that of X-ray astronomy about 30 years ago. The sensitivities achievable are such that a few hundred sources in the sky are detectable, and the angular resolution attainable is of the order of one degree.

Gamma-ray astronomy has suffered from limitations which have constrained its development compared with other observation bands. It has never been possible to concentrate flux from a large collecting area onto a small detector. Furthermore, except at the highest energies, imaging has been possible only by indirect means such as offered by coded-mask and Compton telescopes. Finally, with current techniques there is little prospect of improving angular resolution beyond 0.1-1°. The poor angular resolution is particularly ironic in view of the fact that the ultimate physical limit is imposed by diffraction and this is least constraining in the gamma-ray part of the spectrum where wavelengths are shortest. Current instrumentation falls short by a factor $\sim 10^9$ of the diffraction limit of even a modest size instrument.

The problems arise from the present lack of optics which can focus and image gamma radiation. If radiation from a moderately large collecting area could be concentrated onto a small, low background detector, an enormous improvement in sensitivity would be possible and if a diffraction limited optical system of even one metre diameter could be made, the angular resolution would be better than that currently achieved in any other waveband. The prospects of using the Fresnel lenses to achieve these objectives are reviewed here. More details of some aspects of this work are presented in [1].

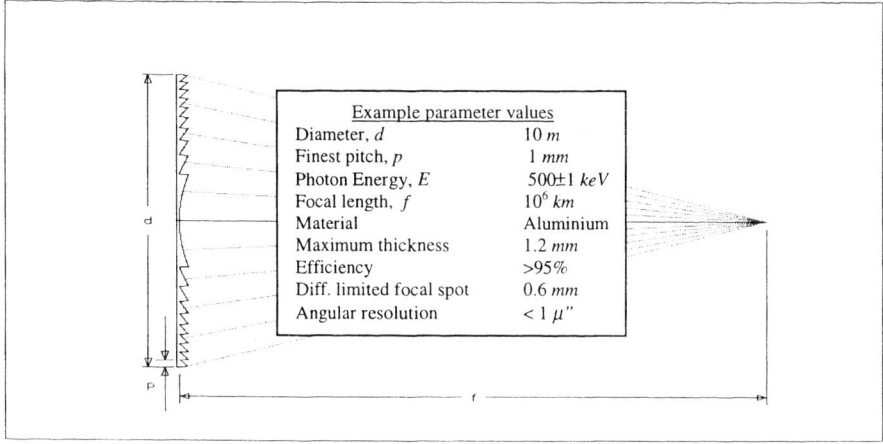

FIGURE 1. A gamma ray Phase Fresnel Lens and some example parameters

DIFFRACTIVE/REFRACTIVE GAMMA-RAY OPTICS

Fermat's principle demands that radiation taking different routes through an optical system should have the same effective path length (modulo the wavelength) and so arrive in phase. In a conventional lens using only refraction, this implies a lens whose thickness $t(r)$ as a function of radius r is

$$t(r) = t(0) - \frac{r^2}{(\mu-1)f}, \qquad (1)$$

where μ is the refractive index and f is the focal length.

Well above absorption edges, the real part of the refractive index of any material is less than unity by an amount delta which depends only on the photon energy E and the electron density n_e in the material:

$$\delta = 1 - \mu = \left(\frac{hc}{E}\right)^2 \frac{n_e r_e}{2\pi} \sim 1.9 \times 10^{-10} \left(\frac{E}{1\,MeV}\right)^{-2} \left(\frac{\rho}{1\,g\,cm^{-3}}\right) \qquad (2)$$

where r_e is the classical electron radius and ρ is the mass density. The numerical approximation assumes A/Z=2.1, typical of most elements.

The refractive index is so close to unity that grazing incidence mirrors are impracticable and that refractive lenses of a useful size would be so thick that absorption would be prohibitive, even if a very long focal length is accepted.

The most basic element of the family of devices considered here is the Fresnel Zone Plate (FZP) in which radiation which would arrive at the focal point with the wrong phase is blocked by an opaque band. The efficiency of such a device is limited to $\pi^{-2}=10.1\%$. By shifting the phase of the radiation by π rather than blocking it, the maximum efficiency is raised to 40.4%. By modifying the phase as a continuous function of radius so that all radiation arrives at the focal point with exactly the same phase one arrives at the Phase Fresnel Lens (Figure 1); all of the incident energy can in principle be directed towards the focal point.

The maximum thickness needed is that which gives a phase-shift of 2π and is given by

$$t_{2\pi} = 6.5 \times 10^{-3} \left(\frac{E}{1\,MeV} \right) \left(\frac{\rho}{1\,g\,cm^{-3}} \right)^{-1} m.. \qquad (3)$$

The thickness profile required is simply that of Equation 1, reduced modulo this distance. For a wide range of materials and over a broad band of gamma ray energies, the absorption of such a lens is low (Figure 2).

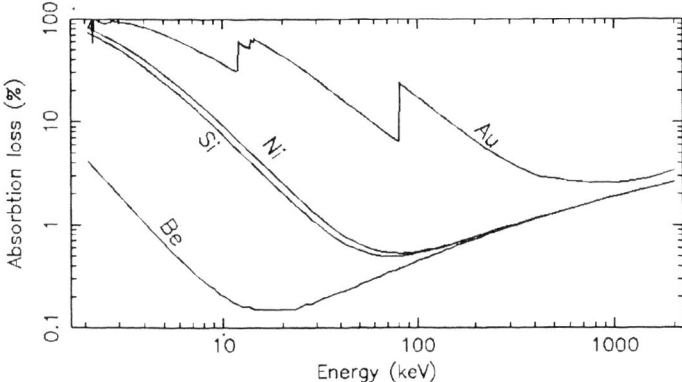

FIGURE 2. The mean absorption in a Phase Fresnel Lens as a function of gamma ray energy for various materials. The curve for aluminium falls slightly below that for Silicon.

PARAMETERS GOVERNING THE DESIGN

The diffraction limited angular resolution of a lens of diameter d and the focal length of a lens with finest groove pitch p are given by

$$\vartheta_D = \frac{1.22}{d}\left(\frac{hc}{E}\right) = 0.31 \left(\frac{E}{1\,MeV}\right)\left(\frac{d}{1\,m}\right) \quad micro\ seconds\ of\ arc,. \qquad (4)$$

$$f = \left(\frac{E}{hc}\right)\frac{pd}{2} = 0.403 \times 10^6 \left(\frac{E}{1\,MeV}\right)\left(\frac{p}{1\,mm}\right)\left(\frac{d}{1\,m}\right) \quad km \quad (5)$$

An implication of these two equations is that the linear size in the detection plane of the diffraction limited focal spot size is simply

$$s_D = f\vartheta_D = 0.61\,p. \quad (6)$$

Blurring due to chromatic aberration will exceed the diffraction limited focal spot size if the bandwidth $\Delta E/E$ is less than about

$$\frac{d^2}{6f\lambda}, \quad \text{i.e. less than about} \quad 135 \left(\frac{E}{1\,MeV}\right)\left(\frac{f}{10^6\,km}\right)^{-1}\left(\frac{d}{1\,m}\right)^2. \quad (7)$$

so the system is essentially a narrow band one. It may, however, be used with close to optimal performance over a wide band by varying the lens detector spacing f in proportion to the energy [1].

NON-PROBLEMS AND POSSIBLE PROBLEMS

The very long focal lengths implied by Equation 6 for a gamma ray PFL of useful diameter and practical pitch are clearly only possible if the optic and the detector are on two separate spacecraft. But such configurations have been extensively investigated for numerous projects in recent years. The 'formation flying' of spacecraft is a recognised branch of space engineering which has been extensively studied recently, often for projects with constraints much more stringent than those of a PFL telescope.

The spacecraft separation needed is greater than that required for most other missions, though it is less than that for the gravitational wave mission LISA which will have 3 spacecraft separated by 5×10^6 km. The line joining the two spacecraft needs to remain fixed in celestial co-ordinates. Retaining a 'formation' in a fixed orientation in such an inertial reference frame is also a requirement for interferometry missions such as SIM, Maxim and Darwin.

The type of system proposed here is exceptional only in requiring the *combination* of large separations and stability of orientation in inertial space. We assume here that like LISA the telescope is placed in orbit around the sun at 1 a.u., leading or trailing the earth by sufficient distance that it is not adversely affected by the earth-moon system. If one spacecraft is passive, then when pointing at a particular object the second must be controlled to be always separated from it by a fixed vector. If the lens spacecraft is passive, the detector spacecraft must track the focal point in 3-dimensional space. The accelerations of the two spacecraft will thus be identical; those of the former will be produced by the gravity field of the solar system, while the second will have to be forced to follow the same acceleration pattern despite experiencing a slightly different gravitational field.

The effect of the gradient of the solar gravitational field is that a spacecraft of mass m will require a motor continuously providing a thrust given by

$$F = \frac{M_{sun}mGf}{r^3}\left(3\cos^2(\beta)+1\right)^{1/2} \sim 4\left(\frac{m}{100\,kg}\right)\left(\frac{f}{10^6\,km}\right)\left(3\cos^2(\beta)+1\right)^{1/2} \quad mN. \quad (8)$$

where $r=1$ a.u. and β is the angle between the vectors **r** and **f**. Such forces are well within with the capabilities of existing continuous thrust ion engines.

Repointing the telescope to a different target will be non-trivial but does not seem to present an overriding problem. A reasonable objective might be to observe 10 targets per year for 2×10^6 s each, leaving about 1/3 of the time for manoeuvres. A typical repointing will require a translation in inertial space of the order of $f/2$. For $f=10^6$ km and $m=100$ kg, this requires a thrust of about 150 mN, which is not out of the question.

Modest errors or variations in the distance between the two spacecraft are not important; even a 1000 km error in 10^6 km can be tolerated. The requirements on pointing in the conventional sense, that is to say on attitude control, are also extremely lax. The orientation of the spacecraft carrying the lens can change by up to 1° without significant effect on the imaging. Similarly the focal plane spacecraft need only keep the detector normal roughly directed towards the lens.

The manufacture of the lens is not a major issue. It can be made from a disk of aluminium and maintaining $\lambda/10$ precision involves tolerances no tighter than 100 μm.

What is critical is the direction of the line joining the detector centre to the lens centre, which must be controlled, or at least determined, with a precision consistent with the extremely high angular resolution.

ANTICIPATED CAPABILITIES

The sensitivity achievable with a PFL telescope could readily be that corresponding to an effective collecting area of 10 m^2 with a background event rate corresponding to a detector volume (per resolution element) of the order of 1 cm^3 or less. Taking as an example a narrow line at 847 keV, the sensitivity will be better than 2×10^{-9} photons cm^{-2} s^{-1} (5σ in 10^6 s). This is 4 orders of magnitude better than the present generation of instrumentation (e.g. INTEGRAL) and sufficient to detect ^{56}Co emission from Type II supernovae at up to 70 Mpc.

Despite the narrow bandpass, the continuum sensitivity is also extremely good. For example, 4×10^{-9} photons cm^{-2} s^{-1} keV^{-1} (5σ) should be achievable at 500 keV in 10^6 s, representing more than a thousand fold improvement over INTEGRAL.

With angular resolution better than a microsecond of arc it will be possible to:-
- Image and study accretion disks, jets and the deformation of space in the region of the event horizon of the supermassive black holes in AGNs (one Schwarzschild radius subtends an angle of an micro arc second or more in many AGN)
- Image the disks of stars at kpc distances, study their activity and observe directly the instabilities and interactions in the wind systems surrounding both single massive stars, and those in interacting binary systems
- Watch the expansion of the radioactive ejecta of supernovae at up to 50 Mpc

REFERENCES

1. Skinner, G. K., *Astron. Astrophys.* To be published.

B-MINE, The Balloon-Borne Microcalorimeter Nuclear Line Explorer

E. Silver[a], H. Schnopper[a], C. Jones[a], W. Forman[a], S. Bandler[a], S. Murray[a], S. Romaine[a], P. Slane[a], J. Grindlay[a], N. Madden[b], J. Beeman[b], E. E. Haller[b], D. Smith[c], M. Barbera[d], A. Collura[d], F. Christensen[e], B. Ramsey[f], S. Woosley[g], R. Diehl[h], G. Tucker[i], J. Fabregat[j], V. Reglero[j] and A. Gimenez[k]

[a]*Smithsonian Astrophysical Observatory,* [b]*Lawrence Berkeley National Laboratory,*
[c]*University of California at Berkeley,* [d]*Osservatorio Astronomico G.S. Vaiana,*
[e]*Danish Space Research Institute,* [f]*George C. Marshall Space Flight Center,*
[g]*University of California at Santa Cruz,* [h]*Max Planck Institut für Extraterrestriche Physik,*
[i]*Brown University,* [j]*Universidad de Valencia,*
[k]*INTA/Laboratorio De Astrofísica Espacial Y Física Fundamental*

Abstract. *B-MINE* is a concept for a balloon mission designed to probe the deepest regions of a supernova explosion by detecting ^{44}Ti emission at 68 keV with spatial and spectral resolutions that are sufficient to determine the extent and velocity distribution of the ^{44}Ti emitting region. The payload introduces the concept of focusing optics and microcalorimeter spectroscopy to nuclear line emission astrophysics. *B-MINE* has a thin, plastic foil telescope multilayered to maximize the reflectivity in a 20 keV band centered at 68 keV and a microcalorimeter array optimized for the same energy band. This combination provides a reduced background, an energy resolution of 50 eV and a 3σ sensitivity in 10^6 s of 3.3×10^{-7} ph cm^{-2} s^{-1} at 68 keV. During the course of a long duration balloon flight, *B-MINE* could carry out a detailed study of the ^{44}Ti emission line centroid and width in CAS A.

INTRODUCTION

Nuclear line astrophysics provides a direct probe of the details of one of the most violent events in the universe -- a supernova explosion that expels the heavy elements into the ISM from the nuclear furnace in which they were created. Because of its production deep in the stellar core and 60 year half-life, ^{44}Ti provides a key diagnostic of supernova explosions. Although other elements are produced at high density and temperature in SN explosions, the key feature of ^{44}Ti is that its half life is longer than the few years required for the overlying strata to become optically thin at high energies and sufficiently short that the ^{44}Ti remains localized around the SN site while it emits intensely. Thus, the nuclear properties of ^{44}Ti make it unique among all elements of nature for probing deep into the cauldron of a SN explosion. Model calculations of nucleosynthesis yields for Type Ia and Type II supernovae show that the conditions for ^{44}Ti production are significant only within the central cores of massive stars or in white dwarfs following a surface detonation.

Hence, the mass in ^{44}Ti is an important diagnostic of the most extreme densities and temperatures.

Despite the importance of ^{44}Ti as a diagnostic of supernova explosions, observational evidence remains scarce. Only recently have instruments of sufficient sensitivity become available. COMPTEL detected ^{44}Ti emission from Cas A in the 1.156 MeV line produced in the final decay of the ^{44}Ti chain as ^{44}Sc decays to stable ^{44}Ca [1]. ^{44}Ti emission also has been reported from GRO J0852-4642 [2], possibly a young supernova remnant RX J0852.0-4622 [3] in the direction of the older Vela remnant.

We have developed a concept for a Balloon-borne Microcalorimeter Nuclear Line Explorer (*B-MINE*) to study the 68 keV ^{44}Ti emission from young supernova remnants. It is designed to measure the structure and centroid of the ^{44}Ti line as well as its flux from Cas A. It will also determine the mass of ^{44}Ti and the velocity of the ejecta, both critical to understanding the nucleosynthesis and the explosion in core collapse supernovae.

B-MINE

B-MINE, the Balloon-Borne Microcalorimeter Nuclear Line Explorer, combines the high spectral resolution of a microcalorimeter array with modest focusing, multilayered optics to reduce the background and enhance the effective area in a balloon-borne experiment. The telescope has a focal length of 7.8 m and is a thin plastic foil, conical approximation to a Wolter I design that is multilayered to maximize the reflectivity in a 20 keV band centered at 68 keV[4]. Multilayers of platinum/carbon enhance the reflectivity at 68 keV resulting in a peak effective area of 60 cm^2 and a FOV that is 5' in diameter.

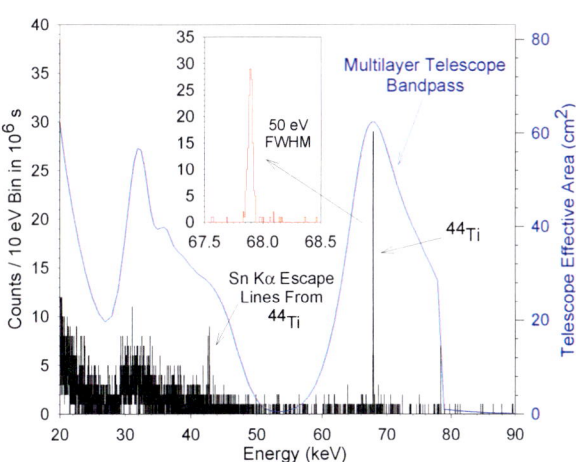

Figure 1. The blue curve shows effective area of the telescope (without detector). The black curve is a simulation of a 10^6 s observation of the Cas A supernova remnant. It includes the detector efficiency. The 68 and 78 keV (unredshifted) ^{44}Ti lines are both seen. The red curve in the inset is the 68 keV line broadened by the 50 eV detector resolution.

The primary energy range for *B-MINE*, centered at 68 keV, is defined by the range of velocities due to the outflow of the ^{44}Ti ejecta. Velocities are presently unknown but could reach 4,000 km s^{-1} in Cas A. A secondary energy range of 30 - 50 keV includes the cyclotron resonance features observed in Her X - 1 and other binary pulsars. These two ranges can be achieved with a novel multilayer design. Figure 1 shows the effective area as a function of energy.

The telescope consists of 191 multilayered shells ranging from 12 cm to 45 cm in diameter. The individual cones are 15 cm long each. The multilayer d-spacing gradation is divided into two narrow ranges; Pt/C multilayers applied to the inner 103 shells (with a grazing angle ≤ 0.25 deg) are for the band centered at 68 keV, and W/Si multilayers applied to the outer 88 shells (0.25 - 0.40 deg) are for the 30 - 50 keV band. We have carefully matched the range of d-spacings in the 68 keV band to accommodate the size of our primary science target (Cas-A) so that the effective FOV is about 5 arcmin.

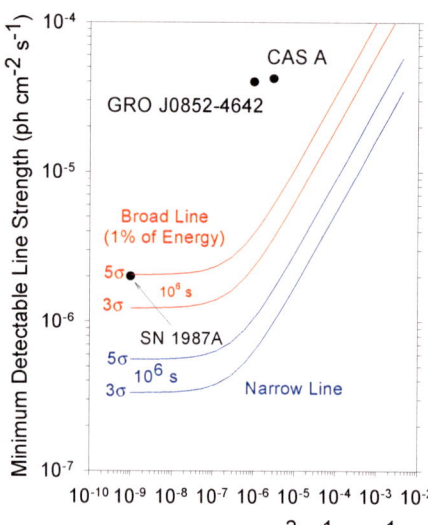

Figure 2. Minimum detectable line strength needed by *B-MINE* for 3σ and 5σ statistical precision.

The microcalorimeter array is optimized for the same energy bands. Each pixel consists of a tin (Sn) absorber and a germanium thermistor, providing a quantum efficiency of 15% and energy resolution of 50 eV at 68 keV. Since a large area array of microcalorimeters is small compared to traditional hard x-ray detectors, the volume induced background is several orders of magnitude smaller than that in more conventional hard x-ray spectrometers. With an energy resolution of 50 eV at 68 keV, the sensitivity for nuclear line detection is unprecedented ; the 3 σ sensitivity in 10^6 s for narrow line detection is 3.3×10^{-7} ph cm^{-2} s^{-1} at 68 keV. During the course of a long duration balloon flight, *B-MINE* would carry out a detailed study of the ^{44}Ti emission line centroid and width in Cas A. Figure 2 shows the sensitivity to line detection.

The improved signal-to-background ratio of the microcalorimeter over past experiments on Cas A would provide a precise measure of the flux, line shape, and centroid of ^{44}Ti emission. As a simple model one can assume a uniform outflow at a constant ejection velocity from the center of the supernova explosion. The radius of the ^{44}Ti shell is given by the time since the explosion and the ejecta velocity. Limb brightening will enhance the flux from the limb (zero velocity in the line-of-sight) compared to the flux from the center. Thus the observed line emission will peak at zero velocity, with flux throughout the energy range corresponding to the maximum and minimum velocities. Figure 3 shows spectra that would be obtained for Cas A if the ^{44}Ti ejecta had velocities of 300, 1000, or 4000 km s^{-1}. If ejecta have a range of velocities, the spectra would be a sum of these simulations, with the line broadened.

With *B-MINE*'s spectral resolution of 50 eV, one could detect broadening if the velocity of the expanding ^{44}Ti shell exceeds 300 km s^{-1}. Detecting velocity broadening would provide clear evidence for mixing of the ejecta layers, thus introducing new constraints on instabilities and transport mechanisms in massive stellar explosions.

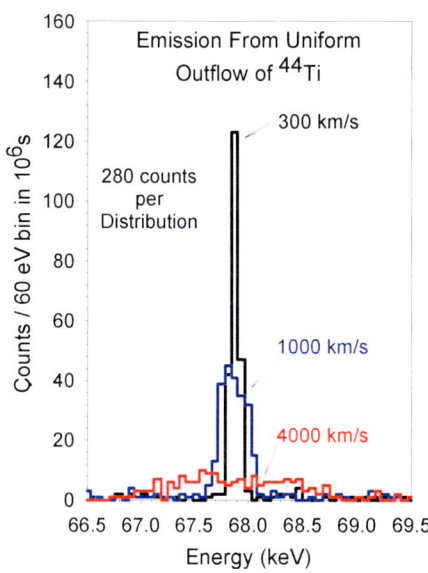

Figure 3. Spectra modeled for ejecta created at the same time and moving with a constant velocity in all directions from the source center. X-ray emission comes from a shell of radius given by the velocity and the age of the remnant. Three different outflow velocities are modeled. The spectra are dominated by the zero redshifted emission from the source limb. The higher the velocity, the broader the distribution.

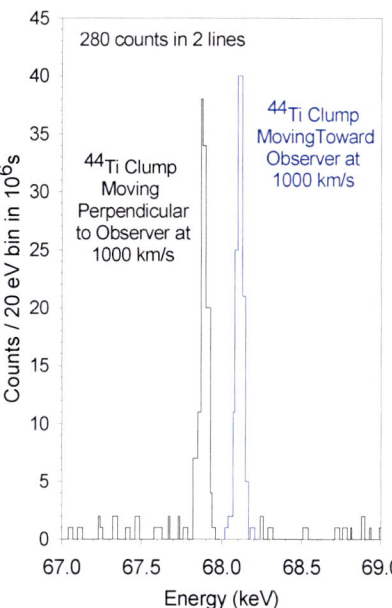

Figure 4. Spectra modeled for the case of directed emission from the center. In one case, the emission appears to come from the center of the source at maximum blue shift, in the other it comes from the limb at zero redshift. For the zero redshift case, emission from the entire limb cannot be distinguished spectroscopically from directed beams of particles striking the limb at one or more places. Imaging is required to resolve the emission site(s).

If the ^{44}Ti source is composed of a few bright knots, the spectrum will have features that correspond to each knot. Figure 4 shows the simulated *B-MINE* spectrum of Cas A if the ^{44}Ti ejecta were in two knots, one moving toward the observer at 1000 km s^{-1} and the other moving perpendicular to the line of sight.

Focal Plane Microcalorimeter Array

The *B-MINE* hard-X-ray microcalorimeter is based on a laboratory prototype that has a 25 µm thick tin X-ray absorber operating at a base temperature of 80 mk [5]. The prototype has a quantum efficiency (Q.E.) of 40% at 35 keV, 8% at 68 keV and an energy resolution of 50 eV. The resolving power at 68 keV is $E/\Delta E = 1360$. A typical X-ray spectrum using ^{241}Am and ^{55}Fe sources is shown in Figure 5. The *B-MINE* spectrometer will use a tin absorber that is 50 µm thick to provide a Q.E. of 15% at 68 keV. Since the energy resolution scales as the square root of the heat capacity, one could expect the energy resolution to degrade to 66 eV by increasing the absorber thickness to 50µm. By decreasing the temperature of the detector and increasing the responsivity of the thermistor by choosing an NTD-germanium thermistor with a slightly different doping, it is very feasible to build a microcalorimeter with 64% Q.E. at 35 keV and 15% Q.E. at 68 keV without degrading the resolution beyond 50 eV. An array of 20 × 20 pixels is required to cover the 5' × 5' angular size of Cas A with the 7.8 m *B-MINE* telescope. The two-dimensional array is is shown in Figure 6 and the details may be found in the literature [5].

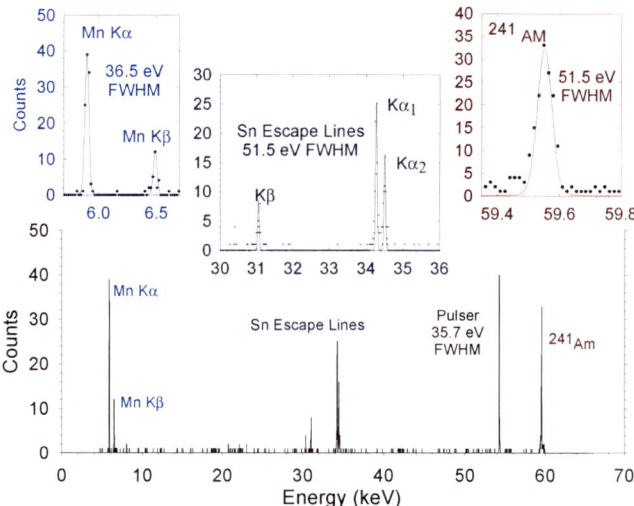

Figure 5. The spectrum of emissions from ^{241}Am and ^{55}Fe sources. The Np L lines between 13.7 keV and 20.8 keV produced in the ^{241}Am decay have been absorbed by a copper foil.

ACKNOWLEDGEMENTS

The authors wish to thank G. Austin, R. Ingram, G. Nystrom, J. Gomes, D. Boyd, and B. Podgorski for their technical input to the *B-MINE* concept. This work was performed in part by NASA Grant NAG5-5104.

REFERENCES

1. Iyudin, A. F. et al., A&A, **284**, L1, 1994.
2. Iyudin, A. F. et al., *Astrop. Letters and Communications*, **38**, 383, 1999.
3. Aschenbach, B., *Nature*, **396**, 141, 1998
4. Schnopper, H. W. et al., *Proc. SPIE* **3766**, 350, 1999
5. Silver et al., *SPIE*, **4140**, 397, 2000

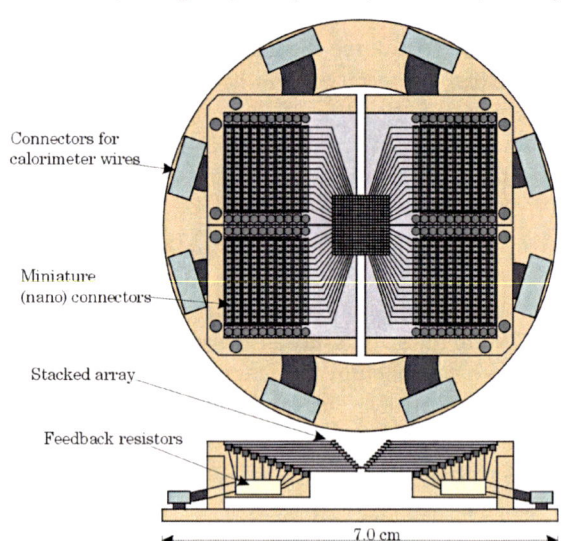

Figure 6. The B-Mine microcalorimeter array design.

Mission Concepts: Advanced Compton Telescopes

Progress towards an Advanced Compton Telescope

J.D. Kurfess and R.A. Kroeger

Naval Research Laboratory
Washington, DC 20375

Abstract. With the termination of the *Compton Observatory* mission and the forthcoming launch of ESA's *INTEGRAL* mission in 2002, plans for a follow-on low/medium-energy gamma ray mission are underway. NASA's Gamma Ray Program Working Group has endorsed an Advanced Compton Telescope (ACT) as the next major mission for gamma ray astronomy. A Compton telescope instrument appears to be the best approach for meeting the sensitivity demands of a broad range of scientific objectives within acceptable cost and weight constraints. With a narrow line sensitivity of a ~ 10^{-7} γ cm^{-2} s^{-1}, gamma ray observations of many supernovae and novae will be realized for the first time. ACT will also provide detailed maps of the Galaxy in several nuclear lines associated with radioactive decays and cosmic ray interactions with the ISM. ACT will enable high sensitivity gamma ray studies of galactic compact objects, active galactic nuclei, and solar flares. The several components of the diffuse low-energy gamma ray background will be investigated. Alternative approaches for the instrument are presented, including the use of time projection chambers, position-sensitive solid-state detectors, and electron tracking in arrays of thin silicon strip detectors. A novel concept for using multiple Compton interactions to achieve high efficiency is discussed.

INTRODUCTION

Following termination of the *Compton Observatory* (CGRO) mission, many high priority scientific objectives remain that require a substantial improvement in sensitivity beyond that achieved with CGRO or the capabilities of the upcoming *INTEGRAL* mission. The Gamma Ray Program Working Group (GRAPWG) has recommended an Advanced Compton Telescope as the next major mission in gamma ray astrophysics following GLAST [1]. This mission will have a factor of 20-50 improved line sensitivity compared to previous missions. The Compton telescope technique is judged the best approach for meeting the broad objectives associated with the line and continuum emissions from discrete and diffuse sources, both galactic and extragalactic. Several instrumental options are being pursued or are candidates for development. These include the use of position-sensitive solid-state detectors, thin silicon strip detectors for tracking the scattered electrons coupled with a position-sensitive calorimeter, and liquid or gaseous time projection chambers.

CURRENT STATUS OF LOW/MEDIUM ENERGY GAMMA RAY ASTROPHYSICS

Low/medium-energy gamma ray astronomy (100 keV – 30 MeV) advanced considerably with the launch and extended mission of the *Compton Observatory*. Major accomplishments include (1) evidence for the extragalactic origin of gamma ray bursts, (2) extragalactic gamma ray astrophysics, with the identification of two classes of high-energy AGN: blazers and Seyfert galaxies, (3) extensive studies of galactic compact objects, including the discovery of several unique systems (e.g. the bursting pulsar and soft gamma-ray repeaters), (4) first maps of the Galaxy in ^{26}Al and positron annihilation radiation, (5) improved observations of the MeV diffuse gamma ray background, (6) extensive solar flare observations, (7) timing and spectral characteristics of galactic black holes

The near future lies with ESA's *INTEGRAL* mission which includes two coded-aperture instruments: SPI dedicated to high spectral resolution observations, and IBIS dedicated to improved imaging. The SWIFT mission, dedicated to gamma ray burst observations, will also undertake a hard X-ray sky survey, and monitor galactic and extragalactic sources in the hard X-ray band.

For many low/medium-energy objectives, very significant improvements in sensitivity beyond *Compton Observatory* and *INTEGRAL* are required. This is the case for (1) detection of many Type Ia supernovae other than those rare events in the Galaxy or nearby galaxies within a few Mpc. (2) first detection of several novae in line gamma rays, (3) high sensitivity detections of active galactic nuclei to support broadband observations with high-energy gamma ray, X-ray and longer wavelength observations, (4) first gamma ray polarization measurements, (5) observations of unique spectral features from galactic black holes and neutron stars.

ACT SCIENCE/PRIORITIES

The ACT science capabilities and likely priorities are discussed below. Several of these relate to improved capabilities for line gamma ray emissions. The GRAPWG narrow and broad lines sensitivity goals for ACT are compared with the capabilities of CGRO and *INTEGRAL* in Figure 1.

Type Ia Supernovae

Type Ia SN are the thermonuclear explosions of white dwarf stars. A significant fraction of the star is synthesized into ^{56}Ni. It has been a long-standing goal of gamma ray astronomy to detect this prompt radioactive debris which includes the gamma ray lines associated with the decay of ^{56}Ni and ^{56}Co, along with the 511 keV emissions associated with positron annihilation. With the current

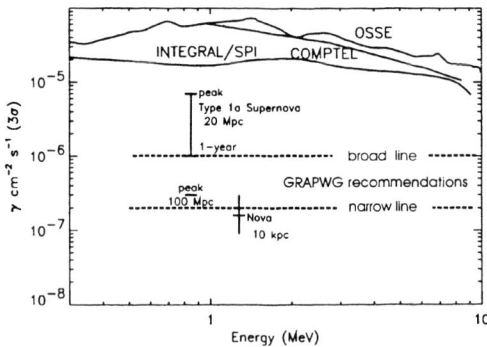

Figure 1. GRAPWG sensitivity recommendations

levels of sensitivity, this requires relatively nearby supernovae, within 5-10 Mpc. None have been observed with the exception of a 3σ observation of SN1991t by COMPTEL [2]. The sensitivity of the SPI instrument on *INTEGRAL* for broad lines from Type Ia SN will be comparable. Therefore, the systematic investigation of a large number of Type Ia SN in gamma rays requires a significant improvement in sensitivity. The GRAPWG baseline goal of 10^{-6} γ cm^{-2} s^{-1} for broad lines typical of Type Ia SN is adequate to detect theses event to ~ 50 Mpc. To that distance 10-20 events per year will be detectable [3]. This will be adequate to distinguish between several models, and to determine the existence of sub-Chandrasekhar models in which a surface detonation is presumed to be the explosion mechanism Understanding the detailed nature of Type Ia SN could also relate to their use as cosmological distance indicators.

Novae

Galactic novae are also believed to be sources of gamma radiation associated with the explosive hydrogen burning on a white dwarf star. Hernanz et al. [4] have made the latest estimates of the gamma ray production in the two classes of events: CO and ONe novae. Both events produce 511 keV emission associated with the annihilation of positrons following production and decay of ^{13}N, ^{14,15}O, and ^{18}F. The former may also produce ^{7}Be (478 keV; $\tau_{1/2}$ = 53d) while the latter is expected to produce copious amounts of ^{22}Na (1275 keV , 511 keV; $\tau_{1/2}$ = 2.6 yr.). It is estimated that several novae per year should be detectable in each class [5].

Active Galactic Nuclei

The Compton Observatory opened the field of extragalactic gamma ray astronomy and identified two classes of sources: blazers and Seyfert galaxies. The former have jets of relativistic matter oriented toward the observer, while the latter appear to be massive black holes lacking jets either observed directly (Seyfert 1s) or observed through a torus that absorbs most of the low energy X-ray emission (Seyferts 2s). ACT will advance the studies of AGN by: providing high sensitivity to low/medium energy gamma ray observations in coordination with other multi-wavelength measurements, searching for the evidence of positron annihilation features (blue-shifted?) that may indicate the nature of the jets (pair jets vs. electron-baryon jets), evidence for the jet emission mechanism from polarization observations (see below), determination of the gamma ray spectra for many individual Seyfert galaxies, and the contribution of AGN to the low/medium energy cosmic diffuse background.

Galactic Compact Objects

Many questions regarding galactic compact objects will be addressed with ACT. These include: the thermal vs. non-thermal nature of the emission in the various states of galactic black hole candidates, observation of annihilation radiation as an indicator of non-thermal activity in black hole accretion, and red-shifted lines (e.g. 2.223 n-p capture) from the surface of neutron stars that can be used in a determination of the equation of state of neutron star material.

Polarization

A Compton telescope is an excellent polarimeter. The compact instrument designs under consideration for the ACT, with high detection efficiency for Compton scatters near 90° will have excellent sensitivity for polarization observations. McConnell [6] has estimated the sensitivity for several configurations, and indicates that a 3σ polarization sensitivity for some ACT options should be about 1%. With this capability, it is likely that important results from polarization measurements will be derived [7]. Candidate sources for which polarization observations may be important for distinguishing emission mechanisms include: blazers, gamma ray bursts, pulsars, and galactic black hole candidates.

Galactic Mapping

COMPTEL has provided initial maps of the galaxy in the ^{26}Al line and in several continuum energy bands from 700 keV to 30 MeV. OSSE has provided the first maps of positron annihilation radiation from the galactic center region. With the dramatically improved sensitivity planned for ACT, highly detailed maps for these most prominent emissions will enable detailed understanding of important galactic features. Furthermore, maps of many additional lines will be realized. These include ^{60}Fe, ^{12}C, ^{16}O, ^{44}Ti (recent supernovae), ^{22}Na (recent novae).

Gamma Ray Bursts

ACT should provide important information on the sources of gamma ray bursts. Temporal and spectral characteristics of the primary gamma ray burst emission may elucidate the likely source mechanism for gamma ray bursts. Very sensitive observations will be available in the MeV region for the first time, permitting the spectral evolution within individual burst pulses to be measured with high precision. With its excellent sensitivity to both continuum and line emissions, ACT will investigate the relationship between gamma ray bursts and nearby supernovae (e.g. SN1998bw).

ADVANTAGES OF A COMPTON TELESCOPE FOR THE NEXT LOW/MEDIUM ENERGY GAMMA RAY MISSION

The next gamma ray mission should meet several requirements. First and foremost, the mission should provide a dramatic improvement in sensitivity, sufficient to meet the objectives outlined in the previous section. Ideally, it should also maintain the excellent energy resolution provided by the *INTEGRAL* SPI, and provide an angular resolution comparable to the *INTEGRAL* IBIS. Finally, it should provide a wide field-of-view for mapping observations and detection of transient sources.

The OSSE instrument on CGRO and the coded-aperture SPI and IBIS instruments on *INTEGRAL* approach reasonable limits of capability for collimated and coded-aperture techniques. Improving the sensitivity by an order of magnitude with these techniques would require approximately two orders of magnitude increase in size.

The capabilities of the Compton telescope technique, on the other hand, can be significantly improved without a corresponding increase in size. For example, the typical efficiency of COMPTEL was about 1%, driven in large measure by the separated D1-D2 configuration and the relatively small solid angle for events scattered from D1 to D2. Much higher efficiencies will be achieved with techniques under development. Also, replacing the scintillation detectors with detectors that provide improved energy and position resolution will result in significantly improved performance for an instrument of comparable scale. Therefore, an ACT appears to be the best approach to meet the broad scientific objectives discussed above.

ACT OPTIONS

Two workshops have been organized in the last year to discuss the science, technology and relevant issues concerning ACT. Proceedings from these are available at http://gamma.nrl.navy.mil/ngram. Implementation options discussed at the workshops include an all germanium instrument (NRL and UC Berkeley); all silicon (NRL), thin silicon tracker with CsI absorber (UC Riverside, MPE/Garching with US collaborators); liquid xenon (Columbia Univ.); and gaseous xenon or argon.

The germanium Compton telescope has evolved from the concept of simply substituting position sensitive germanium strip detectors for the liquid scintillator D1 and NaI(Tl) D2 detectors in used COMPTEL. The merit of this idea was that the superior energy resolution of germanium would provide better background rejection for narrow line emission, but also that the combination of position and energy resolution would provide a sharper point spread function (i.e. narrower rings), providing improved background rejection for point sources. To first order, background rejection efficiency for a narrow-line point source scales as the square of energy resolution in a generic Compton telescope, thus sensitivity scales proportionally to energy resolution until limiting effects of position resolution and Doppler broadening become significant. A gamma ray must scatter exactly once in the D1 detector, then be totally absorbed in the D2 detector to be properly detected. The efficiency in this configuration is rather low. The ATHENA mission concept developed by NRL in the mid 1990's [8] is the first example of an all germanium configuration. ATHENA is based on two arrays of germanium strip detectors, each with 1 m^2 in collecting area (Figure 2). The two arrays are separated by 1 m, provide 2 mm spatial resolution, and ~2 keV FWHM energy resolution. ATHENA's energy resolution is about 30 times better than COMPTEL, thus a sensitivity improvement on this same order was anticipated. Significant technical challenges include cooling two large detector subsystems (and possibly electronics)

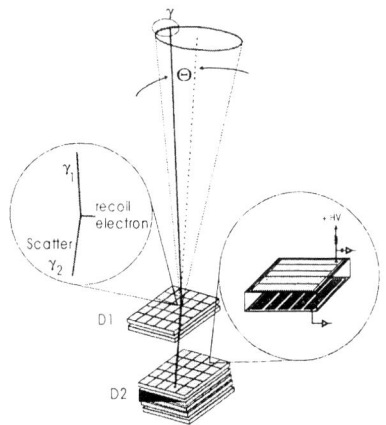

Figure 2. ATHENA, a germanium Compton telescope in a D1/D2 configuration similar to COMPTEL.

to 80 K and achieving the required ~2 ns time-of-flight resolution with germanium detectors on a relatively low power budget. The cost of the 1000-3000 germanium strip detectors required would also be high.

More recent work in the germanium configurations replaces the traditional D1/D2 configuration with a single detector subsystem with three-dimensional position readout. The key advantages of this compact configuration are significantly improved detection efficiency, and the possibility that a somewhat smaller total volume of germanium may be needed. The increased efficiency is largely the result of a higher probability of detecting the scattered gamma ray since all scatter angles are potentially viable events. The compact configuration can utilize gamma rays that scatter through large angles. The main requirement is that the detectors provide good spatial resolution in all three dimensions. This has been demonstrated in germanium strip detectors [9] [10]. Further, scatter angles in a range around 90 degrees provide optimum sensitivity to polarization, thus the compact configuration will be a sensitive gamma ray polarimeter [11] [12].

The compact configuration depends on the ability to use large scatter angles in the data analysis, and the ability to reject background through means other than time-of-flight. In principle, backgrounds from beta-gamma and multiple gamma decays can be distinguished from true gamma rays by the pattern of energy losses. Further, the sequence of the interactions in good events must be properly determined. Probabilistic and Compton scatter consistency arguments can be used to determine interaction orders [13] [14]. These concepts need further study throughMonte-Carlo simulations.

The NRL group has studied the sensitivity a 1 m^2 all-germanium telescope [15], while Boggs and Jean have evaluated a similar concept [16]. The UC Berkeley group has recently proposed a small version for a balloon flight [17]. The proposed Berkeley balloon instrument consists of 12 germanium strip detectors (Figure 3), and provides 14 cm^2 effective area at 1 MeV, similar to the much larger COMPTEL instrument on CGRO. The Berkeley instrument would be surrounded below and on the sides by active shielding to reduce background.

The groups at MPE/Garching and UC Riverside are developing instruments using a D1 detector composed of a stack of thin silicon strip detectors in order to track the recoil electron from the first interaction. The D2 detector would surround D1 to the sides and below with a higher-Z detector such as CsI, CdZnTe or germanium. The MEGA concept (Figure 4) studied by MPE/Garching [18] [19] utilizes a stack of 0.5 mm thick silicon strip detectors with 0.5 mm pitch, and arrays of CsI logs with an area of 0.5x0.5 cm^2 and lengths of 4 and 8 cm on the sides and bottom respectively. Tracking the electron provides full knowledge of the momentum transfer in the first interaction, thus the direction of the

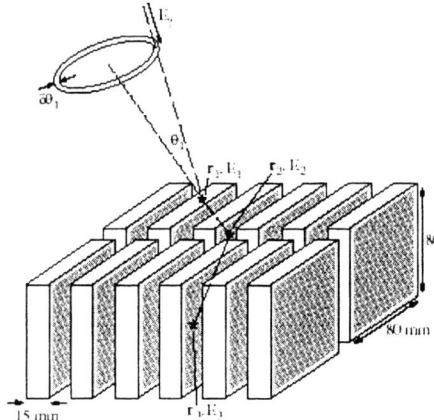

Figure 3. Germanium Compton telescope for a balloon mission proposed by U.C. Berkeley.

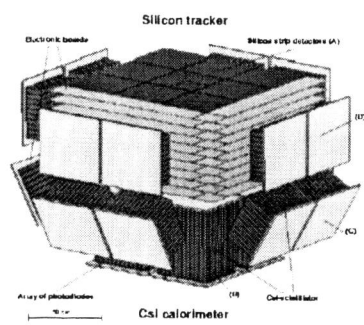

Figure 4. MEGA, with a D1 silicon tracking detector and CsI D2 detector.

incoming gamma ray is uniquely determined within the uncertainties of the measurements. The direction of the recoil electron is inferred from the positions and relative energy losses in each of the silicon detectors that it penetrates. A minimum of two silicon detectors must trigger to determine the electron's direction. However, secondary scattering of the electron and an up vs. down determination require several consecutive detectors in the stack to trigger. The energy threshold for MEGA electron tracking is in the range of 1-2 MeV.

The TIGRE concept developed by UC Riverside [20] is similar to MEGA in many respects, differing in the use of 0.3 mm- thick silicon strip detectors for D1, and also considering CdZnTe detectors for D2 [21].

Liquid nobel gas detectors also provide a promising alternative to solid-state detectors (Figure 5). The Columbia University group has flown a xenon time projection chamber on two balloon flights [22] [23]. The detector provides 1 mm spatial resolution in the x- and y-directions through a series of wire readouts, and 0.3 mm in the z-direction by timing the charge collection. LXeGRIT has a homogeneous detection volume maintained at a temperature of −100 C and a density of 3.1 g/cm^3. A large detector system would most probably be composed of multiple modules.

The energy resolution achieved by LXeGRIT is ΔE/E ~ 8.8% at 1 MeV, which is somewhat poorer than NaI. This appears to be a fundamental limit in the liquid phase [24]. An alternative under consideration is a high-pressure gas time projection chamber with a gas density <0.55 g/cm^3. High pressure gas detectors have achieved an energy resolution of 1.7% at 1 MeV [25]. The gas detector also has the potential of tracking the recoil electron down to about 1.5 MeV, thus having capability similar to MEGA.

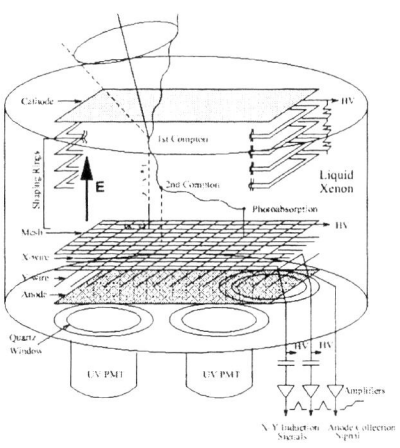

Figure 5. LXeGRIT, a liquid Xenon time projection chamber.

The position-sensitive, high-resolution detectors being developed enable a new capability in gamma ray detection. The "3-Compton" technique uses events that undergo two Compton scatters, followed by a third interaction. The concept can be realized, in principle, in germanium, silicon, gas, or potentially CdZnTe detectors, which offer 3-dimensional readout of large volumes, characteristics common to all of the mission concepts listed above. The energy of the incoming gamma ray is determined by measuring the energy loss in the first two interactions, and the scatter angle of the second interaction [26]. Full energy absorption is not required. The uncertainty in the reconstructed energy is a function of the energy resolution of the detectors, the uncertainty in the scatter angle at the second interaction, and the initial momentum state of the recoil electron. The latter is referred to as Doppler broadening, and is a characteristic of the scattering medium and is manifest as a small uncertainty in the Compton scattering angle and energy [27].

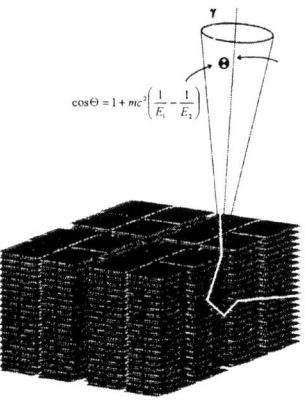

Figure 6. 3-Compton telescope proposed by NRL composed of Si(Li) strip detectors.

The advantage of the 3-Compton approach is that it is possible to measure energy without requiring total absorption, thus higher efficiencies are possible with thinner detectors, or with lower-Z detectors such as silicon. 3-Compton energy spectra do not have the characteristic Compton shelf that is typical of a simple detector. Moreover, it is possible to identify which events are fully absorbed by adding the individual energy losses and comparing with the estimated energy using 3-Compton. Simulations or flight data are required to demonstrate that consistency checks such as this may be powerful tools to reject instrumental backgrounds from internal radioactive decays.

The NRL group is developing silicon detectors for the ACT mission. Silicon has two key advantages over other detector materials: 1) the low-Z minimizes Doppler broadening, and 2) silicon detectors may be operated at temperatures of -40 C or warmer. Segmented Si(Li) detectors with thickness ~7 mm and area 7x7 cm^2 are under development. The thickness is optimized to assure a single scatter in a detector, contain the recoil electron, and minimize detector count. The Si(Li) ACT concept, shown in Figure 6, has 1 m^2 frontal area and an active thickness of 42 g/cm^2. Two-mm thick intrinsic silicon detectors are also being pursued as an alternative to Si(Li). Detector requirements are <2 mm spatial resolution in three-dimensions, and <2 keV FWHM energy resolution. The energy resolution of the 3-Compton energy determination is non-gaussian with a characteristic width ~10 keV in the MeV region.

SENSITIVITY ISSUES

Achieving a significant improvement in sensitivity will require a combination of increased detection efficiency, background rejection efficiency, exposure, and source localization (imaging).

Detection efficiency of a traditional Compton telescope that depends on a single Compton scatter followed by a total absorption is generally less than a similar-sized detector used in full-energy absorption mode. This is because only a specific sequence of interactions within a certain range of scatter angles can produce valid Compton events. For example, COMPTEL had an efficiency on the order of one percent at 1 MeV. Several of the ACT concepts address this issue by collapsing the D1 and D2 detectors into a single volume, with the intent of using all or most events that are absorbed in the detector volume. This way, the efficiency may approach that of the full absorption detector, but with certain caveats. One of these is that gamma rays that initially scatter through large angles are a major part of this increased efficiency. These events trace out wide diameter Compton cones passing through the true source, but also through a larger portion of the sky, and potentially the spacecraft or the Earth. Tracking of the recoil electron will restrict the gamma ray direction to a small segment of the Compton cone at higher energies. However, at lower energies or with instruments without tracking or shielding, a portion of the large scatter events may need to be rejected because they could be consistent with local backgrounds.

The ACT must measure the positions and energies of at least the first two interactions, and properly determine the correct order of the interactions. The ordering problem lends itself to a simple consistency analysis of event ordering called Compton Kinematic Discrimination (CKD) developed by Aprile et al. [13]. Boggs and Jean [14] applied CKD to a Monte-Carlo simulation of a germanium ACT and found efficiencies of ~70% for the total energy events at 200 keV. The event ordering efficiency drops slowly to ~60% at 5 MeV and 50% at 20 MeV. In independent work, van der Marel and Cederwall [28] developed the 'backtracking' method to determine event order for nuclear spectroscopy of radioactive beams and targets. Backtracking applies a similar

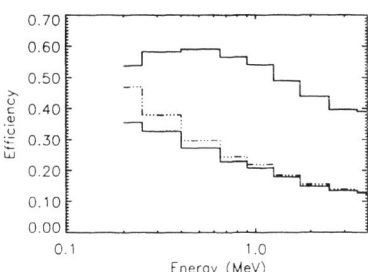

Figure 7. Silicon ACT efficiencies: 3-Compton (upper curve), total energy absorption (dashed curve), and 3-Compton with total energy absorption (lower curve). Fully active volume.

consistency check as CKD and achieves a similar efficiency of ~60%, also in germanium. The primary difference is the CKD test is applied to all possible event orders, whereas backtracking searches for the most likely photoelectric interaction, and then searches for the previous interaction, in sequence, until the full order is determined.

The 3-Compton approach has the potential to provide efficiencies that are higher than that of the full absorption detector in the MeV region. Events with partial energy loss are not lost to the Compton background as in a calorimeter, provided there are at least three interactions, and that the interaction order is properly determined. A large, fully active silicon calorimeter (80x80 cm^2 x 43 g/cm^2) at 1 MeV has a total absorption efficiency 27% and a 3-Compton efficiency of 54%, assuming a fully active volume and not considering the ordering efficiency (Figure 7). The combined total energy and 3-Compton efficiency is 55%, suggesting that almost all total energy events interact at least three times in this silicon detector concept. A similar sized germanium array has efficiencies of 57% total energy absorption, 54% 3-Compton events, and 66% either total energy or 3-Compton. A full 46% of these events are both 3-Compton *and* total absorption. Efficiencies drop with the introduction of passive materials in a realistic detector system. The loss of efficiency scales between the second and third power of the active mass fraction for the 3-Compton technique, so it is important to achieve high active detector fractions in the detector volume.

We note that the 3-Compton/total energy events are unique in that there are two independent means to determine the incoming gamma ray energy. This is one example of a powerful consistency test that will efficiently discriminate between real gamma rays and radioactive backgrounds. Also, any Compton detector with excellent spatial and spectral resolutions can screen events where any combination of energy losses adds up to a known internal background feature. Boggs and Jean [14] have investigated this powerful background discrimination capability.

It is difficult to predict the capabilities and sensitivities of space gamma-ray instruments due to the complex response of the instrument to the space radiation environment. This has been improving as better environmental models and simulation tools have become available. A major emphasis must be given to adequate simulations for the several alternative concepts for ACT as the definition of the mission evolves.

ACKNOWLEDGMENTS

The authors wish to acknowledge the contributions and interest of the participants at the Advanced Compton Telescope workshops held in May 2000 and April 2001.

REFERENCES

1. Recommended Priorities for NASA's Gamma Ray Astrophysics Program, NP-1999-04-072-GSFC
2. Morris, D.J., et al. AIP Conf. Proc. **410**, 1084 (1997)
3. Milne, P.A., et al., Proc. 2nd Chicago Conf. on Thermonuclear Astrophysical Explosions, eds. Niemeryer and Truran, in press (2000)
4. Hernanz, M., et al., ApJL **526**, L97 (1999)
5. Hernanz M., private communication
6. McConnell, M., see paper at http://gamma.nrl.navy.mil/ngram
7. Baring, M., private communication
8. Johnson, W.N., et al., SPIE **2518**, 74 (1995)
9. Momayezi, M., Warburton, W.K. and Kroeger, R.A., SPIE **3768**, 530 (1999)
10. Amman, M., and Luke, P.N., Nucl. Instr. Meth. A, **452**, 155 (2000)
11. Kroeger, R.A., et al., Nucl.Instr. Meth. A **436**, 165 (1999)
12. Aprile, E., et al., ApJS **92** 689 (1994)
13. Aprile, E., et al., Nucl.Iinstr. Meth. A **327**, 216 (1993)
14. Boggs, S.E., and Jean, P., Astron. and Astrophys Supp. **145**, 311 (2000)
15. Kroeger, R.A., et al., AIP Proc **510**, 794 (1999)
16. Boggs, S.E., and Jean, P., astro-ph001331, to be publ. in Proc. 4th INTEGRAL Workshop
17. Boggs, S.E., et al. these proceedings
18. Kanbach, G., et al., see paper at http://gamma.nrl.navy.mil/ngram
19. Kanbach, G., et al. these proceedings
20. Tumer et al., 1995, IEEE Trans. Nuc. Sci. **42**, No. 4, 907
21. O'Neill, T., et al., these proceedings
22. Aprile, E., et al., SPIE **4140**, 344 (2000)
23. Aprile, E., et al., see paper at http://gamma.nrl.navy.mil/ngram
24. Bolotnikov, A.E., et al., Instrum & Exp. Techn., **29**, 809 (1986)
25. Dmitrenki, V.V., et al., IEEE Trans. on Nucl. Sci, **47**,, No 3, 939 (2000)
26. Kurfess, J.D., et al., AIP Conf. Proc. **510**, pp 789 (1999)
27. Du, Y.F., et al., Nucl. Instr. Meth. A **457**, pp 203 (2001)
28. van der Marel, J., and Cederwall, B., Nucl. Instr. Meth. A **437**, 538 (1999)

The Nuclear Compton Telescope: a balloon-borne soft γ-ray spectrometer, polarimeter, and imager

S. E. Boggs[*], P. Jean[†], R. P. Lin[*], D. M. Smith[*], P. vonBallmoos[†], N. W. Madden[**], P. N Luke[**], M. Amman[**], M. T. Burks[**], E. L. Hull[**], W. Craig[‡] and K. Ziock[‡]

[*]*Space Sciences Laboratory, University of California, Berkeley CA, 94720, USA*
[†]*Centre d'Etude Spatiale des Rayonnements, UPS-CNRS, Toulouse, France*
[**]*Lawrence Berkeley National Laboratory, Berkeley CA, 94720, USA*
[‡]*Lawrence Livermore National Laboratory, Livermore CA, 94550, USA*

Abstract. Our collaboration has begun the design and development of a prototype high resolution Compton telescope utilizing 3-D imaging germanium detectors. The Nuclear Compton Telescope (NCT) is a balloon-borne soft gamma-ray (0.2-15 MeV) telescope designed to study astrophysical sources of nuclear line emission and polarization. NCT is a prototype design for the Advanced Compton Telescope, to study gamma-ray radiation with very high spectral resolution, moderate angular resolution, and high sensitivity. The instrument has a novel, ultra-compact design optimized for studying nuclear line emission in the critical 0.5-2 MeV range, and polarization in the 0.2-0.5 MeV range. We have proposed to develop and fly NCT on a conventional US balloon flight in Summer of 2004. This first flight will perform gamma-ray polarization measurements the Crab nebula, Crab pulsar, and Cyg X-1, and ^{26}Al emission from the Cygnus Region. This flight will critically test the novel instrument technologies and analysis techniques we have developed for high resolution Compton telescopes, and qualify the payload to begin a series of ~10-day long duration ballon flights from Alice Springs, Australia starting in Spring 2005.

INTRODUCTION

The *Nuclear Compton Telescope* (NCT) is a balloon-borne soft γ-ray (0.2-15 MeV) telescope designed to study astrophysical sources of nuclear line emission and polarization, and is being developed as a prototype for the *Advanced Compton Telescope* (ACT). It employs a novel Compton telescope design (Figure 1), utilizing twelve 3-D imaging, high spectral resolution germanium detectors (GeDs). The Compton imaging serves three purposes: imaging the sky, measuring polarization, and very effectively reducing background. An overview of the performance characteristics is presented in Table 1. NCT is designed to optimize sensitivity to nuclear line emission over the crucial 0.5-2 MeV range (Table 2), and sensitivity to polarization in the 0.2-0.5 MeV range. NCT's guiding principle is that high efficiency and excellent background reduction are critical for advances in soft γ-ray sensitivity.

Between our NASA *Supporting Research and Technology Program* (SR&T) at UC Berkeley and *Department of Energy* programs at LBNL and LLNL, we have developed and tested large volume, 3-D, cross-strip GeDs using LBNL's amorphous Ge contact

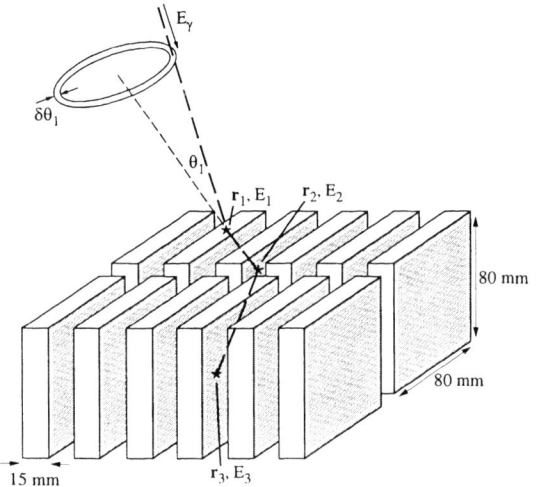

FIGURE 1. The heart of NCT is an array of 12 crossed-strip GeDs with 3-D position resolution.

technology [1]. Timing techniques for measuring the third dimension (depth) have been verified [2], and we have prototyped, tested, and begun fabrication of a 3-D electronics chain, including a novel charge sensitive preamplifier utilizing predominantly surface mount components. This preamp achieves excellent spectroscopic performance in a small footprint, and at modest cost and low power [3]. In parallel, we have actively developed Compton reconstruction techniques for high-resolution instruments [4], and have extensively simulated Compton telescope performance [5]. Our technology developments and instrument designs have progressed to the point where construction of NCT can begin.

We have proposed to NASA's SR&T program to build NCT and fly it on a first continental US (ConUS) flight in Summer 2004. This program will test event reconstruction and imaging techniques unique to high resolution Compton telescopes, as well as demonstrate the excellent background rejection capabilities and high sensitivity of our compact, GeD design. NCT is designed to be fully long duration balloon flight (LDBF) capable, and the Summer 2004 ConUS flight will qualify NCT to begin a series of LDBFs from Alice Springs, Australia starting in Spring 2005. On an LDBF, NCT will achieve nuclear line sensitivities comparable to SPI (the Spectrometer on INTEGRAL), and polarization sensitivities to a level of $<10\,\text{mCrab}$. Primary nuclear science goals for NCT/LDBF and ACT are presented in Table 2.

DESIGN OVERVIEW

The NCT detector configuration is presented in Figure 1, and the full scientific instrument including active shields, cryostat, and dewar in Figure 2. NCT is an array of twelve 15-mm thick planar GeDs of active area 76 *mm* × 76 *mm*. The 15 mm thickness has been

TABLE 1. NCT Performance at 1 MeV (except Polarimetry at 0.2-0.5 MeV).

Performance	NCT/ConUS 1.5×10^4 s	NCT/LDBF 2×10^5 s	NCT/ACT 10^6 s
# Detectors	12	12	120
Effective Area [cm^2]	13.9	13.9	194.
Angular Res. Meas. 1σ [deg]	2.03	2.03	2.03
ΔE FWHM [keV]	2.49	2.49	2.49
FOV [str]	0.5	0.5	0.5
Narrow Line Sensitivity, 3σ [$10^{-5} ph/cm^2/s$]	5.10	1.21	0.076
Broad Line Sensitivity, 3σ, 5000 km/s [$10^{-5} ph/cm^2/s$]	13.1	3.13	0.20
Continuum Sensitivity, 3σ, $\Delta E = E$ [$10^{-4} ph/cm^2/s/MeV$]	6.03	1.43	0.10
100% Polarization Sensitivity [$mCrab$] (0.2-0.5 MeV)	32	7.5	0.53

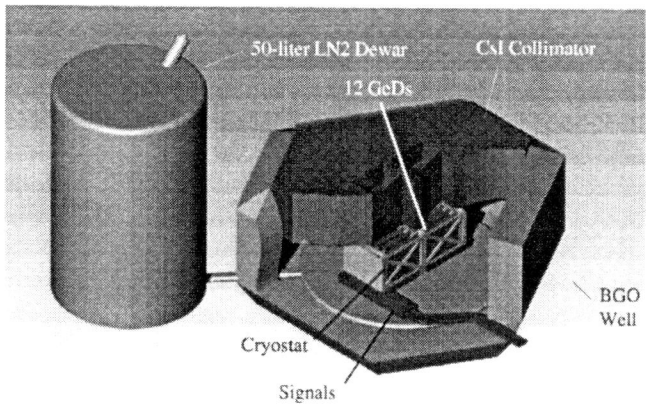

FIGURE 2. The NCT GeD, cryostat, and shield configuration.

chosen as a compromise between producing thick detectors to minimize the number of electronics channels per volume, and keeping the detectors thin to optimize the scattering efficiency between layers in the γ-ray line energy range (0.5-2 MeV). Orthogonal 2-mm electrode strips on the opposite faces, combined with signal timing, provide full 3-D position resolution to 2 mm.

Here we have arranged the GeDs in a somewhat unconventional manner for a Compton telescope: the detectors are set vertically with the edge of the detector facing the collimator opening, and hence the astrophysical sources. This configuration optimizes the effective area per unit GeD volume in the 1-2 MeV range. (For example, the effective area increase by 30% at 1 MeV over the case where the identical GeDs are arranged horizontally in three 2×2 arrays.) Also, this detector configuration is most efficient to

TABLE 2. NCT & ACT Nuclear Science Goals.

Source	Origin	Energy [MeV]	Diffuse
SNe II/Ib	^{26}Al	1.809	y
	^{60}Fe	1.173, 1.333	y
SNe	^{44}Ti	1.154	n
Galactic e$^+$	e^+e^-	0.511	y
BH	e^+e^-	≤ 0.511	n
NS	H(n,γ)D	≤ 2.223	n
Novae	^{22}Na	1.275	n
	^{7}Be	0.478	n
SNe Ia	^{56}Co	0.847, 1.238	n
	^{56}Ni	0.158, 0.812	n

photons which Compton scatter at right angles in the instrument, optimizing sensitivity to polarized radiation. This compact design allows enclosure in an active shield. All of the stopping power required vertically is present, so the design can be easily scaled by adding more "modules" horizontally. The gap between detectors was set at 15 mm. Such a compact geometry can only be achieved by using detectors with fine 3-D imaging capabilities.

The GeDs will be housed in a common cryostat, which is attached to a 50 liter liquid nitrogen dewar to provide ≥ 20 days of cooling at 85 K. The GeD array is enclosed on the sides and bottom by a 5-cm thick active bismuth germanate (BGO) anticoincidence shield. A 10-cm thick CsI front shield collimates the detector field-of-view (FOV) to $40° \times 40°$. The active shield is another unconventional component for Compton telescope designs, but plays a crucial role in helping reduce the background induced by the atmospheric γ-ray emission to a level where the relatively small GeD volume achieves high sensitivity.

In the compact NCT geometry and with GeD timing performance, time-of-flight (TOF) measurements are not able to distinguish the photon interaction order. We have developed several techniques for reconstructing this interaction order independently without the TOF [4]. We also showed that the Compton telescope performance depends on the choice of "event cuts," such as the minimum permitted distance between the first and second scatter (lever arm), and whether 2-site interactions and photons which backscatter ($\theta_1 > 90°$) are accepted [5]. An overview of the NCT performance characteristics are given in Table 1. Detailed analysis of the event cuts, instrument performance, and background calculations will be presented elsewhere.

SCIENTIFIC GOALS

Nuclear astrophysics studies the lifecycle and evolution of matter in our Universe: stellar evolution ending in supernovae, with the ejection of heavy nuclei back into the galaxy to be reborn in new stars. Radioactive nuclei produced through this cycle of creation emit characteristic photons that fingerprint the isotopes themselves, and quantify their abundance, speed, temperature, and attenuation. High spectral resolution measurements of these line emissions probe deep into the center of a supernova explosion, revealing the nuclear burning and dynamics in the core. Nuclear excitations and positron annihilations also reflect the extreme environment on the surface of neutron stars and white dwarfs, and near the event horizon of black holes. CGRO made dramatic achievements in the investigation of nuclear γ-ray sources. INTEGRAL will build upon these discoveries with improved sensitivities and much higher spectral resolution with SPI, and angular resolution with IBIS. NCT is an important advance over SPI in two regards: its development and flight will provide a testing ground for novel event analysis, background reduction, and imaging techniques crucial to ACT, and break new ground in the measurement of polarized γ-ray emission from astrophysical sources. NCT's ability to Compton image 0.511 MeV photons will provide an unprecedented annihilation emission map of our Galaxy. In addition, NCT will supplement SPI's core science with detailed studies of nuclear processes in our Galaxy.

SUMMARY

The NCT compact geometry achieves high photopeak efficiencies, increasing the effective area per unit volume by a factor of $\sim$100 over COMPTEL [6]. This is possible through the use of high 3-D spatial resolution detectors. For a photon or background event, the ability to spatially resolve interactions and measure the energy depositions with high resolution is an powerful new tool for increasing the sensitivity of γ-ray instruments. Compared to SPI, we have decreased the background per unit volume by factors $\geq$30. However, the phase-space for improvement in background rejection techniques is largely unexplored. NCT will provide a crucial testing ground for developing novel background techniques to their full potential. The NCT/ACT sensitivities we presented in Table 1 are conservative upper limits. We believe these sensitivities can be significantly improved through further development of background rejection techniques utilizing the high spectral and spatial resolution of our 3-D GeDs.

ACKNOWLEDGMENTS

This work was supported by NASA grant NAG5-5285 and the California Space Institute.

REFERENCES

1. Luke, P. N., et al., *IEEE Nucl. Sci. Symp. Conf.*, **39** (1992).
2. Amman, M., and Luke, P. N., *NIM*, **A452** (2000).
3. Fabris, L., Madden, N., and Yaver, H., *NIM*, **A424** (1999).
4. Boggs, S. E., and Jean, P., *A&AS*, **145** (2000).
5. Boggs, S. E., and Jean, P., *Proc. 4th INTEGRAL Workshop* (2000), in press.
6. Schoenfelder, V., et al., *ApJS*, **86** (1993).

The TIGRE Gamma-Ray Telescope

T.J. O'Neill[*], A. Akyüz[†], D. Bhattacharya[*], M. Polsen[*], J. Samimi[**] and A. Zych[*]

[*]*Institute of Geophysics and Planetary Physics, University of California, Riverside, CA 92521*
[†]*Dept. of Physics, Cukurova University, Adana, Turkey*
[**]*Dept. of Physics, Sharif University, Tehran, Iran*

Abstract. TIGRE is an advanced telescope for gamma-ray astronomy. From 0.3 to 10 MeV it is a Compton telescope. Above 1 MeV, its multi-layers of double sided silicon strip detectors allow for Compton recoil electron tracking and the unique determination for incident photon direction. From 10 to 100 MeV the tracking feature is utilized for gamma-ray pair event reconstruction. The acceptance of large scattered gamma-rays also makes TIGRE a good polarimeter below 2 MeV. Here we present the status of the TIGRE Instrument.

TIGRE Instrument Description: The TIGRE instrument (O'Neill et al. 1995, Akyüz et al. 1995, Tümer et al. 1995), shown in Figure 1a, features multilayers of silicon strip detectors (SSD) as both the Compton converter and recoil electron tracker. Double-sided SSDs provide submillimeter x and y spatial resolutions as the recoil electron is tracked through successive layers until it is fully absorbed. CsI(Tl)-Photodiode detector arrays on the bottom and sides are used as a calorimeter for the scattered photons and the electron-positron pairs that escape the silicon detectors. The electron track in the silicon strips allows for Up/down photon determination. The calorimeter also serves as a low energy gamma-ray shield. This is important at energies below 1 MeV where up-down discrimination of gamma-rays cannot be determined due to the short electron track length. The side CsI(Tl) arrays also serves as a polarimeter for energies below 2 MeV. A particle anticoincidence plastic scintillator surrounds the entire sensitive material. Below the detectors are the support electronics and the entire assemble is contained in a pressure vessel. Figure 1b shows the kinematics of a Compton scattered event. The width of the event circle (also refered to as the Angular Resolution Measurement ARM) is dictated by the position and energy resolutions of the silicon strips and CsI(Tl) crystals. Figure 2a shows the simulated point spread function (PSF) or ARM ($\Delta\theta$) for 1.8 MeV incident gamma-rays for the TIGRE instrument. The event arc, ψ (see Fig. 1b), is heavily dependent on the scattered electron energy and is determined by the electron track and the event circle. Figure 2b shows the spread in the event arcs for 1.8 MeV incident gamma rays. Table 1 shows the dependence of ψ on the electron energy.

Gamma-ray pair events are also detected in the traditional manner by tracking the electron and positron individually through successive layers of silicon strip detectors until at least one particle or annihilation photon exits and interacts in the calorimeter. Both the energy losses and positions of the pair particles are measured in each Si layer as these particles are tracked through the array. For Compton events, the detailed energy loss and tracking record that the silicon strip provide for each recoil electron together

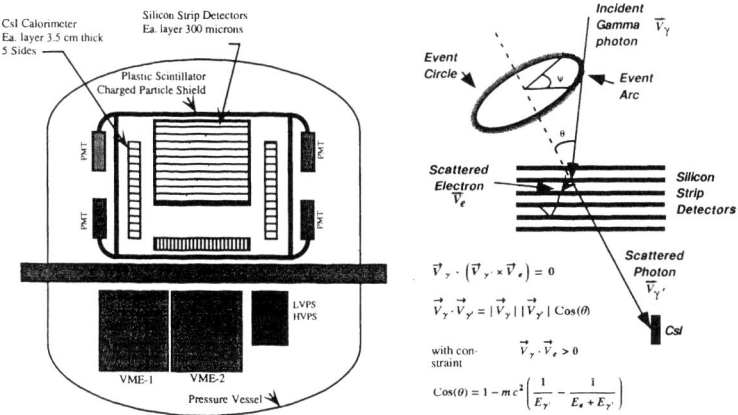

FIGURE 1. Baseline TIGRE instrument. The Compton electron tracking kinematics is shown on the right.

with its forward scatter direction in the Compton collision has led to another Compton telescope innovation. Previous Compton telescopes use the time-of-flight of the scattered gamma ray between two scintilator arrays to determine its direction of motion and discriminate against unwanted background. This necessitates a large separation between the two arrays and a smaller subtended solid angle thereby producing a much lower efficiency. For the TIGRE instrument, Compton scattered electrons that traverse two or more successive layers allow us to determine a direction-of-motion (DOM) parameter from the energy losses and angle changes via multiple scattering. The percentages of events with multiple track hits (i.e. hits in more than one silicon layer) are 57%, 82%, 94% and 94% at 1, 2, 6 and 10 MeV, respectively (Oneill et al. 1996).

Silicon Strip Detectors: TIGRE uses double sided silicon strip detectors to measure the direction and energy of Compton electrons and electron-positron pairs. Each detector is read out using two 128 channel TA-1 ASIC chips from IDE-AS. Triggers from the ohmic and junction sides (x and y strips) are AND'ed on each board.

The right side of figure 3 shows a completed Si detector board containing a 10 cm × 10 cm double-sided 300μm thick Si detector with one TA-1 hybrid wire-bonded to the 128 ohmic strips and one wire-bonded to the 128 junction strips. The hybrid consists of a custom ceramic carrier and a TA-1 chip. Figure 4 shows a ^{57}Co energy spectra for this silicon detector to be 4.6keV (1σ). The TA-1 trigger thresholds were set at about 60 KeV. The left side of Figure 3 shows our stack of 7 layers where each layer contains 4

Scattered Electron Energy (MeV)	Point Spread Function Ψ (Deg.)
E < 0.5	40
0.5 < E < 1.0	30
1.0 < E < 2.0	15
E > 2.0	8

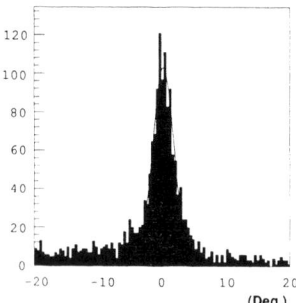
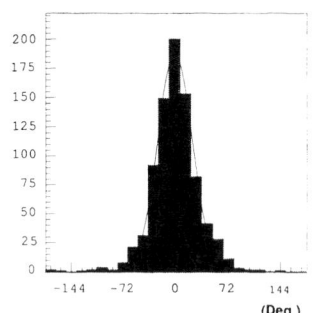

Radial component of the point spread function Δθ for 1.8 MeV incident gamma rays averaged over all scattered electron energies. The 1σ fit is 1.6°

Point spread function ψ along the event circle for 1.8 MeV incident gamma rays averaged over all scattered electron energies. The 1σ fit of 23° reduces the event circle and associated background by a factor of 15 (=360/23).

FIGURE 2. Point Spread Functions.

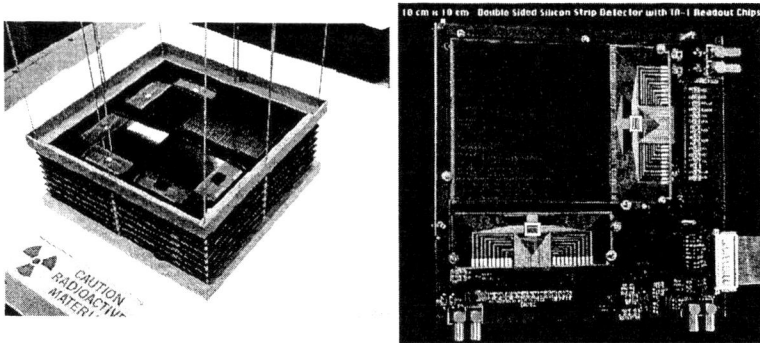

FIGURE 3. A silicon stack with an area of 400 cm². A single detector is shown on the right.

detectors. Triggers from each board are then or'ed producing the D1 trigger. The stack is being enlarged to 16 layers. All of the silicon layers are read out in parallel.

CsI Arrays: TIGRE uses thousands of 1 cm × 1 cm × 3.5 cm CsI crystals, individually wrapped in teflon tape and bonded to photodiodes to measure the energy and position of the scattered gamma ray. Single crystal spectra connected to discrete electronics give 5% FWHM at 662 keV. Arrays consisting of 256 crystals each are read out using the identical ASIC chip and ceramic carrier. Obtaining the energy spectra using the ASIC chip is in progress. The triggers for all crystals are or'ed producing the D2 trigger. The D1 and D2 triggers are then AND'ed producing the good event trigger. Ten arrays are now being constructed and they too are read out in parallel. Figure 5 shows one of these arrays.

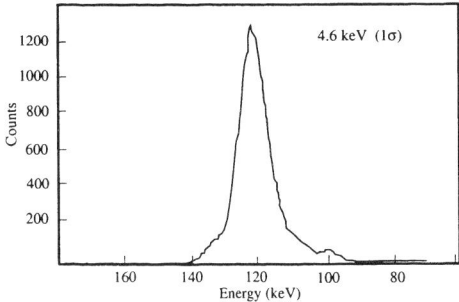

FIGURE 4. Silicon energy spectra for a ^{57}Co source (122 keV).

Simulation: We have extensively modeled the TIGRE instrument described above with the general-purpose MCNP code developed at LANL. Energy and spatial resolutions for each individual Si pixel are taken as a conervative 3 keV (1σ) and 0.75 mm. Thresholds of 30 keV for the Si and 100 keV for the CsI(Tl) and plastic anticoincidence scintillators 100 keV were used.

For pair events, either the pair particle or an annihilation photon needs to be detected in the calorimeter to generate the trigger. Event reconstruction is used to identify gamma-rays with incident directions within the instrument's FOV. The DOM parameter was determined for each track for up/down discrimination. The TIGRE angular resolutions and sensitivities are presented in (Bhattacharya et al. 1999).

FIGURE 5. A single CsI crystal array.

TIGRE has a broad maximum efficiency of 5% and a high effective area of 80 cm^2 in the Compton regime above 0.5 MeV. All Compton events are either "tracked" or "non-tracked" depending on the number of silicon layers the recoil electron traverses. The percentage of tracked events increase from 50% at 0.5 MeV to >90% above 2 MeV. In the pair regime the efficiency remains constant with a value of 5% up to 100 MeV. The effective combination of shielding and kinematics reduces the background contribution from outside the FOV ($\sim \pi$ sr).

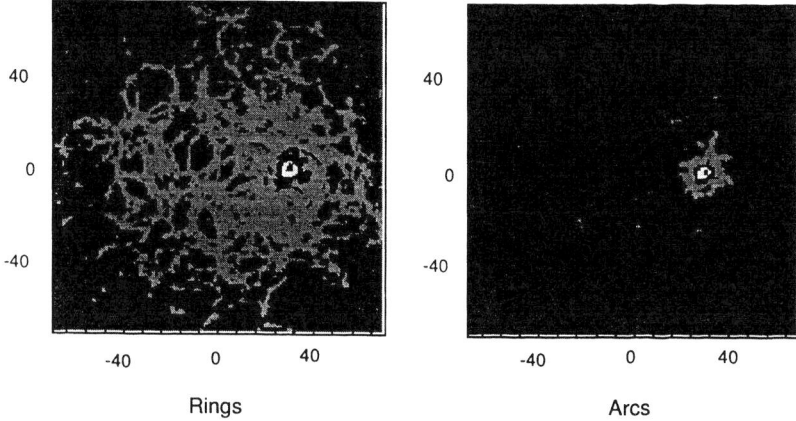

FIGURE 6. Images of a Crab-like source at 30° zenith with an istropic background using rings (left Fig.) and arcs (right Fig.). The axes are in units of degrees.

Three million source events with a Crab-like spectra from 0.5 to 30 MeV were generated at 30 degrees zenith along with 20 million isotropic background events with the same energy spectrum. The left side of Figure 6 is an image of the source and background events when the tracking information is not utilized. Here the event rings were added. The right side of Figure 6 shows the improvement using the tracking information thereby reducing the event rings to event arcs. Three million source events are expected to be incident on TIGRE for a 9 hour observation of the Crab region. During this period the total incident background photons are expected to be $\sim$ 1 billion at an altitude of 3 mbar. Using this background we expect to see the Crab at the 9σ level using just the tracked events (with arcs) and 6σ for all of the events when the tracking information is not used. The 50% increase in sensitivity correlates to a factor of 2.25 reduction in the observing time.

REFERENCES

Akyüz A. et al. 1995, *Experimental Astronomy*, 6, 274.
Bhattacharya, D. et al. 1999, *26th ICRC Conf. Proc.*, 5, 72.
O'Neill, T. J. et al. 1995, *IEEE Trans. Nuc. Sci.*, 42, 933.
O'Neill, T. J. et al. 1996, *A&A Suppl.*, 120, 661.
Tümer et al. 1995, *IEEE Trans. Nuc. Sci.*, 42, 907.

MEGA - A Next Generation Mission In Medium Energy Gamma-Ray Astronomy

Gottfried Kanbach[a] for the MEGA Collaboration[b]

[a] *Max-Planck-Institut für Extraterrestrische Physik, Postfach 1312, 85741 Garching, Germany*
[b] *France: CESR, Toulouse; U.S.A:. UNH, Durham, NH; GSFC/NASA, Greenbelt, MD; NRL, Washington, DC; Columbia U., NY; IGPP, UCR, Riverside, CA; UA, Huntsville, AL*

Abstract. A Medium Energy Gamma-Ray Astronomy (MEGA) detector is being developed and proposed for a small satellite mission. MEGA intends to improve the sensitivity at medium γ-ray energies (0.4-50 MeV) by at least an order of magnitude with respect to past instruments. Its large field of view will be especially important for the discovery of transient sources and for conducting all-sky surveys. Key science objectives for MEGA are the investigation of cosmic high-energy accelerators and of nucleosynthesis sites with γ-ray lines. The large-scale structure of the galactic and cosmic diffuse background is another important goal for this mission. MEGA records and images γ-ray events by completely tracking Compton and pair creation interactions in a stack of double sided Si-strip track detectors and 3-D resolving CsI calorimeters.

INTRODUCTION

After a decade of high-energy astronomy with CGRO and nearly complete spectral coverage extending from several 10 keV to above 30 GeV the next generation of missions are in preparation. INTEGRAL with a launch in 2002, the Italian AGILE project in ~2003, and the GLAST mission in 2005 will take up astronomy from space in this exciting and important energy range of astrophysics. At medium energies, above several 100 keV and up to ~ 50 MeV, no sensitive new telescope with survey capabilities as a follow up of COMPTEL has yet been defined. Figure 1 shows the sensitivity for past and future gamma-ray instruments and the goal for the next step for a combined Compton- and Pair-creation telescope in the range 0.4 to 50 MeV.

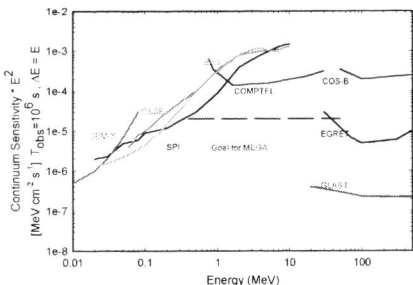

FIGURE 1. Typical sensitivity levels for current telescopes for continuum sources observed for 10^6 s

The improvement for a 2nd generation instrument in the medium energy range would be comparable to the steps of improvement at higher energies that have now led to the 3rd generation telescope GLAST.

INSTRUMENT DESIGN AND PROTOTYPE DEVELOPMENT

Combined imaging and spectroscopy of photons with MeV energies is a difficult task. The interaction cross-section at these energies goes through a minimum (masks become transparent) and the primary interaction of these photons produces secondary, scattered photons of considerable range. Around 8 MeV (e.g. in Si) the type of interaction changes from Compton scattering to pair-creation. In order to make full use of all interacting photons in a telescope one requires not only sufficient material depth to record a good number of events (efficiency) but also spatial tracking of secondary ~MeV electrons and the resolved measurement of all energy deposits. The construction of such instruments is based on solid-state tracking detectors and scintillators (e.g. the TIGRE concept [1]) or liquid Xenon time projection chambers [2]. For MEGA we have chosen a design shown in figure 2.

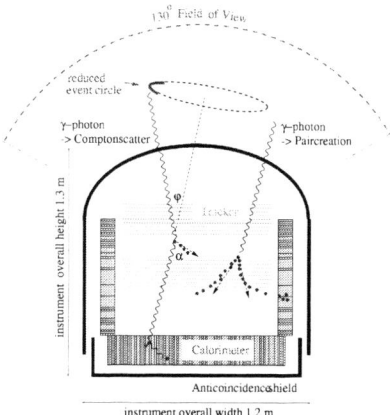

FIGURE 2. Schematic design of the MEGA detector with typical photon interactions of Compton scattering and pair-creation. The tracking of the Compton recoil electron allows the restriction of the full event circle to an arc of possible photon origin. From the combination of energy deposits and scattering characteristics along the electron tracks the direction of motion can also be estimated. Both parameters lead to a very efficient suppression of background events.

The tracker contains 32 layers of double sided Si-strip detectors (6×6 wafers of 6×6 cm^2 each, thickness 500μm, with strip-pitch ~470μm), and is enclosed by a calorimeter made of CsI cells (5×5×40(80) mm^3) read out with Silicon PIN diodes. The detector assembly is surrounded by an anticoincidence shield.

A prototype detector for MEGA has been assembled with 10 layers of 3×3 Si wafers and 20 modules of CsI detectors (8, 4, and 2 cm deep with 120 pixels each). Readout of the detectors occurs via an ASIC front-end (TA1 chip by IDE, 128 channels), custom made front-end control units, and laboratory VME electronics. Figure 3 shows this prototype at a preliminary stage of completion and imaging results using three ^{137}Cs sources (662 keV) of different intensities. Further calibrations with sources and at accelerators are planned for the near future.

Simulations for the full MEGA detector based on the presently available prototype properties indicate an effective area of about 100 cm^2, angular resolution of ~2.5° and energy resolution of 3-4% (FWHM at 2 MeV). A very exciting prospect for an instrument based on Compton scattering through large angles is its sensitivity to the polarization of the detected photons. Simulations carried out by [1] have shown that a MEGA like telescope will be sensitive to polarization levels of 10% from a source of Crab intensity after about 100 hours of observation.

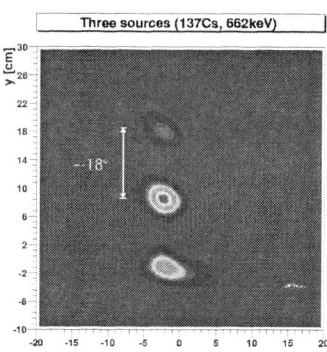

FIGURE 3: Preliminary set-up of the MEGA prototype detector and imaging of three ^{137}Cs sources.

MISSION CONCEPT AND SURVEY SENSITIVITY

The baseline MEGA detector has been considered in a pre-phase A study for a small satellite mission. The detector will have a mass of about 650 kg and dimensions of 1.3m diameter by 1.1m length. Placed on a standard small satellite platform the launch payload mass is about 950 kg, has a diameter of 2.0 m and 2.35 m length. The electrical power requirement will be ~400 W and the average telemetry rate about 50 kbit/s. The development time of MEGA to launch could be about 5.5 years and an orbital mission of 3-5 years should be foreseen. Operations overlapping with the GLAST mission would be of very high scientific value because of the complementary energy bands of both projects.

MEGA should be placed in a low-earth orbit (~550 km) with low inclination, to provide an environment with lowest charged-particle background and improved telescope sensitivity. MEGA is planned to perform continuous all-sky scans with the axis always pointed close to zenith. The large field-of-view describes a wide path of exposure during each orbit and allows to monitor most of the sky continuously for transient sources. Real-time satellite telemetry is planned through the TDRSS Demand Access System. The data will be analyzed promptly for bursts and transients and appropriate alert messages will be sent out to initiate follow-up observations.

We have estimated the survey sensitivity of MEGA with simulations based on GEANT 3. The assumed orbital background for MEGA is taken as 3 times the COMPTEL background at 5GV cut-off rigidity. The realism of this assumption will have to be investigated in more detailed background simulations, once the mass and material compositions of MEGA on a satellite and the orbit are defined more accurately. The sensitivities for MEGA in 3 and 5 years of operation compared to the average sensitivity of the COMPTEL all-sky survey are shown in figure 4.

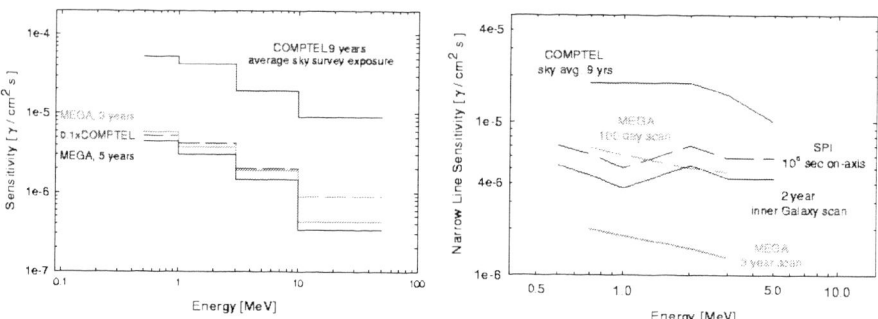

FIGURE 4: Sensitivity for MEGA for sources with continuum spectra (left). The goal of improvement of a factor 10 compared to COMPTEL is shown as the dashed line and compared to 3 and 5 years of a MEGA survey. The sensitivity for narrow spectral lines is shown in the right panel and compared to the values for SPI on INTEGRAL.

SIMULATIONS AND EXPECTED RESULTS

Synthetic observational data for MEGA in a scanning survey mode were generated in Monte Carlo simulations. The inversion of these data, which include statistical and systematic errors of energy and spatial resolution, into images is not straightforward and iterative algorithms have to be used. The complex MEGA data space is efficiently handled with the application of novel list-mode algorithms (adapted from [3]). The image of calibration sources in figure 3 was derived with this method. For a realistic sky distribution of sources in the range 1-10 MeV we chose a region of the galactic anticenter with the bright COMPTEL sources Crab and the blazar PKS 0528+134. Further sources were extrapolated from the 3rd EGRET catalog. The EGRET spectra were assumed to extend down to 10 MeV and continued below that energy with a power law harder by 0.5 in index. Figure 5 (left panel) shows the input map of sources. After an exposure, corresponding to six weeks of orbital scan mode, the image shown in the right panel of figure 5 can be reconstructed. Most of the 'soft' spectrum EGRET sources are clearly detected, although also a 'ghost' image between the two brightest sources appears. An improved model of system response should reduce this kind of artefact. It should be noted that during an orbit time of six weeks a large part of the sky will be imaged with comparable quality and many sources can be monitored continuously.

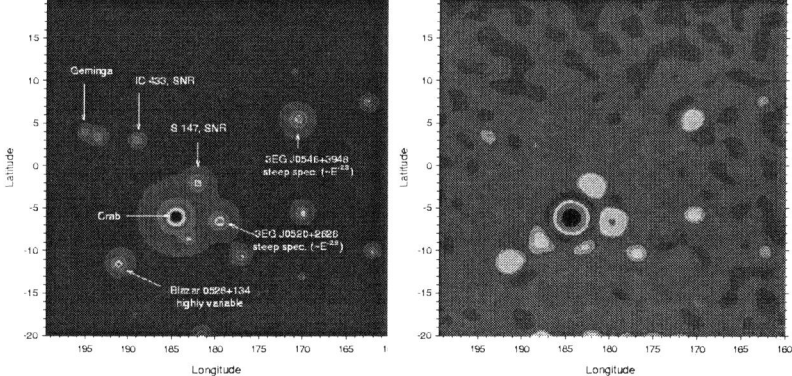

FIGURE 5: Simulation of sources in the galactic anticenter. The 3EG catalog, except the COMPTEL detected sources Crab and PKS 0528+134, was extrapolated with a broken power law to the 1-10 MeV range (left panel). Reconstruction of a simulated MEGA observation after 6 weeks of orbital scan mode (right panel) shows that most 'soft' spectrum 3EG sources are detected.

A preliminary estimate of the number of sources detectable in a MEGA sky survey predicts that about 100 unidentified EGRET sources will be seen. The number of known pulsars at MeV energies should rise to about 10. About a dozen compact galactic binary systems containing black holes like Cyg X-1 will be detected. In extragalactic space about 100 blazars and more than 10 radio and Seyfert galaxies will be visible. A gamma ray burst should occur and be imaged in the large field of view of MEGA about once every two days.

The observation of sites of nucleosynthesis, either from explosive events like Novae and Supernovae, or through their radioactive debris in SNRs and the galactic diffuse radioactivity are the other important key science objective for MEGA. We expect to detect ~5 Novae and 2-3 SNe each year. About 5 young galactic SNRs should be discovered through their ^{44}Ti emission. Detailed mapping of galactic star formation regions will allow to investigate the production of elements in massive stars.

The origin and composition of the diffuse cosmic background at MeV energies is still mysterious and not well measured. At about 5 MeV a transition from predominantly thermal (Seyfert galaxies) to non-thermal ('EGRET Blazars') sources seems to be indicated. However radioactivity from distant SNe, radio galaxies and new 'MeV' blazars, or clusters of galaxies could be important new components of the background with wide ranging cosmological implications.

REFERENCES

1. O'Neill, T.J., et al., *Astron. Astrophys. Suppl. Ser.* **120**, 661-664 (1996).
2. Aprile, E., et al., *Astron. Astrophys. Suppl. Ser.* **120**, 649-652 (1996).
3. Wilderman, S.J., et al., *IEEE Trans. Nucl. Sci.*, **45**, 957-961 (1998).

Development of CdTe Imaging Detectors for a Compton Telescope

K.-L. Giboni, E. Aprile & U. Oberlack

Columbia Astrophysics Laboratory, 550 W. 120th Street, NY, NY 10027

Abstract. Imaging arrays of thick CdTe detectors, with millimeter spatial resolution and better than 10 keV energy resolution at 1 MeV, are very promising for the calorimeter section of an Advanced Compton Telescope. Measurements with small CdTe detectors with Schottky contacts, and advances in fabrication technology, have led us to the design of an 8×8 pixel array, with 2×2 mm^2 pixel size, as a first prototype module for large arrays. The detectors are arranged with the electric field direction perpendicular to the incoming gamma-ray direction, allowing several centimeters deep arrays, with the minimum amount of passive material. With the development of the 8×8, 10 mm deep array, we aim to demonstrate the feasibility of the concept for gamma-ray imaging.

INTRODUCTION

We have proposed the development of an imaging array of high resolution Cadmium Telluride (CdTe) detectors to test the feasibility of this room temperature solid-state detector technology for an Advanced Compton Telescope (ACT). A schematic concept for an ACT is shown in Fig. 1. We envision an array of Si-strip detectors for gamma-ray conversion and Compton electron tracking, surrounded by a modular CdTe calorimeter with 3D position sensitivity. The high energy and spatial resolution of the Si and CdTe arrays allow to place the two detector layers close by, to gain detection efficiency without loss of angular resolution. Large effective area, excellent spectral and imaging performance, and efficient background rejection, are key requirements to achieve the sensitivity level desired for a next generation MeV astrophysics mission [1].

Imaging gamma-rays with a Compton telescope requires two distinct tasks which call for differently optimized detectors, the converter and the calorimeter. Si strip detectors are a prime candidate for the converter. Large area (10 cm × 10 cm), thin (300 μm) Si detectors with fine segmentation (<1 mm) are available. The calorimeter preferably consists of a dense, high Z material to totally absorb the remaining gamma-ray energy, with good position and energy resolution. Most gamma ray interactions in the MeV range are multiple Compton interactions, with the first one assumed in the converter, and the remaining ones in the calorimeter. It is necessary to separate the second from the following interactions, although they can be close by. Thus, besides a good position resolution, also a good double interaction resolution is needed. The energy resolution has to be good as it determines the angular resolution and background discrimination capability.

Different materials and designs are being studied or proposed, among them scintillator crystals (e.g. CsI with pin diode read out), Ge strip detectors, LXe or high pressure Time

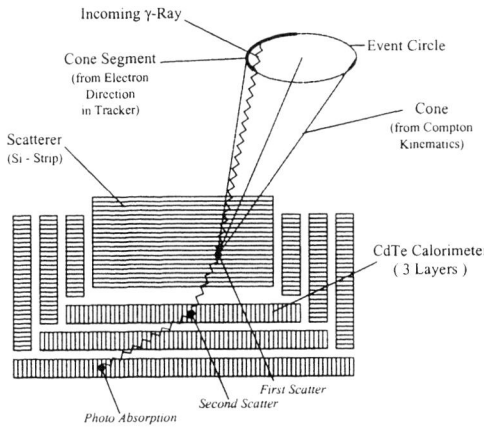

FIGURE 1. Schematic of a Compton telescope concept with solid state detectors.

Projection Chambers, and CZT or CdTe planar arrays. The past few years have seen large progress in CZT detectors for low energy gamma-rays. A number of approaches have been demonstrated for mitigating the effect of the low hole mobility and improving the spectral performance. The tailing effect caused by hole trapping is either compensated by electronic pulse shape correction schemes, or counteracted by using only the electrons signal, with the introduction of virtual grids in the crystal [2]. However, for gamma rays of energy above a few hundreds keV, a good photofraction efficiency would require many layers of these thin planar arrays, each with some dead material in between to support the crystals.

In the following section, we discuss our approach with Schottky CdTe detectors. The design takes advantage of the particular requirements in a Compton calorimeter. Since most scattering events are in forward direction, the Z coordinate, in the direction of the incoming radiation, contributes less to the final angular resolution than the X-Y directions. A calorimeter with three or more layers of 10 mm thick arrays, with 2×2 mm^2 pixel size, would still provide sufficient position resolution.

SCHOTTKY CdTe DETECTORS

CdTe is one of the semiconductors materials which have a sufficiently high bandgap (1.52 eV) for operation at or around room temperature. It has a high density (6.06 g/cm^3), and a high atomic number (Z_{av} = 50) for good efficiency in gamma ray detection. Together with CZT, it is the only compound semiconductor extensively used as nuclear radiation detector. The efficiency per unit thickness is higher for CdTe than for Ge, and the photofraction for an equivalent volume also will be larger. Compared with Ge, however, the charge carriers liberated by an ionizing event drift much slower, with a difference of a factor 10 between the values for electrons and holes. Also the lifetime of the carriers before trapping is much shorter than in Ge. With hole-trapping, the

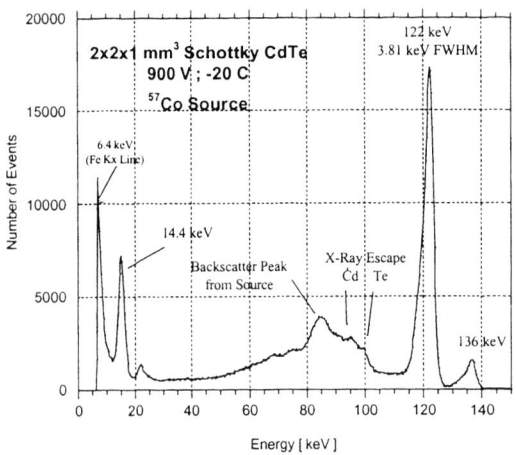

FIGURE 2. ^{57}Co spectrum obtained with a 1 mm thick CdTe detector at 900 V bias voltage and $-20°$.

number of photons in the photopeak is reduced and the spectrum is distorted by a tail towards lower energies. To minimize hole-trapping effects, practical detectors are limited in thickness to 1 or 2 mm. The detectors are typically oriented such that the gamma rays enter through the cathode contact, in a direction parallel to the field. With the higher probability for gamma rays to interact close to the cathode, the drift path for holes is smaller, on average. A shaping constant shorter then the maximum drift time of holes, further discriminates against long hole drift times, and the observed energy spectra show photopeaks with good resolution. Reversing the detector, however, shows much worse energy distributions, indicating that only a fraction of the detector thickness is used to measure the photopeak energy. Interactions occuring close to the anode are misinterpreted and contribute background events at lower energies.

Single CdTe detectors can be made very long, in excess of 30 mm, without loosing performance. The limit of the thickness of order 1 mm, however, is very close to the desired pixel size in X and Y. In the CdTe detector configuration which we propose, the detectors are therefore arranged with the drift field along X or Y instead of Z. With the incoming gamma ray direction perpendicular to the field, interactions close to the cathode or the anode have equal probability. This arrangement, originally discussed in [3], requires that the photopeak efficiency is constant throughout the detector volume. Schottky CdTe [4] [5] [6] detectors are preferred, because they can withstand much higher bias voltage with a leakage current orders of magnitude lower than detectors with ohmic contacts. The higher field strength translates directly into shorter drift times and, thus, a reduction of hole trapping. We have studied the performance of this type of CdTe detectors, and shown that they are fully efficient with good energy resolution [7].

Fig. 2 shows, for example, the energy spectrum of a ^{57}Co source measured with a 1 mm thick Schottky CdTe detector at 900 V bias. At this operating point, the 122 keV photopeak is symmetric.

Most of the activity below the photopeak is identified as X-ray escape peaks, due to the small size of the detector, or as scattering in surrounding material. Besides the good energy resolution, we also note the low minimum energy threshold, with the 14.4 keV peak clearly identified.

Much better resolution has been obtained with 0.5 mm thick detectors in a stack of 12 layers [8]. For a Compton telescope, however, the enhanced energy resolution with very thin planar detectors, would not appreciably add to the ultimate sensitivity, but it would add significant technical complications and cost. In an earlier study [9], detectors of different thickness were evaluated, and 1 mm appeared as most appropriate for an instrument of practical size.

A concern with CdTe detectors is the long time stability of their performance. Early CdTe detectors with ohmic contacts exhibited a decrease in signal with time due to charge build up. With Schottky CdTe detectors, we observed that this polarization effect completely disappears when operating around $0°$ or below. A similar finding was independently reported in [5].

PROPOSED DESIGN OF A CdTe IMAGING DETECTOR

The design of a prototype CdTe array, with 2×2 mm^2 pixel size, is shown in Fig. 3. The depth of the 8×8 pixel array is 1 cm. Each pixel is made of two 2 mm wide, 1 mm thick detectors mounted back to back. The 16 planar arrays are only segmented on the cathode contact, with a single anode contact spread over the full crystal. A 25 μm thick brass foil extends the contacts to one side where they are soldered to a printed circuit board. On the circuit board, the detectors are wired to form an X-Y read out, requiring 16 electronics channels. The 128 detectors which make the array appear like a 1 cm thick block with 8 strips in both X and Y directions. The interconnections are shown on the right side of Fig. 3. The multitude of thin foils still warrants the mechanical stability, while keeping the dead material between pixels to a minimum. For future large arrays, more detectors can be connected together to form larger X-Y patterns, and only the pulseheight of a shaped signal and the start time of the pulse will be significant. The number of detectors in each X-Y strip is limited by the input capacitance. In previous tests [11], the effect of detector capacitance on resolution was studied. Results showed that even 32 detectors in parallel would still provide good energy resolution.

The appropriate size of the basic module will finally depend on the yield of producing the array, with all detectors working according to specifications. Crystal production capabilities would already now allow to choose a larger depth of 3 cm, and possibly 5 cm [10]. Segmenting also the anodes, perpendicular to the cathode strips, can provide a higher granularity in the third coordinate, of course, at the expense of more electronics read out channels.

CONCLUSIONS

The measured characteristics of small volume Schottky CdTe detectors show very promising spectral performance, detection efficiency and stability of operation. To

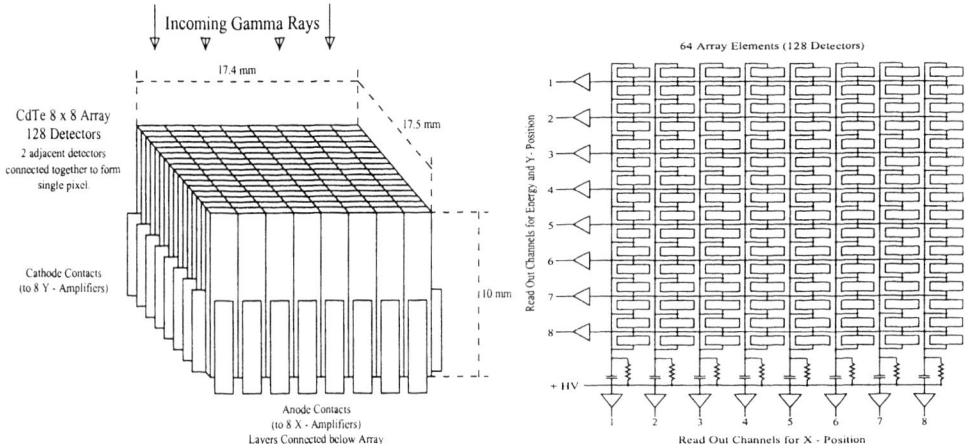

FIGURE 3. Left: Schematic view of an 8×8 pixel array, with 2×2 mm^2 pixel size. Right: Interconnection of the 128 detectors which make the array. 16 read out channels are connected to the 8 columns and 8 rows of the array.

demonstrate the application of these detectors for gamma-ray imaging, we have proposed to develop an initial prototype 8×8 pixel array, with a thickness of 1 cm and a pixel size of 2×2 mm^2. If approved, our research plan is to fully characterize the array with gamma rays, both in the laboratory and in near space, with a balloon flight. The small CdTe array can be easily integrated with the LXeGRIT payload [12], providing important information on the expected performance of these detectors in a future Compton telescope mission.

REFERENCES

1. GRAPWG,'Recommended priorities for NASA's gamma ray astronomy program 1999 – 2013', http://universe.gsfc.nasa.gov
2. Luke, P.N.. *IEEE Trans. Nucl. Sci*, **NS-42 (4)**, p. 2077, 1995.
3. Dusi W., et al., *SPIE*, **1734**, p. 122, 1992.
4. Matsumoto M., et al., presented at the IEEE Nucl. Science Symposium in Denver, November 1997.
5. Takahashi T., et al., *SPIE*. **3446**, p. 29, 1998.
6. ACRORAD, Tokyo, Japan.
7. Giboni, K.-L, and Aprile, E., *NIM A*, **416**, p. 319, 1998.
8. Takahashi, T. et al., *NIM A* **436**, p. 111, 1999.
9. Giboni, K.-L., et al., Contribution to the International Symposium on Solid State Detectors for the 21st Century, Nara, Japan, December 1998.
10. Funaki, M., et al., *NIM A*, **436**, p. 120, 1999.
11. Giboni, K.-L., Aprile, E., and Rochwarger, I., *SPIE*, **3446**, p. 228, 1998.
12. Aprile, E. et al., *SPIE*, **4140**, p. 35, 2000.

Mission Concepts: Hard X-ray Telescopes

EXIST: The Ultimate Spatial/Temporal Hard X-ray Survey

J. Grindlay[1], L. Bildsten[2], R. Blandford[3], D. Chakrabarty[4], M. Elvis[1],
A. Fabian[5], F. Fiore[6], G. Fishman[7], N. Gehrels[8], C. Hailey[9], F. Harrison[3],
D. Hartmann[10], C. Kouveliotou[7], T. Prince[3], B. Ramsey[7],
R. Rothschild[11], G. Skinner[12], and S. Woosley[13]

1. CfA, 2. UCSB, 3. Caltech, 4. MIT, 5. IoA, 6. Rome Obs., 7. MSFC, 8. GSFC, 9. Columbia Univ., 10., Clemson, 11. UCSD, 12. CESR, 13. UC Santa Cruz (EXIST Science Team institutions)

Abstract. The Energetic X-ray Imaging Survey Telescope (EXIST) is a proposed mission to conduct an all-sky imaging hard x-ray (HX) survey (~5-600 keV) with ~0.05mCrab sensitivity (5σ; 6mo.; ~5-100keV) comparable to the ROSAT soft x-ray survey, and to provide the maximum sensitivity and resolution (spatial and temporal) HX imager as the Next Generation GRB mission. Its *primary* science goals are to i) identify and measure obscured AGN and constrain the accretion luminosity of the universe as well as the cosmic IR background from Blazar spectra coincident with GeV-TeV observations, ii) measure spectra, variability and locations for the faintest GRBs to study the most energetic events in the universe and the earliest epoch of star formation, and iii) study black holes on all scales, from x-ray transients to luminous AGN. EXIST would incorporate a very large area (~8m^2) imaging Cd-Zn-Te detector and coded aperture telescope array with nearly half-sky instantaneous view which images the full sky each orbit. With fixed zenith pointing, it could be mounted on the ISS or a free flyer and would complement both GLAST and Constellation-X science if launched before 2010, as recommended by the Astronomy and Astrophysics Decadal Survey.

REVEALING THE OBSCURED HARD X-RAY UNIVERSE

At energies >5-10 keV, the buried nuclei of active galaxies, which provide most of the cosmic x-ray background, become increasingly clear (if not Compton thick) and reveal the accretion power of the universe [1]. The hard x-ray (HX) band provides the most direct view of the central regions of AGN and near-horizon views of black holes; lower energies are often absorbed and γ-ray emission is likely converted to pairs in the most compact regions. The energy band from ~5-600 keV, which covers the transition from the thermal x-ray universe of hot gas traced by ubiquitous Fe line emission (6.4-6.7keV) to the extremes of pair plasmas (511keV) and the non-thermal universe, is rich in phenomena and yet remains relatively obscure in realization. The sky has not yet been surveyed in this band with the sensitivity or imaging resolution achieved in 1991 with ROSAT for the soft x-ray band. The last full-sky HX survey, carried out with HEAO-1 in 1979 [2] was sensitive to sources only down to ~1/20 of the brightest (Crab, CygX-1) for uncrowded high latitude fields, and had only coarse angular resolution (~3°) and energy resolution (~20%). An imaging (17' resolution) HX survey with ~10-30X the HEAO sensitivity will be conducted with the upcoming Swift mission [3] and over a similar energy band (~15-150keV) as the HEAO-A4 survey.

The EXIST (Energetic X-ray Imaging Survey Telescope) mission, proposed first as a New Mission Concept in 1994 [4] and recently recommended for implementation within the current decade by the Astronomy/Astrophysics Decadal Review, would conduct the ``ultimate" HX survey: it would extend the energy band to cover the full HX to soft γ-ray range (~5-600 keV) with the broadest possible temporal coverage (microsec to months) to measure the inherently variable HX sky in both space and time. Such broad energy and temporal coverage requires use of the coded aperture imaging technique (cf. Caroli et al [5]), whereby position-sensitive detectors record shadows of the distribution of sources cast by a coded mask and which then achieves sensitivity for minimum fluxes decreasing with total detector area A, integration time T and recorded background B as $(T/A \cdot B)^{0.5}$. EXIST would extend the coded aperture technique to the practical limits, with total detector area A~8m² and effective exposure time (per source) T maximized so that any source is observed for >20% of the time and the full sky is imaged each 95min orbit. This combination of very large area and field of view, and thus source temporal coverage, makes EXIST also the core of the ``Next Generation Gamma-Ray Burst (GRB) Mission".

We present an overview of the EXIST mission concept and then describe the major scientific objectives that are made possible by the proposed detector-telescope configuration. The technical implementation of EXIST, as currently envisioned [6] for the International Space Station (ISS), is then outlined along with a brief description of the mission development plan. Studies for a free flyer implementation will be conducted, along with continuing programs for detector development and balloon flight tests.

OVERVIEW OF EXIST CONCEPT

The possible implementation of EXIST on the ISS is shown in Figure 1. Mounted on the P3 Attached Payloads site, EXIST is configured as two large-area (4m² each) and large field of view (FOV; 80° x 80° each) coded aperture telescopes that are mounted back-to-back along the main truss of the ISS. The combined FOV[1] of 160° x 80° is aligned along the truss, or perpendicular to the orbital ram direction, so that the combined fan beam sweeps out the entire sky each 95min orbit. The instantaneous FOV of almost ~60% of the sky un-occulted by Earth gives good sensitivity and coverage for bursts and fast transients.

The primary driver for the detector and telescope design is the desired survey sensitivity: the science goals described below are based on a desired all-sky survey sensitivity of 0.05mCrab (5σ; 6 months) over the ~5-100 keV band, yielding flux

---

[1] The EXIST-ISS concept is a contiguous 2 x 4 array of 40° (FWHM) fields. Thus the response is flat over 120° x 40°, reduced to 0.5 at 160° x 80° and 0.25 at 180° x 100°, yielding relative (sensitivities, half-sky fractions) of: (1, 0.23), (0.7, 0.62), (0.5, 0.88).

sensitivities of $F_X = 5 \times 10^{-13}$ cgs (5-10keV) or 9×10^{-13} cgs (20-100 keV). This matches the ROSAT all-sky soft x-ray (0.5-2.5 keV) survey sensitivity, allowing the first extension into the HX band of a survey able to detect and locate the same objects for a (typical) Crab-like spectrum. The corresponding optical magnitude for an object

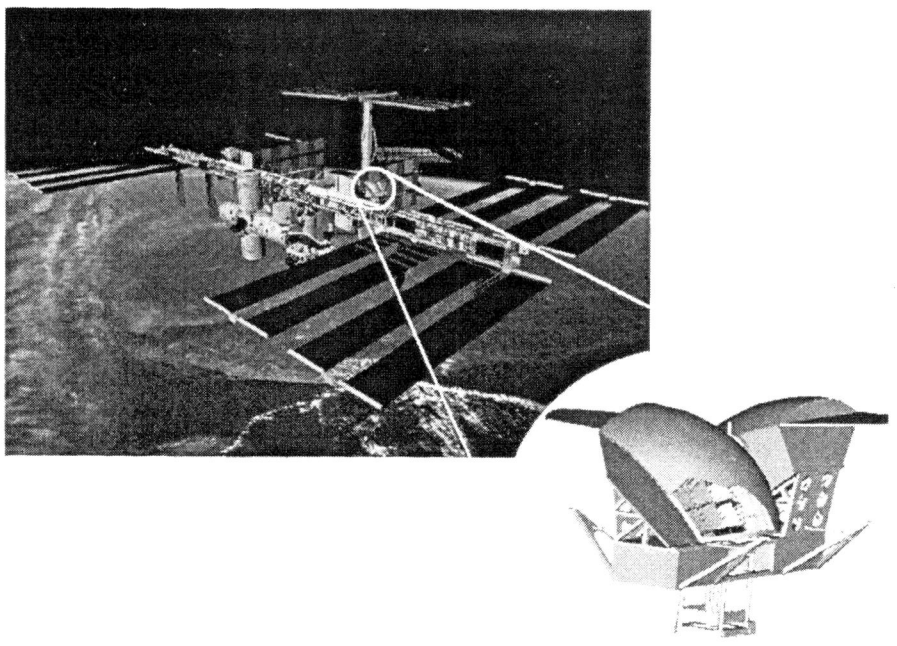

FIGURE 1. EXIST concept for ISS: back-to-back 80° x 80° coded aperture telescopes image (5') the full sky each orbit. Each telescope contains 2 x 2 contiguous 1m² arrays of CZT imaging detectors (partly visible), each actively collimated to 40° x 40°, to view the sky through an extended hemispherical coded aperture.

with x-ray/optical flux ratio $F_X / F_V = 1$ (such as for many CVs and AGN) is V = 20.9, using the relation between F_V and V magnitude, $\log F_V = -3.96 - 0.4V$, which combines the near-uv and optical flux. The sensitivity goal requires a large total detector area (8m²), which is realized by 8 x 1m² telescopes each with 40° x 40° FOV (cf. Fig. 1).

The detectors are Cd-Zn-Te (CZT) crystals, each perhaps 20mm x 20mm and 5-10mm thick, and fabricated with anode pixels (1.3 - 2.5mm pixel pitch) which contact onto a coupling board with VLSI/ASIC readout to record the peak pulse height pixel and nearest neighbors for each x-ray event. A part of the total CZT array (~1/4 total area) may be apportioned to a low energy (<100 keV) optimzed array (0.6mm pixels; 2mm thick CZT) with reduced FOV (10° x 40°) to reduce bright source backgrounds. The CZT crystals could be grouped 2 x 2 onto a common ASIC board (cf. Fig. 8), which are then in turn close-tiled into a large area array (possibly 32cm x 32cm) with common digital readout and control as well as surrounding collimating shield (1cm CsI). Relatively thick CZT ($\geq$5mm) is needed for the primary detector to extend the response to 600 keV, and depth sensing would be used to reduce detector background. For 5mm thick CZT, the ~100-600 keV sensitivity is ~0.5mCrab (cf. Fig. 2).

Source confusion is avoided at the projected sensitivity (cf. Fig. 2), which should detect $\geq$1 AGN/square degree (see below), by imaging with 5' resolution (a factor of 2 better than the usual criteria of ~1/40 source per "beam"). A source on the orbital equator is in the FOV for a fraction 80/360 = 0.22 of the time, whereas those near the orbital poles are nearly continuously observed. The collimation provided by the active shields gives a FOV of 40° x 40° FWHM (cf. footnote 1) for each 1m² telescope is, so that in fact 1/4 of the area of the telescopes is available at the ±90° limits of the combined FOV to cover the orbital poles. Likewise, in the scan direction the FOV is similarly extended from 80° to 100°, at 1/4 response, thereby increasing the total minimum source exposure fraction to 100/360 = 0.28. Thus the full sky is imaged and monitored each 95min orbit.

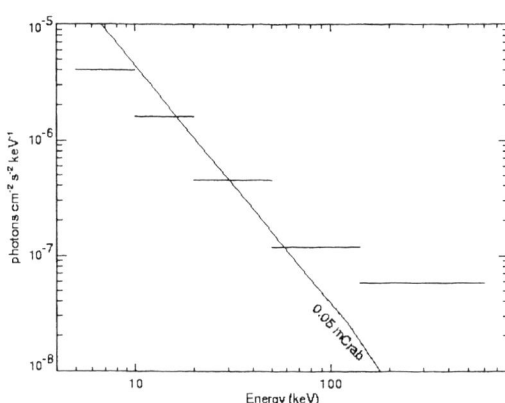

FIGURE 2. Approximate continuum sensitivity (5σ, 1year) for EXIST-ISS configuration. Narrow line sensitivities (phot/cm²-sec) are reduced by factor of ~10.

EXIST SCIENTIFIC OBJECTIVES

Obscured AGN

EXIST would achieve a sensitivity for AGN that would allow the logN-logS distribution of obscured objects to be measured over the full sky for the first time. As pointed out by Fabian [7], the 3 closest AGN (Cen A, NGC 4945, and the Circinus galaxy) are all obscured. What is not known from optical surveys is how many more highly obscured Seyfert II galaxies are relatively nearby. These would be revealed by the EXIST survey and could then be studied in detail by followup observations with higher sensitivity and resolution. BeppoSAX spectra [8] of two optically selected and relatively nearby Seyfert II systems, NGC1068 and NGC6240, are shown in Figure 3 and are indicative of what the x-ray selected survey sample might look like.

An unbiased sample of the number of obscured AGN is essential for testing models of the cosmic x-ray background and the likely contribution of obscured AGN. The relative contribution of obscured AGN (with log NH $\geq$ 23.5) to the total number counts is predicted [9] to exceed that of unabsorbed AGN by factors of $\geq$2-4 at flux levels of log F(5-10keV) ~ -12.5 (cgs), which is readily measured with EXIST.

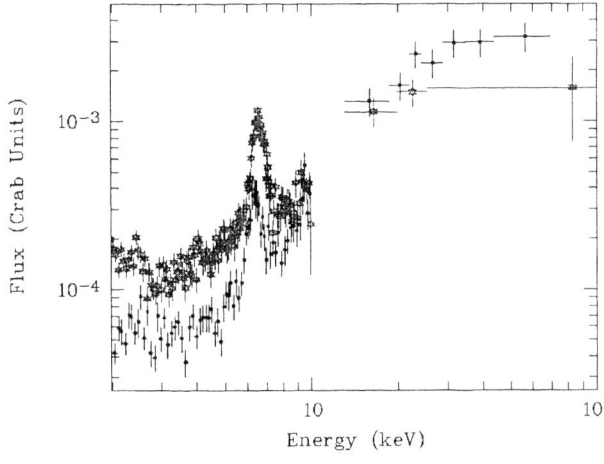

FIGURE 3. BeppoSAX spectra [8] of optically selected Seyfert II's: NGC1068 (top) & NGC6240. EXIST would go ~20X deeper, full sky.

With its sensitivity for AGN that are highly obscured (and becoming Compton thick), and with the overall AGN sample expected (from logN-logS) to be at least ~30,000, EXIST will provide a large sample of both nearby Seyfert II's and more distant and luminous Type 2 QSOs. The luminosity and redshift distributions of the obscured AGN will measure the accretion vs. nuclear (starlight) power in the universe.

Blazars and the Diffuse IR Background

A related objective is to measure the intrinsic HX spectra of Blazars, highly beamed and variable AGN, in order that (near-)simultaneous γ-ray observations can be combined to constrain the diffuse IR background at ~300μ. The recent detections of several blazars as TeV sources (cf. Weekes, these proceedings) shows that these "extreme" BL Lac objects (e.g. Mkn 501) are detected at distances (z ~ 0.03) where their TeV photons could have undergone photon-photon absorption on the cosmic IR background photons (converting to electron pairs).

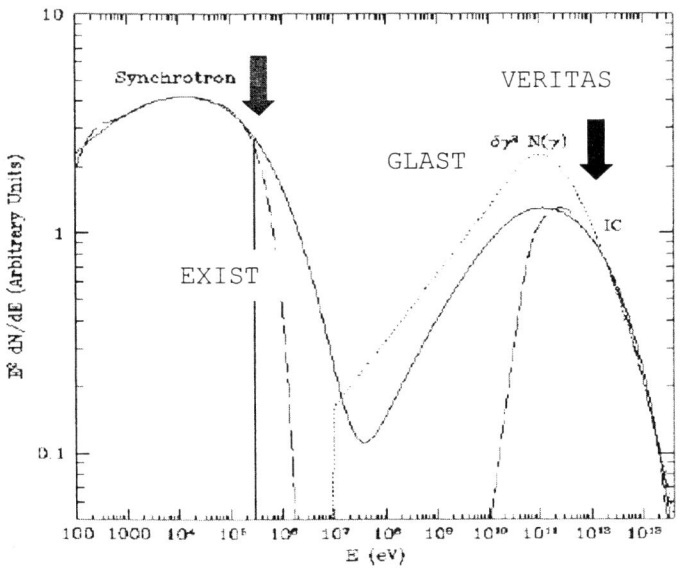

FIGURE 4. Adapted from [10]: HX spectral break (EXIST) constrains the GeV-TeV break (GLAST-VERITAS) and thus the cosmic diffuse IR.

These "blue blazars" have spectral energy distributions (SEDs) with a broad peak in the x-ray/hard x-ray and a second peak in the GeV-TeV gamma-ray band (cf. Figure 4). The natural interpretation (e.g. Coppi & Aharonian [10]), increasingly favored, is that the emission is synchrotron self-Compton (SSC). Thus a simultaneous measurement of the synchrotron spectral cutoff (likely to be time variable) with EXIST allows the observed inverse Compton TeV spectral break to constrain the diffuse IR flux. Interestingly, this IR background is likely produced in part by the same obscured AGN (and starburst galaxies) responsible for the HX background and directly detected by EXIST.

Gamma-Ray Bursts

The capability of EXIST for high sensitivity/resolution hard x-ray imaging over a wide field and energy band renders it particularly suitable for GRB science, which together with the HX survey objectives are the *primary science drivers* of the mission. Indeed, the study of GRBs at the faintest flux levels, as well as high time resolution spectral studies of brighter GRBs, make EXIST the natural hub and primary trigger of a Next Generation GRB Mission (which might also include a separate x-ray afterglow telescope, as a coordinated small mission, and optical/IR rapid-response capability).

Numerous key questions for the origin and nature of GRBs can be addressed. As but one example, the possible determination of GRB redshifts directly from the GRB data using the possible correlation of GRB luminosity and temporal lags over a broad energy band, as recently suggested by Norris et al [11] (cf. Figure 5) is well suited to the broad band energy coverage and high time resolution of EXIST.

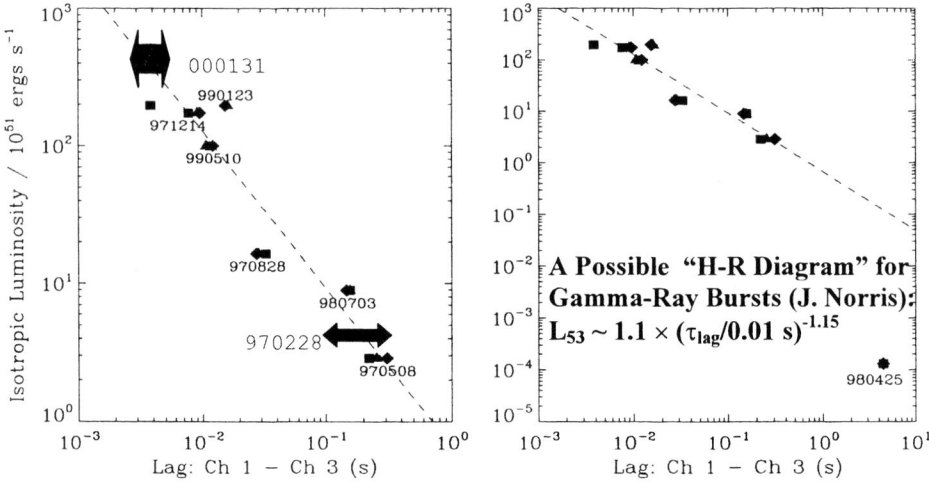

FIGURE 5. Adapted from [11], with 2 bursts added (blue): GRB luminosity vs. measured time lag (~50–300 keV) for BATSE data on GRBs with measured redshifts. The broad energy band and high sensitivity/temporal resolution of EXIST yield such "photometric" redshifts for a large GRB sample.

EXIST would increase the sensitivity for GRB detection by factors of ~5X (vs. Swift) and ~20X (vs. BATSE). As such, it will measure the very faintest GRBs, locate them within ~10sec to ≤1' (GRBs above the BATSE threshold would have positions limited only by the absolute aspect uncertainties, probably ~5"), and provide high time resolution spectra that can be constructed from the un-binned photon arrival times (~1μsec resolution) that would be brought down in the full telemetry for followup analysis. With ~60% of the BATSE instantaneous sky coverage but ~20X increased sensitivity, the GRB detection rate should be ~3-5X that of BATSE so that comparable GRB samples (but for much fainter bursts) can be achieved in ~2 years. The spectral resolution expected for the large CZT array (~1-3 keV (FWHM) at ≤200 keV; ~3-6 keV at higher energies) enables searches for fast-time spectral variations which could constrain fireball and internal shock models as well as the possibility of 511 keV (blue or red-shifted) in the GRB-afterglow transition spectrum.

These capabilities allow EXIST to carry out the most sensitive search for and measurements of GRBs at high redshifts, z ~5 – 20, and thus realize the full potential of GRBs as probes of the high redshift universe [12]. The primary objectives include:

- *extending logN-logS to the limit*, with a large sample of faint GRBs which both maximizes the measurement of the GRB luminosity function (e.g. searches for more 980425 events) and the redshift distribution of GRB sources.
- *measurement of PopIII star formation*, with GRBs at the putative z ~10-20 epoch of PopIII star formation expected if these earliest stars were massive and likely to produce hypernova or collapsar systems as expected [13] for GRBs.
- *constrain metallicity at high z*, since detection of even a moderate sample of high z (>7) GRBs with early (<1d) afterglow spectra (near and mid-IR) would enable Ly-α forest measures of the history of metallicity enhancement, and growth of structure and galaxy formation, after the initial PopIII starburst era.

Black Holes on All Scales

Black holes are ubiquitous hard x-ray sources. Both BHs in galactic x-ray binaries and AGN of all types are detected with a significant fraction of their total luminosity in the ~10-200 keV (rest) band and with spectral distributions remarkably similar [14; cf. Figure 6] for similar source types (BH-LMXBs/Seyfert I and II; microquasars/radio loud QSOs). Thermal Comptonization processes dominate in the former, with spectral breaks and characteristic electron temperatures typically at ~100 keV, and non-thermal power laws in the latter. The high-sensitivity and spatial resolution of the EXIST survey would allow spectral measurements of AGN some 30X fainter than the mean OSSE spectrum shown in the lower right panel of Figure 6. Fundamental questions about accretion onto super-massive BHs can be investigated: i) how do Comptonization temperatures and optical depths scale with luminosity? ii) how do Compton reflection components (~10-30 keV) vary in time and on what timescales (for the brighter AGN) which in turn restrict the emission geometry? and iii) how do these same components scale with relativistic-broadened Fe lines from the inner disk which could be measured in detail with simultaneous high resolution observations from Constellation-X? New clues to these questions are enabled by the all-sky, all-time measuring/monitoring of AGN possible for the first time with EXIST.

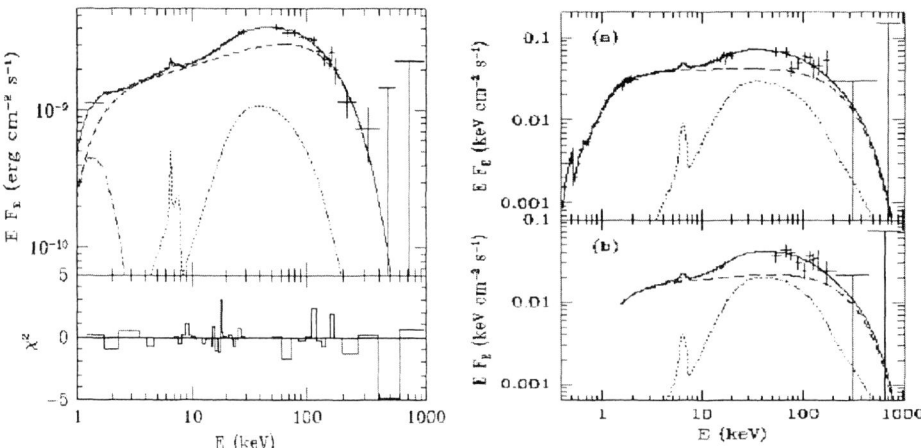

FIGURE 6. Spectra of BH x-ray binary (GX339-4), left, vs. AGN (IC4329A, top; average of 5 Seyfert Is detected by both Ginga and OSSE, bottom), right, showing how similar the thermal Comptonization models (dashed lines) and reflection/fluorescent components (dots) fit both classes (cf. Zdziarski [14]).

Galactic BHs are thus far found predominantly in soft x-ray transients (SXTs) in which the BH accretes from a low mass secondary companion star when mass transfer is triggered by an ionization instability in the accretion disk. Dynamical masses (~3-7 solar masses) have been measured for 13 of these when the secondary star becomes detectable between outbursts, and a very much larger galactic population (≥3000) is expected depending on recurrence times. EXIST would detect and locate these SXTs anywhere in the Galaxy for outbursts either ≥100X fainter and/or shorter than typical values (~100-300mCrab, ~10-30d) so that the full population of galactic BHs in low mass binaries could be studied. BHs accreting from high mass companions must be much more numerous than the one dynamically confirmed case (Cyg X-1); a large population may be lurking among the Be-binary transients for which HX discovery is needed given their expected obscuration in the galactic plane.

Finally, the EXIST survey is uniquely able to find, or limit, the number of possible intermediate mass BHs (with ≥30-100Msun) which would be detected anywhere in the Galaxy if they pass through giant molecular clouds with typical densities ≥10^3 cm^{-3} at their expected velocities (~10km/s) by the Bondi-Hoyle accretion expected. Thus, measurement of the expected hard X-ray spectra from isolated black holes offers a unique probe of the BH content of the Galaxy. If the Milky Way contains ~10^9 neutron stars produced by ~10Msun progenitors, and if BHs form from >30 Msun stars, then ~ 2 x 10^8 BHs are expected within ~200pc of the plane and ~10^3 would be within, and accreting from, dense giant molecular clouds to be discovered by EXIST.

Neutron Stars and SGRs

Accretion-powered pulsars (e.g. Her X-1), with either low or high mass companions, are readily detected and monitored by EXIST throughout the Galaxy. Thus, the studies initiated with BATSE for pulsar timing and accretion torques [15] can be extended to a much larger neutron star (NS) population of both disk-fed and wind-fed systems, which will include the large population of relatively faint x-ray pulsars with Be companions expected as NS transient systems. The recent BeppoSAX discovery of a significant new population of faint (~10mCrab) NS transients/bursters among the LMXBs [16] suggests a larger reservoir to be discovered with EXIST. These may include more sources like SAX J1808.4-3658, the only known accretion-powered millisecond pulsar and direct link to the millisecond pulsar population. EXIST can measure the magnetic fields for virtually the entire x-ray pulsar population from cyclotron lines at energies ~10 keV x B_{12} for fields of strength B_{12} x 10^{12} gauss.

Soft Gamma-ray Repeaters (SGRs) are the extremes of the NS population and likely are magnetars with magnetic fields $\geq$100X that (~10^{12} gauss) of ordinary accretion-powered pulsars. Only three are confirmed in the Galaxy, and yet their highly irregular duty cycles for their sporadic ~10msec, ~30 keV bursts suggests there are many more, which could constrain the evolution of magnetic fields on NSs. The rare super-bursts observed so far from two SGRs (one in the LMC) are so luminous that EXIST could measure them out to the Virgo cluster for the first survey of isolated NSs in external galaxies and possible links to the short-duration GRB population.

Supernova and Nova Rates in the Galaxy

The death rate of both massive and few solar mass stars can be constrained with EXIST. Type II SNe eject ^{44}Ti from a layer close to the mass cut within which material in the core collapses. The decay of this isotope with 96y half-life is marked by HX lines at 68 and 78 keV, easily resolved by EXIST. The full galactic plane survey of EXIST will reveal the supernova remnants (SNR) totally obscured by dust in the galactic plane. Scaling from the 2 SNR detected by the accompanying 1.18MeV line (cf. Diehl, these proceedings), EXIST should detect and locate ($\leq$2') all SNR within ~8kpc for the past 10^3 years, or ~15±4 for a nominal SN rate of ~3/century. The observed sample thus measures the galactic SNII rate and can be studied with Con-X.

Stellar novae are expected to be signalled by a flash of 511 keV emission from ^{18}F, which emits positrons as it decays with 158min half-life. The scattering of these same photons produces a detectable emission through the entire EXIST energy range, down to X-ray energies. After this initial explosion, which could only be detected with an all-sky 511keV imager like EXIST, other characteristic line emission from elements like ^{7}Be (478 keV energy, half-life of 77 days) can be detected for new measures of the nova rate in the Galaxy and the origin of the diffuse 511keV emission.

POSSIBLE IMPLEMENTATION OF EXIST CONCEPT

The very large area imaging CZT array presents the greatest technical challenge for the development of EXIST. The detector-telescope packaging concept for an ISS implementation of EXIST is shown in Figure 1 and in more detail in Figure 7. A CZT

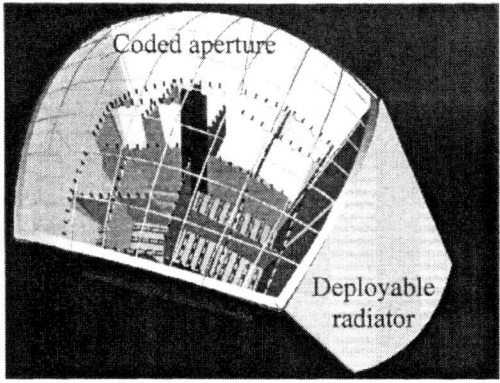

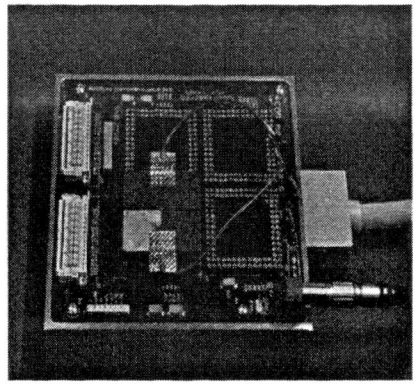

FIGURE 7. Partial detail of concept EXIST-ISS telescope (1 of 2; cf. Fig. 1) showing CZT arrays (16 x 2500cm²), CsI collimators, and extended coded aperture mask and supporting structures.

FIGURE 8. Prototype tiled CZT array [17] module (8cm x 8cm) with 2 x 2 subarray of 2cm CZT crystals with test cathode contacts (3 Au + 1 In, showing).

imager development program is underway at 4 of the institutions of the EXIST Science Team (Harvard, GSFC, Caltech and UCSD) and elsewhere. An example prototype module with ASIC readout (2.5mm pixels; 256 channels) on each of 4 close-tiled 4cm x 4cm CZT sub-arrays is shown in Figure 8. An EXIST imager module could be fabricated from scaling this tiling up a factor of 4 to 32cm x 32cm.

EXIST-ISS is currently estimated to be 5500kg and require 1.5Mb telemetry and 1.5kW power. Design studies for a Free Flyer version of EXIST, as well as a balloon-borne demonstration imager (for a ULDB proto-survey mission), are in progress.

REFERENCES

1. Fabian, A. and Iwasawa, K., *MNRAS*, **303**, L4 (1999).
2. Levine, A. et al, *Ap. J. Suppl.*, **54**, 581 (1984).
3. Gehrels, N., *B.A.A.S.*, **31**, 1512 (1999).
4. Grindlay, J. et al, *Proc. SPIE*, **2518**, 202 (1995).
5. Caroli, E. et al, *Sp. Sci. Rev.*, **45**, 349 (1987).
6. Grindlay, J. et al, *AIP Conf. Proc.*, **510**, 784 (2000).
7. Fabian, A., preprint (astro-ph/0103439) (2001).
8. Vignati, P. et al, *A & A*, **349**, L57 (1999).
9. Comastri, A. et al, *MNRAS*, in press (astro-ph/0105525) (2001).
10. Coppi, P. and Aharonian, F., *Astroparticle Phys.*, **11**, 35 (1999).
11. Norris, J.P., Marani, G.F. and Bonnell, J. T., *Ap. J.*, **534**, 238 (2000).
12. Lamb, D. and Reichert, D., *Ap. J.*, **536**, 1 (2000).
13. MacFadyen, A., Woosley, S. and Heger, A., *Ap. J.*, **550**, 410 (2001).
14. Zdziarski, A., *ASP. Conf. Ser.*, **161**, 16 (1999).
15. Bildsten, L. et al, *Ap. J. Suppl.*, **113**, 367 (1997).
16. In 't Zand, J., preprint (astro-ph/0104299) (2001).
17. Narita, T., Grindlay, J. et al, *Proc. SPIE*, in press (2001).

The Development of Coplanar CZT Strip Detectors for Gamma-Ray Astronomy

M. L. McConnell[†], L.-A. Hamel[‡], J. R. Macri[†], M. McClish[†], and J. M. Ryan[†]

[†]*Space Science Center, University of New Hampshire, Durham, NH 03824*
[‡]*Physics Department, University of Montreal, Montreal, Quebec, Canada*

Abstract. The development of CdZnTe detectors with an orthogpnal coplanar anode structure has important implications for future astrophysical instrumentation. As electron-only devices, like pixel detectors, coplanar anode strip detectors can be fabricated in the thickness required to be effective for photons with energies in excess of 500 keV. Unlike conventional double-sided strip detectors, the coplanar anode strip detectors require segmented contacts and signal processing electronics on only one surface. This facilitates the fabrication of closely-packed large area arrays that will be required for the next generation of coded aperture imagers. These detectors provide both very good energy resolution and sub-millimeter spatial resolution (in all three spatial dimensions) with far fewer electronic channels than are required for pixel detectors. Here we summarize results obtained from prototype detectors having a thickness of 5 mm and outline a concept for large area applications.

INTRODUCTION

Cadmium Zinc Telluride (CZT) detectors are well suited for fabrication of large area, high performance X-ray and γ-ray imaging spectrometers. They have the desirable properties of high stopping power, low thermal noise, room-temperature operation, excellent energy resolution and unsurpassed spatial resolution. With current technologies, high performance CZT imaging spectrometers can be adapted for use in a variety of imaging techniques to image photons at energies up to a few hundred keV. In the future, as the quality of CZT material improves to the point where the use of thicker detector substrates become feasible, the useful energy range of CZT detectors will likely be extended into the MeV energy range.

The spectroscopic value of CZT is exemplified by its use in NASA's upcoming SWIFT mission [1]. Related CdTe detectors are also being deployed on the INTEGRAL mission of ESA [2]. Neither of these applications take advantage of the properties of CZT that allow for the determination of a photon interaction site *within* a detector. Instead, small detector elements serve to isolate the location of the photon interaction site to within several millimeters. Outfitting large areas with numerous small detector elements so that the improved (sub-mm) spatial resolution could be achieved would be prohibitively complex and expensive. The only practical solution would be to employ large-area position-sensitive detectors, drastically reducing the number of detectors and associated electronic channels.

An image plane fabricated using closely-packed arrays of CZT detector modules is a candidate for the central detector of several proposed hard X-ray (30-600 keV) coded aperture telescopes, including HEXIS [3], AXGAM [4], EXIST [5], and MARGIE [6,7]. CZT strip detectors are also under investigation for use as calorimeter detectors in some Compton telescope designs for the 1-100 MeV energy range (e.g., TIGRE [8]). These designs will require image planes with areas exceeding 1000 cm^2 and spatial resolutions on the order of 1mm or better.

THE ORTHOGONAL COPLANAR ANODE STRIP DETECTOR

Good efficiency, energy resolution and position resolution have been demonstrated in pixellated CZT detectors up to 10 mm thick [9]. These detectors are *electron-only* devices that avoid the deleterious consequences of poor hole transport. Unfortunately, the anode pixel contact geometry requires an electronic signal channel for each of the N^2 pixels. Strip detectors, on the other hand, provide N^2 pixels with only 2N electronic channels, one for each row and one for each column. This is important in space flight applications where channel count greatly affects the complexity of the instrument design. Double-sided CZT strip detectors, if carefully designed, can address many of the limitations associated with poor hole transport [10], but they will still be limited in terms of detector thickness (~3 mm) and, consequently, the effective energy range (<300 keV).

We have been developing a novel CZT detector concept: an *electron-only* device featuring *orthogonal coplanar anode strips* [11-14]. Each row takes the form of N discrete interconnected anode pixels, while each column is a single anode strip. Figure 1 illustrates the design of an 8 × 8 *pixel* orthogonal coplanar anode strip detector. The opposite side has a single uniform cathode electrode. The anode pixel contacts, interconnected in rows, are biased to collect the electron charge carriers. The orthogonal anode strips, surrounding the anode pixel contacts, are biased between the cathode and anode pixel potentials. The strips register signals from the motion of electrons as they migrate to the pixels. Since electrons are much more mobile than holes in CZT, signals from photon interactions at all depths in the detector are detected. Given the published results with pixel detectors [9], we expect our coplanar anode strip approach to be effective in CZT detectors at least 10 mm thick. This will permit thicker, more efficient, CZT imaging planes than are practical with double-sided strip detectors and will extend the effective energy range to >1 MeV. More compact packaging is also possible since all electrical connections for processing are on one side of the detector.

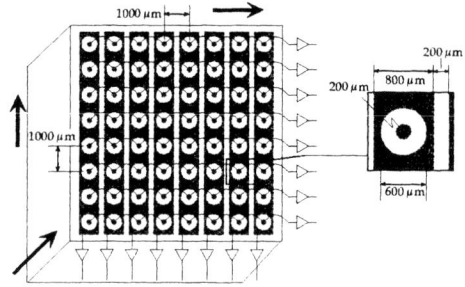

FIGURE 1. A schematic diagram showing the layout of the orthogonal coplanar anode design as used in our prototype detectors. Strip columns (X) are read out on the bottom. Pixel rows (Y) are read out on the right. The CZT thickness is 5 mm.

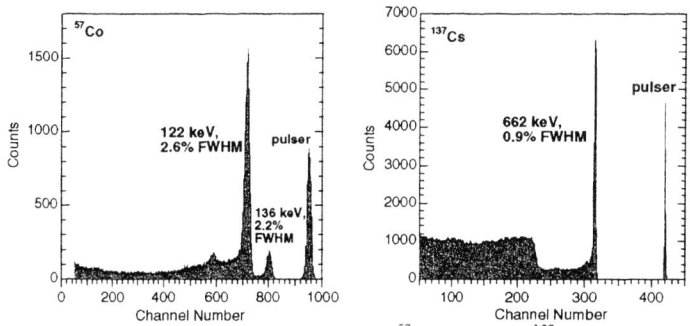

FIGURE 2. Single pixel spectra from ^{57}Co (left), and ^{137}Cs (right).

PROTOTYPE PERFORMANCE

We have fabricated several prototype detectors based on the orthogonal coplanar anode strip design. Each prototype module consists of a patterned CZT substrate that has been polymer flip-chip (PFC) bonded to a ceramic (LTCC) substrate or multi chip module (MCM) [15]. Reliable connections to all strips and pixel rows have been achieved with this fabrication approach. The spectroscopic performance of a single pixel element is demonstrated in Figure 2. The test pulse data shown for two of these spectra indicate that electronic noise presently limits the energy resolution and suggests that further improvements may still be possible.

Some level of charge sharing between adjacent strips/pixels is required in order to infer locations from relative signal measurements. Some charge sharing occurs in both directions, although the level of charge sharing in the X-direction is somewhat higher due to the larger surface area of the strips. This results in somewhat better spatial resolution in the X-direction. Alpha particle scans across the surface of the detector have yielded 1σ spatial resolutions of ~100 μm in X and ~300 μm in Y. The interaction depth (Z) can be inferred from the ratio of the amplitudes of the cathode and anode signals. We have demonstrated a 1σ FWHM position resolution of 650 μm in the Z-coordinate [13, 14].

While the cathode signal can be used for the depth of interaction measurement, the need to feed the cathode signals from the front surface of the detector to front-end electronics

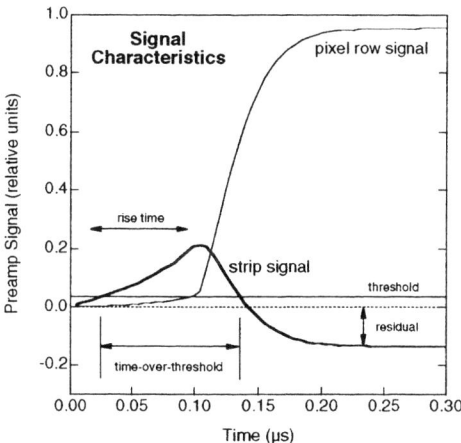

FIGURE 3. Signal characteristics associated with the pixel rows and the orthogonal strips. The strip signal parameters are related to the depth of interaction.

located behind the detector will interfere with the fabrication of closely packed arrays of detector modules. We have therefore been studying approaches that will permit the measurement of the interaction depth using only those signals that are available on the back (anode) side of the detector [16]. These signals are shown in Figure 3.

The pixel signals, rising in only the positive direction, are typical of small-pixel anodes in CZT detectors. The initial slope of the pixel signals is small but increases rapidly when electrons reach the anode region. These signals provide a measure of the energy deposit and identify the Y-coordinate of the photon interaction location. The strip signals identify the X-coordinate of the photon interaction location, but are not used for an energy measurement. The strip signals are bipolar in nature. They have faster initial rises than the pixel signals due to the larger strip areas. They reach a maximum shortly before electron transit time and decrease as the electrons approach the pixel. Of particular interest are three features of the anode strip signal (risetime, time-over-threshold, and residual; see Figure 3) that can be used to measure the depth of interaction, independent of any signal from the cathode. We have been working to design and implement a circuit that can directly measure the interaction depth, using the time-over-threshold of the bipolar strip signal [16].

FABRICATION CONCEPT FOR LARGE-AREA ARRAYS

The baseline approach for packaging a single detector module is illustrated in Figure 4. Here we envision a module with 16×16 logical pixels formed from 16 strips and 16 pixel rows, each on a 1 mm pitch. The CZT substrate in this case is roughly 16×16 mm^2.

The PFC bonding process establishes the mechanical and electrical interconnection between the CZT and LTCC substrates. The multi-layer LTCC substrate establishes the interconnection of the anode pixels in rows and the routing of the pixel row and strip signals (total of 32) plus the guard ring contact to flat gold contact pads on the underside of the module. The fact that 256 *pixels* are achieved with only 32 signal processing channels leaves ample space on the underside of the LTCC substrate for the necessary signal processing electronics

Figure 5 shows the concept of an image plane board supporting a 20×20 CZT module array having 1024 cm^2 active area. The image plane is a large-area, mechanically reinforced circuit board that supports an array of closely packed CZT detector modules.

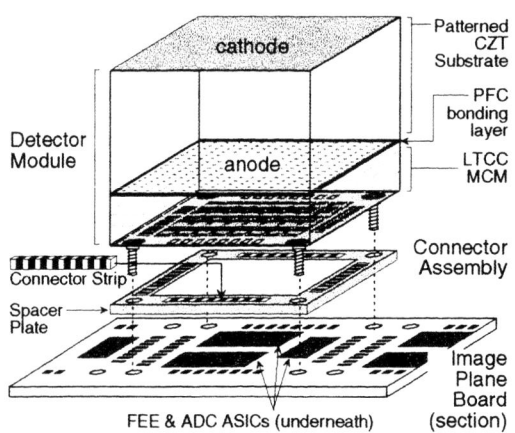

FIGURE 4. Concept for a detector module to be used in fabrication of large area arrays.

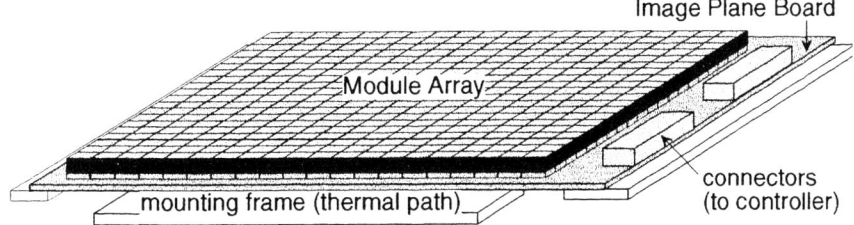

Figure 5. Image plane concept. A 20 × 20 array of imaging CZT modules.

The estimated packing fraction may be as high as 94%. The board supports detector bias and the front-end electronics. These electronics are on the underside of the board. The bonding and connector assembly technologies provide good thermal isolation between the image plane board and the CZT. The cathode bias will be provided using a thin flexible circuit located between rows of modules (not shown).

Assuming 2 mW/channel for the pixel and strip FEE ASICs, the total FEE power dissipation for the array shown in Figure 4 would be 26 W. A pixel detector array of the same size would require 205 W. In our design, the heat dissipated on the image plane-board is conducted to the experiment structure via its mounting frame. The electrical interfaces between the image plane and the image-plane controller are digital, thereby permitting flexibility in the location of the controller electronics.

ACKNOWLEDGMENTS

This work is supported at UNH by grants from NASA's High Energy Astrophysics SR&T program and at UM by the Natural Sciences and Engineering Research Council (NSERC) of Canada.

REFERENCES

1. Barthelmy, S. D., *Proc. SPIE* **4140**, 50-63 (2000).
2. Lebrun, F., et al., *Proc. SPIE* **2806**, 258-268 (1996).
3. Matteson, J. L., et al., *Proc. SPIE* **3445**, 445-457 (1998).
4. Tumer, T. O., et al., *IEEE Trans. Nucl. Sci.* **44 (3)**, 572-576 (1997).
5. Grindlay, J., et al., in *Fifth Compton Symposium*, edited by M.L. McConnell & J.M. Ryan, AIP Conference Proceedings 510, New York, 2000, pp. 784-788.
6. McConnell, M. L., et al., *Proc. SPIE* **2806**, 349-360 (1996).
7. Cherry, M. L., et al., *Proc. SPIE* **3765**, 539-550 (1999).
8. Bhattacharya, D., et al., Proc. 26th Internat. Cosmic Ray Conf., Salt Lake City, **5**, 72-75 (1999).
9. He, Z., et al., *NIM* **A422**, 173-178 (1999).
10. Matteson, J. L., et al., *Proc. SPIE* **3446**, 192-201 (1998).
11. Hamel L.-A., et al., Proceedings of the 1997 Fall Meeting of the Material Research Society, Boston, December (1997).
12. Mayer, M., et al., *NIM* **A422**, 190-194 (1999).
13. Tousignant, O., et al., *Proc. SPIE* **3768**, 38-48 (1999).
14. McConnell, M. L., et al., *Proc. SPIE* **4141**, 157-167 (2000)
15. Jordanov, V. T., et al., *NIM* **A458**, 511-517 (2001).
16. Larson, K., et al., *Proc. SPIE* **4141**, 336-341 (2000).

Ground-Based High-Energy Astronomy

Gamma Ray Astronomy with Air Shower Arrays

A.I. Mincer[1]

New York University, New York, NY 10003, USA

Abstract.
Detectors such as Milagro and ARGO-YBJ reconstruct the energy and arrival direction of very high energy (VHE) photons incident on the earth's atmosphere from measurements of the secondary particles in the air shower which the primary photon generates. These detectors operate continuously and with a wide field of view, allowing them to study all VHE gamma ray sources that pass overhead on a daily basis. This makes them ideal for monitoring time varying sources such as active galactic nuclei and for searching for transient sources such as gamma ray bursts. The Milagro detector is currently operating in the Jemez Mountains near Los Alamos, New Mexico and ARGO is under construction in Yangbajing,Tibet. The current status and future plans of both Milagro and ARGO are discussed and recent results from Milagro are presented.

INTRODUCTION

Astrophysical gamma rays have been directly detected to energies up to tens of GeV. Since the gamma ray flux typically falls as the inverse of energy squared or cubed, satellite and balloon experiments to date have had too small an area and time exposure to be sensitive above these energies. When a primary energetic photon enters the earth's atmosphere, it interacts producing secondary particles, which in turn interact and produce more particles. The result of this process is an atmospheric "air shower" of particles. At energies of about 100 Gev to 10 TeV, the shower particles, mainly electrons, positrons, and photons, increase in number down to a depth of a few hundred g/cm^2 after which they decrease exponentially. By the time this shower reaches sea level, at a depth of about 30 radiation lengths, it contains only a small remnant of shower particles which, depending on the primary energy, may be spread over an area of about a square kilometer. Ground level detectors which are sensitive to these products of the shower are used to indirectly measure properties of primaries whose flux is too small for direct detection.

Ground level air shower characteristics vary widely from event to event for primaries with identical properties. On an event by event basis, it usually not even possible to tell whether the shower was generated by a primary photon or a much more abundant cosmic ray nucleus. Cosmic rays, whose arrival is isotropic due to interstellar magnetic fields, are therefore the major source of background in such measurements. The success of ground based experiments thus depends strongly on the ability to find characteristics which differentiate between these two types of primaries.

Air cerenkov telescopes operate on cloudless moonless nights and view the cerenkov

---

[1] For the Milagro Collaboration

light created throughout the atmosphere by relativistic shower particles. The cerenkov image is thus a projection of the full shower which, as first shown by the Whipple experiment in 1989 [1], by virtue of the clumpiness of hadron showers compared with photon showers, can be used to reject a large fraction of cosmic ray generated showers.

In order to continuously monitor the full overhead sky, to search for transient sources such as gamma ray bursts or to study the time variation of "steady" sources, air shower arrays are used. In these methods only the air shower particles which survive to detection altitude are measured by detectors which are shielded from low energy photons. These detectors can thus monitor the full overhead sky at all times. Since they only sample a small surviving part of the shower, it has been more difficult in these methods to discriminate between primary types.

Because it is expensive to cover large areas with particle detectors, such experiments usually sparsely cover with counters a small fraction of a large area. Since the number of surviving particles goes down with decreasing primary energy, these experiments typically have been sensitive to photons with energy above 50 to 100 TeV. A few experiments claimed detection of sources at these energies during the 1980s, but these have not been confirmed. Possible time variation of sources makes definite interpretation of these results difficult.

A new generation of experiments uses a combination of high altitude and large area coverage to lower the threshold of the method enough to overlap with direct detection measurements. The principles of such VHE photon detectors are discussed in the following section. The Milagro experiment, which is currently taking data, and the ARGO-YBJ experiment, which will begin a run with one segment later this year, will be discussed in this paper. Milagro has already demonstrated the power of this technique. Both experiments are developing gamma hadron separation methods which will, if successful, make this technique even more powerful. The rest of this paper discusses the status of both these experiments, discusses their future plans, and presents some recent results from the Milagro experiment.

THE AIR SHOWER ARRAY METHOD

Individual counters in an air shower array will be triggered by the large flux of low energy muons resulting from the abundant GeV cosmic rays. In order to trigger on air showers, a minimum number of counters are therefore required to detect particles within a short time window, usually up to a few hundred nanoseconds. A large central core of particles typically follows a trajectory which is a continuation of that of the primary particle. Extending laterally from the core is an almost planar cone of particles of thickness on the order of 10 ns near the core but with time spread increasing with distance from the core.

The measured counter hit times are fit to a cone to determine the incidence direction of the primary particle. Measurement of the core location, as determined from the lateral signal size distribution, allows fitting the conical shape and improves angular resolution. The lateral distribution of the deposited energy also gives a measure of the primary particle energy. Knowledge of the core location is vital for this measurement, since without this it is difficult to tell the difference between the distribution of a low energy shower near the core and a high energy one far from it.

Since the isotropic TeV cosmic ray background is large, there are several basic methods used to search for a photon signal from a source:

- Angular resolution: Since the background is isotropic, the better the detector angular resolution, the smaller the region over which one looks for signal, and therefore the smaller the average background contribution.

- Gamma - hadron separation: Hadron induced air showers tend to be clumpier than those due to photons. They are also more likely to include hadrons and muons. Although shower fluctuations make it impossible to identify the primary on an event by event basis, one can find parameters on which to select in order to decrease the number of hadrons in any sample relative to the number of photons.

  For a given bin size with a mean number of expected background events B, the significance (in number of standard deviations) of M measured events is approximately given by $(M - B)/\sqrt{B}$. (A more precise formulation takes into account the the fact that the measurement, regardless of its source, has in this case fluctuated to the value M [2].) The gain from imposing some cut can then be quantified by the quality factor Q defined as $Q = [(M_C - B_C)/\sqrt{B_C}]/[(M - B)/\sqrt{B}]$ where the subscript C identifies the same parameters with the selection cuts imposed.

  For purpose of comparison with numbers presented below, air cerenkov experiments allow elimination of over 99% of hadrons while keeping over 50% of photons for a quality factor of about 9.

- Triggring with other experiments: A detector which covers a large region of the sky with a large number of small angular resolution bins will suffer from statistical upward background fluctuations in random bins. If another experiment is used to fix a location or time bin, the probability of such fluctuations are reduced by the number of bins that would have had to be searched without this information. For example, for gamma ray bursts we refer to the triggered versus untriggered searches, where the sensitivity will be better in the former case.

Because only a small remnant of the shower is detected, these experiments rely heavily on Monte Carlo computer simulations of shower development and detector response. Although accelerator measurements for proton - proton interactions at these energies exist, they do not include the forward going particles, which go unmeasured down the accelerator beam-pipe, but which carry the energy forward and determine the major properties of the air shower. In addition, nucleus - nucleus interactions at these energies still remain largely unmeasured. In spite of these problems, the uncertainties due to particle interaction models are often small compared with the uncertainties in the details of detector properties. It is therefore especially important here to compare data and simulation predictions of event properties.

THE MILAGRO EXPERIMENT

The main part of the Milagro experiment is an instrumented, water-filled, man-made 60m x 80 m x 8m pond located at an altitude of 2630 m in the Jemez Mountains near Los Alamos, New Mexico. The pond is instrumented with two layers of photomultiplier tubes (PMTs): 450 PMTs on a 25 by 18 grid of 2.8 meter squares at a depth of 1.4 meters below the water surface, and 273 PMTs on a 19 by 12 grid at a depth of 6 meters. When an air shower hits the detector, relativistic charged particles emit cerenkov radiation along a 41 degree cone. Because of the counter depth and cerenkov angle, a single PMT is sensitive to charged particles anywhere in the local grid square. Photons, which are typically about 5 times as numerous as charged particles in these showers, either pair produce or Compton scatter in the 1.4 m of water, transferring energy to detectable charged particles. The net result is that light from about half of the particles entering the the pond is detected, in contrast to a fraction of a percent for standard air shower arrays.

Because a large fraction of showers which trigger the experiment have cores outside the detector, energy resolution of the existing experiment is poor. This will improve with the addition of counters outside the pond currently in progress and described below. In absence of such knowledge, one simulates detector rates for assumed primary spectra and thereby find bounds on possible primary energy distributions.

A Results from Milagrito

The Milagro prototype, Milagrito, consisted of 228 PMTs whose depths ranged during the run from 1.0 to 2.0 meters and took data from February 1997 to May 1998. This allowed study of the water cerenkov method, comparison of computer calculations with measured data, and was also an experiment in its own right.

Lessons learned from the experimental method include:

- The optimum depth at which to set the counters to maximize time resolution. This was done by comparing Δ_{EO} for data recorded at different water depths.
- The necessity of blocking almost horizontal moving photons. Such cerenkov photons can induce late signals in far away PMTs which have not been previously hit by other shower particles. These were eliminated from Milagro by putting baffles around the PMTs.
- Comparisons of the monte carlo with data showed cosmic trigger rates [3] and zenith angle distribution [4] could be accurately calculated. The angular resolution as seen by Δ_{EO}, the difference between fit arrival angle calculated using two independent sets of PMTs interleaved like the light and dark squares of a checkerboard, is also well simulated [4]. The resulting point spread function is close to that determined from the cosmic ray shadow the moon casts [5].

The Milagrito measurements included:

- Study of the shadowing of cosmic rays by the moon [5] and the sun: This measurement allows study of the detector resolution, study of the earth's magnetic field, study of the solar magnetic field, and a search for anti-particles in cosmic rays.

- Study of solar energetic particle emission events [6]: This showed the sensitivity of the detector's individual PMT rates to changes in the GeV cosmic ray flux and its resulting usefulness as a large area neutron monitor type detector.
- Detection of Markarian 501 [3] which showed the usefulness of this method for studying time varying sources.
- The possible detection of a TeV gamma ray burst seen by BATSE [7]: This showed the power of this method which allows uninterrupted study of the full overhead sky.

Some additional studies, including an untriggered gamma ray burst search and study of other sources are in progress. Still other analyses, although possible with Milagrito data, will probably not be performed because data from the full detector is now available.

STATUS OF MILAGRO

The full Milagro pond detector began an engineering run in July of 1999 and commenced regular data taking in December of 1999. The trigger, a hardware sum requiring at least 50 PMTs above 1/4 equivalent photoelectrons in a 150 ns time window, fires at a rate of ~ 1.5 KHz. Except for a down period during the Los Alamos fire, it has run with a duty cycle better than 95%.

A fit to the shower cone is performed off line using PMTs with pulse height greater than 2 equivalent photoelctrons. PMTS whose timing is far from the fit are removed in an iterative procedure. The number of PMTS surviving in the fit, N_{FIT} is typically required to be greater than 20. The median energy of cosmic ray triggers is 1.8 TeV and the angular resolution for showers whose core is inside the pond is $0.75°$.

Photon hadron separation currently uses the clumpiness of hadronic showers versus that of purely electromagnetic ones as observed in the bottom layer PMTs. The parameter X_2 is defined as the ratio of the number of bottom layer PMTs with a pulse height greater than 2 photoelectrons ($NHIT_{BOT}$) divided by the maximum pulse height in equivalent photoelelectrons of a bottom layer PMT ($PEMAX_{BOT}$). A large value of X_2 indicates a less spiked distribution and cutting on $X_2 > 2.5$ provides a quality factor of 1.8 as determined from the monte carlo. The simulation currently does not completely match the data (Figure 1) and the cut appears to be more effective than predicted, as can be seen in the Crab data presented below.

The fraction of the sky to which Milagro is sensitive can be seen in Figure 2, which presents the exposure for 60 days of data.

The efficacy of the X_2 cut can be seen in the preliminary Milagro Crab signal (Figure 3) based on 1.35 years of exposure, which has an excess of 1.0 σ above background (8,749,562 events with an expected background of 8,746,621) for no gamma hadron cuts, increasing to 4.8σ (787,503 events with an expected background of 783,059) when the $X_2 > 2.5$ cut is applied. This would correspond to Q=4.8 if statistical fluctuations could be ignored. Milagro accumulates approximately 10 Crab photons per day.

The recent outburst of Markarian 421 is still being analyzed by Milagro. The preliminary analysis of data from 17 January 2001 to 26 April 2001 shows a signal of 5.2σ significance (See Figure 4).

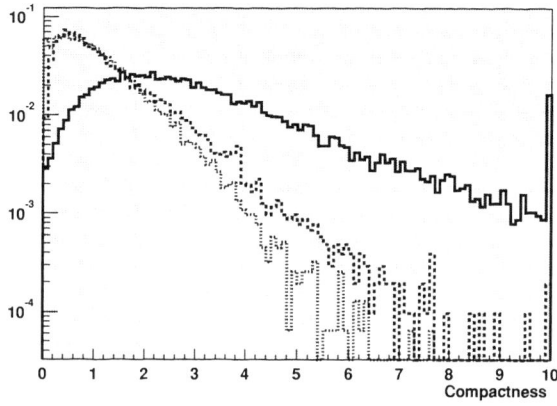

FIGURE 1. Data X_2 distribution (small dots, botom curve) and simulated distribution for photons (solid) and hadrons (large dots)

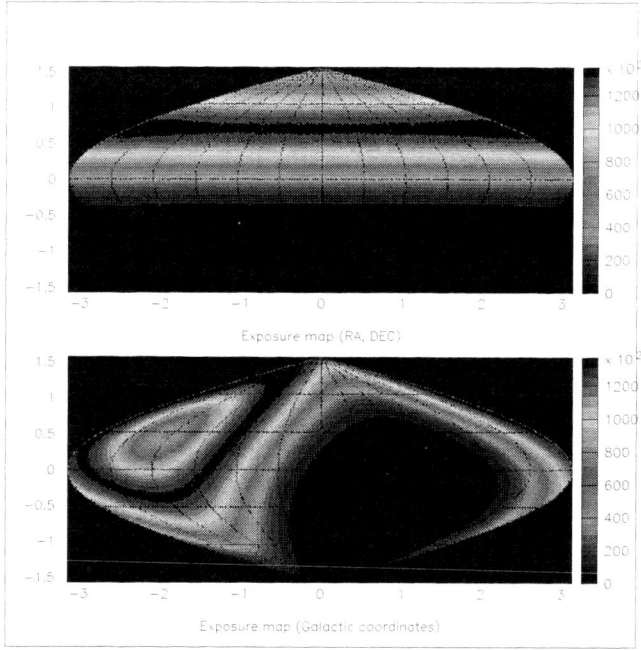

FIGURE 2. Milagro exposure for 60 days of data in bins of 0.5 degrees in declination by 1.0 degrees in -(right ascension) times cos(declination) (top), and in similar sized bins in galactic coordinates (bottom).

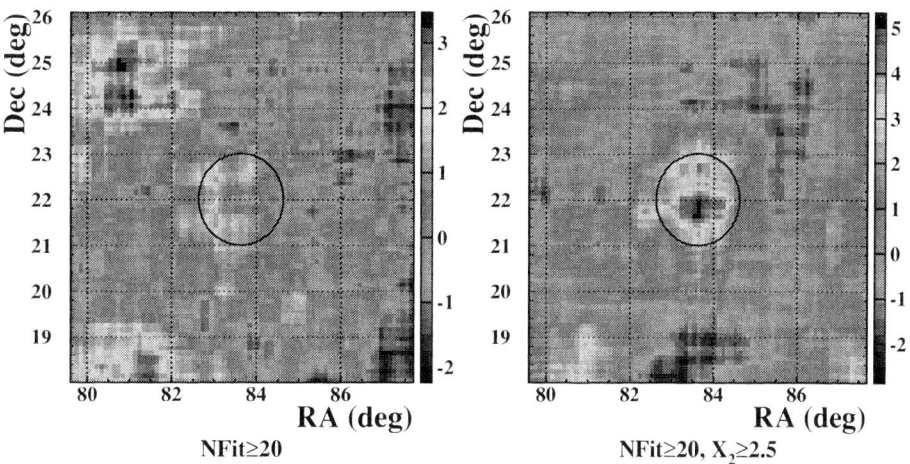

FIGURE 3. Preliminary Milagro background subtracted sky map in the region of the Crab for events with $N_{FIT} > 20$ and no X_2 cut (left), and with $X_2 > 2.5$ (right).

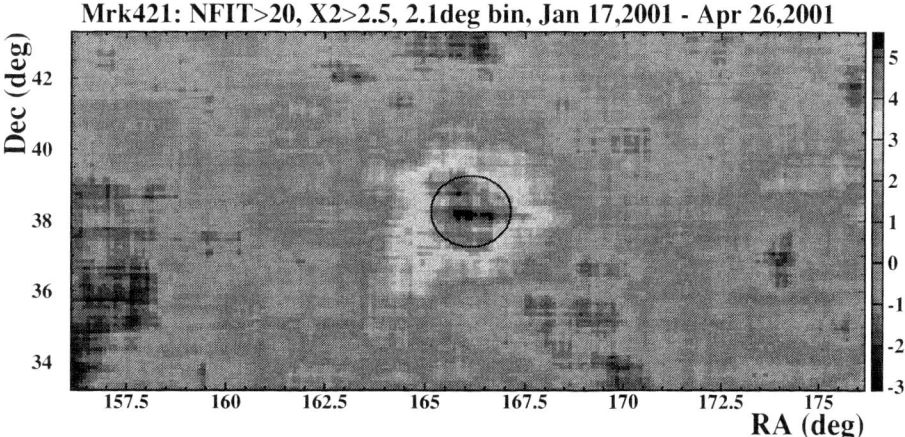

FIGURE 4. Preliminary Milagro background subtracted sky map in the region of Markarian 421.

As Milagro continuously monitors the full overhead sky, studies of the Crab and other galactic sources, studies of AGNs and search for gamma ray bursts are all in progress, as is a search for VHE emission from the galactic plane. Milagro sensitivity to gamma ray bursts is similar in fluence to that of BATSE (Figure 5). Note that this figure does not include any gammma hadron separation or the improvement with the planned outriggers, discussed below. Other cosmic ray studies currently in progress include those of the sun and moon shadows, of cosmic rays by measuring surviving unaccompanied hadrons,

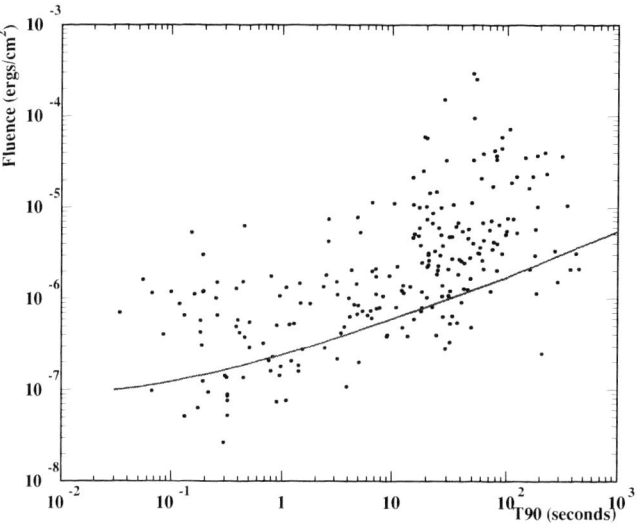

FIGURE 5. Milagro sensitivity to gamma ray bursts compared with BATSE detected events.

and of solar energetic particle events. Taking advantage of the fact that, unlike optical detectors, Milagro can monitor the sun and its vicinity, a search is also underway for VHE photons from annihilating relic neutralinos which could be trapped by the solar gravitational field.

FUTURE MILAGRO PLANS

The Milagro energy threshold is currently determined by the muon background rate. If the N_{HIT} cut is lowered, the trigger rate is overwhelmed by non-vertical muons whose cerenkov light can fire many PMTs. The Milagro angular resolution, about 0.75 degrees for on-pond showers which allow the cone shape and sampling corrections to the fit, is appreciably worse (Figure 6) for showers whose core is off the pond. Milagro can tell little about primary energy from the shower it measures without knowledge of the core position. For all of these reasons, a array of 176 counters is being deployed in an area about 10 times that of Milagro centered on the Milagro pond. The construction began in the summer of 2000 and is on-going. With these counters in place, angular resolution is expected be better than 0.9 degrees for showers as far as 200 meters from the pond center, energy resolution is expected to be about 50% for showers of energy above ~ 2 TeV, and the air shower size threshold can be lowered significantly.

The gamma hadron separation capabilities of Milagro are just beginning to be understood and exploited. In addition to X_2, the two dimensional $NHIT_{BOT}$ and $PEMAX_{BOT}$ space is being studied. Additional parameters which can be used include the shower signal rise time and the lateral signal distribution for both top and bottom PMT layers. Early studies already show the possibility of significant improvement in quality factor.

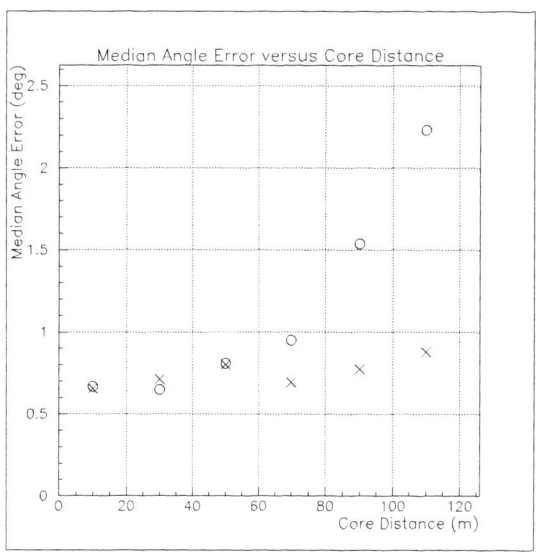

FIGURE 6. Milagro angular resolution as a function of core distance from the pond center with (X's) and without (circles) outriggers.

The future of Milagro is thus a bright one.

ARGO-YBJ

The ARGO-YBJ experiment [8], a Chinese - Italian collaboration, uses resistive plate chambers (RPCs) to provide a 6700 m^2 active area at an altitude of 4300 m in Yangbajing, Tibet. This allows detection of primaries with energies as low as 20 GeV and brings the sensitivity down to about 1/10 the Crab flux from 100 GeV to 20 TeV. An overall trigger rate of 20 KHz (with a 15 pad trigger) and an angular resolution of about 0.4 degrees for 100 pad multiplicity events is expected.

Each RPC is constructed of 10 pads of size 0.60 m by 0.56 m. The plan is to cover 92% of a 78m by 74 m central carpet, and an additional 20 % of the remaining area inside a 111m by 99 m rectangle. Photons will be converted by a layer of 0.5 cm lead covering the RPCs.

Gamma hadron separation using the the radial distribution, expected to be steeper for photons, local lateral fluctuations in the signal, expected to be larger for protons, and a neural network analysis [9] is expected to give a quality factor of about 1.8 retaining 80% of photons.

From February to May of 1998 a 51 m^2 area with 91% active area was tested with the following results [10], all consistent with expectations for the final detector listed above:

- 1.3 ns time resolution
- Δ_{EO} about 2 degrees for 100 pad multiplicities.

- Resolution for 35 pad multiplicity events goes down from 8 degrees to 5 degrees with the addition of the lead layer.

Construction of an 800 m^2 area began in October of 2000 and is expected to begin taking data sometime in 2001. The full central carpet is expected to be completed by the end of 2003 and the outer ring by 2004 [11].

CONCLUSIONS

The current generation of VHE gamma ray air shower detectors is coming into its own. Milagro has shown that the method works and the advantages of a method which can continuously monitor the full overhead sky. ARGO will provide significant overlap with existing satellite detector energies. A qualitative change in the power of air cerenkov experiments occurred when a powerful gamma-hadron separation technique was introduced. Milagro is already using some separation techniques, but improvements are possible in both experiments and should lead to major improvements.

ACKNOWLEDGMENTS

The Milagro experiment is funded by the National Science foundation (Grant Numbers PHY-9722617, -9901496, -0070927, -0070933, -0070968, -0096256), The US Department of Energy Office of High Energy Physics, The US Department of Energy Office of Nuclear Physics, The LDRD Program at Los Alamos National Laboratory, Los Alamos National Laboratory, The University of California, The Institute of Geophysics and Planetary Physics, The Research Corporation, and The California Space Institute.

REFERENCES

1. Weekes, T.C., *Astrophys. J.*, **342**, 379-395 (1989).
2. Li, T,. and Ma, Y., *Astrophys. J.*, **272**, 317-324 (1983).
3. Atkins, R. et. al., *Astrophys. J.*, **525**, L25-L28 (1999).
4. Atkins, R. et. al., *Nucl. Instr. Meth.*, **A449**, 478-499 (1999).
5. Wascko, M.O., *Study of the Shadow of the Moon in Very High Energy Cosmic Rays with the Milagrito Water Cerenkov Detector*, Ph.D. thesis, The University of California at Riverside, CA 92521 (2001).
6. Falcone, A. and Ryan, *J. Astropart. Phys.*, **11**, 283-285 (1999).
7. Atkins, R. et. al., *Astrophys. J.*, **533**, L119-L122 (2000).
8. Aloisio, A. et. al., *Proc. of the "Chacaltaya Meeting on Cosmic Ray Physics", La Paz (Bolivia)*, July 23-27 (2000), to be published in Nucl. Instrum. Meth.
9. Bussino, S. and Mari, S.M., *Astropart. Phys.*, **15**, 65-77 (2001).
10. Bacci, C. et. al., *Nucl. Instrum. Meth.*, **A456**, 121-125 (2000).
11. Vallania, P., *Private Communication*.

CANGAROO-II and CANGAROO-III

Masaki Mori* and the CANGAROO collaboration[†]

*Institute for Cosmic Ray Research, University of Tokyo, Kashiwa, Chiba 277-8582, Japan
E-mail: morim@icrr.u-tokyo.ac.jp
[†]Institute for Cosmic Ray Research, Univ. of Tokyo; Univ. of Adelaide, Ibaraki Univ.; Institute of Space and Astronautical Science; National Astronomical Observatory of Japan; Tokai Univ.; Tokyo Institute of Technology; Kyoto Univ.; STE Laboratory, Nagoya Univ.; Yamagata Univ.; Yamanashi Gakuin Univ.; Osaka City Univ.; Konan Univ.; Ibaraki Prefectural Univ. Health Sciences; Australian National Univ.; Shinshu Univ.

Abstract. Preliminary results from CANGAROO-II, a 10 m imaging Cherenkov telescope, in Woomera, South Australia are presented. They include the confirmation of detections of TeV gamma-ray sources we have reported using a 3.8 m telescope, CANGAROO-I. Also the status of the construction of an array of four 10 m telescopes, called CANGAROO-III, is reported. The first telescope of the array was upgraded from the 7 m telescope and the second one is being constructed in this year. The full array will be operating in 2004.

INTRODUCTION

CANGAROO is an acronym for the Collaboration of Australia and Nippon (Japan) for a GAmma Ray Observatory in the Outback. After successful operation of the 3.8m imaging Cherenkov telescope (CANGAROO-I) [1] for 7 years, which was the first of this kind in the southern hemisphere, we constructed a new telescope of 7m diameter (CANGAROO-II) in 1999 [2] next to the 3.8m telescope in Woomera, South Australia (136°47'E, 31°06'E, 160m a.s.l.). Then the construction of an array of four 10m telescope (CANGAROO-III) was approved and as a first step the 7m telescope was upgraded to 10m diameter, which is the first telescope of the CANGAROO-III array [3, 4, 5, 6, 7]. The major parameters of the CANGAROO telescopes are summarized in Table 1.

TABLE 1. Properties of the CANGAROO telescopes. ([†]: CFRP mirrors of 80cm in diameter)

	3.8m telescope	7m telescope	10m telescope
Focal length	3.8m	8m	8m
Number of mirrors	7 ($11m^2$)	60 ($30m^2$) [†]	114 ($57m^2$) [†]
Number of PMTs	256 (3/8")	512 (1/2")	552 (1/2")
Readout	TDC & ADC	TDC	TDC & ADC
Point image size (FWHM)	0.1°	0.15°	0.20°
Operation	1992–1998	May 1999–Feb 2000	Mar 2000–

FIGURE 1. The 10m imaging Cherenkov telescope in Woomera, South Australia. The huts contain the electronics and the telescope power.

PRELIMINARY RESULTS FROM CANGAROO-II

Initial performance of the 7m telescope is reported in ref.[8]. Observations with the 7m and 10m telescope were carried out in 1999 and 2000, respectively. The target objects were selected from our list of TeV gamma-ray sources: Crab, PSR 1706-44, Vela, SN1006, RXJ 1713.7-3946, in order to confirm our previous detections with the 3.8m telescope. Also nearby X-ray selected BL Lacs were observed: PKS 2005-489, PKS 2155-304 and PKS 0548-322 along with multiwavelength campaigns. Here we give brief description of some preliminary results obtained so far.

Crab. As the standard candle in the TeV gamma-ray astronomy, we observed the Crab repeatedly, although it is visible only at large zenith angles ($53° \sim 56°$) and thus at higher threshold energies (~ 6 TeV for 1999 observations) compared with northern Cherenkov telescopes. The flux obtained from observation by the 7m telescope is consitent with our previous report by the 3.8m telescope [9] (Fig. 2).

RXJ 1713.7−3946 (G347.3−0.5). This is a supernova remnant detected with the 3.8m telescope [10]. Figure 3 shows the alpha distribution for the data taken in 2000 with the 10m telescope after the standard imaging analysis, showing we have confirmed the detection. The threshold energy is ~ 400 GeV. The peak near alpha of zero is broader than that for point sources and may indicate the emission is extended. The details will be given elsewhere [11].

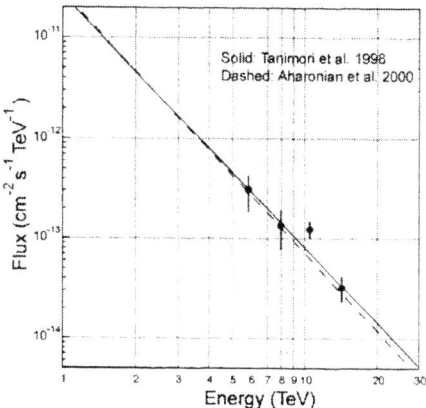

FIGURE 2. The flux of the Crab obtained from observation with the 7m telescope based on 43 hours of on- and 40 hours of off-source data. (Preliminary)

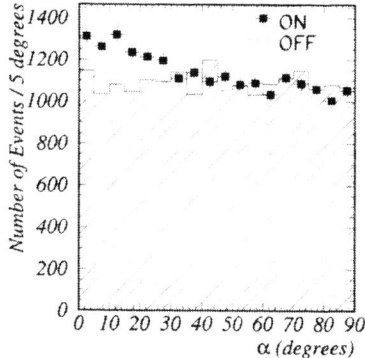

FIGURE 3. The alpha plot of RXJ1713.7−3946 data in 2000. Plots are for on-source data and histograms are for off-source data (11 hours each). The excess events near alpha of zero indicate a gamma-ray signal. (Preliminary)

PKS 2005−489 and PKS 2155−304. Although we spent a lot of time observing these southern nearby BL Lac objects, we could only set upper limits for these sources. From observations in 2000 Jul/Aug/Sep, they are 6.4×10^{-12} cm^{-2}s^{-1} above 450 GeV and 1.2×10^{-11} cm^{-2}s^{-1} above 400 GeV, respectively (2σ level, preliminary). The details will be given elsewhere [12].

Mrk 421. Following the alert of TeV flaring activities by the HEGRA group, we started observation of Mrk 421 at large zenith angles (69 ∼ 72°) in Feb/Mar 2001. Although the observation time is limited, preliminary analysis shows a gamma-ray signal at 5σ level above 9 TeV. A detailed analysis is continuing [13].

Other sources. Results on SN1006 and PSR 1706-44 will be presented elsewhere [14, 15].

STATUS OF CANGAROO-III

We will start the construction of the second 10m telescope at the end of 2001. The full array of four telescopes, set at the corners of a diamond with sides of about 100m, will be operational in 2004. The performance as a system of telescopes will be described elsewhere [7]. With experience from the construction and operation of the CANGAROO-II telescope, we are making following efforts to improve the sensitivity of the telescopes [5].

Reflector. The reflector design is the same as the first 10m telescope. The mirrors, made of CFRP [16], are light and have proven to be durable, but they are under further improvement, especially to obtain better optical quality by refining the production process. The mirror attitude adjustment system has been redesigned to match our needs and save cost.

Telescope control. Each alt-azimuth telescope is controlled by a PC running Linux with a realtime extention (KURT). A master PC issue directives to each control PC via network and tracking modes can be flexibly changed. Clocks are synchronized by NTP software to a GPS receiver.

Camera. The new design of an imaging camera at the prime focus is hexagonal shape to minimize the dead space between PMTs. The total field-of-view is about 4 degrees covered with 427 PMTs of 3/4" diameter [7]. The light guides have been redesigned to maximize photon collection for the new hexagonal arrangement. High volages are supplied to PMTs individually. Each PMT base is included a preamplifier and signals are transmitted via twisted cables to the electronics which will be installed at the verandah of the telescope.

Electronics. The new electronics are all based on the VME specification. The frontend module amplifies signal and feeds to an ADC, discriminates it and feeds to a TDC, an internal scaler and a trigger circuit. The ADC is an improved version of the module used in the first 10m telescope and includes an internal delay of 150ns which eliminates a long external delay cable. A pattern trigger circuit using a Programmable Logic Device is under developement to decrease accidental triggers due to night-sky background photons. Details will be given elsewhere [17].

Monitor. Cloud monitors detect infrared radiation from clouds making use of a thermopile module and supply useful information on data quality [18]. Weather monitors can record temperature, humidity and wind speed. These data are read out via serial line connection and stored for offline analysis.

SUMMARY

The CANGAROO-II 7/10m telescope has been in operation since 1999 and we have begun to produce preliminary results which confirm detections made with the CANGAROO-I 3.8m telescope. This is the first telescope of an array of four telescopes, called CANGAROO-III, which will be in operation in 2004. The final goal will be an energy threshold of 100 GeV and an angular resolution of less than 0.1 degee.

ACKNOWLEDGMENTS

We thank Communication Systems Center, Mitsubishi Electric Corporation, and DSC Woomera for their assistance in constructing the telescopes. This project is supported by a Grant-in-Aid for Scientific Research of Ministry of Education, Culture, Science, Sports and Technology of Japan, and the Australian Research Council.

REFERENCES

1. Hara, T. et al., *Nucl. Inst. Meth. Phys. Res.*, **A332**, 300 (1993).
2. Tanimori, T. et al., "Construction of New 7m Imaging Air Čerenkov Telescope of CANGAROO", in *Proc. 26th ICRC*, edited by Kieda, D. et al., 5, 1999, p. 203.
3. Mori, M. et al., "The CANGAROO-III Project", in *GeV-TeV Gamma Ray Astrophysics Workshop*, edited by Dingus, B. L. et al., AIP Conference Proceedings 515, American Institute of Physics, New York, 2000, p. 485.
4. Mori, M. et al., "Status of the CANGAROO-III Project", in *High Energy Gamma-ray Astronomy*, AIP Conference Proceedings 558, American Institute of Physics, New York, 2001, p. 578.
5. Mori, M. et al., "The CANGAROO-III Project: Status Report", in *Proc. 27th ICRC (submitted)*, 2001.
6. Tanimori, T. et al., "Recent Status of CANGAROO-III Project", in *High Energy Phenomena in Universe*, edited by T. T. Vanh, Rencontre de Moriond, 2001.
7. Enomoto, R. et al., *Astropart. Phys. (in press)* (2001).
8. Kubo, H. et al., "Initial Performance of CANGAROO-II 7m Telescope", in *GeV-TeV Gamma Ray Astrophysics Workshop*, edited by Dingus, B. L. et al., AIP Conference Proceedings 515, American Institute of Physics, New York, 2000, p. 313.
9. Tanimori, T. et al., *Astrophys. J.*, **492**, L33 (1998).
10. Muraishi, H. et al., *Astron. Astrophys.*, **354**, L21 (2000).
11. Enomoto, R. et al., "Likelihood Analysis of sub-TeV Gamma-rays from RXJ1713-39 with CANGAROO-II", in *Proc. 27th ICRC (submitted)*, 2001.
12. Nishijima, K. et al., "Very High Energy Gamma-Ray Observation of Southern AGNs with CANGAROO-II", in *Proc. 27th ICRC (submitted)*, 2001.
13. Okumura, K. et al., "Search for Gamma-ray Above 10 TeV from Markarian 421 in High State with CANGAROO-II Telescope", in *Proc. 27th ICRC (submitted)*, 2001.
14. Hara, S. et al., "Observation of TeV Gamma rays from NE-rim of SN1006 with CANGAROO-II 10m Telescope", in *Proc. 27th ICRC (submitted)*, 2001.
15. Kushida, J. et al., "Observation of PSR1706-44 with CANGAROO-II Telescope", in *Proc. 27th ICRC (submitted)*, 2001.
16. Kawachi, A. et al., *Astropart. Phys.*, **14**, 261 (2001).
17. Kubo, H. et al., "Development of Data Aquisition System of CANGAROO-III Telescope", in *Proc. 27th ICRC (submitted)*, 2001.
18. Clay, R. W. et al., *Publ. Astron. Soc. Aust.*, **15**, 332 (1998).

The Current Status and Future Plans of the STACEE Observatory

R. Mukherjee[1,2], L. M. Boone[3], D. Bramel[2], E. Chae[4], C. E. Covault[5], P. Fortin[6], D. M. Gingrich[7,8], J. A. Hinton[4], D. S. Hanna[6], C. Mueller[6], R. A. Ong[9], K. Ragan[6], R. A. Scalzo[4], D. R. Schuette[9], C. G. Theoret[6], and D. A. Williams[3]

1 *Barnard College, Columbia University, New York, NY 10027*
2 *Department of Physics, Columbia University, New York, NY 10027*
3 *Santa Cruz Institute for Particle Physics, University of California, Santa Cruz, CA 95064*
4 *Enrico Fermi Institute, University of Chicago, Chicago, IL 60637*
5 *Department of Physics, Case Western Reserve University, Cleveland, OH 44106*
6 *Department of Physics, McGill University, Montreal, Quebec H3A 2T8, Canada*
7 *Centre for Subatomic Research, University of Alberta, Edmonton, Alberta T6G 2N5, Canada*
8 *TRIUMF, Vancouver, British Columbia V6T 2A3, Canada*
9 *Deptartment of Physics & Astronomy, University of California, Los Angeles, CA 90095*

Abstract.
The Solar Tower Atmospheric Cherenkov Effect Experiment (STACEE) represents a new type of atmospheric Cherenkov detector that achieves a low energy threshold for γ-ray detection by using heliostat mirrors in a pre-existing solar research facility. STACEE is designed to study astrophysical sources of γ-rays in the energy range of 50 to 500 GeV. A prototype of the experiment using 32 heliostats (STACEE-32) has previously detected the Crab nebula at high significance, demonstrating the viability of the technique. The completed version of STACEE will use 64 heliostats, and will have a total collection area of ~ 2300 m^2. Astrophysics in the 10 to 300 GeV regime has proved to be elusive to both ground-based and satellite experiments and STACEE has the potential of filling an important niche in high energy astrophysics. Here we describe the current status and future goals of STACEE.

I INTRODUCTION

STACEE is an experiment that studies sources of 50-500 GeV γ-rays by detecting the Cherenkov radiation produced in extensive air showers. STACEE is located at the National Solar Thermal Test Facility (NSTTF) at Sandia National Laboratories in Albuquerque, New Mexico, USA. STACEE is designed to use 64 of the heliostats at the NSTTF to focus the Cherenkov light onto an array of photomultiplier tubes (PMTs) on a central tower, using secondary mirrors. Figure 1 is a photograph of the NSTTF, showing the 212 heliostat mirrors, and the central solar tower in the facility. The central tower houses the STACEE secondary mirrors and cameras of photomultiplier tubes, as well the electronics for data acquisition.

STACEE is a wavefront-sampling Cherenkov telescope, which is able to achieve a lower energy threshold than any of the existing imaging atmospheric Cherenkov telescopes by taking advantage of the large collection area of the heliostat mirrors. The energy range 50 to 250 GeV is particularly important for understanding the emission mechanisms in many high energy astrophysical objects (such as pulsars and active galaxies). This is also an energy range that has proved to be difficult to reach by previous satellite- and ground-based experiments (e.g. see Ong 1998 for a review). Other solar tower Cherenkov experiments that use similar techniques as STACEE to detect high energy γ-rays are as follows: CELESTE in the Pyrenees (de Naurois et al. 2000); GRAAL (Arqueros et al. 1999); and Solar Two Gamma-Ray Observatory near Barstow, California (Zweerink et al. 1999).

FIGURE 1. Aerial view of NSTTF at Sandia National Laboratories, the facility in which STACEE is located. The STACEE heliostats direct Cherenkov radiation generated in extensive air showers onto secondary mirrors and cameras of photomultiplier tubes, located on the solar tower shown in the above picture. The smaller building located at the far end of the field houses the heliostat control computers.

The construction of STACEE began in 1997, and the first stage of the experiment, STACEE-32, used a total of 32 heliostats to measure instrument performance and conduct initial observations of a few astrophysical sources. STACEE-32 was operated successfully, and unambiguously detected the Crab nebula with a high statistical significance, at an energy threshold of 190 ± 60 GeV (Oser et al. 2001). A description of STACEE-32 may be found elsewhere (Ong 2000; Covault 2000).

During 1999 and 2000 considerable work has been done towards the completion of STACEE. The experiment now uses 48 heliostats, with substantial improvements made to the optics and electronics systems. In this article we summarize the work done with STACEE-48 and discuss the construction of the full STACEE experiment which will use 64 heliostat channels.

II STACEE-48

Figure 2 shows a plan view of NSTTF indicating the heliostats used by STACEE in the different stages of the experiment. The full STACEE instrument will have 64 heliostats and 5 secondary mirrors, and is expected to operate as a γ-ray observatory by Fall 2001 (Ong et al. 2001). STACEE is currently operating with 48 heliostats. A detailed review of the status of STACEE-48 is given elsewhere (Covault et al. 2001). Here we summarize some of the significant features of STACEE-48.

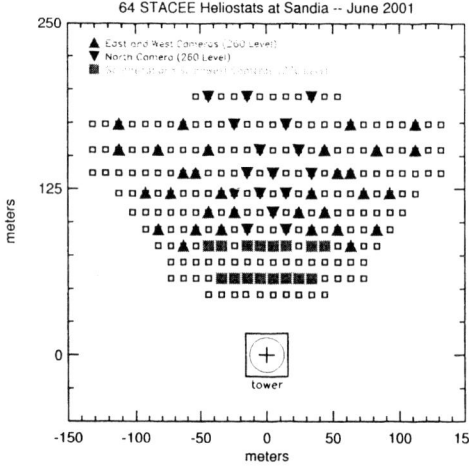

FIGURE 2. Plan view of NSTTF showing the heliostats used by STACEE. STACEE-32 used 32 heliostats (marked in red), and the East and West cameras. STACEE-48 uses an additional 16 heliostats (shown in blue), as well as the North camera. The 16 other heliostats that will be needed for the full STACEE instrument are shown in green (Covault et al. 2001).

Besides adding the 16 heliostats, which will obviously improve the light collection, the key improvements in STACEE-48 are in the areas of optics and electronics. These are listed below.

- The focusing and alignment of the heliostat facets have been improved significantly, leading to an increase in the amount of light reaching the secondary mirrors.

- The optical alignment of the secondary mirrors and cameras are monitored regularly using a set of alignment lasers mounted on the secondary optic structures.

- A black sealant has been applied to the ground underneath the heliostats to reduce the background light due to albedo.
- A new VME-based custom dynamic delay and trigger system, using FPGA technology (Martin & Ragan 2000), has been designed and installed.
- High speed (1 GHz) 8-bit Flash ADCs (Acqiris DC 270) have been installed on PMT channels to achieve better time and pulse amplitude resolution. A description of the key features of the STACEE FADC system is given in Covault et al. (2001).

In addition, considerable effort has been put into Monte Carlo studies, and on-site calibration of the experiment. The angular reconstruction and energy determination of the high energy γ-rays depend on the accurate determination of the arrival times and pulse amplitudes of the Cherenkov signals seen by each channel. Calibrating and characterizing the timing resolution and optical response of STACEE is thus very important. STACEE now has a laser calibration system (Hanna & Mukherjee 2001) that is used *in situ*. An event reconstruction algorithm has been developed for the STACEE data, and extensive work is in progress to compare the data collected with simulations in order to determine the energy threshold of the experiment. Figure 3 shows an estimate of the angular resolution of STACEE, obtained by dividing the array into two overlapping sub-arrays, and using each sub-array to separately reconstruct the arrival direction. Details on the timing measurement, angular resolution and trigger rate of STACEE-48 are given elsewhere (Covault et al. 2001).

Since 2000 November, STACEE-48 has been observing several astrophysical sources. STACEE observed the BL Lac object Markarian 421, for a total of 40 hours on-source, during the period 2001 February to May when the source was reported to be in a high flux state (Boerst et al. 2001). Results on the observations of Mrk 421 with STACEE-48 are described in Hinton et al. (2001). In addition to Mrk 421, STACEE-48 also observed the Crab (for diagnostic tests) and the BL Lac object Markarian 501. These observations are described by Ong et al. (2001).

III SUMMARY AND FUTURE PROSPECTS

With the completion of the STACEE instrument in Fall 2001, STACEE will be operating as a γ-ray observatory engaged in a comprehensive observing program. STACEE's future observing program includes both Galactic and extragalactic sources, such as, pulsars, supernova remnants and blazars. In particular, with an energy threshold of 50 GeV, STACEE should be able to see active galaxies to redshifts beyond 1.0, in comparison to current imaging Cherenkov telescopes that see only the closest blazars (redshift < 0.3). In addition, STACEE also plans to observe the unidentified EGRET sources, that have no known counterparts at other wavebands. A description of STACEE's scientific potential and future observing plans is given in Ong et al. (2001).

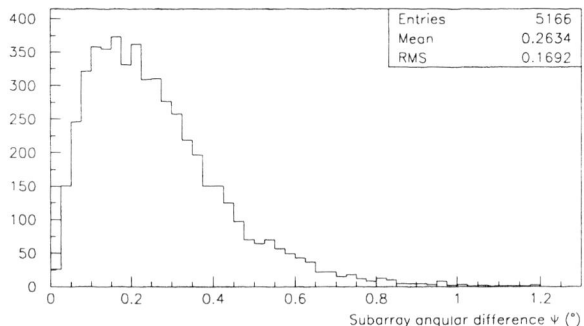

FIGURE 3. An estimate of the angular resolution of STACEE on reconstructed showers, using the difference of the reconstructed direction for each of the two sub-arrays (Covault et al. 2001). The angular resolution is estimated to be $\leq 0.2°$.

We are grateful to the staff at the NSTTF for their excellent support. We acknowledge the contributions of M. Lewandowska, G. Mohanty, S. Oser, T. Tumer, F. Vincent, and J. Zweerink. This work was supported in part by the National Science Foundation (under Grant Numbers PHY-9983836, PHY-0070927, and PHY-007095), the Natural Sciences and Engineering Research Council, FCAR (Fonds pour la Formation de Chercheurs et l'Aide à la Recherche), the Research Corporation, and the California Space Institute. CEC is a Cottrell Scholar of Research Corporation.

REFERENCES

Arqueros, F., et al., *Proc. 26th ICRC, Salt Lake City*, **5**, 215, 1999.
Boerst, H. G., et al., IAU Circ. 7568, 3, 2001.
Covault, C. E. et al., *Proc. 27th ICRC, Hamburg*, 2001.
Covault, C. E. et al., *Towards a Major Atmospheric Cherenkov Detector VI: (Snowbird, UT) conf. proc.* AIP Press, 411–415, 2000.
de Naurois, M., et al., *Proc. 26th ICRC, Salt Lake City*, **5**, 211, 1999.
Hanna, D. S. and Mukherjee, R., *Nuclear Instruments and Methods A*, submitted, 2001.
Hinton, J. A., et al., *Proc. 27th ICRC, Hamburg*, 2001.
Martin, J.-P. and Ragan, K., *Proc. IEEE Nuclear Science Symposium, Lyon, France*, 2000.
Ong, R. A., et al., *Proc. 27th ICRC, Hamburg*, 2001.
Ong, R. A., et al., *Towards a Major Atmospheric Cherenkov Detector VI: (Snowbird, UT)* AIP Press, 401–410, 2000.
Ong, R. A., *Phys. Rep.* **305**, 93 (1998)
Oser, S. et al., ApJ., 547, 949–958, 2001.
Zweerink, J. A. et al., *Proc. 26th ICRC, Salt Lake City*, **5**, 223, 1999.

The Keck Solar Two Gamma-Ray Observatory

M. Tripathi*, D. Bhattacharya†, J. Lizarazo*, G. Mohanty†, U. Mohideen†,
P. Murray*, H. Tom†, T. Tümer†, G. Xing* and J. Zweerink†

Department of Physics, University of California, Davis, CA, USA
†IGPP, University of California, Riverside, CA, USA

Abstract. Ground-based telescopes play a role that is complementary to satellite based gamma ray detectors and are expected to provide a rich data set in the 50 GeV–1 TeV region. Heliostat arrays provide a way of making sensitive observations in this energy regime at an economical cost. With close to 2000 heliostat mirrors, the Keck Solar Two Gamma Ray Observatory near Barstow, CA, is the largest such facility in the world, and thus has the potential of being one of the most sensitive gamma ray detectors. A 32-channel camera is now operational: we discuss the current status and future prospects for the instrument.

INTRODUCTION

Of the 271 gamma ray sources in the third EGRET catalog [1], only about five independently verified detections have been made by ground-based detectors that operate above an energy of about 250 GeV. Furthermore, the ground-based obeservations to date have surveyed only about 1% of the sky. Thus, there is a need to continue observations in this high energy regime and extend explorations into the intermediate region from 10–250 GeV, that is above EGRET's reach but currently inacessible to ground-based detectors. The ground-based efforts include multiple-telescope projects, namely, HESS [2], VERITAS [3], and CANGAROO-III [4], as well as large single-dish telescopes such as MAGIC [5]. The future GLAST satellite [6] is expected to observe in an energy range of 20 MeV to 300 GeV, albeit, with a relatively poor sensitivity at the high-energy end. Heliostat-array detectors such as STACEE [7, 8], CELESTE [9, 10], and the somewhat similar GRAAL [11] project have been active for a few years and have observed high-energy gamma-ray emission from the Crab Nebula. The Keck Solar Two Gamma-ray Telescope which uses a heliostat-array facility located near Barstow, CA, has now been commissioned: we have successfully detected Cherenkov radiation (CR) produced in extensive air showers initiated by high-energy particles.

The gamma ray sky provides for a rich observational program: sources of interest include galactic ones like supernova remnants (plerionic and shell type) and pulsars, as well as extragalactic ones such as active galactic nuclei (AGN) and gamma-ray bursters. Improved source-location capabilities offered by Solar Two should help in identifying the large number of unidentified EGRET sources. Exploration of the previously unobserved energy region should help in understanding the physical mechanisms underlying the emission, and in distinguishing between various emission models. For example, coordinated observations, at multiple wavelengths, of highly-variable AGN will help discriminate between the leptonic and hadronic classes of models. Lately, it has also been realized that observations of high-energy gamma-ray sources can address issues in cosmology, such as estimating the epoch of galaxy formation based on the absorption of gamma rays from distant AGN [12, 13, 14]. In addition, gamma ray measurements can contribute to more fundamental experiments, such as testing quantum gravity theories using the temporal structure of gamma-ray bursts [17] and searching for supersymmetric dark matter via their annihilations into photons.

THE SOLAR TWO TELESCOPE

The telescope presently consists of 32 individual heliostats, each with a mirror area of $40m^2$, that are spread over a circle of about 200 m diameter. The central receiver tower has been equipped with a detector composed of a spherical secondary mirror and a camera with 32 photo-multiplier tubes (PMTs) located at the focal plane. This technique was conceived by Danaher et al [15], followed by Tümer's proposal [16] to use secondary optics in order to resolve the individual heliostat images. For gamma-ray showers, the CR wavefront is roughly spherical with a radius of curvature of about 8km, a temporal width of a few ns, and an extent of about 300m diameter at ground level. Given the large available mirror area in heliostat arrays, such telescopes can provide significant improvements in signal-to-noise ratio for CR detection, thereby lowering the operating threshold to about 50 GeV. While this technique is highly cost-effective in terms of collection area, the relatively poor optical properties of the heliostats do not allow for the use of a fine-grained camera and hence degrade the overall ability of using imaging techniques to differentiate between proton and photon showers. On the other hand, Solar Two is the only telescope, either existing or planned, that has an array of mirrors large enough to contain the entire CR pool of photons. We are currently in the midst of adding a second 32 channel camera to the tower which will increase the instrumented heliostat array diameter to about 300 m.

The heliostats are individually steerable in elevation and azimuth, and thus the telescope functions as a directed instrument that can be used to track a celestial source. They are controlled by a doubly-redundant system running on a mixed network of OpenVMS and Linux computers. The pointing of the heliostats is checked periodically by measuring and correcting for biases observed in tracking the sun, and in tracking individual stars. The cumulative tracking and pointing error for each individual heliostat is estimated to be about $0.1°$. The spherical secondary mirror, located inside the central tower, is made of 13 hexagonal facets, each with a diameter of 1m and radius of curvature of about 6m. The composite secondary has an aperture of 4.4m × 2.6m and reduces the 3m spot-size from the heliostats to much smaller diameters that match entrance apertures of light-collecting parabolic Winston cones which further focus the photons onto PMTs. Thus, each PMT views a single heliostat: the field-of-view in the sky being about $1°$ across, defined by the sharp angular cutoff imposed by the Winston cones. The secondary mirror facets were individually aligned using a laser-pointer fixed to the centre of the appropriate heliostat mirror, and double-checked by imaging the full moon onto the camera.

The readout and trigger system used at Solar Two is largely based on the STACEE-32 design. The trigger electronics are intended to solve two main problems: First, the uniformity and extent of the gamma-ray Cherenkov light pool makes it advantageous to utilize narrow coincidence intervals and impose predetermined majority logic within the various PMTs. The signal from each PMT is discriminated, and digital logic is used to demand a coincidence of 5 out of 8 PMTs in a cluster, and a further coincidence of 3 out of 4 clusters. The coincidence trigger also affords significant rejection of less uniform, background cosmic-ray showers. Second, relative delays need to be added in between channels in order to compensate for the different times of flight from different heliostats. Because this depends on the position of the source in the sky, the delays need to be changed as the source moves across the sky. This is achieved by a combination of programmable delays at two levels, and by low-loss analog delay cables. The delays can be adjusted in steps of 1 ns to a maximum of 1 μs, allowing observation of sources to a zenith angle of 45°.

Two data-recording paths are available. The first are the time-delay counters (TDCs) that record the time of arrival of the discriminated pulse on each channel. The second, more sophisticated, path is the 1GSamples/s digitizers that are continuously sampling a separate analog output on each channel. When a trigger occurs, acquisition is stopped and the data corresponding to a 1μs slice of time on each side of the trigger are read out. The available memory segmentation, and the high-speed cPCI bus allows accurate characterization of the Cherenkov pulse on each channel with small deadtime even at high trigger rates. The digitizers also provide an absolute time-stamp with the records so that both amplitude and time information is preserved in the data stream.

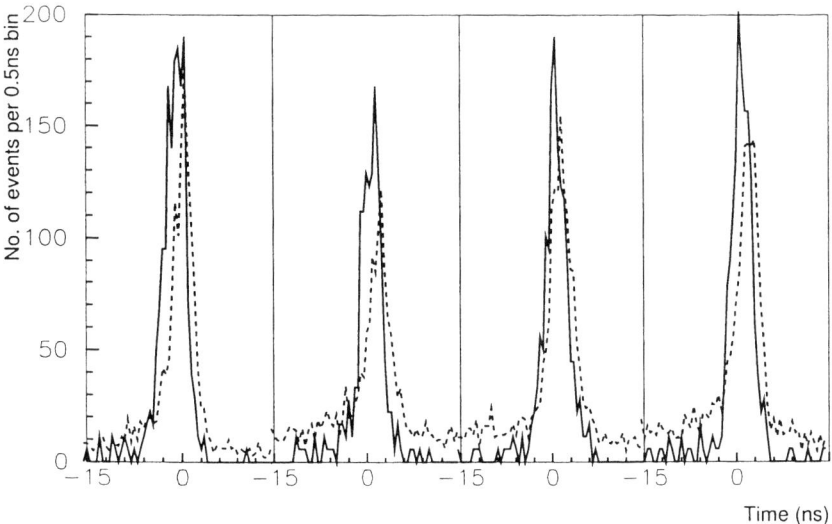

FIGURE 1. Distributions of the arrival times of discriminated pulses. The solid line is for zenith data and the dashed line for Crab data.

DATA ANALYSIS AND RESULTS

The typical gamma-ray signal is small compared to the preponderance of background cosmic-ray showers. Thus, data are taken in an ON/OFF mode, where a putative gamma-ray source is tracked for a given amount of time, and then the heliostats are slewed back to track an OFF-source region corresponding to the same range of elevation and azimuth angles. The simplest mode of analysis is to subtract the two TDC distributions, revealing any excess from the source. Slightly more sophisticated analyses have been used by STACEE and CELESTE to improve background rejection, e.g., by imposing further post-acquisition time-coincidence requirements, selecting showers where the arrival times are well-fit by a circular wavefront expected for gamma-ray showers. We are in the process of developing a more detailed technique that uses a semi-analytical model for the development of a gamma-ray shower [18, 19] to predict temporal as well as amplitude distributions for the Cherenkov light at each heliostat as a function of various parameters such as the energy of the primary particle, the impact point of the shower core, etc. This makes full use of the detailed information from the digitizers, and is expected to provide significantly better background rejection.

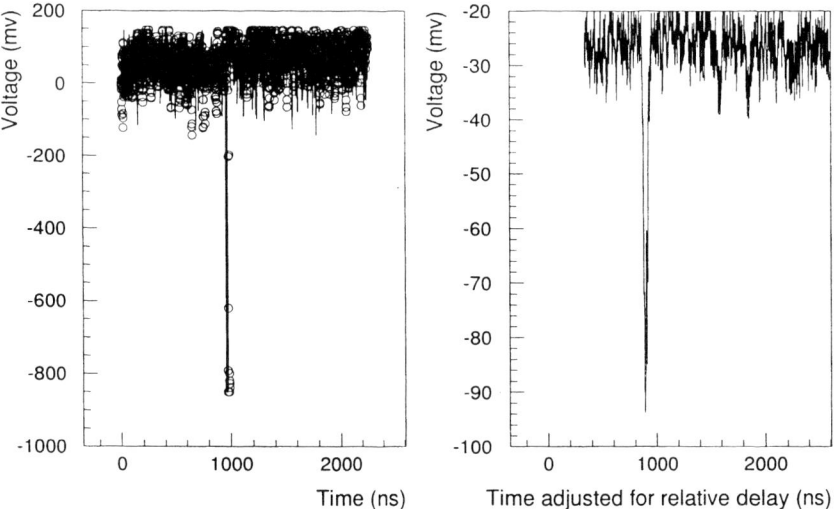

FIGURE 2. (a) Two pulses recorded at zenith on a single digitizer channel (shown by the line and dots, respectively). (b) A simple average of 300 pulses, at zenith, on the same channel.

We started source observations in late 2000, focussing initially on calibration zenith data, and on the Crab Nebula. While we have recorded some ON-source data on the Crab Nebula, the experiment is still in a debugging phase. Hence, here we restrict ourselves to only presenting preliminary evidence that we are indeed detecting a Cherenkov signal. Fig. 1 shows the TDC distributions for triggered events from four representative channels out of the 32. The solid line is the distribution at zenith and the dashed line was obtained when tracking the Crab. All channels show similar, sharply-peaked distributions with a typical FWHM of 8 ns, which demonstrates that most of these events do indeed come from air-shower CR, and that the Cherenkov signal is seen both when the heliostats are static at the zenith, and when they are tracking a source. Fig. 2 shows data from the digitizers. Fig. 2(a) shows two successive acquisitions at the zenith, indicating that the large signal pulses are indeed in time. Fig. 2(b) is a simple average for one channel over 300 acquisitions at the zenith. The overall width is consistent with the expected time jitter due to the angular width within our field of view and due to fluctuations within the cosmic ray showers.

FUTURE DEVELOPMENT

We are in the process of upgrading the electronics so as to maintain a bandwidth of about a GHz through the entire chain. Amplifiers with a 1-GHz bandwidth have already been built and installed. In order to further lower the energy threshold, we plan to employ a new trigger system capable of real-time temporal pattern recognition, also at a 1 GHz bandwidth. The optical throughput is also expected to increase due to improvements in the camera box and Winston cones. We are planning to complete the scheduled expansion to 64 heliostats by early 2002. The heliostats for the second set of 32 channels have been selected, and the design of the second port in the tower has been completed. The secondary mirror facets are already on-site and construction of the port, secondary mirror and camera is expected to begin shortly. When this expansion is completed, the Keck Solar Two observatory will be the largest gamma-ray astronomy facility and will have the unique capability of containing the gamma-ray Cherenkov light pool in one direction. Longer-term plans include increasing the number of heliostats to upto about 250.

REFERENCES

1. Hartman, R. C. et al, *Astrophys. J. Suppl. Series*, **123**, 79–202 (1999).
2. G. Hermann et al, in *Proc. XXXII Rencontres de Moriond, Gif-sur-Yvette, France*, 1997, p. 141.
3. Weekes, T. C., in *Proc. XXXII Rencontres de Moriond, Gif-sur-Yvette, France*, 1997, p. 161.
4. M. Mori et al, in *Proc. intl. symp. on high energy gamma-ray astronomy*, Heidelberg, Germany, 2000.
5. E. Lorenz et al, in *Proceedings of the workshop on TeV astrophysics*, Snowbird, UT, 1999.
6. Gehrels, N., and Michelson, P., *Astroparticle Phys.*, **11**, 277–282 (1999).
7. Chantell, M. et al, *Nucl. Instrum. Methods Phys. Res., Sect. A*, **408**, 468–504 (1998).
8. Oser, S. et al, *Astrophys. J.*, **547**, 949–958 (2001).
9. E. Paré et al, "", in *Towards a major atmospheric Cherenkov detector-V, Kruger Park*, 1997.
10. M. de Naurois et al, *Astrophys. J.* (2001), submitted.
11. F. Arqueros et al, in *Proc. 26th Int. Cosmic Ray Conf., Salt Lake City, UT*, 1999, vol. 5, pp. 215–218.
12. Gould, R. P., and Schréder, G. P., *Phys. Rev.*, **155**, 1408 (1967).
13. Stecker, F. W., and de Jager, O. C., *Astron. Astrophys.*, **334**, L85–L87 (1998).
14. Biller, S. D. et al, *Phys. Rev. Lett.*, **80**, 2992–2995 (1998).
15. Danaher, S. et al, *Solar Energy*, **28**, 335 (1982).
16. Tümer, T. O. et al, in *Nucl. Phys. B (Proc. Suppl.)*, Elsevier, 1990, vol. 14A, pp. 351–355.
17. Amelino-Camelia, G. et al, *Nature*, **393**, 763–765 (1998).
18. Hillas, A. M., *Nucl. Phys. B, Part. Phys.* (1985).
19. Bohec, S. L. et al, *Nucl. Instrum. Methods Phys. Res., Sect. A*, **416**, 425–437 (1998).

The Next Generation of Ground-based Gamma-ray Telescopes

Trevor C. Weekes

*Whipple Observatory, Harvard-Smithsonian Center for Astrophysics,
P.O. Box 97, Amado, AZ 85645-0097.
e-mail: tweekes@cfa.harvard.edu*

Abstract. The status of ground-based gamma-ray astronomy, in terms of current and planned observatories, is reviewed.

I INTRODUCTION

Since the launch of EGRET, imaging atmospheric Cherenkov telescopes have progressively improved their flux sensitivities and lowered their energy thresholds. In the next few years there will be a dramatic improvement in the 50 GeV to TeV energy range with the completion of the new telescope arrays (CANGAROO-III, HESS and VERITAS) and the large telescopes (MAGIC and MACE). These telescopes will be complemented by newly completed "solar farm" Cherenkov telescopes and mountain arrays of particle detectors. Energy thresholds as low as 30 GeV and flux sensitivities as low as 0.5% of the Crab Nebula are anticipated. The geographical distribution of the proposed arrays is such that both the northern and southern skies will be covered.

It has become increasingly politically correct to regard ground-based astronomy as an adjunct to space astronomy. Although ground-based astronomy was once the proud parent of space astronomy, it is now more appropriate to see it as the poor cousin. This is unfortunate for a number of reasons. The concentration of observations in one or two sophisticated, sensitive, space-based instruments with finite lifetimes leads to an unnatural growth pattern with progress coming in fits and starts seperated by long droughts. The rather slow but steady development of ground-based facilities, coupled with their long lifetimes, is a more natural progression and one that lends itself to teaching, research and orderly longterm planning. Taken together, the parallel programs in ground- and space-based gamma-ray astronomy can provide a diverse and stable base for research.

The scientific objectives of space-based and ground-based high energy gamma-ray astronomy are very similar, but the methodologies are very different. This may account for the fact that rarely do we find individuals or groups with overlapping interests in both space and ground-based gamma-ray astronomy. This is particularly noticeable in the frequent lulls in space activity which are not characterized by increases in ground-based activity. The growth in activity of the latter in recent years has been due largely to the influx from high energy physics where the sociology, if not the science, is seen to be more compatible. Ground-based astronomy collaborations are generally large. As the complexity of the techniques increases, this trend is unlikely to reverse. As in high energy physics, these teams of scientists are generally drawn from the universities (rather than from national laboratories).

The nature of the research in ground-based gamma-ray astronomy lends itself to the training of graduate students where the ability to design, build, test and use an instrument is regarded as essential. The broad diversity of physical disciplines: particle physics, cosmic rays, atmospheric physics, is also attractive as an educational project.

One of the fundamental differences between ground- and space-based gamma-ray astronomy is that VHE facilities are seen as observatories, not as missions or experiments. The ground-based telescopes steadily improve with time whereas those in space gradually deteriorate. There is an inherent versatility in earth-based telescopes; the telescopes are readily accessible and easily modified.

II CURRENT STATUS

When EGRET ceased to operate in 2000, it left a rich archive of data to be analyzed; however observational high energy (HE) gamma-ray astronomy effectively came to a standstill and this will be so until there is a new HE telescope in orbit (AGILE in 2003). In contrast there has never been so much activity in very high energy (VHE) astronomy with large teams of physicists using the many fine instruments that have recently come into operation. The techniques are steadily improving with greater understanding of shower parameters, atmospheric effects, statistical red herrings, and instrumental eccentricities. One can have considerable confidence in results reported from this generation of experiments.

Observatories: Although the efforts of many in the ground-based community are now directed towards third-generation systems (below), there are more than ten active observatories which will probably only be phased out when the new generation come on-line. We can, thus, expect some continuity in the field. There are currently at least eleven atmospheric Cherenkov observatories, seven using the imaging technique and four using existing or extant solar facilities to push to lower energies [1].

Status of Observations: In contrast to HE gamma-ray astronomy where the catalog of 271 sources [2] has undergone some downward revisions [3], the TeV catalog has grown steadily in recent years. While still very small, it contains a pleasing variety of sources. Many of these sources are not in the 3rd EGRET Catalog; hence VHE astronomy is not merely an adjunct to HE astronomy but a unique field of high energy astrophysics in its own right.

Recent additions to the TeV source catalog [1] are 1H1426+428 [4] and BL Lac [5]. 1H1426+428 (plus the confirmation of 1ES2344+514 [6]) are important because, together with Markarian 501, they constitute the class of "extreme" galaxies identified by Ghisellini [7]. Plerions are another class of object that seem to emit strongly at TeV energies. The recent detections of SN1006, RJX1713.7-3946 and Cas A constitute the strongest evidence at any gamma-ray energy for emission from shell supernova remnants.

III NEXT GENERATION TELESCOPES

A Requirements

It is generally agreed that the next (third) generation of ground-based telescopes should have the following properties as a minimum goal:

- *Better Flux Sensitivity*: detection of sources which emit γ-rays at levels of 0.5% of the Crab Nebula flux at energies of 300 GeV in 50 hours of observation.
- *Reduced Energy*: an effective peak energy sensitivity <100 GeV with significant sensitivity at 50 GeV or below.
- *Improved Energy Resolution*: an RMS spectral resolution of $\Delta E/E < 0.10 - 0.15$ for an individual shower over a broad energy range (E $>$ 300 GeV).
- *Increased Angular Resolution*: $<0.05°$ for individual showers and source location better than $0.005°$ ($>$ 100 photons).
- *Large Effective Area*: $>0.1 \, \text{km}^2$ to provide sensitive measurements of short variability time-scales.
- *Large Field of View:* at least $3°$ diameter as is used in many current atmospheric Cherenkov imaging telescopes.

B Observatories

Here we briefly describe the characteristics of some of the next generation ground-based telescopes. The MACE project [8] is sufficiently similar to MAGIC that it will not be described separately. It will consist of two telescopes and will use the same telescope design as MAGIC.

MAGIC: MAGIC (Major Atmospheric Gamma Imaging Cherenkov) is an extension of the single large telescope concept pioneered by Whipple, CAT and CANGAROO-I but with a significant increase in aperture and with the introduction of several new technologies. The MAGIC collaboration is headquartered at the Max-Planck-Institute for Physics in Munich, Germany with significant partners in Germany, Spain, and Italy. Like HEGRA, MAGIC will be located on La Palma, Canary Islands, Spain at an elevation of 2.3 km [9].

The most formidable parameter of MAGIC is its large aperture (17m) with 234 m^2 of mirror area made up of 1000 square facets of side 50 cm. The f/number

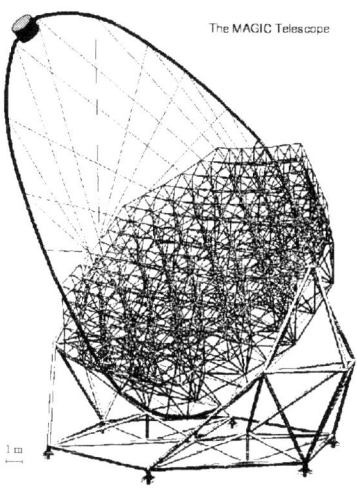

FIGURE 1. The MAGIC Telescope

is 1.03. The facets are diamond-ground and polished aluminium. They are under active control and will be realigned to compensate for gravitational deflections as the telescope moves. The optical support structure is made of carbon fiber to minimize weight (40 tonnes). It can be slewed at the extraordinarily high rate of 3° per second. The telescope design is illustrated in Figure 1.

The camera has pixels of two sizes: 397 of 2.5 cm diameter and 180 of 3.7 cm diameter. Analog signals are transmitted over optical fibers and are processed remotely using 300 MHz flash ADCs.

The flux sensitivity of MAGIC depends critically on its ability to suppress the signal from single muons. It will achieve the lowest peak energy (30 GeV) of any of the atmospheric Cherenkov experiments currently under construction. Construction of MAGIC is well under way and first light is scheduled in 2002.

HESS: The HESS (High Energy Stereoscopic System) collaboration has its headquarters at the Max-Planck-Institut fur Kernphysik in Heidelberg, Germany. The major partners in this large international collaboration come from Germany, France, Namibia, South Africa, and the U.K. The facility will be located at Goellschau, Namibia, Africa at an elevation of 1.8 km [10]. The project has the distinction of being the first gamma-ray telescope to have its conceptual design featured on a postage stamp (Figure 2).

The array will consist of four large telescopes at the corners of a square of side 120m. The design of the optical support structure and mount is somewhat similar to that of MAGIC; both are based loosely on previously constructed solar reflector designs. The azimuth table rotates on a circular rail at ground level. The rotation

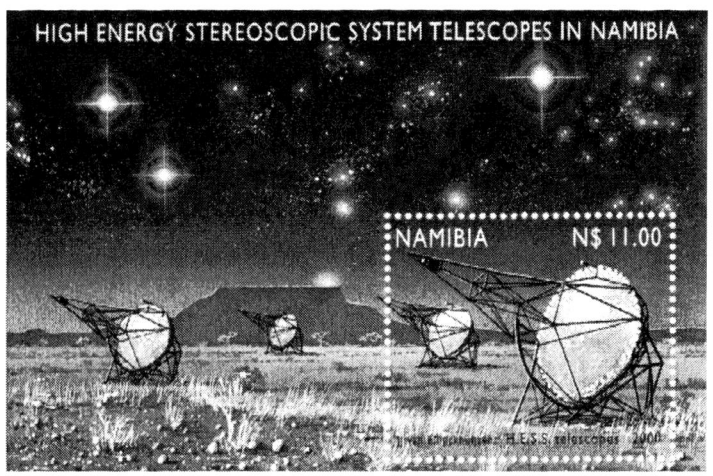

FIGURE 2. The HESS Array as depicted on the postage stamp from Namibia

rate is 1.6° per second. Each complete telescope weighs 50 tonnes. The aperture is 12m and focal length 15m (f/number = 1.2). The optical design is Davies-Cotton with 382 circular facets of diameter 60 cm (total mirror area = 108 m^2). The facets will be made of glass with front-aluminized coatings and quartz overcoating; they can be adjusted remotely.

Each camera will have 960 pixels (diameter 0.16°) giving a field of view of 5°. The readout and trigger electronics are within the camera box which has a total weight of 600 kg. The signal is measured using a dual-range 1GHz Analog Ring Sampling ASIC.

The schedule for HESS is the most advanced of the next generation projects with total funding approved and first light scheduled for 2002. A second phase envisages the addition of up to 12 more telescopes.

CANGAROO-III: CANGAROO-III (Collaboration of Australia and Nippon for a GAmma Ray Observartory in the Outback) is an international collaboration involving Japan and Australia. The III indicates the third stage of development, not the number of reflectors which is four. The Project Office is at the Institute of Cosmic Ray Research, Tokyo, Japan and the facility is sited at Woomera, Australia near sea level [11].

The four reflectors, each of aperture 10m and focal length 8m (f/number 0.8), will be located on the corners of a square 100m by 100m. The optical support structure will be parabolic and made of steel. The mirrors facets are circular (80 cm diameter) and made of composite plastic.

The cameras will have 552 pixels and will have many similarities with the cameras used in CANGAROO-I.

Funding for the complete project has been approved and the first telescope,

FIGURE 3. The first of the four CANGAROO telescopes which was installed in 2000 at Woomera, Australia. It is shown here before all the mirrors were installed.

installed in 2000, is now operational (Figure 3). It is planned to add one telescope each year (2001, 2002, 2003) with first light for the full array early in 2004. Potential problems include manmade lights at the Woomera site (a nearby detention center and rocket facility) but a cooperative agreement is being worked out to maximize shielding. The array is unusual in having reflectors with small f/numbers, composite mirrors and sea level elevation.

VERITAS: VERITAS (the Very Energetic Radiation Imaging Telescope Array System) is a collaboration involving groups in the U.S.A., U.K., and Ireland. Its Project Office is at the Whipple Observatory, Arizona, U.S.A. and the collaboration membership includes the groups that were part of the former Whipple Collaboration. The designated site for VERITAS is in Montosa Canyon, Arizona, a dark site at an elevation of 1.3 km. It will be less than 2km from the Whipple Observatory Administrative Complex, a major logistical advantage [12].

The seven identical telescopes in VERITAS, each of aperture 10m, will be deployed in a filled hexagonal pattern of side 80 m. The reflectors will be rectangular (10m by 11m) so that at a later stage they can be expanded to carry 315 facets, giving a mirror area equivalent to a 12m aperture (Figure 4). The Optical Support Structure will be made of tubular steel with a total weight of 30 tonnes. The mounts will have a slew speed of $>1°$/s. The telescopes will be of Davies-Cotton design with focal length 12m (f/number 1.2); the larger f/number (compared with the existing 10m reflector) gives a significant improvement in optical resolution (Figure 5). The 250 mirror facets will be made of glass with anodized aluminum coatings. A feature of the reflectors is that they will have provision for retractable covers which will extend the life of the mirror coatings.

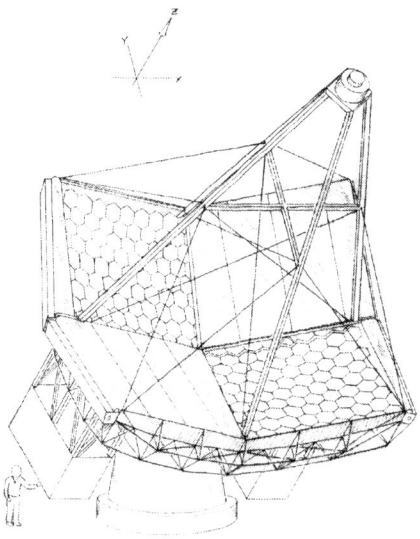

FIGURE 4. Sketch of one of the VERITAS telescopes (with partially opened covers).

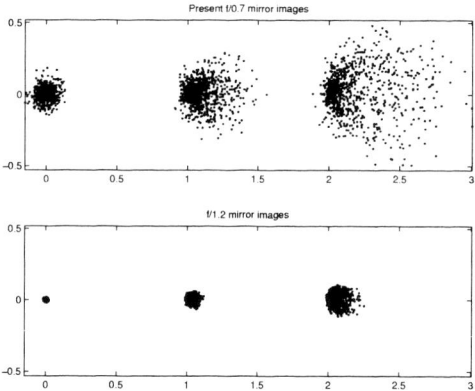

FIGURE 5. Point spread images in the focal plane corresponding to a source at 0°, 1°, and 2° off axis for f/numbers of 0.7 (top, Whipple telescope) and 1.2 (bottom, VERITAS).

The cameras will have 499 pixels consisting of PMTs of angular diameter 0.15° with FOV 3.5°. They are designed so that at a later stage they can be expanded to give a larger FOV with the addition of more pixels. There will be four levels of trigger: individual pixels, pattern trigger within individual telescopes, array trigger involving two or more telescopes and array pattern trigger. The signals from all the pixels will be analyzed with flash ADCs which have a sample rate of 500 MHz and a dynamic range of 10^3. The array will be designed to operate as seven individual telescopes, as sub-arrays of three and four telescopes or as a full array of seven telescopes.

Although VERITAS was the first array of large telescopes proposed, it is the most ambitious and has not yet achieved funding approval (despite receiving endorsements from all scientific review panels). Construction of a prototype telescope and camera are now underway and it is planned to have the full array completed in 2005, hopefully in time to complement the GLAST mission.

C Sensitivity: energy and flux

The three next generation array observatories are remarkably similar and can be expected to have similar flux and energy sensitivities. Differences are more in the definition of parameters than in inherent sensitivities. Detailed Monte Carlo simulations by the HESS and VERITAS groups are in good agreement ([13]; [10]). Published estimates by the CANGAROO group [11] are extrapolations and probably overestimate the sensitivity at low energies (M.Mori, private communication).

Two points must be made:

- *Collection Area:* The collection area of the array reflects a combination of the area over which the array can detect γ-ray showers and the efficiency for retaining those events after analysis. The effective collection area of VERITAS as a function of photon energy is shown in Figure 6.

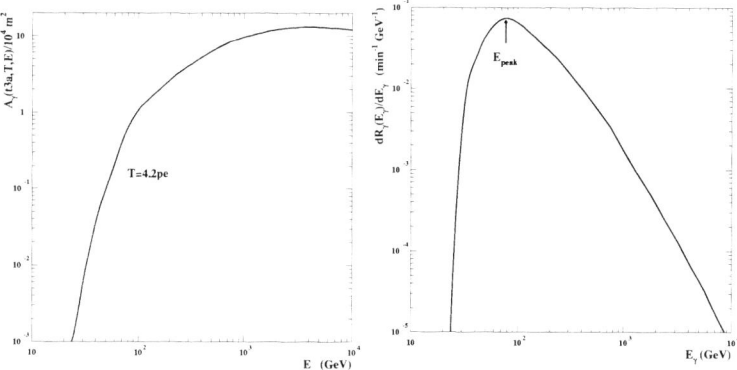

FIGURE 6. *Left figure:* Collection area of VERITAS versus energy. *Right figure:* The arrow indicates the Peak Energy. Actually there is considerable sensitivity down to energies of 50 GeV and below.

- *Energy Threshold and Peak Energy:* The term *energy threshold* is often used to define the point where the differential rate of γ-rays per unit energy interval from the Crab Nebula is the highest; this is more accurately called the peak energy since the actual threshold may be considerably below this level as shown for VERITAS in Figure 6.

- Flux Sensitivity: The minimum detectable integral flux of γ-rays is that which yields a 5σ excess of γ-rays above the background, or a minimum of 10 photons in 50 hours of observations on an object with a differential spectrum given by $dN/dE \propto E^{-2.5}$, (e.g., the Crab Nebula). This conservative estimate of the γ-ray integral flux sensitivity of VERITAS for point sources as a function of energy is shown in Figure 7. The relative integral flux sensitivities of the next generation of space and ground-based telescopes are compared with existing or previous telescopes in Figure 8.

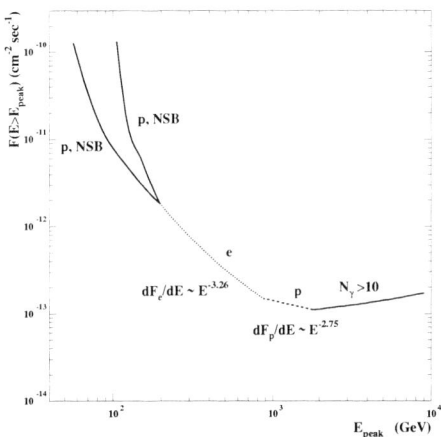

FIGURE 7. Flux sensitivity is limited by night-sky background (NSB), cosmic electrons (e), cosmic hadrons (p), and statistics N> 10. The lower curve indicates a relatively dark observation region (near the zenith with no bright stars in the FOV) while the upper curve indicates a region where the NSB light is approximately 4 times brighter (as in some regions of the Galactic plane).

Acknowledgements: This research in VHE gamma-ray astronmomy is supported by the U.S.D.O.E. I am grateful to Stephen Fegan, Tony Hall and Deirdre Horan for a critical reading of the paper.

REFERENCES

1. Weekes, T.C. in Proc. of International Symposium on High Energy Gamma-ray Astronomy, (Heidelberg, June, 2000), AIP Conf. Proc. 558 15.
2. Hartman, R.C., et al. ApJS, 123, (1999), 79
3. Thompson, D. et al. this symposium (2001)

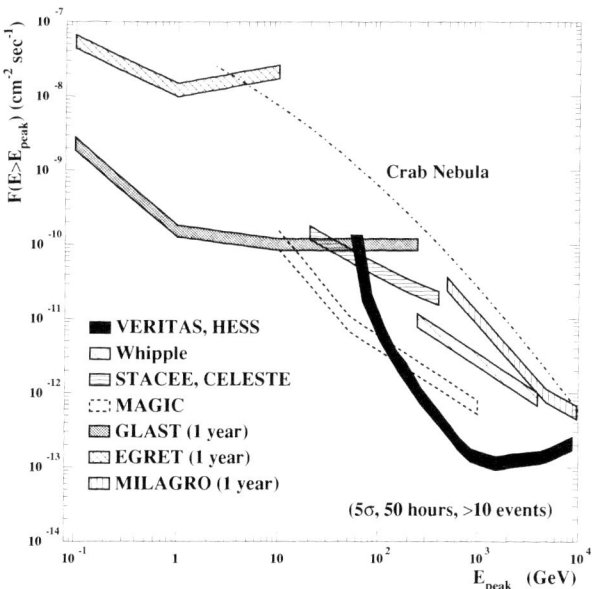

FIGURE 8. Comparison of the point source sensitivity of VERITAS (comparable with HESS and CANGAROO-III) to Whipple (comparable to HEGRA, CAT and CANGAROO-II), MAGIC, CELESTE/STACEE, GLAST, EGRET and MILAGRO. EGRET, GLAST and MILAGRO are wide field instruments.

4. Horan, D. this symposium (2001)
5. Neshpor, Yu.I. et al. Astron. Rep., 45 (2001) 291
6. Badran, H. this symposium (2001)
7. Ghisellini, G. in TeV Astrophysics of Extragalactic Sources, Astropart. Phys., 11, (1999), 11
8. Sharma, M. et al. in Proc. of Symposium on Very High Gamma-ray Astronomy, (Heidelberg, June, 2000) AIP Conf. Proc. 558. (2001) 621
9. Lorenz, E. in GeV-TeV Astrophysics: Towards a Major Atmospheric Cherenkov Detector VI (Snowbird), (1999), 510
10. Hofmann, W. in GeV-TeV Astrophysics: Towards a Major Atmospheric Cherenkov Detector VI (Snowbird), (1999), 500
11. Mori, M. et al. n GeV-TeV Astrophysics: Towards a Major Atmospheric Cherenkov Detector VI (Snowbird), (1999), 485
12. Weekes, T.C. et al. Astropart. Phys. (in press) 2001
13. Vassiliev, V.V. Astropart. Phys., 11 (1999) 247

LIST OF PARTICIPANTS

Andritschke, Robert
 MPE
 randrits@mpe.mpg.de

Aprile, Elena
 Columbia University
 age@astro.columbia.edu

Apt, Jay
 Orbit Experience
 Jay@OrbitExperience.com

Arzoumanian, Zaven
 NRC/NASA-GSFC
 Zaven.Arzoumanian.1@gsfc.nasa.gov

Badran, Hussein
 Smithsonian Institution
 badran@egret.sao.arizona.edu

Baganoff, Frederick K.
 MIT
 fkb@space.mit.edu

Bahcall, John
 Institute for Advanced Study
 jnb@ias.edu

Band, David
 UMBC/NASA/GSFC
 dband@lheapop.gsfc.nasa.gov

Barbiellini, Guido
 INFN TS
 barbiellini@ts.infn.it

Baring, Matthew
 Rice University
 baring@rice.edu

Barnes, Sandy
 USRA/NASA/GSFC
 sbarnes@lheapop.gsfc.nasa.gov

Barthelmy, Scott
 NASA/GSFC
 scott@lheamail.gsfc.nasa.gov

Bazzano, Angela
 Istituto di Astrofisica Spaziale

Bednarek, Wlodek
 University of Lodz
 bednar@krysia.uni.lodz.pl

Bergstrom, Lars
 Stockholm University
 lbe@physto.se

Berrington, Robert
 NRL/ASEE
 rberring@gamma.nrl.navy.mil

Bertsch, David
 NASA/GSFC
 dlb@mozart.gsfc.nasa.gov

Bhattacharya, Dipen
 Univ. of California, Riverside
 dipen@tigre.ucr.edu

Bignami, Giovanni
 ASI
 bignami@asi.it

Blandford, Roger
 Caltech
 rdb@tapir.caltech.edu

Bloom, Elliott
 SLAC/Stanford University
 elliott@slac.stanford.edu

Bloom, Steven
 Hampden-Sydney College
 sbloom@hsc.edu

Bloser, Peter
 MPE
 bloser@mpe.mpg.de

Boettcher, Markus
 Rice University
 mboett@spacsun.rice.edu

Boggs, Steven
    Univ. of California, Berkeley
    boggs@ssl.berkeley.edu

Bonnell, Jerry
    USRA/NASA/GSFC
    bonnell@grossc.gsfc.nasa.gov

Bouchet, Laurent
    CESR, Toulouse
    bouchet@sigma-0.cesr.cnes.fr

Braga, Joao
    INPE - Brazil
    braga@das.inpe.br

Brennan, Don
    Orbital Sciences Corp.
    brennan.don@orbital.com

Bretthauer, Joy
    NASA/GSFC
    joy.w.bretthauer.1@gsfc.nasa.gov

Bridgman, William
    GSFC/GST
    bridgman@wyeth.gsfc.nasa.gov

Buckley, James
    Washington University
    buckley@wuphys.wustl.edu

Buesching, Ingo
    Ruhr-Universitat-Bochum
    ib@tp4.ruhr-uni-bochum.de

Bunner, Alan
    NASA HQ
    alan.bunner@hq.nasa.gov

Burnett, Thompson
    University of Washington
    tburnett@u.washington.edu

Caraveo, Patrizia
    IFC "G.Occhialini", Milano
    pat@ifctr.mi.cnr.it

Carlson, Per
    KTH Stockholm
    carlson@msi.se

Carosso, Paoloc
    NRL/Swales
    pcarosso@swales.com

Chang, Heon-Young
    Korea Institute for Advanced Study
    hyc@ns.kias.re.kr

Chekhtman, Alexandre
    NRL/George Mason Univ.
    chehtman@gamma.nrl.navy.mil

Chen, Andrew
    C.I.F.S./I.F.C.
    chen@ifctr.mi.cnr.it

Chenette, David
    Lockheed Martin
    david.chenette@lmco.com

Chester, Margaret
    Penn State Univ.
    chester@astro.psu.edu

Chiang, James
    UMBC/GSFC
    jchiang@milkyway.gsfc.nasa.gov

Chuang, Kuan-Wen
    Univ. of California, Riverside
    chuang@ucrph0.ucr.edu

Cline, Thomas
    NASA/GSFC
    cline@apache.gsfc.nasa.gov

Cocco, Veronica
    Univ. di Roma "Tor Vergata"
    veronica.cocco@roma2.infn.it

Colafrancesco, Sergio
    Osservatorio Astronomico di Roma
    cola@coma.mporzio.astro.it

Collmar, Werner
    MPE, Garching
    wec@mpe.mpg.de

Cominsky, Lynn
    Press Officer/Sonoma State Univ.
    lynnc@charmian.sonoma.edu

Connors, Alanna
    Eureka Scientific
    aconnors@frances.wellesley.edu

Conte, Dominick
    Spectrum Astro
    Dominick.Conte@specastro.com

Corbel, Stephane
    CEA Saclay/Universite Paris 7
    corbel@discovery.saclay.cea.fr

Cravens, James
    Southwest Research Institute
    jcravens@swri.edu

D'Amico, Flavio
    Brazil

D'Amico, Nichi
    Bologna Astronomical Observaory
    damico@bo.astro.it

De Cesare, Giovanni
    IAS-CNR

Del Santo, Melania
    IAS-CNR
    delsanto@ias.rm.cnr.it

Del Sordo, Stefano
    IFCAI-CNR

Dermer, Charles
    Naval Research Lab
    dermer@gamma.nrl.navy.mil

Di Biasi, Lamont
    L. Di Biasi Associates
    ldibias@attglobal.net

Di Cocco, Guido
    ITESRE/CNR
    dicocco@tesre.bo.cnr.it

Diehl, Roland
    MPE, Garching
    rod@mpe.mpg.de

Digel, Seth
    USRA/NASA/GSFC
    digel@blazar.gsfc.nasa.gov

Dingus, Brenda
    University of Wisconsin
    dingus@physics.wisc.edu

Djannati-Atai, Arache
    PCC/College de France
    djannati@in2p3.fr

Dorman, Lev I.
    Tel Aviv University
    lid@physics.technion.ac.il

Dubois, Richard
    Stanford Linear Accelerator Center
    richard@slac.stanford.edu

Durham, Ian
    University of New Hampshire
    durham@usna.edu

Dyks, Jaroslaw
    N. Copernicus Astronomical Center
    jinx@ncac.torun.pl

Edwards, Philip
    ISAS
    pge@vsop.isas.ac.jp

Ellison, Don
    NCSU
    don_ellison@ncsu.edu

Fan, Junhui
    Guangzhou University
    jhfan@guangztc.edu.cn

Fegan, Stephen
    Smithsonian/Univ. of Arizona
    sfegan@egret.sao.arizona.edu

Feng, Yuxin
    Purdue University
    fengyx@physics.purdue.edu

Fishman, Gerald
    NASA/MSFC
    fishman@msfc.nasa.gov

Fossati, Giovanni
    Univ. of California, San Diego
    gfossati@ucsd.edu

Fruchter, Andrew S.
 STScI
 fruchter@stsci.edu

Gehrels, Neil
 NASA/GSFC
 gehrels@lheapop.gsfc.nasa.gov

Genzel, Reinhard
 MPE, Garching
 genzel@mpe.mpg.de

Goldwurm, Andrea
 Service d'Astrophysique/CEA-Saclay
 agoldwurm@cea.fr

Gonthier, Peter
 Hope College
 gonthier@physics.hope.edu

Grenier, Isabelle
 University Paris VII & CEA Saclay
 isabelle.grenier@cea.fr

Grindlay, Jonathan
 Harvard Observatory
 josh@cfa.harvard.edu

Grove, J. Eric
 Naval Research Lab
 grove@gamma.nrl.navy.mil

Grunsfeld, John
 NASA/Johnson Space Center
 john.m.grunsfeld1@jsc.nasa.gov

Guessoum, Nidhal
 American University of Sharjah, UAE
 nguessoum@aus.ac.ae

Hardee, Philip
 University of Alabama
 hardee@athena.astr.ua.edu

Harding, Alice
 NASA/GSFC
 harding@twinkie.gsfc.nasa.gov

Harris, Michael
 USRA/GSFC
 harris@tgrs2.gsfc.nasa.gov

Harrison, Fiona
 Caltech
 fiona@srl.caltech.edu

Hartman, Robert
 NASA/GSFC
 rch@egret.gsfc.nasa.gov

Hartmann, Dieter
 Clemson University
 hdieter@clemson.edu

Hays, Liz
 University of Maryland
 ehays@umdgrb.umd.edu

Henriksen, Mark
 Univ. of Maryland, Baltimore Cty.
 mark@jca.umbc.edu

Hermsen, Willem
 SRON-Utrecht
 w.hermsen@sron.nl

Hernanz, Margarida
 IEEC/CSIC (Barcelona)
 hernanz@ieec.fcr.es

Horan, Deirdre
 Smithsonian/N.U.I.D., Dublin
 horan@egret.sao.arizona.edu

Hunter, Stanley D.
 NASA/GSFC
 sdh@gamma.gsfc.nasa.gov

Iyudin, Anatoli
 MPE, Garching
 ani@mpe.mpg.de

Johnson, W. Neil
 Naval Research Lab
 johnson@gamma.nrl.navy.mil

Jones, Frank
 NASA/GSFC
 frank.c.jones@gsfc.nasa.gov

Kamae, Tuneyoshi
 SLAC
 kamae@slac.stanford.edu